33rd UIT (Italian Union of Thermo-Fluid-Dynamics) Heat Transfer Conference 2015

Journal of Physics: Conference Series Volume 655

L'Aquila, Italy
22 – 24 June 2015

Editors:

Domenica Paoletti
Dario Amrosini
Stefano Sfarra

ISBN: 978-1-5108-1729-6
ISSN: 1742-6588

Printed from e-media with permission by:

Curran Associates, Inc.
57 Morehouse Lane
Red Hook, NY 12571

Some format issues inherent in the e-media version may also appear in this print version.

Printed by Curran Associates, Inc. (2016)

For permission requests, please contact the Institute of Physics at the address below.

Institute of Physics
Dirac House, Temple Back
Bristol BS1 6BE UK

Phone: 44 1 17 929 7481
Fax: 44 1 17 920 0979

techtracking@iop.org

Additional copies of this publication are available from:

Curran Associates, Inc.
57 Morehouse Lane
Red Hook, NY 12571 USA
Phone: 845-758-0400
Fax: 845-758-2634
Email: curran@proceedings.com
Web: www.proceedings.com

Table of contents

Volume 655

33rd UIT (Italian Union of Thermo-fluid dynamics) Heat Transfer Conference
22–24 June 2015, L'Aquila, Italy

Accepted papers received: 16 October 2015
Published online: 16 November 2015

Preface

011001

33rd UIT (Italian Union of Thermo-fluid dynamics) Heat Transfer Conference OPEN ACCESS D Paoletti, D Ambrosini, S Sfarra

011002

Peer review statement OPEN ACCESS

Papers

Keynote Lectures

012001

On the horizontal throughflow across a vertical porous wall OPEN ACCESS Antonio Barletta pg. 1

012002

Infrared image processing devoted to thermal non-contact characterization-Applications to Non-Destructive Evaluation, Microfluidics and 2D source term distribution for multispectral tomography OPEN ACCESS Jean-Christophe Batsale and Christophe Pradere pg. 11

012003

On sustainable and efficient design of ground-source heat pump systems OPEN ACCESS W Grassi, P Conti, E Schito and D Testi pg. 29

Technical Seminars

012004

Fundamentals and recent advances in X-ray micro computed tomography (microCT) applied on thermal-fluid dynamics and multiphase flows OPEN ACCESS Maurizio Santini pg. 47

012005

Schlieren visualization of water natural convection in a vertical ribbed channel OPEN ACCESS M Fossa, M Misale and G Tanda pg. 56

Heat transfer and efficiency in energy systems, environmental technologies, and buildings

012006

Validation of quantitative IR thermography for estimating the U-value by a hot box apparatus OPEN ACCESS I Nardi, D Paoletti, D Ambrosini, T de Rubeis and S Sfarra pg. 67

012007

A comparison between thermographic and flow-meter methods for the evaluation of thermal transmittance of different wall constructions OPEN ACCESS I Nardi, D Ambrosini, T de Rubeis, S Sfarra, S Perilli and G Pasqualoni pg. 76

012008

Space Heating Load Estimation Procedure for CHP Systems sizing OPEN ACCESS P Vocale, G Pagliarini and S Rainieri pg. 86

012009

Experiments and simulations on a thermosyphon solar collector with integrated storage OPEN ACCESS P Toninelli, A Mariani and D Del Col pg. 94

012010

CO_2 post-combustion capture in coal-fired power plants integrated with solar systems OPEN ACCESS R Carapellucci, L Giordano and M Vaccarelli pg. 105

Measurement of thermophysical properties coupled with LCA assessment for the optimization of a historical building retrofit OPEN ACCESS A Bortolin, P Bison, G Cadelano, G Ferrarini and S Fortuna pg. 115

012012

Beneficial effects of retroreflective materials in urban canyons: results from seasonal monitoring campaign OPEN ACCESS F Rossi, E Morini, B Castellani, A Nicolini, E Bonamente, E Anderini and F Cotana pg. 125

012013

A novel approach to CFD analysis of the urban environment OPEN ACCESS F Nardecchia, F Gugliermetti and F Bisegna pg. 137

012014

The performance of a mobile air conditioning system with a water cooled condenser OPEN ACCESS Davide Di Battista and Roberto Cipollone pg. 148

012015

Development of an Organic Rankine Cycle system for exhaust energy recovery in internal combustion engines OPEN ACCESS Roberto Cipollone, Giuseppe Bianchi, Angelo Gualtieri, Davide Di Battista, Marco Mauriello and Fabio Fatigati pg. 158

012016

Uncertainty evaluation of EnPIs in industrial applications as a key factor in setting improvement actions OPEN ACCESS G D'Emilia, D Di Gasbarro, A Gaspari and E Natale pg. 168

012017

Thermal and optical characterization of natural and artificial marble for roof and external floor installations OPEN ACCESS F Asdrubali, G Baldinelli, F Bianchi, A Presciutti, F Rossi and S Schiavoni pg. 178

012018

Experimental investigation of a packed bed thermal energy storage system OPEN ACCESS Mario Cascetta, Giorgio Cau, Pierpaolo Puddu and Fabio Serra pg. 188

012019

Analysis of the temperatures measured in very thick and insulating roofs in the vicinity of a chimney OPEN ACCESS M Neri, D Luscietti, S Bani, A Fiorentino and M Pilotelli pg. 197

012020

Analysis and clustering of natural gas consumption data for thermal energy use forecasting OPEN ACCESS Alessandro Franco and Fabio Fantozzi pg. 208

012021

A note on calculation of efficiency and emissions from wood and wood pellet stoves OPEN ACCESS D Petrocelli and A M Lezzi pg. 218

012022

Design and experimental analysis of a screened heat pipe for solar applications OPEN ACCESS D Jafari, S Filippeschi, A Franco and P Di Marco pg. 228

012023

Numerical simulation of radiant ceiling panels for indoor cooling OPEN ACCESS Giuliano Cammarata, Giuseppe Petrone and Filippo Masi pg. 238

012024

Thermal design of spacecraft solar arrays using a polyimide foam OPEN ACCESS N Bianco, M Iasiello and V Naso pg. 245

012025

Some critical remarks about the radiative heat transfer in air frame cavities according to EN ISO 10077-2 OPEN ACCESS G Cuccurullo and L Giordano pg. 253

012026

Magnetic refrigeration: an eco-friendly technology for the refrigeration at room temperature OPEN ACCESS C Aprea, A Greco, A Maiorino and C Masselli pg. 261

012027

Mitigating the cooling need and improvement of indoor conditions in Mediterranean educational buildings, by means of green roofs. Results of a case study OPEN ACCESS F Ascione, N Bianco, R F De Masi, F de Rossi and G P Vanoli pg. 274

Micro and nano scale thermo-fluid dynamics

012028

Determination of droplet contours in liquid-liquid flows within microchannels OPEN ACCESS B Pulvirenti, B Rostami, G Puccetti and G L Morini pg. 287

012029

Simulation of copper–water nanofluid in a microchannel in slip flow regime using the lattice Boltzmann method with heat flux boundary condition OPEN ACCESS A D'Orazio, Z Nikkhah and A Karimipour pg. 297

012030

Nano-PCMs for passive electronic cooling applications OPEN ACCESS L Colla, L Fedele, S Mancin, B Buonomo, D Ercole and O Manca pg. 307

Multi-phase fluid dynamics, heat transfer and interface phenomena

012031

An unsteady model for the simulation of the rapid depressurization of vessels containing two-phase mixtures in non-equilibrium conditions OPEN ACCESS R Ricci, V D'Alessandro, S Montelpare, L Binci and A Zoppi pg. 317

012032

Vaporization inside a mini microfin tube: experimental results and modeling OPEN ACCESS A Diani and L Rossetto pg. 327

012033

Venturi flow meter and Electrical Capacitance Probe in a horizontal two-phase flow OPEN ACCESS G Monni, M Caramello, M De Salve and B Panella pg. 337

012034

Influence of temperature on oxygen permeation through ion transport membrane to feed a biomass gasifier OPEN ACCESS T Antonini, P U Foscolo, K Gallucci and S Stendardo pg. 347

012035

Modelling heat and mass transfer in a membrane-based air-to-air enthalpy exchanger OPEN ACCESS S Dugaria, L Moro and D Del Col pg. 357

012036

Air entrainment in the primary impact of single drops on a free liquid surface OPEN ACCESS E Salerno, P Levoni and G S Barozzi pg. 367

012037

Energy dissipation and transfer processes during the breaking of modulated wave trains OPEN ACCESS F De Vita, R Verzicco and A Iafrati pg. 378

012038

Experimental and numerical analyses on a plate heat exchanger with phase change for waste heat recovery at off-design conditions OPEN ACCESS Roberto Cipollone, Giuseppe Bianchi, Davide Di Battista and Fabio Fatigati pg. 388

012039

Influence of electrolyte concentration on holdup, flow regime transition and local flow properties in a large scale bubble column OPEN ACCESS Giorgio Besagni and Fabio Inzoli pg. 398

012040

Experimental study of critical heat flux of refrigerant R1234yf in a multi-minichannel heat sink at medium saturation temperatures OPEN ACCESS W Imparato, R Mastrullo, A W Mauro and L Viscito pg. 408

012041

Thin Film Breakup and Rivulet Evolution Modeling OPEN ACCESS Nicola Suzzi, Giulio Croce and Paola D'Agaro pg. 419

012042

Pulsating Heat pipe Only for Space (PHOS): results of the REXUS 18 sounding rocket campaign OPEN ACCESS F Creatini, G M Guidi, F Belfi, G Cicero, D Fioriti, D Di Prizio, S Piacquadio, G Becatti, G Orlandini, A Frigerio, S Fontanesi, P Nannipieri, M Rognini, N Morganti, S Filippeschi, P Di Marco, L Fanucci, F Baronti, M Mameli, M Manzoni and M Marengo pg. 429

012043

Effect of the application of an electric field on the performance of a two-phase loop device: preliminary results OPEN ACCESS F Creatini, P Di Marco, S Filippeschi, D Fioriti and M Mameli pg. 439

012044

Electric field effects on the dynamics of bubble detachment from an inclined surface OPEN ACCESS P Di Marco, N Morganti and G Saccone pg. 449

Computational fluid dynamics and heat transfer

012045

Backcoupling of acoustic streaming on the temperature field inside high-intensity discharge lamps OPEN ACCESS J Schwieger, B Baumann, M Wolff, F Manders and J Suijker pg. 460

012046

Numerical validation of a κ-ω-κ_θ-ω_θ heat transfer turbulence model for heavy liquid metals OPEN ACCESS D Cerroni, R Da Vià, S Manservisi, F Menghini, G Pozzetti and R Scardovelli pg. 470

012047

Numerical and experimental study of local heat transfer enhancement in helically coiled pipes. Preliminary results. OPEN ACCESS F Bozzoli, L Cattani, S Rainieri and A Zachár pg. 480

012048

Parameter estimation approach to the thermal characterization of intumescent fire retardant paints OPEN ACCESS L Calabrese, F Bozzoli, G Bochicchio, B Tessadri, P Vocale and S Rainieri pg. 490

012049

CFD study on electrolyte distribution in redox flow batteries OPEN ACCESS S Bortolin, P Toninelli, D Maggiolo, M Guarnieri and D Del Col pg. 500

012050

A preliminary investigation of the growth of an aneurysm with a multiscale monolithic Fluid-Structure interaction solver OPEN ACCESS D Cerroni, S Manservisi and G Pozzetti pg. 510

012051

CFD Simulation of Melting and Solidification of PCM in Thermal Energy Storage Systems of Different Geometry OPEN ACCESS S Arena, G Cau and C Palomba pg. 520

012052

Numerical investigation of the flow in axial water turbines and marine propellers with scale-resolving simulations OPEN ACCESS Mitja Morgut, Dragica Jošt, Enrico Nobile and Aljaž Škerlavaj pg. 530

012053

Scaled models in the analysis of fire-structure interaction OPEN ACCESS A Andreozzi, N Bianco, M Musto and G Rotondo pg. 541

012054

A fast algorithm for Direct Numerical Simulation of natural convection flows in arbitrarily-shaped periodic domains OPEN ACCESS D Angeli, E Stalio, M A Corticelli and G S Barozzi pg. 550

012055

Heat transfer along the route to chaos of a swaying thermal plume OPEN ACCESS D Angeli, M A Corticelli, A Fichera and A Pagano pg. 561

012056

Analysis of thermal field within an urban canyon with variable thermophysical characteristics of the building's walls OPEN ACCESS A De Lieto Vollaro, G Galli, A Vallati and R Romagnoli pg. 571

Heat transfer in nuclear plants

012057

Characterization of the thermal conductivity for ceramic pebble beds OPEN ACCESS R Lo Frano, D Aquaro, L Scaletti and N Olivi pg. 581

012058

An anisotropic numerical model for thermal hydraulic analyses: application to liquid metal flow in fuel assemblies OPEN ACCESS F Vitillo, D Vitale Di Maio, C Galati and G Caruso pg. 588

012059

Transient analysis of "2 inch Direct Vessel Injection line break" in SPES-2 facility by using TRACE code OPEN ACCESS S D'Amico, C Lombardo, I Moscato, M Polidori and G Vella pg. 598

Natural, forced and mixed convection

012060

Heat transfer characteristics in forced convection through a rectangular channel with broken V-shaped rib roughened surface OPEN ACCESS D Fustinoni, P Gramazio, L Colombo and A Niro pg. 608

IR thermography

012061

Passive and active infrared thermography: An overview of applications for the inspection of mosaic structures OPEN ACCESS P Theodorakeas, E Cheilakou, E Ftikou and M Koui pg. 616

012062

The use of infrared thermography to detect the skin temperature response to physical activity OPEN ACCESS G Tanda pg. 626

012063

Stream temperature estimated in situ from thermal-infrared images: best estimate and uncertainty OPEN ACCESS F Iezzi and M T Todisco pg. 636

Conduction and radiation

012064

Spectral BRDF measurements of metallic samples for laser processing applications OPEN ACCESS L Vitali, D Fustinoni, P Gramazio and A Niro pg. 645

012065

Accuracy of lumped-parameter representations for heat conduction modeling in multilayer slabs OPEN ACCESS Paola Gori, Claudia Guattari, Roberto de Lieto Vollaro and Luca Evangelisti pg. 654

33rd UIT (Italian Union of Thermo-fluid dynamics) Heat Transfer Conference – L'Aquila (Italy) 22-24 June 2015

Giovan Paolo Cardone, ***Gonfalon of the city of L'Aquila*** **(1579), still the most accurate depiction of the XVI century city, as it must have appeared from the hills of Roio. Painting on silk, 442 cm x 335 cm, Museo Nazionale d'Abruzzo.**

Preface

The 33rd UIT (Italian Union of Thermo-Fluid Dynamics) Heat Transfer Conference was organized by the Dept. of Industrial and Information Engineering and Economics, University of L'Aquila (Italy) and was held at the Engineering Campus of Monteluco di Roio, L'Aquila, June 22-24, 2015. The annual UIT conference, which has grown over time, came back to L'Aquila after 21 years.
The scope of the conference covers a range of major topics in theoretical, numerical and experimental heat transfer and related areas, ranging from energy efficiency to nuclear plants. This year, there was an emphasis on IR thermography, which is growing in importance both in scientific research and industrial applications.
2015 is also the International Year of Light. The Organizing Committee honored this event by introducing a new section, Technical Seminars, which in this edition was mainly devoted to optical flow visualization (also the subject of three different national workshops organized in L'Aquila by UIT in 2003, 2005 and 2008).
The conference was held in the recently repaired Engineering buildings, six years after the 2009 earthquake and 50 years after the beginning of the Engineering courses in L'Aquila.
Despite some logistical difficulties, 92 papers were submitted by about 270 authors, on eight different topics: heat transfer and efficiency in energy systems, environmental technologies and buildings (32

papers); micro and nano scale thermo-fluid dynamics (5 papers); multi-phase fluid dynamics, heat transfer and interface phenomena (16 papers); computational fluid dynamics and heat transfer (15 papers); heat transfer in nuclear plants (6 papers); natural, forced and mixed convection (6 papers); IR thermography (4 papers); conduction and radiation (3 papers).

The conference program scheduled plenary, oral and poster sessions. The three invited plenary Keynote Lectures were given by Prof. Antonio Barletta (University of Bologna, Italy), Prof. Jean-Christophe Batsale (Arts et Metiers Paris Tech, Talence, France) and Prof. Walter Grassi (University of Pisa, Italy). The two invited Technical Seminars were given by Dr. Maurizio Santini (University of Bergamo, Italy) and Prof. Giovanni Tanda (University of Genova, Italy). There were also 13 oral sessions and three poster sessions.

This special issue collects the five papers presented in the plenary sessions (keynote lectures and technical seminars) plus 60 papers selected from those presented and discussed during the congress.

The UIT 2015 conference has been a useful occasion to stimulate discussion, further the understanding of heat transfer and related phenomena, present the state-of-the-art of some topics, discuss emerging trends and promote collaborations. We hope this issue will maintain and extend some of these features.

A special thank you is due to the Organizing and Scientific Committees, to the sponsors and to all the participants.

Domenica Paoletti, Dario Ambrosini, Stefano Sfarra, editors

Journal of Physics: Conference Series **655** (2015) 011001 doi:10.1088/1742-6596/655/1/011001

33rd UIT Conference, L'Aquila 2015 – Scientific Committee

33rd UIT Conference, L'Aquila 2015 – Organizing Committee

UIT Steering Committee

Gian Piero Celata	*ENEA Roma*
Davide Del Col	*University of Padova*
Paolo Di Marco	*University of Pisa*
Vincenzo Naso (President)	*University of Napoli*
Alfonso Niro	*Polytechnic of Milano*
Sara Ranieri (Scientific Secretary)	*University of Parma*
Giuseppe Vella (Vice-President)	*University of Palermo*

UIT Accounting Auditors

Mario De Salve	*Polytechnic of Torino*
Marco Spiga	*University of Parma*
Luca Tagliafico	*University of Genova*

UIT Conferences

I	1983, Trieste	XVIII	2000, Cernobbio (CO)
II	1984, Bologna	XIX	2001, Modena
III	1985, Palermo	XX	2002, Maratea (PZ)
IV	1986, Genova	XXI	2003, Udine
V	1987, Torino	XXII	2004, Genova
VI	1988, Bari	XXIII	2005, Parma
VII	1989, Firenze	XXIV	2006, Napoli
VIII	1990, Ancona	XXV	2007, Trieste
IX	1991, Pisa	XXVI	2008, Palermo
X	1992, Genova	XXVII	2009, Reggio Emilia
XI	1993, Milano	XXVIII	2010, Brescia
XII	1994, L'Aquila	XXIX	2011, Torino
XIII	1995, Bologna	XXX	2012, Bologna
XIV	1996, Roma	XXXI	2013, Moltrasio (CO)
XV	1997, Torino	XXXII	2014, Pisa
XVI	1998, Siena	XXXIII	**2015, L'Aquila**
XVII	1999, Ferrara		

Journal of Physics: Conference Series **655** (2015) 011002 doi:10.1088/1742-6596/655/1/011002

Peer review statement

All papers published in this volume of *Journal of Physics: Conference Series* have been peer reviewed through processes administered by the proceedings Editors. Reviews were conducted by expert referees to the professional and scientific standards expected of a proceedings journal published by IOP Publishing.

33rd UIT (Italian Union of Thermo-fluid-dynamics) Heat Transfer Conference IOP Publishing
Journal of Physics: Conference Series **655** (2015) 012001 doi:10.1088/1742-6596/655/1/012001

On the horizontal throughflow across a vertical porous wall

Antonio Barletta

Department of Industrial Engineering, Alma Mater Studiorum Università di Bologna, Viale Risorgimento 2, 40136 Bologna, Italy

E-mail: antonio.barletta@unibo.it

Abstract. The stationary two-dimensional mixed convection flow in a vertical porous wall subject both to a temperature and to a pressure gap across its boundaries is studied. The instability of this flow is investigated by adopting a linear modal analysis. The two-dimensional transverse rolls are shown to be the preferred pattern at onset of instability. Transverse rolls turn out to be travelling modes. They are driven by the net vertical flow predicted by the basic solution. A numerical procedure based on the joint use of the Runge-Kutta method and on the shooting method is employed to solve the perturbation eigenvalue problem. This numerical solution yields both the neutral stability curves, defining the threshold of linear instability, and the critical Darcy-Rayleigh number yielding the onset condition of instability.

1. Introduction

When an infinite vertical porous layer saturated by a fluid is subject to a boundary temperature difference, a parallel vertical flow establishes under steady-state conditions. This flow is one of purely free convection, it is characterised by a zero net flow rate and by a linear velocity profile. Such a simple flow can be considered as a single convection cell with an infinite height developing along the porous layer. A pioneering study published by Gill [1] proved that this single-cell flow is linearly stable. This means that random perturbations of this flow having a very small amplitude are unable to change it significantly and permanently. Extensions of Gill's work to a domain of perturbations with larger amplitude, namely to the nonlinear domain, were carried out later on. A significant finding on this issue was achieved by Straughan [2]. In fact, this author was able to prove that a nonlinear stability analysis, based on the energy method, leads to the same response declared by Gill [1]: the stationary single-cell flow in a vertical porous slab is stable.

The practical interest of Gill's stability proof relies on the peculiar heat transfer characteristics of the single-cell flow in a vertical porous layer. In fact, as far as the vertical porous layer has an infinite height, the free convection flow does not modify the temperature profile and, hence, the heat transfer rate with respect to a case of pure heat conduction. The stability of this flow demonstrates the impossibility to enhance heat transfer beyond heat conduction. As pointed out by Gill [1], this information is definitely significant with respect to the design of thermal insulation in buildings.

A step further on this subject was taken in a recent paper [3] where the role of the momentum boundary conditions is pointed out. In particular, the author proved that the stability of the single-cell flow in a vertical porous layer strongly depends on the velocity boundary conditions. If the boundaries are assumed to be perfectly impermeable, then Gill's proof applies and the single-cell flow is stable. If the conditions of impermeability are relaxed and replaced by conditions of open boundaries, then Gill's proof does not hold and instability is possible. The instability means transition from the single-cell flow regime to a multiple-cells flow pattern, occurring when the Darcy-Rayleigh number exceeds the threshold of 197.081 [3].

The aim of this contribution is to investigate a setup where the basic single-cell flow studied by Gill [1] is superposed to a prescribed horizontal throughflow across the vertical porous layer, so that the resulting stationary basic flow is two-dimensional instead of one-dimensional and parallel. This generalisation is a sensible one, having in mind applications to the building insulation industry such as the breathing walls [4].

2. Problem formulation

A porous vertical layer is considered. Its width is L, so that the parallel vertical planes bounding the layer are at $x = -L/2$ and $x = L/2$ (see the sketch in Fig. 1). These boundaries are kept at different temperatures T_1 and T_2, respectively. On denoting by p the excess pressure over the hydrostatic pressure distribution, namely the hydrodynamic pressure, we assume a pressure gap acting across the layer and causing a forced horizontal throughflow. Hence, the hydrodynamic pressure on the boundary planes at $x = -L/2$ and $x = L/2$ is considered as uniform with asymmetric values p_1 and p_2, respectively. Hereafter, for the sake of brevity, p will be called pressure field.

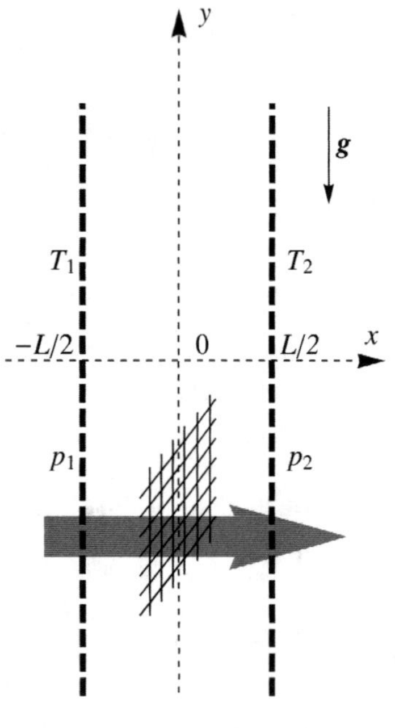

Figure 1. A sketch of the vertical porous layer and of the boundary conditions.

The Cartesian coordinates x, y and z define the horizontal direction perpendicular to the boundaries, the vertical direction, and the horizontal direction parallel to the boundaries, respectively. The seepage velocity field $\boldsymbol{u}$ has components (u, v, w) along the (x, y, z)−axes, time is denoted by t and the temperature field by T.

By introducing the scaling,

$$\frac{1}{L}\,(x,y,z) \to (x,y,z), \quad \frac{\alpha}{\sigma L^2}\,t \to t, \quad \frac{K}{\mu\alpha}\,(p - p_0) \to p,$$
$$\frac{L}{\alpha}\,\boldsymbol{u} = \frac{L}{\alpha}\,(u,v,w) \to (u,v,w) = \boldsymbol{u}, \quad \frac{T - T_0}{T_2 - T_1} \to T, \tag{1}$$

one defines the dimensionless coordinates, time, and fields, so that the local mass, momentum and energy balance equations can be written in a dimensionless form as

$$\boldsymbol{\nabla} \cdot \boldsymbol{u} = 0, \tag{2a}$$

$$\boldsymbol{u} = -\boldsymbol{\nabla} p + R\,T\,\hat{\boldsymbol{e}}_y, \tag{2b}$$

$$\frac{\partial T}{\partial t} + \boldsymbol{u} \cdot \boldsymbol{\nabla} T = \nabla^2 T. \tag{2c}$$

In Eq. (1), α is the average thermal diffusivity of the saturated porous medium, $p_0 = (p_1 + p_2)/2$ is the reference pressure, $T_0 = (T_1 + T_2)/2$ is the reference temperature, μ is the dynamic viscosity, K is the permeability, and σ is the heat capacity ratio of the saturated porous medium. The latter quantity is defined as the ratio between the average volumetric heat capacity of the saturated medium and the volumetric heat capacity of the fluid.

The Darcy-Rayleigh number, R, and the Péclet number, Q, associated with the horizontal throughflow are defined as

$$R = \frac{g\beta(T_2 - T_1)KL}{\nu\alpha}, \qquad Q = \frac{(p_1 - p_2)K}{\mu\alpha}, \tag{3}$$

where ν is the kinematic viscosity of the fluid. Physically, it is not restrictive to assume $R > 0$, provided that Q can be either positive or negative. Thus, all the forthcoming results will be discussed by tacitly assuming a positive value of R.

The dimensionless boundary conditions are given by

$$x = \pm\frac{1}{2}: \qquad p = \mp\frac{Q}{2}, \quad T = \pm\frac{1}{2}. \tag{4}$$

3. Basic solution

There exists a stationary solution of Eqs. (2) and (4) expressed as

$$u_b = Q, \quad v_b = R\,\frac{e^{Qx} - \cosh(Q/2)}{2\sinh(Q/2)}, \quad w_b = 0,$$
$$T_b = \frac{v_b}{R} = \frac{e^{Qx} - \cosh(Q/2)}{2\sinh(Q/2)}, \quad p_b = -\,Q\,x. \tag{5}$$

One may easily check that the basic solution given by Eq. (5) coincides with that assumed in the analysis by Gill [1] when $Q \to 0$. In fact, Gill's basic flow is a parallel vertical flow with a vanishing mass flow rate, such that $u_b = w_b = 0$ and $v_b = R\,x$. On the other hand, Eq. (5) implies a two-dimensional velocity field with a non-vanishing vertical flow rate when $Q \neq 0$. Indeed, the average vertical flow rate is

$$Q_v = \int_{-1/2}^{1/2} v_b\,dx = -\frac{R}{Q}\left[\frac{Q}{2}\coth\left(\frac{Q}{2}\right) - 1\right]. \tag{6}$$

The right hand side of Eq. (6) is positive, meaning a bulk upward flow, when Q is negative. On the other hand, the average vertical flow rate is negative, *viz.* the bulk flow is downward, if Q is positive.

Figures 2 and 3 illustrate the qualitative patterns of the streamlines for $R = 10$ and either positive (Fig. 2) or negative (Fig. 3) Péclet numbers, Q. These figures reveal that an increasing value of $|Q|$ gradually changes the streamlines from an almost straight vertical shape to an almost straight horizontal shape. The former shape corresponds to Gill's parallel flow, while the latter suggests a forced convection regime where buoyancy is almost negligible and the horizontal throughflow prevails. In fact, Fig. 4 reveals that things are not exactly so since, with $Q = \pm 50$, a very thin buoyant boundary layer develops either close to $x = -1/2$ (for $Q = -50$) or close to $x = 1/2$ (for $Q = 50$).

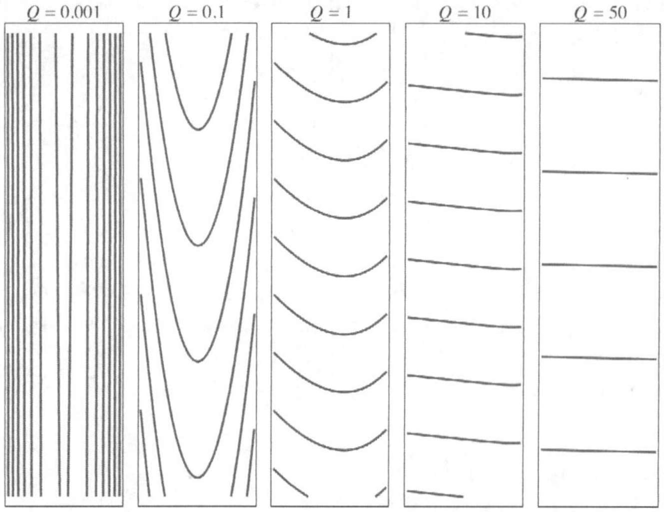

Figure 2. Streamlines of the basic flow, with $R = 10$ and positive values of Q

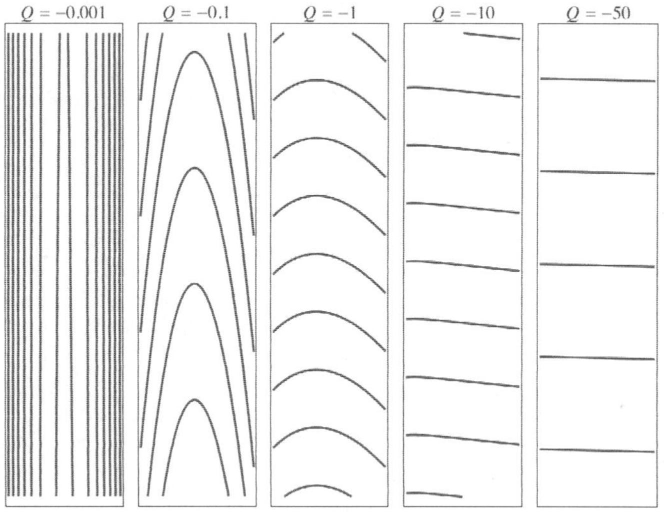

Figure 3. Streamlines of the basic flow, with $R = 10$ and negative values of Q

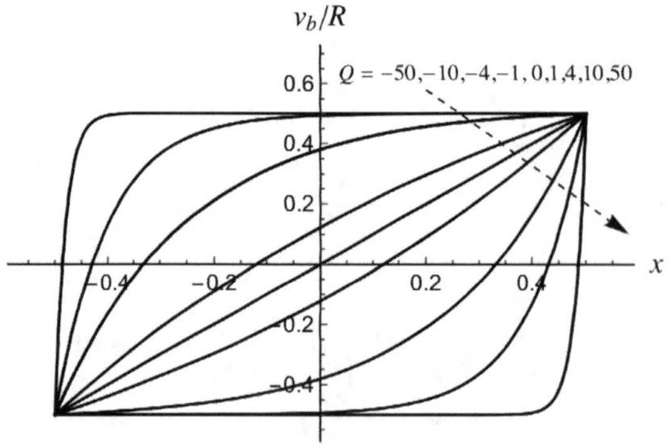

Figure 4. Plots of v_b/R with different values of Q

There is a symmetry underlying the basic solution given by Eq. (5). In fact, the solution (5) is left invariant by the transformation,

$$\boldsymbol{u} \to -\boldsymbol{u}, \quad T \to -T, \quad x \to -x, \quad Q \to -Q. \tag{7}$$

Figures 2 and 3 show that the streamlines of the basic flow conform to this symmetry.

4. Linear stability analysis

We perturb the basic state (5) with small-amplitude disturbances given by

$$\boldsymbol{u} = \boldsymbol{u}_b + \varepsilon\, \tilde{\boldsymbol{u}}, \quad p = p_b + \varepsilon\, \tilde{p}, \quad T = T_b + \varepsilon\, \tilde{T}. \tag{8}$$

Here, the perturbation parameter ε is such that $\varepsilon > 0$ and $\varepsilon \ll 1$, while $\tilde{\boldsymbol{u}} = (\tilde{u}, \tilde{v}, \tilde{w})$, $\tilde{p}$ and $\tilde{T}$ are the perturbation fields. Substitution of Eq. (8) into Eqs. (2) and (4), by taking into account Eq. (5) yields

$$\boldsymbol{\nabla} \cdot \tilde{\boldsymbol{u}} = 0, \tag{9a}$$

$$\tilde{\boldsymbol{u}} = -\boldsymbol{\nabla}\tilde{p} + R\tilde{T}\,\hat{\boldsymbol{e}}_y, \tag{9b}$$

$$\frac{\partial \tilde{T}}{\partial t} + Q\,\frac{\partial \tilde{T}}{\partial x} + v_b(x)\,\frac{\partial \tilde{T}}{\partial y} + T_b'(x)\,\tilde{u} = \nabla^2 \tilde{T}, \tag{9c}$$

$$x = \pm\frac{1}{2}: \qquad \tilde{p} = 0, \quad \tilde{T} = 0, \tag{9d}$$

where prime means derivative with respect to x. We may adopt a pressure-temperature formulation, by evaluating the divergence of Eq. (9b) and by taking into account Eq. (9a). Thus, we obtain

$$\nabla^2 \tilde{p} = R\,\frac{\partial \tilde{T}}{\partial y}, \tag{10a}$$

$$\frac{\partial \tilde{T}}{\partial t} + Q\,\frac{\partial \tilde{T}}{\partial x} + R\,T_b(x)\,\frac{\partial \tilde{T}}{\partial y} - T_b'(x)\,\frac{\partial \tilde{p}}{\partial x} = \nabla^2 \tilde{T}, \tag{10b}$$

$$x = \pm\frac{1}{2}: \qquad \tilde{p} = 0, \quad \tilde{T} = 0. \tag{10c}$$

By employing Eq. (5), one may show that Eqs. (10) are invariant under the transformation,

$$x \to -x, \quad y \to -y, \quad Q \to -Q, \quad \tilde{p} \to -\tilde{p}. \tag{11}$$

A stability analysis versus normal modes is carried out by writing

$$\tilde{p} = f(x)\, e^{i(k_y y + k_z z)}\, e^{\lambda t}, \quad \tilde{T} = h(x)\, e^{i(k_y y + k_z z)}\, e^{\lambda t}, \tag{12}$$

where (k_y, k_z) is the wave vector, and $k = (k_y^2 + k_z^2)^{1/2}$ is the wave number. While k_y and k_z are real, λ is a complex parameter whose real part, $\Re(\lambda)$, expresses the time-growth rate of the disturbance. On the other hand, the angular frequency is given by $\omega = -\Im(\lambda)$, namely by the negative imaginary part of λ. On account of Eq. (12), Eqs. (10) give rise to an ordinary differential eigenvalue problem

$$f'' - k^2 f - i k_y R\, h = 0, \tag{13a}$$

$$h'' - Q\,h' - \left[k^2 + \lambda + i k_y R\, T_b(x)\right] h + T_b'(x) f' = 0, \tag{13b}$$

$$x = \pm\frac{1}{2}: \qquad f = 0, \quad h = 0. \tag{13c}$$

Journal of Physics: Conference Series **655** (2015) 012001 doi:10.1088/1742-6596/655/1/012001

We note that the eigenvalue problem (13) is endowed with a symmetry. In fact, on account of Eq. (5), Eqs. (13) are left invariant by the transformation,

$$x \to -x, \quad Q \to -Q, \quad k_y \to -k_y, \quad f \to -f. \tag{14}$$

The change $k_y \to -k_y$ is equivalent to $\omega \to -\omega$ as its effect is just reversing the direction of the wave vector. Therefore, we can reformulate the symmetry of the eigenvalue problem (13) as an invariance under the transformation,

$$x \to -x, \quad Q \to -Q, \quad \omega \to -\omega, \quad f \to -f. \tag{15}$$

Longitudinal rolls are normal modes with $k_y = 0$, oblique rolls are such that $0 < k_y < k$, while transverse rolls are defined by $k_y = k$. It is immediately verified that oblique or longitudinal rolls may be mapped onto equivalent transverse rolls by a suitable rescaling of the Rayleigh number,

$$S = \frac{k_y}{k} R, \tag{16}$$

so that Eq. (13) can be rewritten as

$$f'' - k^2 f - ik\, S\, h = 0, \tag{17a}$$

$$h'' - Q\, h' - \left[k^2 + \lambda + ik\, S\, T_b(x)\right] h + T_b'(x) f' = 0, \tag{17b}$$

$$x = \pm\frac{1}{2}: \qquad f = 0, \quad h = 0. \tag{17c}$$

The Squire-like transformation given by Eq. (16) was previously adopted by Rees [5]. An immediate consequence of this transformation is that longitudinal rolls ($k_y = 0$) are equivalent to transverse rolls with a zero Darcy-Rayleigh number. This means that no convective instability can be induced by longitudinal rolls, since no buoyancy force exists when the Darcy-Rayleigh number is zero. Moreover, oblique rolls are equivalent to transverse rolls having a smaller Darcy-Rayleigh number. This implies that transverse rolls are selected at onset of convective instability, being the modes which magnify the action of the buoyancy force.

5. Numerical solution

Equations (17) are homogeneous. Hence, they obviously admit the trivial solution $f = h = 0$. Besides this solution, nontrivial solutions are allowed for special choices of the parameters (k, λ, S, Q). In other words, Eqs. (17) define an eigenvalue problem. Its solutions can be found numerically by the combined use of an initial value solver for ordinary differential equations, based on an explicit Runge-Kutta method, and the shooting method. The steps to be made are well-known in the literature (see, for instance, § 19.1 of [6]). A way to formulate them is the following:

(i) Complete Eqs. (17) so that the boundary value problem is transformed into an initial value problem. In other words, we replace Eq. (17c) with

$$f(-1/2) = 0, \quad f'(-1/2) = \eta_1 + i\eta_2, \quad h(-1/2) = 0, \quad h'(-1/2) = 1, \tag{18}$$

where $h'(-1/2) = 1$ is a scale-fixing condition intended to break the scale-invariance of any solution of Eqs. (17), and (η_1, η_2) are unknown real constants to be determined by the shooting method.

(ii) Solve the initial value problem given by Eqs. (17a), (17b) and (18) for prescribed values of (k, S, Q), and guessed values of $(\lambda, \eta_1, \eta_2)$.

(iii) Determine the actual values of the four real parameters $\Re(\lambda)$, $\Im(\lambda) = -\omega$, η_1 and η_2 by employing the shooting method to satisfy the target conditions, $f(1/2) = 0$ and $h(1/2) = 0$, implied by Eq. (17c). As f and h are complex-valued functions, these target conditions are, in fact, four real equations to be satisfied by the four unknowns $\Re(\lambda)$, $\Im(\lambda) = -\omega$, η_1 and η_2.

Journal of Physics: Conference Series **655** (2015) 012001 doi:10.1088/1742-6596/655/1/012001

This method can be efficiently implemented within the computational software program *Mathematica 10* (© Wolfram Research), by using the built-in functions `NDSolve` and `FindRoot`. The main aim is the determination of the neutral stability curves in the parametric plane (k, S), for different prescribed values of Q. This task is accomplished by setting $\Re(\lambda) = 0$, *viz.* by determining the locus of vanishing growth rate, and thus the curve dividing stable ($\Re(\lambda) < 0$) from unstable ($\Re(\lambda) > 0$) conditions.

6. Discussion of the results

The neutral stability curves are plotted, for different positive values of Q in Fig. 5. The essential meaning is that unstable conditions, for a given Q, lie in the region above the corresponding neutral stability curve. The region below the curve is that defining the parametric conditions of linear stability. As a consequence, the absolute minimum of the neutral stability curve yields the so-called critical values of k, S and ω. These values are denoted as k_c, S_c and ω_c.

The value of S_c is specially important. In fact, convective instability takes place whenever $S > S_c$. Thus, Fig. 5 discloses an important feature: the basic flow is stabilised by increasing the value of Q. In other words, the basic horizontal throughflow has a stabilising effect. A similar feature was pointed out, in the case of a horizontal porous layer with a vertical throughflow, by Homsy and Sherwood [7], and by Jones and Persichetti [8].

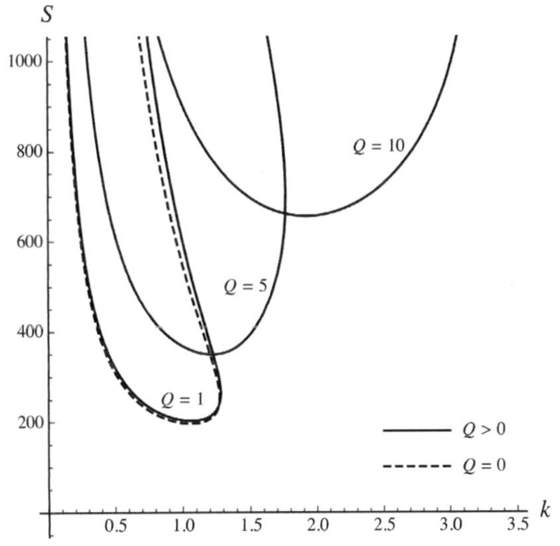

Figure 5. Plots of the neutral stability curves in the (k, S)−plane with $Q = 0$ (dashed line) and $Q > 0$ (solid lines).

Numerical values of k_c, S_c and ω_c versus Q are reported in Table 1. The data reported in this table corroborate the behaviour revealed by Fig. 5, namely that S_c monotonically increases with Q. There is an additional information concerning the values of ω_c. The neutral stability condition with $Q = 0$ is marked by $\omega = 0$, meaning that instability is driven by non-travelling modes [3]. This feature is altered by the presence of a horizontal throughflow. In fact, with $Q > 0$, Table 1 reports negative values of ω_c. Having in mind transverse rolls, this means a negative vertical phase velocity, $\omega_c/k_c < 0$, so that unstable transverse rolls travel in the downward direction. Physically, this behaviour is reasonable since Eq. (6) predicts a net basic vertical flow velocity Q_v whenever $Q > 0$. Thus, it is the net basic downflow which determines the direction of the travelling perturbation modes. With reference to transverse rolls, a comparison between the plots of $|Q_v|$, evaluated with $R = R_c$, and $|\omega_c/k_c|$ versus Q is displayed in Fig. 6. This figure reveals that the values of the phase velocity are slightly larger than those of the basic vertical flow velocity, so that unstable transverse rolls travel faster than the basic net flow rate in the vertical direction. One must be careful on converting this information from dimensionless velocities to dimensional velocities. In fact, in Eq. (1), the scale adopted for velocity is not consistent with the ratio between the scales of length and time.

Journal of Physics: Conference Series **655** (2015) 012001 doi:10.1088/1742-6596/655/1/012001

Table 1. Critical values of k, S and ω for different values of Q.

Q	k_c	S_c	ω_c
0	1.059498	197.0812	0.000000
0.1	1.059540	197.1448	-2.241053
0.5	1.060546	198.6713	-11.24815
1	1.063750	203.4443	-22.76430
2	1.077505	222.5464	-47.67209
3	1.103842	254.1847	-76.86275
4	1.147907	297.3569	-112.4635
5	1.216155	349.7842	-156.5670
6	1.314163	408.1566	-211.1184
7	1.442516	469.2641	-277.4873
8	1.593426	531.2067	-355.7898
9	1.755564	593.5152	-445.1874
10	1.922379	656.2993	-545.1334
11	2.092200	719.6239	-655.6562
12	2.264884	783.4235	-776.9483
13	2.440254	847.5836	-909.1300
14	2.617907	911.9994	-1052.227
15	2.797377	976.5929	-1206.219

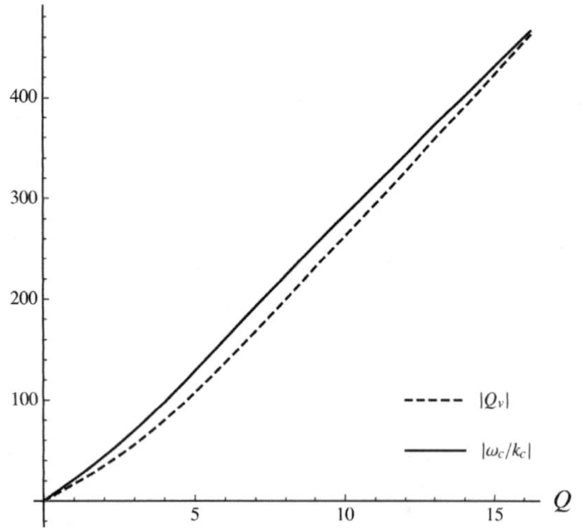

Figure 6. Transverse rolls: plots of $|Q_v|$ with $R = R_c$ (dashed line) and of the modulus of the phase velocity $|\omega_c/k_c|$ (solid line).

Thus, our conclusion that the phase velocity is larger than the basic vertical flow velocity holds in dimensional terms if $\sigma \leqslant 1$, while it may hold or not if $\sigma > 1$.

Having established the neutral stability condition when $Q > 0$, one may question on the conditions for the onset of convective instability when $Q < 0$. The answer readily comes from the symmetry (15) governing the stability eigenvalue problem. When $Q < 0$, the neutral stability curves coincide with the corresponding curves obtained for $Q > 0$. This means that S_c does not change if $Q \to -Q$. Besides that, Eq. (15) implies that the angular frequency ω change its sign at neutral stability, when $Q \to -Q$. Thus, the same values of $|\omega_c|$ reported in Table 1 are retrieved by changing the sign of Q. One may conclude that, when $Q < 0$, unstable modes travel in the upward direction, since $\omega > 0$.

Figure 7 is relative to transverse rolls ($k_y = k$). This figure displays the streamlines and the isotherms of transverse rolls at onset of convection under critical conditions ($k = k_c, R = R_c$), with either $Q = 0$, $Q = 1$ or $Q = 5$. Displaying explicitly the orientation of the streamlines is unnecessary as corotating and counterrotating cells are alternately piled up within the vertical layer, as one may devise from Eq. (12). The comparison between the three cases is definitely interesting as the perturbation patterns gradually change from perfectly symmetric when $Q = 0$, to slightly asymmetric when $Q = 1$ and, eventually, to markedly asymmetric when $Q = 5$.

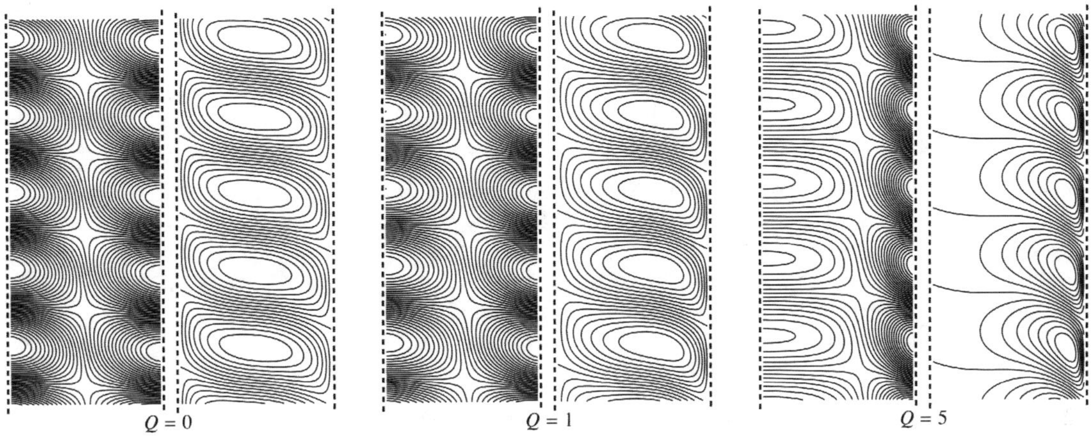

Figure 7. Transverse rolls: plots of the streamlines (left) and isotherms (right) at a fixed time and critical conditions ($k = k_c$, $R = R_c$) for $Q = 0$, $Q = 1$ and $Q = 5$.

Trying to figure out the corresponding perturbation patterns when $Q = -1$ and $Q = -5$ is relatively easy if one applies the transformation (11). This means that reflections of the horizontal $x-$axis and of the vertical $y-$axis are needed, so that a gradual compression of the isotherms to $x = -1/2$, instead of $x = 1/2$, occurs as $|Q|$ increases.

The behaviour of the isotherms of the secondary flow, superposed to the basic single-cell flow, suggests that the Dirichlet boundary conditions for the temperature may become harder and harder to be sustained when $|Q|$ increases. In fact, the strong temperature gradient at the left/right boundary of the porous layer might induce deviations from the isothermal boundary conditions assumed in our model. To address this issue, it seems reasonable a future improvement of the model employed for the temperature boundary conditions. One can assume Robin temperature conditions instead of the Dirichlet boundary conditions, and introduce suitable Biot numbers to modulate the heat transfer rate to the external fluid environments.

7. Conclusions

The effects of an imposed horizontal pressure gradient on the buoyant flow in a plane vertical porous layer have been studied. The plane boundaries of the layer have been modelled as isobaric and isothermal, with both a pressure gap and a temperature gap imposed across the layer. The basic stationary mixed convection flow is two-dimensional, consisting of a parallel vertical buoyant flow superposed to a uniform horizontal throughflow. The governing parameters are the Darcy-Rayleigh number, R, and the Péclet number, Q, associated with the horizontal throughflow. Without any loss of generality, the study has been carried out by keeping $R > 0$, and by allowing Q to be either positive or negative. It has been shown that, when $Q > 0$, a net vertical downflow occurs with non-dimensional average velocity $Q_v < 0$. On the other hand, when $Q < 0$, a net vertical upflow occurs with non-dimensional average velocity $Q_v > 0$.

Suitable symmetries of the governing equations have been established allowing one to gather information on the regime $Q < 0$ from the results obtained when $Q > 0$.

A modal linear stability analysis has been performed to determine the neutral stability condition for the onset of convective instability. A Squire-like transformation has been defined in order to map the analysis of any normal mode onto that of two-dimensional transverse rolls. According to this transformation, three-dimensional normal modes are always more stable than transverse rolls. Thus, convective instability is triggered by two-dimensional perturbation patterns superposed to the two-dimensional basic flow. Neutral stability curves for some values of Q have been drawn, showing that the effect of an increasing value of $|Q|$ is stabilising. Transverse rolls turned out to be travelling in the same direction of the net vertical downflow/upflow, having modulus $|Q_v|$, associated with the basic solution. However, the dimensionless phase velocity of the transverse rolls has a modulus slightly larger than $|Q_v|$.

The presented stability analysis is pivotal to a fully nonlinear analysis of the supercritical regime which is the natural future development of this research. In particular, expected results from the nonlinear study regard the stationary or non-stationary flow patterns spawned when $R > R_c$, as well as the evaluation of the heat transfer rate across the wall. The latter can be a crucial information for an efficient design of building elements such as the porous dynamic insulation media, well-known as breathing walls.

References

[1] Gill A E 1969 *Journal of Fluid Mechanics* **35** 545–547
[2] Straughan B 1988 *Geophysical & Astrophysical Fluid Dynamics* **42** 269–275
[3] Barletta A 2015 *Journal of Fluid Mechanics* **770** 273–288
[4] Imbabi M S E 2006 *Renewable Energy* **31** 729–738
[5] Rees D A S 1988 *International Journal of Heat and Mass Transfer* **31** 1529–1534
[6] Straughan B 2004 *The Energy Method, Stability, and Nonlinear Convection* 2nd ed (New York, NY: Springer)
[7] Homsy G M and Sherwood A E 1976 *AIChE Journal* **22** 168–174
[8] Jones M C and Persichetti J M 1986 *AIChE Journal* **32** 1555–1557

Infrared image processing devoted to thermal non-contact characterization-Applications to Non-Destructive Evaluation, Microfluidics and 2D source term distribution for multispectral tomography

Jean-Christophe Batsale[1], **Christophe Pradere**[2]

[1] Arts et Metiers ParisTech, I2M, UMR 5295, F-33400 Talence, France
[2] CNRS, I2M, UMR 5295, F-33400 Talence, France

Corresponding author: jean-christophe.batsale@ensam.eu

Abstract. The cost of IR cameras is more and more decreasing. Beyond the preliminary calibration step and the global instrumentation, the infrared image processing is then one of the key step for achieving in very broad domains.
Generally the IR images are coming from the transient temperature field related to the emission of a black surface in response to an external or internal heating (active IR thermography). The first applications were devoted to the so called thermal Non-Destructive Evaluation methods by considering a thin sample and 1D transient heat diffusion through the sample (transverse diffusion). With simplified assumptions related to the transverse diffusion, the in-plane diffusion and transport phenomena can be also considered.
A general equation can be applied in order to balance the heat transfer at the pixel scale or between groups of pixels in order to estimate several fields of thermophysical properties (heterogeneous field of in-plane diffusivity, flow distributions, source terms).
There is a lot of possible strategies to process the space and time distributed big amount of data (previous integral transformation of the images, compression, elimination of the non useful areas…), generally based on the necessity to analyse the derivative versus space and time of the temperature field. Several illustrative examples related to the Non-Destructive Evaluation of heterogeneous solids, the thermal characterization of chemical reactions in microfluidic channels and the design of systems for multispectral tomography, will be presented.

1. Introduction

The processing of space and time temperature fields is more and more necessary because devices are now currently available in order to quickly measure, store and process thermal information. Infrared thermography devices are the most usual in the field of heat transfer analysis. But a lot of other possibilities with contact or non-contact sensors (with optical or mechanical scans) will appear in the future. A lot of questions about "how to process" and "how to estimate" thermophysical properties from such fields are posed.
A 2D temperature field can be the signature of a lot of heat transfer phenomena at the surface of a solid or a liquid (heat conduction or diffusion through homogeneous or heterogeneous media, convective transport in complex systems…). One of the main intuitive way to process the signal is to study the time or space derivatives of such fields, in order to link the temperature observation to a heat

transfer model. Several strategies for the temperature field processing in space and in time will be illustrated on a suitable analytical temperature field. The application examples devoted to Non-Destructive Evaluation, microfluidic systems and multispectral tomography subsequently presented are based on a derivation of the experimental signal. Such examples are also illustrating the main domains approached in our team in Bordeaux University.

2. Thermal processing of a 2D transient *T(x,y,t)* field-Derivation

From a 2D transient *T(x,y,t)* field, and a heat transfer model (even simplified) is it often possible to estimate a resultant thermophysical properties field. The diffusion and convection transport modes can be considered in order to set out an identification model. The simple observation of the evolution of the temperature field can give the intuition of such phenomena. The processing of such fields needs of course to be aware of the noise and perturbations considerations about the temperature field recording. The random noise with a zero mean and a uniform and diagonal covariance matrix will generally be taken into account. Several aspects related to the "bad pixels", the space or time correlation and the non-regular time steps will here be assumed to be previously circumvented. Maybe, the first "natural" processing of such data is to try to set out the derivative (versus space or time) of such a discrete temperature field. The space or time derivation of noisy fields is a difficult task, because such operator is amplifying the random measurement noise, if no precaution is taken. A "filtering" is then necessary, but the risk is to lose a part of the original information. Several strategies will be here examined (the finite differences, the polynomial fit, the orthogonal basis decomposition and the convolution), in order to numerically implement the derivation of a discrete signal.

In order to test and illustrate the processing possibilities, it is here proposed to consider the 1D temperature relaxation of a solution to a system such as:

$$a\frac{\partial^2 T}{\partial x^2} = \frac{\partial T}{\partial t}$$

with: $T(x,t=0)=1 \text{ if } 0<x<b$ and $T(x,t=0)=0 \text{ if } b<x<L$

and at $x=0$ $\left.\frac{\partial T}{\partial x}\right|_0 = 0$ and at $x=L$ $\left.\frac{\partial T}{\partial x}\right|_L = 0$

After a time t, the temperature field is relaxed (or filtered) by diffusion, and an approximation of the field is then:

$$T(x,t) = b/L + 2/L\sum_{n=1}^{N}\frac{\sin(\alpha_n b)}{\alpha_n}\exp(-a\alpha_n^{\,2}t)\cos(\alpha_n x)$$

with: $\alpha_n = n\pi/L$ and n, N finite integers.

Rigorously, the previous expression is a serie with N tending to infinity. A physical filter (due to the exponential term) is acting and limiting the space-frequency content of the signal. The parameter which allows to control the filter is here the observation time. This observation time will be fixed in this section and only the space field will be considered.
The observed temperature is a vector obtained from the previous expression at regularly spaced space steps and a white gaussian noise is added.

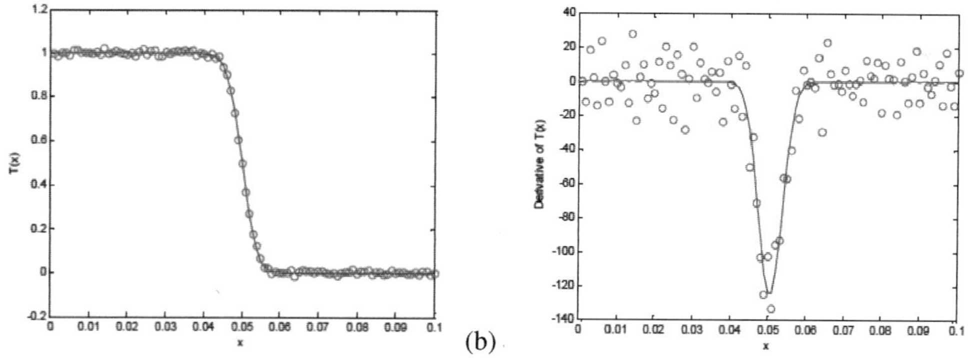

(a) (b)

Figure 1. (a) Temperature field from the previous analytical expression at time *t=0.5s;* $a=10^{-5}\ m^2\ s^{-1}$; *b=L/2; L=0.1m;* (continuous line: real signal, 'o': discrete noisy signal); (b) Derivative of the previous noisy signal by finite differences.

The initial gaussian noise is with zero mean and $\sigma = 0.01$. The resulting noise is amplified by the derivation operation.

2.1 Finite differences

The finite difference approach used on figure 1 is generally presented at each space step *i*, as:

$$\hat{T}'_i = \frac{\hat{T}_{i+1} - \hat{T}_i}{\Delta x}$$

If it is assumed, that the relation between the observed temperature $\hat{T}_i$ and the real temperature T_i is:

$$\hat{T}_i = T_i + e_{T_i}\ ,$$

with e_{T_i} representing the random variable related to the gaussian noise, uniform whatever the position x_i. The asymptotic expansion around x_i is such as:

$$\hat{T}'_i = \frac{T(x_{i+1}) - T(x_i)}{\Delta x} + \varepsilon(x_{i+1}) + \frac{e_{Ti+1} - e_{Ti}}{\Delta x}$$

With:

$$\lim_{x \to x_i} \varepsilon(x) = 0$$

Two kinds of errors have then to be considered: the *approximation error* $\varepsilon(x_i)$ related to the rest of the asymptotic expansion and the *random error* related to the random variable e_{T_i}. Unfortunately, when the space step Δx is tending to zero, the approximation error is effectively tending to zero, but the random error is tending to infinity! The difference of two random variables is a linear operation which amplify the initial noise. In order to avoid such a difficulty, it is necessary to "filter" the signal or to project the discrete observed information in a basis of functions. One of the simplest basis of function can be a polynomial basis.

Journal of Physics: Conference Series **655** (2015) 012002 doi:10.1088/1742-6596/655/1/012002

2.2 Polynomial fitting

From the same previous measurements, a polynomial fitting can be implemented such as:

$$T(x) = \sum_{n=0}^{N} \beta_n x^n$$

The estimation of the β_n parameters in vector: $\mathbf{B} = [\beta_1, \beta_2, \beta_3 .... \beta_n]^T$ can then be obtained by a linear least-square relation such as :

$$\mathbf{B} = \left(\mathbf{X}^t\mathbf{X}\right)^{-1}\mathbf{X}^t\hat{\mathbf{T}}$$

With:

$$\mathbf{X} = \begin{bmatrix} 1 & x_1 & x_1^{\ 2} \cdots \\ 1 & x_2 & x_2^{\ 2} \cdots \\ . & & \\ 1 & x_m & x_m^{\ 2} \cdots \end{bmatrix}$$

In order to minimize the distance:

$$\left\|\hat{\mathbf{T}} - \mathbf{XB}\right\| = \left(\hat{\mathbf{T}} - \mathbf{XB}\right)^T\left(\hat{\mathbf{T}} - \mathbf{XB}\right)$$

The construction of the derivative will then consist in considering the derivative of the polynomial function. The Matlab software is convenient in order to implement such calculations because the $\mathbf{X}^t\mathbf{X}$ matrix can be bad conditioned (Wandermonde matrix) and must be inversed with special precautions. One example of such processing is shown on the figure 2.

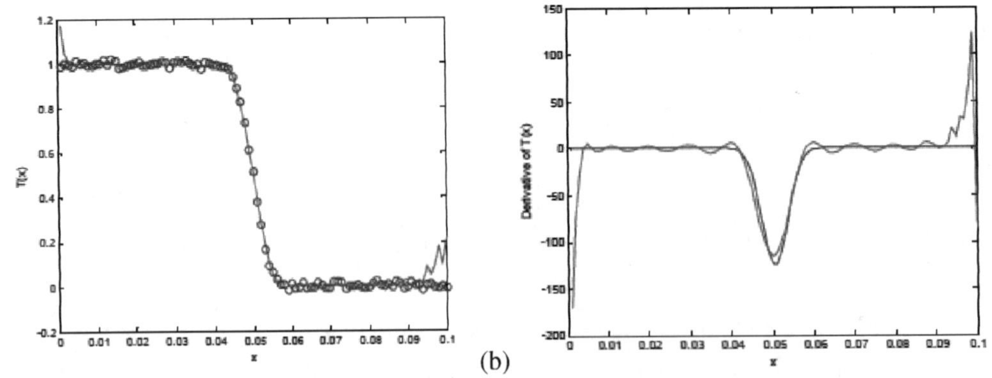

(a) (b)

Figure 2. (a) Polynomial fitting of the signal shown on figure 5; (b) Derivation. The chosen degree of the polynomial fitting is here: *N=30*. The number of observation points is *100*.

It is shown here that the direct derivation of the estimated polynomial expression is giving a suitable continuous approximation in the considered domain. Approximation errors are occurring at the boundary of the domain if no precautions or assumptions are taken. One other strategy is to chose a basis "near from the considered physical phenomenon". Here, the ideal basis is the basis made of the

Fourier cosine functions, because the cosine vectors are here verifying the boundary conditions (null derivative at $x=0$ and $x=L$) and are also the eigenvectors of the diffusion phenomenon (eigenvectors of the Laplacian operator in Cartesian coordinates and adiabatic boundaries).

2.3 Fourier cosine basis

The same processing as the polynomial fitting can be considered with such an expression:

$$T(x)=\sum_{n=0}^{M}\beta_n\cos(\alpha_n x)\ \text{ with: }\ \alpha_n=n\pi/L$$

The estimation of the parameter vector: $\mathbf{B}=[\beta_1,\beta_2,\beta_3....\beta_M]^T$ is then obtained by the same expressions as previously (in section 2.2), excepted that the $\mathbf{X}^t\mathbf{X}$ matrix is orthogonal, and then very easy to be inverted. The covariance matrix related to the parameters is then as diagonal as the observable vector. The result of the estimation is then, with 30 terms for the series, illustrated on figure 3.

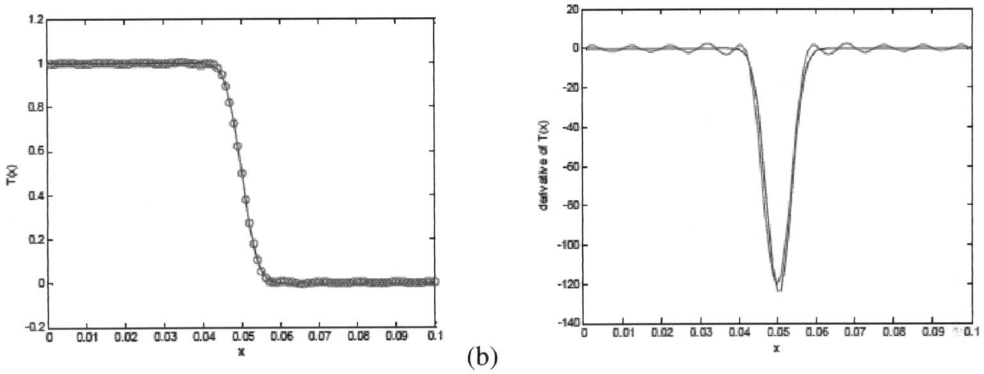

(a) (b)

Figure 3. (a) Fourier fitting of the signal shown on figure 1; (b) Derivation. The number of terms is $N=30$. The number of observation points is *100*.

2.4 Filtering with a convolution kernel

The filtering is a usual operation in signal processing, which consist in weighting the signal with a moving average. It must be noticed that the observable signal himself is maybe previously filtered by the proper instrument. A new approximation of the signal can then be considered by $\tilde{T}(x)$ such as :

$$\tilde{T}(x)=\int_0^L p(\chi)T(x-\chi)d\chi$$

$p(x)$ must be normed such as : $$\int_0^L p(\chi)d\chi=1$$

The derivative of $\tilde{T}(x)$ is then conveniently considered by the commutability of the convolution product such as :

$$\frac{d\tilde{T}}{dx}(x)=\int_0^L p(\chi)\frac{dT(x-\chi)}{dx}d\chi=\int_0^L T(\chi)\frac{dp(x-\chi)}{dx}d\chi$$

The discrete approximation of the derivative is then conveniently considered by a convolution with a « derived » kernel.
One very simple illustration is given on figure 4. The discrete convolution kernel of the filter is for example [1 2 1]/4 and an approximation of the convolution kernel for the derivation is then [1/2 0 -1/2]. It can be noticed that this "slight" convolution (affecting only a few number of neighbours and very similar to the finite difference method) allows to obtain good results with less effort. The main problem is to find a compromise between the bias caused by the filtering and the perturbation caused by the noise.

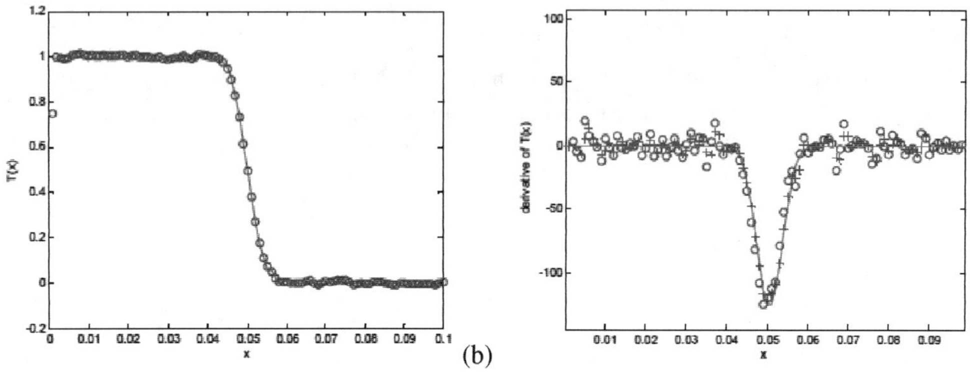

(a) (b)

Figure 4. (a) Filtering of the signal shown on figure 3, with a convolution kernel: [1 2 1]/4; (b) Derivation with a convolution kernel: [1/2 0 -1/2] (plot'o') and comparison with the finite difference derivative of the previously filtered signal (plot'+').

2.5 Singular Value Decomposition (SVD) of the whole space and time signal

The previous methods consists in finding a compromise between the "approximation error" and the "filtering". The number of terms of the serie (or the rank of basis), or the width of the convolution kernel are biasing the signal if they are "used too far". A lot of other methods can be considered (for example more sophisticated regularisation techniques, see [1]). When a large field must be processed the choice of the compromise between the filtering and the bias is made by trial and error. The key point is the knowledge of the random noise (at the minimum the standard deviation). Often, the experiencer does not know the characteristic of the noise in his proper experience. It is then difficult to implement an optimal filtering of the signal. One way to separate the "available signal" from the "random noise", when the experiencer has a great amount of space and time information, consists in implementing the singular value decomposition (SVD) of the $T(x,t)$ field, or the $\hat{T}(x_i, t_j)$ discrete observable matrix. Applying the SVD to temperatures matrix $\hat{\boldsymbol{T}}$ yields (with $n>m$):

$$\hat{\boldsymbol{T}}_{n,m} = \boldsymbol{U}_{n,n} \Sigma_{n,n} \boldsymbol{V}^T_{n,m}$$

Where $\Sigma_{n,n}$ is a diagonal matrix

So that finally the following truncated formulation of the SVD is commonly accepted

$$\hat{\boldsymbol{T}}_{n,m} = \boldsymbol{U}_{n,n} \Sigma_{n,n} \boldsymbol{V}^T_{n,m} = \sum_{k=1}^{n} \gamma_k \left(U_k V_k^T\right)$$

Where the modes U_k and V_k are the column vectors of matrices $\boldsymbol{U}$ and $\boldsymbol{V}$ respectively, and the singular values γ_k are the diagonal elements of Σ arranged in descending order.

Such a decomposition of the global space/time field is offering a lot of advantages. First, the examination of the singular values γ_k allows to select the really available signal. Thermal phenomena are often related to diffusion problems (naturally filtered), with only a few available singular values. The decomposition obtained with the reduced number of singular values is then offering a reduced representation of the field which allows a lot of possibilities (reduced computational effort for the further parameters estimations, low memory storage, new orthogonal basis and projection possibilities, see [2] to [5]). In the case illustrated on figure 5, the SVD is allowing an optimal filtering without any previous knowledge about the random noise at any time. The finite differences derivative is then available without precautions.

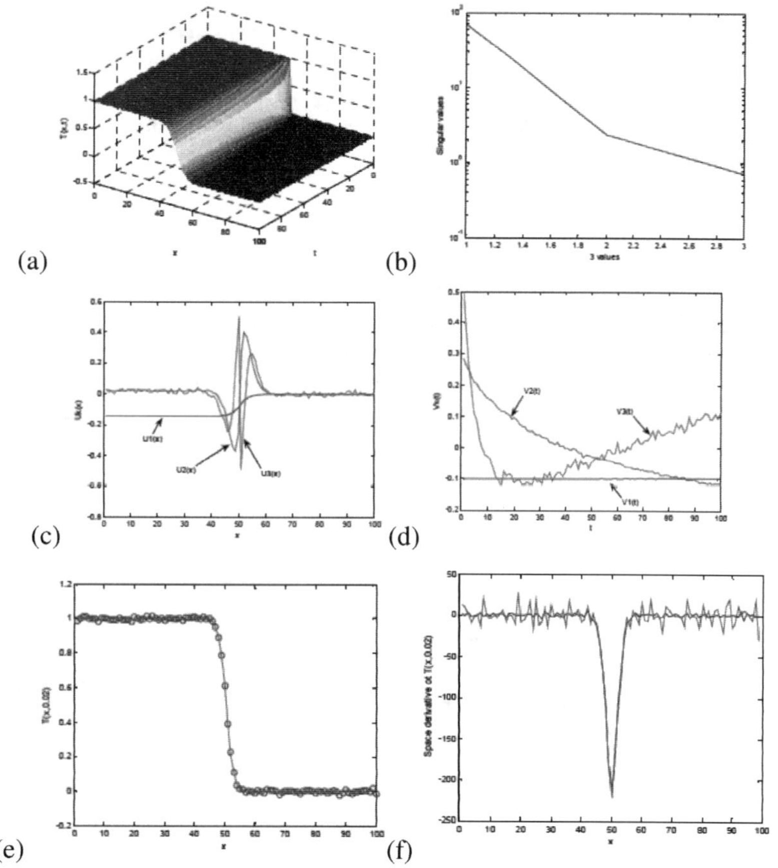

Figure 5. (a) *T(x,t)* noisy temperature field (time extension from figure 5); (b) 3 first singular values; (c) and (d) First ***U*** and ***V*** vectors, coming from the SVD decomposition; (e) Reconstruction of the previous signal at time *t=0.02s*, with 3 term of the SVD decomposition; (f) "Noisy" derivation of the previous signal (initial noisy observation) compared the finite differences derivative of the SVD reconstruction.

The previous example is showing that the transient situation even non-stationary, is giving a greater amount of data than a unique image. The SVD decomposition is a mean to squize and to process the great amount of data quite with the same effort as in the stationary case. It must be noticed that the orthogonal basis ***U*** is more reduced than the basis of cosine functions used in section 2-4.

3. Applications

The approaches previously cited (finite differences, polynomial decomposition, Fourier analysis, Singular Value Decomposition, convolution …) are not only a range of tools for the derivation of the temperature fields but also widely used in classical applications for the processing of the fields recorded with infrared cameras.

In the case of Non-destructive evaluation with the 1D flash experiment, the time-logarithmic derivative of the transient signal is related to a transverse diffusivity variation which gives a suitable decomposition of the signal. Very often, the "temperature contrast" related to a defect detection can be considered with a derivation operator.

More recently, the transmission of wide-band multispectral infrared sources through heterogeneous objects can be detected by simple thin opaque screens where the temperature field must be processed in order to avoid the in-plane diffusion. The time and space derivatives of the temperature field will then be considered.

In the case of microfluidic devices characterization, it is necessary to discriminate and calibrate the convective transport in micro-channels from the purely diffusive in-plane heat transport. The comparison between a first derivative (convective transport) and a second derivative (diffusive transport) must be considered.

3.1 Application to NDE

Non Destructive Evaluation (NDE) with infrared cameras consists generally to apply a heat pulse on a non-homogeneous parallelepipedic slab and to process the temperature response $T(x,y,t)$ from one face of the sample (the front face or the rear face) (see [6]). The aim of such processing is to estimate some characteristics of the heterogeneities (structure, size, nature, position inside the sample…). Instead of a 3D heterogeneous geometry, it is often convenient to consider thin samples with thin heterogeneities such as the transfer is locally 1D (versus z direction). One asymptotic expansion assuming that the heterogeneities fluctuations are small compared to the mean value of one thermophysical property of the sample yields a linear relationship, such as:

$$T(x,y,z,t) = q(x,y)\left(f(z,t,\beta_0) + \Delta\beta(x,y)\left.\frac{\partial f}{\partial \beta}\right|_{(z,t,\beta_0)} \right)$$

Where the functions $q(x,y)$ is the spatial distribution of energy coming from a flash excitation and $\Delta\beta(x,y)$ is the spatial thermophysical property variation(diffusivity, conductivity, thickness…). $q(x,y)$ and $\Delta\beta(x,y)$ have to be specified with a finite number of parameters which will be the object of the estimation procedure. Such procedure will then consists in processing the weighted sum of the images $T(x,y,z,t)$ where the weighting functions are the sensitivity functions $f(z,t,\beta_0)$ and $\left.\frac{\partial f}{\partial \beta}\right|_{(z,t,\beta_0)}$.

The temperature response of the front or the rear face of the sample will be recorded by a camera in order to estimate a map or a field of thermophysical properties. In the ideal case, the sample (a plane plate of small thickness L) is assumed to be thermally insulated and with a temperature field initially uniform ($T(x,y,z,t=0)=0$). If the heat transfer is supposed 1-D, then, the temperature response related to a unique location (x,y) corresponding to a pixel, to an instantaneous thermal pulse, is given (See [7]) on the front face $(z=0)$:

$$T(z=0,t)=\frac{Q}{\rho c L}\cdot\left(1+2\cdot\sum_{n=1}^{\infty}\exp\left(-\frac{n^2\cdot\pi^2\cdot a\cdot t}{L^2}\right)\right)=\frac{Q}{\rho c L}f\left(at/L^2\right)$$

From the previous expression, the estimation problem of several parameters can be considered. Instead of the thermal diffusivity, the estimation problem of the sample thickness L, the thermal conductivity λ and the volumic heat capacity ρc can be considered. In each case, a reference approximated value of the parameter must be known (and noted: L_0, λ_0 and ρc_0).

In many cases such reference values can be obtained by a previous global measurement. The following first order asymptotic expansions can be for example written for the front face, at each time t_i: For example, if a thermal conductivity variation is to be estimated (see[8]):

$$T(0,t_i)\approx\frac{Q}{\rho c L}f\left(\lambda_0 t_i/\rho c L^2\right)+\frac{Q}{\rho c L}\frac{\Delta\lambda}{\lambda_0}t\left.\frac{\partial f}{\partial t}\right|_{(\lambda_0 t_i/\rho c L^2)}$$

It is very important to notice that it is possible to replace the sensitivity functions $f(z=0,t,b_0)$ and $\left.\frac{\partial f}{\partial\beta}\right|_{(z,t,\beta_0)}$ by a linear combination of $f(z=0,t,b_0)$ and the time logarithmic derivative $t\left.\frac{\partial f}{\partial t}\right|_{(z=0,t,\beta_0)}$. It is then possible to implement a linear relationship such as:

$$T(0,t_i)\approx\beta_1 X_{\beta_1}(t_i)+\beta_2 X_{\beta_2}(t_i)$$

The functions $X_{\beta_j}(t_i)$ are the sensitivity functions of $T(0,t_i)$ to parameters β_j. If other parameters combination estimation (such as thermal diffusivity estimation) is considered, the resulting sensitivity function will be a linear combination of functions f and $t\ (\partial f/\partial t)$.

The parameters β_j are defined in each estimation situation such as: $\beta_1=\frac{Q}{\rho c L}$ and $\beta_2=\frac{Q}{\rho c L}\frac{\Delta\lambda}{\lambda_0}$

That is to say with matrix notation, considering the vectors and matrices:

$$\mathbf{T}=\left[T(0,t_1)\quad\cdots\quad T(0,t_N)\right]^t,\ \mathbf{X}=\begin{bmatrix}X_{\beta_1}(t_1) & \ldots & X_{\beta_1}(t_N)\\ X_{\beta_2}(t_1) & \ldots & X_{\beta_2}(t_N)\end{bmatrix}^t,$$

and under matrix notations: $\mathbf{T}=\mathbf{X}\begin{bmatrix}\beta_1\\ \beta_2\end{bmatrix}$

If the measurement noise on each component of the experimental temperature vector $\hat{\boldsymbol{T}}$ is assumed to be related with a constant standard deviation and not correlated, in the domain of validity of the previous asymptotic expansions, then, the optimal estimator of the parameters vector $[\beta_1\ \beta_2]$ is obtained by :

$$\begin{bmatrix}\hat{\beta}_1\\ \hat{\beta}_2\end{bmatrix}=\left(\mathbf{X}^t\mathbf{X}\right)^{-1}\mathbf{X}^t\hat{\mathbf{T}}$$

The linear approximation allows not only the estimation of the parameters, but also the confidence interval of this estimation to be studied.

The method can be used whatever the length of the vector $\hat{\mathbf{T}}$. These expressions can be implemented simultaneously with all of the pixels of the image. Thus, the matrix product $\mathbf{X}^t\hat{\mathbf{T}}$ can be incremented and consists in a real time weighting. The choice of the weighting is linked to the choice of the estimation strategy (estimation of L, ρc or λ). The terms of the sensitivity matrix are theoretically calculated with the references values or the averaged images. This sequential method considerably eases the problems of storage and images manipulation. It is very suitable for a simple Non Destructive Evaluation.
Moreover, $f(t)$ can be "measured" on the data, because at each time step, the average of the image is a filtered approximation of $f(t)$. The computation of the logarithmic derivative of $f(t)$ would then give a suitable method applicable without any idea about the knowledge of the nominal values of the thermophysical properties of the considered sample.
Unfortunately, the logarithmic derivative of this experimental signal is not easy with simple finite differences methods. In order to overcome such difficulties Shepard et al [9] proposed intuitively a logarithmic time-fitting each pixel signal and Rajic[2] to consider the SVD of the global data cube.
In order to conveniently process the great amount of data provided by the flash NDE experiment, Shepard proposed to decompose the signal with a polynomial fitting, such as:

$$Ln(T(x,y,z=0,t))=\beta_0(x,y)+\beta_1(x,y)Ln(t)+\beta_2(x,y)Ln^2(t)+\ldots$$

Such decomposition has no physical meaning because the new parameter vector is not related to a physical model, but the time-logarithmic derivative (considered by Shepard) appears to be very well correlated with the depth or thermophysical properties changes of the tested samples. The calculation of the logarithmic derivative is then taking the advantages related in section 2-2. It is also a very efficient and convenient method in order to reduce and manipulate the great amount of data (only N images corresponding to the degree of the polynomial expressions are to be manipulated). An other way to reduce the amount of data consists in considering the SVD of the information cube. Rajic [2] proposed the SVD decomposition (explained in 2.5) such as:

$$T(x,y,z=0,t)=\sum_{k=1}^{K} u_k(x)\gamma_k v_k(t)$$

Such data obtained from NDE experiments appears to be nicely reduced by 2 or 3 terms of the previous series. Generally, the $\mathbf{U_1}$ vector (or $u_1(x)$ function) is giving a good approximation of the spatial energy distribution. The $\mathbf{V_1}$ vector (or $v_1(t)$ function) is related to $f(t)$. The $\mathbf{U_2}$ vector (or $u_2(x)$ function) is giving a good approximation of the defects localisation. Bamford et al [4] proposed to compare the asymptotic expansion firstly explained in this section to the previous SVD decomposition. The slight differences are coming from the non-orthogonality of the functions. In fact all the methods presented in section 3.1 are very similar. They consist in projecting the data cube in a suitable basis of functions (in space or in time, physical or not) and then to try to process the signal with a physical model. Such methods are suitable because every pixel the whole image related to the analysed field are of interest. Such asymptotic approaches are very convenient in the case where the thermophysical properties mapping is slightly varying for example versus mechanical stresses (see [10]).

3.2 Applications to thermal multispectral tomography

The transmission of wide-band multispectral infrared sources through heterogeneous objects can be detected by simple thin opaque or semitransparent thin screens. Such screens are then acting as bolometers where the transient temperature field can be analysed by a unique IR thermography device. Our team is developing at this moment a large range of instruments based on terahertz radiation (or long wave infrared radiation (see [11]) or wide band infrared multispectral radiation, coupled with thermal imaging systems (see [12]). One of the key obstacle is to detect the thermal effect of very low-energy sources absorbed in a short time by a thin screen.

One illustration of this estimation problem is corresponding to the 1D observation of the transient temperature response $T(x,t)$ to a pulse heating of a laser spot on a beam graduated following the *x*-direction. Thanks to infrared thermography, the instantaneous temperature fields can be measured by infrared thermography at N_x different space steps x_i (lying between *0* and *L)* and at N_t different time steps t_j at N_t different time steps (lying between *0* and t_{max}). It is classically assumed that the measurement errors at each space and time step are uncorrelated, of zero mean and on uniform variance σ^2 or standard deviation σ. The processing of the observable temperature response: $\hat{T}(x,t)$ will be illustrated in these examples. The objective, is to use the great amount of data (N_x X N_t data) in order to filter the estimation of the initial temperature field $T(x,t=0)$.

The general 1-D problem of diffusion with initial condition, without internal source, in an infinite medium can be presented under the following system

$$\frac{\partial T}{\partial t} = a\frac{\partial^2 T}{\partial x^2}$$

with at initial time : $T(x,t=0) = f(x)$

It is here assumed that the 1D medium is infinite and the temperature is zero for $T(x,t)$ and $f(x)$ when x tend to infinite.

A direct solution of this system is obtained by implementing a space Fourier transform (see [13], [14]). Once again the Fourier transform is a suitable way to consider the Laplacian or second derivative of the temperature versus space:

$$\theta(\alpha_n,t) = \int_0^L T(x,t)\cos(\alpha_n x)dx \qquad \text{with} \qquad \alpha_n = n\pi / L \;;\; n \text{ >or=0}$$

L is here assumed as a "large" bound of the *x*-domain, where the temperature is zero and no gradient is to be considered. The previous system becomes an ordinary differential equation, after Fourier transform:

$$-\alpha_n^2\,\theta(\alpha_n,t) = \frac{1}{a}\frac{d\theta}{dt}$$

The initial condition becomes: $\theta(\alpha_n,0) = F(\alpha_n)$

The solution of the forward problem is then:

$$\theta(\alpha_n,t) = \theta(\alpha_n,0)\exp(-a\alpha_n^{\;2}t) \qquad \text{for every } n;$$

Or in the real space: $$T(x,t)=\frac{1}{L}\left[\theta(0,0)+\tfrac{1}{2}\sum_{n=1}^{\infty}\cos(\alpha_n x)\theta(\alpha_n,0)\exp(-a\alpha_n^{\,2}t)\right]$$

From these direct expressions, the estimation of the initial temperature field yields with a linear least square estimator:

$$\hat{\hat{\theta}}(\alpha_n,0)=\frac{\sum_{j=0}^{N_t}\hat{\theta}(\alpha_n,t_j)\exp(-a\alpha_n^{\,2}t_j)}{\sum_{j=0}^{N_t}\exp(-2a\alpha_n^{\,2}t_j)}$$

Such an estimator is decreasing the influence of the measurement noise on the initial estimator: $\hat{\theta}(\alpha_n,t_j)$ the terms of the covariance matrix of the previous estimator are then:

$$\operatorname{cov}\left(\hat{\hat{\theta}}(0,0)\right)=\frac{\sigma^2L^2}{N_xN_t}\ \text{and}\ \operatorname{cov}\left(\hat{\hat{\theta}}(\alpha_n,0)\right)=\frac{\sigma^2L^2}{N_x}\left(\frac{1}{\sum_{j=0}^{N_t}\exp(-2a\alpha_n^{\,2}t_j)}\right)\ \text{if } n \text{ non zero.}$$

Then, the standard deviation of the estimator in the *x*-space (σ_E) are in the following interval:

$$\sigma/\sqrt{N_t}<\sigma_E<\sigma$$

It is necessary to remark that the processing of such amount of data is efficient if the time steps related to the observation are near from *0*. It becomes then a convenient way to avoid the blurring effects due to diffusion. Unfortunately, the key parameter is here the perfect knowledge of the thermal diffusivity: *a*. Very often such parameter is unknown and is not suitably measurable directly on the instrumental device. Once again a Singular Value Decomposition is a quite similar decomposition as Fourier, but does not require the knowledge of a physical model and the of the diffusivity parameter. An extension of theses remarks to a 2D separable case are then even more efficient. If the initial condition is then depending on two space variables: x and y and that such initial condition is separable such as:

$$f(x,y)=f_x(x)f_y(y)$$

With the Fourier transform method, the solution of the 2D problem is then a product of two series depending on *x* and *t* and *y* and *t*, such as :

$$T(x,y,t)=\frac{1}{L}\left[\theta(0,0,0)+\tfrac{1}{2}\sum_{n=1}^{\infty}\cos(\alpha_n x)\theta(\alpha_n,0,0)\exp(-a\alpha_n^{\,2}t)\right]\frac{1}{l}\left[\theta(0,0,0)+\tfrac{1}{2}\sum_{m=1}^{\infty}\cos(\beta_m y)\theta(0,\beta_m,0)\exp(-a\beta_m^{\,2}t)\right]$$

β_m is a spatial frequency related to the *y*-direction, equivalent to α_n *related to the* x-direction.

The analogous estimator in the 2D Fourier space is also separable and gives:

$$\hat{\hat{\theta}}(\alpha_n,\beta_m,0)=\frac{\sum_{j=0}^{N_t}\hat{\theta}(\alpha_n,0,t_j)\exp(-a\alpha_n^{2}t_j)}{\sum_{j=0}^{N_t}\exp(-2a\alpha_n^{2}t_j)}\,\frac{\sum_{j=0}^{N_t}\hat{\theta}(0,\beta_m,t_j)\exp(-a\beta_m^{2}t_j)}{\sum_{j=0}^{N_t}\exp(-2a\beta_m^{2}t_j)}$$

In such a particular case where each instantaneous temperature field is separable versus the x-direction and y-direction, it is suitable to consider preliminarily a SVD, versus x and y at each time step such as:

$$\hat{T}(x_i,y_j,t)=\sum_k U_k(x_i,t)W_k(t)V_k(y_j,t)$$

Only the first term of such a decomposition is then to be taken into account. The order of magnitude of the second term must be much lower than the first one. It is then suitable to consider a secondary decomposition versus the x and y directions such as:

$$U_1(x_i,t_j)=\sum_{r=1}^{r0x}U_r^x(x_i)W_k^xV_r^x(t_j) \qquad \text{and} \qquad V_1(y_i,t_j)=\sum_{r=1}^{r0y}U_r^y(y_i)W_k^yV_r^y(t_j)$$

The advantage of such a truncated double-decomposition is to operate a strong data reduction. For example, if $N_x=N_y=N_t=1000$; the global number of data to be processed is 10^9. But the successive truncations allow to reduce the number of data to be considered to 12000 (if $r0x=r0y=3$). The measurement noise attenuation is than more than 100 times higher (see [13]).

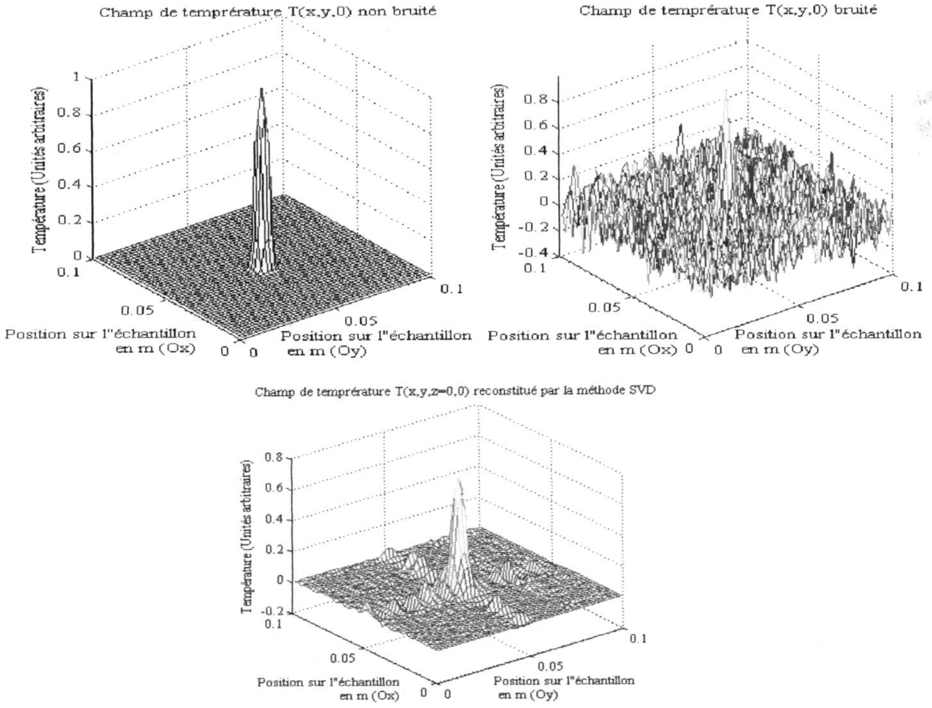

Figure 6. Initial temperature field (upper left without noise; upper right with a high noise/signal ratio; above filtered and reduced by SVD with r_0=3).

The images in figure 6 are illustrating the 2D temperature response to a punctual heat pulse lost in the noise. It shows that even when the noise level is of the same order of magnitude as the signal, the SVD decomposition allows the detection of a very localised point source.
This argue in favour of the bolometric detection of localised radiative source of very low amplitude such as terahertz radiations (see [11]), or in the case of multispectral imaging where the low-energy of each radiative band must be considered (see [12]). From such detection method it is possible to set out tomography instruments with very large possibilities.

3.3 Application to microfluidic devices

The micro fabrication techniques originally devoted to electronics have reach spectacular advances in more general domains such as chemical micro reactors with micro fluidic channels (lab on chip see [15], [16]). Such systems are very profitable because the low fluid or solid mass requirements and small volumes induce low cost fabrication. It leads new ways to design and intensify the experiments such as screening for catalyst testing related to chemical engineering see [17]). Micro-devices also have a lot of interesting characteristics in the field of heat transfer (high speed of the diffusive transfer, possibility of implementing localised heat sources, high surface/volume ratio and intensive local heat flux with low global energy consumption). One of the main drawbacks is related to the instrumentation. Particularly, for temperature measurement, the IR thermography is a non intrusive, way for the recording and processing of a large amount of data. The main challenge is then to set out processing methods of the surface temperature field of such systems through a heat transfer model. Once again, the main characteristic of such data processing is related to temperature differences or derivatives versus time and space. This will be illustrated with the characterization of a simple transient in plane diffusion on a thin plate and then the characterization of convective effect caused by a micro-channel stuck on a thin conductive plate (see [18]).

The simple in plane diffusion with source term is met usually in microelectronics. The objective can be to estimate the space and time distribution of energy given on a silicon plate from electronic chips from the observed transient temperature fields. It can also illustrate the case where a single droplet of reactant is deposited on the homogeneous conductive plate (a glass or silicon thin layer). A classical model of the temperature field with simplified assumptions, is given with the following expression:

$$\frac{\partial T}{\partial t} = a\left(\frac{\partial^2 T}{\partial x^2} + \frac{\partial^2 T}{\partial y^2}\right) + Q(x,y,t)/\rho c$$

The estimation method of the source term field : $Q(x,y,t)$ *in* Wm^{-3} from a sequence of temperature images illustrated by figure 7 consists in considering the Laplacian of the temperature field : $\Delta T = \left(\frac{\partial^2 T}{\partial x^2} + \frac{\partial^2 T}{\partial y^2}\right)$ and the time derivative $\frac{\partial T}{\partial t}$ at each time step and each pixel location, with a previous calibration for the knowledge of the thermal diffusivity a.

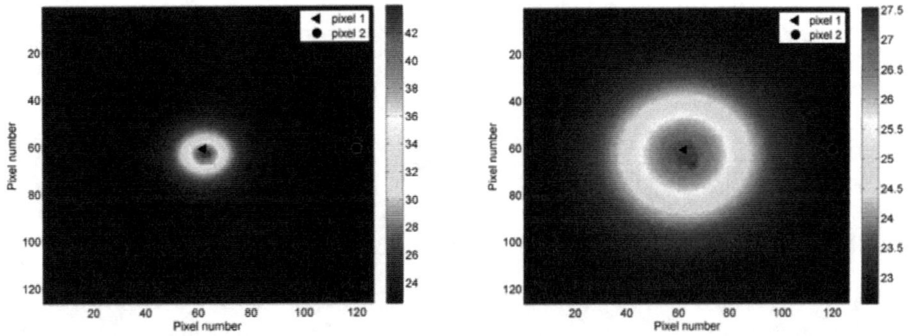

Figure 7. Temperature response field obtained with a step heating made with a laser spot of 200 micrometers of diameter on a glass plate of 0.2 mm thick, at two different times; (left) Image obtained at a short time ($t=0.8s$ after the beginning of the heating); (right) Image obtained at a longer time ($t=1.6s$). The pixel size is 10μm.

One illustration of a possible processing is to study the linear relation between Laplacian and time derivative at each pixel of the image (see figure 8). Such derivatives are obtained from finite differences versus space or time of the temperature information related to the pixels regularly spaced versus time and space.

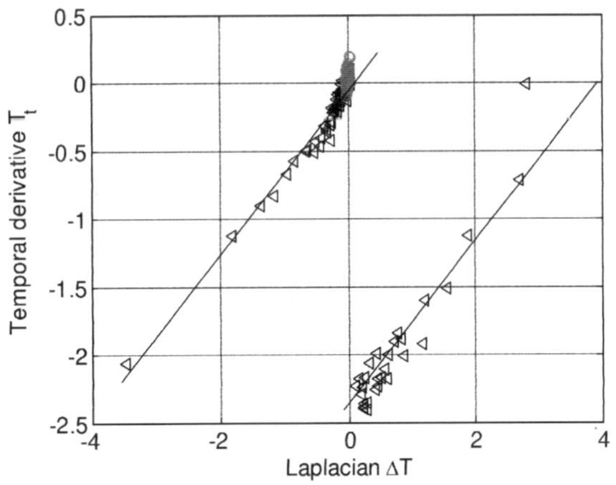

Figure 8. Time derivative versus Laplacian of temperature; a- for an irrelevant pixel out of the heating zone (in red), b-for a pixel inside the heating zone (in black).

On figure 8, the straight line with origin ordinate is corresponding to pixel where a constant heating occurs. The straight line passing through the origin is corresponding to a case of pure diffusion (when the temperature field is relaxing after the heating). The source term can then be estimated versus space independently of the transport property (the thermal diffusivity: a). A criterion suitable to detect such zones is to consider the local correlation between the laplacian ΔT and the time derivative $\partial T = {}^{\partial T}/_{\partial t}$:

$$\rho_{i,j}^{FF_t} = \frac{\sum_{F_t} \Delta T_{i,j}^k \partial T_{i,j}^k}{\sqrt{\sum_{F_t} \Delta T_{i,j}^{k\,2}} \sqrt{\sum_{F_t} \partial T_{i,j}^{k\,2}}}$$

where F_t is a temporal window such as: $F_t = [k, k+lt] with\ k \in [1, N-lt]$, k is the time step number and lt the width of the time window. If such a coefficient is near from 1, a diffusivity parameter is then estimable. If the correlation coefficient is near from zero the pixel or the area is irrelevant and must be discarded.

This example is demonstrating that the great amount of thermographic data coupled with computer facilities is offering large possibilities to estimate not only one source term but also a regularly spaced distribution of sources. It offers then the possibility to design thousands of simultaneous independent experiments. It is then also possible to extend the experimental configurations to microchannels.

The idea which consists in implementing a slow in-plane transfer in a conductive plate can be also used to estimate the convective transport effect due to a flow in a micro channel (of diameter from 10 to 100 micrometres) stuck to a thermally conductive plate. The local laplacian of the temperature field is compared to the local temperature gradient versus the direction of the channel in order to estimate a non dimensional velocity field (Peclet Field). The details of such processing are described in [19]. From these considerations, the microfluidic applications have been largely improved. It is now possible to estimate the enthalpy of chemical reactions in a lot of channel configurations and micro or millifluidic devices (see [18] to [22]). A special attention was paid to droplet flows and the related analysis of the heat release of such complex systems (see [23]). More recently, a complementary step was devoted to the detection of the transmission of wide-band multispectral infrared sources in such chemical devices with the processing related in section *3.2*. There is now a challenge to combine heat transfer and chemical analysis is order to estimate simultaneously thermophysical and thermochemical parameters related to coupled heat and mass transport in such devices.

4. Conclusion

The time-space thermal signal is offering a great amount of data which must be processed from several points of views. Several strategies are available in order to process the signal in relation with a physical model. Often it will consist in projecting the signal in a function basis which can come from intuition, statistical processing or physical analysis. This projection is advantageous in order to filter the random noise, to reduce the great amount of data and to conveniently manipulate and estimate the parameters. But the projection is not always the most suitable strategy. The direct consideration of a physical model will allow to eliminate or discriminate the data correlated or not with a chosen physical phenomenon (one example has been evocated with the estimation of a thermal diffusivity field).

Such considerations were also a common thread in order to present the IR thermography activities of our team at Bordeaux University.

The perspective of these works is now clearly devoted to the exploration of multispectral instrumentation with IR cameras. Especially, the spectral derivative of temperature or luminance fields will be a new direction for the signal processing of an amount of data always bigger with 4 dimensions to be considered.

References

[1] Thikonov A and Arsenine V 1977 *Solutions for ill posed problems* (Washington: Winston & Sons)

[2] Rajic N 2002 Principal component thermography for flaw contrast enhancement and flaw depth characterisation in composite structures *Composite Structures* **58** 521-28

[3] Bamford M, Batsale J-C and Fudym O 2009 Nodal and Modal Strategies for longitudinal Thermal Diffusivity Profile Estimation. Application to the non-destructive Evaluation of SiC/SiC composites under uniaxial tensile tests *Infrared Physics and technology* **52** 1-13

[4] Bamford M and Batsale J-C 2008 Analytical singular value decomposition of infrared image sequences : microcrack detection on ceramic composites under mechanical stresses *Comptes Rendus - Mecanique* **336** 440–47

[5] Bamford M, Florian M, Vignoles GL, Batsale J-C, Cairo CAA and Maillé L 2009 Global and local characterization of the thermal diffusivities of sicf/sic composites with infrared thermography and flash method *Composites Science and Technology* **69** 1131–41

[6] Balageas D, Delpech P, Boscher D and Deom A 1991 New developments in stimulated infrared thermography applied to non destructive evaluation of laminates *Review on Progress in Quantitative Non-Destructive Testing*, Eds. Thompson and Chimienti (New York: Plenum Press) **10 A** 1073-81

[7] Parker WJ, Jenkins W and Abott J 1961 Flash method of determining thermal diffusivity, Heat capacity and thermal conductivity *Journal Appl. Phys.* **32** 1679-84

[8] Mourand D and Batsale J-C 2001 Sequential method for thermal diffusivities discrimination by infrared thermography *High Temperatures - High Pressures* **33** 127-34

[9] Shepard SM, Hou YL, Ahmed T and Lhota JR 2006 Reference-free interpretation of Flash Thermography Data *Insight* **48** 298-307

[10] El Yagoubi J, Lamon J, Batsale J-C, Dhote J and Le Flem M 2015 Multiscale Thermal charactérization of Méchanically loaded Ceramic Matric Composites, Estimation of the Diffusivity at Micro/Mésoscale *Experimental Mechanics* **55** 783-94

[11] Pradere C, Caumes JP, Salort S, Balageas D, Chassagne B and Batsale J-C 2010 Photothermal converters for quantitative 2D and 3D real-time terahertz imaging *Qunatitative infrared Thermography* **7** 217-35

[12] Romano M, Ndiaye C, Duphil A, Sommier A, Morikawa J, Mascetti J, Batsale J-C, Servant L and Pradere C 2015 Fast infrared imaging spectroscopy technique, (FIIST) *Infrared Physics and Technology* **68** 152-8

[13] Avayzyan V 2012 Etude de champs de temperature séparables avec une double decomposition en valeurs singulières Quelques applications à la caracterisation de proprieties thermophysiques de matériaux et au contrôle Non-Destructif *Doctorate Thesis, Université de Bordeaux (France)*

[14] Maillet D, André S, Batsale J-C, Degiovanni A and Moyne C 2000 *Thermal Quadrupoles – Solving the Heat Equation through Integral transforms* (Chichester: John Wiley & Sons)

[15] Jensen KF 2001 Microreaction engineering - is small better? Chemical Engineering Science **56** 293-303

[16] Möllmann KP, Lutz N and Vollmer M 2004 Thermography of microsystems *Inframation Proceedings*, ITC 104 A 2004 07-27

[17] Senkan S, Krantz K, Ozturk S, Zengin V and Onal I 1999 High-throughput testing of heterogeneous catalyst libraries using array microreactors and mass spectrometry *Angewandte Chemie- International Edition* **38** 2794-99

[18] Pradere C, Morikawa J, Toutain J, Batsale J-C, Hayakawa E and Hashimoto T 2009 Microscale thermography of freezing biological cells in view of cryopreservation *QIRT Journal* **6** 37–61

[19] Pradere C, Joanicot M, Batsale J-C, Toutain J and Gourdon C 2006 Processing of temperature

Journal of Physics: Conference Series **655** (2015) 012002 doi:10.1088/1742-6596/655/1/012002

field in chemical microreactors with infrared thermography *QIRT Journal* **3** 117-35

[20] Ravey C, Pradere C, Regnier N and Batsale J-C 2012 New temperature field processing from IR camera for velocity, thermal diffusivity and calorimetric non-intrusive measurements in microfluidics systems *QIRT Journal* **9** 79-98

[21] Hany C, Pradere J, Toutain J, Batsale J-C and Joanicot M 2008 A millifluidic calorimeter without contact for reaction enthalpy and kinetics measurements *Proc. QIRT 2008 (Krakow, Poland, 02-05 July 2008)*

[22] Hany C, Pradere J, Toutain J and Batsale J-C 2008 A millifluidic calorimeter with infrared thermography for the measurement of chemical reaction enthalpy and kinetics *QIRT Journal* **5** 211-29

[23] Romano M, Pradere C, Sarrazin F, Toutain J and Batsale J-C 2015 Enthalpy, kinetics and mixing characterization in droplet flow millifluidic device by infrared thermography *Chemical Engineering Journal* **273** 325-32

On sustainable and efficient design of ground-source heat pump systems

W Grassi[1], P Conti, E Schito, D Testi
BETTER (Building Energy Technique and Technology Research group), DESTEC (Department of Energy, Systems, Territory and Constructions Engineering), University of Pisa, Largo Lucio Lazzarino – 56122 Pisa, Italy

E-mail: walter.grassi@unipi.it

Abstract. This paper is mainly aimed at stressing some fundamental features of the GSHP design and is based on a broad research we are performing at the University of Pisa. In particular, we focus the discussion on an environmentally sustainable approach, based on performance optimization during the entire operational life. The proposed methodology aims at investigating design and management strategies to find the optimal level of exploitation of the ground source and refer to other technical means to cover the remaining energy requirements and modulate the power peaks. The method is holistic, considering the system as a whole, rather than focusing only on some components, usually considered as the most important ones. Each subsystem is modeled and coupled to the others in a full set of equations, which is used within an optimization routine to reproduce the operative performances of the overall GSHP system. As a matter of fact, the recommended methodology is a 4-in-1 activity, including sizing of components, lifecycle performance evaluation, optimization process, and feasibility analysis. The paper reviews also some previous works concerning possible applications of the proposed methodology. In conclusion, we describe undergoing research activities and objectives of future works.

1. Introduction

Heat pump (HP) systems are a widely used technology for thermal energy generation in buildings, capable of efficiently supplying heating, cooling, and sanitary hot water. Particularly, ground-source heat pumps (GSHPs) are potentially able to reach higher performances with respect to their traditional alternatives, albeit special attention must be paid to the design of the overall system (heat pump equipment, ground heat exchangers (GHEs), connecting ductwork, and end-user loop). A general layout of a heat pump installation consists of the following main elements: A) external source; B) piping connecting the external source with the external heat pump heat exchanger (i.e. *ground-coupled loop*); C) ground-coupled heat pump (GHP); D) piping connecting the internal heat pump heat exchanger with the internal heating (or cooling) components (radiators, fan coils, …); E) building. Each element is, in turn, made of a number of components interplaying with one another, and directly interacts with the closer elements. This chain of strictly connected components has to be carefully analyzed as a whole. It is absolutely inappropriate to give prominence only to some specific apparatus.

On the contrary, current design methodologies (see, for instance, [1–4]) are typically based on a sequential and hierarchical logic. The first step concerns the analysis of building needs to identify a

[1] To whom any correspondence should be addressed.

reference thermal power (typically the peak load [3,5]) and select the heat pump capacity. The second step deals with the characterization of the ground source in terms of thermo-physical properties and undisturbed temperature level (i.e. thermal response test or pumping test). Finally, ground heat exchangers (GHEs) are sized to match the thermal requirements of the ground-coupled heat pump unit. In this perspective, each subsystem aims to meet the demands of the previous one. The latter considerations, together with traditional engineering "precautionary principle", often lead to oversized systems, with related energetic and economic disadvantages.

Moreover, as also stressed in [6], most of traditional design standards calculate the number and the size of ground heat exchangers on the basis of several design parameters decided *a priori* (e.g., operative temperatures and flow rates, generators capacity, heating/cooling load share assumed by the ground source, reference design period), without analyzing and simulating the operational life of the global system. Therefore, once the system is operational, the interactions among GSHP subsystems may result in unexpected coupling effects, possibly decreasing the overall operative performance. As a matter of fact, the full design process is mainly based on the personal evaluations of the designer(s) [6,7] and a high level of experience is needed to obtain a sound project. However, there is no guarantee that the final design is the most cost-effective.

Our considerations seem to be confirmed by experimentally monitored seasonal performance factors (SPF): several works have shown that actual SPFs of real GSHP systems vary significantly (from 1 to 6) also for very similar configurations [8–11]. Consequently, despite their theoretical potential in energy and economic savings, high installation costs and the uncertainty on final performances limit GSHPs attractiveness with respect to alternative technologies.

In these circumstances, the establishment of innovative design and management approaches seems necessary to increase the viability of GSHP technology. The installation design must be the product of the complete view of the building needs, the system for energy production (HP and back-up generators), the distribution system and controls, and the characteristics of the ground source. Both equipment size and control have to be optimized according to proper cost-benefit considerations between operative performances and installation costs.

Several works have already dealt with GSHPs simulation and optimization methods (see, for instance, [6,12–17]). However, these performance calculation methodologies have not yet been employed by professionals at the initial design level. It is worth stressing that, in this work, we consider an "optimal design" the one that identifies the best design solution among current market alternatives. We deal with optimal design parameters (e.g., GHEs number and depth, HPs number and capacity) and control variables (temperature set points, speed of compressor, pumps, and fans) with given equipment characteristics. In other words, we do not deal with any technological development.

Herein, the following subjects will be discussed: general behavior of the ground source with regard to the influence of climatic conditions, soil nature, and thermo-physical properties, extraction of energy and degradation on thermal evolution, GSHP equipment modeling, sizing of BHE field, sizing of GSHP unit and back-up generators, and share of the required thermal load between GSHP and back-up generators (control strategy) during the operational life. In order to clarify and exemplify the proposed methodological approach, we built a simple test case, in which we illustrate how to apply the method and obtain positive effects in terms of savings of primary energy and installation and operative costs with respect to traditional design practices. The interconnected effects of BHE sizing and control strategy will be exposed, as evidence of the need for an all-inclusive methodology

2. The design process as an optimization of operative performances

Design methodologies based on matching a reference thermal power demand under nominal conditions are not suggested for a ground-coupled system. As above-mentioned, the final goal of the design method is not to guarantee that a certain peak thermal power is delivered by the GSHP, but that the energetic and economic performances of the overall system (back-up generators included) are maximized with reasonable installation costs. A proper design is based on two main compromises:

- Depending on local ground source characteristics, GSHP operation alters the initial state of the heat source. In other words, the heat exchanges between soil and GHEs decrease/increase the source temperature, reducing the ground-coupled heat pump efficiency and risking several environmental issues (e.g., ground freezing or overheating).
- Installing large size components (e.g., large GHEs surfaces) limits the alteration of the heat source, as we are reducing the "density" of exploitation. However, this strategy increases investment costs without ensuring sufficient economic savings to repay the initial expenditure. This issue has prompted the development of the so-called *hybrid systems*, which share the thermal load among ground source and other back-up technologies in order to reduce size and cost of ground-coupled equipment [1,18].

Willing to highlight the thermodynamic bases of the above-mentioned trade-off, which affects GSHPs optimal design, we built a simple test case, discussed in the following section

2.1. Thermodynamic features of GSHPs operation: an illustrative analytical example

The existence of an optimal level of exploitation of the geothermal source, corresponding to the best synergy among GSHP unit and back-up generators, can be shown by means of a simple test case. In this section, we will investigate the electrical energy use of a vertical GCHP system, depending on the BHEs size and the share of the building thermal load due to the ground source (control strategy). For the sake of simplicity, we do not consider all the necessary elements of a real design, but we deal with a plain analytical model, in order to highlight the main thermodynamic mechanisms that determine a minimum value of energy consumption.

As already mentioned, GSHPs involve different subsystems, viz. ground reservoir, ground heat exchangers (i.e. vertical BHEs), ground-coupled loop and connecting ductwork, HP unit, back-up generators, and building end-user loop or destination thermal source (see Fig. 1). Each of them operates in strict connection with the others, creating a reciprocal influence on their own performance. Therefore, to apply the proposed design methodology, based on performance simulation, we have to employ a comprehensive set of equations, including at least the physical models of each element involved in the energy conversion process, namely: GSHP unit and back-ups, BHE field, and ground source.

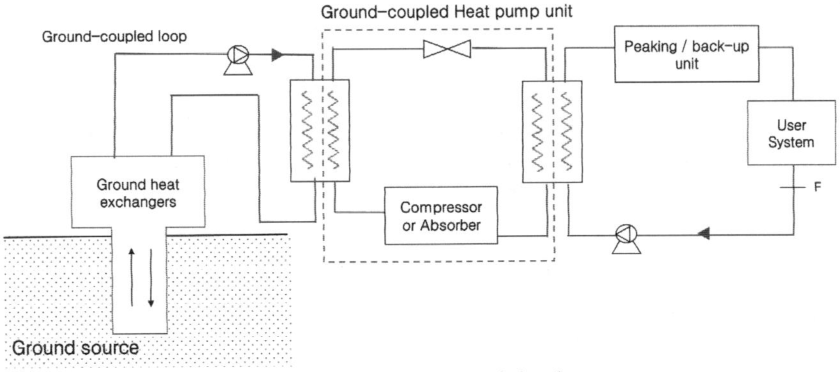

Figure 1. Scheme of the model subsystems.

In this example, we considered a heating system with a GSHP unit and an air heat pump (AHP) as back-up. Thermal performances of the two generators were calculated assuming a constant second-law efficiency value, η^{II} , for each generator (Eqs. 2.a and 2.b): in other words, we took into account only the effects of the different temperature evolution of the two sources. The ground temperature was evaluated by means of the infinite-line source model [19] and the time-superposition technique (i.e.

Duhamel's principle, Eq. 2.e). For sake of simplicity, both external air temperature and building load profiles were assumed sinusoidal (Eqs. 2.c and 2.d).

Similarly to [6], we used a coefficient p_l to represent the fraction of the building heating load delivered by the geothermal heat pump (Eq. 2.g). As above-mentioned, we aimed at analyzing the overall system performance depending on the different load share between air and ground systems: to do that, we performed a sensibility analysis of the overall system energy use (Eq. 1) at various p_l values. In this simple example, p_l was assumed as constant over the system lifetime, leaving a more accurate treatment of the problem to the next section. All the input parameters, including the thermo-physical properties of the soil, are reported in Table 1.

The total electrical energy use, E_{in} (Wh), of the two heat pumps after a period of time τ (yr) was calculated as:

$$E_{in} = \int_0^{\tau} L(t)\left(\frac{p_l}{COP_{GSHP}(t)} + \frac{1-p_l}{COP_{AHP}(t)}\right)dt \tag{1}$$

where:

$$\begin{cases} COP_{GSHP}(t) = \dfrac{T_l}{T_l - T_g(t)}\eta^{II}_{GSHP} & \text{(2.a)} \quad \text{(GSHP unit performance)} \\ COP_{AHP}(t) = \dfrac{T_l}{T_l - T_a(t)}\eta^{II}_{AHP} & \text{(2.b)} \quad \text{(Back-up performance)} \\ L(t) = \max\left[A_l \cos\left(2\pi/\omega \cdot t\right);0\right] & \text{(2.c)} \quad \text{(Building thermal load profile)} \\ T_a(t) = \langle T_a \rangle - A_a \cos\left(2\pi/\omega \cdot t\right) & \text{(2.d)} \quad \text{(External air temperature profile)} \\ T_g(t) = T_g^0 + \int_0^{\tau} W(t-\beta)\dfrac{d\dot{q}}{dt}(\beta)d\beta & \text{(2.e)} \quad \text{(Ground temperature evolution)} \\ W(t) = \dfrac{1}{2\pi\lambda_g}\displaystyle\int_{\frac{R_{BHE}}{2\sqrt{\alpha_g t}}}^{+\infty} \dfrac{e^{-\beta^2}}{\beta}d\beta & \text{(2.f)} \quad \text{(Infinite line source model)} \\ \dot{q}(t) = \dfrac{p_l L(t)}{N_{BHE} H}\left(\dfrac{COP_{GSHP}(t)-1}{COP_{GSHP}(t)}\right) & \text{(2.g)} \quad \text{(Energy balance of the BHE field)} \end{cases}$$

The set of Eqs. 1 and 2 was solved numerically adopting a time-step of 700 h. The latter value was investigated through a convergence analysis on final E_{in} value: it was seen that a shorter time step provides the very same results in spite of higher computational costs. We can conclude that, for the selected α_g and R_{BHE}, monthly-averaged values of simulation variables (e.g., T_g and COP_{GSHP}) do not differ significantly from an actual integration of the instantaneous values.

The total electric energy consumption after 20 years of operational life is shown in Fig. 2, as a function of p_l and BHEs number. In particular, we used a dimensionless efficiency parameter (ε) to normalize and compare the electric energy consumption of the different cases. In the present work, ε reads:

$$\varepsilon = \frac{E^*_{in}}{E_{in}} \tag{3}$$

where E_{in} is the actual electric energy use calculated through Eqs. 1 and 2 and E^*_{in} is the electric energy use of an ideal case, in which the only GSHP unit is used (i.e., $p_l = 1$) and the ground temperature remains always constant (i.e., N_{BHE} approaches infinity). In other words, ε evaluates the gap between the actual performance and the theoretical minimum energy consumption (i.e. $\varepsilon = 1$). This latter concept, also known as "task efficiency", has already been discussed and applied in a previous work [20].

Table 1. Input parameters applied to the set of Eqs. 1 and 2.

Parameter	Value	Unit
λ_g	1.7	W/(m-K)
α_g	0.68	mm²/s
R_{BHE}	7.5	cm
H	100	m
$\langle T_a \rangle$	15	°C
A_a	15	°C
$2\pi / \omega$	1	yr
A_l	20	kW
η_{GSHP}^{II}	0.55	-
η_{AHP}^{II}	0.45	-
τ	20	yr
T_l	45	°C
T_g^0	15	°C

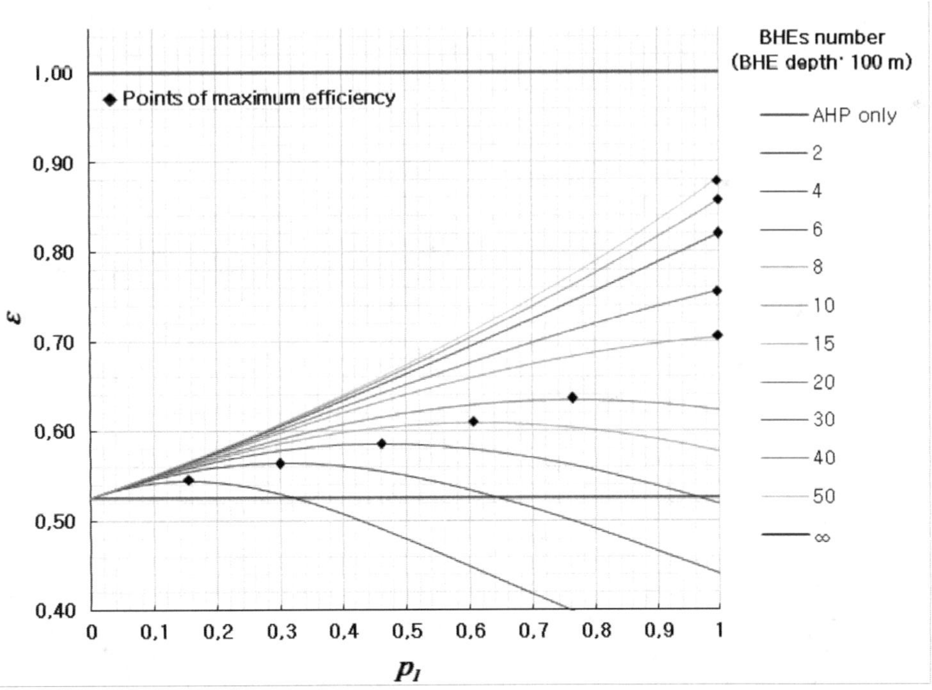

Figure 2. Dimensionless efficiency parameter (ε) as a function of the building load share provided by the GSHP unit p_l and BHEs number.

The results in Fig. 2 show how the ground source is not always convenient with respect to air. Nonetheless, an optimal share of the building load between the two sources can be found. For a given BHEs total depth, the minimum energy use is the result of an optimal compromise between two impairing effects:

- at high p_l, the soil temperature at the borehole surface decreases and can even become lower than air temperature (e.g., when p_l is greater than 0.65 and 4 BHEs are used);
- at low p_l, we are not fully exploiting the ground thermal storage.

We note also that points of maximum efficiency are quite insensitive to a small change in value, which makes the implementation of the control system easier. Regarding BHEs size, we can observe how maximum ε points monotonically increase with borehole number, as a consequence of a reduced alteration of the ground temperature; however, energy savings show a saturation trend, hinting that an oversized system is not going to be cost effective.

The conclusions of this small example confirm our previous considerations for a proper design of GSHP systems. In particular, energetic synergy between ground and back-up technologies can be optimized according to the local external climate, building thermal load, BHEs depth and soil thermo-physical properties. Besides, boreholes number and depth have to be chosen as the optimal tradeoff between savings in operative costs and installation investment.

In this simple test case, we employed a single coefficient η^{II} to take into account all the characteristics and the intrinsic efficiencies of the technological equipment: more complex models are needed to analyze the optimal design of real systems (see section 3). In particular, performance under partial loads strongly depends on the generator capacity, operational limits of components, and on the energy required for the auxiliary systems (e.g., pumping energy in the ground-coupled loop). Besides, significant improvements should be obtained by introducing a time-dependent control law for p_l within the optimization process. To achieve that, a comprehensive physical model of the overall system (including both thermodynamic and technological aspects) has been developed, together with the mathematical formulation of the optimal-design problem (see section 4) and a suitable solution strategy, described in section 5.

3. GSHP model definition and discussion

The previous considerations highlight how the GSHP design process is not a simple calculation of the size of every component, but it consists in a comprehensive procedure based on the evaluation of the overall performance of the system during its operational life and cost-benefit considerations. In other words, the final goal of the designer(s) should be aimed at identifying the best tradeoff among system performance and installation costs.

Current design methodologies do not provide an explicit focus on the optimal share of load delivered by the GSHP with respect to the total thermal need of the building; moreover, the criteria for selecting GHP capacity is not clearly established [5,6].

The optimization of generators capacity and GHEs dimension is a key step for increasing GSHP viability/diffusion. In particular, we are interested in those formulations aimed at investigating the operative performance of the project through the set-up of a physical-based model. This optimization approach is named *"simulation-based optimization procedure"*, as we are dealing with an optimization problem based on the simulation of the operative behavior of the analyzed system [21]. It follows that GSHP design can be seen as an optimization process, in which both design and control variables have to be concurrently evaluated according to their impact on final system performances. With the proposed approach, feasibility study, sizing process, performance analysis and design optimization converge into the very same activity.

As above-mentioned, GSHPs are made of several subsystems (see Fig. 1), with ground source, ground heat exchangers (GHEs) and ground-coupled heat pump unit(s) being the key ones. For each of them, we will discuss the main physical phenomena governing their energetic behavior and a tailored simulation model will be proposed. These expressions will be finally coupled to a full set of equations to simulate the operative performances of the overall GSHP system (section 3.4).

3.1. External source

The external source of a HP system may be air, surface water, deep water, or ground. The main appeal of the ground source is its rather constant temperature at a certain depth, which should ensure, at least in theory, higher performances with respect to the outdoor air. However, it is worth recalling that transient heat conduction is the main heat transfer mechanism between the ground and GHEs (especially in dry soils); consequently, at equal heat exchangers dimension, the system capacity is lower than air or water systems where the mechanism of heat transfer between the heat source and the heat pump heat exchangers is forced convection (typical thermal conductivity and diffusivity are generally low, varying within a range of 1 – 3 W/(m K) and 10^{-7} – 10^{-6} m^2/s, respectively [1–4]).

Once convection places a role, mass transfer is the main responsible of the heat exchange, jointly with the fluid thermo-physical properties as grouped in Prandtl and Reynolds numbers. The heat transfer coefficient is given by correlations of the form:

$$Nu = C\,\mathrm{Pr}^m\,\mathrm{Re}^n \tag{4}$$

Even if we suppose this correlation (same C, m, and n) to hold both for air and water, for a given heat exchanger geometry, at the same Reynolds number, we get:

$$\frac{Nu_{Air}}{Nu_{Water}} = \left(\frac{\mathrm{Pr}_{Air}}{\mathrm{Pr}_{Water}}\right)^m \tag{5}$$

with the ratio between the two Prandtl numbers roughly around 0.1. In addition, the type of heat exchanger used for air is quite different from that used for water (e.g., cross flow battery for air and plate for water). As very well known, water is a more suitable source than air for two main reasons: it is more effective in terms of heat transfer performances (Eq. 5) and its temperature is higher/lower, on an average basis, and much more constant, thus allowing for a better SCOP and SEER.

With regard to ground source, it is worth stressing that the characterization of the ground source can be divided into two different steps. The first one concerns the characterization of the initial/undisturbed state of the source medium in terms of temperature distribution (varying with time and depth); the second one deals with the evaluation of the ground thermal response when the coupled GSHP system operates.

3.1.1. Characterization of the ground source.

Soil temperature is determined both by the energy exchanges with the above air, which are affected by average outdoor air temperature, solar radiation and surface cover, evapo-transpiration of the soil (both connected to the soil moisture content and surface botany) and those from the surrounding ground [22]. Grossly, we can divide the area affecting the heat transfer to the geothermal exchanger into two main regions: the upper zone, mainly influenced by the climatic conditions above the ground level, and a lower one unaffected by them, but only by the surrounding ground thermo-physical properties.

The typical model to reproduce the underground temperature evolution in undisturbed conditions is the transient heat conduction in a semi-infinite solid. This temperature distribution versus time t and depth z can be also calculated, for instance, by using the model proposed by Labs [23]:

$$T_g(z,t) = \langle T_{air}\rangle - A_s \exp\left(-z\cdot\sqrt{\frac{\pi}{365\alpha_g}}\right)\cos\left[\frac{2\pi}{365}\left(t - t_0 - \frac{z}{2}\cdot\sqrt{\frac{365}{\pi\alpha_g}}\right)\right] \tag{6}$$

where α_g is the average annual (apparent) thermal diffusivity of undisturbed ground, m^2/s; t_0 is the phase of air temperature wave, days; $\langle T_{air} \rangle$ is the average annual air temperature, °C; z is the depth, m; A_s is the amplitude of annual average air temperature wave (i.e. $T(0,t) = A_s \cos\left[2\pi/365\left(t-t_0\right)\right]$), °C. [24,25] take also into account surface conditions through a proper coefficient k_v,which depends on surface vegetation and other correction constants. They adopt the following formula:

$$T_g(z,t) = \left(\langle T_{air} \rangle \pm \Delta T_m\right) - 1{,}07 \cdot k_v A_s \exp(-0{,}00031552 z a_g^{-0.5}) \cdot \cos\left[\frac{2\pi}{365}\left(t - t_0 - 0{,}018335 z a_g^{-0.5}\right)\right] \quad (7)$$

with the same meaning of symbols. According to Eqs. 6 and 7, it is possible to define a *thermal penetration depth* $d_p \approx \sqrt{2\alpha_g / \omega}$ as the depth at which the amplitude of the temperature oscillation, A_s, decreases of a factor equal to $1/e$. Typical d_p values vary from 1 to 10 meters for annual oscillations.

In addition, during the design process, effective thermo-physical properties of the ground have to be determined. For a preliminary feasibility study, one can use reference values or data from previous nearby projects; however, *in-situ* evaluation methods are generally suggested [1,2]. The main site-investigation technique for closed-loop systems is the thermal response test (TRT), which is able to determine the effective thermal conductivity and diffusivity of the soil, together with the so-called "*borehole thermal resistance*" (see section 3.2). Reference standards, procedures and recommendations for TRTs have been published in literature and engineering handbooks [1,2,26].

3.1.2. Evaluation of ground response during GSHP operation

A first conclusion of the previous discussion is that the effect of the external climate does not propagate below, say, 10 meters from the surface, depending on many factors like: nature of soil (thermo-physical properties), amount of rainfalls and solar radiation, type of soil cover surface, etc. Actually, [27] proved to exert some influence on the outlet temperature of the heat exchanging fluid, also for boreholes 55 m deep, however, since the typical BHEs depth is about 100 m, the temperature oscillation on the top of the borehole is assumed to be negligible on the global heat transfer process during the GSHPs operation [28,29].

Many works present methods for the evaluation of ground temperature evolution at the BHEs surface in a *purely conductive medium* [19,29,30] or in *saturated porous medium* [30–32]. For the sake of simplicity, let us refer to the former case, where Fourier number controls the entire process. According to [6,31], axial effects become significant at very long times (about 1 year); hence, in multi-year simulations, the finite line-source (FLS) [19] model seems the most appropriate one, at least among analytical formulations. The latter reads:

$$\Theta_g(x,t) = \frac{1}{4\pi}\int_0^1 \left[\frac{1}{d/L}\, erfc\left(\frac{d/L}{2\sqrt{Fo_H}}\right) - \frac{1}{d'/L}\, erfc\left(\frac{d'/L}{2\sqrt{Fo_H}}\right)\right] dH' \quad (8)$$

$$Fo_H = \frac{\alpha_g t}{H^2} \qquad R = \frac{r}{H} \qquad Z = \frac{z}{H} \qquad \Theta_g = \frac{\left(T_g - T_g^0\right)\lambda_g}{\dot{q}_{BHE}}$$

$$d/L = \sqrt{R^2 + (Z - H')^2} \qquad d'/L = \sqrt{R^2 + (Z + H')^2}$$

where H is the borehole depth, m; $\dot{q}_{BHE}$ is the linear heat flux at the BHE surface, W/m.

As heat transfer $\dot{q}_{BHE}$ is not stationary during the operational time, time superposition can be applied to calculate the actual temperature distribution. If we can assume that the radial dimension of the BHE is negligible compared to the size of the geothermal field, also space superposition can be applied to evaluate interferences among the BHEs.

3.2. Ground-coupled borehole heat exchangers model

We propose to analyze the boreholes through the classical heat exchanger theory (see, for instance, [33]). In such a way, we want to ensure that the heat transferred at the evaporator/condenser is coherent with the temperature variation of the fluid in the BHE field. This model can be easily coupled to another concept, widely used in the analysis of BHEs: the *borehole thermal resistance* (R_b) defined as [34]:

$$R_b = \frac{\bar{T}_w - T_g}{\dot{q}} \tag{9}$$

where:

- $\bar{T}_w$ is the average temperature of the fluid in the U-loops of the BHE (°C);
- $\dot{q}$ is the linear heat flow through the BHE perimeter in steady-state conditions (W/m).

Following this definition, R_b is a stationary parameter that relates the mean fluid temperature to the BHE average boundary temperature: dynamic and axial effects are thus neglected. Furthermore, R_b conventionally refers to the BHE depth and not to the length of the ducts. Reference formulas to evaluate R_b can be found in [34,35].

We propose to evaluate the fluid outlet temperature from the BHEs through the classical effectiveness method [33] for heat exchanger analysis. Eqs. 10.a and 10.b represent the energy balance and the heat transfer equation for a single BHE

$$\dot{m}_w c_w \left(T_{w,in} - T_{w,out}\right) = \dot{q}H = \dot{Q}_g \tag{10.a}$$

$$\dot{Q} = KA\left(\bar{T}_w - T_g\right) \tag{10.b}$$

Combining Eqs. 10.a and 10.b, we obtain:

$$\frac{H}{R_b} = KA \tag{11}$$

This simple relation illustrates the meaning of R_b within the heat exchanger theory. Again, we stress that the characteristic length is the BHE depth H_{BHE} and not the length of the ducts buried in the borehole. The arrangement and the number of pipes only affect the value of R_b.

Since the borehole surface temperature T_g is assumed to be uniform and constant during each time step, we can write the expression of heat transfer effectiveness, ε_{BHE}, as [33]:

$$\varepsilon_{BHE} = \frac{T_{w,in} - T_{w,out}}{T_{w,in} - T_{w,out}} = 1 - \exp(-NTU) \tag{12}$$

where $NTU = \dfrac{KA}{\dot{m}_w c_w}$ is named “Number of Transfer Units”. Other ε_{BHE} expressions for BHEs are described in [35]. Similar correlations for energy piles are currently undergoing a research activity through a regression analysis of the results of several transient FEM simulations, in order to identify the most relevant parameters, characteristic lengths, and time scales [36]. Once $T_{w,out}$ is obtained for every BHE, the overall energy transferred to the ground during the time step Δt is:

$$Q_g = \sum_{N_{BHE}} \dot{m}_w c_w \left(T_{w,in} - T_{w,out}\right)\Delta t \tag{13}$$

3.3. Ground-coupled heat pump unit model and heat generators

Different approaches have been proposed to simulate the heat pump performance as a function of the operative conditions. For design purposes, *black-box* based models seem to be the most effective ones, in terms of quality of result and implementation efforts. The overall performance of the heat pump unit is predicted by appropriate interpolation of manufacturers' data at the operative sources temperature and capacity ratio (CR) [7,12,37,38]. In this work, we refer to CR as the useful thermal output of the HP in heating or cooling mode, divided by its maximum capacity, when operating at the actual temperatures of the thermal source [39]. We stress that CR refers to the maximum thermal output of a HP unit at given source conditions; it should not be confused with the load ratio, which is the ratio between seasonal average thermal power demand and building peak load. The effect of CR on operative coefficients of performance depends on the choice of the HP unit size and on its modulation capability in response to the evolution of the thermal load.

3.4. The full set of equations describing the overall system

In this section, we illustrate the proposed full set of equations necessary for simulating the behavior of a vertical GCHP system during its operational life. The latter one collects the above-discussed modeling formulations and it will be employed in the optimization procedure illustrated in section 4.

Rigorous dynamic simulation methods would need dynamic models of each component and all parameters and boundary conditions should be available with sufficient accuracy at a much shorter time scale. Such a level of detail is often not available, especially at the earliest stages of a design process. Moreover, according to the aims of this work, we need to avoid complex formulations that would be impractical within the above-mentioned optimization procedure.

Consequentially, we decided to adopt a quasi-steady-state method, calculating energy balances over a sufficiently long time step, Δt (i.e. a month), which allows us to neglect internal energy variations (except for the ground) and employ simpler models. All the time-dependent inputs and unknown variables of the system have to be considered as constant for the entire duration of the time step, assuming average values. In section 2, we verified the applicability of this strategy: indeed, we observed that there is a negligible deviation between a monthly-average evaluation and the actual ground temperature evolution.

The choice of a monthly time scale seems appropriate also for BHEs modeling. As above-mentioned, evaluating the heat transfer process within the boreholes through a stationary equivalent thermal resistance (i.e. R_b) is a widely accepted technique (see, for instance, [6,12,34]), as BHEs heat capacity can be neglected after a sufficiently long period of continuous operation [40,41]).

The two coefficients, f_H and f_C, were introduced to represent the control strategy of the system. They are defined as the heat delivered/removed to/from the end-user loop divided by the total energy load during a given time step. In other words, selecting a $f_{H/C}$ value corresponds to change the thermal load at the ground source. Such a control can be achieved, for instance, by varying the CR value of the heat pump unit through its capacity control system. Although their similarity, we stress that f_H and f_C do not represent the integral-average of the instantaneous share of the building thermal load delivered by the ground heat pump (i.e. p_l in Eqs. 2).

The full set of equations reads:

$Q_{eva/cond} = Q_g$	[14.a]	
$Q_{eva/cond} = \dot{m}_w c_w \left\| T_{w,in} - T_{w,out} \right\| \Delta t$	[14.b]	Energy balance for the fluid of the ground-loop at the evaporator/condenser section
$Q_{eva/cond} = F\left(T_{w,out}; T_l; CR\right)$	[14.c]	GHP unit performance
$Q_g = H\left(\dot{m}_w; c_w; T_{w,in}; T_g; R_b; R_{BHE}; H; N_{BHE}\right)$	[14.d]	BHE field
$T_g = S\left(T_g^0; Fo_g; Pe_g; Q_g; R_{BHE}; H; N_{BHE}; D_{BHE}\right)$	[14.e]	Ground source
$f_H = \frac{Q_{eva}}{Q_l}\left(\frac{COP}{COP-1}\right) \quad f_C = \frac{Q_{cond}}{Q_l}\left(\frac{EER}{EER+1}\right)$	[14.f]	Shares of the building load for heating and cooling at GSHP unit
$Q_{bk} = B(T_l; CR_{bk})$	[14.g]	Back-up generator(s)
$Q_{bk} = (1 - f_{H/C})Q_l$	[14.h]	

Eqs. 14 include both equipment models (Eqs. 14.b-14.g) and two auxiliary equations (Eq. 14.a and Eq. 4.h). The former imposes that heat the exchanged between the BHE field and the ground (Q_g) is equal to the heat transferred in the evaporator/condenser ($Q_{eva/cond}$), in accordance with the quasi-steady-state method); the latter imposes that the building thermal load (Q_l), up to the end-user distribution system, is given by the sum of the thermal energies delivered/removed in heating/cooling mode by GSHP unit and back-up generators.

For the sake of generality, the equations are voluntarily presented in an implicit form, but possible expressions for functions F, H, S and B have been provided in sections 3.1, 3.2 and 3.3. Anyway, any proper formulation can be used accordingly to the specific design project (e.g., Eqs. 2).

4. Statement of the optimal-design problem

As already mentioned, we propose to simulate GSHP systems by means of a quasi-steady-state approach; therefore, our goal is to minimize the finite series of a proper "*return/cost function*" over the operative time under consideration. The decisions about the management of the system (f_H and f_C values) have to be made at each stage of the simulation (i.e. at each time step). This kind of optimization problems are typically called *multistage decision problems* [42]. The mathematical formulation of the problem can be stated as:

$$\textit{find the control sequence } \mathbf{U} = \left\{\mathbf{u}^0, \mathbf{u}^1, \mathbf{u}^2, \mathbf{u}^3 \ldots \mathbf{u}^N\right\}$$

that minimizes:

$$J(\mathbf{U}) = \sum_{n=1}^{N} R_{H/C}\left(\mathbf{x}^n, \mathbf{u}^n, n\Delta t\right) \tag{15.a}$$

subject to:

$$\mathbf{x}^{n+1} = f\left(\mathbf{x}^n, \mathbf{u}^n, n\Delta t\right) \tag{15.b}$$

$$\mathbf{h}\left(\mathbf{x}^n\right) = 0 \qquad \mathbf{g}\left(\mathbf{x}^n\right) \leq 0 \tag{15.c}$$

$$u_{p,\min} \leq u_p \leq u_{p,\max} \tag{15.d}$$

where:

- $\mathbf{U}=\{\mathbf{u}^0,\mathbf{u}^1,\mathbf{u}^2,\mathbf{u}^3\ldots\mathbf{u}^N\}$ is the set containing all the $\mathbf{u}^n$;
- $\mathbf{u}^0=\{u_1^0,u_2^0,u_3^0\ldots u_k^0\}$ is the vector of design variables; the superscript *0* indicates that these variables are not time-dependent;
- $\mathbf{u}^n=\{u_1^n,u_2^n,u_3^n\ldots u_k^n\}$ is the vector of control variables at the *n-th* stage;
- $\mathbf{x}^n=\{x_1^n,x_2^n,x_3^n\ldots x_p^n\}$ is the vector of state variables at the *n-th* stage;
- $J(\mathbf{U})$ is the so-called performance index:
- $R_{H/C}(\mathbf{x}^n,\mathbf{u}^n,n\Delta t)$ is the so-called *"return/cost function"*; it represents the contribution of the *n-th* stage to the total performance index;
- $f(\mathbf{x}^n,\mathbf{u}^n,n\Delta t)$ is the mathematical model of the system, relating the state variables of a stage to the control variables and the state variables of the previous stage (set of Eqs. 14);
- $\mathbf{h}(\mathbf{x}^n)$ is the vector of the equality constraints;
- $\mathbf{g}(\mathbf{x}^n)$ is the vector of the inequality constraints.

4.1. Return function

The objective function can be chosen both in thermodynamic, energetic, economical, and thermo-economical terms; multi-objective formulations can be used, too. If we are interested in primary energy uses, we can express $R_{H/C}(\mathbf{x}^n,\mathbf{u}^n,n\Delta t)$ as:

$$R_H(\mathbf{x}^n,\mathbf{u}^n,n\Delta t)=\frac{Q_{cond}}{COP_{GSHP}}f_{ep,GSHP}+\frac{Q_{bk}}{\eta_{bk}}f_{ep,bk}\quad\text{(heating mode)}\tag{16.a}$$

$$R_C(\mathbf{x}^n,\mathbf{u}^n,n\Delta t)=\frac{Q_{eva}}{EER_{GSHP}}f_{ep,GSHP}+\frac{Q_{bk}}{EER_{bk}}f_{ep,bk}\quad\text{(cooling mode)}\tag{16.b}$$

For an economically-driven optimization, the return function could be the net present value (NPV), defined as the net cash flow at a given time step, resulting from the algebraic sum of present values of: incentives, savings on operational costs with respect to benchmark technologies, installation costs, and maintenance costs. The operational costs of the energy production in heating (C_H) and cooling (C_C) modes are given by:

$$C_H=\frac{Q_{cond}}{COP_{GSHP}}c_{e,GHP}+\frac{Q_{bk}}{\eta_{bk}}c_{e,bk}\quad\text{(heating mode)}\tag{17.a}$$

$$C_C=\frac{Q_{eva}}{EER_{GSHP}}c_{e,GSHP}+\frac{Q_{bk}}{EER_{bk}}c_{e,bk}\quad\text{(cooling mode)}\tag{17.b}$$

where $c_{e,GSHP}$ and $c_{e,bk}$ are the unitary costs (possibly undergoing inflation) of the energy needed for running, respectively, GSHP unit and back-up generators. As a matter of fact, the optimal sequence of $\mathbf{u}^n$ depends on the choice of the specific energetic or economic goal; this is particularly true when the ratio between $f_{ep,GSHP}$ and $f_{ep,bk}$ is different from the one between $c_{e,GSHP}$ and $c_{e,bk}$.

Journal of Physics: Conference Series **655** (2015) 012003 doi:10.1088/1742-6596/655/1/012003

4.2. Control variables

The overall optimization problem involves two types of control variables: the variables related to the sizing of the system (design variables) and the ones related to the control strategy. The main control strategy variable is the capacity ratio (CR) of the heat pump unit. At each time step, we can change the thermal output of the GSHP unit and, consequently, the share of the building load delivered by the GSHP system $(f_{H/C})$. At some stages, we can also decide to turn off the heat pump and match the thermal load only with the back-up generators. Hence, the optimization solution will also include the schedule of the HP operation.

The design variables $(\mathbf{u}^0)$ are also "*control variables*" in the terminology common to optimization theory, but they are not time dependent. This notwithstanding, they have to be evaluated within the optimization process. In addition, the design variables can be divided into two sub-groups: the *continuous variables* and the *discrete variables* (also called *design parameters*). The former can assume every value within the allowed range, the latter are instead limited to certain discrete values, such as integer numbers; therefore, different and specific optimization techniques have to be applied. Besides, the possible GSHP system layouts and the size of the components have to be considered as *discrete parameters* for the optimization process, even if they cannot be associated to a numerical variable. The main variables, parameters and constraints for the optimization of GSHP systems are summarized in Table 2.

Table 2. List of variables considered in the proposed optimization procedure.

	Symbol	Name and comments	Lower and upper bounds
Control-strategy variables			
	CR	Capacity Ratio	*Min:* Minimum value specified by manufacturer *Max:* 1
Design variables			
Continuous variables	H	Borehole depth	*Min:* 0 *Max:* 100-150 m (suggested values)
	$\dot{m}_w$	Mass flow rate of ground-coupled loop	*Min:* Hydraulic considerations: e.g. the flow rate must be large enough to guarantee a fluid velocity higher than 0.3 m/s or a turbulent regime within the ducts (suggested values).
Discrete variable or parameters	N_{BHE}	Number of BHEs	Integer value
	$F(\ )$	Function describing HP capacity and performance (Eq. 14.c)	
	$H(\ )$	Function describing BHEs design (Eq. 14.d)	
	$B(\ )$	Function describing back-ups capacity and performance (Eq. 14.g)	

The proposed optimization methodology starts creating a set of possible design alternatives (indicated in Fig. 3 as "*conf*"). As said in section 4.2, the system layout is considered as a discrete variable, thus all the reasonable combinations in terms of BHEs number, generators models and equipment arrangement are included in this "*conf*" set. The latter is investigated through an "*exhaustive enumeration method*"; in other words, we evaluate the $J(\mathbf{U})$ value of each configuration to find the optimal one. In this context, a physical insight of the problem can significantly reduce the number of tests, saving computational time and efforts.

We investigate continuous design variables and control strategy for each *"conf"* element by an iterative procedure:

- at first, design variables are set to an initial set of guesses and the corresponding control strategy is optimized;
- then, using the previously-found control strategy, design variables are optimized;
- finally, the procedure is iterated, till a convergence criterion is satisfied.

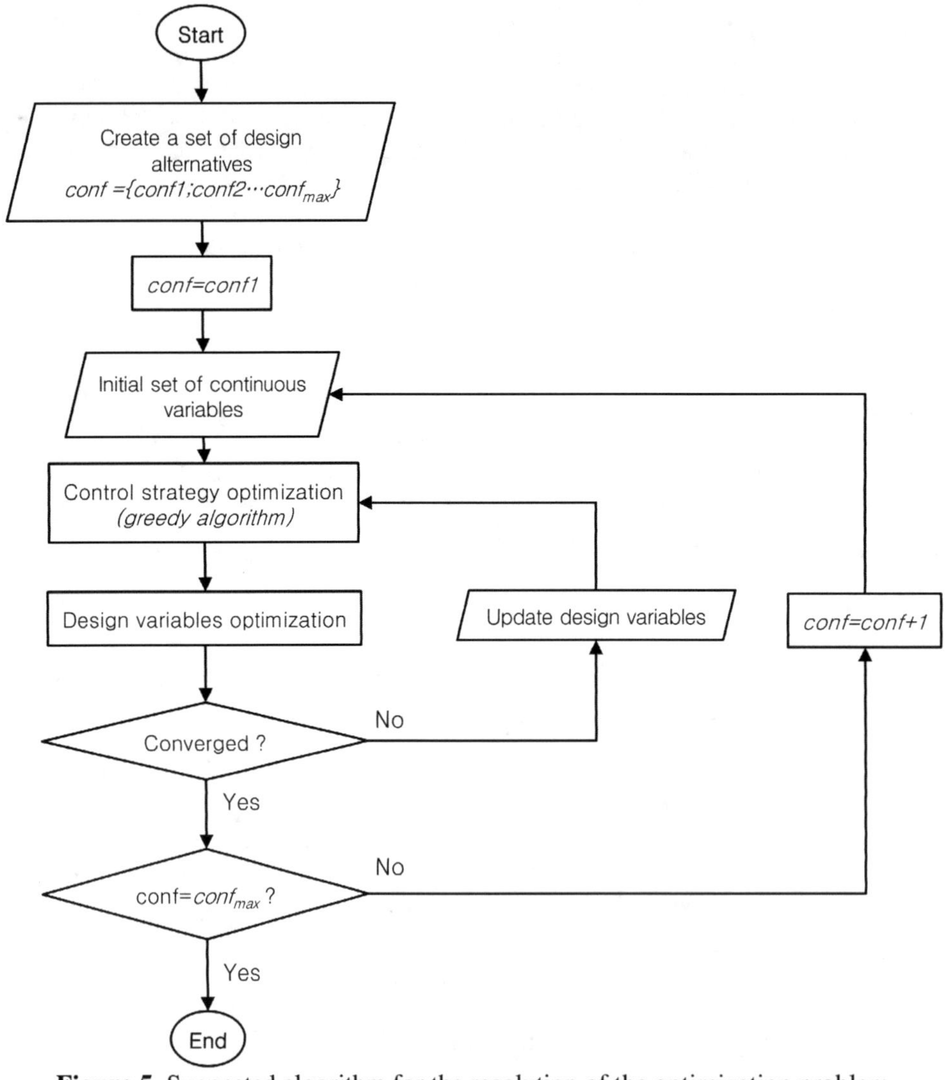

Figure 5. Suggested algorithm for the resolution of the optimization problem.

The optimal control in multistage decision problems is generally determined by means of dynamic programming (DP) techniques (i.e. backward induction). This latter, when applicable, reduces the dimension of the solution space, decomposing the original problem in a sequence of N single decision problems [42]. However, in ground source applications, DP cannot be applied, as both next state vector, $\mathbf{x}^{n+1}$, and return function, $R\left(\mathbf{x}^n,\mathbf{u}^n,n\Delta t\right)$, depend on all previous steps. Indeed, as above said, the value of the ground temperature is the result of the entire history of heat exchange. The other common strategy is the employment of the so-called evolutionary algorithms (i.e. genetic algorithms), however, the large dimension of the solution space (corresponding to the sum of time steps and design variables number) makes their application impractical and time-consuming.

A *greedy algorithm* is a rapid and straightforward alternative: this resolution strategy finds the optimal choice at each individual time step "*...in the hope that this choice will lead to a globally optimal solution*" [43]. The effectiveness and the limits of this algorithm in GSHP control problems will be discussed in future works, though promising research is still ongoing on the subject.

5. Conclusions and examples of application

In the present work, we proposed an innovative approach to the design of GSHPs, based on energetic/economic optimization on the entire lifecycle of the system. Traditional design methods determine the size of the equipment on the basis of some critical reference conditions; on the contrary, we propose to investigate the possible design alternatives in terms of operative performance. In section 2, we analyzed the main thermodynamics features that affect an optimal design of GSHP systems.

We showed that increasing BHEs number does not always entail a significant increase of system performances: therefore, system oversizing should be avoided. Besides, optimal ground exploitation corresponds to a proper share of the delivered heat among GSHP and back-up system. This optimal share depends on the efficiencies of the involved generation technologies and, for GSHPs, to an optimal thermal level of the ground source.

We propose to evaluate both design and control variables of real GSHPs systems by means of an optimization algorithm, coupled to a comprehensive set of equations representing each subsystem of GSHPs (back-up generators included). The main outputs of the method are: optimal capacity of the ground heat pump unit, optimal number and depth of BHEs, and optimal control strategy.

The proposed set of equations can be promptly adapted to different technological solutions, maintaining the overall structure of the method (e.g., GWHPs, energy piles, solar-assisted HP systems). Besides, the overall approach can be applied in several contexts with different aims: designers can be supported during their professional activity, political authorities can investigate proper assessments and criteria for specific incentives, energy efficiency operators (e.g., energy service companies) can evaluate the investment profitability of any specific GSHP project, industrial operators (e.g., drilling companies and GHP manufacturers) can analyze maximum operative savings to properly decide drilling and equipment prices, researchers can investigate current GSHP systems to seek technological developments and room for improvement. In short, possible applications concern numerous professional, political, economic, and research activities.

For instance, we employed the proposed design methodology in [17], to deal with a typical professional design case: both energetic and economic benefits with respect to traditional design methods were fully illustrated and discussed. In [44], we analyzed a real case study (a primary school located in the north of Italy characterized by high heating loads), to investigate the achievable benefits of possible design alternatives (GS-GAHP and condensing boiler) and different control strategies (lower supply temperature to the end-user loop). In [16], we investigated the optimal BHEs number and the investment profitability as a function of BHEs drilling fees: we showed how an accurate evaluation of maximal GSHPs performances, covering a proper set of benchmark buildings and loads, may help authorities to assess amounts and access criteria for financial incentives, encouraging GSHPs diffusion, but avoiding market distortions or speculations. Finally, in [20], we illustrated as the proposed methodology can be applied to investigate the technological room for improvement of GSHP technology: in other words, we figured out the subsystem on which technological development should be focused, the expected benefits and some hints about a possible strategy for research activities.

Future works will be aimed at improving simulation models and methods, together with optimization algorithms. Besides, we are currently investigating the sensibility of GSHP performances on a wide range of technological, economic, and environmental factors (e.g., generators technology and efficiencies, thermo-physical properties of the ground, groundwater effects, external air temperature, GHEs number and size, ground-coupled loop design and arrangement, thermal load profile, control strategies, installation costs, and energy prices), in order to investigate the possibility of building reference performance maps relating the mentioned parameters. The final goal is to develop some simplified and straightforward guidelines, in order to extend the practical applicability of the proposed methodology to GSHP operators and stakeholders.

Nomenclature

Symbols and Acronyms

A	Amplitude of a periodic variable
BHE	Borehole heat exchanger
c_e	Energy price per unit (€/Wh)
C	Operational costs (€)
CR	Capacity ratio
D	Dimensionless BHEs distance
d_p	Penetration depth (m)
E_{in}	Energy use (Wh)
E_{in}^*	Energy use of an ideal case, in which the only GSHP unit is used and the ground temperature remains always constant (Wh)
f_{ep}	Primary energy factor
$f_{H/C}$	GSHP share of building load in heating/cooling mode
Fo	Fourier number
GHP	Ground-coupled heat pump unit
H	Borehole depth (m)
$J(\mathbf{U})$	Performance index
K	Overall heat transfer coefficient (W/m²/K)
k_v	Surface vegetation coefficient (Eq. 7)
L	Building thermal load profile (W)
$\dot{m}$	Mass flow rate (kg/s)
N_{BHE}	Number of boreholes
Nu	Nusselt number
Pe	Péclet number
Pr	Prandtl number
p	Fraction of the instantaneous building load delivered by the geothermal heat pump
$\dot{q}$	Heat flow per unit length (W/m)
Q	Thermal energy (Wh)
$\dot{Q}$	Thermal power (W)
R	Dimensionless radius
$R_{H/C}$	Return function in heating or cooling period
R_b	Borehole thermal resistance (m K/W)
R_{BHE}	Borehole radius (m)
Re	Reynolds number
r	Radial coordinate (m)
t_0	Phase of a periodic variable (h)
$\mathbf{U}$	Set containing all the u^n
W	Infinite line source model (Eq. 2.f)
$\mathbf{x}^n$	Vector of state variables
Z	Dimensionless depth
z	Depth (m)

Greek Letters

α	Thermal diffusivity (m²/s)
β	Auxiliary variable
ε	Task efficiency (Eq. 3)
ε_{BHE}	Heat exchanger effectiveness
η	First-law efficiency
η^{II}	Second-law efficiency
Θ	Dimensionless temperature
λ	Thermal conductivity (W/m/K)
τ	Reference operational period (h)
ψ	Exergy efficiency
ω	Angular frequency

Subscripts

AHP	Air HP unit
bk	Back-up generator
C	Cooling mode
$cond$	Condenser
eva	Evaporator
H	Heating mode
g	Ground
in	Inlet/supply
l	Building thermal load
out	Outlet/return
w	Water circulating in the ground-coupled loop

Superscripts

0	Initial time
$*$	Ideal conditions

References

[1] ASHRAE 2011 *ASHRAE Handbook - HVAC Applications* 31.1-34.34
[2] GEOTRAINET 2011 *Geotrainet training manual for designers of shallow geothermal systems*
[3] CEN 2007 EN 15450
[4] UNI 2012 UNI 1466
[5] Alavy M, Nguyen H V., Leong W H and Dworkin S B 2013 *Renew. Energy* **57** 404–12
[6] Robert F and Gosselin L 2013 *Appl. Therm. Eng.* **61** 481–91
[7] Pardo N, Montero Á, Sala A, Martos J and Urchueguía J F 2011 *Appl. Therm. Eng.* **31** 391–8
[8] Bakirci K 2010 *Energy* **35** 3088–96
[9] Hwang Y, Lee J-K, Jeong Y-M, Koo K-M, Lee D-H, Kim I-K, Jin S-W and Kim S H 2009 *Renew. Energy* **34** 578–82
[10] Michopoulos a., Zachariadis T and Kyriakis N 2013 *Energy* **51** 349–57
[11] Karabacak R, Güven Acar Ş, Kumsar H, Gökgöz A, Kaya M and Tülek Y 2011 *Int. J. Refrig.* **34** 454–65
[12] Nagano K, Katsura T and Takeda S 2006 *Appl. Therm. Eng.* **26** 1578–92
[13] Arteconi A, Brandoni C, Rossi G and Polonara F 2013 *Int. J. Energy Res.* **37** 1971–80
[14] Li F, Zheng G and Tian Z 2013 *Energy Build.* **58** 27–36
[15]S Retkowski W and Thöming J 2014 *Appl. Energy* **114** 492–503
[16] Conti P, Grassi W and Testi D 2013 Proposal of a Holistic Design Procedure for Ground Source Heat Pump Systems *European Geothermal Congress 2013* (Pisa, IT)
[17] Conti P, Grassi W and Testi D 2015 Proposal of Technical Guidelines for Optimal Design of Ground-Source Heat Pump Systems *World Geothermal Congress 2015* (Melbourne, AU)
[18] Preene M and Powrie W 2009 *Géotechnique* **59** 261–71
[19] Ingersoll L R, Zobel O J and Ingersoll A C 1954 *Heat conduction with engineering, geological and other applications* (New York: McGraw-Hill)
[20] Casarosa C, Conti P, Franco A, Grassi W and Testi D 2014 *J. Phys. Conf. Ser.* **547** 011001
[21] Nguyen A-T, Reiter S and Rigo P 2014 *Appl. Energy* **113** 1043–58
[22] Banks D, Scott H and Ogata C 2012 *Acque Sotter.* **130** 9–18
[23] Labs K 1982 *Undergr. Sp.* **7** 37–65
[24] Popiel C ., Wojtkowiak J and Biernacka B 2001 *Exp. Therm. Fluid Sci.* **25** 301–9
[25] Baggs S A 1983 *Sol. Energy* **30** 351–66
[26] Raymond J, Therrien R, Gosselin L and Lefebvre R 2011 *Ground Water* **49** 932–45
[27] Zarrella A and Pasquier P 2015 *Appl. Therm. Eng.* **78** 591–604
[28] Eskilson P 1987 *Thermal analysis of heat extraction boreholes* (University of Lund (S))
[29] Philippe M, Bernier M and Marchio D 2009 *Geothermics* **38** 407–13
[30] Carslaw H S and Jeager J C 1959 *Conduction of heat in solids* ed C Press (Clarendon Press)
[31] Molina-Giraldo N, Blum P, Zhu K, Bayer P and Fang Z 2011 *Int. J. Therm. Sci.* **50** 2506–13
[32] Sutton M G, Nutter D W and Couvillion R J 2003 *J. Energy Resour. Technol.* **125** 183
[33] Lavine A S, DeWitt D P, Bergman T L and Incropera F P 2011 *Fundamentals of Heat and Mass Transfer* (Hoboken (NJ): John Wiley & Sons, Inc.)
[34] Lamarche L, Kajl S and Beauchamp B 2010 *Geothermics* **39** 187–200
[35] Conti P 2015 *Sustainable design of ground-source heat pump systems: optimization of operative life performances,* PhD Thesis (University of Pisa)
[36] Batini N, Rotta Loria A F, Conti P, Testi D, Grassi W and Laloui L 2015 *Appl. Therm. Eng.* **86** 199–213
[37] CEN 2008 EN 15316-4-2
[38] UNI 2012 UNI/TS 11300-4
[39] CEN 2012 UNI EN 14825
[40] Bauer D, Heidemann W, Müller-Steinhagen H and Diersch H-J G 2011 *Int. J. Energy Res.* **35** 312–20

[41] Ruiz-Calvo F, De Rosa M, Acuña J, Corberán J M and Montagud C 2015 *Appl. Energy* **140** 210–23

[42] Rao S S 1996 *Engineering Optimization: Theory and Practice* (Hoboken (NJ): John Wiley & Sons, Inc.)

[43] Cormen T H, Leiserson C E, Rivest R L and Stein C 2009 *Introduction to Algorithms* (Cambridge (MA): The MIT press)

[44] Ghisleni M, Pellegrini G, Conti P, Testi D and Grassi W 2015 Analysis of monitoring data and simulation of seasonal energy performance of the GS-GAHP system installed in the kindergarten building of Oulx, Turin *L'impiantistica per i climi estremi: tecnologie per i nuovi mercati della climatizzazione* (Padova (IT))

Fundamentals and recent advances in X-ray micro computed tomography (microCT) applied on thermal-fluid dynamics and multiphase flows

Maurizio Santini

Department of Engineering and Applied Sciences, University of Bergamo, Italy

Email: maurizio.santini@unibg.it

Abstract. X-ray computed tomography (CT) is a well-known technique nowadays, since its first practical application by Sir. G. Hounsfield (Nobel price for medicine 1979) has continually benefited from optimising improvements, especially in medical applications. Indeed, also application of CT in various engineering research fields provides fundamental informations on a wide range of applications, considering that the technique is not destructive, allowing 3D visualization without perturbation of the analysed material. Nowadays, it is technologically possible to design and realize an equipment that achieve a micrometric resolution and even improve the sensibility in revealing differences in materials having very radiotransparency, allowing i.e. to distinguish between different fluids (with different density) or states of matter (like with two-phase flows). At the University of Bergamo, a prototype of an X-ray microCT system was developed since 2008, so being fully operative from 2012, with specific customizations for investigations in thermal-fluid dynamics and multiphase flow researches. A technical session held at the UIT International Conference in L'Aquila (Italy), at which this paper is referring, has presented some microCT fundamentals, to allow the audience to gain basics to follow the "fil-rouge" that links all the instrumentation developments, till the recent applications. Hereinafter are reported some applications currently developed at Bergamo University at the X-ray computed micro-tomography laboratory.

1. X-ray computed micro-tomography (microCT)

The pervasive use of NDT (non-destructive technologies) in research and industrial applications are currently in their best development opportunities, especially the high resolution X-ray applications. A comprehensive recent review by T.J. Heindel [1] reports the applications on flow visualization and characterization of multiphase flows.

At the *Department of Engineering and Applied Science* of the University of Bergamo, in 2012, began a new research focused on the investigation of interfaces and multiphase systems at the microscale. A first prototype of an X-ray computed micro-tomography (microCT) was designed and built that overcomes the limitations of the very few currently available semi-commercial devices, generally unsuitable for laboratory researches, with low density materials and at micrometric scale.

Technically, the microCT is a sophisticated equipment that allows to obtain the digitized three-dimensional scan of an irradiated sample. In fact, it can be described as a non-destructive 3D microscope, whose spatial resolution is a few micrometers (i.e. consider that a human hair is circa 100 microns and a human red blood cell is almost 8 microns). The currently available prototype is fully operational at the laboratories of the University of Bergamo, and it is rather unique nationwide and competes with the few synchrotrons equipped in the world. The microCT setup is used, but not limited at, in research in the subject area of Technical Physics; its particularities have allowed to promote publications of scientific interest (here, in part, cited) and activates contracts for industrial research and technology transfer that have actually co-financed the construction. The research fields are multidisciplinary, so the technical session held at the UIT International Conference in L'Aquila (Italy), at which this paper is referring, reports only some of the fields of application among many others that are under development.

2. MicroCT applied to diagnostics of cavitation in fuel injection systems

Cavitation plays an important role in fuel injector systems as it alters the injector's internal flow structure, discharge coefficient and can contribute to injector damage and component wear. Increasing injection pressure and decreasing nozzle hole sizes and injection duration cause an acceleration of the fuel as it travels from the sac through the restricting orifice and can lead to rapid drop of the static pressure and as a consequence to local cavitation in the entrance of the nozzle hole. Cavitation has been extensively studied using transparent large scale [2] and real size injectors [3, 4]. Flows with cavitation can be extremely three-dimensional and requires experimental approaches capable of measuring such structures. Visible light is scattered at the vapor-liquid interface and so that X-ray diagnostics have been used over the past few years to obtain quantitative information of the void fraction [5, 6]. Radiography and fluorescence diagnostics are reported and provide time-averaged vapor distributions from raster-scanning approach. X-ray phase contrast technique allows higher temporal resolution in respect to both previous techniques but is still showing a projected two dimensional cavitation images. CT measurements are presented by Bauer et al. [6] using a clinical scanner and reported three-dimensional time-averaged void-fraction distribution in quasi-steady pipe flow with the limited resolution of clinical scanners. At the University of Bergamo, in collaboration with the International Institute for cavitation Research (ICR) for Prof. M. Gavaises CITY University London (UK), the microCT facilty is applied to cavitation diagnostics can deliver three-dimensional (time-averaged) information with noteworthy higher resolution. Figure 1 shows the time-averaged vapor distribution in high-resolution.

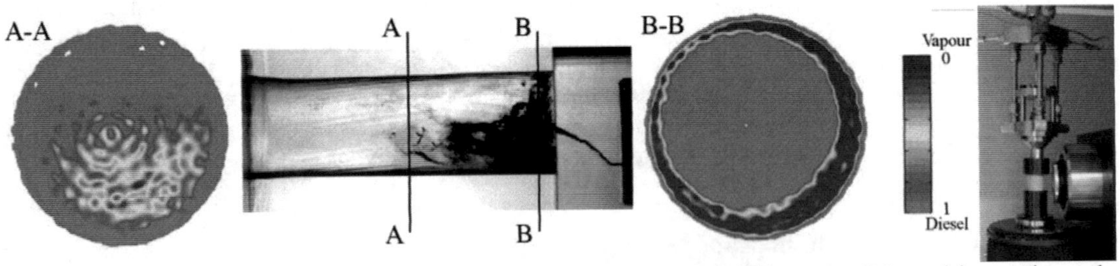

Figure 1. Vapor distribution within an Diesel injector at two different axial positions; the color represents the vapor fraction; blue corresponds to 100% vapor and red to 100% Diesel (liquid).

3. MicroCT applied to inspection and measurements of fuel injectors

MicroCT allows three-dimensionally profiling of the inner of complex injector parts and comparing this profile with nominal CAD geometries with a resolution of few microns. By doing this, CAD overlays tolerance bands could be generated and/or CAD overlays can be aligned with the inspected part to allow best fit analysis. Figure 2 show such comparison between nominal CAD drawings and fabricated new (not used) injector. Producer companies of microCT scanners often commercially offer such metrology analysis, and literature data are sparsely regarding fabrication tolerances of fuel injectors.

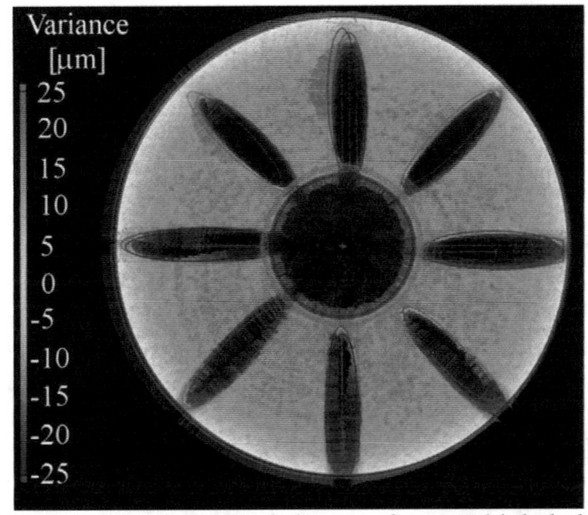

Figure 2. MicroCT volume of a multiple-hole injector with nominal CAD geometry (red lines) best-fitted on sac and resulting tolerance band.

Instead, several investigations on deposit formation and erosion in fuel injectors can be found due to two motivations: (a) continued legislative pressure to reduce exhaust emissions and as a consequence the development of advanced injection systems using higher temperatures and pressures at the injector tip, where deposit formation and erosion is initiated; and (b) the use of renewable and fuels with additives which are capable of providing good engine performance in short term engine operation but during long term operation degradation of engine performance, excessive carbon and lacquer deposits and actual damage to the engine were observed [7]. The formation of deposition and erosion within the holes of the injector affects the injection pattern and fuel flow rate and may have an adverse effect on the overall system performance [8]. Inspections of deposition were done with scanning electron microscopy, field emission scanning electron microscopy and energy dispersive X-ray spectroscopy [7, 9]. MicroCT allows a full three-dimensional inspection of such deposition layer and separation of this layer from the nozzle material is possible due to their differences in X-rays attenuation (see figure 3). Such a carbon layer after segmentation is shown figure 4 for the first time.

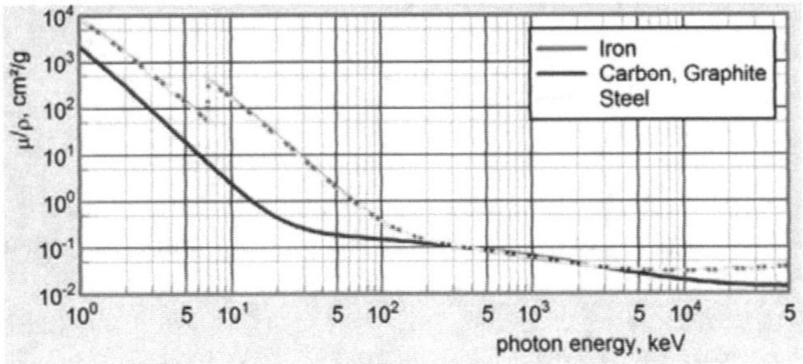

Figure 3. X-ray attenuation profile for carbon, steel and iron, a suitable energy range allows to reveal fouled surfaces.

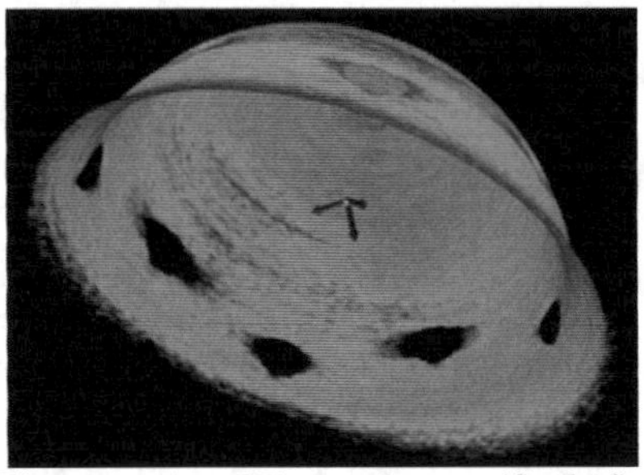

Figure 4. 3D carbon (fouled) deposition layer of injector tip after segmentation

4. MicroCT applied to the estimation of phase transition rate of arbitrary shaped droplets

The experimental investigation on fluid evaporation or sublimation are generally complex, even more in highly deformed droplets, nevertheless available models needs to be improved in some extent. In general, the evaporation rate is a key parameter directly linked to the feasibility to measure the droplet volume variations (eventually in 3D) under non-stationary boundary conditions. The effect of particle deformation on phase transition was investigated by [10, 11]. These previous studies concluded that the transition rates (per unit area) are higher for deformed particles. The dependency of transition rate from the local curvature (an arbitrary shaped particle) can be studied starting from microCT investigations. 3D volume variation due to sublimation is presented here for the first time. Figure 5 shows the local change of the radius of a camphor sphere in 24 hours for different side views. These first results demonstrating microCT as a very promising technique for estimating phase change rates with adequate 3D accuracy and high sensitivity and accuracy. MicroCT allows studying any shaped geometry. Further, this technique can even used with imposed and oriented external convective flux, laminar or turbulent.

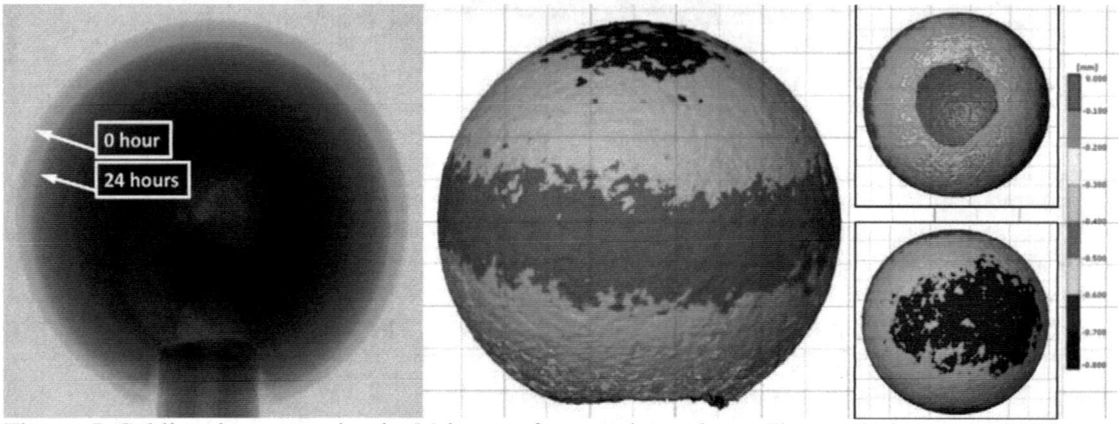

Figure 5. Sublimation regression in 24 hours of a camphor sphere. The resolution corresponds to a voxel of 7 microns, on the left 2 overlapped radiography, on the right the 3D contour plot of local ΔR.

5. MicroCT applied on wetting behaviour

Wetting of ideal surfaces is characterized by the equilibrium contact angle given by the well-known Young Equation [12]. Instead, wetting of rough surfaces is characterized by an apparent contact angle (APCA), which is defined as an equilibrium contact angle measured macroscopically [13], since the detailed microscopic topography cannot be viewed using classical optical methods. Therefore, APCA is the angle between the tangent to the liquid-vapor interface and the apparent solid surface. Two important models for describing the wettability behavior of rough surface are commonly used to explain the effect of roughness on the APCA: the Wenzel [14] model for homogeneous and the Cassie-Baxter [15] model for heterogeneous wetting. Further wetting states are reported by [16-19].

The APCA is governed by the wetting regime occurring in the vicinity of the triple line [14]; hence, visualization of this area is crucial and classical optical methods cannot be applied. Several advanced techniques were used for visualization the triple line such as reflection interference contrast microscopy [20], laser scanning confocal microscopy [21] or microCT [22-25] as shown in figure 6 and figure 7.

MicroCT allows analyzing the wetting behavior in terms of drop shape, wetting state, real and apparent wetted area; both enabling to estimate the Wenzel ratio, real and apparent contact angles, and the amount of drop liquid penetrating in texture fibers [22-25]. There, the drop and the substrate volumes can be extracted from the total scanned volume by means of volume segmentation on the basis of the X-ray attenuation of the different substances. The analysis of the drop and surface contact area gives the information about the contact angle applying of the axisymmetric drop shape analysis (ADSA) [26]. With respect to the commonly used ADSA; using microCT the drop/surface side pictures are replaced with real slices of the drop and substrate volumes in the chosen direction and at the chosen location [22-25]. Thus local information about both the real and the apparent contact angle along the triple line can be obtained, overcoming the limitation of the conventional optical approach based on side projections. For further discussion of applying microCT to wetting analysis, it is referred to [22,25].

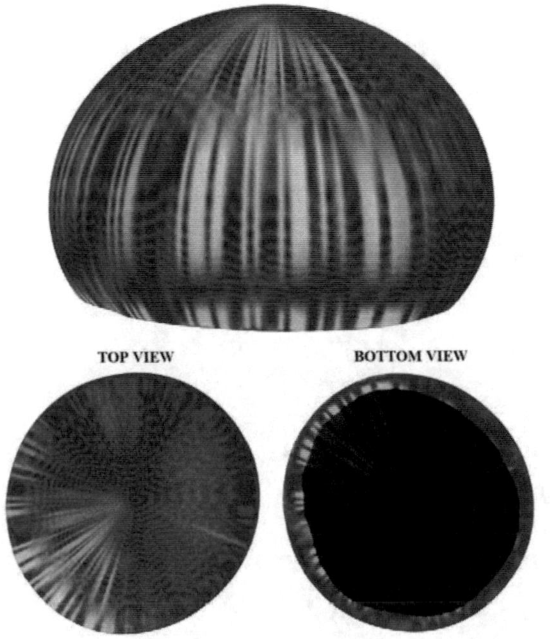

Figure 6. 3D orthographic, top and bottom view of the microCT droplet after the reconstruction based on the Laplace-Young fitting of 240 half slices of the droplet. Color is proportional to the dimensionless error $\Delta R/R_{max}$ between the Laplace-Young fitted surface and the original surface (maximum error 3.85%) [22].

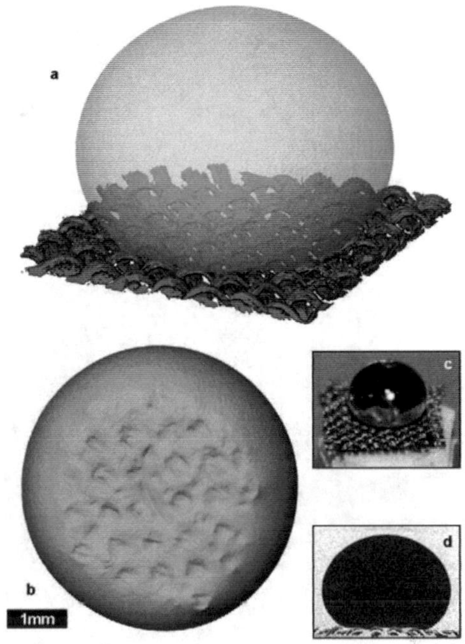

Figure 7. MicroCT of complex artificial substrate and a sessile water drop: segmented and rendered volume [23].

6. MicroCT applied to biofilm morphologies

Power output limitation is one of the main concerns that need to be addressed for full-scale applications of the microbial fuel cell technology. Fouling and biofilm growth on the electrodes in single chamber microbial fuel cells (MCF), the most likely design of microbial fuel cells, affects their performance in long-term operation with wastewater. These performances losses are caused by the direct exposure of the electrodes to the solution enriched with the inorganic/organic substrate and bacteria which induces the growth of an electroactive biofilm [27-30].

Previously reported techniques for biofilm structural characterization include optical sectioning, fourier transform infrared spectroscopy analysis, and other microscopy methods, where the most challenging problem is to determine morphological features of thick biofilms (i.e., having a thickness of a few millimeters). Scanning electron microscopy (SEM) requires sample pretreatment, increases chances of artifacts and morphological distortions [31, 32]. The developed environmental SEM allows the imaging of the biofilm also in wet conditions reducing before mentioned problems [33]. But, this technique is quite complex and requires a dedicated instrumentation with a special detector. Confocal laser scanning microscopy (CLSM) imaging is arguably the most common nondestructive method in use for imaging biofilms [31, 32]. However, CLSM cannot be used to image inorganic fouling layers such as the carbonate deposits commonly observed in MFC electrodes, as they are impenetrable to photons. Recently, microCT was successfully used for volumetric investigation of MFC electrodes without the above described limitations of other established techniques in this sector for the first time [34]. MicroCT was used to analyze both the surface and the deep layers of biofilms and inorganic fouling in their native state, without any pretreatment.

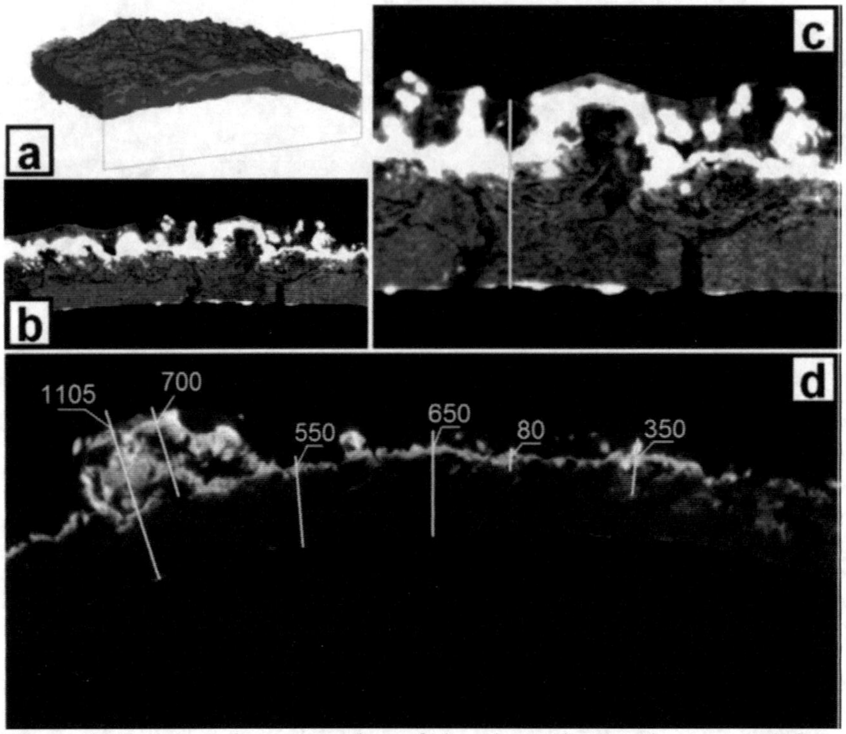

Figure 8. Orthogonal cross-section of the electrode sample extracted from the reconstructed microCT volume and post-processed to extract estimated thickness information (dimensions in micrometer); in (c) the line is equal to 650 μm [34].

In figure 8, such analysis is shown reporting also quantitative measure of distance (and volume) available due to proper calibration. Unveil details are reported like surface defects, density of the fouling layers, and the interface between biological and inorganic fouling, Finally, being a nondestructive method, microCT can be used *in-situ*, thus allowing experiments at different time on the same sample.

7. MicroCT applied to biological tissues: human heart valve disease visualizations in foetuses

The most frequent causes of fetal death are represented by chromosomal or genetic abnormalities, structural congenital malformations, and other vascular abnormalities. Currently in the prenatal period, the gold standard for clinical evaluation of fetal abnormalities is represented by ultrasound-based diagnoses. Numerous studies have shown that a nuchal scan during the first trimester screening performed by experienced operators can identify of major structural and cardiac abnormalities [35-39]. However in case of fetal death or miscarriage, the prenatal ultrasound information is unavailable in the majority of cases. Here, the traditional autopsy techniques provide only partial clinical responses, especially before 16 weeks of gestation due to the small size of fetal organs. Consequently in recent years, scientific research has focused on the study of alternative methods of investigation traditional postmortem autopsy, such as the use of magnetic resonance imaging (MRI) and post-mortem computed tomography (CT), which allow performing an virtual non-destructive autopsy and obtaining as result high resolution images of fetal organs in their entirety. The main limitations of these methods are represented by limited resolution. Recently, Lombardi et al. [40] presented a feasibility study on the use of microCT using a commercial system and this technique - also with quite low resolution - was able to more fully and accurately the presence and type of anatomic abnormality in comparison with the mentioned established methods.

Recently, the Author was authorized to start large post-mortem study in cooperation with Policlinico Milano, Clinica Mangiagalli on human foetuses with low gestational age (first and second trimester) affected by severe cardiac pathologies. The study is aimed to create the first 3D anatomical atlas for human heart foetuses for different gestational ages. Figure 9 shows the feasibility test on a murine heart (provided by Istituto di ricerche farmacologiche "Mario Negri", Dr. A. Remuzzi).

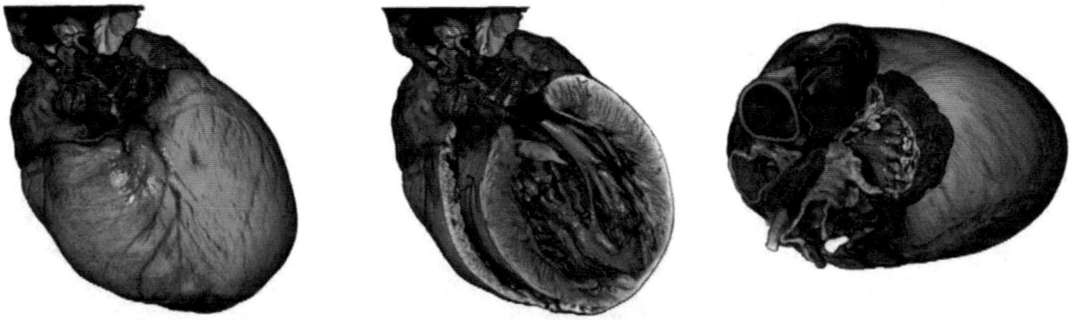

Figure 9. Visualization of a millimetric heart size from a reconstructed microCT volume, voxel 7 microns.

Acknowledgments
This research was partially supported by the projects: ITALY® 2013 "Microtomografia a raggi X: strumento interdisciplinare per le scienze" (University of Bergamo) and the Regional Project "Fluids/porous X-Ray μ-tomography", within the frame of Regione Lombardia call for Technological and Scientific Cooperation Agreements in cooperation with the Institute of Aerospace Thermodynamics (ITLR), Prof. B. Weigand University of Stuttgart.
Cavitation researches are done in cooperation with the International Institute for cavitation Research (ICR), Prof. M. Gavaises City University London (UK), partially financed by the EU project FP7 Marie Curie Actions and grant number IOF 300410.
The Author would like to thank M.Sc. M. Lorenzi for his precious help for tomographic acquisitions, Prof. M. Guilizzoni for his unconditioned support on wettability analysis and Prof. G.E. Cossali for his contributions on several ideas and researches.

References

[1] Heindle T J 2011 A review of X-ray flow visualization with apllication to multiphase flows *J. Fluids Eng.* **133** 074001

[2] Gavaises M, Andriotis A, Papoulias D, Mitroglou N and Theodorakakos A 2009 Characterization of string cavitation in large-scale Diesel nozzles with tapered holes *Phys. fluids* **21** 052107

[3] Blessing M, König G, Krüger C, Michels U and Schwarz V 2003 Analysis of Flow and Cavitation Phenomena in Diesel Injection Nozzles and Its Effects on Spray and Mixture Formation *SAE Technical Paper* 2003-01-1358

[4] Watanabe H, Nishikori M, Hayashi T, Suzuki M, Kakehashi N and Ikemoto M 2015 Visualization analysis of relationship between vortex flow and cavitation behavior in diesel nozzle *Int. J. Engine Res.* **16** 5-12

[5] Duke D Swantek A, Kastengren A, Fezzaa K and Powell C 2015 Recent developments in X-ray diagnostics for cavitation *SAE Technical Paper* 2015-01-0918

[6] Bauer D, Chaves H and Arcoumanis C 2012 Measurements of void fraction distribution in cavitating pipe flow X-ray CT *Meas. Sci. Technol.* **23** 055302

[7] Liaquat A M, Masjuki H H, Kalam M A and Rizwanul Fattah I M 2014 Impact of biodiesel blend on injecto deposit formation *Energy* **72** 813-23

[8] Birgel A, Ladommatos N, Aleiferis P, Zülch S, Milovanovic N, Lafon V, Orlovic A, Lacey P and Richards P 2008 Deposit Formation in the Holes of Diesel Injector Nozzles: A Critical Review *SAE Technical Paper* 2008-01-2383

[9] Hagos F Y, Aziz A R A and Sulaiman S A 2015 Investigation of deposit formation in direct-injection spark-ignition engine powered on syngas *Int. J. Automot. Techn.* **16** 479-85

[10] Mashayek F 2001 Dynamics of evaporating drops. Part I: formulation and evaporation model *Int. J. Heat Mass Tran.* **44** 1517-26

[11] Jeng S M and Deng Z T 1996 Numerical simulation of deformed droplet dynamics and evaporation *Recent advances in spray combustion: Spray combustion measurements and model simulation* **2** 305-28

[12] Young T 1805 An essay on the cohesion of liquids *Phil. Trans.* **95** 65–87

[13] Marmur A 2009 A guide to the equilibrium contact angles maze *Contact Angle Wettability and Adhesion* **6** 3–18.

[14] Wenzel R N 1936 Resistance of solid surfaces to wetting by water *Ind. Eng. Chem.* **28** 988–94

[15] Cassie A B D and Baxter S 1944 Wettability of porous surfaces *Trans. Faraday Soc.* **40** 546–51

[16] Miwa M, Nakajima A, Fujishima A, Hashimoto K and Watanabe T 2000 Effects of the surface roughness on sliding angles of water droplets on superhydrophobic surfaces *Langmuir* **16** 5754–60

[17] Marmur A 2003 Wetting on hydrophobic rough surfaces: to be heterogeneous or not to be? *Langmuir* **19** 8343–8

[18] Bico J, Thiele U and Quéré D 2002 Wetting of textured surfaces *Colloids Surf. A* **206** 41–6

[19] Ishino C, Okumura K and Quéré D 2004 Wetting transitions on rough surfaces *Europhys. Lett.* **68** 419–25

[20] Moulinet S and Bartolo D 2007 Life and death of a fakir droplet: impalement transitions on superhydrophobic surfaces *Eur. Phys. J. E* **24** 251–60.

[21] Papadopoulos P, Mammen L, Deng X, Vollmer D and Butt H J 2013 How superhydrophobicity breaks down *Proc. Natl Acad. Sci. USA* **110** 3254–8

[22] Santini M, Guilizzoni M and Fest-Santini S 2013 X-Ray computed microtomography for drop shape analysis and contact angle measurements *J. Colloid Interface Sci.* **409** 204-10

[23] Santini M and Guilizzoni M 2014 3D X-ray micro computed tomography on multiphase drop interfaces: from biomimetic to functional applications *Colloids Interface Sci. Commun.* **1** 14-7

[24] Guilizzoni M, Santini M, Lorenzi M, Knisel V and Fest-Santini S 2014 Micro computed tomography and CFD simulation of drop deposition on gas diffusion layers *J. Phys.: Conf. Ser.* **547** 012028

[25] Santini M, Guilizzoni M, Fest-Santini S and Lorenzi M 2015 A novel technique for investigation of complete and partial anisotropic wetting on structured surface by X-ray microtomography *Rev. Sci. Instrum.* **86** 023708

[26] del Rio O I and Neumann A W 1997 Axisymmetric Drop Shape Analysis: Computational Methods for the Measurement of Interfacial Properties from the Shape and Dimensions of Pendant and Sessile Drops *J. Colloid Interface Sci.* **196** 136-47

[27] Cristiani P, Franzetti A, Gandolfi I, Guerrini E and Bestetti G 2013 Bacterial DGGE fingerprints of biofilms on electrodes of membraneless microbial fuel cells *Int. Biodeterior*

Biodegrad. **84** 211-9

[28] Cristiani P, Carvalho M L, Guerrini E, Daghio M, Santoro C and Li B 2013 Cathodic and anodic biofilms in Single Chamber Microbial Fuel Cells *Bioelectrochem.* **92** 6-13

[29] Ishii S, Suzuki S, Norden-Krichmar T M, Phan T, Wanger G, Nealson K H, Sekiguchi Y, Gorby Y A and Bretschger O 2014 Microbial population and functional dynamics associated with surface potential and carbon metabolism *ISME J.* **8** 963-78

[30] Ishii S, Suzuki S, Norden-Krichmar T M, Wu A, Yamanaka Y, Nealson K H and Bretschger O 2013 Identifying the microbial communities and operational conditions for optimized wastewater treatment in microbial fuel cells *Water Res.* **47** 7120-30

[31] Resat H, Renslow R S and Beyenal H 2014 Reconstruction of biofilm images: combining local and global structural parameters *Biofouling* **30** 1141-54

[32] Beyenal H, Lewandowski Z and Harkin G 2004 Quantifying biofilm structure: Facts and fiction *Biofouling* **20** 1-23

[33] Biffinger J C, Pietron J, Ray R, Little B and Ringeisen B R 2007 A biofilm enhanced miniature microbial fuel cell using Shewanella oneidensis DSP10 and oxygen reduction cathodes *Biosens. Bioelectron.* **22** 1672-9

[34] Santini M, Guilizzoni M, Lorenzi M, Atanassov P, Marsili E, Fest-Santini S, Cristiani P and Santoro C 2015 Three-dimensional X-ray microcomputed tomography of carbonates and biofilm on operated cathode in single chamber microbial fuel cell *Biointerphases* **10** 031009

[35] Sonek J 2007 First trimester ultrasonography in screening and detection of fetal anomalies *Am. J. Med. Genet. C Semin. Med. Genet.* **145** 45-61

[36] Syngelaki A, Chelemen T, Dagklis T, Allan L and Nicolaides K H 2011 Challenges in the diagnosis of fetal non-chromosomal abnormalities at 11-13 weeks *Prenat. Diagn.* **31** 90-102

[37] Bellotti M, Fesslova V, De Gasperi C, Rognoni G, Bee V, Zucca I, Cappellini A, Bulfamante G and Lombardi C M 2010 Reliability of the first-trimester cardiac scan by ultrasound-trained obstetricians with high-frequency transabdominal probes in fetuses with increased nuchal translucency *Ultrasound Obstet. Gynecol.* **36** 272-8

[38] Persico N, Moratalla J, Lombardi C M , Zidere V, Allan L and Nicolaides K H 2011 Fetal echocardiography at 11-13 weeks by transabdominal high-frequency ultrasound *Ultrasound Obstet. Gynecol.* **37** 296-301

[39] Lombardi C M, Bellotti M, Fesslova V and Cappellini A 2007 Fetal echocardiography at the time of the nuchal translucency scan *Ultrasound Obstet. Gynecol.* **29** 249-57

[40] Lombardi C M, Zambelli V, Botta G, Moltrasio F, Cattoretti G, Lucchini V, Fesslova V and Cuttin MS 2014 Postmortem microcomputed tomography (micro-CT) of small fetuses and hearts *Ultrasound Obstet. Gynecol.* **44** 600-9

Schlieren visualization of water natural convection in a vertical ribbed channel

M Fossa, M Misale, G Tanda[1]
DIME, Università degli Studi di Genova, Genova, Italy

Abstract. Schlieren techniques are valuable tools for the qualitative and quantitative visualizations of flows in a wide range of scientific and engineering disciplines. A large number of schlieren systems have been developed and documented in the literature; majority of applications involve flows of gases, typically air. In this work, a schlieren technique is applied to visualize the buoyancy-induced flow inside vertical ribbed channels using water as convective fluid. The test section consists of a vertical plate made of two thin sheets of chrome-plated copper with a foil heater sandwiched between them; the external sides of the plate are roughened with transverse, square-cross-sectioned ribs. Two parallel vertical walls, smooth and unheated, form with the heated ribbed plate two adjacent, identical and asymmetrically heated, vertical channels. Results include flow schlieren visualizations with colour-band filters, reconstructions of the local heat transfer coefficient distributions along the ribbed surfaces and comparisons with past experiments performed using air as working fluid.

1. Introduction

The schlieren technique introduced by Foucault [1] and Toepler [2] has long been an attractive diagnostic tool for the study of fluid flow fields. Since the schlieren effect is related to gradients of the refractive index, density/temperature variations in a fluid flow can be easily visualized by means of this optical method, basically requiring a white light source, an optical setup made of mirrors or lenses, a camera and a special filter to detect the light deflections in the observed field. A comprehensive and exhaustive description of the schlieren technique, including qualitative and quantitative applications, optical principles and devices to detect light deflections, is provided in the book by Settles [3].

Due to the simple relationship between the refractive index and the density of gases, the schlieren technique has been mainly employed for the investigation of flow and thermal fields using a gas as flowing medium [3]. Conversely, schlieren was not a popular method in the past to visualize liquid flows. Liquids usually induce strong light refractions and this limits the use of schlieren devices, which do not typically need or want high sensitivity, to phenomena characterized by very small changes in temperature or a very small width of the test section in the light beam direction. In addition, quantitative results can be gained if the refractive index variations with the liquid temperature and pressure are precisely known *a priori*.

The Authors of the present work were apparently the first to apply a quantitative schlieren technique to the study of heat transfer in liquid flows [4-5]; in particular, local heat transfer

[1] To whom any correspondence should be addressed. E-mail: giovanni.tanda@unige.it

coefficients along a vertical plate in the presence of laminar, natural convective water flow have been determined. Results have been compared with numerical simulations, theoretical relationships and literature experimental data obtained in air by the same technique and under similar conditions in terms of dimensionless groups. The experiments performed proved the validity of the experimental method and its applicability to the study of thermal fields not only in gases but even in transparent liquid flows. This work aims at gaining qualitative and quantitative visualization of natural convection in water channels having repeated ribs on the heat transfer surfaces. These surfaces are commonly encountered in several engineering applications (electronic equipment, solar collectors, passive heating of buildings, etc.). The roughening of a heat transfer surface is a common practice to increase the convective heat transfer [6-10]. Roughness elements may significantly affect the free convection heat transfer in vertical channels owing to the following circumstances: (i) the blockage effect associated with the presence of protrusions could provoke a weaker induced flow, potentially reducing the heat transfer; (ii) the roughness could induce disturbances in the overlying laminar boundary layer, thus causing premature transition to turbulence; (iii) the roughness elements, if heated, add an extra heat transfer surface area. Moreover, there are situations (surfaces of buildings, electronic circuit boards) in which the roughness occurs naturally and is not added for the specific purpose of modifying heat transfer performance. Thus, understanding the thermal behaviour of these systems is essential for their correct design.

2. Theoretical background

The angular deflection of a light ray due to a density/temperature gradient in a fluid layer can be estimated from geometrical optics.

Let us consider a light ray, propagating along the *z*-axis of a Cartesian *x-y-z* coordinate system, which passes through a fluid layer, of depth *L*, bounded by two surfaces lying onto *x-y* parallel planes, as shown in Fig.1. The deflection angle of the light ray in the *y* direction as it reaches the *x*=*L* plane is given by

$$\alpha_y = \int_0^L \frac{1}{n}\frac{\partial n}{\partial y} dz \tag{1}$$

where *n* is the refractive index of the fluid and the integration is performed over the entire length of the light ray in the fluid layer. Eq. (1) is derived under the assumption of very small angles of deflection.

If the phenomenon under study is two-dimensional, i.e. density/temperature of the fluid does not depend on *z*-coordinate, as the entering angle is zero, the angle at *z*=*L* is

$$\alpha_y = \frac{L}{n_0}\frac{\partial n}{\partial y} \tag{2}$$

where the term $1/n$ within the integral has been assumed not to change greatly through the test section and set equal to $1/n_0$, n_0 being a reference value for the fluid refractive index. As the gradient of the refractive index with respect to the temperature is introduced, Eq.(2) becomes

$$\alpha_y = \frac{L}{n_0}\frac{\mathrm{d}n}{\mathrm{d}T}\frac{\partial T}{\partial y} \tag{3}$$

For gases, $\mathrm{d}n/\mathrm{d}T$ can be easily deduced by applying the Gladstone-Dale and the ideal gas laws to give $\mathrm{d}n/\mathrm{d}T = -K\,P/(R\,T^2)$, where *K* is the Gladstone-Dale constant, *R* is the ideal gas constant, *P* is the pressure and *T* is the absolute temperature. For liquids, $\mathrm{d}n/\mathrm{d}T$ has to be experimentally determined. In the case of water, Harvey et al. [11] reported the refractive index at different wavelength in the visible spectrum, versus pressure and temperature. A fitting of data, at the atmospheric pressure and averaged over the visible wavelength range, gave n_0 = 1.335 (at 20°C) and the following relationship of dn/dT with T in the 15-40°C range

$$dn/dT = -1.1869 \times 10^{-5} - 4.4407 \times 10^{-6}\, T + 2.5 \times 10^{-8} T^2 \tag{4}$$

where the fluid temperature T is expressed in °C. It is worthy of note that averaging dn/dT with the wavelength results in a 2% maximum deviation from Eq.(4) if any single wavelength in the visible spectrum is considered.

In the case of a water layer, enclosed by glass walls (plane and at uniform thickness) and surrounded by ambient air (at refractive index n_{air}), an additional angular deflection occurs according to Snell's law. If α'_y is the angle of the light ray as it emerges into the surrounding air after it has passed through the water layer, it follows that

$$\alpha'_y = \alpha_y \frac{n_0}{n_{air}} \tag{5}$$

where, due to the small values of the angles, sin α'_y and sin α_y have been replaced by α'_y and α_y, respectively, while (n_0/n_{air}) is the ratio between refractive indices of water and air at a standard condition (20°C, atmospheric pressure).

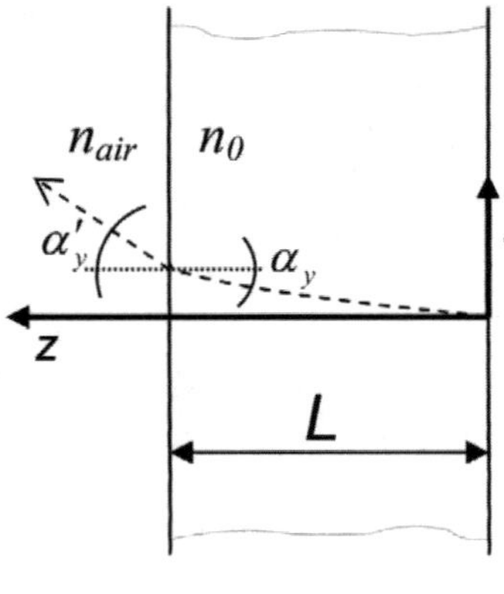

Figure 1 Deflection of a light ray due to optical disturbances in a fluid layer of depth L. Snell's law applies at the fluid/air interface if the fluid refractive index n_0 significantly differs from the air refractive index n_{air}

3. The experiment

3.1 The optical setup

The Z-type schlieren arrangement used in the present experiment is shown in Fig.2; it has been employed for quantitative heat transfer experiments in buoyancy-induced air flows for almost two decades (see, for instance, refs.[12-14]) and applied here to the study of water flows.

A noncoherent light beam coming from a 30μm-wide vertical slit source is collimated by the concave mirror M_1 to form parallel light rays crossing the test section, located inside a water chamber, whose details will be provided in Section 3.2. Without inhomogeneities of the water refractive index in the test section, the parallel light rays, focused by the concave mirror M_2, converge to a thin vertical line on the primary focal plane, representing the primary image of the light source and called primary focal line. All rays are then projected onto the screen (image plane) where a real image of the test section is formed. When inhomogeneities are present in the test section, the light rays undergo angular deflections. All rays (deflected and non-deflected) are recombined on the screen, where a non-distorted image of the test section is still formed.

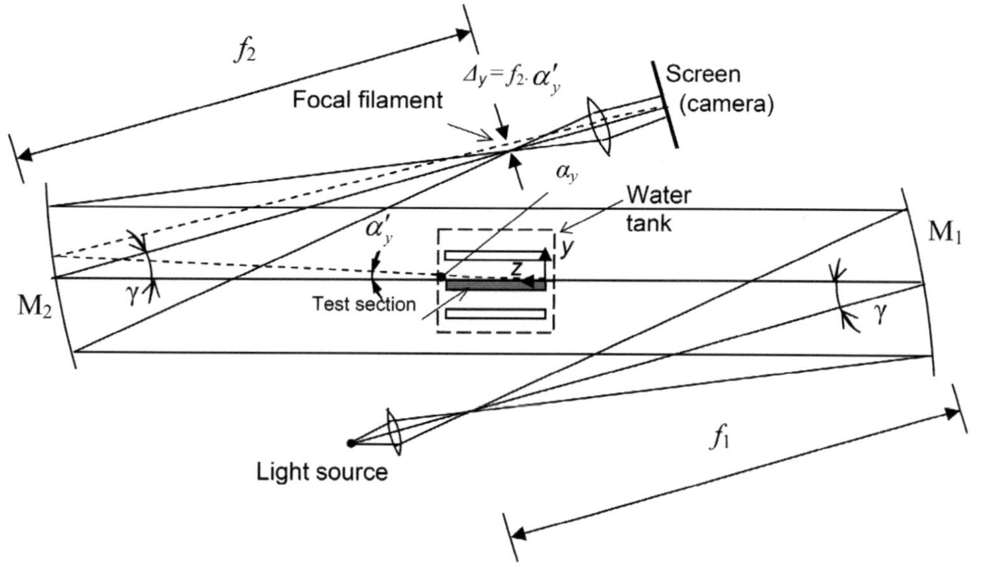

Figure 2 Schematic layout of the Z-shaped schlieren system (top view); M_1 and M_2 are 380 mm diameter concave mirrors (with focal lengths f_1 and f_2, respectively), α_y is the *y*-component of the light ray angular deflection within the test section, α'_y is the angular deflection of the light ray emerging into the ambient air, and Δ_y is the related shifting at the focal plane of mirror M_2 ($f_1 = f_2 =$ 1.9 m, $\gamma = 9$ deg, distance between mirrors about 8 m)

Regions of the test section that give rise to the same light angular deflection along the *y* direction can be identified by using a vertical filament (for instance a thin opaque strip or wire) in the primary focal plane of mirror M_2 (also called the *schlieren head*). On this plane, the deflected rays are shifted by a quantity Δ_y given by

$$\Delta_y = f_2 \alpha'_y = f_2 \alpha_y \frac{n_0}{n_{air}} \tag{6}$$

where f_2 is the distance between mirror M_2 and the primary focal line, n_{air} (=1.0003) is the refractive index of ambient (laboratory room) air and n_0 (=1.335) is the refractive index of water at a standard condition. It is useful to remind that α_y is the angular deflection, along the *y* direction, induced by thermal gradients in the water according to Eq.(3). In order to detect the light shifting Δ_y (and the related angular deflection), the focal filament is first placed in its reference position (position 1 in Fig.3). This position, for which the primary focal line is completely intercepted by the filament, can be clearly identified since it corresponds, without inhomogeneities in the fluid, to a uniformly dark image on the screen. Then, moving the vertical filament in the focal plane, parallel to itself, allows one to measure the values of the light ray deflections in the whole optical field. For instance, when the focal filament is shifted from position 1 to position 2 (Fig.3) by a known quantity Δ_y, a shadow is formed on the image plane. This shadow defines the locus of the images of points deflecting the light by the same angle (termed curve of equal light deflection angle or, more simply, *iso-deflection* line). When deflections of light rays passing in the vicinity of the heated wall are of interest (in order to deduce the heat transfer coefficient), the focal filament has to be shifted until its shadow, projected onto the image plane, intercepts the vertical surface profile at the desired location.

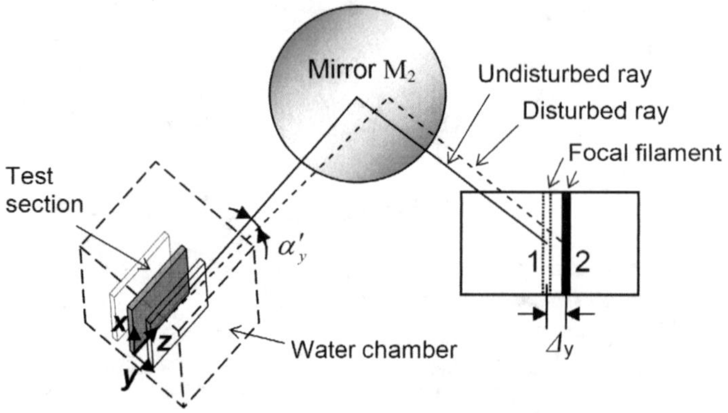

Figure 3 Measurement of the shifting Δ_y of the deflected ray in the focal plane of schlieren mirror M_2 by means of the focal filament, moved from position 1 to position 2

This schlieren technique, called *focal filament method* by Vasil'ev [15], in special conditions allows a direct measurement of the local heat transfer coefficient. For this purpose, the local convective heat transfer coefficient is now introduced

$$h = -\frac{k_w}{(T_w - T_f)}\left(\frac{\partial T}{\partial y}\right)_w \tag{7}$$

where $(\partial T/\partial y)_w$ is the fluid temperature gradient, in the direction y normal to the heated plate, evaluated at the wall, T_w is the wall temperature, k_w is the thermal conductivity of the fluid at the wall temperature and T_f is the fluid temperature.

From the combination of Eqs.(3), (5), (6) and (7), it follows that

$$h = -\frac{k_w}{(T_w - T_f)}\left(\frac{n_{air}\Delta_w}{f_2 L(\mathrm{d}n/\mathrm{d}T)_w}\right) \tag{8}$$

where the subscript w denotes quantities to be evaluated at the wall. Therefore, assuming the water properties k and $\mathrm{d}n/\mathrm{d}T$ as known, provided that T_w and T_f are independently measured, Eq.(8) states the direct relationship between the optical quantity Δ_w and the local heat transfer coefficient for a 2-D thermal field. In particular, Δ_w represents the shifting of the light ray passing in the vicinity of the heated wall and is detected by the schlieren apparatus according to the procedure illustrated in Fig.3.

A filter, operating according to the principles of the focal filament previously described, has been obtained by photographing a violet strip using a slide film. The slide produced in this manner, consisting of a 370μm-wide, transparent, violet band, was directly used as focal filament filter. As the filter is vertically mounted in the focal plane of mirror M_2, light rays deflected by the same angle pass through the violet band, and the corresponding points of the test section appear, in the image plane, coloured in violet. The width of the violet band was properly selected in order to minimize light diffraction effects and to provide a relatively high sensitivity of the technique. The use of a transparent colour band instead of an opaque band makes it easier to identify the correct location of the intersection between the iso-deflection line (appearing violet in the image plane) and the surface wall profile (appearing black in the image plane). To facilitate the description of successive steps, the violet band will be simply termed *focal filament*.

As an alternative to the focal filament method, an insight of the fluid flow/thermal field can be obtained at a glance by using a colour filter (made of a set of thin transparent coloured bands) mounted in the focal plane of mirror M_2. The colour schlieren method does not require any displacement of the filter. This method has the advantage of giving a whole-field image of the phenomenon without having to take a number of images recorded at different time steps, as occurs with the focal filament method.

Therefore, light angular deflection associated with any points framed inside the optical field can be extracted in principle from a single colour image. However, if the filter is made of discrete coloured bands, the measurement sensitivity is limited because of the finite number of bands that can be mounted on the same filter and the resulting colour images cannot typically be used for quantitative purposes.

In the present work, a colour filter with discrete bands was used only for the purpose of qualitative visualization, while the focal filament filter was used for the direct measurement of local heat transfer coefficient from Eq.(8). As previously mentioned, the present schlieren setup was originally designed for the study of convective air flows. When water is considered as convective fluid, the refraction index derivative with temperature is two orders of magnitude larger than that for air; this results in an extra-sensitivity of the optical apparatus as the fluid is changed. This problem was alleviated by properly setting the length L of the test section (in the light beam direction) and the wall-to-fluid temperature difference T_w-T_f, in order to achieve the best compromise among the sensitivity of the optical device (which increases with L and T_w-T_f), the uncertainty in the heat transfer coefficient measurement (increasing as T_w-T_f decreases), and the respect of a nearly 2-D thermal phenomenon (controlled by the extent of L).

3.2 The test section

The description of the test section used in the experiments is facilitated by the schematic view presented in Fig.4. The thermally active component of the apparatus was a vertical plate (termed *heated plate*) made of two thin sheets of chrome-plated copper with a 0.5mm-thick electric foil heater sandwiched between them. The two copper sheets were sealed with a water-resistant cement to prevent any contact between the heater and the water. When electrical power was delivered to the heater, the heated plate was expected to attain a uniform surface temperature at the steady state owing to the high thermal conductivity of copper. The dimensions of the heated plate were the following: height H = 87 mm, length L = 48 mm, overall thickness t = 8 mm. Each side of the heated plate exposed to the water flow was roughened with five transverse, square-cross-sectioned ribs, made integral with the baseplate to guarantee the absence of contact resistance. The square ribs had a height e of 2.42 mm and were regularly spaced at intervals of P = 17.4 mm, resulting in a pitch-to-height ratio P/e =7.2.

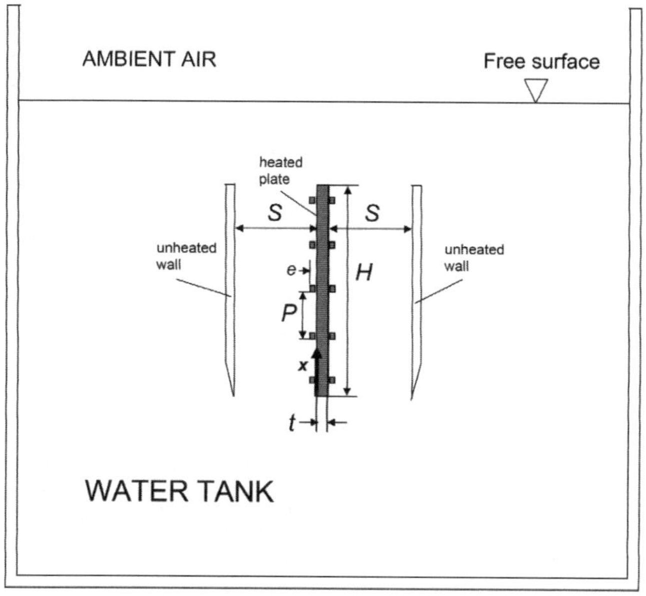

Figure 4 Schematic layout of the vertical ribbed channels inside the water tank. Drawing not to scale

Two parallel vertical walls, smooth and unheated, formed with the heated plate two adjacent, identical, and asymmetrically heated, channels. The spacing S between each unheated wall and the heated plate, set equal on both sides, was 17.4 mm, corresponding to a channel aspect ratio S/H equal to 0.2. The symmetrical arrangement of the heated plate/shrouding wall assembly permitted the optical measurements to be repeated on both sides and, owing to the symmetry, averaged at the same elevation, thus reducing the experimental error.

The heated and unheated walls were suspended, by using a supporting frame, inside a tank with inner dimensions 180x65x390 mm (width x length x height) filled with distilled water. Special care was taken in the design of the supporting frame to permit the correct positioning of the channel walls and to avoid flow obstructions in the vicinity of the channel openings. The tank was open on the top side to provide ambient pressure conditions at the air/water interface. The exit of the channel was situated 75 mm below the water surface and the entrance at 190 mm above the tank inner floor. The vertical sides of the tank normal to the light beam were made of 6mm-thick, high quality glasses so as to permit the schlieren measurements. The remaining sides and the bottom of the tank were made of 10mm-thick chrome-plated copper and finned on the ambient air side to facilitate the dissipation of the input power to the laboratory ambient air and to reduce, as much as possible, the thermal stratification in the fluid.

The heated plate and the water tank were instrumented with 0.5-mm-dia sheathed thermocouples, calibrated to $\pm$ 0.05 K. Six thermocouples were placed into the heated plate at different locations through small diameter holes drilled into the material as close to the exposed surfaces as possible. The wall temperature T_w was obtained by averaging the readings of the six thermocouples embedded in the plate material. The fluid temperature T_f was measured by a thermocouple located at the inlet section of one of the two identical channels; an additional thermocouple, able to travel vertically within the tank and outside the channels, was used to check the presence of undesirable water fluid temperature stratification.

3.3 Experimental procedure

Experiments were performed according to the following procedure:

(i) the heated plate was placed in the tank and the two unheated walls were set to provide a channel aspect ratio of 0.2;

(ii) the plate was heated by a given input of electrical power delivered to the heater, in order to achieve the desired uniform temperature over the heated plate;

(iii) at the steady state, measurements of wall and fluid temperatures were obtained by the readings of the thermocouples deployed in the material and of the thermocouple at the channel inlet, respectively;

(iv) for each run, several schlieren images were taken by moving the violet band (i.e., the focal filament) from its reference position as shown in Fig.3; the displacement impressed to the filament, directly measured by a micrometer, allows one to obtain the local heat transfer coefficients at several locations along the vertical surfaces of the heated plate by applying Eq.(8).

Typical examples of photographs taken by the schlieren apparatus using the focal filament method are reported in Fig. 5. As explained in Section 3.1, the thin line visible in the ribbed channel identifies the zones of fluid that give rise to a known light ray angular deflection (related to the displacement of the focal filament from its original reference position). In order to detect angular deflections of rays passing in the vicinity of the heated wall, the focal filament has to be moved (as shown in the sequence of pictures displayed in Fig.5) until the corresponding iso-deflection line generated in the image plane intercepts the wall profile at a given point.

Moreover, qualitative information on the thermal field around the heated plate can be inferred from the schlieren image, recorded using the colour filter, shown in Fig. 6; the artificially coloured fluid regions give, at a glance, a visualization of the growing thermal boundary layer. The same figure reports a schlieren colour image obtained for the same experiment, conducted in air under thermal similarity conditions (same Ra_H) in a geometrically similar system (same number of ribs and P/e) [13]. The colour patterns show only minor differences due to the different Pr numbers in the regions immediately downstream of each rib, where the stagnation zone, inactive from the heat transfer point of view, appears to be wider for the Pr=0.7 (air) case.

3.4 Data reduction

The experimentally determined heat transfer results have been recast in terms of dimensionless groups. Local Nusselt number Nu_x and Rayleigh number Ra_x were respectively evaluated from

$$Nu_x = \frac{hx}{k} \tag{9}$$

$$Ra_x = \frac{\beta g x^3 Pr}{\nu^2}\left(T_w - T_f\right) \tag{10}$$

where k, β, ν, and Pr represent the thermal conductivity, coefficient of expansion, kinematic viscosity and Prandtl number of water, respectively, and g is the acceleration of gravity. All the thermophysical properties were evaluated at the film temperature $\frac{1}{2}(T_w + T_f)$. A single experimental run has been identified by the Rayleigh number Ra_H, based on the channel height H

$$Ra_H = \frac{\beta g H^3 Pr}{\nu^2}\left(T_w - T_f\right) \tag{11}$$

Experiments were conducted for Ra_H in the 1.4×10^7-1.7×10^7 range.

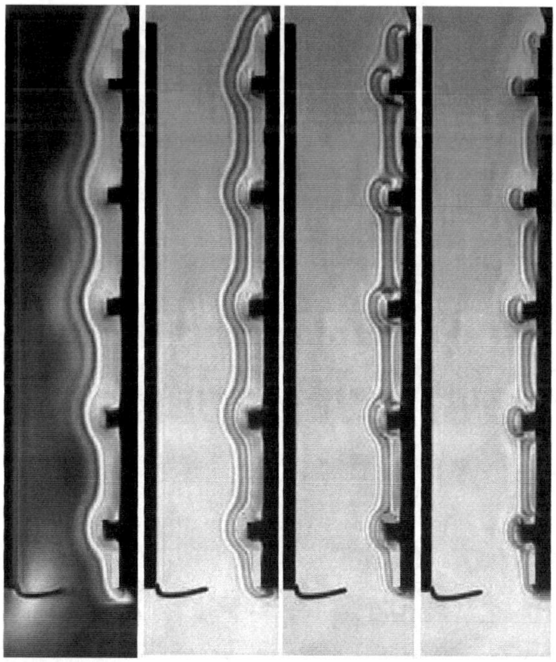

Figure 5 Typical schlieren images recorded with the focal filament progressively moved from its reference position (left-hand image) in order to show the images of points of the optical field deflecting the light by the same known amount (central and right-hand images)

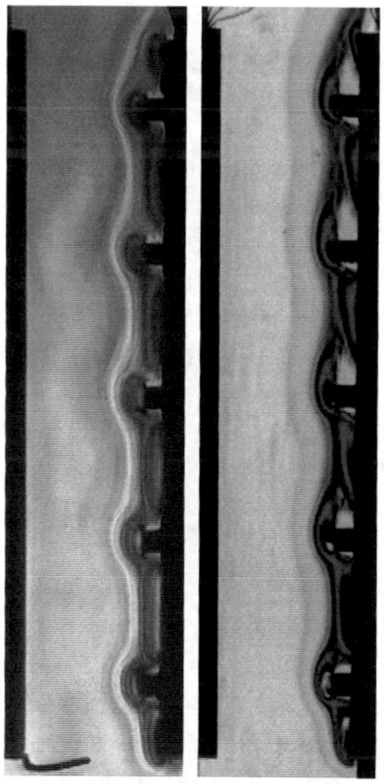

Figure 6 Typical schlieren images recorded with a colour filter. Left-hand: water channel, present work; right-hand: air channel at the same P/e and Ra_H values [13]

3.5 Experimental uncertainty

The uncertainty in the results (at the 95% confidence level) was evaluated by using a root-sum-square of the contributions made by the uncertainties in each of the individual measurements, according to the procedures outlined in [16]. The uncertainty in the local heat transfer coefficient h as expressed by Eq.(8) is generally sensitive to the errors associated with the temperature measurements, with the dn/dT relationship assumed for water, and with the schlieren Δ_w readings. As stated by Tanda [12], the accuracy in schlieren Δ_w readings improves with increasing Δ_w, whose value was, for the present experiments, typically between 3 and 6 mm, with associated errors of less than 10%. As errors in the wall-to-fluid temperature difference and dn/dT relationship are accounted for, the uncertainty in h (and Nu_x) values falls into the 15-18% range, while the uncertainty in Ra_x (and Ra_H) is estimated to be 7%.

4. Results and discussion

As mentioned in the Introduction, the aim of the present work was to apply the focal filament schlieren method to measure the local heat transfer coefficient for the buoyancy-induced water flow in an asymmetrically heated, ribbed vertical channel having an aspect ratio S/H equal to 0.2. Heat transfer experiments were performed by keeping the input power, delivered to the heated plate, in the 3 - 4 W range, giving a wall-to-fluid temperature difference of 1.4 to 1.8 K. For values lower than 1.4 K the h-uncertainty related to temperature measurement errors becomes too large; conversely, values higher than 1.8 K give rise to light ray angular deflections too large to be detected by the focal filament (or colour) filter. Depending on the laboratory room temperature, steady-state values of T_w and T_f were in the 19-21°C and 17-20°C ranges, respectively. These experimental conditions were selected in order to achieve the best compromise between schlieren sensitivity and measurement accuracy, and to reproduce, in terms of dimensionless groups, the same conditions considered in a previous experimental investigation conducted using air as convective fluid.

Schlieren images, taken at the steady-state with the colour filter, revealed that the water flow in the channels was always laminar. Under the above mentioned experimental conditions, a large mass of water between the exterior sides of the channels and the tank sides was seemingly featured by no motion; here, the fluid was thermally stratified. Measurements performed by the thermocouple travelling through the fluid showed that the temperature difference between two points, taken in the stagnant fluid and aligned to the inlet and outlet sections of the channels, respectively (i.e. at a vertical distance equal to the channel height H), was typically 0.3-0.4 K, corresponding to about 20-25% of the mean wall-to-fluid temperature difference.

Figure 7 shows a typical distribution of the local heat transfer coefficient h along the heated plate measured for a wall-to-fluid temperature difference of 1.5 K. Different symbols for the inter-rib and the rib (at the center of each vertical side) regions have been adopted. For comparison purposes, the h-distribution obtained in [4] for a smooth vertical channel, by using the same experimental technique, aspect ratio and wall-to-fluid temperature difference, is reported. Inspection of the figure reveals that along the inter-rib regions heat transfer coefficient distribution is characterized by low values close to the ribs, due to the presence of fluid stagnation zones just upstream and downstream of each rib. The space between ribs is sufficient for the main flow to wash the inter-rib surfaces before being deflected by the next rib downstream; this explains the presence of heat transfer coefficient peaks approximately at the midpoint of each inter-rib region. The developing thermal field is responsible for the progressive reduction of h values registered at points separated by an elevation equal to the rib pitch. Relatively high heat transfer coefficients are found on the vertical sides of ribs since they are washed by cool water flow. The heat transfer performance of the ribbed surface is typically lower than that obtained for the smooth surface, with the exception of points located on the ribs close to the channel exit.

Heat transfer coefficient distributions obtained for a larger wall-to-fluid temperature difference (up to 1.8 K) were qualitatively similar in shape, with individual values slightly higher, in line with the typical proportionality between h and $(T_w - T_f)^{0.25}$ occurring in laminar natural convection heat transfer.

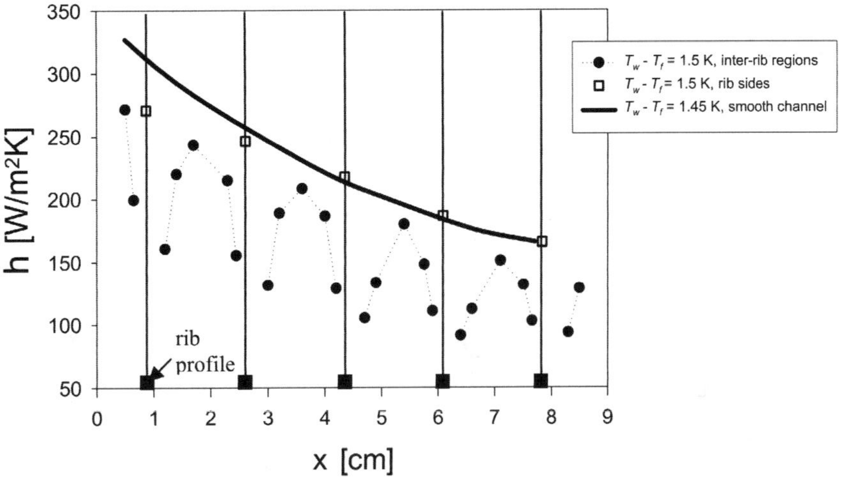

Figure 7 Local heat transfer coefficients along the heated plate with ribs (symbols) and without ribs (continuous line, [4])

Figure 8 reports experimental results presented in the traditional form of local Nusselt number against local Rayleigh number. The same figure displays the experimental results obtained using air (Tanda, [8]) as convective fluid; for the sake of clarity, only the inter-rib regions are considered. The two series of experiments were performed by using the same schlieren technique (focal filament method) and under similarity conditions (almost the same Rayleigh number Ra_H based on the heated plate height H, the same channel aspect ratio S/H, the same number of ribs for each side, the same rib pitch-to-height ratio P/e).

The agreement between dimensionless data recorded for two different Pr-number fluids is excellent and only minor differences are observed. As previously noted in the comment to Fig.6, the highest Pr number fluid (water) produces more effective heat transfer in the recirculation zone downstream of each rib. Moreover, in the case of water, the inter-rib Nu distribution tends to be more uniform and less peaked. This Pr number effect is qualitatively consistent with results obtained in [17] for the forced convection heat transfer in water and air ribbed channels.

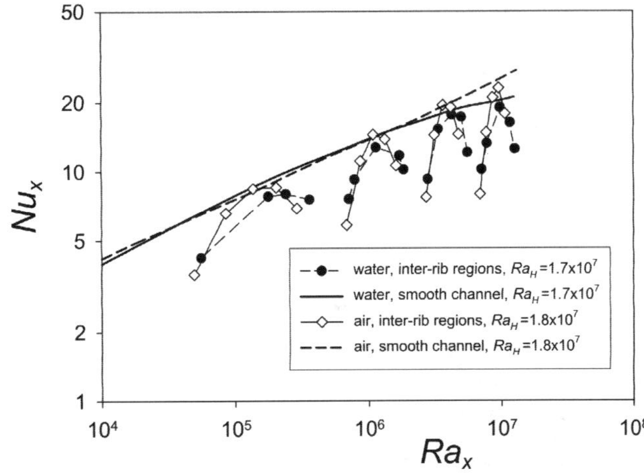

Figure 8 Local Nusselt number versus local Rayleigh number. Full symbols: present experiments for the water ribbed channel, open symbols: experiments for the air ribbed channel (Tanda [8]). Lines: smooth channel experiments for water [4] and air [8]

5. Conclusions

A schlieren technique for the study of convective heat transfer inside vertical channels, with water as working fluid, was presented. This experimental technique, typically employed for quantitative investigations in gases, has been successfully applied here for a liquid.

The investigated channel geometry consisted of an isothermal ribbed heated wall delimited by two, symmetrically arranged, unheated walls. The channel aspect ratio was kept fixed and equal to 0.2.

Local heat transfer coefficients, recorded along the heated side of the ribbed channel, displayed a distribution affected by thermally inactive regions just upstream and downstream of each rib, with heat transfer performance lower than that of the corresponding smooth channel. The same result was obtained using air as working fluid in a previous literature study.

The measurements performed confirm the validity of the schlieren method as quantitative diagnostic tool and encourages its use for the study of thermal fields not only in gases but even in transparent liquid flows.

References

[1] Foucault L 1859 Mémoire sur la construction des télescopes en verre argenté *Annales de l'Observatoire Impérial de Paris* **5** 197-237
[2] Toepler A 1864 *Beobachtungen nach einer neuen optischen Methode – Ein Beitrag zur Experimentalphysik* (M.Cohen & Son)
[3] Settles GS 2001 *Schlieren and Shadowgraph Techniques* (Springer)
[4] Tanda G, Misale M and Fossa M 2014 Heat transfer measurements in water using a schlieren technique *Int. J. Heat Mass Transfer* **71** 451-458
[5] Misale M, Fossa M and Tanda G 2014 Investigation of free convection in a vertical water channel *Exp. Thermal Fluid Science* **59** 252-257
[6] Bhavnani SH and Bergles AE 1990 Effect of surface geometry and orientation on laminar natural convection heat transfer from a vertical flat plate with transverse roughness elements *Int. J. Heat Mass Transfer* **33** 965-981
[7] Aydin M 1997 Dependence of the natural convection over a vertical flat plate in the presence of the ribs *Int. Comm. Heat Mass Transfer* **24** 521-531
[8] Tanda G 1997 Natural convection heat transfer in vertical channels with and without transverse square ribs *Int. J. Heat Mass Transfer* **40** 2173-2185
[9] Onbasioglu SU and Onbasioglu H 2004 On enhancement of heat transfer with ribs *Applied. Thermal Engineering* **24** 43-57
[10] Imbriale M, Panelli M and Cardone G 2012 Heat transfer enhancement of natural convection with ribs *Quantitative Infrared Thermography J.* **9** 55-67
[11] Harvey AH, Gallagher JS and Levelt Sengers JMH 1998 Revised formulation for the refractive index of water and steam as a function of wavelength, temperature and density *J. Physical Chemical Ref. Data* **27** 761-774
[12] Tanda G 1993 Natural convection heat transfer from a staggered vertical plate array *ASME J. Heat Transfer* **115** 938-945
[13] Tanda G 1996 *Application of Optical Methods to the Study of Convective Heat Transfer in Rib-Roughened Channels* Ph.D. Thesis (The City University, London, UK)
[14] Tanda G 2008 Natural convective heat transfer in vertical channels with low-thermal-conductivity ribs *Int. J. Heat Fluid Flow* **29** 1319-1325
[15] Vasil'ev LA 1971 *Schlieren Methods* (Keter Inc)
[16] Moffat RJ 1988 Describing the uncertainties in experimental results *Exp. Thermal Fluid Science* **1** 3-17
[17] Webb BW and Ramadhyani S 1985 Conjugate heat transfer in a channel with staggered ribs *Int. J. Heat Mass Transfer* **28** 1679-1687

Journal of Physics: Conference Series **655** (2015) 012006 doi:10.1088/1742-6596/655/1/012006

Validation of quantitative IR thermography for estimating the U-value by a hot box apparatus

I Nardi, D Paoletti, D Ambrosini, T de Rubeis and S Sfarra

DIIIE, University of L'Aquila, Piazzale Pontieri, 1 Monteluco di Roio, I 67100, L'Aquila, Italy

E-mail: iole.nardi@graduate.univaq.it

Abstract. Energy saving plays a key role in the reduction of energy consumption and carbon emission, and therefore it is essential for reaching the goal of the 20-20-2020 policy.
In particular, buildings are responsible of about 30% of the total amount of Europe energy consumption; the increase of their energy efficiency with the reduction of the thermal transmittance of the envelope is a point of strength with the actions and strategies of the policy makers.
Currently, the study of energy performance of buildings is based on international standards, in particular the Italian one allows to calculate the U-value according the ISO 6946 or by in-situ measurements, using a heat flow meter (HFM), following recommendations provided in ISO 9869.
In the last few years, a new technique, based on Infrared Thermography (IRT) (also referred to as Infrared Thermovision Technique – ITT), has been proposed for in situ determination of the thermal transmittance of opaque building elements. Some case studies have been reported.
This method has already been applied on existing buildings, providing reliable results, but also revealing some weaknesses.
In order to overcome such weak points and to assess a systematic procedure for the application of IRT, a validation of the method has been performed in a monitored environment.
Infrared camera, the heat flow meter sensors and a nearby meteorological station have been used for thermal transmittance measurement. Comparison between the U-values measured in a hot box with IRT as well as values calculated following international standards and HFM results has been effected.
Results give a good description of the advantages, as well as of the open problems, of IR Thermography for estimating the U-value.
Further studies will help to refine the technique, and to identify the best operative conditions.

1.Introduction

Today's in Eu Countries total buildings energy consumptions are 30-40% of the final energy and the demand is raising in the residential sector. Some recent European directives encourage policy actions to reduce the energy need for heating and cooling by 20% by 2020 and 80% in 2050 compared to 1990 levels in order to mitigate the environmental impacts of the built environment. It appears very important to underline that the knowledge of the thermal behaviour of the building envelope is an essential component of the strategies of the policy makers. Among the thermal parameters, the thermal transmittance (U-value $W/m^2{\cdot}K$) is very important, since it is used to define the total energy consumption and to determine the optimal thermal comfort of occupants.

The thermal transmittance is a measure of how well an envelope resists to heat transfer and its knowledge is essential for retrofitting and energy saving measures.

Nowadays the U evaluation is based on international standards, in particular it is possible to calculate the design value with ISO 6946 [1] or by in situ measurements, using a flow meter following recommendations foreseen in ISO 9869 [2].

The real energy performance of an envelope can be accurately estimated only for well-characterized systems; in particular the U-value depends on the materials, their age, thickness, density, thermal mass.

For this reason significant inaccuracies in these parameters may be great source of uncertainty: especially in historic buildings in some conditions the actual wall U-value does not always agree with the design value, therefore it is necessary to measure and analyze in situ the U-value.
Currently, for in situ measurements the heat flux method is used. The U value is determined by measuring the temperatures on the surfaces of the wall and the heat flux through it, determined by a sufficient temperature gradient between indoor and outdoor. At least 72 h of measurements are necessary to evaluate the transmittance, and the progressive average method is commonly used to process data.
Several authors [3-9] have demonstrated that some errors are due to non-accurate measurements of surface temperatures, location, number and type of sensors and probes, thermal bridges, shielding or variation of temperature gradient; it is clear, however, that measurement is less precise if the temperature gradient is low.
Many studies have been effectuated for increasing the measurement accuracy, for example Cesaratto *et al* [4] found that the location of sensors causes an error up to 46% related to theoretical value.
Recently an alternative approach has been proposed, the quantitative infrared thermography (namely IRT) for in situ measurements of thermal transmittance, and different studies have been performed [10-17]
This method can be applied in a quasi-steady state condition and may be considered rapid and non-destructive.
However despite the IR inspection seems to be an easy task, it is not free from error; in fact, the neglecting of some factors of influence can lead to erroneous evaluation of the object under analysis.
Such factors can be characterized as procedural (camera calibration, focus, placement), technical or environmental factors.
A validation of the method is proposed in a controlled environment.
In fact, in this new method the most sensitive variables proved are the exact measure of the reflected apparent temperature, the wall surface temperature, and the emissivity of the surface of the envelope.
Today a detailed study on the influence of the several environmental variables on the determination of the U-transmittance with quantitative IR thermography compared with HFM method and theoretical value in a controlled environment is missing.
The aim of this work is to validate the use of IRT for measuring the U-value of a test wall with well-known characteristics in an environment with controlled parameters for comparison with HFM method. A hot-box has been employed to perform such test.
Results are compared and the possible causes of differences between them are also analyzed.

2. Guarded Hot Box description

The importance of a hot box stands in its capability to analyze and investigate the thermal transmittance of homogeneous and big samples of walls, glazing, ceilings or floors or thermal bridges [18], in a quite simple way: two rooms, a hot and cold one, are kept at a desired temperature, while a specimen (object of the study) separates them. By measuring the power provided to the hot chamber to keep it at a constant temperature, and knowing the imposed temperatures on the two sides, it is possible to obtain the thermal transmittance.
Starting from this concept, several hot box designs have been developed and realized all over the world, and different guidelines have been emanated through the years, varying from a country to another. A comprehensive overview of them is described in [19].
The hox box employed at the Department of Industrial Engineering and Information and Economics (DIIIE) of the University of L'Aquila to perform tests is built according to the UNI EN ISO 8990 [20] and it is a Guarded Hot Box (GHB). This means that the hot chamber is composed by a smaller chamber, called *metering chamber* or *metering* box, whose task is to further confine the heat of the hot chamber, and to host sensors and probes, and by a guard chamber. It is well known in literature that the measure errors are reduced while increasing the perimeter of the metering box that, according to UNI EN ISO 8990, should measure at least 100x100 cm.
To further minimize the heat loss through the lateral sides of the sample, there could be a *tempering ring* that helps to strengthen the hypothesis of mono-dimensional heat flow through the specimen.
Figure 1, rearranged from the abovementioned norm, helps to understand how the fluxes are scattered in a GHB.

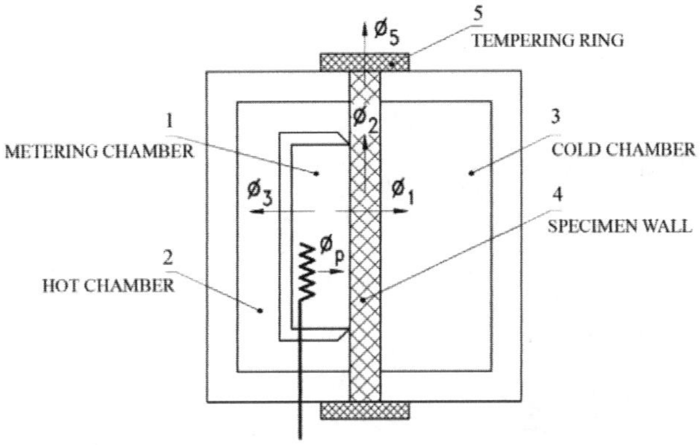

Figure 1. Heat fluxes in a guarded hot box.

Standing on figure 1, the heat flux Φ_1 through the sample equals the thermal power Φ_P inlet minus the flanking flux Φ_2 (that might exit the apparatus through the tempering ring causing a flux Φ_5) minus the heat flux Φ_3 through the walls of the metering box, i.e.:

$$\Phi_1 = \Phi_P - \Phi_2 - \Phi_3 \quad (1)$$

Therefore, it is possible to know the flux that interests the specimen by subtracting from the power Φ_P (that is required from the heating devices to keep constant the hot chamber temperature, and which can be easily measured) the power lost by the apparatus itself, that is through the specimen (causing a bidimensional flux) and through the envelope of the metering chamber. Of course, it is important to minimize as much as possible Φ_2 and Φ_3, and to measure these “unbalances”, for an appropriate evaluation of the flux through the wall.

The GHB used to perform our tests, and depicted in figure 2, allows to study specimen whose net sizes can be larger than 180 x 180 cm (sizes of the metering box) and up to 300 x 300 cm (sizes of the chambers). Thanks to mobile metal supports, it is possible to move separately the two chambers and the specimen, and to clamp it to the chambers; moreover, the metering chamber moves on a rail: at the end of the run, the seal of the chamber is perfectly adherent to the specimen.

The wall of the metering box and of the two chambers are made of galvanized steel sheet (thickness 0.1 cm) filled with 10 cm of expanded polyurethane.

Inside the chambers, groups of electric resistances (er), each one of 1000 W, provide the thermal power; near them, fans (f) are mounted to allow air circulation and forced convection, while grid fans (gf) split and direct the air flow.

The voltage regulator that controls the electric resistances, and all the devices needed to set the parameters in the chambers, are mounted on the sides of the chambers themselves (sw1 and sw2), as the reciprocating water-cooled compression refrigerating unit, that cools through copper pipes the cold chamber (cc).

Figure 3 represents the overall view of the hot box when it is joined to the specimen with its support.

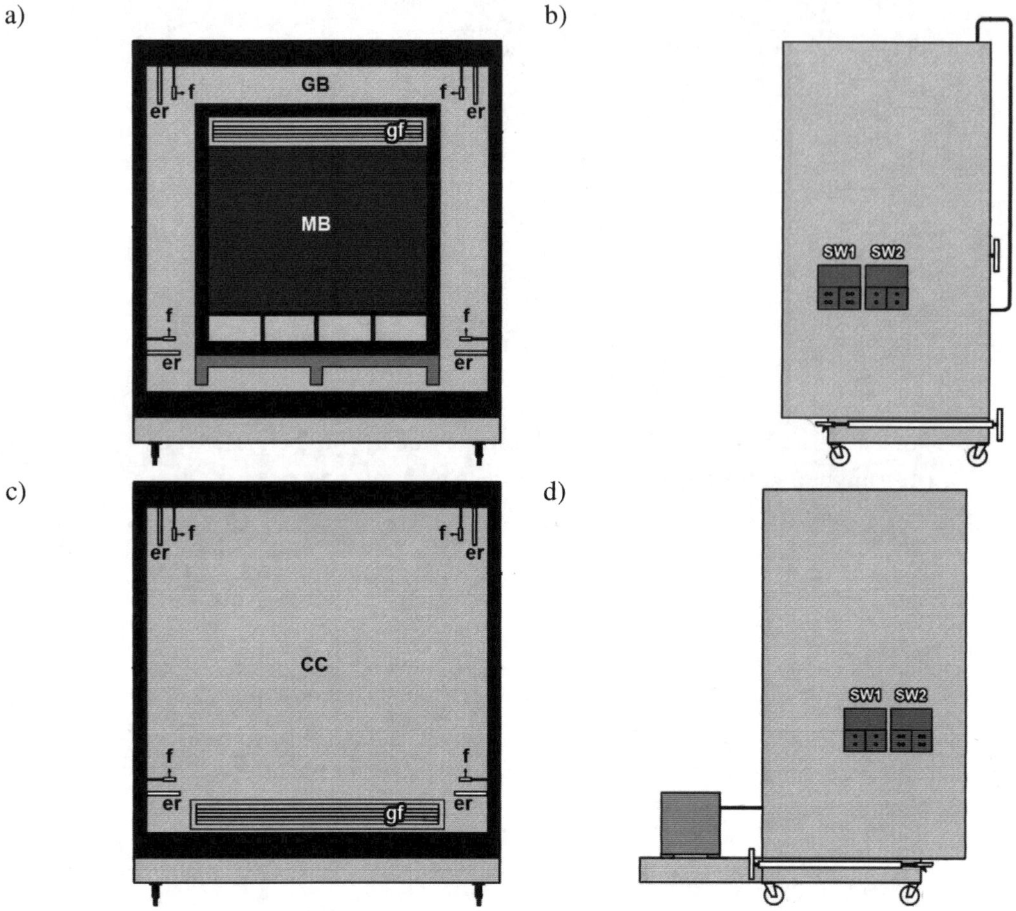

Figure 2. Drawing of the guarded hot box: (a) hot chamber - inside view; (b) hot chamber - side view; (c) cold chamber – inside view; (d) cold chamber – side view.

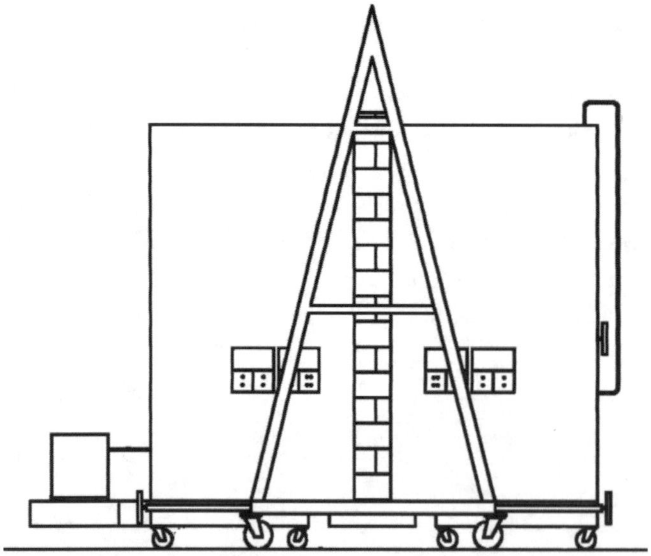

Figure 3. Overall view of the guarded hot box: (1) cold chamber, (2) hot chamber, (3) specimen and its support (4).

Journal of Physics: Conference Series **655** (2015) 012006 doi:10.1088/1742-6596/655/1/012006

3. Equipment employed

In order to perform the tests, two air temperature probes have been employed to monitor the temperature inside each chamber. Also a heat flux meter has been employed, together with three surface temperature probes (two for the cold side, and one for the hot one) placed using a thermal compound.
Finally, an infrared camera has been used: the technical specifications of the equipment are listed in table 1.

Table 1. Technical specifications of the equipment used.

	Instrument	*Brand and type*	*Input*	*Measuring range*	*Resolution*	*Error (@25°C)*
1)	Fluxmeter	Hukseflux HFP01		-2000÷+2000 W/m^2	50 μV/W/m^{-2}	5% on 12 h of measurement
2)	Temperature probes	LSI Lastem EST124 - Pt100		-40 ÷ 80°C	0.01°C	0.15°C
3)	Data logger	LSI Lastem M-Log ELO008	*Potential difference*	-300 ÷ 1200 mV	40 μV	±100 μV
			Pt100	-50 ÷ 125°C	0.003°C	±0.05°C
4)	Temperature probe	Kimo HD100		-20 ÷ 70°C	0.1°C	±0.4% of reading ±0.3°C
5)	Infrared camera	FLIR ThermaCAM® S65 SH		-40 ÷ 1500 °C	0.05° C	± 2° C or ± 2%
				FOV	24° x 18° / 0.3 m	
				IFOV	1.3 mrad	
				Spectral range	7.5 -13 μm	
				Sensor	Focal plane array uncooled microbolometric; 320 x 240 pixels	

4. InfraRed Technique (IRT)

In the last years, the efforts of different research groups aim to find the best way to measure the thermal transmittance using an infrared camera.
Therefore, different procedures have been proposed, and a recent work [21] compares them proposing a kind of sensitivity analysis.
According to that study, the methodology proposed by Albatici et al. is the most variable, but it is the only one that involves the use of an IR camera to evaluate all the quantities needed for the calculation of the thermal transmittance. Hence, in this work, we decided to use and validate that approach.
Assuming that ε_v is the emissivity of the wall on the IR camera spectral range, v is the wind velocity, and T_{in} T_{out} and T_w are respectively the indoor environment, the outdoor environment temperature and the surface temperature of the element expressed in Kelvin, the thermal transmittance can be calculated as follows:

$$U = \frac{5.67\,\varepsilon_v\left[\left(\frac{T_w}{100}\right)^4 - \left(\frac{T_{out}}{100}\right)^4\right] + 3.8054\,v\left(T_w - T_{out}\right)}{\left(T_{in} - T_{out}\right)} \tag{2}$$

The previous equation can be eased in the following expression in case of absence of ventilation:

$$U = \frac{5.67\,\varepsilon_v\left[\left(\frac{T_w}{100}\right)^4 - \left(\frac{T_{out}}{100}\right)^4\right]}{\left(T_{in} - T_{out}\right)} \tag{3}$$

Equation 2 derives from the heat balance of a wall, assuming a simplification of the Jurge's equation for the expression of the convective heat transferred.
The procedure explained by Albatici et al is easily managed. Using approximating black bodies, even with simple devices whose length is 5-6 times bigger than their opening, it is possible to calculate T_{in} and T_{out}. During our tests, we used a black PVC pipe. As an alternative, for the measure of the indoor temperature, Albatici et al propose to frame from the outside with an IR camera, the inside of the room during a quick opening of a window.
Then, T_w can be derived in a piece of wall, after setting in the IR camera the wall emissivity in the spectral range of the camera. The latter can be evaluated, once again, with the IR camera, using the procedure [22] or using a black tape [13] or using the procedure of [23].

5. Results and discussion

The validation of the IRT method has been performed comparing the U-value obtained by comparison with the results of the heat flow method and of theoretical calculation, on a specimen wall, by means of a guarded hot box.
The comparison is needed in order to assess the reliability of the IRT technique, and to stand out if there are limits to its applicability. Therefore, a hot box has been used to test this method at different temperature levels and different temperature gradients, aimed to simulate multiple boundary conditions of a building, since the strength of the IRT technique could consist in the possibility to quickly investigate buildings' envelope in a quite simple way, and therefore to ease the work of energy auditors.
Although in a recent work [21] the IRT technique has been applied to different small samples (0.5 x 0.5 m) of homogeneous material, the approach followed in this work has been chosen to replicate in-situ measurement conditions, of a bigger and multi-layered sample.
To proceed with the theoretical calculation, norm UNI EN ISO 6946:2008 has been followed, therefore the thickness and the conductivity (derived form of UNI 10351 [24]) of each layer of the wall have been collected.
Table 2 shows the data, listed from the inside to the outside.

Table 2. Stratigraphy of the sample.

	Material	Thickness [m]	Superficial mass [kg/m²]	Conductivity [W/m·K]
1	Plaster	0.003	4.2	0.70
2	Concrete blocks	0.20	183.0	0.62
3	Plaster	0.025	35.0	0.70
4	EPS (UNI 7819 (15 kg/m³))	0.05	0.8	0.04
5	Plaster	0.003	4.2	0.70

The overall thermal transmittance calculated equals 0.56 $W/m^2{\cdot}K$
In order to determine the U-value following ISO 9869:1994 [2], three thermoresistances and a flux meter have been employed on the specimen wall of the GHB. The acquisition rate of data by the data logger has

been 10 minutes. The two chambers simulated the indoor and outdoor environments: furthermore, different temperature gradients have been imposed in the GHB to simulate different boundary conditions.
The temperature gradients remained quite constant over the measurements campaigns, due to the set point imposed, unless the low-oscillating temperature in the cold chamber due to the repetitive on-offs of the chiller. It is possible to state, however, that steady-state conditions occurred during the measurement campaigns.
The results obtained, shown in table 3, confirm what already assessed in literature: the HFM method values depend on the temperature gradient. This raises a new issue: which of the three values obtained should be considered for the following evaluation. We decided to consider a mean value, of about 0.62 W/m^2·K.

Table 3. HFM Results.

Campaign	Mean heat flux [W/m^2]	Mean inside wall temperature [°C]	Mean outside wall temperature [°C]	U-Value [W/m^2·K]
1	10.85	19.79	3.15	0.59
2	6.5	20.45	11.64	0.66
3	10.1	28.55	13.45	0.60

Finally, the IRT technique has been employed, with a little variation of the method; in fact the measurements of temperatures cannot be performed without splitting the hot box.
Therefore, the two chambers of the hot box have been separated, but the specimen wall remained clamped to the hot chamber; the warehouse where the GHB is located has been assumed as the outdoor, and since there isn't forced convection, it has been possible to apply equation 3.
With this set up, it was possible to measure T_{out} by an approximating black body realized with a black PVC pipe internally finned with a closed end, and it was also possible to frame the wall with the IR camera to measure T_w. The last temperature needed, T_{in}, has been acquired with a temperature probe (the fourth of table 1). Data and results are shown in table 4.

Table 4. Results obtained with IRT technique.

Campaign	T_{in} [°C]	T_{out} [°C]	ΔT [°C]	T_{refl} [°C]	T_{wall} [°C]	U-Value [W/m^2·K]
1	32.20	11.30	20.90	11.88	13.48	0.51
2	32.50	12.27	20.23	12.75	14.68	0.59
3	24.60	12.27	12.33	12.57	13.59	0.53
4	23.30	12.24	11.06	13.26	13.98	0.78

Campaigns 1 and 2 differ in the outdoor temperature, which affects the reflected temperature and the U-value, that is increased of 15.7%.
Keeping constant the outdoor temperature and decreasing the indoor temperature of about 8°C (as in campaigns 2 and 3) the measured thermal transmittance decreases of 10.2%.
The wall temperature being similar (as in campaigns 1 and 3), and by varying the reflected temperature and the air temperature gradient, the thermal transmittances obtained are comparable.
The result obtained during the fourth campaign is much higher than the others, and it seems an outlier; in this test the air temperature gradient is the minimum and the reflected temperature is the maximum.
From table 4 it appears clear that even the IR techniques provides oscillating results, depending on the boundary conditions imposed. This drawback, typical of the IR method, has been already outlined in literature, even by applying other techniques [15, 21, 25, 26]; this weak point, however, is common to all methods that involve in-situ measurements, and the IRT technique is not an exception in this sense.
The mean U-value of the first three campaigns is 0.54 W/m^2·K, while the mean value between all the campaigns is 0.60 W/m^2·K; both these values have been considered in the following discussion.
Despite the variations already discussed, the IR thermography results show good agreement to the design value and to HFM method.
In table 5 the percentage differences between the results of the three methods are shown.

Table 5. Results and their percentage difference.

U [$Wm^{-2}K^{-1}$]			Percentage difference		
Design Value	HFM	IRT	Design-HFM	HFM-IRT	Design-IRT
0.56	0.62	0.60	10.71%	3.23%	7.14%
		0.54		12.90%	3.70%

The percentage differences between the methods range from 3.23% to 12.9%, according to which U- value from IRT is considered; however, these results can be considered acceptable, even though they are gathered in steady-state condition, never achieved in in-situ campaigns; furthermore, in real tests, errors in the estimation of variables like wind speed, mean reflected temperature or wall emissivity, or the wall exposure, can lead to mistaken results.

The reflected temperature in our tests didn't vary significantly, since they were performed in a warehouse, but further experiments will be performed to assess the possible sensitivity to this parameter

The tests performed have confirmed that an optimum temperature difference between the inside and the outside must be almost 20°C, so it occurs a good thermal flux through the envelope, in order to reduce uncertainty in the measurements.

In conclusion, the main goal of our research is a validation of the IR method for quick in-situ determination of the energy performance of buildings, very important for planning energy refurbishment processes.

It is clear that a more extensive study is necessary, based on different walls typologies, and in order to assess the sensitivity to the above mentioned parameters.

Currently some improvements of the hot box are in progress: measurements will be performed with a fundamental logic based on software and hardware components, setting and control devices.

References

[1] EN ISO 6946: 2007. Building components and building elements. Thermal resistance and thermal transmittance Calculation method.

[2] ISO 9869: 1994 Thermal insulation. Building elements. In-situ measurement of thermal resistance and thermal transmittance.

[3] Peng C and Wu Z 2008 *Energ. Buildings* **40** 2076-2082.

[4] Cesaratto PG, De Carli M and Marinetti S 2011, *Energ. Buildings* **43** 1792-1801.

[5] Cesaratto PG and De Carli M 2013 *Energ. Buildings* **59** 29-36.

[6] Ahmad A, Maslehuddin M and Al-Hadhrami L, 2014 *Energ. Buildings* **84** 132-141.

[7] Asdrubali F, D'Alessandro F, Baldinelli G and Bianchi F 2014 *Case Studies in Construction Materials* **1** 53-59.

[8] Meng X, Yan B, Gao Y, Wang J, Zhang W and Long E 2015 *Energ. Buildings* **86** 754-765.

[9] Desogus G, Mura S and Ricciu R 2011 *Energ. Buildings* **43** 2613-2620.

[10] Wrobel A and Kisilewicz T 2008 *Proc. 9th International Conference on Quantitative Infrared Thermography*, Krakow, PL.

[11] Albatici R, Tonelli AM 2008 *Proc PLEA 2008 - 25th Conference on Passive and Low Energy Architecture* (Dublin, Ireland).

[12] Albatici R, Tonelli AM 2010 Infrared thermovision technique for the assessment of thermal transmittance of opaque buildings elements on site *Energ Build* **42** 2177.

[13] Fokaides PA and Kalogirou SA 2011 *Appl Energ* **88** 4358.

[14] Dall'O G, Sarto L, Panza A 2013 *Energies* **6** 3859 www.mdpi.com/journal/energies.

[15] Tzifa V, Papadakos G, Papadopoulou AG, Marinakis V and Psarras J, 2014 *International Journal of Sustainable Energy,* DOI: https://dx.doi.org/10.1080/14786451.2014.982119.

[16] Nardi I, Sfarra S and Ambrosini D 2014 *J. Phys.* **547**

[17] Albatici R, Tonelli AM and Chiogna M 2015 *Appl. Energ.* **141** 218-228.

[18] Martin K, Campos-Celador A, Escudero C, Gomez I and Sala J M 2012 *Energ. Buildings* **50** 139-149.

[19] UNI EN ISO 8990:1999, Thermal insulation – Determination of steady-state thermal transmission properties – Calibrated and Guarded Hot Box.

[20] Asdrubali F and Baldinelli G 2011 *Energ. Buildings* **43** 1618-1626.

[21] Simoes I, Simoes N, Tadeu A and Riachos J 2015 *Qirt 2014* .

[22] Albatici R, Passerini F, Tonelli AM and Gialanella S 2013 *Energ. Buildings* **66** 33-40

[23] Marinetti S, Cesaratto PG 2012 *NDT&E Int* **51** 127-34

[24] UNI 10351:1994 Conduttività termica e permeabilità al vapore.

[25] Lehmann B, Ghazi Wakili K, Frank T, Vera Collado B and Tanner C 2013 *Appl Energ* **110,** 29-43

[26] Van De Vijver S, Steeman M, Van Den Bossche N, Carbonez K and Janssens A 2015 *Qirt 2014.*

33rd UIT (Italian Union of Thermo-fluid-dynamics) Heat Transfer Conference IOP Publishing
Journal of Physics: Conference Series **655** (2015) 012007 doi:10.1088/1742-6596/655/1/012007

A comparison between thermographic and flow-meter methods for the evaluation of thermal transmittance of different wall constructions

I Nardi, D Ambrosini, T de Rubeis, S Sfarra, S Perilli and G Pasqualoni

DIIIE, University of L'Aquila, Piazzale Pontieri, 1 Monteluco di Roio, I 67100, L'Aquila, Italy

E-mail: iole.nardi@graduate.univaq.it

Abstract. One of the key parameters that "meter" the energy performance of the whole structure of buildings is the thermal transmittance.
This parameter can be evaluated with a theoretical approach, regulated by standard ISO 6946, once the stratigraphy of the envelope and the properties of the constituent materials are known, or by using a heat flow meter (HFM), following the recommendations provided in standard ISO 9869.
Recently, the use of quantitative IR Thermography (IRT) has been proposed by several researchers; this method allows to determine the overall transmittance of an envelope in a short time (especially in comparison with HFM method). However, the theoretical or experimental transmittance, measured on real buildings having walls composed by different materials, can be rather distant from those calculated or measured with different procedures.
For this reason, for a correct certification of the thermal performance of a building envelope, it is necessary the availability of experimental procedures for a direct and reliable evaluation of the thermal transmittance, suitable for different walls.
Research has found that, especially in historical constructions, faults in the building envelope and the age of the materials can greatly affect the HFM measurements.
The aim of this work is to analyze in situ the thermal performance of three different walls which have been selected according to: different materials, different age, and different construction. On each envelope, a comparison between U-values, measured by HFM and IRT, and computed according the standard procedure, has been effected.

1. Introduction

In the European context, the existing buildings are responsible for more 30% of the final energy consumption, 65% of which is due to residential buildings.
To adopt proper energy policy actions, it is important to know the energy performance of this sector; clearly, it is not possible to base on a particular building but the attention must be addressed to a set of constructions belonging to a specific territorial context [1-2].
This need is particularly strong in Italy, where the building stock has a large range from historical to modern buildings, and almost 75% of the residential houses were built prior to 1991. Regarding the thermal performance of an envelope, the difficulties arise because of lack of information about important parameters that affect the thermal transmittance (U-value) of the components (age, materials, wall thickness, humidity and so on).

Unfortunately, the building energy data used for energy planning at national or local scale, or for re-qualification of ancient buildings, rarely are based on measured values; the laws propose standard data of U-value suitable for Italian context and particularly for new constructions. The thermal transmittance can be calculated with a theoretical approach, through specific software, regulated by standard ISO 6946 [3], once the stratigraphy of the envelope and the properties of the constituent materials are known or by using several non destructive and destructive methods; in particular currently the most used is a Heat Flow Meter (HFM) following the recommendations provided in standard ISO 9869 [4].
However, to evaluate the energy performance of a building stock, in situ measurements are needed, but they are often expensive, time consuming, or require expert knowledge. Recently, several researchers have proposed the use of quantitative thermography for a rapid in situ measurement of the overall transmittance of an envelope building in a short time [5-12].
Often some differences have been found between theoretical and/or experimental data obtained on real buildings [13-15], especially for ancient buildings, with these different procedures [16-18]; for this reason it is difficult to decide which technique to use.
The aim of this work is to present a series of experimental measurements of the thermal transmittance of different buildings belonging to different historical periods, comparing different non invasive techniques: infrared technology, heat flow meter method and calculated value following the standards mentioned above.
The experimental methods have been applied at the same time on an ancient masonry, a well insulated house, and a public building. The differences among the calculated and U value measured with both methods have been outlined. These data might constitute a base for a correct energy diagnosis of the opaque envelope, especially of ancient buildings with a rapid, economic and appropriate method.

2. Description of the test buildings
The test buildings used for in situ measurements of thermal transmittance are located in L'Aquila, a small city in the Abruzzo Region, in the centre of Italy. Detailed climatic data for this city are shown in table 1.

Table 1. Climatic data of L'Aquila city.

Altitude (a.s.l)	714 m	Average annual temperature	12.1°C
Geographical coordinates	42°21'14.43"N 13°23'31.17"E	Indoor Design temperature (heating season)	20°C
Degree Days	2514*	Indoor Design temperature (cooling season)	26°C
Climatic zone	E*	Outdoor Design temperature (heating season)	-5°C

* according to Italian Law DPR 26/08/1993, n.412

The buildings selected as case studies are an historical building, a private house, realized with innovative and sustainable materials, and a public building characterized by large volumes and high ceilings. Figure 1 shows the main characteristics, the urban context, the construction materials and the IR images of the case studies together with the weather conditions during measurements, while the stratigraphy of the walls and their thermal transmittances are displayed in table 2.

Case study **A**
- Historical building
- city centre
- Second half of 1800
- Rough-cut stones
- Weather condition: slightly overcast

Case study **B**
- Private house
- Suburb
- 2011
- Well-insulated envelope
- Weather condition: slightly overcast

Case study **C**
- Public building
- suburb, above a hill
- Early '90
- Internally insulated
- Weather condition: overcast

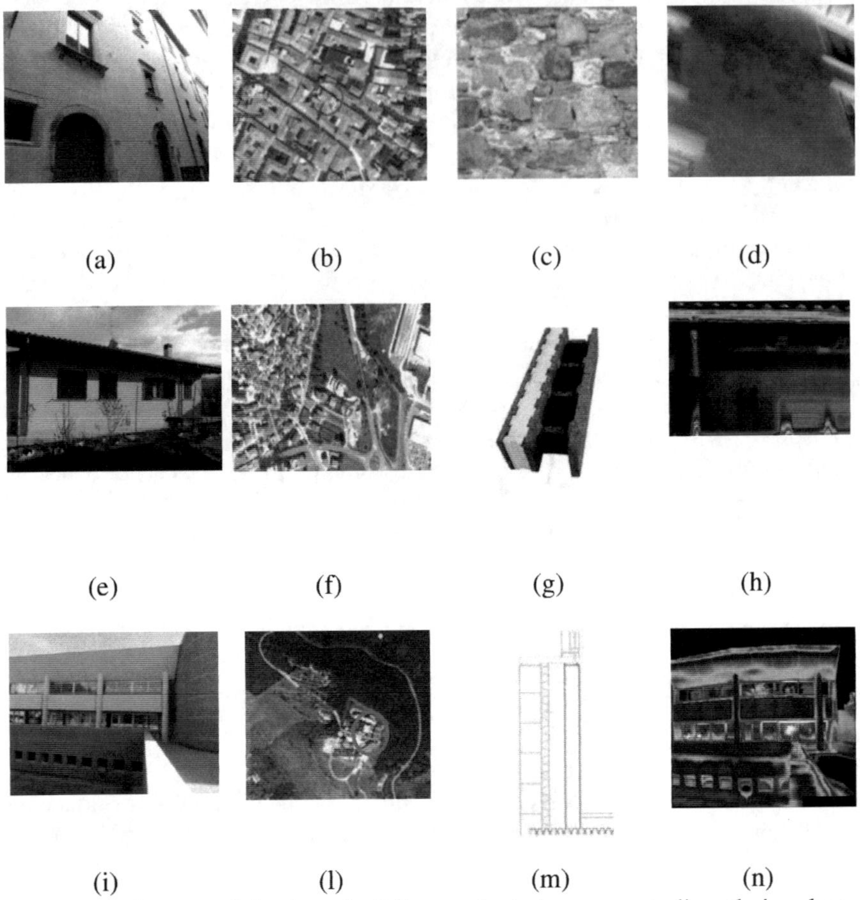

Figure 1. Main characteristics and pictures of the three buildings selected as case studies, their urban context from aerial view, the type of stratigraphy and an IR image: (a)-(d) historical building; (e)-(h) private house; (i)-(n) public building.

2.1 Case study A: historical building

The selected case study is representative of the houses (second half of the 18th century) generally present in the historical centre of L'Aquila and surroundings, and linked to the local typology and technological traditions.

In fact, about 60% of buildings' masonry in L'Aquila is realized with small stone elements (15-20 cm), of irregular shapes and setting, with possible lacks among the stones, sometimes filled by stone wedges. Typical masonry width is about 80 cm.

The earthquakes that hit the city during past centuries have determined the development of several construction techniques, and the reuse of recovered materials with different characteristics.

The façade of the selected building is 80 cm width, and it is made of irregular ashlars of rough-squared, randomly arranged with other materials detected by preliminary thermographic inspections, and covered with plaster. Generally there is no information about materials' characteristics, therefore it is important to use thermography for revealing decay and construction typologies.

2.2 Case study B: private house

This case study is a private house recently built in the suburb of the city with a cement-wood brick, internally insulated with 10 cm of high density EPS (expanded polystyrene).

This building material is widespread in the North of Europe, but still not diffused in Italy, especially in the Abruzzo Region. The building material used for this house is made up of wastes of wood processing techniques, in particular fir fibers, mixed with iron oxide, water and Portland cement. The production technologies, as well as the raw materials used for these bricks, are ever-evolving.
Bricks are produced by extrusion, and a part of the inner cavity is filled with an insulating layer; in this case, an EPS layer has been used.

2.3 Case study C: public building
The facility was built in the early of '90, and it hosts the Faculty of Engineering.
The bearing structure is made with reinforced-concrete frames, while the masonry walls are made with clay bricks, an air layer, an insulating panel and a single or double curtain wall (depending on the façade). After the earthquake that hit L'Aquila in 2009, the building had undergone to structural damages, repaired in 2011. The interventions did not include the energetic refurbishment of the building.

Table 2. Thermophysical properties of the investigated walls, and their thermal transmittance obtained by theoretical calculus.

	Material	Thickness [m]	Superficial mass [kg/m²]	Conductivity [W/m·K]	Resistance [m²·K/W]	Thermal transmittance[a)] [W/m²·K]
Case Study A	Plaster	0.020	28.0	0.66	0.03	1.25
	Rough-cut stones	0.800	1520.0	-	0.56	
	Mortar	0.020	36.0	1	0.02	
Case Study B	Plaster	0.015	27.00	1.00	0.015	0.23
	Wood-cement (density 550 kg/m³)	0.04	22.00	0.13	0.308	
	Concrete	0.18	432.00	1.87	0.096	
	Wood-cement (density 550 kg/m³)	0.025	13.75	0.13	0.192	
	High density EPS	0.10	4.20	0.031	3.387	
	Wood-cement (density 550 kg/m³)	0.03	16.5	0.13	0.231	
	Plaster	0.015	27.00	1.00	0.015	
Case Study C	Plaster	0.02	28	0.66	0.66	0.51
	Hollow brick	0.08	48	-	-	
	Air layer	0.10	-	-	-	
	Fiber glass panel	0.04	0.6	0.04	0.04	
	Concrete blocks	0.15	147.1	0.48	0.48	
	Plaster	0.02	28	0.66	0.66	

a) The thermal transmittance has been calculated assuming as internal and external liminar resistances respectively 0.13 and 0.04 m²·K/W

3. U-value measurement: procedures and results

To evaluate the U-value of the envelopes, three different approaches have been used: a theoretical calculus, according to EN ISO 6946:2007 [3], the heat flow meter method following ISO 9869:1994 [4], and the IRT technique, a recently proposed technique already used by the Authors in a previous work [19].

For the theoretical calculus, the thickness and conductivity of each layer are needed, in order to calculate the thermal resistance of each component of the stratigraphy using the lumped- parameter electrical analogy. The sum of the resistances provides the wall thermal resistance, that, added to the internal and external liminar resistances, gives the total thermal resistance of the wall, whose reciprocal represents the overall wall transmittance.
The HFM provides the U-value by reciprocating the sum of the standard liminar resistances (both internal and external) and the wall thermal resistance, as in equation (1). The reciprocate of the wall thermal resistance is the thermal conductance C, that is calculated by the HFM by dividing the instantaneous measured flow rate through the wall by the superficial temperature difference measured on the inside and the outside of the wall, according to equation (1).

$$U = \frac{1}{\frac{1}{h_i} + \sum R + \frac{1}{h_e}} \qquad C = \frac{1}{R} = \frac{\Phi}{\Delta T} \tag{1}$$

A heat flow meter (Hukseflux HFP01) and three thermal- resistances (LSI Lastem Pt100) were used in our campaigns; data were acquired every 10 minutes with the help of a data-logger(LSI Lastem M-Log ELO008), and the series of the instantaneous thermal conductance values were processed by using the progressive average procedure, in order to obtain the asymptotic thermal conductance true value [20]. In addition, at the same time, outdoor and indoor air temperatures were recorded with two thermo-hygrometers (OnSet HOBO H8).
Anyway, a preliminary thermographic inspection of the outside walls has been performed for identifying the best position of the heat flow meter probes. As IR camera, a FLIR ThermaCAM® S65, operating in the range of 7.5 -13 μm and whose IFOV is 1.3 mrad, was employed, even to apply the IRT technique, following the procedure explained in [6].
According to this technique, by using an infrared camera it is possible to detect the heat flow through the envelope and its emissivity; the outdoor temperature is measured by using an approximating black body, namely using a black PVC pipe whose diameter is 6 times smaller than its length, while the indoor temperature is calculated approximating the black body with the opening of a window. The mean apparent reflected temperature is measured by means of a lambertian surface, like an aluminum foil crumpled and then stretched, taped on the surface and framed by the IR camera together with the wall itself.
The U-value is calculated by using the following Equation 2, proposed in [8]:

$$U = \frac{5.67\,\varepsilon_v \left[\left(\frac{T_w}{100} \right)^4 - \left(\frac{T_{out}}{100} \right)^4 \right] + 3.8054\,v\,(T_w - T_{out})}{(T_{\mathrm{int}} - T_{out})} \tag{2}$$

where T_{int} is the indoor environment temperature [K], T_{out} is the outdoor environment temperature [K], T_w is the surface temperature of the element [K], ε_v is the emissivity of the material in the measurement range of the IR camera, and v is the wind velocity, neglected in our application.
This technique has to be fully exploited, and the continuous updates of the procedure demonstrates that several researches on the matter are still ongoing; the variety of the building typologies proposed in this work can help to widen the knowledge of limits and strengths of this innovative application of the quantitative thermography.
Hereafter are shown, for each building analyzed, the main features of the façade, together with the elaborations of the data acquired by the HFM after a measurement campaign of 72 hours for case

study B and C, and 144 hours for the historical building, performed with the progressive average method using a simple tool provided by the HFM manufacturer (InfoFlux by LSI Lastem). Moreover, the U-value calculated with the IRT is proposed.

3.1 Case study A: historical building

The thermophysical properties of the materials that constitute the stratigraphy of this ancient building are not known; therefore, values foreseen by the UNI ISO 10351 [21] have been chosen, while the wall thickness has been deduced by similar buildings of the same historical period and then checked with an endoscope.

The building has heavy walls, whose stratigraphy is shown in table 2, and the overall thermal transmittance equals 1.25 $W/m^2{\cdot}K$.

Data acquired with the HFM for 144 hours have been processed using the progressive average procedure, gaining an asymptotic thermal conductance. The value of the thermal transmittance achieved by taking into account even the laminar coefficients is 1.17 $W/m^2{\cdot}K$.

Finally, the thermal transmittance has been measured using IRT technique. In this case, the U-value recorded is 1.14 $W/m^2{\cdot}K$.

3.2 Case study B: private house

The stratigraphy of this house recalls the composition of the cement-wood bricks used and reported in table 2. Using the procedure explained before, it is possible to calculate the overall thermal transmittance, that equals 0.23 $W/m^2{\cdot}K$.

The measurements with HFM were carried out on the north-west wall for three days. During the in-situ survey the internal surface temperature is almost constant although the building is lived, while the external temperature shows a periodic trend, that recall the alternation of days and nights. The U-value obtained using the progressive average procedure is 0.42 $W/m^2{\cdot}K$. This value is much higher than the previous.

The IRT technique provided U= 0.22 $W/m^2{\cdot}K$.

3.3 Case study C: public building

The stratigraphy of this building, listed in table 2 together with the conductivity, leads to an overall transmittance equal to 0.51 $W/m^2{\cdot}K$.

Following the rules of ISO 9869:2014, in situ thermal transmittance measurements with HFM have been performed, obtaining once again an asymptotical trend of the conductance. The U value is 0.75 $W/m^2{\cdot}K$ during the monitoring campaign (72 hours) the internal and external surface temperature have the trend shown in figure 2.

Using the IRT, a similar value, 0.76 $W/m^2{\cdot}K$ has been obtained.

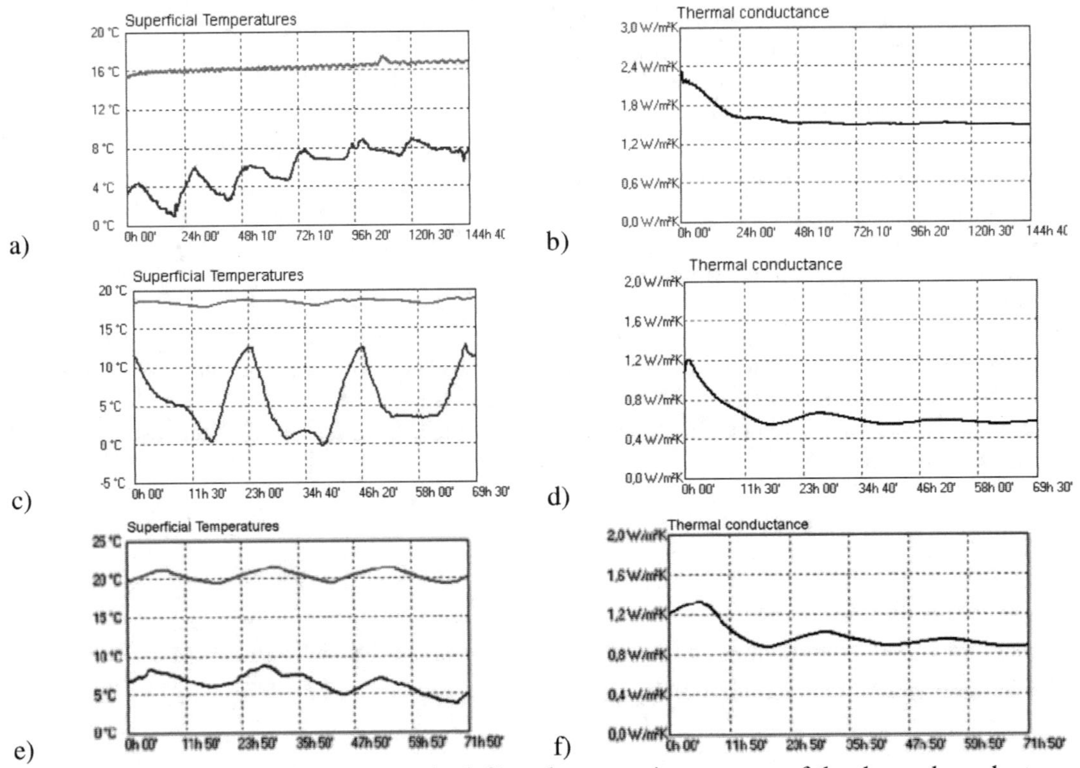

Figure 2. Superficial temperatures (on the left) and progressive average of the thermal conductance (on the right) for case study A (a), b)), case study B (c),d)), and case study C (e), f)).

Results are summarized in table 3.

Table 3. U-values obtained with the different methods and their percentage differences.

	U [$Wm^{-2}K^{-1}$]			Percentage differences		
	Design	HFM	IRT	Design-HFM	HFM-IRT	Design-IRT
Case study A	1.26	1.17	1.14	7.14%	2.56%	9.52%
Case study B	0.23	0.42	0.22	82.6%	47.62%	4.35%
Case study C	0.52	0.75	0.76	44.23%	1.33%	46.15%

For the case study A, the in situ measurements suggest that the thermal performance of the envelope is better than the one revealed by the calculation based on thermo-physical properties of the components of the wall. This might depend on the uncertainty related to the building construction and the ratio of mortar to stone in this particular wall construction. During the HFM acquisition the inside surface temperature is almost constant (being the house unoccupied) but the difference between internal and external conditions does not seem to strongly affect the results.

The HFM and IRT values are near, and their percentage difference is 2.56%, confirming the validity of the IRT technique even for heavy walls, if the measurement campaign follows the procedure mentioned and IR images are acquired during a proper heat release phase, that is after important heat storage during the day.

For case study B, the thermal transmittances obtained by theoretical approach and IRT are similar (difference of 4.35%), but results don't merge with the value obtained with the HFM. This mismatch can be due to several reasons; first of all, the flux meter might have been exposed to direct solar radiation since it has not been possible to install it to a completely north-exposed wall. This condition, although inadvisable, is unfortunately common in the practice, unless the probe is sheltered. Moreover, as seen in figure 2, the wall temperature difference that occurred during the measuring campaign has not been greater than 10°C, but it swung through the days and this can significantly influence the final results [22-23].

In this case, the envelope can be considered as made of light wall, and it has been assessed that there is a great sensitivity to small variations of the wall temperature measured along the investigated wall. For case study C, the design U-value is smaller than the ones obtained experimentally. Considering that the wall has a complex stratigraphy, the calculated value can be considered as the less reliable, whilst the U-value measured with HFM can be considered trustworthy since the surface temperatures of figure 2e) had quite the same trends, maintaining a constant gradient along the measurement campaign. The results provided after the post process with the average method is near the one obtained with the IRT (difference of 1.33%), demonstrating the validity of the new technique.

4. Conclusion

Based on the results mentioned above, the following conclusions can be outlined.

Three envelopes of buildings built in L'Aquila with different techniques were investigated. The U-values obtained in situ by using HFM and IRT have been compared with the one obtained knowing the thermal properties of the materials of each stratigraphy.

Firstly climatic conditions suggest when thermographic measurements should or should not be undertaken [24-25]; the recommendation is a cloud covered sky, low wind speed and a temperature gradient of at least 10°C between internal and external rooms. However, as demonstrated by a study of Fox [26] climatic changes can also have an impact on the thermal properties of the materials on a transient basis. Consequently, thermographic measurements should be performed for multiple days in different environmental conditions for having more accurate results, since

Furthermore, it has been noticed that temperature changes in the investigated areas cause an uncertainty in the U-value measurement: an IR image shows a temperature distribution by a pixel pattern, and by selecting different pixel of the wall IR image it is possible to gather different temperatures and, therefore, different U-value. A confirmation of these results, obtained from experimental data, can be found in the study of Albatici *et al* [8], that shows a sensitivity analysis of the expected U-value by simulating percentage differences on the input parameters. As recommendation for future works, the investigations on the same wall but on different positions could be furthered and the average U value could be evaluated. Also for data calculated following ISO 6946 important differences between the results can be noticed, due to the uncertainty on the conductivity and thickness values of materials.

HFM measurements were performed in compliance with standard ISO 9869 and results seem to be in accordance to the data obtained with IRT method. In two case studies similar values of thermal transmittance are obtained from calculations and in situ measurements, with the exception of case study B.

Anyway it is important to notice that the errors and uncertainties are present in all methods; the influence factors of the in situ measurement are very complex: the U-value obtained with HFM depends on the variability of the climatic conditions during the monitoring, as well as on the method used for the data process (progressive average method, black-box method) [27-28].

Measurement uncertainty is reduced if the temperature of the internal surface is kept constant as much as possible, and the difference between indoor and outdoor temperature is higher than 10°C. As a general remark, according to norm ISO 9869 the uncertainty of in situ measurement performed by HFM ranges from 14% to 28%.

The U-value obtained by IRT depends on local conditions, on the wall emissivity, the air temperature gradient between indoor and outdoor (which has to be at least 15 °C), while in our campaigns it didn't seem to depend on the thermal capacity of the walls (light or heavy ones).

IRT provides better results if the measurement is performed on the North wall, since there isn't direct solar irradiation.
Finally, it is important to note that IRT method is non-invasive and allows the knowledge of the thermal transmittance of a large area, while the HFM provides only a punctual measurement.
In conclusion, basing on the case studies proposed, it is difficult to establish which method gives the best result, the one that is more representative of the real thermal transmittance of the envelope, but it is not possible to establish a common method, that has to be chosen case by case. In scientific literature, however, it is stated that U-value of walls measured in situ with different methods can vary up from 20% to 50% [29-30].
However the results proposed in this paper seem to show that IRT method might be used in alternative to HFM especially for historical buildings, where faults in the building and the ageing of the materials can greatly affect HFM measurements, although the errors sources are not still identified and quantified, as explained in a previous work.

References

[1] De Lieto Vollaro R, Guattari C, Evangelisti L, Battista G, Carnielo E and Gori P 2015 *Energ Buildings* **87** 87-94.
[2] Meng X, Yan B, Gao Y, Wang J, Zhang W and Long E 2015 *Energ Buildings* **86** 754-65.
[3] EN ISO 6946: 2007. Building components and building elements. Thermal resistance and thermal transmittance Calculation method.
[4] ISO 9869: 1994 Thermal insulation. Building elements. In-situ measurement of thermal resistance and thermal transmittance.
[5] Albatici R and Tonelli AM 2008 *Proc PLEA 2008 - 25th Conference on Passive and Low Energy Architecture* (Dublin, Ireland).
[6] Albatici R and Tonelli AM 2010 *Energ Build* **42** 2177.
[7] Dall'O G, Sarto L and Panza A 2013 *Energies* **6** 3859.
[8] Albatici R, Tonelli AM and Chiogna M 2015 *Appl Energ* **141** 218-28.
[9] Fokaides PA and Kalogirou SA 2011 *Appl Energ* **88** 4358-65.
[10] Van De Vijver S, Steeman M, Van Den Bossche N, Carbonez K and Janssens A 2015 *Qirt 2014.*
[11] Belussi L, Danza L, Meroni I and Salamone F 2015 *Opto-Electron. Rev* **23 (1)** 83-87.
[12] Kylili A, Fokaides PA, Christou P and Kalogirou SA 2014 *Appl Energ* **134** 531-49.
[13] Walker R and Pavìa S 2015 *Build Environ* **94** 155-65.
[14] Peng C and Wu Z 2008 *Energ Buildings* **40** 2076-82.
[15] Baker P 2011 Historic Scotland Technical Paper 10.
[16] Adhikari R S, Lucchi E and Pracchi V 2012 *PLEA – 28th Conference, Opportunities, Limits & Needs Towards an environmentally responsible architecture.* (Lima, Perù).
[17] De Berardinis P, Rotilio M, Marchionni C and Friedman A 2014 *Energ Buildings* **80** 415-423.
[18] Simoes I, Simoes N, Tadeu A and Riachos J 2015 *Qirt 2014.*
[19] Nardi I, Sfarra S and Ambrosini D 2014 *J Phys* **547** 012016.
[20] Evangelisti L, Guattari C, Gori P and De Lieto Vollaro R 2015 *Proceedings of the 15th CIRIAF National Congress.*
[21] UNI 10351:1994 Conduttività termica e permeabilità al vapore.
[22] Desogus G, Mura S and Ricciu R 2011 *Energ Buildings* **43** 2613-2620.
[23] Asdrubali F, D'Alessandro F, Baldinelli G and Bianchi F 2014 *Case Studies in Construction Materials* **1** 53-59.
[24] Lehmann B, Wakili KG, Frank T, Collado BV and Tanner C 2013 *Appl Energ* **110,** 29-43.
[25] Jeonggook K, Junghun L, Jonghun K, Cheolyong J, Hakgeun J and Doosam S 2015 *Building Serv. Eng. Res. Technol.* **0(0)** 1-16 doi: 10.1177/0143624415600701.
[26] Fox M, Coley D, Goodhew S and De Wilde P 2015 *Energ Buildings* **92** 95-106.
[27] Ahmad A, Maslehuddin M and Al-Hadhrami LM 2014 *Energ Buildings* **84** 132-41.
[28] Cesaratto PG and De Carli M 2013, *Energ. Buildings* **59** 29-36.
[29] Evangelisti L, Guattari C, Gori P and De Lieto Vollaro R 2015 *Sustainability* **7** 10388-98.
[30] Cesaratto PG, De Carli M and Marinetti S 2011, *Energ. Buildings* **43** 1792-1801.

Journal of Physics: Conference Series **655** (2015) 012008 doi:10.1088/1742-6596/655/1/012008

Space Heating Load Estimation Procedure for CHP Systems sizing

P Vocale, G Pagliarini, S Rainieri
Department of Industrial Engineering, University of Parma, Parco Area delle Scienze 181/A, 43124 Parma – Italy

E-mail: pamela.vocale@unipr.it

Abstract. Due to its environmental and energy benefits, the Combined Heat and Power (CHP) represents certainly an important measure to improve energy efficiency of buildings. Since the energy performance of the CHP systems strongly depends on the fraction of the useful cogenerated heat (i.e. the cogenerated heat that is actually used to meet building thermal demand), in building applications of CHP, it is necessary to know the space heating and cooling loads profile to optimise the system efficiency. When the heating load profile is unknown or difficult to calculate with a sufficient accuracy, as may occur for existing buildings, it can be estimated from the cumulated energy uses by adopting the loads estimation procedure (h-LEP). With the aim to evaluate the useful fraction of the cogenerated heat for different operating conditions in terms of buildings characteristics, weather data and system capacity, the h-LEP is here implemented with a single climate variable: the hourly average dry-bulb temperature. The proposed procedure have been validated resorting to the TRNSYS simulation tool. The results, obtained by considering a building for hospital use, reveal that the useful fraction of the cogenerated heat can be estimated with an average accuracy of ± 3%, within the range of operative conditions considered in the present study.

1. Introduction

Combined heat and power (CHP) is a highly efficient method of utilising the energy resources (fossil or renewable), therefore its use has become increasingly widespread in the construction sector in order to achieve primary energy savings and consequently reduce the greenhouse gas emissions.
The increase in energy efficiency relating to the cogeneration with respect to separate fossil-fired generation of heat and electricity is mainly affected by the fraction of the cogenerated heat which is actually used to meet building thermal demand.
As a consequence, the matching between the cogenerated heat and the heat load demand plays a leading role in the CHP system optimization [1,2].

An attractive solution to improve the efficiency of the cogeneration systems is offered by the usage of thermal energy storage which allows to store the cogenerated heat when building heat demand is low thus making it available when load is high [3,4].
In absence of such devices, it is mandatory to perform a comparison between the cogenerated heat and the heat load demand to properly size the CHP systems.

Unlike what happens in industrial sector, where the time scheduling of heat load demand is generally known from the process characteristics, in building sector heating and cooling loads profile depends on many factors, such as building characteristics, climatic conditions, building usage, ventilation and air conditioning equipment used and so on.

As highlighted by Pagliarini e Rainieri [5], when the CHP system is planned to meet buildings energy uses, the energy consumption on hourly basis have to be known. The hourly heating and cooling loads can be obtained by using monitored data or can be evaluated by resorting to a suitable simulation software.

For existing buildings it could be hard to perform hourly simulation of the space heating and cooling due to the possible lack of whole-building metering and detailed information about the building envelope.

On the other hand, for this kind of buildings average monitored data are usually available from utility bills, which can be used for buildings energy analysis.

By using a multivariate regression model, Pagliarini and Rainieri proposed an hourly Loads Estimation Procedure (h-LEP) which allows to restore the heating and cooling pattern of existing buildings on an hourly time scale from the monthly averaged energy consumption, the dry bulb air temperature and the global solar radiation on the horizontal plane [6].

More recently, the h-LEP based on a multivariate regression has been modified to dump the effect of the hourly solar radiation on the reconstruction of the hourly loads profile [7].

When the daily averaged energy consumption is regressed against the dry bulb air temperature only, the h-LEP does reduce to the single-variate two-parameter model [8].

In this paper the h-LEP is implemented with a single climate variable (i.e. the hourly average dry temperature) with the aim to evaluate the useful fraction of the cogenerated heat for different operating conditions in terms of buildings characteristics, weather data and system capacity.

Although the single variable regression analysis has been widely applied in building energy research and several inverse model have been proposed [8-10] (i.e. linear, change-point linear, variable-based degree-day and multivariate inverse models), it has to be pointed out that these inverse models have been developed and applied for establishing the baseline model which is used for identify energy savings in energy retrofitting projects.

To validate the proposed estimation procedure, a comparison between the restored heating loads profile and the hourly profile obtained by using TRNSYS software is performed. A building for hospital use has been considered as case study.

2. Problem statement

The overall energy efficiency of the CHP system, based on the first principle of thermodynamics, can be defined as the ratio of useful energy output to the energy input [11]:

$$\eta_{CHP} = \frac{W_{CHP} + Q_{CHP,u}}{E_{CHP,in}} \tag{1}$$

where W_{CHP} and $Q_{CHP,u}$ are the electricity and the useful thermal energy produced by the cogeneration unit, respectively, and $E_{CHP,in}$ is the energy supplied to the CHP system.

Therefore by sizing the CHP for the thermal requirements of the facility, the highest CHP system efficiency can be reached.

Unfortunately, for existing buildings the hourly space heating profile is seldom known. However, it can be estimated from the cumulated energy consumption by adopting an inverse approach. Pagliarini and Rainieri [5] have suggested an interesting inverse model (h-LEP), based on a multivariate regression approach, which allows to restore the space heating pattern by regressing the monthly energy consumption against the dry-bulb temperature of the external air and the global solar radiation on the horizontal plane.

When considering the dry bulb temperature only, then h-LEP reduces to the single-variate two-parameter model [8], which expresses the daily net heat load of the building, as a linear function of the dry bulb temperature, averaged in the time period for which the cumulated energy use is known:

$$Q = K\left(T_{bal} - T_e\right) \tag{2}$$

being K (W/°C) the overall sensible heat transfer coefficient accounting for both transmission and ventilation, T_e (°C) the outdoor dry bulb temperature and T_{bal} (°C) the balance temperature, which accounts for both internal and solar heat gains [7].

In this condition, the estimation procedure becomes straightforward and it can be implementable on a spreadsheet.

It has to be highlighted as the assumptions made in the present analysis are consistent with the results of previous studies, which show that in the regression analysis the outdoor dry-bulb temperature is the most important weather variable, also when the hourly time scale is considered [12].

3. Case study

The inverse procedure here presented has been adopted to restore the hourly space heating load profile of a building for hospital use (Figure 1). This kind of building has been chosen because in hospitals a good simultaneity between the electric and thermal loads occurs, which is the best condition for consideration of cogeneration plant [13-15]. In particular, in winter the thermal energy can be used for space heating, water heating and for the process (sterilization, laboratory requirements, and so on). Since only the space heating is dependent on the climate data, in this work other contributions have been disregarded.

To evaluate the influence of climatic conditions, the building has been supposed located in two different climates, namely Berlin (Germany) and Stockholm (Sweden).

Since the inverse model here adopted considers only the influence of the external air temperature on the space heating demand, it is important to analyse different operating conditions in terms of heat gains and building thermal losses to better evaluate the accuracy of the proposed model.

For this purpose, the building behaviour has been studied by considering two values of internal heat generation per unit of floor surface area, namely q_g= 0 W/m^2 and q_g= 5 W/m^2 and three values of air change rate (i.e. n = 0, 0.3 and 1 h^{-1}). The internal temperature has been assumed equal to 20 °C.

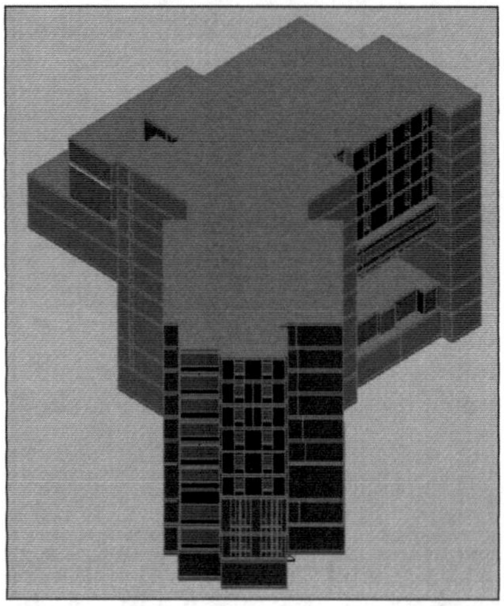

Figure 1. Case study building.

The thermal behaviour of the case study building has been analysed by using the *Multi-Zone Building* component within the TRNSYS environment.

Figure 2 shows the monthly daily averaged energy use as a function of the mean daily temperature in the period, as required by the application of the single-variate two-parameter model, together with the interpolating straight, for two different operating conditions and for all the months included in the heating season (i.e. from November to March).

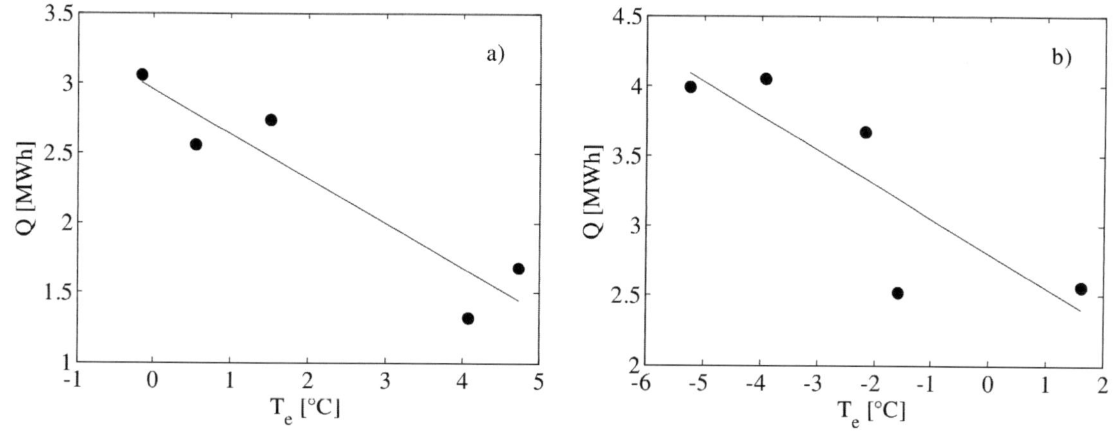

Figure 2. Daily average energy use as a function of the average temperature, for q_g=5 W/m^2 and n=0.3 h^{-1}: a) Berlin; b) Stockholm.

By regressing the daily average values of the energy use against the average temperature, the values of the overall heat transfer coefficient (K) and of the balance temperature (T_{bal}) can be estimated.

Once the unknown parameters are estimated, the hourly space heating profile can be restored, by means of Eq.(2).

The comparison between the heating load profile obtained by using the TRNSYS tools and the restored hourly building load is shown in Figure 3, for two months of heating season. Since in the inverse procedure here adopted the effect of solar radiation is disregarded, the model leads to an overestimation of the peaks of the heating load, as it can be observed in Figure 3.

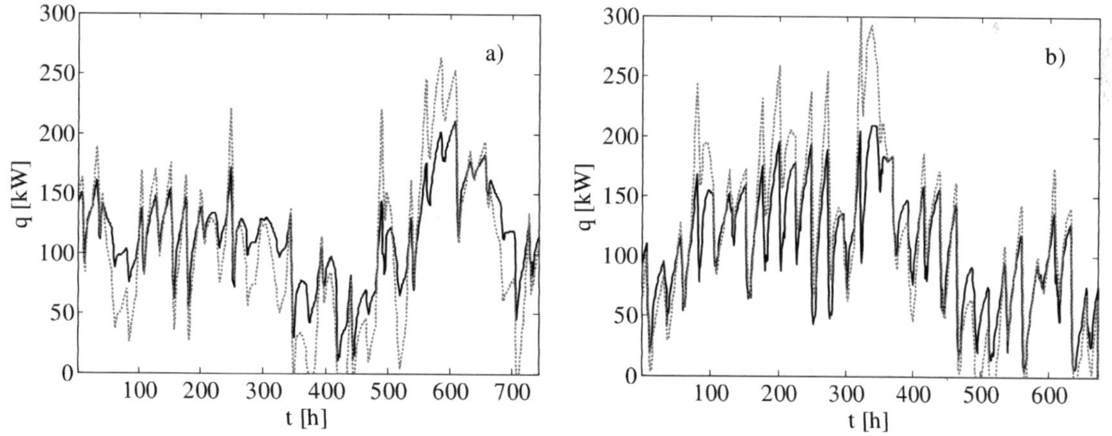

Figure 3. Comparison between the simulated heating load (continuous black line) and the restored one (dashed red line). Berlin, q_g =5 W/m^2, n =0.3 h^{-1}. a) December; b) February.

Therefore it is expected that the accuracy of the h-LEP with a single climate variable can be improved by considering only the months characterised by low values of both external air temperature and solar radiation. In fact if the unknown parameters (i.e. K and T_{bal}) for the case study building are evaluated by considering only four months of the heating season (i.e. from November to February), the estimated data fit the simulated ones much better.

Journal of Physics: Conference Series **655** (2015) 012008 doi:10.1088/1742-6596/655/1/012008

Figure 4 shows a comparison between the simulated profiles (continuous black line) and the estimated ones obtained by regressing the monthly energy consumptions for the whole heating season (dashed red line), under the h-LEP with a single climate variable, and by excluding the consumption of March from the regression analysis (dashed blue line).

To put in evidence the difference between the multivariate regression models and the single variable ones, in Figure 4 the restored heating profiles obtained by applying the modified h-LEP [7] are presented as well.

It can be observed that if only four months are considered, both the h-LEP with a single climate variable and the modified h-LEP provide a satisfactory fit of the simulated load profiles.

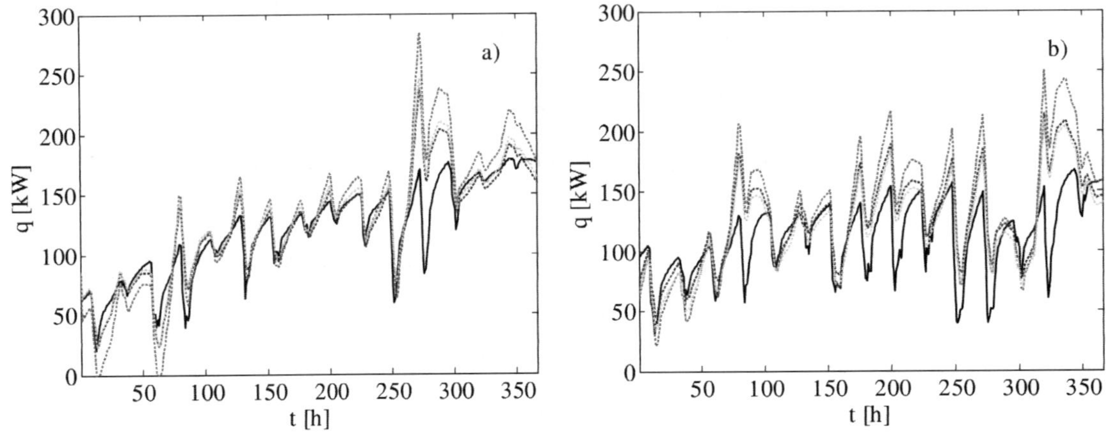

Figure 4. Comparison between the simulated heating load (continuous black line) and the restored ones (dashed red line for h-LEP with a single climate variable applied by considering 5 months, blue line for h-LEP with a single climate variable applied by considering 4 months and green line for the modified h-LEP). Berlin, q_g =0 W/m^2, n =0 h^{-1}. a) First and second week of January; b) first and second week of February.

As expected, the inverse model here adopted guarantees better accuracy if the thermal losses are relevant compared to the heat gains. This effect it is evident in Figure 5 where the influence of the air change rate (n) on the restored heat loads profile is presented.

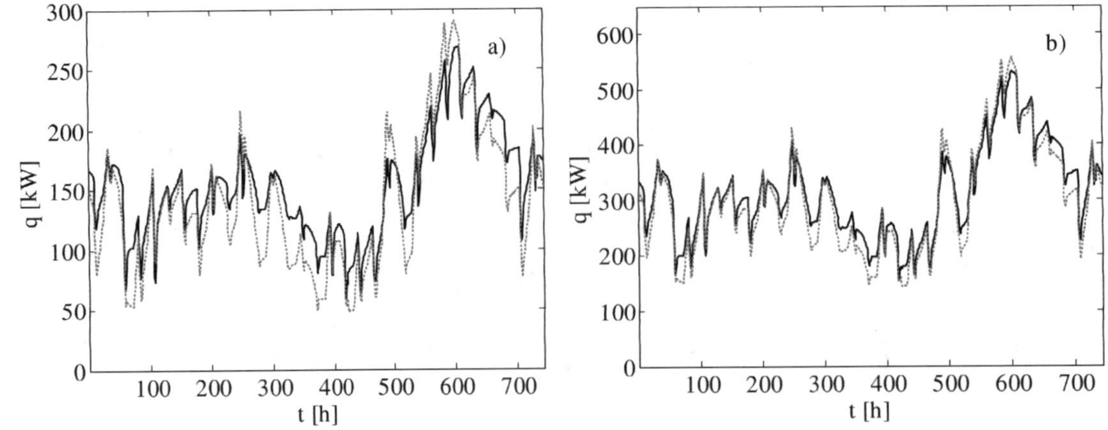

Figure 5. Comparison between the simulated heating load (continuous black line) and the restored one (dashed red line). Stockholm, q_g=5 W/m^2. a) n =0.3 h^{-1}; b) n =1 h^{-1}.

The quality of the h-LEP with a single climate variable is presented in Tables 1 and 2 where the values of coefficient of determination R^2 [12] are indicated for each operating condition here analysed.

Table 1. Coefficient of determination R^2. Berlin.

	Present model (5 months)	Present model (4 months)	Modified h-LEP [4]
q_g =0 W/m^2, n =0 h^{-1}	0.33	0.56	0.69
q_g =5 W/m^2, n =0.3 h^{-1}	0.63	0.77	0.80
q_g =5 W/m^2, n =1.0 h^{-1}	0.87	0.93	0.94

Table 2. Coefficient of determination R^2. Stockholm.

	Present model (5 months)	Present model (4 months)	Modified h-LEP [4]
q_g =0 W/m^2, n =0 h^{-1}	0.58	0.70	0.71
q_g =5 W/m^2, n =0.3 h^{-1}	0.76	0.86	0.84
q_g =5 W/m^2, n =1.0 h^{-1}	0.91	0.96	0.94

To assess the effectiveness of the proposed approach in the evaluation of energy efficiency of the CHP systems, both the restored profiles and the simulated ones have been used to estimate the useful heat obtainable by the cogeneration unit.

With the aim of optimising the energy performance of the CHP systems, the analysis has been carried out by considering several values of the CHP unit capacity.

Figure 6 shows the useful fraction of the cogenerated heat α_u versus the normalized cogeneration unit capacity γ (i.e. the ratio between the CHP capacity and the average heating demand of the case study building) for both space heating profiles (i.e. the load pattern simulated with TRNSYS software and the profile restored by applying the h-LEP with a single climate variable).

It can be observed that the regression approach here adopted allows to evaluate the useful thermal energy with a good accuracy, especially for Stockholm which is characterized by values of the dry-bulb air temperature and solar radiation lower than those of Berlin.

The robustness of the h-LEP with a single climate variable has been quantified by calculating the relative error between the useful fraction of the cogenerated heat α_u estimated by considering the restored load profile and the corresponding one evaluated by means of the simulated load profile:

$$\varepsilon_r = \frac{\alpha_{u,est} - \alpha_{u,\mathrm{sim}}}{\alpha_{u,\mathrm{sim}}} \tag{3}$$

The restored profiles lead to an underestimation or overestimation of the useful cogenerated heat, depending on the operating conditions and on the CHP unit capacity. The maximum and minimum relative errors that occur by using the estimated load pattern are +2.51% and -6.87%, respectively, within the CHP capacity range considered in the present analysis. Nevertheless, the relative error, averaged over the operating conditions considered in the present study, ranges between -3% and +3%; therefore the h-LEP with a single climate variable performs quite satisfactorily.

Journal of Physics: Conference Series **655** (2015) 012008 doi:10.1088/1742-6596/655/1/012008

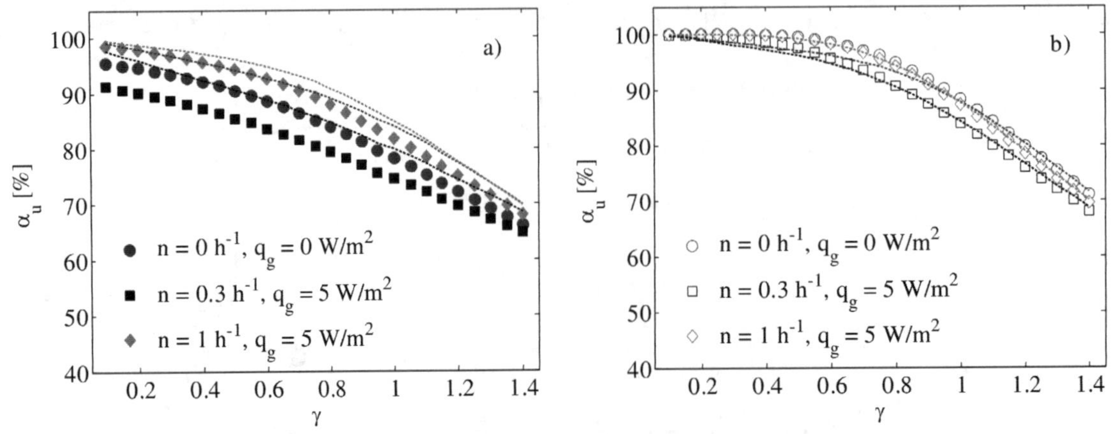

Figure 6. Comparison between the useful fraction of the cogenerated heat obtained by the simulated load profile (dashed lines) and the corresponding values evaluated by means of the space heating pattern restored by considering the whole heating season (markers): a) Berlin climate data; b) Stockholm climate data.

As discussed below, if the h-LEP with a single climate variable is applied by considering only the months characterized by low values of external air temperature and solar irradiation (i.e. the period from November to February for the here considered case study), the estimated profiles fit better the simulated ones and consequently the fraction of useful heat achievable by the CHP system can be estimated with better accuracy, as it can be observed in Figure 7.
The maximum and minimum relative errors on the useful fraction of the cogenerated heat, within the here considered operating conditions, are +0.46% and -4.38%, respectively.

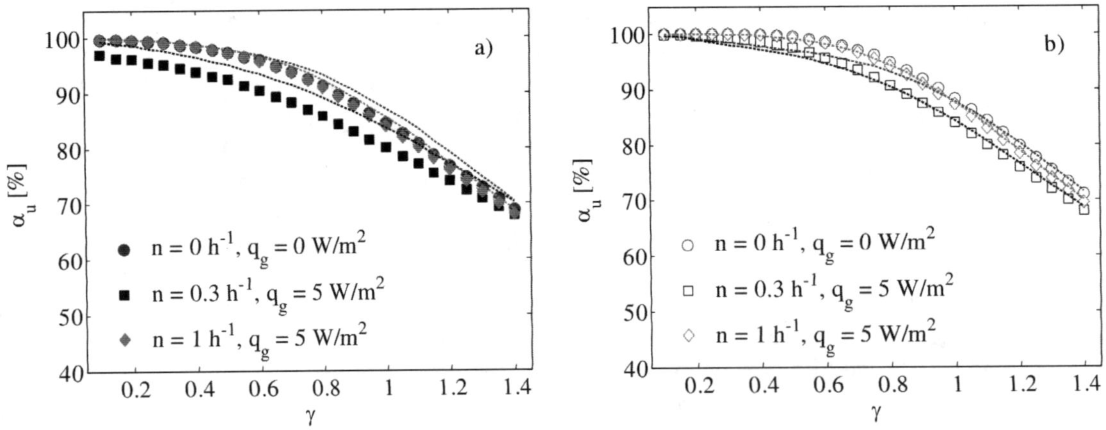

Figure 7. Comparison between the useful fraction of the cogenerated heat obtained by the simulated load profile (dashed lines) and the corresponding values evaluated by means of the space heating pattern restored by considering four months of the heating season (markers): a) Berlin; b) Stockholm.

4. Conclusions

With the aim to restore the hourly load profiles of existing buildings from the monthly energy consumption, the hourly loads estimation procedure (h-LEP) has been implemented by using as input variable only the dry-bulb air temperature.

The validity of the here adopted model has been assessed by comparing the estimated space heating load of a case study building (i.e. a building for hospital use) with the heating demand obtained by using the TRNSYS tool.

To evaluate the influence of the building characteristics and weather data on the accuracy of the h-LEP with a single climate variable, several operating conditions have been analysed by changing the value of the internal heat generation, the air change rate and the building location.

The results of the simulations point out that the h-LEP with a single climate variable works quite satisfactorily; however it has to be highlighted that the here adopted procedure works better for buildings characterized by constant internal heat gains and air ventilation rates and for locations characterized by low solar radiation.

By using the restored heating load, the useful fraction of the cogenerated heat can be estimated with an average relative error ranging between -3% and +3%, for the operative conditions here considered.

5. References

[1] Kavvadias K C, Tosios A P and Maroulis Z B 2010 *Energy Conversion and Management* **51** 833–845
[2] Mago P J and Smith A D 2012 *Building and Environment* **53** 74-82
[3] Pagliarini G and Rainieri S 2010 *Applied Thermal Engineering* **30** 1255-1261
[4] Smith A D, Mago P J and Fumo N 2013 *Sustainable Energy Technologies and Assessments* **1** 3-12
[5] Pagliarini G, Rainieri S and Vocale P 2014 *Energy & Environment* **25** 1423-1438
[6] Pagliarini G and Rainieri S 2012 *Energy and Buildings* **49** 348-355
[7] Pagliarini G, Rainieri S and Vocale P 2015 *Proc. ASME-ATI-UIT 2015 Conference on Thermal Energy Systems: Production, Storage, Utilization and the Environment (Napoli, Italy)* in press
[8] American Society of Heating, Refrigerating and Air-Conditioning Engineers 2001 *ASHRAE Handbook – Fundamentals,* SI Edition (Atlanta: ASHRAE) pp. 31.1 – 31.34
[9] Reddy T A, Deng S and Claridge D E 1999 *ASME Journal of Solar Energy Engineering* **121** 40–46
[10] Reddy T A and Claridge D E 1994 *Energy and Buildings* **21** 35–44
[11] Ertesvag I S 2007 *Energy* **32** 2038–2050
[12] Reddy, T A 2011 *Applied Data Analysis and Modeling for Energy Engineers and Scientists* (New York: Springer)
[13] Kayo G, Hasan A and Siren K 2014 *Energy and Buildings* **82** 685-695
[14] Teke A and Timur O 2014 *Renewable and Sustainable Energy Review* **33** 224-235
[15] Pagliarini G, Corradi C and Rainieri S 2012 *Applied Thermal Engineering* **44** 150-158

Experiments and simulations on a thermosyphon solar collector with integrated storage

P Toninelli[1], **A Mariani**[2,*] **and D. Del Col**[1]

[*] Corresponding Author: mariani_a@enea.it
[1] Università di Padova, Dipartimento di Ingegneria Industriale, Padova, Italy
[2] ENEA, C.R. Casaccia, Roma, Italy

Abstract. This paper deals with the thermal behaviour of a new type of flat solar collector that integrates the fluid storage tank. Often the main limitation of the solar thermosyphon installations is the prohibition to adopt external storage tanks due to their impact, especially for historical centres of particular architectural significance. To avoid this issue, a new system, that includes the collector and the storage, has been developed. This new apparatus works as a thermosyphon: it is possible to take advantage of the natural convection to avoid using a pump. Experimental tests have been conducted in such a collector with and without the absorbing plate. Furthermore, CFD simulations are reported to analyze in detail the dynamic thermal performance of the innovative solar collector and a good-agreement with the experimental tests has been found. Finally, both in numerical simulations and in experimental data the thermosyphon effect has been verified, obtaining the desired water temperature for domestic applications.

1. Introduction

The solar thermal system is a developed technology that aims at converting the solar radiation into thermal energy available to user. There are numerous designs of solar collector on the market to meet the needs of different locations and areas of applicability, i.e. water and air heating, cooling and electrical power production and so on.

For domestic applications the solar collectors can be, mainly, divided in three groups: forced-circulation, natural convection and integrated collector storage.

The advantage of integrated collector storage solar water heaters (ICSSWHs) is, primarily, an isolated, simple and cheap system with no moving parts, allowing the user to be power grid independent. Furthermore, the collector and the thermal storage are a single unit, so the esthetic impact of the system is reduced as compared to common thermosyphon devices that are prohibited in some particular areas, e.g. old towns, for the presence of the external thermal storage. The ICSSWH configurations are reviewed in detailed by Smyth and coworkers [1] from the first technologies to the recent ones. The main disadvantages, instead, are due to high top heat losses and hot water retention during the night and the not-insolation periods. These performance problems can be limited or overcome with appropriate modifications.

About solar radiation capturing, more systems have been analyzed and optimized from an optical point of view with the concentrating device installation; for example a ICSSWH with compound parabolic concentrator has been studied in detail by Devanarayanan and Kalidasa Murugavel [2] when varying geometry of reflectors and number of tanks. A similar study was performed by Tripanagnostopoulos and Souliotis [3] when varying the position of two tanks inside the solar water heater to analyze its thermal performance.

The key-element of the solar collector is the absorber plate that is designed to maximize the solar radiation absorption with special coating. The absorber plate can be coated with black, semi-selective or selective layer. The selective coating on the absorber plate guarantees a high degree of absorption (>0.95) in the range of ultraviolet and visible region and at the same time a low emissivity factor (<0.1) in the infrared region having low heat losses by radiation to ambient air. A comparison between different coatings has been experimentally analyzed by Del Col *et al.* [4] for the flat solar collector and Ge and Xie [5] for ICSSWH. In addition, Kumar and Rosen [6] analyzed a ICSSWH corrugated absorber plate to increase the solar radiation collection and improve the heat transfer coefficient. Beside the reflectors application and the treatments on the absorber plate, there are other possibilities to increase the water temperature:

- reducing heat losses by convection (transparent honeycomb layer, Kaushika and Reddy [7] or single/multi glass cover on the top, Kuman and Rosen [8]), by conduction (insulation on the back) and by radiation (type of glass cover and absorber plate);
- choosing the proper ICSSWH geometry, i.e. rectangular size when varying aspect ratios (Henderson *et al.* [9]), triangular (Kaushik *et al.* [10]) or trapezoidal one (Smyth and Norton, [1]);
- use of baffles layer inside the ICSSWH.

It is also fundamental to be able to maintain for a long time the water at high temperature inside the ICSSWH avoiding too high heat losses and reversal thermosyphon flow. To retain the fluid temperature at user operating condition the hot water could be withdrawn continuously from the top of the system through constant/intermittent forced flow rate (Kumar and Rosen [6,8]) or through natural one with a ICSSWH geometry subdivided in two parts A and B, exposing part A to solar radiation and well-insulating part B (Kumar and Rosen [11]).

Furthermore, in case of forced flow there has to be a suitable draw-off regime making better use of the water thermal stratification. For this reason, Smyth and co-workers [1] suggest different manifold designs to improve the flow inlet to the ICSSWH.

A lower system efficiency can be due to a reversal thermosyphon flow. This phenomenon takes place when the solar radiation is low or null, i.e. during cloudy sky or night, but it can be prevented by using mechanical devices as reported by Faimal *et al.* [12].

Another option to maintain a high water temperature is the application of PCM (Phase Change Material) around the thermal storage. Chabaane *et al.* [13] has performed a comparison between two ICSSWH systems with and without PCM technology. It is worth noting that during the day for the PCM case the water temperature remains almost constant at the maximum value which is nearly equal to the melting temperature of the PCM, allowing a better performance than the sensible heat storage unit.

The objectives of this paper are:

- to verify the natural convection occurring inside the thermosyphon presented here and the achievement of the desired water temperature for domestic applications;
- to analyze in detail the dynamic thermal performance of the innovative solar collector and to validate the CFD simulations with experimental data.

As compared to the past studies, this paper presents an innovative solar thermosyphon which permits:

- to avoid the installation of the external tank;
- to store hot water in the well-insulated side reducing the heat losses to external air and improving thermal stratification;
- to obtain hot water in a compact device with modular configurations, adaptable to the user needs.

2. The new solar panel

A new ICSSWH design is reported in this paper. As compared to the common solar thermosyphon, to compensate the absence of the external tank, the idea is to build a solar collector with the single modular element composed by a small tank, placed below a receiver tube welded to an absorber plate, as shown in figure 1. In this way the big external tank was replaced by a sufficient number of modules in order to satisfy the needs of hot water.

The simple and cheap design of an ICSSWH can be here observed. In fact, the working fluid is tap water and there are not moving parts since the fluid moves under natural convection due to the thermosyphon effect.

This new ICSSWH design allows to heat the water in the receiver tube by the solar radiation and to withdraw the hot water from the receiver tube to the small tank place below it. In this way, the hot water is driven by natural circulation to the small tank where the heat losses are mitigated by a good thermal insulation. This configuration may be similar to that reported by Kumar and Rosen [11].

One disadvantage of the system may be the higher depth (about 6 cm) as compared to traditional solar flat plate collectors, but this aspect should be optimized during industrialization.

During the experimental tests, the section was composed by a single modular element, that was filled with water until the air was completely eliminated. Hereinafter, the experimental tests were performed in no flow conditions, so the device runs as a closed-loop.

In this innovative ICSSWH the tanks are connected directly to the tap water at the inlet and the hot water at the outlet. Hot water is taken directly from the top of the tank, when there is a demand for it, and at the same time the same tank is refilled with cold tap water.

Figure 1. The design of the new solar thermosyphon.

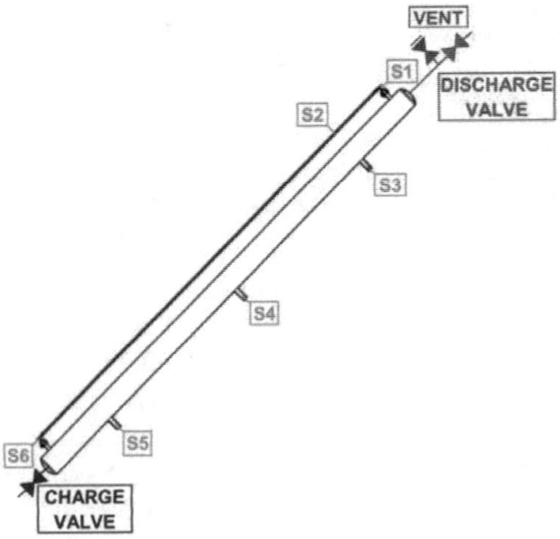

Figure 2. Thermocouples position in the test section.

3. Experimental data

Three different test sections have been run during the experimental campaign and each one consists of a single solar panel module, properly instrumented. The first two cases were preliminarily tested and then the third one.

The solar irradiance has been simulated through Joule effect by applying an electrical resistance: this allows to carry out indoor tests with simple and fast procedures. In fact, in case of outdoor tests, since the apparatus is exposed directly to solar radiation, the experimental data depend on the climate dynamics, so it becomes fundamental to measure solar radiation and air temperature during the whole test. For the first two tests a cylindrical electric heater of 4 mm diameter and length equal to 1600 mm has been used. This one was applied directly over the receiver pipe, centred along the total length and for its installation the receiver plate has been removed. The receiver tube and the resistance were wrapped with aluminium tape to ensure good thermal contact. The external diameter of the channel is

8 mm for the first test and 10 mm for the second one. The 10 mm tube was chosen because it is more widely available on the market and therefore cheaper to purchase.
For the third test section a heating strip of length equal to 1600 mm has been placed on the absorbent plate using a glue. This configuration, as compared to the one performed for the preliminary tests, ensures a better agreement with the operating conditions of a solar collector.
Since the absorber plate is welded on a 10 mm channel, as depicted in figure 1, a conductive paste was used to avoid contact thermal resistances due to air presence between the electrical strip and the absorber plate. With this operation a good thermal contact is guaranteed because the paste will fill the spaces that, otherwise, would be filled by air.
All the test sections were instrumented with six K-type thermocouples (figure 2): S1 at the end of the received tube, S2 about 200 mm from the end of the electrical resistance, S3, S4, S5 respectively at 250 mm from the top, at the centre and at 250 mm from the bottom of the tank and, finally, S6 at the beginning of the received tube. Since all the thermocouples are placed in the fluid, they measure the local water temperature in the point where they are placed, except S2 that is applied on the adsorbed plate. The measurement uncertainty of the K-type thermocouples was ±0.2 °C.
The supplied power by the electrical resistance is measured using a power meter LAUGHLIN DMG 700 while the data acquisition system is National Instruments Compact Point File. The measurement uncertainty of the electrical power was ±10 % due to the disturbances in the electrical grid during the experimental tests. The device was insulated with neoprene on any surface to limit heat losses to external air; precisely, the thickness was about 5 mm around the receiver pipe and about 10 mm around the collector tank for the preliminary tests. For the third test section, instead, it has been preferred to insulate only the top of electrical strip and the whole tank with a thickness of 35 mm and of 15 mm respectively. The insulation were made by applying self-adhesive layers of neoprene (thermal conductivity is in the range of 1 ÷ 2.5 W m^{-1} K^{-1}). In addition, to limit the radiation heat transfer, an aluminium tape has been applied on the last neoprene layer.
It is worth highlighting that, during the experimental test, the top side of absorber plate has been insulated to input the source term avoiding heat losses directly to the ambient; in this way it is possible to know the real source inputted to the solar collector. Furthermore, it has been decided to not insulate the bottom side of absorber plate to work closer to the real performance of the solar panel, where the heat losses to the ambient, primarily, occur through the glass cover.
Since the thermosyphonic phenomenon is driven by the natural convection the fluid velocity is very low, i.e. few centimetres per second, for this reason any sensor would affect the measurements and the flow itself, making the data collection very difficult.
The thermal power has been supplied in two different ways. For the first test the heat source has been performed with: an increasing ramp from 0 to 200 W for two hours, a constant value of 200 W for three hours and, finally, a decreasing ramp from 200 to 0 W for two hours. For the second and third test, instead, a constant heating power of 130 W for three hours has been supplied.
The first heating power profile has been chosen to study the thermosyphon performance under not-constant supplied power. The second heating power profile has been adopted to analyze the temperature trends inside the solar collector, with a constant power of 800 W m^{-2} over the absorber plate.
The experimental data were obtained at varying device tilt angle, precisely, from 15° to 45° for the preliminary tests and 45° for the third one.
The solar thermosyphon was filled by water and at the beginning of each test the fluid and device temperature was around 27°C.
This experimental apparatus, due to its simplicity, allows to modify its configuration quickly. In fact, it is possible to change the inclination of the circuit and the power supplied to the electrical resistance, simulating different situations: exposure to direct solar radiation during peak hours (power kept constant at a value close to the maximum solar radiation available in Northern Italy latitudes) or the evolution of solar radiation during a typical day, with the characteristic bell-shaped curve of solar power.

4. Experimental results

As depicted in figures 3-5 in all the cases the establishment of natural convection is confirmed.

In fact, the temperatures increasing can be observed in time inside the receiver tube, showing that natural circulation causes the water flow and the good operation of the whole system.

Furthermore, it can be confirmed that the average temperature in the collector tank is compatible with the domestic demand of hot water.

From figures 3-4 the effect of collector tilt angle can be evaluated. At higher inclination the temperatures in the upper part of collector, primarily T_S2, T_S1 are lower and the other temperatures, i.e. T_S3 and T_S4, are higher than those at lower inclination. The reason of this behaviour is due to a different buoyancy effect at varying the tilt angle of thermosyphon. Infact, at higher device inclination the buoyancy effect is high and it improves the natural circulation having a better temperature distribution inside the storage tank.

In all the cases it can be noted that although the thermocouple T_S6 is placed near the heated area its value does not increase much in time, because it is more influenced by the cold fluid coming from the collector tank rather than by the conduction on the metal parts of the tube.

From figures 3-4, it can be observed as the temperature trends are strictly influenced by the heating mode. In fact for the first test all the temperatures follow the bell-shaped function of heating power, so three steps occur:

- in the first step the temperatures increase at similar rate of the input power (0-2 h), primarily for T_S1 and T_S due to the short distance from the thermal source;
- in the second step the temperature trends are flatter due a constant heating power (2-5 h);
- finally, when the electrical power decreases the thermocouples records a maximum value, that occurs a different locations inside the thermosyphon, and then the temperatures decrease, except for the T_S6. The temperature trend of T_S6 could be due to a reversal flow of natural circulation inside the device.

In the second and third test the trends are different as compared to the first one except for the third step, where the temperature trend are quite similar. In fact, in figures 4-5 it is worth noting that, at the test beginning, when a constant value of electrical power is supplied, the temperatures increase very rapidly in the first 450 seconds and then the temperature increasing becomes much flatter. This could be due to natural circulation behaviour: when the density gradient between the thermosyphon ends is enough high to improve the flow, i.e. around 450 seconds after the beginning, then the temperature slopes are lower.

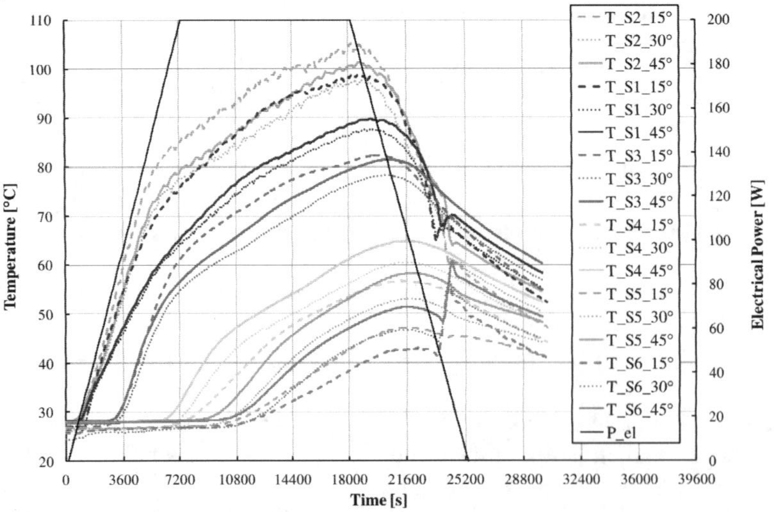

Figure 3. Temperatures in the thermosyphon with a 8 mm diameter channel at different tilt angles (Test 1).

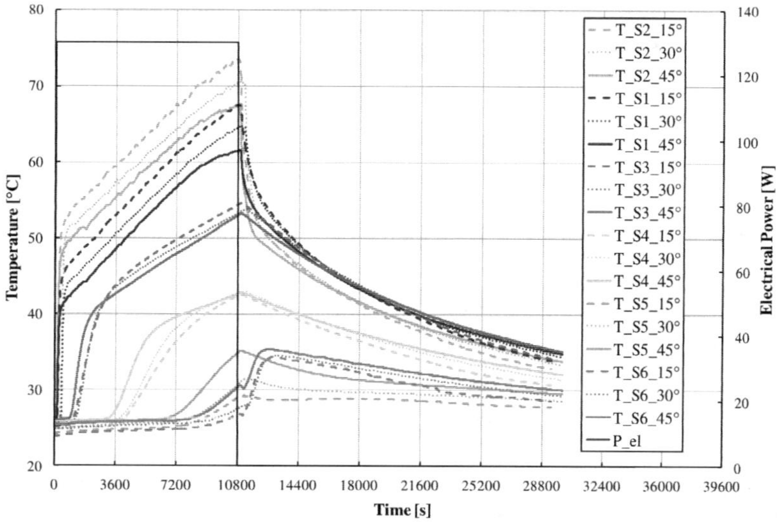

Figure 4. Temperatures in the thermosyphon with a 10 mm diameter channel at different tilt angles (Test 2).

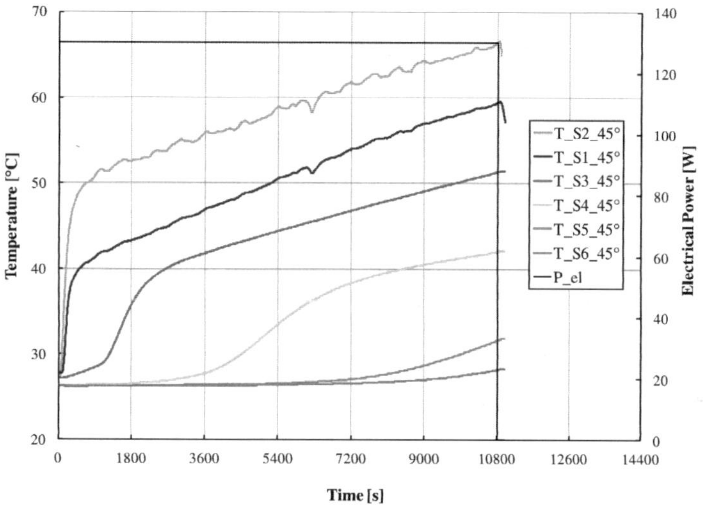

Figure 5. Temperatures in the thermosyphon with a 10 mm diameter channel and the absorber plate at 45° of tilt angle (Test 3).

5. Numerical simulations

In this section the comparison between the CFD simulations and the experimental data is reported. The goal of CFD simulations is the detailed description of thermosyphon performance, obtaining the temperature distribution and the velocity field during its operating conditions.

The setting of the CFD simulations is formed by three parts:

- the definition of thermosyphon geometry;
- the setting of the mesh, i.e. type and number of elements, so the mesh refinement;
- the setting of models (i.e. single/multi-phase, turbulence, transient/steady-state…), boundary conditions, materials and solution methods.

The dimensions of the absorber plate were 2037 mm, 106 mm and 1 mm, respectively for the length, the width and the thickness. The material of the absorber plate and the pipe was copper, whereas the material of tank was steel. The external diameter was 8÷10 mm for the pipe and 89 mm for the tank.

Three-dimensional and time-dependent simulations of solar thermosyphon are performed at the same operating condition of test 3. The working fluid is considered incompressible.

To set the thermosyphon effect it is important to apply three conditions to improve the natural convection: a source term (i.e. a constant wall heat flux), a temperature-dependent density of the fluid and the presence of gravity with orientation that depends by the thermosyphon tilt angle.
For this reason, all fluid properties (i.e. water) are considered to be temperature-dependent and are interpolated by polynomials computed from NIST Refprop Version 9.0.
The initial conditions of simulations are a fluid temperature of 27°C and a mass flow rate equal to zero.
Since the convection is natural and not forced the fluid velocity is very low (i.e. centimeters per second), so the fluid flow has been simulated as laminar.
The constant value of the heat flux imposed on the absorber plate is 800 W m^{-2}.

5.1. Spatial Mesh
The tridimensional domain is discretized into about 4.5 millions of elements. Due to computational efforts and costs two mesh types are chosen and the symmetry condition has been applied.
The set up of the mesh is not simple since the domain size is wide. For this reason, in the tank domain the mesh is unstructured and much coarser, while in the copper pipe and absorber plate domains the mesh is structured and much finer, as depicted in figure 6.
In this way it is possible to have:

- better accuracy in the zones where the fluid dynamics calculations are more complex;
- computational saving where the fluid properties are almost constant and the fluid dynamics calculations are simpler.

Furthermore, the mesh independence test has been performed. The new mesh was composed by about 5.8 millions of elements. The maximum deviation between the temperature profiles obtained with the two meshes was about 0.5 °C.

5.2. Temporal Mesh
As mentioned above, the simulations are in transient configuration since the phenomenon studied is time-dependent. Each simulation analyzed the thermosyphon performance for a time period equal to 2 hours. The time domain is discretized with two methods:

- for the first ten minutes the time step is five seconds and the results are recorded every minute;
- hereinafter the time step is thirty seconds and the results are recorded every ten minutes.

This procedure is done because the applied heat flux is immediately equal to 800 W m^{-2} and therefore the temperature and velocity profiles of fluid vary rapidly.

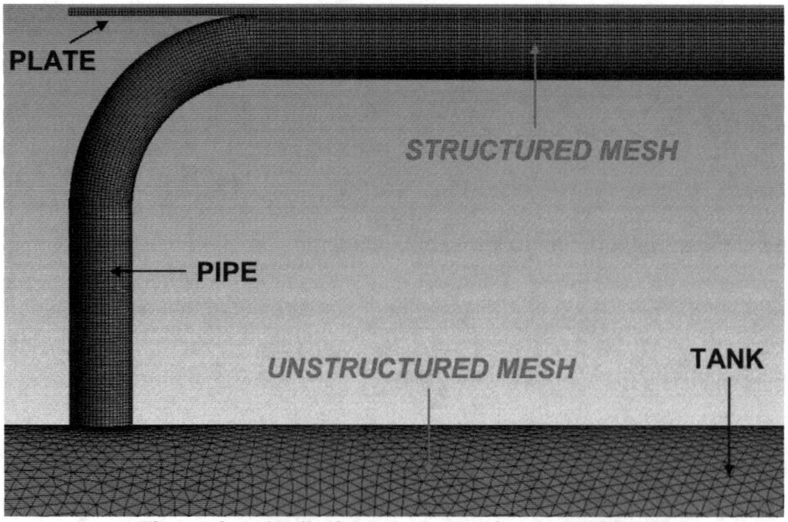

Figure 6. A detail of the mesh used for simulations.

5.3. Boundary Conditions

During the experimental campaign the most of heat losses occurs at the bottom side of the absorber plate, as described in the previous section, while the upper side is well-insulated with neoprene. So, to evaluate a similar amount of heat losses a heat transfer coefficient of 8 W m^{-2} K^{-1} has been applied over the absorber plate to take into account the forced convection with air (i.e. a wind velocity of 1 m s^{-1}) and the radiation transfer with sky. The air temperature is equal to the initial temperature of the fluid, 27°C.

The heat flux over the absorber plate has been defined as a body source term, i.e. 800 kW m^{-3}, for a volume of 1600 mm x 53 mm x 1 mm (since the symmetry conditions have been applied the width of the volume is equal to the half width of the absorber plate), yielding a heating power of about 130 W. This method has been adopted since in CFD simulations it is not possible to define two different boundary conditions on the same surface.

Furthermore the dimensions used for the body source term are related to the length of the heating resistance, the width (symmetry condition) and the thickness of the absorber plate.

During the experimental tests the solar collector was well-insulating with neoprene, so for numerical simulations the heat losses on the other external surfaces of the thermosyphon are imposed equal to zero (adiabatic boundary).

When the heat losses from the absorber plate to ambient are taken into account, two effects in different periods can be noted:

- in the short term, since the heat flux is not transferred totally to the fluid, the density gradient is lower at the start, and in the first 10 minutes the maximum velocity is lower than that one under completely adiabatic condition;
- in the long term, the heat losses affect on the slope of the temperature profiles.

5.4. Setting solver

The commercial CFD code Ansys FLUENT release 14.0 has been used for the unsteady simulations to define the thermosyphonic behavior inside the new apparatus presented here.

An implicit first order time scheme is used to solve the momentum and energy equations and second order upwind schemes are applied for their spatial discretization.

The standard scheme is used for the pressure interpolation, while the pressure-velocity coupling was handled by means of the SIMPLE algorithm. The under-relaxation factors used are the default values presented in the FLUENT code.

6. Experimental data vs. numerical simulations

In this section the validation of numerical simulations with experimental data is performed. In figure 7 the temperature trends of both numerical simulations and experimental data using thermocouples readings are reported.

It is worth noting that, at the beginning, the temperature at the end of the copper pipe increases rapidly since it is the nearest to the thermal source term. The heat is applied as a step function at the start of the test, and then the slope of the temperature curve becomes lower and its value is defined by the heat losses to air. In fact, in adiabatic conditions the temperature slope is higher than that one in the simulations presented here, since the heat flux would be completely transferred to the working fluid. A similar trend can be noted for the other temperatures, precisely for T_S3 and T_S4, and it is important note that the rapid increase of temperature occurs at different times with different slope for the different positions of thermocouples in agreement with the experimental data. This behavior of temperature trends highlights the start of natural convection due to buoyancy effects and the presence of a thermal stratification inside the tank, as depicted in figures 7-8.

The different slope of temperature profiles is due to the amount of fluid volume and the fluid velocity in the zone where the thermocouple is placed. In fact, the amount of fluid near the S1 thermocouple (in the copper pipe) is much lower than that one around the S3 thermocouple (at the top of the tank). Furthermore, the fluid velocity in the tank is much lower than that one in the pipe due to the different

external diameter. For these aspects the temperatures in the tank increase more gradually as compared to the temperatures in the pipe.
It can be observed that the T_S5 and the T_S6 trends are flat, so a time of two hours is not enough to heat the volume of water at the bottom of tank.

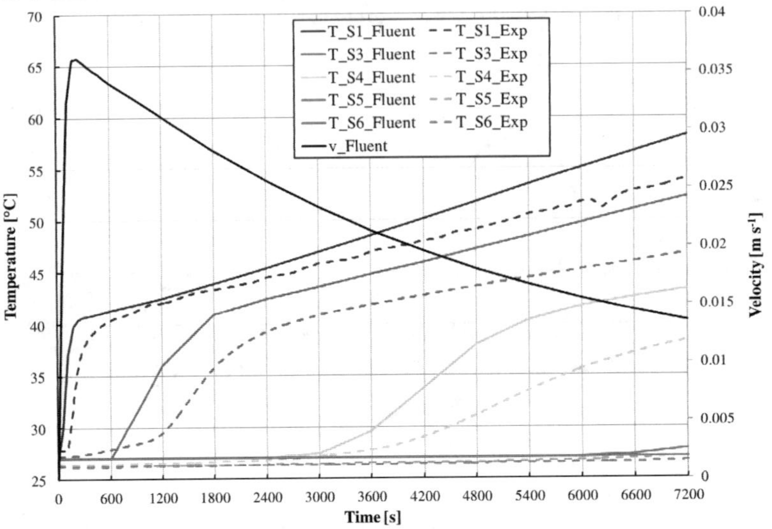

Figure 7. Comparison between the measured temperatures (Test 3) and numerical simulations.

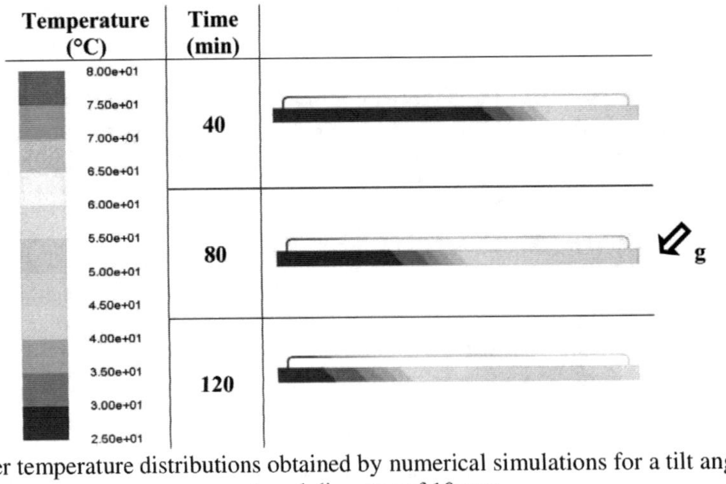

Figure 8. The water temperature distributions obtained by numerical simulations for a tilt angle of 45° and for an external diameter of 10 mm.

The good agreement between numerical simulations and experimental data can be confirmed.
Furthermore, it is worth noting that the numerical simulations tend to slightly overpredict the experimental trends. This aspect could be due to different reasons, such as:

- a complex insulation operation of the thermosyphon during the experimental tests due to the short distance between the tank and the copper pipe: this causes more heat losses both from tank and from copper pipe to external;
- difficulties to guarantee a costant heat flux over the absorber plate due to disturbances in the electrical grid during the experimental test;
- a non-optimal thermal contact between the electrical resistance strip and the absorber plate despite of the conductive paste application, so it could create some air bubbles that decrease the local conduction.

- Finally, in figure 8 the temperature field inside the thermosyphon is depicted at different times, i.e. at 40, 80 and 120 minutes after the test start. In this figure the thermosyphon phenomenon due to the temperature gradients created in the first moments by the heat sink and the gravity effect can be seen. Over time, this process increases the amount of volume heated and a thermal stratification occurs at top side of the tank.

As depicted in figure 7, the deviations between the experimental temperatures and the numerical ones, that have been calculated at 3600 s, are 1.6 °C, 3.0°C, 1.8°C, 0.5°C, 0.7°C for T_S1, T_S3, T_S4, T_S5 and T_S6, respectively.

7. Conclusions

Experimental tests were carried out on three different test sections at varying operating conditions and the following conclusions can be drawn.

- In all the test sections the presence of the natural convection is confirmed. It is verified with the experimental temperature trends and also by the numerical simulations in figure 7 where the velocity trend inside the pipe is reported.
- A good agreement between the experimental data collected in test 3 and the numerical simulations is found. This validation is very important to define in detail the performance of the solar thermosyphon, obtaining more informations about the temperature distribution and the velocity field inside the solar collector. It is worth highlighting that the flow rate is not measured because the fluid velocity is very low and any sensor would affect the measurement and the flow itself, but this parameter, as mentioned above, can be obtained for the numerical simulations.
- The effect of device inclination has been investigated. It can be observed that at low tilt angle of the solar thermosyphon the temperature values are higher and the thermal stratification is more pronounced. This means a better performance at high tilt angle due to the influence of buoyancy effect with a higher circulation of working fluid and with a higher fraction of heated fluid inside the tank storage. For this reason it would be interesting to analyze experimentally the solar thermosyphon at 90° tilt angle, that is the typical configuration on vertical facades.
- When the heating system is switched off the temperature at the bottom of the pipe increases; this phenomenon could be due to the reversal flow of the fluid. For this reason, it may be fundamental to investigate this problem that would be verified when solar irradiance decreases.

8. References

[1] Smyth M, Eames P C and Norton B 2006 Integrated collector storage solar water heaters *Renew. Sust. Energy Rev.* **10**, 503-38

[2] Devanarayanan K, Kalidasa Murugavel K 2014 Integrated collector storage solar water heater with compound parabolic concentrator - Development and progress *Renew. Sust. Energy Rev.* **39**, 51-64

[3] Tripanagnostopoulos Y and Souliotis M 2006 ICS solar systems with two water tanks *Renew. Energy* **31**, 1698-717

[4] Del Col D, Padovan A, Bortolato M, Dai Pre M and Zambolin E 2013 Thermal performance of flat plate solar collectors with sheet-and-tube and roll-bond absorbers *Energy* **58**, 258-69

[5] Ge H and Xie G 2001 A study on the evacuated-tubes integral-collector-storage solar water heater with two different coated absorbers *Proc. Intern. Conf. on Energy Conver. Appl.*, 1227-30

[6] Kumar R and Rosen M A 2010 Thermal performance of integrated collector storage solar water heater with corrugated absorber surface *Appl. Therm. Eng.* **30**, 1764-68

[7] Kaushika N D and Reddy K S 1999 Thermal design and field experiment of transparent honeycomb insulated integrated-collector-storage solar water heater *Appl. Therm. Eng.* **19**, 145-61

[8] Kumar R and Rosen M A 2011a Comparative performance investigation of integrated collector-storage solar water heaters with various heat loss reduction strategies *Int. J. Energy Res.* **35**, 1179-87

[9] Henderson D, Junaidi H, Muneer T, Grassie T and Currie J 2007 Experimental and CFD investigation of an ICSSWH at various inclinations *Renew. Sust. Energy Rev.* **11**, 1087-116

[10] Kaushik S C, Kumar R and Garg H P 1995 Effect of baffle plate on the performance of a triangular built-in-storage solar water heater *Energy Convers. Manage.* **36**, 337-42

[11] Kumar R and Rosen M A 2011b Integrated collector-storage solar water heater with extended storage unit *Appl. Therm. Eng.* **31**, 348-54

[12] Faiman D, Hazan H and Laufer I 2001 Reducing the heat loss at night from solar water heaters of the integrated collector-storage variety *Sol. Energy* **71**, 87-93

[13] Chaabane M, Mhiri H and Bournot P 2014 Thermal performance of an integrated collector storage solar water heater (ICSSWH) with phase change materials (PCM) *Energy Convers. Manage.* **78**, 897-903

Journal of Physics: Conference Series **655** (2015) 012010 doi:10.1088/1742-6596/655/1/012010

CO_2 post-combustion capture in coal-fired power plants integrated with solar systems

R Carapellucci [1], L Giordano [2], M Vaccarelli [1,*]

[1] Department of Industrial and Information Engineering and Economics, University of L'Aquila (Italy), Via Giovanni Gronchi 18, 67100 L'Aquila

[2] Laboratoire Réactions et Génie des Procédés (LRGP) (UMR 7274) Université de Lorraine, ENSIC, 1, rue Grandville - BP 20451.54001 Nancy Cedex, France

[*] Corresponding author E-mail:mauravaccarelli@gmail.com

Abstract. The majority of the World's primary energy consumption is still based on fossil fuels, representing the largest source of global CO_2 emissions. According to the Intergovernmental Panel on Climate Change (IPCC), such emissions must be significantly reduced in order to avoid the dramatic consequences of global warming. A potential way to achieve this ambitious goal is represented by the implementation of CCS (Carbon Capture and Storage) technologies.
However, the significant amount of energy required by the CCS systems still represents one the major barriers for their deployment. Focusing on post-combustion capture based on amine absorption, several interesting options have been investigated to compensate the energy losses due to solvent regeneration, also using renewable energy sources. One of the most promising is based on the use of concentrating solar power (CSP), providing a part of the energy requirement of the capture island.
In this study the integration of a CSP system into a coal-fired power plant with CO_2 post-combustion capture is investigated. Basically, a CSP system is used to support the heat requirement for amine regeneration, by producing saturated steam at low temperature. This allows to reduce or even eliminate the conventional steam extraction from the main power plant, affecting positively net power production and efficiency. The energy analysis of the whole system is carried out using the GateCycle software to simulate the coal-fired power plant and ChemCad platform for the CO_2 capture process based on amine absorption.

1. Introduction

Human activities, since the pre-industrial era, have contributed enormously to increase the overall level of greenhouse gas emissions in the atmosphere. According to the statistics of International Energy Agency, CO_2 emissions from fuel combustion have reached almost 32 Gt in 2012, corresponding to an year on year increase of around 1.2%. Electricity and heat generation have contribute to around 40% of such emissions, heavily relying on the use of coal, the most carbon-intensive fossil fuel [1].
The integration of technologies based on renewable energy into power generation system can offer the possibility to drastically reduce CO_2 emissions. Despite their huge potential, the use of renewable technologies is still hindered by their high capital costs and their low density and discontinuity, threatening the security and the reliability of supply [2]. In this context, CCS (Carbon Capture and

Storage) is regarded as an interesting option to obtain a short term reduction of CO_2 emissions, until renewable energies will reach the technical and economic feasibility.
Currently, amine-based post-combustion capture represents the most suitable technology for the retrofitting of existing power plants. Research efforts in this field are focused on reducing the considerable efficiency penalty and losses in power plant capacity, mainly arising by the need to extract a considerable fraction of medium-low pressure steam for the stripper reboiler heating.
Several solutions have been proposed, including the optimization of the operating parameters of absorber/stripper system, the modification of CO_2 scrubbing process and improvement of heat integration by modifying the original power plant layout [3].
An alternative option to mitigate performances penalties induced by the CO_2 capture island is the use of a solar thermal energy system, assisting the solvent regeneration process.
A few number of studies have addressed the integration of solar energy into power plants retrofitted with CO_2 capture. The concept was first proposed by Wibberley et al. [4], to study the retrofit of a coal-fired power plant. They investigated the partial replacing of steam extraction with steam generated by a solar collector field, observing a considerable reduction of power plant derating . Later Cohen [5] carried out a preliminary study of supplying the overall reboiler duty, by combining mid-to high temperature solar collectors with a molten salt thermal energy storage. Considering the same technology, Mokhtar et al. [6] evaluated the techno-economic feasibility of solar assisted post-combustion capture, using actual weather and wholesale electricity price data to assess the impact of regional and climatic variables. Considering a 600 MW coal-fired power plant with a 80% CO_2 capture rate, Zhao et al. [7] compared, from the energy point of view, solar assisted solvent regeneration with the solar feed-water repowering operated through the closure of high pressure steam bleedings.
This paper aims to investigate several options to retrofit a coal-fired power plant with a MEA-based absorption system, capturing 90% of CO_2 from exhaust flue gas.
Assuming a conventional steam extraction to provide heat for solvent regeneration, the study first assesses the potential reduction of power losses attainable by operating the coal fired power plant to various degrees of high-pressure steam turbine overload. The retrofit options based on steam extraction are then compared with the case of integrating a CSP system, providing part of steam for the stripper reboiler operation. The performance improvement of hybrid system is assessed, evaluating the range of efficiency and power production increase on a mean seasonal and annual basis.

2. Model of the solar field

A linear Fresnel collector with direct steam generation has been selected as reference technology for the solar field. Performances have been evaluated based on technical specifications of a commercially available solar collector of Novatec Solar technology [8], NOVA-1. The solar field involves the use of several base modules connected in series in order to obtain a solar collector with the required temperature increment. Besides, the required steam flow rate is attained by connecting in parallel several solar collectors.
Figure 1 shows a vertical cross-section of the base module, while Table 1 summarizes the product data sheet.
Neglecting the power losses from the receiver, the thermal power output of the solar collector is the product of primary reflector surface, Direct Normal Irradiation (DNI) and the optical efficiency

$$P_{th} = A \cdot (DNI \cdot \eta_{opt} - P_{loss}) \tag{1}$$

The specific thermal loss of primary reflector is calculated as

$$P_{loss} = u_0 \Delta T + u_1 \Delta T^2 \tag{2}$$

where $u_0 = 0.056$ W/(m^2K) and $u_1 = 0.000213$ W/(m^2K^2).
Finally, the optical efficiency is evaluated by multiplying the baseline value ($\eta_0 = \eta(\Theta_\perp = \Theta_{II} = 0) = 0.67$) by the transversal $K_\perp(\Theta_\perp)$ and longitudinal $K_{II}(\Theta_{II})$ correction factors:

$$\eta_{opt} = \eta_0 \cdot K_\perp(\Theta_\perp) \cdot K_{II}(\Theta_{II}) \quad (3)$$

Table 2 summarizes the thermal power output of the solar collector under reference conditions.

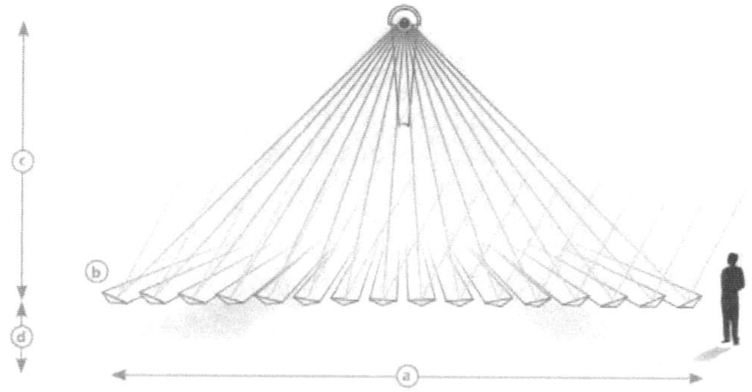

Figure1 Cross section view of the base module [8]

Table 1 Technical data of the base module[8]

Geometrical parameter	*Value*
Width (a)	16.56 m
Length (b)	44.8 m
Absorber tube height above primary reflector level (c)	7.4 m
Height of primary reflector level (d)	0.75 – 1.05 m above ground level
Recommended minimum clearance between parallel rows	4.5 m
Aperture surface of primary reflectors	513.6 m^2
Orientation	±20° longitudinal deviation from north-south axis
Minimum row length	5 modules, 224 m in length
Maximum row length	22 modules, 985.6 m in length

2.1. Site characterization

In order to assess the feasibility of integrating a CSP system into a coal fired power plant with CO_2 capture based on amine absorption, a potential installation in a site of the south of Italy (Brindisi) has been considered.
The availability of solar energy source and the air temperature have been evaluated on a hourly basis, using a software developed at the Department of Mechanical, Energy and Management Engineering (University of L'Aquila), as a part of a doctoral thesis [9].
In this respect, Figure 2 shows the monthly values of solar radiation incident on a horizontal surface for the site of interest.

Table 2 Thermal output under reference conditions [8]

Reference conditions
External temperature=40°C
Inlet flow temperature=100°C
Outlet flow temperature=270°C
No wind
900 W/m^2 direct normal radiation
30° transversal angle $\Theta_\perp$
10° incidence angle Θ_{II}
Performance
278 kW per control unit
541 W/m^2 of aperture surface area for primary reflectors

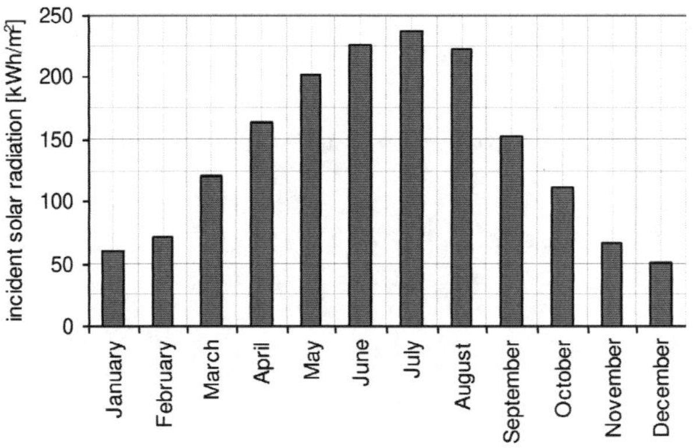

Figure 2 Monthly values of monthly solar radiation on a horizontal surface in the city of Brindisi

For the purposes of designing a concentrating solar power plant, the information of greatest interest is related to the direct radiation on a normal surface to the Sun (DNI, Direct Normal Irradiation), being such systems unable to collect the diffuse component.

Conventionally, the design of CSP systems is based on the radiation detected on June 21 at 12:00, or at the summer solstice. In this study, a radiation of 870 W/m^2 has been considered, being representative of the highest values of DNI for the site of Brindisi; moreover, it has been assumed a solar height of 77.8° (maximum solar height of 21 June), an azimuth angle of zero (solar noon), and an ambient air temperature of 30°C.

In order to assess the variation of CSP performances, due to the daily fluctuation of solar energy source, hourly data have been properly averaged on a seasonal and annual basis. Hence, five diurnal profiles of DNI and air temperature have been evaluated, four representative of a typical day of each season and one describing the mean annual daily behavior.

With respect to the city of Brindisi, Figure 3 shows the summer, winter and annual diurnal profiles of DNI, while Table 3 summarizes the overall daily radiation of typical days and the corresponding average air temperature.

Table 3 Solar insolation data in typical days

Typical day	Period	Day numbers	Daily radiation (kWh/m^2)	Average temperature (°C)
Winter	I trimester	90	2.82	10.31
Spring	II trimester	91	6.51	17.93
Summer	III trimester	92	6.66	23.82
Autumn	IV trimester	92	2.49	14.44
Year	year	365	4.63	16.66

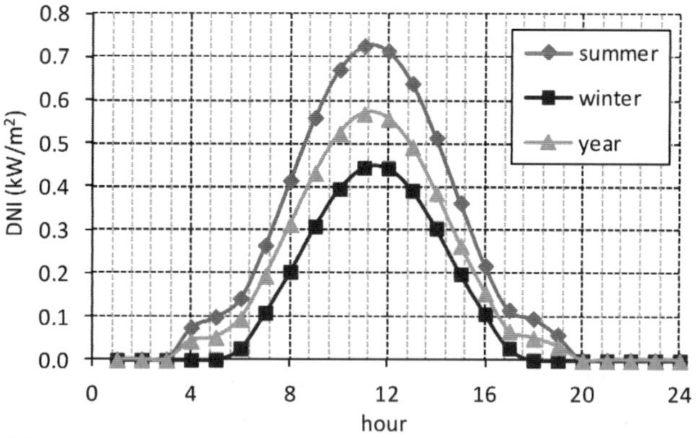

Figure 3 Summer, winter and annual diurnal profiles of DNI

3. Energy analysis of integrating a CSP system into a coal fired power plant with CO_2 capture

The study investigates the retrofit of a coal-fired power plant with a post combustion capture system based on MEA absorption. At first, several options to provide heat to the reboiler through the steam extracted from the main plant are compared. Then, the integration of a CSP system providing part of steam for solvent regeneration is analyzed, with the aim to assess the potential improvement of energy performances.

3.1. Retrofit options based on conventional steam extraction

Figure 4 shows the layout of coal fired-power plant, modeled using the GateCycle software [10]. At design conditions, it produces a net power output of 100 MW, with a LHV efficiency of 42%.

Post-combustion capture of CO_2 is accomplished by a chemical absorption system using monoethanolamine (MEA) as solvent. The CO_2 capture island has been simulated using the software ChemCAD 6.3, by Chemstations [11]. Assuming a capture ratio (φ) of 90% and a lean loading of 0.25 kg_{CO2}/kg_{MEA}, the specific reboiler duty states at 3.5 MW/kg_{CO2} [12], corresponding to a thermal power requirement for solvent regeneration of around 65.7 MW_{th}. This thermal power can be provided using saturated steam at about 3 bar.

Table 4 allows to compare the coal-fired power plant energy performances at design conditions (*Case 0*) and after CO_2 capture retrofit, assuming that the heat duty for amine regeneration is fully met by extracting steam at crossover pipe between the intermediate and low pressure steam turbines.

Two retrofit options have been investigated, operating the steam power plant at off-design without (*Case 1*) or with (*Cases 2 and 3*) high-pressure overload. Assuming the same fuel flow rate of design

conditions (*Case 1*), that is avoiding an overload of the high pressure steam turbine, the steam extracted along the expansion for solvent regeneration produces a significant derating of the coal-fired power capacity, that reduces from 100 to 79.1 MW (-21%), the electricity consumption of CO_2 capture and compression processes accounting for 35% ($P_{CO2\ CAPT\&COMPR}$=7.4 MW). As a result, the efficiency reduces to 33.1%, corresponding to an efficiency penalty of around 9% pts.
In *Case 2*, the coal flow rate has been properly increased in order to produce an overload at high-pressure steam turbine of 10%. Hence, with the same steam extraction of *Case 1* (25 kg/s), the power plant derating reduces to 16% (84.0 MW), while the net efficiency remains almost unchanged.
Considering the maximum capability of boiler [13] and steam turbine [14,15] compared to the design rating, the HPST overload has been increased up to a maximum value of 15% (*Case 3*). Hence, the power plant capacity reaches 85.5 MW, against a slight decrease of net efficiency, that states at 32.9% (- 9% pts). With a capture ratio of 90%, specific CO_2 emissions reduces from 752 kg/MWh to less than 96 kg/MWh.
Comparing the retrofit options, it is noted that *Case 2* allows to provide a good compromise between net power output of the steam power plant with CO_2 capture and the overload of its high pressure turbine. Hence, the integration of CSP system will be investigated with reference to a steam power plant operating at off-design conditions with a 10% HPST overload.

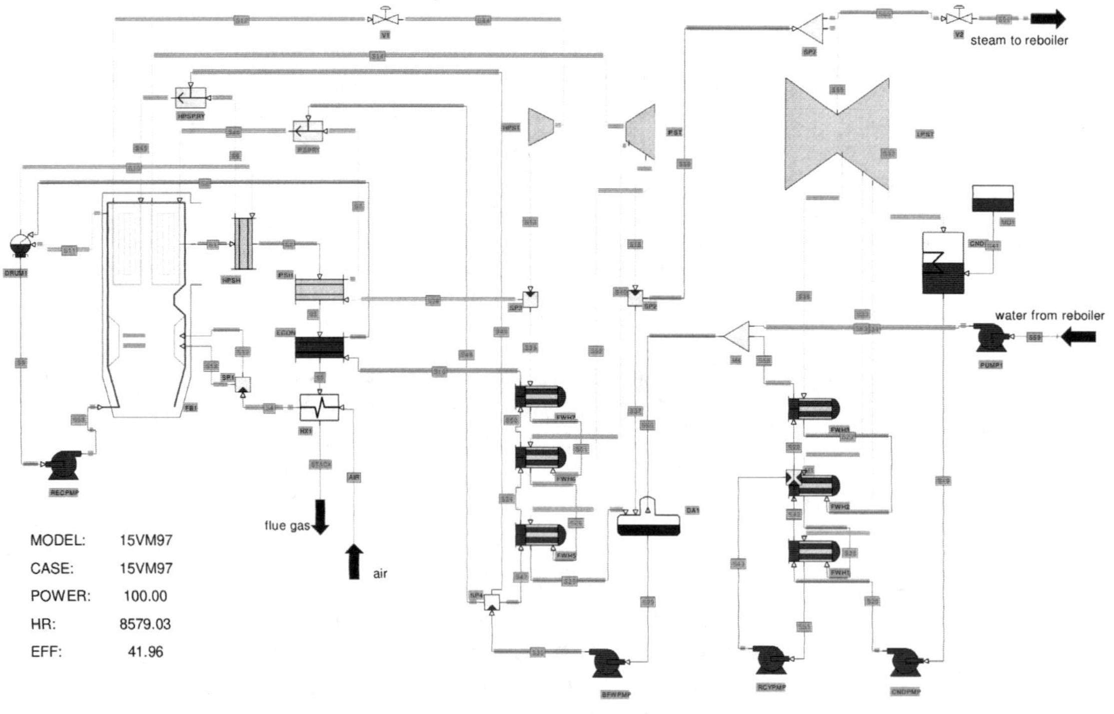

Figure 4 Layout of pulverized coal-fired power plant

3.2. Retrofit options based on CSP system integration

As outlined in paragraph 2, the CSP system based on linear Fresnel technology has been sized considering the greatest DNI achieved all year round (June 21 at 12:00: DNI=870 W/m^2). In this regard, Table 5 summarizes the main project and performance parameters of the solar field, while

Figure 4a shows the trend of steam flow rate production in typical days of summer and winter, as well as the mean annual diurnal trend.

Table 4 Energy performance of the power plant integrated with the CO_2 removal system

	Case 0 **On-design** φ= 0%	*Case 1* **Off-design** φ= 90%	*Case 2* **Off-design** φ = 90%	*Case 3* **Off-design** φ = 90%
High pressure overload	0%	0%	+10%	+15%
P_{NET} [MW]	100.0	79.1	84.0	85.5
$P_{CO2\ CAPT\&COMPR}$ [MW]	-	7.4	7.8	7.9
η_{NET} [%]	42.0	33.1	33.0	32.9
$CO_{2,\ em}$ [kg/s]	20.9	2.1	2.2	2.3
$CO_{2,\ em}$ [kg/MWh]	752.0	95.3	95.5	95.8
m_{FUEL} [kg/s]	9.4	9.4	10.0	10.2
m_{SH} [kg/s]	79.3	79.3	87.2	91.1
m_{RH} [kg/s]	74.8	74.7	81.7	84.7
m_{CD} [kg/s]	55.0	33.1	37.8	39.8
p_{CD} [kPa]	5.0	4.4	4.5	4.6
$m_{steam\ to\ reb}$ [kg/s]	-	24.9	24.9	24.9

Table 5 Main project and performance parameters of the solar field [16]

Modules in series by line	10
Line length [m]	448
Line width [m]	16.56
Line area catchment [m^2]	5136
Line temperature input [°C]	110
Line temperature output [°C]	140
Available solar power [kW]	2795.7
Power lost [kW]	37.2
Line net power output [kW]	2758.5
Number of Lines	24
Total power solar field [MW]	66.2
Net efficiency solar field [%]	61.7
Available steam flow [kg/s]	29.1

The steam available from the solar field, produced at 133°C and 3 bar, allows to reduce the steam extraction from the main plant. In this respect, Figure 4b shows the effect of CSP system integration on steam extraction, assuming an HPST overload of 10%.

As expected, the steam extraction achieves the highest value (25 kg/s) at night, due to the lack of solar energy source, while it progressively decreases with the increase of solar energy availability. During the day, the total steam flow requirement is slightly higher than in the night, due to the lower valuable thermodynamic properties of solar steam production with respect to steam extracted from the turbine of the power plant. During summer the solar field is able to produce almost all the steam required by the reboiler in the middle of the day (24.1 kg/s at 11:00). Hence, the corresponding steam extraction from the main power plant reaches a minimum value of 4.2 kg/s (Table 6). Otherwise, during winter, the CSP system reaches a maximum steam production of 14.5 kg/s, thus requiring a steam extraction of 12.4 kg/s.

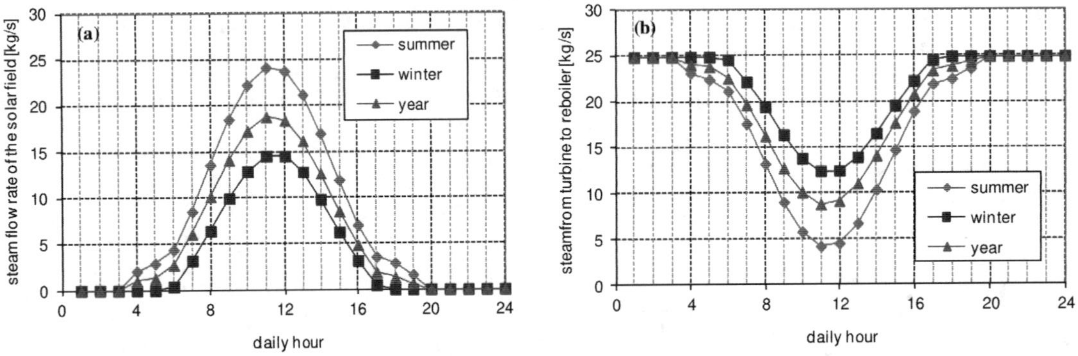

Figure 5 Steam flow production of solar field and corresponding steam extraction from the power plant for typical days of summer and winter and mean annual diurnal trend

Figures 6 and Table 6 show the effect of CSP system integration on the energy behavior of the retrofitted coal-fired power plant. Due to the solar thermal power production, the net power plant capacity reaches a maximum value of 95.1 MW in the middle (11:00) of a typical summer day, thus decreasing the capacity derating to less than 5% pts. Moreover, Figure 6d shows that, due to the off-design operating conditions of HPST (m_{HPST}=+10%), LPST operates with a slight overload (about +2%), in the case of minimum steam extraction (4.2 kg/s).
On the other hand, during winter, the power plant derating remains lower than the case of steam extraction only, reaching in the middle of day a minimum value of -10% (90.2 MW) (Table 6).
It is also noteworthy to note that, due to the steam extraction at crossover pipe between IPST and LPST sections, the steam sent to condenser, as well as its operating pressure, reduces compared to design conditions. In this regard, Figure 6c shows that such deviation decreases with the rise of the solar steam contribution, being entirely eliminated in the middle of a typical summer day.
As regard to the net efficiency, the maximum values range between 35.5% (winter) up to 37.4% (summer), corresponding to an efficiency penalty of 6 and 5% pts (Fig. 6b). As a result, the specific CO_2 emissions further decrease, reaching a minimum value of 77.8 kg/MWh.
Figure 7a shows that integrating the CSP system also produces an increase of the average daily net plant efficiency, corresponding to 0.7% pts in winter and 1.4% pts in summer. Moreover, the increase of the average daily electricity production reaches around 2.9% on annual basis (Fig. 7b).

Table 6 Energy results of the integration

	Off-design φ = 90% Steam extr.	**Off-design φ = 90% Steam extr. + CSP summer**	**Off-design φ = 90% Steam extr + CSP winter**
P_{NET} [MW]	84.0	95.1	90.2
$P_{CO2\ CAPT\&COMPR}$ [MW]	7.8	7.8	7.8
η_{NET} [%]	33.0	37.4	35.5
$CO_{2,\ em}$ [kg/MWh]	87.3	77.8	81.7
$m_{steam\ from\ CSP}$ [kg/s]	-	24.1	14.5
$m_{steam\ from\ extr}$ [kg/s]	24.9	4.2	12.4
LP overload [%]	-31.0	2.2	-11.1
HP overload [%]	10.0	10.0	10.0

Journal of Physics: Conference Series **655** (2015) 012010 doi:10.1088/1742-6596/655/1/012010

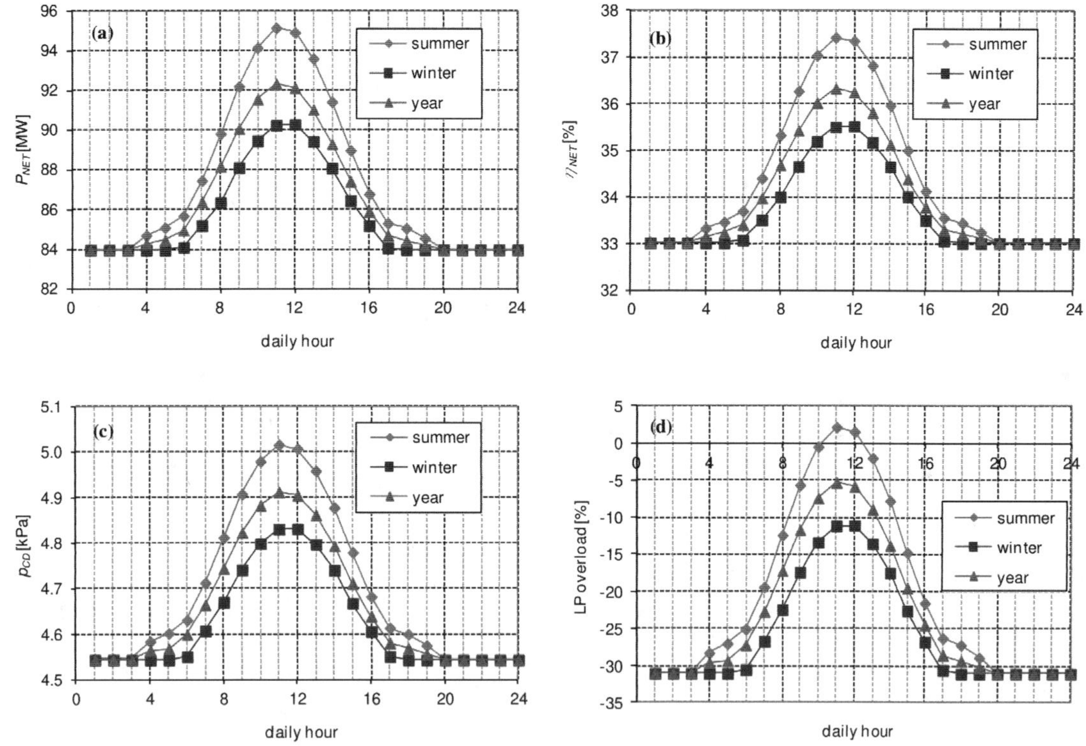

Figure 6 Trend of the net power output (a), net efficiency (b), condenser pressure (c) and LPST overload (d) for typical days of summer and winter and mean annual diurnal trend

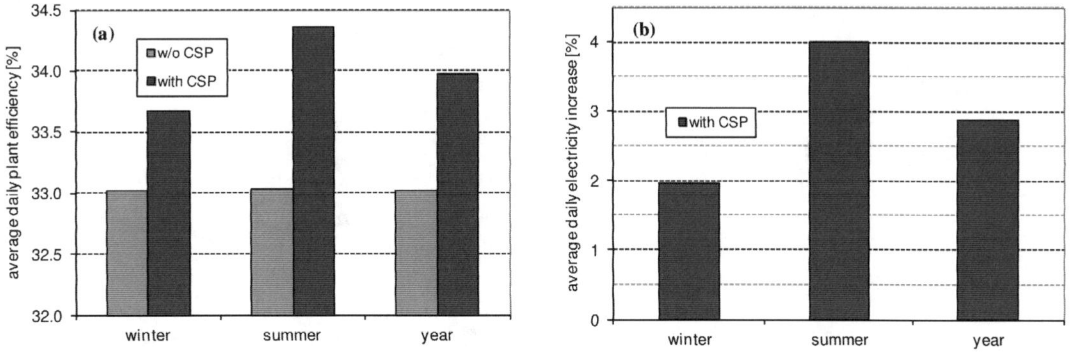

Figure 7 Average daily plant efficiency without and with CSP integration and increase in average daily electricity production thanks to the CSP integration

4. Conclusions

The aim of this paper has been the use of solar energy source in a coal-fired power plant retrofitted with a post-combustion amine-based CO_2 capture island. A concentrated solar collector based on Fresnel technology has been chosen as auxiliary unit, in order to provide a fraction of steam requirement for MEA regeneration.

The energy analysis has highlighted that CSP system enables to reduce the steam extraction from the

main plant, to an extent that depends on the availability of solar energy resource. The maximum solar steam production occurs in the middle of a typical summer day (11:00 a.m.), drastically reducing the steam extraction (-83%) compared to the retrofit plant without CSP system integration (25 kg/s). Conversely, during winter, the maximum contribution of solar energy source reduces to less than 50% of the overall steam requirement (26.9 kg/s).
As a result, performances of retrofitted power plant enhance compared to the case of steam extraction only. Assuming an HPST overload of 10%, the maximum power plant capacity states at 90.2 MW during a typical winter day, thus reducing the capacity derating from 16% to 10%; on the other hand, the net efficiency gains more than 5% pts, stating at 35.5%. Considering a typical summer day, the CSP system allows to reach in the middle of day a maximum power output of 95 MW, with a corresponding net efficiency of 37.4% and specific CO_2 emissions lower than 80 kg/MWh.
The improvement of energy performances is noticeable even on a mean daily basis. For instance, the increase of average daily net plant efficiency ranges from 0.7% pts (winter) up to 1.4% pts (summer). In addition, the mean annual increase of daily electricity production states at around 3%.

References

[1] International Energy Agency (IEA) 2014 CO_2 emissions from fuel combustion - Highlights Paris.
[2] Parvareh F, Sharma M, Qadir A, Milani D, Khalilpour R, Chiesa M, Abbas A 2015 Integration of solar energy in coal-fired power plantsr etrofitted with carbon capture: A review, *Renew Sust Energ Rev*, **48** 49-61.
[3] Rubin E S, Mantripragada H, Marks A, Versteeg P, Kitchin J 2012 The outlook for improved carbon capture technology, *Progr Energ Combust*, **38** 630-671.
[4] Wibberley L J, Scaife P 2008 Retro-fitting post combustion capture, Cooperative Research Centre for Coal in Sustainable Development (Australia), *Technology Assessment Report* **75**.
[5] Cohen S M W M, Rochelle G T 2010 Utilizing solar thermal energy for solvent regeneration in post-combustion CO_2 capture, *ASME 4th International Conference on Energy Sustainability, Phoenix*.
[6] Mokhtar M, Ali M T, Khalilpour R, Abbas A, Shah N, Hajaj A A et al 2012 Solar-assisted post-combustion carbon capture feasibility study, *Appl Energy*, **92** 668-76.
[7] Zhao Y, Hong H, Zhang X, Jin H 2012 Integrating mid-temperature solar heat and post-combustion CO_2-capture in a coal-fired power plant, *Sol Energy*, **86** 3196-3204.
[8] Novatec Solar 2010 Nova 1 - Turnkey solar boiler, mass produced in industrial precision, Novatec Solar GmbH, Karlsruhe.
[9] Giordano L 2010 Studio ed ottimizzazione di un'isola energetica da fonti rinnovabili con sistema di accumulo ad idrogeno, PhD Thesis, University of L'Aquila.
[10] General Electric Company 2005, GateCycle™, Getting started & installation guide, General Electric Company.
[11] Chemstations, Process Flowsheet Simulator, User Guide Chemstations Inc., 2008, Houston, USA.
[12] Vaccarelli M, Carapellucci R, Giordano L 2014 Energy and economic analysis of the CO_2 capture from flue gas of combined cycle power plants, *Energy Procedia*, **45** 1165-1174.
[13] Collins G, Naval Engineering Plants, 2012, AuthorHouse (Bloomington).
[14] Raja A K, Shriwastav A P, Dwivedi M, Power Plant Engineering, 2006, New Age international Ltd, Publishers (New Delhi).
[15] Deidewig F, Wechsun M, Thermodynamic Aspects of Designing the New Siemens High Pressure Steam Turbine With Overload Valve for Supercritical Applications, *Proceedings of ASME 2006 Power Conference*.
[16] Cau G, Cocco D, Tola V 2013 Analisi dell'integrazione di impianti di cattura post-combustione della CO_2 con fonti rinnovabili, ENEA - Ricerca di Sistema Elettrico.

Measurement of thermophysical properties coupled with LCA assessment for the optimization of a historical building retrofit

A. Bortolin[1], P. Bison[1], G. Cadelano[1], G. Ferrarini[2], S. Fortuna[1]

[1] CNR-ITC, Corso Stati Uniti, 4, 35127, Padova, Italy
[2] DII-UNIPD, Via Venezia 1, 35131, Padova, Italy

E-mail: alessandro.bortolin@itc.cnr.it

Abstract. Historical buildings are a significant part of the Italian building stock and, in most cases, need deep refurbishment interventions to reach the energy criteria required by the current standards. A workflow that integrates on-site surveys and building modeling is mandatory to obtain effective energy saving measures. This work describes the analysis and modeling of the San Vito alla Rivera church, a XIV century building that was damaged during 2009 L'Aquila earthquake, suffering a partial collapse of the façade and of the roof. The latter was selected for a complete restoration that could improve its thermal performance while maintaining, as much as possible, the original structure. Several elements of the roof were collected in situ in order to measure, in laboratory, its thermophysical properties applying standard techniques and alternative methods based on infrared thermography. The accurate characterization of the materials was the starting point for the estimation of the environmental impact of the retrofit aimed to reach a defined thermal transmittance. A model of the building was created with TRNSYS software to calculate the energy consumption before and after the intervention. A Life Cycle Assessment (LCA) analysis was conducted on different insulation materials to determine the one with the lowest impact.

1. Introduction

The energy efficiency policies, both at an European and national level, are stressing the impact of buildings on the overall energy consumption. Economical and legislative actions are promoted to increase the energy savings of newly built and existing edifices.
Historical and listed buildings are a wide part of the Italian building stock both in terms of quantity and value [1]. Frequently these edifices are residential or host institutional or commercial activities, hence they are required to comply with the current standards on energy performance, thermal comfort and safety, even if they were planned and built without these goals [2]. At the same time, especially in the case of listed buildings, they are subject to strict codes that strongly limit even small modifications to the original building structure. A further challenge is that usually few information are available on the design and material specification of these buildings and, even if data are present, the effect of aging on the structure must be taken into account.
Nowadays several historical buildings must undergo a refurbishment process in order to comply with the energy performance prescribed by European and Italian standards and codes. The knowledge of the building condition is a primary necessity in order to plan an adequate retrofit strategy. For this reason, on-site surveys and laboratory testing should be performed before any other analysis. When

dealing with cultural heritage, the diagnostic instruments that are non-invasive and non-destructive are preferred or even mandatory. From this standpoint, infrared thermography [3] (IRT) is a well-established technique for the survey of buildings and the measurement of material properties [4, 5]. The data obtained from the survey should be utilized for the creation of a mathematical model of the building, where the availability of real data should significantly increase the accuracy and the reliability of the model itself. The model is necessary to compare the impact of different retrofit solutions both in terms of applicability and expected impact of the proposed measures. After the comparison of different solutions, only the most effective ones should be implemented and their impact should be measured. The complete process, summarized in figure 1, suggests that different and complex tasks should be completed by researchers and technicians with various and complementary skills.

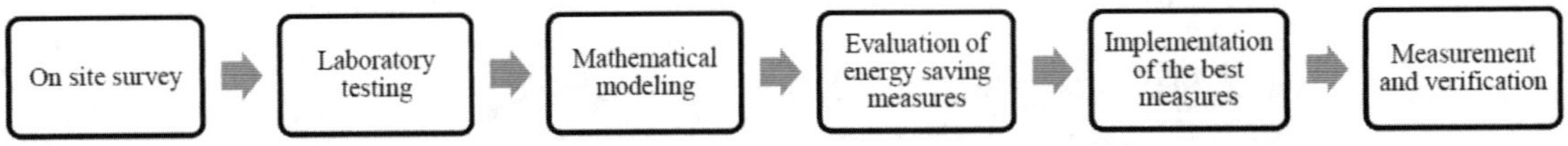

Figure 1. Workflow of the retrofit process of a historical building

This study is focused on the first part of the retrofit process. In this framework, the goal of the work is underlining the limits and possibilities of the application of Life Cycle Analysis (LCA) during the modeling and evaluation of energy savings phases. This method has been applied more and more frequently as a planning tool for new buildings, however it is not commonly chosen for building retrofit and in particular for the renovation of historical buildings. In Italy the LCA method is not mandatory, therefore it has been applied only in few cases (e.g. planning competitions) and only under direct request of the customer [6, 7]. The stock of Italian historical and listed buildings that needs retrofit interventions would benefit from the application of LCA, as it would lead to the use of materials with a lower environmental impact. The main barriers to the diffusion of LCA for historical building retrofit are the lack of a standardized approach and the difficulty of obtaining accurate data about the manufacturing processes and the material properties of construction elements.

2. Building evaluation

2.1. On-site survey

The building studied in this work is the San Vito alla Rivera church, located inside the historical city center of L'Aquila (Italy). This church dates back to the 14th century and was damaged by the 2009 earthquake, suffering a partial collapse of the roof and the façade, as shown in figure 2. The on-site survey on the building started with a thermographic campaign that was aimed to identify hidden structural elements and to investigate the thermal behavior of the building. The experiment were repeated in summer and winter conditions with the use of advanced scanning techniques. Other measurements were performed with video endoscopy and temperature sensors. The roof was the thinnest and most damaged element, therefore it was easier to identify it as the primary component that should be retrofitted.

2.2. Laboratory testing – Thermal diffusivity

Few material samples were collected from the remains of ceiling tiles and from walls. The most important material parameters from a thermal standpoint are thermal conductivity (λ) and thermal diffusivity (α), defined in equation (1) as:

$$\alpha = \frac{\lambda}{\rho\, c} \tag{1}$$

where λ is the thermal conductivity (W m^{-1} K^{-1}), ρ is the volumic mass (kg m^{-3}) and c is the specific heat (J kg^{-1} K^{-1}). Conductivity is related to the building behavior in steady state conditions while diffusivity influences the dynamic performance of the building. Several techniques based on infrared thermography are available to measure the thermal diffusivity of materials and have been applied in the literature [8]. In this work firstly a transmission method was used [9], were one side of the sample was heated with a pulse produced with a 1000 W lamp and the temperature increase was measured with the thermal camera on the opposite side. A different measurement was carried on in a reflection scheme, where the specimen was heated with a laser on a localized spot and the temperature increase and distribution was measured with the thermal camera on the same side. This technique, with an adequate setup, could be transferred from the laboratory to the on-site survey. The third method is derived from the classical Ängstrom method that applies two thermoelectric devices to create a sinusoidal heating on one end of the specimen while the thermal camera measures the propagating thermal wave along it. These techniques and the analysis of results are described in details in [10] and the final results are resumed in table 1.

Table 1. Comparison of thermal diffusivity measurements obtained with different procedures

	α [m^2 s^{-1}] average value	α [m^2 s^{-1}] standard deviation
Transmission method	4.96 E-07	± 2.01 E-08 (± 4%)
Reflection method	5.07 E-07	± 9.85 E-08 (± 19.4%)
Wave method	4.67 E-07	± 3.80 E-08 (± 8.1%)

2.3. Laboratory testing – Thermal conductivity

For small samples of construction materials, thermal conductivity could be measured directly on the specimen sandwiched between two thermoelectric modules that generate a heat flux through it, with an infrared camera monitoring the temperature gradient [11]. In this work also specific heat (with differential scanning calorimetry) and volumic mass were measured, making possible to derive the conductivity value from the diffusivity, as shown in table 2.

Table 2. Comparison of thermal conductivity measurements obtained with different procedures

	λ [W m^{-1} K^{-1}] average value	λ [W m^{-1} K^{-1}] standard deviation
Direct measurement	0.76	± 0.01 (± 1%)
Indirect (transmission method)	0.64	± 0.06 (± 9.3%)
Indirect (reflection method)	0.66	± 0.16 (± 24.7%)
Indirect (wave method)	0.61	± 0.08 (± 13.4%)

3. Mathematical modeling

3.1. Building simulation

Several simulation strategies are available to model the thermal behavior of a building. In this work the San Vito church was studied with the commercial software TRNSYS version 16 [12]. This

software was chosen for the easy and reliable implementation and comparison of different energy savings measures. The creation of the model was based on a geometrical model, with respect to the sizes and orientations of walls, and on the measured thermal properties of the walls. When data could not be obtained directly, typical values derived from the literature were used. The TRNSYS software was used to determine hourly values of heating demand over one year of simulation. The weather data for the city of L'Aquila were selected from the standard weather libraries.
The first simulation was performed to determine a baseline, with the hypothesis of a retrofit aimed to restore the building exactly as it was before the earthquake. Then the impact of energy saving measures was assessed. Usually the first retrofit option is the complete insulation of the external walls but in this case aesthetical and practical reasons discouraged this solution. From a structural standpoint the first element that should be retrofitted was the ceiling, where an insulating layer could be added inside the existing stratigraphy and the estimated thermal transmittance value was higher than 3 W m^{-2} K^{-1} [13]. A realistic goal was reaching a transmittance equal to 0.3 W m^{-2} K^{-1}, value suggested by the current standards on energy efficiency. The second option was the replacement of the existing windows (single pane, metal frame) with new efficient glazing systems (triple pane, wooden frame) that could decrease the windows thermal transmittance to a value close to the one of the walls (around 1.3 W m^{-2} K^{-1}). The replacement of windows has also a positive effect on building leakage. An estimation based on literature values led to quantify the decrease in thermal leakages equal to 25%. With the combination of both energy savings measures, the yearly energy demand was lower than 40%, going from the baseline value of 65000 kWh to the post-retrofit value of 38000 kWh. More simulations were performed with a parametric thickness of the insulation material on the roof, combining the energy output with the LCA method as described in the next chapter.

3.2. LCA method

The LCA method [14] was applied in two successive steps with different goals. Initially, three different insulating materials were compared to determine the one with the lowest impact. A multiple search conducted on the ECOshopping database of insulation materials [15] and on the EcoInvent 2.2 database [16, 17] led to group all the materials in three different categories. For each category one material was chosen on the basis of an estimated on-site availability. The first group was made by traditional material with very good thermal properties and low cost, and the selected material was extruded polystyrene (XPS). The second group was made by natural materials with good thermal properties and high cost, and the selected material was cork. The last group was made by intermediate thermal properties and cost, and the selected material was wood wool. The material properties are described in the next table, including the quantity needed to reach the target thermal transmittance of the renovated wall equal to 0.3 W m^{-2} K^{-1}.

Table 3. Properties of the three insulation materials compared in the LCA properties. The quantity is the amount of material required to reach the target thermal transmittance of the wall.

	Thickness [cm]	Volumic mass [kg m^{-3}]	Thermal conductivity [W m^{-1} K^{-1}]	Quantity [kg m^{-2}]
Cork	11.6	120	0.039	13.9
XPS	10.4	35	0.034	3.6
Wood wool	26.9	400	0.090	107.6

The software used for the LCA calculation was SimaPro 7.3 [18]. The study considered the impact due to transportation from the manufacturer to the building site and also the impact of the end-of-life disposal of the materials. The available solutions for the end-of-life are various, including recycling; this work considered the material will be incinerated after use.

The LCA study could be conducted with different methodologies. The results shown in the next figures 2-5 were respectively obtained with five methods: IMPACT 2002+ v2.11, EDIP 2003 v1.04, EPS 2000 v2.07, Re.Ci.Pe. Endpoint (E) v1.09 / Europe ReCiPe E/A, and IPCC 2007 GWP 100a v1.02 [19,20,21]. All methods have different impact categories and different weighting factors. The obtained results are represented in figures 2-6 as environmental damage indexes.

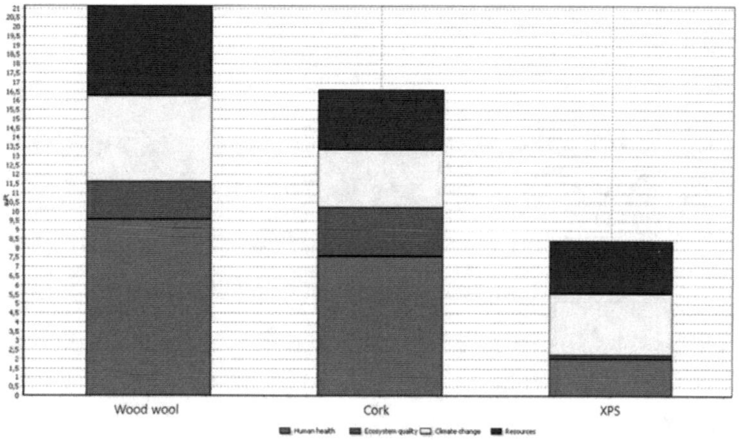

Figure 2. Comparison based on IMPACT 2002+ method.

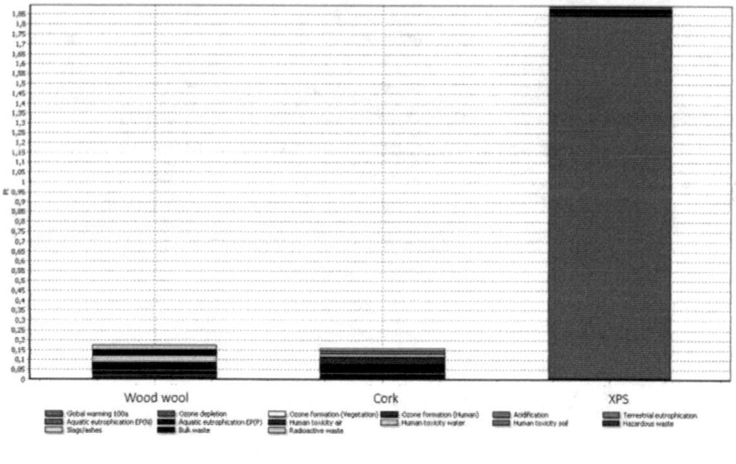

Figure 3. Comparison based on EDIP 2003 method

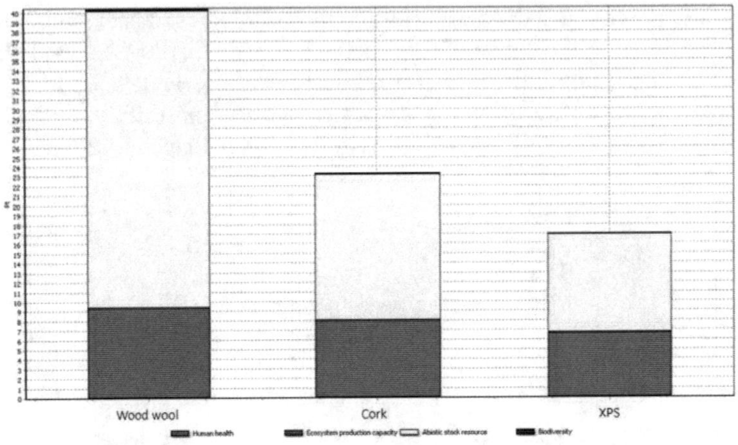

Figure 4. Comparison based on EPS 2000 method.

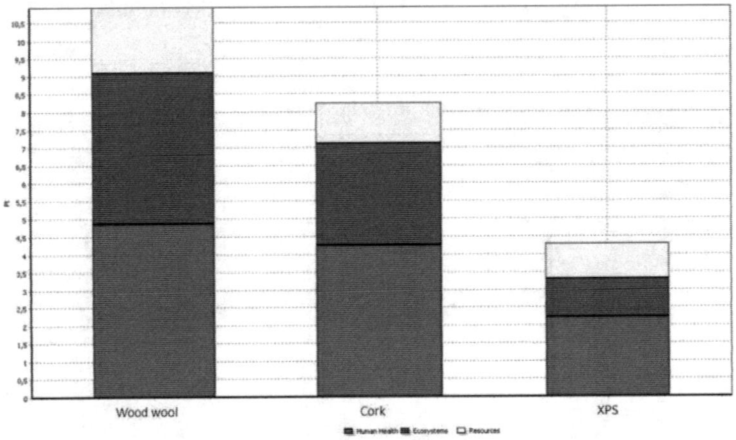

Figure 5. Comparison based on Re.Ci.Pe method.

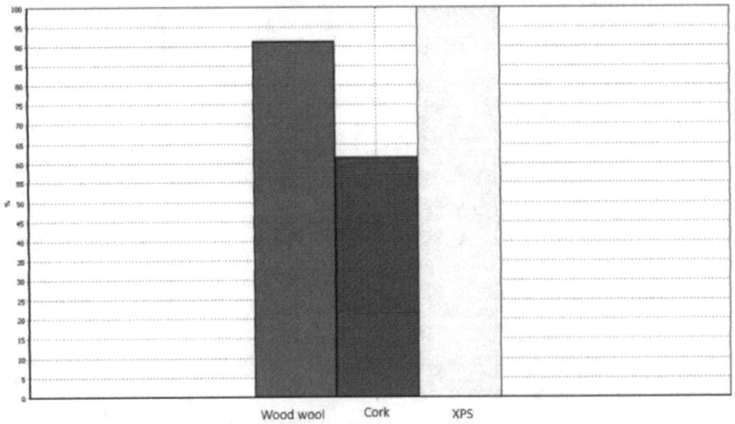

Figure 6. Comparison based on IPCC method.

Journal of Physics: Conference Series **655** (2015) 012011 doi:10.1088/1742-6596/655/1/012011

The comparison between materials is not straightforward, as different methods give different results. Moreover the evaluation should rely only on relative comparison between materials analyzed with the same methods, as the absolute values cannot be compared between different methods. The best material for the majority of methods was Extruded Polystyrene (on the third column) while EDIP 2003 and IPCC gave different indications. For the EDIP method, the high impact is due to the category "Ozone depletion" that represents the 96.8% of the total impact. This is caused by the use of CFC-113, a foaming agent with a very high atmospheric lifetime, during the production process. Also for the IPCC method the greatest part (81.7%) of the environmental damage is due to the presence of foaming agents, in this case HFC-134a, and to Carbon dioxide.
After the choice of XPS as the insulating material for the roof, several parametric simulations were made in TRNSYS software. The thickness of the insulating layer was changed from 0 to 35 cm with a step of 5 cm. The obtained annual energy consumptions were integrated with further LCA analysis, in order to determine if there was an optimum thickness with respect to energy consumption and material impact. The results obtained with five different LCA methods are shown in the next figures 7-11.

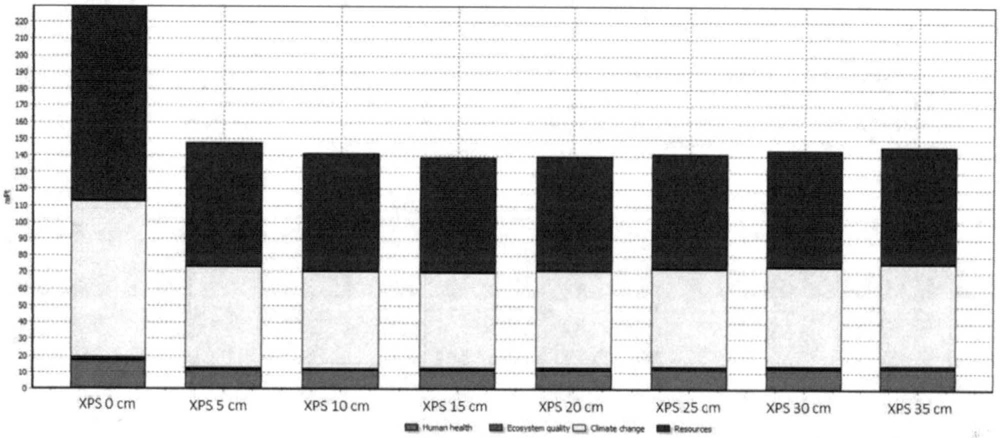

Figure 7. Comparison of different insulation thicknesses with the IMPACT 2002+ method

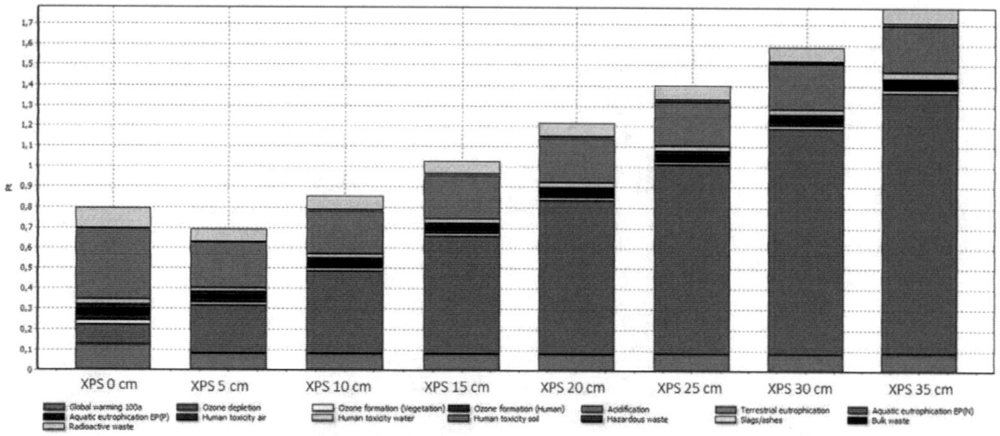

Figure 8. Comparison of different insulation thicknesses with the EDIP 2003 method

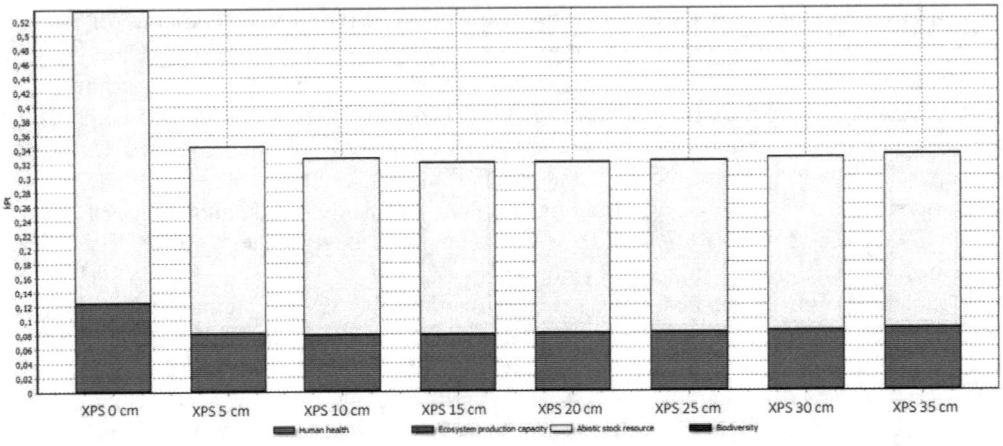

Figure 9. Comparison of different insulation thicknesses with the EPS 2000 method

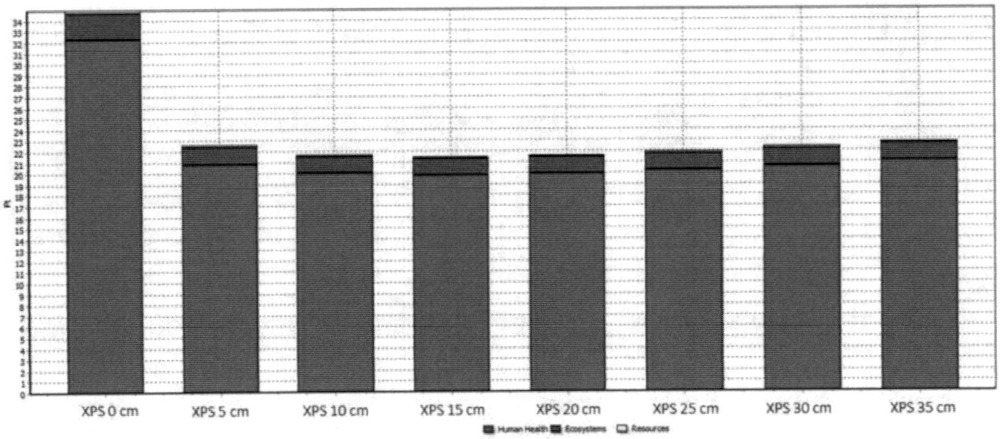

Figure 10. Comparison of different insulation thicknesses with the Re.Ci.Pe. method

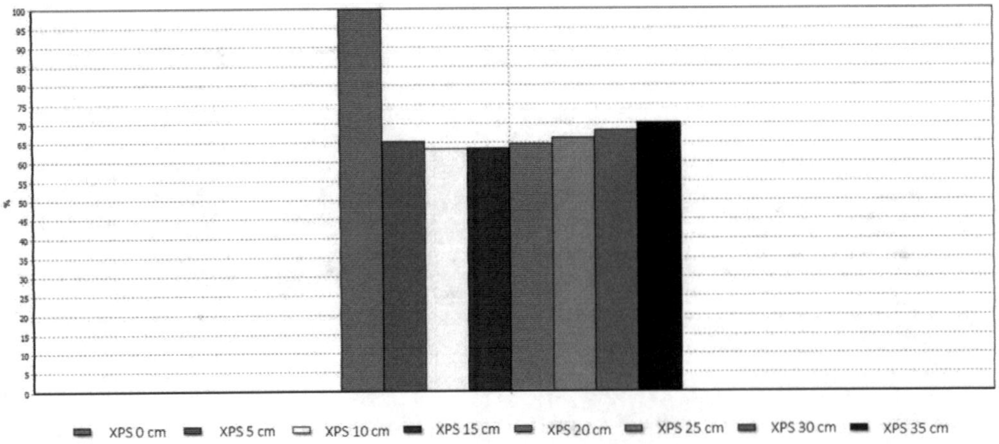

Figure 11. Comparison of different insulation thicknesses with the IPCC method

The effect of insulation is always positive, as even the thinner insulation is always better than the case without insulation, but an excess of insulation could be detrimental. Each method shows that the reduction of energy consumption, obtained increasing the insulation thickness, is counterbalanced by the increase of the environmental impact. Therefore an optimum thickness value could be identified and it is generally located between 5 and 15 cm. This result is shown in figure 12, where the parametric results for all the methods are reported.

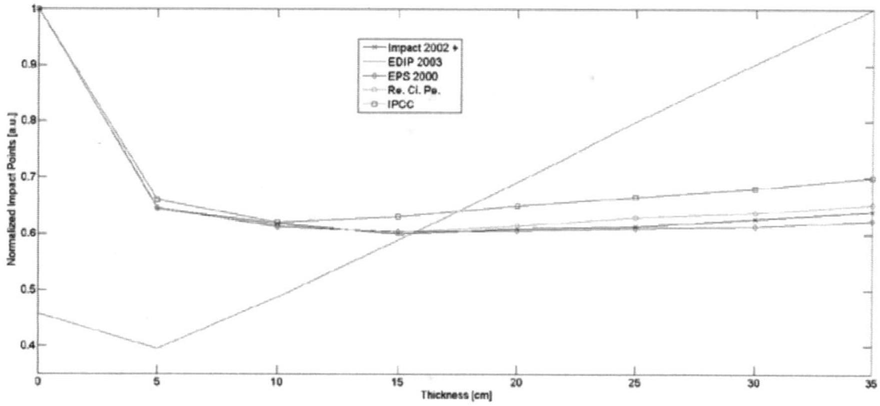

Figure 12. The comparison of different insulation thicknesses with all methods allows to identify the optimum thickness value.

4. Conclusions

The study of a historical building aimed to an energy retrofit is a complex task, that should be arranged in multiple steps. The preliminary operations should be focused on the evaluation of the current state of building, from an energetic and structural point of view, gathering data with on-site surveys. In these phase the use of non-destructive evaluation techniques (e.g. infrared thermography) is always advisable and, in some cases, mandatory. This work showed how IRT could be successfully applied to characterize the main thermal parameters of a construction material.
The measured data are the basis or the benchmark value of building mathematical models that are necessary to evaluate the impact of different energy savings measures. In this case, for energetic and structural reasons, the best energy saving measure was the insulation of the roof. The combination of energy simulation and LCA to compare different insulating material and find the best solution is driven to a holistic approach, that should be considered when dealing with a multi-criterion optimization. The benefits of the insulation with respect to the baseline standard are as clear as the damages due to an excess of insulation. The results showed that different LCA methods find an optimum insulation thickness in the range between 5 and 15 cm, that contains the value of 10.4 cm that is suggested by the energy savings standards.
Future work on this building may regard the final part of the retrofit process, including the evaluation of the energy savings obtained with the implemented energy measures, or a more in-depth analysis of other possible energy saving strategies. More work should focus also on the transfer of different data types between each phase of the process, with the aim of integrating different sources into a single streamlined procedure.

5. References

[1] Mazzarella L 2015 Energy retrofit of historic and existing buildings. The legislative and regulatory point of view *Energy and Buildings* **95** 23–31

[2] Camuffo D, Pagan E, Bernardi A and Becherini F 2004 The impact of heating, lighting and people in re-using historical buildings: a case study *Journal of Cultural Heritage* **5** 409–16

[3] Maldague X 2001 Theory and practice of infrared technology for nondestructive testing (Wiley)

[4] Meola C 2013 Infrared Thermography in the Architectural Field *The Scientific World Journal* **2013**

[5] Cadelano G, Bison P, Bortolin A, Ferrarini G, Peron F, Girotto M and Volinia M 2015 Monitoring of historical frescoes by timed infrared imaging analysis *Opto-Electronics Review* **23** 102–8

[6] Fortuna S, Neri P and Peron F. Social and energy redevelopment of an old building, BSA 2013 conference, Bozen

[7] Ruggieri G, Dotelli G, Melià P and Sabbadini S 2013 Life cycle assessment of refurbishment strategies for historic buildings, in Retrofitting the Built Environment (John Wiley & Sons)

[8] Philippi I, Batsale J C, Maillet D and Degiovanni A 1995 Measurement of thermal diffusivities through processing of infrared images *Review of Scientific Instruments* **66** 182–92

[9] Parker W J, Jenkins R J, Butler C P and Abbott G L 1961 Flash Method of Determining Thermal Diffusivity, Heat Capacity, and Thermal Conductivity *Journal of Applied Physics* **32** 1679–84

[10] Bison P, Bortolin A, Cadelano G, Ferrarini G and Grinzato E 2012 Comparison of some thermographic techniques applied to thermal properties characterization of porous materials *Proc of 11th international conference on quantitative infrared thermography-QIRT2012. Naples, Italy*

[11] Bison P and Grinzato E 2010 Fast estimate of solid materials thermal conductivity by IR thermography *Quantitative InfraRed Thermography Journal* **7** 17–34

[12] TRNSYS, A transient simulation program (Version 16). Solar Energy Laboratory, University of Wisconsin, 1994. http://sel.me.wisc.edu/trnsys/index.html

[13] UNI/TR 11552, 2014. Abaco delle strutture costituenti l'involucro opaco degli edifici - Parametri termofisici, UNI Italian Organization for Standardization.

[14] EN/ISO 14040, 2000. Environmental management - life cycle assessment - principles and framework (ISO 14040:2000), CEN European Bureau for Standardisation

[15] ECOShopping Database of insulation materials. Retrieved from www.ecoshopping-project.eu

[16] Frischknecht R, Jungbluth N et al. 2003 *Implementation of Life Cycle Impact Assessment Methods, Final report ecoinvent 2000* (Swiss Centre for LCI Dübendorf Switzerland)

[17] Frischknecht R, Jungbluth N, Althaus H J, Doka G, Dones R, Hischier R, Hellweg S, Humbert S, Margni M, Nemecek T and Spielmann M 2007 *Implementation of Life Cycle Impact Assessment Methods: Data v2.0. ecoinvent report No. 3* (Swiss centre for Lyfe Cycle Inventories Duebendorf Switzerland)

[18] Goedkoop M, Oele M., De Schryver A and Vieira M, 2008 *SimaPro Data Base Manual* (PRé Consultants The Netherlands)

[19] Goedkoop M 2008 *ReCiPe 2008: A life cycle impact assessment method which comprises harmonised category indicators at the midpoint and the endpoint level* (Ministry of Housing Spatial Planning and the Environment The Hague Netherlands)

[20] Steen B 1999. *A systematic approach to environmental strategies in product development (EPS). Version 2000 - General system characteristic* (Centre for Environmental Assessment of Products and Material Systems Chalmers University of Technology)

[21] Wenzel H and Hauschild M 2001. *Environmental Assessment of Products: Methodology, Tools and Case Studies in Product Development* (Kluwer Academic Publisher)

Beneficial effects of retroreflective materials in urban canyons: results from seasonal monitoring campaign

F Rossi, E Morini, B Castellani[1], A Nicolini, E Bonamente, E Anderini, F Cotana
University of Perugia, Department of Engineering, CIRIAF, Via G. Duranti 67, Perugia, Italy, IT

E-mail: beatrice.castellani@unipg.it

Abstract. Urban Heat Island (UHI) is a phenomenon that happens in urban areas consisting in higher temperatures than those in the surrounding rural areas. Cool materials and urban forestry have been identified as the mean countermeasures to this phenomenon. The specific research target was to investigate the benefit that can be obtained by the application of new retro-reflective (RR) cool materials on building envelopes in urban canyons. An ideal RR is a high reflective material that reflect incident radiation backward to the same direction of incidence. The test facility is made of two twin arrays resembling urban canyons with different aspect ratios. On the East array a cool, white, diffusive material is applied. On the West array a cool, white Retro-Reflective (RR) material is applied. The result of a seasonal (summer) monitoring campaign is discussed, with a particular focus on the temperature trends inside canyons with the same geometry but different envelope materials. While temperatures measured by the vertical sensors reveal a higher surface temperature of the RR, air temperatures and pavements' superficial temperatures are lower when RR are applied. These results suggest that RR materials have a cooling potential as coatings in urban canyons, and can thus improve urban climate conditions during summer.

1. Introduction

Urban Heat Island phenomenon (UHI) is related to higher temperatures in urban areas with respect to the suburban ones [1]. UHI is mainly given by urban facilities and human activities which contribute to urban temperature increase: buildings absorb daytime heat and release it at night-time; air conditioning, vehicles and industry release thermal heat; buildings decrease air flow; the lack of green spaces and the city pavements also increase absorption and storage of solar radiation [2]. UHI intensity depends on: geography, topography, climate, morphological and geometrical properties, materials, land-use and land-cover configuration, emission profiles and energy use [3]. Urban warming has negative effects on environmental and social aspects and affects the energy balance of the cities [4]. The main consequences are a significant increase of the peak and total electricity demand for cooling [5], a greater energy consumptions of several urban facilities as outdoor lighting [6] the worsening in thermal comfort in outdoor spaces and air quality conditions [7-10] and human health risks and increase of mortality [11]. Several researches also showed that the use of appropriate materials on the building envelope can have positive effects on the phenomenon mitigation and on the improvement of outdoor air quality and comfort conditions [12-14]. These materials are called "cool": cool roofs [15], cool pavements [16] and other innovative solutions including phase change materials [17] and thermochromic materials [18] are object of research. They are typically characterized by high solar reflectance (albedo) and high infrared emittance. Many studies show the positive effects of this relatively simple and cost-effective technique on urban heat island. It is well known and documented that a large scale change of albedo has a serious impact on the local peak ambient temperature [19]. It has been

[1] To whom any correspondence should be addressed.

demonstrated that increasing the albedo by 0.15 can reduce peak summertime temperatures by up to 1.5 °C in urban areas like Los Angeles [20]. In [21] through MM5 climate model two modified albedo scenarios have been compared to assess the effect of cool materials on the reduction of UHI in Athens: results show that temperature decreases up to 2.2 K. With a global climate model [22] it has been assessed the effects of albedo control on urban areas. The annual mean heat island decreased by 33%, the daily maximum and minimum urban temperature decreased respectively of 0.6° C and 0.3°C. Numerous studies also focused on the reduction of building energy consumption due to cool materials use [23-26]. During the DESIREX Campaign 2008 [27], it was stated that a higher albedo leads to about 5% reduction in energy consumption for air conditioning during summertime periods for the area of Madrid.
It has also been estimated that an increase of urban albedo (roofs and pavements) by 0.1 give a negative radiative forcing equivalent to offsetting 44Gt of CO_2 emission globally [28]. In [29,30] it has been demonstrated that 10 m^2 of high albedo surface can compensate 1 tCO_{2eq} at Mediterranean latitudes.
Retro-Reflective (RR) materials are here proposed as new cool materials for UHI mitigation. RR is a high reflective material that reflects mostly the incident radiation in the same direction of the incoming radiation. RR materials have been recently proposed as cool materials [31]. The purpose of this paper is to investigate the benefits that can be obtained by the application of RR materials on building envelopes in urban canyons.
The test facility, located at the University of Perugia (Italy) is made of two twin arrays resembling urban canyons with different aspect ratios [32]. On the east array a white diffusive material was applied while on the west array a white RR material is applied. The monitored temperatures in a cloudless, sunny summer day are discussed. The objective of the monitoring campaign is to compare the effect of two different types of cool materials, diffusive and RR, simultaneously during the same environmental conditions (solar irradiation, wind). Such analysis is based on the temperatures that are measured by the sensors of the facility and has a purely comparative significance. The improvement of cooling potential of retroreflective materials with respect to the white diffusive envelope is assessed.

2. The test field

The facility is composed of two twin arrays (West field and East field), each one representing two sets of canyons (Canyon 1, Canyon 2 and Canyon 3). Canyon 1, 2 and 3 are characterized by a H/D ratio of 1, 0.5 and 2 respectively. The facility, shown in Figure 1, is located on a roof enclosed by a 1-meter high stone parapet.

Figure 1. Experimental facility located at the University of Perugia.

In both the fields a bituminous membrane stands for a typical asphalt road on the horizontal surface. On the vertical surfaces of the canyons a white diffusive membrane (East field) and a RR membrane (West field) are installed. Optical characteristics of the investigated materials were assessed by spectrophotometric analysis. They are shown in Figure 2. Diffusive and RR materials basically differ from each other for their optical properties, in particular for the angular distribution of the reflected incident rays. An ideal diffusive surface is defined as a surface that reflects incident light in a hemisphere (a half disk in 2D). In this case thus energy is reflected by the surfaces towards all the directions, with an isotropic trend. An ideal RR surface, instead, is defined as a surface that reflects all the radiation toward the direction of incidence. The light is thus reflected in a ellipse whose major axis is in the direction of the incident radiation and its length depends on a $\cos^n$

mathematical relation [33]. In light of this, the materials that were applied on the vertical surfaces of the facility have different behaviours and influences on the nearby surfaces.

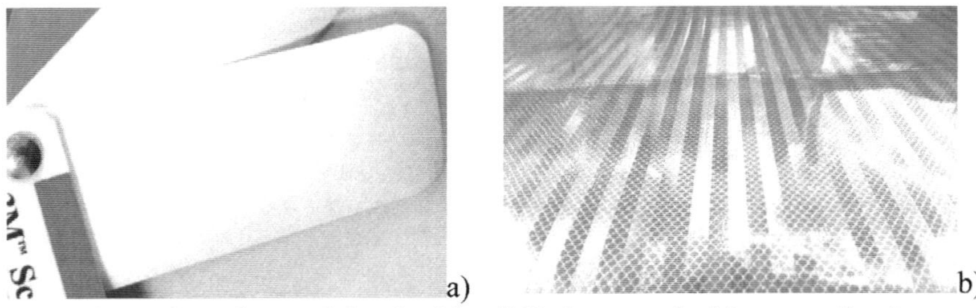

Figure 2. Investigated materials: a) white diffusive sample, b) retro reflective sample.

Temperature sensors are represented in Figure 3. "P" sensors measure the surface temperature of the pavement, "A" sensors measure the temperature of the air and V_S and V_N sensors measure the surface temperature of the vertical walls exposed to South and to North respectively. The same scheme is adopted for Canyons 1, 2, and 3 for both the West and East field. Temperature sensors are constituted by K-type thermocouples. The test field is also equipped with a meteorological station and an upward-facing pyranometer. The facility is not thought for dynamic similarity studies since fluid dynamic phenomena that take place inside the canyon are not under control. Therefore, results from the experimental investigation are just comparative but not scalable.

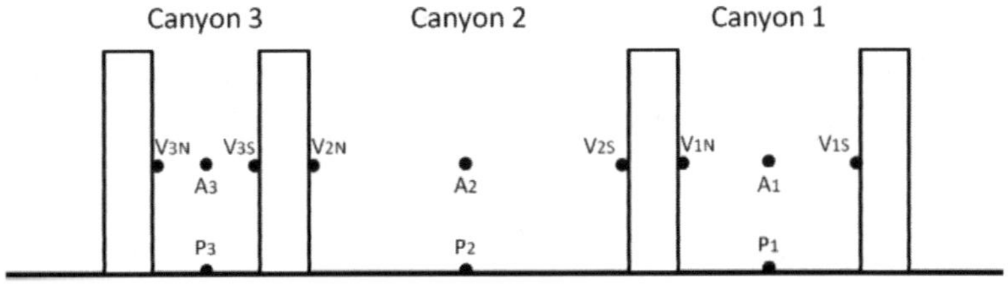

Figure 3. Scheme of the sensors inside the canyons. The same scheme is adopted for the East (white) and West (RR) field.

2.1. Spectrophotometric analysis of materials

Global reflected radiation profiles of RR and diffusive samples over the wavelength are reported in Figure 4.

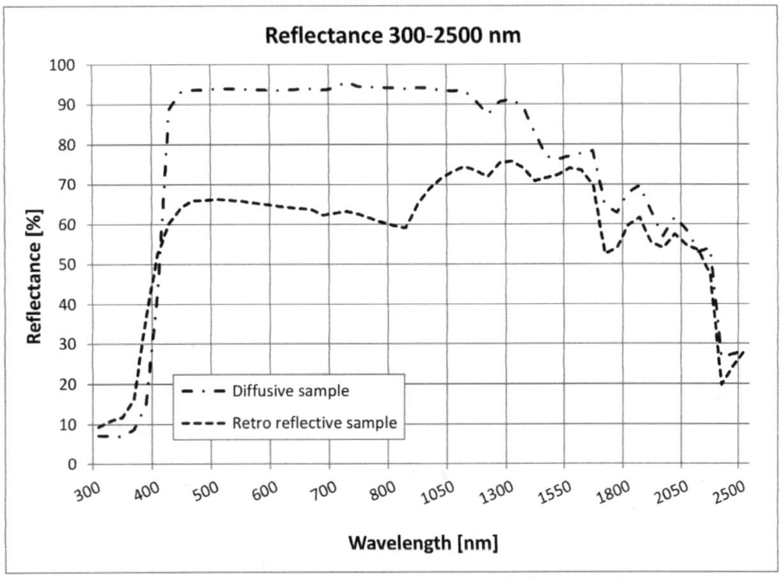

Figure 4. Reflectance spectrum of the samples.

Reflectance of the investigated materials in the solar spectrum is measured by Shimatzu SolidSpec 3700 spectrophotometer equipped with 60 mm integrating sphere. The spectrophotometer measurement range is 280–2500 nm, which includes the 99% of the solar energy. The value of solar reflectance is calculated as described in ASTM Standard G 173 [34,35]. The purpose of these measurements is the evaluation of the global hemispherical solar reflectance even in terms of spectral distribution. Solar reflectance measurements by the spectrophotometer show that the RR material has an overall reflectance equal to 62.8%, lower than that of the diffusive sample, equal to 86.5%. RR global reflectance in the range from 300 to 400 nm is higher than that of the diffusive film. In the infra red range RR global reflectance is much lower than that of diffusive material as well as in the visible spectrum.

3. Discussion of results

3.1. Data analysis

A three-month monitoring campaign during summer 2014 was carried out in the test facility, located in Perugia. July 20th was chosen as reference hot sunny day to evaluate the effect of the RR material on the energy balance in the canyons. The maximum, minimum and average measured temperatures are about 34° C, 26° C, and 18° C respectively and levels of irradiation measured by the pyranometer are higher than 0.8 kW/m^2 for many consecutive hours throughout the day. Wind velocity measurements were lower than 2 km/h during the entire reference day. Wind intensity has surely an influence on temperature trends, also because fluid dynamic phenomena that take place inside the canyon are not under control. Optical properties of the investigated materials are not affected by outdoor conditions. In addition, both RR and diffusive materials are studied simultaneously and under the same climatic conditions, therefore the related temperature profiles are influenced to the same extent. This does not invalidate the experimental campaign characterized by a comparative approach. Temperature difference ΔT was calculated as the difference between temperatures recorded by sensors in the East field and the West field respectively in each "n" position (ΔT_n). Since in the East field the vertical surfaces are white and in the West field the vertical surfaces are RR, a value of $\Delta T>0$ reveals that RR allow to reduce the energy absorbed, trapped and circulating in the canyons. The trend of ΔT has been graphed for the n measurement points and the subtended areas have been calculated ($A_{\Delta T,n}$). Since data are collected every 10 minutes, temperatures between two consecutive records were linearly interpolated. $A_{\Delta T,n}$ was thus calculated as the sum of trapezoids, whose parallel sides are represented by the recorded temperatures. As for the ΔT_n itself, a positive value of the subtended area means that energy in that point of the RR canyon is less than energy in the corresponding point of the white canyon. $A_{\Delta T,n}$ values have been then compared with the areas subtended by the curves that represent the temperatures at the respective n position in the East field ($A_{E,n}$). The percentage of energy trapped within the canyons saved by the RR material is thus showed in comparison with the energy trapped within the white canyons during six hours, to distinguish the daytime and the night-time savings. To consider the effect of the energy trapped during July 20th on night-time temperatures, the evaluations are carried out from July 20th at 00:00 to July 21st at 06:00. These evaluations have been treated separately for each measurement point in the canyons (pavement, air, vertical points) and finally a combined evaluation of the points belonging to the same canyon is presented. The results are shown below.

3.2. Air temperature

Air temperature profiles inside canyons are shown in Figure 5. The represented profiles show very low differences between RR and diffusive material in all the three canyons. To better comment such results, temperature differences ΔT_{A1}, ΔT_{A2}, and ΔT_{A3}, related to Canyon 1, Canyon 2 and Canyon 3 respectively were calculated according to procedure in Paragraph 3.1. Profiles are shown in Figure 6.

As can be seen from the graphs, the maximum value of ΔT occurs during the central part of the day, from 11:00 to 15:00, when solar irradiation is higher. Nevertheless, temperature differences oscillate rapidly from positive to negative values. Data are therefore organized and analysed on a 6-hourly basis and subtended areas are calculated. For the 6-hourly trends, Table 1 shows increases or reductions of trapped energy calculated as ratios between subtended areas of ΔT profiles and subtended areas of diffusive materials' temperature profiles ($A_{\Delta T,A}/A_{E,A}$). Negative values of ΔT occur during night (from 00:00 to 05:50) in Canyon 1 and Canyon 3, both characterized by an higher H/D ratio than Canyon 2. This means that RR materials tend to increase nocturnal air temperatures in narrow canyons if compared to diffusive cool materials.

Journal of Physics: Conference Series **655** (2015) 012012 doi:10.1088/1742-6596/655/1/012012

a)

b)

c)

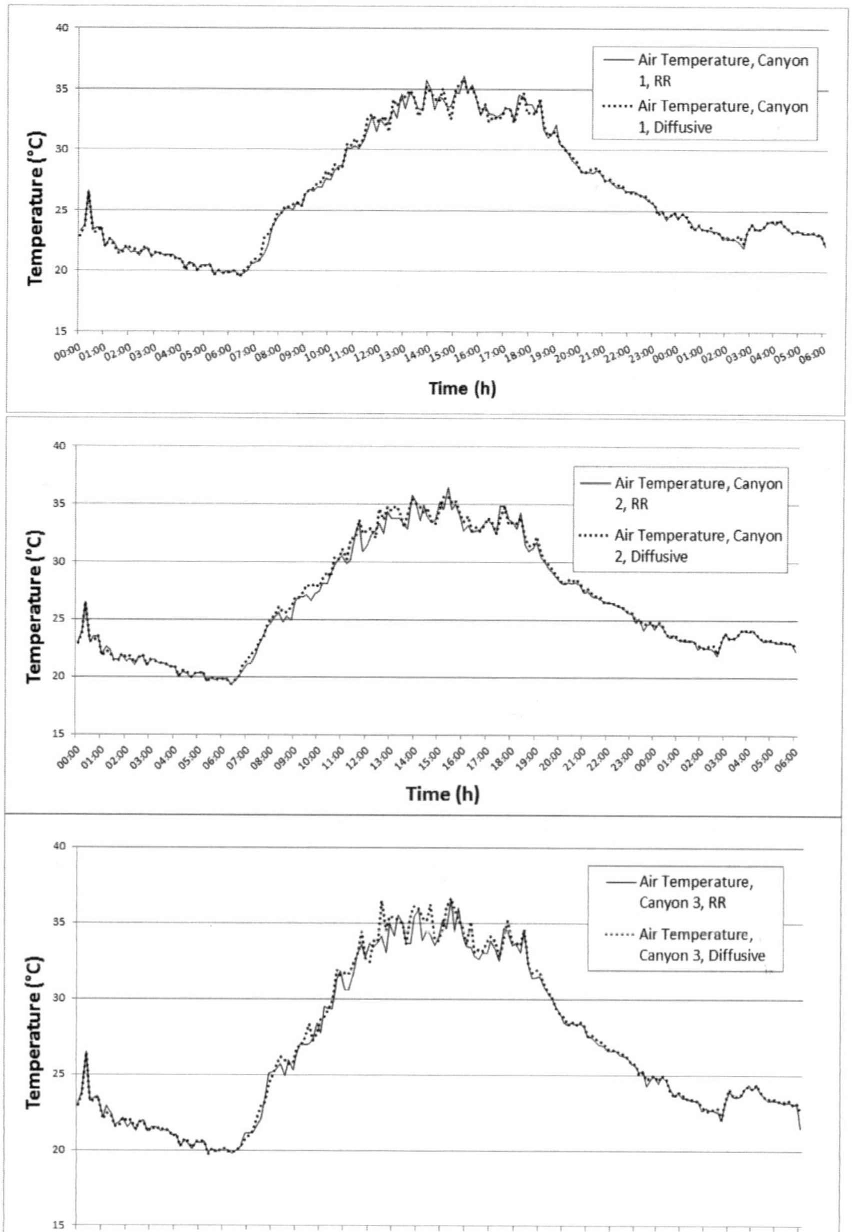

Figure 5. Air temperature profiles (A1, A2, A3): a) Canyon 1, b) Canyon 2, c) Canyon 3.

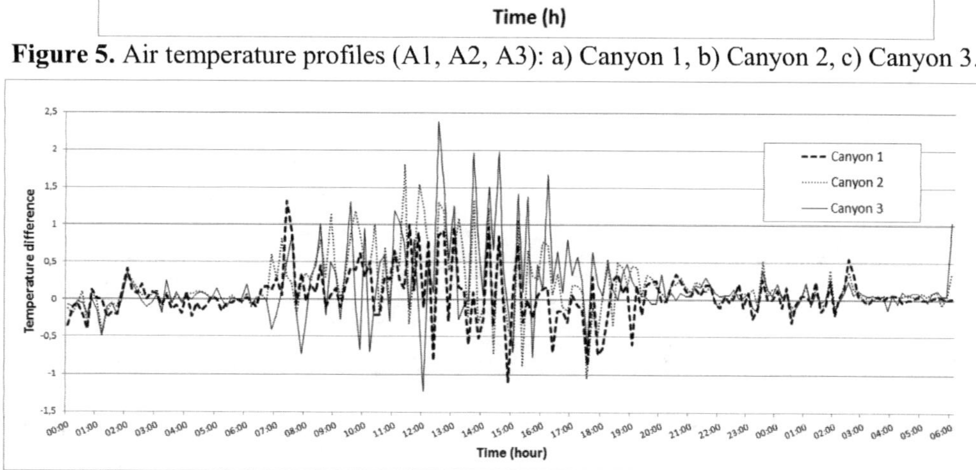

Figure 6. Differences of air temperature between RR and diffusive materials for the three canyons.

It is important to underline however that this negative effect is practically negligible since the increase of trapped energy is lower than 0.2%.
A cooling effect of RR materials can be observed during the central part of the day in all the three canyons with a reduction of energy trapped in the canyon equal to 1-2%.

Table 1. *6-Hourly trend of* $A_{\Delta T,A}/A_{E,A}$ *(%)*

hours	$A_{\Delta T,A1}/A_{E,A1}$ (%)	$A_{\Delta T,A2}/A_{E,A2}$ (%)	$A_{\Delta T,A3}/A_{E,A3}$ (%)
from 00:00 to 05:50	-0.179	0.024	-0.048
from 06:00 to 11:50	1.013	1.751	0.177
from 12:00 to 17:50	-0.0501	0.721	1.535
from 18:00 to 23:50	0.002	0.606	0.436
from 24:00 to 06:00	0.162	0.305	0.329

The global effect of RR materials during the entire reference day (24 hours) is quantified equal to 0.23% (for A_1), 0.71% (for A_2) and 0.68% (for A_3) avoided energy. This reveals a lower energy trapped in the RR canyons at temperature air level, with respect to energy trapped in the white diffusive canyons.

3.3. Pavement temperature

Pavement superficial temperature profiles are shown in Figure 7. Temperature differences on pavements are more marked, especially for Canyon 2 and Canyon 3 in the central part of day, showing a pronounced cooling effect of RR materials. In the same period, RR and diffusive materials in Canyon 1 show the same effect on temperatures. During night time, temperatures of diffusive pavements are slightly higher than those of RR pavements. Differences of pavement superficial temperature between RR and diffusive materials measured at P1, P2 and P3 points are shown in Figure 8. It can be noticed that temperatures of RR pavements are always lower than temperatures of diffusive pavements during the entire reference day. The maximum difference in the range of [3-7] °C occurs during the hottest hours of the day. Temperatures of the pavements, lower in the RR Canyons can be explained recalling what have been stated before about the optical properties of the vertical surfaces. The RR material reflects backward to the incidence direction the most of the incoming radiation and just a little radiation towards the other directions. The diffusive one, instead, diffuses the reflected radiation on its typical hemisphere and in an isotropic way. This justifies the higher temperatures of the pavements in the diffusive Canyons, that are struck by the direct radiation and also by reflected waves.
Data are then analyzed in accordance with procedure in Paragraph 3.1 on 6-hourly basis and shown in Table 2. The hourly comparison reveals savings up to 6.77%, 7.57% and 6.74% in P_1 and P_2 from 06:00 to 11:50 am, and in P_3 from 12:00 to 17:50 pm respectively. As shown in Figure 7, two negative percentages are found in P_2 from 00:00 to 05:50 am and from 18:00 to 23:50 pm. The daily comparison reveals savings of 5.15%, 2.41% and 5.27%, showing an RR global cooling effect on canyon's pavements.

a)

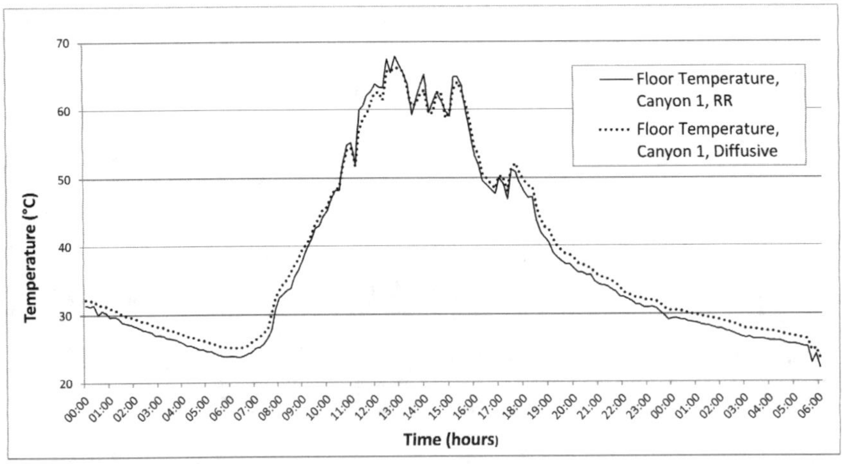

b)

c)

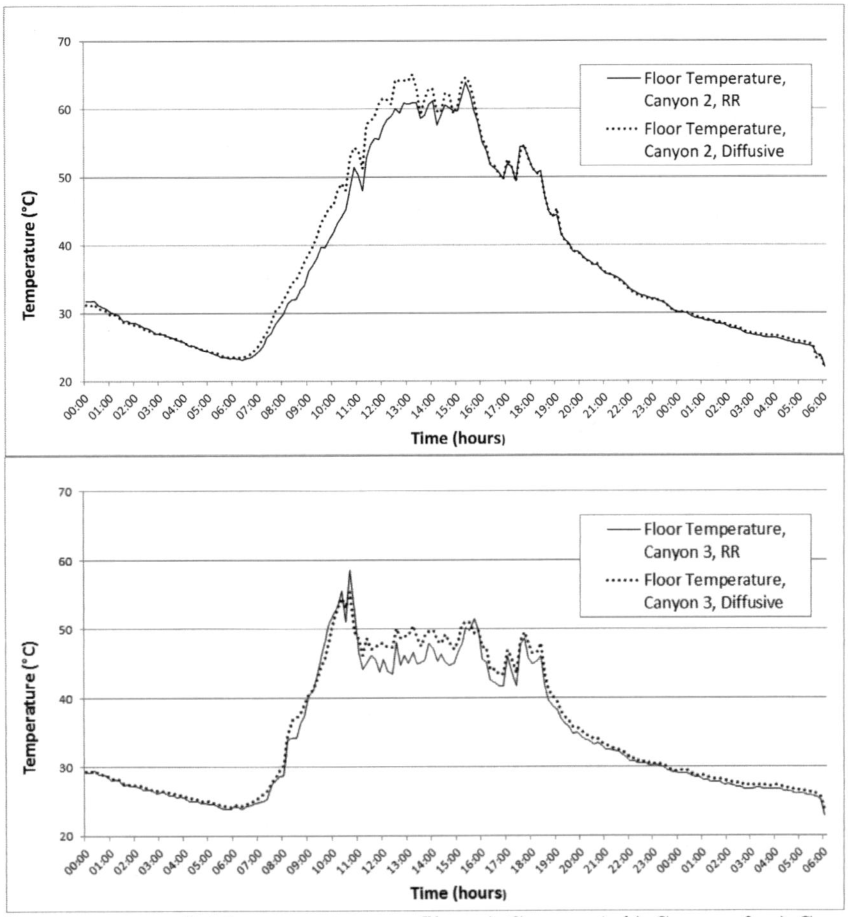

Figure 7. Pavement superficial temperature profiles: a) Canyon 1, b) Canyon 2, c) Canyon 3.

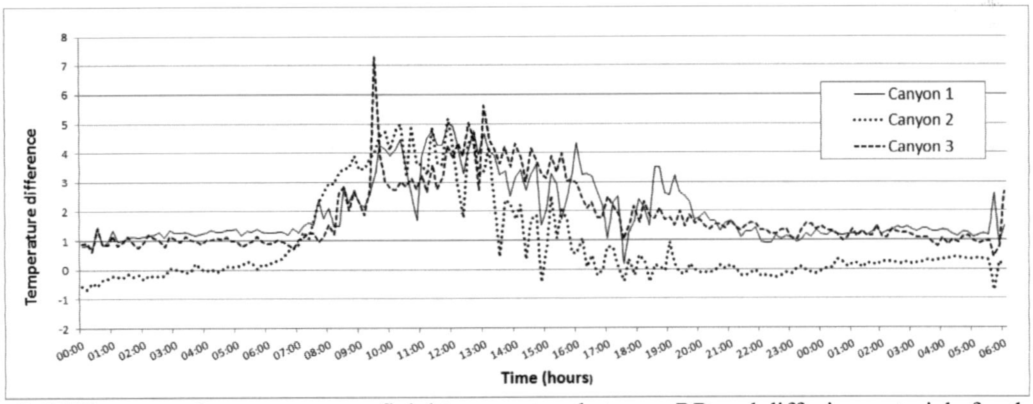

Figure 8. Differences of pavement superficial temperature between RR and diffusive materials for the three canyons.

Table 2. *6-Hourly trend of $A_{\Delta T,P}/A_{E,P}$ (%)*

hours	$A_{\Delta T,P1}/A_{E,P1}$ (%)	$A_{\Delta T,P2}/A_{E,P2}$ (%)	$A_{\Delta T,P3}/A_{E,P3}$ (%)
from 00:00 to 05:50	4.205	-0.369	3.733
from 06:00 to 11:50	6.771	7.572	1.638
from 12:00 to 17:50	5.103	2.530	6.739
from 18:00 to 23:50	0.045	-0.065	4.444
from 24:00 to 6:00	4.635	0.825	4.163

Journal of Physics: Conference Series **655** (2015) 012012 doi:10.1088/1742-6596/655/1/012012

3.4. Façades temperatures

Façades have been treated separately, according to their exposition.

3.4.1. South-oriented façades

Superficial temperature profiles on south-oriented façades are shown in Figure 9. For all the three canyons, temperatures during the central part of the day are higher in case of RR materials. During night time, results of comparison are not uniform: in Canyon 1 temperature trends overlap, in Canyon 2 RR materials show higher temperatures while in Canyon 3 they have lower temperatures.

Differences of south-oriented façades temperature between RR and diffusive materials measured at V_{1s}, V_{2s} and V_{3s} points are shown in Figure 10. Profiles reveal many positive values for ΔT but also many large negative values, with peak negative values at 10:50 (ΔT= - 5.82 °C), 12:20 (ΔT= - 7.68 °C), and 12:20 (ΔT= - 5.66 °C) for V_{11}, V_{2s}, and V_{3s} respectively.

According to this trend, the percentage of energy reveals that up to a 3.8% extra of energy is absorbed by the vertical surfaces exposed towards South in the RR canyon than in the white one.

a)
b)
c)

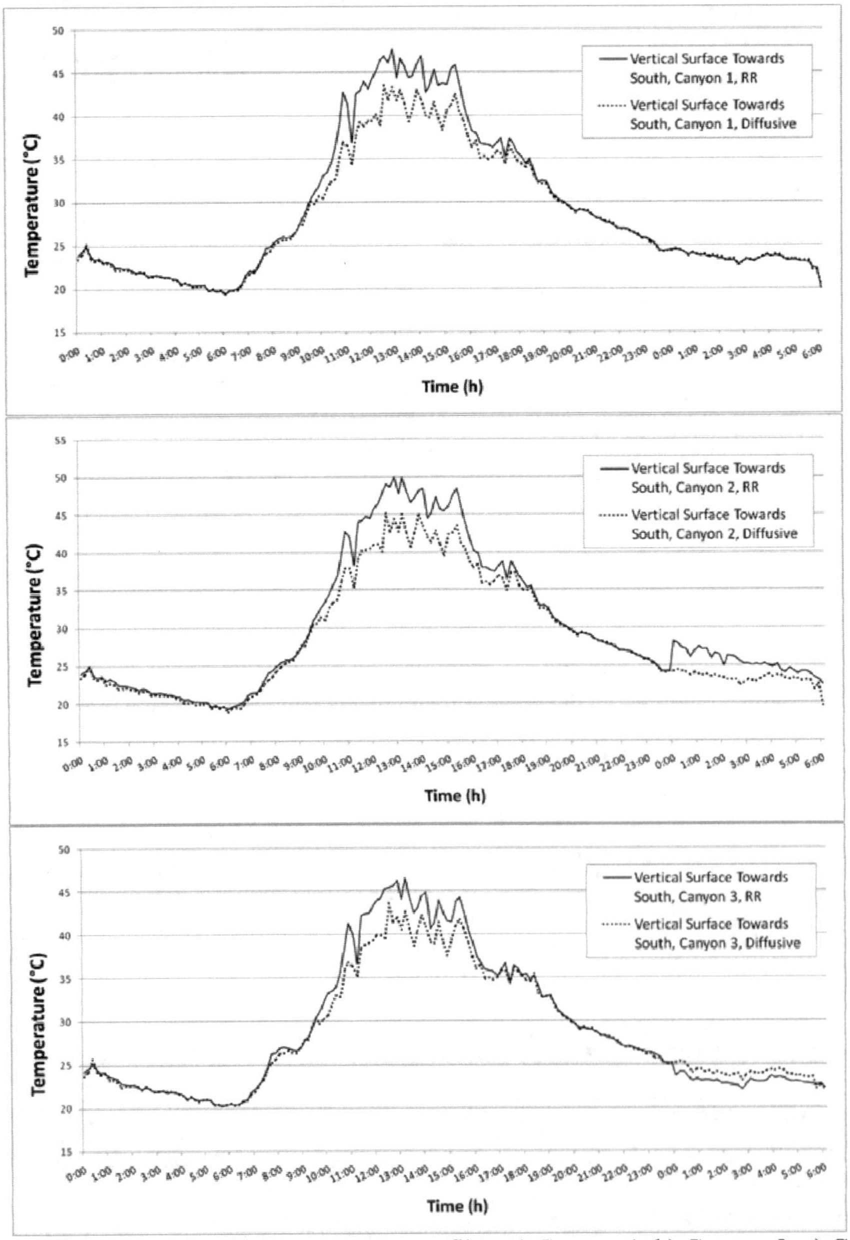

Figure 9. South-oriented façades' temperature profiles: a) Canyon 1, b) Canyon 2, c) Canyon 3.

Values of AΔT,VS/AE,VS are -3.070 for V1s, -3.794 for V2s and -2.130 for V3s. On 6-hourly basis, the percentage of extra energy kept into the canyon is up to about 7%, 9% and 6% in the middle hours of the day (Table 3). Data show a negative effect of RR materials on energy absorbed by south-oriented façades. The worst behaviour is observed in Canyon 2, the flattest one.

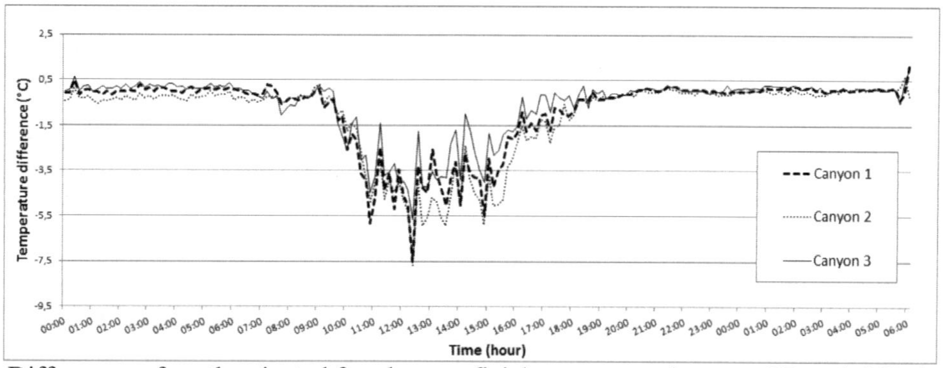

Figure 10. Differences of south-oriented façade superficial temperature between RR and diffusive materials for the three canyons.

***Table 3.** 6-Hourly trend of $A_{\Delta T,VS}/A_{E,VS}$ (%)*

hours	$A_{\Delta T,VS1}/A_{E,VS1}$	$A_{\Delta T,VS2}/A_{E,VS2}$	$A_{\Delta T,VS3}/A_{E,VS3}$
from 00:00 to 05:50	0.190	-1.262	0.863
from 06:00 to 11:50	-5.149	-4.966	-1.082
from 12:00 to 17:50	-7.653	-9.042	-5.587
from 18:00 to 23:50	-0.002	-0.437	0.027
from 24:00 to 6:00	0.526	0.214	0.682

3.4.2. North-oriented façades

Superficial temperature profiles on north-oriented façades are shown in Figure 11. Differences between RR and diffusive materials are not as sharp as in the south-oriented vertical surfaces analysed before.

Temperature differences between RR and diffusive materials measured at V_{1N}, V_{2N} and V_{3N} points are shown in Figure 12. Profiles reveal many positive values for ΔT and many large negative values. The peak positive values are 1°C, 0.5°C and 1.8°C in canyon 1, canyon 2 and canyon 3 respectively. The peak negative values are -3.9°C, -2.9°C, -2.4°C in in canyon 1, canyon 2 and canyon 3 respectively. The overall daily effect of RR is an increase, although minor (up to 1%), of the energy that is hold in the canyon. Values of $A_{\Delta T,VN}/A_{E,VN}$ are -0.594 for V_{1N}, -1.093 for V_{2N} and -0.045 for V_{3N}.

On 6-hourly basis, this increase is up to 2%, 2.7% and 0.7% in canyon 1, 2, and 3 respectively from 12:00 to 17:50 (Table 4). The increase of superficial temperature is due to an overall lower level of reflectance of RR materials with respect to a white diffusive material.

a)

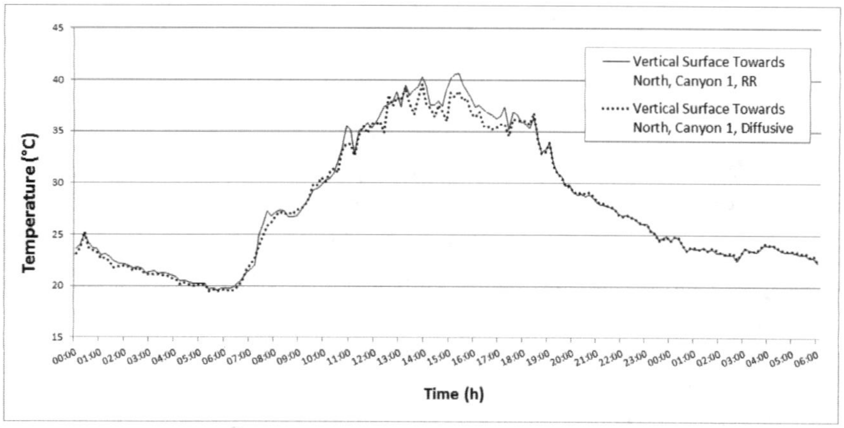

b)
c)

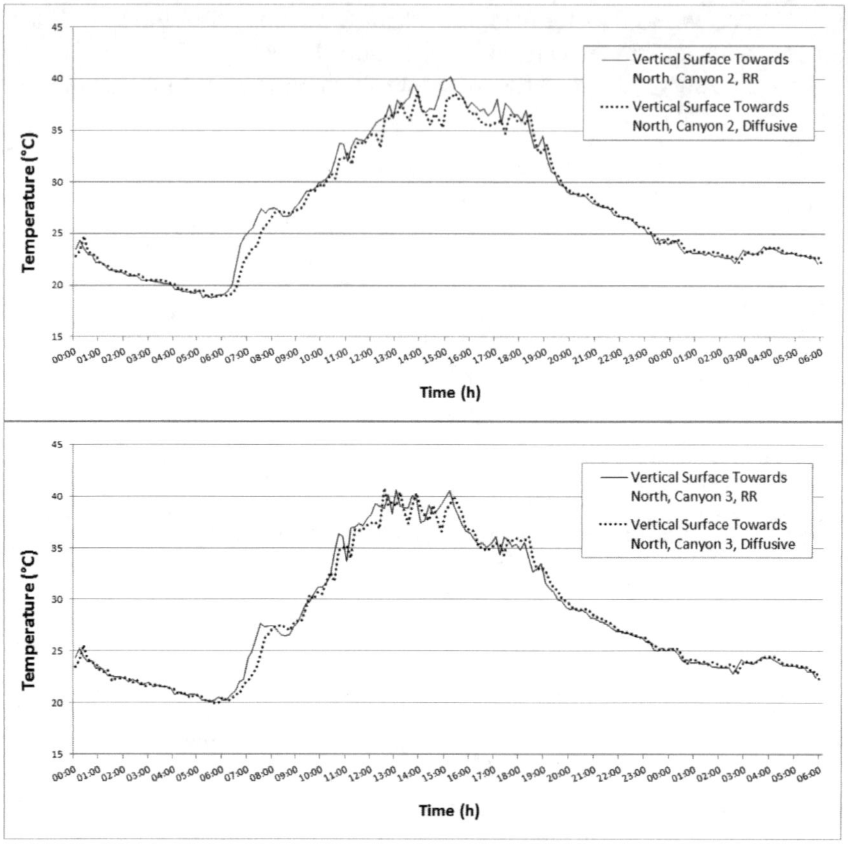

Figure 11. North-oriented façades' temperature profiles: a) Canyon 1, b) Canyon 2, c) Canyon 3.

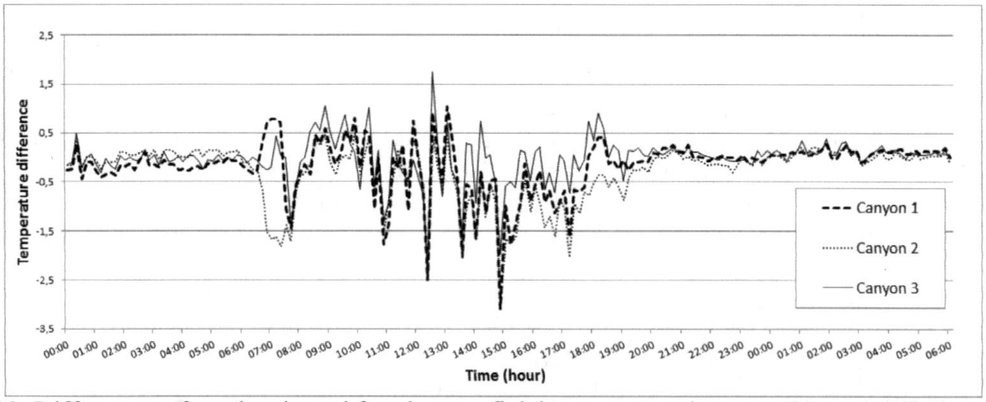

Figure 12. Differences of north-oriented façade superficial temperature between RR and diffusive materials for the three canyons.

Table 4. *6-Hourly trend of* $A_{\Delta T,VN}/A_{E,VN}$ *(%)*

hours	$A_{\Delta T,VN1}/A_{E,VN1}$	$A_{\Delta T,VN2}/A_{E,VN2}$	$A_{\Delta T,VN3}/A_{E,VN3}$
from 00:00 to 05:50	-0.787	0.207	-0.168
from 06:00 to 11:50	-0.216	-1.515	0.005
from 12:00 to 17:50	-1.984	-2.764	-0.686
from 18:00 to 23:50	0.001	-0.623	0.315
from 24:00 to 6:00	0.468	0.270	0.551

The main disadvantage of RR materials is related to the temperatures measured by the vertical sensors, that reveal a higher surface temperature of the RR on the façades. On the other hand, the decrease of the air and ground temperature reveal its UHI mitigation potential. In addition, the effect of RR materials on energy trapped in the canyon depends strictly on the geometry of the canyon itself. In particular the cooling potential of the investigated RR coating on air temperature is evident in Canyon 2, that is the flattest one. A lower air temperature is associated to higher façades' superficial temperatures as shown in Table 4.

4. Conclusions

The trend of temperatures in three urban canyons on a scale test facility with H/D ratio equal to 1, 0.5 and 2 has been analyzed. A comparative analysis about the influence of a retro-reflective (RR) material with respect to a white diffusive cool material applied on vertical surfaces of urban canyons has been treated.
The comparative analysis is based on temperatures' trend: a positive ΔT and a positive area subtended by ΔT reveals a benefit from the RR compared to the white material. An individual evaluation for air, vertical surfaces and horizontal surfaces reveal benefits and disadvantages from RR materials. The main disadvantage is related to the temperatures measured by the vertical sensors, that reveal a higher surface temperature of the RR. Solar reflectance measurements by the spectrophotometer show indeed that the RR material has an overall reflectance equal to 62.8%, lower than that of the diffusive sample, equal to 86.5%. The temperature measured by air temperature sensors and the temperature measured by pavement sensors, and thus the urban heat island effect, decreased in the RR canyons. As for the air temperature a decrease up to 1% is measured. As for the pavement temperature a much higher reduction of temperature is registered: the reduction is between 5% and 22% in canyon 1, between 1% and 2% in canyon 2, and between 5% and 10% in canyon 3 at night-time (from 00:00 to 06:00). These results suggest that RR materials have a cooling potential with respect to traditional white diffusive coatings in urban canyons, and can thus improve urban climate conditions during summer. The evaluations provided in this paper are based on a comparative analysis between the performance of an RR material and a white diffusive material. The validation for real scale application will be object of the next steps of the research. The extension of the present evaluations to real scale is foreseen. With this purpose, a dimensional analysis and similitude will be elaborated to correlate the above experimental results to real scale applications.

5. References

[1] Santamouris M 2001 James and James Science Publishers
[2] Oke TR, Johnson DG, Steyn DG, Watson ID 1991 *Boundary-Lay Meteorol* **56** 339
[3] Mihalakakou G, Flocas HA, Santamouris M, Helmis CG 2002 *J. Appl. Meteorol.* **41** 519
[4] Santamouris M 2014 On the energy impact of urban heat island and global warming on buildings *Energy and Buildings* **82** 100
[5] Asimakopoulos DA, Santamouris M, Farrou I, Laskari M, Saliari M, Zanis G, Giannakidis G, Tigas K, Kapsomenakis J, Douvis C, Zerefos SC, Antonakaki T, Giannakopoulos C 2012 *Energy Build.* **49**
[6] Rossi F, Bonamente E, Nicolini A, Anderini E, Cotana F 2015 *Energy and Buildings* In press doi:10.1016/j.enbuild.2015.04.054
[7] Pantavou K, Theoharatos G, Mavrakis A, Santamouris M 2011 *Building and Environment* **2** 339
[8] Pantavou K, Theoharatos G, Santamouris M, Asimakopoulos 2013 *Building and Environment* **66** 82
[9] Sarrat C, Lemonsu A, Masson V, Guedalia D *Atmos. Environ.* **40** 1743
[10] Stathopoulou E, Mihalakakou G, Santamouris M,. Bagiorgas HS 2008 *J. of Earth Syst. Scie.* **117** 227
[11] Santamouris M, Kolokotsa D *Energy and Buildings* **98** 125
[12] Akbari H, Matthews HD *Energy and Buildings* **55** 2
[13] Santamouris M, Synnefa A, Karlessi T 2011 *Solar Energy* **85** 3085
[14] Pisello AL, Rossi F, Cotana F 2014 *Energies* **7** 2343
[15] Joudi A, Svedung H, Bales C, Rönnelid M 2011 *Appl Energy* **88** 4655
[16] Levinson R, Pan H, Ban-Weiss G, Rosado P, Paolini R, Akbari H *Appl Energy* **88** 4343
[17] Castellani B, Morini E, Filipponi M, Nicolini A, Palombo M, Cotana F, Rossi F *Sustainability* **6** 6815
[18] Doulos L, Santamouris M, Livada I 2004 *Solar Energy* **77** 231
[19] Synnefa A, Santamouris M, Livada I 2006 *Solar Energy* **80** 968
[20] Oleson KW, Bonan GB, Feddema J 2010 *Geophysical Research Letters* L **37** 03701
[21] Kolokotroni M, Ren X, Davies M, Mavrogianni A 2012 *Energy and Buildings* **47** 302
[22] Kolokotsa D, Diakaki C, Papantoniou S, Vlissidis A *Energy Build* **55** 85
[23] Santamouris M *Solar Energy* **103** 682

[24] Zinzi M, Carnielo E, Agnoli S 2012 *Energy Build* **50** 111
[25] Salamanca F, Martilli A, Yague C 2008 *International Journal of Climatology* **32** 2372
[26] Campra P, Garcia M, Canton Y, Palacios-Orueta A 2008 *J.of Geophysical Research* D**113** 18109
[27] Taha H 1997 *Energy and Buildings* **25** 99
[28] Akbari H, Menon S, Rosenfeld A 2009 *Climatic Change* **95** 3
[29] Rossi F, Cotana F, Filipponi M, Nicolini A, Menon S, Rosenfeld A 2013 *Adv. Build. En.Res.***7** 254
[30] Cotana F, Rossi F, Filipponi M, Coccia V, Pisello AL, Bonamente E, Petrozzi A, Cavalaglio G 2014 *Applied Energy***130** 641
[31] Rossi F, Castellani B, Presciutti A, Morini E, Filipponi M, Nicolini A, Santamouris M 2015 *Applied Energy* **145** 8
[32] Bonamente E, Rossi F, Coccia V, Pisello AL, Nicolini A, Castellani B, Cotana F, Filipponi M, Morini E, Santamouris M 2013 *Advances in building Energy Research* **7** 222
[33] ASTM E903-12. Standard test method for solar absorptance, reflectance, and transmittance of materials using integrating spheres.
[34] ASTM G 173-03. Standard tables for reference solar spectral irradiances: direct normal and hemispherical on 37° tilted surface; 2012.
[35] Rossi F, Pisello AL, Nicolini A, Filipponi M, Palombo M 2014 *Applied Energy* **114** 621

A novel approach to CFD analysis of the urban environment

F Nardecchia[1], F Gugliermetti[1], F Bisegna[1]

[1] SAPIENZA University of Rome, Department of Astronautical, Electrical and Energetic Engineering (DIAEE), Via Eudossiana 18 - 00184 Rome, Italy

E-mail: fabio.nardecchia@uniroma1.it

Abstract. The construction of cities, with their buildings and human activities, not only changes the landscape, but also influences the local climate in a manner that depends on many different factors and parameters: weather conditions, urban thermo-physical and geometrical characteristics, anthropogenic moisture and heat sources. Land-cover and canopy structure play an important role in urban climatology and every environmental assessment and city design face with them.
Inside the previous frame, the objective of this study is both to identify both the key design variables that alter the environment surrounding the buildings, and to quantified the extension area of these phenomena. The tool used for this study is a 2D computational fluid dynamics (CFD) numerical simulation considering different heights for buildings, temperature gaps between undisturbed air and building's walls, velocities of undisturbed air. Results obtained allowed to find a novel approach to study urban canopies, giving a qualitative assessment on the contribution and definition of the total energy of the area surrounding the buildings.

1. Introduction

With further understanding of human health, nature and environment protection, people are more concerned about the relationship between buildings and the surrounding environment [1]. Due to rapid urbanization in developing countries, environmental issues have gained increasing attention in cities with all kind of climates, especially for the metropolis [2]. Therefore, the interest in the microclimate around buildings in urban areas has increased as it affects, among other things, outdoor and indoor thermal comfort [3-5], energy use for heating and cooling [6], the dispersion of air pollution and all the other parameters that can affect the anthropogenic heat [7-8]. In the urban environment, a comfortable climate is important for well-being [9-10].
Great problems in the assessment of urban climatology are due to heterogeneity, especially in land-cover and canopy structure and so many problems are due to the building design [11-15].
The objective of this study is to evaluate the influence of a single isolated building to identify the key urban design variables that influence the urban canopy, and evenly to obtain a correlation that allows giving indications about influence that a building could produce in the surrounding environment. To do this, 2D computational fluid dynamics (CFD) numerical simulations, using standard k–ε turbulence model of individual buildings have been performed, varying alternately some geometric parameters and environmental conditions.
This approach allows both to identify, in a simple and unique way, the extension of the area influenced by the building and to calculate the difference in total energy between disturbed and undisturbed areas surrounding the building.

Journal of Physics: Conference Series **655** (2015) 012013 doi:10.1088/1742-6596/655/1/012013

The novel aspect of this approach is mainly due to the introduction of only one energy parameter to take into account the building influence on its surrounding environment.
In this way, it is also possible a comparison among different urban settlements characterized with different environmental microclimate, in terms temperature, in order to define possible actions to reduce energy consumption and to find out the weight that each of them can have on the urban microclimate change, to highlight urban critical issues. This can lead to defining a strategy of urban actions (creation of green areas, restricted traffic areas, etc.) and architectural (different materials applications, cool roofs, green roofs, cool pavements, etc.) to improve the microclimate of the considered area; by this way, qualitative results can be reached in a quicker time respect to current approaches.
The study also reports a correlation between the average height of the buildings and the extension of the relative influenced area that could be easily improved, following the proposed approach, also to groups of buildings.

2. A background on Computational Fluid Dynamics for urban areas
In the last years a lot of studies were developed [16-18] on external building ventilation with CFD models, as this approach could allow to deepen phenomena that are objectively difficult to investigate experimentally, thus lowering significantly the costs of the study. However, almost all those are unsteady analyses with macro models of 1km×1km mesh and steady analyses at daytime with micro models [19-21]. Unsteady analysis with a micro model is not so spread because the calculation for the solar radiation requires a lot of computational time and resources [22].
The state of the art of microclimate residential studies led to the definition of two main categories for its approach: lumped parameters [23-24] and distributed parameters method [25–26]. Currently, most calculations cases mesh the surfaces and air into the same size, which lead to long calculating time and few practical applications some other researches give assumed surface temperatures as boundary conditions of CFD simulation, instead of calculating both the quantities with the equations in CFD simulation, which may lead to serious inaccuracy. Using measured surface temperatures for CFD calculation may be reliable, but it may also require many detailed boundary conditions. Therefore, the research about how to reduce the calculating time without reducing the accuracy for numerical simulation of the thermal environment around buildings is needed.

3. Model Description
Computational fluid dynamics (CFD) modelling is based on the numerical solution of the governing fluid flow, which is derived from basic conservation and transport principle: (a) the mass conservation (continuity) equation and (b) the three momentum conservation (Navier–Stokes) equations in x, y, z. The air around the buildings can be regarded as an incompressible turbulent inert flow, and the air density is assumed to be constant. These assumptions are reasonable for most lower atmosphere environments as described by Sini et al. [27]. Besides, the turbulence produced by the buoyancy effect is not included because the convective thermal effect around the building is not taken into consideration in the present study. So it has been considered only the turbulence provided by the forced convection. For the two-dimensional problem of an isolated building, the standard k–ε turbulence model governing equations (continuity (1) and momentum (2)) can be expressed as follows:

$$\frac{\partial \overline{U}_i}{\partial x_i} = 0 \tag{1}$$

$$\overline{U}_j \frac{\partial \overline{U}_i}{\partial x_j} = -\frac{1}{\rho}\frac{\partial \overline{P}}{\partial x_i} + \frac{\partial}{\partial x_j}\left(\upsilon \frac{\partial \overline{U}_i}{\partial x_j} - \overline{u_i u_j} \right) \tag{2}$$

k and ε transport equations in the standard k–ε turbulence model:

$$\overline{U}_j \frac{\partial k}{\partial x_j} = \frac{\partial}{\partial x_j}\left(\frac{\nu_t}{\sigma_k}\frac{\partial k}{\partial x_j}\right) + \upsilon_t\left(\frac{\partial \overline{U}_i}{\partial x_j} + \frac{\partial \overline{U}_i}{\partial x_i}\right)\frac{\partial \overline{U}_i}{\partial x_i} - \varepsilon, \tag{3}$$

$$\overline{U}_j \frac{\partial \varepsilon}{\partial x_j} = \frac{\partial}{\partial x_j}\left(\frac{\upsilon_t}{\sigma_\varepsilon}\frac{\partial \varepsilon}{\partial x_j}\right) + \frac{\varepsilon}{k}\left(C_{\varepsilon 1}\upsilon_t\left(\frac{\partial \overline{U}_i}{\partial x_j} + \frac{\partial \overline{U}_j}{\partial x_i}\right)\frac{\partial \overline{U}_i}{\partial x_j} - C_{\varepsilon 2\varepsilon}\right) \tag{4}$$

where

$$\overline{u_i u_j} = \frac{2}{3}k\delta_{ij} - \upsilon_t\left(\frac{\partial \overline{U}_i}{\partial x_i} + \frac{\partial \overline{U}_j}{\partial x_j}\right) \tag{5}$$

$\upsilon_t = C_\mu (k^2/\varepsilon)$, k is the turbulent kinetic energy, ε denotes the turbulent dissipation rate. The constants for k–ε turbulence model are summarized in Table 1.

Table 1. Constants for k–ε turbulence model

C_μ	σ_k	σ_ε	C_{z1}	C_{z2}
0.09	1.00	1.22	1.44	1.92

Figure 1 shows the computational building block configuration, where a two-dimensional computational domain is considered. The inlet boundary is placed at a distance ten times that of the building height (H) upstream of the building, where the velocity inlet boundary condition is applied, and similarly the outlet boundary is placed at a distance ten times that of the building height (H) downstream of the building, where pressure outlet boundary condition is applied. The definition of the extension for the control fluid domain was done by following the guidelines for CFD simulations already known in the literature [28].

The velocity inlet boundary condition is chosen at the inflow boundary with a constant profile and a static pressure of 0 Pa with respect to the reference pressure is imposed as outlet boundary condition. It is used to force the flow in the direction normal to the outlet without any backflow. A free-slip wall condition is applied at the upper boundary.

The domain is discretized by using an orthogonal grid. Since the geometry is relatively simple, a block-based hexahedral mesh is used to enhance the quality of the mesh. The grid lines are refined near the solid surfaces (ground, rooftops, and building walls) and the spacing between grid lines is gently stretched at increasing distances from the solid surfaces, where all inflation ratios of the mesh are kept about 1.05.

Extensive tests of the independence of the meshes have been developed with increasing mesh numbers until further refinement is shown to be less significant, to ensure an independence of the results from grid size. The physical domain and cells number were changed according to the size of the building. A typical computational domain when H = 10 m is 300x100 m, to which it corresponds a computational grid of about 50 000 cells. The sketch of the grid mesh is reported in Figure 1: the grid chosen was fine when closer to building and ground and then expanded further away. The expansion ratio in this non-uniform grid system is about 1.1.

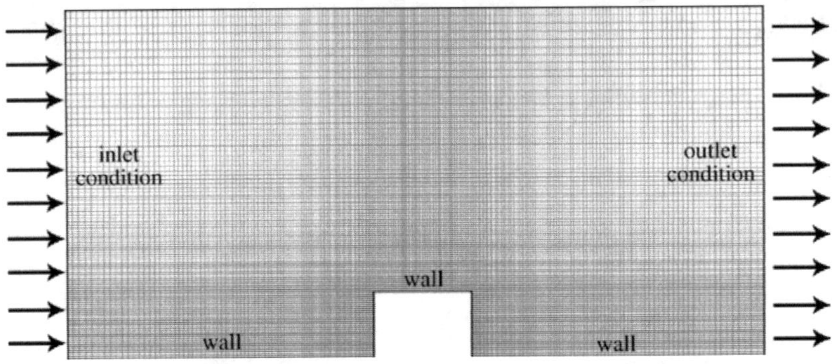

Figure 1. Computational domain and grid

The commercial CFD code ANSYS 14.5 was used in this research. The 2D steady RANS equations are solved in combination with the realizable k-ε turbulence model. The realizable k-ε turbulence model is chosen because of its generally good performance for wind flow around buildings [29]. The governing equations were discretized using a finite volume method and the Density Based method was used to solve the equations. The numerical solver employs a structured, non-orthogonal, fully collocated, cell-centered, finite volume approach for the discretization of the computational domain, and the velocity vector is decomposed into its Cartesian components. Physical diffusive cell fluxes are approximated using a conventional second-order central differencing scheme, and the SIMPLE algorithm [30] is used for pressure correction. Pressure interpolation is standard and second order discretization schemes are used for both the convection terms and the viscous terms of the governing equations [30].
The convergence target based on root mean square is set to 10^{-7}. The quantities of interest, such as velocity and turbulent kinetic energy were monitored at several grid points during the solving process to check stable levels before convergence is met.

4. Methodology

There is not a clear set of variables to study the urban morphology and climatology [31-32]. Therefore, the choice of variables is influenced by the following factors: those identified by the researchers as critical to understanding the urban canopy environment [31-35]; factors ignored in previous researches; variables that can be readily manipulated by designers at the concept design stage.
In this work three critical variables has been identified as major variables to be analyzed: wind speed (V_∞); temperature difference between the undisturbed air and the building surface temperature (ΔT); height of the buildings.
This study assumes the width (W) of the building constant as it is, in real cases, very small respect to the extension of the zone influenced by the building.
The temperatures of ground and building surfaces play an important role in the thermal environment around buildings and so in the canopy layer. They influence the air distribution significantly and the simulation needs them as thermal boundary conditions. The following assumptions are adopted to develop the heat balance model of ground and building surfaces:

- The walls of the buildings have uniform thermal characteristics (windows are neglected for simplification);
- The temperature inside buildings and the ground temperature has been set to be always constant and equal to 303 K, thus the simulation of the thermal environment around buildings focuses on the outdoor air temperature.

Two different temperature differences between the undisturbed air and the building surface temperature (ΔT) have been imposed, to represent summer, winter and spring conditions. In this study,

the contribute of radiation was not considered, also if it is known to be an important factor in the study of the urban environment. For this reason, to emphasize the presence of the temperature, it has been chosen $\Delta T = +5$ K and -5 K.
Even though the geometry of the flow configuration is rather simple, the flow is physically quite complex, with multiple separation regions and vortices, as widely described in the ASHRAE [36]. The study on the height of the buildings H and of its influence on urban canopy layer was conducted by simulating gradually increasing heights. Starting from a minimum height of buildings up to H = 5 m, we analysed buildings up to H = 30 m height, with variations of H in each simulation of about 5 m. As shown below, the parameter H is proved crucial to assessing the extent of the urban environment extension.
Moreover, in this work have been analysed three different speed regimes, 0.1, 0.5, 1 m/s, chosen to represent typical speed that can be found in a typical urban environment.
Summarizing, using the model described in the previous sections, we carried out a series of CFD simulation studies to assess the impact of the previously defined parameters on the building influence extension. Three macro cases have been simulated:

- Case 1: fixing the building height H and the temperature difference ΔT (free stream temperature-building temperature), the undisturbed velocity V_∞ has been varied from 0.1 m/s to 1 m/s;
- Case 2: fixing the building height H and the velocity V_∞, the temperature difference (free stream temperature-building temperature) ΔT has been varied from –5 K to +5 K;
- Case 3: fixing the temperature difference ΔT (free stream temperature-building temperature) and the velocity V_∞, the buildings height H has been varied from 5 m to 30 m.

5. The urban scale analysis

Since the velocity and temperature difference seem to be the major responsible for the energy balance in an urban area, these variables have been used as parameters to evaluate the energy balance, during the numerical simulations. An energy indicator defined as a function of the velocity and the temperature difference has been introduced:

$$E_{ENV} = f(E_K, E_P, Q) \quad [J] \tag{6}$$

where E_K is the kinetic energy, E_P is the potential energy and Q is the total heat. From a theoretical point of view, it would be more accurate to distinguish this index into two parts, one for the summer and one for the winter period. But in this paper a practical approach has been chosen and the part relating to the temperature difference has been considered in its absolute value (6) and so a unique energy index can be used. More, in the calculation of E_{ENV} the term relating to the potential energy, has not been considered, as its influence contribution appears to be negligible.
The environment energy (E_{ENV}), that is the energy in the urban canopy layer, is calculated on a surface with origin in the center of the building and has a certain extension that to have to be determined. This extension is calculated as follows:

- (for the zone) above the building, the point in which the variation of the undisturbed velocity with the height does not exceed 0.03% has been identified;
- for the downstream zone of the building, however, a distance calculated between the center of the building and the point at which the variation of speed undisturbed with the height remains constant (0.5%) as one moves away from the body in a direction parallel to the current (see Curve B in Figure 2) is defined. Figure 2 shows the velocity profiles as you move away from the building (for example, V1, V2, ...) and a point in which the speed (V4) is approximately equal (0.5%) to the previous one (V3). This trend is any way shown precisely in Figure 2, where velocity profiles as a function of distance from the building are presented.

The environment energy should be compared with a value calculated in an undisturbed area, where it is not affected from the presence of buildings. By this way, it is possible to have a quantification of the influence of the urban building considered. Thus, this quantification would be considering the variation of the E_{ENV}, and then the change of total energy.

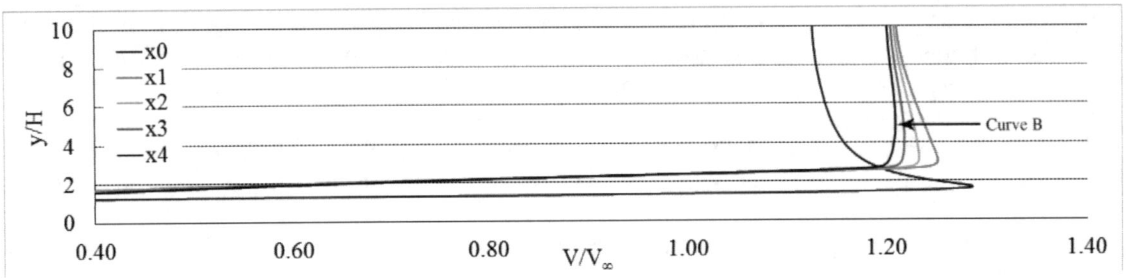

Figure 2. Velocity variation with height

6. Results

6.1. Case 1

Building height H and temperature difference ΔT have been fixed, while varying the free stream velocity V_∞ between 0.1 m/s and 1 m/s. In this case, the building height H is set at 5 m and then at 10 m, its width is 15m, and the temperature difference ΔT between the free stream air and the walls of the building is equal to 5 K (summer situation). In particular, the air temperature was set at 298 K and the temperature of the building at 303 K. As shown in Figure 3, the variations of the ratio E_{ENV}/H with velocity are minimal and scarcely appreciable. Further testing has been done by varying the height H of the building, from 5 m to 10 m. Also in this case the variations of the ratio E_{ENV}/H with velocity are minimal and slightly appreciable.

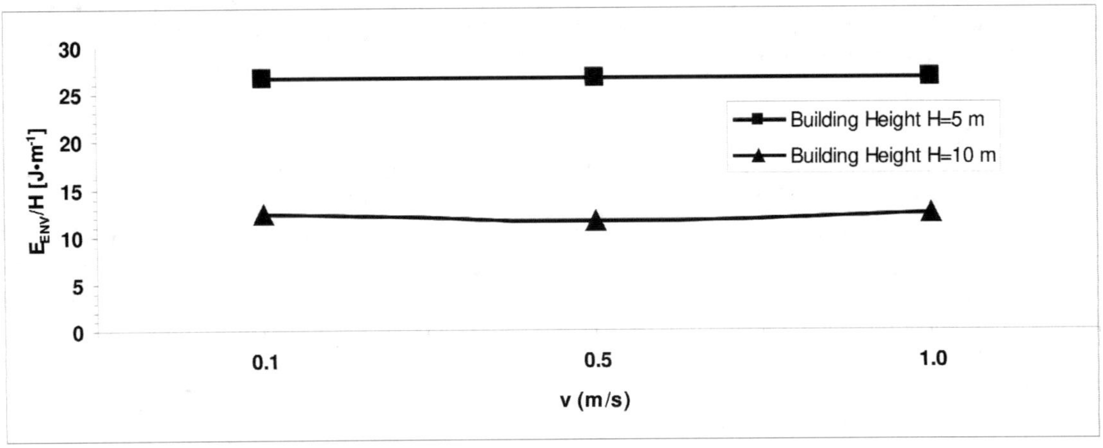

Figure 3. Evolution of E_{ENV}/H as function of free stream velocity (H = 5 m)

In Table 2 the values of the E_{ENV}/H ratio in function of the free stream velocity are reported, for both the cases in which there is the presence of buildings, called Disturbed (indicated with D_s), and in which it is not affected by the presence of buildings, called Undisturbed (indicated with U_n). The Undisturbed area was calculated considering the influence of the Disturbed area, but eliminating any kind of influence due to the building. So the simulation was developed with the same boundary conditions of the one with the building, but eliminating the presence of the building.

Table 2. E_{ENV}/H values in function of V_∞

	H = 5 m		H = 10 m	
V_∞ [m/s]	E_{ENV}/H [J/m] D_s	E_{ENV}/H [J/m] U_n	E_{ENV}/H [J/m] D_s	E_{ENV}/H [J/m] U_n
0.1	26.56	21.58	12.31	10.04
0.5	26.56	21.58	11.54	10.04
1	26.57	21.58	12.34	10.04

The difference in values between the D_s area and the U_n one are evident: as expected, in the presence of buildings, the value of E_{ENV} grows significantly, as the total energy in this area is higher. So, we can conclude that as the free stream velocity V_∞ increases, the parameter E_{ENV}, and hence the extension of the building influence area, does not vary substantially

6.2. Case 2

In this case, the building height H and the velocity V_∞ have been fixed, while varying the temperature difference ΔT from –5 K to +5 K. The building height H is set at 5 m, its width is 15 m and the undisturbed air velocity is equal to 0.1 m/s.

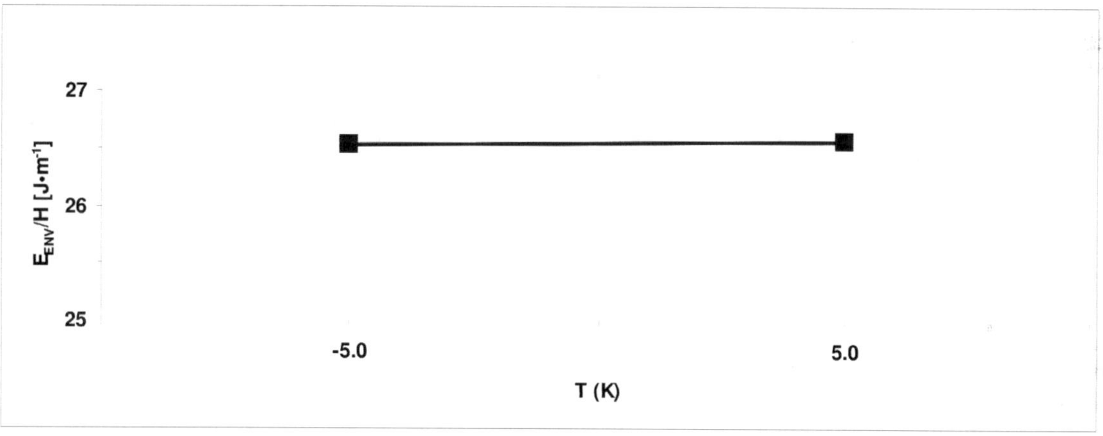

Figure 4. Evolution of E_{ENV}/H as a function of ΔT.

As shown in Figure 4, the variations of the ratio E_{ENV}/H in function of the ΔT are minimal and slightly appreciable, and we can conclude that the parameter E_{ENV} does not vary substantially with the variation of the temperature difference between undisturbed air and the walls of buildings, and hence the extension of the urban heat island does not vary. Even in this case (Table 3) the values of the ratio E_{ENV}/H as a function of the temperature difference, both for the area where there is the presence of buildings, called D_s, and for the area without buildings, called U_n, are reported.

Table 3. E_{ENV}/H values in function of ΔT

ΔT [K]	E_{ENV}/H [J/m] D_s	E_{ENV}/H [J/m] U_n
+5	26.56	21.58
-5	26.53	21.58

Also, in this case, the difference in values between the D_s area and the U_n area is clearly visible. As expected, in the presence of buildings the value of E_{ENV}/H increases significantly, as the total energy present in this area surrounding the building is higher.

6.3. Case 3

The temperature difference ΔT and the velocity V_∞ have been fixed, while varying the buildings height H from 5 m to 30 m. The temperature difference ΔT between the undisturbed air and the walls of the building is equal to 5 K, its width is 15 m and the undisturbed air velocity is equal to 0.1 m/s.

Table 4. A comparison of E_{ENV} values for a city/rural condition

H [m]	E_{ENV}/H [J/m] D_s	E_{ENV}/H [J/m] U_n
5	26.56	21.58
10	12.31	10.04
15	5.60	4.58
20	3.88	3.42
25	2.56	2.35
30	1.12	1.02

As expected, even in this case the difference is clearly visible between the values of the D_s area and the U_n area: in presence of buildings, the value of E_{ENV}/H increases significantly, as in previous cases. Figure 5 and Table 4, shows that increasing the building height, the value of E_{ENV} decreases with some potential law, found in:

$$E_{ENV} = a \cdot H^{(b+1)} \tag{7}$$

being a numerical coefficient equal to 470.6 and b an exponent equal to -1.66. A comparison between the analytical and numerical correlation found is reported in Figure 5.

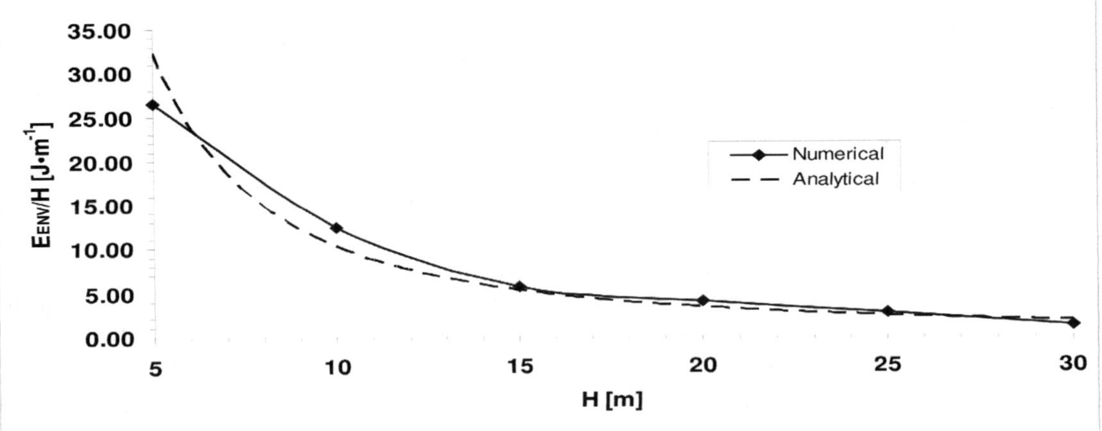

Figure 5. Analytical and numerical correlation comparison for the E_{ENV}/H

This correlation shows that an increase of the height of the buildings produces a decrease of the factor E_{ENV}/H with a potential law. Being E_{ENV} = f (D) a function of the extension (D) of the area surrounding the building, increasing the height of building, the extension of the influenced area increases. This means that it is also possible to find a correlation that links D and the average height of the buildings. In this case the correlation becomes:

$$D = c \cdot H^{d} \tag{8}$$

with c e d coefficients respectively 38.9 e 0.48. A comparison between the analytical and numerical correlation found is reported in Figure 6.

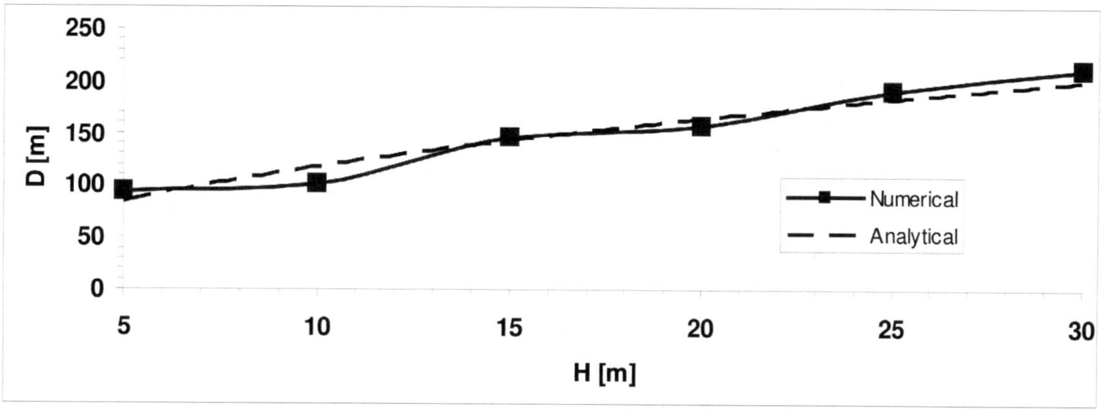

Figure 6. Analytical and numerical correlation comparison for D.

In this way, it is possible to have both a qualitative value, using E_{ENV}, and a quantitative value linked to the extent, by D, of the disturbed urban environment, considered.

7. Discussion and conclusions

Results obtained allowed to find a correlation between the average height of buildings H and the external building environment total energy E_{ENV}. This parameter allows to give a qualitative assessment on the contribution of the total energy of the urban environment. The next step was then to give a quantitative assessment, tying this parameter to the extension of the influence of the single building, D, and also to find in this case a correlation linking the D to the average height of buildings. With these correlations, it is possible to characterize quickly and easily an influence due to the buildings, in terms of total energy and geometric extension.

With this approach it is also possible a comparison among different urban environment in order to define possible actions to reduce energy consumption and to find out the weight that each of them can have on the urban microclimate change, to highlight urban critical issues.

This could lead to defining a strategy of urban actions (creation of green areas, restricted traffic areas, etc.) and architectural (different materials applications, cool roofs, green roofs, cool pavements, etc.) to improve the microclimate of the considered area.

Then next step will be to create a complete model that, in the frame of the presented approach, could take into account other parameters like radiation, evapotranspiration, transmission and absorption of the buildings, etc.. in order to develop a guide line of possible actions to mitigate city impact on environment.

8. References

[1] Sorrentino G, Scaccianoce G, Morale M and Franzitta V 2012 The importance of reliable climatic data in the energy evaluation *Energy* **48** (Amsterdam: Elsevier) 74-79

[2] Sun R, Lü Y, Chen L, Yang L and Chen A 2013 Assessing the stability of annual temperatures for different urban functional zones *Building and Environment* **65** (Amsterdam: Elsevier) 90-98

[3] Reichrath S, Davies T W 2002 Using CFD to model the internal climate of greenhouses: past, present and future *Agronomie* **22** 3-19.

[4] Norton T, Grant J, Fallon R and Sun D W 2010 Optimising the ventilation configuration of naturally ventilated livestock buildings for improved indoor environmental homogeneity *Build Environ* **45** 983-95

[5] Salata F, Golasi I, de Lieto Vollaro E, Bisegna F, Nardecchia F, Coppi M, Gugliermetti F and de Lieto Vollaro A 2015 Evaluation of different urban microclimate mitigation strategies through a PMV analysis *Sustainability* **7** 9012-9030

[6] Chow W K 1996 Application of Computational Fluid Dynamics in building services engineering *Building and Environment* **31** 425-436

[7] Tominaga Y, Mochida A, Yoshie R, Kataoka H, Nozu T, Yoshikawa M and Shirasawa T 2008 AIJ guidelines for practical applications of CFD to pedestrian wind environment around buildings *Journal of Wind Engineering and Industrial Aerodynamics* **96** 1749–1761

[8] Salata F, Nardecchia F, de Lieto Vollaro A and Gugliermetti F 2015 Underground electric cables a correct evaluation of the soil thermal resistance *Applied Thermal Engineering* **78** 268-277

[9] Mochida A and Lun I Y F 2008 Prediction of wind environment and thermal comfort at pedestrian level in urban area *Journal of Wind Engineering and Industrial Aerodynamics* **96** (Amsterdam: Elsevier) 1498-1527

[10] Gugliermetti F, Passerini G and Bisegna F 2004 Climate models for the assessment of office buildings energy performance *Build Environment* **39** (Amsterdam: Elsevier) 39-50

[11] Huang H, Ooka R and Kato S 2005 Urban thermal environment measurements and numerical simulation for an actual complex urban area covering a large district heating and cooling system in summer *Atmospheric Environment* **39** (Amsterdam: Elsevier) 6362–6375

[12] Gugliermetti F and Bisegna F 2003 Meteorological days for HVAC system design in Mediterranean climate *Building and Environment* **38** (Amsterdam: Elsevier) 1063-1074

[13] Bottillo S, de Lieto Vollaro A, Galli G and Vallati A 2014 Fluid dynamic and heat transfer parameters in an urban canyon *Solar Energy* **99** pp 1-10

[14] Hunt G R and Linden P F 1999 The fluid mechanics of natural ventilation displacement ventilation by buoyancy driven flows assisted by wind *Building and Environment* **34** 707–720

[15] Etheridge D W 2002 Non dimensional methods for natural ventilation design *Build Environment* **37** 1057-72

[16] Kato S, Murakami S, Takahashi T and Gyobu T 1997 Chained analysis of wind tunnel test and CFD on cross ventilation of large-scale market building *Journal of Wind Engineering and Industrial Aerodynamics* **67-68** 573-87

[17] de Lieto Vollaro A, de Simone G, Romagnoli R, Vallati A and Botillo S 2014 Numerical study of urban canyon microclimate related to geometrical parameters *Sustainability* **6** 7894-7905

[18] Ambrosini D, Galli G, Mancini B, Nardi I, Sfarra S 2014 Evaluating mitigation effects of urban heat islands in a historical small center with the ENVI-Met® Climate Model *Sustainability* **6** 7013-7029

[19] van Hooff T and Blocken B 2010 On the effect of wind direction and urban surroundings on natural ventilation of a large semi-enclosed stadium *Computer & Fluids* **39** 1146-55

[20] Ashie Y and Kono T 2011 Urban-scale CFD analysis in support of a climate-sensitive design for the Tokyo Bay area *International Journal of Climatology* **31** 174–188

[21] Kolokotsa D, Psomas A and Karapidakis E 2009 Urban heat island in southern Europe: The case study of Hania Crete *Solar Energy* **83** 1871–1883

[22] Bottillo S, de Lieto Vollaro A, Galli G and Vallati A 2014 CFD modeling of the impact of solar radiation in a tridimensional urban canyon at different wind conditions *Solar Energy* **102** 212-222

[23] Swaid H and Milo E H 1990 Prediction of urban air temperature variations using the analytical CTTC model *Energy and Buildings* **20** (Amsterdam: Elsevier) 313–24

[24] Elnahls M M and Williamson T J 1997 An improvement of the CTTC model for predicting urban air temperatures *Energy and Buildings* **27** (Amsterdam: Elsevier) 41–49

Journal of Physics: Conference Series **655** (2015) 012013 doi:10.1088/1742-6596/655/1/012013

[25] Mochida A, Murakami S and Ojima T 1997 CFD analysis of mesoscale climate in the center Tokyo area *Journal of Wind Engineering and Industrial Aerodynamics* **67-68** (Amsterdam: Elsevier) 459–477

[26] Urano A, Ichinose T and Hanaki K 1999 Thermal environment simulation for three dimensional replacement of urban activity *Journal of Wind Engineering and Industrial Aerodynamics* **81** (Amsterdam: Elsevier) 197–210

[27] Sini J F and Anquetin S 1996 Pollution dispersion and thermal effects in urban street canyon *Atmospheric Environment* **30** (Amsterdam: Elsevier) 2659–2677

[28] Franke J, Hellsten A, Schlünzen H and Carissimo B 2007 Best practice guideline for the CFD simulation of flows in the urban environment *COST Office* University of Hamburg Meteorological Institute Centre for Marine and Atmospheric Sciences.

[29] Lomax H, Pulliam T H and Zingg D W 1999 Fundamentals of Computational Fluid Dynamics

[30] Ansys Inc., ANSYS FLUENT 12.0/12.1 Documentation.

[31] Chandler T J 1976 Urban climatology and its relevance to urban design *World Meteorological Organisation* Technical notes no. 149 Geneva.

[32] Golany G S 1996 Urban design morphology and thermal performance *Atmospheric Environment* **30** (Amsterdam: Elsevier) 455–65

[33] Landsberg H E 1981 *The urban climate* (New York: Academic Press)

[34] Oke T R 1987 *Boundary layer climates* (USA: Methuen)

[35] Sawid H and Hoffman M E 1990 Climatic impacts of urban design features for high and mid latitude cities *Energy and Buildings* **14** (Amsterdam: Elsevier) 325–36

[36] American Society of Heating, Refrigeration and Air-Conditioning Engineers 2005 *ASHRAE Fundamentals Handbook* SI Edition Atlanta USA

Journal of Physics: Conference Series **655** (2015) 012014 doi:10.1088/1742-6596/655/1/012014

The performance of a mobile air conditioning system with a water cooled condenser

Davide Di Battista[1], **Roberto Cipollone**
Department of Industrial and Information Engineering and Economics,
University of L'Aquila, v. G. Gronchi, 18 – L'Aquila, Italy

E-mail: davide.dibattista@univaq.it

Abstract. Vehicle technological evolution lived, in recent years, a strong acceleration due to the increased awareness of environmental issues related to pollutants and climate altering emissions. This resulted in a series of international regulations on automotive sector which put technical challenges that must consider the engine and the vehicle as a global system, in order to improve the overall efficiency of the system. The air conditioning system of the cabin, for instance, is the one of the most important auxiliaries in a vehicle and requires significant powers. Its performances can be significantly improved if it is integrated within the engine cooling circuit, eventually modified with more temperature levels.
In this paper, the Authors present a mathematical model of the A/C system, starting from its single components: compressors, condenser, flush valve and evaporator and a comparison between different refrigerant fluid. In particular, it is introduced the opportunity to have an A/C condenser cooled by a water circuit instead of the external air linked to the vehicle speed, as in the actual traditional configuration. The A/C condenser, in fact, could be housed on a low temperature water circuit, reducing the condensing temperature of the refrigeration cycle with a considerable efficiency increase.

Introduction

The use of mobile A/C system for cabin cooling and drying is about 7 billion liters of fuel per year, which is the 3,2% of the global fuel consumption of vehicles [1, 2]. Similar studies evaluate the vehicles overconsumption due to the A/C system is between 1,8 and 2 l/100 km, equivalent to 20-25% of a mean sedan car fuel consumption on the homologation cycle [3].
The importance of A/C system in terms of vehicle energy balance and environmental issue invites to consider new technologies and opportunities to improve its efficiency. On the thermodynamic point of view, the most significant opportunity is to decrease the condensing temperature. However, this temperature is strictly related to the external cooling source that usually is the external air: actually, the condenser is placed in the front end of the vehicle and its heat removal rate is linked to the external air temperature and speed. Therefore, if the condenser was cooled by a liquid circuit [4], it could be managed at a lower temperature; the mean temperature difference could be reduced thanks to the higher specific heat of the liquid (i.e. water/glycol) and, also, with a suitable flow rate. An interesting opportunity is related to the possibility to integrate the condenser in the engine cooling system, particularly if it is split into two separate circuits [5-7]. In all the cited papers [4-7] the idea of integrating the A/C vehicle's system in the cooling circuit was discussed making reference to a double

1 To whom any correspondence should be addressed.

temperature circuit. In [4] the attention was focused to the heat exchanger integration inside the engine bay without considering the A/C performances as a function of the cooling power requested by the cabin. A more oriented systematic approach was discussed in [5] verifying the experimental performances achieved on a small vehicle. In [6,7] the performances of a double temperature cooling system were deeply analyzed, considering the A/C condenser as a lumped thermal load, represented by a specific power, without entering in the A/C modeling.
The thermodynamic equilibrium of the A/C system is not straightforward to determine and depends on several factors: ambient temperature, solar irradiation, vehicle speed or fan operation, amount of the refrigerant and, above all, components design. The frequent operation at partial load or in off-design conditions makes the A/C equilibrium even more complex.
In this paper, the Authors present and discuss a comprehensive mathematical model of the A/C system, which includes the off design conditions of the evaporator and of the condenser, the flush valve and the compressor according to conservation equations (mass, momentum and energy).
The paper discusses the behavior of the A/C system at different loads. The benefits related to a liquid cooled condenser inserted in the low temperature circuit in a double temperature level engine cooling layout are discussed in comparisons with a traditional A/C system with an air cooled condenser, considering a new interesting fluid (R1234yf [8]) instead of the traditional R134a. The two configurations have been studied in term of thermodynamic equilibrium and, in particular, the compressor energy saving have been assessed and compared to the effects of the coolant circuit modification.

2. Cabin air conditioning system model description

In air conditioning systems, a working fluid flows in a thermodynamic circuit composed by a compressor, a condenser, an evaporator, a flush valve and a little tank (Figure 1). The modelling of the A/C system has an inner complexity, due to the continuous matching of the condenser and evaporator temperatures with the external conditions and cabin cooling requests.

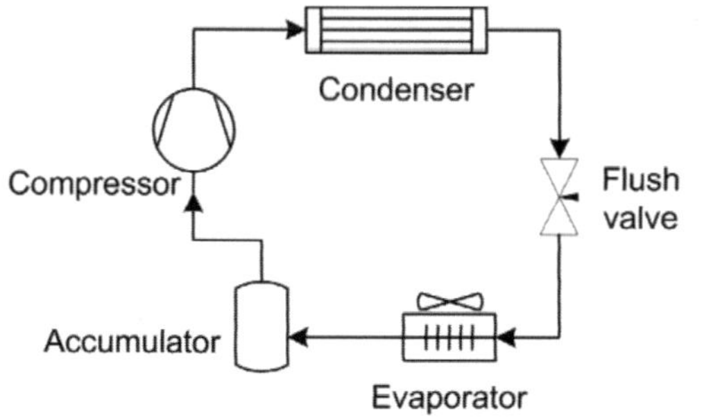

Figure 1: traditional A/C circuit layout: evaporator is placed in the cabin side, having the role to transfer the heat from the cabin to the refrigerant fluid (evaporating it), while the condenser is in the front end of the vehicle, having to carry out the condensation heat of the fluid towards the environmental air.

2.1 Compressor

The compressor usually used in mobile A/C is a volumetric one. In ideal case, the pressure is fixed by the circuit and the mass flow rate depends on the revolution speed ω of the compressor.

$$\dot{m} = \eta_{vol} \cdot \rho_{refr} \cdot V \cdot \omega \qquad (1)$$

Volumetric efficiency η_{vol} is slightly dependent on pressure delivered and speed, which is imposed by the electric motor. Refrigerant density ρ_{refr} refers to the inlet section of the compressor and depends on the thermodynamic conditions in the accumulator. The isentropic compressor efficiency, assumed by manufactures data, allows the calculation of outlet fluid enthalpy and mechanical power requested [9].

2.2 Accumulator

The accumulator is placed on the low pressure side of the A/C circuit, downstream the evaporator. It is represented by a capacity which separates the evaporator outlet and the compressor inlet, storing the extra-liquid during the refrigerant low request conditions. Moreover, it avoids that the liquid enters in the compressor, condition that can damage the compressor itself and strongly reduces its efficiency. The accumulator acts, also, as a dryer and as a filter.

2.3 Evaporator and condenser

The evaporator is the heat exchanger which cools down the cabin. In this component, the refrigerant fluid usually enters as a two-phase mixture (downstream the flush valve) and exits as saturated vapor or slightly superheated. Once the exchanger geometry and the entering conditions of both fluids are known, outlet condition can be evaluated. Thermodynamic conditions of the refrigerant fluid at the inlet are known in terms of pressure (saturation pressure) and quality.

The condenser is the heat exchanger used to extract heat from the refrigerant fluid and to exchange toward the air coming from the external environment. In this component, the refrigerant usually enters as superheated vapor (conditions fixed by the compressor) and exits as saturated liquid (or as a mixture with saturated vapor or also in some conditions as sub-cooled fluid) depending from the heat exchanger behavior whose geometry is fixed. Its thermal balance gives the temperature of the fluid that crossed it (hotter than the inlet one): usually, it is represented by environmental air whose flow rate is fixed by the vehicle or by auxiliary fan.

Condenser and evaporator are represented in one dimensional way (uniform conditions in the cross section area) and divided in subvolumes: the geometry of the fluids passages is known and, therefore, the heat exchange surface. Continuity, momentum and energy equations are solved step by step following the fluid motion inside a 1D passage, considered in steady state conditions.

Forced convective heat exchange coefficients are evaluated considering proper correlations: Dittus-Boelter for single phase, Shah's and Klimenko's equations [10, 11] for two phase region:

$$\begin{cases} h_{1phase} = 0.0243\,\mathrm{Re}^{0.8}\,\mathrm{Pr}^{n}\,\dfrac{k}{D} \\ h_{cond} = 0.023\,\mathrm{Re}_{sl}^{0.8}\,\mathrm{Pr}_{sl}^{0.4}(1-x)^{0.8}\,\dfrac{3.8x^{0.76}(1-x)^{0.04}}{p/p_{cr}}\,\dfrac{k}{D} \\ h_{evap} = 0.087\,\mathrm{Re}^{0.8}\,\mathrm{Pr}^{1/6}\left(\dfrac{\rho_{sv}}{\rho_{sl}}\right)^{0.2}\left(\dfrac{k_{wall}}{k_{sl}}\right)^{0.09}\dfrac{k_{sl}}{D} \end{cases} \quad (2)$$

n is equal to 0.3 for heating fluid and 0.4 for cooling fluid, k is the fluid thermal conductivity while D the equivalent cross section diameter, the quality x is currently evaluated for each subvolumes of the heat exchanger pipe, considering homogeneous flow; sl subscript refers to saturated liquid, sv to saturated vapor, p_{cr} is the critical pressure of the fluid, k_{wall} the thermal conductivity of the exchanger walls. Previous equations consent to solve the heat exchange in operating conditions, calculating the outlet thermodynamic conditions of the fluids by knowing the mass flow rates (and, so, fluid velocities) and thermodynamic inlet conditions.

2.4 Flush valve

The flush valve is considered as a reduction of the fluid passage section (orifice). This produces an expansion from high pressure liquid (almost saturated liquid) to low pressure liquid (two-phase conditions) downstream the orifice. At this point, a partial pressure recovery is realized when the fluid fills again the whole section of the duct, just upstream the evaporator.

When the refrigerant fluid crosses the valve, the following processes and conditions must be represented and constraints respected:

- inlet and outlet pressure (p_{max} and p_{min}) must match temperature levels at condenser and evaporator;
- orifice inlet conditions (in particular density ρ_{in} and enthalpy h_{in}) are fully determined by knowing the condenser outlet conditions;
- the fixed orifice geometry imposes the mass flow rate according to (eq. 3):

$$\dot{m} = C_D A_{throat} \sqrt{2\rho_{in}(p_{\max} - p_{throat})} \tag{3}$$

the discharge coefficient C_D is function of the section ratio and Reynolds. It has been calculated according to the Lichtarowitcz's correlation [12];

- since pressure recovery downstream the orifice is negligible for short orifices, in a first approximation, p_{throat} can be assumed equal to the outlet pressure (i.e. evaporation pressure p_{min}). However, in a more precise way, pressure recovery Δp_{rec} can be evaluated to Borda-Carnot formula;
- the momentum equation (in steady condition), continuity and energy equation are applied from the inlet to the outlet section of the valve (eq. 4), considering the flow as homogeneous:

$$\begin{cases} \Delta p_{rec} = \dfrac{A_{throat}}{A}\left(1 - \dfrac{A_{throat}}{A}\right)\rho_{in} u_{throat}^2 \\ (p_{\max} - p_{\min})A + \dot{m}(u_{in} - u_{out}) = 0 \\ (h_{in} - h_{out}) = \dfrac{1}{2}(u_{out}^2 - u_{in}^2) \end{cases} \tag{4}$$

A is the full pipe cross section, u_{out} is the fluid speed downstream the orifice where the fluid fill the overall pipe section; knowing orifice geometry and considering momentum and continuity equations through it, fluid-dynamic conditions in the throat section (in particular, fluid density ρ_{throat} and fluid speed u_{throat}) are evaluated and, therefore, the outlet enthalpy h_{out} of the fluid;

- once evaporation pressure and valve outlet enthalpy are known the outlet thermodynamic conditions are fully determined; in particular, fluid quality x.

Definitely, when the pressure drop through the valve is known from both condensing and evaporating temperatures, the valve imposed the mass flow rate (eq. 3) and the momentum equation gives the fluid speed at the valve outlet, being known the inlet conditions. Once also momentum equation is solved, enthalpy downstream the valve can be evaluate according to energy conservation (eq. 4) and, so, fluid quality. This procedure is, therefore, an iterative method: in fact, pressure levels (p_{max} and p_{min}) have to be adjusted at each iteration, in order to match the flow rate imposed by the orifice and that one of the volumetric compressor.

2.5 Model integration

The sequence of the equations representing the A/C system is solved according to an iterative procedure as in Figure 2. Heat exchangers are sized and their geometry is chosen; at the same time, the valve geometry is fixed (orifice dimensions) and A/C ducts defined. Refrigerant fluid is chosen too.
Knowing the temperature of the fluids which behave as source at two heat exchangers and their flow rates, condensing and evaporating temperature can be hypothesized. Then, knowing the condenser outlet conditions, in particular fluid density ρ_{in}, equation (3) gives a flow rate value of the refrigerant fluid that must correspond to the one imposed by the compressor (eq. 1). At the same time, vapor quality is evaluated from equations (4). At this point, the heat exchangers model of the evaporator gives the thermodynamic conditions of the fluid at its outlet, i.e. compressor inlet. So, the flow rate given by eq. (1) can be verified and compared to the one given by the eq. (3): if their difference is within a given tolerance ε, the solution of all the equations is found. If the two flow rates do not correspond, the two pressure levels (p_{max} and p_{min}, condensation and evaporation) are adjusted and the procedure is re-iterated (Figure 2). The procedure is closed by the condenser behavior: once the compressor outlet conditions are known, the heat exchanger model of the condenser gives the fluid conditions at its outlet. In particular, fluid density ρ_{in} must match that of equation (3).

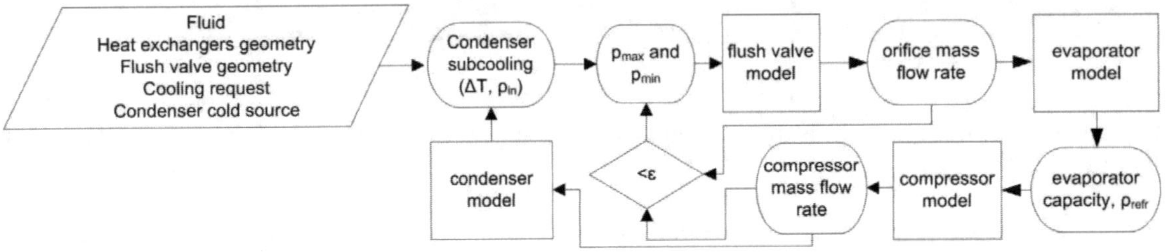

Figure 2: procedure for the A/C system model. Rounded shapes represent the outputs of the models.

3. Results

The A/C model can be considered as a virtual platform of the conditioning system and it can be used to study different layouts and arrangements. The availability of low temperature water flow is necessary to fulfill a liquid cooled condenser: an interesting solution is to split the traditional cooling system into two circuits each of them operating at different temperature [7].

The most common operating fluid for automotive A/C is R134a, but its use is forbidden in new vehicles and will be replaced by R1234yf, which represents the actual technological standard [13, 14]. Figure 3 show the thermodynamic cycles of A/C system with R134a and R1234yf as operating fluids. Both cycles have been compared at the same cooling power (about 4 kW) required by the cabin and exchanged within the evaporator. They are evaluated in steady conditions: the heat surface of the evaporator is about 6 m^2 and that one of the condenser is 10 m^2; air flow rate at evaporator is 190 g/s at a temperature of 45 °C, while at condenser it is 530 g/s at 35°C, corresponding to a vehicle speed of about 60 km/h. The revolution speed of the volumetric compressor is 1650 RPM. Tubes connecting the exit of the condenser have 10 mm diameter while the orifice is 1.183 mm. In this condition, the two pressure levels of R1234yf cycle are 2.6 bar and 22 bar, while R134a cycle had a maximum pressure of about 24 bar and similar minimum pressure (Figure 3).

The mathematical model demonstrates that the new fluid has an average pressure level slightly lower than the R134a, but with a maximum temperature about 10°C higher. The condensing temperatures are similar, but the condensing heat of the R1234yf is lower and, so, it is necessary to overheat more the fluid in order to keep the same power exchanged in the evaporator. Moreover, the R1234yf has a higher specific heat at constant pressure which leads to a lower mass flow rate inside the A/C circuit [15]. The cycles have been reported making reference the absolute entropy with respect to specific mass flow rates of the fluids (42 g/s for R134a and 39 g/s for R1234yf).

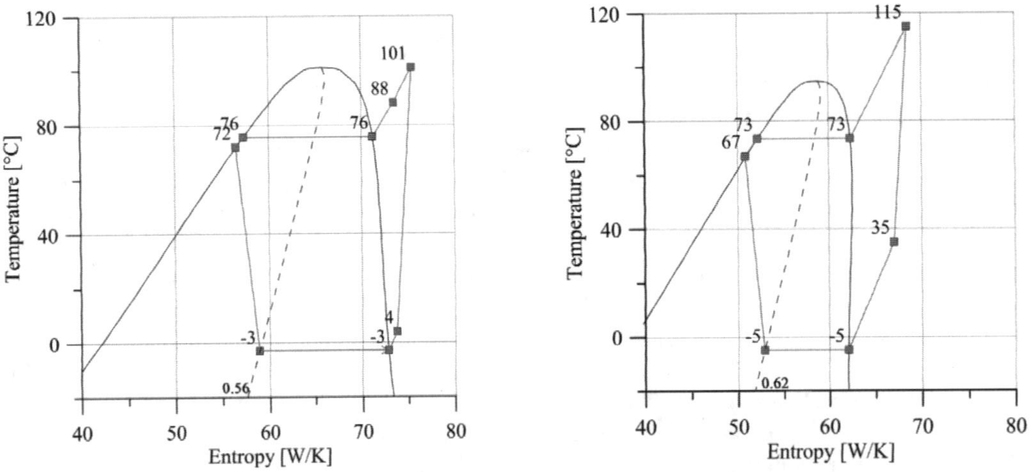

Figure 3: comparison between thermodynamic cycles of A/C system with R134a (on the left) and R1234yf (on the right).

Journal of Physics: Conference Series **655** (2015) 012014 doi:10.1088/1742-6596/655/1/012014

Figure 4 shows the performance of the A/C system with R1234yf as working fluid and air cooled condenser: the air mass flow rate which crosses the condenser is strictly related to the vehicle speed (it is placed in the front end of the vehicle) and, so, modifies the heat rejection capacity of the condenser itself. When the air mass flow rate growths, its power rejected increases till to a saturation value due to the exchanger's fixed geometry; the evaporator power rises as well, increasing the cooling power at cabin side. At low vehicle speed, instead, this cooling capacity is limited and COP value is very low: compressor power increases as well as the refrigerant mass flow rate. Figure 5 shows the pressures reached in the circuit with different condenser air inlet flow rate: orifice throat pressure is slightly higher than evaporating pressure, indicating the low pressure recovery downstream the orifice. Condensing pressure decreases when external air flow rate increase, confirming the COP increase.

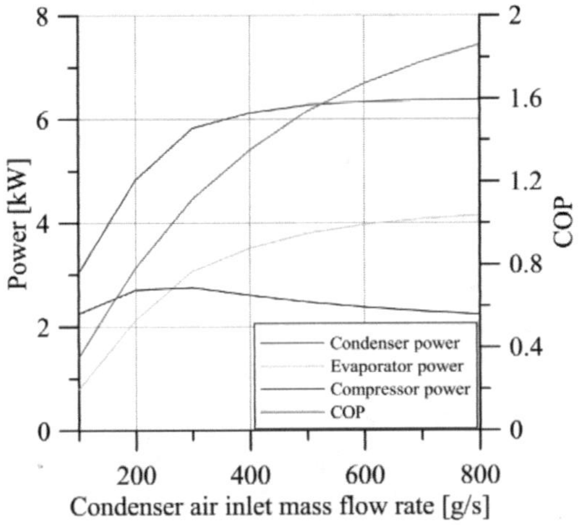

Figure 4: A/C performances operating with R1234yf and different air mass flow rate at condenser side

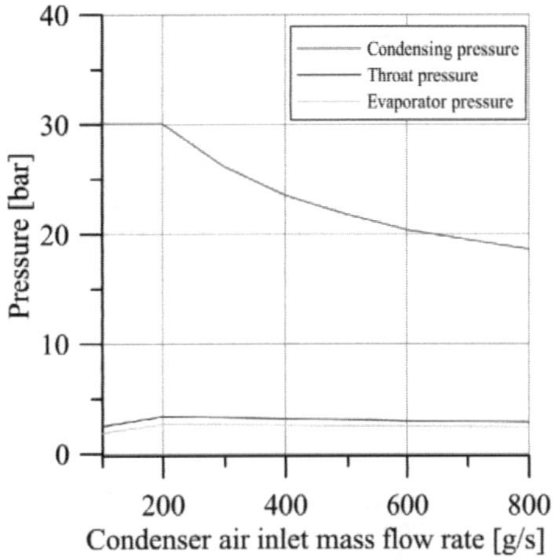

Figure 5: pressure levels in the circuit with reference to orifice

A significantly changing parameter is the air inlet temperature at the condenser, which actually is the external air temperature. When the air inlet temperature increases (for instance, in summer), COP decreases because of the higher mean temperature of the thermodynamic cycle. However, the system is able to keep the cooling capacity of the evaporator increasing the compressor work (Figure 6).

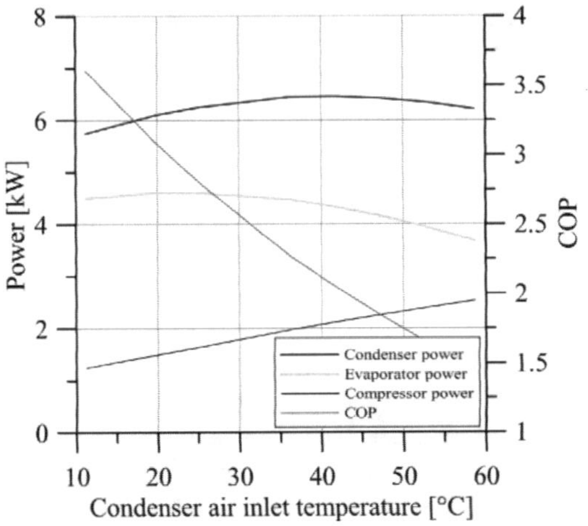

Figure 6: A/C performances operating with R1234yf and different air inlet temperature at condenser side

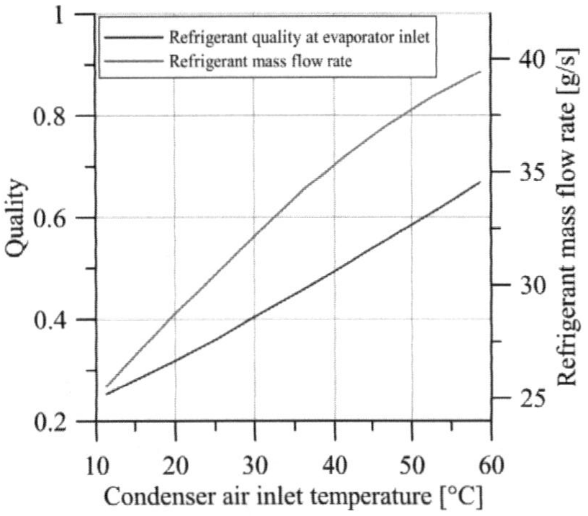

Figure 7: refrigerant mass flow rate and evaporator inlet quality of the A/C system for different air temperature at condenser

The performances variation on Figure 4 and Figure 6 is due to a re-arrangement of the A/C parameters, in particular of refrigerant mass flow rate and its quality at the evaporator inlet. When the external air temperature increases, the quality of the refrigerant at the very inlet of the evaporator increases as well and the cooling capacity would decrease; therefore, the A/C system reacts increasing the mass flow rate of the working fluid, in order to satisfy the cabin cooling request (Figure 7).

The opportunity to cool-down the condenser by water can be very interesting: the heat transfer coefficient is higher and the overall equilibrium of the A/C system could be modified (Figure 8).

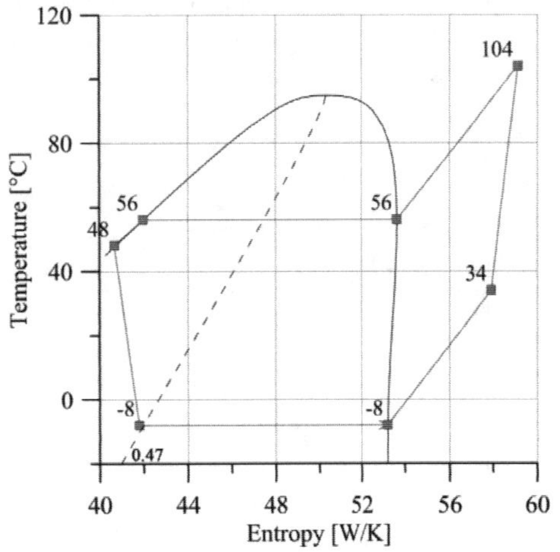

Figure 8: thermodynamic cycle of A/C system with R1234yf as working fluid and water cooled condenser (WCC). The cabin cooling request is fixed at 4 kW (evaporator side) and the water flow rate is 35 l/min, entering in the condenser at 42°C

The higher heat transfer coefficient of the water produces a better heat exchange within the condenser and a lower condensing temperature with respect to the air cooled one: the maximum pressure decrease (it is about 15 bar), leading to a lower compression ratio, with benefit in term of specific work required by the compressor.

The effects of the variation of the water flow rate at the condenser are showed in Figure 9. Such flow rate can be easily changed using an electrical pump on a secondary engine cooling circuit. A higher water flow rate leads to a lower condensing temperature and, so, a lower condensing pressure. The performances are, however, more stable than in the case of air cooled condenser.

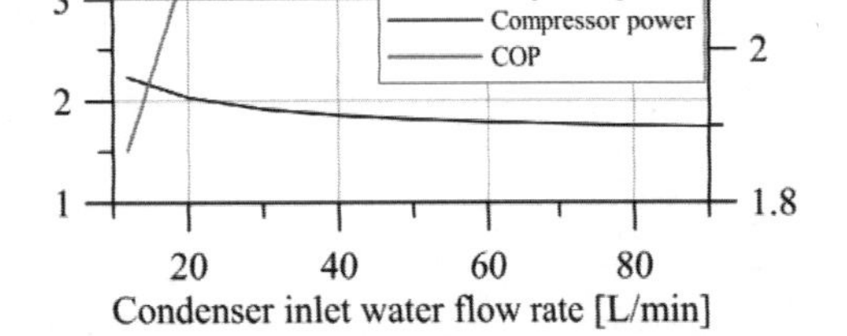

Figure 9: A/C performances with liquid cooled condenser at different water flow rate

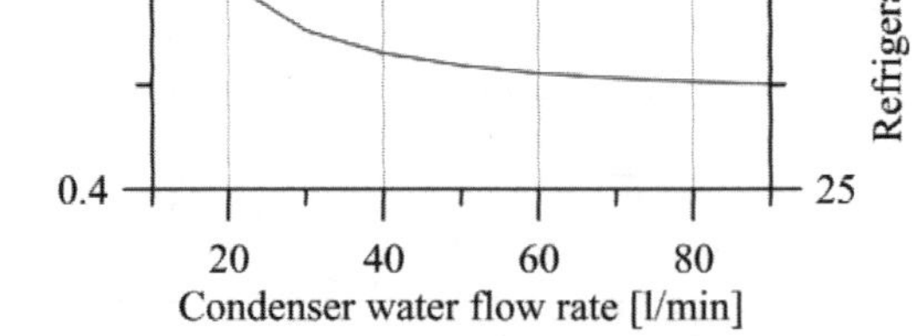

Figure 10: R1234yf mass flow rate and quality at evaporator inlet for A/C equipped with liquid cooled condenser at different flow rate

The saturation value of the power is noticeable also in this case and it is consequence of the fixed geometry of the heat exchanger. COP increases when the water flow rate growths, indicating that the two pressure levels are approaching each other. The higher COP produces a decrease of the working fluid flow rate, which is compensated by a reduction of the quality at the evaporator inlet in order to preserve the cooling request by the cabin (Figure 10).

The benefits of a liquid cooled condenser are clearly in Figure 11: the power absorbed by the compressor decreases and COP increases with a water cooled condenser.

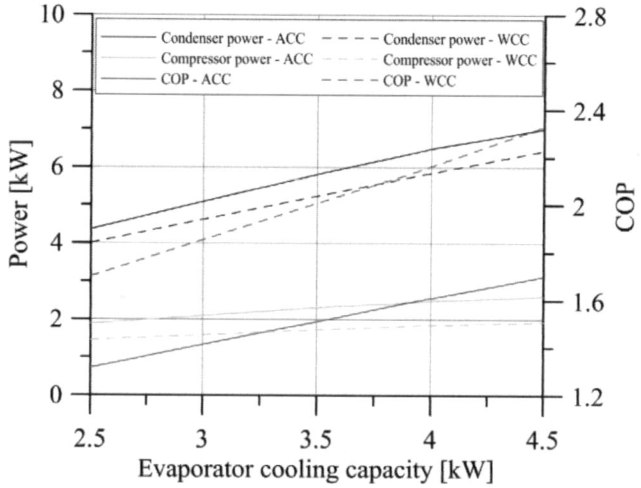

Figure 11: comparison between performances of the A/C system varying the cabin cooling request. Straight lines are referred to ACC, while dashed lines to WCC

In Figure 12 and Figure 13 are represented the A/C parameters that rule the performances highlighted in Figure 11. A water cooled condenser slightly reduces the minimum pressure, (at the evaporator) with respect to the air cooled one, because of the higher heat exchange; so, a higher degree of subcooling is reached in the condenser outlet. The maximum pressure level (in the condenser) decreases as well, producing a higher thermodynamic efficiency.

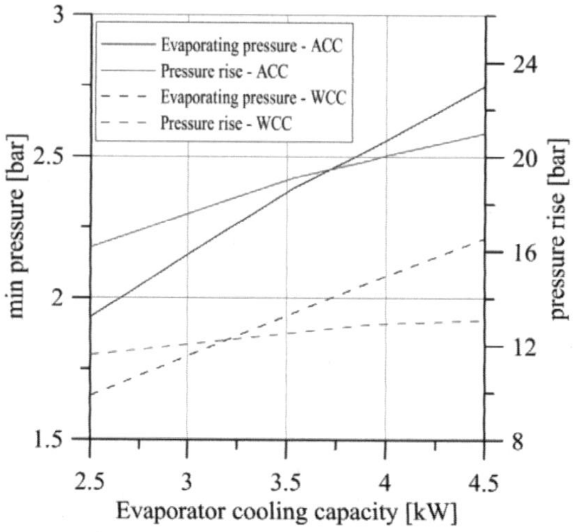

Figure 12: comparison between air cooled condenser (ACC) and water cooled condenser (WCC A/C system):evaporating pressure and pressure rise

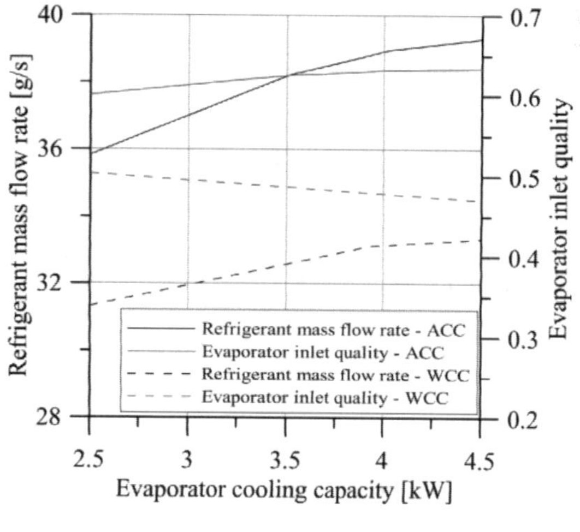

Figure 13: comparison between air cooled condenser (ACC) and water cooled condenser (WCC) A/C system: refrigerant mass flow rate and evaporator inlet quality

Figure 14 shows a quantitative comparison between a mobile A/C system equipped with air cooled condenser and a water cooled one, keeping the same condensing power (6 kW): with a water cooled condenser, the cooling capacity at the cabin side increases as demonstrated by the 12 % higher power exchanged in the evaporator; the most important results is, however, the decrease of the mechanical power absorbed by the compressor of about 22%. These results produce a really higher COP (40%).
The refrigerant mass flow rate is lower, indicating a further reduction of the compression work (Figure 14). When the cabin cooling request increases, the quality of the working fluid at the evaporator inlet is almost constant, but it is really lower when the condenser is water cooled: this demonstrates a higher potential in terms of cooling capacity.

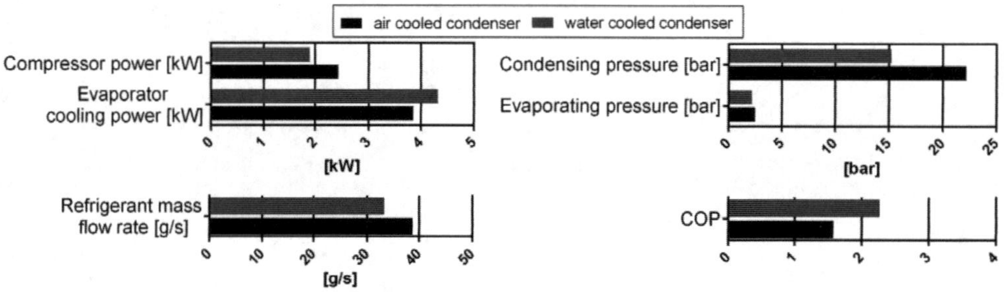

Figure 14*:* comparison between A/C systems with air cooled and water cooled condenser

Further benefits are represented by the reduction of the pressure levels of the system, with lower leakages probability and saving on the refrigerant mass flow rate (that can lead to a lower overall amount of the refrigerant on-board); moreover, with a water cooled condenser, the A/C components can be re-allocated within the engine bay, reducing the ducts length; a WCC, in fact, can be placed closer to the evaporator and it has not to be placed in the front end of the vehicle, as in air cooled configuration.
All this benefits lead to a more reliability of the whole A/C system in terms of fluid leakages and environmental aspects. Finally, a water cooled condenser, having higher heat transfer coefficient, can be more compact of an air cooled one, but it is necessary to introduce a secondary liquid circuit in the engine bay. This, however, can fulfill also other thermal issues, like the charge air cooling, integrating different engine and vehicle thermal needs.

Conclusions

Mobile air conditioning systems have some critical operating issues due to the very often off-design and transient working conditions. Heat transfer coefficients at the two heat exchangers, in fact, strongly depend on the vehicle speed and on the temperatures of the two sources with which heat is exchanged (in particular outside temperature).
The paper presents a comprehensive mathematical model of the A/C system, conceived to represent specific operating conditions of the mobile units. The model catches the typical off-design conditions of the single components (compressor, heat exchangers, flush valve) when external parameters (cabin cooling request, external air temperature and vehicle speed) are significantly changed. This required an iteration procedure in which condensing and evaporating temperature are unknown and they are the result of the model.
Thanks to the model, which behaves as a virtual platform of the unit, some technological improvements for the sector have been studied regarding to:

a) the performances of a new refrigerating fluid (R1234yf), which will replace by laws the conventional R134a, shows slightly lower operating pressures and lower mass flow rates;
b) a water cooled condenser fed by a secondary engine cooling circuit, on which low temperature vehicle's needs could be fulfilled (intercooler, cabin heater, A/C condenser as well), realizes a lower and steadier condensing temperature, because it is not related to the changing of external air temperature and flow rate. Considering R1234yf as working fluid, a reduction of about 0.5 kW

(22%) of the compressing work is demonstrated, with respect to the ACC case. Very interesting is, also, the stronger increase of the COP (about 40%), which leads to a more efficient system. This power reduction is, in any case, higher than the one needed by the secondary cooling circuit [7], which is in the order of 0.1 kW and can be easily compensated by the other benefit related to the double temperature cooling system [6].

c) a reduction of the mean pressure level and the refrigerant mass flow rate can be achieved, implying a lower leakages probability and lower amount of refrigerant on board, with savings on management costs and environmental safety.

REFERENCES

[1] Roujol S and Joumard R 2009 Influence of passenger car auxiliaries on pollutant emission factors within the Artemis model *Atm. Envir.* **43** 1008–1014

[2] Rugh J P, Hovland V and Andersen S O 2004 Significant Fuel Savings and Emission Reductions by Improving Vehicle Air Conditioning *15th Annual Earth Technologies Forum and Mobile Air Conditioning Summit (Washington, D.C, USA)*

[3] Bharathan D, Chaney L, Farrington R B, Lustbader J, Keyser M and Rugh J (NREL) 2007 An Overview of Vehicle Test and Analysis from NREL's A/C Fuel Use Reduction Research *Vehicle Thermal Management Systems Conf. & Exh. (VTMS-8) (Nottingham, UK)*

[4] Mathur G., Hara J, Iwasaki M and Meguriya Y 2012 Development of an Innovative Energy Efficient Compact Cooling System "SLIM" *SAE Tech. Paper 2012-01-1201*

[5] Malvicino C, Di Sciullo F, Ferraris W, Vestrelli F and Beltramelli F 2012 Advanced Dual Level Heat Rejection System for Passenger Cars *SAE Tech. paper 2012-01-1204*

[6] Cipollone R and Di Battista D 2012 Performances and opportunities of an engine cooling system with a double circuit at two temperature levels *SAE Tech. Paper 2012-01-0638*

[7] Cipollone R, Di Battista D and Gualtieri A 2013cA novel engine cooling system with two circuits operating at different temperatures *En. Conv. and Manag.* **75** 581-592.

[8] Janković Z, Sieres Atienza J and Martínez Suárez J A 2015 Thermodynamic and heat transfer analyses for R1234yf and R1234ze(E) as drop-in replacements for R134a in a small power refrigerating system, *Appl. Th. Eng.* **80** 42-54

[9] Bianchi G and Cipollone R 2015 Theoretical modeling and experimental investigations for the improvement of the mechanical efficiency in sliding vane rotary compressors *Ap. En.* **142** 95-107

[10] Klimenko V V 1990 A generalized correlation for two-phase forced flow heat transfer—second assessment, *Int. J. of Heat and Mass Tr.*, **33** 10 pp 2073-2088

[11] Shah M M 2009 An Improved and Extended General Correlation for Heat Transfer during Condensation in Plain Tubes *HVAC&R Res.* **15** N. 5, pp. 889-913

[12] Lichtarowicz A, Duggins R K and Markland E 1965 Discharge Coefficients for Incompressible Non-Cavitating Flow Through Long Orifices *J. of Mech. Eng. Sc.* **7** no. 2 210-219

[13] Navarro-Esbrí J, Mendoza-Miranda J M, Mota-Babiloni A, Barragán-Cervera A and Belman-Flores J M 2013 Experimental analysis of R1234yf as a drop-in replacement for R134a in a vapor compression system *Int. Jour. of Refr.* **36** Issue 3 pp 870-880

[14] Lee Y and Jung D 2012 A brief performance comparison of R1234yf and R134a in a bench tester for automobile applications *Appl. Th. Eng.* **35** 240-242

[15] Zilio C, Brown J S, Schiochet G and Cavallini A 2011 The refrigerant R1234yf in air conditioning systems *Energy* **36** Issue 10 pp 6110-6120

Development of an Organic Rankine Cycle system for exhaust energy recovery in internal combustion engines

Roberto Cipollone, Giuseppe Bianchi, Angelo Gualtieri, Davide Di Battista, Marco Mauriello, Fabio Fatigati

University of L'Aquila, via Giovanni Gronchi 18, L'Aquila 67100, Italy

E-mail: roberto.cipollone@univaq.it

Abstract. Road transportation is currently one of the most influencing sectors for global energy consumptions and CO_2 emissions. Nevertheless, more than one third of the fuel energy supplied to internal combustion engines is still rejected to the environment as thermal waste at the exhaust. Therefore, a greater fuel economy might be achieved recovering the energy from exhaust gases and converting it into useful power on board. In the current research activity, an ORC-based energy recovery system was developed and coupled with a diesel engine. The innovative feature of the recovery power unit relies upon the usage of sliding vane rotary machines as pump and expander. After a preliminary exhaust gas mapping, which allowed to assess the magnitude of the thermal power to be recovered, a thermodynamic analysis was carried out to design the ORC system and the sliding vane machines using R236fa as working fluid. An experimental campaign was eventually performed at different operating regimes according to the ESC procedure and investigated the recovery potential of the power unit at design and off-design conditions. Mechanical power recovered ranged from 0.7 kW up to 1.9 kW, with an overall cycle efficiency from 3.8% up to 4.8% respectively. These results candidate sliding vane machines as efficient and reliable devices for waste heat recovery applications.

1. Introduction

The current worldwide trend of increasing energy demand in road transportation sector remarkably contributes to the growing share of fossil fuel usage and to the release of harmful greenhouse gas emissions. To date, road transportation sector is currently responsible for 17% of the total share of global CO_2 emissions [1]. Nevertheless, the multitude of irreversible processes occurring in Internal Combustion Engines (ICEs) limit their capability to achieve a high efficiency. Indeed, a large amount of fuel energy is rejected from the engine to the environment as waste heat, with a significant fraction (30 - 40%) through the exhaust gases, whose outlet temperature makes them thermodynamically attractive. Hence, a conversion of exhaust heat into useful power would not just bring measurable advantages for improving fuel consumptions but also increase engine power density, further reducing CO_2 and other harmful exhaust emissions correspondingly.

Several direct and indirect energy recovery strategies have been developing by scientists and the automotive industry: the direct ones use gas enthalpy to drive an auxiliary turbine [2], while the indirect ones rely on the high gas temperature either to perform a direct conversion of heat into electric power using thermoelectric generators [3] or applying innovative recovery approaches such as the realization of an additional expansion phase in a six-stroke engine cycle [4]. A more conventional methodology is the usage of exhaust gases as upper thermal source of a thermodynamic bottoming cycle such as Brayton, Stirling, Kalina and Rankine ones [5]. Among the indirect Waste Heat Recovery (WHR) methodologies, Rankine bottoming cycles that use an organic working fluid are characterized by high recovery potential as well as reduced investment costs [6]. However, although this idea of heat recovery was conceived in the 1970s, the industrial applicability of the technology still lacks of mature and reliable components, especially the pump and expander of the recovery system [7].

Recent literature reveals that an Organic Rankine Cycle (ORC) system equipped on an ICE might lead to a fuel economy up to 10-15% [8] and an overall efficiency increase up to 12% [9]. Besides, in reference studies [8, 10, 11] up to 20% of the waste heat from a heavy-duty diesel engine was recovered. As concerns light duty gasoline engines, an experimental analysis stated that at high engine powers, maximum efficiency of the ORC exhaust energy recovery system was up to 14%, and 3-8% under general vehicle operating conditions [12]. Researches on a heavy duty diesel engine using R141b as working fluid of the ORC system showed that engine efficiency increased by 3% and CO_2 emissions decreased by 150 tons/year. On the other hand, working fluids like R123 and water performed slightly worse [13].

Dual loop ORC systems to recover waste thermal energy both from engine exhaust gases and coolant were also developed: in a study on a gasoline engine as thermal source, it was concluded that engine efficiency increased up to 14-16% at peak power and up to 30-50% at lower loads [14]. For engine cooling water waste heat recovery, the use of regenerative ORC systems using R236fa and reheat regenerative ORC systems using R134a allowed to achieve an increase of ICE electrical efficiency up to 4.9% and a net efficiency of 6.6% [15].

Drawbacks due to vehicle weight increase and back-pressure effects along the exhaust gas path were recently outlined by the Authors to state the limits of a bottoming cycle energy recovery system in passenger cars and light-duty vehicles [16, 17].

In the current work, the development of an ORC-based energy recovery system for internal combustion engines is presented with respect to its design, measurement setup and experimentation. The novelty of the study relies on the usage of sliding vane machines for the expander and pump technologies. The recovery performances of sliding vane devices and ORC system were assessed through a test campaign at different engine operating regimes that led the recovery plant to work at design and off-design conditions.

2. Design Methodology

2.1. Exhaust gas mapping

The operating point of the engine, that can be expressed in terms of revolution speed (ω) and torque ($\widehat{T}$), highly affects mass flow rate and temperature of exhaust gases. Although the flexible nature of the internal combustion engine offers wide ranges of variation, in truck applications such as the ones for freight transportation, the vehicle mostly operates at cruise regime. For this reason, in order to design the energy recovery system, an experimental campaign at different stationary operating points was carried out on the IVECO F1C engine instrumented on the AVL APA 100 Transient dynamometer engine test bench complete with automation system based on Puma management software.

Exhaust gases were monitored downstream the particle trap to let the catalytic converter work in proper conditions. Main quantities of interest to design the recovery system were the gas mass flow rate and temperature. The first quantity resulted from the measurements of air and fuel mass flow rates while for the latter one a K-type thermocouple was used. Tests were carried out combining operating points at constant load (torque) with the ones at constant revolution speed. The region investigated is displayed in Figure 1 and covers those engine operating points

Journal of Physics: Conference Series **655** (2015) 012015 doi:10.1088/1742-6596/655/1/012015

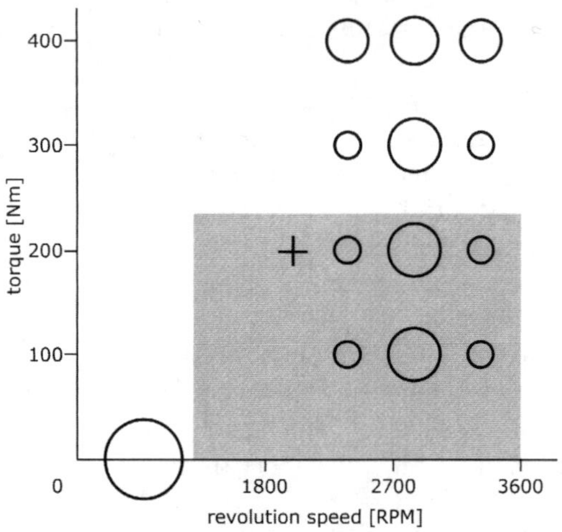

Figure 1. ESC and investigation area (grey region) - design point marked with a cross

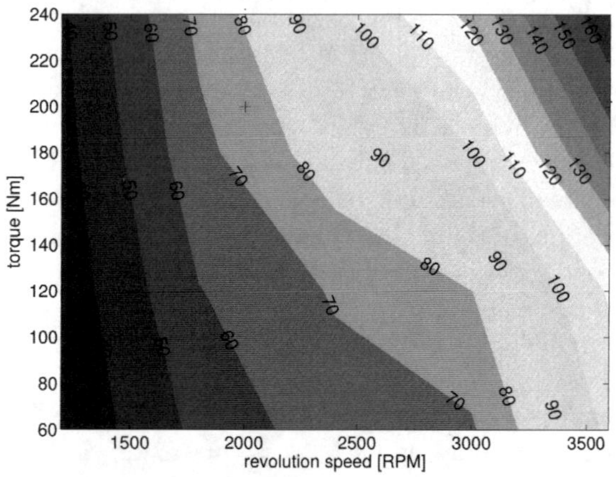

Figure 2. Exhaust gas mass flow rate [g/s]

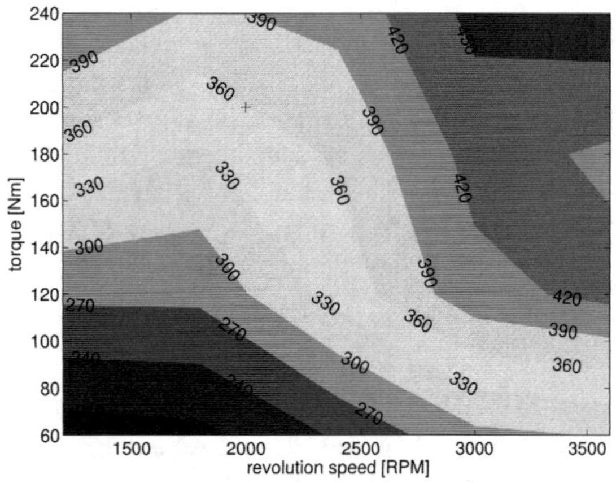

Figure 3. Gas temperature downstream the particle trap [°C]

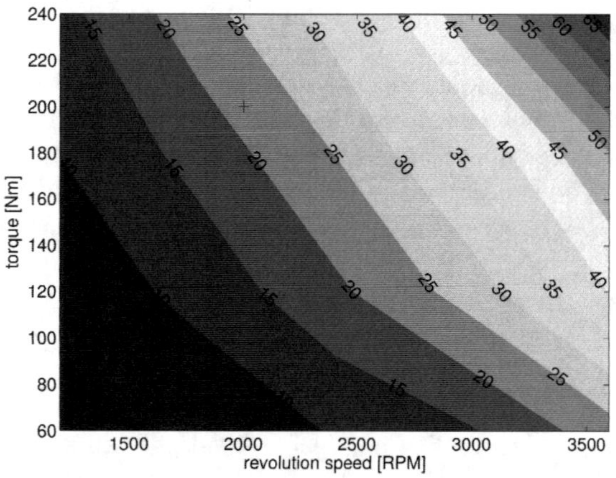

Figure 4. Available thermal power [kW] (discharge gas temperature 100 °C)

which commonly occur during driving conditions of freight transportation vehicles; it further embeds the bottom operating points of the ESC procedure (engine load at 25% and 50%).

As shown in Figures 2 and 3, there is a proportional dependence of exhaust gas mass flow rate and temperature with revolution speed and load. At high loads, gas temperature reached up 450 °C with a mass flow rate of 0.16 kg/s. With reference to an ideal heat transfer between gases and working fluid of the ORC-based recovery plant, assuming a specific heat at constant pressure for the gas of 1.1 $\mathrm{kJ\,kg^{-1}\,K^{-1}}$ and an outlet gas temperature of 100 °C, the thermal power that could be theoretically recovered is reported in Figure 4. The measurement uncertainty on the experimental characterization of the upper thermal source of the energy recovery plant did not exceed 1 % of the calculated value, as it resulted from the uncertainty propagation of the quantity measured.

Figure 3 shows higher gas enthalpy when load and revolution speed of the engine increase. Although from a pure thermodynamic viewpoint the design of the recovery system should be oriented to the maximum exergy of the upper thermal source, those regimes seldom occur during the vehicle operation. Therefore, in order to maximize the energy recovery, the system was designed with reference to a usual operating point of the engine, namely 200 Nm at 2000 RPM.

2.2. Working fluid selection

The proper selection of the working fluid for the recovery plant is a fundamental step to achieve the best trade-off in terms of performance, costs and environmental requirements. In literature, several comprehensive studies provide suitable selection criteria [18]: fluids were mainly analyzed through a screening procedure based on thermodynamic performance [10, 19–21], preselected according to an idea of the operating map of the expansion device [22] or according to thermo-economic considerations [23]. Since the current research aimed at the development of machines for WHR, the thermodynamic approach was preferred.

Table 1. Input data for the thermodynamic analysis at design point

ICE at 200 Nm, 2000 RPM			Assumptions		
gas mass flow rate	$\dot{m}_g$	0.077 kg/s	cooling water inlet temperature	$T_{w,in}$	60 °C
gas inlet temperature	$T_{g,inl}$	356 °C	superheating	ΔT_{sh}	10 K
gas outlet temperature	$T_{g,outl}$	100 °C	subcooling	ΔT_{sc}	5 K

Among the assumptions and constrains listed in Table 1, the role of the superheating deserved a deeper analysis since the technological implications to enhance this parameter are rather challenging. A slight superheating always prevents liquid phase at the expander inlet. However, as demonstrated by the theoretical analyses reported in Figure 5, for a given available thermal power (like the one provided by the ICE at design conditions) cycle efficiency is not significantly affected. This fact, in disagreement with the steam power plant theory, is due to the shape of the saturation curve for wet and isentropic fluids. Indeed, the saturated vapor curve has a slope which always leads to a superheated state at the expansion end, even if expansion starts in saturated conditions. From that point on, since the slope of the upper and lower isobars is almost the same, an increase of superheating at the expansion inlet does not lead to benefits on the cycle efficiency.

The effects of superheating on the actual target performance parameters for waste heat recovery applications, i.e. the product between mass flow rate and net specific work, are further reported in non dimensional terms in the right chart of Figure 5. Unlike the other sensitivity analyses in which the cycle pressure ratio did not contribute to qualitatively modify the effects of superheating, the net power recovered is influenced both by the superheating and the cycle pressure ratio. Thermodynamic cycles with manometric pressure ratios (β_p) below 4 perform better if not superheated. On the other hand, maximum net power recovery would require superheating of the working fluid if the thermodynamic cycle operated at high pressure ratios. However, considering the theoretical approach of the analysis carried out and the minimal variation in percentage terms, it can be concluded that superheating does not contribute to enhance the recovery but it is highly advisable for technological reasons. In the current application, it was limited to 10 K.

On the other hand, a trade-off value of 5 K was assumed for the degree of subcooling at the condenser outlet in order to prevent cavitation phenomena at the pump without oversizing both condenser and evaporator because of a larger thermal power to reject or provide respectively (for a given maximum cycle temperature). Hence, the design choice of a large subcooling would have been twice disadvantageous in terms of weight and costs of the overall recovery system.

Main parameter investigated in the thermodynamic analysis was the cycle pressure ratio: since minimum pressure was constrained by the engine cooling system ($p_{min} > p_{sat}(T_{w,in})$),

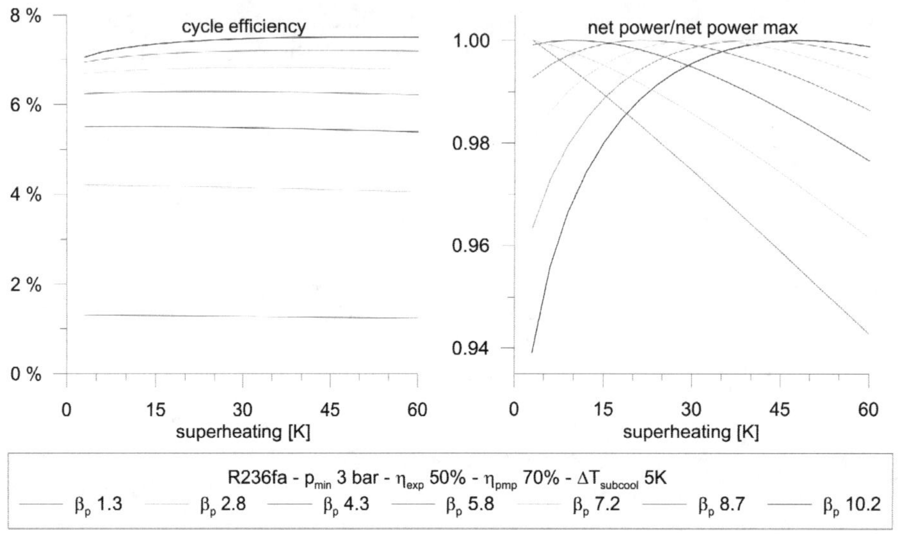

Figure 5. Effects of superheating on theoretical cycle performances

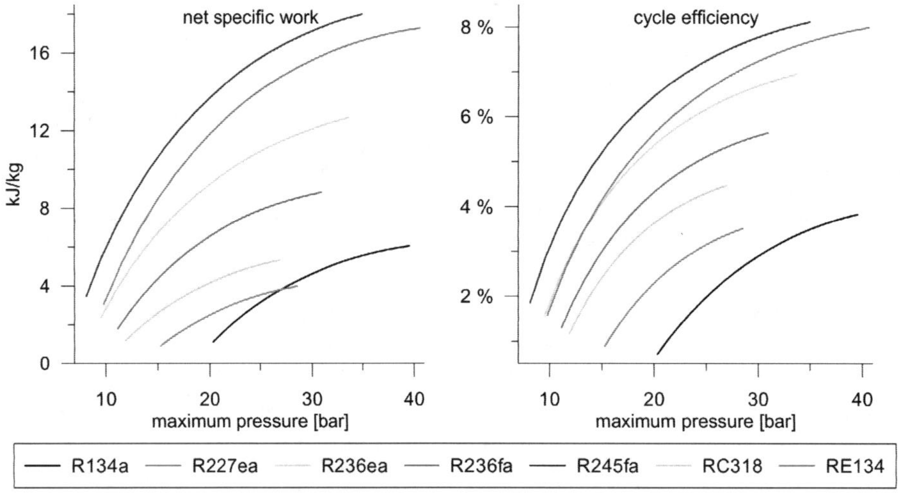

Figure 6. Working fluid selection

the maximum cycle pressure of the organic fluid was varied until 95 % of the critical threshold was reached. Results reported in Figure 6 show that R134a, R227ea and RC318 offer very low performance for the present application: they would require high flow rates and low pressure ratios to produce low net specific work and cycle efficiencies not exceeding 4 %. On the other hand, R245fa demonstrates the best theoretical performance as well as flexibility in the operating point (pressure ratio can vary from 1.5 to 5.5). At 30 bar the cycle efficiency would reach 7.7 % with a net specific work of 17.2 kJ kg^{-1} computed with a pump isentropic efficiency of 50 % and an expander isentropic efficiency of 70 %.

When cost variable was additionally taken into account as selection criterion, at this preliminary stage of the research the availability of R245fa became not affordable. Hence, the working fluid that was adopted for the development of the recovery system was R236fa (Hexafluoropropane with a GWP_{100} of 6300, an atmospheric lifetime of 209 years and a nil

Table 2. Theoretical design approaches (ICE at 200 Nm, 2000 RPM)

			vehicle-scale design	lab-scale design
fluid			R245fa	R236fa
maximum pressure	p_{max}	bar	30	15
minimum pressure	p_{min}	bar	6.2	3
working fluid mass flow rate	$\dot{m}_{WF}$	kg/s	0.098	0.109
expander volumetric flow rate	$\dot{V}_{WF,exp}$	L/min	31.38	64.38
pump volumetric flow rate	$\dot{V}_{WF,pmp}$	L/min	4.80	4.78
heat gain at the evaporator	$\dot{Q}_1$	kW	21.4	21.4
expander output power	$\dot{W}_{exp}$	kW	2.04	1.98
pump input power	$\dot{W}_{pmp}$	kW	0.38	0.19
cooling water mass flow rate	$\dot{m}_w$	kg/s	0.537	0.955
overall cycle efficiency	η_{ORC}	%	7.74	8.32

ODP).

2.3. Vehicle and lab-scale designs

In lab-scale tests, the availability of water at lower temperature ($T_{w,in}$ = 20 °C), further allowed to decrease the pressure levels of the overall system at the same operating point of the internal combustion engine. Table 2 summarizes the operating parameters of the recovery system at design conditions with both the vehicle-oriented and lab-scale approaches: in both cases, cycle pressure ratio and mass flow rate are essentially the same while volumetric flow rate at the expander in vehicle-scale conditions is halved with respect to the laboratory ones. Therefore, the similarity study that led to the establishment of the design specifics for pump and expander of the recovery system led to the following conclusions:

- from a structural point of view, machines were designed with reference to the real operating conditions, whereas stresses are higher because of greater pressure values in the thermodynamic cycle;
- since inlet conditions at the pump led to the same volumetric flow rate, operating conditions in lab-scale tests actually simulate the vehicle application;
- at this stage of development, the reference specifics for the expander design were the laboratory ones. To restore the volumetric flow rate at design point in vehicle operating conditions, the expander should work at 50 % of the revolution speed. If the machine was coupled with a DC generator (or an AC generator with AC/DC converter) to store the energy recovered in the batteries, this requirement would be effortlessly achievable.

3. Experimental Setup

The power unit displayed in Figure 7 is able to recover heat from the exhaust gases of the IVECO F1C engine and converts it into electrical energy. The working fluid (a mixture of R236fa and 5% of POE oil) accomplishes a series of thermodynamic transformations according to a slightly superheated Rankine cycle. The heat recovery takes place downstream the particle trap of the engine and allows pre-heating, vaporization and superheating of the working fluid that further expands in the sliding vane expander. The mechanical energy recovered is eventually converted into electric energy through an asynchronous generator. After the expansion process, the organic fluid is de-superheated and condensed using water as lower thermal source. The recovery system

Figure 7. Engine and ORC system test benches

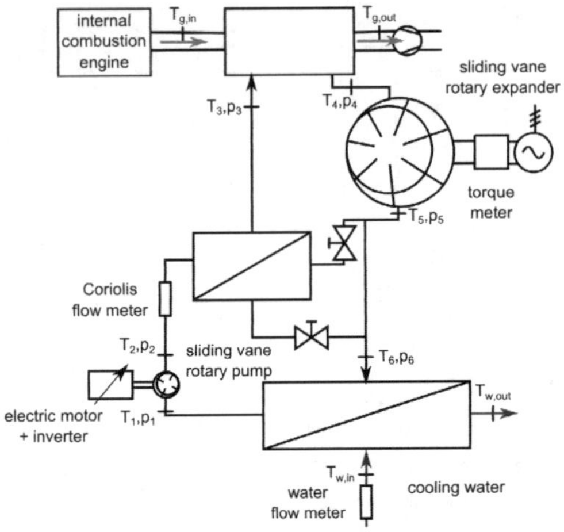

Figure 8. Plant layout and instrumentation

layout can also take into account regeneration. At the condenser outlet, the working fluid reaches a sub-cooled state to be properly pressurized by the sliding vane rotary pump. An inverter on the electric motor of the pump allowed to vary the mass flow rate circulating in the plant while the expander rotated at revolution speeds slightly grater than 1500 RPM, constrained by the electric grid frequency.

Pressure and temperature transducers were installed across all the plant components, as reported in Figure 8. Due to their limited operating range, gas temperatures across the hot side of the evaporator were monitored using K-type thermocouples instead of T-type ones used elsewhere. The sliding vane expander performances were assessed through a direct measurement of mechanical power as product of torque and revolution speed measured by a torque meter. A Coriolis mass flow meter with a HART loop converter was installed downstream the pump and allowed a direct measurement of the working fluid mass flow rate as well as its temperature and density at the same location.

Pressure and temperature data acquired during the tests were linked to the Coolprop LabVIEW library to retrieve all the thermophysical properties of the working fluid in the measuring points. In this way, the real-time entropy diagram was graphed to provide immediate information on the recovery system behavior. At the same time, energy balances were implemented in a graphical interface.

4. Results and Discussion

The test campaign was performed in stationary conditions for the internal combustion engine and the recovery power unit in order to investigate the behavior of both systems and their reciprocal influence. The reference test conditions were chosen according to the operating points of the European Stationary Cycle (ESC), a 13-mode steady state procedure used for emission measurement from heavy-duty diesel engines (Directive 1999/96/EC). However, in order to assess an effective heat recovery from the engine, the idle operating point shown in Figure 1 was not considered. In particular, since operating regimes between engine loads at 25% and 50% are the typical driving conditions of freight transportation vehicles, thus the ones in which the energy recovery would effectively impact on the overall fuel savings in absolute terms, results are presented for six test cases and summarized in Table 3.

Figure 9 details the test case closest to the design point (200 Nm, 2175 RPM) with absolute

Journal of Physics: Conference Series **655** (2015) 012015 doi:10.1088/1742-6596/655/1/012015

Table 3. Test matrix

$\widehat{T}_{ICE}$	Nm	100	200	100	200	91	182
ω_{ICE}	RPM	2175	2175	2750	2750	3325	3325
$\dot{m}_g$	kg/s	0.063	0.080	0.073	0.100	0.077	0.114
$T_{g,inl}$	°C	304.2	350.2	346.4	364.2	422.9	431.2
$T_{g,outl}$	°C	71.4	90.8	85.5	106.4	101.5	137.0
$\dot{m}_{WF}$	kg/s	0.101	0.139	0.127	0.179	0.167	0.208
p_{max}	bar	9.8	13.7	12.5	17.4	16.9	21.9
β_{exp}	-	2.5	2.8	2.9	3.2	3.3	3.4
T_{max}	°C	72.7	89.0	84.8	101.0	100.1	119.5
T_{min}	°C	29.0	36.8	33.5	41.5	40.1	48.4
$\widehat{T}_{exp}$	Nm	4.25	6.68	6.27	8.66	8.24	11.24
ω_{exp}	RPM	1535	1569	1553	1573	1572	1606

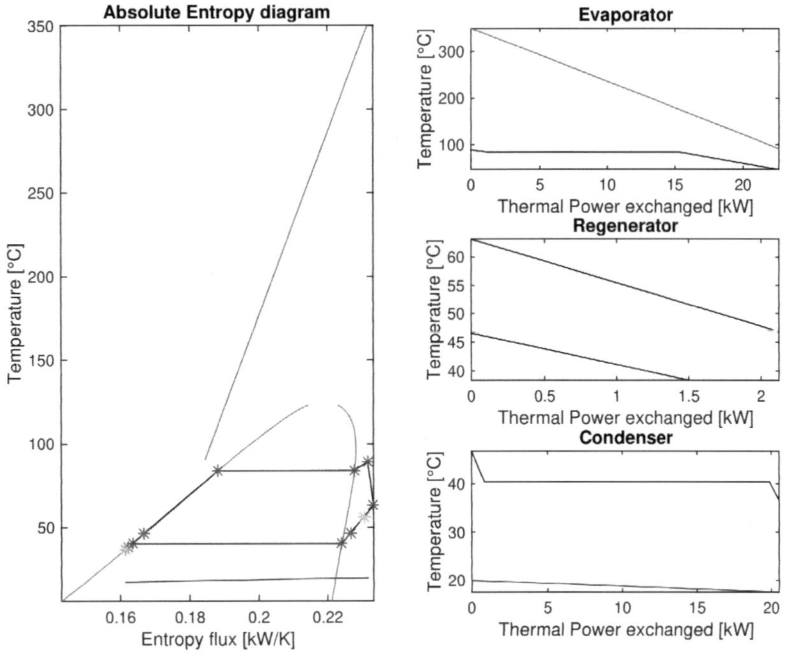

Figure 9. Thermodynamic summary - 200 Nm, 2175 RPM

entropy diagram and heat transfer curves at evaporator, condenser and recuperator. Since the first two devices were carefully insulated, the efficiency of the heat transfers from the exhaust gases to the working fluid and from the working fluid to the cooling water reveals almost unitary. Conversely, in the recuperator, thermal losses to the environment reduced the amount of regeneration. In particular, the share of regenerated power over the total gain at the regenerator and the evaporator was limited between 5% and 10%. In all the experiments, the pinch point at the evaporator occurred between working fluid inlet and gas outlet. Furthermore, the significant gap between red and black curves reveals high irreversibilities in the heat transfer.

Figures 10 and 11 provide a summary of the recovery performance of the ORC system achieved in the 6 ESC points. Energy recovery increases at higher loads and engine revolution speeds

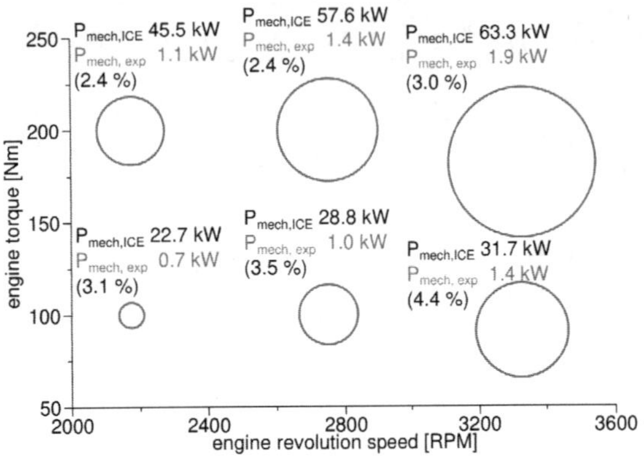

Figure 10. Summary of the energy recovery performances

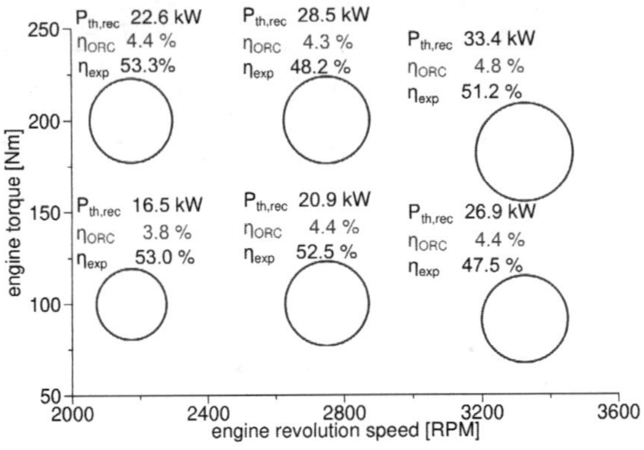

Figure 11. Summary of the energy recovery efficiencies

due to greater exhaust gas mass flow rates and temperatures. At 3225 RPM and 182 Nm, the sliding vane expander was able to provide up to 1.9 kW in mechanical form (3% of the engine power). Additionally, overall cycle efficiency reached 4.8%. Although the availability of the thermal waste decreased at low regimes, the recovery system was able to perform efficiently also at off-design conditions. At 3225 and 91 RPM, mechanical power at the expander shaft reached 4.4% of the engine shaft power. In all the experiments, overall efficiency of the power unit was always around 4%.

As concerns the global expander efficiency, computed as ratio of the mechanical power over the isentropic power, best performances occurred in the test case closest to the design point (η_{exp} 53.3%). At off-design conditions with higher heat gain, mechanical power recovered increased due to a higher working fluid mass flow rate and cycle pressure ratio. However, the large superheating that characterized these operating points worsened the expander efficiency because of greater discharge pressures at the condenser.

A more extensive investigation of the fluid behavior within the machine might be provided retrieving the expander indicator (pressure-volume) diagram from high frequency pressure data. This activity was previously carried out by the Authors in [24] with the methodology developed for sliding vane compressors [25] and it will be performed in the near future on this application using piezoresistive pressure transducers rather than piezoelectric ones in order to minimize the measurement uncertainty due to the reconstruction procedure of the pressure-volume diagram [26, 27].

5. Conclusions

Bottoming organic Rankine cycles for the energy recovery from exhaust gases of internal combustion engines recently gained the attention of scientific community and automotive industry thanks to their remarkable theoretical performances and economic feasibility. In this context, the current research work dealt with the design, development and testing of an ORC-based recovery power unit. The novelty of the study was the usage of the sliding vane technology for pump and expander of the recovery system in order to face the lack of mature, efficient and reliable devices for waste heat recovery in automotive applications.

Power unit design was carried out after an experimental campaign on an IVECO F1C engine that assessed the available thermal power at the engine exhaust. Two design configurations were compared to reproduce vehicle operating conditions in laboratory tests. The experimental test bench of the ORC recovery system equipped with sliding vane machines was characterized by a high-accuracy measurement chain and a data acquisition routine that also provided the real-time entropy diagram as monitoring tool during the experiments. Recovery system performances were experimentally assessed through a steady test campaign where the ORC power unit was coupled with the internal combustion engine. Mechanical power recovery reached up to 1.9 kW which corresponded to 3% of the engine mechanical power at that given operating conditions. This share increased at low engine loads and high revolution speeds. The overall cycle efficiency ranged between 3.8% and 4.8%, while the overall expander efficiency values were between 47.5% and 53.3%.

Improvements on the overall cycle efficiency are expected thanks to an upgraded plant layout and thermal insulation of the components. Greater isentropic and volumetric efficiencies of the sliding vane expander are going to be achieved optimizing intake and exhaust processes. These activities will be supported by the analysis of the expander indicator diagram using high frequency pressure sensors to investigate the fluid pressure evolution within the machine.

Acknowledgement

Ing. Enea Mattei S.p.A., in the persons of Dr. Giulio Contaldi and Dr. Stefano Murgia, is gratefully acknowledged for the technical support provided throughout the development of sliding vane expander and pump prototypes.

References

[1] IEA 2014 *Key world energy statistics* (International Energy Agency)
[2] Weerasinghe W, Stobart R and Hounsham S 2010 *Applied Thermal Engineering* **30** 2253 – 2256
[3] Riffat S and Ma X 2003 *Applied Thermal Engineering* **23** 913 – 935
[4] Conklin J C and Szybist J P 2010 *Energy* **35** 1658 – 1664
[5] Sprouse C and Depcik C 2013 *Applied Thermal Engineering* **51** 711 – 722
[6] Saidur R, Rezaei M, Muzammil W, Hassan M, Paria S and Hasanuzzaman M 2012 *Renewable and Sustainable Energy Reviews* **16** 5649 – 5659
[7] Lopes J, Douglas R, McCullough G, O'Shaughnessy R and et al 2012 *SAE Technical Paper* **2012-01-1942**
[8] Teng H, Klaver J, Park T, Hunter G L and van der Velde B 2011 *SAE Technical Paper* **2011-01-0311**
[9] Vaja I and Gambarotta A 2010 *Energy* **35** 1084 – 1093
[10] Teng H, Regner G and Cowland C 2007 *SAE Technical Paper* **2007-01-0543**
[11] Teng H and Regner G 2009 *SAE Technical Paper* **2009-01-2913**
[12] Wang T, Zhang Y, Zhang J, Shu G and Peng Z 2013 *Applied Thermal Engineering* **53** 414 – 419
[13] Wang T, Zhang Y, Zhang J, Peng Z and Shu G 2014 *Energy Conversion and Management* **84** 97 – 107
[14] Arias D A, Shedd T A and Jester R K 2006 *SAE Technical Paper* **2006-01-1605**
[15] Peris B, Navarro-Esbr J and Mols F 2013 *Applied Thermal Engineering* **61** 364 – 371
[16] Di Battista D, Mauriello M and Cipollone R 2015 *Applied Energy* **152** 109 – 120
[17] Di Battista D, Mauriello M and Cipollone R 2015 *SAE Technical Paper* **2015-01-1608**
[18] Bao J and Zhao L 2013 *Renewable and Sustainable Energy Reviews* **24** 325 – 342
[19] Wang Z, Zhou N, Guo J and Wang X 2012 *Energy* **40** 107 – 115
[20] Tian H, Shu G, Wei H, Liang X and Liu L 2012 *Energy* **47** 125 – 136
[21] Liu B T, Chien K H and Wang C C 2004 *Energy* **29** 1207 – 1217
[22] Quoilin S, Declaye S, Legros A, Guillaume L and Lemort V 2012 *Working fluid selection and operating maps for Organic Rankine Cycle expansion machines*
[23] Quoilin S, Declaye S, Tchanche B F and Lemort V 2011 *Applied Thermal Engineering* **31** 2885 – 2893
[24] Cipollone R, Bianchi G, Di Battista D, Contaldi G and Murgia S 2014 *Energy Procedia* **45** 121 – 130
[25] Bianchi G and Cipollone R 2015 *Applied Energy* **142** 95 – 107
[26] Bianchi G and Cipollone R 2015 *Applied Thermal Engineering* **84** 276 – 285
[27] Bianchi G, Cipollone R, Murgia S and Contaldi G 2015 *International Journal of Refrigeration* **52** 11 – 20

Journal of Physics: Conference Series **655** (2015) 012016 doi:10.1088/1742-6596/655/1/012016

Uncertainty evaluation of EnPIs in industrial applications as a key factor in setting improvement actions

G D'Emilia, D Di Gasbarro, A Gaspari and E Natale

Dipartimento di Ingegneria Industriale e dell'Informazione e di Economia, University of L'Aquila, 67100 L'Aquila, Italy

E-mail: giulio.demilia@univaq.it

Abstract. A methodology is proposed assuming high-level Energy Performance Indicators (EnPIs) uncertainty as quantitative indicator of the evolution of an Energy Management System (EMS). Motivations leading to the selection of the EnPIs, uncertainty evaluation techniques and criteria supporting decision-making are discussed, in order to plan and pursue reliable measures for energy performance improvement. In this paper, problems, priorities, operative possibilities and reachable improvement limits are examined, starting from the measurement uncertainty assessment.
Two different industrial cases are analysed with reference to the following aspects: absence/presence of energy management policy and action plans; responsibility level for the energy issues; employees' training and motivation in respect of the energy problems; absence/presence of adequate infrastructures for monitoring and sharing of energy information; level of standardization and integration of methods and procedures linked to the energy activities; economic and financial resources for the improvement of energy efficiency.
A critic and comparative analysis of the obtained results is realized. The methodology, experimentally validated, allows developing useful considerations for effective, realistic and economically feasible improvement plans, depending on the specific situation. Recursive application of the methodology allows getting reliable and resolved assessment of the EMS status, also in dynamic industrial contexts.

1. Introduction

Energy management requires a systematic and continuous approach, whose final goal is using energy as efficiently and effectively as possible.

Companies often make simply isolated attempts to reduce consumptions, rather than adopt a systemic approach to the problem; even when the organizations intend to face the problem in a systematic way, it is not easy for them to self-assess their possibilities for improvement.

The International standard ISO 50001:2011 [1] provides a framework of requirements for organizations to develop a policy for: more efficient use of energy, fix targets and objectives to meet the policy, use data to better understand and make decisions about energy use, measure the results, review how well the policy works, and continually improve energy management.

Although the process-based PDCA (Plan-Do-Check-Act) methodology still holds, the general requirements and procedures are not particularly focused on how an organization can understand and quantitatively assess its current position within this operating path to reach the targets. Being aware of the position of an organization in the operating sequence for the implementation of an Energy Management System (EMS) allows identifying and reaching strategic objectives more effectively and efficiently.

In literature, some approaches to evaluate the maturity in the energy management field exist. Carbon Trust [2] presents two tools that organizations can use to self-assess their performance across the areas of energy: the Energy management matrix [3], helping organizations assessing their strengths and weaknesses across the evaluation of six areas of energy management (policy, organizing, training, performance measurement, management); the Energy Management Assessment (EMA) tool, providing a more comprehensive and detailed self-assessment. Both consist of a simple quantitative evaluation, based on a numerical assignment to different areas examined in a range of values. Other approaches can be found, at different levels of complexity [4-7]; they are all based on qualitative evaluations of different aspects and even quantitative analysis are always based on a judgment: therefore, these assessments are inevitably characterized by elements of subjectivity.

Furthermore, even pushing the analysis at a major level of detail, the resolution of these methods is often insufficient. As a matter of fact, by means of these approaches, the ability of distinguishing little but useful improvements along the growing and progressive path of an EMS development, is strongly limited.

Usually, the energy performance is evaluated through different Energy Performance Indicators (EnPIs), which are general or depending on the specific situation examined [8]. The uncertainty evaluation of EnPIs becomes relevant to its own ability to represent correctly the aspects of interest for the scenario considered, even though just a few cases can be found in literature, where the role of the raw data uncertainty is stressed [9].

The uncertainty evaluation of the indicators is not a trivial task, especially if they are aggregate: if so, many aspects should be carefully considered. In other words, when high-level aggregate EnPIs have to be evaluated, also productive, organizational and plant aspects and features should be taken into account, and considered among the others.

The EnPIs uncertainty assume characteristic interest also for further needs and motivations, summed up as follows:

- Reliability of information:

The EnPI uncertainty is directly linked to the accuracy of the information carried out by the indicator itself. In this sense, the EnPIs uncertainty and its monitoring can be intended as a part of the data check procedure, useful to assure adequacy of data to provide reliable information.

- Tailoring the actions, since each context has its own peculiarity:

Carrying out the EnPIs uncertainty analysis and setting up actions supporting its reduction leads to a deeper understanding and knowledge of the processes under control, allowing focusing on the specific improvement action, which depends on the specific context analysed.

- Dynamic, changing and variable industrial operating contexts:

Measurement uncertainty of physical quantities of interest for the evaluation of the EnPIs strictly depends on the specific situation and it changes, if the conditions to be examined change. As a consequence, the EnPIs uncertainty involves and brings with itself also the dynamic aspects of the scenarios examined.

Only an EMS which is carried out in a complete and comprehensive manner allows a deeper knowledge able to consider all these factors, singularly and in their mutual interactions. In this sense, the EMS development path seems to support strategies for uncertainty reduction. Therefore, assuming the uncertainty as a quantitative parameter to take a picture of the current status of the inherent activities, appears interesting and promising. In fact, the EnPI uncertainty involves the information about quality and amount of actions realized to build and maintain the EMS, being the EnPI one of the possible indirect measurement of the EMS itself.

In authors' opinion, measurement methodologies and uncertainty analysis are able to return reliable information about improvement actions even if their application is referred to different situations (i.e. different industrial scenarios), at different levels of awareness, and in different stages of progress in the implementation of an energy management system. Since the actions related to an EMS are general, systematic and systemic (i.e. they are not carried out according to a spot approach focused on single machine), they require a widespread managerial ability.

In this paper, in order to endorse the systematic approach of an EMS, high-level synthetic and enough meaningful EnPIs will be considered. The possibility to evaluate their uncertainty will be studied in order to get a quantitative parameter (the EnPI uncertainty) able to return:

- the position the company achieved along the EMS life-cycle;
- a support to the designing of action plans for energy improvement depending on the definition of intervention capability;
- the reliability of the improvement actions with respect to the predicted targets.

In this way, a more resolved and accurate method could be set, aiming to overcome a mere classification into development levels.

In summary, any decision making process based on measured EnPIs could be compromised if uncertainty on EnPI is too large. The level of confidence of the decision making process depends on the weight of uncertainty on EnPI relative to the target improvement in EnPI which one aims to achieve.

Section 2 describes the methodology proposed and the industrial contexts analysed in order to test the methodology. The opportunities deriving from the test cases analysed are also discussed. Section 3 shows the obtained results with reference to the actual situations examined. Considerations of practical interest are discussed and proposed. Conclusions and future works end the paper and show also some perspectives in order to develop the concept discussed in this paper.

2. Methodology

The methodology provides for the use of the EnPI uncertainty as an indicator of:

- level of awareness of issues and opportunities linked to the development of an EMS, in the specific situation;
- ability to set the strategies referred to the energy efficiency improvement and to the evaluation of the action effectiveness;
- energy saving that can be guaranteed in advance, in relation to a pre-defined improvement target.

In particular, the proposed methodology pivots on two main conceptual steps (figure 1):

1. Step 1: EnPI identification

   The selection of the EnPI is based on its information capability and synthesis, meaningfulness and accuracy. These characteristics will assure that it will represent exhaustively the energy processes it is called to indicate. Many energy contributions should be included, in order to have a more accurate indicator providing the possibility to take into account wide ranges of actions for energy performance improvement.
2. Step 2: EnPI uncertainty evaluation in terms of standard deviation.

   The relationship in equation (1) is the simple criterion used to assess if the predicted percentage energy saving (t%, in the following) can be guaranteed on the basis of the percentage EnPI uncertainty (u(EnPI)%, in the following) which is based on the current experimental data.

$$u(EnPI)\% \leq t\% \tag{1}$$

The criterion in (1) is in accord with the common practice [10] assuming the standard deviation value, as detectability threshold.

If condition (1) is positively fulfilled, and if the action has been already defined, it is suggested that the decision-maker adopts the measure for energy savings. Else, any improvement actions to reduce measurement uncertainty should be considered. Once these latter are carried out, then the uncertainty of the selected EnPI should be evaluated again (step 2) and the flow in figure 1 repeated.

If the action has not been defined, a target according to the detectability criterion in (1) can be set. The relationship in (2), aims to provide a conceptual idea of the size of the improvement action to be carried out, in order to let the same EnPI show the target is reached (being equal the other conditions):

$$t\% = u(EnPI)\% \tag{2}$$

The technical and economic feasibility of the action should be checked (according to the specific energy policy). If the technical and economic feasibility is confirmed, it is suggested that the decision-maker adopt the measure for energy savings. Else, any measure to reduce the EnPI uncertainty should be set, and the cycle (starting from step 2) of figure 1 begins again.

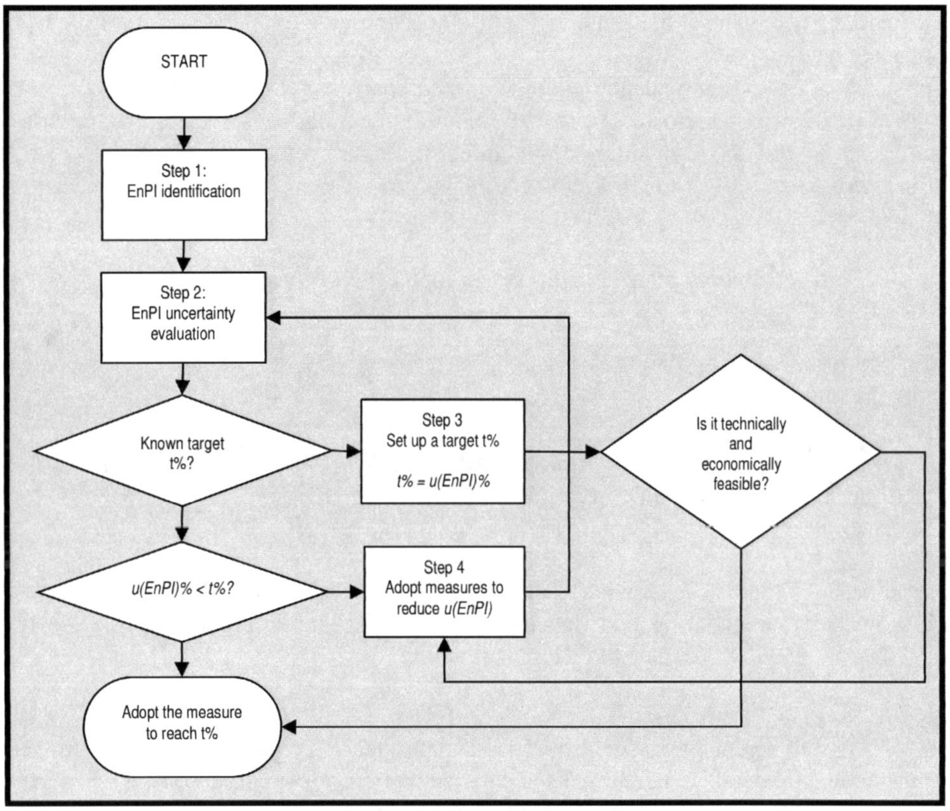

Figure 1. Methodology flow diagram

In order to verify generality and effectiveness of the methodology proposed, this latter is applied to two industrial cases, different for level of awareness in energy management and measurement issues. The differences between the two situations (named "Test case A" and "Test case B", in the following) are qualitatively highlighted using the Energy management matrix [3], able to establish a starting point for the analysis.

The test case A consists in the production of carbon fibre components for avionic industry. The main part of production process from an energetic point of view consists of certified thermal curing in autoclave.

The main utilities referred to the whole plant can be summarized as follows: autoclaves, air compressors connected to the autoclaves, industrial CNC cutting plotters, devices for mechanical treatments, refrigerating cells for Pre-preg components storage, lighting network, ventilation unit for clean room air supply.

Meters for gas, electrical power and water are available only with reference to the whole plant. There are also measuring instruments for process control with manual data recording.

A part of the recorded data is not electronically stored but it is available only on paper.

The test case B consists in the production of small plastic components through injection moulding.

The main utilities referred to the plant section involved in the analysis can be summarized as follows: injection moulding machines, chillers for mould cooling, evaporation towers and pumps for fluid handling.

Meters for electrical power are available with reference to each utility involved in the analysis, with automatic data acquisition and recording. There are also measuring instruments for process control with automatic data recording. All these data are stored in shared databases.

Further aspects to be highlighted are the following:

- formal energy management policy, but not continuously active commitment from top management;
- presence of an energy manager leading an interdepartmental team;
- diffusion of energy performance results (monthly meetings, posters, specific reports,…).

With reference to the above-mentioned situations, the Energy management matrix can be filled showing framework according to table1. For clarity is recalled that the minimum value is 0 and the maximum is 4.

Table 1. Energy management matrix for Test-case A and Test-case B.

	Test-case “A”	Test-case “B”
Energy management policy	1	3
Organizing	1	2
Staff motivation	0	2
Tracking, monitoring and reporting systems	0	3
Staff awareness/training and promotion	1	2
Investment	1	2

Table 1 underlines that the examined cases are quite different as for the current status of each company’s effort in implementing and developing an EMS. This could be useful also to test the proposed methodology with reference to different situations.

3. Results

Depending on the different level of awareness, data accuracy is a key-factor in setting the improvement actions and reaching the related targets. In fact, depending on the specific situation, paths, solutions and quality of the obtained results are different.

According to the methodology described in section 2, the EnPIs chosen for the analysis are described.

3.1. Methodology application - Step1: EnPI identification

3.1.1. Test case A. Specific Energy Consumption (SEC) is defined as follows:

$$SEC = \frac{E}{P} \tag{3}$$

where E is the total energy consumption [kWh] and P is the number of pieces produced. Both quantities refer to the same time interval. Generally this indicator is synthetic, meaningful and accurate, provided that the energy processes it is called to indicate are satisfactorily monitored. It is usually considered for energy performance evaluations [8,11].

3.1.2. Test case B. Energy per Working Hour (EWH) is defined as follows:

$$EWH = \frac{E_m}{\sum_i h_i} \tag{4}$$

where E_m is the energy consumption of moulding plant section [kWh] and h_i are the machines working hours. Both quantities refer to the same time interval. This indicator is significant for test case B because the process is time-driven, the machines are general purpose and the produced components are similar one to each other.

3.2. Methodology application - Cycle 1

*3.2.1. Test Case A.*As for the Test case A, figure 2. As an example, shows the normalized SEC behaviour during one week of production, which has been selected being a full time production week. The SEC percentage uncertainty calculated as standard deviation of the SEC value, in percentage of the mean SEC value, during the examined week (5 days, 1 value for each day), is about 20%.

Target t% has not been defined. SEC uncertainty suggests to identify an improvement action so that the SEC reduction is in the order of 20%. Because of the remarkable amount of the requested improvement, the management considers this target technically and economically not feasible. Thus, some actions to improve the EnPI accuracy should be provided.

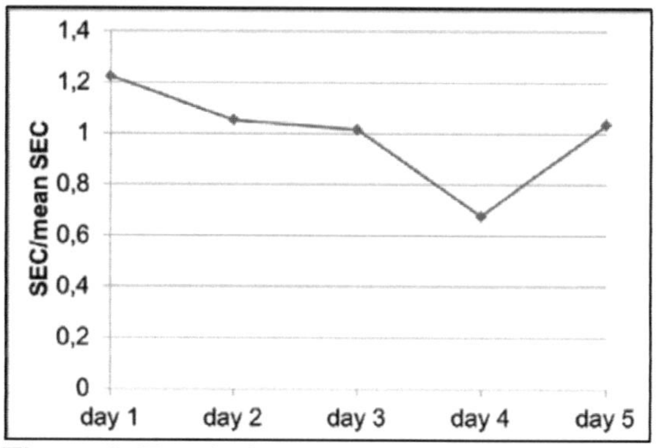

Figure 2. Trend of normalised SEC_Cc for Test-case A

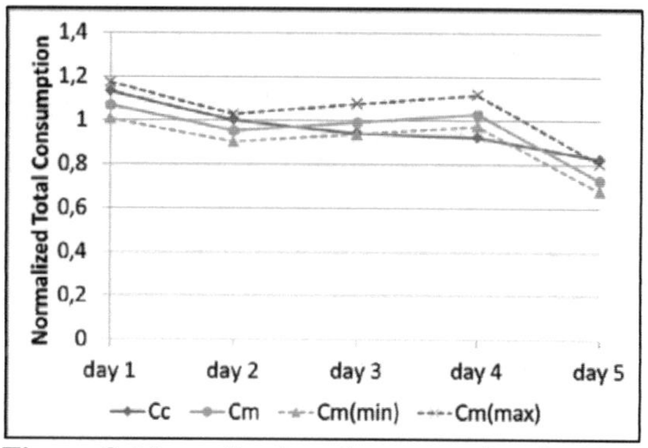

Figure 3. Normalized total consumption for Test-case A

In order to reduce the EnPI uncertainty, a consumption model considering the energy-intensive utilities based on the electric size of the utilities themselves has been set up. The model has been validated using experimental data. Figure 3 shows the obtained behaviour, together with the variability bands of the model, obtained considering the uncertainty of estimation of contributions to the whole consumption. The subscript Cc refers to the measured consumption, the subscript Cm refers to the consumption model.

3.2.2. Test Case B. As for the Test case B, figure 4, as an example, shows the normalized EWH behaviour during one year of production.

The percentage uncertainty calculated as standard deviation of the EWH value, in percentage of the mean EWH value, during the examined year is about 11%.

The management could allow some actions to improve the efficiency of the injection moulding machines, extending a spot action to more machines. The extension criterion is that the specific efficiency increase should be not less than 15%, based on the payback period consideration. It should be noted that the specific efficiency increase depends on the machine technology which is different among the machines potentially interested.

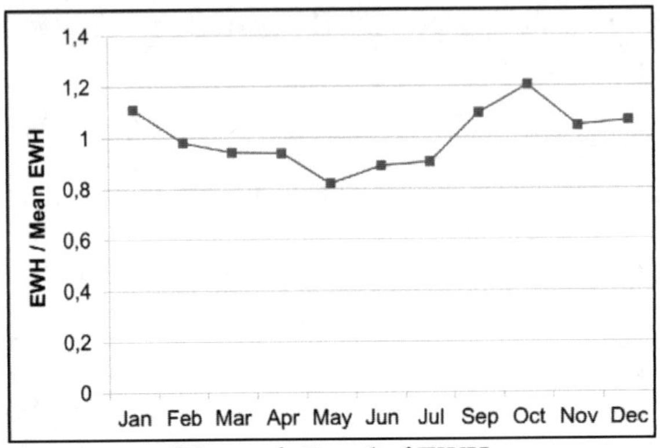

Figure 4. Trend of EWH

Table 2 shows the entity of the predictable saving corresponding to the reduction of the adopted EnPI.

Considering the EnPI uncertainty, and the target (t% = 4.2%) the criterion (1) is not satisfied, therefore the success of the action cannot be demonstrated by the EnPI itself. Thus, some action to improve the EnPI accuracy should be provided.

Table 2. Evaluation of the savings for single machine and plant section.

Injection molding machine				Plant section	
Quantity	**Type**	**Single machine saving**		**Consumption**	**Saving**
1	1	17%		5,2%	0,9%
2	2	15%		2 x 5,5% = 11%	1,5%
3	3	15%		3 x 4% = 12%	1,8%
4	4	12%		4 x 3% = 12%	No upgrade
4	5	12%		4 x 4% = 16%	No upgrade
12	6	9%		12 x 3% = 36%	No upgrade
1	7	No possible upgrade		7,8%	No possible upgrade
			Total	100%	4,2 %

In order to reduce the EnPI uncertainty, some actions to improve data reliability have been identified, which are summarized as follows:

- In field check of power meters, in terms of:
  - compliance of installation connections to the electric diagrams;
  - compliance of setting parameters in respect to the current transformers (e.g. turn ratio).
- Activation of a more reliable IT architecture, specific for energy data storage and management, such as:
  - development of alarm systems in case of anomalies to avoid data loss;
  - comparison of data to assure that stored data are the same of the power meters;
  - comparison of data stored on the previous data base to the last one.
- Reallocation of consumptions after a more detailed mapping of energy flows (e.g. compressed air consumption).

3.3. Methodology application - Cycle 2

3.3.1. Test case A. Figure 5 shows the normalized SEC behaviour during the same week of production as figure 4, evaluated on the basis of the consumption model, (SEC_Cm).

The percentage uncertainty calculated as standard deviation of the SEC_Cm value, in percentage of the mean SEC_Cm value, during the examined week (5 days, 1 value for each day) is about 16%.

The definition of a consumption model allowed to reduce the SEC uncertainty with respect to the previous estimation and it also supplied an indication about the relevance of contributions of different energy consuming utilities, underlining the contribution of the autoclave which is about 50% of the whole consumption. Because of the remarkable amount of the requested improvement, the management considers this target is still too hard to achieve. Thus, the EnPI accuracy has to be strongly improved by more diffused measurement of the energy flows for most relevant utilities and improved production monitoring and data registration.

3.3.2. Test case B. Figure 6 shows the normalized EWH behaviour before (11 months, 1 value for each month) and after (8 months, 1 value for each month) the interventions described in previous section.

The percentage uncertainty calculated as standard deviation and in percentage of the mean EWH value, during the examined period after the interventions is about 4%.

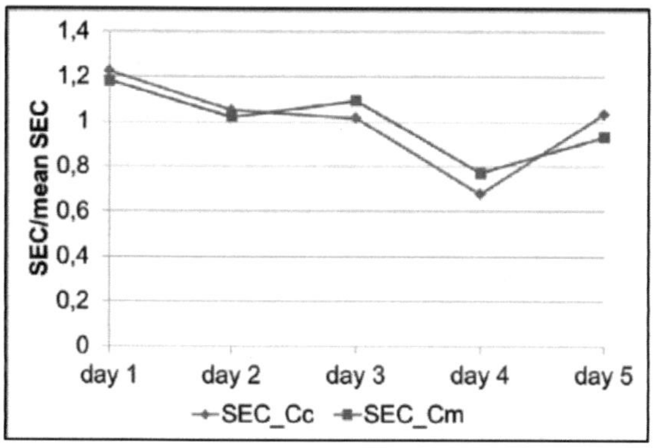

Figure 5. Trend of SEC_Cc and SEC_Cm

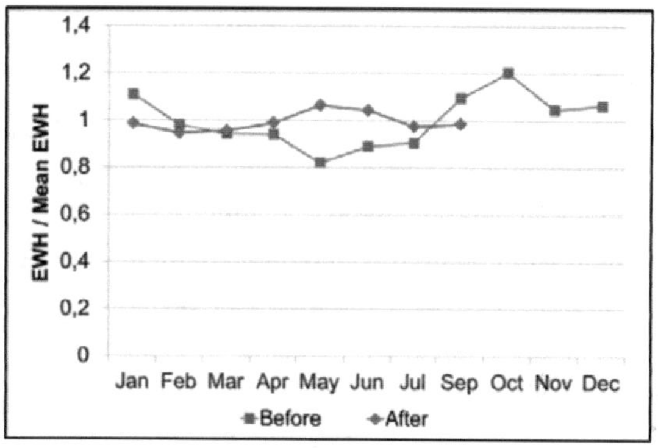

Figure 6. Trend of EWH before and after the interventions

Having reduced the EnPI uncertainty from the previous value of 11% to 4% the improvement action proposed by the management (summarized in table 2) can be done. Considering the new EnPI uncertainty, and the target (t% = 4.2%) the criterion (1) is now satisfied, therefore the success of the action can be demonstrated through the EnPI itself.

In figure 6 the data "before" and "after" the interventions cross each other; this is linked to the reduced EnPI uncertainty. In fact a higher EnPI uncertainty corresponds to a high fluctuation of the indicator; if EnPI uncertainty is reduced the time behaviour of the indicator is much more stable.

3.4. Discussion of results

In both cases, the applied methodology allowed providing some useful indications about reachable savings and action areas to be considered to achieve the improvements targets.

About the test case A, the measurement uncertainty allows validating the consumption model, which is set up in order to compensate the lack of energy data. The possibility to set the target as the minimum of the variability band of the consumption model, allows identifying further improvement areas, even though relevant investments are required.

The actions leading to a reduction of the EnPI uncertainty allows identifying the most energy-intensive utilities, i.e. the autoclaves. Suggestions for a further reduction of the EnPI uncertainty are also given.

Ultimately, the level of the target definition results coherent with the quality of the available measurements. In this sense, the EnPI uncertainty can be seen as a support of a higher awareness and knowledge of the analysed situation.

About the test case B, a more organic strategy involving the data validation within the whole measurement chain has been set up. As a consequence, being the EnPI uncertainty decreased, the EnPI meaningfulness results increased. Therefore, the reduced EnPI uncertainty allows demonstrating the success of the proposed improvement action, and eventually the success of further improvement actions in the same plant section.

4. Conclusions and future works

In this work, the possibility of evaluating and controlling the EnPIs uncertainty has been discussed in order to demonstrate that it allows increasing the transparency level of the energy performance in terms of the aspects affecting the energy consumption. The high level EnPI uncertainty is not easy to evaluate but it offers the possibility of supporting the decision-making activities, since the measurement uncertainty involves the ability of modelling and knowing the actual energy flows and their interactions, which the indicator itself should represent.

As more the EnPI uncertainty is reduced, as more the possibility of choosing among many suitable actions is increased, maintaining a high level of confidence of obtaining the predicted improvement. The increased awareness of the actual situation is also a positive result.

In this paper a methodology has been developed and applied for validation purposes to two different cases regarding different industrial sectors and level of evolution of the EMS life-cycle. In particular, the uncertainty evaluation of high-level EnPIs has been studied.

The recursive application of the approach to the same test case allowed to demonstrate that the EnPI uncertainty is a parameter resolved and accurate enough to detect different positions along the EMS life-cycle. The same result could not be obtained by means of other qualitative methods. The same result has been achieved for both cases examined.

Furthermore, the discussion about the reached level of uncertainty supplied indications about the design of action plans for energy improvement and their reliability with respect to the predicted targets. In fact, depending on the different quality of measurements available, effective guidelines to carry on the EMS in terms of establishing, implementing, maintaining and improving coherent action plans have been suggested.

The methodology allowed also to highlight technical and economic feasibility of the improvements taking into account the possibility of guaranteeing the achievement of the predicted target with a satisfactory level of reliability.

Future works:

- testing the reproducibility of the methodology taking into account more industrial sectors and more different energy vectors typologies (compressed air, steam, renewable energies,...);
- experimentally assessing the measurement uncertainty ability to create more diffused awareness;
- evaluating the possibility of using portable plug-in measurement systems able to support the requirements of an EMS, to set demonstrator apparatuses for awareness diffusion of achievable advantages of implementing an EMS;
- examining the possibilities offered by integrating the proposed methodology in IT and informative systems business modelling, taking into account the present and future high connectivity scenario.

References

[1] International Standard Organization ISO 50001:2011 Energy Management Systems – Requirements with Guidance for Use
[2] Carbon Trust CTG054 2011 Energy Management - a Comprehensive Guide to Controlling

Energy Use (Great Britain)

[3] Gordic´ D, Babic´ M, Jovicˇic´ N, Šušteršicˇ V, Koncˇalovic´ D, Jelic D 2010 Development of energy management system – Case study of Serbian car manufacturer *Energ. Convers. Manage.* **51** 2783

[4] Curry E, Conway G, Donnellan B, Sheridan C, Ellis K 2012 A maturity model for Energy efficiency in mature data centres *Proc. 1st International Conference on Smart Grids and Green IT Systems* (Porto) (SciTePress Digital Library) pp 263-267

[5] Ngai EWT, Chau DCK, Poon JKL 2013 Energy and utility management maturity model for sustainable manufacturing process *Int. J. Prod. Econ.* **146** 453

[6] Introna V, Cesarotti V, Benedetti M, Biagiotti S, Rotunno R 2014 Energy Management Maturity Model: an organizational tool to foster the continuous reduction of energy consumption in companies *J. Clean. Prod.* **83** 108

[7] Antunes P, Carreira P, Silva MM 2014 Towards an energy management maturity model *Energ. Policy* **73** 803

[8] Doty S, Turner WC 2013 *Energy management Handbook* (Lilburn: Fairmont Press)

[9] Perotto E, Canziani R, Marchesi R, Butelli P 2008 Environmental performance, indicators and measurement uncertainty in EMS context: a case study *J. Clean. Prod.* **16** 517

[10] D'Aponte F, D'Emilia G, Lupinetti S, Natale E, Pasqualoni P 2015 Detection threshold of a measurement method as a key indicator for diagnostic ability in converting machinery for diapers production, accepted for publication on *Int. J.Metrol.Qual. Eng.*

[11] Morvay ZK, Gvozdenac DD 2008 *Applied Industrial Energy and Environmental Management* (Chichester: John Wiley & Sons)

Thermal and optical characterization of natural and artificial marble for roof and external floor installations

F Asdrubali[1], G Baldinelli[1], F Bianchi[1], A Presciutti[1], F Rossi[1], S Schiavoni[1]
[1] University of Perugia, Department of Engineering, Via Duranti 67, Perugia, Italy

E-mail: francesco.asdrubali@unipg.it

Abstract. Some types of buildings need to use certain materials for aesthetic reasons, like churches or mosques. Marble is one of the most common materials usually installed on roofs and floors. The measurement of the thermal and optical characteristics can be useful to understand its behaviour when it is subjected to thermal loads such as solar radiation or high temperature winds. The paper shows a comparison study between natural and artificial types of marble, to investigate the thermal characteristics both in steady-state and transient conditions. Optical properties and surface emissivity were evaluated, in order to calculate the Solar Reflectance Index (SRI); the specific heat, the thermal conductivity and the density were measured to define the thermophysical properties useful for the dynamic analysis. Finally, a test bench was created to check the marble behaviour under known artificial irradiation.

1. Introduction

The constructions external layers, being exposed to solar radiation, are recently investigated with raising attention ([1], [2], [3], [4]) especially for applications in hot climates. For instance, the application of a layer of growing vegetation on the roof or on the walls of a building can led to advantages from both thermal and acoustic point of view [5]. A different approach is to increase the ratio of the energy reflected and emitted from the building envelope through the application of cool coatings [6]. All the efforts directed to reduce the cooling load in the built environment must take into account the heat load deriving from the sun [7], [8]. Within this framework, marble plays the double role of fine and cooling material, if chosen with suitable thermal and optics characteristics [9], [10]. Many studies have been conducted to evaluate the change of aesthetic features of the marble caused by chemical reactions [11], the consequences of high temperatures and thermal cycling on structural properties [12], [13], [14]. The marble thermophysical performance assessment needs the knowledge of various parameters [15] whose combination could give a complete information also in dynamic conditions, the most common situation encountered in summer season [16].
Natural and an artificial marble samples were analysed in order to define the thermophysical properties describing the transient thermal behaviour and the marble response to solar irradiation. A set of several tests were performed on the two types of marble, according to the specific standard that rules each measurement. The characterization of the samples was useful to compare the natural and artificial composition, as well as to study two models for describing the transient thermal behaviour: an analytical approach and finite volume simulations; results were then compared with measured data, obtained by a laboratory experimental setup.

2. Thermal and optical characterization of natural and artificial marble

2.1. Thermal conductivity

The homogeneous materials thermal conductivity is commonly measured with a guarded hot plate equipment, where two large square-shape samples are used: 500 x 500 x 30 mm for both natural and artificial marbles. The device creates a temperature gradient on the sample surfaces with a known heat flux that passes through the sample itself. The apparatus can use single or double samples, according to the Standard EN 12664 [17]; for the analysed marble, a single specimen setup was chosen. The equipment is composed by a heating and a cooling unit (figure 1), in order to create the desired temperature gradient.

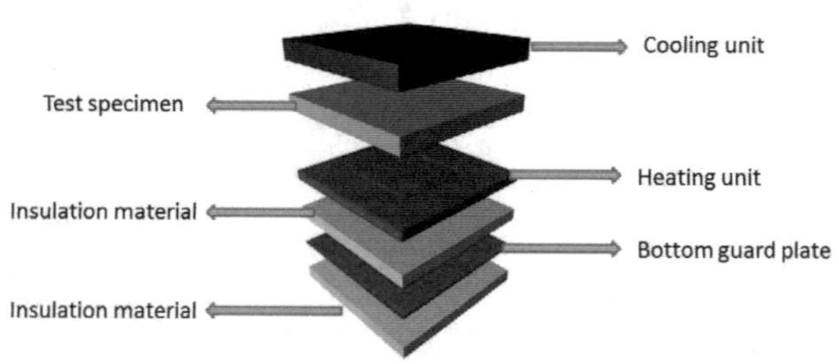

Figure 1. Guarded hot plate equipment.

The heating unit (figure 2) is composed by a metering section (250 x 250 mm) and a guard section, aimed at avoiding lateral losses; besides, another bottom guard hot plate is positioned to prevent the creation of vertical fluxes in the direction opposite to the specimen.

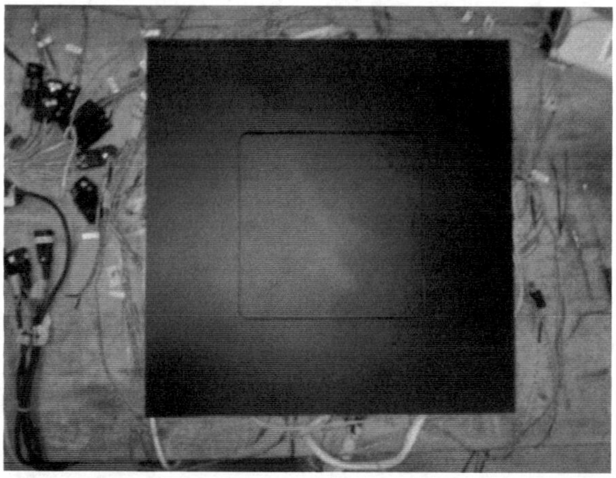

Figure 2. Guarded hot plate: metering section and guarding section.

The control and acquisition system were developed in the LabView environment [18], by a specific software that allows the control and acquisition of the various parameters, thanks to the reading and the recording of the 43 temperature probes and the current and voltage probes, installed to evaluate the heat flux. During the tests, when the system reaches the steady state conditions, the specimen thermal

conductivity is calculated by the one-dimensional Fourier expression, with the temperatures and the heat flux of the measuring zone.
Table 1 shows the thermal conductivities of the natural and artificial marbles, with the measured uncertainties, calculated according to the Standard ENV 13005 [19].

Table 1. Results of thermal conductivity tests.

	Thermal conductivity (W/m K)
Natural marble	1.46 ± 0.16
Artificial marble	1.08 ± 0.08

2.2. Specific heat
The specific heat of solid materials such as marble can be measured by a Hot Disk [20] setup, an instrument based on the Transient Plane Source (TPS) method [21], [22]. During the measurement, the hot disk sensor is sandwiched between two identical samples and it constitutes at the same time the heater and the recorder of the resistance increase as a function of time. An electrical current flows in the sensor spiral to increase the temperature between a fraction of a degree up to several degrees. With a specific algorithm based on the transient behaviour of the system, the instrument measures the thermal conductivity, the diffusivity and the volumetric specific heat. The volumetric specific heat was obtained and the specific heat in [J/kg K] has been calculated with the density measurement. Respect to the thermal conductivity assessment, different samples were used for this measurement: 100 x 100 x 10 mm square-shape for both types of marble. The natural marble has a density of 3020 kg/m^3, with a specific heat of 773 J/kg K; the artificial one has a density equal to 2730 kg/m^3 with the specific heat equal to 706 J/kg K. The uncertainty of the measurements is equal to 5% according to the Hot Disk specifications [20].

2.3. Emissivity and optical properties
2.3.1. Emissivity. The method for the measurement of the surface total emissivity is regulated by the Standard ASTM C1371 [23], that defines the procedures and the suitable instruments. The device consists of a circular head with a diameter of 50 mm, heated by an electric power supplier up to a temperature of 355 K. During the measurement, the sample surface and the measurement head surface remain separated by an air layer of about 4.3 mm, confined by the circular crown of the plastic cylinder that encases the entire head and facilitates its use. The instrument includes a heat flow differential gauge made of two pairs of sensors, that returns the difference between the voltage signals produced by the sensors themselves: one couple with the same surface treatment of the head (black-opaque, high-emissive), the other couple with a golden low-emissive treatment (figure 3).

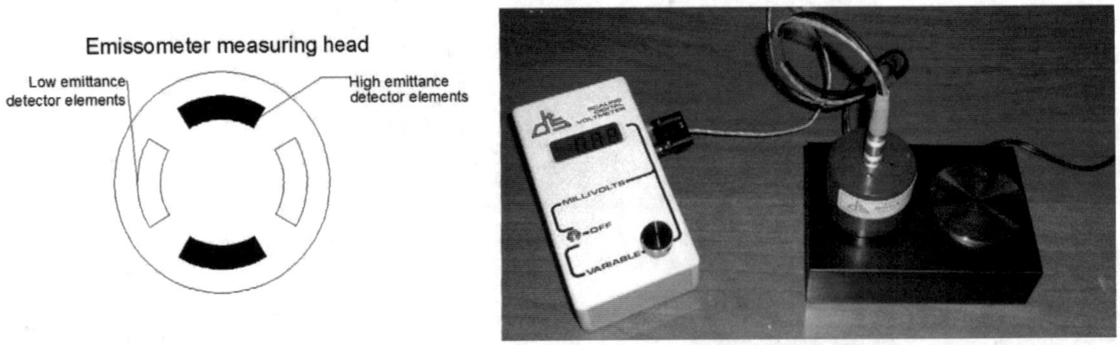

Figure 3. Calibration phase of the emissometer.

By means of this measurement it is possible to assess the radiation heat exchange, since the heat flux registered by the first pair includes both the conduction and radiation heat transfer, while the second pair evaluates contribution of the the only conduction; the small thickness of the cavity does not allow the activation of convection.
From the knowledge of the radiation heat flow and the surfaces temperatures (the sample temperature must be kept close to the ambient temperature), the emissivity of the sample can be determined.
The two tested samples (square-shape 100 x 100 x 10 mm) showed similar values of total emissivity. The emissivity for the natural marble is 0.88 and for the artificial one is 0.85, both with an uncertainty equal to ±0.03.

2.3.2. Optical properties. The optical characterization of marble samples (square-shape 100 x 100 x 10 mm) were carried out through specular gloss and reflectance. Specular gloss is used to evaluate the tendency of an object to reflect light in a specular (mirror-like) direction. This optical property was measured in terms of Gloss Unit (GU) using a ETB-0833 triple angle (20°, 60° and 85°) Gloss Meter in compliance with the ASTM D523-14 standard [24]. The instrumentation detects gloss between 0 GU (perfectly matte surface) and 200 GU (high gloss behaviour). Reflectance was evaluated using a double beam spectrophotometer in the wavelength range 300-2500 nm. The instrumentation allows to evaluate the amount of energy reflected by a solid sample, in comparison to a standard material. In this case, a $BaSO_4$ flat plate was used, which absolute reflectance coefficient is certified in the analyzed wavelength range. An estimation of the absolute reflection coefficient of the assessed samples was then calculated multiplying the absolute reflectance coefficient of the $BaSO_4$ reference sample with the measured reflection coefficient of the assessed samples. These elaborated data were used to calculate the single rating parameters solar reflectance ρ_s (according to ASTM E903-12 [25]) and light reflectance ρ_v (according to EN 410:2011 [26]). The first one allows to easily evaluate the sample reflection between the whole examined wavelength range, whereas ρ_v is calculated considering only the wavelengths perceivable to human eyes, fixed [26] in the range 380-780 nm. Several measurements were performed for each sample type to evaluate the uncertainty. The results are reported in table 2.

Table 2. Optical characterization of artificial and natural marble

Marble type	Parameter	Value
	20° Specular Gloss	101.0 ± 7.0 GU
	60° Specular Gloss	95.7 ± 8.4 GU
Artificial	85° Specular Gloss	91.9 ± 11.9 GU
	Solar reflectance ρ_s	82.00 ± 0.14 %
	Light reflectance ρ_v	86.90 ± 0.12 %
	20° Specular Gloss	0.9 ± 0.1 GU
	60° Specular Gloss	1.7 ± 0.2 GU
Natural	85° Specular Gloss	0.8 ± 0.3 GU
	Solar reflectance ρ_s	84.39 ± 0.41 %
	Light reflectance ρ_v	87.26 ± 0.18 %

Data reported in table 2 shows that artificial and natural marble reflect practically a similar quantity of light, but not in the same way. Solar reflectance and light reflectance values of the two assessed samples are quite similar, whereas specular gloss values are deeply different. Natural marble has a matte surface avoiding specular reflections; this aspect is crucial in some practical application, such as pavements of long corridors artificially illuminated. In this case, a specular gloss value close to 0 GU is advisable, since it avoids uncomfortable luminaire reflections on people walking on the corridor. As far as reflection properties, both materials absorb less than the 20% of the energy in the 300-2500 nm wavelength range and reflect only above the 13% of the radiation perceivable to human eyes.

2.4 Solar Reflectance Index (SRI). According to ASTM E1980-11 [27], it is possible to calculate the Solar Reflectance Index from the measurements of total emissivity and solar reflectance. These parameters are related to the surface temperature of the specimen, since they represent respectively the capacity of the surface to emit energy to the environment and the effect of solar radiation on the surface itself. The Standard reports two different approaches for the calculation of *SRI*. The index in standard solar and ambient conditions (solar flux equal to 1000 W/m^2, ambient air temperature equal to 310 K and sky temperature equal to 300 K) is evaluated with the following equation:

$$SRI = 123.97 - 141.35\chi + 9.655\chi^2 \tag{1}$$

where:

$$\chi = \frac{((1-\rho_s) - 0.029\varepsilon)(8.797 + h_c)}{9.5205\varepsilon + h_c} \tag{2}$$

ρ_s and ε represent respectively the measured solar reflectance and thermal emissivity. The term h_c is the convective coefficient that, according to the Standard, could assume three different values: 5, 12 and 30 W/m^2 K, to take into account of different wind conditions. Table 3 shows the results of SRI for the two marble samples with the final uncertainties calculated according to the Standard ENV 13005 [19].

Table 3. Results of the calculation of the Solar Reflectance Index (SRI).

Solar Reflectance Index (SRI)	**h_c = 5 W/m^2 K**	**Uncertainty**	**h_c = 12 W/m^2 K**	**Uncertainty**	**h_c = 30 W/m^2 K**	**Uncertainty**
Natural marble	105	± 0.79	105	± 0.70	105	± 0.63
Artificial marble	101	± 0.65	101	± 0.48	102	± 0.35

3. Models

3.1. Experimental setup
Thermal measurements were carried out in order to evaluate the surface temperature of marble under constant solar irradiation. The solar spectrum was provided using the Sun simulator Solar 2000, (ABET Technology, Model 11018) which generates a standard solar spectrum according to the IEC60904-9 (solar simulator performance requirements).
The sample (100 mm x 100 mm x 10) was placed into an insulated cylinder under the sun simulator lamp. The sample distance from the lamp was 25 cm, to obtain a radiation equal to the maximum value of 1200 W/m^2 on the sample surface.

Three thermocouples were attached on the sample surface, as shown in figures 4 and 5. At the same time, a fourth thermocouple was used to measure the air temperature at 10 cm from the marble surface.

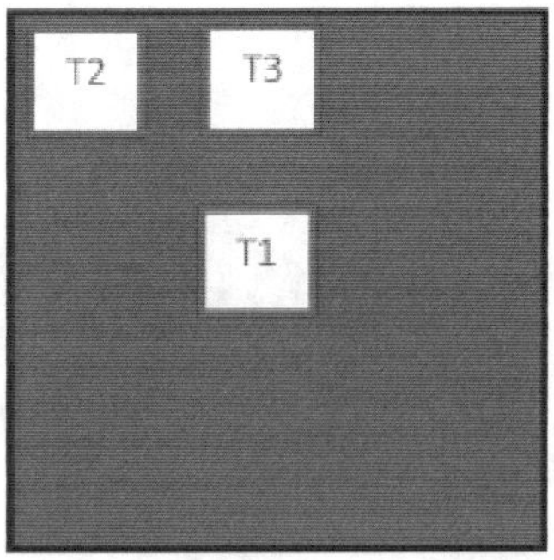

Figure 4. Positions of temperature probes on the surface sample.

Figure 5. Temperature probes on the surface sample (with the thermocouple for air temperature measurement).

As soon as the shutter was opened, the radiation value was measured with a global radiation probe. Hence, the four values of temperatures were acquired each 10 seconds, until the average surface temperature reached a value that remained constant for at least 20 minutes. Afterwards, the radiation was verified at the end of measurement, in order to check its steadiness during the acquisition. Lower radiation values were obtained inserting a filter into the light beam.

3.2. Analytical model

The marble tile irradiated with a constant heat flux could be modelled, in first approximation, with the Lumped Capacitance Method [28]; the equations driving the phenomenon could be written as follows:

$$\frac{dU}{dt} = \frac{dQ}{dt} + \alpha W_i \tag{3}$$

$$\rho CV \frac{dT}{dt} = -hA[T(t) - T_\infty] + \alpha W_i \tag{4}$$

Where U is the internal energy of the entire tile, Q is the heat transferred to the external environment, t is the time, α is the solar absorbance, W_i is the incident radiation, ρ is the marble density, C is the marble specific heat, V is the tile volume, h is the heat transfer coefficient between the upper surface of the marble tile and the external air, hypothesizing that there are no thermal losses downwards, A is the tile upper surface, T is the tile temperature, T_∞ is the air temperature (supposed constant) and T_0 is the tile temperature at the initial instant.

The initial boundary condition of $T = T_\infty$ for $t = 0$, brings to the solution:

$$T(t) - T_\infty = (T_0 - T_\infty - \frac{\alpha W_i}{hA})e^{-\frac{hAt}{\rho CV}} + \frac{\alpha W_i}{hA} \tag{5}$$

It is therefore possible to follow the transient of the marble tile temperature, once the parameters above described are known.

3.3. CFD Analysis

The marble transient behaviour was also studied by a finite volumes method by means Ansys Fluent [29]. The analysis was carried out with the definition of the 2-D geometrical model of the marble and a gridding process with hybrid meshes (figure 6). The thermophysical properties of the two types of marble, measured by the methods showed in the previous paragraph, were inserted in the material properties of the virtual domain.

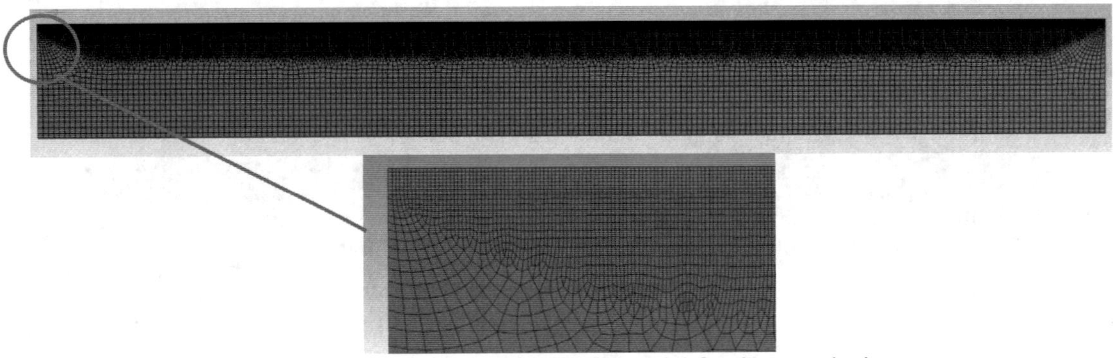

Figure 6. Grid of the 2-D marble model for CFD analysis.

At the aim of simulating the heat flow density (W/m^2) that hits the surface of the model, a thin layer of marble was created on top of the domain. An internal heat source was generated in this thin layer with the same value of the radiation heat flow, reduced by the solar reflectance index of the considered marble sample.

The initial temperature of the simulation was chosen by the values retrieved from the experimental analysis, to make the temperature transient comparable. Furthermore, the convective boundary conditions on the top surface of the marble were assigned (and considered constant during the simulations), hypothesizing that the marble in the CFD model reaches the same asymptotic temperature of the experimental data after two hours of simulation. Adiabatic boundary conditions were given for the other three sides. It is necessary to underline that for the comparison between the numerical analysis and the experimental setup, the density of the marble in the CFD model was increased to take into account of the thermal inertia of the polystyrene and the metal structure positioned under the test sample. The time-step of the simulations was setup to 1 second, with a complete transient analysis of around 6300 seconds.

4. Comparative analysis of the different approaches

The two models were compared to assess the transient thermal behaviour of the marbles. In particular, on the basis of the experimental approach, the aim of the work consists of evaluating the agreement of the models with measured data. Figures 7 and 8 show the trend of temperature on the top surface of the marbles, according to the three approaches, when a heat flow density equal to 1100 W/m^2 hits the sample. The boundary conditions of the two marbles were defined by the experimental analysis in the final steady state conditions and they were assigned to the analytical and numerical models. The two models need to define the same boundary conditions, but they have different approaches to the phenomenon: the analytical model - Lumped Capacitance Method [28] applies the heat transfer coefficient h to the whole marble perimeter and it considers the entire solid at the same temperature each time-step. On the other hand, the numerical model applies the heat transfer coefficient to the top surface of the marble, according to the real behaviour of the experimental test, creating a temperature gradient inside the solid. Therefore, the numerical analysis needs to define the boundary conditions in the other three sides, which, according to the experimental setup, were considered adiabatic. Table 4 shows the convective boundary condition parameters.

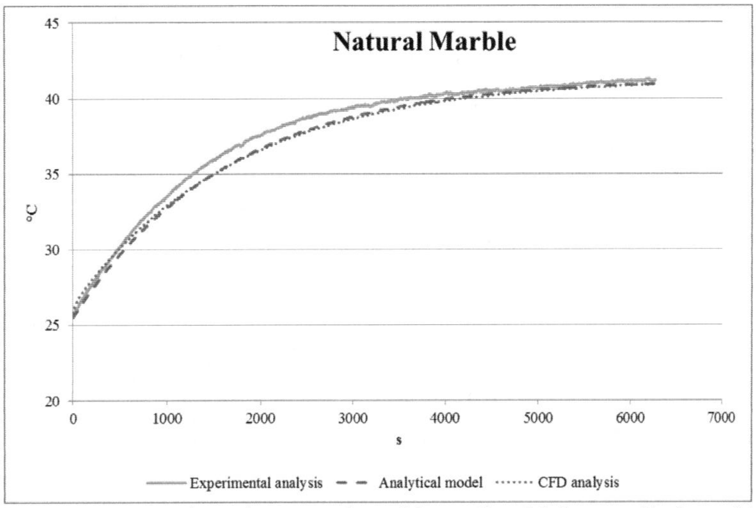

Figure 7. Temperature trend on the natural marble surface hit by a radiation of 1100 W/m^2.

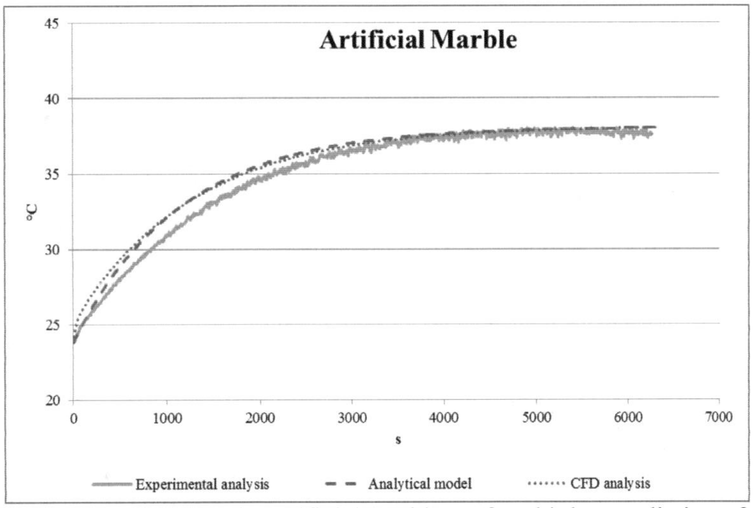

Figure 8. Temperature trend on the artificial marble surface hit by a radiation of 1100 W/m^2.

Table 4. Boundary conditions measured during the experimental analysis.

	Heat transfer coefficient (W/m^{2}K)	Free steam temperature (°C)
Natural marble	27.58	34.90
Artificial marble	33.67	32.19

The trend of the analytical and numerical models are generally comparable with the experimental results for both types of marble. The slight differences are linked to the models approximations and to the definition of the intrinsic materials properties. In particular, the analytical and numerical models consider a constant heat transfer coefficient value for each time-step of the analysis, whereas in actual conditions the heat transfer coefficient varies with time. The other factor that may contribute to the trend differences is the definition of the material thermal properties that, as previously seen, are evaluated with tests affected by a level of uncertainty.

5. Conclusions

An experimental and numerical comparative analysis on natural and artificial types of marble was conducted. The characterization of the samples properties was performed with several methods, in order to evaluate the main parameters that can affect the thermal behaviour of the materials hit by solar irradiation. The samples steady-state and transient thermal response depend on the thermal conductivity, the specific heat and the density; all these properties were measured according to the corresponding Standards. By means of these parameters it was possible to calculate the thermal diffusivity, which resulted equal to 6.25 x 10^{-7} m^2/s for the natural marble and 5.60 x 10^{-7} m^2/s for the artificial one, highlighting a quicker propagation of the thermal wave in the natural sample.
The calculated SRI (the main parameter that describes the energy balance of a surface hit by solar radiation) of tested marbles resulted higher than 100, proving its high performance as reflective material for roof coverings. SRI values higher than 100 testify that the marble cooling capacity is better than the standard white surface [30]. Although the experimental measurements did not follow exactly the SRI standard conditions, the excellent behaviour of the tested marbles is confirmed by the measured surface temperatures (almost 39 - 41°C), values lower than the white standard surface temperature (44,7°C). This feature makes the marble suitable for cool roofs or coating of outdoor surfaces, where people can walk barefoot during the summer season, such as swimming pools or religious places (mosques and churches). The comparison of the SRI between the two typologies of marble indicates that the natural marble, when subjected to the same solar irradiation of the artificial one, reaches a slightly lower temperature. The comparative analysis of a theoretical model (Lumped Capacitance Method) and a finite volume method with experimental data showed encouraging results on the thermal transient assessment, with small differences mainly due to the simplifying hypotheses and to the uncertainty of the marble thermophysical properties measurements. Thus, the analytical and the numerical model could constitute an easy and quick tool to evaluate the transient and stationary response of materials subjected to solar irradiation.

6. References

[1] Asdrubali F, Baldinelli G, Bianchi F and Sambuco S 2015 A comparison between environmental sustainability rating systems LEED and ITACA for residential buildings *Build. Environ.* **86** 98–108

[2] Baldinelli G 2010 A methodology for experimental evaluations of low-e barriers thermal properties: Field tests and comparison with theoretical models *Build. Environ.* **45** 1016-24

[3] Pisello A L, Castaldo V L, Pignatta G, Cotana F and Santamouris M 2015 Experimental in-lab and in-field analysis of waterproof membranes for cool roof application and urban heat island mitigation *Energy Build.* Article in Press doi:10.1016/j.enbuild.2015.05.026

[4] Asdrubali F, Cotana F, Pisello A L , Mencarelli N, D'Alessandro F and Schiavoni S 2015 Acoustic properties of stone aggregates used in cool roofs *Proc. 22st ICSV* 12-16 July Florence Italy

[5] Asdrubali F, Schiavoni S and Horoshenkov K V A 2012 review of Sustainable Materials for Acoustic Applications *Build. Acoustics* **19 (4)** 283-312

[6] Synnefa A, Santamouris M and Akbari H 2007 Estimating the effect of using cool coatings on energy loads and thermal comfort in residential buildings in various climatic conditions *Energy Build.* **39 (11)** 1167-1174

[7] Rossi F, Castellani B, Presciutti A, Morini E, Filipponi M, Nicolini A and Santamouris M 2015 Retroreflective façades for urban heat island mitigation: Experimental investigation and energy evaluations *Appl. Energy* **145** 8–20

[8] Baldinelli G, Bonafoni S, Anniballe R, Presciutti A, Gioli B and Magliulo V 2015 Spaceborne detection of roof and impervious surface albedo: Potentialities and comparison with airborne thermography measurements *Sol. Energy* **113** 281–94.

[9] Careddu N and Marra G 2013 The effects of solar UV radiation on the gloss values of polished stone surfaces *Constr. Build. Mater.* **49** 828–834
[10] Krimpalis S and Karamanis D 2015 A novel approach to measuring the solar reflectance of conventional and innovative building components *Energy Build.* **97** 137-145
[11] Bams V and Dewaele S 2007 Staining of white marble *Mater. Charact.* **58** 1052–62
[12] Rodríguez-Gordillo J and Sáez-Pérez M P 2006 Effects of thermal changes on Macael marble: Experimental study *Constr. Build. Mater.* **20** 355–65
[13] Ferrero A M, Migliazza M and Spagnoli A 2009 Theoretical modelling of bowing in cracked marble slabs under cyclic thermal loading *Constr. Build. Mater.* **23** 2151–9
[14] Ozguven A and Ozcelik Y 2013 Investigation of some property changes of natural building stones exposed to fire and high heat *Constr. Build. Mater.* **38** 813–21
[15] Amaral P, Correia A, Lopes L, Rebola P, Pinho A and Carrilho Lopes J 2013 On the Use of Thermal Properties for Characterizing Dimension Stones *Key Eng. Mat.* **548** 231-238
[16] Gobakis K, Kolokotsa D, Maravelaki-Kalaitzaki N, Perdikatsis V and Santamouris M 2015 Development and analysis of advanced inorganic coatings for buildings and urban structures *Energy Build.* **89** 196–205
[17] EN 12664 2001 Thermal performance of building materials and products - Determination of thermal resistance by means of guarded hot plate and heat flow meter methods - Dry and moist products of medium and low thermal resistance
[18] NI LabView Environment – Virtual Instruments – National Instruments 2015
[19] ENV 13005 1999 - ISO Guide to the expression of uncertainty in measurement
[20] Hot Disk, Thermal Constants Analyzer – Instruction Manual, Revision date 2014-03-27
[21] ISO 22007-2 2008 Plastics - Determination of thermal conductivity and thermal diffusivity - Part 2: Transient plane heat source (hot disc) method
[22] Gustafsson S E 1991 Transient plane source techniques for thermal conductivity and thermal diffusivity measurements of solid materials **Rev. Sci. Instrum. 62(3)** 797
[23] ASTM C1371 - 04a 2010 Standard Test Method for Determination of Emittance of Materials Near Room Temperature Using Portable Emissometers *ASTM International* West Conshohocken PA
[24] ASTM D523-14 2014 Standard Test Method for specular gloss *ASTM International* West Conshohocken PA
[25] ASTM E903-12 2012 Standard Test Method for Solar Absorptance, Reflectance, and Transmittance of Materials Using Integrating Spheres *ASTM International* West Conshohocken PA
[26] EN 410 2011, Glass in building - Determination of luminous and solar characteristics of glazing
[27] ASTM E1980 – 11 2011 Standard Practice for Calculating Solar Reflectance Index of Horizontal and Low-Sloped Opaque Surfaces *ASTM International* West Conshohocken PA
[28] Rohsenow W M, Hartnett J P and Cho Y I 1998 *Handbook of heat transfer* (United States of America: McGraw-Hill)
[29] Fluent version 14.5 2014 User's guide USA
[30] Akbari H, Levinson R, and Berdahl P 1996 ASTM Standards for Measuring Solar Reflectance and Infrared Emittance of Construction Materials and Comparing their Steady-State Surface Temperatures *Energy and Environment Division* American Council for Energy Efficient Economy Summer Study

Experimental investigation of a packed bed thermal energy storage system

Mario Cascetta[1], Giorgio Cau[1], Pierpaolo Puddu[1], Fabio Serra[2]

[1]Dep. of Mechanical, Chemical and Materials Engineering (DIMCM), University of Cagliari, via Marengo, 2, 09123 - Cagliari, Italy
[2]Solar Concentration Technologies and Hydrogen from RES Laboratory, Sardegna Ricerche - Z.I. Macchiareddu 09010 Uta (CA) - Italy

E-mail: fabio.serra@sardegnaricerche.it

Abstract. In this work experimental investigations on a thermal energy storage system with a solid material as storage media and air as heat transfer fluid will be presented. The experimental test rig, installed at the DIMCM of the University of Cagliari, consists of a carbon steel tank filled with freely poured alumina beads that allows investigations of heat transfer phenomena in packed beds. The aim of this work is to show the influence of the operating conditions and physical parameters on thermocline formation and, in particular, the thermal behaviour of the thermal energy storage for repeated charging and discharging cycles. Better charging efficiency is obtained for lower values of mass flow rate and maximum air temperature and for increasing aspect ratio. A decreasing influence of the metal wall with continuous operation is also highlighted. In conclusion, the analysis focuses on the thermal hysteresis phenomenon, which causes degradation of the thermocline and the reduction of the energy that can be stored by the accumulator as the repeated number of cycles increases.

1. Introduction

Energy storage is assuming a fundamental role in power generation from renewable energy sources owing to their intermittent and non-programmable nature. Energy storage can be effectively used to decouple electricity production and demand, meet random fluctuations in demand and reduce part-load generation in fossil fuel power plants [1, 2].

In particular, thermal energy storage (TES) systems allow storage of heat for later use and are useful in managing the decoupling between the power required by users and that produced by a solar field [3, 4]. Moreover, TES can be used in numerous commercial and industrial applications, often integrated with conventional energy sources to achieve major reliability by reducing peaks of electricity generation, increasing generation capacity and reducing production costs.

There are several methods for storing thermal energy that can be divided into physical and chemical processes [5]. In every type, a sequential process of charge, storage and discharge is present, and this needs to be reversible to recover the energy stored. Physical processes involve sensible heat and latent heat. The former is by far the most common way of TES: the energy is stored as sensible heat of a liquid or solid material and represents the simplest and least expensive form of thermal storage [6]. When the energy is stored in the form of latent heat, a phase change of the storage material occurs [5, 7]. The chemical type of storage exploits the heat of reaction of a reversible reaction; it has not yet been extensively studied but it will become an attractive option at relatively low cost in the future [6, 8]. Although technologies based on latent heat and chemical energy are considered the most promising, some technological and economic aspects make sensible heat storage superior [9, 10].

The thermal storage system investigated herein consists of a tank filled with solid beads (packed bed) crossed by air used as heat transfer fluid (HTF). In past decades, flow and heat transfer in packed beds, including random and structured packings, were extensively investigated by many researchers [11]. A packed bed with rocks generally represents the most suitable energy storage solution for air-based solar systems, with the aim to provide thermal comfort inside residential buildings [12].

The most common solution for TES section in concentrating solar power (CSP) plants is today represented by two tank systems with thermal oil or molten salts employed as storage medium [3, 4, 7]. To reduce the cost of the TES section, a thermocline system, based on a single-tank packed bed containing a low-cost filling material can be used [13-16]. In thermocline systems, during the charging phase, the hot HTF flowing from the top of the tank gradually heats the filling material before exiting from the bottom of the tank. A high temperature zone, located in the upper part of the tank, is separated from the low temperature one located in the lower part of the tank by a thermal gradient (thermocline) that moves downwards. During the discharging phase, the flow is reversed and the cold HTF is pumped from the bottom of the tank so that the thermocline moves upwards. The use of low-cost filling materials in a single-tank system reduces the cost of the TES section and the volume of HTF required [15, 16].

Beds packed with rocks are particularly suitable in CSP plants when air is used as HTF [17]. Indeed, these systems allow elimination of operating temperature or pressure constraints due to chemical instability of the HTF or solid material or their interaction and elimination of the need for a heat exchanger between the HTF and the storage medium. On the other hand, owing to the lower volumetric heat capacity and thermal conductivity of air compared to thermal oil or molten salts, a higher air mass flow rate and surface area are required, with higher pressure drops and energy losses.

However, owing to the temperature gradient inside the tank, thermocline systems are less efficient than two-tank systems because of the presence of an unexploited zone of the tank and the progressive reduction of the useful energy during continuous operation due to thermal hysteresis [14, 17, 18].

In this work, an experimental study was carried out to explore how operating conditions, such as mass flow rate or maximum HTF temperature, and physical parameters, such as the aspect ratio of the tank, influence the thermocline formation and, in particular, the thermal behavior of the TES in the case of repeated charging and discharging cycles.

2. Experimental setup and instrumentation

The experimental setup built at DIMCM is shown schematically in Figure 1. It consists of an open circuit where, by operating some valves, it is possible to perform single phases and complete cycles of charging and discharging of the thermal storage.

The TES system consists of an insulated steel tank filled with small alumina spheres. The experimental set-up, equipped with a variable speed screw compressor and an electric heater, allows investigation of heat transfer performance of packed beds using air as heat transfer fluid. An inverter allows changing of the compressor speed to modify the delivered flow rate from 255 to 940 m^3/h of free air. The mass flow rate is evaluated by an orifice plate (Q in Figures 1 and 2) with an uncertainty of about 2.5% of the read value. A PID controller allows modulation of the electrical heater mean power and setting of the air flow temperature to the desired value up to 300 °C. Table 1 provides the most relevant data of the rig.

The storage material is made of commercial sintered alumina beads with a diameter in the range 7-9 mm (Al_2O_3 89.5 wt %) freely poured into the tank with a bed void fraction in the range 0.385-0.395. The net volume of the porous bed inside the reservoir is about 0.5 m^3, 0.58 m in diameter and 1.8 m in height. A layer of 100 mm mineral wool covers the carbon steel tank to minimize thermal losses to the outside. Two flow distributors placed at the top and bottom of the reservoir spread the air uniformly in the porous bed. The temperature profile in the storage tank is carefully investigated using nineteen T-type thermocouples placed along the vertical axis 100 mm from each other in a metallic rack as shown in Figure 3. Five more thermocouples are placed along a radius, with a decreasing distance from each other close to the wall to investigate the influence of the wall on the temperature profile. Two homemade feed-throughs link the sensors to the outside. The external wall temperature is measured by ten K-type thermocouples along the axial direction and by five K-type thermocouples in the circumferential direction at about 600 mm from the top of the bed. The sensors are fastened using a special high-temperature cement. Other thermocouples are used to measure flow temperature at different locations of the plant to handle the valves before starting the charging or discharging phases.

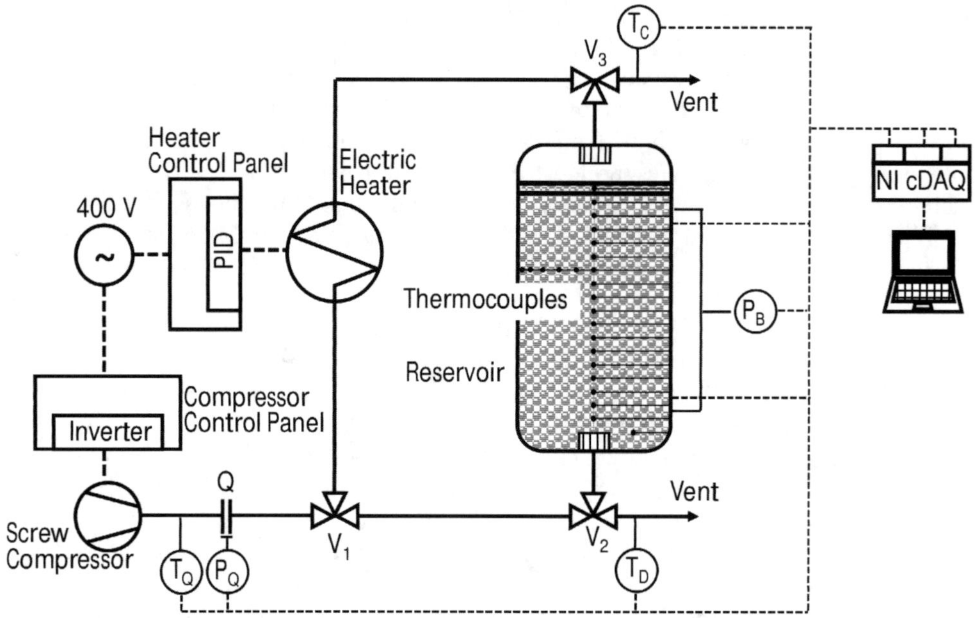

Figure 1. Schematic of the experimental setup.

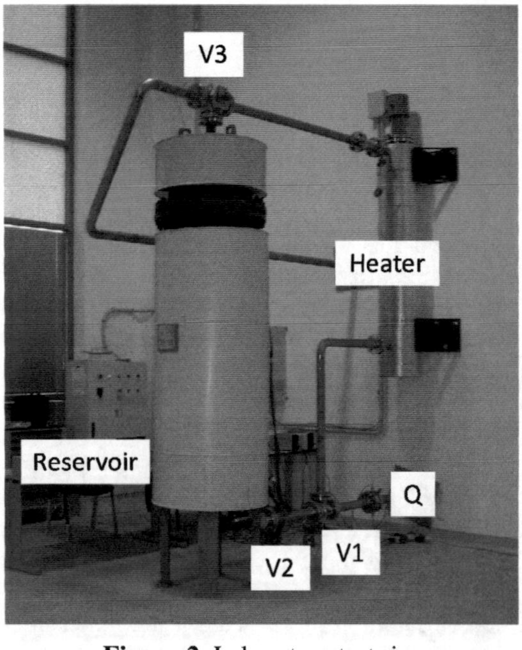

Figure 2. Laboratory test rig.

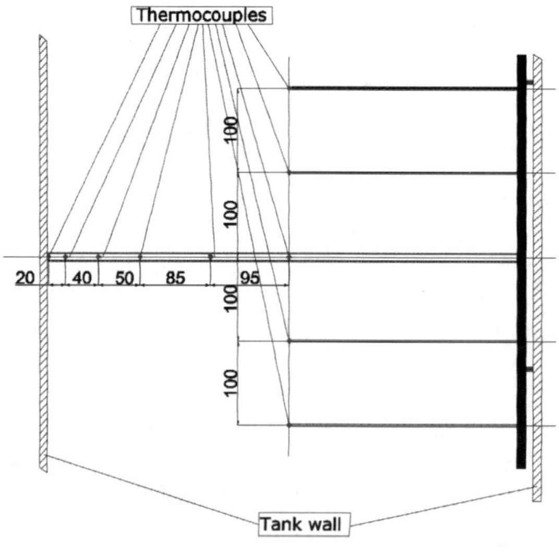

Figure 3. Thermocouple rack.

Table 1. Technical data of rig components.

Compressor delivered flow	255-940 m^3/h, FAD
Electrical heater max. power	70 kW
Maximum air temperature	300 °C
Bed height	1800 mm
Bed diameter	584 mm
Alumina bead diameter	7-9 mm
Alumina bead density	3550 kg/m3
Alumina bead heat capacity (100 °C)	902 J/kgK
Bed void fraction	0.385-0.395
Tank coating thickness (Mineral wool)	100 mm

All the thermocouples are calibrated in the range of 25-300 °C with an uncertainty of ±1°C. A differential pressure transmitter detects the pressure drop along the bed. All signals are acquired by a National Instruments Compact DAQ USB chassis with several modules for thermocouples and analog inputs. The most important recorded signals are monitored with a Graphical User Interface developed using LabVIEW software.

The experimental set-up is suitable for investigating and evaluating performance of the storage system under different operating conditions obtained by varying the mass flow rate, aspect ratio, particle dimension and temperature range during the charging and discharging phases.

Three 3-way stainless steel full-bore valves allow performance of a single phase of charge or discharge as well as repeated sequences of charge and discharge cycles. At the beginning of the first charging phase, the initial temperature of the bed is stabilized sending the air directly from the compressor to the tank. The air flows through valves V1 and V2 from the bottom to the top of the tank before being sent into the atmosphere through valve V3. After this first preliminary phase, the piping connecting the heater to the accumulator is heated until a stable temperature is reached. Subsequently, valve V1 connects the compressor to the heater to allow an increase in air temperature to the desired value, while the air is sent to the vent through valve V3. When the temperature detected by thermocouple TC (Figure 1) reaches the desired value, valve V3 is actuated to start the charging phase. The air moving along the porous bed from the top to the bottom heats the solid material before being vented through V2. During the discharging phase the valves are set as in the preliminary phase to recover the thermal energy stored in the accumulator, sending a flow of cold air from the bottom to the top of the bed. The minimum fluid temperature is determined by the compressor delivery.

3. Experimental characterization and discussion

Defining T_{max} and T_{min} respectively as the maximum and the minimum temperature detected in the porous bed, the temperature distribution can be reported in the dimensionless form (eq. 1):

$$\theta = \frac{T - T_{min}}{T_{max} - T_{min}} \tag{1}$$

Figure 4 (a) depicts the time evolution of the temperature profile along the bed during a typical charge-discharge cycle. During the charging phase, the temperature increases from the top to the bottom of the bed. This phase was stopped when the temperature at the bottom of the tank reached a threshold value given by the tolerance with respect to the minimum temperature allowed by user's needs. Likewise, the discharging phase ended when the temperature at the top of the tank reached a threshold value defined in the same manner with reference to the maximum temperature. In this case a threshold tolerance of about 2.5% (θ=0.025 and 0.975) of the temperature range considered (about 200°C) was assumed. It is possible to observe that the temperature at the top (x/L = 0) changes in time to the maximum value due to the thermal inertia of the upper part of the tank. The discharging phase ended when the temperature at the outlet of the accumulator reached the threshold value. The area between the curves of the temperature distribution at the end of charging and discharging phases is indicative of the useful energy actually recovered during the entire cycle, and shows that most of the bed is not exploited. The amount of stored energy depends on several process parameters, such as mass flow rate and temperature thresholds, inlet–outlet temperature difference and physical parameters, such as aspect ratio AR (length to diameter ratio L/D), bed void fraction and particles diameter. The influence of some of these parameters will be presented below.

Figure 4 (b) shows the time evolution of the temperature distribution along the radial direction. In about 40% of the radius the temperature falls from an almost constant value to the external wall temperature value. This behavior leads to a further reduction of useful energy. The wall effect along the radius obviously depends strongly on the constructive characteristics of the reservoir and on the working fluid properties, with particular disadvantages for gases with respect to liquids [14, 19].

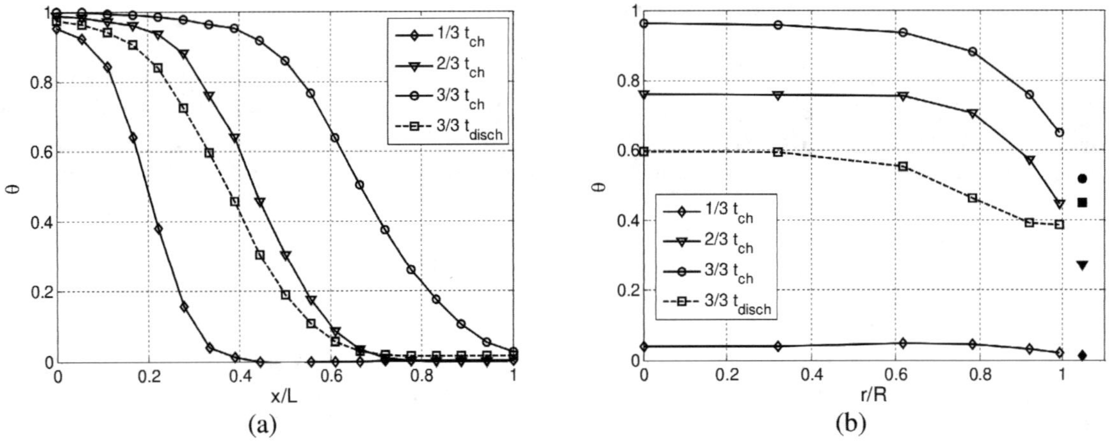

Figure 4. Temperature distribution during a typical charging-discharging cycle (T_{max} 237°C, T_{min} 38°C, mass flow rate 0.15 kg/s) (a) axial distribution (b) radial distribution (markers placed beyond r/R=1 refers to external wall).

A more effective way to improve the accumulator storage capacity is to increase the threshold tolerance. This solution obviously depends on the value of the temperature threshold that can be accepted by the user. Figure 5 shows a noteworthy increase in the energy stored which is obtained when the outlet dimensionless temperature is equal to 0.2 and 0.3. As shown below, this condition gives benefits also in continuous operation. The same figure reports the effect of mass flow rate on the energy stored in the tank. By reducing the mass flow rate the energy charged increases, and this effect is more evident at lower temperature thresholds. A higher mass flow rate means higher flow velocities, higher convective heat transfer and lower charging time. However, in the case of low mass flow rate the lower velocities determine a lower flux of energy through the bed thus enhancing the release of energy in a smaller portion of the bed, with higher temperature gradients and a steeper temperature profile.

The aspect ratio influence was evaluated by considering three different heights of the bed during a charging phase: L=D (AR=1), L=2D (AR=2) and L=3D (AR=3). Figure 6 shows the temperature distribution along the corresponding heights of the bed during the charging phase, when the threshold values (θ=0.01 and θ=0.2) are reached. A higher aspect ratio leads to higher energy stored with a most significant improvement at lower temperature thresholds. This behavior can be explained by taking into account that the thermal process is characterized by the time constant τ, defined as the ratio between the thermal energy transferred to the storage material and the flux of thermal energy exchanged, i.e. the ratio between the thermal capacity of the storage material and the convective heat transfer of the fluid.

$$\tau = \frac{M_b \cdot c_b}{h_{fb} \cdot A} = \frac{\rho_b \cdot (1-\varepsilon) \cdot L \cdot c_b}{h_{fb}} \tag{2}$$

In eq. 2 M_b, c_b and ρ_b are respectively the mass, the specific heat capacity and the density of the bed, h_{fb} is the heat transfer coefficient between fluid and bed, ε is the bed void fraction, A and L are respectively the cross section area and the length of the bed.

Finally, Figure 7 shows the influence of the inlet temperature on the temperature distribution along the bed during the charging phase. By keeping the minimum temperature constant, a lower inlet temperature allows better exploitation of the bed. The slight difference in the temperature distribution shown in Figure 7 can be justified by considering the variation of the thermal properties of fluid and bed with the temperature.

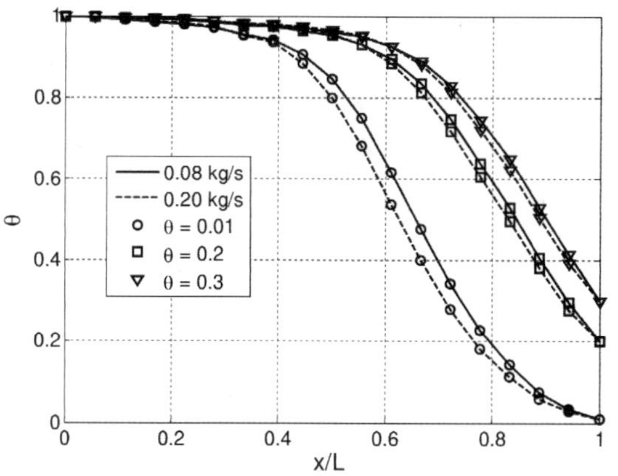

Figure 5. Axial temperature distribution for different mass flow rate and temperature threshold (mass flow rate 0.08 kg/s: T_{max} 230°C, T_{min} 45°C; mass flow rate 0.2 kg/s: T_{max} 239°C, T_{min} 42°C).

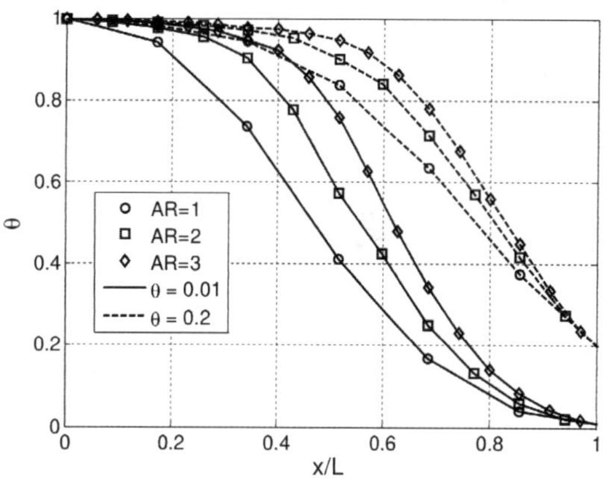

Figure 6. Influence of the aspect ratio on temperature distribution ($T_{max,AR1}$ 226°C (θ=0.01) and 232°C (θ=0.2), $T_{max,AR2}$ 236°C, $T_{max,AR3}$ 239°C, T_{min} 42°C, mass flow rate 0.2 kg/s).

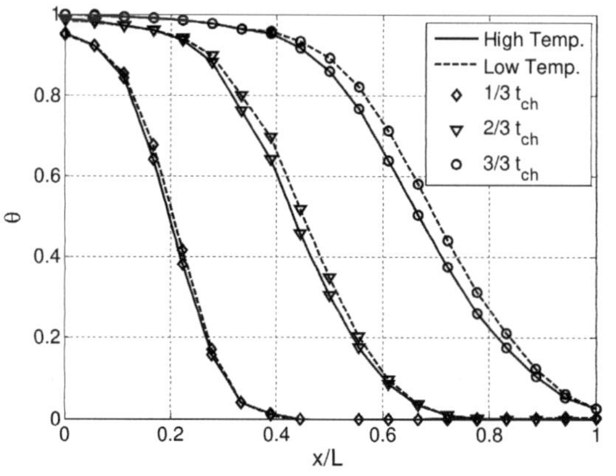

Figure 7. Temperature distribution during the charging phase as a function of the inlet temperature (high temp. T_{max} 237 °C, low temp. T_{max} 158°C, T_{min} 38°C, mass flow rate 0.15 kg/s).

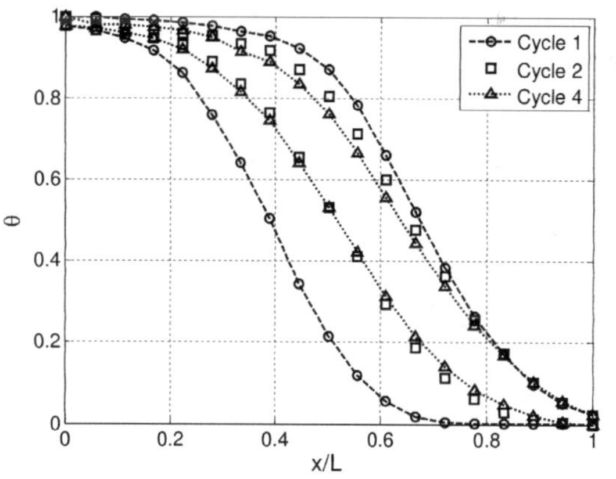

Figure 8. Temperature distribution during continuous operation (T_{max} 191°C, average T_{min} 56°C, threshold tolerance 3°C, mass flow rate 0.2 kg/s).

4. Continuous operation

The TES unit was continuously charged and discharged at a constant mass flow rate. At the beginning of the first charging cycle the temperature along the bed was uniform, while the second cycle started with the temperature profile obtained in the previous phase of discharge. Figure 8 shows the temperature distribution at the end of both charging and discharging phases for four consecutive cycles. As shown in Figure 8, continuous operation affects TES performance negatively, with a gradual and progressive reduction of the energy that can be stored and recovered. In fact, after four cycles the useful energy is more than halved. Performance improvement of the thermal energy storage system was investigated during continuous operation for different threshold values. Figure 9a reports the temperature distributions at the end of each charging and discharging phase for seven consecutive cycles with a threshold tolerance of 10% (θ=0.1 and 0.9). Also in this case Figure 9a shows the continuous reduction of the energy that can be stored and recovered from the accumulator. It is evident that after seven cycles no appreciable differences in the temperature distribution are recorded and the bed appears to reach a constant behavior, with no more degradation in the energy transfer between the bed and the HTF. For the same experiment Figure 9b shows the temperature distributions in the external tank surface at the end of the charging and discharging phases for the first and last cycle. The temperature profile of the wall appears to follow the one of the bed until it reaches an almost stationary state, with a reduction of its influence during TES operation.

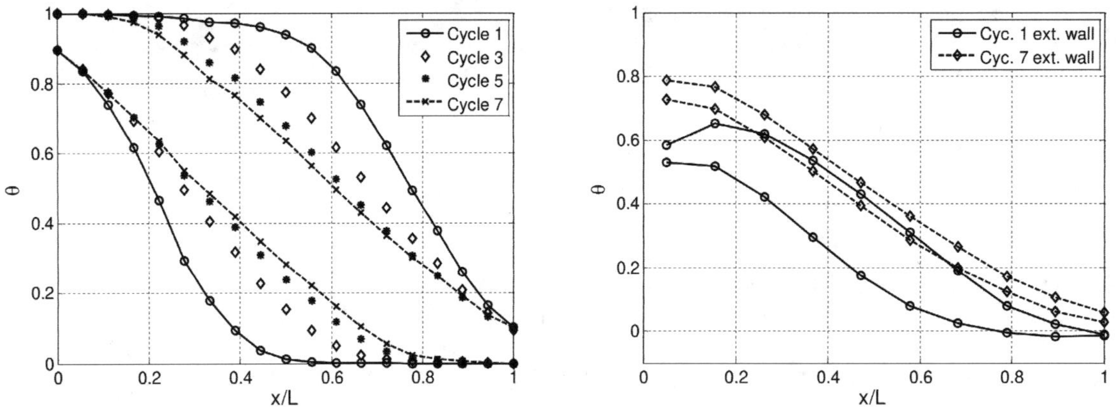

(a) (b)

Figure 9. Temperature evolution during continuous operation with threshold tolerance equals to 10% (θ=0.1 and 0.9) (T_{max} 194°C, average T_{min} 57°C, mass flow rate 0.2 kg/s).

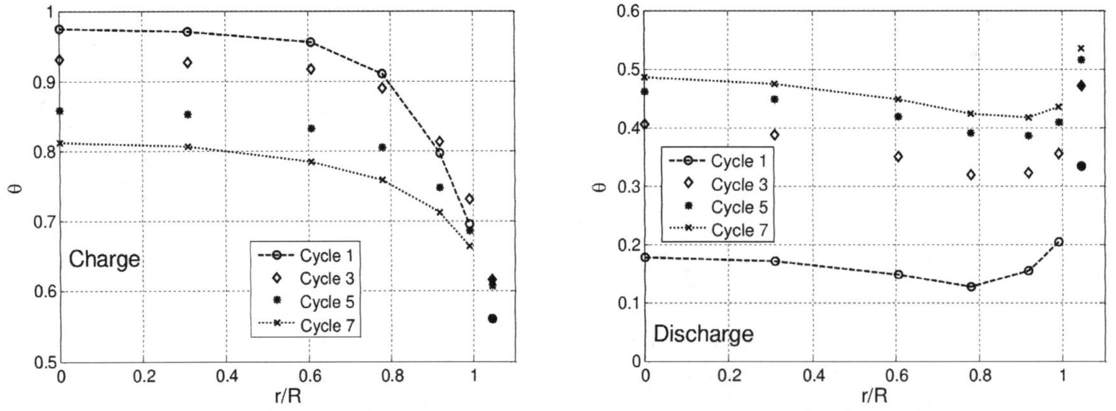

(a) (b)

Figure 10. Radial temperature evolution during continuous operation with threshold tolerance equals to 10% for charging (a) and discharging (b) phase (T_{max} 194°C, average T_{min} 57°C, mass flow rate 0.2 kg/s, markers placed beyond r/R=1 refers to external wall).

Figure 10 depicts how continuous operation affects the radial temperature distribution during the charging and discharging phases. The radial temperature distribution at the end of each charging and discharging phase during continuous operation shows a progressive decreasing difference between the maximum and minimum temperature along the radius. Moreover the difference in the distribution progressively decreases approaching almost constant behavior as the number of cycles increases (Figures 10a and 10b).

A further improvement can be obtained by increasing the threshold tolerance, as shown in Figure 11a, where a constant behavior seams reached after only four cycles. In this case the reduction of energy stored and recovered during continuous operation is more contained if compared to the previous cases considered. As shown in Figure 10a, also in Figure 11b the radial temperature distribution during the charging phase tends to assume a steady-state value, but with a progressively less marked influence of the wall.

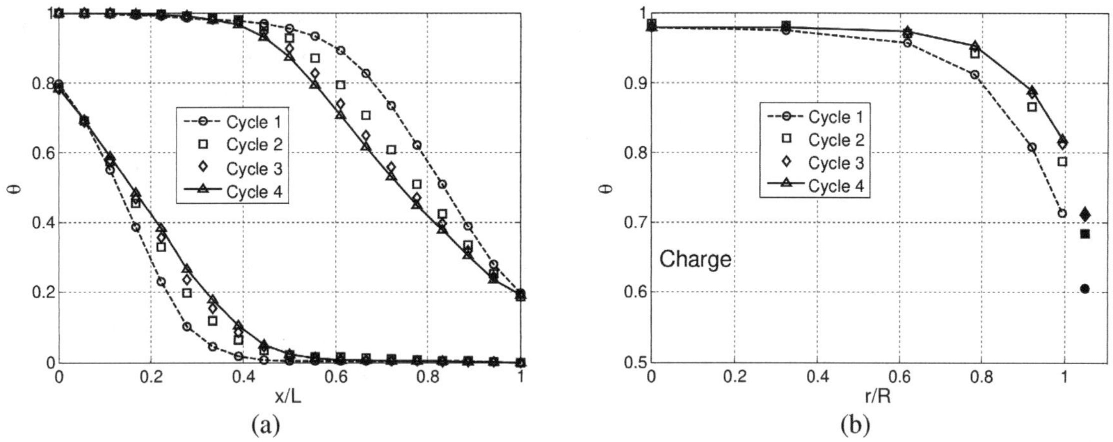

Figure 11. Temperature distribution during continuous operation with threshold tolerance of 20% (θ=0.2 and 0.8) along the vertical axis (a) and the radius (b) (T_{max} 246°C, average T_{min} 56°C, mass flow rate 0.2 kg/s, markers placed beyond r/R=1 refers to external wall).

5. Conclusions

This work presents some results of an experimental investigation on a thermal energy storage system carried out at the Department of Mechanical, Chemical and Materials Engineering of the University of Cagliari. The test facility consists of an open circuit with an insulated steel tank filled with small spheres of alumina, representing the storage media and uses hot air as heat transfer fluid. The influence of some operating conditions and physical characteristics of the system on the formation of the temperature profile "thermocline" was analyzed, as well as the thermal behavior of the TES for repeated cycles of charge and discharge was investigated. The following results from the experimental investigations were obtained:

- energy stored in the bed can be increased increasing the temperature threshold (higher value of the dimensionless temperature θ), the aspect ratio and reducing the mass flow rate;
- slightly higher energy can be stored in the bed if the charging phase is performed at a lower maximum temperature;
- radial temperature distribution shows that metal wall influence cannot be neglected. Radial temperature gradient in observed for at least of 40% of the tank radius from the wall;
- thermal hysteresis is evidenced during repeated cycle operation, highlighting the reducing capacity of the energy stored in the bed (after 4 cycles a reduction of about 60% with a threshold tolerance of 3°C is observed in Figure 8);
- performance of the thermal energy storage system are improved increasing the threshold tolerance between the exit air temperature and the minimum temperature during the charging phases and the maximum gas temperature during the discharging phases.

- a threshold tolerance of 20% of the temperature range considered, shows a considerable improvement of the net useful energy that can be stored after many cycles of continuous operation and a less influence of the thermal hysteresis (only a reduction of about 20% is obtained with a threshold tolerance of 20% in Figure 11).

References

[1] Kousksou T, Bruel P, Jamil A, ElRhafiki T and Zeraouli Y 2014 Energy storage: Applications and challenges *Sol. Energ. Mat. Sol. C.* **120** 59–80

[2] Hadjipaschalis I, Poullikkas A and Efthimiou V 2009 Overview of current and future energy storage technologies for electric power applications *Renew. Sust. Energ. Rev.* **13** 1513–22

[3] Tian Y and Zhao C Y 2013 A review of solar collectors and thermal energy storage in solar thermal applications *Appl. Energ.* **104** 538–53

[4] Medrano M, Gil A, Martorell I, Potau X and Cabeza LF 2010 State of the art on high temperature thermal energy storage for power generation. Part 2–Case Studies *Renew. Sust. Energ. Rev.* **14** 56–72

[5] Mehling H and Cabeza LF 2008 *Heat and cold storage with PCM* (Berlin: Springer)

[6] Pilkington Solar International GmbH 2000 Survey of thermal storage for parabolic trough power plants *National Renewable Energy Laboratory Report*

[7] Gil, Medrano, Martorell, Lazaro, Dolado, Zalba and Cabeza 2010 State of the art on high temperature for power generation. Part 1–Concepts, materials and modellization *Renew. Sust. Energ. Rev.* **14** 31–55

[8] Cot-Gores J, Castell A and Cabeza LF 2012 Thermochemical energy storage and conversion: A-state-of-the-art review of the experimental research under practical conditions *Renew. Sust. Energ. Rev.* **16** 5207–24

[9] Fernandes D, Pitié F, Cáceres G and Baeyens J 2012 Thermal energy storage: How previous findings determine current research priorities *Energy* **39** 246-57

[10] Kuravi S, Trahan J, Goswami DY, Rahman MM and Stefanokos EK 2013 Thermal energy storage technologies and systems for concentrating solar power plants *Prog. Energ. Combust.* **39** 285-19

[11] Singh H, Saini RP, Saini JS and April 2010 A review on packed bed solar energy storage systems *Renew. Sust. Energ. Rev.* **14** 1059–69

[12] Coutier JP and Faber EA 1982 Two applications of a numerical approach of heat transfer process within rock beds *Sol. Energy* **29** 451–62

[13] Yang Z and Garimella SV 2010 Thermal analysis of solar thermal energy storage in a molten-salt thermocline *Sol. Energy* **84**, 974-985

[14] Bruch A, Fourmigué JF and Couturier R 2014 Experimental and numerical investigation of a pilot-scale thermal oil packed bed thermal storage system for CSP power plant *Sol. Energy* **105** 116–25

[15] Kolb GJ 2011 Evaluation of annual performance of 2-tank and thermocline thermal storage systems for trough plants *J. Sol. Energ. – T. Asme* **133**

[16] Cocco D and Serra F 2015 Performance comparison of two-tank direct and thermocline thermal energy storage systems for 1 MWe class concentrating solar power plants *Energy* **81** 526-36

[17] Hänchen M, Brückner S and Steinfeld A 2011 High-temperature thermal storage using a packed bed of rocks and heat transfer analysis and experimental validation *Appl. Therm. Eng.* **31** 1798-06

[18] Cascetta M, Cau G, Puddu P and Serra F 2014 Numerical Investigation of a Packed Bed Thermal Energy Storage System with Different Heat Transfer Fluids *Energy Procedia* **45** 598-07

[19] Wen D and Ding Y 2006 Heat transfer of gas flow through a packed bed *Chem. Eng. Sci.* **61** 3532-42

Journal of Physics: Conference Series **655** (2015) 012019 doi:10.1088/1742-6596/655/1/012019

Analysis of the temperatures measured in very thick and insulating roofs in the vicinity of a chimney

M Neri[1], D Luscietti[2], S Bani[3], A Fiorentino[2] and M Pilotelli[2]

[1,2] University of Brescia, via Branze, Brescia, Italy
[3] Associazione ANFUS, via della Cascina Pontevica, Folzano, Brescia, Italy

E-mail: manuelaneri.brescia@gmail.com

Abstract. Chimneys convey exhaust gas produced in domestic heat appliances to the external environment and to do this they have to pass through elements such as roofs and floors. If these elements are made up of flammable materials the fire hazard may occur. In some European countries the number of roof fire is very high and they affect also certified chimneys, that is, tested following the prescription of the related standards. The aim of this paper is to highlight that the certification procedure does not allow to test chimneys in the worst condition, therefore, chimney installed following the manufacturer prescriptions may in some cases cause the overheating and subsequent roof fire. To do this, experimental tests have been performed for measuring the temperature on roofs in the vicinity of a certified chimney. The results show that the certification procedure should be revised.

1. Introduction

The overheating of flammable materials due to the presence of a chimney has been analyzed in the past for determining the temperature of the outer surface of the chimney [1-4] and to define safety distances for preventing the burning of structural elements [4-7]. Nowadays, despite of the chimney heat performances are tested following the EN ISO 1859 standard [8], the number of roof fires in some European countries do to the presence of a chimney is very high [9-14].

The standard [8] prescribes the heat stress test and the thermal shock test by means of which the distance to be respected between chimney and flammable materials is determined. The heat stress test, analyzed in this paper, consists in installing the chimney to be certified in a structure following the manufacturer prescriptions, in producing exhaust gas at a predetermined temperature and in measuring the temperatures on the test structure until the achievement of the final test condition. The finale test condition is defined as an increase of the temperature on the test structure lower than 2°C in 30 minutes. If the measured temperatures on the flammable materials do not exceed 85°C, the chimney is certified and a label is attached on it. The label reports the permitted exhaust gas temperature and the required distance from flammable materials, but no information is given about the way of sealing the eventual clearance between chimney and roof. The test structure is composed of two walls at right angle of thermal resistance equal to 2.73 m^2K/W and two floors positioned at different height. The floors are made up of three layers: a lower wooden layer 12 mm thick, an upper wooden layer 20 mm thick and an intermediate layer of mineral wool that is 200 mm thick for the lower floor and 100 mm thick for the upper floor.

Recent studies investigated the differences between the conditions in the certification procedure prescribed by the standard [8] and those in real installations, but all of them analyzed configurations in which a chimney was installed in the corner of a roof and near two walls. These studies highlighted that in real installations the condition can be much more critical. Real roofs can be thicker than the floors prescribed by the standard [8], for example in the countries of the north of Europe their thickness can reach 1000 mm. In the label no information is given about how to seal the space between chimney and roof (clearance); then, it may be sealed in a different way with respect to that adopted in the certification procedure. For example, some chimney producers prescribe to test chimneys with open clearance, but this installation can not be performed in real installations because the entering of atmospheric agents must be prevented. In energy-saving buildings the clearance is usually filled with insulating material to avoid condensation and thermal bridges and this represents a very dangerous practice. The influence of the clearance sealing mode and the roof characteristics on the temperature in the roof was experimentally shown by [15]: this study analyzed the configuration prescribed by the standard [8] and it demonstrated the importance to specify in the label reported on the chimney the clearance sealing mode adopted in the certification. Weaknesses related to the way of measuring the exhaust gas temperature in the certification procedure were experimentally highlighted: Leppänen et al. [16] showed the flue gas temperatures of fireplaces is higher than temperature indicated in CE marking [16], Inha et al. [17] showed that for light-weight metal chimneys one fire cause is the high flue gas temperatures of sauna stoves. In [15] it was shown that the temperatures measured on flammable materials are related to a lower exhaust gas temperature than that used to define the class temperature of the chimney, and the final test condition does not allow to measure the stationary temperature on flammable materials. Experiments were performed to assess the influence of the smouldering combustion of the binder of rockwool used to manufacture chimneys [18]: this phenomenon determines an escalation of temperature that could cause the ignition of flammable materials in the vicinity of the chimney. In [19] a 3D numerical model was designed to reproduce the condition of the heat stress test, and a lumped element model for estimating the stationary temperature from the temperature-time curves measured in the heat stress test was also proposed.

The aim of this paper is to highlight, by means of experimental tests, that the heat stress test proposed by the standard [8] does not allow to test chimneys in the worst conditions. To do this, a certified chimney has been installed at the center of three roofs following the manufacturer prescriptions. By installing a chimney at the center of a roof, it is not affected by the presence of the walls that, having a lower thermal resistance than the roofs, facilitate the heat dispersion. One roof is representative of the more insulating roof prescribed by the standard [8], while the other two are representative of roofs in energy-saving buildings. The clearance between the chimney and the roofs has been sealed in four different ways to reproduce the conditions that may occur in real installations, that is, open, sealed with metal sheets, sealed with insulating panels, and filled with insulating material. Despite the exhaust gas temperature has been kept lower than that for which the chimney is certified, in the majority of the tests, temperature of flammable materials has been higher than 85°C. The results of these have confirmed that one possible cause of roof fires is a weak certification procedure.

2. Materials and methods

The tests described in this paper consisted in the following steps: installation of a certified chimney in the center of a roof following the scheme in Figure 1 ((a) and (b)), production of exhaust gas at a predetermined temperature T_{ch}, measurement of the temperatures in the roof at the positions indicated in Figure 1 ((b) and (c)).

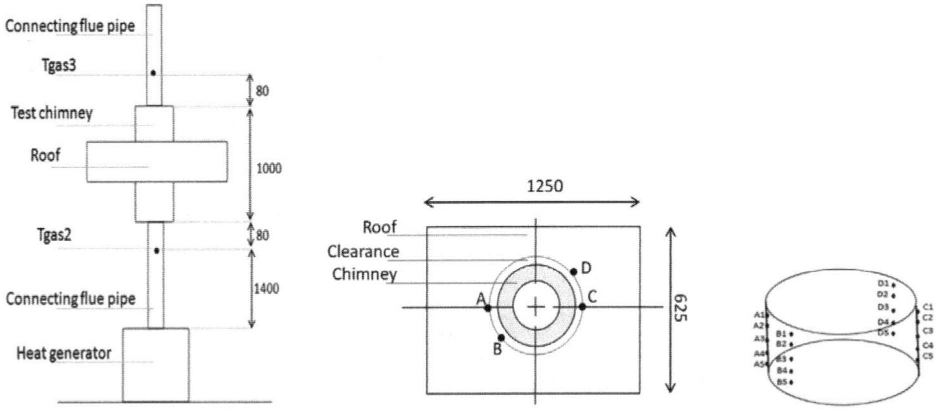

a) b) c)

Figure 1. Test structure: a) front view, b) top view, c) thermocouples positioning. Letters A, B, C and D indicates the lines on which the thermocouples were posed. Dimensions are in millimeters.

2.1 Test structure

The test structure is represented in Figure 1 (a) and it consists of a roof, a chimney and a heat generator. The chimney was installed at the center of the roof, and connected to the heat generator by means of a flue pipe, and an additional one was also installed to convey the exhaust gas far from the test structure. The chimney used in the test is certified T600 when it is installed 20 mm far from flammable materials. This means that during the certification procedure the exhaust gas temperature was 700°C and the maximum temperature measured on flammable materials did not exceed 85°C. In our tests, the chimney was installed respecting the manufacturer prescriptions reported in the label attached on it, that is, 20 mm from the roof and for safety reasons, the exhaust gas temperature was maintained at 500°C instead of 700°C.

The roofs in which the chimney was installed are shown in Figure 2. Roof R1 is similar to the lower floor prescribed by the standard [8] but it is a little thicker and more insulating, roofs R2 and R3 have same thickness and same thermal resistance but different layers position and are representative of roofs in energy-saving buildings. By performing tests on roof R2 and R3 has allowed to understand if the layers position affects the maximum temperature on the roof. The chimney was installed 20 mm far from the roof as prescribed by the manufacturer and the clearance was sealed in four ways that are represented in Figure 3. The clearance was left open (*op*), sealed by means of metal sheets (*ms*), sealed with insulating materials (*ad*) and filled with insulating materials (*fi*). The sealing with metal sheet (*ms*) allowed air movement in the clearance and heat transfer through the sealing, the sealing with insulating panels (*ad*) allowed air movement in the clearance but prevented heat transfer through the sealing and it has been realized to approach an adiabatic sealing. Finally, the filling (*fi*) of the clearance reproduced the installation usually adopted in energy-saving buildings. Each roof was tested with each sealing mode, for a total of 12 tests, which are reported in Table 1. The tests are identified by acronyms: the first two letters represent the roof and the last two the clearance sealing mode. The standard [8] prescribes to stop the test when the increase in temperature is lower than 2°C in 30 minutes, or after 8 hours. However, it was not possible to reach this condition in all the tests as specified in Table 1.

Journal of Physics: Conference Series **655** (2015) 012019 doi:10.1088/1742-6596/655/1/012019

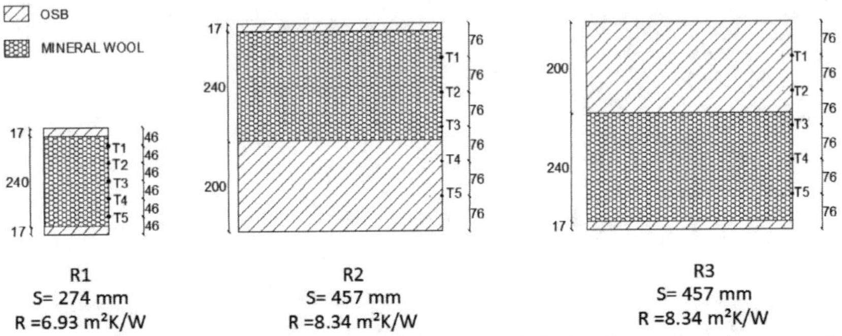

Figure 2. Tested roofs. *S* represents the thickness and *R* the thermal resistance of the roof. T1, T2, T3, T4 and T5 indicate the positions where temperatures have been measured. Dimensions are in millimetres.

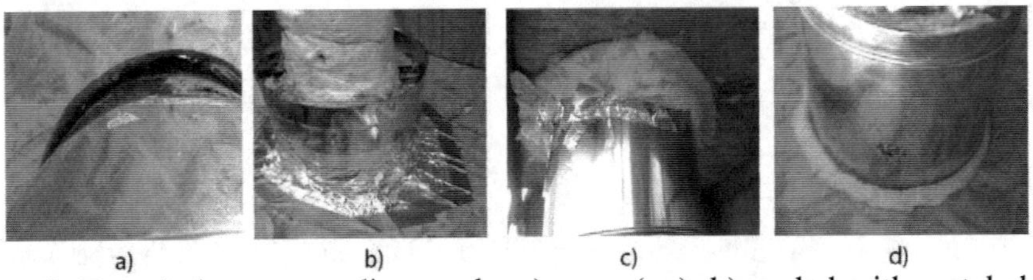

Figure 3. Tested clearance sealing mode: a) open (*op*), b) sealed with metal sheet (*ms*), c) sealed with insulating material (*ad*), d) filled with insulating material (*fi*).

Table 1. Experimental tests and their characteristics. The acronym a.a.t. means aluminum adhesive tape.

ROOF	TEST	CLEREANCE SEALING MATERIAL	T_{ch} [°C]	T_{env} [°C]	Time length [hours]	Final condition achiev.
Roof 1	$R1_{op}$	Open	525.8	8.12	3.6	yes
	$R1_{ms}$	Sealed with aluminum sheets	499	6.8	7.5	yes
	$R1_{ad}$	Sealed with ceramic fiber	494.5	8.3	6.8	yes
	$R1_{fi}$	Filled with super wool 96	498.4	9.8	8	no
Roof 2	$R2_{op}$	Open	504	5.7	3.7	yes
	$R2_{ms}$	Sealed with aluminum sheets	501.2	6.6	7.2	no
	$R2_{ad}$	Sealed with ceramic fiber	507.7	6.2	7.3	yes
	$R2_{fi}$	Filled with super wool 96	503.4	6.1	8	no
Roof 3	$R3_{op}$	Open	498.8	5.2	4.5	yes
	$R3_{ms}$	Sealed with aluminum sheets	504.7	9.1	6.8	yes
	$R3_{ad}$	Sealed with ceramic fiber	500.5	7.2	7.4	no
	$R3_{fi}$	Filled with Super wool 96	500.2	6.3	8	no

2.2. *Temperatures measurement and elaboration*

Temperatures were measured by means of 24 thermocouples of type K and two DAQ NI 9213 connected to a PC. K thermocouples have an uncertainty of ±2.8°C for temperatures ranging between 0°C and 350°C and of ±0.75\% for temperatures ranging between 350°C and 1260°C. The uncertainty of the DaQ is ±0.02°C. Temperatures were recorded every 10 seconds. Exhaust gas temperatures were measured at the chimney inlet (T_{gas2}) and at the chimney outlet (T_{gas3}) as shown in Figure 1, and also the environment temperature (T_{env}) was measured. Flammable materials temperatures were measured according to the schemes in Figures 1 and 2. Thermocouples were positioned vertically and equidistantly and then they were spaced 4.6 cm in tests performed on *Roof 1* and 7.6 cm in tests performed on *Roof 2* and *Roof 3*. The tests were stopped as indicated in Table 1 and for each thermocouple, the final temperature on the roof (T_{end}) was determined as average of the temperatures recorded in the last five minutes of the test. In this way, it was possible to filter final results from instantaneous fluctuations.

Since the conditions in the clearance can be considered axial-symmetrical, a unique temperature value (calculated as average of them) was determined for the four corresponding thermocouples (positioned at the same height). Five temperatures were calculated: for example, from thermocouples A1, B1, C1 and D1, a value T1 was determined. Due to the length of their duration, it was not possible to repeat the tests but, by averaging four values, it was possible to obtain a rough estimate of their internal variability due to the installation mode of the chimney and of the thermocouples. The temperature inside the chimney (T_{ch}) in correspondence to the roof was calculated as $T_{ch}=(T_{gas2}+T_{gas3})/2$.

2.3. *Uncertainty and internal variability*

Uncertainty of temperatures was determined as the sum of the uncertainty related to the measuring instruments and that related to the data elaboration. The uncertainty is ±2.9°C for gas temperatures (T_{gas2}, T_{gas3}) and the environment temperature (T_{env}), and the uncertainty of the flammable materials temperatures is 3.7°C. Final flammable materials temperatures have been determined as average of four values and the related variability as been determined as the difference between the maximum and the minimum values.

3. Results and discussion

The analysis has been done by comparing measured maximum temperatures, and comparing the temperature profiles measured on roofs *R2* and *R3*.

3.1. *Roof maximum temperatures*

Figure 4 shows the maximum temperature measured in the tests. It can be noted that, despite the exhaust gas temperature was 200°C lower than in the certification procedure, in the majority of the tests, the maximum temperature exceeds the limit of 85°C. The only configuration that allows to respect the limit temperature is with open clearance ($R1_{op}$, $R2_{op}$ and $R3_{op}$) and for the others configurations ($R2_{ms}$, $R2_{ad}$, $R2_{fi}$, $R3_{ms}$, $R3_{ad}$, $R3_{fi}$) the more insulating the clearance sealing mode, the higher the temperature in the roof. In the test $R2_{fi}$ the measured maximum temperature is lower than that measured in the test $R2_{ad}$: this is due to the fact that $R2_{fi}$ test was interrupted at an instant farther from the stationary condition with respect to $R2_{ad}$. However, by estimating the stationary temperature by means of the lumped element model proposed in [19], also for roof R2 the higher temperature has been reached in the case with filled clearance. In roof R3, the clearance has been sealed with aluminum adhesive tape in order to reduce air infiltration.

The fact that the measured temperatures are higher than 85°C can be due to three causes: the roofs used in this experimental campaign are thicker than those prescribed by the standard [8], the clearance has been closed in different ways than those adopted in the certification procedure (that it is unknown) and, moreover, installing the chimney at the center of a roof represents a worst condition than that occurs when the chimney is installed in the corner of it. However, the installations were performed following the indications reported in the label of the chimney. Then, it can be stated that the information reported in the label is not enough to ensure safety installations: information about how the clearance was sealed in the test procedure would allow chimney installers to reproduce the same conditions tested in the certification procedure.

It must be remarked that in last years, the thickness of the roofs has incremented so much that those prescribed by the standard [8] do not represent the worst condition that can occur in real installations.

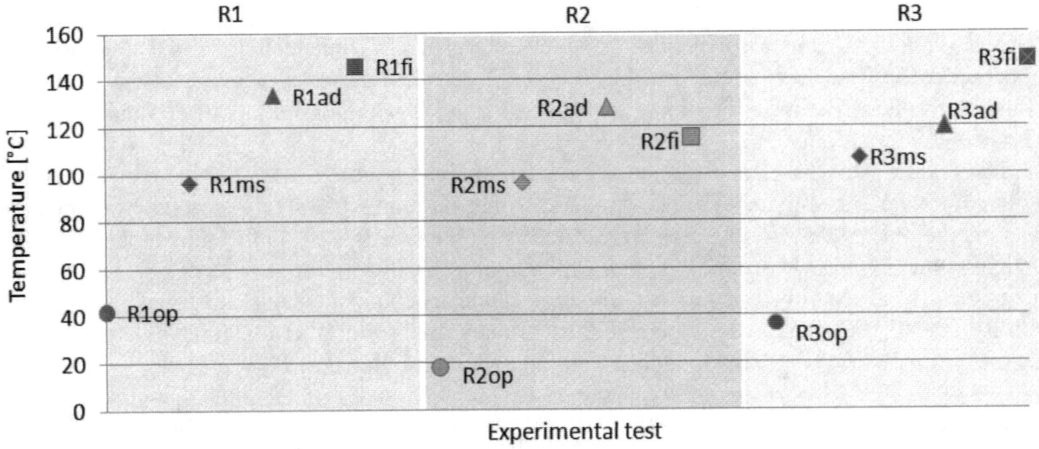

Figure 4. Maximum temperatures measured in the experimental tests performed on roofs R1, R2 and R3 with the four clearance sealing modes (open (*op*), sealed by means of metal sheets (*ms*), sealed with insulating panels (*ad*) and filled with insulating materials (*fi*)).

3.2. *Roof layers position influence*

In order to understand whether the roofs layers position affects the position of the maximum temperature on roofs, temperature profiles measured on roofs R2 and R3 are reported in Figures 5, 6, 7 and 8. Roofs R2 and R3 have same thickness and thermal resistance but different layers position.

Figure 5 shows the temperatures measured on the roofs when the clearance was left open. It can be noted that for this installation mode the mean temperatures are quite homogeneous but for the roof R3 they are affected by high variability. High variability can be attributed to the loss of the axial-symmetric condition due to environmental conditions (speed and direction of the wind), and the presence of the heat generator.

Figures 6 and 7 show the temperatures measured on roof R2 and R3 when the clearance was sealed by means of metal sheets (ms), and by means of insulating material (ad) respectively. It can be noted that with respect to the cases with open clearance (Figure 5) the temperatures are less homogeneous: in both cases the higher temperature have been measured on the insulating layer, that is, in the upper part of roof R2 and in the lower part of roof R3.

Journal of Physics: Conference Series **655** (2015) 012019 doi:10.1088/1742-6596/655/1/012019

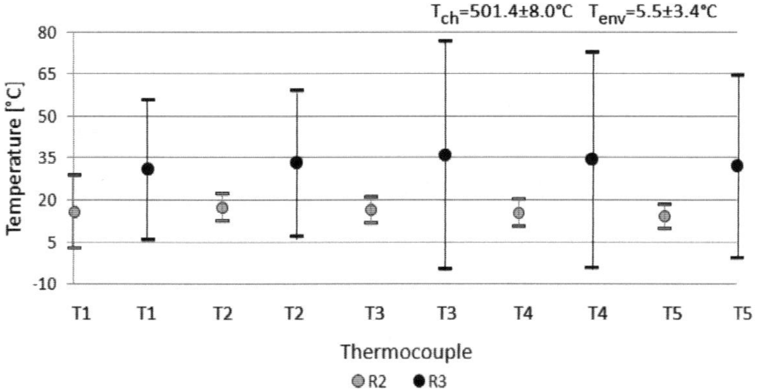

Figure 5. Temperatures on roofs R2 and R3 measured with open clearance. T1, T2, T3, T4 and T5 indicate the thermocouple positions, where T1 is the highest one.

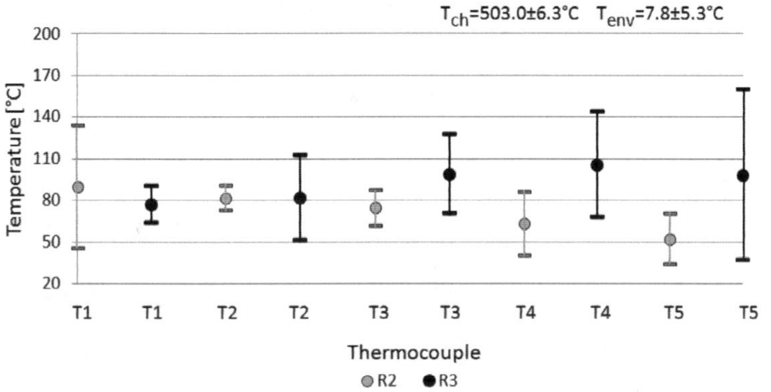

Figure 6. Temperatures on roofs R2 and R3 measured with clearance sealed by means of metal sheets. T1, T2, T3, T4 and T5 indicate the thermocouple positions, where T1 is the highest one.

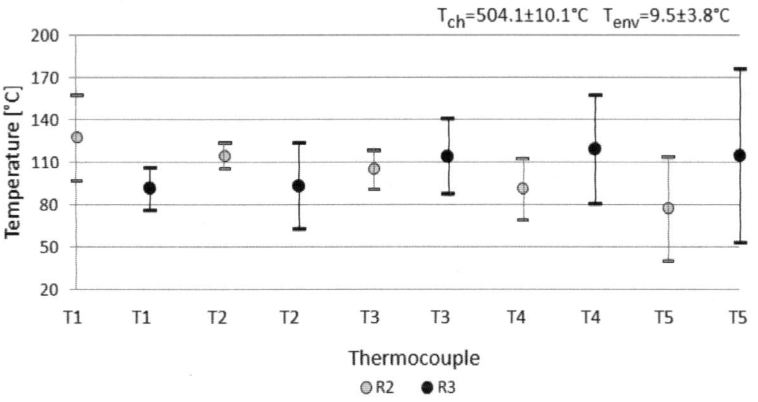

Figure 7. Temperatures on roofs R2 and R3 measured with clearance sealed by means of insulating panels. T1, T2, T3, T4 and T5 indicate the thermocouple positions, where T1 is the highest one.

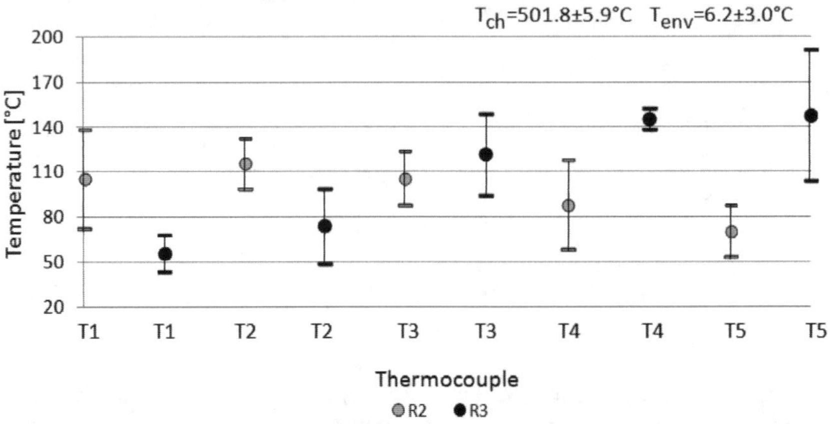

Figure 8. Temperatures on roofs R2 and R3 measured with clearance filled with insulating material. T1, T2, T3, T4 and T5 indicate the thermocouple positions, where T1 is the highest one.

Figure 8 shows the temperatures measured when the clearance was filled with insulating material. Also in this case the maximum temperature was measured on the insulating layer, but the difference in temperature between the upper and the lower zones of the roofs is up to 100°C for roof R3 and it is higher than in the tests performed with other clearance sealing modes. When the clearance is sealed in the upper and lower parts the natural convection push hot air upward but the radiation tends to make uniform the horizontal heat flux in the vertical direction: on roof R2 the effect of the natural convection is accentuated by the presence of the insulating layer in the upper part. When the clearance is filled and only conduction takes place, the temperature increases on the more insulating layers. If thermocouples are installed following the standard [8] prescription - i.e. on the center line of the roof - the maximum temperature on flammable materials may not be measured.

3.3. Test final condition

The standard [8] prescribes to stop the heat stress test when the increase in temperature on the test structure is less than 2°C in 30 minutes or after 8 hours. In Figure 9, four temperature time traces measured in tests with different clearance sealing modes are reported. All of them represent temperatures whose increase at the end of the test was less than 2°C in 30 minutes except for the thermocouple related to $R1_{fi}$. It can be noted that if the tests had continued, the temperatures would have increased. Then, the standard [8] should be modified in order to allow the stationary temperature detection. Since the achievement of the stationary temperature may take long time, the stationary temperature could be estimated by means of the lumped element model proposed by [19] applied to the temperatures measured during the heat stress test.

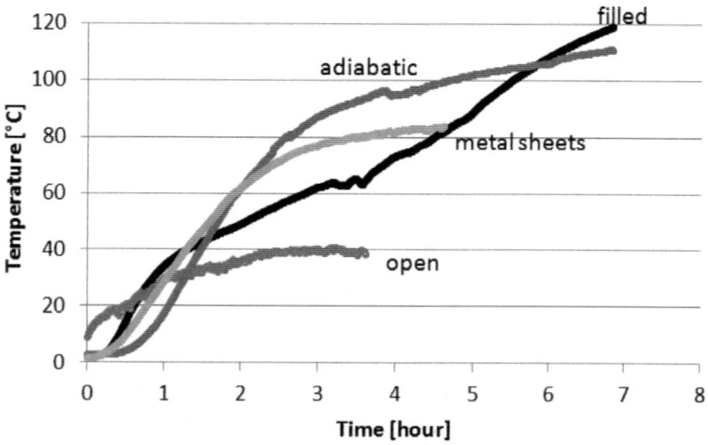

Figure 9. Temperatures time traces measured in the tests performed on roof R1 for different clearance sealing modes.

3.4. Influence of the thermophysical characteristics of the materials

By analyzing the temperatures in Figure 4 it can be noted that the roof characteristics do not affect strongly the maximum temperature in the roof, but rather the temperature distribution. Higher temperatures occur in the more insulating layer especially in the configurations with filled clearance. In the configurations with blocked air between chimney and roof (clearance sealed either with metal sheets or sealed with insulating panels) temperature distribution is more homogeneous because the effect of the layers arrangement is counterbalanced by radiation and convection.

In the configurations with air in the clearance, hot air rises because of natural convection and it heats the upper layer, even if radiation tends to homogenize the temperature distribution: then, if in the upper part of the roof there is a more conductive layer (in this case the wooden layer), the heat flow from the roof to the environment is facilitated because on the upper surface of the roof a greater convective heat transfer coefficient occurs.

In the clearance, the standard [8] prescribes a wooden lath 5 cm thick attached to the roof to contain the insulating material: the presence of the lath flatters the temperature distribution and, having a higher thermal conductivity with respect to that of the material in the roof, it acts as a thermal bridge and determines a lowering in temperature. Then, the presence of a conductive element in this position is suitable for keeping the temperature low, but at the same time, its presence in energy-saving buildings could represent a critical aspect.

By filling the clearance the increase in temperature is slower with respect to the configurations with air in the clearance, but it leads to higher temperature on flammable materials, as can be seen in Figure 9.

4. Conclusions

In this paper differences between the condition in the certification procedure prescribed by the standard [8] and the condition in real installations have been experimentally shown. The study focuses on the influence of the roof characteristics and the clearance sealing mode on the flammable materials in the vicinity of a chimney.

The tests have been performed by installing a certified chimney at the center of three roofs and the clearance between chimney and roof has been sealed in four ways. The configuration with a chimney installed at the center of a roof is more critical than the condition described in [8] because the walls of the test structure can act as a thermal bridge.

Despite the chimney has been installed following the manufacturer prescriptions reported on the label attached on it, and the exhaust gas temperature has been kept 200°C lower than that for which the

chimney is certified, the temperature of flammable materials has exceeded the limit of 85°C in the majority of the tests. The limit temperature has been respected only in the configuration with open clearance. Since in real installations the clearance can not be left open, especially in roof penetrations, the certification procedure should not be performed with open clearance. In any case, the clearance sealing mode adopted in the certification should be specified in the label, as was already suggested for the installations in the test structure by [15]. Since higher temperatures have been measured in roofs representative of roofs in energy saving buildings in which the clearance is usually filled with insulating materials, tests ad hoc should be performed for these configurations. The layers position in the roof affects the temperature distribution, and higher temperatures are measured on more insulating layers. Then, to detect the actual maximum temperature, thermocouples should be installed vertically instead of horizontally. Temperature-time curves measured in the tests have shown that the final test condition does not allow to reach stationary temperatures. Therefore, they should be estimated by means of the lumped element model described in [19] and compared with the maximum temperature at which materials can be exposed. By comparing temperatures measured in these tests with those measured in the tests performed in the test structure and reported in [15], it can be stated that installations at the center of a roof is more critical because for the same configuration higher temperatures have been measured when the chimney was at the center of the roof, despite the environment temperature was lower.
In conclusion, the position of the chimney in the roof strongly affects the flammable materials temperature and to test chimneys in favour of safety, the certification procedure should be modified and a new structure should be designed. In this new structure, the chimney should be installed at the center of a roof and enough far from the border of the roof in order to avoid edge effects.

Acknowledgments

This research has been done in collaboration with Centro Studi ANFUS (Associazione Nazionale Fumisti e Spazzacamini) of Brescia. We also thank Legno Camuna of Brescia (Italy) and Bolletta srl (Perugia).

References

[1] National Fire Protection Association (1984) Standard for Chimneys, Fireplaces, Vents and Solid Fuel Burning Appliances, NFPA 211, Quincy
[2] Voigt G Q 1933 Fire Hazard of Domestic Heating Installations, National Bureau of Standard **11** 353-372
[3] Lawson D I, Fox L L and Webster C T 1952 The Heating of Panels by Flue Pipes, Fire Research Special Report Fire Protection Association
[4] National Bureau of Standards 1941-1945 Prefabricated Metal Chimneys, Fire Research Section, Unpublished Report to the Federal Public Housing Authority More references
[5] Peacock R D 1987 Wood Heating Safety Research: an Update, *Fire Technology* 292-312
[6] Lawson D I and Simms D L 1952 The ignition of wood by radiation *Brit. J. Appl. Phys.* **3** 288-292
[7] Kline L M and Witte H 1959 Performance of type B gas vent for gas-fired appliance *Bulletin of research*
[8] UNI EN 1859 2009 Camini - Camini metallici - Metodi di prova, UNI, Milano
[9] Buffo S and Dadone P N (2007) Studio Statistico Vigili del Fuoco di Brescia sulle Cause dell'Incendio Tetto, Vigili del Fuco di Brescia, Brescia (in Italian)
[10] International Partnership for the Investigation of Fires Explosions and Other Major Incidents

www.burgoynes.com, Accessed 21 July 2015

[11] Dadone P N 2009 Analisi 00. Casi Verificatisi sul Territorio della Provincia di Brescia e Statistica delle Cause, Incendi tetto e Canne Fumarie Proceeding Conference, Brescia (in Italian)

[12] Leppänen P 2010 Fire Safety of Metal Chimneys. Master's Thesis. Tampere University of Technology, Department of Civil Engineering, p 51 (in Finnish)

[13] Hakala V M 2013 Fire Investigations, Thematic Investigation 3: Chimney and Fireplace as Causes of Fire, 14 (in Finnish)

[14] Ministry of the Environment of Finland 2011 Finnish Concern on Fire Safety Risk Due to CE Marking of Appliances Fired by Solid Fuel and Chimney Products

[15] Neri M, Luscietti D, Fiorentino A and Pilotelli M 2015 Experimental Analysis of Chimneys in Wooden Roofs, *Fire Technology* DOI 10.1007/s10694-015-0525-7.

[16] Leppänen P, Inha T and Pentti M 2015 An Experimental Study on the Effect of Design Flue Gas Temperature on the Fire Safety of Chimneys, *Fire Technology*, p 847-866

[17] Inha T, Leppänen P, Peltomäki M and Pentti M 2011 Fire Safety of Light-weight Metal Chimneys. Tampere University of Technology, Department of Civil Engineering. Research report no. PALO 1950/2011 (in Finnish)

[18] Leppänen P, Neri M and Mäkinen J 2015 Heat release caused by the smouldering combustion of the binder of rockwool, *Rakenteiden Mekaniikka (Journal of Structural Mechanics)* **48**, No 1 2015 68-82

[19] Neri M, Luscietti D, Bani S, Fiorentino A and Pilotelli M 2015 Chimneys in Wooden Roofs: a 3D Steady Numerical Model for the Prediction of the Temperatures, Conf. ASME-ATI-UIT, Napoli

Analysis and clustering of natural gas consumption data for thermal energy use forecasting

Alessandro Franco, Fabio Fantozzi

Department of Energy, Systems, Territory and Constructions Engineering
University of Pisa, Largo Lucio Lazzarino, 56126 PISA

E-mail: alessandro.franco@ing.unipi.it

Abstract. In this paper, after a brief analysis of the connections between the uses of natural gas and thermal energy use, the natural gas consumption data related to Italian market are analyzed and opportunely clustered in order to compute the typical consumption profile in different days of the week in different seasons and for the different class of users: residential, tertiary and industrial. The analysis of the data shows that natural gas consumption profile is mainly related to seasonality pattern and to the weather conditions (outside temperature, humidity and wind chiller). There is also an important daily pattern related to industrial and civil sector that, at a lower degree than the previous one, does affect the consumption profile and have to be taken into account for defining an effective short and mid term thermal energy forecasting method. A possible mathematical structure of the natural gas consumption profile is provided. Due to the strong link between thermal energy use and natural gas consumption, this analysis could be considered the first step for the development of a model for thermal energy forecasting.

1. Introduction

The development of instruments for the short and mid term prevision of the patterns that govern the energy demand can permit significant energy savings [1]. Accurate models for electric and thermal energy short and mid term forecasting are essential to the operation and planning of a utility company and to increase the efficiency in the energy production and distribution, [2].

Forecasting of the thermal energy use is a quite complex task: an important amount of the thermal energy consumed is caused by the heating of civil/tertiary sector, but the part relating to industrial uses, such as can be for example the process steam, is important too. While the first component can be directly correlated to climatic variables (mainly temperature, humidity and wind speed) and seasonality pattern, the second one is more directly related to the planning of production (day of week) and less dependent on meteo data. The geographic aggregation of the data is important too. The development of forecasting method for electricity can be based on available data of consumption and electricity load, [3-4] the same cannot be so simple for the thermal energy. Peak energy demand is surely a critical element in the short, mid and long term which for a country like Italy, consists of electricity (the mix and spatial distribution of generation technologies, transmission, and distribution) and natural gas (production, transmission, distribution, and storage), [5-6]. Some important drivers like the use of natural gas for thermal energy production and the diffused use of combined cycle power plants using natural gas have altered the traditional cyclical demand of natural gas. Another important factor is the cost and availability of natural gas, which is itself related to many political and

economical factors, many times not fully predictable. The economic factor has an influential impact on the demand for natural gas in both the residential and in the industrial sector, and in both cases it is not simple to quantify the short, mid and long term impact in gas consumption. The analysis of the drivers of natural gas consumption is currently object of research, as suggested by [7], but it is clear that in a particular country like Italy, the use of thermal energy is the most important driver. Furthermore forecasting of natural gas consumption has been already object of analysis in the literature, for example by [8] and [9] with a general perspective. The residential and non residential consumption of natural gas in Italy has been recently analyzed in the literature. In particular Bianco et al. in [10-11] proposed a model to forecast residential and non-residential consumption of natural gas. Consumption drivers are identified and discussed and a single equation demand model is identified. Anyway the object of the two papers is the obtainment of long term scenario and the effect of economic drivers is considered of primary importance. It is important to remark that the renovation of the natural gas wholesale and retail markets together with a technological review of current smart energy practices is a key step to accomplish an optimal energy consumption pattern, to avoid energy wastes and to implement a sustainable energy system. In this vision smart meters installed at the utility endpoints to monitor usage could be an important component of the smart grid. As discussed in the previous section, high resolutions automated meter reading system for residential gas meters, which can be used to record gas consumption for each appliance have been already experimentally tested and applied, even if in most cases they are not in the commercial development phase yet ([12]). Research on smart meters appears to be interesting from a technological point of view, but the development and application of smart metering is a particularly complex task mainly for aspects related to data treatment and communications. The true challenge and difficult task in the field of smart meters' development, which also occurs in the use of Automatic Meter Reading (AMR) devices, appears to be the fact that monitoring energy usage data poses privacy risks that might not be easy to solve, [13-14]. On the other hand, fine-grained energy consumption data collected by AMR could reveal sensitive information coming from home (number of occupants, vacancies, typical energy usage, etc.). While there is a general consensus that energy monitoring devices can be an opportunity to "*provide consumers with a more comprehensive and nuanced understanding of their usage patterns*", the only real perspective of considering demand side management is the gas flow monitoring, [15]. Even if there has recently been a lot of research in this topic, [12] and the italian distributors were required to install Smart Gas Meters at 100% of all non-domestic consumers' premises by the end of 2012 and at 80% of domestic consumers' premises by the end of 2016, [16], the availability of data is limited.
This work represents the first step of the construction of the model for the definition of thermal energy use consumption in complex and largely extended energy systems. Results can be extended in some directions that are currently under investigation by the authors; for example to other countries to check if the features that are informative in the Italian case are equally important in other countries.

2. The link between natural gas consumption and thermal energy use: the Italian case

Natural gas has been used since the '70s to satisfy the needs of many commercial and residential users throughout the world through a huge and complex network. Demand for natural gas has traditionally been cyclical, simply following the characteristic seasonal patterns. As a consequence, the cyclical nature of the demand was used in the past to accurately predict the required natural gas: a direct correlation with average medium environmental temperature could be observed. In addition, until a few years ago, the connection between the gas sector and the electricity sector was marginal. Things began to change at the beginning of the '90s, when the commercial development of combined cycle power plants led to a shift towards the use of natural gas for the generation of electricity as well. This had a strong impact on the traditional cyclical gas demand. Obviously, while demand of natural gas heating decreases during the summer months, the demand for air conditioning/cooling is relevant during the same period. Nowadays electricity provides the primary source of energy for residential and commercial cooling, thus causing a significant increase in electricity demand especially in summer months. As a further complication to this scenario, in the last few years there has been a consistent increase in the penetration of intermittent renewable energy production, particularly in countries like Italy, Denmark and Germany. In turn, this has implied that gas is being used as a back-up for electricity production when renewables are not available (e.g., to back-up solar plants at night time), thus contributing to tighten the connections between gas and electricity. Due to the particular development in the 80 and 90s of thermal energy systems based on the use of natural gas, Italy represents a particular case where the use of thermal energy could be correlated with the consumption of natural gas. Only in recent years, the gas sector was heavily influenced by new constraints emerged with this new integrated vision system of the energy systems. The significant penetration of RES in electricity production has imposed specific logic to the consumption of natural gas in thermoelectric plants, mainly natural gas combined cycle power plants. The topic is diffusely discussed in [17].

2.1. Database definition

For the Italian case, daily gas flows of 2012 are available from the database of SNAM network national splitted to intended use (Fig. 1), [18]; the raw data set, containing data as-is from the Snam Rete Gas database has been used. The weather is represented by the average temperature data set, obtained from a specific database, [19]. The specific year has been selected because in the year between 2011 and 2014 Italy has been interested in a great development of renewable energy systems (PV, Wind, Bioenergy). Both the use of heat pumps and of thermal energy systems based on biomass has been largely increased. As additional element, we can also observe that in 2013 and 2014 the effect of economic crisis has influenced the consumption of natural gas, so that the year 2012 must be the last year in which it is exactly clear the connection between thermal energy use and natural gas consumption. Fig. 1 provides the volume of gas natural consumption in Italy in 2012, to support the previous discussion regarding the difficulty of predicting natural gas demand. The natural gas consumption data are represented in aggregated form and in the three main customer components (gas for thermoelectric plants, gas for industrial sector and gas for civilian/urbane sector). The effect of season is represented by an average daily value of the temperature registered in the observed day in a particular town of Central Italy. Fig. 1 shows the strong fluctuation of the natural gas demand, expressed in Mm^3 ($1\ Mm^3 = 10^6\ m^3$), taking into account the main sectors of use. As can be seen in Fig. 1, gas for heating varies in a significant way depending on the time of the year (cold vs. warm seasons); but gas for thermoelectricity remains of the same order of magnitude over the year, and fluctuates according to other dynamics (e.g., weekly patterns). Overall, it is not trivial to predict accurately the natural gas demand. Residential consumption of natural gas is determined by the demand for heating, sanitary water and cooking facilities in residential buildings.

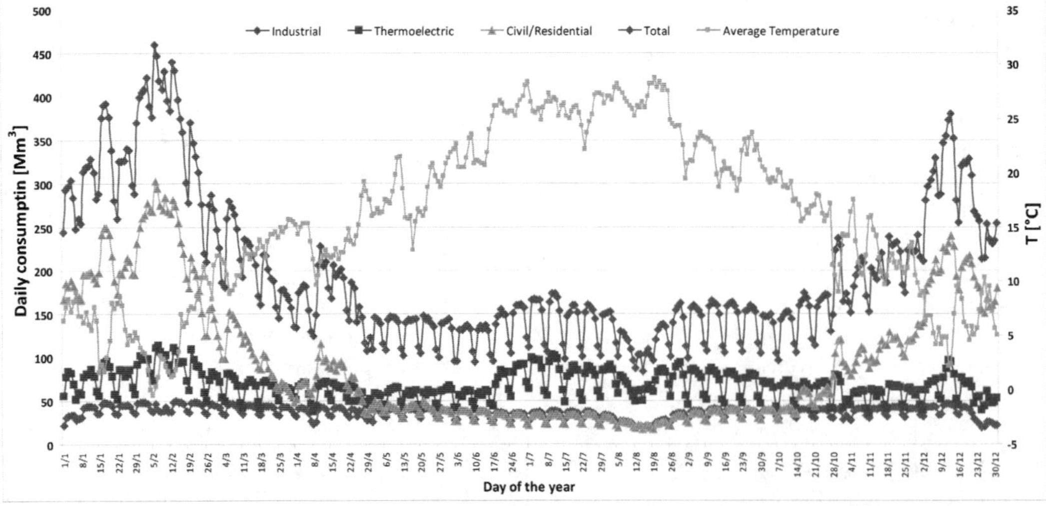

Figure 1. Consumption of natural gas in Italy (reference year 2012)

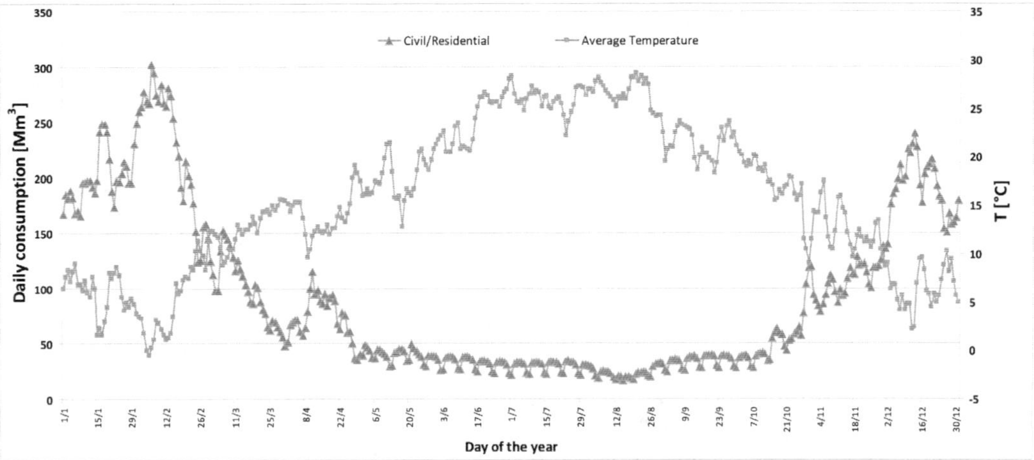

Figure 2. Consumption of natural gas in civil sector (reference year 2012)

In this case the main consumption drivers are the external temperature during the heating season, the population and the buildings characteristics (i.e. insulation, facilities, etc.). Non-residential natural gas consumption represents the use related to economic activities, particularly industry and services (i.e. offices, shops, etc.). Concerning natural gas consumed in the industrial sector, this is mainly used in production processes, even though a substantial share is also utilized for heating purposes (i.e. heating of very large industrial buildings), whereas the consumption in the service sector is substantially due to heating demand. Natural gas consumption in power plants is linked to the electricity market, because it is directly referred to the electricity generated in gas fuelled power plants and is not dependent on thermal energy use. A clear link between natural gas consumption and average outside temperature is evidenced. But the influence of other elements, like the day of the week (in particular working days and holidays) cannot be neglected. By making an initial clustering of the data between the various sector, analyzing the trend of the civilian consumption during the year (Fig. 2) it is possible to clearly identify that the natural gas demand depends not only on the outside temperature.

2.2. Preliminary analysis of the data

Fig. 1 shows clearly that the natural gas consumption, directly related to heat load and thermal energy consumption is clearly correlated to seasonality patterns. By visual analysis of the data if is possible to propose two possible clusters related to two different meteorological seasons: the "cold" season (starting from the middle of October to the end of April) in which the natural gas consumption can be clearly linked to the average external temperature and the "warm" season (from May to middle of October), in which the consumption is not particularly influenced by the outdoor temperature. Analyzing the aggregated data of natural gas consumption it can be also evidenced that the gas consumption data, both in the "cold" season and in the "warm" season rely on the specific type of the day, e.g., working days, vs. holidays, vs. pre-holidays, two weeks of August. The gas consumption is smaller on Saturdays (some industries and some activities are closed) and even smaller on Sundays.

3. Data analysis and clustering of natural gas consumption data: research of peculiar patterns

Classification and clustering of time series date is recognized as an important area of research for energy data both considering demand and production. Clustering refers to the ability to aggregate similar data together and the basic clustering operation corresponds to take a set of a given number n of objects and/or data and group them into k clusters. A good clustering method has predictive power; in this case: clusters can be used to predict future trend of the data, for example concerning energy consumption. Moreover clusters allow compressing a great number of information into a single or a reduced number of information. In addition clusters permits to identify the "outliers", i.e., the cases in which clusters fail to accurately represent particular data. The objective of this section is to propose possible clustering of natural gas consumption data. Analyzing natural gas consumption data, the seasonality effects connected with the average temperature can be clearly observed with sensibly higher consumption in winter, lower consumption values in autumn and spring and the lowest values in summer. But it is also interesting to distinguish the various components of natural gas consumption, in particular referred to industrial load, civil/residential load and thermoelectric load: the correlation with the temperature is different in the three cases. Also, three main types of energy consumption users can be identified. The first type (majority of users) is characterised by the domestic consumers who consume a significant amount of energy also during the night time. The second type of consumers is represented by the tertiary activities, having a quasi-flat consumption profile during the day. The third type involves industrial use of natural gas, considering thermal energy and other industrial uses.

3.1. Detailed analysis of the data

Concerning the residential usage of natural gas, the analysis of the consumption data shows that the consumption can be particularly referred to seasonality pattern: this as can be evidenced by the strong link with the average external temperature, mainly when the temperature is below 15 °C. However different profiles in different days of the week can be also evidenced. It can be easily seen by visual inspection that natural gas consumption data rely on the specific type of the day, e.g., working days, vs. holidays, vs. preholidays. The consumption is smaller on Saturdays, due to the fact that some activities (like offices and schools) are closed and even smaller on Sundays (a lot of commercial activities are closed too). This suggests that the load can be splitted into two or three clusters corresponding to the category of the day of the week (e.g., working days, preholidays like Saturdays and holidays). In case of industrial users, by classifying load consumption into two categories (working days and holidays), it is possible to identify two clusters while the natural gas consumption appears to be less connected with the outside temperature. Visual inspection of the aggregated natural gas consumption data concerning the civil sector, reported in Fig. 2, suggests that the consumption, for a well defined value of the outside temperature is different depending on whether the day is a working day or not. More generally, it is possible to recognise whether a given datum belongs to a holiday (i.e., Sundays, other Italian holidays, Christmas day), to a working day, or to a pre-holiday. With the term preholidays it is defined the set of Saturdays, the days within two holidays when most offices are closed, and the working days during conventional Italian summer and Christmas holidays (i.e., the two weeks around August 15 and the two weeks between Christmas and January 6). Fig. 3 and 4 represents the link of daily consumption of natural gas and average outside temperature in the weekdays and in all the other days, grouping preholidays, holidays and special days, defined in Fig. 5.

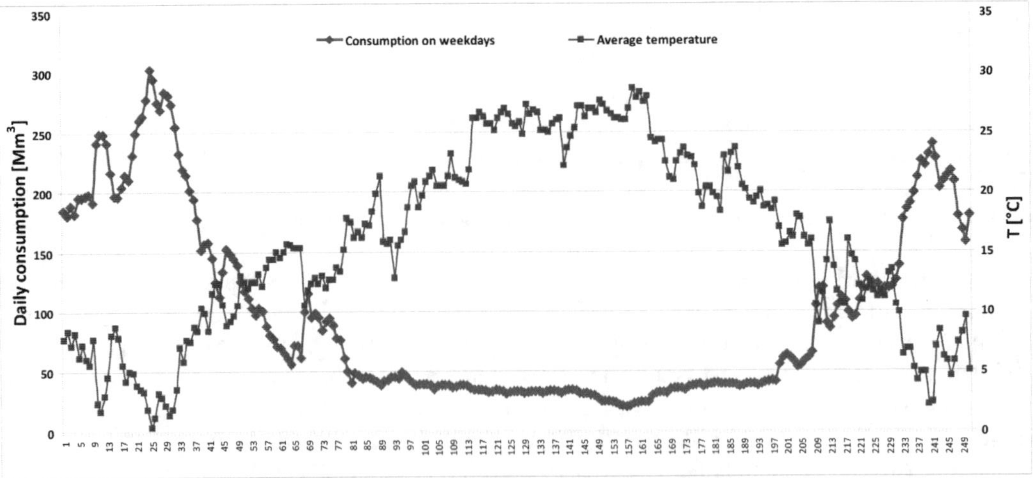

Figure 3. Natural gas consumption data during the days of the week

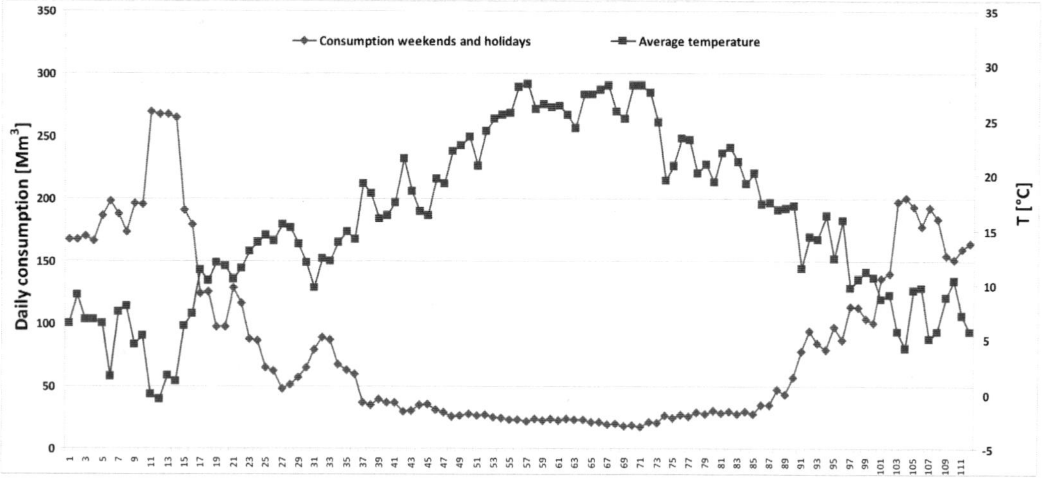

Figure 4. Natural gas consumption data during preholidays, holidays and special days

Mon	1	2	3	4	5	6	7	8	9	10	11	12	13	14	15	16	17	18	19	20	21	22	23	24	25	26	27	28	29	30	31
Jan	S	S	S	S	S	S	H	H	W	W	W	W	W	P	H	W	W	W	W	W	P	H	W	W	W	W	W	P	H	W	W
Feb	W	W	W	P	H	W	S	S	S	W	P	H	W	W	W	W	W	P	H	W	W	W	W	W	P	H	W	W	W		
Mar	W	W	P	H	W	W	W	W	W	P	H	W	W	W	W	W	P	H	W	W	W	W	W	P	H	W	W	W	W	W	P
Apr	H	W	W	W	W	P	H	S	S	W	W	W	W	P	H	W	W	W	W	W	P	H	W	W	P	W	W	P	H	P	
May	S	W	W	W	P	H	W	W	W	W	W	P	H	W	W	W	W	W	P	H	W	W	W	W	W	P	H	W	W	W	W
Jun	P	S	H	W	W	W	W	W	P	H	W	W	W	W	W	P	H	W	W	W	W	W	P	H	W	W	W	W	W	P	
Jul	H	W	W	W	W	W	P	H	W	W	W	W	W	P	H	W	W	W	W	W	P	H	W	W	W	W	W	P	H	W	W
Aug	W	W	W	P	H	W	W	W	W	W	S	S	S	S	S	S	S	S	S	W	W	W	W	W	P	H	W	W	W	W	W
Sep	P	H	W	W	W	W	W	P	H	W	W	W	W	W	P	H	W	W	W	W	W	P	H	W	W	W	W	W	P	H	
Oct	W	W	W	W	W	P	H	W	W	W	W	W	P	H	W	W	W	W	W	P	H	W	W	W	W	W	P	H	W	W	W
Nov	S	P	P	H	W	W	W	W	W	P	H	W	W	W	W	W	P	H	W	W	W	W	W	P	H	W	W	W	W	W	
Dec	P	H	W	W	W	W	W	P	H	W	W	W	W	W	P	H	W	W	W	W	W	P	H	S	S	S	S	S	S	S	S

Figure 5. Working days (W), preholidays (P), holidays (H), special days (S) during 2012 in Italy.

The data on the horizontal axis represents the progressive number of the days according to Fig. 5, in which all holidays, pre-holidays, working days and special days in Italy during 2012 are defined. The methods of short and mid term forecasting can be various: among them similar-day approach, regression methods, time series, neural networks, expert systems and fuzzy logic. Considering the aggregated data analyzed, a multiplicative model seems to be useful in order to predict the natural gas consumption in the civil sector in Italy. The focus of the study is in particular the winter data. For predicting the consumption data, two different types of models can be proposed. An additive model that takes the form of predicting consumption as the function of four components:

$$C(t) = C_N + C_w + C_s + C_r \tag{1}$$

where C is the total consumption, C_N represents the "normal" part of the consumption, which is a set of standardized load shapes for each "type" of day that has been identified as occurring throughout the year, C_w represents the weather sensitive part, C_s is a component that creates a deviation from the usual pattern, C_r is a random term. As alternative model, a multiplicative model may be of the form

$$C(t) = C_N \cdot f(w) \cdot f(s) \cdot f(r) \tag{2}$$

where C_N is the normal (base) consumption and the correction factors, f(w), f(s), and f(r) are positive numbers. These corrections are based on current weather, f(w), special events, f(s) and random fluctuation, f(r). Factors such as pricing, f(p), can also be included. After analysis of the raw data and comparison of several models the conclusion is that the following model, that combines additive and multiplicative model can be the most accurate. The mathematical formulation can be:

$$C(t) = F(d(t) \cdot f(w(t))) + R(t) \tag{3}$$

where C(t) is the actual consumption at time t, d(t) is the day of the week, F(d) is the daily component, w(t) is a function of the weather data that include the temperature, humidity and wind chill, f(w) is the weather function, and R(t) is a term of correction (random). The term $w(t)$ is a vector that consists of the current and lagged weather variables. This last term reflects the fact that thermal energy uses depends not only on the current weather conditions but also on the weather during the previous days. In particular special conditions occur when the cold weather continues for several days or when a cold day arrives after some "hot" days.

Analysing the consumption of natural gas referred to industrial sector (Fig. 6), it is possible to evidence the reduced dependence on the temperature and a most important influence of the day of the week: a remarkable difference between working days and holidays can be clearly evidenced. Elasticity values with respect to the temperature have been estimated, for the natural gas consumption in industrial sector. Fig. 7 reports the link between the daily consumption and the temperature in the weekdays, while Fig. 8, 9 and 10 provides the same data for the three different types of days: pre-holidays, holidays and special days of the year 2012. The days are defined using the reference data of Fig. 5. As it can be evidenced from the various figures, there is a basic consumption level and an additional amount directly linked to the temperature. It is possible to conclude this analysis putting in evidence that if in the case of natural gas consumption in the civil/residential sector, the link with weather conditions and temperature is clearly evident (the difference between the minimum level is remarkable, going from 20-30 Mm^3 for each day up to 300 Mm^3 for each day in the coldest period of the year), the industrial sector appears to be different. As it can be evidenced in Figs 7-9, a minimum level of 20 Mm^3 for each day of the year is consumed, while an additional amount of maximum 30 Mm^3 for each day can be related to the temperature and to the day of the week. Analyzing the data of natural gas consumption in the industrial sector, an additive model can be proposed for prediction:

$$C = C_N + C_w + C_d + C_R \tag{4}$$

where C is the total consumption, C_N represents the "baseload" component, which is a set of standardized load shapes for each "type" of day that has been identified as occurring throughout the year, C_w represents the weather sensitive part, C_d is a special day component that create a substantial deviation from the usual load pattern and C_R is a completely random term, the typical noise term.

3.2. Discussion, use of the present analysis and possible structure of thermal energy use forecasting

The analysis of the aggregated natural gas data consumption permits some interesting considerations on the use of thermal energy and for the definition of models for short and mid term forecasting.

In principle it seems that both for natural gas consumption in Italy (strongly linked to the thermal energy end uses) and for thermal energy use in general, a combination of end-use model and econometric model can be proposed as the more appropriate one. The end-use approach directly estimates energy consumption by using extensive information on end use and end users. Concerning the civil/residential sectors information about sizes of houses, appliances, the customer use, their age, and so on, are required. Statistical information about customers along with dynamics of change is another additional element for the forecast. In general, weather conditions influence in maximum part both the load and the natural gas consumption. It is clear that forecasted weather parameters are the most important factors for defining each kind of model. So, the accuracy of load forecasting depends on the accuracy of forecasted weather scenarios. But the time factors including the time of the year and the day of the week (mainly the day type) and in principle the hour of the day are important too.

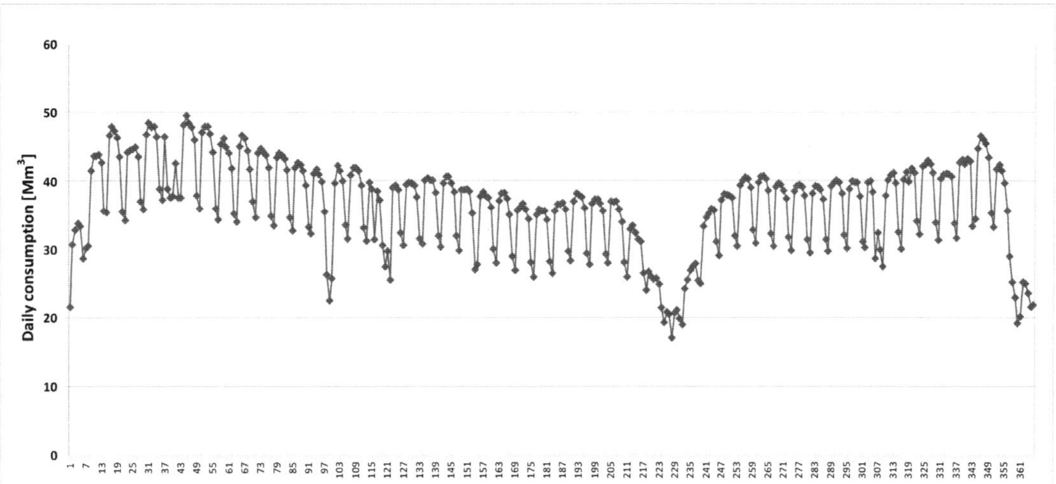

Figure 6. Natural gas consumption in the industrial sector

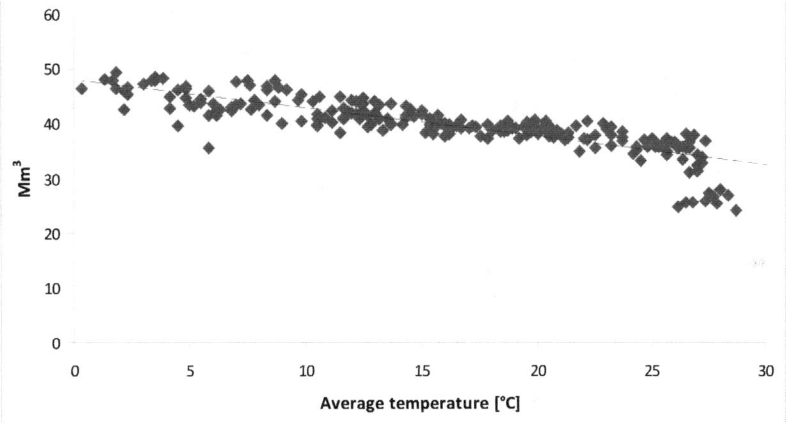

Figure 7. Natural gas consumption trend in industrial sector during the weekdays

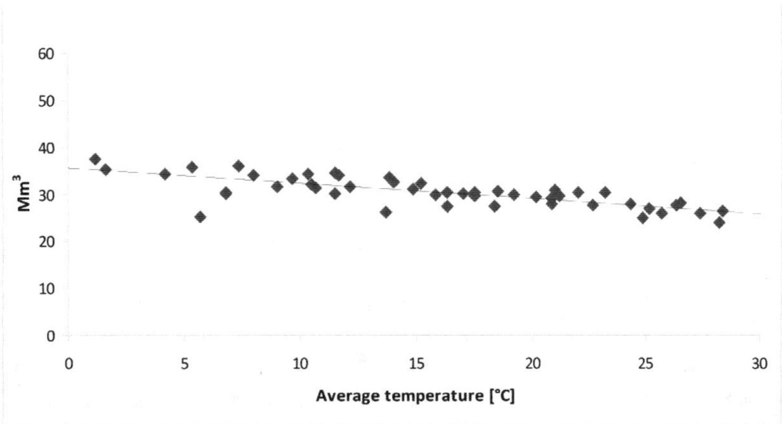

Figure 8. Natural gas consumption trend in industrial sector during the holidays

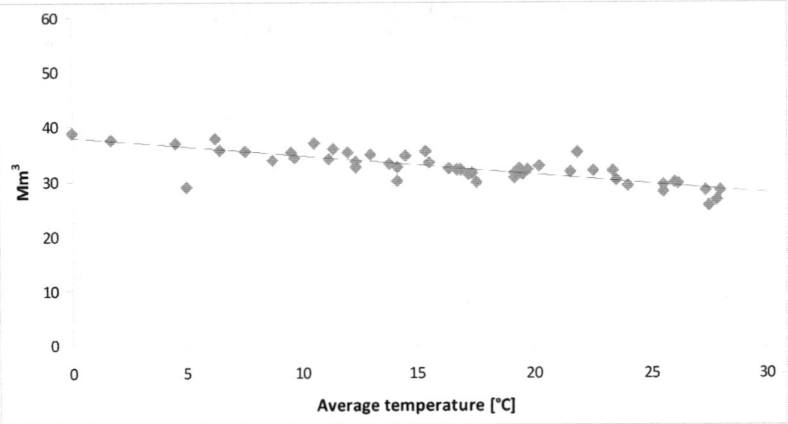

Figure 9. Natural gas consumption trend in industrial sector in the pre-holiday days

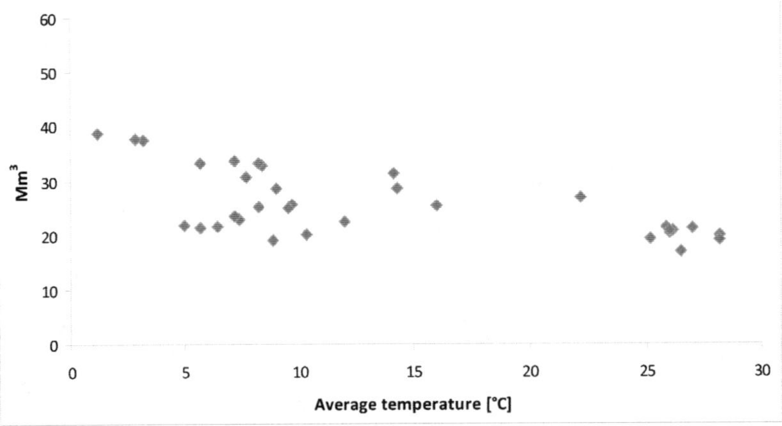

Figure 10. Natural gas consumption trend in industrial sector in the pre-holiday days

While it is really difficult to elaborate hourly models, remarkable differences can be evidenced between weekdays and weekends, mainly in the industrial sector. For this reason, a combination of additive and multiplicative models can be used to develop a model for thermal energy use and heat load forecasting. Economic factors such as per capita incomes and natural gas prices can influence the thermal energy use and natural gas consumption and can be included in econometric models.

4. Conclusions

This paper reports the first step of the development of a method for the forecast of thermal energy use of a complex system. The first part provides a preliminary analysis of the forecasting of energy uses. Considering the particular link of thermal energy use and natural gas consumption the analysis is then focused on a specific objective: to analyze and describe the structure of natural gas consumption and the correlation of those data with physical and statistical elements like a given period of the year, a given day in the week and a given customer category. Then the consumption data of one year, in particular 2012, are completely analyzed. The research of some peculiar patterns in natural gas consumption data started from a visual inspection of the aggregated national data. The natural gas consumption data, are organized and clustered; the analysis provides interesting information regarding a typical energy consumption in a given day of the week or in a given season. It also shows the particular link of the natural gas consumption with the weather conditions, mainly represented by the temperature. The consumption in civil/residential sector represents the main component during cold season and is directly connected with the temperature, while the natural gas consumption in the industrial sector is only partially connected with the temperature and part with the day of the week. The authors believe that this information can be used both to summarise the thousands of profile data in a single daily data (e.g., the centroid of the corresponding cluster), and also to support energy providers with some information that can be used to tailor bills according to actual energy consumption. This work represents just the first step of the construction of a model, conceived for the definition of thermal energy use consumption in complex and largely extended energy systems. It can be extended in some directions that are currently under investigation by the authors. The results of the present analysis will be useful for the development of a model for thermal energy forecasting, whose basic elements are briefly introduced in section 3.2.

References

[1] Suganthi L and Samuel Anand A 2012 Energy models for demand forecasting—A review. *Renew. and Sust. Energy Reviews* ***16***, 1223–40.
[2] Huang Z Yu H Peng Z and Zhao M., 2015. Methods and tools for community energy planning: A review. *Renew. and Sust. Energy Reviews* **42**, 1335–48.
[3] Sandels C Widén J and Nordström L 2014 Forecasting household consumer electricity load profiles with a combined physical and behavioral approach. *Appl. Energy* **131**, 267–78.
[4] Ardakani F J and Ardehali M M 2014 Long-term electrical energy consumption forecasting for developing and developed economies based on different optimized models and historical data types. *Energy* **65**, 452-61.
[5] Schellong W 2011, *Chapter 5: "Energy Demand Analysis and Forecast*", Energy Management Systems, Dr Giridhar Kini (Ed.), ISBN: 978-953-307-579-2.
[6] Hernandez L, Baladron C, Aguiar J M, Carro B, Sanchez-Esguevillas A and Lloret J, 2014. Artificial neural networks for short-term load forecasting in microgrids environment. *Energy* **75**, 252-64.
[7] Dilaver O, Dilaver Z and Hunt L C 2014 What drives natural gas consumption in Europe? Analysis and projections. *Journal of Natural Gas Science and Engineering* **19**, 125-36.
[8] Sabo K, Scitovski R, Vazler I and Zekic´-Sušac M., 2011. Mathematical models of natural gas consumption. *Energy Conv. and Management* **52**, 1721–27.
[9] Soldo B, 2012. Forecasting natural gas consumption. *Appl. Energy* **92**, 26–37.
[10] Bianco V, Scarpa F and Tagliafico L A, 2014 Analysis and future outlook of natural gas consumption in the Italian residential sector. *Energy Conv. and Management* **87**, 754–64.
[11] Bianco V, Scarpa F and Tagliafico L A, 2014 Scenario analysis of nonresidential natural gas consumption in Italy. *App. Energy* 113, 392–403.
[12] Tewolde M, Longtin J P, Das S R and Sharma S, 2013 Determining appliance energy usage with a high-resolution metering system for residential natural gas meters. *App. Energy* **108**, 363-72.
[13] NIST U.S., 2010 Guidelines for smart cyber security: vol. 2, privacy and the Smart Grid The Smart Grid, NISTIR 7628.
[14] Rouf I Mustafa H Xu M Xu W Miller R and Gruteser M 2012. Neighborhood watch: security and privacy analysis of automatic meter reading systems, In: Proceedings of the 2012 ACM conference on computer and communications security. Raleigh, North Carolina: 462–73.
[15] Kerr R and Tondro M 2012 Residential feedback devices and programs: opportunities for natural gas. US Department of Energy, Energy Efficiency & Renewable Energy – Report 2012.
[16] Di Castelnuovo M and Fumagalli E 2013 An assessment of the Italian smart gas metering program. *Energy Policy* **60**, 714-21.
[17] Franco A and Salza P 2011 Strategies for optimal penetration of intermittent renewables in complex energy systems based on techno-operational objectives. *Ren. Energy* **36**, 743-53.
[18] Snam Rete Gas 2015 The gas transportation in Italy: day ahead capacity booking, http://www.snamretegas.it/en/services/Thermal_Year_2012_2013/Transportation_capacity/
[19] Il Meteo. 2015 http://www.ilmeteo.it/portale/archivio-meteo/ [in Italian].

Journal of Physics: Conference Series **655** (2015) 012021 doi:10.1088/1742-6596/655/1/012021

A note on calculation of efficiency and emissions from wood and wood pellet stoves

D Petrocelli[1,2], **A M Lezzi**[2]

[1] Ravelli S.r.l., Via Kupfer 31, Palazzolo sull'Oglio, BS 25036, Italy
[2] Dipartimento di Ingegneria Meccanica ed Industriale,
Università degli Studi di Brescia, via Branze 38, 25123 Brescia, Italy

E-mail: d.petrocelli@ravelligroup.it; domenico.petrocelli@unibs.it

Abstract. In recent years, national laws and international regulations have introduced strict limits on efficiency and emissions from woody biomass appliances to promote the diffusion of models characterized by low emissions and high efficiency. The evaluation of efficiency and emissions is made during the certification process which consists in standardized tests. Standards prescribe the procedures to be followed during tests and the relations to be used to determine the mean value of efficiency and emissions. As a matter of fact these values are calculated using flue gas temperature and composition averaged over the whole test period, lasting from 1 to 6 hours.

Typically, in wood appliances the fuel burning rate is not constant and this leads to a considerable variation in time of composition and flow rate of the flue gas. In this paper we show that this fact may cause significant differences between emission values calculated according to standards and those obtained integrating over the test period the instantaneous mass and energy balances. In addition, we propose some approximated relations and a method for wood stoves which supply more accurate results than those calculated according to standards. These relations can be easily implemented in a computer controlled data acquisition systems.

1. Introduction

Wood is one of the most ancient sources of energy for residential heating. In some areas it is still the most largely used fuel, mainly in the form of wood logs. In the last twenty years use of other forms, such as wood chips and pellets, has increased very much.

Although wood is a renewable energy source, old and inefficient appliances are a source of ambient air pollution in many urban areas. For this reason several European countries are trying to promote the development and diffusion of wood-burning appliances that ensure the best utilization of fuel. This goal is pursued through enforcement of more strict limits on efficiency and emissions from appliances and through subsidies to the more efficient models. The evaluation of efficiency and emissions is made during the certification process that consists in a prescribed series of tests made according to national or international standards. Standards on wood-burning appliances have been frequently revised, lately. In most of the revisions only standard's sections regarding construction and safety requirements are updated. Sections reporting methods and equations to compute or to determine efficiency, emissions and related quantities are usually unchanged.

In woody biomass appliances the fuel burning rate is not constant and this leads to a considerable variation in time of concentration and temperature of flue gas. In appliances such as wood stoves in particular, the burning rate at the beginning of the test period is very different from that at the end. Also combustion in pellet stoves and boilers is not strictly stationary. In order to take this intrinsically unsteady behavior into consideration, standards and certification procedures prescribe to calculate efficiency and emissions using as input data average values of temperature and composition. The average is taken over the entire test duration, ranging from 1 up to 6 hours depending on the type of appliance. However, it seems more correct to determine efficiency and emissions using instantaneous temperature and composition values, and then averaging the results. Differences between values obtained by the two approaches are not negligible and affect, in our opinion, the correct evaluation of appliance performance.

In this paper, in Sec. 2 we recall the main relations to calculate efficiency and emissions on the basis of flue gas data collected during a typical testing session. In Secs. 3 and 4 we present characteristic features of combustion and testing of wood and pellet stoves, respectively. In Sec. 5 we propose approximated relations and methods that can be used for the calculation of efficiency and emissions for both wood and pellet appliances. Finally in Sec. 6 experimental data for pellet stoves and wood stoves post-processed with equations introduced in Sec. 5 are compared and discussed with those obtained according to standards' methods.

2. Combustion of wood

Wood is mainly composed by carbon, hydrogen, oxygen and nitrogen, plus traces of other elements that will be neglected here. Elements are combined in complex molecules as such cellulose, hemicellulose, lignin. The products of complete combustion are carbon dioxide and water. In real combustion other products are present: CO, NOx, hydrocarbons, dust, etc. They are the result of a non-ideal combustion and they are often unwanted because pollutant. Emissions are expressed as the quantity – either total volume or total mass – of pollutant emitted, referred to the quantity of burned fuel. In standards, emission values are usually supplied in $\mathrm{mg/Nm^3}$ referred to a fixed quantity of $O_{2,\mathrm{ref}}$, i.e. the mass of substance present in 1 $\mathrm{m^3}$ of flue gas at normal conditions (0 °C, 101.325 kPa) and characterized by oxygen concentration equal to $O_{2,\mathrm{ref}}$. Actually combustion does not occur in such a way that concentration of oxygen in flue gas is $O_{2,\mathrm{ref}}$, but reporting data in this way it is necessary for a meaningful comparison of different appliances. Another way to report the concentration of a pollutant is in mg/MJ, i.e. the mass of pollutant emitted referred to the input energy $m\,\mathrm{LHV}$. The analysis of the combustion can be done starting from the composition of fuel from elemental analysis and the composition of flue gas. Common gas analyzer can measure the volume concentration in the dry part of flue gases of O_2, CO_2, CO, NO_x, HC, etc. For this reason, flue gas is ideally divided in dry part and water vapor. The dry part of flue gas for appliances that burn wood is mainly composed by N_2, O_2, CO_2, CO, NO_x, HC. Since CO, NO_x and HC concentrations are typically two or three order of magnitude lower than those of N_2, O_2 and CO_2, they are neglected for calculation of volume and mass flow rate. Here we consider the CO, so that we can make a comparison with relations of standards. The goal of the following analysis is to recall some useful relations for the evaluation of emitted substances and energy content of flue gas.

Treating flue gas as an ideal mixture of ideal gases, the volume flow rate of dry flue gas at normal conditions is

$$\dot{V}_f = \dot{m}\frac{22.4\,C}{12\,(y_{\mathrm{CO}} + y_{\mathrm{CO_2}})} \tag{1}$$

where $\dot{m}$ is the burning rate, y_{CO} and $y_{\mathrm{CO_2}}$ are the volume concentration of CO and CO_2, respectively, and C the mass fraction of carbon in the fuel; all quantities are in SI units. It's worthwhile to mention that in standards the ratio 12/22.4 is expressed as 0.536. The CO mass flow rate can be expressed as

$$\dot{m}_{\mathrm{CO}} = \dot{m}\frac{7C}{3}\frac{y_{\mathrm{CO}}}{y_{\mathrm{CO}} + y_{\mathrm{CO_2}}} \tag{2}$$

Regarding the water vapor, it derives from the evaporation of moisture of the fuel and oxidation of hydrogen, hence if W is the mass fraction of water in the fuel and H the mass fraction of oxygen the volume flow rate at normal conditions is

$$\dot{V}_w = \frac{22.4}{18}\dot{m}\left(9H + W\right) \tag{3}$$

In standards the ratio 22.4/18 is explicitly indicated as 1.245. Here, we want to represent the emission of a substance, CO for example, as the ratio between the mass of CO which is emitted during a test and the total volume of dry flue gas at normal conditions and at $O_{2,\text{ref}}$. Denoting t_{test} the test duration, the total mass of CO which is emitted can be calculated from the instantaneous values collected during the test,

$$m_{\text{CO}} = \int_0^{t_{\text{test}}} \dot{m}_{\text{CO}}\, dt = C\frac{7}{3}\int_0^{t_{\text{test}}} \dot{m}\frac{y_{\text{CO}}}{y_{\text{CO}} + y_{\text{CO}_2}} dt \tag{4}$$

Since the volume flow rate of flue gas at normal conditions at $O_{2,\text{ref}}$ is

$$\dot{V}_{f\text{ref}} = \dot{V}_f\left(\frac{0.21 - y_{\text{O}_2}}{0.21 - O_{2,\text{ref}}}\right), \tag{5}$$

the total volume of flue gas is given by the following relation

$$V_{f\text{ref}} = \int_0^{t_{\text{test}}} \dot{V}_{f\text{ref}} dt = \frac{C}{0.536}\int_0^{t_{\text{test}}} \frac{\dot{m}}{(y_{\text{CO}_2} + y_{\text{CO}})}\left(\frac{0.21 - y_{\text{O}_2}}{0.21 - O_{2,\text{ref}}}\right) \tag{6}$$

It follows,

$$\frac{m_{CO}}{V_{f\text{ref}}} = \frac{\dfrac{7}{3}\displaystyle\int_0^{t_{\text{test}}} \dot{m}\frac{y_{\text{CO}}}{y_{\text{CO}} + y_{\text{CO}_2}} dt}{\dfrac{1}{0.536}\displaystyle\int_0^{t_{\text{test}}} \frac{\dot{m}}{y_{\text{CO}} + y_{\text{CO}_2}}\left(\frac{0.21 - y_{\text{O}_2}}{0.21 - O_{2,\text{ref}}}\right) dt} \tag{7}$$

A similar equation can be obtained for all substances measurable with a continuous analyzer, i.e. NOx, THC, SOx, etc.

Efficiency of appliances which release heat to the surroundings by convection and radiation is determined through an indirect method by the relation

$$\eta = \frac{Q_{\text{in}} - Q_a - Q_b}{Q_{\text{in}}} \tag{8}$$

where Q_a represents the total thermal losses, while Q_b the chemical losses in flue gas. They are defined by

$$Q_a = \int_0^{t_{\text{test}}} \dot{Q}_a dt \qquad Qb = \int_0^{t_{\text{test}}} \dot{Q}_b dt \tag{9}$$

$\dot{Q}_a$ is the sum of the thermal losses in the dry part of flue gas and in water vapor

$$\dot{Q}_a = \dot{V}_f\, c_{p,\text{dry}}\left(T_f - T_{0,i}\right) + \dot{V}_w\, c_{p,\text{w}}\left(T_f - T_{0,i}\right) \tag{10}$$

where $c_{p,\text{dry}}$ and $c_{p,\text{w}}$ are the volumetric specific heat at constant pressure of the dry part of flue gas and of the water vapor, measured in kJ/m^3 K. $\dot{Q}_b$ is the power that can be obtained from combustion of CO in flue gas, i.e.

$$\dot{Q}_b = \dot{m}_{\text{CO}}\, LHV_{\text{CO}} = \dot{m}\, C\frac{7}{3}\frac{y_{\text{CO}}}{y_{\text{CO}} + y_{\text{CO}_2}} LHV_{\text{CO}} \tag{11}$$

Since $Q_{\text{in}} = m\,LHV$, the expression of efficiency is

$$\eta = 1 - \frac{\int_0^{t_{\text{test}}} \dot{Q}_{a,i}dt}{m\,LHV} - \frac{\int_0^{t_{\text{test}}} \dot{Q}_{b,i}dt}{m\,LHV} = 1 - q_{a\,\%,i} - q_{b\,\%,i} \tag{12}$$

For example, $c_{p,\text{dry}}$ depends on T_f and y_{CO_2} and is represented by a polynomial; $c_{p,\text{w}}$ is a function of T_f. The other contribution is chemical losses. Since they are related to the concentration of CO in flue gas, and concentration of CO is limited by standards and laws at 400 ppm, its contribution to total efficiency is of the order of magnitude of 0.1%.

3. Wood stoves

Wood stoves are simple appliances for combustion of wood, usually in logs. In a wood stove combustion takes place in the combustion chamber. A door permits the charge of wood logs in combustion chamber. Often, in modern wood stoves, on the door is present a glass window that allows to see the flame and contributes to the release of heat to the room. Combustion air is usually supplied from the bottom grid (primary air) and from the top of the chamber (secondary air). The residues of combustion pass through the bottom grid and are collected in an ash box. Some firebricks cover the combustion chamber. They maintain a high temperature in the combustion chamber and, at the same time, regularize the heat flux by acting as thermal inertia. Hot gases formed in combustion process release heat to the body of the stove, usually steel or cast iron, and to the flue pipe that in turn releases heat to the room by natural convection and radiation. At the flue exit of a traditional wood stove it is necessary a natural draught. Regulation of combustion is done by varying the section of the primary air. The draught of the chimney, can be modified with a damper at the flue exit. These regulations are usually based on user's experience. When the appliance is heated up and an adequate fire bed is formed, the wood introduced in combustion chamber ignites after a few seconds receiving heat from embers and wall of the combustion chamber. A similar appliance is the wood-cooker, in which part of the heat power is used to heat the cooking plate and the oven.

3.1. Combustion in wood stoves

Wood log stoves are intermittent burning appliances, i.e. appliances in which refueling of the combustion chamber is done after partial or full consumption of the previous charge of fuel. The combustion of wood logs is characterized by some consecutive phases. Just after the insertion of the wood log in the combustion chamber, it dries with the evaporation of water. Then log temperature increases, volatile gases form from decomposition of wood, and a first flame appears. The decomposition of wood reaches the maximum rate: wood mass rapidly decrease (high burning rate) and CO2 concentration in flue gas is due to the high rate of consumption of oxygen. Finally, during the burn-out phase, burning rate diminishes, tending to zero, and CO2 concentration in flue gas decreases.

The behavior of CO concentration depends on the characteristics of combustion chamber: usually at the beginning of the combustion phase increases with CO2 (low excess air), then when CO2 decreases, it reaches the lowest values and finally increases during the burn-out phase, when combustion takes place with high excess air (see figure 4).

3.2. Testing methods

Tests for determination of efficiency and emissions of wood burning appliances are made according to EN 13240 [1] for wood stoves and according to EN 12815 [2] for wood cookers. They consist in a pre-test period and one or more test periods. During pre-test the appliance heats-up and on the grate an adequate fire bed has to form, only measurements of test period are taken into account. A test period begins when a specified quantity of fuel is loaded in the combustion chamber and finishes when the weight of the appliance return at the value measured just before the insertion of the fuel charge. The test is valid only if it lasts at least 45 minutes (for wood log). In order to perform test, the appliance must be placed on a platform scale. Temperature and composition of flue gas must be measured at the flue exit. During the test the following quantities must be measured: flue gas temperature, flue gas composition, i.e. the concentration of CO2, CO, NO, THC, with a frequency of at least one sample per minute, room temperature. For stoves with an integrated boiler, inlet and outlet water temperature and water mass flow rate must be measured in order to determine the power transferred to water.

4. Pellet stoves and boilers

Pellet stoves and pellet boilers are modern appliances, which are designed for the combustion of wood pellet. The dimension of pellet particle is suitable for automatic feeding systems that ensure a relatively regular provision of fuel in combustion chamber, usually in a burning pot. Product of combustion are evacuated by a fan placed at the flue exit. All modern pellet stoves and pellet boiler are equipped by an electronic control unit. In simple models, the user sets the power level and a control unit regulates the rotation speed of flue gas fan and of the screw auger for the delivery of fuel in the burning pot; In complex models, the control unit regulates the combustion on the basis of feedbacks from transducers like thermocouples in combustion chamber, lambda sensors, flow meters, etc.

4.1. Combustion in pellet stoves

Combustion in pellet stoves and pellet boilers takes place continuously. For this reason phases of combustion are not distinguishable, because at the same time in the burning pot there are particles just inserted that dries, particles that are in the phase of pyrolysis and gasification and other particles at the end of combustion. However, it is not possible to obtain a strictly constant burning rate, so composition and temperature in flue gas fluctuate in time.

4.2. Test methods

Tests for pellet burning appliances, with automatic feeding system, according to EN 14785 [3] for pellet stoves or EN 303-5 [4] for pellet boilers, consist in a pre-test period, during which the stove reaches steady state condition of operation, and a test period lasting 6 hours for boilers or two test periods lasting 3 hours for pellet stoves. As in tests on wood stoves, also in this kind of test flue gas temperature, flue gas composition, and room temperature, must be measured and recorded at least once every minute.

5. Emission and efficiency calculation

5.1. Method of standards

According to standards, the concentration of a pollutant substance X referred to the reference concentration of oxygen $O_{2,\mathrm{ref}}$, on the basis of data measured during a test is

$$X \text{ content} = \overline{X}\frac{0.21 - O_{2,\mathrm{ref}}}{0.21 - \overline{y_{O_2}}} \tag{13}$$

where $\overline{X}$ and $\overline{y_{O_2}}$ are, respectively, the time average of X and O_2 concentration over the whole test period.

Efficiency is determined by equation (12), where $q_{a\%,i}$ and $q_{b\%,i}$ are calculated using the average CO and CO_2 concentration and the average flue gas temperature. For both pellet and wood appliances, standards take into account only the average over the whole test period of all data measured. For pellet boiler, only heat transferred to water is considered useful, therefore the efficiency is calculated with the "direct method" , i.e.

$$\eta_{\text{direct}} = \frac{Q_{\text{water}}}{m \, \text{LHV}} \tag{14}$$

where the power transferred to water is calculated from the measure of water mass flow rate and of water temperature increment in the boiler.

5.2. Emissions of pellet appliances

As mentioned in section 4.1, burning rate is not strictly constant, then also composition and flow rate of flue gas varies in time. CO emission is strongly correlated to excess air [5]: in particular, it results low in an optimal range of excess air, that is to say in an optimal range of oxygen concentration, while for higher or lower values of excess air it increases rapidly [6]. Concentration of a gas in flue gas depends on the concentration of other gases. For example, when combustion take place with low excess air, the concentration of oxygen is low and, consequently, the concentration of all other gases – pollutants included – is higher. Conversely, when combustion takes place with high excess air, concentration of oxygen is high, therefore the concentration of other substances results lower. Therefore, the same concentration of CO at different O2 concentrations corresponds to different total emissions: the mass of CO emitted when oxygen level is low is less than the mass of CO emitted when oxygen level is high. Equation (13) does not distinguish from the two cases, because only the overall average of O_2 and CO concentration are used. Equation (7) is not affected by this and it can be used to determine a more accurate value for pollutant concentration. However, for pellet appliances is not possible to measure the burning rate $\dot{m}\,(t)$ accurately, because the resolution of the platform scale, that measure the overall weight of the appliance, is not enough, often 10 or 50 grams. For the accurate measurement of pollutant emissions, Win et al. [7] measure directly the flue gas flow rate. In our opinion, this method is too complicated to be proposed for the test procedure of the Standards. What we can do is to consider $\dot{m}$ approximately constant, since it fluctuates randomly around its mean value. Furthermore, since $y_{\text{CO}} \ll y_{\text{CO}_2}$, we can approximate $y_{\text{CO}} + y_{\text{CO}_2} \sim y_{\text{CO}_2}$. Then, the denominator in Eq. (7) can be simplified as follows

$$\int_0^{t_{\text{test}}} \frac{\dot{m}}{y_{\text{CO}_2}} \left(\frac{0.21 - y_{\text{O}_2}}{0.21 - O_{2,\text{ref}}} \right) dt = \dot{m} \int_0^{t_{\text{test}}} \frac{1}{y_{\text{CO}_2}} \left(\frac{0.21 - y_{\text{O}_2}}{0.21 - O_{2,\text{ref}}} \right) dt = \dot{m} \int_0^{t_{\text{test}}} v_{\text{f,ref}} \, dt = \dot{m} \, v_{\text{f,ref}} \, t_{\text{test}} \tag{15}$$

where $v_{\text{f,ref}}$ represents the volume of flue gas at normal conditions when concentration of oxygen is equal to $O_{2,\text{ref}}$ per mass of fuel burned. Eq. (7) can be rewritten

$$\frac{m_{\text{CO}}}{V_{\text{f,ref}}} = \frac{1.25}{t_{\text{test}}} \int_0^{t_{\text{test}}} \frac{y_{\text{CO}}}{y_{\text{CO}_2}} \frac{1}{v_{\text{f,ref}}} dt = \frac{1.25}{t_{\text{test}}} \int_0^{t_{\text{test}}} y_{\text{CO}} \left(\frac{0.21 - O_{2,\text{ref}}}{0.21 - y_{\text{O}_{2,\text{i}}}} \right) dt \tag{16}$$

Therefore the CO content is the time average of instantaneous CO concentration referred at $O_{2,\text{ref}}$. The CO emission measured in mg/MJ, can be obtained dividing Eq. (4) by $m \, \text{LHV}$:

$$\frac{m_{\text{CO}}}{Q_{\text{in}}} = \frac{7\,C}{3\text{LHV}} \frac{1}{t_{\text{test}}} \int_0^{t_{\text{test}}} \frac{y_{\text{CO}}}{y_{\text{CO}} + y_{\text{CO}_2}} dt \tag{17}$$

5.3. *Efficiency of pellet appliances*

The efficiency of pellet appliances, can be determined using Eq. (12). Even in this case, if $\dot{m}$ can be considered constant, the terms $q_{a\%}$ and $q_{b\%}$ become:

$$q_{a\%} = \frac{1}{t_{\text{test}}\,\text{LHV}} \int_0^{t_{\text{test}}} \left[\frac{C}{0.536\,(y_{\text{CO}} + y_{\text{CO}_2})} c_{p,\text{dry}} + c_{p,\text{w}}\,(9H + W) \right] (T_F - T_{0,i})\, dt \quad (18)$$

$$q_{b\%} = \frac{12644}{t_{\text{test}}\,\text{LHV}} \int_0^{t_{\text{test}}} \frac{C\, y_{\text{CO}}}{0.536\,(y_{\text{CO}} + y_{\text{CO}_2})} dt \quad (19)$$

Hence, terms $q_{a\%}$ and $q_{b\%}$, are the time average of their instantaneous values.

5.4. *Emissions of wood appliances*

With reference to Eq. (2) the time history of the emission rate (in mass) of CO, as of any other pollutant, depends on the time histories of CO concentration in flue gas and of the burning rate. In intermittent wood appliances, like wood stoves, the burning rate decreases in time rather smoothly [8, 9]. CO emissions can be determined accurately upon use of Eq. (7). For the measurement of burning rate, Calvo et al. [8] measure directly the weight of the combustion grid of a wood stove by means a load cell, but this implies that the appliance must be modified. Eq. (13) prescribed by standards leads to a value of CO emissions which only approximates the result of Eq. (7). For wood appliances, we suggest here a better approximation that can be obtained by the following method:

(i) divide the test period in n sub-periods, so that for each sub-period the consumption of wood is m/n of the total mass m of the batch;

(ii) apply Eq. (13) proposed by standards, not to the entire test period, but to each single sub-period, that means to calculate the value of CO emission at $O_{2,\text{ref}}$ for each sub-period;

(iii) calculate the value of CO as the mean of the n CO emission values calculated for each sub-period: this corresponds to a weighted average where the weights are the fuel mass burned in the sub-period.

The number of sub-periods n must be chosen univocally in order to prevent many interpretations. For wood stoves we have obtained good results upon setting $n = 3$. In general, n should be chosen on the basis of the weight of the batch of fuel used for each test period. In fact, in each interval the burned wood must be large enough with respect to the resolution of the platform scale, usually between 10 and 50 grams.

5.5. *Efficiency of wood appliances*

At the same way, for efficiency calculation it is possible to apply the same procedure of emissions calculation, i.e. calculate the efficiency as proposed by standard for each of n sub-periods, then calculate the value of efficiency as the average of the n values.

6. Examples and discussion

In this section we want to show that there may be significant differences between results obtained from exact Eqs. (7), (10), (11), and (12) and the approximate relations, like (13), prescribed by standards. In addition we show that the approximate Eqs. (16), (18), and (19) and the method proposed in Sec. 5.4 seem to supply more accurate results than standards.

6.1. *Pellet stove*

We consider two different tests, A and B, performed on pellet stove at Ravelli's Lab according to standard EN 14785. Composition of wood is C = 46.9%, H = 6.3%, W = 6.0%. Its heating

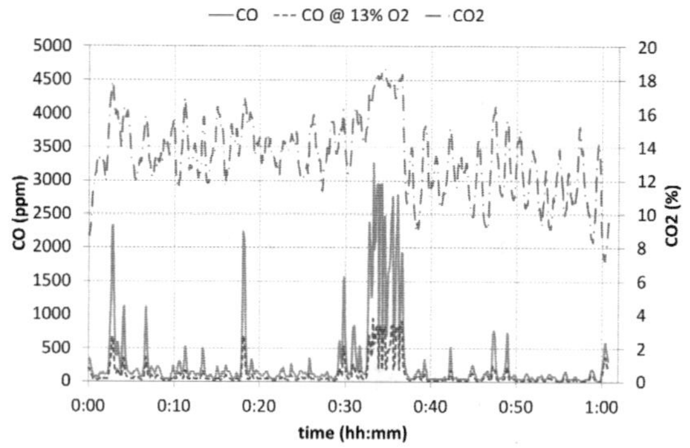

Figure 1. Pellet stove, case A. Concentrations of CO, CO at 13% of O_2, and CO2 vs. time.

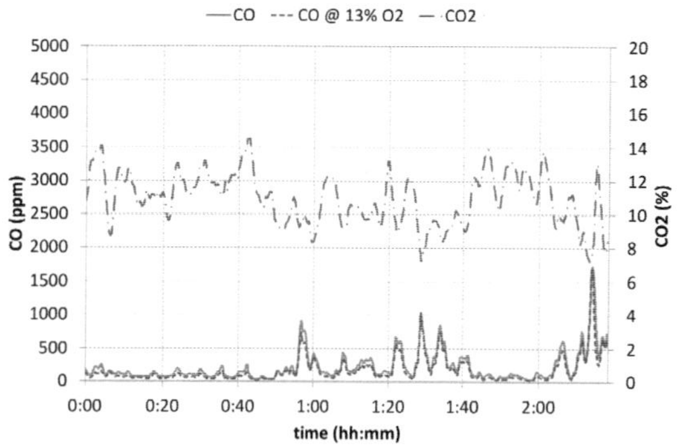

Figure 2. Pellet stove, case B. Concentrations of CO, CO at 13% of O_2, and CO2 vs. time.

Table 1. Measured data (time average) for pellet stoves.

	O_2 %	CO_2 %	CO ppm	T_f °C	T_0 °C
A	6.7	13.4	298	133.5	18.2
B	9.4	10.9	223	185.9	22.3

Table 2. Efficiency and emissions calculated according standards and equations (18) and (19), (16) and (17).

	Efficiency						Emissions			
	Standard			Eqq. (18) and (19)			Standard CO 13% O_2		Eqq. (16) and (17) CO 13% O_2	
	q_a %	q_b %	η %	q_a %	q_b %	η %	mg/Nm^3	mg/MJ	mg/Nm^3	mg/MJ
A	5.95	0.14	93.91	6.11	0.11	93.77	208.4	139.2	183.5	122.0
B	10.12	0.12	89.76	10.29	0.14	89.99	191.3	127.9	213.5	142.3

value is LHV= 17.4 MJ/kg. Time histories of CO and CO2 concentration during tests are shown in figures 1 and 2. Time average values over the entire test are reported in table 1.

Test case A is characterized by a CO peak at low level of excess air. CO concentration at 13% O2, calculated by Eq. (13), i.e. according to standards, is 167 ppm that correspond to 208 mg/Nm^3. The concentration of CO at 13% O2, calculated by Eq. (16) is 147 ppm that corresponds to 184 mg/Nm^3. In test case B, peaks of CO are recorded in correspondence of high levels of excess air. The concentration of CO at 13% O2 calculated according to standards is 153 ppm that corresponds to 191 mg/Nm^3. The concentration of CO at 13% O2 calculated by Eq. (16) is 171 ppm that corresponds to 214 mg/Nm^3. Differences between the two methods are not negligible. For instance, if the limit of CO emission is fixed at 200 mg/Nm^3, as for the German regulation BImSchV for pellet stove with integrated boiler [10], for one method law requirements are satisfied, whereas for the other are not.

Table 2 reports also the efficiency calculated by two methods. For this quantity results do

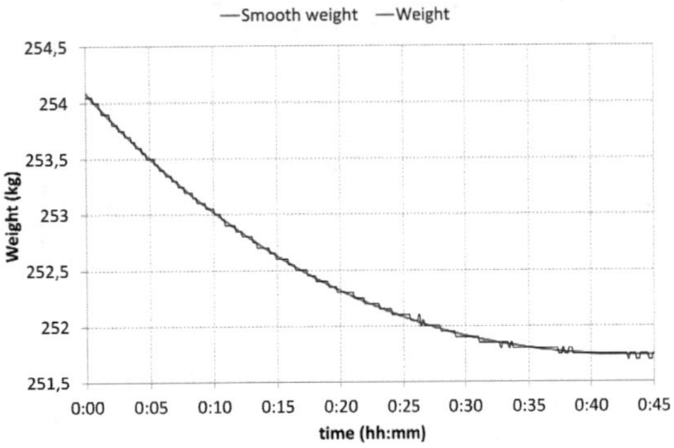

Figure 3. Weight of wood stove during the test.

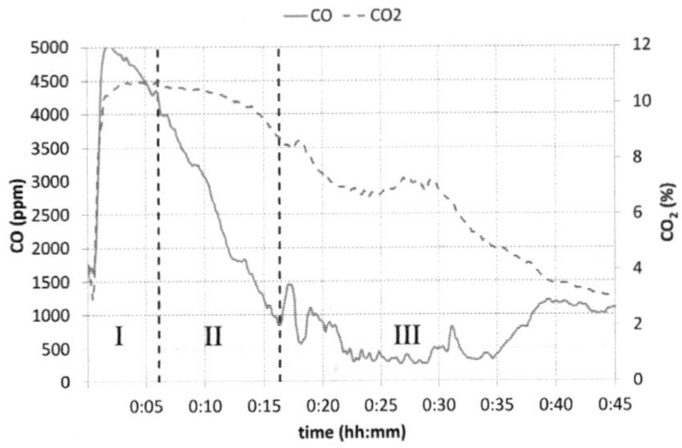

Figure 4. CO and CO_2 concentration in flue gas of wood stove during a test.

agree. In both methods, $c_{p,\text{dry}}$ and $c_{p,\text{w}}$ are calculated using relations proposed by standards.

Although time dependence of burning rate induces time changes of properties that contribute to efficiency, i.e. temperature, specific heats, and volume flow rates, these variations do not seem significant. For instance, taking into account that $c_{p,dry}$, in the ranges of $y_{CO_2} = [0.01, 0.19]$ and $T = [25, 250]$°C varies between 1.303 kJ/m^{3}K and 1.407 kJ/m^{3}K, the variations of $c_{p,dry}$ during a test is small because temperature of flue gas at the exit of a pellet stove at steady state condition of operation varies within a range of 10-15°C around the mean value. From a comparison between the results of thermal losses calculated with expression supplied by standards and thermal losses calculated using the Peng–Robinson model implemented in the software AspenPlus, we found that in the range $T_f = [80, 250]$°C and $y_{CO_2} = [0.03, 0.17]$ thermal losses are overestimated by standards and the error is between 0.4% to 1.3%.

6.2. Wood stove

In this example we report data recorded during a test of a wood stove performed at Ravelli's Lab. Composition of wood is C = 43.2%, H = 5.2%, W = 12.8%. Its heating value is LHV= 15.0 MJ/kg. Some properties related to flue gas and the weight measured during the test are reported In figure 3 stove weight vs. time is shown: the absolute value of time derivative is the burning rate. In figure 4 measured CO and CO2 concentration vs. time are plotted. The weight of the fuel charge was 2.30 kg. The number of sub-divisions of test period is $n = 3$. Duration of test was 45 minutes. In Table 3 we report time average of data over the test period and over each single sub-period. As expected, the first sub-period is the shortest. Consequently, the high CO level reached during the first sub-period will contribute to estimate of overall CO emission more than in the standards' method.

Table 3. Measured data (time average) for wood stove.

	Duration mm:ss	Consumption kg/h	O_2 %	CO_2 %	CO ppm	T_f °C	T_0 °C
1	6:05	7.56	10.6	7.3	4220	248.9	31.9
2	10:05	4.56	10.4	10.1	2449	254.4	32.9
3	28:30	2.15	14.5	5.8	699	226.8	31.8
Overall	45:00	3.06	13.0	7.3	1588	236.1	31.8

Journal of Physics: Conference Series **655** (2015) 012021 doi:10.1088/1742-6596/655/1/012021

Table 4. Results for wood stove.

	Efficiency			Emissions	
	q_a	q_b	η	CO 13% O_2	
	%	%	%	mg/Nm^3	mg/MJ
Real	19.39	1.71	78.89	2345	1694
Standard	21.16	1.51	78.14	1997	1375
Sec.5.5	19.11	1.76	79.11	2482	1708

Table 4 reports the results of test, calculated according to standards and with the procedure proposed in Sec. 5.4. Real emission and real efficiency are computed upon use of Eqs. (7) and (12), where $\dot{m}$ is estimated from the weight history represented in figure 3. There are significant differences between the real values and the ones calculated according to the standards. Differences reduce when applying Eq. (13) to sub-periods and then taking the mean value.

7. Conclusions

In this work we made a review of the methods proposed by European Standards for the calculation of efficiency and emission of wood and wood pellet appliances. We propose some modification to equations prescribed in standards in order to improve the accuracy of calculated data. The methods for the calculation of efficiency and pollutant concentration (CO, NOx, HC) presented here allow to calculate the mean concentration of these substances taking into account both the variation of the flue gas mass flow rate and composition during the test which is a peculiar characteristic of these appliances. In fact, we show that for pellet stoves the method of standard can overestimate or underestimate emissions, while for wood stove there can be considerable differences between the calculated values and real data.

For pellet stove and pellet boilers, we propose equations for the calculation of pollutant concentrations only. As a matter of fact efficiency calculated according to standards is quite accurate.

For wood appliances we propose to apply the same method proposed by standards, but dividing the test period in sub-periods. This method does not introduce new relations and it can be easily implemented in software or spreadsheets.

References

[1] EN 13240+A2 2004 *Roomheaters fired by solid fuel. Requirements and test methods*
[2] EN 12815 2006 *Residential cookers fired by solid fuel. Requirements and test methods*
[3] EN 14785 2006 *Residential space heating appliances fired by wood pellets. Requirements and test methods*
[4] EN 303-5 2012 *Heating boilers - Part 5: Heating boilers for solid fuels, manually and automatically stoked, nominal heat output of up to 500 kW - Terminology, requirements, testing and marking*
[5] D Petrocelli 2013 *Indicazioni sperimentali per la progettazione di stufe a pellet ad alta efficienza* Ph.D. thesis University of Brescia
[6] Petrocelli D and Lezzi A 2014 *J. Phys.: Conf. Ser.* **501** 012036
[7] Win K, Kaung M, Persson T and Bales C 2012 *Atmospheric Environment* **59** 320–327
[8] Calvo A, Tarelho L, Alves C, Duarte M and Nunes T 2014 *Fuel Processing Technology* **126** 222–232
[9] Vicente E, Duarte M, Calvo A, Nunes T, Tarelho L and Alves C 2015 *Fuel Processing Technology* **131** 182–192
[10] 1 BImSchV (Germany) 2010 *Verordnung uber kleine und mittlere Feuerungsanlagen*

Design and experimental analysis of a screened heat pipe for solar applications

D Jafari, S Filippeschi, A Franco, P Di Marco

Università di Pisa, DESTEC, Largo Lucio Lazzarino 2, 56122 Pisa, Italy

E-mail: d.jafari@studenti.unipi.it

Abstract. This paper summarizes the design, the construction and the preliminary results of a transient and steady state investigation of the heat transfer mechanisms of a horizontal heat pipe (HP). The experiments are performed using a custom-made HP constituted by copper tube with outer diameter and length as 35 mm and 510 mm, respectively, with the inner surface covered by three layers stainless steel mesh wick (100 mesh/inch). Water is used as a working fluid. The evaporator section is heated by electrical resistances wrapped around the tube and the cooling system consists of an insulated water manifold with inner diameter of 39 mm, connected to chilled water bath to maintain the inlet temperature of the circulating cooling water at 25 °C for various heat loads (30-100 W). The aims of this activity is to obtain data to verify the steady state HP analytical model already presented by authors at a fixed filling volume and to determine the effect of the heat transfer load on the heat transfer performance of screen mesh HPs. The heat transfer coefficients are determined using thermocouples on the outer wall and within the core of the HP. The agreement between the analytical results and the preliminary experimental data appears to be very good.

1. Introduction

Phase change heat transfer is suitable for transferring large amount of heat because of the high heat transfer coefficients associated with the evaporation and condensation processes. Heat pipe (HP) is one of such a passive device, which utilize phase change of the working fluid inside of a sealed container (generally a pipe), efficiently transporting large amount of heat from evaporator section to the condenser section; The condensed liquid is drawn back to the evaporator by the capillary force in the wick structure laid at the internal periphery of the device [1-3].

HPs have been widely applied for several engineering application such as electronics cooling, air conditioning and power generation [4-7]. HPs can be also used for thermal control, in particular in connection with solar energy systems like solar collectors [8-11], solar walls [12, 13] or photovoltaic systems [14, 15].

The wide use of HPs has led to increased demand to study its thermal performance in order to efficiently designing them, mainly in connection with solar application for medium temperature (50-150°C). Many theoretical studies of the operation of low temperature and high temperature HPs have been reported such as simple lumped model [16] and analytical model [17-19]. However, the focus of the present study is on experimental investigations of the transient and steady state operation of HPs,

Journal of Physics: Conference Series **655** (2015) 012022 doi:10.1088/1742-6596/655/1/012022

with emphasis on obtaining data to verify an existing steady state HP model presented by the authors in [20].

The thermal behaviour in a HP is determined by many parameters, such as heat input, filling ratio of the working fluid, geometry, orientation, characteristics of the capillary structure, and the thermophysical properties of the working fluid. The heat transfer characteristics of the HP have been experimentally investigated by several researchers. Among the others, El-Genk and Huang [21] focused the attention on the transient response of a HP (screen mesh) at different input powers and cooling rates. Wang and Vafai [22] experimentally investigated the transient thermal performances of flat plate HPs. They introduced the concept of HP time constants to describe its transient behaviour and to understanding the start-up and shut-down phenomena of HPs. Kempers et al. [23] considered the effect of the number of mesh layers and filling ratio on the heat transfer performance of copper HPs (screen mesh wicks) using water as working fluid. It was found that the effective thermal resistance decreases with an increase in heat flux, and the thermal resistance is larger for fewer mesh layers. Further, Kempers et al. [24] analysed the heat transfer mechanisms in the condenser and evaporator sections of a 3 layers of screen-mesh HP. Wong and Kao [25] investigated HPs consisting of a two-layer copper mesh wick of 100 and 200 mesh/inch screens using ether as working fluid. They discussed the effects of different wicks and fluid filling ratios on the evaporation/boiling characteristics.

For the design of HPs for particular applications several theoretical models have been developed based on the solution of conservation equations in two phase systems, but they must be joined with an experimental characterization of the heat and mass transfer mechanisms. In fact, the link between fluid-dynamics, thermal behaviour and gravitational effects are not completely known and obtained data are difficult to be used for general cases or extended to particular cases like medium temperature applications. For this reason the authors first developed a model for sensitivity analysis of HP [20] and then designed and constructed an experimental facility to validate the model. This paper describes and summarizes the progress and the preliminary results of an investigation into the heat transfer process of a screen mesh HP for solar thermal application. The test rig allows measurements of the wall temperature and the vapour temperature in the different regions of the HP (evaporator, adiabatic, and condenser sections) as well as vapour pressure and also the thermal power removed by means of the cooling manifold. The purpose of the present study is to characterize the effect of heat load on the transient and steady state operation of copper HPs. The evaporation heat transfer coefficient of the HP is evaluated and the working characteristics and its heat transfer performance are discussed.

2. Principles and operation of HP for the specific application

Figure 1 shows the operation of a cylindrical HP. The components of a HP are pipe wall and end caps, a wick structure and a small amount of working fluid. The length of a HP could be divided into three parts: the evaporator, adiabatic and condenser sections. Heat applied to the evaporator section is conducted through the pipe wall and wick structure, where the working fluid evaporates. The resulting vapour pressure drives the vapour to the condenser, where it releases its latent heat of vaporization to the provided heat sink [2]. The capillary pressure created by the wick pumps the condensed fluid back to the evaporator section. This process will continue as long as there is a sufficient capillary pressure to drive the condensate back to the evaporator. The selection of the wick structure could be an important parameter which depends on several factors: working fluid, pipe material and working conditions.

For designing a HP for specific application like solar collectors, one of the important elements is the characterisation of the wick structure, in particular its permeability, the developed capillary pressure and the effective conductivity [3]. In order to have a large capillary pressure difference, capillary radius must be small. Permeability must be large in order to have a lower liquid pressure drop, a better distribution of the liquid and therefore, higher heat transport capability. A large value of effective thermal conductivity gives a small temperature drop across the wick, which is a favourable condition for the HP performance.

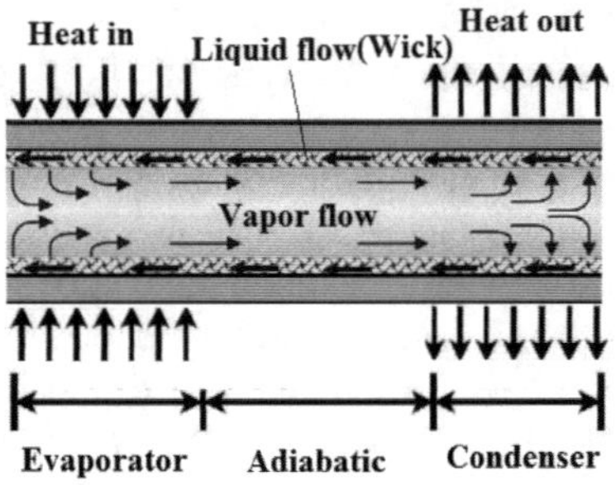

Figure 1. Cylindrical HP: a schematic view.

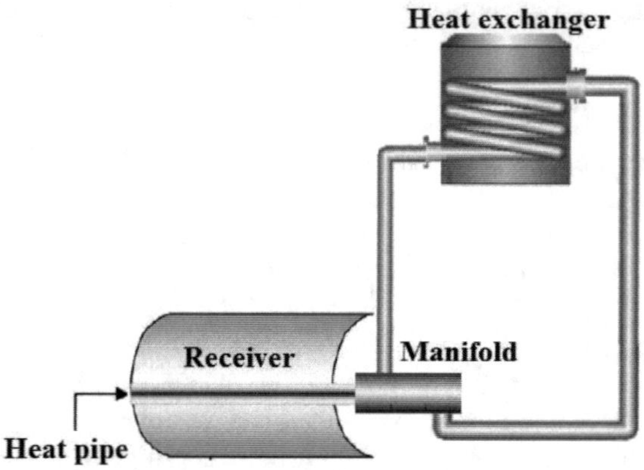

Figure 2. Schematic view of concentrating systems.

Table 1. Expressions for HP wick design (screen mesh)

Parameter	Expression
Effective thermal conductivity	$k_{eff} = k_l\left[(k_l + k_w) - (1-\varepsilon)(k_l - k_w)\right] / (k_l + k_w) + (1-\varepsilon)(k_l - k_w)$
Effective capillary radius	$R_{eff}=1/2N$
Permeability	$K = d_w^2 \varepsilon^3 / 122(1-\varepsilon)^2$
Porosity	$\varepsilon = 1 - S\pi N d_w / 4$

The wick thickness also plays an important role in the determination of the thermal resistance and heat transfer capacity. The screen mesh detailed parameters are calculated according to references [1-3] as evidenced in table 1. Wang and Peterson [26] found that increasing the thickness of the porous layer would correspondingly increase the maximum heat flux based on a capillary limit. Kempers et al. [23] showed that there is a small increase in thermal resistance when increasing the number of mesh screens; however, the maximum heat transfer also increases. The idea pursued in this paper is to conduct an analysis on the operating HP in order to be able to identify criteria for the analysis of thermal performances of such systems in configurations typical of concentrating solar collectors (ranged from 200-10000 W/m^2). The concentrating solar systems using HP are composed a concentrator and a receiver and the receiver is the key component to transfer the energy collected by the concentrator to the heat storage systems. Two particular aspects, that can also be joined together, could be considered for concentrating collectors: the concentration of solar radiation and the use of a fluid with phase change as energy vector. The use of a two-phase fluid flow allows high thermal fluxes to be transferred with very low change in the average temperature, allowing a higher concentration ratio. Figure 2 shows the proposed system including a reflector and a HP.

3. Experimental apparatus

Basing on the objective of the present study and previous discussion and considering the results of the main experimental activities available in the literature the authors have designed, a specific experimental setup aimed to analyse the thermal behaviour of HP. The pipe wall is made of copper with an outer diameter of 35 mm, a thickness of 1 mm and a length of 510 mm. The lengths of evaporator and condenser section are 300 mm and 160 mm, respectively. The working fluid is distilled water. Three layers of stainless steel wick with 100 mesh/inch size is prepared, formed and rolled onto a rod, inserted through the pipe and by means of a specially built device, in order to ensure that the wick and the wall are in good thermal contact. Table 2 and table 3 list the detailed HP and wick parameters. After having assembled, cleaned and sealed the HP using brazing techniques, the pipe is initially evacuated. This process is started at ambient temperature, and then the pumping is continued

to 10^{-4} mbar while the pipe is heated, to favor outgassing. The HP is then filled with the desired amount of degassed, ultra-pure water. Figure 3 shows a schematic diagram of the experimental setup.
The rig consisted of the HP, electrical heating with a voltage controller, cooling flow circuit, vacuum device and measuring system. The evaporator section is uniformly heated using four silicon type thermofoil heaters (model MINCO HK5488R17.2L12A) clamped to the HP using Teflon type insulation. The power input is supplied by a DC Power supply (Agilent DC6575A) which has an accuracy of ±1 percent of reading. In the condenser section, heat is convectively removed by water extracted from a cooling bath (HAKKE F-3C DIN 58966), by means of a 160 mm long copper manifold mounted around the HP, with outside diameter and wall thickness of 42 and 1.25 mm. During the thermal performance tests the temperature of the cooling bath is maintained at an imposed value and the mass flow rate can be exactly controlled. An electromagnetic flow meter (Siemens SITRANS F M MAGFLO5000) allows measurement the mass flow rate of the cooling water with an accuracy of about 1%. The thermocouples in the manifold inlet and outlet and mass flow measurement allow to calculate the power output from condenser section and to compare it with the input electrical power.
All the HP vapours and wall temperatures are measured with T-type thermocouples, which have been calibrated with an accuracy of ±0.2°C. The external surface temperatures of the HP are measured with eight wire thermocouples (T_{w1} to T_{w8}) mounted along the wall in the evaporator, adiabatic and condenser sections. The vapour temperatures are measured by additional five T-type stainless steel thermocouples (T_{v1} to T_{v5}) mounted axially within the HP vapour space. This handmade multipoint thermocouple are attached to the outer surface a PEEK probe (outer diameter of 3 mm) which exists in the condenser end cap and is swaged to provide a well leak seal. The location of wall and vapour thermocouples is shown in figure 5. In the evaporator end cap a pressure transmitter (PTX7511 accuracy of ±0.15%) is installed to measure vapour pressure. All the signals to monitor HP temperature and pressure and cooling mass flow rate are acquired by the Agilent HP32790 data acquisition system, and stored in a computer.

3.1. Experimental procedure
The experimental procedure began after charging the working fluid in the HP. Following evacuation, 42 cm^3 (filling ratio of 10%) of degassed and distilled water are inserted into the pipe through a special valve arrangement, to avoid air contamination. The amount of water is enough in order to ensure that the wick completely saturated. Figure 4 shows the rig used for HP evacuation and charging in the present study. The cooling water supplied to the water jacket has a constant flow rate of 0.02 kg/s and a temperature of 24.8 °C. At the beginning of the test, cooling water is pumped in the condenser jacket before power is supplied. The transient is initiated by increasing the electric power in the evaporator section from 30 to 100 W. The experiment continues until steady state is reached and then a further heat power is applied. The system is considered to have attained steady state when each thermocouple temperature fluctuation is no more than 0.5 °C within 5 min. At the steady state and transient operation, the power input, the water flow rate in the cooling jacket, and the temperature at the various locations are recorded. Energy balance between the heat input by electrical heaters and the heat removed by sink is monitored to ensure an energy balance of at least 90 percent.

3.2. Data reduction and uncertainty analysis
The measured variables in the experimental analysis are: heat rate applied to the evaporator section (Q_{in}), heat output of the condenser section (Q_{out}), inlet of coolant water temperature (T_{in}), outlet temperature (T_{out}), mass flow rate of the coolant water ($\dot{m}$), wall surface temperatures (T_w), vapour temperature (T_v) and vapour pressure. The heat transfer capacity of evaporator section for a HP is determined by heat transfer coefficient (h_e). From the measured data of wall temperature and vapour temperature (equivalent to the wall temperature of adiabatic section), the heat transfer coefficient in the evaporator can be evaluated using the following equation:

$$h_e = Q_{av} / \pi d_i L_e (T_{w,e} - T_v) \tag{1}$$

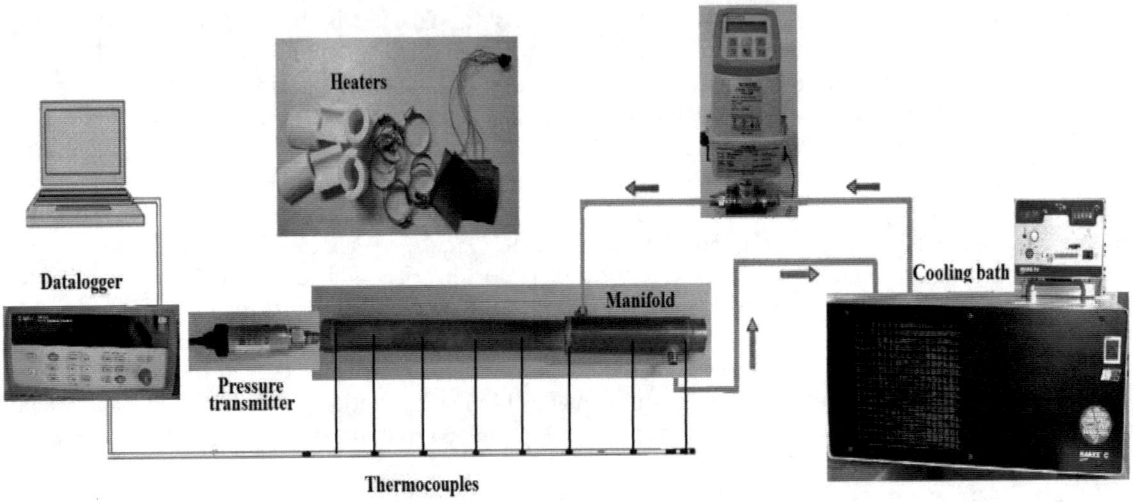

Figure 3. Overall experimental setup.

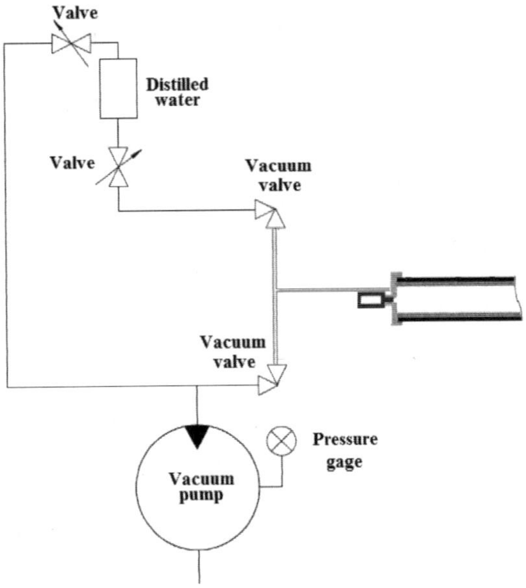

Figure 4. HP filling facilities.

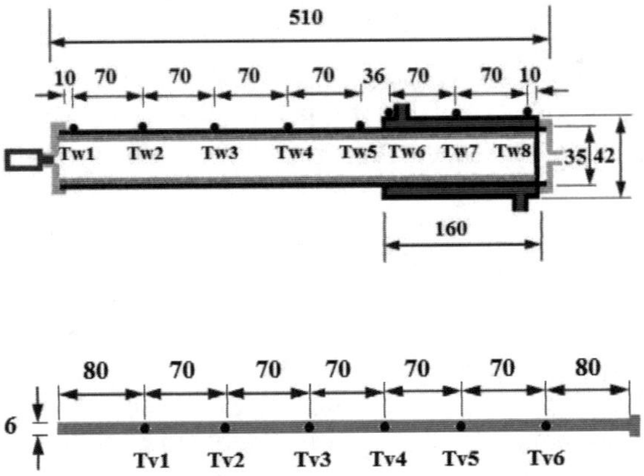

Figure 5. Thermocouple positions along the HP

Table 2. Design summery of HP

Pipe material	Copper
Working fluid	Water
Fluid charge (cm^3)	42
Total length (mm)	510
Evap. length (mm)	225
Cond. length (mm)	160
Outside diameter (mm)	35
Inside diameter (mm)	33

Table 3. Design summery of screen mesh wick

Wick material	Stainless steal
Mesh number (mesh/inch)	100
Effective capillary radius (mm)	0.127
Screen wire diameter (mm)	0.11
Screen wick thickness (mm)	0.66
Porosity	0.63
Permeability	2.01×10^{-10}
Effective thermal conductivity	1.38

where $T_{w,e}$ is average evaporator wall temperature defined considering four temperature as described in figure 5 (T_{w1}, T_{w2}, T_{w3} and T_{w4}), T_v is average vapour temperature (T_{v1}, T_{v2} and T_{v3}) and Q_{av} is average heat transfer rate

$$Q_{av} = Q_1 + Q_2/2 \tag{2}$$

The rate of heat transfer to the evaporator section is obtained from the following relation:

$$Q_1 = VI \tag{3}$$

where V is voltage and I is current. The rate of heat removal from the condenser section is calculated from the following relation:

$$Q_2 = \dot{m}C_p(T_{out} - T_{in}) \tag{4}$$

The experimental uncertainties of the various parameters: heat input, thermal resistance and heat transfer coefficients are calculated. The relative experimental uncertainty can be described for the variable $R = R(x_1, x_2, \ldots, x_n)$ as:

$$U_R/R = \sqrt{\sum_{i=1}^{n} [(x_i/R)(\partial R/\partial x_i)]^2 (P_i/x_i)} \tag{5}$$

where P is the bias limit of the variable R. From above equations and mentioned uncertainty of apparatus, the relative uncertainties of the output heat flux from condenser section and evaporation heat transfer coefficient are estimated 4.8% and 5.5%, respectively for the temperature and heat load ranged 25-40°C and 30-100W, respectively.

4. Experimental results and analysis

The objective of the experiments is to characterize the heat loading on the transient and steady state operation of a screen mesh HP. Further, to obtain data to verify an existing steady state HP model previously presented by authors [20].

4.1. Transient behaviour of HP

Before test, the entire HP is set at ambient temperature, and the interface between the liquid in the wick and the vapour is at saturation. Figure 6 shows the results of the transient response of the HP to a step function increase in electric power input up to 100 W. The HP reached steady-state approximately 600 s after the initiation of the transient, for an input heat load of 30 W. At steady state, the output heat flux is only about 2.1 W lower than the electric power input (see table 3). Figure 7 shows the measured axial distributions of the wall and vapour temperatures at different times during the transient reported in figure 6. As can be seen in figure 7a, the measured wall temperatures in the evaporator and condenser sections are almost uniform. The vapour temperatures are nearly uniform along the entire length of the HP, except for the condenser section (figure 7b): variations are less than 1K. The reason for this low vapour temperature in the condenser section for lower heat loads will be discussed in section 4.2.

4.2. Steady state temperature distribution along the outside wall surface

Experimental results of the temperature distribution along the wall surface of HP at horizontal orientation are compared in figure 8a. The tests are conducted at different heat inputs and constant cooling water temperature (24.8°C). The results indicate that, as expected, the evaporator wall temperature rises with the increase of heat flux, and the temperature distribution of the HP is nearly uniform and decreases slightly towards the adiabatic section. The temperature reaches the minimum value at the condenser section. The wall temperature of the condenser is approximately uniform for all heat transfer rates, with an average value of 25 °C. It should be noted that the condenser temperature is the temperature of outer manifold wall. The axial vapour temperature distribution corresponding to figure 8a is shown in figure 8b. The vapour core temperatures in the evaporator section are nearly equal and increase with the heat input as expected. The core temperatures in the condenser section at the lowest heat inputs of 30 and 40 W are significantly lower than that in the evaporator section. This low vapour temperature in the condenser section could be due to the presence of residual non-condensable gases which tend to accumulate at the condenser end of the heat pipe. At low heat fluxes, the non-condensable gases occupy a larger part of the condenser due to the lower vapour pressure. As the heat flux increases (for heat input above 40W), the vapour core temperatures in the condenser section increase. The incondensable gases seem to have a minor role on these conditions and the active region of the condenser is increased. The vapour temperatures are also nearly uniform and closer to the adiabatic wall temperature (T_{w5}).

4.3. Evaluation of evaporation heat transfer coefficient

The thermal resistance of the evaporator section can be expressed in terms of a heat transfer coefficient. The measurements of heat flux and of average wall and vapour temperature, temperature in the evaporator and are used to calculate the heat transfer coefficients in the evaporator section of the HP by using Eq. (1). The heat transfer coefficients for the HP are plotted versus average temperature of evaporator section as shown in figure 9. As can be seen, the evaporation heat transfer coefficient is non-linear and increases as the temperature (heat flux) increases. At higher heat fluxes (1850 W/m^2 to 2100 W/m^2), the heat transfer increases rapidly, indicating the thermal resistance of the HP is decreasing.

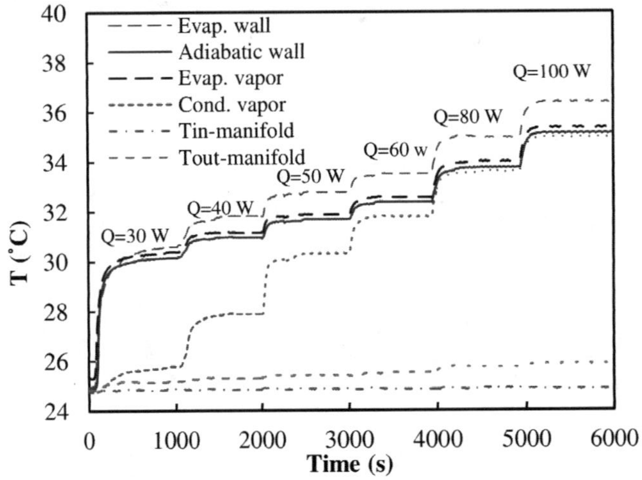

Figure 6. Transient temperature behaviour of HP for different heat input (T_c = 24.8 °C).

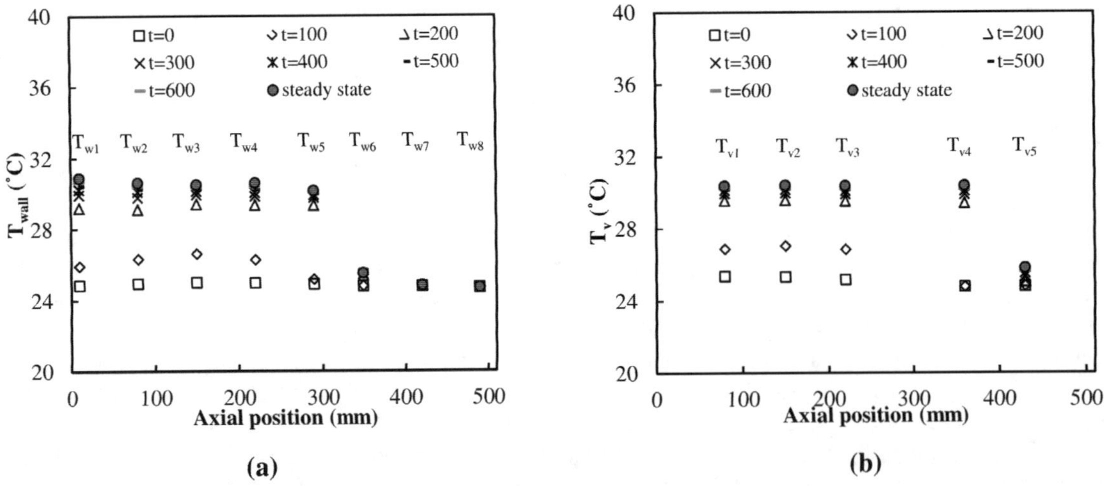

Figure 7. Transient axial distributions of wall temperatures (a) and vapour temperatures (b), Q=30 W.

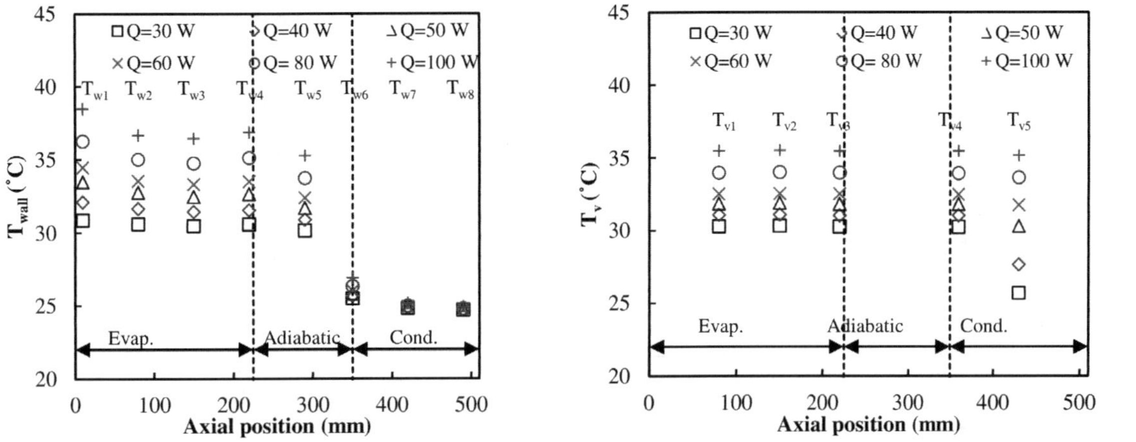

Figure 8. Steady state axial distribution of wall temperatures (a) and (b) vapour temperatures (b) different heat input.

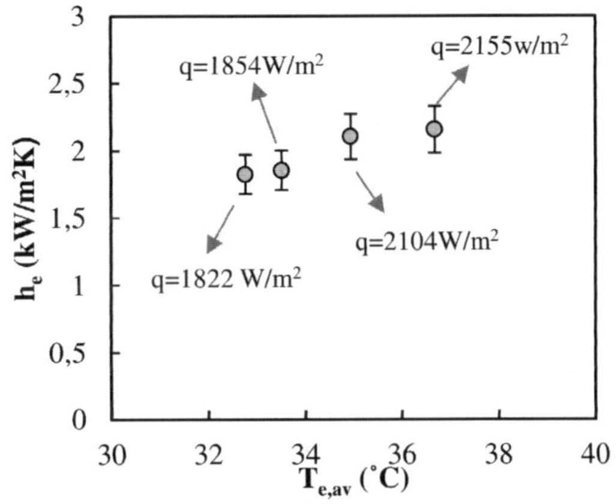

Figure 9. The heat transfer coefficients against average temperature of evaporator section

5. Comparisons between the experimental data and the expected analytical results

As already mentioned in previous section, a steady–state analytical model has been developed and presented in [20] to predict the performance of HPs. The model couples two–dimensional heat conduction in the HP wall with the liquid flow in the wick and the vapour hydrodynamics. The experimental data obtained in the present analysis have been compared with the results expected with the model for different heat input. In theoretical analyses, it is assumed that the condenser is cooled by water, according to the experimental apparatus. Thus, a convection heat transfer coefficient is defined as boundary condition on the condenser wall. The condenser heat transfer coefficients (h_{inf}) are determined based on the experimental data (output heat flux in the condenser section Q_{out}, and average cooling water temperature $T_{c,av}$) are shown in table 4.

Figure 10a and figure 10b show the comparison between the HP wall and vapour temperatures obtained from analytical model with the experimental data, respectively. The agreement of the results with experimental data is very good. A difference less than about 1K is observed between the evaporator vapour and wall temperatures predicted by theoretical models and experimental data, which can be acceptable at this step of analysis considering the experimental error and approximations in the analytical model. It should be noted that the condenser temperature is not reported in the current experimental study because the temperature of condenser section is obtained based on wall manifold temperature which is different from HP condenser section.

Table 4. Condenser heat transfer coefficients for different heat inputs.

Evaporator	Condenser			
Q_{in} (W)	Q_{out} (W)	T_{in}	$T_{c,av}$ (°C)	h_{inf}(W/m^{2}K)
60	56.5	24.8	25.18	476.9
80	74.59	24.8	25.3	531.33
100	92.81	24.8	25.41	571.88

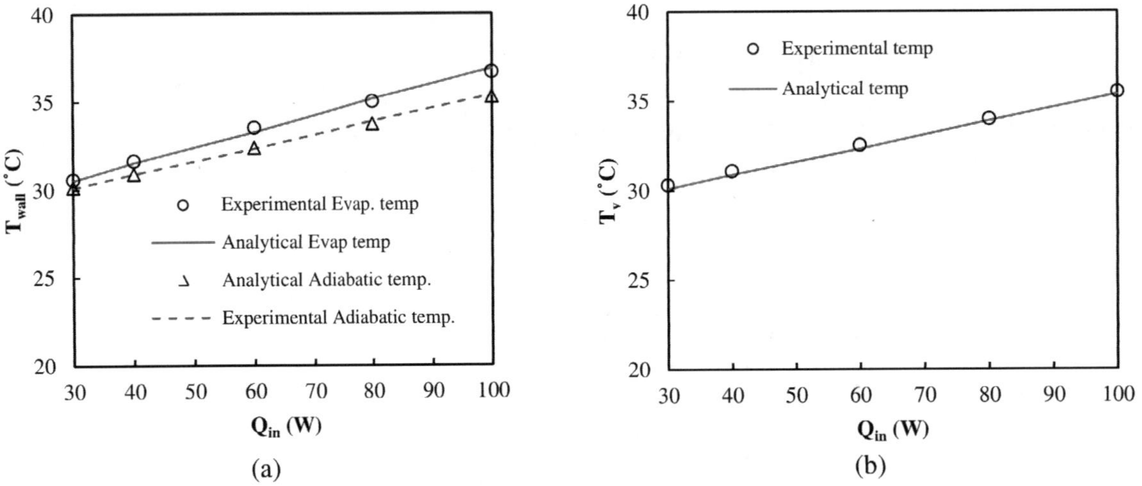

Figure 10. Variation of wall temperature (a) and vapour temperature (b) with heat input obtained from the analytical and experimental studies for the evaporator and adiabatic section.

6. Conclusions

The study proposed in this paper presents a detailed experimental apparatus and preliminary experimental results about the heat transfer performances of a HP designed for a particular solar application. The experimental apparatus has been developed to investigate the thermal performance of a cylindrical copper HP using water as a working fluid. Transient experiments were performed using a uniformly heated evaporator section and a convectively cooled condenser section (temperature of 24.8°C). The effect of the heat load (30-100 W) on the heat transfer performance of HPs was experimentally investigated by testing a HP with 3 layers of wire screen mesh (100 mesh/inch). The axial distribution of the wall and vapour temperatures were measured and discussed in detail. By experimenting on different heat loads under steady-state and transient conditions, it was observed that the temperature variation along the wall surface of the evaporator section of the HP was nearly uniform and decreased slightly along the adiabatic section. In the experiment, the vapour temperature was uniform along the HP, except at low heat loads (30 and 40 W) where it was significantly lower than that in the evaporator section, presumably due to the presence of residual non-condensable gases. The results obtained are in very good agreement with data obtained by a numerical HP model previously presented by the authors. After these preliminary results, the authors will further develop the experimental analysis to evaluate the maximum heat transfer capacity of HP for different filling ratios (10-50%), inclination angles (0-90°) and low-to-medium cooling water temperatures (25-90°C).

7. References

[1] Peterson GP 1994 *Heat Pipes: Modelling, Testing and Applications* (USA: John Willey & Sons)
[2] Faghri A 1995 *Heat Pipe Science and Technology* (Washington, DC: Taylor and Francis)

[3] Reay DA and Kew PA 2006 *Heat Pipes, Theory Design and Applications* (New York: Elsevier)

[4] Vasiliev LL 2005 *Heat Pipes in Modern Heat Exchangers* Appl. Therm. Eng. **25**(1) 1-19.

[5] Wan JW, Zhang JL and Zhang WM 2007 *The effect of heat-pipe air-handling coil on energy consumption in central air-conditioning system* Energy Build. **39** 1035–40.

[6] Mochizuki M, Nguyen T, Mashiko K, Saito Y, Nguyen T and Wuttijumnong V 2011 *A Review of Heat Pipe Applications Including New Opportunities* Frontiers in Heat Pipes **2** 013001.

[7] Remeli MF, Tan L, Date A, Singh B and Akbarzadeh A 2015 *Simultaneous power generation and heat recovery using a heat pipe assisted thermoelectric generator system.* Energy Convers. Manage. **91** 110–19.

[8] Riffat S, Zhao X and Doherty PS 2005 *Developing a theoretical model to investigate thermal performance of a thin membrane heat-pipe solar collector* Appl. Therm. Eng. **25** 899–915.

[9] Azad E 2008 *Theoretical and experimental investigation of heat pipe solar collector* Exp. Thermal Fluid Sci. **32** 1666-72

[10] Arab M and Abbas A 2013 *Model-based design and analysis of heat pipe working fluid for optimal performance in a concentric evacuated tube solar water heater* Solar Energy **94** 162–76.

[11] Brahim T, Dhaou MH and Jemni A 2014 *Theoretical and experimental investigation of plate screen mesh heat pipe solar collector.* Energy Convers. Manage. **87** 428–38.

[12] Wang Z, Duan Z, Zhao X and Chen M 2012 *Dynamic performance of a facade-based solar loop heat pipe water heating system* Solar Energy **86** 1632–47.

[13] He W, Hong X, Zhao X, Zhang X, Shen J and Ji J 2015 *Operational performance of a novel heat pump assisted solar façade loop-heat-pipe water heating system.* Appl. Energy **146** 371–82.

[14] Akbarzadeh A and Wadowski T 1995 *Heat pipe-based cooling systems for photovoltaic cells under concentrated solar radiation* Appl. Therm. Eng. **16**(1) 81–7.

[15] Gang P, Huide F, Tao Z and Jie J 2011 *A numerical and experimental study on a heat pipe PV/T system* Solar Energy **85** 911–21.

[16] Zuo ZJ and Faghri A 1998 *A network thermodynamic analysis of the heat pipe* Int. J. Hcat Mass Trans. **41**(11) 1473-84.

[17] Huang XY and Liu CY 1996 *The pressure and velocity fields in the wick structure of a localized heated flat plate heat pipe* Int. J. heat mass trans. **39** 1325–30.

[18] Zhu N and Vafai K 1999 *Analysis of cylindrical heat pipes incorporating the effects of liquid – vapor coupling and non-Darcian transpor –a closed form solution* Int. J. heat mass trans. **42** 3405–18.

[19] Shabgard H and Faghri A 2011 *Performance characteristics of cylindrical heat pipes with multiple heat sources* Appl. Therm. Eng. **31** 3410–18

[20] Di Marco P, Filippeschi S, Franco A and Jafari D 2014 *Theoretical Analysis of Screened Heat Pipes for Medium and High Temperature Solar Applications* J Phys Conf Ser **547** 012010.

[21] El Genk MS and Huang L 1993 *An experimental investigation of the transient response of a water heat pipe* In. J. Heat Mass Trans. **36**(15) 3823-30.

[22] Wang Y and Vafai K 2000 *An experimental investigation of the thermal performance of an asymmetrical flat plate heat pipe* Int. J. Heat Mass Trans. **43** 2657-68.

[23] Kempers R, Ewing D and Ching CY 2006 *Effect of number of mesh layers and fluid loading on the performance of screen mesh wicked heat pipes* Appl. Therm. Eng. **26** 589–95.

[24] Kempers R, Robinson AJ, Ewing D and Ching CY 2008 *Characterization of evaporator and condenser thermal resistances of a screen mesh wicked heat pipe* Int. J. Heat Mass Trans. **51**6039–46.

[25] Wong SC and Kao YH 2008 *Visualization and performance measurement of operating mesh-wicked heat pipes.* Int. J. Heat Mass Trans. **51** 4249–59.

[26] Wang YX and Peterson GP 2003 *Analytical model for capillary evaporation limitation in thin porous layers* J. Thermophys. Heat Transfer **17**(2) 145–9.

Numerical simulation of radiant ceiling panels for indoor cooling

Giuliano Cammarata, Giuseppe Petrone*, Filippo Masi

University of Catania, Department of Industrial Engineering, Viale A. Doria, 6 – 95126 Catania, ITALY

*Corresponding author: gpetrone@dii.unict.it

Abstract. The aim of this work is to analyse the thermal performance in cooling of two different models of radiative ceiling panels, by varying the discharge temperature and the velocity of heat-carrying fluid. The first one geometrical configuration refers to a coil- embedded panel, while the second one is made by a matrix of tubes. Numerical simulations were carried-out by using a FE-approach to solve governing equations for the physical system. Thermal performances were analysed as a function of the inlet temperature and the inlet velocity of the fluid. Then interpolation functions are proposed in order to assess thermal performances for both geometrical configurations against several working conditions.

1. Introduction

The idea of using the pavement as an endpoint for giving way to heat dates back to more than two thousand years. The Chinese, Egyptians and Romans achieved heating systems inspired by such an idea. In the last decades the heating and cooling radiative panel systems are spreading. In these last years the heating and cooling radiative panel systems are spreading. Practically the usual indoor terminal elements are replaced with panels built by means of pipes appropriately inserted into floors, walls or ceiling installed in order to shape a radiative surface. The ceiling panels are able to provide higher performances than those of the floor during both heating and cooling. The trend of reduction of the water temperature in the heating systems has proceeded to go to input temperatures in heating around 35° C; such a result has been reached both from the point of view of the generation and from the point of view of the elements of exchange coupled with these technologies.

Several studies were carried-out on this topic. Among those, Su et al. [1] studied heat transfer and cooling characteristics of the concrete ceiling radiant cooling panel. Two-dimensional mathematical model of steady state heat transfer was developed by using finite difference method. Heat transfer in the concrete panel was numerically simulated. Miriel et al. [2] experimentally and numerically investigated performances, thermal comfort and energy consumptions of radiant ceiling panel heating–cooling systems. Diaz et al. [3] exploited some results of an experimental study to develop a computational model of cooling ceiling systems. Tye-Gingras et al. [4] developed a CFD model of a room coupled with a semi-analytical radiant panel model to study a heating ceiling and wall hydronic radiant panel system in a typical residential building. They also performed a multi-objective optimization investigation for energy consumption and comfort. Causone et al. [5] investigated in order to assess the heat transfer coefficients between radiant ceiling and room in typical conditions of occupancy of an office or residential building. Internal gains were therefore simulated using heated cylinders and heat losses using cooled surfaces. Evaluations were developed by means of experimental tests in an environmental chamber. A simplified calculation method for estimating heat flux from ceiling radiant panels was proposed by Okamoto et al. [6] using pipe density on panels and the temperature difference between the room air and the supply water. Their proposed method was validated by experimental data. Using of different energy source as feeding for ceiling radiant panels was recently studied by Bojić et al [7] that highlighted the opportunity of energy saving and environmental benefit related to that opportunity. Li et al [8] shared some experimental detected results, providing useful heat transfer coefficients for this kind of application. In this framework of studies, the present work is devoted to numerically analyse the thermal performance in cooling of two different models of radiative ceiling panels. Models were built-up exploiting a FE-based commercial software and results were carried-out for several working conditions considered for both geometrical configurations. As final product of our investigation, some interpolation functions are proposed in order to assess thermal fluxes with respect to different functional conditions concerning the working fluid.

2. Modelling

The geometry of the studied systems are presented in Fig. 1a-b. Panels are made of multiple layers of which the back one, in both cases, is made of expanded sintered polystyrene (EPS), a material assuring good thermal insulating properties (λ=0.034 W/(m K)). In general, its use finds justification because of the opportunity of avoiding the space above the suspended ceiling also being cooled, limiting the final efficiency in that way. The back-side insulating panel turns out to be, in both cases, carefully shaped sa as to host the active element of the thermal exchange with a different geometry for the two models. In the first case (Fig. 1a) it consists in a coil ("coil panel"), in the second one (Fig. 1b) it is made by an array of radiating tubes joining the inlet to the output collector ("array panel").

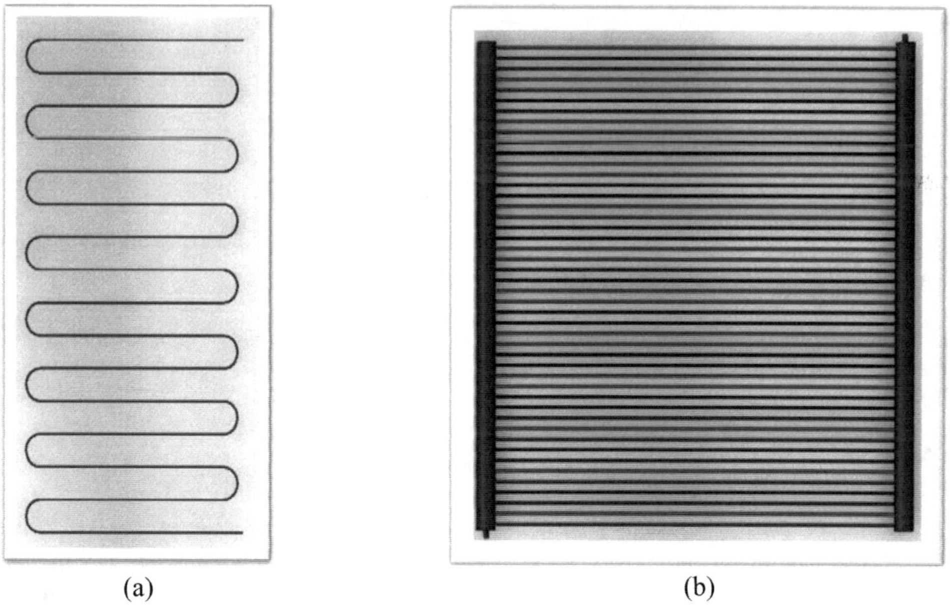

(a) (b)

Figure 1. Geometry of numerical models: "coil panel" (a) and "array panel" (b).

Another significant constitutive element is the aluminum diffuser, that is present in both analysed geometries, even that different thickness and arrangements were considered. The high thermal conductivity of this element allows a remarkable increasing in panels performance. The last layer constituting the panels is a "finishing" layer guarantying closing and protection with respect to the external environment. It also assure the overall aesthetics of the panels. In our geometrical models we choose prefabricated plasterboard for the first panel and a plaster of lime and gypsum in the second case. The details concerning dimensions and shape of the two investigated models of panel, as well as the arrangement of the different layers that make it up, are graphically reported in Fig. 2, where cross-sections of the two panel configurations are shown.

In order to solve fluid-dynamical and thermal distribution on the 3D numerical models built-up for simulating the cooling panels, the Navier-Stokes and energy equations were numerically integrated in Comsol Multyphysics environment, a FE-based software allowing to solve coupled PDE's equations [9].

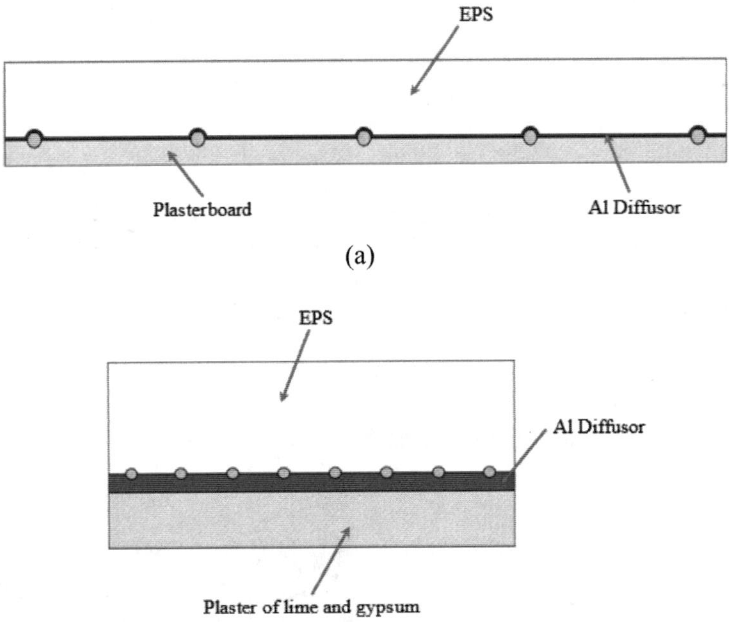

Figure 2. Longitudinal cross-section (portion) of "coil panel" (a) and "array panel" (b).

Under assumptions of Newtonian fluid and incompressible flows, Navier-Stokes and energy equations were discretized on no-structured and no-uniform numerical grids made of tetrahedral second order elements. A standard turbulence k-ε method [10, 11] was applied to solve the flow by an eddy-viscosity approach. Solutions were achieved by applying an iterative procedure based on a modified Newton-Raphosn method [12]. At each computational step, a direct solver is applied to solve linear systems coming from numerical discretization. Governing equations were solved by considering the following boundary conditions: entry velocity (U_{IN}) and entry temperature (T_{IN}) for inlet fluid, logarithmic wall function for fluid at solid walls and null pressure at the out-coming sections, thermal insulation on lateral and back side on the panels, thermal flux on the front-face of the panels (facing to the ambient). Considering a conjugate convective and radiating flux to the ambient, a convective heat transfer coefficient ($h_c = 13$ W/(m^2 K)) and a surface emissivity value ($\varepsilon = 0.95$) were set in the models. Simulations were carried-out for parametric value of $U_{IN} = 0.1$-0.5 m/s and $T_{IN} = 16$-25 °C. A constant ambient temperature ($T_{AMB} = 26$ °C) was considered in simulations. Some test grids were preliminarily carried out in order to adequately refine the computational grid to obtain mesh-independent results.

3. Results

Simulation results are now presented reporting the thermal analysis for both geometries. The performances were analysed according to the variation of the temperature and the velocity of entry of the working fluid. Fig. 3a-b shows the thermal distribution in a front view of the "coil panel" (faced to the ambient) in two different working conditions. It can be noticed as increasing mass flow rate allows to minimize water temperature difference between inlet and outflow section. This item is due to the minor residence time of the fluid vector in the heat exchanger. Otherwise, this result was attended because of in our simulations we assumed constant heat flux coefficients at the panel/air-room interface. As a consequence, we can deduce that increasing in mass flow rate not only assures increasing in heat flux transferred to the indoor ambient, but also better thermal comfort conditions, related to a much more uniform radiant temperature distribution with respect to the panel surface. In analogy, in Fig. 4a-b thermal maps are reported for the "array panel". In that configuration, non-homogeneous thermal distribution related to low mass flow rate is less evident. This can be considered a direct consequence of the water distribution system.

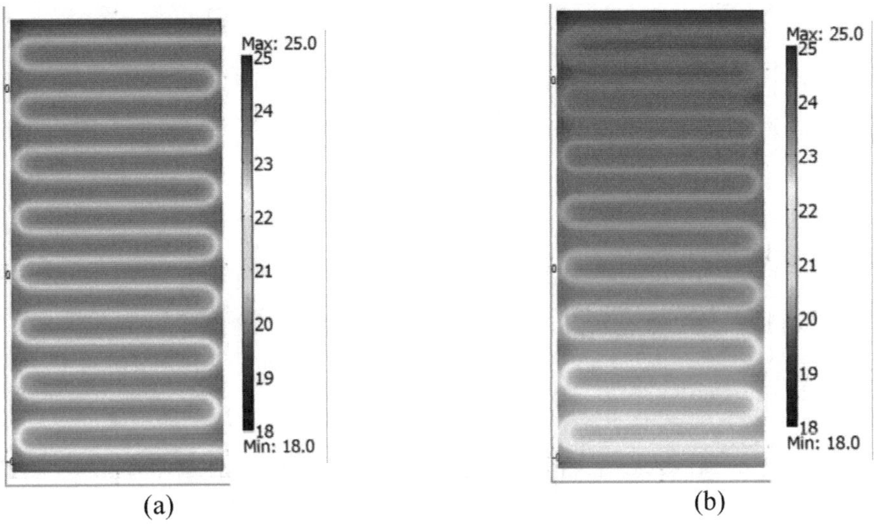

Figure 3. Thermal distribution in front-view (ambient-faced) for "coil panel" and T_{IN}=16 °C, considering U_{IN} =0.5 m/s (a) and U_{IN} = 0.1 m/s (b).

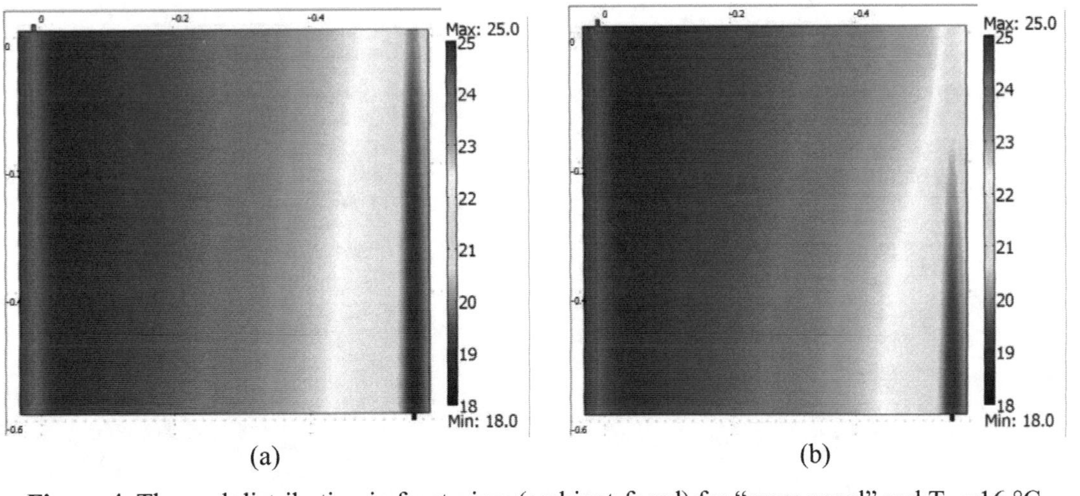

Figure 4. Thermal distribution in front-view (ambient-faced) for "array panel" and T_{IN}=16 °C, considering U_{IN} =0.5 m/s (a) and U_{IN} = 0.1 m/s (b).

Starting from temperature distributions solved by the FE models, values of the specific thermal flux supplied by the panel to the indoor environment were evaluated. A quantitative post-processing analysis of the obtained results was carried-out by determining thermal flux of cooling panels as a function of the inlet velocity and difference between inlet and ambient temperature. Graphics plotted in Figs. 5-6 reports computed values for the "coil panel", while results derived from "array panel" simulations are shown in graphics of Figs. 7-8. From those, some interpolation functions were proposed, allowing to deduce thermal efficiency for both configurations by using as input data the inlet velocity or the inlet temperature of the working fluid.

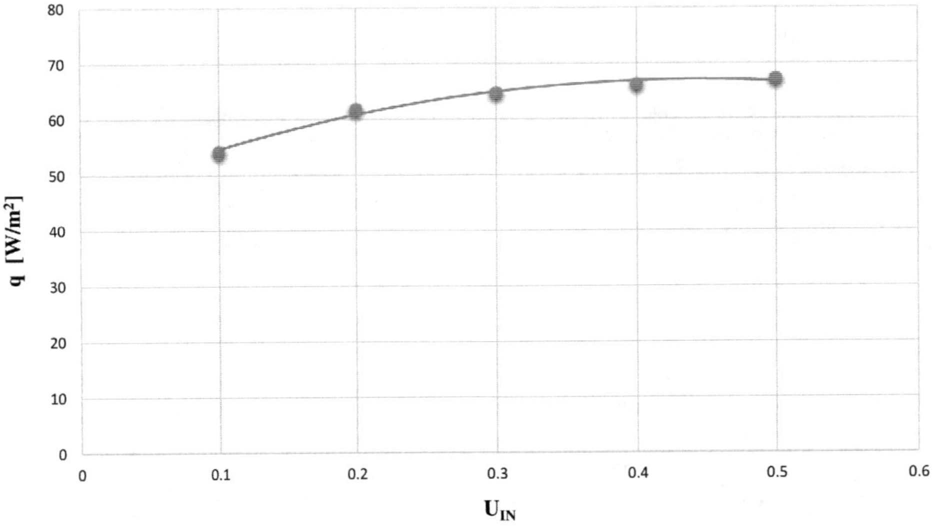

Figure 5. Supplied thermal flux to the ambient by the "coil panel" as a function of the inlet velocity of the working fluid.

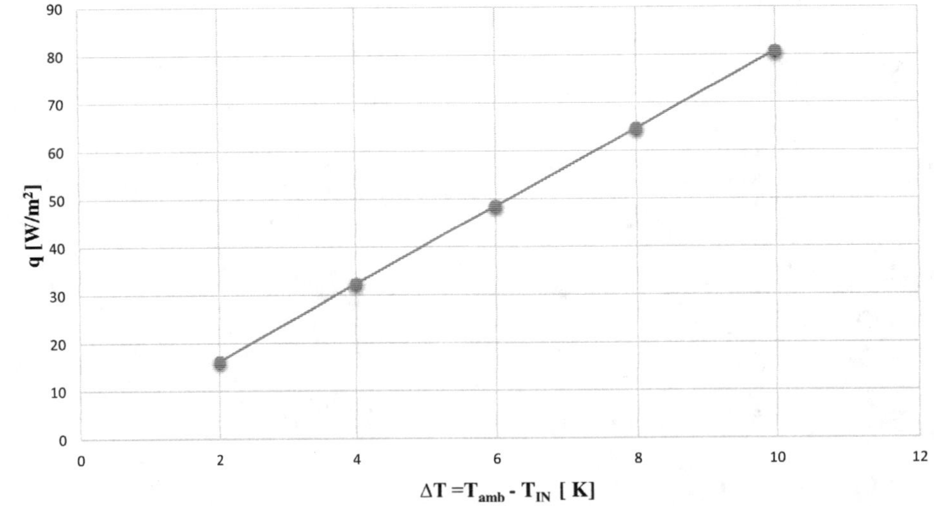

Figure 6. Supplied thermal flux to the ambient by the "coil panel" as a function of the difference between ambient and inlet temperature of the working fluid.

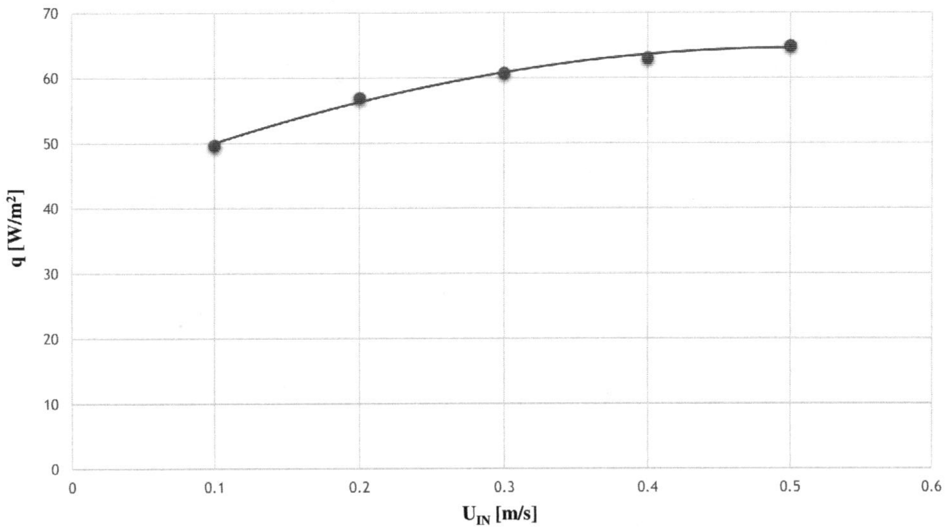

Figure 7. Supplied thermal flux to the ambient by the "array panel" as a function of the inlet velocity of the working fluid.

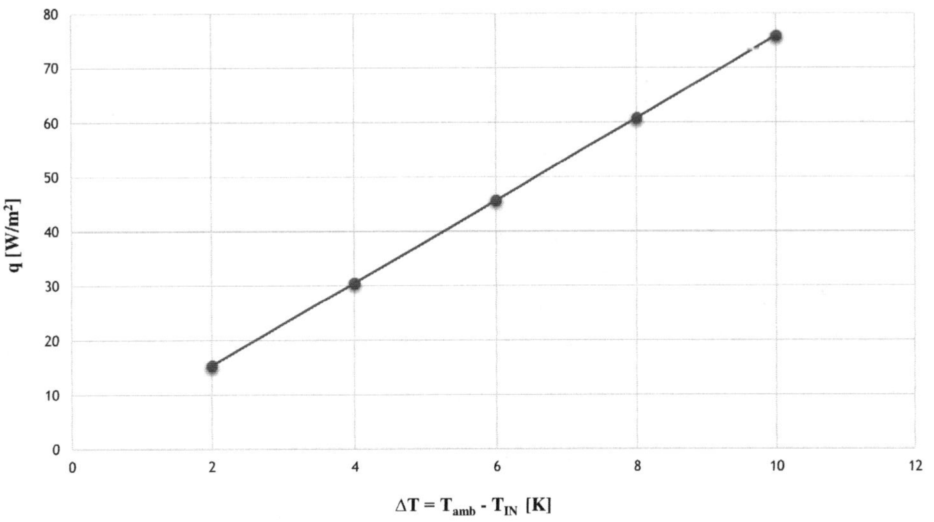

Figure 8. Supplied thermal flux to the ambient by the "array panel" as a function of the inlet velocity of the working fluid.

Interpolation functions used for "coil panel" data set are proposed below, reporting also the obtained regression coefficient (R) for each one:

$$q = -\,106.26\, U_{IN}^2 + 94.11\, U_{IN} + 46.25 \quad (R=0.984) \tag{1}$$

$$q = 8.0749\, \Delta T + 0.0668 \quad (R=1.000) \tag{2}$$

In a similar manner, interpolation functions used for “coil panel” data set are proposed below:

$$q = -\,88.23\, U_{IN}^2 + 89.34\, U_{IN} + 41.97 \quad (R=0.993) \tag{3}$$

$$q = 7.5622\, \Delta T + 0.2408 \quad (R=1.000) \tag{4}$$

From the proposed correlation laws it is possible to assess the thermal performance of the studied configurations as a function of the feeding fluid inlet conditions.

4. Conclusions

In this paper a numerical analysis is proposed in order to investigate on thermal performance of ceiling panels for indoor cooling in two geometrical configurations. The first one refers to a coil shape of the working fluid tube, the second one refers to an array and collector configuration. From parametric analyses carried-out as a function of the fluid inlet temperature and mass flow rate it was possible to point-out some interesting aspects concerning both configurations. In particular, it was highlighted as the “coil” layout assures good thermal distribution for the higher mass flow rate analysed; otherwise, it appears that this kind of system could determinate non homogeneous thermal distribution if the mass flow rate is low. This item could be cause of potential indoor thermal discomfort. From a quantitative point of view, thermal flux transferred by the “coil” configuration appears slightly higher than this offered by the “array” configuration. As exploitable result of our investigation, we propose a set of interpolation functions in order to assess the influence of the inlet temperature and mass flow rate of the working fluid on the thermal performance of the studied component in both geometrical configurations considered.

References

[1] Su L, Li N, Zhang X, Sun Y, Qian J 2015 *Appl. Therm. Eng.* **84** 170
[2] Miriel J, Serres L, Trombe A 2002 *Appl. Therm. Eng.* **22** 1861
[3] Diaz N F, Lebrun J, André P 2010 *Int. J. Refrigeration* **33** 793
[4] Tye-Gingras M, Grosselin L 2012 *Build. and Environment* **54** 1
[5] Causone F, Corgnati S, Filippi M, Olesen BW 2009 *Energy and Build.* **41** 622
[6] Okamoto S, Kitora H, Yamaguchi H, Oka T 2010 *Energy and Build.* **42** 29
[7] Milorad Bojić M, , Dragan Cvetković D, Ljubiša Bojić L 2015 *Appl. Energy* **138** 404
[8] Li R, Yoshidomi T, Ooka R, Olesen B W 2015 *Energy and Build.* **86** 58
[9] Comsol Multiphysics Version 4.3 User's Guide 2012
[10] Ignat L, Pelletier D, Ilinca F 2000 *Comp. Meth. in Appl. Mech. and Eng.* **189** 1119
[11] Launder BE and Spalding DB 1974 *Comp. Meth. in Appl. Mech. and Eng.* **3** 269
[12] Deuflhard P 1974 *Num. Math.* **22** 289

Thermal design of spacecraft solar arrays using a polyimide foam

N Bianco[1], **M Iasiello**[1,2] **and V Naso**[1]

[1] Dipartimento di Ingegneria Industriale (DII) – Università degli studi di Napoli Federico II – P.le Tecchio, 80 – 80125 – Napoli -Italy,

Abstract. The design of the Thermal Control System (TCS) of spacecraft solar arrays plays a fundamental role. Indeed, the spacecraft components must operate within a certain range of temperature. If this doesn't occur, their performance is reduced and they may even break. Solar arrays, which are employed to recharge batteries, are directly exposed to the solar heat flux, and they need to be insulated from the earth's surface irradiation. Insulation is currently provided either with a white paint coating or with a Multi Layer Insulation (MLI) system [1]. A configuration based on an open-cell polyimide foam has also been recently proposed [2]. Using polyimide foams in TCSs looks very attractive in terms of costs, weight and assembling. An innovative thermal analysis of the above cited TCS configurations is carried out in this paper, by solving the porous media energy equation, under the assumption of Local Thermal Equilibrium (LTE) between the two phases. Radiation effects through the solar array are also considered by using the Rosseland approximation. Under a stationary daylight condition, temperature profiles are obtained by means of the finite-element based code COMSOL Multiphysics®. Finally, since the weight plays an important role in aerospace applications, weights of the three TCS configurations are compared.

Nomenclature

$\boldsymbol{G}$	Incident heat flux (W/m^2)
i	Current layer
H	Length of the computational domain (m)
k	Thermal conductivity (W/m·K)
L	Length (m)
n	Effective index of refraction
N	Layers number
$\boldsymbol{q}$	Heat flux (W/m^2)
s	MLI thickness (m)
T	Temperature (K)
x, y	Rectangular coordinates (m)
V	Volume (m^3)

Greek letters

α	Absorptivity
β	Extinction coefficient (1/m)
ε	Emittance
ρ	Density (kg/m^3)
σ	Stefan-Boltzmann constant (W/m^2 K^4)

Subscripts

al	aluminum
c	conduction
eff	effective
i	layer
MLI	Multi Layer Insulation
p	polyimide
r	radiation
s	solar
sc	cover glass
wp	white paint

Other symbols

< >	Volume average of a variable
∇	Nabla operator

[2] To whom any correspondence should be addressed. E-mail: marcello.iasiello@unina.it

1. Introduction

Thermal management of a space solar array is a primary task during its design phases. An example of the importance of this task is that internal electronic components of the spacecraft need to operate in certain temperature ranges. Solar arrays are employed to catch energy from the sun and convert into electricity. The solar cell efficiency, the ratio of the output electrical power to the incident radiation, is inversely proportional to its temperature; then, a correct thermal management guarantees the cells to operate into acceptable efficiency ranges [3].

The thermal heat rates involved on a satellite is sketched in figure 1. The absorbed fraction of the solar irradiation enters the satellite across its boundary facing the sun while the absorbed fraction of the sum of the earth albedo and the heat rate emitted by the earth enters the satellite across its boundary facing the earth. Heat rates emitted by both boundaries exit the satellite. Other heat rates can be neglected [2]. The control of the cell temperature requires to reduce the heat rate entering the boundary facing the earth. This is currently attained either coating it with a white paints [1] or affixing a multi layer insulation (MLI) system on it. The white paint coating AZW-LA-II has the best performances in terms of high emittance and low absorptance [4]. However, this solution has a high impact on costs of the insulation system [5]. Less expensive white paints are also employed. It is worth noticing that degradation of the coating optical properties can occur after many hours of operation [6]. On the other side, multilayer insulations allow to increase the thermal resistance, thus lowering the temperature of the solar cell. However, difficulties in assembling multilayer insulations are a drawback [2].

The above mentioned limitations can be overcome by using a polyimide foam insulation material on the rear of the solar array [2]. A high temperature polyimide foam for space applications, named LaRC TEEK, is manufactured by NASA Langley Research Center [7]. It has excellent flammability resistance, thermal insulation, cryogenic capability and a low weight. Compared to traditional systems, performances in terms of costs, easiness of assembly and weight can be also improved. In the present paper, a solar array configuration with a polyimide foam is proposed, and its thermal analysis is carried out. Results are compared with the thermal analysis of traditional Thermal Control Systems (TCS) of solar arrays.

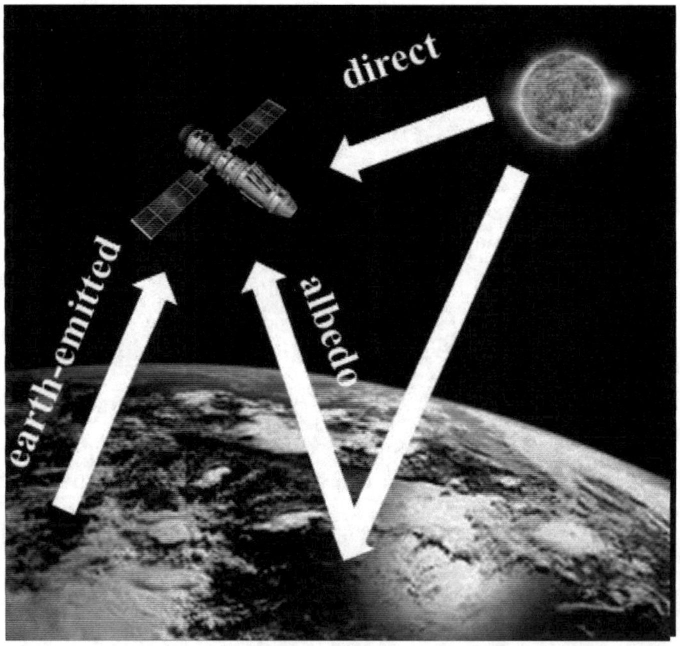

Figure 1. Thermal heat rates on a satellite.

2. Solar array design

The configurations of traditional space solar arrays are sketched in figures 2(a) [8] and 2(b) [1], for the white paint and MLI configurations, respectively. In both cases, solar cells are covered with a thin protective glass. Carbon-Fiber Reinforced Plastic (CFRP) is used to link the cell with the rest of the array. A honeycomb substrate is used for mechanical purposes and because of the high thermal resistance provided by its porosity. The bottom of the array can be coated with a white paint. In the multi-layer configuration (figure 2(b)), Kapton external sheets and aluminized Kapton internal sheets are employed. Sheets are separated by spacers, and the heat conduction between the sheets can be neglected. It is worth remarking that the MLI configuration is thicker than the white painted one. The configuration proposed by Zaglauer and Pitz [9], with an insulating open cell polyimide foam inserted between the honeycomb and the solar cell, is sketched in figure 2(c). The above mentioned configurations will be referred to in this paper.

3. Mathematical model

3.1. Governing equations

The thermal analysis of the TCS of solar arrays reported in figure 2 is carried out by solving the energy equation in steady-state condition.

For non-porous layers one can write:

$$\nabla^2 T = 0 \tag{1}$$

where T is the temperature.

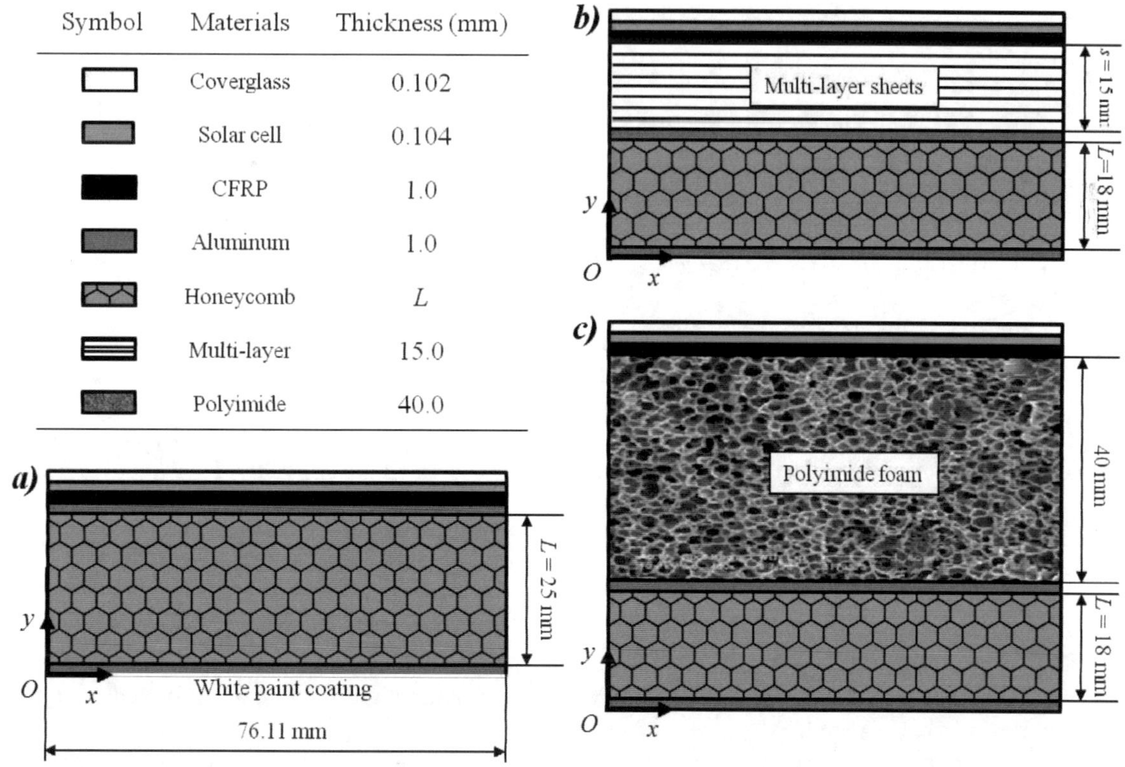

Figure 2. Space array configurations: a) with white paint; b) with MLI system; c) with polyimide foam.

Journal of Physics: Conference Series **655** (2015) 012024 doi:10.1088/1742-6596/655/1/012024

For the open cell polyimide foam and the honeycomb layers, because of their porous nature, the energy equation is volume-averaged over a Representative Elementary Volume (REV) of the foam [10] and then solved for the whole porous domain. The averaged temperature is:

$$\langle T \rangle = \frac{1}{V}\int_V T\, dV \tag{2}$$

where V is the REV volume. The volume-averaged energy equation, under Local Thermal Equilibrium (LTE) between the two phases [11], is:

$$k_c \nabla^2 \langle T \rangle - \nabla \cdot \boldsymbol{q}_r = 0 \tag{3}$$

where k_c is the effective thermal conductivity of the porous material and $\boldsymbol{q}_r$ is the radiative heat flux. The radiative heat flux in porous media is usually described by solving the Radiative Transfer Equation (RTE) [12]. However, when the medium is optically thick, Rosseland approximation for radiation can be employed [13]. The radiative heat flux $\boldsymbol{q}_r$ can be calculated by means of the Fourier law:

$$\boldsymbol{q}_r = -k_r \nabla T \tag{4}$$

where the radiative thermal conductivity k_r is defined as [13]:

$$k_r = -\frac{16\sigma n^2 T^3}{3\beta(T)} \tag{5}$$

with σ the Stefan-Boltzmann constant, n the effective index of refraction, equal to 1 since porosity of foams is usually high [13], and $\beta(T)$ the mean extinction coefficient in whole wavelength range.

Equations (4) and (5) can be combined as:

$$\boldsymbol{q} = \boldsymbol{q}_c + \boldsymbol{q}_r = -(k_c + k_r)\nabla T \tag{6}$$

where $\boldsymbol{q}$ is the total heat flux, due to conduction and radiation, and $\boldsymbol{q}_c$ is the conduction heat flux. Combining equations (4) and (6), equation (3) for porous layers becomes a temperature volume-averaged Laplace equation:

$$\nabla^2 \langle T \rangle = 0 \tag{7}$$

Governing equations (1) and (7) are solved with reference to the rectangular system of coordinate *Oxy*, depicted in figure 2.

3.2. Boundary conditions

Boundary conditions for equations (1) and (7) are presented in the following. All the domains are considered to be grey for the radiation. Side walls of the computational domains reported in figure 2 are assumed to be adiabatic.

On the top of the computational domain, namely the cover glass of the solar cell, the following net heat flux $|\boldsymbol{q}_s|$ boundary condition holds:

$$|\boldsymbol{q}_s| = \alpha_{sc}|\boldsymbol{G}_s| - \varepsilon_{sc}\sigma T_{sc}^4 \tag{8}$$

with $|\boldsymbol{G}_s|$ the solar incident heat flux, equal to 1,367 W/m^2, α_{sc} the effective absorptivity of the solar cell, ε_{sc} the emittance of the outer cover glass surface, T_{sc} its temperature. Reference will be made to α_s = 0.696 [8] and ε_{sc} = 0.899 [8] in the following.

On the bottom of the computational domain, namely the satellite cover facing the earth, the boundary condition is:

$$|\boldsymbol{q}_{bs}| = \alpha_{bs}|\boldsymbol{G}_{bs}| - \varepsilon_{bs}\sigma T_{bs}^4 \tag{9}$$

where $|\boldsymbol{G}_{bs}|$ is the radiative heat flux incident on the satellite bottom surface, because of the earth albedo and the infrared emission by the earth, equal to 650 W/m^2. For the MLI configuration in figure 2(b) and for the polyimide foam configuration in figure 2(c) reference will be made to $\alpha_{bs} \equiv \alpha_{al}$ = 0.15 [15] and $\varepsilon_{bs} \equiv \varepsilon_{al}$ = 0.030 [15]; for the white paint configuration in figure 2(b), $\alpha_{bs} \equiv \alpha_{wp}$ = 0.39 [8] and $\varepsilon_{bs} \equiv \varepsilon_{wp}$ = 0.88 [8] are assumed.

At the interface of each layer, the continuity of temperature and heat flux between two adjacent layers i and $i+1$ holds:

$$T_i\big|_{y_-} = T_{i+1}\big|_{y_+} \tag{10}$$

$$-\left[(k_c + k_r)\right]_i \frac{\partial T}{\partial y}\bigg|_{y_-} = -\left[(k_c + k_r)\right]_{i+1} \frac{\partial T}{\partial y}\bigg|_{y_+} \tag{11}$$

3.3. Thermophysical properties

As far as non-porous materials are concerned, the following values of conduction conductivity, k_c, are assumed: 1.0 W/m K for the coverglass, 100.0 W/m K for the solar cell, 54.4 W/m K for the CFRP, 235.0 W/m K for the aluminum.

As far as porous materials are concerned, reference will be made to a total conductivity, made up by the sum of the conduction conductivity, k_c, and the radiation conductivity, k_r.

As to the multilayer sheets, a 0.12 W/m K [15] effective conduction conductivity of an N = 10 layers material, is assumed. Accounting for the rather low maximum difference in the temperatures of the cover glass and the bottom surface of the solar array, the effective radiation conductivity of the MLI can be expressed by:

$$k_r = 4\sigma T^3 \frac{s}{N\left(\dfrac{2}{\varepsilon_{MLI}} - 1\right) + 1} \tag{12}$$

with s the MLI thickness and ε_{MLI} = 0.14 [15].

The effective total conductivity of the honeycomb structure is assumed equal to 116.0 W/m K [1].

The effective total conductivity of the polyimide foam is made up by the sum of the conduction contribution, 0.001 W/m K [14] for a 0.965 porosity polyimide foam, and the radiation contribution, given by equation (5) with the mean extinction coefficient given by [14]:

$$\beta(T) = \rho_p\left(17.4 + 0.229T - 2.27 \cdot 10^{-4} T^2\right) \tag{13}$$

where ρ_p is the polyimide foam relative density.

3.4. Numerical solution

Governing equations (1) and (7), with boundary conditions described in the previous subsections, were solved by using the finite-element commercial code COMSOL Multiphysics. A triangular mesh was used for the discretization of the domain, with less than 25,000 elements in each simulated case. Grid

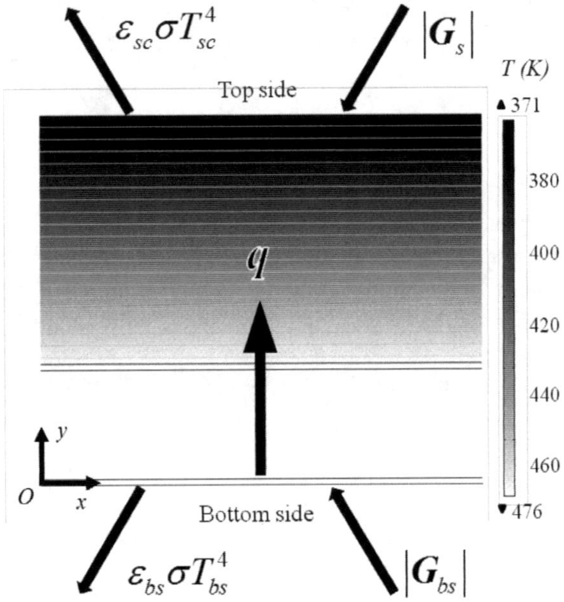

Figure 3. Temperature field in the TCS with polyimide foam.

independence was checked by performing simulations with 50,000 and 100,000 elements, that showed negligible differences in solar cell temperature fields. A convergence criterion of 10^{-4} was used, while the accuracy of energy balances was verified with a 10^{-10} order error.

4. Results

The temperature field in the TCS with the polyimide foam is shown in figure 3. The maximum temperature (about 470 °C) is attained on the back cover of the array. The temperature of the solar cell is about 370 K.

Temperature profiles in the TCSs of solar arrays in figure 2 are reported in figures 4 – 6. They show that the temperature of the solar cell is about 330 K for the white paint configuration and nearly 370 K for the MLI and the polyimide foam configurations. We can also notice that temperature gradients in the MLI and in the polyimide foam are negligible. Temperature is uniform in the honeycomb, because of its higher thermal conductivity than that of other materials.

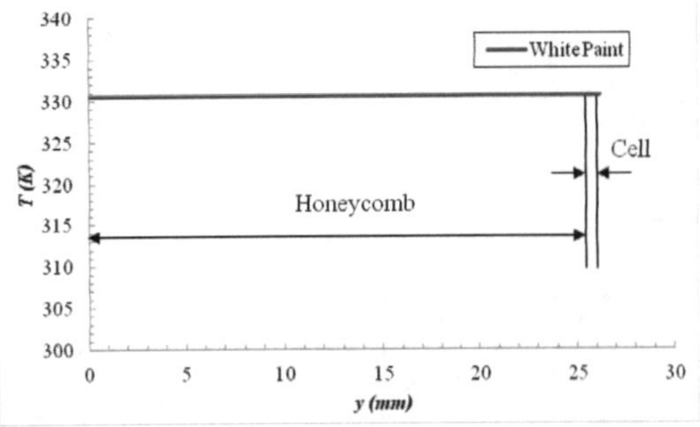

Figure 4. Temperature profile in the white paint configuration.

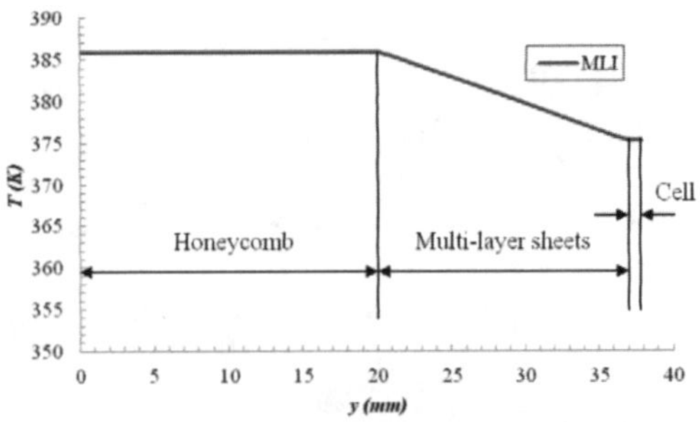

Figure 5. Temperature profile in the MLI configuration.

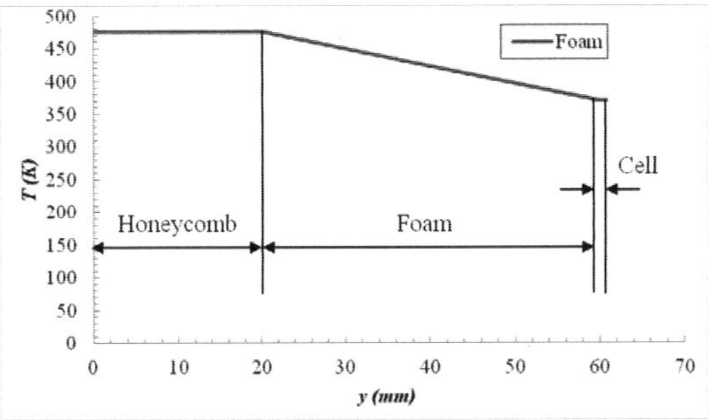

Figure 6. Temperature profile in the polyimide foam configuration.

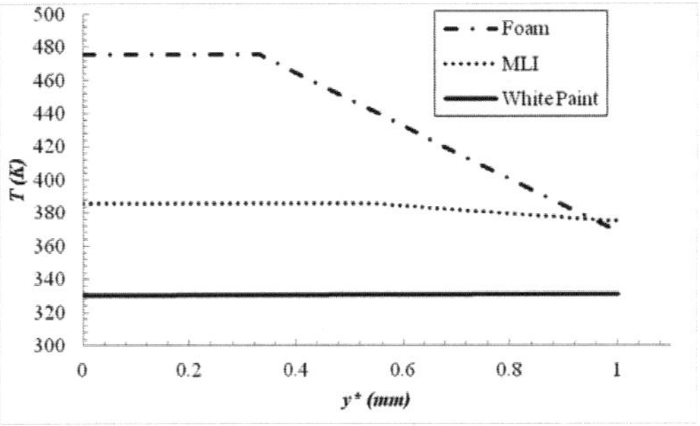

Figure 7. Comparisons of temperature profiles in different configurations.

For the sake of comparison, temperature profiles as a function of the dimensionless coordinate $y^* = y/H$, with H the length of the computational domain, that is the distance between the outer surfaces of the cover glass and the bottom cover of the satellite, are presented in figure 7. Even if the white paint configuration provides the lowest value of the cell temperature, it is worth reminding that the white paint has a high influence on the TCS overall costs. Besides, it is the configuration for which the weight becomes higher than the other two configurations, and the weight has a primary role in aerospace applications.

The weight of the three configurations is reported in table 1. Because, of structural reasons, the honeycomb of the latter device is thicker (25 mm) than in other two cases (18 mm). Therefore, one can notice that the polyimide foam configuration has the lowest weight, and the white paint configuration has the highest one. This implies that the polyimide foam has an advantage in terms of weight.

Table 1. Weights of space solar arrays.

Configuration	Weight (kg)
White paint	0.202
MLI	0.177
Polyimide foam	0.166

5. Conclusions

An innovative thermal analysis of some Thermal Control System (TCS) of spacecraft solar arrays has been presented. Temperature profiles have been compared by using a finite-element method. Results showed that the temperature reached on a polyimide foam based TCS is low enough to allow a proper operation of the solar cell. A comparison between the three configurations in terms of weight showed that the polyimide foam can be used to insulate a space solar array by reducing the weight of the solar array, in addition to the other mentioned advantages of costs, assembling and longevity.

6. References

[1] Gilmore D G *et al.* 2002 Thermal design examples *Spacecraft Thermal Control Handbook vol. 1 Fundamental Technologies* ed D G Gilmore (El Segundo, CA: The Aerospace Press) chapter 3 pp 71 - 137

[2] Wolters R 2001 Structure of thermal insulation of satellites US6318673 (Patent)

[3] Skoplaki E and Palyvos J A 2009 On the temperature dependence of photovoltaic module electrical performance: A review of efficiency/power correlations *Sol. Energy* **83** 614-24

[4] Choi M K 2004 AZW-LA-II White paint on swift: lessons learned from first time flying on spacecraft radiators *Proc. 2nd International Energy Conversion Engineering Conference (Providence, RI, 16 – 19 August 2004)*

[5] Anvari A, Farhani F and Niaki K S 2009 Comparative study on space qualified paints used for thermal control of a small satellite *IJChE* **6** 50-62

[6] Ding Y, Feng W and Yan D 2009 Degradation of optical-properties of thermal control coatings under space low energy electrons Proc. 11th International Symposium On Materials in a Space Envionment (Aix-en-Provence, France, 15 – 18 September 2009)

[7] Pater R H and Curto P A 2007 Advanced materials for space applications *Acta Astronaut.* **61** 1121-9

[8] Kim H K and Han C Y 2010 Analytical and numerical approaches of a solar array thermal analysis in a low-earth orbit satellite *Adv. Space Res.* **46** 1427-39

[9] Zaglauer A and Pitz W 2003 Champ-The first flexbus in orbit *Acta Astronaut.* **52** 747-51

[10] Nakayama A and Kuwahara F 2008 A general bioheat transfer model based on the theory of porous media *Int. J. Heat Mass Tran.* **51** 3190-99

[11] Amiri A and Vafai K 1994 Analysis of dispersion effects and non-thermal equilibrium, non-Darcian, variable porosity incompressible flow through porous media *Int. J. Heat Mass Tran.* **37** 939-54

[12] Coquard R, Rochais D and Baillis D 2010 Conductive and radiative heat transfer in ceramic and metal foams at fire temperatures *Fire Technol.* **48** 699-732

[13] Baillis D and Coquard R 2008 Radiative and conductive thermal properties of foams *Cellular and Porous Materials: Thermal Properties Simulation and Prediction* ed A. Öchsner, G E Murch and M. J. S. de Lemos (Weinheim: Wiley-VCH) chapter 11 pp 343-384

[14] Caps R, Heinemann U, Fricke J and Keller K 1997 *Int. J. Heat Mass Tran.* **40** 269-80

[15] Donabedian M, Gilmore D G, Stultz J W, Tsuyuki G T and Lin E I 2002 Thermal design examples *Spacecraft Thermal Control Handbook vol. 1 Fundamental Technologies* ed D G Gilmore (El Segundo, CA: The Aerospace Press) chapter 5 pp 161-205

Some critical remarks about the radiative heat transfer in air frame cavities according to EN ISO 10077-2

G Cuccurullo[1], L Giordano
Department of Industrial Engineering, University of Salerno,
Via Giovanni Paolo II 132, 84084 Fisciano (SA), Italy

E-mail: gcuccurullo@unisa.it

Abstract. Thermal performances of windows frames are established, in Europe, by the international standard UNI EN ISO 10077-2:2012. The standard introduces an equivalent thermal conductivity for air frame cavities thus simplifying the original combined heat transfer problem to a merely two-dimensional conductive one. The equivalence is referred to a rectangular cavity and is not able to fully recover the same radiative heat flux involved in the original problem. In view of that, the paper is focused on the radiative heat transfer taking place in the air cavities and aims to check if different equivalence criteria could lead to improved results. Thus, numerical tests involving an accurate description of radiative heat transfer in air cavities are compared to the simplified fully-conductive one provided by the standard. Results show that different criteria lead to quite different results. The optimal criterion turns out to depend on both geometrical and surface radiative parameters. It is also shown that, in any case, a proper radiative resistance but not the one suggested by the ISO 10077 should be adopted.

1. Introduction

Actually, energy efficiency is one of the main goals of building design being required to reduce energy consumption and face the gradual exhaustion of natural resources. Moreover, high insulation degrees affect thermal comfort since they influence radiative exchange between inner people and outer walls. In the above connection, the study of the thermal performance of glass doors and windows assumes a primary role, since these items represent a weak area in terms of heat exchange with the external environment [1].

In the last two-decades, the need to contain thermal dispersions through fenestration components has triggered the development of new highly-performant glazing with results that were unimaginable just few years ago. Performances have been greatly enhanced mainly by lowering natural convection effects inside multiple pane windows or selecting special glasses with suitable radiative properties. Since the technological asymptote seems to be very close by now, in the last years researchers have shifted their attention to consider heat transfer through window frames. Consider that, the area occupied by the frame may even weight as much as 20-30% of the total windows area while its impact on the total window heat transfer may be much larger because of the large difference in transmittance featuring other building enclosures. This is the case for aluminum alloys used for windows frames where a 160 W/(m K) thermal reference conductivity is assumed. The latter is at least two magnitude

[1] To whom any correspondence should be addressed.

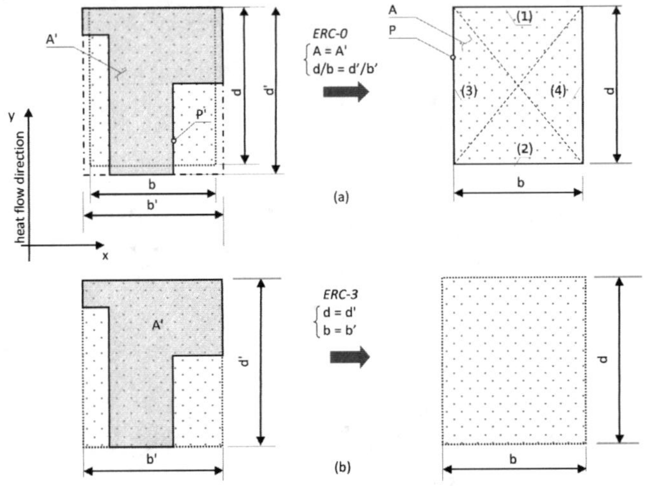

Figure 1. Equivalent rectangular cavities, ERC-0, ERC-1.

order higher than the one related to wood or PVC. This explains why other joining materials (e.g. polyamide) are used than aluminium to realize intermediate junctions. As a matter of fact, reducing the transmittance values for window-frames seems to exhibit wide profit margins wishing to reduce heat loss through the building envelope.

At the same time, determining thermal performances of windows frames in an accurate and repeatable fashion is a primary need triggered by market reasons. It is then advisable to check the accuracy of the procedures involved in featuring the product performance, i.e. the thermal transmittance. To this purpose, numerical procedures based on international standards are often employed whereas some niche options based on the hot box calorimeter can still be adopted (e.g. according to EN ISO 12567 [2]). It is well known and stated by the ISO EN 15099 itself that two different numerical methods of calculating the thermal transmittance are available: the ISO linear method [3-4] and the ASHRAE standard 142P edge method [5]. They treat in a different way the effect of the glazing spacer on the heat transfer through the frame and the glazing unit near the frame [6]. The two methods give sometimes slightly different results [7-8] which are almost contained within the 3% difference. A substantial agreement among 2D-numerical and experimental procedures is found but some improvements can be searched [9-10]. Three dimensional effects are somewhat described [7].

Actually, in Europe, the thermal characteristics of windows, doors and shutters are established by the international standard UNI EN ISO 10077-2:2012. Thus, in the present work attention is focused on such a standard. Nevertheless, with some advices, it is confirmed in what follows that ISO15099 is more accurate than 10077 as claimed by [11, 12]. Also other methods are available, e.g. the one suggested by the National Fenestration Rating Council (NFRC) [13, 14].

Generally speaking, the method proposed by different standards for window frames is based on numerical approaches simplifying the heat transfer problem to a merely two-dimensional conductive one; for such a problem computational efforts strictly required by the combined heat transfer effectively taking place in cavities are strongly reduced. The key parameter for realizing the simplified procedure is the equivalent thermal conductivity which accounts for the radiative-convective heat transfer in the internal frame cavities. The equivalent thermal conductivity doesn't allow to fully recover the radiative heat transfer of the original cavities, thus its introduction is not painless. Then, this paper focuses the attention on the criteria adopted by the standard for realizing the above equivalence. It is aimed to check if new possibilities can be identified in order to attain improved results. To this purpose, numerical tests, carried on by a commercial FEM code (Comsol), act as a reference since it has been involved an accurate description of radiative heat transfer in air cavities. Numerical runs are compared to the corresponding simplified fully-conductive one provided by the

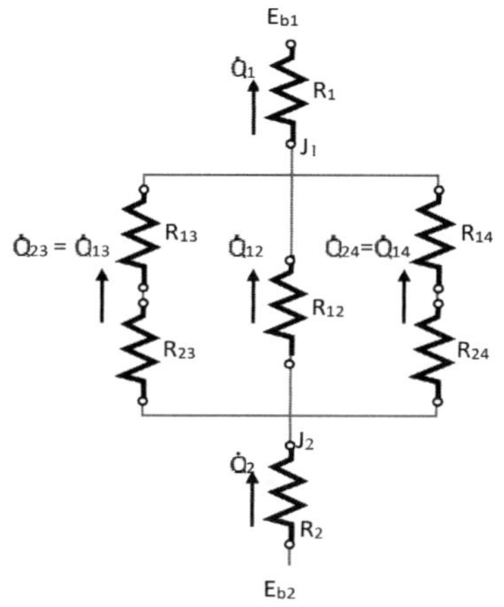

Figure 2. Radiative resistances.

standard. In what follows, it is shown that both the former and the latter approach lead to comparable results in terms of heat fluxes but meaningful improvements can be attained provided that proper corrections are considered with respect to the simplified correlation recommended by the EN ISO10077-2.

2. Basic equations for radiative heat transfer

In order to evaluate the heat flux flowing across the frame, the standard UNI EN ISO 10077-2 states that unventilated non-rectangular air cavities have to be transformed into rectangular cavities filled with a material exhibiting a suitable equivalent thermal conductivity accounting both for radiative and convective contributions.

This approach was early introduced by Standaert [15] and later both by Jonsson [16] and Carpenter [17] which formulated the effective conductivity referred to the rectangular geometry, area $A = d$ x b, perimeter $P = 2(b + d)$, see figure 1. Based on the early approach, several standards, such as ASHRAE 96, ISO 2000a and CEN 2001a, introduce an equivalent thermal conductivity defined as:

$$k_{eq} = (h_a + h_r)d \; ; \; h_r = 4 \cdot f \cdot \sigma \cdot T_m^3 \tag{1}$$

where h_a and h_r are the convective and radiative heat transfer coefficients, d is the length of the cavity in the heat flow direction, σ is the Stefan-Boltzmann constant, and f is a correcting factor with respect to the black body heat transfer; it has to depend on the cavity geometry and the radiative surface properties assumed as grey ones. In particular, according to the 10077 standard the correcting factor f is given by:

$$\mathrm{f} = \mathrm{E\,F}; \quad F = \frac{1}{2}\cdot\left(1+\sqrt{1+(d/b)^2} - d/b\right); \quad E = \left(\frac{1}{\varepsilon_1}+\frac{1}{\varepsilon_2}-1\right)^{-1} \tag{2}$$

where ε_1 and ε_2 are the emissivities of the hot and cold surfaces normal to the heat flow.

In order to apply the previous equations, the transformation of a generic irregular cavity of area A' and perimeter P' into an equivalent rectangular cavity (ERC) has to be performed. After identifying the smallest circumscribing rectangle, dim. b' x d' in figure 1 (a), the transformation can occur according to different criteria. The CEN 2001 a and lately the 10077 standards follow the rule illustrated in figure 1 (a): the ERC saves the same area, $A = A$', and the same aspect ratio, $d/b = d'/b'$, in order to feature the equivalent rectangle (dim. b x d). In addition to the previous one, identified below with the label "ERC-0", other equivalences can be proposed, for instance:

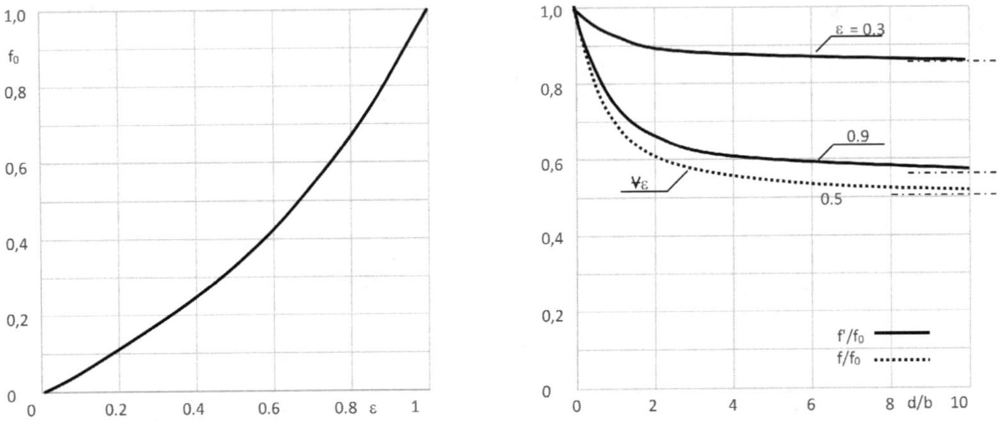

Figure 3. The correcting factors according to the 10077 standard (f) and to this paper (f').

- ERC-1: $A/P = A'/P'$, $d/b = d'/b'$, i.e. the ratio of area to perimeter and aspect ratio are saved;
- ERC-2: $d/d' = A/A'$, $d/b = d'/b'$; the scaling evidences the role of the depth d along which the heat transfer mainly takes place.
- ERC-3→ $d = d'$ and $b = b'$; the smallest circumscribed rectangle is selected, as suggested by the North American National Fenestration Rating Council, figure 1 (b)

After realizing the transformation into a rectangle of the cavity at hand by one of the available criteria, its view factors can be evaluated, as shown in the next paragraph.

3. View factors for rectangular cavity

In order to explicitly evaluate the coefficient h_r for a rectangular cavity, the network method, based on the electrical network analogy by A. K. Oppenheim, has been applied. According to the 10077, the heat transfer essentially takes place along the *y*-direction since the lateral surfaces (3) and (4) , see figure 1, are sought as adiabatic; for the remaining two surfaces, first-type boundary conditions are given, i.e. $T_1 = 0$ °C and $T_2 = 20$ °C.

Due to the geometry and load symmetry along the *y*-axis, the radiosities inherent to the lateral surfaces are equal, $J_3 = J_4$, therefore the related cross flow vanishes. The equivalent electrical network is shown in figure 2; here, the $R_i = (1-\varepsilon_i)/(\varepsilon_i\ w_i)$ are the surface resistances and $R_{ij} = (w_i\ F_{ij})^{-1}$ are the space resistance to radiation, w_i and ε_i being the width and the emissivities of the i-th surface, F_{ij} being the related view factors. Because of symmetry constraints, the view factors between the longitudinal and transverse rectangle surfaces turn out such as $F^* = F_{14} = F_{13} = F_{24} = F_{23}$ and their corresponding resistances are equal.

The radiative heat fluxes can be written in terms of the equivalent resistance, R_{eq}, of the network: $\dot{Q} = \dot{Q}_1 = \dot{Q}_2 = (E_{b2} - E_{b1})/R_{eq}$, where the potentials $E_{b1} = \sigma \cdot T_1^4$ and $E_{b2} = \sigma \cdot T_2^4$ are known, since the corresponding surface temperatures are prescribed.

The only unknown parameters required to evaluate the heat flow along the *y*-direction are the view factors. These have been calculated using the well known Crossed-Strings Method by Hottel which, in particular, yields:

$$F_{12} = \sqrt{1+\left(\frac{d}{b}\right)^2} - \frac{d}{b}; \qquad F^* = \frac{1}{2}\left[1 + \frac{d}{b} - \sqrt{1+\left(\frac{d}{b}\right)^2}\right] \tag{3}$$

Finally, after linearizing, the radiative heat flux between the hot and cold surfaces becomes:

$$\dot{Q} \approx h_r' \cdot A \cdot (T_2 - T_1) \tag{4}$$

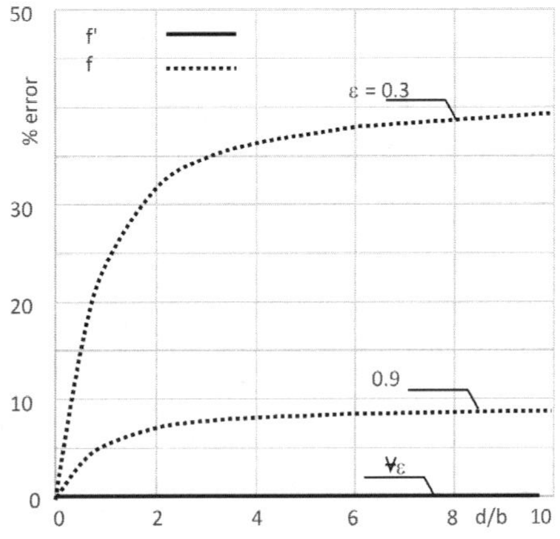

Figure 4. Percentage error for the correcting factors.

where the corrected radiative heat transfer coefficient $h_r' = 4 \cdot \sigma \cdot f' \cdot T_m^3$ is expressed in a similar fashion to eq. (1), the new correcting factor f' being defined as

$$f' = \frac{1}{E^{-1} + F_{34}}; \quad F_{34} = \sqrt{\left(\frac{b}{d}\right)^2 + 1} - \frac{b}{d} \tag{5}$$

The factor f' turns out to be the same of that given by ISO 15099 and formerly introduced by Roth [8], but here it has been expressed in a more compact way.

Finally, by comparing eq. (1) and the (5), it clearly appears that the use of the correction factor f' as an alternative to the corresponding f proposed by the EN ISO 10077-2 gives different results. Both the correcting factors turn out to be decreasing functions of the ratio d/b while selecting the same emissivity ε for both hot and cold surfaces. Considering that both the correcting factors attain the same value $f_0 = \varepsilon/(2-\varepsilon)$ for the ratio d/b being zero, the corresponding normalized curves f/f_0 and f'/f_0 are shown in figure 3 for $\varepsilon = 0.3$ and 0.9. The two selected values for emissivities are typical extreme values of aluminium frames: the former is prescribed by ISO 10077-2, if no other information is available, and is near the one featuring a painted or anodized frame; the latter is characteristic of galvanized or chromed untreated aluminium 6060, not far from the one related to shiny metallic surfaces.

All the normalized curve decrease monotonically attaining different asymptotic values: $f(\varepsilon, d/b \to \infty)/f_0 = 1/2$, independently of ε, while $f'(\varepsilon, d/b \to \infty)/f_0 = (2-\varepsilon)/2 > 1/2$. Thus, it can be inferred that, for a selected d/b, the factor f turns out to be always lower than f'.

4. Comparison between numerical and simplified results for rectangular cavities

In order to check the analytical expression obtained for the factor f', the radiative heat power leaving the hot surface has been numerically calculated using the software Comsol Multyphisics 5.0, based on the Finite Element Method (FEM). The commercial code allows to evaluate the radiative heat exchange in the cavities according to the diffuse opaque surface model. Thus, a rectangular cavity is singled out and, after building the geometry, the required boundary conditions were imposed: that is, with reference to figure 1, temperatures T_1 and T_2 on (1) and (2) and adiabatic conditions on (3) and (4). The convergence of the solution was verified by increasing the number of elements using the physics controlled mesh. Assuming the numerical solution "to be exact", figure 4 shows the percentage errors committed on the radiative power with respect to the numerical solution: as expected, the correcting factor f' allows to recover numerical results in any case, while the error made

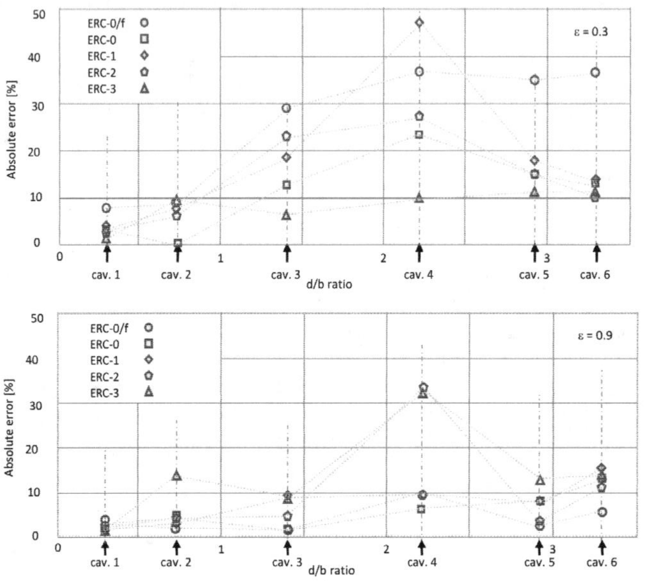

Figure 5. Percentage absolute error on the heat flux.

using the factor f increases with decreasing emissivity and increasing the cavity aspect ratio. Note that for typical emissivity of 0.9 the error made using f is sensibly lower than the one related to low emissivities.

It is then proven that the factor f given by UNI 10077-2 doesn't recover the exact value of radiative heat power even in the most favourable case of a singled out rectangular cavity for which the boundary conditions are strictly applied.

5. Analysis of the equivalence criterion

Six cavities taken from commercial profile (namely, mod. Planet 62 TT (Z), Meral S.p.A, Italy) were analysed in order to check the approximation made by the ERC criteria, see table 1. In view of the conclusion drawn in the previous paragraph, the factor f' is adopted to calculate the equivalent thermal conductivity with the noticeable exception of the case ERC-0 for which, in addition, the correction factor f provided by the 10077 standard is considered as a reference and labelled as ERC-0/f.

The procedure outlined below was followed. After applying the selected ERC criterion to the cavity at hand, its equivalent thermal conductivity was determined, assuming only radiative heat transfer to happen. Then, according to the procedure outlined by the 10077 standard, the original cavity was filled with a fictitious material whose thermal conductivity was the equivalent one. After that, the original shape was surrounded by an additional rectangular frame, the latter being identified by the smallest circumscribing rectangle whose uniform thickness was arbitrarily chosen to be equal to $b'/10$. Proceeding in such a way, it was allowed to encompass irregularly shaped cavities too by uniquely assigning boundary conditions as already done in the scheme of figure 1. The adjoining frame was assumed to exhibit the already determined equivalent conductivity. Thus, because of the symmetry of the load and geometry, a nearly 1D - heat flux was realized which was easily calculated. Such results were compared with that obtained by using the FEM code to solve the original radiative problem.

Results are summarized in figure 5, where the percentage absolute error on the heat flux leaving the boundary 1 is shown for the two already selected emissivities, i.e. 0.3 and 0.9. As a consequence of what observed in the previous paragraph about the correction factors, lower errors are realized for incresing emissivity. In particular, for $\varepsilon = 0.3$, it can be stated that the criterion ERC-0/f, i.e. the one suggested by the 10077 standard, gives very nearly the worst results for each cavity, whereas the best

Journal of Physics: Conference Series **655** (2015) 012025 doi:10.1088/1742-6596/655/1/012025

results are given by the ERC-3 with the noticeable exception of the most irregular "cavity 2" where ERC-0 prevails. None of the ERC-1 and ERC-2 criteria seem to give advantages. On the contrary, with reference to $\varepsilon = 0.9$, the ERC-3 seems to be the worst case, whereas the ERC-0 is almost the best one for the first four cavities at hand; surprisingly, for the two higher d/b ratio taken under consideration, the criterion proposed by the 10077 standard allows to realize lower errors. In any case, a lower sensitivity to the selected criterion can be appreciated when emissivity decreases.
Finally, it seems that the present work almost entirely confirms that more precise results can be obtained for the heat flux occurring in the cavity by employing the expressions here presented rather than the one suggested by the 10077 standard. Such results were confirmed, for instance, by several authors [6, 11, 12] both from experimental and theoretical point of view. Whilst, due to unpredictable geometrical and emissivity combinations, the criterion suggested by the standard seems to give better results.

Table 1. The six cavity shapes under test.

cav. n.	d/b	shape	cav. n.	d/b	shape
1	0.3		4	2.2	
2	0.72		5	2.92	
3	1.4		6	3.3	

6. Conclusions

In this paper the different methods for calculating the equivalent thermal transmittance for window frame cavities are compared. The reference method is sought to be the one given in the ISO EN 10077. With respect to the latter, both view-factor and geometrical equivalence criteria are compared by processing six typical frame cavities. It has been found that different criteria combinations give different results from the 10077 standard, they almost giving better results depending on the geometry and surface emissivity under consideration. It has been proven that a better description of the heat transfer across the cavity can be realized by using a proper equivalent radiative resistance calculated by the Crossed-Strings Method for the rectangular equivalent shape. Such a result is coherent with that reported by ISO EN 15099 standard, "Thermal Performance of Windows, Doors and Shading Devices Detailed Calculations". Moreover, it is shown that using the geometrical equivalence to a rectangular shape given from the 10077 standard is the best criterion available among that under test for higher emissivities, whereas for lower emissivities the NFRC criteria prevails.

7. References

[1] Giordano L, Cuccurullo G and Fiore P 2014 An approximate solution for the inner-glass longitudinal temperature profile in double glazed windows *Proc. 40th IAHS World Congress on Housing , Sustainable Housing Construction* (December 16-19, Funchal – Portugal)
[2] European Committee for Standardization, CEN. 2000. EN ISO 12567-1, *Thermal Performance of Windows and Doors—Determination of Thermal Transmittance by Hot Box Method—Part 1: Complete Windows and Doors* (Brussels)
[3] ISO TC163/SC. ISO/DIS 15099.5 1999 *Thermal Performance of Windows, Doors and Shading Devices –Detailed Calculations.*

[4] CEN 1998 *Thermal performance of windows, doors and shutters – Calculation of thermal transmittance – Part 2: Numerical methods for frames*. CEN 10077-2

[5] American Society of Heating, Refrigerating and Air Conditioning Engineers ASHRAE.1998 *Standard Method for Determining and Expressing the Heat Transfer and Total Optical Properties of Fenestration Products*, Public Review Draft of Standard 142P (Atlanta)

[6] Gustavsen A 2008 *State-of-the-Art Highly Insulating Window Frames –Research and Market Review* (Lawrence Berkeley National Laboratory, University of California)

[7] Weitzmann P, Jensen C F and Svendsen S 2000 Comparison of calculations of thermal transmittance of windows using two and three-dimensional models, *Proc. of the EuroSun Conference* (Department of Buildings and Energy, Technical University of Denmark)

[8] Roth H 1998 *Comparison of Thermal Transmittance Calculation Methods Based on ASHRAE and CEN/ISO Standards*, Master of Science Thesis (Dept. of Mechanical and Industrial Engineering ,Univesity of Massachusetts Amherst)

[9] Gustavsen A, Goudey H, Arasteh D, Uvsløkk S, Talev G, Jelle B. P and Kohler C 2010 Experimental and Numerical Examination of the Thermal Transmittance of High Performance Window Frames, *Proc. ASHRAE Buildings XI Conference*

[10] Blanusa P, Goss W P, Roth H, Weitzmannn P, Jensen C. F, Svendsen S and Elmahdy H 2007 Comparison between ASHRAE and ISO thermal transmittance calculation methods, *Energy and buildings* **39** 374-384

[11] Gustavsen A 2001 *Heat transfer in window frames with internal cavities* Ph.D. Thesis (Norwegian University of Science and Technology)

[12] Noyé P A, Laustsen J B and Svendsen S 2004 Calculating the heat transfer coefficient, of frame profiles with internal cavities, *Nordic Journal of Building Physics* **3**

[13] Mitchell R, Kohler, Arasteh D, Carmody J, Huizenga C and Curcija D 2006 *THERM 5.2/WINDOW 5.2 NFRC Simulation Manual* (Berkeley, CA: Lawrence Berkeley National Laboratory)

[14] MD: National Fenestration Rating Council, NFRC. 2001. NFRC 100:2001, *Procedure for Determining Fenestration Product U-factors* (Silver Spring).

[15] Standaert P 1984 Thermal evaluation of window frames by the finite difference method, *Proc. Proc. Windows in building design and maintenance* (Stockholm: Swedish Council for Building Research)

[16] Jonsson B 1985 *Heat transfer trough windows. During the hours of darkness with the effect of infiltration ignored* (Stockholm: Swedish Council for Building Research)

[17] Carpenter S C and McGowan A G 1989 Frame and spacer effects on window U-value *ASHRAE Transactions* **95** 604-608

Journal of Physics: Conference Series **655** (2015) 012026 doi:10.1088/1742-6596/655/1/012026

Magnetic refrigeration: an eco-friendly technology for the refrigeration at room temperature

C Aprea[1], A Greco[2], A Maiorino[1], C Masselli[1]

[1]DII, Università di Salerno, via Giovanni Paolo II 132, 84084 Fisciano, Salerno, Italia
[2]DII, Università degli Studi di Napoli Federico II, P.le Tecchio 80, 80125 Napoli, Italia

Email: adriana.greco@unina.it

Abstract. Magnetic refrigeration is an emerging, environment-friendly technology based on a magnetic solid that acts as a refrigerant by magneto-caloric effect (MCE). In the case of ferromagnetic materials MCE is a warming as the magnetic moments of the atom are aligned by the application of a magnetic field, and the corresponding cooling upon removal of the magnetic field. There are two types of magnetic phase changes that may occur at the Curie point: first order magnetic transition (FOMT) and second order magnetic transition (SOMT). The reference cycle for magnetic refrigeration is AMR (Active Magnetic Regenerative cycle) where the magnetic material matrix works both as a refrigerating medium and as a heat regenerating medium, while the fluid flowing in the porous matrix works as a heat transfer medium. Regeneration can be accomplished by blowing a heat transfer fluid in a reciprocating fashion through the regenerator made of magnetocaloric material that is alternately magnetized and demagnetized. In this paper, attention is directed towards the near room-temperature range. We compare the energetic performance of a commercial R134a refrigeration plant to that of a magnetic refrigerator working with an AMR cycle. Attention is devoted to the evaluation of the environmental impact in terms of a greenhouse effect. The comparison is performed in term of TEWI index (Total Equivalent Warming Impact) that takes into account both direct and indirect contributions to global warming. In this paper the AMR cycle works with different magnetic refrigerants: pure gadolinium, second order phase magnetic transition ($Pr_{0.45}Sr_{0.35}MnO_3$) and first order phase magnetic transition alloys ($Gd_5Si_2Ge_2$, $LaFe_{11.384}Mn_{0.356}Si_{1.26}H_{1.52}$, $LaFe_{11.05}Co_{0.94}Si_{1.10}$ and $MnFeP_{0.45}As_{0.55}$). The comparison, carried out by means of a mathematical model, clearly shows that $GdSi_2Ge_2$ and $LaFe_{11.384}Mn_{0.356}Si_{1.26}H_{1.52}$ has a TEWI index always lower than that of a vapor compression plant. Furthermore, the TEWI of the AMR cycle working with FOMT materials is always better than that of SOMT materials. $Gd_5Si_2Ge_2$ is the best FOMT material.

1. Introduction

Worldwide, about 15% of the overall energy consumption originates from refrigeration. Most modern refrigeration units are based on vapor compression plants, whose development is strictly related to the characteristics of the working fluids, since the very beginning of their commercial diffusion. The traditional refrigerant fluids, i.e. CFCs and HCFCs, have been banned because of their contribution to the disruption of the stratospheric ozone layer (Ozone-Depleting substances ODs).

Journal of Physics: Conference Series **655** (2015) 012026 doi:10.1088/1742-6596/655/1/012026

Human activities have increased the concentration of greenhouse gases in the atmosphere, thus resulting in a substantial warming of both earth surface and atmosphere that adversely affect the natural ecosystem. The impact of greenhouse gases on global warming is quantified by their GWP (Global Warming Potential). The GWP is defined as the mass of CO_2 that would result in the same net impact on global warming as the release of a single unit (kg) of the component.

The Kyoto Protocol (1987), pursuant to the United Nations Framework Convention on Climate Change (UNFCCC), sets binding targets for greenhouse gas emissions.

A vapor compression plant produces both a direct and an indirect contribution to global warming. The former depends on the GWP of refrigerant fluids and on the fraction of refrigerant charge which is either directly released in the atmosphere during operation and maintenance, or is not recovered when the system is scrapped. The indirect contribution is related to energy-consumption of the plant. In fact, a vapor compression refrigerator requires electrical energy produced by a power plant that typically burns a fossil fuel, thus releasing CO_2 into the atmosphere. The amount of CO_2 emitted is a strong function of the COP of the vapor compression plant.

Magnetic refrigeration is an emerging new technology. It is based on the magneto-caloric effect taking place in solid-state refrigerants (MCE). Compared to conventional vapor compression systems, magnetic refrigeration can be an environment-friendly and efficient technology. The magnetic refrigerant is a solid, has essentially zero vapor pressure, and therefore is ecologically sound with no direct Ozone Depletion Potential (ODP) and zero direct Global Warming Potential (GWP). The Active Magnetic Regenerator (AMR) is the core of a magnetic refrigerator system. It is a special kind of thermal regenerator made of magnetic material which works both as a refrigerating and as a heat regenerating medium.

The present paper compares a commercial R134a refrigerator and a 1.5T AMR cycle using different magnetic materials and water as a secondary fluid. The higher efficiency of the magnetic refrigeration unit resulted in a reduced CO2 release and in a lower electric energy consumption. Thus, the development of magnetocaloric air conditioners would provide a relevant contribution to energy conservation and to the reduction of global warming. Magnetic refrigeration is not yet a commercial technology because of a number of challenges that have still to be overcome. Improved engineering to overcome are: limitations of the currently available magnetic refrigerant materials, developing new magnetic materials with giant magneticaloric effect, producing magnetic refrigerants on large scale, developing processes to inexpensively fabricate these materials into useable forms for regenerators (spheres, wires, foils, plates, etc.) without losing the magnetocaloric effect, increase of the magnetic field strength of the permanent magnets while reducing the size, mass and costs.

2. Magnetic materials and magnetocaloric effect

The performances of an AMR refrigerator are mostly influenced by the particular solid materials employed, which is for sure a magnetocaloric material (MM) [1-6]. In fact, only magnetocaloric materials experience magnetocaloric effect (MCE) which is a coupling between MM entropy and the variation of an external magnetic field applied to the material, providing an amount of cooling that needs to be transferred out of the regenerator by a secondary fluid, used only as heat vector. The explanation of the MCE lies in the coupling between the magnetic spin and the crystal lattice. When a ferro- or paramagnetic material is magnetized there will be some ordering of the magnetic spins, forcing them towards the direction of the applied field. If this is done isothermally, this will lower the material's magnetic entropy by the isothermal entropy change (ΔS_M). However, if the magnetization is done adiabatically, the total sample entropy remains constant and the decrease in magnetic entropy is countered by an increase in the lattice and electron entropy. This causes a heating of the material and a temperature increase given by the adiabatic temperature change (ΔT_{ad}). The inverse procedure also applies: under adiabatic demagnetization the magnetic entropy increases, causing a decrease in lattice vibrations and by that a temperature decrease. At its Curie temperature, where is located its own magnetic phase transition, a MM shows the peak of MCE, in terms of ΔT_{ad} and ΔS_M. The two possible magnetic phase changes that one can observe at the Curie point are first order magnetic transition (FOMT) and second order magnetic transition (SOMT).

At the Curie point a magnetic transition has FOMT characteristics when the material exhibits a discontinuity in the first derivative of the Gibbs free energy (G.f.e.), whereas has a SOMT behavior when the gap is detected in the second derivative of G.f.e. while its first derivative is a continuous function.

The most employed material for the refrigeration at room temperature is gadolinium, a rare-earth metal which exhibits a second order paramagnetic to ferromagnetic transition at its Curie temperature of 294 K. It exhibits excellent magnetocaloric properties that are difficult to improve upon. Nowadays a huge part of researchers is orienting its attention on some new alloys of magnetocaloric materials. The most promising are $Gd_5Si_{4-x}Ge_x$ compounds (where $0 < x < 0.5$), alloys of gadolinium, silicon and germanium which show a FOMT characterized by a peak of ΔT_{ad} and ΔS_M much greater than gadolinium ones, but the whole function is quite sharper in $Gd_5Si_{4-x}Ge_x$ alloys. In particular, $Gd_5Si_2Ge_2$ exhibits the larger MCE among $Gd_5Si_{4-x}Ge_x$ compounds, named GIANT magnetocaloric effect of Pecharsky [7-10]. $Gd_5Si_2Ge_2$ presents two different phase transitions: at 276 K one can observe a FOMT which constitute the MCE highest temperature peak, whereas at 299K it is possible to appreciate a SOMT where, according to it, the material orders paramagnetically. MnAs alloys also assume the role of candidates for new magnetic refrigeration materials at room temperature because those compounds have a giant MCE exhibiting a first order magnetic transition; by varying alloy's composition, the Curie temperature could be switched in the range 220÷318 K. Among Mn-based compounds, $MnFeP_{1-x}As_x$ compounds that are stable for $0.15 < x < 0.66$ and exhibit interesting magnetic properties associated with a first order metamagnetic transition. The Curie temperature of the alloy increases linearly with the As contents. In the present paper compound $MnFeP_{0.45}As_{0.55}$ [11] has been investigated; it undergoes a FOMT from paramagnetic to ferromagnetic at 307 K (on heating) according to a rapid decrease of the material parameters which changes its Debye temperature and its electronic structure. The adiabatic temperature change for these compounds is relatively low and the thermal conductivity is significantly lower than that of gadolinium and other magnetic materials.

Other interesting compounds for magnetic refrigeration are the rare-earth transition-metal $La(Fe_xSi_{1-x})_{13}$ because of the large MCE exhibiting around room temperature (ΔS_T of up to 30 J/kgK under a 0-5T magnetic field variation). The main advantages of employing $La(Fe_xSi_{1-x})_{13}$ alloys are cheap, readily available and easy preparation. They present a first-order itinerant electron metamagnetic transition producing a giant magnetocaloric effect. In this paper $LaFe_{11.384}Mn_{0.356}Si_{1.26}H_{1.52}$ and $LaFe_{11.05}Co_{0.94}Si_{1.10}$ [12-14] have been considered. Both compounds have a first order transition at 290 and 287 K, respectively.

In the present work also $Pr_{0.45}Sr_{0.35}MnO_3$ [15,16] is considered. It belongs to $Pr_{1-x}Sr_xMnO_3$ compounds, which are an easy synthetizable series, and it presents a second order phase transition at 295 K where it shows a modest MCE.

Table 1 summarizes the characteristics of the presented magnetic materials in the room temperature range. The main parameters of the different materials allow a fast comparison in terms of cost vs performance analysis. ΔT_{ad} and ΔS_M reported are the peak value with a magnetic field variation ΔH of 1.5 T.

Table 1 allow a quick comparison between the different magnetic materials. The FOMT materials show peak values of ΔT_{ad} and ΔS_{MT} always higher than that of Gd. $Gd_5(Si_xGe_{1-x})_4$ shows the higher peak values, but out of the room temperature range. The greater thermal conductivity is that of Gd, but also LaFeSi alloys show high values. From a commercial point of view, $Gd_5(Si_xGe_{1-x})_4$ and Gd are too expensive and the magnetic transition metals are more adequate than the rare earths for an industrial production of magnetic cooling engines. La is the cheapest element of the rare-earths series, and Fe, Si, Mn are available in large amounts. The material costs of $MnFeP_{0.45}As_{0.55}$ are quite low, but processing of As is complicated due to its toxicity.

Table 1. The different magnetic materials.

Materials	**Tc**	**ΔH**	**ΔT_{ad}**	**ΔS_M**	**ρ**	**k**	**Cost**
	(°C)	**(T)**	**(°C)**	**(J/kgK)**	**(kg/m³)**	**(W/mK)**	**(€/kg)**
Gd	294	1.5	6	5	7900	10.9	3000
$Gd_5(Si_xGe_{1-x})_4$	276	1.5	7.8	14	7205	5.8	9000
$LaFe_{11.384}Mn_{0.356}Si_{1.26}H_{1.52}$	290	1.5	5	10.5	7100	9	1200
$LaFe_{11.05}Co_{0.94}Si_{1.10}$	287	1.5	3.17	5.5	7290	8.9	1200
$MnFeP_{0.45}As_{0.55}$	307	1.5	4	12	7300	2.5	1500
$Pr_{0.45}Sr_{0\text{-}.35}MnO_3$	295	1.5	1.5	2.5	5800	1.8	1050

In figure 1 and 2 are reported ΔS_{MT} and ΔT_{ad} as a function of temperature for the above mentioned magnetic materials for a magnetic field change of 1.5 T.

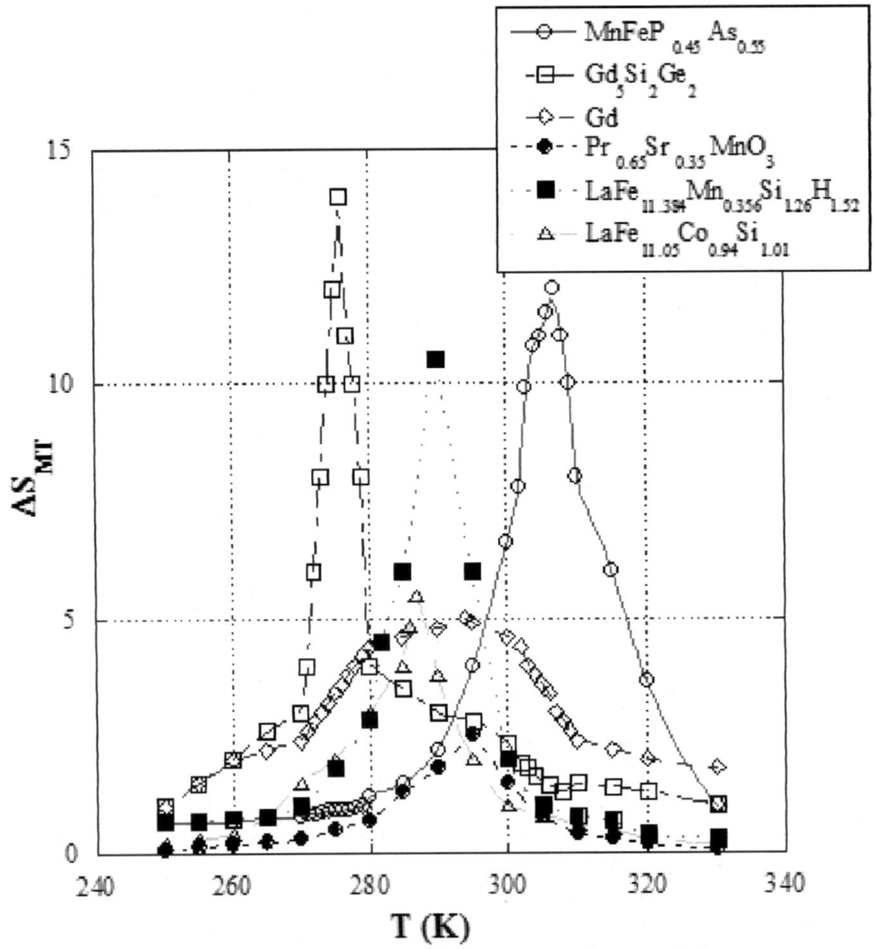

Figure 1. ΔS_{MT} vs T for the different magnetic materials.

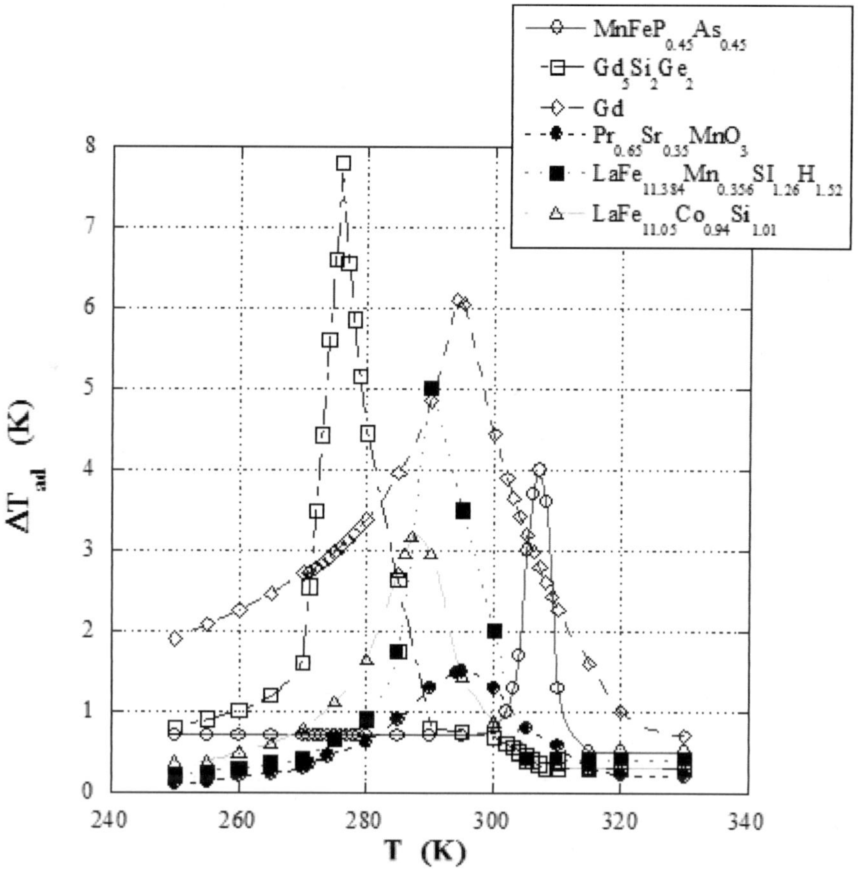

Figure 2. ΔT_{ad} vs T for the different magnetic materials.

Figure 1 clearly shows that $Gd_5(Si_xGe_{1-x})_4$, $LaFe_{11.384}Mn_{0.356}Si_{1.26}H_{1.52}$ and $MnFeP_{0.45}As_{0.55}$ materials shows the greater values of ΔS_{MT}. These materials are FOMT and therefore the magnetic entropy variation is centered in a restricted temperature interval. Whereas Gd shows lower values of ΔS_{MT} but over a broad temperature range. $Pr_{0.45}Sr_{0.35}MnO_3$ shows a very low magnetic entropy variation in the whole temperature range. $Gd5(SixGe_{1-x})_4$ and Gd show high values of ΔT_{ad}. Whereas, $MnFeP_{0.45}As_{0.55}$ has the disadvantage that the adiabatic temperature change is not very large due to the relatively high heat capacity. The main goal of this paper is to investigate the effect of the different magnetic materials on the energetic performances of an AMR cycle and to compare the energetic performance of a commercial R134a refrigeration plant to that an AMR cycle. Attention is devoted to the evaluation of the environmental impact in terms of a greenhouse effect. To this hope, a practical 2D model for predicting the refrigeration capacity and the efficiency of an AMR cycle in the room temperature range has been developed. Different magnetic materials have been considered as refrigerant: pure gadolinium [17], second order phase magnetic transition $Pr_{0.45}Sr_{0.35}MnO_3$ and first order phase magnetic transition alloys $Gd_5(Si_xGe_{1-x})_4$, $LaFe_{11.384}Mn_{0.356}Si_{1.26}H_{1.52}$, $LaFe_{11.05}Co_{0.94}Si_{1.10}$ and $MnFeP_{0.45}As_{0.55}$.

3. The TEWI concept

Human activities have increased the concentration of greenhouse gases in the atmosphere. This resulted in a substantial warming of earth surface and atmosphere that adversely affected the natural ecosystem. The impact that greenhouse gases on global warming is quantified by their GWP (Global

Journal of Physics: Conference Series **655** (2015) 012026 doi:10.1088/1742-6596/655/1/012026

Warming Potential). The GWP is defined as the mass of CO2 that would result in the same net impact on global warming as the release of a single unit (kg) of the atmospheric component in question [18].

Vapor compression plants produce both a direct and an indirect contribution to global warming. The former depends on the GWP of refrigerant fluids and on the fraction of refrigerant charge released in the atmosphere during operation and maintenance, or not recovered when the system is scrapped. The indirect contribution consists in the so-called energy-related contribution. In fact, a vapor compression refrigerator requires electrical energy produced by a power plant that typically burns a fossil fuel releasing CO2 to the atmosphere. The amount of CO2 emitted is a strong function of the COP of the vapor compression plant. The Total Equivalent Warming Impact (TEWI) index takes into account both contributions to global warming of the system.

The concept of total equivalent warming impact (TEWI) was developed to combine the effect of direct refrigerant emission with those due to energy consumption and the related combustion of fossil fuels for the electric energy production. TEWI provides a measure of the environmental impact of greenhouse gases originating from operation, service and end-of-life disposal of the equipment. TEWI is the sum of the direct contribution of the greenhouse gases used to make or operate the systems and the indirect contribution of carbon dioxide emissions resulting from the energy required to run the systems over their normal lifetimes.

The TEWI is calculated as :

$$\begin{aligned} TEWI &= CO_{2,dir} + CO_{2,indir} && [kg\,CO_2] \\ CO_{2,dir} &= RC\left[P_L + \left(\frac{1-P_R}{V}\right)\right] V \cdot GWP && [kg\,CO_2] \\ CO_{2,indir} &= \alpha \cdot \frac{\dot{Q}_{ref}}{COP} \cdot H \cdot V && [kg\,CO_2] \end{aligned} \quad (1)$$

The direct global warming effect of refrigerant fluids, stemming from the absorption they produce of long-wave radiations, depends on their GWP and on the fraction of refrigerant charge released in the atmosphere. The last is mainly due to leakage during the operational plant life time (PL) and to the residual amounts which, according to the current state of technology, are not recyclable and thus are released to the atmosphere when taking the plant out of operation (1-PR). In the simulation PL is assumed as 5%, whereas PR has not been considered. As already stated, the indirect contribution to TEWI consists in the so-called energy-related contribution. Indeed, an electrical refrigerator requires electrical energy from a CO_2 releasing power plant that typically burns a fossil fuel. The amount of CO2 emitted is a function of the refrigerator COP, of the power plant efficiency and of the fuel used in the conversion plant that affect the emissions per unit energy converted. The typical power-plant technology adopted varies from one country to another. The literature provides some indicative, average levels of CO_2 release per KWh of electrical energy for various countries. For Italy, the value is 0.6 kg CO_2/kWhe. The annual power consumption in the TEWI simulation is 290 kWh per year that corresponds to a commercial medium size wine cooler. R134a is an HFC with zero ODP and a GWP of 1300.

4. The AMR cycle

Barclay firstly introduced the Active Magnetic Regenerative refrigeration cycle, well known as AMR, in 1982 [20]. It considers a magnetic Brayton cycle.The main innovation consists of introducing the AMR regenerator concept, i.e. the employ of the magnetic material itself as refrigerant and as regenerator. A secondary fluid is used to transfer heat from the cold to the hot end of the regenerator. Substantially every section of the regenerator experiments its own AMR cycle, according to the proper working temperature. Through an AMR one can appreciate a larger temperature span between the regenerator material and the auxiliary fluid.

The principle of operation of the presented magnetic refrigerator is based on the AMR thermodynamic cycle. The working principle of an AMR is presented in figure 3. Let us assume that the bed is at a steady state condition with the hot heat exchanger at Th whereas the cold heat exchanger is at Tc. Four processes are simultaneously present in the AMR cycle: (a) adiabatic magnetization: each particle in the bed warms up; (b) isofield cooling: the high field is present, the fluid is blown from the cold end to the hot end, and absorbs heat from the bed and expels heat at a temperature higher that Th in the hot heat exchanger; (c) adiabatic demagnetization: each particle in the bed cools again; (d) isofield heating: the field is zero, the fluid is blown from the hot end to the cold one, and it expels heat to the particles of the bed and absorbs heat at a temperature lower than Tc in the cold heat exchanger. In Figure 3 the dashed line represents the initial temperature profile of the bed in each process, while the solid line depicts the final temperature profile of the process.

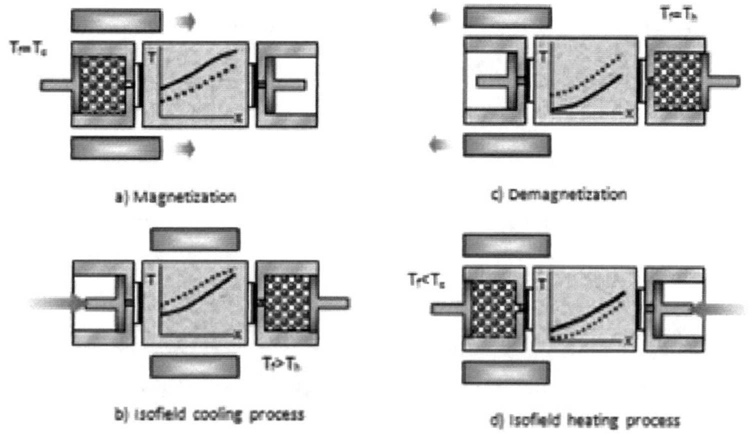

Figure 3. Four schematic processes of an AMR cycle: solid and dashed lines referred to final and initial temperature profile respectively.

5. Model description

The model presented in this paper is of a two-dimensional porous regenerator operating at room temperature. The regenerator has a rectangular shape with a height of 20mm and a length of 45 mm. The area of the regenerator is filled with a regular matrix of 3600 circles that constitute the porous media; every circle has a diameter of 0.45mm and the amount of the area occupied by all of the circles is 63% of the total rectangular area. A group of channels is formed by stacking particles in the regenerator area: the fluid flows through these interstitial channels.

In this paper the 2D simulations of the numerical model are performed according to operate in the same conditions of a Rotary Permanent Magnet Magnetic Regenerator (RPMMR) [21-23], so that a comparison between the model and the RPMMR could be done. The rotary permanent magnet magnetic refrigerator (RPMMR) presents a rotating magnet while the magnetocaloric material (MCM) is stationary and is made of 8 regenerators.

In this model are taken into account both the fluid flow and heat transfer between the solid and the fluid. The regenerator of the proposed model has a porous media geometry: it is made of a porous matrix of gadolinium circles and the fluid flows through interstitial channels paragraphs are indented.

The fluid flow in the positive x direction during the isofield cooling process and in the opposite direction during isofield heating.

The assumption for the porous region of the AMR are:

(1) the magnetic material is isotropic and the magnetic particles are rigid spheres with the same diameter;

(2) heat radiation is neglected;
(3) the porosity of porous medium is constant;
(4) the bed is adiabatic.

The mathematical formulation that describes the AMR cycle includes several distinct groups of equations according to the different processes of the AMR cycle that the regenerator experience. paragraphs are indented (BodytextIndented style).

The equations that rule the regenerative fluid flow processes, in both directions, are: the Navier-Stokes equations for the fluid flow and the energy equations for both the fluid and the solid particles. With the assumptions that the fluid is incompressible, the viscous dissipation in neglected, due to low mass flow, the above equations are as follows:

$$\begin{cases} \frac{\partial u}{\partial x}+\frac{\partial v}{\partial y}=0 \\ \frac{\partial u}{\partial t}+u\frac{\partial u}{\partial x}+v\frac{\partial u}{\partial y}=-\frac{1}{\rho_f}\frac{\partial p}{\partial x}+\upsilon\left(\frac{\partial^2 u}{\partial x^2}+\frac{\partial^2 u}{\partial y^2}\right) \\ \frac{\partial v}{\partial t}+u\frac{\partial v}{\partial x}+v\frac{\partial v}{\partial y}=-\frac{1}{\rho_f}\frac{\partial p}{\partial y}+\upsilon\left(\frac{\partial^2 v}{\partial x^2}+\frac{\partial^2 v}{\partial y^2}\right) \\ \frac{\partial T_f}{\partial t}+u\frac{\partial T_f}{\partial x}+v\frac{\partial T_f}{\partial y}=\frac{k_f}{\rho_f c_{fp}}\left(\frac{\partial^2 T_f}{\partial x^2}+\frac{\partial^2 T_f}{\partial y^2}\right) \\ \frac{\partial T_s}{\partial t}=\frac{k_s}{\rho_s c_{sp}}\left(\frac{\partial^2 T_s}{\partial x^2}+\frac{\partial^2 T_s}{\partial y^2}\right) \end{cases} \tag{2}$$

The fluid is sufficiently low speed to be considered laminar. The solid and the fluid temperature are strictly related and they are constantly evolving during the whole AMR cycle.

The equations that model the magnetization and the demagnetization process of the cycle must take into account the MCE which elevates or lowers the temperature of the solid by the variation of the external magnetic field applied to the regenerator. Hence the MCE temperature variation Is converted into a heat source Q:

$$Q=Q(H,T_S)=\frac{\rho_S c_{sp}(H,T_S)\Delta T_{ad}(H,T_S)}{\Delta t} \tag{3}$$

which has the dimensions of a power density and included in the solid energy equation, only for magnetization and demagnetizations phases, as follows:

$$\rho_s c_{sp}\frac{\partial T_s}{\partial t}=k_s\left(\frac{\partial^2 T_s}{\partial x^2}+\frac{\partial^2 T_s}{\partial y^2}\right)+Q \tag{4}$$

The term Q is positive during magnetization, negative during demagnetization. Δt is the period of the magnetization/demagnetization process. Therefore, the equations that govern the magnetization and the demagnetization phases are:

$$\begin{cases} \rho_f C_{fp}\frac{\partial T_f}{\partial t}=k_f\left(\frac{\partial^2 T_f}{\partial x^2}+\frac{\partial^2 T_f}{\partial y^2}\right) \\ \rho_s C_{sp}\frac{\partial T_s}{\partial t}=k_s\left(\frac{\partial^2 T_s}{\partial x^2}+\frac{\partial^2 T_s}{\partial y^2}\right)+Q \end{cases} \tag{5}$$

The magnetic properties of the magnetocaloric materials have been evaluated interpolating experimental data available from the literature. From the construction of a table function, who describes Q in a range of temperature and magnetic field intensity, and by the help of a mathematical function finder software, a mathematical expression for Q has been found both for magnetization and demagnetization processes. The intensity of the magnetic field varies from 0 to the maximum value of 1.2 T during magnetization and from 1.2T to 0 during demagnetization.

The coupled equations that govern the AMR cycle, imposed on this model, are solved using Finite Element Method. The AMR cycle is modelled as four sequential steps. The same time step Δt has been chosen for the resolution during all the four periods of the cycle. The cycle is repeated several times with constant frequency until the regenerator reaches steady state operation.

The refrigeration energy and the energy supplied in the environment are calculated according to the following equations:

$$Q_{ref} = \int_{t_M+t_{CF}+t_D}^{t_M+t_{CF}+t_D+t_{HF}} \dot{m}_f C_f \left(T_C - T_f(0,y,t)\right) dt \quad (6)$$

$$Q_{rej} = \int_{t_M}^{t_M+t_{CF}} \dot{m}_f C_f \left(T_f(L,y,t) - T_H\right) dt \quad (7)$$

The work of the secondary fluid circulation pump is:

$$W_p = \frac{\dot{m}\left(\Delta p_{CF} + \Delta p_{HF}\right)}{\eta_p \rho_f}\left(t_{CF} + t_{HF}\right) \quad (8)$$

The Coefficient of Performance is:

$$COP = \frac{Q_{ref}}{Q_{rej} - Q_{ref} + W_p} \quad (9)$$

The COP of the vapor compression plant working under the same operating conditions of the above mentioned AMR cycle was predicted with the DOE/ORNL Heat Pump model (Rice 2006). The vapour compression plant used in the simulation is a commercial, small size, R134a refrigerator with a semi-hermetic compressor, air cooled condenser, forced air circulating evaporator [24].

6. Results and discussion

Several AMR cycles with different magnetocaloric materials employed as refrigerant, were simulated. The MCE materials kept under investigation have been Gd, $Gd_5Si_2Ge_2$ and alloys of $LaFe_{11.384}Mn_{0.356}Si_{1.26}H_{1.52}$, $LaFe_{11.05}Co_{0.94}Si_{1.10}$, $MnFeP_{0.45}As_{0.55}$, $Pr_{0.45}Sr_{0.35}MnO_3$. In all of the cases, the secondary fluid is pure water. The simulations were performed with fixed flow rate (5 l/min), AMR cycle frequency (1.25 Hz) and cold heat exchanger temperature T_C (288 K). The hot heat exchanger temperature TH was varied in the range 295÷302 K to characterize the performance sensitivity of the heat rejection temperature in proximity of the refrigerant Curie temperature. The results presented were generated for a magnetic induction, which varies from 0 to 1.5 T.

TEWI is the sum of the direct and indirect contribution to global warming. The direct contribution of the AMR cycle is zero because the refrigerant is solid and therefore has essentially zero vapor pressure and zero GWP. The direct contribution of the vapor compression plant accounts about 10 % of the whole value. The parameter that affects the indirect contribution of the TEWI is the COP of the plant. In Figure 4 is shown the Coefficient of Performance (COP) as a function of hot heat exchanger temperature for all the materials presented in this work and for the vapor compression plant.

The Figure clearly shows that $Gd_5Si_2Ge_2$ and $LaFe_{11.384}Mn_{0.356}Si_{1.26}H_{1.52}$ have the highest values of COP and always over performs a vapor compression plant. The results of the simulation clearly show that $Gd_5(Si_xGe_{1-x})_4$ is the best magnetic material with a COP that is always greater than that of a traditional vapor compression plant in the same operating conditions (from a minimum of + 26 to a maximum of +44 %). Gd show energetic performances similar to a vapor compression plant. Instead $LaFe_{11.05}Co_{0.94}Si_{1.10}$, $MnFeP_{0.45}As_{0.55}$ and $Pr_{0.45}Sr_{0.35}MnO_3$ always underperfom vapor compression.

Therefore the COP of the AMR cycle working with these materials needs an improvement

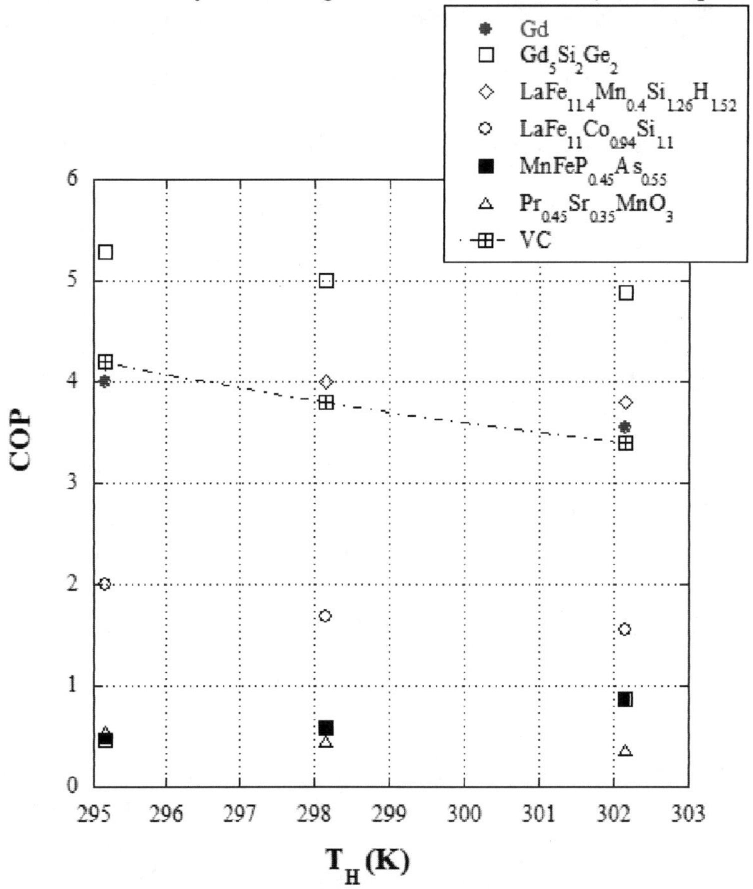

Figure 4. COP as a function of the hot heat exchanger temperature.

In order to make a comparison ΔTEWI has been defined according to the following equation:

$$\Delta TEWI = \frac{TEWI_{AMR} - TEWI_{VC}}{TEWI_{VC}} \tag{10}$$

In figure 5 is reported ΔTEWI of Gd, $Gd_5(Si_xGe_{1-x})_4$ and $LaFe_{11.384}Mn_{0.356}Si_{1.26}H_{1.52}$ as a function of T_H.

The figure clearly shows that the AMR cycle working with these magnetic materials shows a lower greenhouse effect with respect to a vapor compression plant. In particular with $Gd_5Si_2Ge_2$ ΔTEWI varies between a minimum of -44 to a maximum of -38 %. In figure 6 is reported ΔTEWI of $LaFe_{11.05}Co_{0.94}Si_{1.10}$, $MnFeP_{0.45}As_{0.55}$ and $Pr_{0.45}Sr_{0.35}MnO_3$ as a function of T_H.

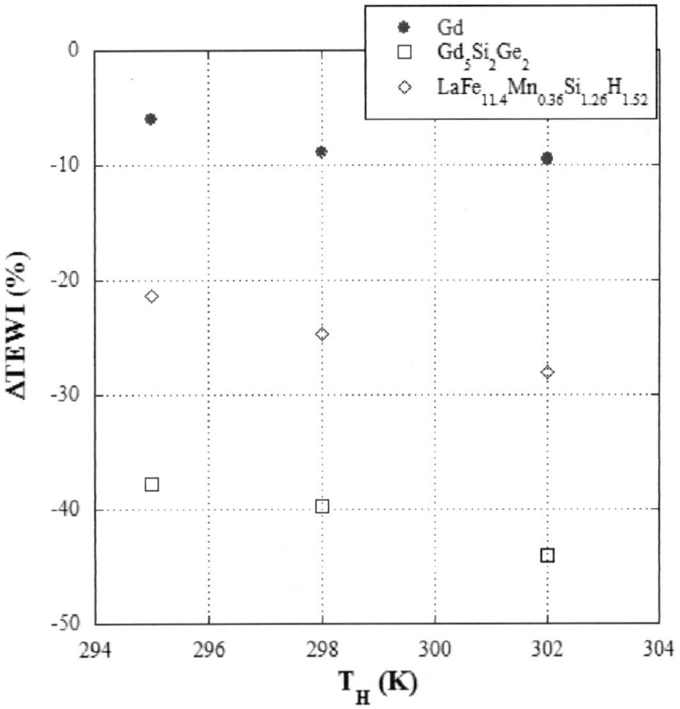

Figure 5. ΔTEWI of Gd, $Gd_5(Si_xGe_{1-x})_4$ and $LaFe_{11.384}Mn_{0.356}Si_{1.26}H_{1.52}$ as a function of the hot heat exchanger temperature.

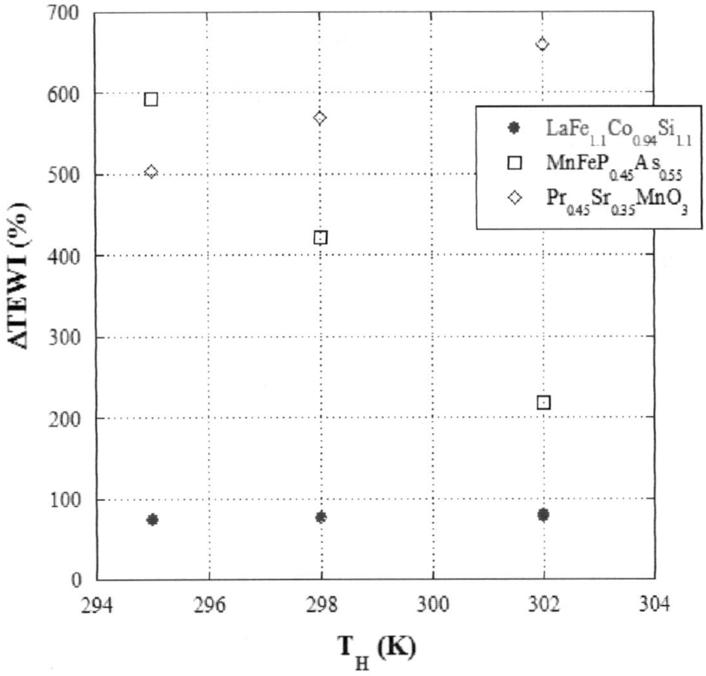

Figure 6. ΔTEWI of $LaFe_{11.05}Co_{0.94}Si_{1.10}$, $MnFeP_{0.45}As_{0.55}$ and $Pr_{0.45}Sr_{0.35}MnO_3$ as a function of the hot heat exchanger temperature.

The Figure clearly shows that the AMR cycle working with these magnetic materials strongly under performs a vapor compression plant. Therefore, these magnetic materials cannot be considered for a commercial use because show a contribution in term of greenhouse effect greater than that of a vapor compression plant.

7. Conclusions

In the present paper, a practical model for predicting the performance and efficiency of an AMR refrigerator system has been introduced. The model is able to simulate an AMR made of FOMT and SOMT materials. In particular: Gd, $Gd_5Si_2Ge_2$, $LaFe_{11.384}Mn_{0.356}Si_{1..26}H_{1.52}$, $LaFe_{11.05}Co_{0.94}Si_{1.10}$, $MnFeP_{0.45}As_{0.55}$ and $Pr_{0.45}Sr_{0.35}MnO_3$ have been considered as solid magnetic refrigerants. In order to make a comparison , the energetic performances of a vapor compression plant working in the same operating conditions have been evaluated by means of a computer program. Attention has been devoted to the evaluation of the greenhouse effect of both AMR and vapor compression cycle. The greenhouse effect has been evaluated by means of TEWI index. The latter takes into account both direct and indirect contribution to global warming.

The simulations clearly show that:

- $Gd_5Si_2Ge_2$ and $LaFe_{11.384}Mn_{0.356}Si_{1.26}H_{1.52}$ has a TEWI index always lower than that of a vapor compression plant. In particular GdSi2Ge2 is the best magnetic material because an AMR cycle working with the latter material shows a greenhouse effect lower than that of a traditional plant of around -40%.
- Gd, that is the benchmarck material for magnetic refrigeration, shows a greenhouse effect similar to that of a vapor compression plant.
- $LaFe_{11.05}Co_{0.94}Si_{1.10}$, $MnFeP_{0.45}As_{0.55}$ and $Pr_{0.45}Sr_{0.35}MnO_3$ cannot be considered because show a contribution in term of greenhouse effect greater than that of a vapor compression plant.

These results indicate that magnetic refrigeration is a promising refrigeration technology that will be used in chiller applications. An AMR cycle can be an eco-friendly technology if used with magnetic materials that show a significant magnetocaloric effect. As expected, from an efficiency point of view, the best candidates to magnetic refrigeration is $Gd_5(Si_xGe_{1-x})_4$ which exhibits the lower values of TEWI index, but on the other side is a very expensive material, disadvantage which makes it impractical for every economic plan of magnetic refrigerator commercialization. From a global point of view (performances and cost), the most promising material are LaFeSi compounds which are really cheaper than rare earth compounds and they give a contribution to global warming sufficiently lower than that of a vapor compression plant.

References

[1] Brown G V 1976 Magnetic heat pumping near room temperature *J. Appl. Phys.* **47** 3673
[2] Pecharsky V K, Gschneidnerr K A jr 2006 Advanced magnetocaloric materials: What the future hold? *Int. Journ. of Refrig.* **29** 1239
[3] Aprea C, Greco A, Maiorino A 2014 Magnetic refrigeration: a promising new technology for energy saving *International Journal of Ambient Energy* DOI:10.1080/01430750.2014.962088.
[4] Aprea C, Greco A, Maiorino A 2011 A numerical analysis of an Active Magnetic Regenerative Cascade system *International Journal of Energy Research* **35** 177
[5] Aprea C, Greco A, Maiorino A 2011 A numerical analysis of an Active Magnetic Regenerative Refrigerant system with a multi-layer regenerator *Energy Conversion and Management* **52** 97
[6] Aprea C, Greco A, Maiorino A 2013 The use of the first and of the second order phase magnetic transition alloys for an AMR refrigerator at room temperature: a numerical analysis of the energy performances *Energy Conversion and Management* **70** 40
[7] Guillou F, Porcari G, Yibole H, van Dijk N, Brück E 2014 Taming the First-Order Transition in Giant Magnetocaloric Materials *Adv. Mater.* **26** 2671

[8] Franco V, Blázquez J S, Ingale B, Conde A 2012 The magnetocaloric effect and magnetic refrigeration near room temperature: materials and models *Annual Review of Materials Research* **42** 305

[9] Pecharsky V K, Gschneidner K A jr 1997 Giant magnetocaloric effect in $Gd_5Si_2Ge_2$ *Phys. Rev. Lett.* **78** 4494

[10] von Moos L, Bahl C R H, Nielsen K K and Engelbrecht K 2015 The influence of hysteresis on the determination of the magnetocaloric effect in Gd5Si2Ge2 *J. Appl. Phys.* **48** 1

[11] de Oliveira N A, von Ranke P J 2005 Theoretical calculations of the magnetocaloric effect inMnFeP0.45As0.55 a model of itinerant electrons *J. Phys. Condensed Matter* **17** 3325

[12] Burriel R, Sazatornil M, Palacios E, Barcza A, Katter M, 2014 Accurate determination of the magnetocaloric parameters of LaFeCoSi compounds *6th IIF-IIR Int. Conf. on Magnetic Refrigeration*, Victoria, BC, 7-10 September 2014.

[13] Morrison K, Sandeman K G, Cohen L F, Sasso C P, Basso V, Barcza A, Katter M, Moore J D, Skokov K P, Gutfleisch O 2012 Evaluation of the reliability of the measurement of key magnetocaloric properties: A round robin study of La(Fe,Si,Mn)H_δ conducted by the SSEEC consortium of European laboratories *Int. J. of Refr.* **35-36** 1528

[14] Bjørk R, Bahl C R H, Katter M 2010 Magnetocaloric properties of $LaFe_{13-x-y}Co_xSi_y$ and commercial grade Gd *J. Magn. Magn. Mater.* **322** 3882

[15] Guillou F, Legait U, Kedous-Lebouc A, Hardy V 2012 Development of a new magnetocaloric material used in a magnetic refrigeration device *EPJ Web of Conferences EMM-FM2011 – First Euro Mediterranean Meeting on Functionalized Materials* **29**

[16] Phan M-H, Yu S-C 2007 Review of the magnetocaloric effect in manganite materials *J. of Magn. and Magn. Mat.* **308** 325

[17] Ponomarev B.K 1986 Magnetic properties of gadolinium in the region of paraprocess *J. Magn. Magn. Mater.* **61** 129

[18] Wuebbles D J 1995 Weighing functions for ozone depletion and greenhouse gas effects on climate *Ann. Rev. Energy Env.* **20** 45

[19] Intergovernmental Panel on Climate Change (IPCC) of the World Meteorological Organization (WMO) and the United Nations Environment Programme (UNEP). Climate Change 2001: The Scientific Basis – *Contribution of Working Group I to the IPCC, Third Assessment Report, Cambridge University, Cambridge, 2001.*

[20] Barclay J A 1982 The theory of an active magnetic regenerative refrigerator *NASA STI/Recon Technical Report* N 83, 34087.

[21] Aprea C, Greco A, Maiorino A, Mastrullo R, Tura A 2014 Initial experimental results from a rotary permanent magnet magnetic refrigerator *Int. J. of Refr.* **43** 111

[22] Aprea C, Cardillo G, Greco A, Maiorino A, Masselli C 2015 A comparison between experimental and 2D numerical results of a packed-bed active magnetic Regenerator *Appl. Therm. Eng.* **90** 376

[23] Aprea C, Greco A, Maiorino A, Masselli C 2015 A comparison between rare earth and transition metals working as magnetic materials in an AMR refrigerator in the room temperature range *Appl. Therm. Eng.* **91** 767

[24] Rice K 2006 DOE/ ORNL Heat Pump model, *Mark VI Version.*

Mitigating the cooling need and improvement of indoor conditions in Mediterranean educational buildings, by means of green roofs. Results of a case study

F Ascione[1], N Bianco1, R F De Masi[2*], F de Rossi[2], G P Vanoli[2]
[1] Università degli Studi di Napoli Federico II, DII – Dipartimento di Ingegneria Industriale, Piazzale Tecchio 80, 80125 Napoli, Italy
[2] Università degli Studi del Sannio, DING – Dipartimento di Ingegneria Industriale, Piazza Roma 21, 82100, Benevento, Italy

*E-mail: rfdemasi@unisannio.it

Abstract. Indoor overheating risk and increased energy demand for cooling are becoming more and more frequent in the building sector of the Mediterranean area. In detail, for the reduction of the energy consumption of educational buildings, characterized by high endogenous gains, the particular boundary conditions affecting their use should be taken in consideration, and thus schedules of occupancy, wide necessity of air-changes for air quality. This paper, with reference to a case study, proposes deep investigations aimed at optimizing the annual energy performance of an educational building of the University of Sannio, located in the Southern Italy. A numerical model of the building has been designed and validated according to monitored data. Starting from the present scenario, after a complete refurbishment of the building envelope, the potentialities of several typologies of green roofs - by considering also the implementation of the adaptive approach in the comfort standard - have been tested. The scope is the optimization of the energy demand for the annual microclimatic control, by avoiding an energy-intensive operation of the air-conditioning devices during the warm season.

1. Introduction

Inappropriate design of buildings, combined with the heat island effect in dense urban areas, as well as the global warming and the necessity of improvement of living standards, have led to an increase (around 46%) of buildings with air-conditioning systems [1]. In the last decennia, peak loads are moving from winter to summer and it implied criticalities for the electricity grids, with increasing costs of energy. For this reason, priority should be given to strategies which enhance the thermal performance of buildings during the summer period, by encouraging, especially, an energy-oriented refurbishment of the existing stock.

Several studies have shown the positive effects of innovative technologies as cool material, cool roofing system and integration of phase change materials (PCM) in different envelope elements. Cool roofs represent an innovative and relatively inexpensive technique proper for reducing the building cooling energy requirements and for improving the indoor thermal comfort conditions. These applications primarily consist of high-reflectance and high-emissivity coatings or membranes, largely investigated, in recent studies, by Cotana *et al.* [2-4]. On the other hand, PCMs are widely investigated technologies and there is a fast developing research area concerning these. Pomianowski *et al.* [5]

have presented a review of PCM technologies developed to serve the building industry. Recently, different solutions have been investigated to increase the efficiency of PCM integration, in terms of improvement of indoor thermal comfort and reduction of energy consumption for cooling [6-8].
Also green roofing systems have shown good potentialities in terms of reduction of the air-conditioning requests, improvement of comfort sensation in the outdoor and indoor environment, better sound insulation and urban water management [9-12]. However, high construction and maintenance costs often do not allow to justify their economic feasibility [13], that is strictly influenced by the chosen type of plants, specific climatic conditions and characteristics of the buildings [14, 15]. Moreover, the coupling with traditional energy efficiency measures should be taken into account for a global optimization of the annual performances.
Here, the main motivations of this paper: it discusses the optimization of the annual energy performance of an educational building in the South Italy. After a complete refurbishment of the building envelope, several typologies of green roof have been tested. After the description of the case study and calibration of the building numerical model, the proposed study consists of three main phases:

a) evaluation of "traditional" insulation measures for the building envelope;
b) energy and environmental analysis of three green roof typologies and economic profitability;
c) assessment overheating risk and indoor thermal conditions according to adaptive approach.

2. The case study: an educational building in Mediterranean climate

The investigated building is located just close to the ancient center of Benevento, a southern Italian city with moderate Mediterranean climate. It is an educational building hosting offices and classrooms of the Economics of the University of Sannio. All the information necessary for the building performance assessment have been obtained through in-situ surveys, interviews with managers and occupants, in-field measurements. In order to simulate the real energy performances, hourly simulations, by means of EnergyPlus [16], have been performed according to procedures of "tailored ratings" [17]. In this section, audit procedure aimed at the numerical modeling are described.

2.1. Building simulation energy modeling

The building (figure 1) has a quite rectangular shape, being placed on a sloped area, so that the articulation of volumes provides a large amount of sun- and wind- exposed roof. It has a global elevation of about 17 m, with four usable floors above the ground. The net conditioned building area is equal to 4'707 m^2. The "surface to volume ratio" (S/V) is equal to 0.41 m^{-1}.

Figure 1. The investigated building of University of Sannio

2.1.1. Characterization of building uses and definition of thermal zones

The numerical model has been defined by combining all detected information about occupancy schedules and installed equipment. A typological floor is shown figure 2. Some areas of the basement are presently interested by renovations; however some classrooms (figure 2a) and a bar/lunch area are already operative. The ground and second floors host mainly classrooms, libraries and offices for teachers. At the first floor, there are mostly offices, a conference room and some laboratories. In the modeling phase, even if diversified depending on the room use, the average air change rate has been defined around 1.5 AHC, for guaranteeing the required comfort conditions established by the standard UNI EN 15251 [18]. In every space, an additional air change rate equal to 0.5 ACH has been considered too, because of the lack of air-tightness.

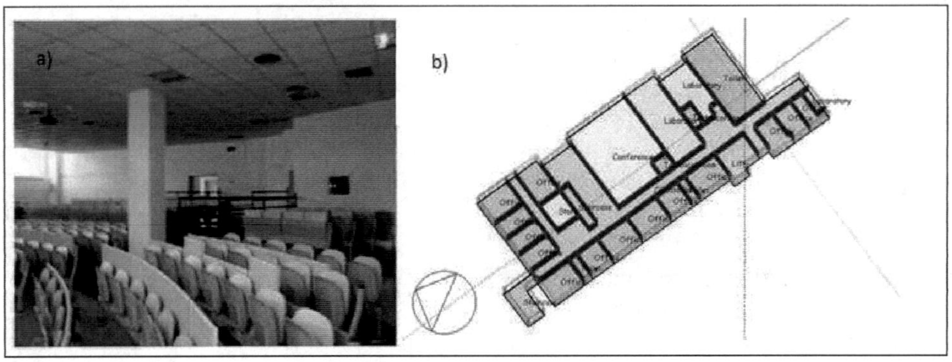

Figure 2. Thermal zones: a) classroom at basement floor; b) geometrical model of 1st floor

2.1.2. Building envelope and technical plant audit

By means of the audit of the building envelope, it is known that the building has a reinforced concrete structure with insulated external walls. The average wall thickness is 0.25 m and the reliable thermal transmittance (U_{wl}) of the vertical wall is 0.55 $W{\cdot}m^{-2}{\cdot}K^{-1}$. Ceiling, basement and roofs have mixed structures, given by the parallel presence of concrete beams, joists and interposed hollow bricks, with insulating layers (i.e., 3.0 cm of expanded polystyrene). The thermal transmittance is around 0.76 $W{\cdot}m^{-2}{\cdot}K^{-1}$. The window-wall ratio is 27%, with double glazed systems, air filling, aluminum frame and external shading. The evaluated U_W is 2.7 $W{\cdot}m^{-2}{\cdot}K^{-1}$. Table 1 summarizes the building HVAC system.

Table 1: Thermal zones and HVAC system

	Thermal zones	HVAC System	Generation system
Basement floor	Education activities	Mixed air/water system	Condensing gas boiler (P_n=388 kW_{th}) Air-cooled chiller (P_n=406 kW_{th}).
	Bar/ dining hall	Autonomous direct expansion systems	Various installed power
I, II and ground floor	Classrooms and offices	Hydronic air-conditioning systems	Air-cooled electric heat pump: heating: P_n= 91.2 kW cooling P_n= 85.5 kW

According to the Italian law, the heating system is turned on from 16th November to 31st March and the building is heated at 20 °C from 8.00 am to 6.00 pm, from Monday to Friday. The cooling set-point is at 26 °C, between 10.00 am and 6.00 pm, also in this case from Monday to Friday, from 1st June - 30th September. A fluorescent lighting system is installed in the whole building.

2.2. Calibration of energy model

The output of simulations have been compared with the billing data of the last five years, according to the "Whole Building Level Calibration with Monthly Data" approach proposed by M&V Guideline [19]. According to this, a model can be declared to be calibrated if:

- ✓ the ERR_{year} (error in the annual energy consumption) is within ±10%;
- ✓ the coefficient of variation of the root mean squared error $CV(RMSE_{month})$ is within ±10%;
- ✓ the mean bias errors (MBE) is within ±5%.

Table 2 shows the evaluated indexes separately for the space heating (E_H) and the electric energy demand for all building uses (E_{el}). A satisfactory convergence is quite evident according to the literature indications, so that the energy model can be considered as well-calibrated. According to the numerical model, by assuming a conventional lower calorific value of natural gas equal to 9.59 kWh/m^3 and by considering the electric efficiency of the Italian system as equal to 0.42 [16], the calculated specific heating demand, in terms of primary energy, is around 32 kWh/m^2, including the energy demand of auxiliaries. Moreover, with reference to the annual electric demand only for the space cooling, the annual primary is equal to 22 kWh/m^2.

Table 2: Calibration indexes evaluation

ERR_{year}		MBE		$CV(RMSE_{month})$	
E_H	E_{el}	E_H	E_{el}	E_H	E_{el}
1.5%	1.5%	2.1%	2.6%	3.7%	10%

3. Optimization of educational building refurbishment

According to the authors, the optimization of refurbishment design should consider at least four issues:

- annual primary energy saving;
- avoided polluting emissions;
- indoor comfort conditions;
- effectiveness under the point of view of the economic feasibility.

These points will be investigated for the proposed case study. More in detail, the annual primary energy saving (DEP expresses in % or kWh/year) will be referred only to the space heating and cooling. With reference to the the avoided emissions of equivalent carbon dioxide (ΔCO_2), the LCA (Life Cycle Assessment) emission factors have been adopted, by assuming for the request of natural gas and electricity respectively 0.237 tCO_{2-eq}/MWh and 0.708 tCO_{2-eq}/MWh [20]. About economic indicators, the discounted pay-back times (SPBs) of efficiency measures and their Net Present Values (NPVs, considering a lifetime equal to 20 years) will be taken into account. Common prices have been assumed for electricity and gas, respectively equal to 0.25 €/kWh_{el} and 0.90 €/m^3 [21].
In order to assess the quality of thermal environments, the discomfort levels have been properly estimated, according to both conventional and adaptive approaches. Thus, firstly the Predicted Mean Vote (PMV) and the Predicted Percentage of Dissatisfied (PPD) have been adopted according to [22-23]. Furthermore, the effect of a less restrictive standard concerning the indoor operative temperature levels has been evaluated. In this regard, several studies underlined that the coupling of adaptive approach and adoption of passive cooling strategies can allow lower cooling consumptions [24].

3.1. Traditional energy saving measures

Presently, the building envelope does not allow levels of thermal resistance completely proper for reducing the energy demand for heating. For this reason, as first step, a complete refurbishment of the building envelope has been studied as summarized in table 3.

Table 3: Interventions and adopted nomenclature

	Technical specification	Nomenclature	C_I [€]
Insulation of the wall (3 cm EPS)	$U_{wl} = 0.37\ W{\cdot}m^{-2}{\cdot}K^{-1}$	[IW]	78’060
Insulation of roof slab (9 cm EPS)	$U_r = 0.29\ W{\cdot}m^{-2}{\cdot}K^{-1}$	[IR9]	74’040
Insulation of roof slab (6 cm EPS)	$U_r = 0.33\ W{\cdot}m^{-2}{\cdot}K^{-1}$	[IR6]	62’934
Low-E glazing (4/16/6 - Argon)	$U_w = 2.20\ W{\cdot}m^{-2}{\cdot}K^{-1}$	[LE]	319’475

Figure 3 shows the expected results. First, the application of 9.0 cm of expanded polystyrene could reduce the heating demand of 9.3% and thus the annual energy demand can be decreased around 6%, meanwhile the polluting emissions of about 5.3%. The reduction of the operational cost is around 1'498 €/year and the DPB (assuming 3% as discounting rate) is greater than 15 years. Conversely, by coupling roof and wall insulations, the annual energy savings and the avoided emissions are around 6% for the solution IW_IR6 and 7% for the second package of energy efficiency measures (EEMs) with wall insulation and the application of 9.0 cm of insulation material for the roof. In both cases, the operational costs are reduced of around 7%.

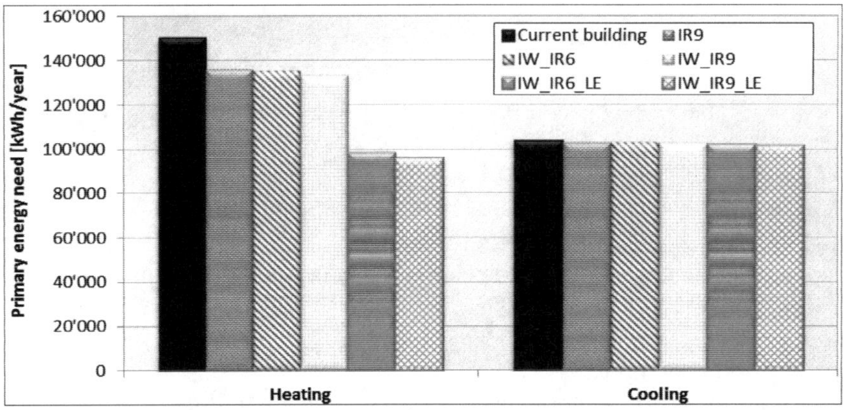

Figure 3. Effect of different refurbishment measures in terms of annual primary energy need

Finally, the previous refurbishment configurations have been coupled with the adoption of low-emissive windows. The overall achievable annual saving is around 21% and 22% when, respectively, 6.0 cm and 9.0 cm are considered for the thickness of the roof insulation. The energy saving implies avoided CO_2 equivalent emissions of about 15.0 tons for the solution IW_IR9_LE and of 14.4 tons for IW_IR6_LE. The operational costs for both configurations are reduced of about 20%. However, the discounted paybacks are always higher than 20 years.

All told, good results can be achieved in terms of energy efficiency and environmental effect. The operational costs can be reduced and it is important for building management. However, the paybacks are quite high and probably suitable incentives' policies are required for making the retrofit feasible.

3.2. Configuration of simulated green roofs

For the case study here analyzed, three configurations of green roofs have been considered:

- HL-LR means High LAI and Low Stomatal Resistance;
- ML-MR means Mean LAI and Mean Stomatal Resistance;
- LL-HR means Low LAI and High Stomatal Resistance.

Each kind of green roof has been considered as equipped with a layer of 12 cm of polystyrene (3 cm already installed + 9 cm of added insulation during the refurbishment), placed just below the drainage layer. Table 4 shows the main characteristics and the unitary installation costs without considering the cost of insulation for the three simulated configuration.

For the building here investigated, considering the roof surface of the second floor, the area potentially suitable for installation of green roof is of 950 m^2.

Each kind of green roof has been considered as equipped with a layer of 12 cm of polystyrene (3 cm already installed + 9 cm of added insulation during the refurbishment), placed just below the drainage layer. Table 4 shows the main characteristics and the unitary installation costs without considering the cost of insulation for the three simulated configuration.

Table 4. Main characteristics of the vegetation modeled on the roofs

	Dimension	LL-HR	ML-MR	HL-LR
Height of plants [m]	[m]	0.1	0.25	0.40
Leaf Area Index (LAI) [m^2/m^2]	[m^2/m^2]	0.8	2.0	3.5
Leaf Reflectivity [-]	[-]	0.3	0.4	0.5
Leaf Emissivity [-]	[-]	0.95	0.95	0.95
Density of Dry Soil [kg/m^3]	[m^2/m^2]	700	1 100	1 000
Thickness of the Soil [m]	[m]	0.10	0.14	0.18
Minimum Stomatal Resistance	(s/m)	300	200	120
Saturation Volumetric Moisture Content of the soil	(-)	0.40	0.40	0.40
Residual Volumetric Moisture Content of the soil	(-)	0.01	0.01	0.01
Initial Volumetric Moisture Content of the soil	(-)	0.20	0.20	0.20
Installation cost per unit of roof area	[€/m^2]	73	78	83

For the building here investigated, considering the roof surface of the second floor, the area potentially suitable for installation of green roof is of 950 m^2.

4. Green roof: potentiality for an educational building in Mediterranean climate

In this section, the green roof installation will be evaluated. Firstly, a deepening about the energy balance of the overall facility, and thus building and HVAC systems, will be presented. Then, the energy and environmental benefits will be evaluated as well as the economic profitability. The indoor air temperature and the comfort conditions will be also analyzed.

The simulations will be performed by using rain data derived by 10 years of monitoring activity (from 2002 to 2012), with a sampling time equal to 10 minutes. These data have been elaborated and a representative reference year has been defined. Figure 4 shows the cumulative monthly rainfall levels.

In this regard, it is very important the evaluation of a realistic daily rainfall. Indeed, this not only influences the green roofs behavior but also it affects greatly the evaluation of the economic expenditure for the artificial irrigations as underlined by Ascione *et al.* [26].

In order to evaluate the effectiveness of the proposed intervention, in the following sections, the analysis of results has been focused on the energy demand of the second floor. Indeed, with reference to this, the green roof has the greater incidence in terms of energy balance.

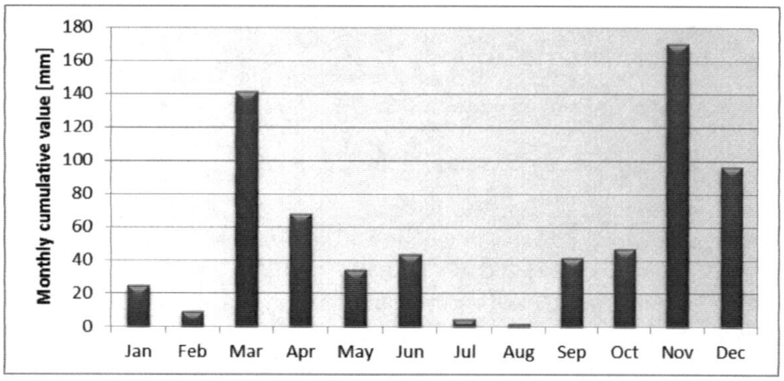

Figure 4. Rainfall for a typical month

4.1. Heat transfer phenomena involved in the building energy balance

The green roof has effects on the energy balance of the building/HVAC system, both during the winter and summer seasons. More in detail, with reference to the heat transfer phenomena, in wintertime, the heat convection of the external side of the roof is greatly reduced by the roughness of the canopy and

the quasi-stable air layer within the foliage. Moreover, in the periods characterized by a dry soil, it contributes to increment the envelope thermal resistance.
Conversely, during the cooling period, an attenuation and time shifting of the thermal wave crossing the roof can be obtained because of the thermal mass of the wet soil layer, characterized by significant thickness, wetness and quite high thermal capacity. Moreover, the evapotranspiration phenomenon (and this is the main effect) that interests both vegetation and soil, allows a significant evaporative cooling. The evapotranspiration is directly related to the solar radiation, the soil moisture contents, the stomatal resistance of the vegetation, the resistance of the boundary limit to the vapor diffusion, the amount of foliage (expressed by the leaf area index).
The green roof model implemented in EnergyPlus is based on the studies of Sailor and this slightly modifies the FASST methodology of Frankenstein and Koenig, as described in [26]. The physical formulation is based on two main equations, concerning the energy balance at the atmosphere/foliage interface (F_f, equation 1) and the heat transfer through the soil (F_g, equation 2).

$$F_f = \sigma_f\left[I_s^{\downarrow}\left(1-\alpha_f\right)+\varepsilon_f I_{ir}^{\downarrow}-\varepsilon_f\sigma T_f^4\right]+\frac{\sigma_f\varepsilon_f\varepsilon_g\sigma}{\varepsilon_g+\varepsilon_f-\varepsilon_f\varepsilon_g}\left(T_g^{\,4}-T_f^{\,4}\right)+H_f+L_f \tag{4.1}$$

$$F_g = \left(1-\sigma_f\right)\left[I_s^{\downarrow}\left(1-\alpha_g\right)+\varepsilon_g I_{ir}^{\downarrow}-\varepsilon_g\sigma T_g^4\right]-\frac{\sigma_f\varepsilon_f\varepsilon_g\sigma}{\varepsilon_g+\varepsilon_f-\varepsilon_f\varepsilon_g}\left(T_g^{\,4}-T_f^{\,4}\right)+H_g+L_g+K\frac{\partial T_g}{\partial z} \tag{4.2}$$

In the equations (4.1) and (4.2), the following terms are:

- H_f: sensible heat flow at the atmosphere/foliage interface, [W m^{-2}];
- H_g: sensible heat flow at the atmosphere/soil interface, [W m^{-2}];
- L_f: latent heat flow at the atmosphere/foliage interface [W m^{-2}];
- L_g: latent heat flow at the atmosphere/soil interface [W m^{-2}];
- s_f: fractional Vegetation Coverage;
- a_f: foliage absorption coefficient;
- a_s: soil absorption coefficient;
- e_f: foliage infrared emissivity;
- e_s: soil infrared emissivity;
- T_f: Temperature of the foliage [K];
- T_s: Temperature of the soil [K];
- z: height or depth [m];
- s:constant of Stefan-Boltzman [W m^{-2} K^{-4}];
- $I_s^{\downarrow}$: total incoming solar radiation [W m^{-2}];
- $I_{ir}^{\downarrow}$: total incoming infrared radiation [W m^{-2}].

Sensible (H_f and H_g) and latent (L_f and L_g) heat flows (where the subscripts "f" and "g" respectively refer to foliage and ground) are calculated as function of LAI, sensible and latent bulk heat transfer coefficients at the ground surface, latent heat of vaporization at the foliage's and soil's temperatures, air velocity and mixing ratio in the canopy. The complete model is available in the technical documentation of EnergyPlus [16].

4.2. Energy-environmental benefits and economic profitability of green roofs

Table 5 shows the comparison between the primary energy requirements of the second floor for the configuration in which only 9.0 cm of insulation has been applied (IR_9) and those with the installation of green roofs as previously described.

Annually, the primary energy saving varies from 2.0% to 3.4% (depending on the type of vegetation), showing the best results for configurations with higher LAI and lower stomatal resistance. Again, with reference to the HL_LR solution, also the emissions are reduced of about 4%. In particular, the savings are mainly due to the reduced cooling need during the cooling season, because the heating savings are always less than 1%; for the summer period, conversely, in case of HL_LR, savings of 7% can be achieved in terms of both primary energy and operating costs. Of course, the HL-LR solution is the one requiring the highest irrigation cost, because of the higher evapotranspiration. Really, the water cost in Italy is quite moderate, around 1.3 €/m^3. By considering the whole year, all solutions are energetically beneficial, even if the costs of artificial irrigation completely nullify the energy savings. In other words, the annual cash flows are disadvantageous.

Table 5. Energy saving, avoided polluting emission and reduction of operational cost (2nd floor)

	Heating [kWh·yr^{-1}]	Cooling [kWh·yr^{-1}]	ΔEP [%]	ΔCO_2 [%]	ΔCE [€·yr^{-1}]
IR9	39'871	26'254	-	-	-
HL_LR	39'545	24'329	3.4	4.0	247
LL_HR	38'889	25'971	2.0	1.7	124
ML_MR	39'290	25'132	2.5	2.3	180

4.3. Thermal comfort analysis

For the whole cooling season, the indexes PMV and PPD have been calculated and it allows the understanding that the solutions ML_MR and HL_LR allow improved comfort conditions at the second floor. More in detail, Figure 5 shows the hourly trends of PPD for an office, facing south-east, for two summer days. It can be seen how the percentage of dissatisfied can be considerably reduced, with indoor operative temperatures more frequently in the range 23-26 ° C. This result is the same for all the investigated thermal zones, by showing that a green roof certainly induces a significant improvement of comfort conditions for the occupants.

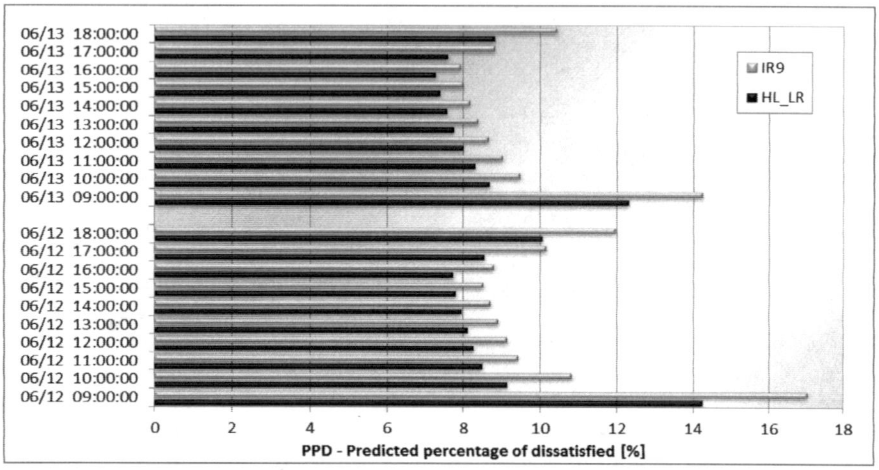

Figure 5. Predicted percentage of dissatisfied for an office with south-west exposure (2nd floor)

5. Indoor thermal conditions according to adaptive approach for traditional slab or green roof

The traditional approach expresses indoor comfort conditions as function of steady temperature levels. Conversely, the adaptive approach takes into consideration temperature levels that are unsteady and follow the variability of the outdoor climate.

The European Committee for Standardization has introduced the adaptive approach in the Standard EN 15251 [18]. The basic equation for the evaluation of the limits of the comfort operative temperature (T_{co}), reliable only for outdoor reference temperatures between 10°C and 30°C, is:

$$T_{co} = 0.33\ T_{rm} + 18 \pm X \tag{5.1}$$

Here, T_{rm} is the running weekly mean temperature, and "X" is a scalar number, ranging between 2 and 4 °C, that provides the acceptability ranges. In this study, the comfort Category I has been considered (i.e., X = 2 °C). It corresponds to the highest level of comfort expectations (PPD < 6%).
According to [18], the adaptive approach should be restricted to the assessment of the summer performances of naturally ventilated and unconditioned buildings.
Thus, firstly, this approach has been applied to the case study, assuming that the HVAC system is turned off at the second floor. In this way, the trend of operative temperatures will be analyzed and the benefits achievable with the installation of green roof will be evidenced.
Then, a hybrid management of HVAC system will be proposed, assuming variable set point temperatures on the basis of the outside temperature. Indeed, since the adaptation takes place also in conditioned buildings provided with "perceived adaptive opportunity" [27, 28], this approach seems reasonable for extending the adaptive criterion to all kinds of buildings.

5.1. Adaptive theory and range of comfort for the climate of Benevento
The definition of reference temperature for Benevento has been applied, and then it was calculated the range of comfort for the operating temperatures, as shown in table 6, for the typical warm season (i.e., from June to September).

Table 6. Running mean temperature and limit values for the operative temperature

Week	T_{rm}	Highest T_{co}	Lowest T_{co}	Week	T_{rm}	Highest T_{co}	Lowest T_{co}
I	23.4	28.5	24.5	X	28.1	30.1	26.1
II	24.5	28.9	24.9	XI	27.9	30.0	26.0
III	25.4	29.2	25.2	XII	27.6	29.9	25.9
IV	26.1	29.4	25.4	XIII	26.9	29.7	25.7
V	27.1	29.7	25.7	XIV	25.5	29.2	25.2
VI	28.0	30.0	26.0	XV	24.8	28.9	24.9
VII	28.8	30.3	26.3	XVI	24.1	28.7	24.7
VIII	28.5	30.2	26.2	XVII	22.9	28.4	24.3
IX	28.4	30.2	26.1				

5.2. Free running building: indoor thermal comfort
The hourly trend of operative temperature has been analyzed, for all rooms of the second floor. By way of example, Figure 6 shows the chart for one day in July for an office with south-west exposure during the occupied hours. The upper and lower limits for the temperatures are respectively 26.2°C and 30.2°C. Each green roof configuration allows lower operative temperatures. More in detail, the configuration HL_LR is able to maintain quite low the oscillations between the hottest hours and the cooler hours, without the use of the air conditioning. Good results are achieved also with the adoption of the ML_MR configuration, even if a high LAI allows much better performance. In the following discussion, the results will concern merely the installation of a green roof with high LAI, because this assures the better energy and environmental performances. Globally, starting from these analyses, it has been noted that the increased inertia induces a reduction of the temperature by 0.3°C to 1.5°C.
In detail, two indices have been introduced to evaluate the hourly temperatures of 2nd floor:

- GTO, global operating temperature index: percentage of hours, in which the operative temperature does not exceed the upper limit (2nd column in table 6) during the working period.
- MTO, mean value for the operating temperature index: this is the percentage of hours in which operative temperature does not exceed the mean value of comfort range (X=0 in equation 9).

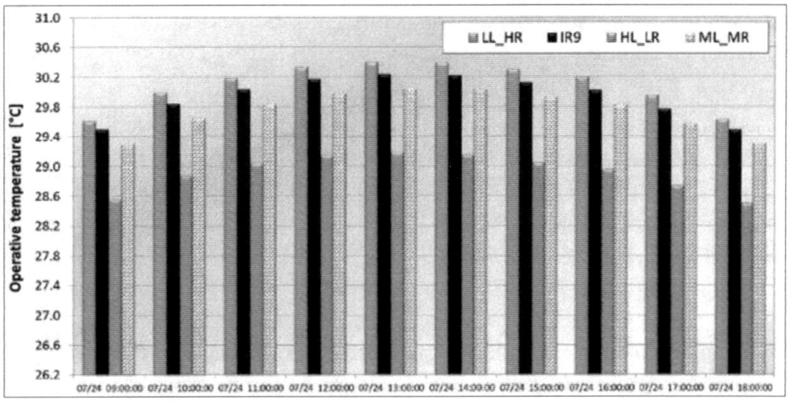

Figure 6. Hourly operative temperatures for an office with south-west exposure (24th July)

Moreover, figure 7 shows that, by referring to the GTO index, the installation of a green roof allows an operating temperature below the maximum allowable temperature, for all occupied hours of the building. This means that - by combining an appropriate strategy of ventilation and the installation of a green roof - the air conditioning could be turned off during the entire summer season. It should be noted that the adaptive theory provides a number of possibilities for the occupants' behavior; a proper natural ventilation must be considered in every case. Furthermore, by considering the MTO index, it is seen that the solution with green roof allows, for about the 80% of the occupancy hours, the achievement of an operating temperature "lower than" or "equal to" the average temperature of the comfort range.

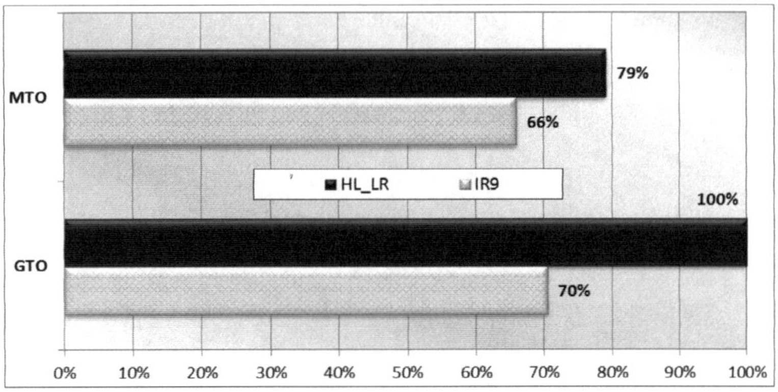

Figure 7. Operating temperature indexes: MTO and GTO

5.3. Energy and environmental benefits and economic profitability by adopting a hybrid approach

At the present time and with reference to the cooling period, the consideration of naturally ventilated office buildings, in the Italian context, is quite unreliable, so that the best potential of the adaptive approach consists in the actual management of air conditioning and cooling systems through variable set-points. For this reason, it has been assumed the consideration of an optimized management of the air-conditioning system, in which the set point temperature is fixed on the average temperature of comfort range defined with an adaptive approach, on a time basis equal to one week. This type of management can be achieved with an automated system that connects the external temperature to the selected set-point. In the following table 7, for the second floor, the comparison between the configuration with roof insulation and traditional management of the HVAC system and green roof installation and adaptive approach for the thermal comfort, is shown. Annually, the primary energy saving with high LAI roof is around 17%. More in detail, the electricity demand during June and September is reduced of around 60% with this configuration and around 30% during the hottest months. A good result is obtained in terms of reduction of polluting emissions with all configurations. The economic analysis starts to get interesting.

Table 7. Energy saving, avoided polluting emission and reduction of operative costs (2nd floor)

	Heating [$kWh \cdot yr^{-1}$]	Cooling [$kWh \cdot yr^{-1}$]	ΔEP [%]	ΔCO_2 [%]	ΔCE [€$\cdot yr^{-1}$]
High LAI-adaptive comfort	39'545	15'014	17	19	1'287
Low LAI-adaptive comfort	38'889	17'105	15	16	1'115
Med. LAI-adaptive comfort	39'290	16'182	16	17	1'180

Table 8. Performance improvement by adopting a High LAI and Low Stomatal Resistance green roof

	ΔEP [%]	ΔCO_2 [%]	ΔCE [%]
Traditional approach (HVAC on)	3.4	4	0.3
Adaptive approach (hybrid HVAC)	17	19	10

Finally, the energy needs have been compared for the proposed solutions, when the adaptive theory is applied to all configurations. The insulation of the roof slab (IR9) and the adaptive management of air conditioning in summer induce a demand for cooling of 18'119 $kWh \cdot yr^{-1}$.
Therefore, an annual (i.e., heating and cooling) energy savings of 9% can be achieved with the configuration high LAI and adaptive approach for thermal comfort, and the electric demand for cooling is reduced by 10% during the July and August and 30% during the less hot months, with positive effects also concerning the emissions. In conclusion, the green roofs surely allow the attenuation of the heat transfer from the outside to the inside environments, in summer conditions and thus energy and environmental benefits. However, there is not a net economic profitability, so that the design must be very careful. The best configuration of green roof is the one with High LAI and Low Stomatal Resistance. The final results for this type of installation, for the 2nd floor, are summarized in table 8, by comparison with the solution of roof insulation and the same HVAC management.

Conclusions

This paper discusses the potentialities of integration of several typologies of green roofs and the management of the microclimatic control with adaptive models. The investigated building hosts department and classrooms of the University of Sannio and it is located in Benevento (South Italy). The first part of the study has been focused on the calibration of a proper energy model and the evaluation of traditional refurbishment actions applied to the building envelope. Moreover, beyond the definition of various eco-roofs, with reference to both properties of the soil and plants, the energy saving, the avoided polluting emissions and the economic profitability have been evaluated. For the case study, by considering only the second floor of the building (where the great incidence of green roof can be perceived), it was found that the primary energy savings in the summer months, compared to a solution in which a mere insulation is applied, ranges from 2.0% to 3.4% depending on the type of vegetation. The best results are for plants with high LAI and low stomatal resistance. The costs of green roof installation and the needs of irrigation, however, do not provide profitable paybacks.
As further analysis, by applying the adaptive theory to evaluate the comfort temperatures, it was found that in case of green roof with high LAI, even with the HVAC system turned off, the thermal comfort constraints are always fulfilled. This leads to the conclusion that an appropriate strategy for natural

ventilation, the application of less stringent comfort temperatures and the installation of a green roof, with high LAI and low resistance stomatal, could completely nullify the demand for the space cooling. By considering a hybrid approach, in which the control temperatures for the active microclimatic are set according to the mean value of the comfort adaptive range, the installation of the aforementioned typology of green roof allows a reduction of primary energy demand for the summer cooling. The saving is of 18% compared to a solution with roof insulation and similar management system, or about 43% compared to a solution with only insulation and a standard management system. In these cases, also the cost-effectiveness of the investment begins to be evident. Although this is not highlighted in this study, also the quality of the external environment can be improved, as investigated in [29-31], not only in terms of reduction of polluting emissions but also relatively to the urban heat island effect. Finally, in order to optimize the building performance in the Mediterranean climate, deep studies are needed about technologies aimed at limiting the energy demand for the air-conditioning in the cooling season.

References

[1] Rodrigues L T, Gillott M and Tetlow D 2013 Summer overheating potential in a low-energy steel frame house in future climate scenarios *Sust. Cities and Society* **7** 1

[2] Cotana F, Rossi F, Filipponi M, Coccia V, Pisello A L, Bonamente E, Petrozzi A and Cavalaglio G 2014 Albedo control as an effective strategy to tackle Global Warming: A case study *Appl. Energy* **130** 641

[3] Pisello A L and Cotana F 2014 The thermal effect of an innovative cool roof on residential buildings in Italy: Results from two years of continuous monitoring *Energy Build.* **69** 154

[4] Pisello A L, Piselli C and Cotana F 2015 Influence of human behavior on cool roof effect for summer cooling *Build. Environm.* **88** 116

[5] Pomianowski M, Heiselberg P and Zhang Y 2013 Review of thermal energy storage technologies based on PCM application in buildings *Energy Build.* **67** 56

[6] Barreneche C, Navarro L, de Gracia A, Fernández A I and Cabeza L F 2016 In situ thermal and acoustic performance and environmental impact of the introduction of a shape-stabilized PCM layer for building applications *Renew. Energy* **85** 281

[7] Castellani B, Morini E, Filipponi M, Nicolini A, Palombo M, Cotana F and Rossi F 2014 Clathrate hydrates for thermal energy storage in buildings: Overview of proper hydrate-forming compounds *Sustainability* **6** 6815

[8] Soares N, Gaspar A R, Santos P and Costa J J 2014 Multi-dimensional optimization of the incorporation of PCM-drywalls in lightweight steel-framed residential buildings in different climates *Energy Build.* **70** 411

[9] Saadatian O, Sopian K, Salleh E, Lim C H, Riffat S, Saadatian E, Toudeshki A, and Sulaiman M Y 2013 A review of energy aspect of green roofs *Renew. Sustain. Energy Rev.* **23** 155

[10] Mentens J, Raes D and Hermy M 2006 Green roof as a tool for solving the rainwater runoff problem in the urbanized 21st century? *Landscape Urban Plan* **77** 217

[11] Versini P A, Ramier D, Berthier E and de Gouvello B 2015 Assessment of the hydrological impacts of green roof: From building scale to basin scale *J Hydrology* **524** 562

[12] Santamouris M 2014 Cooling the cities: A review of reflective and green roof mitigation technologies to fight heat island and improve comfort in urban environments *Sol. Energy* **103** 682

[13] Ascione F, Bianco N, de' Rossi F, Turni G and Vanoli G P 2013 Green roofs in European climates. Are effective solutions for the Energy savings in air-conditioning? *Appl. Energy* **104** 845

[14] Berardi U, GhaffarianHoseini A H and GhaffarianHoseini A 2014 State-of-the-art analysis of the environmental benefits of green roofs *Appl. Energy* **115** 411

[15] Bianchini F and Hewage K 2012 Probabilistic social cost-benefit analysis for green roofs: A lifecycle approach *Build. Environ* **58** 152

[16] U.S. Department of Energy 2013 Energy Plus simulation software *Version 8.1.0*
[17] EN 15603: 2008. Standard: Energy Performance of Buildings – Overall Energy Use and Definition of Energy Ratings
[18] UNI EN 15251: 2008. Indoor environmental input parameters for design and assessment of energy performance of buildings addressing indoor air quality, thermal environment, lighting and acoustics
[19] U.S. Department of Energy Federal Energy Management Program, M&V Guidelines: Measurement and Verification for Federal Energy Projects Version 3.0, 2008, Boulder, CO.
[20] TERNA Statistical Data (in Italian) http://www.terna.it
[21] The Covenant of Mayors, Committed to local sustainable energy. Technical annex to the SEAP template instructions document: the emission factors http://www.eumayors.eu/
[22] European Commission, Eurostat Statistics for Energy Costs http://epp.eurostat.ec.europa.eu/
[23] ASHRAE Standard 55: 2004. Thermal environmental conditions for human occupancy
[24] UNI EN ISO 7730: 2005. Ergonomics of the thermal environment e analytical determination and interpretation of thermal comfort using calculation of the PMV and PPD indices and local thermal comfort criteria
[25] Ferrari S and Zanotto V 2012 Adaptive comfort: Analysis and application of the main indices *Build. Environ* **49** 25
[26] Ascione F, Bianco N, De Masi R F, De Rossi F and Vanoli G P 2015 Technical and economic analysis of green roofs to reduce building cooling needs, *Chapter 13 of Eco-Efficient Materials for Mitigating Building Cooling Needs, Design, Properties and Application 1st Edition* (Woodhead Publishing) p 349
[27] Brager G S and Baker L 2008 Occupant satisfaction in mixed-mode buildings *Proc of Conference: Air Conditioning and the Low Carbon Cooling Challenge, Cumberland Lodge, Windsor (London, United Kingdom)*
[28] ThermCo 2009 Interrelation between different comfort parameters and their importance in occupant satisfaction. Report for the Thermal Comfort in Buildings with Low-Energy Cooling European Project http://www.thermco.org/
[29] Salata F, Golasi I, Vollaro E L, Bisegna F, Nardecchia F, Coppi M, Gugliermetti F and Vollaro A L 2015 Evaluation of different urban microclimate mitigation strategies through a PMV analysis *Sustainability* **7** 9012
[30] Salata F, Golasi I, Vollaro A D L and Vollaro R D L 2015 How high albedo and traditional buildings' materials and vegetation affect the quality of urban microclimate. A case study *Energy Build.* **99** 32
[31] Peri G, Rizzo G, Scaccianoce G and Sorrentino G 2013 Role of green coverings in mitigating heat island effects: An analysis of physical models *Applied Mechanics and Materials* **261-262** 251

Journal of Physics: Conference Series **655** (2015) 012028 doi:10.1088/1742-6596/655/1/012028

Determination of droplet contours in liquid-liquid flows within microchannels

B Pulvirenti[1], B Rostami, G Puccetti and G L Morini

DIN – Alma Mater Studiorum Università di Bologna,
Laboratorio di Microfluidica, Via del Lazzaretto 15/5, Bologna 40131, Italy

E-mail: beatrice.pulvirenti@unibo.it

Abstract. An experimental analysis of the droplet regime with a silicone oil-water two-phase flow within a micro cross-junction, varying the average velocity of the fluids, has been carried out. The micro cross-junction is made as intersection of two glass microchannels with a stadium-shaped cross-section with height H=190 μm and width W_j=195 μm within the junction and W=390 μm before and after the junction. The water flow rate is broken in droplets having spherical shape with dimensions and velocity that depend on the average velocity ratio imposed. Different kinds of intermittent droplets have been observed, in the ranges of average velocity (0.0105-0.042) m/s and (0.0004-0.0050) m/s for oil and water, respectively. The droplet velocity has been calculated starting from the detection of the shape of the droplets and then by the evaluation of the displacement of the droplets in the unit of time. The images of the droplets have been obtained from a high-speed camera, connected to an inverted microscope. The procedure of water phase contour detection is based on Matlab Image Toolbox scripts.

1. Introduction

In the last years, several microdevices have been developed for the production of droplets and emulsions, which have a large use in many fields, especially in the field of drug delivery [1-3]. The most common microdevices for droplet generation are based on T-junctions, flow-focusing junctions and co-flowing devices (a classification of these devices can be found in [4]). Devices based on cross-junction are examples of flow-focusing devices widely used for droplet-based microreactors and chemical synthesis [5].

Based on the experimental observations collected in the last years, the geometry of the microfluidic microchannels used in these devices plays an important role in controlling liquid–liquid flows but this is not the only parameter which influences the droplet production. In fact, a complete control over the two-phase flow behavior, by these devices, requires a detailed understanding of the bubble or droplet formation mechanisms. One physical approach is to analyze the forces involved in the droplet formation process in microchannels [6, 7]. The dynamics of droplet formation is mainly governed by some variables, including channel geometry, structure and property (such as channel type, dimension and hydrophobicity), fluid properties (such as density, viscosity, interfacial tension and contact angle), and operating parameters (such as pressure, flow rate ratio, temperature, electric field, etc.). These parameters can be typically expressed by means of several dimensionless parameters, noting that

[1] To whom any correspondence should be addressed

small geometric length scales typically lead to low Reynolds numbers (*Re*<1) in microfluidic devices. Therefore, the effect of the inertia forces can typically be ignored in this kind of flow which is close to a Stokes flow. In this condition, the Capillary number becomes the most important dimensionless parameter.

In micro-junctions, three typical flow regimes have been distinguished: squeezing, dripping [8, 9], and jetting [10, 11]. Corresponding to these flow regimes, two main dynamical models for droplet breakup in a microfluidic T-junction, namely *rate-of-flow-controlled* breakup [8] and *shear-driven* breakup [5], have been proposed. At low capillary numbers, typically *Ca*<0.01, that is, the shear stress is much smaller than the interfacial force; droplet breakup is dominated by the pressure drop across the emerging droplet or bubble. In this squeezing regime, the mechanism is known as *rate-of-flow-controlled* breakup [8], and it is ubiquitous in the confined T-junction system, and also in flow-focusing geometries. This pressure drop results from the blocking by the droplet or bubble which fills the entire cross-section of the channel. Within this squeezing regime the droplet size becomes a function of the flow ratio of the two fluids [9].

The wetting properties of the fluid–wall interface are extremely important in determining whether ordered or disordered flow patterns occur [12]. When the continuous phase completely wets the microchannels, ordered patterns can be obtained. On the contrary, if the wetting is partial, disordered flow patterns are produced. Typically, water-in-oil (W/O) droplets are produced in hydrophobic devices, and oil-in-water (O/W) emulsions are generated in hydrophilic droplet generators [13]. The hydrophobicity or hydrophilicity property of a solid surface can be obtained by means of specific surface coatings [6].

The aim of this work is to study the water droplet breakup mechanism within a hydrophobic cross-junction in a liquid-liquid (silicone oil-water) two-phase flow (water-in-oil). The micro cross-junction is made as intersection of two glass microchannels with a stadium-shaped cross-section having a restriction within the junction. The water droplet breakup is investigated by varying the average velocity of silicone oil between 0.0105 and 0.042 m/s and water between 0.0004 and 0.0050 m/s.

After the reconstruction of the droplet contour, a droplet average velocity value is computed as well as the main characteristic dimension of the droplet, dimensionless diameter, as a function of the Capillary number and the flow rate ratios (water flow rate/silicone oil flow rate) and a correlation has been proposed.

2. Experimental Setup

The experimental apparatus, shown in figure 1, is based on the use of an inverted microscope (1, *Nikon Eclipse TE2000-U*) illuminated by means of a Mercury lamp (2a, *Nikon C-SHG1*) from the bottom and a tungsten lamp (2b, *Photon Beard PhotonBeam 1000*) from the top of the microdevice.

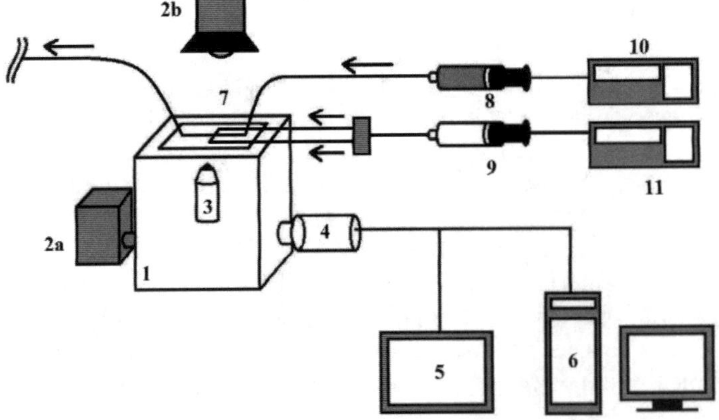

Figure 1. The schematic diagram of experimental setup

An air immersion lens with a numerical aperture NA=0.25 and magnification M=10x is used (3, *Nikon, CFI DS 10X*). The microscope is connected to a high-speed camera (4, *Olympus I-speed II*) in order to acquire a series of images for reconstructing the evolution of water droplets within the cross-junction (7, *Dolomite Junction Chip*). Due to the high frequency values of the droplet breakup, in each experimental analysis up to 5000 frames per second have been acquired by means of the Olympus I-speed II software (6). A LCD monitor (5) directly connected to the high-speed camera enables the real-time visualization of the flow conditions through the microchannel. The micro cross-junction is obtained as intersection of two microchannels, one straight and one curved, as shown in figure 2a. In figure 2a, one can see that the Dolomite chip has also a T-junction, but in this paper only the results obtained by the cross-junction are presented. Figure 2b shows also a top view of the junction.

(a) (b)

Figure 2. Sketch of Dolomite chip (a) and sketch of the cross-junction (b)

The microchannels have a stadium-shaped cross-section (see figure 3a), with the width W=390 µm and the height H=190 µm before and after the junction. In the junction, the cross-section has again the shape of a stadium, but the width W_j=195 µm and the height H=190 µm (see figure 3b). Both microchannels are made by hydrofluoric acid (HF) etching on a substrate of glass and then thermally bonded with the same material. The average roughness of the channel inner walls declared by the manufacturer is 5 nm. The inner walls of the cross-junction have been treated with a self-assembled monolayer of silane groups, rendering the surface as water repelling (hydrophobic behavior). Even if it is known that microchannels made by a similar manufacturing process can show a real cross-section different from the geometry declared by the manufacturer [14], after a qualitative optical check of the cross-junction by microscope, it has been observed that the shape of the cross-section of the cross-junction is coherent with that declared in the data-sheet of the manufacturer.

(a) (b)

Figure 3. Cross-section out of the junction (a) and within the junction (b)

The two liquids adopted in these experimental tests are silicone oil (*SIGMA-ALDRICH*), entering in the two perpendicular and opposite streams, the properties of the silicone oil are shown in table 1, and water, entering in straight direction.

Table 1. The properties of the silicone oil employed for the experiments

Viscosity (cSt) (25 °C)	Density (g/ml) (25 °C)	Vapor pressure (mmHg)		Vapor density
		25 °C	20 °C	
20	0.95	<5	5	>1 (vs air)

The liquids are pumped by means of two syringe pumps (*Harvard Apparatus PHD4400 Programmable* and *Cole-Parmer Version Hills*); the operative ranges of the flow rates for the syringe pumps are reported in table 2 as well as their typical uncertainty values. The range of the volumetric flow rate, both for oil and water, tested in this paper are shown in figure 4a; water volumetric flow rates between 0.1 ml/h and 1.2 ml/h have been tested as well as oil volumetric flow rates between 2.5 ml/h and 10 ml/h. In terms of average velocity, values between 0.0105 and 0.042 m/s are imposed for silicone oil and between 0.0004-0.0050 m/s for water (see figure 4b).

Table 2. Characteristics and uncertainties of the measurement instruments

Labels in figure 1	Instrument	Range		Uncertainty
		Minimum flow rate (μl h^{-1})	Maximum flow rate (ml min^{-1})	
10	Harvard PHD4400	0.0001	220.82	± 0.35%
11	Cole-Parmer Hills	0.2	8.333	± 0.5%

A series of 34 different experimental runs have been made: the complete map of the experimental runs is shown in figure 4a by putting in evidence for each test the imposed values of volumetric flow rates both for oil and water and in figure 4b in terms of imposed average liquid velocity.

(a) (b)

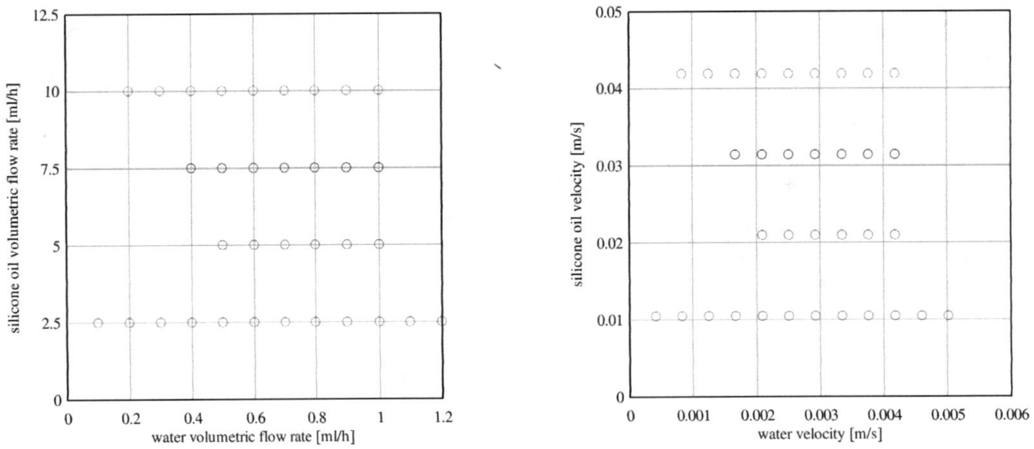

Figure 4. Map of the experiments: volumetric flow rate (a), velocity (b)

3. Processing of the images

A typical sequence of snapshots obtained from the fast camera is shown in figure 5; these snapshots have been obtained by imposing a water volumetric flow rate of 0.6 ml/h and an oil volumetric flow rate equal to 7.5 ml/h.

Figure 5 shows the squeezing of the water liquid phase, followed by the breakup and the formation of a water droplet in oil, in a sequence of 30 snapshots, recorded with a frame rate 5000 fps. Each image has a resolution of 320×240 pixels and the time interval between two images is equal to 200 μs. The snapshots are shown in a sequential order; line by line from top to bottom, in each line from left to right (*i.e.* the 18th snapshot is in the fourth line, the third element from left). Figure 5 shows the good quality of the acquired images with droplet contours well defined and a level of illumination quite uniform in the whole field-of-view. The images allow to recognize with a good resolution the liquid/liquid interface as well as the solid walls of the cross-junction. The central area of the water droplet is characterized by light gray pixels but there is a dark zone around this region which puts in evidence the curved boundary of the droplet.

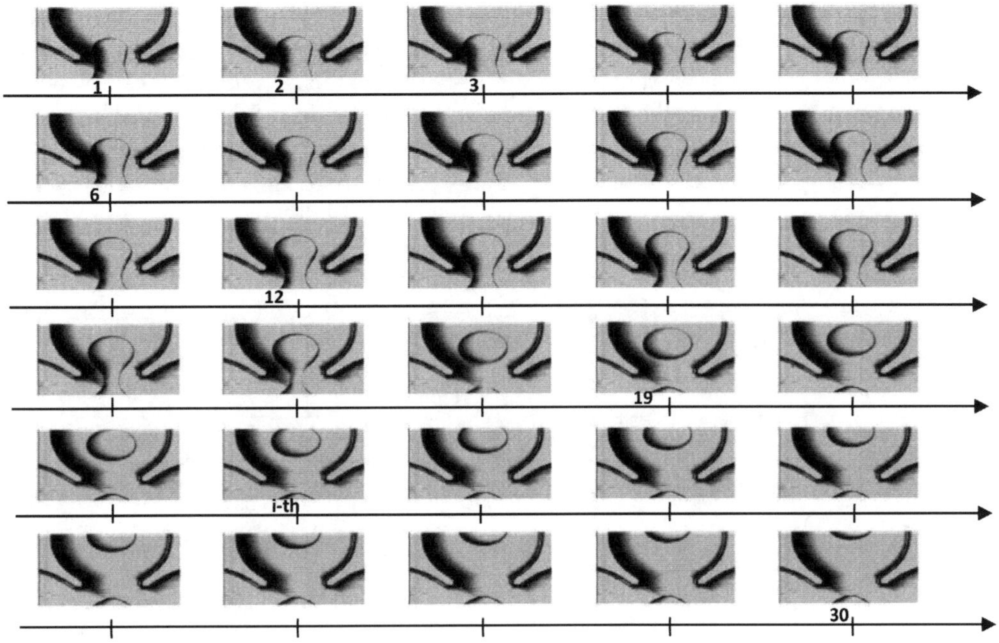

Figure 5. Raw images showing the squeezing and the breakup of water in oil (W/O) for an imposed water volumetric flow rate of 0.6 ml/h and an oil flow rate of 7.5 ml/h

3.1. Analysis of the images

Figure 6a shows an image chosen from the sequence reported in figure 5 (12th snapshot).

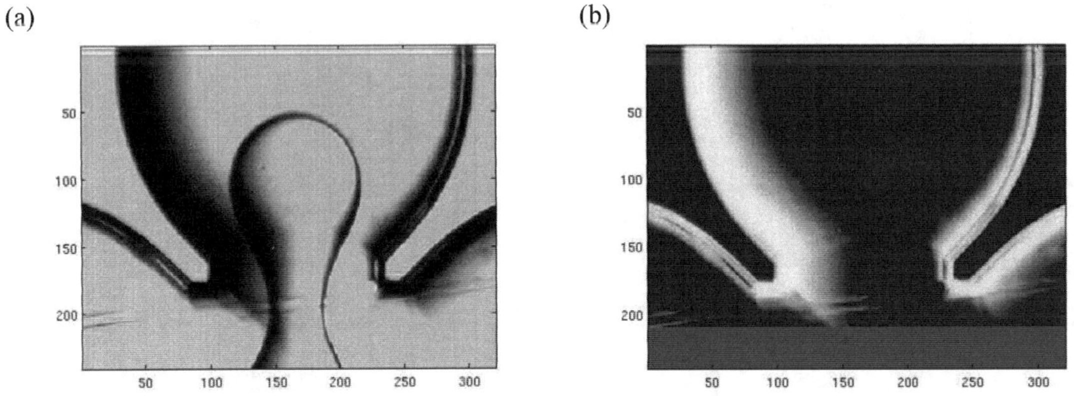

Figure 6. Raw image chosen to explain the processing sequence (a), complement of the image of the cross-junction without droplets (b)

The raw images of the sequence of figure 5 are treated as numerical matrices having 320×240 elements in which each pixel is associated to a number in the range [0-256] which correspond to different gray levels of the pixel. In order to individuate the water-oil interface by avoiding any interference between the water-oil interface and the channel walls, the complement of an image of the cross-junction filled with a single phase liquid (oil or water) (figure 6b) is used with the aim to compute a characteristic threshold t for the image, which is a normalized intensity of the image, evaluated by means of the Otsu method [18]. By this method, the weighted sum of the variances of two classes of pixels in the image, the background and the foreground pixels, can be minimized.

Journal of Physics: Conference Series **655** (2015) 012028 doi:10.1088/1742-6596/655/1/012028

The result is a characteristic threshold of the image that can be used for converting an intensity image to a binary image (by using *graythresh* function in Matlab). For the sequence of images shown in figure 5, by means of this method a threshold t=0.47 has been evaluated. Then, it is possible to transform the intensity values of the gray scale into new values, in order to filter the gray levels and enhance the contrast between background and foreground pixels (by using *imadjust* function in Matlab). The effect obtained by following this pre-processing procedure on the image shown in figure 6a, is shown in figure 7a. It is evident that the background contribution has been neutralized and in the processed images only the presence of the liquid/liquid interface is emphasized. Finally, the obtained gray level maps can be converted into binary images by means of a new threshold, which for the sequence of images shown in figure 5 is t=0.7, by using *im2bw* function in Matlab. The binary images corresponding to the 12th snapshot is shown in figure 7b.

(a)
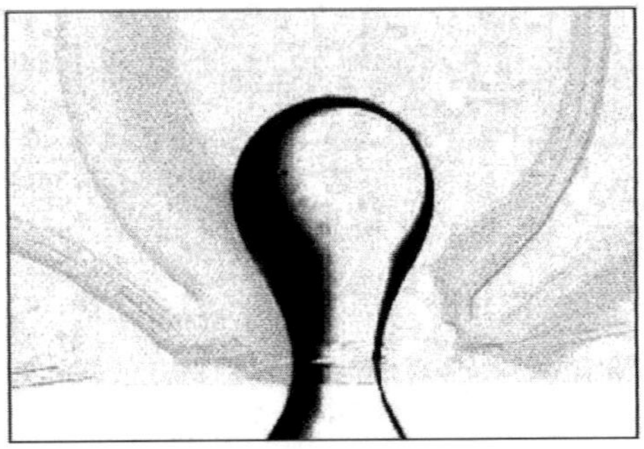
(b)
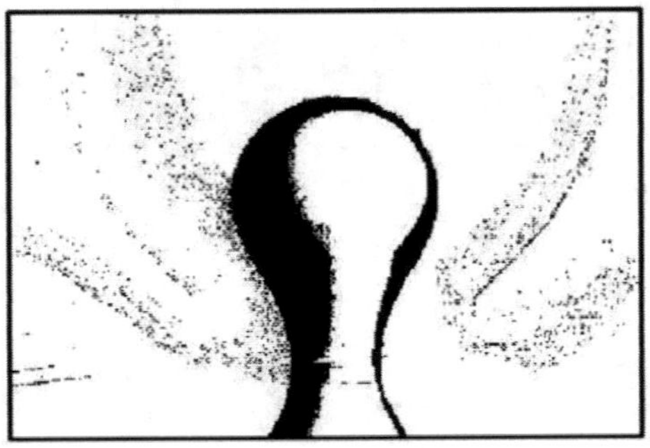

Figure 7. Gray levels image (a) and binary image (b) of the 12th snapshot

3.2. Detecting the edge

In a previous work [17] it has been demonstrated that among different techniques available for finding the edges of an image the Canny method [19] can offer some advantages in terms of compromise between accuracy and computational time. By this technique, one can find edges by looking for local maximal of the gradient of the binary image. To smooth the image, a Gaussian filter is first applied to convolve with the image. Then, the method uses two thresholds, to detect strong and weak gradients, and includes the weak ones in the output only if they are connected to strong gradients. This method is therefore less likely than the others to be fooled by noise, and more likely to detect true weak edges.

(a)
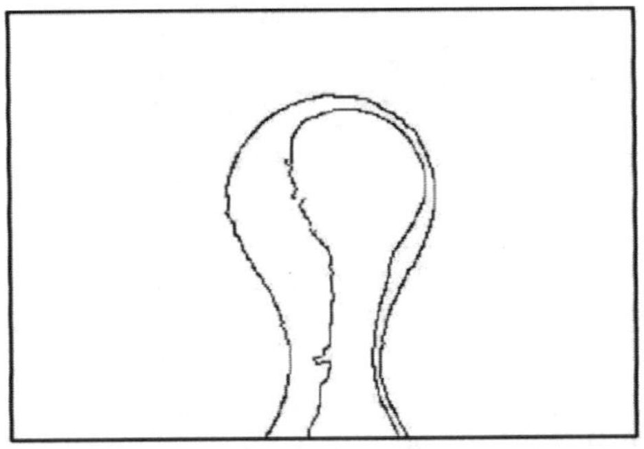
(b)
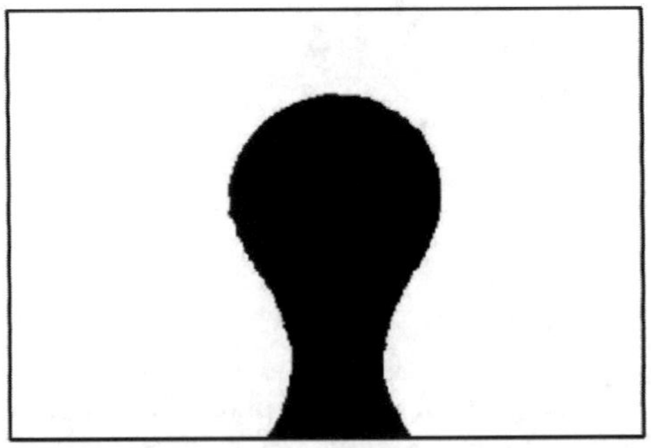

Figure 8. Edges extracted with the Canny method by the binary images (a) and reconstruction of the water domain (dark zone) (b)

The edges individuated by applying the Canny method to the binary image shown in figure 7 are shown in figure 8a. Then the domain occupied by the water phase can be reconstructed by filling the edges with the *imfill* function of Matlab: the results are shown in figure 8b.

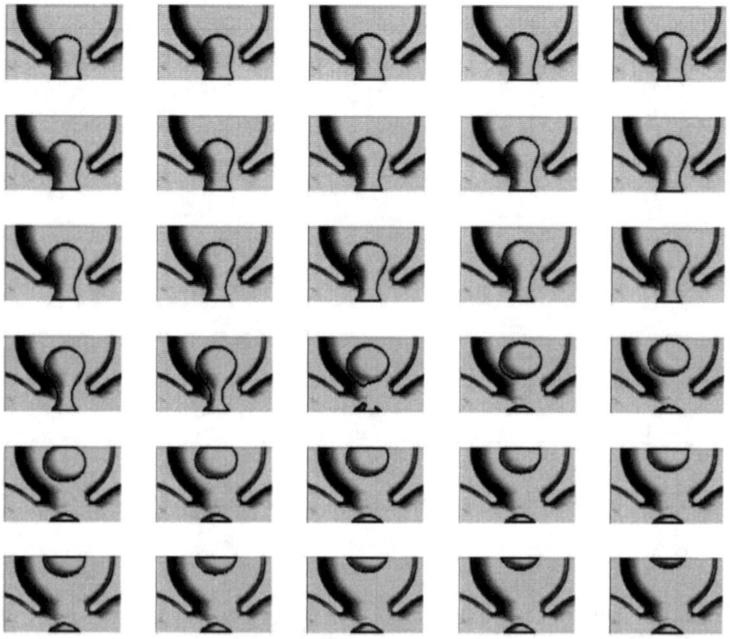

Figure 9. Evolution of the water-phase contours during the squeezing and the breakup of water in oil (W/O) for an imposed water volumetric flow rate of 0.6 ml/h and an oil flow rate of 7.5 ml/h

3.3. Post-processing of the image

Now, it becomes possible to obtain, starting from the oil/water contours, the area A occupied by the water-phase, the perimeter of the water-oil interface and the coordinate of the centre of mass of the droplet (by using the *bwboundaries* function of Matlab). In presence of a single droplet, the centre of mass (x_c, y_c) and also the velocity of the droplet can be calculated. Figure 10 shows the position of the coordinate for the centre of the droplet (red lines) obtained for three consecutive snapshots. The yellow lines show the position of the channel walls within the junction. Considering the first two snapshots in the figure, we can obtain the velocity of the droplet by considering the displacement of the centre of mass along the y-axis and the time interval between two consecutive images ($v = (y_c(2) - y_c(1))/dt$); here dt=1/5000 s.

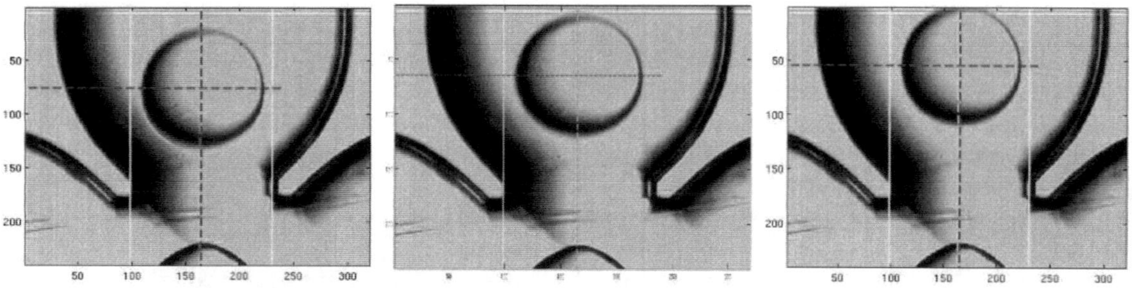

Figure 10. Determination of the droplet centre for three consecutive snapshots

This evaluation can be repeated by considering different couples of images (*i.e.* in figure 10, the evaluation of the droplet velocity can be obtained by considering the images 1-2, or 2-3) in order to obtain an average value of velocity for the droplet.

Finally, from the area A obtained from the droplet contour, it is possible to calculate the ideal radius of a circular droplet having the same area A of the real droplet ($R = \sqrt{A/\pi}$). Then, plotting a perfect circle with radius R, centred in (x_c, y_c), it is possible to find the eccentricity of the droplet. Figure 11 shows an example of comparison between the real shapes of the droplet (blue line) determined experimentally and the circular contours obtained from the ideal radius associated to the droplet (red line). It is evident that the eccentricity of the droplet is very limited in this specific case.

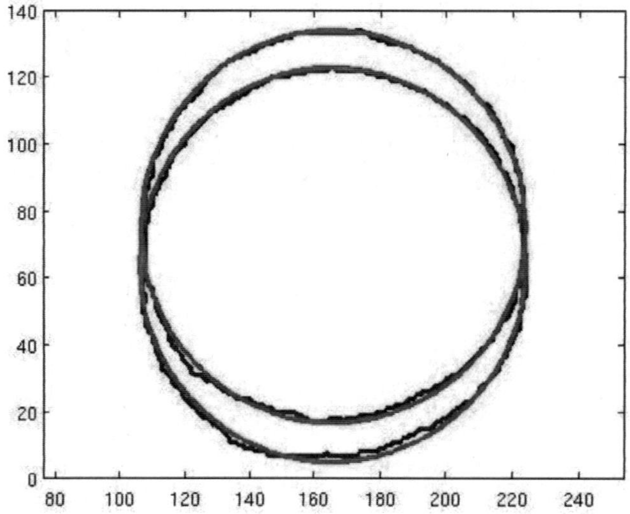

Figure 11. Comparison between the droplet contour obtained by the experimental images (blue line) and the circular contour drawn from the radius R associated to the droplet (red line)

However, figure 11 shows a deviation from the ideal circular profile especially in the bottom of the droplet; in this case the droplet contour can be assumed to be circular with an error less than 2%. The coordinates of the centre show that the droplet is moving only along the y direction (the variation of x_c between two consecutive images is negligible), with a velocity v=8.5 cm/s for the case depicted in figure 11.

4. Result and discussion

In literature two different dynamical patterns have been recognized for explaining breakup in a microfluidic T-junction or cross-junction: the *shear-driven* breakup [5], which is derived from the balance of shear and interfacial stresses, and the *rate-of-flow-controlled* breakup [8], which is proposed for systems operating at low values of the Capillary numbers of the continuous phase ($Ca_c=\mu_c v_c/\sigma$, where μ_c is the dynamic viscosity, v_c the velocity and σ is the interfacial tension between the two fluids) and it is based on the evolution of pressure in the continuous phase upstream of the tip of the dispersed phase.

In *shear-driven* breakup, the predicted size of a droplet under an external shear force is approximated by balancing the interfacial force with the shear force:

$$\frac{\sigma}{R} = \mu_c \varepsilon \tag{1}$$

where R is the final droplet radius and ε is the shear rate. The surface tension (σ) between silicone oil and water has been evaluated by means of the following correlation suggested in [20] as $\sigma = 0.122T + 32.82$, where T is the temperature in °C and σ is in mN/m. Casting equation (1) in order to obtain the dimensionless radius, one can easily find that:

$$R^* = \frac{R}{H} = \frac{1}{Ca_c \frac{\varepsilon H}{v_c}} \tag{2}$$

where H is the height of the microchannel. Around the droplet, a shear rate ε is established in the continuous phase forced to flow in a restricted cross-section region with area S_g, delimited by the walls at one side and the droplet surface at the other side. The shear rate can be estimated as follows:

$$\varepsilon = \frac{v_g}{\delta_g} \tag{3}$$

where δ_g is the distance between the droplet contour and the wall and v_g is the average velocity of the continuous phase within the restricted section S_g. Substituting equation (3) in equation (2), one can reach to a prediction of the dimensionless droplet radius, as a function of the dimensionless shear rate and of the Capillary number of the continuous phase:

$$R^* = \frac{R}{H} = \frac{1}{Ca_c \frac{v_g}{\delta_g}\frac{H}{v_c}} = \frac{1}{Ca_c \varepsilon_d} \tag{4}$$

where $\varepsilon_d = \frac{v_g}{\delta_g}\frac{H}{v_c}$ is the dimensionless shear rate.

Figure 12a shows the comparison of the dimensionless diameter of the droplet D=2R* calculated by using equation (4), with the experimental results. As pointed by other authors [9, 11, 21], the dimensionless diameter of the droplet decreases with Ca_c much more slowly than one would expect from the scaling law (equation (4)). De Menech *et al.* [11] for a T-junction found a good agreement between D^* and $(Ca_c\, \varepsilon_d)^{-0.4}$ while van der Graaf *et al.* [21] demonstrated that D^* scales with $Ca_c^{-0.25}$. In figure 12a the experimental results obtained in this work have been compared with the correlations proposed by De Menech *et al.* [11]; the best agreement is observed when D^* scales with $(Ca_c\, \varepsilon_d)^{-0.3}$. In figure 12b the values of the dimensionless diameter of the droplet has been correlated to the flow rate ratios ϕ (defined as the ratio of water flow rate/silicone oil flow rate). As one can see, D^* for a fixed value of Ca_c does not change significantly with ϕ: the experimental results can be correlated to ϕ and Ca_c by means of the following relationship: $D^* = 0.4(\phi^{0.05} / Ca_c^{0.25})$.

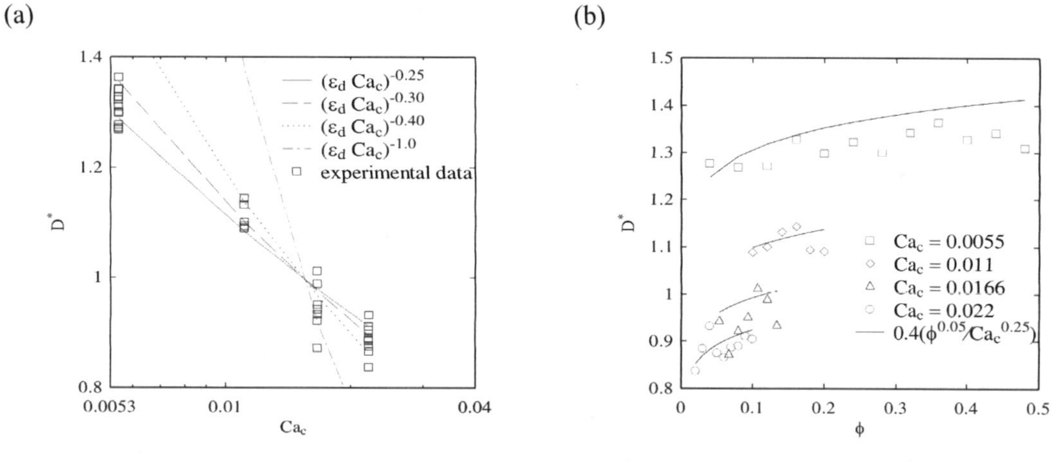

Figure 12. The dimensionless diameter of the droplet: as a function of Ca_c compared with the different correlations (a) as a function of the volumetric flow rate ratios ϕ (b)

It has been demonstrated by figure 12 that the dimensionless diameter of the droplets observed experimentally in this work depends to the Capillary number of the continuous phase and very weakly to the flow rate (water/oil) ratio; this observation seems to confirm that the observed droplet breakup can be considered more similar to a *shear-driven* breakup than a *rate-of-flow-controlled* breakup. However, further investigation is needed in order to clarify these aspects.

5. Conclusions

In this work water-in-oil droplet breakup mechanism within a micro cross-junction with hydrophobic walls has been investigated experimentally by means of optical techniques. By the use of a high-speed camera, connected to a microscope, a sequence of snapshots which capture the squeezing of the water phase, followed by breakup and droplet formation, have been recorded and analyzed. Processing of the raw snapshots provides us useful information for the physical description of the evolution of the squeezing mechanism and of the formation of water droplets. In the paper it has been described the procedure by means of which it is possible to estimate the eccentricity, area, average velocity and the dimensionless diameter of the droplet. The experimental data shows a good agreement with correlation when the diameter of the droplet scales with the Capillary number as $(Ca_c\,\varepsilon_d)^{-0.3}$. The dimensionless diameter has been presented for various flow rate ratios and a correlation to fit the results has been presented.

References

[1] Zhao C X 2013 *Adv. Drug Deliv. Rev.* **65** 1420-1446.
[2] Nisisako T and Hatsuzawa T 2010 *Microfluid Nanofluid* **9** 427-437.
[3] Yang C H, Huang K S, Lin P W and Lin Y C 2007 *Sensor Actuat. B-Chem.* **124** 510-516.
[4] Zhao C X and Middelberg A P J 2011 *Chem. Eng. Sci.* **66** 1394-1411.
[5] Thorsen T, Roberts R W, Arnold F H and Quake S R 2001 *Phys. Rev. Lett.* **86** 4163-4166.
[6] Liu K, Ding H, Chen Y and Zhao X Z 2007 *Microfluid Nanofluid* **3** 239-243.
[7] Tan Y C, Fisher J S, Lee A I, Cristini V and Phillip A 2004 *Lab Chip* **4** 292-298.
[8] Garstecki P, Fuerstman M J, Stone H A and Whitesides G M 2006 *Lab Chip* **6** 437-446.
[9] Xu J H, Li S W, Tau J and Luo J S 2008 *Microfluid Nanofluid* **5** 711-717.
[10] Utada A S, Fernandez-Nieves A, Stone H A and Weitz D A 2007 *Phys. Rev. Lett.* **9** 94502-(1-4).
[11] De Menech M, Garstecki P, Jousse F and Stone H A 2008 *J. Fluid Mech.* **595** 141-161.
[12] Dreyfus R, Tabeling P and Willaime H 2003 *Phys. Rev. Lett.* **90** 144505(1-4).
[13] Barbier V, Williame H and Tabeling P 2006 *Phys. Rev. E* **74** 46306.
[14] Puccetti G, Pulvirenti B and Morini G L 2014 *Energy Procedia* **45** 538.
[15] Puccetti G, Tosi M, Pulvirenti B and Morini G L 2015 *Exp. Therm. Fluid Sci.* **67** 62-69.
[16] Di Marco P 2008 *III Giornata di Studio (L'Aquila, Italy).*
[17] Pulvirenti B, Tosi M, Puccetti G and Morini G L 2014 *XXII UIT Heat Transf. Conf (Pisa)*
[18] Nobuyuki O 1979 *IEEE Trans. Syst. Man Cyber.* **9** 62–66.
[19] Canny J 1986 *IEEE Trans. Pattern Anal. Mach. Intell.* **8** 679–698.
[20] Peters F and Aarabali D 2013 *Colloids Surf.* **426** 1-5.
[21] van der Graaf S, Nisisako T, Schroen C G P H, van der Sman R G M and Boom R M 2006 *Langmuir* **22** 4144-4152.

Simulation of copper–water nanofluid in a microchannel in slip flow regime using the lattice Boltzmann method with heat flux boundary condition

A D'Orazio[1], Z Nikkhah[2], A Karimipour[2]

[1] Dipartimento di IngegneriaAstronautica, Elettrica ed Energetica, Sapienza Università di Roma, Via Eudossiana 18, Roma 00184, Italy
[2] Department of Mechanical Engineering, Najafabad Branch, Islamic Azad University, Isfahan, Iran, Postal Code: 8196848531

E-mail:annunziata.dorazio@uniroma1.it

Abstract. Laminar forced convection heat transfer of water–Cu nanofluids in a microchannel is studied using the double population Thermal Lattice Boltzmann method (TLBM). The entering flow is at a lower temperature compared to the microchannel walls. The middle section of the microchannel is heated with a constant and uniform heat flux, simulated by means of the counter slip thermal energy boundary condition. Simulations are performed for nanoparticle volume fractions equal to 0.00%, 0.02% and 0.04% and slip coefficient equal to 0.001, 0.01 and 0.1. Reynolds number is equal to 1, 10 and 50.The model predictions are found to be in good agreement with earlier studies. Streamlines, isotherms, longitudinal variations of Nusselt number and slip velocity as well as velocity and temperature profiles for different cross sections are presented. The results indicate that LBM can be used to simulate forced convection for the nanofluid micro flows. They show that the microchannel performs better heat transfers at higher values of the Reynolds number. For all values of the Reynolds considered in this study, the average Nusselt number increases slightly as the solid volume fraction increases and the slip coefficient increases. The rate of this increase is more significant at higher values of the Reynolds number.

1. Introduction

The lattice Boltzmann Equation (LBE) is a minimal form of the Boltzmann kinetic equation, which is the evolution equation for a continuous one-body distribution function, wherein all details of molecular motion are removed except those that are strictly needed to represent the hydrodynamic behaviour at the macroscopic scale; it has gained much attention for its ability to simulate fluid flows, and for its potential advantages over conventional numerical solution of the Navier–Stokes equations.

[1] To whom any correspondence should be addressed

The kinetic nature of Lattice Boltzmann Method (LBM) introduces key advantages, including easy implementation of boundary conditions and fully parallel algorithms. In addition, the convection operator is linear, no Poisson equation for the pressure must be resolved and the translation of the microscopic distribution function into the macroscopic quantities consists of simple arithmetic calculations [1]. LBM have met with significant success for the numerical simulation of a large variety of fluid flows, including real-world engineering applications and physical phenomena of various complexities as multiphase flows, complex geometries and interfacial flows [2]. The application to fluid flow coupled with non negligible heat transfer turned out to be much more difficult. The LBE thermal models fall into three categories: the multi-speed approach, the passive scalar approach and the doubled populations approach. In this latter, successful, strategy thermal energy density and heat flux are expressed as kinetic moments of a thermal distribution function, so that no kinetic moment beyond the first order is ever required, thus providing numerical stability, also in case of significant temperature gradient [3]; in addition, with respect to the previous approaches, viscous heat dissipation and compression work done by the pressure were naturally incorporated and the boundary conditions are easily implemented because both populations live in the same lattice, where additional speeds are not necessary.

During the last two decades, much attention has been paid to make and use micro devices. The small sizes as well as high efficiency of micro devices – such as microsensors, microvalves and micropumps – are some of the advantages of using MEMS and NEMS (Micro and Nano Electro Mechanical Systems). To guarantee the performance of such devices and make them cool, many studies have been carried out concerning flow and heat transfer in microchannels. At micro scale level, the surface effects are getting more important which leads to change in the classic boundary conditions. The well known differences of micro flows from the macroscopic ones are the slip velocity and temperature jump on the solid–fluid boundaries. For the gas micro flows, the flow regimes can be slip, transient and free molecular flow regimes; however for liquid micro flows, mainly the slip flow regime can be observed [4]. Therefore, in addition to classic Navier–Stokes (NS), the particle-based methods including direct simulation of Monte Carlo (DSMC), molecular dynamics (MD) and the lattice Boltzmann method (LBM) may be applied [5]. Expensive computation cost and complex mathematical procedure of MD and DSMC, as well as the inability of N–S for simulation of flow in transition and free molecular regimes, have encouraged the researchers to use LBM.

On the other hand, using nanofluids is an innovative way to increase heat transfer which has attracted the researchers' interests who are working on micro flow due to their exciting potential. Nanofluids are a mixture of liquid and dispersed solid nanoparticles. The higher thermal conductivity of nanoparticles leads to the increase of nanofluid heat transfer. Nanofluid's characteristics are different from the traditional solid–liquid mixtures in milli or micro-meter particle's size. There are many studies concerning nanofluid in cavities and tubes which involve its positive effects on the Nusselt number. Some researchers have reported the flow and heat transfer of the nanofluid in microchannels [6-9]. For instance, Raisi et al. [10] simulated the Cu–water nanofluid in a microchannel for both slip and no-slip conditions, ignoring the temperature jump effects and applying the classic Navier–Stokes equations. All in all, theoretical results of fluid flow in slip flow regimes or even simulation of nanofluid flow using LBM (in single phase or multi-phase mixture model) have been presented by several researchers. However, there are few studies concerning nanofluid simulation in microchannels using LBM [11-13]. However, all of them have ignored the slip velocity and temperature jump effects. The open literature suggests that nanofluids are an effective coolant which requires more investigations. In particular, the convection heat transfer of nanofluids in microchannel in the slip flow regime is still not entirely understood.

In the present study, laminar forced convection heat transfer of dilute water–Cu nanofluids in a microchannel is analyzed. The double population Thermal Lattice Boltzmann method (TLBM) is used.

The middle section of the microchannel is heated with a constant and uniform heat flux. This type of boundary condition, representing very usual situations in physical world, is not simple to model in lattice Boltzmann schemes. In effect, the only boundary condition able to simulate imposed temperature and imposed heat flux at a boundary is the counter-slip thermal energy density boundary condition, that has been presented by D'Orazio et al. in [3, 19, 20] (where the boundary were at rest), and used in case of moving wall in [14]. The results obtained in this work are weighed against model validation results found in the literature. Particular attention is paid to the effects of different slip velocities (slip coefficient is equal to 0.001, 0,01 and 0.1) and different solid volume fractions (equal to 0.00%, 0,02% and 0.04%). Reynolds number is equal to 1, 10 and 50. The model predictions are found to be in good agreement with earlier studies.

2. Thermal and hydrodynamic Lattice Boltzmann Method

The lattice Boltzmann equation with a single relaxation time from the BGK model can be expressed as

$$\tilde{f}_i(\vec{x}+\vec{c}_i dt, t+dt) - \tilde{f}_i(\vec{x},t) = -\frac{dt}{\tau_f+0.5dt}\left[\tilde{f}_i - \tilde{f}_i^e\right] \tag{1}$$

$$\tilde{g}_i(\vec{x}+\vec{c}_i dt, t+dt) - \tilde{g}_i(\vec{x},t) = -\frac{dt}{\tau_g+0.5dt}\left[\tilde{g}_i - \tilde{g}_i^e\right] - \frac{\tau_g dt}{\tau_g+0.5dt} f_i Z_i \tag{2}$$

where the populations $\tilde{f}_i$ carry mass and momentum and the populations $\tilde{g}_i$ carry internal energy and heat flux. The discrete distribution functions $\tilde{f}_i$ and $\tilde{g}_i$ are introduced as in [5]:

$$\tilde{f}_i = f_i + \frac{dt}{2\tau_f}(f_i - f_i^e) \tag{3}$$

$$\tilde{g}_i = g_i + \frac{dt}{2\tau_g}(g_i - g_i^e) + \frac{dt}{2} f_i Z_i \tag{4}$$

$$Z_i = (\vec{c}_i - \vec{u}) D_i \vec{u}\ , \quad D_i = \partial_t + \vec{c}_i \cdot \nabla \tag{5}$$

where f_i and g_i are the discrete populations which evolve when a standard first order integration strategy is adopted, the terms Z_i and D_i represent the effects of viscous heating and the material derivative along direction $\vec{c}_i$ respectively, τ_f and τ_g are relaxations times and f_i^e and g_i^e are the equilibrium distribution functions. Throughout of this work, two-dimensional square lattice with the nine speeds, as shown in figure 1, is used.

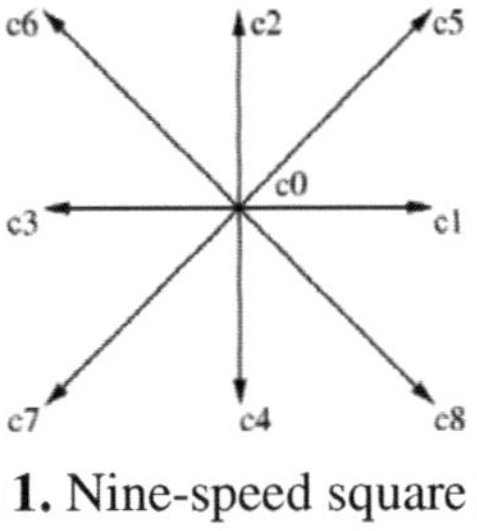

Figure 1. Nine-speed square lattice.

The discrete particle lattice speeds are:

$$c_i = \left(\cos\frac{i-1}{2}\pi, \sin\frac{i-1}{2}\pi\right)c, \qquad i = 1,2,3,4 \tag{6}$$

$$c_0 = (0,0) \qquad c_i = \sqrt{2}\left[\cos\left(\frac{i-5}{2}\pi + \frac{\pi}{4}\right), \sin\left(\frac{i-5}{2}\pi + \frac{\pi}{4}\right)\right]c, \qquad i = 5,6,7,8 \tag{7}$$

where $c^2 = 3RT$ and T is the temperature. The equilibrium density distributions are chosen as follows:

$$f_i^e = \omega_i \rho \left[1 + \frac{3\vec{c}_i \cdot \vec{u}}{c^2} + \frac{9(\vec{c}_i \cdot \vec{u})^2}{2c^4} - \frac{3(u^2+v^2)}{2c^2}\right] \quad (8)$$

$$g_0^e = -\omega_0 \left[\frac{3pe(u^2+v^2)}{2c^2}\right] \quad (9)$$

$$g_{1,2,3,4}^e = \omega_1 pe \left[1.5 + 1.5\frac{\vec{c}_i \cdot \vec{u}}{c^2} + 4.5\frac{(\vec{c}_i \cdot \vec{u})^2}{c^4} - 1.5\frac{(u^2+v^2)}{c^2}\right] \quad (10)$$

$$g_{5,6,7,8}^e = \omega_2 pe \left[3 + 6\frac{\vec{c}_i \cdot \vec{u}}{c^2} + 4.5\frac{(\vec{c}_i \cdot \vec{u})^2}{c^4} - 1.5\frac{(u^2+v^2)}{c^2}\right] \quad (11)$$

where $\vec{u} = (u, v)$ and $\rho e = \rho RT$ (in two-dimensional geometry). The weights of the different populations are

$$\omega_0 = \frac{4}{9} \qquad \omega_i = \frac{1}{9} \;\; i = 1,2,3,4 \qquad \omega_i = \frac{1}{36} \;\; i = 5,6,7,8 \quad (12)$$

The terms enclosed by the square bracket, multiplied by the corresponding weights ω_i, will be called *corresponding form for equilibrium.*

Finally, using f_i^e and g_i^e, hydrodynamic and thermal variables are calculated as follows:

$$\rho = \sum_i \tilde{f}_i \qquad \rho e = \sum_i \tilde{g}_i - \frac{dt}{2}\sum_i f_i Z_i \quad (13)$$

$$\rho\vec{u} = \sum_i \vec{c}_i \tilde{f}_i \qquad \vec{q} = \left[\sum_i \vec{c}_i \tilde{g}_i - \rho e\vec{u} - \frac{dt}{2}\sum_i \vec{c}_i f_i Z_i\right]\frac{\tau_g}{\tau_g + 0.5dt} \quad (14)$$

The kinematic viscosity and the thermal diffusivity in the two-dimensional geometry are given by:

$$\nu = \tau_f R\bar{T}, \qquad \chi = 2\tau_g R\bar{T} \quad (15)$$

3. Nanofluid

Nanofluid is a homogeneous mixture of the liquid and suspended nanoparticles. Its effective density can be obtained by

$$\rho_{nf} = \varphi\rho_s + (1-\varphi)\rho_f \quad (16)$$

where φ is the nanoparticle volume fraction and the subscripts "f", "s" and "nf" refer to base fluid, solid nanoparticles and nanofluid, respectively. Using the heat capacity of nanofluid, the nanofluid thermal diffusivity can be obtained by [15]:

$$(\rho C_p)_{nf} = \varphi(\rho C_p)_s + (1-\varphi)(\rho C_p)_f \qquad \alpha_{nf} = k_{nf}/(\rho C_p)_{nf} \quad (17)$$

The effective dynamic viscosity is expressed by using the Brinkman model [16]:

$$\mu_{nf} = \mu_f/(1-\varphi)^{2.5} \quad (18)$$

Following equation, which was presented by Patel et al. [17,18], is considered to determine the nanofluid thermal conductivity k_{nf} as a function of liquid and solid thermal conductivities

$$k_{nf}/k_f = 1 + \left[\frac{k_s}{k_f}\frac{A_s}{A_f} + Ck_s Pe\frac{A_s}{k_f A_f}\right] \quad (19)$$

in which

$$\frac{A_s}{A_f} = \frac{d_f}{d_s}\frac{\varphi}{1-\varphi} \qquad Pe = \frac{u_B d_s}{\alpha_f} \qquad u_B = \frac{2k_B T}{\pi\mu_f d_s^2} \qquad k_B = 1.3807 \times 10^{-23} J/K \qquad C = 36000 \quad (20)$$

where Pe is the Péclet number with the Brownian motion velocity of particles u_B and k_B is the constant of Boltzmann.

4. Boundary conditions

Non-equilibrium bounce back model, normal to the boundary, is used for inlet and outlet hydrodynamic boundary conditions. In this model, distribution functions are reflected in suitable ways to satisfy the equilibrium conditions and improve accuracy. It results

$$\tilde{f}_1 = \tilde{f}_3 + \frac{2}{3}\rho_i U_i \qquad \tilde{f}_5 = \tilde{f}_7 + \frac{1}{2}(\tilde{f}_4 - \tilde{f}_2) + \frac{1}{6}\rho_i U_i \qquad \tilde{f}_8 = \tilde{f}_6 - \frac{1}{2}(\tilde{f}_4 - \tilde{f}_2) + \frac{1}{6}\rho_i U_i \tag{21}$$

$$\tilde{f}_3 = \tilde{f}_1 - \frac{2}{3}\rho_{out} U_{out} \tag{22}$$

$$\tilde{f}_7 = \tilde{f}_5 - \frac{(\tilde{f}_4-\tilde{f}_2)}{2} - \frac{\rho_{out}U_{out}}{6} - \frac{\rho_{out}V_{out}}{2} \qquad \tilde{f}_6 = \tilde{f}_8 + \frac{(\tilde{f}_4-\tilde{f}_2)}{2} - \frac{\rho_{out}U_{out}}{6} + \frac{\rho_{out}V_{out}}{2} \tag{23}$$

With regard to thermal boundary conditions at inlet and outlet sections, they are obtained by means a thermal counter-slip approach as proposed by [3,19,20], in which the incoming unknown thermal populations are assumed to be equilibrium distribution functions with a counter slip thermal energy density e', which is determined so that suitable constraints are verified. It results

$$\tilde{g}_1 = \frac{6\rho e + 3dt\sum_i f_i Z_i - 6(\tilde{g}_0+\tilde{g}_2+\tilde{g}_3+\tilde{g}_4+\tilde{g}_6+\tilde{g}_7)}{2+3U_i+3U_i^2} \times \left[1.5 + 1.5U_i + 3U_i^2\right]\frac{1}{9} \tag{24}$$

$$\tilde{g}_5 = \frac{6\rho e + 3dt\sum_i f_i Z_i - 6(\tilde{g}_0+\tilde{g}_2+\tilde{g}_3+\tilde{g}_4+\tilde{g}_6+\tilde{g}_7)}{2+3U_i+3U_i^2} \times \left[3 + 6U_i + 3U_i^2\right]\frac{1}{36} \tag{25}$$

$$\tilde{g}_8 = \frac{6\rho e + 3dt\sum_i f_i Z_i - 6(\tilde{g}_0+\tilde{g}_2+\tilde{g}_3+\tilde{g}_4+\tilde{g}_6+\tilde{g}_7)}{2+3U_i+3U_i^2} \times \left[3 + 6U_i + 3U_i^2\right]\frac{1}{36} \tag{26}$$

$$\tilde{g}_3 = \frac{6(\tilde{g}_1+\tilde{g}_5+\tilde{g}_8) - 3dt\sum_i c_{ix} f_i Z_i - 6\rho e u_{out}}{2-3U_{out}+3U_{out}^2} \times \left[1.5 - 1.5U_{out} + 3U_{out}^2 - 1.5V_{out}^2\right]\frac{1}{9} \tag{27}$$

$$\tilde{g}_6 = \frac{6(\tilde{g}_1+\tilde{g}_5+\tilde{g}_8) - 3dt\sum_i c_{ix} f_i Z_i - 6\rho e U_{out}}{2-3U_{out}+3U_{out}^2} \times \left[3 - 6U_{out} + 6V_{out} + 3U_{out}^2 + 3V_{out}^2 - 9U_{out}V_{out}\right]\frac{1}{36} \tag{28}$$

$$\tilde{g}_7 = \frac{6(\tilde{g}_1+\tilde{g}_5+\tilde{g}_8) - 3dt\sum_i c_{ix} f_i Z_i - 6\rho e U_{out}}{2-3U_{out}+3U_{out}^2} \times \left[3 - 6U_{out} - 6V_{out} + 3U_{out}^2 + 3V_{out}^2 + 9U_{out}V_{out}\right]\frac{1}{36} \tag{29}$$

With regard to microchannel walls boundary conditions, the slip boundary condition is applied for hydrodynamic field. More specifically, we refer to the work of Ngoma and Erchiqui [21] who considered the slip length coefficient β and defined the slip velocity u_s for the liquid inside the microchannel on the stationary walls and we write the dimensionless slip velocity U_s as

$$u_s = \pm\beta \left.\frac{\partial u}{\partial y}\right|_{y=0,h} \qquad U_s = \pm B \left.\frac{\partial U}{\partial Y}\right|_{Y=0,1} \tag{30}$$

To determine the slip velocity in LBM, the specular reflective bounce back model (combination of bounce back and specular boundary condition) is applied in this work. For example for the bottom wall, the unknown distribution functions are estimated by

$$\tilde{f}_2 = \tilde{f}_4 \qquad \tilde{f}_5 = r\tilde{f}_7 + (1-r)\,\tilde{f}_8 \qquad \tilde{f}_6 = r\tilde{f}_8 + (1-r)\,\tilde{f}_7 \tag{31}$$

The accommodation coefficient value, r, is chosen appropriately for more accuracy [22].

With regard to thermal boundary conditions at the microchannel walls, the middle section of the microchannel is heated with a constant and uniform heat flux, whereas the final section is insulated. These boundary conditions are obtained by means of the thermal counter-slip approach with imposed heat flux. As an example, for the top wall of the channel, named as "north wall", in which entering heat flux is constant and equal to q_N, the unknown $\tilde{g}_4$, $\tilde{g}_7$ and $\tilde{g}_8$ are chosen as follows.

$$\tilde{g}_i = \rho(e_N + e') \times [corresponding\ form\ for\ equilibrium] \qquad i = 4,7,8 \tag{32}$$

By definition:

$$\sum_i c_{iy}\,\tilde{g}_i = \frac{dt}{2}\sum_i c_{iy} f_i Z_i + \rho e_N V_N + \frac{\tau_g + 0.5dt}{\tau_g} q_y \tag{33}$$

which yields to:

$$\rho e_N + \rho e' = \left[K - \frac{dt}{2}\sum_i \frac{c_{iy}}{c} f_i Z_i - \rho e_N \frac{V_N}{c} - \frac{\tau_g + 0.5dt}{\tau_g}\frac{q_y}{c}\right] \Big/ \left[\frac{1}{3} - \frac{1}{2}\frac{V_N}{c} + \frac{1}{2}\frac{V_N^2}{c^2}\right] \tag{34}$$

where V_N is the component normal to the wall of flow velocity at the wall, K is the sum of the three known populations ($\tilde{g}_2$, $\tilde{g}_5$ and $\tilde{g}_6$) e_N denotes the current value of thermal energy density at the north wall. The numerical value of the heat flux q_N (and similarly for the value q_S at the south wall) is obviously equal to zero for the final section of the microchannel. The corners nodes are treated similarly and the counter-slip procedure can be applied to the five unknown incoming populations at the corner. The same relations are used for the bottom wall (named as "south wall"), in which entering heat flux is constant and equal to q_S and the unknown populations are $\tilde{g}_2$, $\tilde{g}_5$ and $\tilde{g}_6$.
With regard to the initial conditions, the velocities of all nodes inside the cavity are taken as zero initially. The initial density is set to a value of 2.7. The initial equilibrium distribution functions are evaluated correspondingly. The initial distribution functions are taken as the corresponding equilibrium values.

5. Results and discussion

Laminar forced convection heat transfer of a Cu–water nanofluid in a long microchannel is studied numerically using Lattice Boltzmann Method previously described. Length L and height H are shown in Figure 2. The top and bottom walls are partially heated; nanofluid enters the microchannel with inlet velocity and temperature u_i and T_i respectively.

The nanofluid simulated in this work is a dispersion of nanoparticles of copper (Cu) in pure water (as the base liquid). It is assumed that the considered fluid is a Newtonian, incompressible fluid, in laminar flow regime. Nanoparticles are spherical, with diameter as $d_p = 100\text{nm}$. Water and nanoparticles mixture is in the homogeneous mode and the radiation effect is negligible.

Characteristic dimensionless number in the analysis of laminar forced convection problems of nanofluids in microchannels are the Reynolds number and Prandtl numbers, defined as $\mathrm{Re} = u_i h/\nu_f$ and $\mathrm{Pr} = \nu_f/\alpha_f$ respectively, nanoparticle concentration φ and slip coefficient B. As stated before, lattice Boltzmann method is used for near-incompressible flows and therefore Mach number is assumed as $\mathrm{Ma} \ll 1$. More specifically, the characteristic velocity of the flow must be small compared with the fluid speed of sound; in present study the velocity u_i is selected as $u_i = 0.1c$ ($U_i = 0.1$). The Prandtl number is calculated for the nanofluid mixture at $\varphi = 0$ (pure water), $\varphi = 0.02$ and $\varphi = 0.04$ and the and Reynolds number is assumed as $Re = 1$, $Re = 10$ and $Re = 50$. The effects of the slip velocity on the flow and heat transfer are studied; the slip velocity coefficient is assumed as $B = 0.001$, $B = 0.01$ and $B = 0.1$.

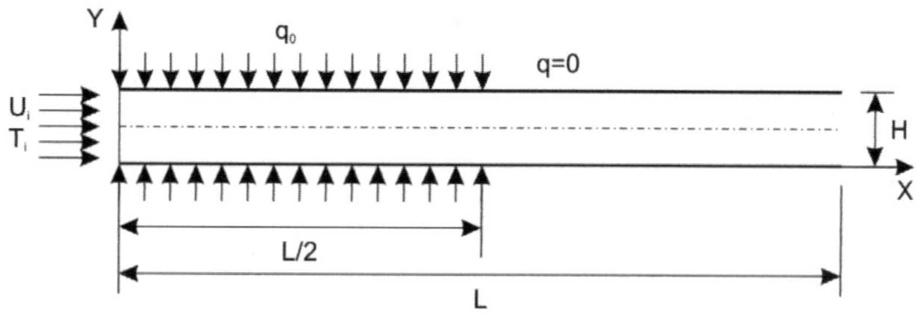

Figure 2. The schematic diagram of the microchannel.

To avoid ambiguity, the boundary of the channel are referred to according to the coordinates shown in Figure 2. In the following the macroscopic variables of fluid flow are made dimensionless as follows, respectively for dimensionless coordinates, velocity components, temperature, time

$$Y = \frac{y}{H}, \quad X = \frac{x}{H}, \quad L = \frac{l}{H}, \quad U = \frac{u}{U_i}, \quad V = \frac{v}{U_i}, \quad \Theta = \frac{T - T_i}{\Delta T}, \quad \Delta T = \frac{q_y H}{k_f}, \quad \tau = \frac{tU_i}{H} \tag{35}$$

Therefore, the local and average Nusselt numbers along the microchannel wall are calculated using following relations

$$Nu_X = \frac{\lambda H}{k_f} = \frac{q_y H}{k_f\,(T_s - T_i)} = \frac{\Delta T}{(T_s - T_i)} = \frac{1}{\Theta_s(X)}, \qquad Nu_m = \frac{1}{L}\int_0^L Nu_X dX \tag{36}$$

where $\lambda = q_y/(T_s - T_i)$ is the convection coefficient. The top wall is the at Y = 1, the bottom wall is at Y=0 and the inlet and outlet sections are at X = 0 and X = 32.

In order to obtain grid independent solution, a grid refinement study is performed for nanoparticle concentration $\varphi = 0.02$ and slip coefficient $B = 0.001$. Grid independence of the results is established in term of average Nusselt number on the wall and dimensionless maximum value of temperature Θ at and Y = 0.1 for three different grid size, namely 40×400, 50×500 and 60×600 lattice nodes; due to small difference between the results of the last two grid sizes, the 50×500 grid is chosen as a suitable one in this work.

To validate the computer code, the comparison with the values obtained by Raisi et al. [10] for the longitudinal velocity U_{out} at the end of the microchannel is examined for the slip coefficient equal to 0.05 and 0.1 and nanoparticles concentration coefficient and Reynolds number equal to 0.03 and 50.0 respectively; in addition, the comparison with the values obtained by Aminossadati et al. [7] for the average Nusselt number Nu_m is examined for the Reynolds number equal to 10.0, 100.0 and 500.0. The values obtained in present work show good agreement with those in [10] and [7], with a maximum error equal to 0.2% and 1.9% respectively.

In order to show the effect of the Reynolds number on the flow field and heat transfer, in Figures 3, 4 and 5 streamlines and isotherms are reported for slip velocity coefficient $B = 0.1$ and nanoparticles concentration coefficient $\varphi = 0.02$ for the cases $Re = 1$, 10 and 50. Nanofluid enters the microchannel from the left and after cooling the walls, it leaves from the right side. So, there will be symmetric and horizontal streamlines along the microchannel. For the same values of Reynolds number in Figure 6, 7 and 8 the slip velocity U_s is reported as a function of the slip coefficient. It can be observed that at the inlet, the slip velocities start from their maximum values and decrease asymptotically along the wall and approach constant values; larger B corresponds to larger slip velocity on the walls.

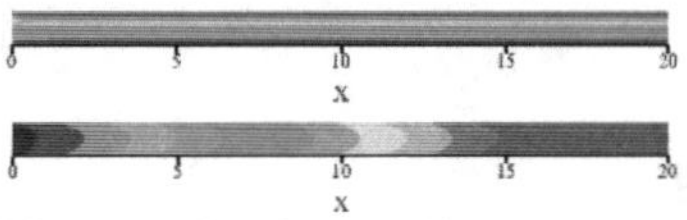

Figure 3. Streamlines (top) and isotherms (bottom) for $B = 0.1$, $\varphi = 0.02$ and $Re = 1$

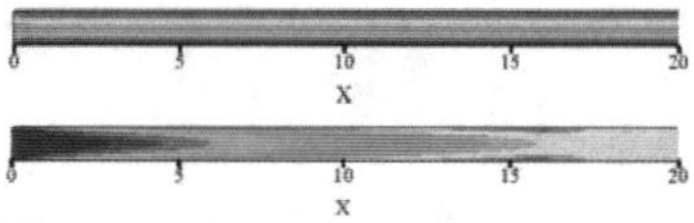

Figure 4. Streamlines (top) and isotherms (bottom) for $B = 0.1$, $\varphi = 0.02$ and $Re = 10$

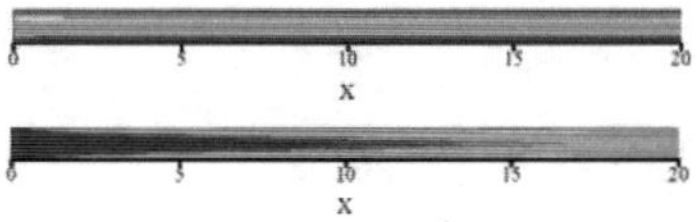

Figure 5. Streamlines (top) and isotherms (bottom) for $B = 0.1$, $\varphi = 0.02$ and $Re = 50$

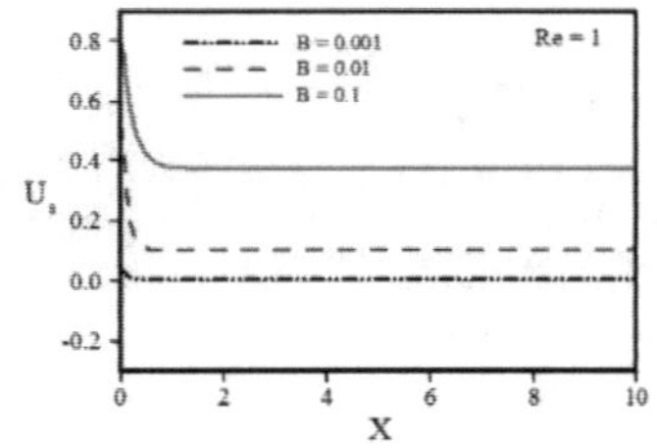

Figure 6. Slip velocity for $\varphi = 0.02$ and $Re = 1$.

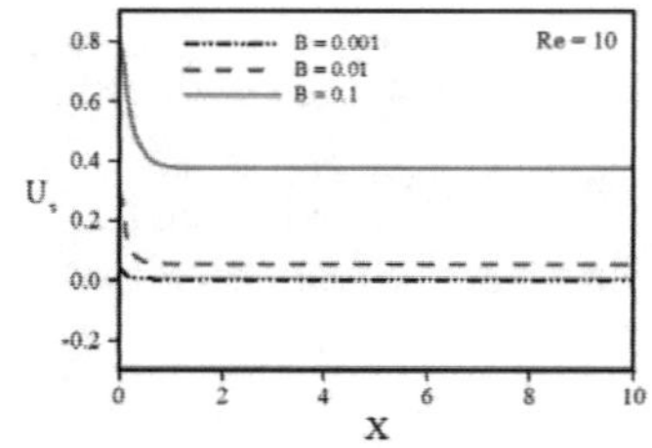

Figure 7. Slip velocity for $\varphi = 0.02$ and $Re = 10$.

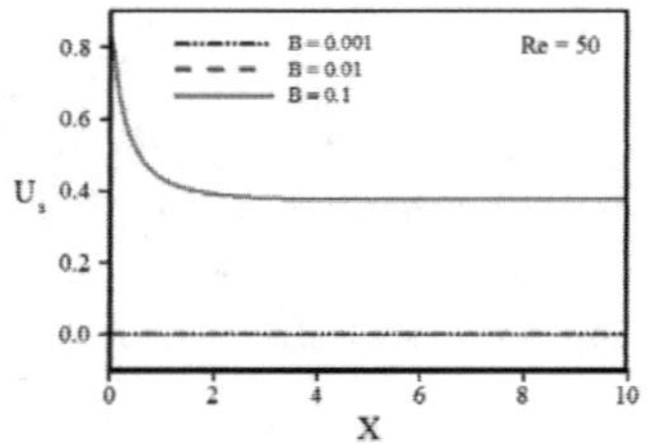

Figure 8. Slip velocity for $\varphi = 0.02$ and $Re = 50$.

The average Nusselt number $\mathrm{Nu_m}$ for different values of φ and B is presented in Figure 9, which indicates that the importance of using nanofluid to increase the heat transfer rate is distinguishable for $Re = 50$. In this case, using 4% of Cu nanoparticles leads to increase almost 11% of the average Nusselt number at $B = 0.1$. This increase is more significant (19.5%) at $B = 0.001$.

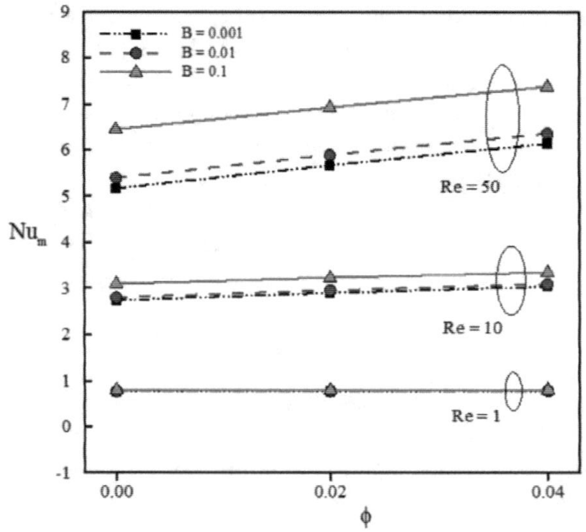

Figure 9. Average Nusselt number $\mathrm{Nu_m}$ as a function of particles concentration φ for different slip velocity coefficient and in case of Reynolds number $Re = 1$, 10 and 50

6. Conclusions

A thermal lattice Boltzmann BGK model with a dedicated boundary condition was used to study numerically laminar two-dimensional laminar forced convection heat transfer of Cu–water nanofluid in a microchannel when he heat transfer rate is imposed at the boundaries. The effects of different volume fractions of copper nanoparticles and slip coefficient were investigated on the slip velocity and Nusselt number for $Re = 1$, 10 and 50. The results show that, as expected, heat transfer rate increases as increases the Reynolds number and the slip velocity at the wall. The effect of particle concentration is detectable for not too low values of Reynolds number and this effect is significant for the lowest slip velocity coefficient. As a result, to increase Nu in micro liquid flows, it is recommended to use nanofluid with $\varphi = 4\%$ and at low values of slip coefficient as like $B = 0.001$. The study shows that LBM together with the counter-slip thermal boundary condition can be effectively used for heat transfer phenomena in case of flows in microchannel in presence of nanofluid.

References

[1] Succi S 2001 *The lattice Boltzmann equation for fluid dynamics and beyond* (Oxford: Oxford University Press)
[2] Wang M and Kang Q 2010 Modeling electrokinetic flows in microchannels using coupled lattice Boltzmann methods *J. of Comp. Phys.* **229** 728-744
[3] D'Orazio A, Succi S and Arrighetti C 2003 Lattice Boltzmann simulation of open flows with heat transfer *Phys. of Fluids* **15** 2778-2780
[4] Nguyen N T and Wereley S T 2006 *Fundamentals and Applications of Microfluidics* (Artech House, Norwood).
[5] Bird G 1994 *Molecular gas dynamics and the direct simulation of gas flows* (Oxford).
[6] Esfe M H, Arani A A A, Karimipour A and Esforjani S S M 2014 Numerical simulation of natural convection around an obstacle placed in an enclosure filled with different types of nanofluids *Heat Transfer Res.* **45** 279–292.
[7] Aminossadati S M, Raisi A and Ghasemi B 2011 Effects of magnetic field on nanofluid forced convection in a partially heated microchannel *Int. J. Non-Linear Mech.* **46** 1373–1382.
[8] Kalteh M, Abbassi A, Saffar-Avval M and Harting J 2011 Eulerian–Eulerian twophase numerical simulation of nanofluid laminar forced convection in a microchannel *Int. J. Heat Fluid Flow* **32** 107–116.
[9] Mital M 2013 Semi-analytical investigation of electronics cooling using developing nanofluid flow in rectangular microchannels *Appl. Therm. Eng.* **52** 321–327.
[10] Raisi A, Ghasemi B, Aminossadati S M 2011 A numerical study on the forced convection of laminar nanofluid in a microchannel with both slip and no-slip conditions *Numer. Heat Transfer A* **59** 114–129.
[11] Guo Y, Qin D, Shen S and Bennacer R 2012 Nanofluid multi-phase convective heat transfer in closed domain: Simulation with lattice Boltzmann method *Int. Commun. Heat Mass Transfer* **39** 350–354.
[12] Ay C, Young C W and Young C F 2012 Application of lattice Boltzmann method to the fluid analysis in a rectangular microchannel *Comput. Math. Appl.* **64** 1065–1083.
[13] Yang Y T and Lai F H 2011 Numerical study of flow and heat transfer characteristics of alumina-water nanofluids in a microchannel using the lattice Boltzmann method *Int. Commun. Heat Mass Transfer* **38** 607–614.
[14] D'Orazio A, Karimipour A, Nezhad A H and Shirani E 2015 Lattice Boltzmann method with heat flux boundary condition applied to mixed convection in inclined lid driven cavity *Meccanica* **50** 945–962
[15] Xuan Y and Li Q 2003 Investigation on convective heat transfer and flow features of nanofluids *ASME J. Heat Transfer* **125** 151–155.
[16] Brinkman H C 1952 The viscosity of concentrated suspensions and solutions *J. Chem. Phys.* **20**

571–581.

[17] Patel H E, Sundararajan T, Pradeep T, Dasgupta A, Dasgupta N and Das S K 2005 A Micro-Convection Model for Thermal Conductivity of Nanofluids *Pramana-J. Phys.* **65** (5) 863-869.

[18] Keblinski P, Phillpot S R, Choi S U S and Eastman J A 2002 Mechanisms of heat flow in suspensions of nano-sized particles (nanofluids) *Int. J. Heat Mass Trans.* **45** (4) 855-863.

[19] D'Orazio A, Corcione M and Celata G P 2004 Application to natural convection enclosed flows of a lattice Boltzmann BGK model coupled with a general purpose thermal boundary condition *Int. J. of Thermal Sc.* **43** 575–586.

[20] D'Orazio A and Succi S 2004 Simulating two-dimensional thermal channel flows by means of a lattice Boltzmann method *Fut. Gen. Comp. Sys.* **20** 935-944.

[21] Ngoma G D and Erchiqui F 2007 Heat flux and slip effects on liquid flow in a microchannel *Int. J. Therm. Sci.* **46** 1076–1083.

[22] Karniadakis G and Beskok A 2002 *Micro flows: fundamentals and simulation* (New York).

[23] Succi S 2002 Mesoscopic modelling of slip motion at fluid-solid interfaces with heterogeneous catalysis *Phys. Rev. Lett.* **89**.

Nano-PCMs for passive electronic cooling applications

L Colla[1], L Fedele[1], S Mancin[2], B Buonomo[3], D Ercole[3], O Manca[3]

[1]Istituto per le Tecnologie della Costruzione - Consiglio Nazionale delle Ricerche, Corso Stati Uniti, 4, 35127 Padova, IT
[2]DTG - Universtà di Padova, Str.lla S. Nicola, 3, 36100 Vicenza, IT
[3]Dip. di Ingegneria Industriale e dell'Informazione, Seconda Università degli Studi di Napoli Real Casa dell'Annunziata, Via Roma 29, Aversa (CE) 81031, IT

E-mail: Oronzio.manca@unina2.it

Abstract. The present work aims at investigating a new challenging use of oxide (TiO_2, Al_2O_3, etc.) nanoparticles to enhance the thermal properties: thermal conductivity, specific heat, and latent heat of pure paraffin waxes to obtain a new class of Phase Change Materials (PCMs), the so-called nano-PCMs. The nano-PCMs were obtained by seeding different amounts of oxide nanoparticles in a paraffin wax having a melting temperature of 45°C. The thermophysical properties such as latent heat and thermal conductivity were then measured to understand the effects of the nanoparticles on the thermal properties of both the solid and liquid PCM. Finally, a numerical comparison between the use of the pure paraffin wax and the nano-PCM in a typical electronics passive cooling device was implemented. Numerical simulations were carried out using the Ansys-Fluent 15.0 code. Results in terms of solid and liquid phase temperatures, melting time and junction temperature were reported. Moreover, a comparison with experimental results was also performed.

1. Introduction

The solid-liquid phase change process is an interesting possibility to reject even high heat loads, especially when they are intermittent. The term Phase Change Materials (PCMs) commonly refers to those materials, which use the solid-liquid phase change process to adsorb and then release heat loads [1]. The use of the latent heat absorption phenomenon associated with melting of a suitable PCM can be considered an effective way to delay or modify the temperature rise of a surface subjected to high and intermittent heat fluxes. The intrinsic advantage of the PCM systems is related to their simplicity and reliability; however, the reversible phase change process must be carefully analysed in order to avoid any unmanageable situations during the real operation of these devices.

As reported in several comprehensive reviews [2-4] published in the open literature, PCMs have been widely suggested for thermal storage applications due to their capability of storing and releasing large amounts of energy with a small PCM volume and a relatively low temperature variation. A large number of PCMs (organic, inorganic, and eutectic) are available in any required temperature range. An ideal PCM should exhibit a suitable phase-transition temperature, high latent heat of fusion, and high thermal conductivity; it should be characterized by high density associated with a small volume change during the melting process and low vapour pressure in the melt; moreover, it should be chemically stable over a long period of time, non-toxic and non-hazardous, and compatible with the constructional materials. Finally, it should be abundant, available, and cost effective.

Paraffin waxes have been found to exhibit many desirable characteristics, such as high latent heat, low vapour pressure in the melt, chemically inert and stable, non-toxic. However, they also have a very low thermal conductivity and a high volume change during the melting process. As such, several researchers have focused their attention to different enhancement techniques proposed to improve the thermal conductivity of the PCMs; Jegadheeswaran and Pohekar [5] and Fan and Khodadadi [6] have reviewed most of these efforts.

A new challenging opportunity to improve the thermophysical properties of the paraffin waxes is represented by the addition of nanoparticles to a base PCM. For example, Khodadadi and Hosseinizadeh [7] seeded PCMs with nanoparticles to improve thermal energy storage by increasing heat conductivity. Recent research (*e.g.* Shin and Banerjee [8], Chieruzzi *et al.* [9]) showed that nanoparticles could enhance the heat capacity of the molten salt PCM. Zhichao *et al.* [10] measured 8% total heat storage capacity enhancement dispersing TiO_2 nanoparticles at 0.2 vol% in erythritol as PCM. Jiang *et al.* [11] studied PCM paraffin microcapsules modified with different percentages of nano-Al_2O_3 obtaining enhanced thermal transfer and phase change properties. Karkri *et al.* [12] used a PCM based on high-density polyethylene (HDPE) mixed with micro-encapsulated paraffin wax. The paraffin wax acted as a high-latent-heat material, whereas the HDPE matrix ensured the compact shape. The authors observed a decrement on the PCM thermal conductivity and diffusivity, and an enhancement on the latent, sensible and total heat of the PCM increasing the paraffin content. He *et al.* [13] studied thermophysical properties of nanofluids as PCM, $BaCl_2$–H_2O PCMs, in low temperature cool storage, using TiO_2 nanoparticles. Thermal conductivity increased up to 16.7% suggesting the possible use of PCM in the industries of low temperature energy storage, as refrigeration and air conditioning systems, solar energy systems and heating and cooling of buildings.

This paper presents some preliminary experimental and numerical results on two nano-PCMs obtained by seeding different amounts, 0.5% and 1.0%wt, of Al_2O_3 nanoparticles in a pure paraffin wax with a melting temperature of 45 °C, called RT45. The thermal conductivity, the specific heat of solid and liquid and the latent heat of the three PCMs are experimentally measured and critically discussed.

A numerical analysis is performed in order to compare the different heat transfer characteristics of the three PCM in a typical passive cooling system for electronic thermal management. In particular, the case study consists of an aluminium heat sink of 40x40 mm and 50 mm high filled up with the PCMs and then heated from the bottom to simulate a hot spot to be cooled. Different heat flow rates were imposed (10, 20 and 30 W); the results are presented in terms of melting time and maximum interfacial temperature.

2. Materials

A commercial paraffin wax with a melting point around 45 °C, RUBITHERM® RT45, was purchased from Rubitherm Technologies GmbH (Germany). Al_2O_3 nanopowder (Nanostructured & Amorphous Materials) has a nominal diameter of 10 nm, 99% purity and 160 m^2/g of Specific Surface Area (SSA). Nanoparticles were added to the liquid paraffin sample, to reach mass fractions of 0.5% and 1.0%, using the analytical balance Mettler PM6100 (FS 6100 g, ± 0.01 g). Corresponding volume fractions are 0.10% and 0.21%. The nano-PCMs were mechanically stirred for 20 min at 600 rpm and then they underwent to an ultrasonic procedure using a probe sonicator (Hielscher UP200s), in order to uniformly disperse nanoparticles. Temperature of the nano-PCMs tends to increase during the sonication process, therefore, in order to maintain the sample temperature under 70 °C, the samples were immersed in a thermostatic bath set at 50 °C, sonicated for at least 150 min with the temperature of the PCMs monitored every 30 min. Then, the samples were quickly solidified and used for thermal conductivity and latent heat measurements.

3. Thermal Conductivity

Thermal conductivity was measured using a ThermTest TPS2500S, an instrument based on the Hot Disk technique. It can measure thermal properties of a wide range of materials (solids, liquids, powders, pastes and foams). The hot disk technique uses the transient plane source mode and is applicable in the transitional regime. The plane sensor (Figure 1) consists of a Nickel thin metal strip wound in a double spiral on a plane support. The spiral is enclosed between two Kapton layers printed on it. Kapton material is a poor electrical conductor, but a good thermal conductor. This disk is then placed between two samples of the same material, as shown in figure 2, and works as both transient heater, to increase the temperature of the sample, and precise temperature reader, to record the time-dependent temperature increase. The heat transfer is assumed purely conductive and the thermal transport properties are determined by the instrument, considering the constant electric current supplied by the sensor and the temperature change as a function of time, by monitoring the total resistance of the hot disk sensor. The declared instrument uncertainty is 5%; however, it has been calibrated with a standard stainless steel sample showing a deviation lower than 1.5%. Pure and nano-PCMs were measured at around 24 °C.

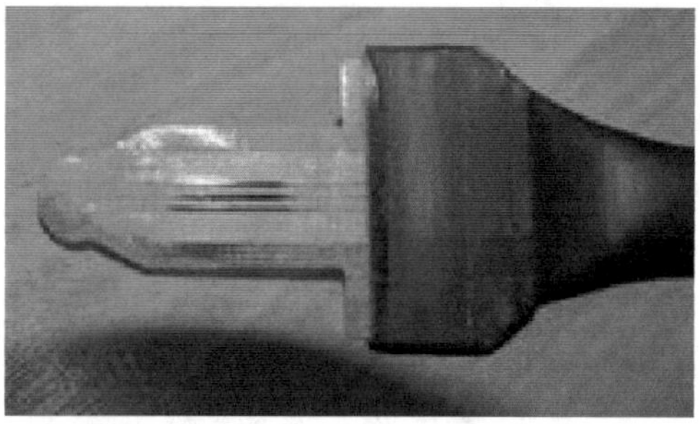

Figure 1. Hot disk sensor.

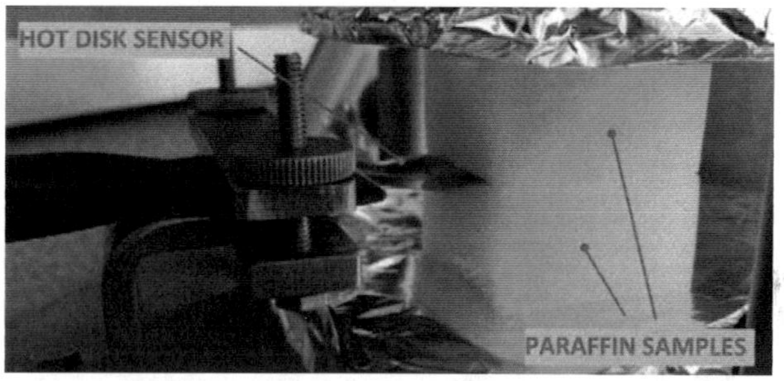

Figure 2. Thermal conductivity setup. The sensor is placed between two paraffin samples.

4. Specific Heat

The measurement of specific heat was performed using a Differential Scanning Calorimeter (DSC), Setaram Instrumentation (Figure 3). The DSC measures specific heat capacity by heating a sample and by measuring the temperature difference between the sample and a reference sample.

Three measurements must be repeated with two vessels, under an identical programmed variation of temperature, between an initial and final value, and with a predefined velocity.

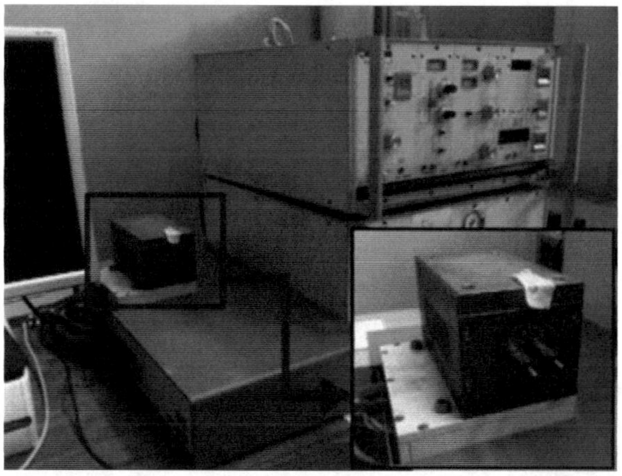

Figure 3. Differential scanning calorimeter, detailed calorimetric cells.

The first run was performed with the two empty vessels and the baseline heat flux was obtained; the second run was performed with one empty vessel and a mass of reference sample in the measuring vessel, providing the reference heat flux; the third run, which imposes the heat flux into the sample, was carried out with one empty vessel and a mass of the unknown sample in the measuring vessel. Sapphire is the reference sample and the instrument uncertainty is 2%.

5. Experimental Results

The values of the thermal conductivity of solid PCM and nano-PCMs were measured at 24 °C, *i.e.* at temperature more than 20 K lower than the reference melting temperature of the base paraffin wax RT45. As listed in Table 1, the experimental value of the thermal conductivity of RT45 was 0.2415 ± 1.5% W/mK, while the measurements carried on the nano-PCM paraffin waxes gave 0.2506 ± 1.5% and 0.2512 ± 1.5% W/mK, at 0.5 and 1.0 wt%, respectively. As an outcome, the addition of Al_2O_3 nanoparticles, up to 1 wt%, seems to lead to an overall enhancement of around 4%.

Table 1. Values of the thermal conductivity measurements.

Type	λ (W/mK)	**Variation** (%)
RT45	0.2415 ± 1.5%	-
RT45 Al_2O_3 0.5wt%	0.2506 ± 1.5%	+3.8
RT45 Al_2O_3 1.0wt%	0.2512 ± 1.5%	+4.0

Latent heat of PCMs and nano-PCMs was measured imposing a programmed temperature variation, between 27 °C and 87 °C, and with a predefined velocity of 30 °C/h. Figure 4 compares the response signals (*i.e.* specific heat) of the three paraffin waxes as a function of temperature. The peaks represent the solid-liquid phase change. At a glance, it clearly appears that the nano-PCM paraffin waxes exhibit higher peaks as compared to the base paraffin wax. The values of the latent heat of fusion of the investigated materials were determined by applying a numerical integration of the recorded curve.

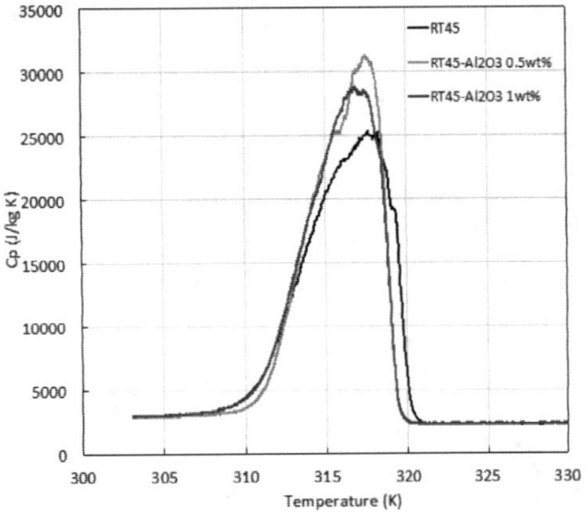

Figure 4. Specific heat capacity of paraffin waxes RT45 and its nano-PCMs, as a function of temperature.

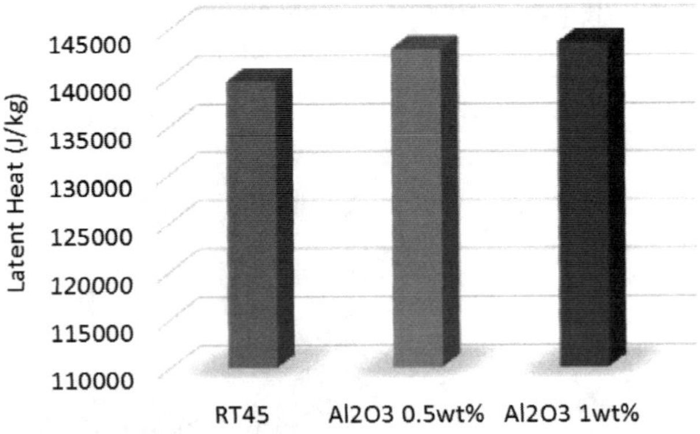

Figure 5. Phase change latent heat of the three studied materials.

In the case of nano-PCMs, only the paraffin wax undergoes a phase change, while the Al_2O_3 nanoparticles remain solid. Thus, in the presence of nanoparticles before integrating the peak area, the contribution of the Al_2O_3 on the measured specific heat capacity ($c_{p,tot}$) was accounted and subtracted from the overall measured one.

Specifically, knowing the specific heat of Al_2O_3 ($c_{p,np}$ = 880 J/kgK) and the mass fraction (ω), specific heat capacity of the only paraffin wax part ($c_{p,wax}$) was calculated from the total measured specific heat capacity ($c_{p,tot}$) as:

$$c_{p,wax} = \frac{c_{p,tot} - \omega c_{p,np}}{1-\omega} \quad (1)$$

The results of the numerical integration are presented in figure 5. It was observed that the latent heat slightly increases when Al_2O_3 nanoparticles are added: 2.3% and 2.8% for 0.5 wt% and 1.0 wt%, respectively. Those enhancements are lower than ones obtained for example by Shaikh *et al.* [14], which measured a maximum increase of around 13% in the fusion heat of paraffin waxes with single wall carbon nanotubes at 1 vol.% and Al_2O_3 nanoparticles at 1.0 and 1.5 wt%, respectively. Moreover, in a previous work (Colla *et al.* [15]), authors measured enhancements in the latent heat of 7.5 and 13.7% with a nanoparticles mass concentration of 1.0% in different paraffin waxes. Assuming that the intermolecular attraction in the paraffin wax/ Al_2O_3 dispersion explains the enhancement of the latent energy, as proposed by Shaikh *et al.* [14] according to the Lennard-Jones theory of intermolecular forces, different paraffin waxes have different surface interaction with the Al_2O_3 nanoparticles. Therefore, the change in latent heat of PCM due to the inclusion of nanoparticles strongly depends on the type of the PCM and the nanoparticles.

6. Physical and Numerical Models

The numerical analysis simulates an electronic passive cooling device, in particular an aluminium box filled up with pure paraffin wax or nano-PCM with 0.5% and 1.0% weight of nanoparticles. The geometry is a rectangular box with height equal to 50 mm and base equal to 40 mm. A two dimensional domain is considered, as depicted in figure. 6.

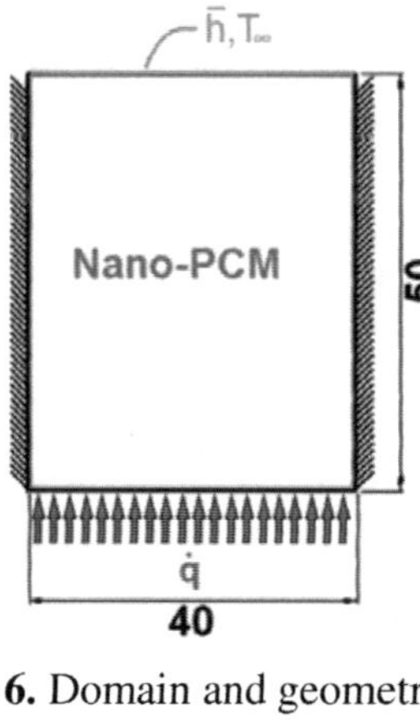

Figure 6. Domain and geometrical size.

The boundary conditions are given by an assigned heat flux at the bottom, to simulate the heat transfer dissipation of an electronic device, and by convective heat transfer on the top, while the sides are adiabatic. The heat flux is related to the heat transfer rate, equal to 10 W, 20 W and 30 W, *i.e.* typical values in an electronic device. The melting of paraffin wax is modelled with the enthalpy-porosity method [16], given that paraffin wax does not melt at a single temperature value, but in a temperature range. In this temperature range, solid and liquid phases coexist, and the "liquid fraction" is defined as a parameter indicating the liquid fraction in a simulation model cell. This parameter is expressed with β and its value changes from 0, when the domain is completely solid, to 1 when it is totally liquid:

$$\beta = 0 \quad for \quad T < T_{SOLIDUS}$$
$$\beta = \frac{T - T_{SOLIDUS}}{T_{SOLIDUS} - T_{LIQUIDUS}} \quad for \quad T_{SOLIDUS} < T < T_{LIQUIDUS} \tag{2}$$
$$\beta = 1 \quad for \quad T > T_{LIQUIDUS}$$

where T is the local temperature of the cell, $T_{liquidus}$ is the temperature upper which the domain is completely liquid and $T_{solidus}$ is the temperature below which it is completely solid. The mixed solid-liquid zone exists in a temperature range between $T_{liquidus}$ and $T_{solidus}$. The gravitational acceleration is taken into account along the y-axis and the Boussinesq approximation is assumed in order to take into account the buoyancy force due to natural convection. The solid part of the domain is modelled by adding a source term in the momentum equation. The single-phase model to simulate the nano-PCM in liquid phase is assumed. In the following, the mass fraction of the nano-PCM is ψ and it indicates the ratio between the mass of nanoparticles and the total mass of the paraffin wax and nanoparticles. The properties of the materials used for the simulations are listed in the table 2.

The specific heat, the thermal conductivity and the latent heat are calculated from the experimental results, while the following properties are calculated.

The density is evaluated by:

$$\rho_{NANOPCM} = (1-\psi)\rho_{PCM} + \psi\rho_{PARTICLES} \tag{3}$$

where $\rho_{NANOPCM}$ is the density of the nano-PCM, ρ_{PCM} is the density of pure paraffin wax, $\rho_{PARTICLES}$ is the density of the nanoparticles (aluminum oxide) and ψ is the mass fraction of the nanoparticles.

The values of specific heat and thermal conductivity are obtained from the experimental results for the liquid phase while for the solid phase it used the value provided by the supplier.

The viscosity is calculated from the Maxwell equation:

$$\mu_{NANOPCM} = \frac{\mu_{PCM}}{(1-\psi)^{2.5}} \tag{4}$$

where μ_{PCM} is the viscosity of the pure paraffin wax, $\mu_{NANOPCM}$ is the viscosity of the nano-PCM.

Table 2. Thermal properties of the materials, *i.e.* pure paraffin wax, nano-PCM at 0.5%wt and nano-PCM at 1.0%wt. [a]) from supplier data sheet, [b]) calculated, [c]) measured.

Properties		Pure PCM	Nano-PCM 0,5%wt	Nano-PCM 1%wt
Density [kg/m^3]	Solid	880 [a]	895 [b]	911 [b]
	Liquid	770 [a]	786 [b]	802 [b]
Specific heat [J/kg·K]	Solid	3028 ± 2% [c]	2943 ± 2% [c]	3032 ± 2% [c]
	Liquid	2333 ± 2% [c]	2322 ± 2% [c]	2376 ± 2% [c]
Thermal Conductivity [W/m·K]	Solid	0.241 ± 1.5% [c]	0.251 ± 1.5% [c]	0.251 ± 1.5% [c]
	Liquid	0.2 [a]	0.203 [b]	0.206 [b]
Viscosity [kg/m·s]		0.0256 [b]	0.0259 [b]	0.0263 [b]
Thermal expansion factor [1/K]		$1.25\cdot10^{-4}$ [a]	$1.22\cdot10^{-4}$ [b]	$1.19\cdot10^{-4}$ [b]
Latent Heat [J/kg]		$1.397\cdot10^5$ ± 2% [c]	$1430\cdot10^5$ ± 2% [c]	$1.436\cdot10^5$ ± 2% [c]
$T_{solidus}$ [K]		308	308	308
$T_{liquidus}$ [K]		322	322	322

The thermal expansion coefficient is calculated from [17]:

$$\beta_{NANOPCM} = \frac{(1-\psi)(\rho\beta)_{PCM} + \psi(\rho\beta)_{PARTICLES}}{\rho_{NANOPCM}} \tag{5}$$

The latent heat is obtained from experimental data. It is worth underlying that with the increasing of mass fraction the latent heat of fusion tends to increase. The thermal conductivity in liquid phase is calculated from the Maxwell equation [17].

The governing equations of the model are:

$$\nabla \cdot (\rho \vec{V}) = 0 \tag{6}$$

$$\rho\left(\frac{\partial \vec{V}}{\partial t} + \left(\nabla \cdot \vec{V}\right)\vec{V}\right) = \mu\left(\nabla^2 \vec{V}\right) - \nabla \vec{p} + \vec{S} \tag{7}$$

$$\frac{\partial(\rho h)}{\partial t} + \nabla \cdot \left(\rho \vec{V} h\right) + L\left[\frac{\partial(\rho\beta)}{\partial t} + \nabla \cdot \left(\rho \vec{V} \beta\right)\right] = k\left(\nabla^2 T\right) \tag{8}$$

where $\vec{V}$ is the velocity vector of the material in the liquid phase, μ is the viscosity of the material, p is the pressure and vector $\vec{S}$ is a source term given as [18]:

$$\vec{S} = \frac{(1-\beta)^2}{\left(\beta^3 + \pi\right)^3} A_{mush} \vec{V} - \rho g \gamma \left(T - T_0\right) \hat{y} \tag{9}$$

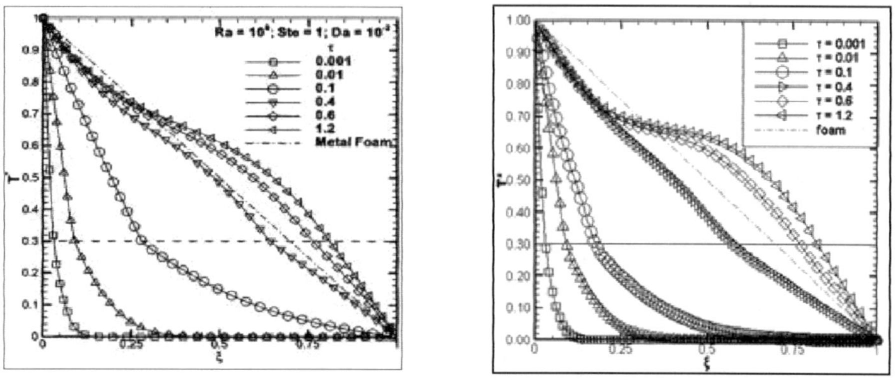

Figure 7 Validation test: a) data from Krishnan *et al.* [21] b) data from present numerical model.

The first term models the presence of the solid part in the mixed (liquid and solid) region, where π is a small number (0.001) to avoid division by zero [18] and A_{mush} is the mushy zone constant which describes the kinetic process in the mushy zone [19]. In these simulations, the value of mushy zone constant is set to 10^7. The second term models the natural convection in the liquid phase of the material according to the Boussinesq approximation. The energy equation uses the enthalpy of the material according to enthalpy-porosity method [20].

The employed grid is a set of quadrilateral control volumes with a coarse grid in the centre and fine grid near the boundary of the domain. The test carried out for the mesh independence solution consists of four different grids: 20 x 25, 40 x 50, 60 x 75, and 80 x 100. The mesh 60 x 75 represents the best trade-off for the present study, as good compromise between the accuracy and the computational time. The validation of the model is made by comparison with the work of Krishnan *et al.* [21]. The same characteristics are employed and then compared. The results are shown in figure 7.

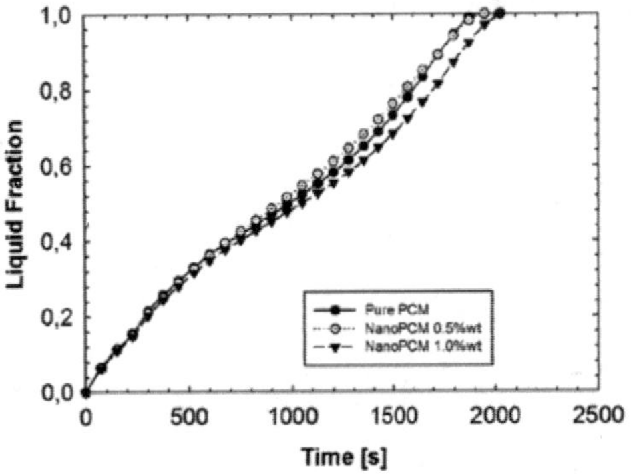

a)

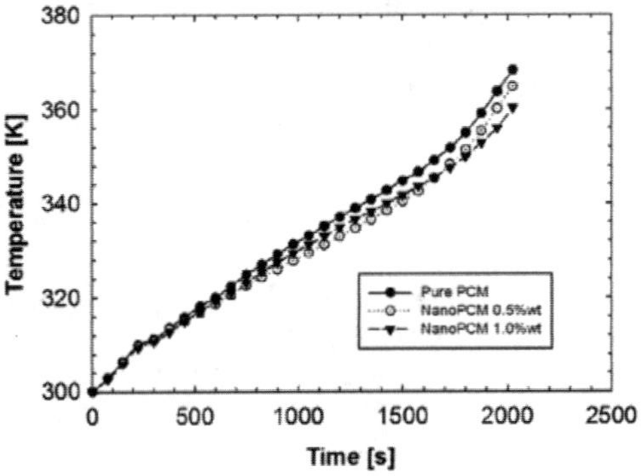

a)

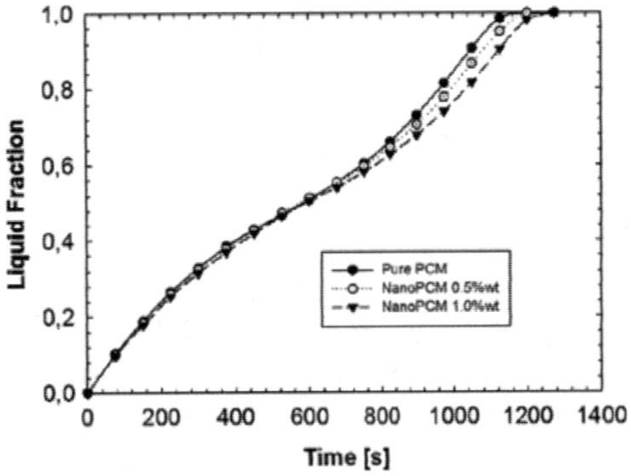

b)

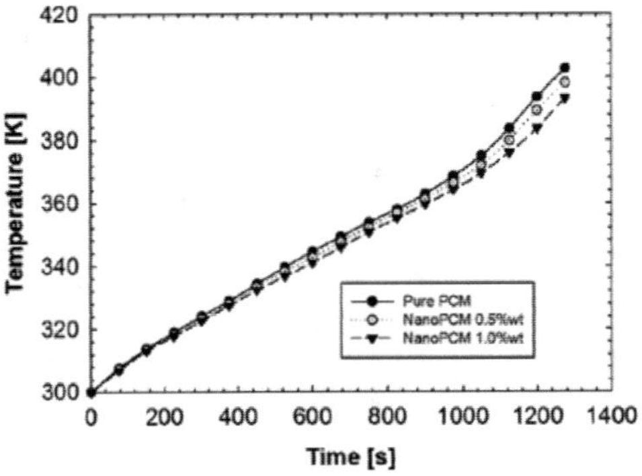

b)

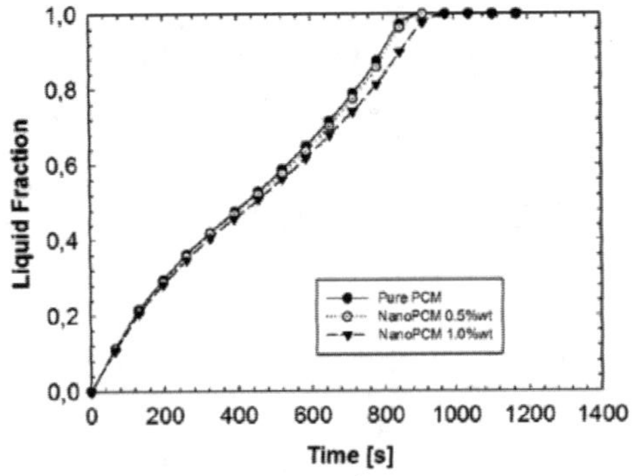

c)

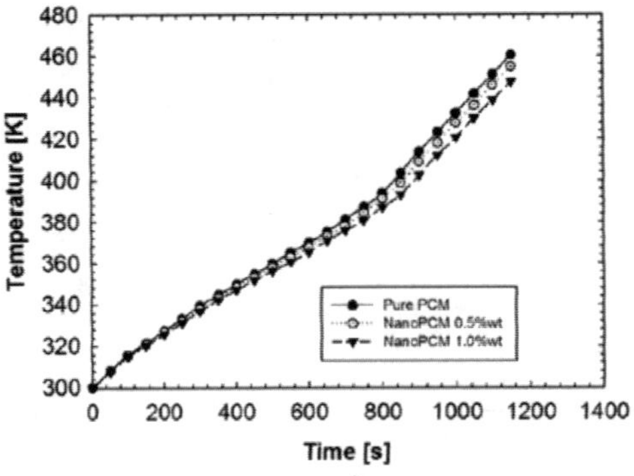

c)

Figure 8. Mean values of liquid fraction vs time, for heat flux equal to: a) $6.25 \cdot 10^3$ W/m^2 (10 W) b) $12.5 \cdot 10^3$ W/m^2 (20 W), c) $18.75 \cdot 10^3$ W/m^2 (30 W).

Figure 9 Mean values of temperature vs time, for heat flux equal to: a) $6.25 \cdot 10^3$ W/m^2 (10 W) b) $12.5 \cdot 10^3$ W/m^2 (20 W), c) $18.75 \cdot 10^3$ W/m^2 (30 W).

7. Numerical Results

The numerical results are shown in terms of average liquid fraction, average temperature of the whole domain, melting time, and the max interfacial temperature at different heat flow rates (10, 20 and 30 W) and mass fractions (pure material, 0.5%wt and 1.0%wt). The maximum interfacial temperature corresponds at the temperature of the bottom taken at time of completed melting.

In figure 8, there are three plots on the average liquid fraction for pure material, 0.5%wt and 1.0%wt concentration, as functions of time, and a), b) and c) refer to 10 W, 20 W and 30 W of imposed power, respectively. When the concentration of nanoparticles increases it seems that the liquid fraction tends to rise more slowly. In fact, the liquid fraction for pure PCM, respect to liquid fraction for the 0.5%wt or 1.0%wt concentrations, moves up more quickly. The only exception happens with a power supplied of 10 W, where the liquid fraction of the material with 0.5%wt of nanoparticles grows faster. This behaviour can be explained by the smaller value of latent heat at decreasing the concentration of nanoparticles into PCM. Table 2 reports the latent heat of the materials that slightly increases as the concentration of nanoparticles is increased.

In figure 9 there are three plots about the average temperature of the pure material, and of the PCM with 0.5%wt and 1.0%wt nanoparticles concentration, as functions of time, and a), b) and c) refer to 10 W, 20 W and 30 W imposed power, respectively.

The temperature path can be divided in three ranges: in the first range the temperature sharply increases in linear way. In the second stage, longer than the first, temperature follows the same trend, even if increasing less suddenly. This stage is important, because the melting occurs in a large part of its domain, leading to a slow increase of the average temperature. In the third range, the melting of whole domain is almost completed and the temperature returns to quickly rise. Table 3 reports the melting time and the corresponding maximum interfacial temperature for the three supplied powers.

Table 3. Melting time and maximum interfacial temperature.

POWER SUPPLIED	10 W		20 W		30 W	
Material	Time [s]	Max interfacial temperature [K]	Time [s]	Max interfacial temperature [K]	Time [s]	Max interfacial temperature [K]
Pure PCM	1906	413.7	1154	477.8	900	541.3
Nano-PCM at 0.5%wt	1935	412.6	1192	476.5	889	528.9
Nano-PCM at 1.0%wt	2020	413.4	1239	480.1	965	554.6

8. Conclusion

This paper presents an experimental and numerical study focused on the possible application of nano-PCM for electronics cooling. A commercial paraffin wax, having the solid-liquid phase transition at 45°C, was selected and used to develop two nano-PCM, by seeding Al_2O_3 nanoparticles at 0.5% and 1.0%wt, respectively. The three PCMs were then experimentally characterized in order to measure their solid thermal conductivity, solid and liquid heat capacity, and latent heat of fusion. The results show that the nanoparticles lead to a very low increase of both the thermal conductivity and latent heat. Furthermore, a numerical analysis was conducted to understand the effects of the nanoparticles during the solid-liquid phase change process in a passive electronics cooling device. In particular, a rectangular box 50 mm high, with a base equal to 40 mm, filled up with the three PCMs, was simulated when heated from the bottom by imposing 10, 20 and 30 W, respectively. The results show that the nano-PCMs are able to delay the melting process with a globally lower mean temperature.

Additional experimental, theoretical, and numerical work is surely needed to understand the heat transfer mechanisms on the basis of the performance of the paraffin waxes, in order to develop rigorous procedure for the nano-PCMs development and characterization.

9. Acknowledgment

The authors thank Mrs. Antonella Barizza and Dr. Paolo Bison for their fundamental help.

10. References

[1] Mancin S, Diani A, Doretti L, Hooman K and Rossetto L 2015 *Int. J. Therm. Sc.* **90** 79-89.
[2] Zalba B, Marìn JM, Cabeza LF and Mehling H 2003 *Appl. Therm. Eng.* **23** 251–283.
[3] Sharma A, Tyagi VV, Chen CR and Buddhi D 2009 *Ren. Sust. Energy Reviews* **13** 318–345.
[4] Agyenim F, Hewitt N, Eames P and Smyth M 2010 *Ren. Sust. Energy Reviews* **14** 615–628.
[5] Jegadheeswaran S and Pohekar SD 2009 *Ren. Sust. Energy Reviews* **13** 2225–2244.
[6] Fan L and Khodadadi JM 2011 *Ren. Sust. Energy Reviews*, **15** (2011), 24–46.
[7] Khodadadi JM and Hosseinizadeh SF 2007 *Int. Commun. Heat Mass Transfer* **34** (5) 534–543.
[8] Shin D and Banerjee D 2011 J. *Heat Transfer* **133** (2) 024501.
[9] Chieruzzi M, Cerritelli GF, Miliozzi A and Kenny JM 2013 *Nanoscale Res. Lett.* **8** (1) 448.
[10] Zhichao L, Qiang Z and Gaohui W 2015, *Int. J. Heat Mass Transfer* **80** 653–659.
[11] Jiang X, Luo R, Peng F, Fang Y, Akiyama T and Wang S 2015 *Appl. Energy* **137** 731–737.
[12] Karkri M, Lachheb M, Nógellovác Z, Boh B, Sumiga B, AlMaadeed MA, Fethi A and Krupa I 2015 *Energy and Build.* **88** 144–152.
[13] He Q, Wang S, Tong M and Liu Y, 2012 *Energ. Convers. Manage.* **64** 199–205.
[14] Shaikh S, Lafdi K and Hallinan 2008 *J. Appl. Phys.* **103** 094302.
[15] Colla L, Fedele L, Mancin S, Danza L and Manca O 2015 *ASME-ATI-UIT Conference 17–20 May, 2015, Napoli, Italy.*
[16] Voller VR and Prakash C 1987 *Int. J. Heat Mass Transfer* **30** 1709-19.
[17] Khodadadi JM and Hosseinizadeh SF 2007 *Int. Comm. Heat Mass Transfer* **34** 534-543.
[18] Nithyanandam K and Pitchumani R 2014 *J. Heat Transfer* **136** (5), 051503.
[19] Al-abidi A, Bin Mat S, Sopian K, Sulaiman MY and Mohammed AT 2013 *Ren. Sust. Energy Reviews* **20** 353-363.
[20] Liu Z, Yao Y and Wu H 2013 *Appl. En.* **112** 1222–32.
[21] Krishnan S, Murthy JY and GarimellaSV 2005 *ASME J. Heat Transfer* **127** 995–1004.

33rd UIT (Italian Union of Thermo-fluid-dynamics) Heat Transfer Conference IOP Publishing
Journal of Physics: Conference Series **655** (2015) 012031 doi:10.1088/1742-6596/655/1/012031

An unsteady model for the simulation of the rapid depressurization of vessels containing two-phase mixtures in non-equilibrium conditions

R Ricci[1], V D'Alessandro[1], S Montelpare[2], L Binci[1] and A Zoppi[1]
[1] Dipartimento di Ingegneria Industriale e Scienze Matematiche
Università Politecnica delle Marche
Via Brecce Bianche, 60100 Ancona (AN), Italy
[2] Dipartimento di Ingegneria e Geologia
Università degli Studi "G. D'Annuzio" di Chieti–Pescara
Viale Pindaro 42, 65127, Pescara (PE), Italy

E-mail: v.dalessandro@univpm.it

Abstract.
This paper describes the development of a simulation tool for the rapid depressurization (blowdown) of vessels containing two-phase mixtures in non–equilibrium conditions. The model adopts the cubic equations of state for fluids mixtures with non–ideal behavior for booth the phases, *i.e.* vapor and liquid, and it is based on a split two fluids model considering internal heat and mass transfer processes, as well as heat transfer with the vessel wall and the external environment. In order to account the mass and energy exchanged between the gas and the liquid phase, in conditions away from the thermodynamic equilibrium, a partial phase equilibrium (PPE) type approach has been introduced.
In this paper the validation of the proposed model with two different literature test cases is addressed and the role of the Peneloux correction for the employed equation of state is also investigated.

1. Introduction

The rapid depressurization of pressure devices is a key phenomenon in the Oil & Gas industry since it can produce a significant heat transfer between the fluid and the vessel walls. In particular the wall temperature shows a dramatic decrease and, if it falls below the ductile–brittle transition temperature of the vessel material, there is a risk for vessel wall rupture. Hence a proper vessel design needs a reliable prediction of the minimum wall and fluid temperatures that can be reached during the blowdown process [1].
It is worth noting that in some cases, starting for example from gas alone inside the vessel, the temperature drop can lead to vapor phase condensation. Moreover several experimental tests showed a significant temperature difference between the phases during and for a period after depressurization. This condition is clearly related to the greater convective heat transfer coefficient for the liquid phase (compared with the vapor one). Consequently the assumption of thermodynamic equilibrium between phases is not appropriate. At this point it is easy to understand as the matter of the numerical models development for this class of problems is a challenging task due to the complexity of the physical phenomena inside the vessel.

Journal of Physics: Conference Series **655** (2015) 012031 doi:10.1088/1742-6596/655/1/012031

In the last years several numerical and experimental works with varying degrees of sophistication have been published. In particular we can find a first class of models based on simplified relations such as in [2, 3] which lead to gross over estimations. On the other hand a second group of models developed up to now, such as those proposed in [4, 5, 6, 7], is based on rigorous analytic procedures. At the time of this writing, the authors are aware only the work published by Haque in [4] as a reliable model able to take into account non–equilibrium conditions between the phases. This model provides results in good agreement with experimental data reported in [5] however is not fully documented. For this reason the main goal of this research work is the development of a reliable code for two–phase vessels blowdown. Our model treats the non–equilibrium conditions between the phases using a revisited version of the Partial–Phase Equilibrium (PPE) approach introduced in [8] and it is suitable for vessels containing one or two-phases fluids mixtures with non–ideal behavior. Peng–Robinson (PR), [9], and Soave–Redlich–Kwong (SRK), [10], equations of state (EoS) with Van der Waals mixing rules are adopted as thermodynamic model. Furthermore the Peneloux correction, [11], is also implemented for the considered EoS. This correction was introduced in order to correct liquid density predictions and it introduces a parameter influencing molar volumes and phase densities without influencing the phase equilibrium.
The code is devoted to the study of vertical and horizontal vessels considering only top vapor venting. The implemented model involves all relevant heat transfer mechanisms: internal convection, conduction through vessel wall, external convection, and it is based on a compositional approach for each phase.
The obtained results have been compared with literature experimental data, [5], as well as numerical results obtained from BLOWDOWN code reported in [1, 4, 5] showing a good agreement.

2. Governing equations

The model implemented in the code here developed is based on the assumption of non–spatial dependence of the thermodynamic quantities. In particular we use two different values of the temperature, one for the vapor, T_G, and one for the liquid, T_L, which represent the bulk temperature of each phase. For what concerns the pressure within the whole vessel it is considered uniform. This hypothesis is supported by evidence from the blowdown process occurs at a time scale which is generally much longer than the adjustment of the pressure within the vessel that occurring at the speed of sound.
Hence the proposed mathematical model is based on a system of ordinary differential equations, (ODEs), devised on the basis of mass and energy conservation laws, that can be written as:

$$\frac{d\mathbf{U}}{dt} = \mathbf{s}(\mathbf{U}) + \frac{\mathbf{D}}{\tau}, \tag{1}$$

where:

$$\mathbf{U} = \begin{bmatrix} n_L \\ n_G \\ n_L u_L \\ n_G u_G \end{bmatrix}, \quad \mathbf{s}(\mathbf{U}) = \begin{bmatrix} 0 \\ -\psi_{G,out} \\ q_{Lw}\mathrm{sgn}(n_L) - q_{LG} \\ q_{Gw} + q_{LG} - h_G\psi_{G,out} \end{bmatrix}, \quad \mathbf{D} = \begin{bmatrix} n_l - n_g \\ n_g - n_l \\ n_l u_l - n_g u_g \\ n_g u_g - u_l n_l \end{bmatrix}. \tag{2}$$

In eq. 1 and in eq. 2 $\mathbf{U}$ is the state vector where n_L and n_G represent the moles number of the bulk liquid and the bulk vapor phases respectively. On the other hand u_L and u_G represent the molar internal energy of each phase. Note that the internal energy is calculated as the sum of an ideal gas contribution and a residual correction depending on the non–ideal behavior:

$$u = u_{res} + \sum_{i=1}^{n_c} z_i u_i. \tag{3}$$

In eq. 3 the residual contribution u_{res} is computed on the basis of the adopted EoS, see [12], while the ideal gas contribution is found using heat capacity data applicable to gas at very low–pressure as in [13]. A similar approach is used for the enthalpy.
The entries of the source terms vector $\mathbf{s}(\mathbf{U})$ are preliminary identified in eq. 2. In particular q_{Lw} and q_{Gw} are the thermal power exchanged by convection between each phase and the vessel walls; q_{LG} is related to the heat transfer between the liquid and the vapor phases. The different heat fluxes are computed using the Newton law as follows:

$$q_{Lw} = h_{Lw} S_{Lw} \left(T_{Lw} - T_L\right), \tag{4}$$

$$q_{Gw} = h_{Gw} S_{Gw} \left(T_{Gw} - T_G\right), \tag{5}$$

$$q_{LG} = h_{LG} S_{LG} \left(T_L - T_G\right) \operatorname{sgn}\left(n_L\right). \tag{6}$$

In the previous equations S_{LG} is the area of the interface between the two phases, while S_{Lw} and S_{Gw} are the areas of the interface between the phases and the wall in contact with themselves.
Moreover the convective heat transfer coefficient between the liquid phase and vessel walls, h_{Lw}, is evaluated using the correlation suggested by Ford in [14]. The h_{Gw} coefficient, instead, is evaluated according to [15, 16] considering as reference length the vertical height of the cylinder for vertical vessels and the vessel inside diameter for horizontal axes cylinders. Lastly to calculate the convective heat transfer coefficient h_{LG}, it has been considered a plane interface between liquid and gas and the equations reported in [15] have been used.
Transport properties, *i.e.* dynamic viscosity and thermal conductivity, are computed using the corresponding states model for both the vapor and liquid mixtures, [17]. The methane transport properties are used as the reference state for the previous models and they are obtained as in [18].
The vapor phase discharge rate, $\psi_{G,out}$, appearing in eq. 2 is calculated as:

$$\psi_{G,out} = \frac{C_D \psi_{is}}{M_W^G} \tag{7}$$

where C_D is the vapor discharge coefficient fixed to a default value of 0.84, M_W^G is molar weight of the vapor while ψ_{is} is mass flow rate of an isoentropic flow through a nozzle, [19].

2.1. Partial Phase Equilibrium

The entries in "daughters phases" vector $\mathbf{D}$ in eq. 1 are evaluated using a PPE type approach which allows to take into account the non–equilibrium between the phases during the blowdown process. More in depth during the blowdown the fluid in the vessel is divided in two different zones: zone G (vapor phase) and zone L (liquid phase). In order to consider the mass and energy exchanged between the liquid and vapor phases, two new phases "g" and "l", called bubbles and droplets, respectively, are introduced. The bubbles derive from the liquid (L) vaporization while the droplets from the vapor (G) condensation. Therefore, it is assumed that the new daughter phase "g" is in phase equilibrium with the parent liquid (bulk) "L" likewise the new daughter phase "l" is in phase equilibrium with the parent gas (bulk) "G". The PPE approach here proposed fixes the instantaneous equilibrium within each partial sub–system, keeping the

initial condition of non–equilibrium between the bulk phases; in other words, it is assumed that the gas formed by vaporization (vapor) from the liquid remains instantly in equilibrium with its "parent phase" (the liquid), before mixing with the gas bulk and homogenizing with it in a time–scale τ. The same happens for the liquid. Hence the PPE approach allows to use the usual equilibrium thermodynamics, for the parent phases handling, without getting involved in the complex thermodynamics of the non–equilibrium phase change. This is certainly the key advantage of this approach.
Therefore the PPE approach allows to compute the number of moles of the droplet daughter phase, n_l, and the number of moles of vapor daughter phase, n_g, as wells as the related internal energies u_l and u_g, appearing in the vector **D**, flashing in each time–step the parents phases.
The code here developed uses, for the sake of computing the entries in **D**, a Vapor-Liquid Equilibria (VLE) solver for two-phase fluids mixtures. A stability analysis method is adopted, according to [20], to get better initial equilibrium coefficients, K_i for the VLE algorithm. The aim of a stability analysis is to check the thermodynamic stability of a phase. In other words if a mixture at given composition has a lower energy remaining in single phase, hence stable, or if the mixture Gibbs free–energy decreases by splitting in two or more phases, hence unstable. The modified tangent plane distance function can be safely used for this purpose and all the details about this procedure can be found in [20].
Starting from the initial estimate of the equilibrium coefficients and the feed molar composition z_i, the liquid and the vapor phases composition, x_i and y_i respectively, are generated for each parent phase introducing the overall vapor fraction β (β_G will be used in the following for vapor parent phase while β_L for liquid parent phase) according to eq. 8.

$$x_i = \frac{z_i}{1-\beta+\beta K_i}, \qquad y_i = \frac{K_i z_i}{1-\beta+\beta K_i}, \tag{8}$$

where the overall vapor fraction for each phase is obtained solving the Rachford–Rice equation:

$$g(\beta) = \sum_{i=1}^{n_c} z_i \frac{K_i - 1}{1-\beta+\beta K_i} \tag{9}$$

with $K_i \doteq y_i/x_i$.
In the VLE solver if the imbalance between the phases fugacity is not less than 10^{-9} then the K_i coefficients for G and L are updated for the iteration $(p+1)$ as follow:

$$K_i^{(p+1)} = K_i^{(p)} \left(\frac{x_i \varphi_i(\mathbf{x})}{y_i \varphi_i(\mathbf{y})}\right)^{(p)}, \tag{10}$$

where the adopted EoS is invoked to get the fugacity coefficients, φ_i. Moreover a General Dominant Eigenvalue Method (GDEM), reported in [21], was also implemented to accelerate successive substitutions method (SSM) iterations of eq. 10.
We have to remark that during the iterations the feed can be unstable even if the Gibbs free–energy increases. In particular $\Delta G/RT > 0$ if either tpd_x or tpd_y are negative:

$$\frac{\Delta G}{RT} = (1-\beta)\underbrace{\sum_{i=1}^{n_c} x_i \ln\left(\frac{x_i\varphi_i(\mathbf{x})}{z_i\varphi_i(\mathbf{z})}\right)}_{\mathrm{tpd}_x} + \beta\underbrace{\sum_{i=1}^{n_c} y_i \ln\left(\frac{y_i\varphi_i(\mathbf{y})}{z_i\varphi_i(\mathbf{z})}\right)}_{\mathrm{tpd}_y}. \tag{11}$$

In our approach the updating of equilibrium coefficient is related to Gibbs free–energy behavior; in other words, if we assume for instance $\mathrm{tpd}_y < 0$, we select K_i for the subsequent iteration as:

$$\ln K_i = \ln\varphi_i(\mathbf{z}) - \ln\varphi_i(\mathbf{y}) \tag{12}$$

otherwise if $\text{tpd}_x < 0$ we use:

$$\ln K_i = \ln\varphi_i(\mathbf{x}) - \ln\varphi_i(\mathbf{z}). \quad (13)$$

The notation used up to now for the feed composition, and for its splitted phases, is obtained from the standards adopted in the Phase–Equilibria community. However in our approach we have two different feed compositions, corresponding to liquid and vapor phase, at the beginning of the specific time–step. Thus the notation is extended using a subscript which refers to a specific parent phase. Therefore the initial value for z_i is obtained from the initial conditions while in each time–step $z_{i,G}$ and $z_{i,L}$ are obtained and updated using the PPE approach as follows.
In particular in each time–step (n) we perform a VLE computation for both the parent phases (L and G) in order to compute n_l and n_g according to the following equations (obtained form eq. 8):

$$n_{l,i} = n_G^{(n)}\left(\frac{z_{i,G}}{1-\beta_G+\beta_G K_{i,G}}\right)^{(n)} \to n_l = \sum_{i=1}^{n_c} n_{l,i}, \quad (14)$$

$$n_{g,i} = n_L^{(n)}\left(\frac{K_{i,L} z_{i,L}}{1-\beta_L+\beta_L K_{i,L}}\right)^{(n)} \to n_g = \sum_{i=1}^{n_c} n_{g,i}. \quad (15)$$

Lastly u_l and u_g are computed invoking the EoS as in eq. 3.

2.2. Wall Temperature

In order to get T_{Lw} and T_{Gw} (the temperature of the inside vessel wall in contact with the liquid phase and the vapor phase, respectively) the 1–D Fourier equation across the vessel wall has been solved with the following boundary conditions:

$$-\lambda\left.\frac{\partial T}{\partial x}\right|_{x=0} = h_{fw}(T_{fw} - T_f) \qquad -\lambda\left.\frac{\partial T}{\partial x}\right|_{x=L_w} = h_{wa}(T_{air} - T_{wf}). \quad (16)$$

where the subscripts f refers to a generic parent phase (L or G) and L_w is the vessel wall thickness. It worth mentioning that we solve the Fourier equation, for the walls, neglecting their curvature (generally cylindrical), as we generally assume the radius of curvature to be very large, compared to the thickness. Hence we assume in first approximation as negligible the heat flux along the wall surface, compared to the heat flux across its thickness.
The Fourier equation is solved using a finite difference method with a second–order accurate central scheme for the spatial derivative while a backward Euler scheme is used for time derivative. Two ghost nodes have been introduced at the domain boundaries in order to use second–order accurate schemes also for the boundary conditions. Moreover the linear systems deriving from the implicit approach used for time integration have been solved with a standard iterative Gauss–Siedel method, [22].
The heat transfer between vessel and surrounding environment has been also considered with the coefficient h_{wa} in eq. 16. It clearly depends on the external conditions and both the natural convection and the forced convection cases have been taken into account as suggested in [16, 23].

3. Numerical Solution

The governing equations, eq. 1, are solved using a 4th accurate explicit Adams-Bashfort scheme, [24], for $\mathbf{s}(\mathbf{U})$ if available three time–steps prior to the actual, otherwise lower order schemes, *i.e.* 1st, 2nd, or 3rd are used, eq. 17. Differently for the sake of stability a first order scheme is used for the vector $\mathbf{D}$:

$$
\begin{aligned}
\mathbf{U}^{(n+1)} &= \mathbf{U}^{(n)} + \Delta t\left[\frac{55}{24}\mathbf{s}^{(n)} - \frac{59}{24}\mathbf{s}^{(n-1)} + \frac{37}{24}\mathbf{s}^{(n-2)} - \frac{3}{8}\mathbf{s}^{(n-3)}\right] + \frac{\Delta t}{\tau}\mathbf{D}^{(n)} \\
\mathbf{U}^{(n+1)} &= \mathbf{U}^{(n)} + \Delta t\left[\frac{23}{12}\mathbf{s}^{(n)} - \frac{4}{3}\mathbf{s}^{(n-1)} + \frac{5}{12}\mathbf{s}^{(n-2)}\right] + \frac{\Delta t}{\tau}\mathbf{D}^{(n)} \\
\mathbf{U}^{(n+1)} &= \mathbf{U}^{(n)} + \Delta t\left[\frac{3}{2}\mathbf{s}^{(n)} - \frac{1}{2}\mathbf{s}^{(n-1)}\right] + \frac{\Delta t}{\tau}\mathbf{D}^{(n)} \\
\mathbf{U}^{(n+1)} &= \mathbf{U}^{(n)} + \Delta t\ \mathbf{s}^{(n)} + \frac{\Delta t}{\tau}\mathbf{D}^{(n)}.
\end{aligned}
\tag{17}
$$

We have to point out that in order to advance the solution in time, starting from the state vector $\mathbf{U}$, it is mandatory to compute the bulk temperatures, T_L and T_G, and the vessel pressure, p. For this purpose the following functions have to be defined:

$$
w_L = f\left(T_L, p_L, Z_L, \mathbf{z_L}\right) - u_L, \quad w_G = f\left(T_G, p_G, Z_G, \mathbf{z_G}\right) - u_G \tag{18}
$$

where Z_L and Z_G are the compressibility factors while $\mathbf{z_L}$ and $\mathbf{z_G}$ are vectors containing the molar compositions of each parent phase in a specific time–step. Hence the computation of T_L, T_G and p consists in the root finding of eq. 18 with the constraints:

$$
V_0 = n_L v_L + n_G v_G, \qquad p_L = p_G. \tag{19}
$$

In our code the solution of eq. 18, 19 has been achieved adopting a Newton method with Aitken acceleration, [22]:

$$
T_L^{(i+1)} = T_L^{(i)} - \omega_L m_L^{(i)} \frac{w_L^{(i)}}{w_L^{'(i)}}, \qquad T_G^{(i+1)} = T_G^{(i)} - \omega_G m_G^{(i)} \frac{w_G^{(i)}}{w_G^{'(i)}} \tag{20}
$$

with

$$
m_L^{(i)} = \frac{T_L^{(i-1)} - T_L^{(i-2)}}{2T_L^{(i-1)} - T_L^{(i)} - T_L^{(i-2)}}, \qquad m_G^{(i)} = \frac{T_G^{(i-1)} - T_G^{(i-2)}}{2T_G^{(i-1)} - T_G^{(i)} - T_G^{(i-2)}} \tag{21}
$$

note that Aitken acceleration can be introduced only when $i > 2$ hence at the beginning of the iterative algorithm a standard Newton method is used. Within each iteration (i) the pressure vessel, needed for the evaluation of w_L and w_G, is obtained by means of a further iterative loop. In particular starting from an initial guess for v_L the molar volume for vapor phase, v_G, is obtained from eq. 19 as:

$$
v_G^{(j)} = \frac{V_0}{n_G^{(n)}} - \frac{n_L^{(n)}}{n_G^{(n)}} v_L^{(j-1)} \tag{22}
$$

where V_0 is the vessel volume. Therefore $p_G = s\left(T_G^{(i)}, v_G^{(j)}\right)$ is computed introducing in the EoS eq. 19 and eq. 22; lastly the EoS in the Z–form is also invoked to get the compressibility factor of the liquid phase Z_L using the constraint reported in eq. 19:

$$
a_3\left(p_G^{(j)}, T_L^{(i)}\right) Z^3 + a_2\left(p_G^{(j)}, T_L^{(i)}\right) Z^2 + a_1\left(p_G^{(j)}, T_L^{(i)}\right) Z + a_0\left(p_G^{(j)}, T_L^{(i)}\right) = 0 \tag{23}
$$

Z_L is the minimum real root of eq. 23 and its coefficients clearly depend from the selected EoS. Thus the outer and inner loop are advanced until the quantities δ, eq. 24, and ϵ, eq. 25, are converged up to the machine precision.

$$\delta = \left| \frac{Z_L R T_L^{(i)}}{p_G^{(j)}} - v_L^{(j-1)} \right|, \tag{24}$$

$$\epsilon = \max \left(\left| \omega_L m_L^{(i)} \frac{w_L^{(i)}}{w_L^{'(i)}} \right|, \left| \omega_G m_G^{(i)} \frac{w_G^{(i)}}{w_G^{'(i)}} \right| \right). \tag{25}$$

The relaxation factors ω_L and ω_G in the following test cases are fixed to 0.1.

4. Results

In this section are presented numerical results obtained for two different test cases named S12 and HSE run3 respectively. In particular, experimental and numerical data available in [4, 5] are used as reference results for S12 problem, while numerical results available in [1] are used as reference solution for HSE run3.
It is really worth noting that time–scale τ related to the mixing between parents and daughters phases is fixed, for all the computations here reported, equal to the time–step size (its value for all the numerical tests here reported is 0.5 sec). In our computational tests stable solutions have been obtained only with this condition thus it is adopted in all the reported tests.

4.1. Case S12

The experiment S12 consists of the blowdown of a cylindrical vessel of 1.130 m diameter, 3.240 m length, 5.9 cm wall thickness and top choke of 1 cm diameter. The fluid used is a mixture of is of 66.5 mole % methane, 3.5 mole % ethane, 30.0 mole % propane. Experimental tests were performed by Haque et. al. and reported in [5]. Measured data are affected by some uncertainties such as: fluid composition and initial temperature. In particular Haque et. al. [5] provided nominal fluid compositions, meaning that, during the experiments, authors found also traces of other higher molecular weight hydrocarbons, in particular of butanes. Differently for what concerns fluid initial temperature in [5] a value ranging from 290 K to 305 K is reported in the text.

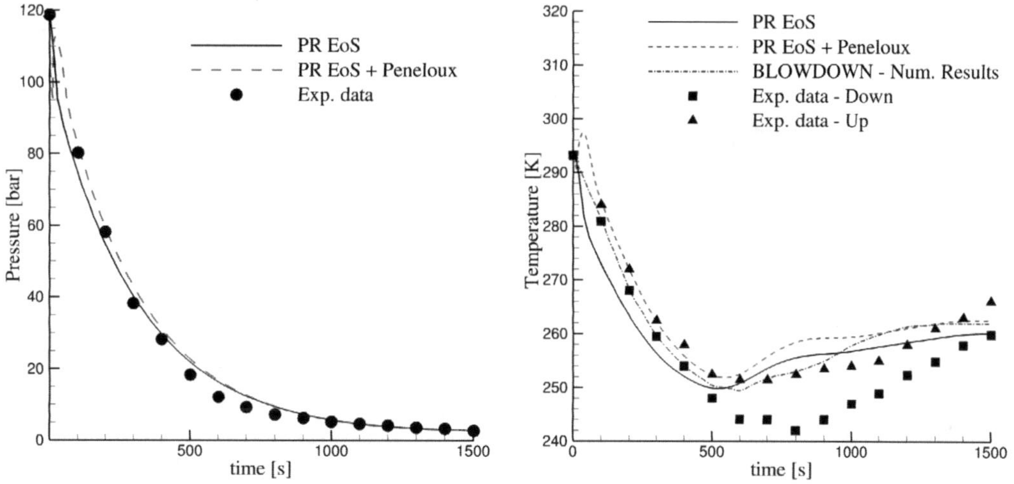

(a) Pressure

(b) Vapor phase temperature

Figure 1. S12 results

In the present work we have considered an initial temperature of 293.15 K as in Fig. 8 of the reference paper [5] and the initial fluid pressure is 118.5 bar. The vessel has been considered immersed in stagnant air at 293.15 K, in equilibrium with the internal temperature as in [5]. All the computations for this test case have been performed using PR EoS with and without Peneloux correction.
Fig. 1 and Fig. 2 show the comparisons between experimental data and numerical predictions. The obtained results for the pressure evolution in the vessel, Fig. 1(a), are in excellent agreement with the experimental data. Similarly the vapor temperature predictions, Fig. 1(b), are very close to the BLOWDOWN code results, [4]. The use of Peneloux correction improves the results in the initial blowdown stage, dominated by the expansion; in the second stage the heat flux from the wall makes the vapor temperature to climb back up. However in this case the use PR equation without Peneloux corrections seems to produce results in better agreement with experimental data.
Fig. 2(a) reports the time behavior of the liquid bulk temperature. The obtained results are compared to the BLOWDOWN code and experimental data reported in [5] showing globally a good agreement. More in depth our results tend to slightly overestimate the temperature while BLOWDOWN tends to underestimates the liquid temperature.

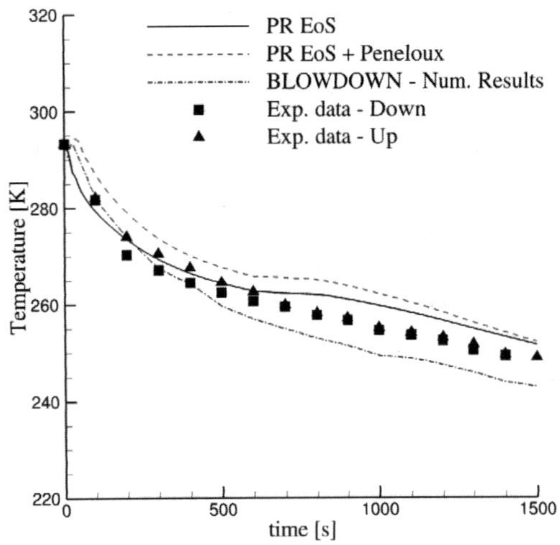

(a) Liquid phase temperature

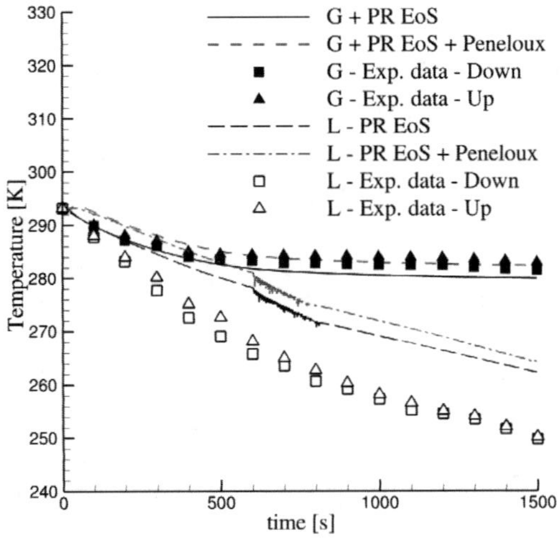

(b) Walls temperature

Figure 2. S12 results

The internal wall temperature in contact with the vapor phase is predicted very well by our code, Fig. 2(b). The use of Peneloux correction in this case seems to improve the prediction of the wall temperature at the end of the blowdown process. Instead, a reasonable agreement is observed for the liquid side which shows the same behavior of the experimental one. In this situation Peneloux volume shift produces a slight overestimation of the wall temperature.

4.2. Case HSE run3

The second test case here presented is related to numerical experiments performed by [1]. These computations were obtained on a horizontal vessel containing a mixture of hydrocarbons using BLOWFIRE code which is a fork of BLOWDOWN. The case RUN3 presented here consists of blowdown of a cylindrical vessel of 1.970 m diameter, 8.429 m length and top choke of 1.4 cm diameter with a wall thickness of 16 mm. The initial composition is of 1.0 mole % carbon dioxide,

25.0 mole % methane, 7.0 mole % ethane, 5.0 mole % propane, 4.0 mole % butane, 2.0 mole % pentane, 7.0 mole % hexane, 7.0 mole % octane, 15.0 mole % nonane and 27.0 mole % decane. As in the previous case the vessel has been considered immersed in air at 323 K and the same initial temperature of the fluid in the vessel is used. An initial molar vapor fraction of 0.83 has been considered in the performed computations while the initial fluid pressure is 22 bar. The computations have been performed using SRK EoS with and without Peneloux correction.
The results obtained using the code developed in this work are good agreement with BLOWFIRE results in terms of pressure time behavior, Fig. 3(a). In this case the adoption of Peneloux correction produces an overshoot in the pressure at beginning of the computations, while globally the pressure is slightly overestimated respect to BLOWFIRE.

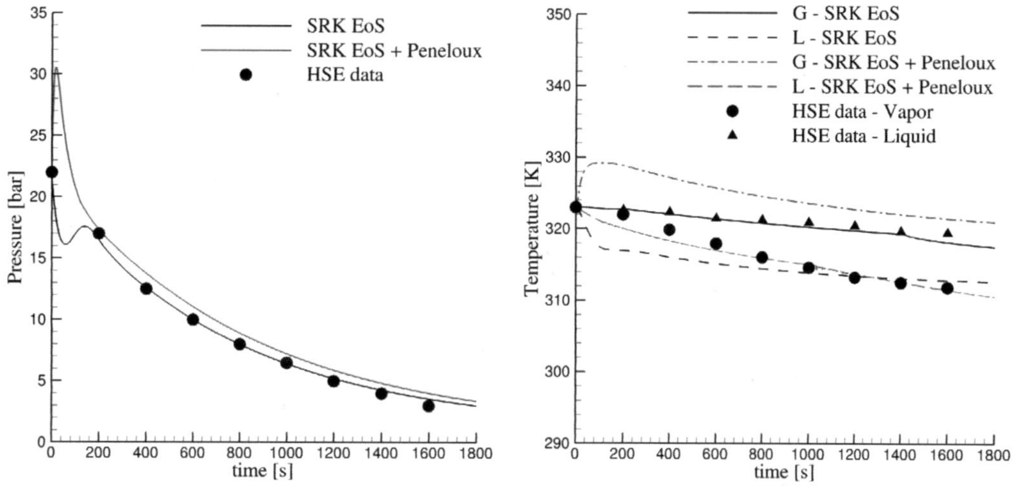

(a) Pressure

(b) Walls temperature

Figure 3. HSE run3 results

Fig. 3(b) shows the predicted temperature of the wall in contact with the vapor and the liquid phases, compared with Roberts et al. [1]. Using SRK EoS without Peneloux correction the agreement is excellent for the temperature trend of the wall in contact with liquid and rather good for the temperature of the wall in contact with vapor. The temperature of the internal wall in contact with liquid phase is perfectly predicted and the temperature of the internal vessel wall in contact with vapor is underestimated at the start of the blowdown. Differently the introduction of Peneloux volume shift produces an improvement in the prediction of the wall temperature in the vapor side while in the liquid side the temperature behavior is overestimated with a different trend in time behavior.

5. Conclusions

In this work a FORTRAN code for the simulation of blowdown process of pressure vessels containing two–phase fluid mixtures has been developed. The code takes into account non–equilibrium effects, *i.e.* temperature differences between phases, using the partial phase equilibrium type approach. The proposed model has been validated comparing its results with numerical and experimental data from literature and it has showed good agreement. In particular pressure–time profiles are predicted with very good accuracy, blowdown times are estimated in line with the experiments and finally temperature–time profiles are obtained with a good accuracy by our code.

Peng–Robinson and Soave–Redlich–Kwong equations of state have been tested on two different test cases. In this work the role of Peneloux correction has been also investigated evidencing its good performance is some cases. This is probably related to the better performance of the EoS, in the liquid density prediction, obtained by introducing Peneloux correction, [25]. Lastly considering the overall performance, the code here developed produces reasonably accurate results requiring a very limited CPU time.

References

[1] Roberts T A, Medenos S and Shirvill L 2000 Review of the Response of Pressurized Process Vessels and Equipment to Fire Attack Tech. Rep. 2000-051 HSE

[2] Reynolds W C and Kays W M 1958 *Trans ASME* **80** 1160–1168

[3] Montgomery G 1995 *Hydr Proc* **4** 85–88

[4] Haque M A, Richardson S M and Saville G 1992 *Trans. Inst Chem Eng Part B* **50** 1–9

[5] Haque M A, Richardson S M, Saville G, Chamberlain G and Shirvill L 1992 *Trans. Inst Chem Eng Part B* **70** 10–17

[6] Mahgerefteh H and Wong S M A 1999 *Comp Chem Eng* **23** 1309–1317

[7] Mahgerefteh H, Gboyega B, Falope O and Oke A O 2002 *AIChE J* **48** 401–410

[8] Speranza A and Terenzi A 2005 *App Ind Math in Italy–Ser Adv Math for App Sci* **69**

[9] Peng D Y and Robinson D B 1976 *Ind Eng Chem Fund* **15** 59–64

[10] Soave G 1972 *Chem Eng Sci* **27** 1197–1203

[11] Peneloux A and Rauzy E 1982 *Fluid Phase Equilibr* **8** 7–23

[12] Michelsen M and Mollerup J M 2004 *Thermodynamic Models: Fundamentals and Computational Aspects* (Tie-Line Publications)

[13] Reid R C, Prausnitz J M and Sherwood T K 1977 *The properties of gases and liquids* (McGraw-Hill)

[14] Ford P E 1955 *4th World Petroleum Congress* (Rome–Italy)

[15] Incropera F P and Witt D D 2007 *Fundamentals of Heat and Mass Transfer* (New York: John Wiley & Sons)

[16] Churchill S W and Chu H H S 1975 *Int J Heat Mass Trans* **18** 1323–1329

[17] Pedersen K S and Fredenslund A 1987 *Chem Eng Sci* **42** 182–186

[18] Hanlet H J M, McCarty R D and Haynes W M 1975 *Cryogenics* **15** 413–417

[19] Kundu P K, Cohen I M and Dowling D R 2011 *Fluid Mechanics* (Accademic Press)

[20] Michelsen M L 1982 *Fluid Phase Equilibr* **9** 1–19

[21] Crowe A M and Nishio M 1975 *AIChE J* **21** 528–533

[22] Quarteroni A, Sacco R and Saleri F 2007 *Numerical Mathematics* (Berlin: Springer–Verlag)

[23] Sparrow E M and Gregg J L 1956 *Trans ASME* **78** 1823–1829

[24] Hairer E, Norsett S P and Wanner G 1993 *Solving ordinary differential Equations I: Nonstiff problems* (Berlin: Springer–Verlag)

[25] Pedersen K and Christensen P L 2006 *Phase Behaviour of Petroleum Reservoir Fluids* (Taylor and Francis)

Vaporization inside a mini microfin tube: experimental results and modeling

A Diani, L Rossetto[1]
Dipartimento di Ingegneria Industriale, Università degli Studi di Padova, Via Venezia 1, 35131 Padova, Italy

E-mail: luisa.rossetto@unipd.it

Abstract. This paper proposes a comparison among the common R134a and the extremely low GWP refrigerant R1234yf during vaporization inside a mini microfin tube. This microfin tube has an internal diameter of 2.4 mm, it has 40 fins, with a fin height of 0.12 mm. Due to the high heat transfer coefficients shown by this tube, this technology can lead to a refrigerant charge reduction. Tests were run in the Heat Transfer in Micro Geometries Lab of the Dipartimento di Ingegneria Industriale of the Università di Padova. Mass velocities range between 375 and 940 kg m^{-2} s^{-1}, heat fluxes from 10 to 50 kW m^{-2}, vapour qualities from 0.10 to 0.99, at a saturation temperature of 30 °C. The comparison among the two fluids is proposed at the same operating conditions, in order to highlight the heat transfer and pressure drop differences among the two refrigerants. In addition, two correlations are proposed to estimate the heat transfer coefficient and frictional pressure drop during refrigerant flow boiling inside mini microfin tubes. These correlations well predict the experimental values, and thus they can be used as a useful tool to design evaporators based on these mini microfin tubes.

1. Introduction

HFCs are greenhouse gases widely used for refrigeration and air conditioning applications. Even though they have a zero ozone depletion potential, they have a non-negligible global warming potential. For instance, R134a, which is one of the most used HFCs, has a GWP of approximately 1400. HFCs must be gradually phased out in order to reduce the emissions of global warming gases. R1234ze(E) and R1234yf have been proposed as replacements of the HFC 134a. R1234ze(E) and R1234yf are hydro fluoro olefins, they have a zero ozone depletion potential and a GWP lower than 1 [1]. Together with the use of low GWP refrigerants, the reduction of greenhouse gases could be achieved with the use of efficient systems, which lead to a refrigerant charge reduction.
Small microfin tubes were initially studied for CO_2 applications, due to its high working pressure. Gao et al. [2] performed experiments on the flow boiling heat transfer of almost pure CO_2 and CO_2-oil mixtures in horizontal smooth and microfin tubes. The smooth tube is a stainless steel tube with an inside diameter of 3 mm, whereas the microfin tube is a copper tube with a mean inside diameter of 3.04 mm. Experiments were carried out at mass velocities from 190 to 1300 kg m^{-2} s^{-1}, heat fluxes from 5 to 30 kW m^{-2}, at a saturation temperature of 10 °C, with oil circulation ratio from <0.01% to 0.72% wt. In case of almost pure CO_2, the flow boiling mechanism was found to be dominated by nucleate boiling. Dang et al. [3] investigated the flow boiling heat transfer of CO_2 inside a small microfin tube with a mean inner diameter of 2.0 mm. Mass flux ranges between 360 and 720

[1] To whom any correspondence should be addressed.

kg m^{-2} s^{-1}, heat flux between 4.5 and 18 kW m^{-2}, at a saturation temperature of 15 °C. Experimental results indicated that the heat flux has a significant effect on the heat transfer coefficient, whereas the coefficient does not always increases with mass velocity: under certain conditions, the heat transfer coefficient at higher mass velocity was lower than that at lower mass velocities, indicating that the convective heat transfer has a suppression effect on nucleate boiling.

R1234yf has a normal boiling temperature approximately 3.15 K lower than that of R134a, thus it is a possible substitute of the common R134a, whose Global Warming Potential is about 1400. McLinden et al. [4] carried out a screening of the possible refrigerants, eliminating those toxic or unstable and focusing on those that presented critical temperatures between 300 K and 400 K; they analysed 62 refrigerants. Two of those were R1234ze(E) and R1234yf, which have recently been matter of research and investigation of the scientific community because they present values of low- GWP and normal boiling temperature close to that of R134a, the traditional fluid commonly used in refrigeration and air conditioning equipment.

Up today, few works experimentally investigated the heat transfer capabilities of these refrigerants during single and two-phase flow inside conventional and enhanced tubes. Grauso et al. [5] presented experimental results for local heat transfer coefficients, adiabatic frictional pressure gradients, and two-phase flow regimes with the low GWP R1234ze(E). The circular smooth tube has an inner diameter of 6 mm. Mass velocities were varied from 146 to 520 kg m^{-2} s^{-1}, heat fluxes from 5.0 to 20.4 kW m^{-2}, at a saturation temperature from -2.9 °C to 12.1 °C. Compared to R134a, R1234ze(E) revealed similar values of local heat transfer coefficients, showing the same trend with vapour quality, but it showed an earlier inception of the dry out phenomenon. The experimental adiabatic frictional pressure drop data of R1234ze(E) result slightly higher than those obtained for R134a at the same operating conditions.

Hossain et al. [6] experimentally measured the heat transfer coefficients during in-tube evaporation of R1234ze(E), R32, R410A, and a mixture of R1234ze(E)/R32 (55/45 mass %). The smooth tube has an inner diameter of 4.35 mm. The experiments have been carried out under the conditions of mass flux from 150 to 445 kg m^{-2} s^{-1}, the saturation temperatures are 5 and 10 °C, over the vapour quality range 0.0-1.0. It was found that the experimental heat transfer coefficients of R1234ze(E) were lower than those of R1234ze(E)/R32, R410A and R32 by 11%, 56%, and 83%, at about 0.48 vapour quality and 300 kg m^{-2} s^{-1}.

Really few works can be found in the open literature regarding low GWP refrigerants flow boiling inside small microfin tubes. Kondou et al. [7] experimentally studied the flow boiling of R1234ze(E), R32, and mixture of R1234ze(E)/R32 in a horizontal microfin tube of 5.21 mm inner diameter. The heat transfer coefficient and pressure drop were measured at a saturation temperature of 10 °C, heat fluxes of 10 and 15 kW m^{-2}, and mass velocities from 150 to 400 kg m^{-2} s^{-1}. The heat transfer coefficients of R1234ze(E) were lower than those of R32, and the heat transfer coefficient of the mixture R1234ze(E)/R32 was even lower than that of R1234ze(E).

One of the few works about R1234yf flow boiling inside microfin tubes is that of Han et al. [8], who experimentally investigated the boiling heat transfer characteristics of R1234yf inside a microfin tube with an outer diameter of 7 mm, at mass fluxes between 100 and 400 kg m^{-2} s^{-1}, heat fluxes of 4, 8 and 12 kW m^{-2}, saturation temperature of 5 °C and 15 °C, with four different oil concentrations: 0%, 1.5%, 3.0%, and 5.0%.

This paper explores the possible viable replacement of R134a with the low GWP R1234yf. Experimental results are given in terms of comparison between R134a and R1234yf heat transfer coefficient and frictional pressure drop. Furthermore, two empirical correlations are proposed and discussed for the estimation of the heat transfer coefficient and frictional pressure drop. The heat transfer model is an updated version of the model by Padovan et al. [9], and it is valid for vapour quality prior to the thermal crisis and for mass velocities from 200 to 940 kg m^{-2} s^{-1}. The model for the estimation of the frictional pressure gradient uses a liquid only two-phase multiplier, which is a modified version of the multiplier suggested by Cavallini et al. [10] for minichannels. These correlations well predict the experimental values, and thus they can be used as a useful tool to design evaporators based on these mini microfin tubes.

2. Experimental set up and test section

The experimental tests were carried out in a facility built at the Heat Transfer in Micro Geometries laboratory of the Dipartimento di Ingegneria Industriale of the University of Padova. This experimental facility permits pressure drop and either flow boiling or condensation measurements to be performed. It has a maximum working pressure of 30 bar, and the mass flow rate can be varied up to 72 kg h^{-1}. A schematic of the experimental set up is depicted in figure 1. The facility consists of three main loops: the refrigerant loop, the hot water loop, and the cold water loop. In the refrigerant loop, the fluid is pumped by means of a magnetic coupled gear pump. An oil free pump was chosen in order to avoid any possible oil contamination in the refrigerant loop. It is connected to an inverter, so that it is possible to vary the refrigerant mass flow rate by controlling an electronic display. The refrigerant is pumped through a Coriolis effect flow meter, which reads the refrigerant mass flow rate with an accuracy of ±0.10% of the reading, and then the fluid reaches an evaporator, at the exit of which the refrigerant is superheated. The evaporator is a brazed plate heat exchanger, where the refrigerant flows in counter current with hot water, which is supplied by the hot water loop. The hot water loop consists of an electric boiler, able to supply up to 5 kW: it is split into three electric resistances, two of which work in on-off mode, whereas the power of the third resistance can be modulated. A controlling valve to adjust the water flow rate, an expansion vessel to cope with water volume variations, and a magnetic flow meter with an accuracy of ±0.25% of the reading, complete the hot water loop. The water temperature difference across the heat exchanger is measured by means of a 4-junction T-type thermopile (accuracy ±0.03 K).

The superheated vapour is then partially condensed in a pre-condenser, which is a tube-in-tube heat exchanger, where the refrigerant flows in the inner tube, and cold water flows in the annulus. The cold water is supplied by the cold water loop, where a stabilized chiller controls the water temperature at the inlet of the pre-condenser. The cold water flow rate can be controlled by means of a valve and it is measured with a magnetic flowmeter (accuracy ±0.25% of the reading). The water temperature across the pre-condenser is measured by means of 4-junction T-type thermopile (accuracy ±0.03 K). It is possible to adjust the heat flow rate by independently controlling the water flow rate and the water temperature, in order to set the vapour quality at the inlet of the test section at the desired value.

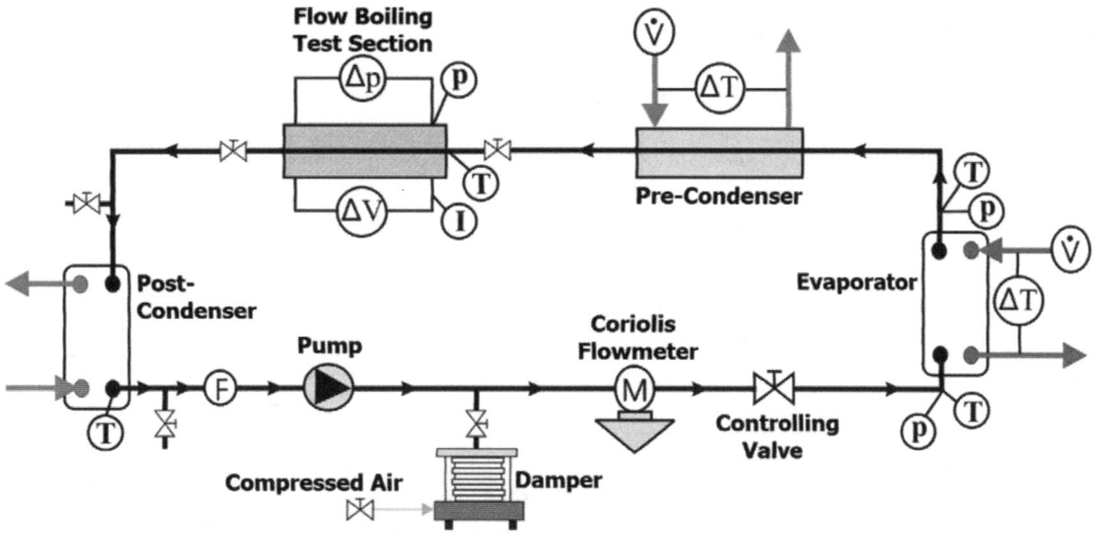

Figure 1. Schematic of the experimental facility.

The refrigerant evaporates in the test section, and is then condensed and subcooled in a post condenser, which is another brazed plate heat exchanger fed with tap water. The saturation conditions can be controlled with a damper, which is connected to the compressed air line.
As highlighted in Figure 1, temperature and pressure measurements are carried out in order to know the refrigerant thermodynamic state throughout the circuit. Calibrated T-type thermocouples, with an accuracy of ±0.05 K, are used to measure the refrigerant temperatures at the inlet of the evaporator, at the inlet of the pre-condenser, at the inlet of the test section, and at the outlet of the post-condenser, whereas absolute pressure transducers, with an accuracy of ±1950 Pa, are used to monitor the refrigerant pressures at the inlet of the evaporator, at the inlet of the pre-condenser, and at the inlet of the test section. A differential pressure transducer (accuracy of ±25 Pa) is used to measure the total pressure drop across the test section.

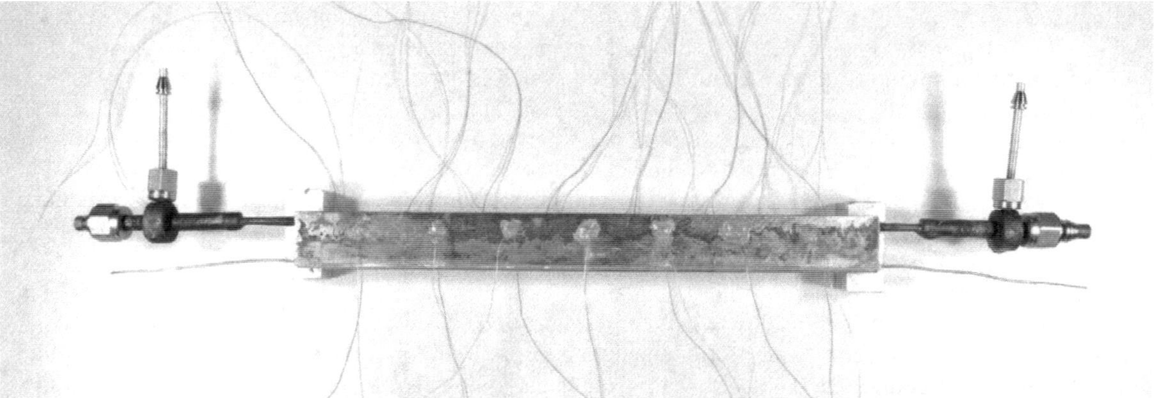

Figure 2. Picture of the test section.

A picture of the assembled test section is reported in figure 2. The microfin tube was brazed in a guide milled on the top surface of a copper plate, which is 225 mm long, 20 mm high, and 10 mm wide. Fifteen holes were drilled just 1 mm beneath the tube, in order to host as many T-type thermocouples (accuracy ±0.05 K) to monitor the wall temperature distribution. Another guide was milled on the bottom surface, where a Nickel-Chrome wire resistance was located. The wire resistance is connected to a DC power supplier which provides the heat flow rate needed to vaporize the refrigerant flowing inside the tube. The electric power can be calculated from two different voltage drop measurements: the first one is across the Ni-Cr wire resistance, and the second one is across a known reference resistance (shunt), which is in series with the Ni-Cr wire resistance and permits to calculate the current flowing in the circuit. The accuracy of the measurements of the electric power is ±0.13% of the reading. Pressure ports are located about 57.5 mm upstream and downstream of the copper plate, and thus the length for the pressure drop measurements is 340 mm. A smooth connection was designated to connect the microfin tube to the tubes of the facility in order to avoid any possible abrupt pressure losses. The microfin tube under investigation has an outer diameter OD of 3 mm, an inner diameter at the fin tip D of 2.4 mm, it has 40 fins along the circumference, and each fin is 0.12 mm high. The apex angle γ is 43°, whereas the helix angle β is 7°.
The test section is located inside an aluminum housing, which contains a 30 mm thick rock wool to limit as much as possible the heat losses due to conduction to the ambient. Preliminary tests were run in order to evaluate these heat losses: tests were carried out in vacuum conditions by supplying the electric power needed to maintain the wall temperature at fixed values. Heat losses q_{loss} were found to have a linear dependence on the mean wall temperature, according to the following equation:

$$q_{loss}[\mathrm{W}] = 0.1235 \cdot \bar{t}_{wall}[^\circ\mathrm{C}] - 2.6114 \tag{1}$$

where $\bar{t}_{wall}$ is the mean wall temperature. During these tests, the wall temperature was maintained between 25 °C and 60 °C, and the ambient temperature was approximately 20 °C.

3. Data analysis

The vapour quality at the inlet of the test section depends on the heat flow rate exchanged in the pre-condenser and on the thermodynamic conditions at the outlet of the evaporator, as:

$$q_{pc} = \dot{m}_{w,pc} \cdot c_{p,w} \cdot \left(t_{w,out} - t_{w,in}\right) = \dot{m}_{ref} \cdot \left(h_{vs} - h_{TS,in}\right) \tag{2}$$

where $\dot{m}_{w,pc}$ is the water mass flow rate flowing in the pre-condenser, $c_{p,w}$ is the water specific heat at constant pressure, and $t_{w,out}$ and $t_{w,in}$ are the water temperatures at the outlet and inlet of the precondenser, respectively. On the right side of equation 2, $\dot{m}_{ref}$ is the refrigerant mass flow rate measured by the Coriolis effect flowmeter, and h_{vs} and $h_{TS,in}$ are the refrigerant specific enthalpies at the outlet of the evaporator (which corresponds to the enthalpy at the inlet of the pre-condenser), and at the inlet of the test section (which corresponds to the enthalpy at the outlet of the pre-condenser). From this thermal balance, it is possible to calculate the refrigerant enthalpy at the inlet of the test section as:

$$h_{TS,in} = h_{vs} - \frac{\dot{m}_{w,pc} \cdot c_{p,w} \cdot \left(t_{w,out} - t_{w,in}\right)}{\dot{m}_{ref}} \tag{3}$$

and thus the inlet vapour quality is given by:

$$x_{in} = \frac{h_{TS,in} - h_L}{h_V - h_L} \tag{4}$$

with h_L and h_V the specific enthalpies of the saturated liquid and vapour, respectively, calculated from the measured values of refrigerant inlet pressure. By controlling the water flow rate at the pre-condenser and the water temperature at the inlet of the pre-condenser, it is possible to set the desired vapour quality x_{in} at the inlet of the test section. All the thermo-physical properties are calculated using Refprop 9.1 [11].

The net heat flow rate exchanged by the microfin tube is calculated as:

$$q_{TS} = P_{el} - q_{loss} = \Delta V \cdot I - q_{loss} \tag{5}$$

where ΔV and I are the voltage and current in the electric wire resistance. Therefore, the refrigerant specific enthalpy at the outlet of the test section can be calculated with the following equation:

$$h_{TS,out} = h_{TS,in} + \frac{q_{TS}}{\dot{m}_{ref}} \tag{6}$$

and the vapour quality as:

$$x_{out} = \frac{h_{TS,out} - h_L}{h_V - h_L} \tag{7}$$

where, in this case, the refrigerant properties are calculated from the knowledge of the outlet pressure. The mean vapour quality can now be calculated. From the error propagation analysis, the mean, maximum, and minimum uncertainties on the mean vapour quality are 0.027, 0.029, and 0.025, respectively.

The thermal performance of the microfin tube is given in terms of heat transfer coefficient *HTC*, referred to the area of an equivalent smooth tube with the same inner diameter at the fin tip A_D, as:

$$HTC = \frac{q_{TS}}{A_D \cdot \left(\bar{t}_{wall} - \bar{t}_{sat}\right)} \tag{8}$$

where $\bar{t}_{wall}$ is the average value of the fifteen temperatures measured by the fifteen thermocouples located 1 mm under the microfin tube, and $\bar{t}_{sat}$ is the mean saturation temperature. The mean,

Journal of Physics: Conference Series **655** (2015) 012032 doi:10.1088/1742-6596/655/1/012032

maximum, and minimum uncertainties on the heat transfer coefficient are 1.8%, 5.4%, and 0.7%, respectively.

The hydraulic performance of the microfin tube is given in terms of frictional pressure gradient, which was calculated from the measured total pressure gradient by subtracting the momentum pressure gradient, as:

$$\left(\frac{dp}{dz}\right)_f = \left(\frac{dp}{dz}\right)_{tot} - \left(\frac{dp}{dz}\right)_a \qquad (9)$$

The model of Rouhani and Axelsson [12] was taken into account to calculate the void fraction, needed to calculate to momentum pressure drop. The gravitational contribution was not considered since the microfin tube is horizontally located.

4. Experimental results

This paragraph proposes a comparison among the thermal and hydraulic perfomances of R1234yf and R134a during vaporization inside a mini microfin tube. The saturation temperature at the inlet of the test section was kept constant and equal to 30 °C, which is suitable for high temperature industrial heat pumps and for electronics cooling applications.

The mass velocity is referred to the cross sectional area of an equivalent smooth tube having an internal diameter equal to the diameter at the fin tip of the microfin tube under investigation, whereas the heat flux is referred to the outer surface area of the mini microfin tube. The vapour quality was varied from 0.1 to 1, the mass velocity was varied from 375 to 940 kg m^{-2} s^{-1}, and the heat flux from 10 to 50 kW m^{-2}: in these operating conditions, the quality variation from the inlet to the outlet of the test section is between 0.04 and 0.39.

Table 1 reports the major thermo physical properties of R1234yf and R134a. As it can be seen, the two fluids exhibits approximately the same properties, which make R1234yf a possible replacement of the more common R134a. R134a shows a slightly lower vapour density, thus at constant mass velocity it might exploit a better performance in case of two-phase forced convection, whereas the liquid densities are similar. Furthermore, reduced pressure of R1234yf is 22% higher than that of R134a, but the latent heat of vaporization is 18% lower: the two opposite trends might balance the nucleate boiling performances of the two fluids.

Table 1. Major thermo-physical properties of R1234yf and R134a. Data from Refprop 9.1 [12].

Property	R1234yf	R134a
p_{sat} [bar]	7.84	7.70
p_{red} [-]	0.232	0.190
ρ_L [kg m^{-3}]	1073	1188
ρ_V [kg m^{-3}]	43.7	37.5
$c_{p,L}$ [J kg^{-1} K^{-1}]	1417	1447
$c_{p,V}$ [J kg^{-1} K^{-1}]	1086	1066
$h_{L,V}$ [kJ kg^{-1}]	141	173
λ_L [W m^{-1} K^{-1}]	0.062	0.079
λ_V [W m^{-1} K^{-1}]	0.014	0.014
μ_L [μPa s]	145	183
μ_V [μPa s]	11.3	11.9
Pr_L [-]	3.32	3.35
Pr_V [-]	0.86	0.88
σ [mN m^{-1}]	5.56	7.38
$(dT/dp)_{sat}$ [K bar^{-1}]	4.7	4.5

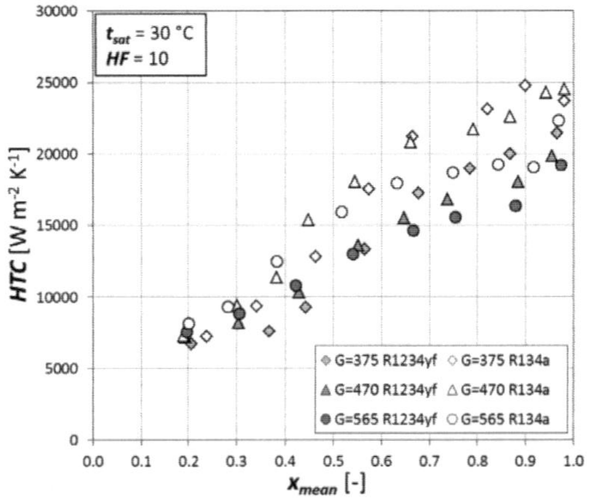

Figure 3. Effect of mass velocity on the heat transfer coefficient. Data at 10 kW m^{-2}.

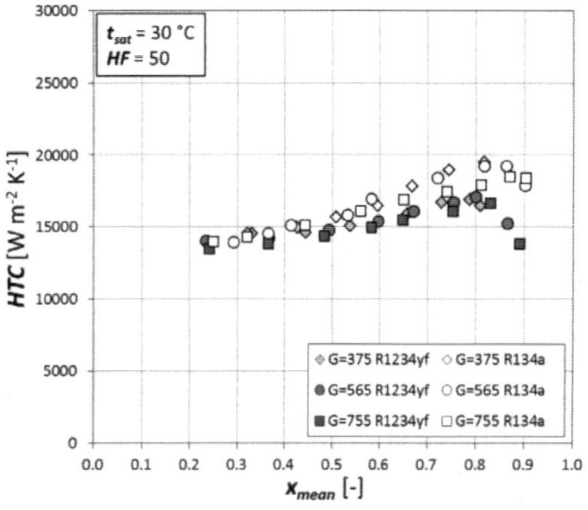

Figure 4. Effect of mass velocity on the heat transfer coefficient. Data at 50 kW m^{-2}.

Figure 3 compares the heat transfer coefficient of R1234yf and R134a. Heat transfer coefficients are plotted against the mean vapour quality as a function of mass velocity with an imposed heat flux of 10 kW m^{-2}. Generally speaking, the heat transfer coefficient always increases as vapour quality increases, meaning that two-phase forced convection mainly controls the phase change process in these operating test conditions. It is interesting to point out that the highest heat transfer coefficient is shown at the lowest tested mass velocity, i.e. 375 kg m^{-2} s^{-1}, for both fluids: this can be linked to the particular geometry of the microfin tube, which enhances the heat transfer mechanism especially at low mass velocities. The higher thermal performances of R134a become more and more evident when the vapour quality increases. This is due to the fact that the contribution of two-phase forced convection increases with increasing vapour quality, i.e. on vapour velocity. Since R134a has a lower vapour density (see table 1), it exploits higher vapour velocity at constant mass velocity, and thus it shows higher heat transfer coefficients than those of R1234yf in these operating test conditions.

Figure 4 collects the same results but with an imposed heat flux of 50 kW m^{-2}. First of all, at low vapour qualities, the heat transfer coefficients are higher than the former case for both fluids, since in this case the heat flux is higher and the contribution of nucleate boiling is stronger, which enhances the heat transfer coefficients. At low vapour qualities, the two fluids show similar values of heat transfer coefficient, due to their thermo-physical characteristics which lead to comparable performances during the nucleate boiling mechanism. When the vapour quality increases, R134a shows slightly higher values of the heat transfer coefficient of R134a, since it has better convective characteristics. In addition, the heat transfer coefficients at high vapour quality are 1.4 times than those at low vapour quality in the case of R134a, whereas they are 1.25 times in the case of R1234yf. A start of the dryout phenomenon can be observed for both refrigerants at mean vapour qualities higher than 0.8.

Figure 5 shows the frictional pressure gradient for R1234yf and R134a as a function of the mean vapour quality with an imposed heat flux of 10 kW m^{-2}. The model of Rouhani and Axelsson [12] was considered to estimate the void fraction to calculate the momentum pressure drops, which were subtracted from the total measured pressure drops. Generally speaking, the results show that, at constant mass velocity, the frictional pressure gradient increases with vapour quality, reaching a maximum value after which it slightly decreases. At constant vapour quality, the frictional pressure gradient increases as the mass velocity increases. R1234yf and R134a show similar values of frictional pressure gradient at low vapour qualities, whereas R1234yf shows slightly lower frictional pressure gradients than those of R134a at higher vapour quality, and the difference increases with increasing the vapour quality: this behaviour can be explained considering the fact that R134a and R1234yf show similar value of liquid density, whereas the vapour density of R134a is lower, which leads to higher vapour velocity at constant mass velocity with consequent higher pressure drops.

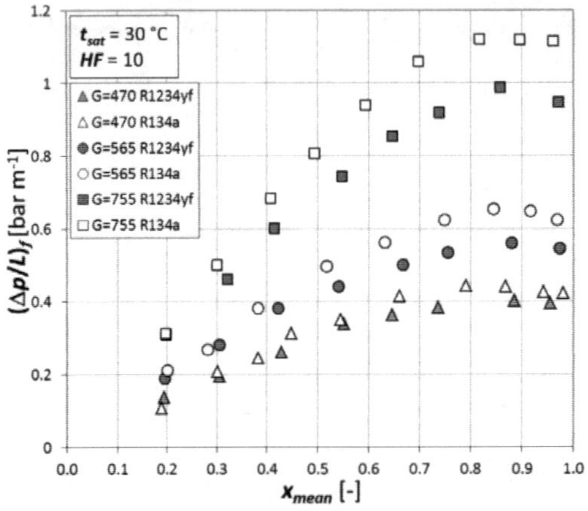

Figure 5. Effect of mass velocity on the frictional pressure gradient. Data at 10 kW m^{-2}.

5. Empirical modelling

Diani et al. [13] recently proposed two empirical equations to estimate the heat transfer coefficient and frictional pressure drop during refrigerant flow boiling inside a mini microfin tube. The models were developed from experimental measurements carried out during R134a and R1234ze(E) flow boiling inside a mini microfin tube with an inner diameter at the fin tip of 3.4 mm, with a fin height of 0.12 mm. The heat transfer model is an updated version of the original one by Cavallini et al. [14], also revised in Padovan et al. [9]. The model is valid for vapour qualities prior to the onset of the dryout and for mass velocities ranging between 200 and 940 kg m^{-2} s^{-1}.

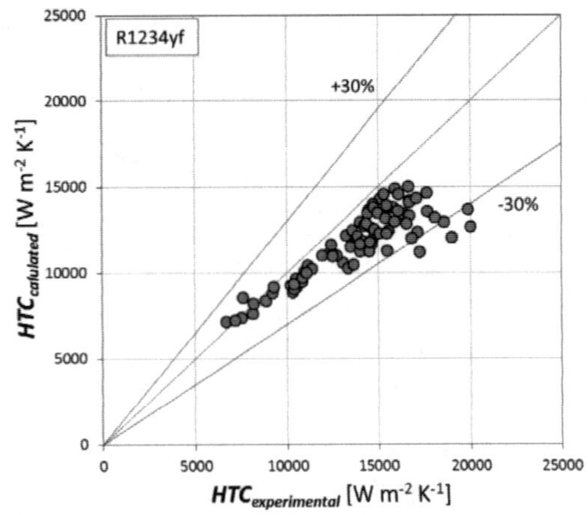

Figure 6. Comparison between experimental and calculated heat transfer coefficient with the model by Diani et al. [13] for R1234yf.

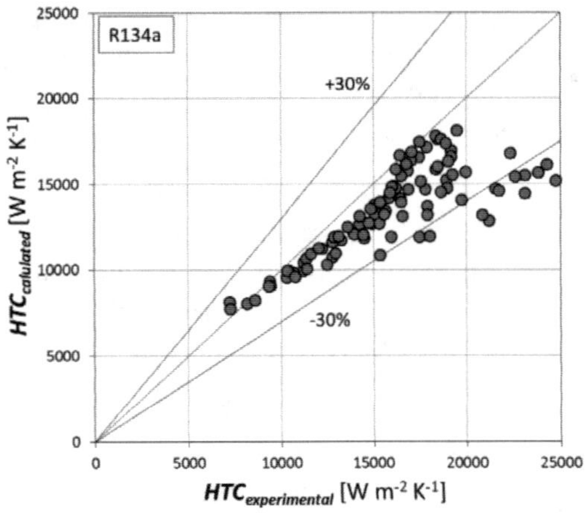

Figure 7. Comparison between experimental and calculated heat transfer coefficient with the model by Diani et al. [13] for R134a.

The comparison of the calculated heat transfer coefficient against the experimental counterpart is shown in figures 6 and 7. The model slightly underestimates the heat transfer coefficient, with relative, absolute, and standard deviations of -13.7%, 14.1%, and 8.8%, respectively, for R1234yf, and -13.7%, 14.1%, and 10.5%, respectively, for R134a.

More in details, the model calculates the heat transfer coefficient as the sum of the contribution of nucleate boiling and of two-phase forced convection. The nucleate boiling heat transfer coefficient can be calculated using the model proposed by Cooper [15] considering a suppression factor, whereas the convective heat transfer coefficient is a revised version of the two-phase heat transfer coefficient

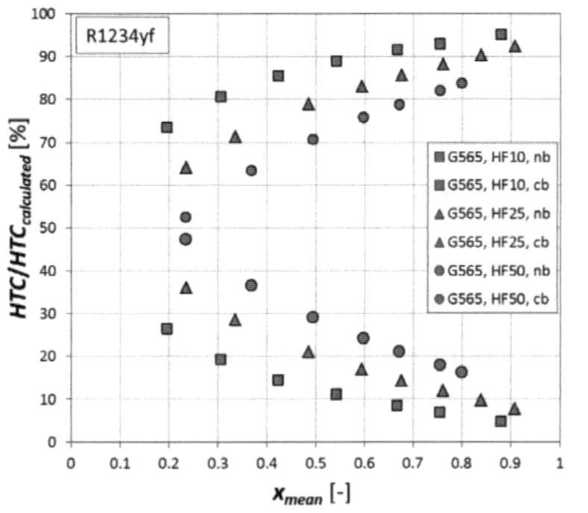

Figure 8. Effect of heat flux with the model by Diani et al. [13] for R1234yf.

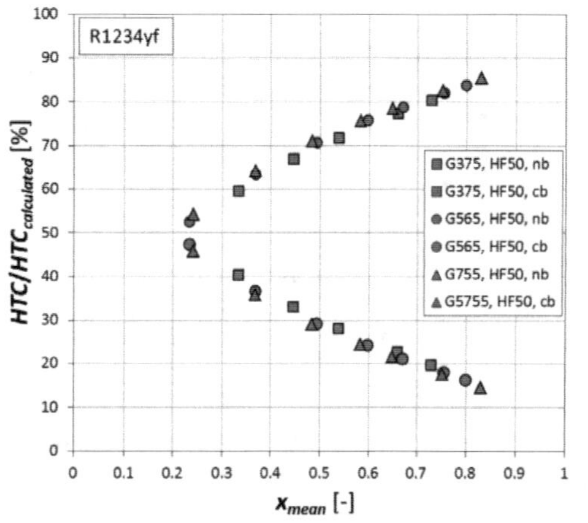

Figure 9. Effect of mass velocity with the model by Diani et al. [13] for R1234yf.

proposed by Cavallini et al. [14]. The effect of heat flux on the heat transfer coefficient calculated with the model is proposed in figure 8 for a case study at 565 kg m^{-2} s^{-1}. As expected, the nucleate boiling contribution decreases with increasing the vapour quality, whereas the two-phase forced convection contribution increases with increasing the vapour quality. It is interesting to point out that there is a strong effect of heat flux on the heat transfer coefficient especially at low vapour qualities, being 26% even at the lowest heat flux of 10 kW m^{-2}. Figure 9 shows the effect of mass velocity on the calculated R1234yf heat transfer coefficient with a heat flux of 50 kW m^{-2}. As it appears, both the nucleate boiling and two-phase forced convection contributions are highly affected by vapour qualities, but are not affected by mass velocity. This behaviour is confirmed by the experimental results collected in figure 4.

Figures 10 and 11 compare the calculated frictional pressure drops against the experimental ones. The model suggests calculating the only two-phase multiplier as proposed by Cavallini et al. [10] for minichannels. The model underestimates the experimental values with a relative and standard deviations of -26.4% and 6.2% for R1234yf, and of -23.8% and 6.2% for R134a. The underestimation can be related to the ratio between fin height to tube diameter, which was 0.035 for the tested tube from which results the correlation was developed, whereas it is 0.050 for the case under investigation.

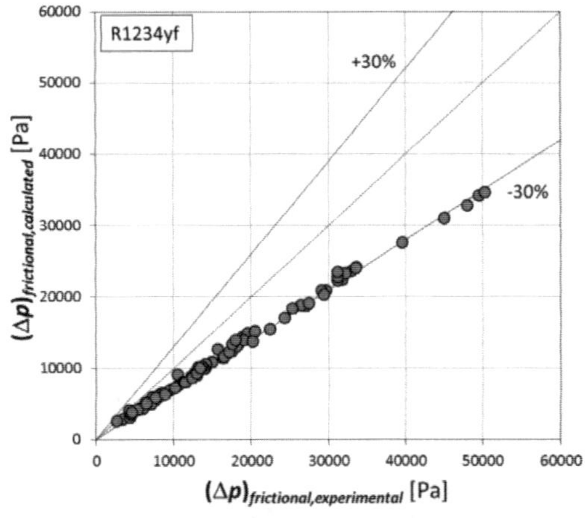

Figure 10. Comparison between experimental and calculated pressure gradients with the model of Diani et al. [13] for R1234yf.

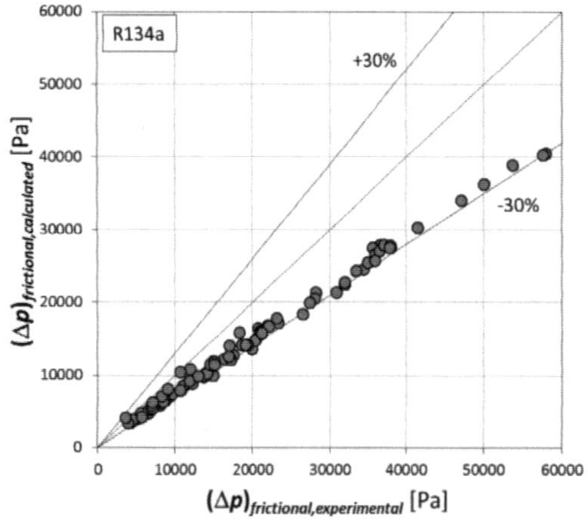

Figure 11. Comparison between experimental and calculated pressure gradients with the model of Diani et al. [13] for R134a.

6. Conclusions

This paper compares the behaviour of R1234yf and R134a during flow boiling inside a mini microfin tube with an inner diameter at the fin tip of 2.4 mm. The microfin tube was electrically heated from the bottom, and the heat flux ranged between 10 and 50 kW m^{-2}. Mass velocity was varied between 375 and 940 kg m^{-2} s^{-1}, and the saturation temperature at the inlet of the test section was kept constant and equal to 30 °C. At low heat flux, R134a shows higher heat transfer coefficients than those of R1234yf, due to its better convective thermo-physical properties, whereas they have comparable performance at high heat fluxes. On the other hand, R1234yf shows lower frictional pressure gradient. Therefore, R1234yf is a possible candidate as replacement of the common R134a.

The experimental results of heat transfer coefficient and frictional pressure drop were compared against the values predicted by the correlation of Diani et al. The model underestimates on average the heat transfer coefficient of -13.7%, and the frictional pressure drop of -25%. The model also permits to discern the effects of nucleate boiling and two-phase forced convection during the calculation of the heat transfer coefficient.

7. References

[1] Hodnebrog O, Etminan M, Fuglestvedt J S, Marston G, Myhre G, Nielsen C J, Shine K P, Wallington T J 2013 *Rev. Geophys.* **51** 300-378
[2] Gao L, Honda T, Koyama S 2007 *HVAC&R Research* **13** 415-425
[3] Dang C, Haraguchi N, Hihara E 2010 *Int. J. Refrig.* **33** 655-663
[4] Mc Linden M O, Kazanov A F, Brown J S, Domanski P A 2013 *Int. J. Refrig.* **38** 80-92
[5] Grauso S, Mastrullo R, Mauro A W, Thome J R, Vanoli G P 2013 *Int. J. Refrig.* **36** 478-491
[6] Hossain Md A, Onaka Y, Afroz H M M, Miyara A 2013 *Int. J. Refrig.* **36** 465-477
[7] Kondou C, BaBa D, Mishima F, Koyama S 2013 *Int. J. Refrig.* **36** 2366-2378
[8] Han X, Hi P, Yuan X, Wang Q, Chen G 2013 *Int. J. Heat Mass Tran.* **67** 1122-1130
[9] Padovan A, Del Col D, Rossetto L 2011 *Appl. Therm. Eng.* **31** 3814-3826
[10] Cavallini A, Del Col D, Matkovic M, Rossetto L 2009 *J. Heat Transfer* **131** 033107-1-8
[11] Lemmon E W, Huber M L, Mc Linden M O 2013 NIST Standard Reference Database 23: Reference Fluid Thermodynamic and Transport Properties - REFPROP, Version 9.1
[12] Rouhani S Z, Axelsson E 1970 *Int. J. Heat Mass Tran.* **13** 383-393
[13] Diani A, Mancin S, Rossetto L 2014 *Int. J. Refrig.* **47** 105-119
[14] Cavallini A, Del Col D, Rossetto L 2006 Flow Boiling Inside Microfin Tubes: Prediction of the Heat Transfer Coefficient Proc. of ECI International Conference on Boiling Heat Transfer Spoleto
[15] Cooper M G 1984 *Adv. Heat Tran.* **16** 157-239

Acknowledgments

The support of Wieland-Werke AG and Dr. Christoph Walther on this research activity is gratefully acknowledged. The MS student Guido Ghion is also gratefully acknowledged.

Venturi flow meter and Electrical Capacitance Probe in a horizontal two-phase flow

G Monni, M Caramello, M De Salve and B Panella
Energy Department, Politecnico di Torino, Corso Duca degli Abruzzi, 24, 10129

E-mail: marco.caramello@polito.it

Abstract. The paper presents the results obtained with a spool piece (SP) made of a Venturi flow meter (VMF) and an Electrical Capacitance Probe (ECP) in stratified two-phase flow. The objective is to determine the relationship between the test measurements and the physical characteristics of the flow such as superficial velocities, density and void fraction. The outputs of the ECP are electrical signals proportional to the void fraction between the electrodes; the parameters measured by the VFM are the total and the irreversible pressure losses of the two-phase mixture. The fluids are air and demineralized water at ambient conditions. The flow rates are in the range of 0,065-0,099 kg/s for air and 0- 0,039 kg/s (0-140 l/h) for water. The flow patterns recognized during the experiments are stratified, dispersed and annular flow. The presence of the VFM plays an important role on the alteration of the flow pattern due to wall flow detachment phenomena. The signals of differential pressure of the VFM in horizontal configuration are strongly dependent on the superficial velocities and on the flow pattern because of a lower symmetry of the flow with respect to the vertical configuration.

1. Introduction

In the framework of integral test facilities for the thermal-hydraulic simulation of nuclear reactors, with particular emphasis to small modular reactors, there is a large interest in the characterization of the behaviour of the system in case of design basis accidents. As an example, the simulation of LOCA accidents can be performed by means of straight horizontal pipes equipped with fast opening valves or rupture disks to connect the external environment with the primary system. Such pipes are also used to locate the proper instrumentation to evaluate the thermal- hydraulic quantities of the flow before the section of the rupture. The final goal of the instrumentation is to estimate the flow rate of the liquid and vapour phases and the energy released from the rupture with time, so to retrieve the behaviour of mass inventory and internal energy inside the primary system during the accidental transient. The flow rates of the phases through the rupture cannot be measured in a direct manner because of the complexity of the flow. It is therefore necessary to retrieve the value of the flow rates on the basis of the characteristic parameters of the two-phase flow, such as the void fraction, the flow pattern, the density and the superficial velocities. These quantities are usually measured by means of a special instrumentation called spool piece (SP). A spool piece is a combination of two or more instruments whose electrical signals are proportional to the characteristic parameters of the two-phase flow. Several instruments can be found in a SP, such as Venturi flow meters (VFM) [1], electrical capacitance probes (ECP) [2], drag disks [3], turbines [3], wire mesh sensors [4].
The correct interpretation of the signals of the instruments requires the development of models able to describe the two-phase fluid-dynamics in the test section, the working principle of the instrument together with the relationships between the electrical signals and the physical parameters of the flow (liquid and gas flow rates, void fraction). The validation of the measurements is often done by comparing the prediction of the instruments with the estimation of theoretical models. The complete characterization of a spool piece for the estimation of the two-phase flow parameters implies the definition of the instruments belonging to the SP from the geometrical point of view, the configuration of the test section and the relationship between the electrical signals and the physical quantities. Other

important values to be estimated in order to complete the characterization of the flow are the absolute pressure, differential pressures along the test section and the temperature of the mixture. During the last years several SPs have been tested and characterized at Politecnico di Torino [5-7] by means of experimental facilities at low pressure using demineralized water and air in horizontal and vertical configuration, with a particular interest in the characterization of annular dispersed and stratified annular flows. In this paper the experimental results obtained with a SP consisting of a Venturi flow meter and an Electrical Capacitance Probe in horizontal configuration are presented. The Venturi flow meter has been widely used for the estimation of the mass flow rate in single phase flow because of its simplicity, the low cost and the low maintenance required. For these reasons, with the aim to characterize its behaviour in two-phase flow, the VFM has been studied by many authors both theoretically and experimentally [8-13]. The Electrical Capacitance Probe has the aim to estimate the void fraction of the fluid and the flow pattern. The working principle of the instrument is based on the fact that the phases constituting the two-phase mixture are characterized by different electrical properties, and that the measured response of the instrument is a function of the void ~~to the~~ fraction of the phases [14-16]. The specific ECP tested during the experimental campaign has been designed and produced by SIET Company.

2. Test facility

A picture of the experimental facility used for the characterization of the SP made of the Venturi flow meter and the Electrical Capacitance Probe is shown in figure 1. It consists of an air blower, the demineralized feedwater system, an air-water mixer, two quick closing valves, a calibrated orifice, an open discharge volume, additional instrumentation and the SP. The piping system is made of Plexiglas for the direct observation of the flow pattern. The internal diameter is 80 mm. The orifice is installed before the air-water mixer and it is used to estimate the air flow rate. It is characterized by a diameter of 65 mm and a minimum section diameter of 30,19 mm. The discharge coefficient has been evaluated according to the UNI standards and has a value of 0,605. The demineralized feedwater flow is imposed by means of two rotameters of different full-scale. The additional instrumentation consists of relative and absolute pressure transducers and K-type thermocouples along the experimental facility. The pressure transducers are installed on the calibrated orifice to estimate the gas flow rate and on the Venturi in order to measure both irreversible and total pressure drops.

Figure 1. The experimental facility.

They are characterized by a linear or parabolic response in which the full scale, the output current (4-20) mA and the conversion resistance of the signal from current to voltage are taken into account. The calibration curves of the differential pressure transducers have been obtained by imposing known water heads on the instruments. The quick closing valves are used to measure the average void fraction in the test section. The estimation of the average void fraction has been done following two different procedures: a) the mass of water between the two quick closing valves is drained from the bottom and measured by means of a weightier (for low void fraction value), b) a mass of water is injected in the intercepted region of the facility until a certain level on the cross section is obtained. From the knowledge of the total volume of the test section the intercepted mass of water and the void fraction (for high void fraction value) can be evaluated. Figure 2 shows a picture of the Venturi flow meter. It is a symmetrical short-type Venturi designed according to ISO 5167-4: 2003 standards. The use of a short-type instrument implies a reduction of the manufacturing costs even though the pressure losses of the instruments are higher compared to the ones of a standard Venturi because of flow detachment phenomena. The symmetrical configuration is an obliged choice if a flow reversal is expected. The geometrical parameters of the instrument are reported in table 1. The conical convergent section is L=2.7*(D-d)=108 mm long.

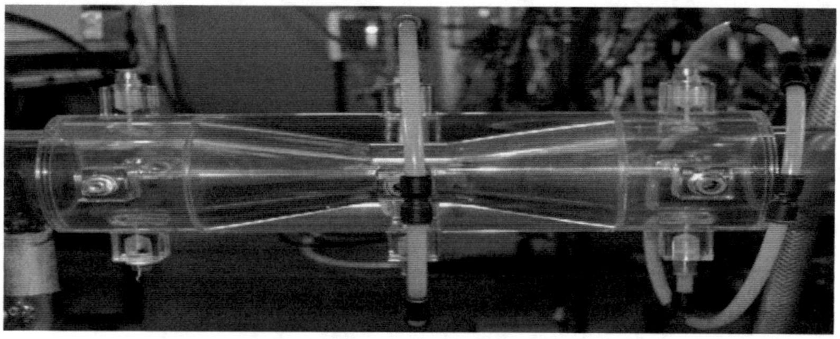

Figure 2. The Venturi flow meter.

Table 1. VFM geometrical data

Parameter	Value	
Diameter	80	mm
Throat diameter	40	mm
Angle	21	°
Total length	340	mm

Figure 3. The Electrical Capacitance Probe.

Figure 3 shows a picture of the ECP. It is constituted by 9 linear electrodes (length of 400 mm and width of 5 mm) set on the external side of the Plexiglas pipe and a central electrode in the middle of the cross section. Figure 4 shows the electrodes position in the horizontal test section set up.

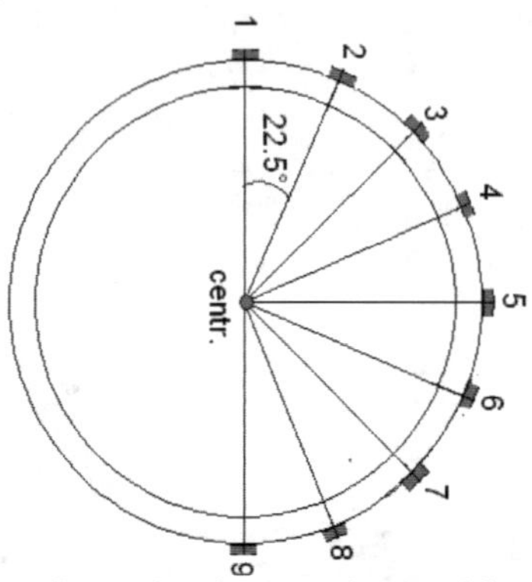

Figure 4. ECP electrodes configuration in experimental horizontal test section set up.

The peripheral electrodes are welded on both sides to the measurement chain, while the central electrode is welded only on one side. The distance between the sockets to which the electrodes are welded is sufficient to reduce the presence of parasitic currents. The geometrical parameters of the ECP and the maximum working conditions are reported in table 2.

Table 2. ECP characteristics

Parameter	Value
Diameter	80 mm
Length	600 mm
Wall thickness	5 mm
Maximum pressure	0.2 Mpa
Maximum temperature	60 °C
Angular distance between the electrodes	22,5°

The input/output electrical signals of the ECP are managed by LabView environment. A NI USB-6259 device is used both as input provider for the electrodes and output receiver. The input signal is sinusoidal with a frequency of 25 kHz and amplitude of 5 Vpp. The sequence of measurement is performed by exciting one electrode at a time and reading the output of the others receiving electrodes. The couple of exciting/receiving electrodes is uniquely defined by a code number. The sampling frequency of the ECP is 250 kHz.

A complete description of the theory of the ECP has been previously presented by the authors here [17]. The working principle is based on the different values of electrical conductivities and relative permittivities of liquid water and air. The RMS value of the voltage drop between an input electrode i and a receiving electrode j when a single phase is present in the test section is proportional to the electrical properties of the medium, its thermodynamic conditions and the angular distance between the electrodes (see equations 1 and 2 for liquid water and air respectively). The same couple of input and receiving electrodes in the presence of a two-phase mixture experiences a voltage drop (V in equation 3) which depends on the thermodynamic conditions, the distance and the electrical properties of the mixture which are dependent on the volume occupied by each phase between the two electrodes. By defining a suitable normalization of the signal in two-phase flow as the one proposed in equation 3 it is possible to correlate the signal of the ECP with the void fraction of the mixture between the electrodes.

$$V_{L,i,j}(T,\theta) = V_L(T,\theta) \quad (1)$$
$$V_{G,i,j}(T,\theta) = V_G(T,\theta) \quad (2)$$
$$V^*_{i,j}(\alpha,\theta,T) = \frac{V_L - V}{V_L - V_G} \quad (3)$$

3. Test procedure and experimental matrix

The test procedure that has been adopted for the experimental activity consists of the following steps:

1. the velocity of the blower and the water flow rate are imposed,
2. once a steady state condition is established, the measures from the calibrated orifice (gas flow rate), the VFM (differential pressures) and the ECP (electrical response) are acquired for a period of 30 seconds, in order to gain the information of RMS and standard deviation,
3. the quick closing valves are closed and both the blower and feedwater system are turned off,
4. the mass of water intercepted by the valves is drained from the system and measured by means of the weight measurement technique.

At the beginning of each session the single phase response of the instruments both for water and air are measured. These values are used in the post-processing of the results to normalize the signals of the ECP. For the present experimental campaign, the range of velocities obtained is between 11 and 18 m/s for air and 0,8 e 8 mm/s for water. The resulting flow patterns are stratified and annular stratified. For the same couple of flow rates of the liquid and gas phases two separate tests have been carried out in order to assess the repeatability of the measurements.

4. Results and discussion

The measurements and the characterization of the Venturi flow meter and the Electrical Capacitance Probe are assessed in the present chapter. Figure 5 shows the experimental relationship between the void fraction and the flow quality for the two experimental sets. The reproducibility of the results is considered satisfactory: the data dispersion is considered to be caused by different entrainment phenomena and the effect of ambient pressure and temperature, that are not easily reproducible.

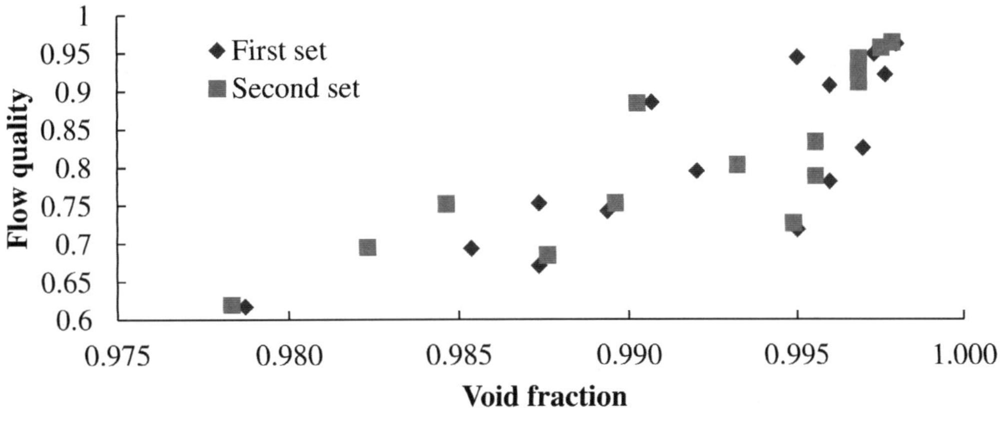

Figure 5. Flow quality as a function of the void fraction.

4.1. Characterization of the Venturi flow meter

Figure 6 shows the dependency between the measured pressure loss inside the VFM and the total flow rate. Three curves, characteristic of the three frequencies of the blower can be observed.

In the tested range the effect of variation of the gas velocity has a greater influence with respect to the liquid phase on the pressure loss inside the VFM. From the values of experimental pressure loss, both

Journal of Physics: Conference Series **655** (2015) 012033 doi:10.1088/1742-6596/655/1/012033

in single-phase and two-phase flow, the relationship between the two-phase multiplier Φ_g^2 and the Lockhart-Martinelli parameter χ, defined as in equation 4, has been derived (figure 7).

$$\chi = \frac{(1-x)}{x}\left(\frac{\rho_g}{\rho_l}\right)^{0.5} \tag{4}$$

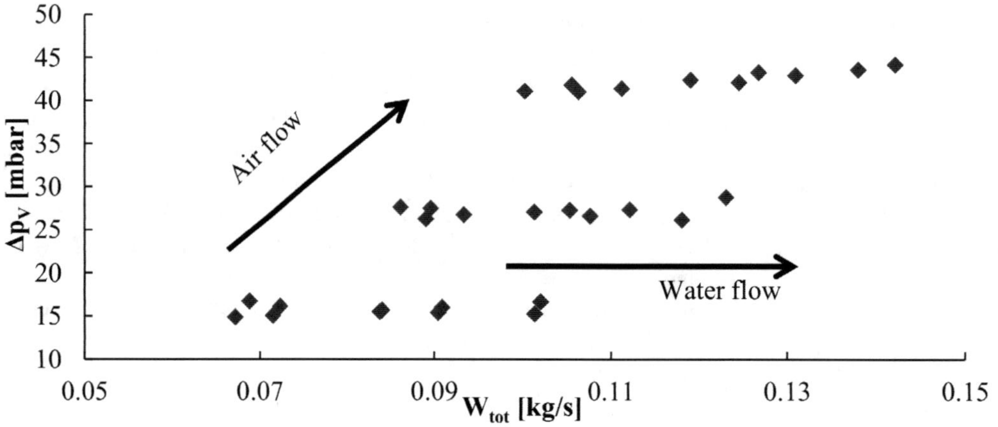

Figure 6. VFM pressure loss vs total flow rate.

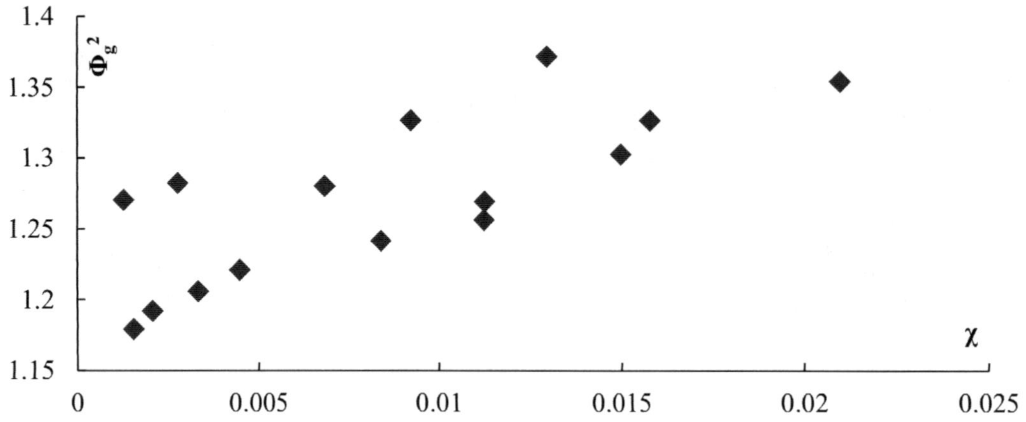

Figure 7. Experimental two-phase multiplier vs Lockhart-Martinelli parameter.

Another important measure obtained from the VFM is the irreversible pressure loss. The irreversible pressure loss shows an higher sensibility to the variation of liquid flow rate with respect to the standard VFM pressure loss (figure 8), and it is particularly influenced by the flow pattern. Three characteristic curves can be observed in figure 8: considering a fixed value of total mass flow rate higher pressure losses are measured in the presence of an higher air flow rate. By analysing the experimental results related to the irreversible pressure losses we found that they are strongly dependent on the density of the gas phase and on the superficial velocities of the phases (j_l, j_g). An empirical correlation has been derived for the estimation of the irreversible pressure loss through the VFM on the basis of the characteristic quantities of the two-phase flow (equation 5).

$$\Delta P_{irr} = 0{,}2705\left(\rho_g J_g^{1{,}91}\right)\left(\frac{J_l}{J_g}\right)^{0{,}13} \quad (5)$$

The accuracy of the relationship in equation 5 with respect to the experimental data is of the order of 12%. In order to reduce the error the dependency of the correlation on the typology of the flow pattern has to be considered.

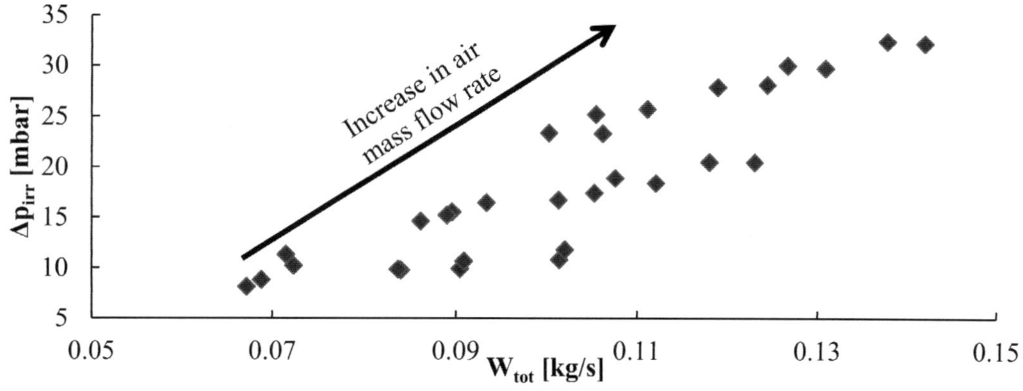

Figure 8. VFM irreversible pressure loss vs total flow rate.

4.2. Characterization of the electrical capacitance probe

Figure 9 shows the behaviour of the normalized signals V^*_{ij} as a function of the position of the electrodes, considering as input electrode i the number 1 in the upper part for the figures on the left side and the number 9 in the bottom part for the figures on the right side. The curves are expressed by using the experimental value of the void fraction as parameter.
Test section void fractions between 0,978 – 0,985 are typical of tick film annular flow regimes. The lower electrode shows a typical response of an annular stratified flow pattern.

In Fig. 9 the different dependency on the void fraction between the two chosen electrodes can be observed. The major contribution to this difference is expected to be the stratification of the liquid phase at the bottom of the cross section, which is clearly visible on the electrode 9. As the velocity of the gas phase increases entrainment phenomena of liquid droplets in the gas phase are more evident and this explains an increase of the dimensionless voltage drop of the electrode 9 as the void fraction increases.
In all of the cases considered in figure 9 an increase in the measured void fraction inside the test section causes a general increase in the output signal of the ECP, leading to the values of the full air case. This is more evident when the input signal is transmitted from electrode 9 which is always covered by a water level characterized by a smaller thickness as the void fraction increases.

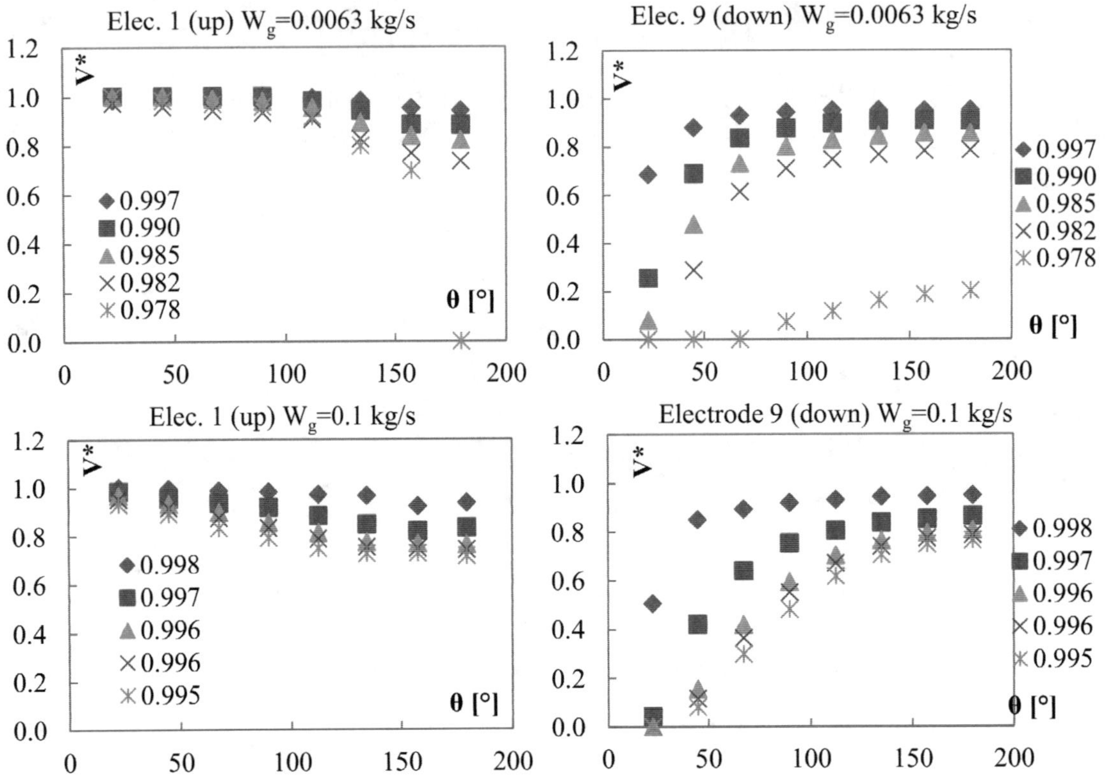

Figure 9. Dimensionless voltage drop for electrodes 1 and 9.

Figure 10 shows the dependency on the void fraction for the electrodes at the minimum angular distance of 22,5° where the electrical signal is maximum for every local void fraction.

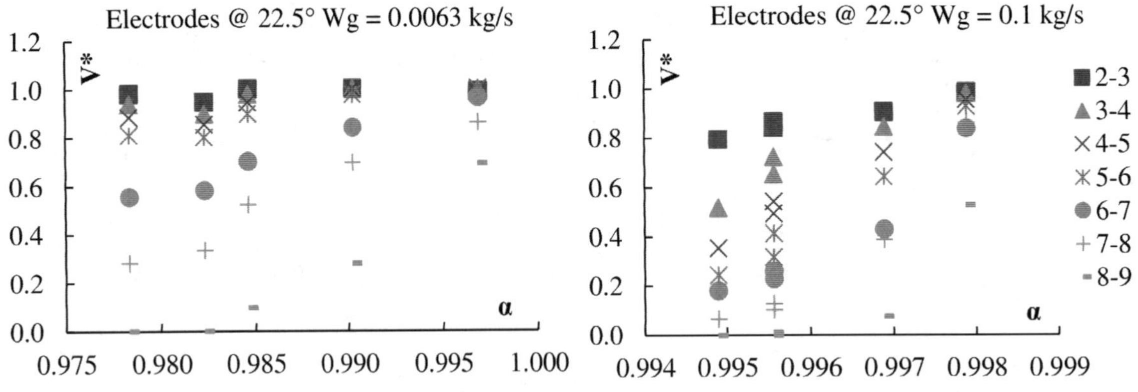

Figure 10. Dependency of the neighbouring electrodes on the void fraction.

The combination of the electrodes 2-3 is located in the upper part of the cross section. For low values of flow rates the behaviour of the signal is independent of the value of the void fraction since the flow pattern is stratified. On the other hand, the electrodes at the bottom of the cross section are largely influenced since they are able to experience the variation of thickness of the liquid film as the average void fraction on the cross section changes. For the couple of electrodes 8-9 figure 10 shows the values of void fraction after which electrode 8 becomes uncovered by the water level (between the second and the third value of experimental void fraction). Figure 11 shows the same results of figure 10 with reference to the central electrode. The signal of the central electrode has a linear behaviour as a function of the void fraction, whose slope depends on the position of the external electrodes on the cross section. As in figure 10, an higher variation in the signal is recorded for the couples lying at the bottom of the cross section.

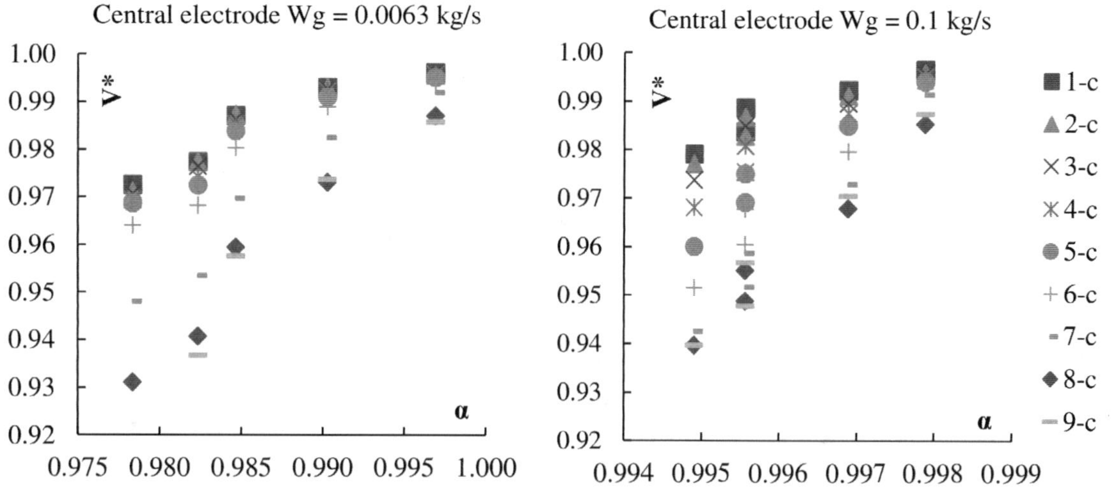

Figure 11. Dependency of the neighbouring electrodes on the void fraction (central electrode).

5. Conclusions

A preliminary methodology and the procedures to characterize a spool piece constituted by a Venturi flow meter and an Electrical Capacitance Probe for the experimental estimation of the flow rates of the phases in a horizontal two phase flow have been presented. The results obtained show the good capability of the spool piece to measure the necessary physical quantities suitable for the estimation of the flow rates. The Electrical Capacitance Probe is useful for the recognition of the flow pattern as it is very sensitive to the characteristic parameters of the two-phase flow. The information of the void fraction distribution along the cross section can be used to define the most appropriate experimental correlation between the pressure loss on the Venturi flow meter and the superficial velocities of the phases, so to decrease the error in the estimation of the flow rates.

The use of the VFM for the estimation of the two-phase flow rate is promising because both irreversible and total pressure drops are highly influenced by the behaviour of the phases. Nevertheless, the information provided by the instrument is not self-standing as the complete characterization of the flow also requires the understanding of the flow regime and of the void fraction, which are provided in this case by the ECP.

With the aim to consolidate the presented methodology and to broaden its applicability for different flow patterns, future work will be devoted to the implementation of deterministic and probabilistic approaches for the analysis of the signals of the SP for higher void fraction conditions and transition flow regimes.

Acknowledgements
The present research has been supported by ENEA and by the ministry of economic development.

List of symbols

L [m]	Length of VFM	*W [kg/s]*	Flow rate
D [m]	Test section diameter	θ	Angle between electrodes
d [m]	Throat diameter of VFM	Φ^2	Two phase multiplier
V [Volt]	Voltage	*J*	Superficial velocity
α	Void fraction	Subscripts	
T [°C]	Temperature	*tot*	Total
χ	Lockhart-Martinelli parameter	*L*	Liquid
x	Flow quality	*G*	Gas
ρ *[kg/m³]*	Density	*i, j*	Indexes

References

[1] De Salve M, Monni G, and Panella B 2013 *ANS Trans.* **108** 1013-16
[2] Monni G, De Salve M, Panella B and Randaccio C 2013 *Sci. Technol. Nucl. Install* **2013** 1-12
[3] De Salve M, Monni G and Panella B 2011 *Proc. UIT 2011* 107-12
[4] De Salve M, Monni G and Panella B 2012 *J. Phys.: Conf. Ser.* **395**
[5] Monni G, De Salve M and Panella B 2014 *Prog. in Nuc. En.* **77** 167-75
[6] Monni G, De Salve M and Panella B 2014 *Proc. HEFAT 2014*
[7] Monni G, De Salve M and Panella B 2014 *Exp. Therm. Fluid Sci.* **5** 213-21
[8] Martinelli R C and Nelson D B 1948 *Trans. ASME* **70** 695
[9] Lockhart R W and Martinelli R C 1949 *Chem. Eng. Prog.* **45** 39
[10] Dukler A E, Wicks M and Cleveland R G 1964 *AIChE J.* **10** 38-43 and 44-51
[11] Baker C R 2000 Flow Measurement Handbook (Cambridge University Press)
[12] Jitschin W 2004 *Vacuum* **76** 89-100
[13] Fang L D, Zhang T and Jin N D 2007 *J. Petroleum Sc. Eng.* **57** 245-56
[14] Huang Z, Wang B and Li H 2003 *IEEE Trans. Instr. and Meas.* **52** 7–12
[15] Wu Y, Li H, Wang M and Williams R A 2005 *Canadian J. Chem. Eng.* **83** 37–41
[16] Warsito W and Fan L S 2001 *Chem. Eng. Sc.* **56** 6455–62
[17] De Salve M, Monni G and Panella B 2010 Report RdS/2010/67, ENEA

Journal of Physics: Conference Series **655** (2015) 012034 doi:10.1088/1742-6596/655/1/012034

Influence of temperature on oxygen permeation through ion transport membrane to feed a biomass gasifier

T Antonini[1], P U Foscolo[1], K Gallucci[1], and S Stendardo[2]
[1] Industrial Engineering Department, University of L'Aquila, Via G. Gronchi, 18-67100 L'Aquila, Italy
[2] ENEA, Casaccia Research Centre, 00100 Roma, Italy

E-mail: tania.antonini@graduate.univaq.it

Abstract. Oxygen-permeable perovskite membranes with mixed ionic-electronic conducting properties can play an important role in the high temperature separation of oxygen from air. A detailed design of a membrane test module is presented, useful to test mechanical resistance and structural stability of $Ba_{0.5}Sr_{0.5}Co_{0.8}Fe_{0.2}O_{3-\delta}$ (BSCF) capillary membrane in the reactor environment. Preliminary experimental results of membrane permeation tests highlight the positive effect of temperature on perovskite materials. This behaviour is also confirmed by a computational model of *char* combustion with oxygen permeated through the membrane module, when it is placed inside a gasifier reactor to provide the necessary input of heat to the gasification endothermic process. The results show that the temperature affects the oxygen permeation of the BSCF membrane remarkably.

1. Introduction

Solid fuel (biomass, coal) gasification-based energy systems offer a reliable, affordable conversion process into fuel gas, with high-efficiency and flexibility in the production of power, heat and a wide range of chemical substances.
Building on current state of the art, gasification-based technologies can be refined and improved via application of advanced gas separation technologies that are being developed [1,2].
Most of the current commercial gasification projects, as well as those planned for the future, consider oxygen-blown gasifiers and require large flow rates of oxygen to establish auto-thermal process conditions. Thus, air separation represents a major area of opportunity for both cost reduction and performance improvement, making it a key part of energy systems. Any technology that can offer a significant reduction in the cost of oxygen will have a substantial impact on the overall economics of gasification-based processes.
Oxygen permeation through ion transport membranes is a novel technology that can be used as a replacement of energy-demanding cryogenic distillation or Pressure Swing Adsorption (PSA)/Pressure Vacuum Swing Adsorption (PVSA) processes. Cryogenic process is expensive and energy intensive, because it operates at very low temperature and at elevated pressures [3], therefore is suitable for very large plants only. In small to medium scale applications PSA or PVSA require compression of a volumetric gaseous stream at least 5 times greater than the oxygen stream effectively utilized in the gasification process with a substantial penalty of the whole energy efficiency.
Ion Transport Membranes (ITM) offer a sound alternative to conventional processes, where the goal is to provide oxygen at high temperature conditions (850-1000°C) in small to medium scale applications.

A considerable number of research works have demonstrated that dense perovskite membranes at elevated temperatures exhibit substantial oxygen ion permeability [4-9].
Ceramic dense membranes must be defect free and gas tight to hinder the gas permeation through pores, only allowing the oxygen ionic transport.
These membranes deliver oxygen ions from the feed side to the permeate side without the aid of any external electric circuit (counter diffusion of electrons preserves the electro-neutrality), which makes the process simple and very attractive.
The driving force is the difference in oxygen partial pressure across the membrane [6], so that it is not needed to pressurize the air feeding stream.
In this paper, the influence of temperature is investigated on the oxygen flux to a biomass fluidized bed gasifier equipped with an ITM module, by means of a simulation model, purposely developed to evaluate partial combustion of *char* with oxygen permeated through a tubular membrane.
In addition, a laboratory test rig has been designed, realized and is presently under operation to investigate experimentally oxygen permeation as a function of temperature exhibited by commercially available capillary membranes [7]. This study will also help identifying technological problems related to the assembly of the oxygen permeating membrane in a self standing module that could be inserted and operated in a gasification reactor.

2. Materials

Dense perovskite membranes exhibit ion permeability at elevated temperatures when subjected to a gradient of oxygen chemical potential [6]. This oxygen semi-permeability is generally attributed to the partial substitution of both A e B cations (in this study A = Ba, Sr and B = Co, Fe) in the perovskite lattice structure $(Ba,Sr)(Co,Fe)O_3$ [8, 9] and the formation of oxygen vacancies.
Their orthorhombic structure makes perovskite materials very stable at high temperature. However, the perovskite itself does not show the capability to conduct oxygen ions; there must be a certain amount of imperfections or atomic defects produced due to the non-stoichiometry of the perovskite materials.
These positively charged vacancies can be disordered at high temperature [10] and are considered to be the most mobile ionic species in these materials [11-13]. The transport of electric charges is compensated by the transport of electron-holes in the reverse direction.
Following Teraoka et al.'s work [8, 9], experimental measurements of oxygen fluxes conducted by Xu and Thomson [6] allowed to conclude that the oxygen permeation at low temperatures (750 °C) is limited by the rate of oxygen-ion recombination but is dominantly controlled by bulk diffusion at high temperatures (> 800 °C).
The cubic perovskite $Ba_{0.5}Sr_{0.5}Co_{0.8}Fe_{0.2}O_{3-\delta}$ (BSCF) is the state of the art ceramic membrane material for oxygen separation technologies above 800 °C making them most interesting for application in the gasification systems. BSCF membranes exhibit one of the highest oxygen permeabilities reported so far for dense oxides combined with excellent phase stability.
The overall oxygen flux is described by the following equation proposed by Tan and Li [14], who adapted to the case of tubular membranes the oxygen transport model originally proposed by Xu and Thomson for flat membranes [6]:

$$J_{O2} = \frac{\left[k_r \cdot \left(P_{O2}^{I\,0.5} - P_{O2}^{II\,0.5}\right)\right]}{\left[\frac{2 \cdot t \cdot k_f}{D_V}\left(P_{O2}^{I} \cdot P_{O2}^{II}\right)^{0.5} + \frac{d_m}{d_{io}} \cdot \left(P_{O2}^{II}\right)^{0.5} + \frac{d_m}{d_o} \cdot \left(P_{O2}^{I}\right)^{0.5}\right]} \quad (1)$$

where P_{O2}^{I} and P_{O2}^{II} are the oxygen partial pressures at the membrane surface, on the feeding-side and on the permeate-side, respectively.
Each of the parameters in the above equation (D_V, k_r and k_f) is expressed as a function of the membrane temperature, T_w, by means of an Arrhenius type equation. Pre-exponential factor and

activation energy values were reported by Xu and Thomson [6] by fitting experimental data of oxygen permeation rates through LSCF ($La_{0.6}Sr_{0.4}Co_{0.2}Fe_{0.8}O_{3-\delta}$) ion-conducting membranes.
More recently, oxygen permeation properties were investigated using a high temperature permeation setup at TU/e (Technical University of Eindhoven, Nederland) [15,16]; two BSCF tubular membranes produced by VITO, with different thickness, were tested. Their specifications are listed in Table 1, together with corresponding parameters characterizing the membrane samples adopted in this study.
Two basic parameters affect positively the oxygen permeation flux: temperature and oxygen partial pressure on the permeate side. High temperature systems improve oxygen permeation rate because of the Arrhenius type dependence law of the parameters D_V, k_r and k_f. On the other hand, by increasing the sweep gas flow rate on the permeate side, the oxygen pressure decreases, as a consequence the driving force across the membrane increases with an improvement in the oxygen permeation flux.

Table 1. Chemical and geometric characteristics of membranes tested at TU/e [15] and in this study.

University	Composition	Acronym	Length [mm]	Outer Diameter [mm]	Thickness [mm]
TU/e	$Ba_{0.5}Sr_{0.5}Co_{0.8}Fe_{0.2}O_{3-}$	$BSCF_{(1)}$	100	2.7	0.20
TU/e	$Ba_{0.5}Sr_{0.5}Co_{0.8}Fe_{0.2}O_{3-}$	$BSCF_{(2)}$	100	3.5	0.50
This study	$Ba_{0.5}Sr_{0.5}Co_{0.8}Fe_{0.2}O_{3-}$	$BSCF_{(3)}$	60	3.5	0.35

A MATLAB® routine was developed to solve a non linear least-squares regression model applied to the permeation data (oxygen fluxes) obtained with the newly developed membranes [1, 15]. In the equation (1) of the oxygen permeation flux, the bulk diffusion coefficient, D_V, and the forward surface exchange rate constant, k_f, appear as their ratio and the contribution of this term is relatively small when compared to the remaining terms. In addition, k_r is a multiplying factor of the whole expression of the oxygen flux. For these reasons, in the regression model k_f and D_V have been maintained those suggested by Tan and Li [14], whereas the pre-exponential factor and the activation energy of k_r have been determined, those minimizing the difference between calculated and experimental values.
As shown in Table 2, it is worth noticing a substantial reduction in the k_r activation energy that is halved compared to the respective value proposed by the previous model [6], confirming that these newly developed membranes are activated at relatively lower temperatures.
In Figure 1, experimental measurements done at TU/e and model predictions of oxygen fluxes are compared in a parity plot: an excellent match is obtained.

Table 2. Pre-exponential factor and activation energy of the parameters in equation (1): values proposed by Xu and Thomson [6] and (last row) the modified k_r according to the state of the art permeation data.

Parameter	Pre-exponential factor	Activation Energy, E_A [kJ/mol]
D_V [cm²/s]	$1.58 \cdot 10^{-2}$	73.6
k_f [cm/atm$^{0.5}$/s]	$5.90 \cdot 10^{6}$	226.8
k_r [mol/cm²/s]	$2.07 \cdot 10^{4}$	241.3
k_r [mol/cm²/s] (state of the art)	3.33	122.0

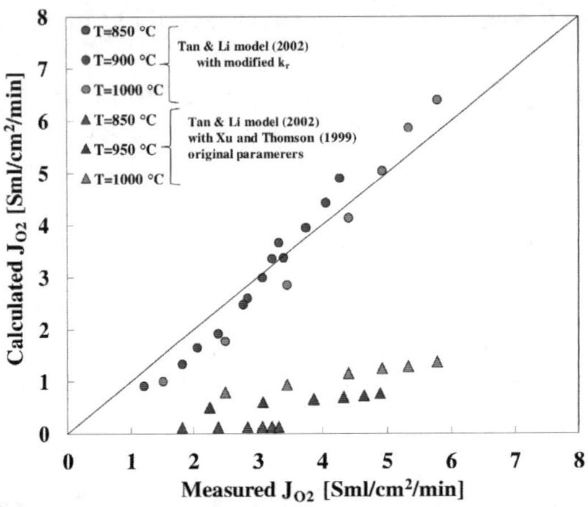

Figure 1. Parity plot of measured vs. calculated O_2 permeability through VITO BSCF tubular membranes, at different temperatures; predictions are performed with the Xu and Thomson model [6] utilizing the original D_V and k_f parameter values and k_r proposed in [1, 15] (Table 2), respectively.

3. Test module to measure oxygen permeation fluxes

Oxygen permeability and structural stability are two key factors of MIEC (Mixed Ionic Electronic Conducting) materials. Generally, material stability is opposite to the oxygen flux value.

In order to seek the tradeoff between the two factors according to the requirements of the environment for a specific application, the design of a module for oxygen generation with a BSCF membrane has been developed.

There are different schools of thought to connect the membrane with the test module.

Fixed-ends configurations, although widely suggested in literature [16] resulted inadequate, causing membranes to break during heat treatment because of different thermal expansion of the perovskite material and the remaining alumina and steel pipes of the reactor module. Instead, a suitable reactor layout is a *once-through* configuration, to reduce thermal stresses by allowing differential dilatation of materials during heating cycles.

An additional goal pursued in this research activity was to minimize the junctions placed inside the hot furnace, limiting them to those involving the membrane itself. The steel/alumina connections to arrange the input of the feeding stream and the output of the permeate stream were located outside the hot region and stainless steel pipes have been used, connected by means of standard SWAGELOK ® junctions. Figure 2 illustrates the whole layout.

The most critical point in assembling the test module is sealing the perovskite capillary membrane to the alumina pipes for adduction and withdrawal of gaseous streams.

The sealing system has been widely studied at TU/e [16], therefore in this work the same technique will be adopted in assembling the membrane reactor.

The inlet gas, helium, see Figure 2, is fed from the bottom of a 99.7% Al_2O_3 capillary (4) placed inside two co-axial 99.7 % Al_2O_3 pipes (1) and (5), where the capillary membrane (3) is housed. Air, at room conditions, flows between the electric furnace inside wall and the membrane outside surface. The test module is shown in Figure 2b and its components are listed in Table 3.

The first step in assembling the module (see Figure 2) is to fix the inner (600 mm long) and outer (400 mm long) Al_2O_3 tubes with silicone adhesive in their respective stainless steel holders (6) and (8), as shown in Figure 3A – 3B.

With reference to Figure 2, the perovskite capillary membrane was connected to alumina pipes by a suitable high temperature gas-tight sealant. Both edges of the membrane surface are protected from chemical reactions by spreading MaTeck gold conducting paste on a capillary segment about 1 cm deep, on each side, see Figure 3C.

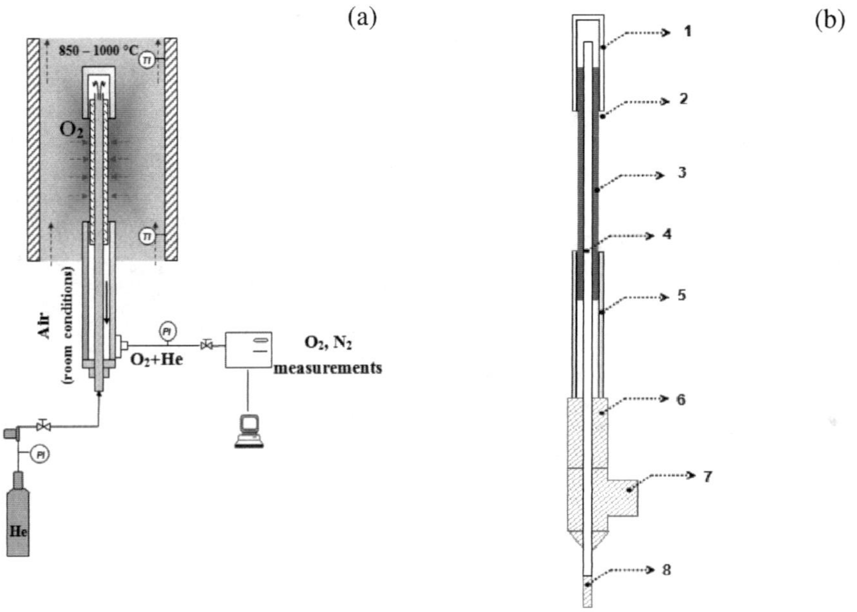

Figure 2. a) Test rig for high temperature oxygen permeation measurements; b) detailed view of the membrane module (components are listed in Table 3).

Table 3. Components of BSCF capillary test module (reference is made to Figure 2b).

1	Al_2O_3 cup (L_1=20 mm, OD_1=6 mm, ID_1=4 mm)
2	Sealing system (gold paste + glass seal)
3	BSCF capillary membrane (as in Table 1)
4	Inner Al_2O_3 tube (L_4=600 mm, OD_4=1.6 mm, ID_4=1 mm)
5	Outer Al_2O_3 tube (L_5=400 mm, OD_5=6 mm, ID_5=4 mm)
6	Stainless steel fitting (Al_2O_3 outer pipe holder)
7	"Tee" tube fitting (permeate output flow, O_2 in He)
8	Stainless steel fitting (He input flow)

The gold paste was dried for 1 h at room temperature and then placed in an oven for the curing procedure (heated up to 1000 °C at a rate of 2 °C/min, kept for 1 h at the maximum temperature, then cooled down to room temperature). The second step consisted of applying the glass seal (Schott 8252) on top of the gold paste, Figure 3D, inserting both membrane heads into the alumina pipes and sealing the joints by heating the sample to 1020 °C at a rate of 2 °C/min, holding it for 1 h at the maximum temperature level and cooling it down to the desired operating temperature (about 850 °C). Gas leaks were monitored by measuring the concentration of nitrogen on the permeate side (by GC analysis).

3.1 Estimated high temperature oxygen permeation fluxes

Preliminary permeability tests were carried out in the range of temperature 850 – 950 °C; the helium flow rate on the sweep side was varied from 80 to 350 Sml/min in order to investigate the influence of the O_2 partial pressure in the permeate. During the experimental campaign, a maximum heating rate of 2 °C/min was used to prevent the sealing and the membrane from cracking due to thermal stresses and

Journal of Physics: Conference Series **655** (2015) 012034 doi:10.1088/1742-6596/655/1/012034

the temperature level was always kept above 800 °C. The oxygen permeation flux estimated from the experimental measurements is reported in Figure 4 as a function of temperature, at different He flow rates.

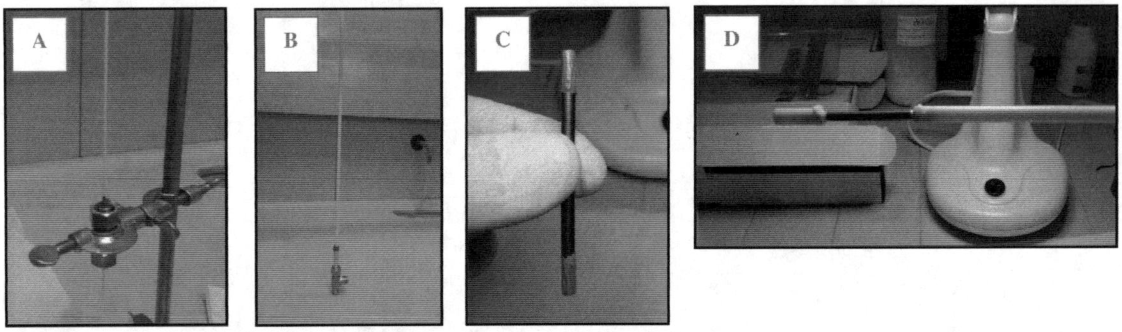

Figure 3. Assembling sequence of the test module for high temperature oxygen permeation measurements: sealing the inner (A) and outer (B) Al_2O_3 tubes in their respective stainless steel holder, (C) MaTeck gold conducting paste at the end of the BSCF capillary, (D) sealing the membrane with the Schott glass paste between the Al_2O_3 cup and the outer tube

The permeation fluxes obtained in this study (Figure 4b) are compared with the corresponding data obtained at TU/e in the experimental study mentioned above [15] (Figure 4a). The permeability shown in Figure 4b is somewhat higher than that in Figure 4a, although of the same order of magnitude; this result can be partially ascribed to the reduced thickness of the membrane employed in this study (Table 1). According to the Arrhenius type dependence law, O_2 permeation rates increase considerably with increasing temperature. A positive effect on the O_2 flux values is also found increasing the sweep gas flow rate, He, because the oxygen partial pressure on the permeate side decreases.

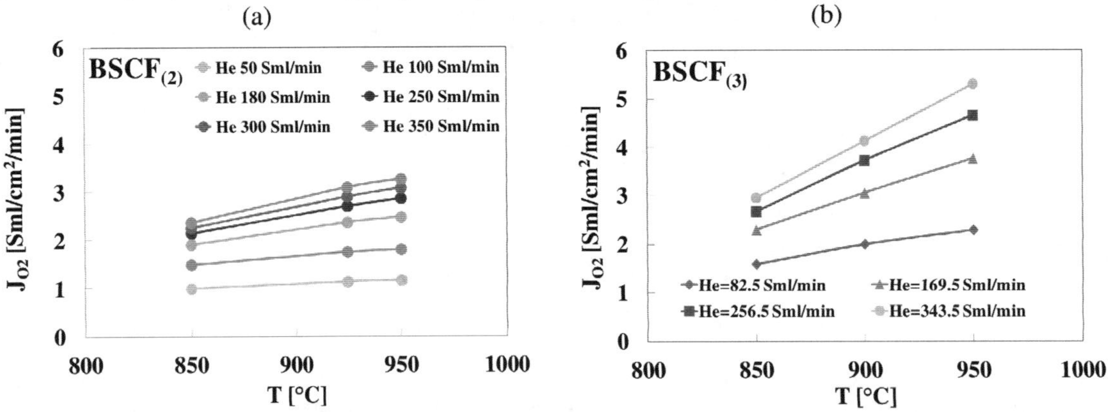

Figure 4. O_2 permeation fluxes as a function of temperature obtained at TU/e (a) and in this study (b) with the BSCF membrane specifications reported in Table 1.

In the experimental procedure, gas samples were analyzed by gas chromatography using a Thermal Conductivity Detector (TCD) and helium as gas carrier (its flow rate is fixed at 20 ml/min). The calibration curve is obtained by performing GC analysis of three standard gas mixtures (1% O_2 - 4% N_2 in He, 10% O_2 - 40% N_2 in He and 30% O_2 - 50% N_2 in He): the relative surface area of peaks is used in the calculations with the unknown gas samples collected during the experiments.

4. Computational model of ITM oxygen permeation in a biomass gasifier

In a previous work [1], a new concept for integration of gasification with oxygen transport membranes and hot gas cleaning in one reactor vessel has been developed in order to obtain a clean fuel gas from biomass that would allow immediate and efficient conversion into power (high temperature fuel cells, gas turbines, bio-fuels or chemicals). A simplified schematization of that concept is illustrated in Figure 5.

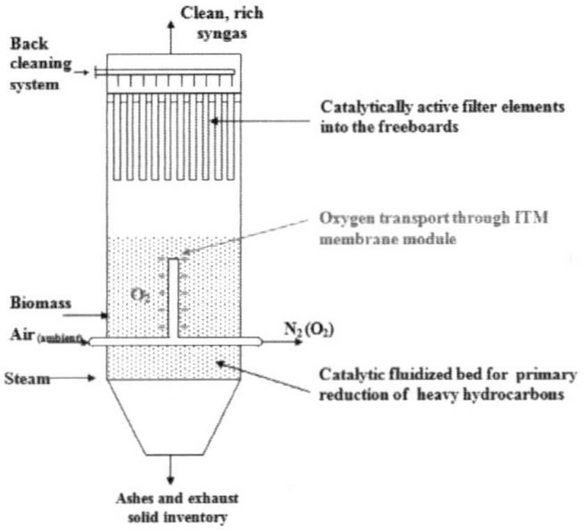

Figure 5. Integration of gasification with ITM membrane module.

In order to investigate the influence of temperature on oxygen permeation flux, numerical evaluations are performed for a single ITM capillary membrane, with the specifications summarized in Table 3, immersed vertically inside the fluidized bed where particles are considered uniformly dispersed in the bed layer surrounding the membrane module and air (oxygen 21% by volume) at atmospheric pressure flows inside the membrane; the air superficial velocity inside the ITM module is fixed (u_{in} = 5 m/s).
In small to medium scale applications (up to 10 MW_{th}) , the knowledge of the gasifier total oxygen demand allows to fix the inlet air flow rate, Q_{air}, and the oxygen transfer system configuration:

$$Q_{air} \frac{P}{R\,T} (y_{O_{2,in}} - y_{O_{2,out}}) = J_{O_2} A_m \, N \, (1 - y_{O_{2,out}}) \tag{2}$$

where the oxygen depletion in the outlet stream $y_{O_{2,out}}$ is kept above a fixed limiting value,
The oxygen demand to assure autothermal behavior of a pilot scale biomass gasifier is of the order of 10 moles per kg of biomass (about 25% of what would be needed for complete combustion); so with a biomass feedstock of 200 kg/h (corresponding to a gasifer capacity of 1 MW_{th}) the oxygen demand would be of the order of 2 kmol/h and the membrane surface area of about 18 m^2, with an expected oxygen permeation flux of 3 $\mu mol/cm^2/s$. The required number of BSCF membrane pipes can be easily arranged inside the gasifier vessel, considering both a reactor inner diameter of the order of 0.3 - 0.5 m. and a reactor height of the order of 4 - 5 m. The membrane bundle configuration could be chosen so to assure optimum oxygen distribution within the reactor; membrane pipes with greater diameter could be also adopted to reduce their number. The *char* fraction inside the fluidized bed is fixed to be on average w_c = 6 % by weight, according to the gasification results reported in [17].

4.1 Description of model equations
In what follows, reference is made to oxygen transfer to the fluidized bed and its consumption by means of *char* combustion: an heterogeneous reaction process. Once the expected oxygen permeation flux is fixed, the geometric specifications of the ITM system are obtained:

$$A_m = \pi \cdot d_m \cdot L, \qquad d_m = \frac{2 \cdot t}{ln[(d_o)/d_{io}]} \tag{3}$$

where t is the membrane thickness, d_o and d_{io} are the perovskite membrane outside and inside diameters, respectively.
Local steady state conditions are imposed; knowing from previous works [18, 1] that the resistance on the air side is negligible allows to disregard it. Two oxygen transfer resistances in series are therefore

considered: the former in the perovskite membrane, the latter related to oxygen mass transfer in the fluidized bed, enhanced by simultaneous *char* combustion.
The combustion process of small *char* particles in the fluidized bed is relatively fast, so that the depth of penetration of the concentration and temperature perturbation is expected very thin in comparison with the fluidized bed scale, of the order of a few millimetres. As a result, a one-dimensional analysis in the tubular membrane radial coordinate is appropriate and resistance to mass and energy transport are assumed to take place in a gaseous layer at the inter-phase transfer surface (sweep side).
The global combustion reaction rate, $\mathbf{r_c}$ (*char* moles burnt per unit time and fluidized bed volume) is assumed to depend linearly on the local oxygen concentration [19]:

$$\boldsymbol{r_c} = k_O \cdot C_{O2}; \qquad 1/k_O = 1/(k_c \cdot a_c) + 1/\left(k \cdot \rho_{M,c}/\rho_{M,g}\right) \tag{4}$$

A diffusion-controlled combustion regime is established because the mass transfer rate constant is small compared to the reaction rate constant at the high temperature level in the gasifier.
If *char* is considered to be pure carbon, the reaction stoichiometry can be expressed as follows:

$$C + \delta O_2 \rightarrow 2(1-\delta)CO + (2\delta - 1)CO_2 \tag{5}$$

where the value of δ is estimated from Arthur empirical correlation [20] as a function of temperature.
The condition of zero oxygen accumulation at the membrane/fluidized bed interface results in the following relation expressing the equality between the oxygen flow transported through the membrane and that diffusing toward the fluidized bed:

$$J_{O_2} \cdot d_m \; = \; -\,\rho_{M,gas} \cdot D^e_{O_2} \cdot \left(\frac{dy_{O_2}}{dr}\right) \cdot 2\,r \;\; at\, r \; = \; d_o/2 \tag{6}$$

In order to evaluate the derivative on the RHS of the above equation, the oxygen concentration profile should be obtained by integration of the differential, stationary mol balance in the fluidized bed layer adjacent to the membrane pipe:

$$\rho_{M,gas} \cdot D^e_{O_2} \cdot \frac{d}{dr} \cdot \left[r \cdot \left(\frac{dy_{O_2}}{dr}\right)\right] = \delta \cdot r \cdot \boldsymbol{r_c} \tag{7}$$

The product $\rho_{M,gas} \cdot D^e_{O_2}$ can be reasonably considered constant in this study because temperature fluctuates no more than 5% of its absolute value in the fluidized bed, T_B, as it will be shown later.
Boundary conditions:

$$r = \frac{d_o}{2} \rightarrow y_{O_2} = P^{II}_{O_2}/P \qquad r = \infty \rightarrow y_{O_2} = y_{O_{2,bulk}} = 0 \tag{8}$$

The above equations (1) and (6) allow to determine the oxygen flux, J_{O2}, and the oxygen partial pressure on the sweep side, P_{O2}^{II} under isothermal conditions.
To estimate the importance of temperature perturbation due to the *char* combustion process, an energy balance is coupled with the oxygen molar balance.
The combustion of *char* should in principle determine a temperature increase in the fluidized bed close to the membrane surface, resulting in a membrane temperature value, T_w, above that in the bulk of the bed and in turn improving the oxygen permeation rate, due to the Arrhenius-type dependence of the parameters in equation (1). In Figure 7 both these effects are illustrated with reference to the case of study examined here.
By considering the effective thermal conductivity of *char* particles, the energy balance is expressed by the following differential equation:

$$k_t^{\,e} \cdot \frac{d}{dr} \cdot \left[r \cdot \left(\frac{dT}{dr}\right)\right] = \Delta H_r \cdot r \cdot \boldsymbol{r_c} \tag{9}$$

Boundary conditions:

$$r = d_o/2 \rightarrow \frac{dT}{dr} = \frac{h_g \cdot (T_w - T_a) \cdot d_{io}}{k_t^{\,e} \cdot d_o} \qquad\qquad r = \infty \rightarrow T = T_B \tag{10}$$

At the membrane boundary, it is imposed that the heat transfer rate from the fluidized bed to the membrane should be equal to that from the membrane to the air stream. On the other hand, with reference to the boundary condition in the bulk of the gasifier, in the numerical integration it has in

fact been assumed that temperature reaches the value T_B at a distance from the membrane corresponding to complete consumption of oxygen. As a matter of fact, fluidized beds are characterized by a strong tendency to smooth temperature differences, over distances comparable with the thickness of one particle layer, as soon as the causing perturbation extinguishes [21].
To estimate the importance of this thermal perturbation quantitatively (Figure 7), it is assumed that the temperature in the bulk of the air stream on the feeding side, T_a, the fluidized bed temperature, T_B, and that of the membrane, T_w, are uniform along the membrane, and the steady state energy balance through the fluidized bed layer surrounding the whole membrane pipe is considered, coupled with the oxygen molar balance corresponding to the oxygen partial pressure at the air inlet section. The contribution of radiative heat transfer is neglected in equation (9) [22].
The system of equations (7) and (9) with corresponding boundary conditions are integrated and iterations are made till the membrane temperature, T_w, assumes a steady value that satisfies both, oxygen transfer and heat transfer constraints.

4.2 Results and discussion
The model described above was implemented in MATLAB®, and calculations were performed with reference to the membrane geometry illustrated in Figure 2b.

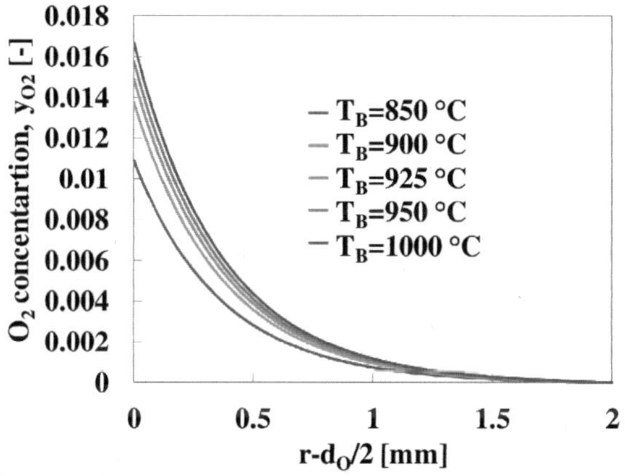

Figure 6. Calculated O_2 concentration in the fluidized bed as a function of the radial distance from the membrane surface in the range of bed temperature 850-1000 °C (w_c=6 %).

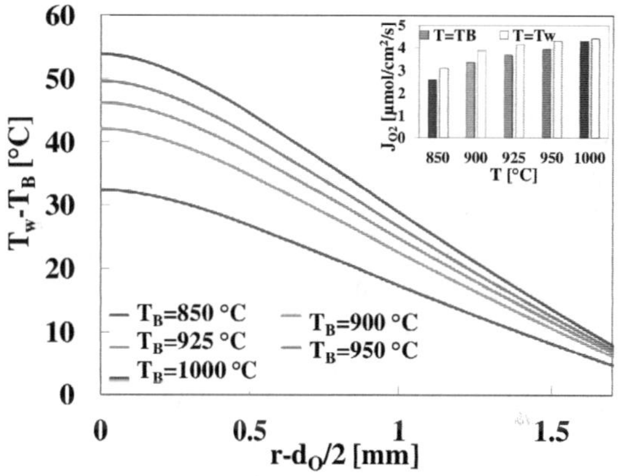

Figure 7. Calculated temperature profile as a function of the radial distance from the membrane surface in the range of bed temperature 850-1000 °C. In the insert, effect of membrane temperature increase on the O_2 permeation flux: comparison at $T=T_B$ (bed temperature) (shaded bar) and $T=T_W$ (membrane temperature) (open bar), respectively (w_c=6 %).

Figure 6 shows the oxygen molar fraction in the fluidized bed as a function of the radial distance from the membrane outer surface, with reference to the air inlet horizontal section of the membrane pipe (z = 0). It is confirmed that the O_2 concentration in the fluidized bed becomes negligible at a distance equal to or less than the membrane external diameter (d_O = 3.5 mm): the greater the *char* concentration, the steeper is the O_2 concentration profile, because the combustion reaction rate is faster.
Finally, Figure 7 allows to estimate quantitatively the temperature perturbation brought about by the exothermic *char* combustion reaction with an appreciable improvement of the membrane permeability. Temperature is imposed to approach the value T_B at a distance of about 1.7 mm from the membrane outer surface corresponding to complete consumption of oxygen, as shown in Figure 6. At T_w= 1054 °C the highest oxygen permeation rate is obtained, J_{O2}=4.4 µmol/cm^2/s.

5. Conclusions

Ion transport membranes integrated in the gasification environment are suitable to separate oxygen from air and feed it to the reactor in order to provide the necessary energy input for the autothermal gasification process, by means of combustion reactions.

Newly developed BSCF-type membranes have been tested to check the influence of temperature and sweeping gas flow rate on the oxygen permeation rates. The values observed are higher than those reported in [6] and slightly above those recently obtained at TU/e [15], with a different lab rig and measurement technique. The expected trends with membrane temperature and sweeping gas flow rate, respectively, have been validated experimentally.

The positive effect of temperature is also confirmed by the computational model. Satisfactory oxygen permeation fluxes are predicted, enough to house the whole membrane module inside the gasifier. The permeability increase due to the exothermic *char* combustion process at the membrane surface is substantial.

Finally, a test module for high temperature oxygen permeation measurements has been realized, which allows to test ITM mechanical resistance, durability and stability in inert and reactive environment.

6. Acknowledgments

The financial support from Italian National Agency for New Technologies, Energy and Sustainable Economic Development (ENEA) to this research activity is acknowledged.

7. References

[1] Antonini T, Gallucci K, Anzoletti V, Stendardo S, Foscolo P U 2014 *Chem. Eng. Process.* doi:10.1016/j.cep.2014.11.009
[2] Heidenreich S, Foscolo P U 2015 *Prog. Energy Combust. Sci.* **46** 72-95
[3] Leo A, Liu S, Diniz Da Costa J C 2009 *Int. J. Greenh. Gas Control* **3** 357-397
[4] Steele B C H 1992 *Mater. Sci. Eng. B* **13** 79-87.
[5] Bouwmeester H J M., Kruidhof H, Burggraaf A J 1994 *Solid State Ionics* **72** 185-194.
[6] Xu S J, Thomson W J 1999 *Chem. Eng. Sci.* **54** 3839–50.
[7] VITO (Vision on technology), Catalogue Materials Technology, www.vito.be.
[8] Teraoka Y, Zhang H M, Furukawa S, Yamazoe N 1985 *Chem. Lett.* 1743-1746.
[9] Teraoka Y, Zhang H M, Okamoto K, Yamazoe N 1988 *Mat. Res. Bull.* **23** 51-58.
[10] Kruidhof H, Bouwmeester H J M, van Doorn R H J E, Burggraaf A J 1993 *Solid State Ionics* **63 (65)** 816-822.
[11] Schaffrin C 1976 *Phys. Status Solidi A* **35** 79-88
[12] Baiatu T, Waser R, Hardtl K H 1990 *J. Am. Ceram. Soc.* **73** 1663-1673.
[13] Smyth D M 1991 *Ferroelectr.* **116** 117-124.
[14] Tan X, Li K 2002 AIChE J. **48** 1469-77.
[15] Anzoletti V 2014 MSc Dissertation in Chemical Engineering, University of L'Aquila.
[16] Di Felice L, Middelkoop V, Anzoletti V, Snijkers F, Van Sint Annaland M, Gallucci F 2014 *Chem. Eng. Process.* doi:10.1016/j.cep.2014.12.004
[17] Rapagnà S, Gallucci K, Di Marcello M, Foscolo P U, Nacken M, Heidenreich S 2009 *Energy Fuels* **23** 3804-3809
[18] Antonini T, Gallucci K, Foscolo P U 2014 *Chem. Eng. Trans.* **37** 91-96.
[19] Di Blasi C 2000 *Chem. Eng. Sci.* **55** 2931-2944
[20] Arthur J R 1950 *Trans. Faraday Soc.* **47** 164-178
[21] Glicksman L R 1984, Heat transfer in fluidized bed combustors, ed P Basu (Toronto: Pergamon Press) 63-100.
[22] Chen J C, Chen K L 1981 *Chem. Eng. Commun.* **9** 255-271.

Modelling heat and mass transfer in a membrane-based air-to-air enthalpy exchanger

S Dugaria, L Moro, D Del Col

Università degli Studi di Padova, *Dipartimento di Ingegneria Industriale*
via Venezia 1, 35131 Padova, Italy.

E-Mail: simone.dugaria@dii.unipd.it

Abstract The diffusion of total energy recovery systems could lead to a significant reduction in the energy demand for building air-conditioning. With these devices, sensible heat and humidity can be recovered in winter from the exhaust airstream, while, in summer, the incoming air stream can be cooled and dehumidified by transferring the excess heat and moisture to the exhaust air stream. Membrane based enthalpy exchangers are composed by different channels separated by semi-permeable membranes. The membrane allows moisture transfer under vapour pressure difference, or water concentration difference, between the two sides and, at the same time, it is ideally impermeable to air and other contaminants present in exhaust air. Heat transfer between the airstreams occurs through the membrane due to the temperature gradient. The aim of this work is to develop a detailed model of the coupled heat and mass transfer mechanisms through the membrane between the two airstreams. After a review of the most relevant models published in the scientific literature, the governing equations are presented and some simplifying assumptions are analysed and discussed. As a result, a steady-state, two-dimensional finite difference numerical model is setup. The developed model is able to predict temperature and humidity evolution inside the channels. Sensible and latent heat transfer rate, as well as moisture transfer rate, are determined. A sensitive analysis is conducted in order to determine the more influential parameters on the thermal and vapour transfer.

1. Introduction

In recent years, forced ventilation has been widely used in large buildings where it is required to provide a comfortable and healthy indoor environment for the occupants. Although exchange between indoor and outdoor air is necessary to maintain the indoor air quality, the energy demand associated with treating the outdoor air can be significant. A smart way to reduce this consumption is to adopt total energy recovery systems. These devices are able to transfer sensible heat and moisture between supply and exhaust airstreams. Since the latent heat of moisture can account for large part of the energy in the exhaust air, recovery of total energy is of great importance. A total energy recovery systems is mainly characterized by the type of air-to-air enthalpy exchanger in use. The two most common types of enthalpy exchangers are membrane based planar plate-type enthalpy exchangers and enthalpy wheels.

A membrane-based enthalpy exchanger is an air-to-air heat exchanger with a semi-permeable membrane used for heat recovery in modern building ventilation system. Supply and exhaust airstreams at different temperature and humidity transfer heat and moisture across the membrane by flowing through the exchanger. The thermal performance of the exchanger are mainly affected by its

Journal of Physics: Conference Series **655** (2015) 012035 doi:10.1088/1742-6596/655/1/012035

geometry, and in particular by the flow configuration of the airstreams. Traditionally the airstreams were in crossflow, but in recent years the quasi-counterflow configuration has been widely used, due to its higher thermal performance. Figure 1 shows the membrane shape of enthalpy exchanger in crossflow and in quasi-counterflow configuration. The red lines highlight the computational discretization.

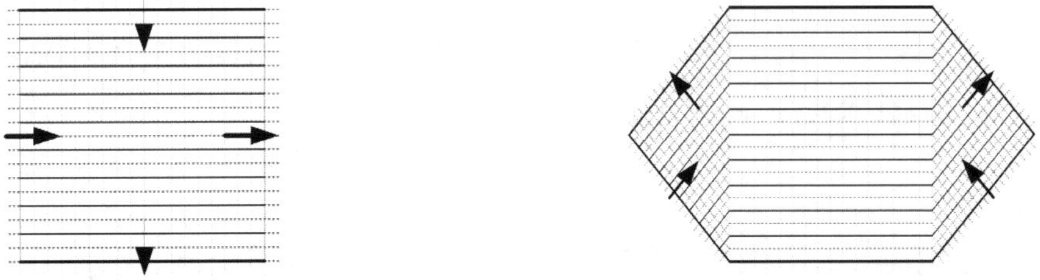

Figure 1. Plate of planar enthalpy exchanger in crossflow (left) and quasi-counterflow (right) arrangements.

In order to promote thermal and moisture transfer, the heat transfer and mass transfer resistance of the membrane should be as lower as possible. On the other hand, the membrane must be stiff enough to withstand the pressure difference and limit the surface warping under both dry and wet conditions. Since the moisture transfer is mainly affected by the moisture permeability of the membrane, it is essential to understand how its physical characteristics affect the exchange performance.

In the last decades, two research groups have performed experimental studies on the characterization of semi-permeable membranes for energy exchange purpose [1,2,3,4,5]. Table 1 summarizes the main results of their works.

Table 1. Typical values of membrane properties

Material	λ_M W m^{-1} K^{-1}	$D_{w,M}$ kg m^{-1} s^{-1}	C_s -	θ_{max} kg$_w$ kg$_M^{-1}$	δ_M mm	*Reference*
Copolimer	-	$2{,}16 \cdot 10^{-8}$	2,5	0,23	0,02	[1]
Paper	0,44	$5{,}33 \cdot 10^{-9}$	6	0,92	0,055	[2]
Cellulose acetate	0,41	$7{,}98 \cdot 10^{-9}$	11.4	0,43	0,005+0,04[a]	[2]
Modified cellulose acetate	0,44	$2{,}5 \cdot 10^{-9}$	8.64	2,5	0,005+0,04[a]	[2]
PES	0,15[b]	$5{,}41 \cdot 10^{-7}$	2.49	0,07	0,095	[3]
PVDF	0,19[b]	$1{,}92 \cdot 10^{-6}$	10.26	0,03	0,082	[3]
Cellulose	0,4[b]	$7{,}10 \cdot 10^{-7}$	8.1	0,25	0,115	[3]

[a] thickness of the support PP net (polypropylene) [b] typical value reported in the literature

The literature contains also studies on modeling membrane-based enthalpy exchangers. Niu and Zhang [1] developed a theoretical model to evaluate the performance of a membrane-based enthalpy exchanger and to investigate the effect of the inlet air conditions. Zhang and Niu [6] developed performance correlations for quick estimation of the sensible, latent and enthalpy effectiveness of the exchanger. They based their further studies [2,7,8] on these works. Min et al. performed similar works by studying heat and mass transfer processes across the membrane [9,10,11]. They analyzed the effect of the heat of adsorption on the process of heat transfer in moisture exchange across a membrane [12,13,14].

In the current work, a detailed model of the coupled heat and mass transfer mechanisms through the membrane between the two airstreams has been developed. The model is able to predict temperature and humidity profile inside the channels. Using this model as a base, the performance of the enthalpy exchanger was evaluated in terms of sensible and latent heat. A sensitive analysis is conducted in order to determine the most influential parameters on the thermal and moisture transfer.

2. Theoretical model

1.1. Physical model

The model consists of supply stream and exhaust stream channels separated by a membrane. Figure 4 shows the schematic of a channel in a typical membrane-based planar plate heat exchanger. The membrane are separated by plastic spacer with the purpose to guide the stream across the axial direction.

In the quasi-counterflow arrangement, supply and exhaust streams in the core-section are in counterflow, whereas in the two head-section the streams are in crossflow (Figure 1, right). Heat and water vapour are transferred transversally to the flow direction through the membrane from the hot and humid stream to the cold and drier one. According to the Solution-Diffusion model [15], when moisture transfers through the membrane, it first adsorbs at high concentration side of the membrane releasing heat. Then, under the driving force of concentration gradients in the membrane, the adsorbed water diffuses through the membrane and finally desorbs from the other (low concentration) side of the membrane absorbing heat.

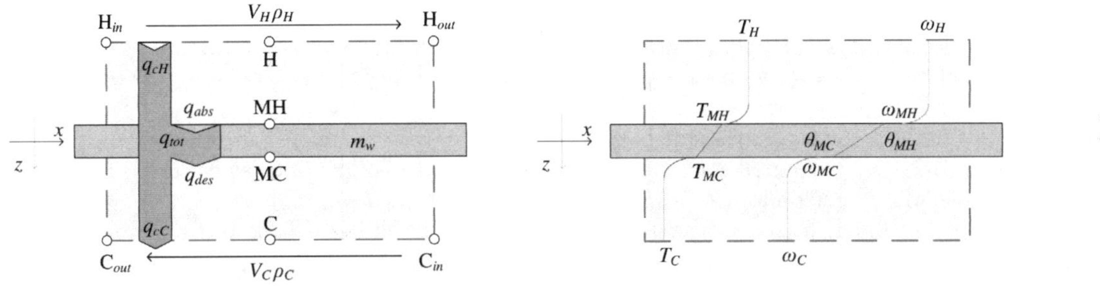

Figure 2. Left: one-dimensional schematic of the heat and mass transfer. Right: temperature T, humidity ω and water uptake θ distribution profiles across the membrane.

During the equilibrium between the membrane and the moisture at its surface, the sorption property of the membrane can be described using sorption isotherms. A popular sorption isotherm equation widely used in the gas separations permits to express the moisture uptake θ ($kg_w\ kg_M^{-1}$) of the membrane as a function of the relative humidity ϕ at the system temperature T [16]:

$$\theta = \theta_{max}\left(1 - C_S + \frac{C_S}{\phi(T)}\right)^{-1} \tag{1}$$

where θ_{max} is the maximum moisture content of the membrane corresponding to the saturation condition (ϕ=100%) and C_S is the membrane sorption coefficient.

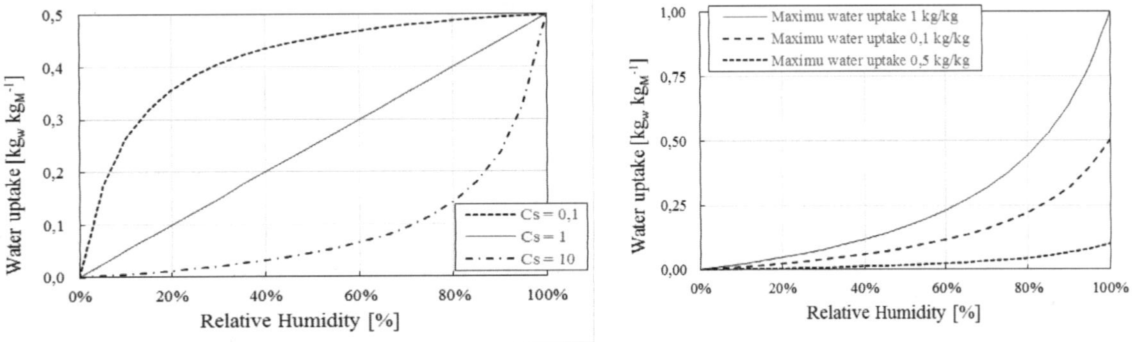

Figure 3. Typical sorption curves when varying the the sorption coefficient C_s (left) and the moisture uptake θ_{max} (right) .

C_S is a variable that represents the shape of the sorption curve and the type of sorption. For C_S =1 (silica gel), the sorption curve is linear with the relative humidity and corresponds to Herny sorption.

C_S <1 (molecule sieve) indicates that the sorption curve is an upward convex curve, which corresponds to Langmiur sorption. C_S >1 (polymers) yields a downward convex curve, which applies to most of the membrane used for enthalpy exchanger as can be seen in Table 1. Figure 3 shows sorption curves for different values of the sorption coefficient C_S (left) and of the maximum moisture uptake θ_{max} (right).

1.2. Modeling assumptions
The mathematical model of coupled heat and mass transfer in a membrane based heat exchanger is based on the following assumptions:

1. Both the heat and mass transfer processes are at steady state.
2. Heat conduction and vapour diffusion in parallel direction to the channel (x, y) are negligible compared to the bulk convection.
3. Heat conduction through the plastic spacer is negligible as compared to the bulk convection.
4. Moisture diffusion in the membrane only occurs in the transvers direction (z) along the thickness
5. The physical properties of the membrane are constant with temperature and the water uptake (profiles of temperature and moisture uptake in the membrane are linear in the transversal direction).
6. Adsorption and desorption of moisture are in equilibrium adsorption state.
7. The heat of adsorption-desorption is assumed constant and equal to the latent heat of condensation-vaporization.
8. Sorption hysteresis is neglected.

Niu and Zhang [1] reported that the effect of axial heat conduction and vapour transfer can be totally neglected for Peclet number Pe ($Pe = Re\ Pr$) greater than *100*.

The effect of the heat of adsorption-desorption on the process of heat and moisture exchanges across a membrane has been studied by Min and Wang [12,13]. They showed that the heat of adsorption-desorption is not a constant but depends on the membrane surface adsorption capacity and temperature, which are affected by the heat and mass transfer characteristics. The heat of adsorption can be considered composed by two parts: the heat of condensation, caused by the interaction among the adsorptive molecules, and the surface energy, caused by the interaction between the adsorptive and adsorbent molecules affected by the system temperature. In the work [12] of Min and Wang has been shown that for small moisture mass flux, the effect of the surface energy can be neglected and the heat of adsorption can be treated as a constant. Furthermore, as shown in their sequent work [13], a variable heat of adsorption-desorption requires the estimation of several membrane physical properties, usually difficult to find in literature. Hu et al. [14] made a similar study on the effect of adsorption heat on the heat transfer. They treated the adsorption heat as a constant and assumed it to be equal to latent heat of condensation. Therefore, in this work the heat of adsorption-desorption is assumed constant and equal to the latent heat of condensation-vaporization, too.

It has been assumed that the temperature along the membrane thickness exhibits a linear distribution. This condition corresponds to ignore the effect of the enthalpy carried by the mass transfer in the membrane on the temperature distribution. This assumption is reasonable because the membrane is usually really thin (less than *0,1* mm) and the mass flux is quite small compared to the bulk flow.

1.3. Governing equations and boundary conditions
The governing equations for studying the coupled heat and mass transfer in a control-volume are based on the energy and mass balances. Referring to Figure 2, the subscript H and C are associated to the hot and cold air stream, while MH and MC refer to the hot and membrane and cold side of the membrane, respectively:

Hot air stream	$\dot{V}_H \rho_H c_{p,H} \dfrac{\partial T_H}{\partial x} + 2\alpha_H (T_H - T_{MH}) = 0$	$\dot{V}_H \dfrac{\partial \omega_H}{\partial x} + 2\alpha_{m,H} (\omega_H - \omega_{MH}) = 0$	(2)
Cold air stream	$\dot{V}_C \rho_C c_{p,C} \dfrac{\partial T_C}{\partial x} + 2\alpha_C (T_C - T_{MC}) = 0$	$\dot{V}_C \dfrac{\partial \omega_C}{\partial x} + 2\alpha_{m,C} (\omega_{MC} - \omega_C) = 0$	

where V, ρ an c_p are the volumetric flow rate (m^3 s^{-1}), the density (kg m^{-3}) and the specific heat (J kg^{-1} K^{-1}) of the airstreams, The symbols α (W m^{-2} K^{-1}) and α_m (kg m^{-2} s^{-1}) are used to identify the heat and mass transfer coefficients, respectively. T (K) and ω (kg$_v$ kg$_{da}^{-1}$) are the temperature and the absolute humidity of the moist air. For the membrane,

$$\dot{m}_w c_{p,w}\frac{\partial T_M}{\partial z}-\lambda_M\frac{\partial^2 T_M}{\partial x^2}-\lambda_M\frac{\partial^2 T_M}{\partial y^2}-\lambda_M\frac{\partial^2 T_M}{\partial z^2}=0 \qquad \dot{m}_w=-D_{wM}\left(\frac{\partial\theta}{\partial z}+\frac{\partial\theta}{\partial x}+\frac{\partial\theta}{\partial y}\right) \tag{3}$$

where the moisture mass transfer flux m_w (kg m^{-2} s^{-1}) is dependent on the moisture diffusion coefficient in the membrane D_{wM} (kg m^{-2} s^{-1}) and the moisture uptake θ (kg$_w$ kg$_M^{-1}$) gradient of the membrane.

Based on the assumptions of the previous paragraph, the boundary conditions for the coupled heat and mass transfer can be expressed as follows:

Hot air stream

$$T_H\big|_{x=0}=T_{H,in} \qquad \frac{\partial T_H}{\partial y}=0$$

$$\omega_H\big|_{x=0}=\omega_{H,in} \qquad \frac{\partial \omega_H}{\partial y}=0$$

Cold air stream

$$T_C\big|_{x=0}=T_{C,in} \qquad \frac{\partial T_C}{\partial y}=0$$

$$\omega_C\big|_{x=0}=\omega_{C,in} \qquad \frac{\partial \omega_C}{\partial y}=0$$

Membrane

$$\frac{\partial T_M}{\partial x}=0 \qquad \frac{\partial T_M}{\partial y}=0$$

$$\frac{\partial \theta}{\partial x}=0 \qquad \frac{\partial \theta}{\partial y}=0$$

where $T_{H,in}$, $\omega_{H,in}$ and $T_{C,in}$, $\omega_{C,in}$ are the temperature and absolute humidity of the two airstreams at the inlet condition

Hot air – membrane interface *Cold air – membrane interface*

$$-\lambda_M\frac{\partial T_M}{\partial z}\bigg|_{\delta=0}=2\alpha_H\left(T_H-T_{MH}\right)+\dot{m}_w\Delta h_{abs} \qquad -\lambda_M\frac{\partial T_C}{\partial z}\bigg|_{z=\delta_M}=-2\alpha_C\left(T_C-T_{MC}\right)+\dot{m}_w\Delta h_{des} \tag{4}$$

where Δh_{abs} and Δh_{des} (J kg^{-1}) represent the isosteric heat of adsorption and desorption, respectively

1.4. Coupled heat and mass transfer

Figure 2 shows a schematic representation of the energy balance in a control-volume membrane. Based on this diagram, the heat transfer described in the previous governing equations can be discretized as follows:

$$2\alpha_H\left(T_H-T_{MH}\right)+\dot{m}_w\Delta h_{abs}=\frac{\lambda_M}{\delta_M}\left(T_{MH}-T_{MC}\right)=2\alpha_C\left(T_{MC}-T_C\right)+\dot{m}_w\Delta h_{des} \tag{5}$$

where δ_M (m) is the thickness of the membrane.

The convective heat fluxes q_c (W m^{-2}) between the bulk flows and the membrane surfaces are defined as,

$$q_{c,H}=\alpha_H\left(T_H-T_{MH}\right) \qquad q_{c,C}=\alpha_C\left(T_{MC}-T_C\right) \tag{6}$$

Substitution of Equation 6 in Equation 5 yield to the expression:

$$q_{c,C}=q_{c,\mathrm{H}}=\left(\left(T_H-T_C\right)-\dot{m}_w\Delta h_{lat}\left(\frac{\delta_M}{\lambda_M}\right)\right)\Bigg/\left(\frac{1}{2\alpha_M}+\frac{\delta_M}{\lambda_M}+\frac{1}{2\alpha_C}\right) \tag{7}$$

Besides the bulk temperatures (T_H and T_C) the convective heat transfer coefficients (α_C and α_H) and the conductivity of the membrane (λ_M), the convective heat is dependent from the moisture flux across the membrane m_w. The moisture mass flux is given by,

$$\dot{m}_w=2\alpha_{mH}\left(\omega_H-\omega_{MH}\right)=\frac{D_{wM}}{\delta_M}\left(\theta_{MH}-\theta_{MC}\right)=2\alpha_{mC}\left(\omega_{MC}-\omega_C\right) \tag{8}$$

According to the sorption isotherm of Equation 1, the moisture uptake, θ_{MH} and θ_{MC}, on the hot and cold side of the membrane can be related to the relative humidity at the membrane surface, hence Equation 2 becomes:

$$\theta_{MH} = \theta_{\max}\left(1 - C_S + \frac{C_S}{\phi(T_{MH})}\right)^{-1} \qquad \theta_{MC} = \theta_{\max}\left(1 - C_S + \frac{C_S}{\phi(T_{MC})}\right)^{-1} \tag{9}$$

In order to obtain an expression of the mass flux similar to that for the convective heat flux, it is useful to relate relative and absolute humidity. Using the well-known Clapeyron equation and assuming atmospheric pressure (*101325* Pa), the relationship between relative to absolute humidity can be approximated as:

$$\frac{\phi}{\omega} = 10^{-6} e^{5294/T} - 1{,}61\phi \tag{10}$$

As reported by [15] neglecting the second term on the right of Equation 10 leads to an error less than 5%. Thus, the mass flux across the membrane assume the following expression:

$$\dot{m}_w = \frac{D_{wM}}{\delta_M}\theta_{\max}\left(\left(1 - C_S + \frac{C_S}{\omega_{MH} 10^{-6} e^{5294/T_{MH}}}\right)^{-1} - \left(1 - C_S + \frac{C_S}{\omega_{MC} 10^{-6} e^{5294/T_{MC}}}\right)^{-1}\right) \tag{11}$$

This results is similar to that obtained from Min and Su [9,10]. It has to be note that the previous equation differs from that given by Zhang and Liang [7]: in the present formulation the moisture gradient across the membrane has been taken in account instead of evaluating the sorption characteristic based only on the humid side of the membrane. Since humidity can change greatly across the membrane the two formulations may differ substantially.

From the above observations, it results that heat transfer and mass transfer in membrane-based enthalpy exchangers are coupled processes that should not be traded separately.

3. Numerical model

A 2-dimensional, finite-difference model has been developed to calculate the heat and moisture exchange in the membrane based planar plate enthalpy exchanger in quasi-counterflow arrangement.

The computational domain has been chosen as a three-dimensional elementary cell taken between two consecutive channels as represented in Figure 4.

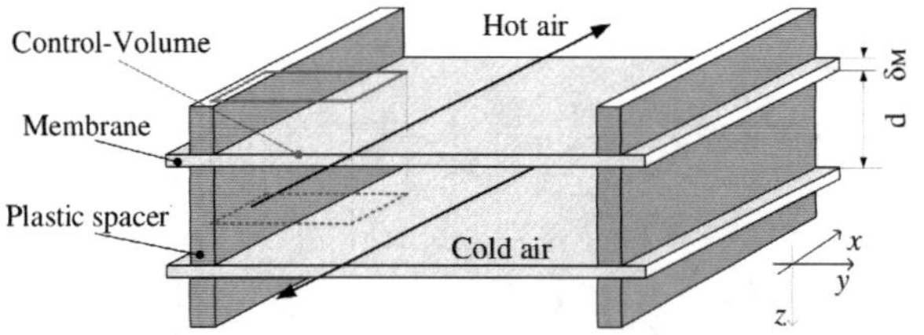

Figure 4. Schematic of a channel in membrane based planar plate heat exchanger and control-volume identification for a counterflow arrangement.

As shown in Figure 1, the geometry of the single control volume depends on its position in the exchanger. The number of elements in the head sections is determined by the number of divisions for each channel, while in the core section the number of element depends on the divisions along the axial direction and the number of division for each channel, as well.

Each control-volume requires iterative calculations to simultaneously solve Equations 5 and 8. The bulk air conditions are taken as average value between inlet and outlet conditions for each element. A main loop updates the inlet conditions for each control-volume until convergence is reached.

3.1. Heat and mass transfer coefficient in the boundary layer

To evaluate the convective heat transfer coefficient, the recent work of Gendebien et al. [17] has been taken as reference. The heat transfer coefficient in the boundary layer can be described by Nusselt correlation.

In the model, laminar flow has been considered for $Re < 2000$. The fully developed laminar flow heat transfer coefficient α (W m^{-2} K^{-1}) in rectangular duct can be calculated as function of the aspect ratio of the cross section only. The Nusselt number correlation developed by Shah and London [18] has been used. The resulting Nusselt number is then corrected to takes account of the Prandtl number and the thermal entry length effects according to Kakaç et al. [19] and to Wibulswas [20], respectively. Generally for turbulent flow, the Nusselt number correlations developed for circular tube can be use also for different cross sectional shape [21]. In this model, the Gnielinski correlation [21] is used to calculate the Nusselt number for $Re > 4000$. The Gnielinski correlation requires to estimate the friction factor, which is calculated with the correlation developed by Petukhov for smooth surfaces [21]. For transitional flow, between $Re = 2000$ and $Re = 4000$, a linear interpolation has been used.

Once the heat transfer coefficient α is known, the mass transfer coefficient α_m (kg m^{-2} s^{-1}) can be obtained from the Chilton-Colburn analogy [21].

4. Results and discussion

Simulations were first conducted to investigate the effects of the membrane characteristics on the coupled heat and mass transfer through the membrane. Calculations were done for one geometrical configuration and one set of inlet air conditions for a variety of membrane parameters.

The parameters of the simulation are summarized in Table 2.

Table 2. Model parameters.

Total length	l_{total}	400 mm	Inlet temperature (H)	$T_{H,in}$	35 °C
Total width	w_{total}	250 mm	Inlet relative humidity (H)	$\phi_{H,in}$	75 %
Total height	h_{total}	150 mm	Inlet mass flow rate (H)	$V_{H,in}$	50 m^3 h^{-1}
Core length	l_{core}	200 mm	Inlet temperature (C)	$T_{C,in}$	20 °C
Spacer thickness	t_{spacer}	1 mm	Inlet relative humidity (C)	$\phi_{C,in}$	50 %
Number of layers	n_{layer}	30	Inlet mass flow rate (C)	$V_{C,in}$	50 m^3 h^{-1}
Number of channels	$n_{channel}$	9			

The results of the different simulations are represented in terms of sensible, latent and total effectiveness, respectively defined as:

$$\varepsilon_s = \frac{Q_s}{\min(\dot{V}\cdot\rho\cdot c_p)_{in}\left(T_{H,in}-T_{C,in}\right)} \qquad \varepsilon_l = \frac{Q_l}{\min(\dot{V}\cdot\rho)_{in}\left(\omega_{H,in}-\omega_{C,in}\right)\Delta h_{lat}}$$
$$\varepsilon_t = \frac{Q_s+Q_l}{\min(\dot{V}\cdot\rho\cdot c_p)_{in}\left(T_{H,in}-T_{C,in}\right)+\min(\dot{V}\cdot\rho)_{in}\left(\omega_{H,in}-\omega_{C,in}\right)\Delta h_{lat}} \tag{12}$$

Calculations were then conducted to estimate the difference between the coupled and the decoupled heat and mass transfer. Decoupling heat and mass transfer can be realized by ignoring the latent heat terms in Equation 9. The results were compared in terms of sensible and latent heat transfer rate and by visualization of the temperature and humidity distributions.

4.1. Parametric analysis

The parametric analysis has been conducted to investigate the performance of the enthalpy exchanger with different membrane characteristics. Diffusion coefficient $D_{w,M}$, sorption coefficient C_s, maximum water uptake θ_{max}, membrane thickness δ_M and membrane thermal conductivity λ_M have been varied in their typical ranges.

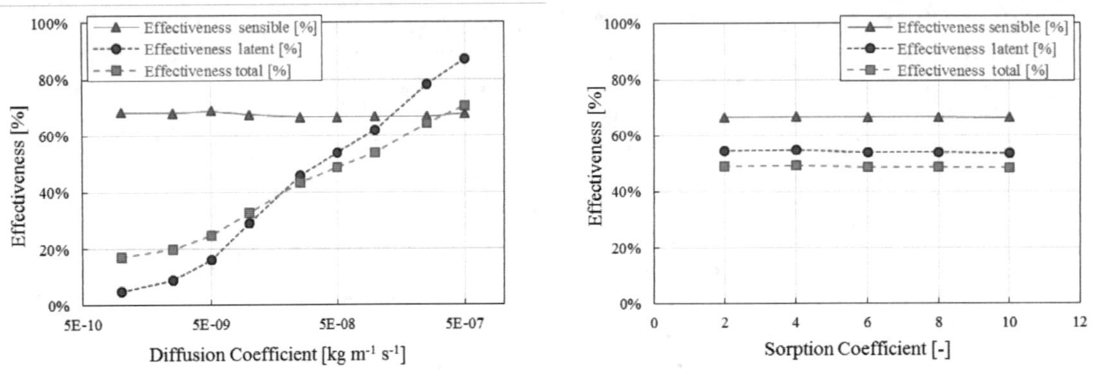

Figure 5. Effectiveness for different values of the diffusion coefficient $D_{w,M}$ of water in the membrane (left) ($C_s = 6$) and for different values of the sorption coefficient C_s (right) ($D_{w,M} = 5{\cdot}10^{-8}\,kg\ m^{-1}\,s^{-1}$). Results are obtained at $\theta_{max}=0{,}3$ kg kg^{-1}, $\delta_M = 0{,}05$ mm, $\lambda_M = 0{,}3$ W m^{-1} K^{-1}.

Figure 5 (left) shows the variations of the sensible, latent and total heat performance with the moisture diffusion coefficient in the membrane. When the moisture diffusion coefficient increases, the moisture flow rate increases as well, causing an enhancement in the latent heat flux. Since a larger moisture diffusion coefficient will cause a larger water uptake gradient in the membrane, it will increase the moisture mass flux between the two sides and the latent heat as a consequence. The total effectiveness, is the results of combining the sensible and the latent heat transfer. Since the sensible effectiveness does not present wide variation, the total effectiveness has a similar variation to that of the latent effectiveness.

The sorption coefficient affects only marginally the heat and the moisture transfers, as plotted in Figure 5 (right) It has to be mentioned that in the considered range, typical for enthalpy exchanging membranes, the variation of the sorption causes only slightly change on the sorption curve. As a consequence, the sensible, latent and total effectiveness remain almost constant with the sorption coefficient.

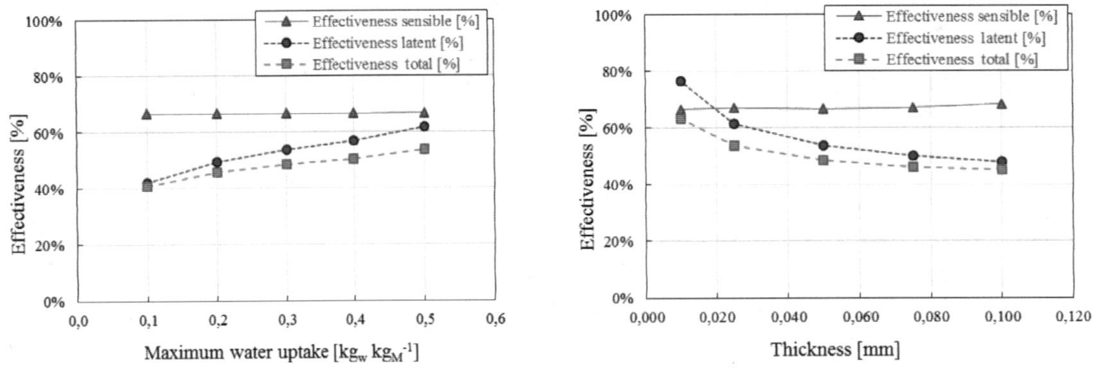

Figure 6. Effectiveness for different values of the maximum water uptake in the membrane θ_{max}.(left) (δ_M = 0,05 mm) and for different values of the membrane thickness δ_M (right) ($\theta_{max}=0{,}3$ kg kg^{-1}). Results are obtained at $D_{w,M} = 5{\cdot}10^{-8}$ kg m^{-1} s^{-1}, $C_s = 6$, $\lambda_M = 0{,}3$ W m^{-1} K^{-1}.

A more influential parameter on the sorption curve is the maximum water uptake in the membrane, as can be seen from Figure 3 (right). The effects of its variation are shown in Figure 6 (left). The increase in the maximum water uptake causes an increase in the maximum potential gradient for the water uptake across the membrane. It results in an enhancement of the moisture transfer flux and of the latent heat transfer as well. As the maximum water uptake varies, the sensible effectiveness remains almost unchanged.

As mentioned in the introduction, in order to minimize the membrane thermal and mass resistances itis fundamental to reduce the membrane thickness. Figure 6 (right) shows the variation of the effectiveness with the variation of the membrane thickness. The sensible effectiveness present only slightly variation as the thickness changes. The reason is that the dominant thermal process is the convective heat transfer in the boundary layer of the airstreams. In the considered conditions, the ratio of conductive (membrane) to convective resistance results less than 1/100. The latent and total effectiveness increases by decreasing the membrane thickness. The trend is super-linear with the thickness decrease. It confirms that with constant sorption characteristics of the membrane, an effective way to enhance the moisture and latent heat transfer is to reduce the thickness of the membrane.

As seen above, the sensible effectiveness is only slightly affected by the variation of the membrane parameters. Even the variation in the membrane thermal conductivity have small influences on the surface temperatures of the membrane, which affect sensible and latent heat transfer. In fact, the performance of the enthalpy exchanger exhibits almost no change with the membrane thermal conductivity varying from *0,1* to *100* W m^{-1} K^{-1} (not shown in the graphs). The reason is that the membrane thickness is really small, leading to a negligible thermal resistance compared to the thermal resistance of the convective processes. For a membrane-based enthalpy exchanger operating under typical conditions, the membrane thermal resistance accounts only for a small fraction of the total thermal membrane resistance. The heat transfer is dominated by the convection from the bulk flow to the membrane surfaces, while the membrane mass resistance dominates the humidity exchange. These present results confirm the observations of Min and Su [8,9] .

4.2. Coupled and decoupled heat and mass transfer comparison

Decoupling the heat and mass transfer consists in ignoring the latent term in Equation 2 and following consequences. The result is a temperature profile independent from the moisture transfer. The dependence of the moisture transfer by the surface temperatures remains in the expression of the relative humidity using Clapeyron equation.

The main effect of decoupling the heat and mass transfer is a reduction in the computational effort to solve the heat and mass transfer problem, which allows to reduce the calculation time.

Coupled and decoupled heat and mass transfer models are compared under the same conditions reported in Table 2. For these simulations, the membrane is characterized by a diffusion coefficient of $5 \cdot 10^{-8}$ kg $m^{-1}s^{-1}$, a sorption coefficient equal to *6*, a maximum water uptake of *0,3* kg kg^{-1}, a membrane thickness equal to *0,05* mm and a membrane thermal conductivity set to *0,3* W m^{-1} K^{-1}.

Minor differences have been registered in the temperature distribution. The calculated sensible heat transfer rate results *160* W and *162* W for the coupled and decoupled model, respectively, with a sensible effectiveness of *65* %. As for the temperature, the absolute humidity presents only slightly difference between the coupled and the decoupled models. The latent heat transfer rate results *409* W for the coupled solution and *410* W for the decoupled model, with a latent effectiveness of *65* %, which leads to a total effectiveness of *47* %.

5. Conclusions

A numerical study was performed to analyze the combined heat and moisture transfer across a membrane. A first set of calculations were performed for a representative air inlet conditions by changing the membrane parameters. A further calculation has been performed to compare the difference between coupled and decoupled heat and mass transfer models.

From the calculations it can be deduced that the sensible heat transfer is practically not affected by the membrane properties. In fact, the heat transfer is dominated by the convection from the bulk flows to the membrane surfaces. The membrane thermal resistance accounts only for a small fraction of the total thermal resistance. The only way to enhance the sensible heat flow rate is to enlarge the exchange surface or increase the convective coefficient. On the contrary, the membrane properties have a strong influence on the latent heat transfer because the membrane mass resistance dominates the humidity exchange. The most influential membrane characteristics result to be the moisture diffusion coefficient and the maximum water uptake in the membrane. When a membrane material has been selected, an effective way to maximize the latent heat transfer is to reduce the membrane thickness as much as possible. At the same time the thickness of the membrane must guarantee enough stiffness to withstand the pressure differential and limit the surface warping under both dry and wet conditions.

Within the assumptions of this study, no evident difference has been registered in the predictions by the coupled and the decoupled heat and mass transfer models. Only slightly differences can be seen in the temperature and absolute humidity profiles. The adoption of the decoupled mass transfer model will lead to spare computational time and simultaneously to predict the performance of a membrane-based enthalpy exchanger with reasonable accuracy.

References

[1] Niu J L and Zhang L Z 2001 Membrane-based enthalpy exchanger: material considerations and clarification of moisture resistance *J. Membr. Sci.* **189** 197-191

[2] Zhang L Z, Liang C and Pei L 2008 Heat and moisture transfer in application scale parallel-plates enthalpy exchanger with novel membrane materials *J Membr. Sci.* **325** 672-682

[3] Min J, Liu X, and Hu T 2008 Experimental studies of transport properties of water vapor through membranes *Int. Ref. Air Cond. Conf.* Paper 989

[4] Min J, Hu T and Liu X 2010 Evaluation of the moisture diffusivities in various membranes *J. Membr. Sci.* **357** 185-191

[5] Zhang L Z, Wang Y, Wang C, Xiang H 2008 Synthesis and characterization of a PVA/LiCl blend membrane for dehumidification *J. Membr. Sci.* **308** 198-206

[6] Zhang L Z and Niu J L 2002 Effectiveness correlations for heat and mass transfer processes in an enthalpy exchanger with membrane cores *ASME J. Heat Trans.* **5** 922-929

[7] Zhang L Z and Liang Y 1999 Heat and mass transfer in a membrane-based energy recovery ventilator *J Membr. Sci.* **163** 29-38

[8] Zhang Y, Jiang Y, Zhang L Z, Deng Y and Jin Z 2000 Analysis of thermal performance and energy savings of membrane based heat recovery ventilator *Energy* **25** 515-527

[9] Min J and Su M 2010 Performance analysis of a membrane-based enthalpy exchanger: Effect of the membrane properties on the exchagner performance *J Membr. Sci.* **348** 376-382

[10] Min J and Su M 2010 Performance analysis of a membrane-based enthalpy exchanger: Effect of membrane spacing and thickness on the ventilator performance *App. Therm. En.* **30** 991-997

[11] Min J and Hu T 2011 Moisture permaeation through porous membranes *J. Membr. Sci.* **379** 496-503

[12] Min J and Wang L 2012 Heat of adsorption and its effects on transmembrane heat transfer *J Membr. Sci* **409-410** 173-179

[13] Min J and Wang L 2013 Coupled heat and mass transfer during moisture exchange across a membrane *J Membr. Sci* **430** 150-157

[14] Hu T, Min J and Song Y 2008 Analysis of the effect of the heat of sorption on the process of heat transfer in moisture exchange across a membrane *Sci. China Ser. E-Tech. Sci.* **51** 2120-2127

[15] Wijimans J G and Baker RW 1995 The Solution-Diffusion Model: a review *J. Membr. Sci.* **107** 1-21

[16] Simonson C J and Besant R W 1999 Energy wheel effectiveness: part I – development of dimensionless groups *I. J. Heat Mass Transfer* **42** 2161-2170

[17] Gendebien S Bertagnolio S and Lemort V 2013 Investigation on a ventilation heat recovery exchanger: Modeling and experimental validation in dry and partially wet conditions *En. and Build.* **62** 176-189

[18] Shah R K and London A L 1978 *Laminar Flow Forced Convection in Ducts* (New York: Academic Press) 205

[19] Kakac S, Shah R K and Aung W 1987 *Handbook of single phase convective heat transfer* (New York: Wisley-Interscience)

[20] Wibulswas P. 1966 *Laminar flow heat transfer in non circular ducts* PhD Thesis, London University

[21] Bergman T L, Lavine A S, Incropera F P and DeWitt D P 2011 *Fundamentals of Heat and Mass Transfer* (New York: John Wiley and Sons)

Air entrainment in the primary impact of single drops on a free liquid surface

E Salerno[1], P Levoni[1,2] and G S Barozzi[1]

[1] DIEF- Dipartimento di Ingegneria "Enzo Ferrari", Università di Modena e Reggio Emilia, Via Vivarelli 10 - 41125 Modena (Italy)
[2] MIMESIS fluid dynamics & energy, Strada Contrada 309 - 41126 Modena (Italy)

Email: elisabetta.salerno@unimore.it

Abstract. Air-bubble entrainment produced by the impact of water drops on a liquid pool is investigated with the use of high-speed imaging. A wide range of drop volumes and impact velocities is considered to determine how the entrainment mechanisms change with varying the impact conditions. Five different entrainment regimes are distinguished on the basis of the observed flow phenomena. Their characteristic features are described in terms of bubble formation, crater evolution, jetting and secondary drop ejection. A regime map is reconstructed in the Froude-Weber space. Results obtained in the present study show good agreement with the phase diagrams reported in the literature and contribute to complete the scenario of the entrainment regimes. Quantitative data about the size and the residence times of the entrained nuclei are also presented.

1. Introduction

The impact of drops on a liquid surface often results in the introduction of gaseous nuclei in the receiving medium. This phenomenon is encountered in a large variety of natural and technological processes. In some applications, such as marine and climate science [1], nucleate boiling [2, 3] and fire suppression systems [4], the entrainment of bubbles can have positive implications since it promotes the transfer of heat and the exchange of gas molecules across the liquid-gas interface. On the contrary, the inclusion of bubbles is an undesirable phenomenon in filling and pouring processes, like the industrial handling of molten metals [5], cosmetic and pharmaceutical preparations [6] and liquid foods [7]. In those fields, highly perturbed filling jets are usually employed, which tend to disrupt into discrete droplets due to instabilities. The consequent entrainment of air resulting from the drops impinging on the free surface is detrimental to the product quality and production rates, as bubbles can cause defects, produce chemical and biological instabilities or give rise to foam.

In any case, it is of crucial importance to define the conditions for which entrainment occurs as well as understand the incidence of the phenomenon in quantitative terms. For these reasons, the impact of drops has been largely investigated in the last decades. In particular, the recent improvements in high-speed imaging techniques have allowed a better insight into these rapidly evolving events, leading to the identification of different mechanisms of entrainment. They can be summarized as follows.

In the case of drops falling on a deep pool of liquid at relatively low impact velocities, Thoroddsen et al. [8] and Tran et al. [9] showed that a thin hemispheric layer of air remains entrained between the two liquid bodies at the very beginning of the impact process. This layer prevents the immediate coalescence of the drop, and its subsequent draining out and rupture lead to the production of hundreds or thousands of micron-sized bubbles within a few of milliseconds. According to the energy of impact, rupture can occur once the entrained layer is completely stretched or can start before it has drained out. The so-called "Mesler entrainment" [2, 3] is obtained in the latter case. When the kinetic energy of the drop is sufficiently high, a thin disk of air remains enclosed rather than a hemispherical cap [10], whose contraction usually results in the formation of a pair of nuclei. These nuclei are often referred to as "Thoroddsen bubbles" and have diameters ranging from tens to hundreds of microns. Accompanying their appearance, a multitude of smaller nuclei arranged in a ring distribution can be observed [10, 11]. As it was first postulated by Oguz and Prosperetti [12], these additional nuclei derive from the evolution of the outer contact line between the drop and the pool surface.

Soon after contact is established, a deep crater may be created in the receiving liquid as the drop impact energy is converted into potential and surface energy of the fluid. When the crater collapses under the action of the surrounding liquid, a portion of air can be left behind and other forms of entrainment are obtained. In some cases a single large bubble detaches from the bottom of the crater. Such a scenario is the result of a delicate balance between the motion reversal of the cavity walls, which starts at different times in different positions, and the convergence of capillary waves at the bottom of the crater [13–15]. With high-energy impacts, surface tension-driven instabilities develop on the base of the crater and several bubbles of smaller size appear.

The conditions of occurrence of the cited entrainment mechanisms are usually reported in terms of non-dimensional parameters. When water is used as the working fluid, the Weber and the Froude numbers are assumed sufficient to describe the impact problem; these are defined as $We = \rho V^2 d / \gamma$ and $Fr = V^2 / gd$, respectively. Here, d is the drop diameter, V the impact velocity, g the gravitational acceleration, ρ the density, and γ is the surface tension of the liquid. The existing studies on air entrainment in deep pools cover a limited number of impact conditions or refer to single stages of the impact process. The majority of data available in the literature concerns the production of single large bubbles by crater collapse. This form of entrainment has been observed to occur repeatedly and predictably in a well-defined region of the Froude-Weber space. For this reason, this mechanism is called "regular entrainment". The lower and upper limits of the regular entrainment regime were determined by Oguz and Prosperetti [13] and are expressed by the power law equations $We = 41.3Fr^{0.179}$ and $We = 48.3Fr^{0.247}$. Out of these boundaries, the process of single bubble formation can still be achieved, though it is no longer reproducible. For Weber and Froude numbers above the upper limit of regular entrainment region, slight differences in the crater dynamics may lead to different forms of entrainment. These types of entrainment regimes are still poorly investigated [11, 14, 16–18]. The definition of the Froude and Weber numbers delimiting the occurrence of the entrainment phenomena due to air film rupture is also quite controversial. According to the observations of Liow [18], the hemispherical cap is not formed when $Fr < 7$ and $We = Fr$. Liow and Cole [11] found values for the Weber number ranging from 6 to 20 for the case of Mesler entrainment, and $We < 6$ for film rupture after complete drainage. At $We > 20$, Thoroddsen bubbles and Oguz-Prosperetti bubble rings may appear [10, 11]. No dependence on the Froude number has been reported relative to these phenomena.

It must be pointed out that the quantitative limits discussed here are valid only for the case of liquids with low viscosity. When viscosity effects cannot be neglected a third dimensionless quantity is required to properly characterize drop impacts. This is usually written in the form of the capillary number Ca ($Ca = \mu V / \gamma$, where μ is the liquid viscosity). For what concerns the production of bubbles by cavity collapse, the global role of viscosity is seen as a damping of the capillary waves propagating along the crater walls, and a weakening of the crater reversal dynamics. This reflects in an increase of the inviscid regular entrainment limits in the Fr-We space [15]. On the other hand, the exact influence of liquid viscosity on the onset of bubble production by air film rupture has not yet been clarified.

From the above considerations it is evident that, despite the progress made so far in the study of the impact process, the global picture is still not exhaustive. The purpose of this research is to obtain a more complete scenario of the entrainment phenomena. To this aim, a series of practical experiments involving water drops impinging on a deep pool of the same liquid was carried out. A wide range of drop volumes and falling heights was explored. High-speed imaging was adopted both to examine the impact dynamics, and to follow the process evolution over long time scales. The recorded image sequences were analysed and a regime map was reconstructed, based on the measured impact conditions. The size and the residence time of the bubbles were also extracted from the images, in order to offer useful indications about the role of bubble entrainment in practical processes.

2. Experimental procedures

2.1. Drop generation and receiving system

As sketched in figure 1, the experimental apparatus used in this work for the study of drop impacts consists of two main sections: i) the drop generation and receiving system, and, ii) the imaging system.

For the generation of the liquid drops, a positive displacement dispensing pipette was adopted. A 12 Volt DC motor, manually activated through a switch, was employed to control drop release and generate drops of different size. Drop volumes ranging from 2 to 23 μl were obtained with this facility. The pipette was suspended above a large rectangular glass tank containing the receiving liquid, and having a base area of 220 x 400 mm and a height of 250 mm. In order to make negligible the influence of the tank walls on the impact dynamics, a minimum lateral distance of 110 mm was always maintained between the syringe tip and the tank sidewalls. Similarly, bottom effects were excluded by filling the tank up to a height of 150 mm. To observe the influence of different falling heights, the pipette was mounted vertically on a traversing arm, capable of 0.1 mm minimum displacements in the three orthogonal directions. Ten different heights for the syringe tips were selected, ranging from 30 to 300 mm relative to the pool free surface. About 16 drop impacts were examined for each falling height, for a total of 176 trials.

Deionized water was used as the working fluid for both the drops and the liquid pool. To avoid surface contamination by large dust particles from the external ambient, in the course of the experiments the tank was covered with a transparent acrylic sheet, just leaving free the passage for the falling drop. At the end of each experimental series, the liquid inside the pool was removed and replaced with fresh fluid. Since the pool temperature could vary in the course of the experiments, due to intense illumination, a resistance thermometer was introduced in the liquid pool. Variations in the physical properties of the liquid, i.e. surface tension and density, could thus be accounted for. Throughout the entire set of experiments, a maximum temperature excursion of 6.8 °C was recorded, from a minimum of 15.6 °C up to 22.4 °C.

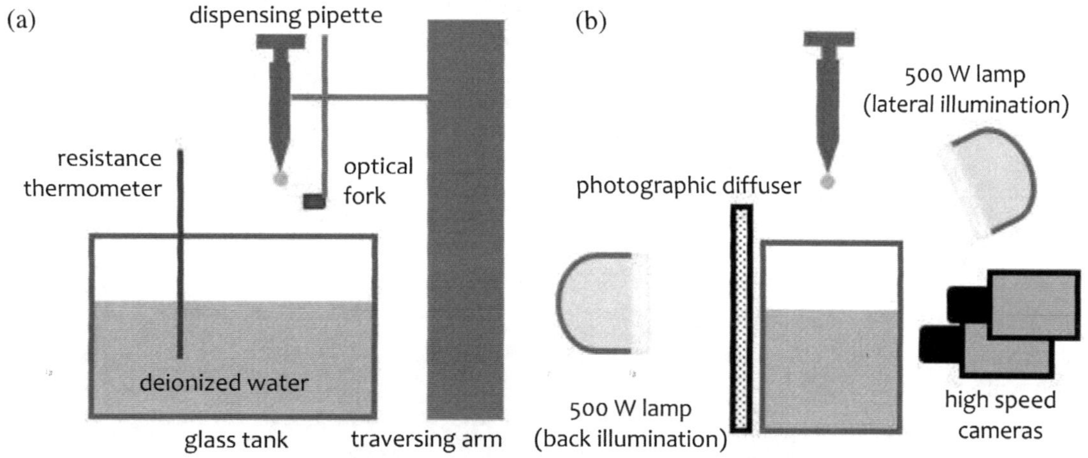

Figure 1. Schematic of the experimental apparatus: front view of the drop generation and receiving system (a), and side view of the imaging system (b).

High-speed imaging was adopted for the recording of the impact events. A pair of NanoSense MkIII CMOS cameras from Dantec Dynamics was arranged in front of the pool, for the simultaneous observation of the flow phenomena below and above the pool free surface. Both the cameras were fitted with a Micro Nikkor 60 mm f/2.8 lens. Magnifications of 36 and 23 μm/pixel were obtained for the camera focused on the falling drops, and the camera focused below the pool free surface, respectively.

The acquisition rate was varied from impact to impact in order to observe phenomena characterized by different time scales. The minimum acquisition rate of 100 fps was set to get information about the residence times of the entrained bubbles. On the other hand, framing rates up to 3600 fps were achieved by reducing the image resolution, so as to follow the impact dynamics with high temporal resolution. The acquisition start was triggered by means of an optical fork sensor, located just below the syringe tip for the detection of the falling drop. When the drop intercepts the light beam, generated by an infrared LED source, the reflected light is detected by a phototransistor and a TTL signal is sent to the cameras to initiate recording. The microcontroller Arduino Uno was employed for the generation of signals, whose rate was verified with the use of an oscilloscope. A maximum synchronization error of 6.6 μs was measured, including the contribution of the cables. Two 500 W halogen lamps were used as the light sources. One of them was placed behind the impact site to provide back illumination. A photographic diffuser was interposed between the lamp and the pool backside, so producing a uniform distribution of the emitted light. The second lamp provided additional lateral illumination.

2.2. Data reduction

The extrapolation of quantitative information on the impact phenomena was performed by processing the acquired images with the software ImageJ (provided by NIH, USA).

The image sequences related to the fly of the drops before impact were analysed to obtain the drop diameter and the impact velocity. In the presence of large three-dimensional drop oscillations after release, the frame exhibiting the less deformed drop shape was processed for the estimate of the drop size. Each drop was approximated with an ellipsoid, from whose volume the equivalent spherical diameter was evaluated. Drop diameters ranging from 1.55 to 3.54 mm were measured. For deformed drops, a maximum relative variation Δd of 9% resulted between the two axes of the ellipsoid. In the majority of cases, however, the shape variation was far below this value.

The impact velocity was derived from the last frames of the drop falling sequence: the coordinates of the centroid were tracked and their shift was then divided by the time interval between frames. The measured horizontal velocity at the impact position was less than 4% of the vertical velocity, indicating that all drops fell almost orthogonally on the pool free surface. The impact velocity range explored in the experiments was estimated to lie between 0.52 and 2.41 m/s.

The measured values of the drop diameter and the impact velocity were later used to calculate the Weber and the Froude numbers for the characterization of the impact conditions. Weber numbers ranging from 10 to 270 and Froude numbers between 8 and 329 were obtained. Considering the maximum uncertainty of drop size and impact velocity, of 1% and 3%, respectively, the accuracy of the Weber and the Froude number was estimated to be of at least 7%.

3. Results and discussion

3.1. Entrainment regimes

The analysis of the impact sequences recorded above and below the pool free surface led to the identification of five different entrainment regimes corresponding to different impact conditions. In the following subsections, the various regimes are presented and discussed for increasing values of We and Fr.

3.1.1. Initial contact I regime. For low Weber and Froude numbers, the sole entrainment mechanism involved in the impact process is the formation of small air bubbles as the drop first comes into contact with the pool surface. Hence, this regime has been denoted as "*initial contact I*". Those bubbles become visible within a fraction of milliseconds after contact is established and are in the form of Thoroddsen bubbles, sometimes accompanied by Oguz-Prosperetti bubble rings. As visualized in figure 2, bubble rings appear as a cloudy line forming an arc around the two larger bubbles.

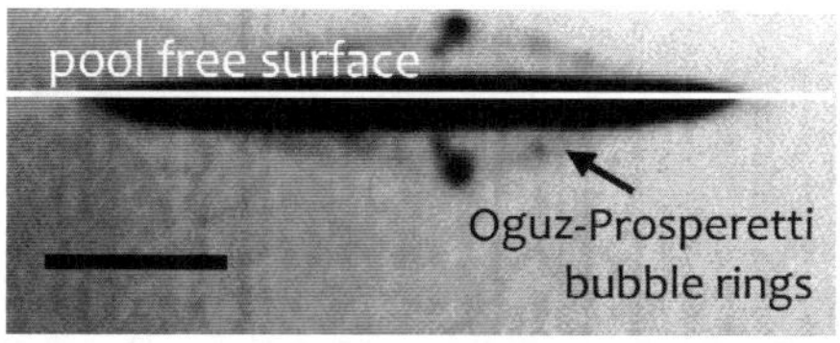

Figure 2. *Initial contact I regime.* Oguz-Prosperetti bubble rings formed in conjunction with Thoroddsen bubbles (Fr = 18, We = 26). Frame acquired at t = 0.8 ms after impact. Scale bar = 1 mm.

The frames reported in figure 3 show the subsequent evolution of the impact process. Following the rupture of the air disk, the liquid of the drop merges with the target fluid and a crater, or cavity, forms. The crater expands changing its initial profile, forced by the flow field induced in the liquid pool: sharp corners develop at the sides of the cavity, while the bottom progressively assumes the shape of a rounded tip as a result of the convergence of a capillary wave. A few milliseconds after impact, the crater reaches its maximum expansion and the surrounding fluid starts to contract the crater. In this regime, the bottom peak rapidly retracts without leaving any bubble. Simultaneously to crater collapse, a smooth bulge rises just above the pool surface and the free surface gradually restores its initial form. Sometimes, a faster rebounding of the crater tip results in the formation of a thicker column of liquid. Small secondary droplets may detach from the liquid column due to capillary instabilities. Those droplets can reach great heights before falling back on the pool surface; however, their size is too small to give rise to any relevant phenomena.

3.1.2. Single bubble I regime. Thoroddsen bubbles are not an exclusive feature of the "*initial contact I regime*". They have been observed for almost the totality of the impact conditions explored in this study. In the regime designated here as "*single bubble I*", their production is followed by the entrainment of one large bubble derived by cavity collapse.

As it can be observed in figure 4.a, in this case the initial profile of the crater is nearly hemispherical as a consequence of the high energy of the impacting drop. When the cavity stops growing downward (t = 19.0 ms after impact), an axisymmetric inward flow develops just above the cavity base [14]. This causes the cavity sidewalls to be pushed inward, while the cavity floor stagnates. Hence, a nipple-like protrusion, visible at t = 21.6 ms, is formed at the base of the cavity. Finally, a capillary wave converging towards the cavity bottom makes the opposite sides of the channel to come into contact before the protrusion can retract. As a consequence, the bottom portion of the cavity is disconnected from the rest of the crater, resulting in the entrainment of a macroscopic bubble in the liquid.

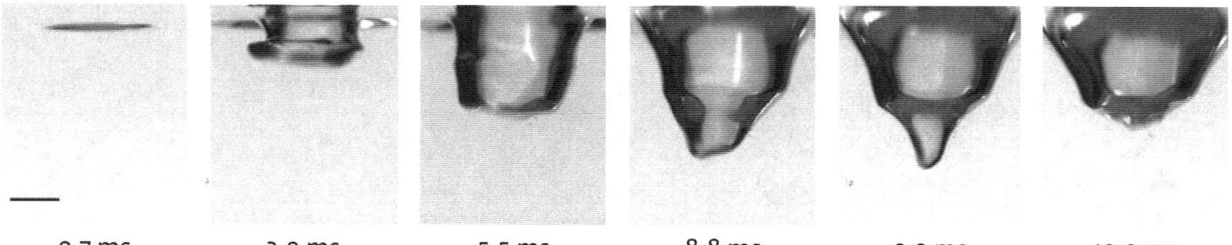

Figure 3. *Initial contact I regime.* Image sequence captured below the pool surface (Fr = 60, We = 48, framing rate = 2740 fps). Scale bar 1.5 mm. Times referred to the instant of the first contact.

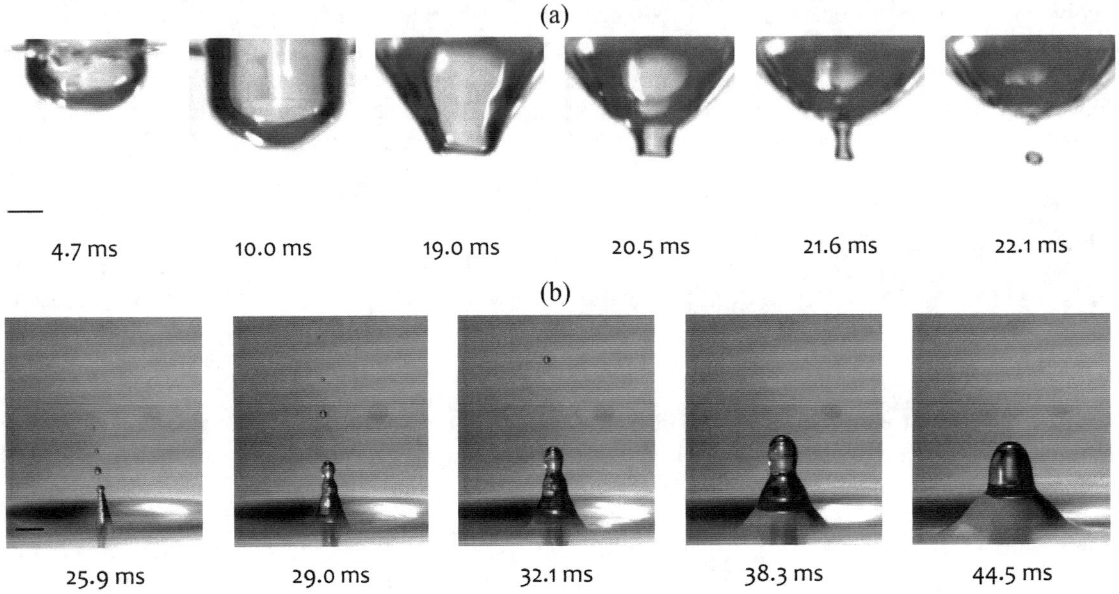

Figure 4. *Single bubble I regime*. Image sequence captured below (a) and above (b) the pool surface (Fr = 87, We = 133, framing rate = 1898 and 967 fps respectively). Scale bar 2 mm.

The inward axisymmetric flow which governs the crater collapse is responsible also for the formation of an upward liquid jet above the liquid surface [19]. In the "*single bubble I regime*", a short and narrow jet appears, together with a series of small droplets ejected from its tip at high velocity. This situation is illustrated in figure 4.b. The jet quickly reaches a maximum height comparable to the diameter of the primary drop, then starts to thicken and finally subsides.

3.1.3. Single bubble II regime. For a narrow range of Weber and Froude numbers just above the region of the "*single bubble I regime*", a different jet behaviour is encountered in conjunction with the entrainment of the large bubble.

The development of the impact crater is similar to the one observed in the previous regime. Surface tension-driven instabilities may be generated at the bottom of the retracting protrusion, which takes the shape of a slightly curved cylindrical stem. Thus, immediately after detachment, the newly-formed bubble appears to be quite stretched as an effect of stem deformations. However, the most distinctive feature of the "*single bubble II regime*" is the subsequent flow dynamics above the free surface.

Supported by the high potential energy stored in the crater, the jet rises well above the pool surface, preceded by a finite number of very small-sized drops (figure 5).

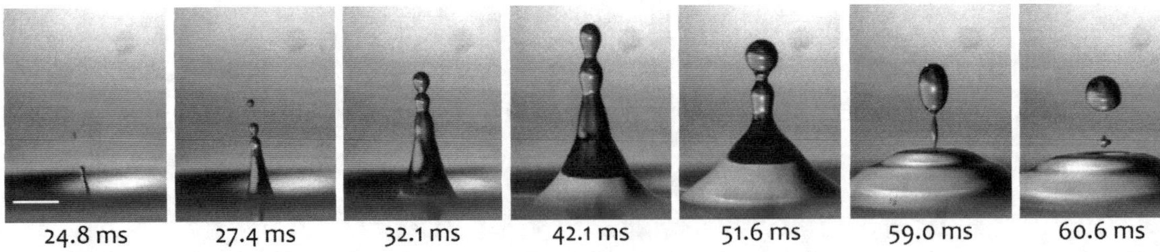

Figure 5. *Single bubble II regime*. Image sequence captured above the pool surface (Fr = 110, We = 158, framing rate = 1898 fps). Scale bar 3 mm.

The jet body becomes thicker while it continues to rise and its upper portion undergoes severe surface distortions, driven by capillary waves travelling down the jet. Those disturbances are generated during the detachment of the first series of droplets. When the liquid column collapses, the fluid progressively necks until a large drop separates. In the specific sequence shown, a smaller satellite droplet also generates. The large secondary drop is comparable in size to the original impacting drop but has a lower velocity. Drops produced with this mechanism can potentially participate in the introduction of gaseous nuclei when falling back into the pool, as from [11, 20]. A more accurate analysis of the role of the secondary drops on entrainment is left for future analysis.

3.1.4. Multiple bubbles regime. At even higher Weber and Froude numbers ($100 < Fr < 220$, $160 < We < 210$), two or more bubbles are entrained in the liquid pool as a consequence of crater collapse. It is worth to point out, however, that we included under the same category regimes characterised by different nature, variable with the size of the entrained bubbles and the shape of the cavity bottom at pinch-off.

These forms of entrainment are obtained when the surface instabilities established at the floor of the cavity are strong enough to break up the stem structure at multiple locations. Consequently, bubbles similar to the ones shown in figure 6 are produced. The development of the upward moving jet, instead, is identical to the one described in section 3.1.3: after the jet has reached its maximum height, one or, more rarely, two large secondary drops detach from the jet body at the point of large curvature.

3.1.5. Initial contact II regime. When the falling velocity and the size of the primary drop are sufficiently high, the production of bubbles by crater collapse is suppressed. On the contrary, Thoroddsen bubbles are still formed at the very beginning of the impact process. Hence, this last class of events is under many aspects similar to the "*initial contact I regime*" for what concerns the entrainment mechanisms involved in the primary impact. The main differences reside in the dynamics followed by the impact crater (figure 7.a) and the jet (figure 7.b).

In the region referred to as "initial contact II regime", the lateral expansion of the primary impact crater is less limited than in previous regimes as inertial forces dominate and surface tension is not able to restrict the cavity growth. The effect of radial focusing of the flow on crater collapse is also weaker: the base of the cavity reverses its motion before any capillary wave travelling downward the crater walls reaches the bottom. Even the smaller disturbances covering its floor do not have sufficient time to develop before contraction occurs. As a result, no bubbles are detached from the base of the primary crater. In addition, the rapid contraction of the crater gives origin to a thicker, but higher and slower jet than in the previous cases. One or more large drops detach while the jet is still growing. These rise further above the free surface before falling back.

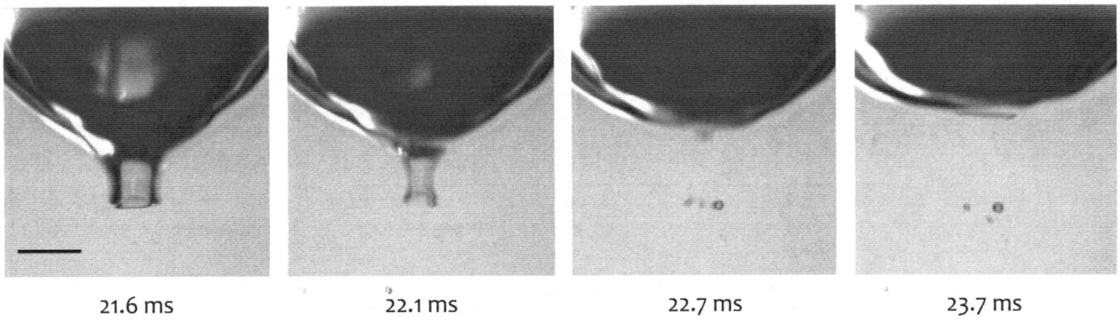

Figure 6. *Multiple bubbles regime*. Image sequence captured below the pool surface ($Fr = 171$, $We = 196$, framing rate = 1898 fps). Scale bar 2 mm.

Journal of Physics: Conference Series **655** (2015) 012036 doi:10.1088/1742-6596/655/1/012036

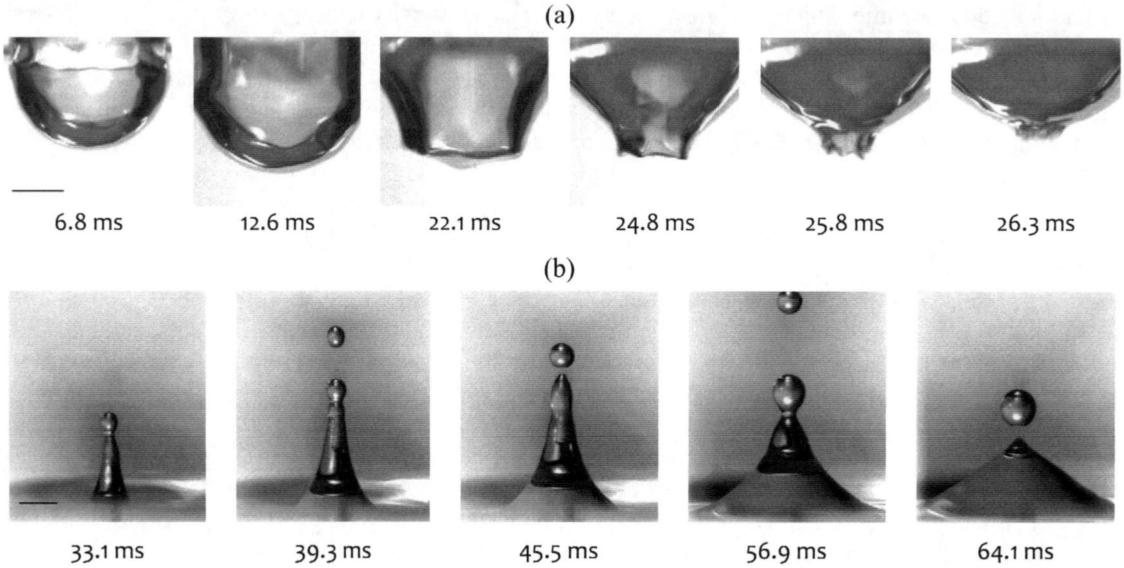

Figure 7. *Initial contact II regime.* Image sequence captured below (a) and above (b) the pool surface (Fr = 172, We = 247, framing rate = 1898 and 967 fps respectively). Scale bar 3 mm.

3.2. Phase diagram

The conditions under which the entrainment regimes observed in the present work occur, are represented in the Froude-Weber phase diagram displayed in figure 8. Boundaries reported by the literature between regions characterised by distinct forms of entrainment and data acquired by different authors [11, 14] are also plotted.

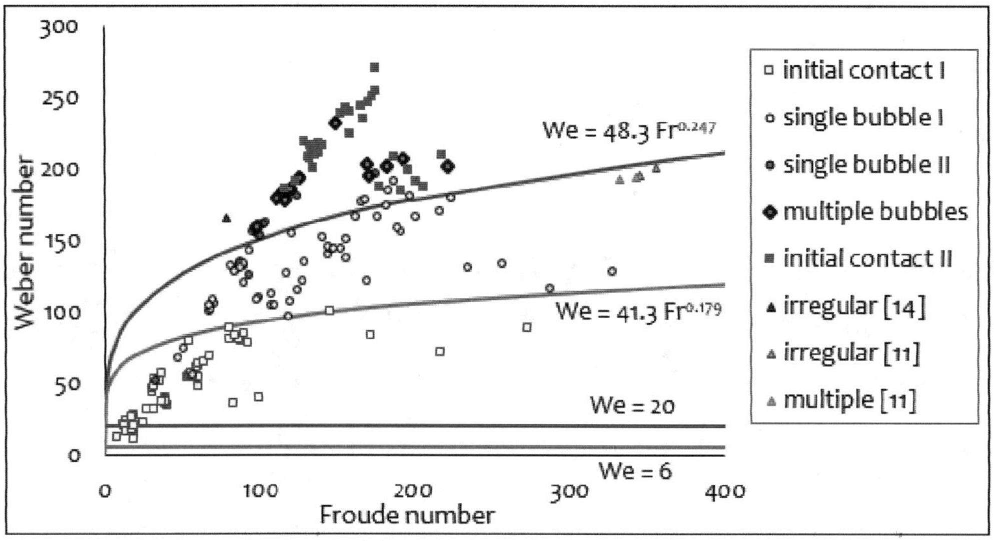

Figure 8. Phase diagram of the entrainment regimes.

The diagram shows that the data acquired herein on *single bubble I regime* fall in the large majority between the curves provided by Oguz and Prosperetti [13], which delimit the region of regular entrainment. Out of this region, results relative to the formation of the single large bubble without large secondary drops are obtained occasionally but are surrounded by points pertaining to different regimes. This implies that a slight change in the impact conditions results in a completely different outcome. Such a behaviour gives a direct confirm of the lack of reproducibility typical of this entrainment mechanism for We and Fr values above and below the two reference curves.

The region below the lower boundary of regular entrainment is dominated by the *initial contact I regime*. The introduction of gaseous nuclei in the form of Thoroddsen bubbles is encountered also for Weber values lower than 20, that is, below the limit determined by Thoroddsen et al. [10] and by Liow and Cole [11]. According to Liow and Cole, Mesler entrainment should, instead, take place for $6 < We < 20$. A possible explanation of this discrepancy may reside in the variability typical of Mesler entrainment in water: at parity of conditions, the presence of surface contaminants or even a small deviation in the shape of the drop may hinder the formation and rupture of the hemispherical air film [2, 21]. For a narrow subset of Froude and Weber numbers in the region of *initial contact I regime* ($8 < Fr < 32$, $10 < We < 32$), Oguz-Prosperetti bubble rings have been observed in conjunction with Thoroddsen bubbles. It is not clear if their disappearance for We and $Fr > 32$ is determined by a suppression of the mechanisms of their production, or if their size becomes so small that they escape detection with the imaging system adopted.

The formation of Thoroddsen bubbles in the absence of other entrainment mechanisms from the primary drop is achieved also at We and Fr numbers above the upper limit of regular entrainment. In that region, large secondary drops are usually pinched-off from the jet emerging after cavity retraction. Their effect in the introduction of gaseous nuclei needs to be ascertained. For $We > 210$, the *initial contact II regime* takes place systematically, but the data collected are too scant to determine the trend of occurrence of this phenomenon. For Weber numbers lower than 210 and higher than the upper limit of regular entrainment, the *initial contact II regime* mixes up with different mechanisms of entrainment, involving the production of one or more bubbles by crater collapse. *Multiple bubbles regime* is, in general, obtained at We and Fr numbers higher than those typical of *single bubble I* and *II regimes*, in the range that we provisionally established to be $100 < Fr < 220$ and $160 < We < 210$. This suggests that, above a certain threshold, surface tension forces are no longer able to contain the crater expansion. However, the definition of the quantitative limits of occurrence of multiple entrainment is rather difficult, since these data are not grouped in a well-delimited region. The present results offer a much richer phase-diagram than the available counterparts in the literature, but further work is required to ascertain if the region exhibiting mixed forms of entrainment is characterised by intrinsic instabilities or if other factors come into play in determining the impact behaviour.

3.3. Size and residence time of the entrained bubbles

As discussed in the previous sections, a large variety of bubbles can be introduced in the liquid pool as a result of drop impacts. Their properties and behaviour depend on the particular mechanism leading to their production.

Thoroddsen bubbles have been detected almost in the totality of impact events, with measured diameters ranging between 0.026 and 0.145 mm. After formation, these nuclei propagate ahead of the expanding impact crater and can be carried deeper into the liquid bulk by vortex rings. Alternatively, they can be drawn into the cavity wake and pulled towards the free surface when the cavity retracts. In both cases, Thoroddsen bubbles are retained in the liquid for long times: increasing the acquisition time of the cameras made it possible to observe that these nuclei were still present about 15 seconds after impact.

As opposite, Oguz-Prosperetti bubble rings disappear a few milliseconds after their formation. It is not clear if this is due to a rapid dissolution of these nuclei in the liquid, or to their outward motion from the impact point, which drives them out of the plane of focus. Bubbles in this category are too small to be resolved with the imaging system adopted in the present experiments, which offers a magnification of 23 μm/pixel. Hence, quantitative information about their exact number and size can not be extracted by the images herein acquired.

Single bubbles entrained by cavity collapse have sizes covering a wide range of values: diameters ranging between 0.07 and 1.22 mm have been found across both the *single bubble I* and *II regimes*. Bubbles of this kind, once released, can move further downward in the liquid, as a consequence of the slight increase in cavity depth before retraction. In this lapse of time, the bubble oscillates radially, until a spherical shape is assumed. As buoyancy becomes dominant, the bubble reverses its motion and rises towards the free surface. The elasticity of the interface can make the bubble bounce on the free surface for one or more times. Eventually, bubble interface coalesces with the pool surface, leading to bubble collapse. This process usually comes to an end within a fraction of second, residence times up to nearly 0.5 s being typical of nuclei produced by this form of entrainment.

In *multiple bubbles regime*, from a minimum of 2 to a maximum of 6 bubbles have been obtained, depending on the peculiar development of capillary instabilities on the cavity floor. In the cases where two bubbles are formed, their diameter is usually of the order of 400 μm. After detachment, the bubbles move deeper into the liquid bulk, until the buoyancy force drives them towards the free surface where they burst nearly immediately (at about 0.6 s after their formation). When three or more bubbles are produced, the size of the gaseous nuclei is usually smaller, and diameters ranging between 26 and 280 μm have been measured. Such small bubbles are less affected by buoyancy and they continue their downward motion into the fluid, until they exit from the field of view and no longer re-appear.

4. Conclusions

Through an extensive experimental campaign, the effect of drop impacts on the population of gaseous nuclei inside a liquid pool was investigated. The analysis of the impact events, performed with the use of high-speed imaging, has revealed the existence of five distinct entrainment regimes, on-setting at different impact conditions. Those regimes have been classified on the basis of the mechanisms taking part in the bubble production process. The fluid dynamics typical of each regime was discussed, after collecting and analysing high-speed image sequences for 176 drop impacts.

The conditions of occurrence of the entrainment regimes were summarized on a Froude-Weber phase diagram. The present data largely confirm the results reported in the literature, showing the existence of well-delimited entrainment regions. In addition, fresh data on *initial contact II*, *single bubble II* and *multiple bubbles regimes* have been provided, thus contributing to complete the scenario of drop-impact regimes. The results also highlight the presence of undefined entrainment regions, whose understanding requires further analysis.

In addition, the size and the behaviour of the entrained nuclei have been characterised in order to assess the incidence of bubble entrainment in practical applications. The data provided help to clarify the role played by bubble entrainment in real-world processes. These, however, require the investigation of additional effects, such as the interaction with solid walls and secondary impacts.

References

[1] Wanninkhof R, Asher W E, Ho D T, Sweeney C and McGillis W R 2009 *Ann. Rev. Mar. Sci.* **1** 213–44
[2] Esamilizadeh H and Mesler R 1986 *J. Colloid Interface Sci.* **110** 561–74
[3] Sigler J and Mesler R 1990 *J. Colloid Interface Sci.* **134** 459–74
[4] Li H, Mei S, Wang L, Gao Y and Liu J 2014 *Int. J. Heat Fluid Fl.* **47** 1–8
[5] Saeedipour M, Schneiderbauer S, Pirker S and Bozorgi S 2014 *Magnesium Technology* ed M Alderman, M V Manuel *et al* (Hoboken: John Wiley & Sons) pp 185–90
[6] Lin T J 1971 *J. Soc. Cosmet. Chem.* **22** 323–37
[7] Ashurst P R 2005 *Chemistry and Technology of Soft Drinks and Fruit Juices* (Oxford: Blackwell Publishing Ltd)
[8] Thoroddsen S T, Thoraval M-J, Takehara K and Etoh T G 2012 *J. Fluid Mech.* **708** 469–79

[9] Tran T, de Maleprade H, Sun C and Lohse D 2013 *J. Fluid Mech.* **726** R3
[10] Thoroddsen S T, Etoh T G and Takehara K 2003 *J. Fluid. Mech.* **478** 125–34
[11] Liow J-L and Cole D E 2007 Bubble entrapment mechanisms during the impact of a water drop *Proc. 16th Australasian Fluid Mechanics Conf.* pp 866–69
[12] Oguz H N and Prosperetti A 1989 *J. Fluid Mech.* **203** 149–71
[13] Oguz H N and Prosperetti A 1990 *J. Fluid Mech.* **219** 143–79
[14] Elmore P A, Chahine G L and Oguz H N 2001 *Exp. Fluids* **31** 664–73
[15] Deng Q, Anilkumar A V and Wang T G 2007 *J. Fluid Mech.* **578** 119–38
[16] Pumphrey H C, Crum L A and Bjørnø L 1989 *J. Acoust. Soc. Am.* **85** 1518–26
[17] Franz G J 1959 *J. Acoust. Soc. Am.* **31** 1080
[18] Liow J-L 2001 *J. Fluid Mech.* **427** 73–105
[19] Rein M 1996 *J. Fluid Mech.* **306** 145–65
[20] Carroll K and Mesler R 1981 *AIChE J.* **27** 853–55
[21] Mills B H, Saylor J and Testik F Y 2012 *AIChE J.* **58** 46–58

Journal of Physics: Conference Series **655** (2015) 012037 doi:10.1088/1742-6596/655/1/012037

Energy dissipation and transfer processes during the breaking of modulated wave trains

F De Vita[1,2], R Verzicco[1] and A Iafrati[2]

[1] University of Rome "Tor Vergata", 1 Politecnico street, 00133 Rome, Italy
[2] CNR-INSEAN Marine Technology Research Institute, 139 Vallerano street, 00128 Rome, Italy

E-mail: Francesco.De.Vita@uniroma2.it

Abstract. In this work numerical simulation of the breaking induced by the modulational instability are presented. The study is focused on the energy and momentum exchanged taking place at the air-water interface during the breaking process. Simulations are performed by a two-fluids Navier-Stokes solver combined with the VOF technique, which provides a detailed description of the flow in both phases. Results obtained by two different initial conditions are presented in terms of free surface shapes, energy in air and water, vertical fluxes of horizontal momentum and vorticity.

1. Introduction

In fully developed sea, ocean and atmosphere are strongly coupled, exchanging mass, momentum and heat through their interface. The ocean waves themselves are the most evident result of the interaction and of the energy input from wind to the wave system [1]. However, even if perhaps less intuitive, there is evidence that the ocean surface cannot be regarded as a flat and passive surface when modelling the wind blowing on top of it [2] and thus the coupled system should be considered for reliable weather and climate forecasting [3].

There are several phenomena taking place at the interface, but the one which is by far the most relevant for the exchange processes at the interface is the wave breaking. The breaking is a strongly nonlinear phenomenon which occurs when the wave grows too much and exceeds some limiting conditions. There are two main mechanisms which make it to happen, which are the linear superposition and the modulational instability, the latter being much more frequent [4]. Regardless the way which leads to its development, the wave breaking occurrence injects horizontal momentum and leads to a strong mixing of the upper ocean layer, with the entrainment of air bubbles [5] and the release of sea droplets which affect the atmospheric boundary layer [6].

Despite its important role, the physics of the breaking phenomenon is not well understood and many points are still obscure. As highlighted in [7], breaking models currently adopted for the breaking in wave forecasting are strongly dependent on their parameters and changing them makes the predictions more or less in agreement with the recorded data.

The development of accurate breaking models is hindered by the difficulties of realizing field measurements, but even in laboratory the situation is not much better. In spite of many detailed experiments which have been performed, there are technical limitations which make it rather difficult to perform detailed non-intrusive investigation of the two-phase flow. For instance, due to the light reflection from the air-water interface, the use of optical measurement techniques is allowed only in a later stage when the largest air bubbles have degassed [8], and thus there are no information for the early stage of the breaking process which is when the most relevant exchange processes take place. Due to the lack of detailed data on the velocity field, the wave energy is computed on the basis of the free surface elevation only [9], thus neglecting the energy amount associated to the rotational flow generated by the breaking process and leading to incorrect estimates of the energy losses. Last but not least, only a little is known for the flow above the free surface and for the air water interaction.

All the above aspects can be investigated by highly detailed simulations of the two-phase flow. Although the numerical modelling is still very expensive from the computationally viewpoint, and thus sometime limited in the scales of the problems which can be faced, it provides a complete picture of the phenomenon and makes it possible to derive all the data which are needed for the analysis [10], [11]. Such models are used here to study the air-water interaction taking place during the breaking event with particular regard to the derivation of quantitative estimates of the exchange processes.

As a specific application, the breaking generated as a result of the modulational instability is here considered being representative of the open ocean conditions. The simulation is started from a fundamental component with sideband perturbations. Under suitable conditions [12], the modulational instability develops giving rise to a steep wave which breaks when a limiting steepness is exceeded [13]. The flow, both in air and water, is simulated by a Navier-Stokes solver for the multiphase system, which uses a Volume of Fluid technique for the interface capturing. Results are presented in term of energy transfer, vorticity generation and momentum flux. Comparisons of the solutions obtained when using different values of the initial steepness are also established.

2. Numerical Setup

The breaking of water waves induced by the classical Benjamin-Feir instability mechanism [14] is simulated numerically by a two-fluid method. The initial solution is composed by a fundamental wave component with two sideband perturbations. The free surface elevation $\eta(x,0)$ is initialized as:

$$\eta(x,0) = \frac{\varepsilon_0}{k_0}[\cos(k_0 x) + \alpha\cos(k^+ x) + \alpha\cos(k^- x)] \qquad 2.1$$

where $k_0 = 2\pi/\lambda_0$ and $\varepsilon_0 = k_0 A_0$ are the wavenumber and the initial steepness of the foundamental component, respectively, λ_0 and A_0 the corresponding wavelength and amplitude, α is the relative amplitude of the sideband perturbations and k^+ and k^- are the wavenumbers of the sideband ($k^+ = k_0 + \Delta k$ and $k^- = k_0 - \Delta k$).

In [13] it is shown that for $\alpha = 0.1$ and $\Delta k = k_0/5$ the breaking occurs when $\varepsilon_0 = 0.12$ and, with the increase of the initial steepness, it gets stronger and starts earlier. Most of the results of this work are presented for $\varepsilon_0 = 0.20$.

The velocity field in water is initialized from the linear theory as:

$$\begin{aligned} u(x,y,0) &= A_0\left[(gk_0)^{1/2}e^{(k_0 y)}\cos(k_0 x) + \alpha(gk^+)^{1/2}e^{(k^+ y)}\cos(k^+ x) + \alpha(gk^-)^{1/2}e^{(k^- y)}\cos(k^- x)\right] \\ v(x,y,0) &= A_0\left[(gk_0)^{1/2}e^{(k_0 y)}\sin(k_0 x) + \alpha(gk^+)^{1/2}e^{(k^+ y)}\sin(k^+ x) + \alpha(gk^-)^{1/2}e^{(k^+ y)}\sin(k^+ x)\right] \end{aligned} \qquad 2.2$$

where g is the gravity acceleration, which is 9.81 ms^{-1}.

Simulation are performed using the open-source Gerris software (http://gfs.sourceforge.net) which solves the incompressible Navier-Stokes equations and has a quad-octree discretization with adaptive mesh refinement. The solver is based on a projection method and use a multilevel Poisson solver (for more details see [15]). The multiphase flow is modelled using the one-fluid approach with the fluids properties depending on the variable $T(\mathbf{x},t)$ representing the volume fraction of the two fluid (*i.e.* the ratio between one fluid in the other so that $T = 1$ represents the air domain and $T = 0$ the water domain). The governing equations are the momentum equation:

$$\frac{d\boldsymbol{u}}{dt} = \frac{1}{\rho(T)}\{-\nabla p + \nabla \cdot [\mu(T)(\nabla \boldsymbol{u} + \nabla \boldsymbol{u}^T)] + \sigma k \delta_s \boldsymbol{n}\} \quad 2.3$$

for the fluid, and the advection equation for the volume fraction field:

$$\frac{\partial T}{\partial t} + \nabla \cdot (T\boldsymbol{u}) = 0 \quad 2.4$$

where $\boldsymbol{u}$ and p denote velocity and pressure, respectively, $\rho(T)$ and $\mu(T)$ the local fluid density and viscosity, k the interface curvature, σ the surface tension coefficient and δ_s the Dirac distribution which is zero out of the interface. The fluid properties are related to the variable T as:

$$\begin{aligned} \rho(T) &= \rho_a T + \rho_w(1-T) \\ \mu(T) &= \mu_a T + \mu_w(1-T) \end{aligned} \quad 2.5$$

The simulation are performed using the real parameters for air and water, which are:

$\mu_w = 1.81\ 10^{-5}$ Nsm^{-2}, $\mu_a = 1.0\ 10^{-3}$ Nsm^{-2}, $\rho_a = 1.125$ kg/m^3, $\rho_w = 1000$ kg/m^3, $\sigma = 0.073$ Nm^{-1}

The wavelength of the foundamental component is $\lambda_0 = 0.6$ m, which is also the main scaling length, the domain width is $5\lambda_0$ and the height is $3\lambda_0$. Periodic boundary conditions are enforced at the right and left boundaries of the domain whereas free-slip boundary conditions are imposed at the top and the bottom boundaries.

The non-dimensional parameters characterizing this problem are the Reynolds and Weber numbers which are defined as:

$$\begin{aligned} Re &= \frac{\rho_w (g\lambda_0)^{1/2}\lambda_0}{\mu_w} = 1455664.8 \\ We &= \frac{\rho_w (g\lambda_0)^{1/2}\lambda_0}{\sigma_w} = 219.95 \end{aligned} \quad 2.6$$

where $(g\lambda_0)^{1/2}$ is the reference velocity, which is related to the phase velocity. It is worth noticing that the Reynolds number based on the phase velocity is not fully representative of the viscous phenomena taking place during the breaking process. A definition of the Reynolds number more representative of the physics should be based on the crest-to-trough velocity jump and on the wave height at the onset of the breaking. With such a definition, the Reynolds number would be approximatively 60000, which is two order of magnitude smaller than that based on the phase speed and makes simulations technically feasible. However, it is difficult to get a unique definition of those parameters as they strongly depend on the criteria adopted to define the onset of the breaking as well as on the points used to extract the velocities at crest and trough. In order to avoid such ambiguities, it was preferred to keep the definition based on the phase velocity despite its limitations.

Due to the high Re numbers, convergence analyses on the energy dissipation (not showed here) were made, and it was found that a maximum resolution of 2048 points per wavelength is enough to resolve the main dissipations scale of the phenomenon. This resolution is not enough to resolve all the dissipative scales but since the main contribution to the energy decay is due to the breaking, which involves larger scales of motion, allows to describe correctly the process. With such a resolution, the highest refinement level corresponds to a total of 10240 x 6144 grid cells. The use of the adaptive refinement capabilities allows to coarsen the grid outside the most relevant regions. The refinement level is based on vorticity and gradient of the interface function. With such choice, the grid about the free surface is always at the highest resolution level.

3. Results

During the evolution of the wave system the two sidebands grow whereas the foundamental component decreases due to nonlinear effects, as shown both experimentally in [12] and numerically in [13]. The growth of the sidebands leads to the formation of one wave in the group which is steeper than others. The steepness of the steepest wave, while increasing due to the nonlinear interaction, oscillates in time with a period which is twice the period of the fundamental component T, the maximum being achieved when the wave passes at the peak of the envelope. When the steepness exceeds a limiting value of 0.32, the breaking commences [16]. It last only a short fraction of the wave period as the modulational process causes a reduction of the steepness but then another breaking event occur in $2T$. In Fig. 1 the free surface profiles at different time instants are shown. The breaking starts with the formation of a bulge at the crest which slides down along the forward face, eventually leading to the entrainment of air cavities which are successively fragmented into smaller bubbles, the largest ones rising back towards the free surface and the finest remaining entrained in the vortical flow. The profiles display the formation of a second breaking event about $2T$ after the first event, which is in agreement with the field measurements in [17].

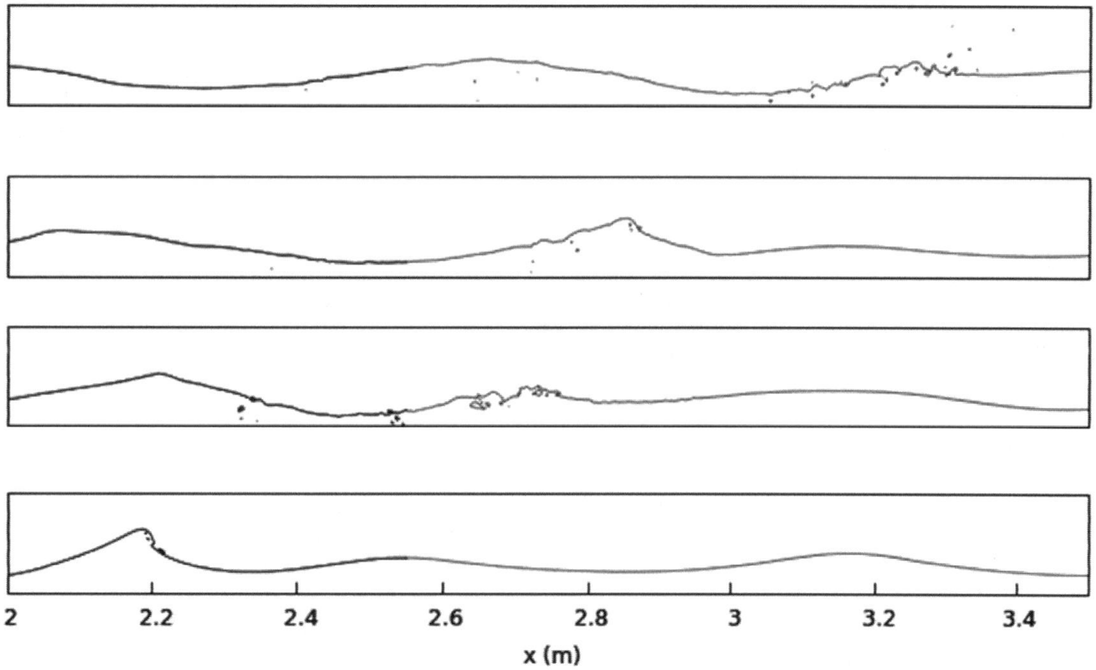

Figure 1. Surface elevation at different instant (the time grows from the bottom to the top) form $t = 14.5s$ to $t = 16$ s. The time interval between two configuration is equal to the wave period, $T = 0.62$ s. The wave is travelling from the left to right.

The important role played by the breaking on the exchange processes at the interface is clearly highlighted by the associated generation of bubbles and droplets. It is worth noting that, in spite of the quite high resolution adopted (the cell size is about 0.3 mm), not all scales are adequately resolved and, furthermore, phenomena like dissolution and aerosol or evaporation are not accounted for by the adopted model. Such phenomena provide significant contributions to the heat and gas transfer at the ocean-atmosphere interface. Due to the above mentioned limits, the results are focused on the analysis of the energy transfer and the vertical flux of horizontal momentum. The latter in particular helps in estimating the thickness of the layer about the interface where advection processes are enhanced by the flow induced by the breaking.

3.1. Mechanical Energy

In order to evaluate the energy transfer between the two fluids, the total energy of water and air is computed as:

$$E_T(t) = \int_V (E_k + E_p)dV = \int_V \left(\rho g y + \frac{1}{2}\rho(u^2 + v^2)\right) dV - E_{p0} \qquad 3.1$$

where E_k is the kinetic energy, E_p is the potential energy and E_{p0} is the potential energy of the still water level. Figure 2 shows the time history of the total energy in air and water normalized with the initial energy content in water. The data refer to the case with initial steepness 0.20. In the early stage there is a rather mild energy reduction in water. This is because, before the onset of the breaking, the flow is essentially irrotational and viscous effects are rather weak [18]. When the breaking starts there is the formation of a plunging jet that impacts onto the free surface, leading to the formation of a rotational flow in water and to air entrainment. After the large air cavity is entrained it starts rotating and it is progressively squeezed and fragmented into small bubbles [5,10]. The bubble fragmentation process generates strong velocity gradients at small scales which significantly enhances the mixing and the dissipation process [10].

The breaking occurrence is also responsible for the transfer of energy to the air (right picture in figure 2). The energy transfer is operated by both tangential and normal stresses at the interface. Although the latter are expected to be dominant, a deeper investigation is needed to achieve a quantitative estimate of the two contributions.

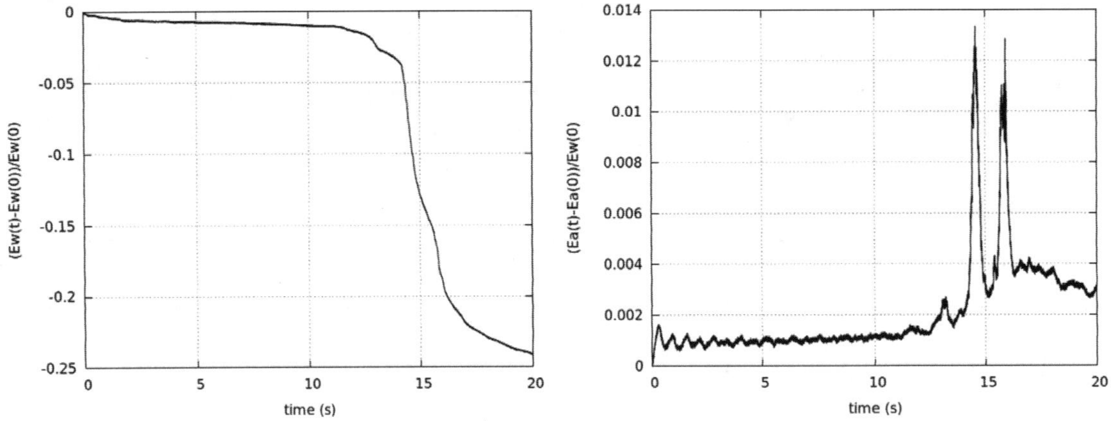

Figure 2. The figure shows the time history of the normalized energy for the case with initial steepness $\varepsilon_0 = 0.20$ in the water domain (left) and the air domain (right).

For the case considered in the present study, the total amount of energy lost by the water during the whole breaking process is about 25% of the initial energy content. For the same case, the energy amount accumulated in the air phase is about 1%. However, as discussed in [19], this is only a small fraction of the energy which is actually dissipated in the air. Indeed, due to the higher value of the kinematic viscosity and to the stronger velocity gradients, the dissipation rate in air is much higher than in water and most of the energy transferred in the air disappears before being accumulated.

As shown in figure 3, the results are essentially the same for the case with initial steepness 0.22. Due to the larger value of the initial steepness, the breaking starts somewhat earlier. The breaking events are more distinguished in this case but there are no significant differences in terms of the total energy fraction dissipated by the breaking and of the energy amount accumulated in air.

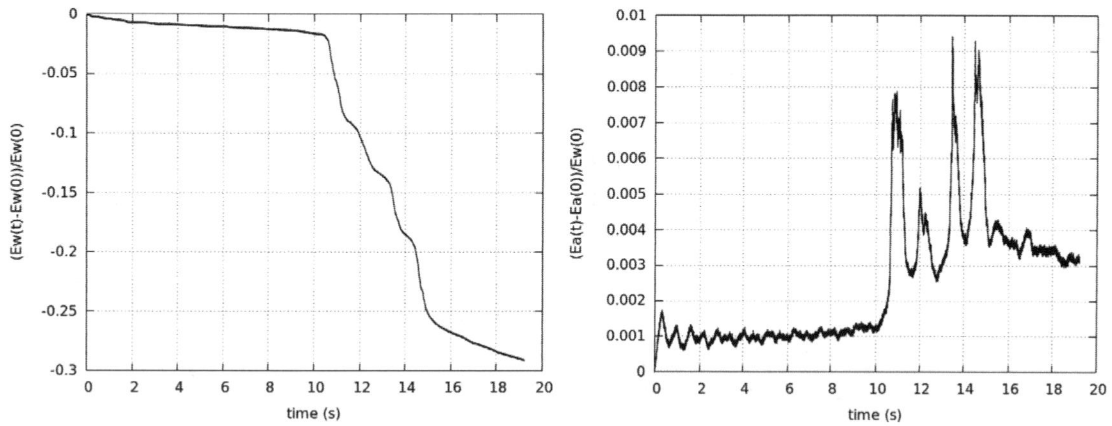

Figure 3. The figure shows the time history of the normalized energy for the case with initial steepness $\varepsilon_0 = 0.22$ in the water domain (left) and the air domain (right).

It is worth noting that, although the total amount is the same, the energy fraction dissipated by the single breaking events exhibits quite large variations from one event to the other, which explains the large scatter found in the experimental data [20].

3.2. Momentum transfer
In order to analyze the vertical transfer of the horizontal momentum, the flux:

$$q(y,t) = \int_x \rho uv \, dx \qquad 3.2.1$$

is computed along several horizontal lines, both in air and water, with a vertical distance of 0.005 m. Hence, the total amount of horizontal momentum transferred across each line at a given time τ is given by:

$$q(y,\tau) = \int_0^\tau q(y,t)dt \qquad 3.2.2$$

Figure 4 shows the total flux $q(y,\tau)$ transferred during the breaking process (i.e. from $t = 12\ s$, which is when the breaking starts, up to $t = 16\ s$). In order to avoid any misleading effect which could be caused by the crossing of the air-water interface through the lines, the fluxes are evaluated only for $y <$ -0.04 m and $y > 0.065$ m. Results indicate that the water layer affected by the momentum transfer is about 0.15 m thick, which is about one-fourth of the wavelength. Much deeper is the propagation of the momentum into the atmospheric layer, with some effects arriving up to one wavelength above the free surface. This has important consequences on the heat and gas transfer processes at the sea surface. It is worth noting that in this case the effects of the wind are missing. In a more realistic windy condition, it is expected that the interaction of the air flow induced by the breaking with the wind further enhances the exchange and transfer processes. The momentum flux is negative in water and positive in air meaning that there is a mean vertical current which is downward in water and upward in air.

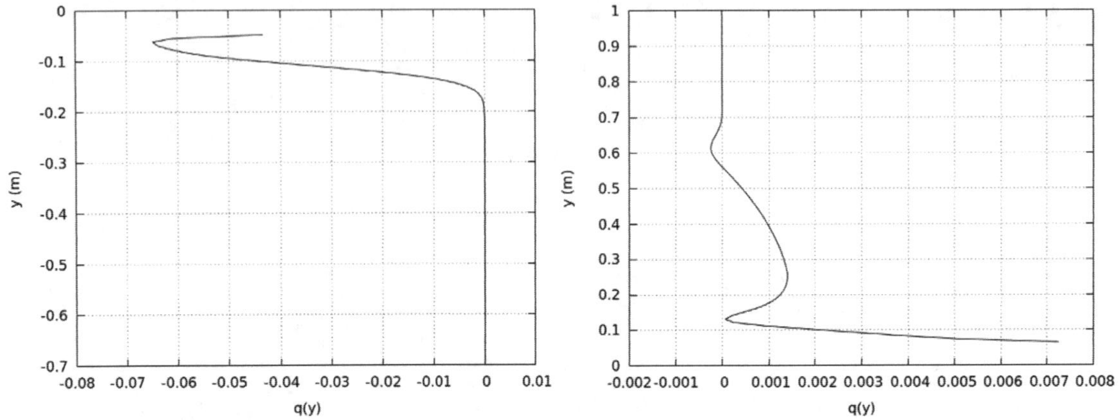

Figure 4. The figure shows the total horizontal momentum transferred in the y direction during the whole breaking process in the water domain (left) and in the air domain (right).

3.3. Vorticity

When the breaking starts, the wave develops a sharp curvature at the crest which induces the separation of the air flow. Due to the flow separation, the normal stresses at the interface at the front and at the rear of the wave crest are no longer balanced, and this contributes substantially to the energy and momentum transfer from the water to the air. Besides the normal stresses, tangential stresses are also quite strong. The flow separation gives rise to the development of a strong vorticity field in the air layer, and the periodicity of the flow associated to the orbital motion of the wave, leads to the shedding of large coherent structures [13].

Figure 5 shows the vorticity field at four different instants. As already indicated by the momentum transfer, the vorticity structures propagate much deeper into the air and some structures arrive even beyond one wavelength above the free surface. The sequence also highlights the vorticity generation in water. The structures in water are much smaller in size and, moreover, due to the lower velocities the downward motion is slower. Partly, such differences are due to the different viscosity value. But the higher velocities in air are also related to the momentum transfer at the interface in combination to the high density ratio of the two fluids.

Figure 6 shows a detailed view of the vorticity field both in water and in air. In the air, structures with different size appear, some of them being generated by the viscous effects at the free surface and others originated by the interaction with the sprays and droplets ejected in the breaking process (e.g. left picture in figure 7). In water, structures are generally much smaller. Also in this case some of them are generated by viscous effects at the interface, see [21], and others are generated by the air entrainment and by the fragmentation process. Aside from the generation mechanism, the vortex structures in water remain in an upper layer, just beneath the free surface.

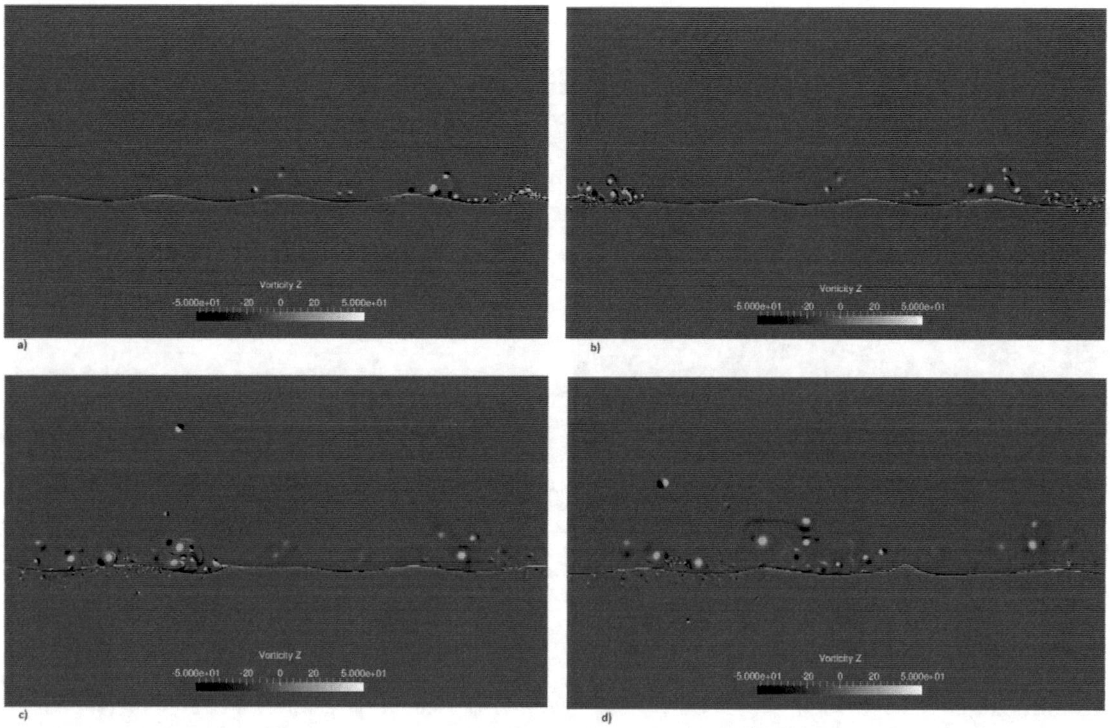

Figure 5. The figure shows the vorticity field distribution at four different instant: a) 14.55 s, b) 15.35 s, c) 17.20 s, d) 18.9 s.

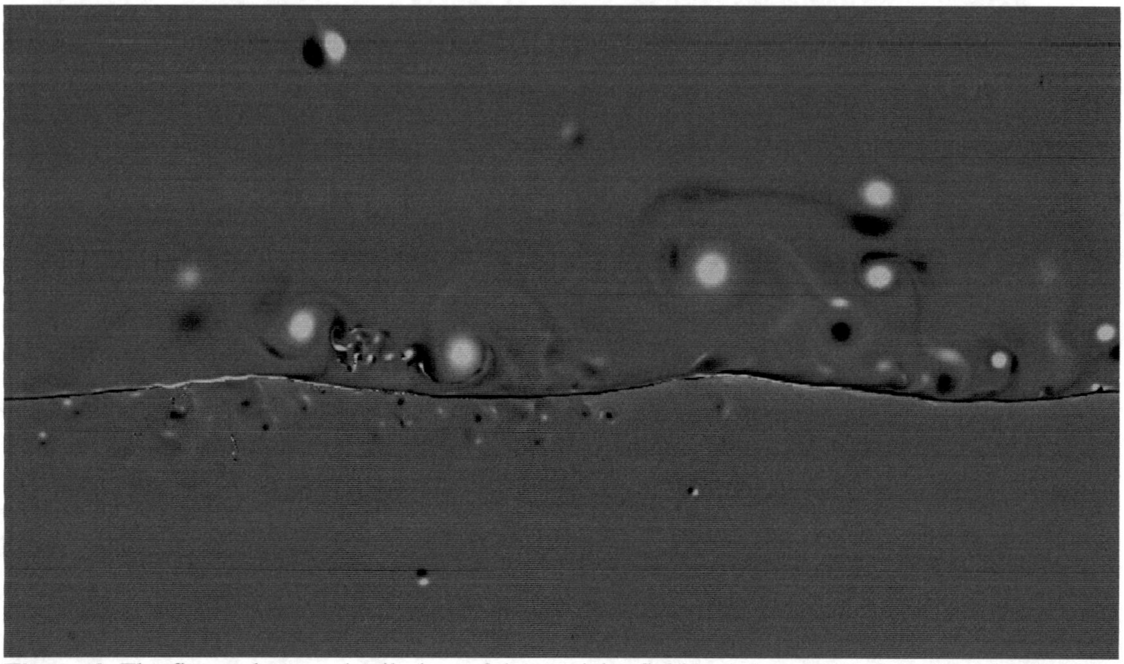

Figure 6: The figure shows a detail view of the vorticity field both in water and air at t = 18.90 s (the colours are the same of figure 5).

By increasing the initial steepness, the vorticity field is more intense both in air and water, the latter mainly related to the augmented level of air entrainment (figure 7).

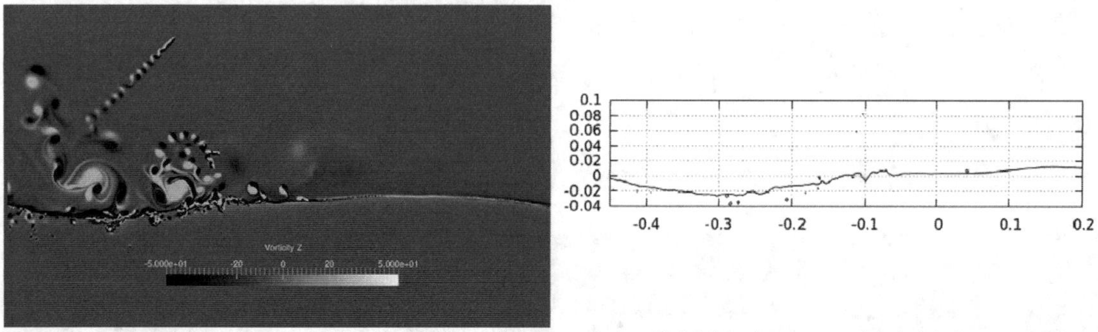

Figure 7. The figure shows a detailed view of the vorticity field for the case with initial steepness ε_0 = 0.22 both in the water domain (left) and in the air domain (right) at time t = 15 s. The picture on the left highlights the Karman vortex sheets generated by droplets propagating in air.

4. Conclusions

The breaking of water waves induced by modulational instability has been numerically simulated by a two-fluid numerical approach. Simulations have been performed for two different values of the initial steepness and quantitative estimates of the energy transfer and momentum flux have been provided. The results show that in correspondence of the breaking events there is an energy transfer to the air, most of which is quickly by the viscous effects. It has been shown that the breaking occurrence enhances the mixing processes, and even without wind, the effects of the breaking arrive up to about one wavelength above the free surface. Further studies are needed in order to evaluate all the fluxes at the interface and to provide a more complete picture of the exchange phenomena at the interface, and thus a better understanding of the role of the breaking on the transfer processes occurring at the ocean-atmosphere interface.

Acknowledgements

The work has been financially supported by the "Flagship Project RITMARE - The Italian Research for the Sea - coordinated by the Italian National Research Council and funded by the Italian Ministry of Education, University and Research within the National Research Program 2014-2015". Authors express their gratitude to the Gerris team and Dr. S. Popinet not only for making the code available but also for the technical help provided (http://gfs.sourceforge.net). The numerical simulations of the Bejnamin-Feir instability have been done at CINECA, the Italian Supercomputing Center. We acknowledge the CINECA award under the ISCRA initiative, for the availability of high performance computing resources within the project IscrC_DIPWAVE. The technical support by Dr. Giorgio Amati is also gratefully appreciated.

References

[1] Janssen P 2009 The interaction of ocean waves and wind, Cambridge, UK, Cambridge Press

[2] Emmanuel K A 1986 An air-sea interaction theory for tropical cyclones. Part I: steady-state maintenance *J. Atm. Scien.* **43** 585-604

[3] Cavaleri L Fox-Kemper B and Hemer M 2012 Wind waves in the coupled climate system *Bull. Amer. Meteor. Soc.* **93** 1651-61

[4] Babanin A 2011 Breaking and dissipation of ocean surface waves Cambridge
[5] Deane G B and Stokes M D 1999 Air entrainment processes and bubble size distribution in the surf zone *J. Phys. Ocean.* **29** 1393-1403
[6] Kudryavtsev V N 2006 On the effects of sea drops on the atmospheric boundary layer *J. Geophy. Res.* **111**
[7] Cavaleri L 2009 Wave modelling – missing the peaks *J. Phys. Ocean.* **39** 2757-78
[8] Drazen D. and Melville W K 2009 Turbulence and mixing in unsteady breaking surface waves *J. Fluid Mech.* **628** 85-119
[9] Banner M L and Peirson W L 2007 Wave breaking onset and strength for two-dimensional deep-water wave group *J. Fluid Mech* **585** 93-115
[10] Iafrati A 2009 Numerical study of the effects of the beaking intensity on wave breaking flows *J. Fluid Mech.* **622** 371-411
[11] Iafrati A 2011 Energy dissipation mechanisms in wave breaking process: spilling and highly areated pluning breaking events *J. Geophy. Res.* **116**
[12] Tulin M P and Waseda T 1999 Laboratory observation of wave group evolution, includin breaking effects *J. Fluid Mech.* **378** 197-232
[13] Iafrati A, Babanin A and Onorato M 2014 Modeling of ocean-atmosphere interaction phenomena during the breaking of modulated wave trains *J. Copm. Phys.* **271** 151-71
[14] Benjamin T R and Feir J E 1967 The disintegratione of wave trains on deep water. Part 1. Theory *J. Fluid Mech.* **27** 471-30
[15] Popinet S 2009 An accourate adaptive solver for surface-tension-drive interfacial flows *J. Comp. Phys.* **228** 5838-66
[16] De Vita F, Verzicco R and A Iafrati 2015 Energy dissipation and spectrum evolution during the breaking of modulated wave trains *Proceeding of the 30th International Workshop Water Waves Floating Bodies,* Bristol, UK
[17] Lamont-Smith T, Fuchs J and Tulin M P 2003 Radat investigation of the structures of wind waves *J. Oceanogr.* **59** 49-63
[18] Landau L D and Lifshitz E M 1959 Fluid Mechanics *Pergamon Press*
[19] Iafrati A, Babanin A and Onorato M 2013 Modulational instability, wave breaking and formation of large-scale dipoles in atmosphere *Phy. Rev. Lett.* **110**
[20] Galchenko A, Babanin A, Chalikov D and Young L R 2010 Modulational instabilities and breaking strength for deep-water wave groups *J. Phys. Ocean.* **40** 2313-24
[21] Iafrati A and Campana E F 2005 Free-surface fluctuations behind microbreakers: space-time behaviour and subsurface flow field *J. Fluid Mech.* **529** 311-47

Experimental and numerical analyses on a plate heat exchanger with phase change for waste heat recovery at off-design conditions

Roberto Cipollone, Giuseppe Bianchi, Davide Di Battista, Fabio Fatigati

University of L'Aquila, via Giovanni Gronchi 18, L'Aquila 67100, Italy

E-mail: roberto.cipollone@univaq.it

Abstract. This paper analyzes the performances of an evaporator for small scale waste heat recovery applications based on bottoming Organic Rankine Cycles with net output power in the range 2-5 kW. The heat recovery steam generator is a plate heat exchanger with oil as hot stream and an organic fluid on the cold side. An experimental characterization of the heat exchanger was carried out at different operating points measuring temperatures, pressures and flow rates on both sides. The measurement data further allowed to validate a numerical model of the evaporator whereas heat transfer coefficients were evaluated comparing several literature correlations, especially for the phase-change of the organic fluid. With reference to a waste heat recovery application in industrial compressed air systems, multiple off-design conditions were simulated considering the effects of oil mass flow rate and temperature on the superheating of the organic fluid, a key parameter to ensure a proper operation of the expansion machine, thus of the energy recovery process.

1. Introduction

Medium and low grade energy recovery might be accomplished in multiple contexts, such as cement and glass industry, heavy duty internal combustion engines as well as solar and geothermal power plants [1]. Among the industrial situations, compressed air systems might highly benefit from an energy recovery system in mechanical or electrical form to lower package specific consumptions of compressors since they typically waste almost 80 % of the electrical energy supplied to the machine through the oil radiator [2]. Indeed, literature studies estimate the saving potential from waste heat recovery as 10 % of the overall energy consumptions [3]. To date, waste heat recovery in oil flooded compressors usually occurs through a thermal approach that uses the hot temperature oil stream (90-120 °C) to heat up water in a heat exchanger and partially fulfill heating or hot water needs of the industrial site. On the other hand, a universally referenced technology which converts thermal energy into mechanical form is based on an Organic Rankine Cycle (ORC).

The evaporator is a key component of the ORC-based energy recovery system since it has to guarantee a correct heat transfer during the pre-heating, vaporization and superheating of the high pressure organic fluid. Furthermore, it should have high thermal performances to maximize the heat recovery as well as low pressure drops on both sides to prevent re-condensation of the organic fluid or excessive pressure loss for the external heat source. Compared with other technologies, plate heat exchangers (PHEs), thanks to the high degree of turbulence established at the flow channels, require lower surface and volume than those needed by shell and tube devices to exchange the same amount of power. Moreover, PHEs can be simply maintained, cleaned and inspected. Their high effectiveness and low cost led to a large use of these devices as evaporators and condensers in industrial applications [4, 5]. Another important consequence of their compactness is the low refrigerant charge [6]. On the other hand, drawbacks of PHEs are large pressure drops due to the restricted flow paths. However, this issue could be overcome using the manifold-multichannel technology [7], which allows to reduce single-phase flow pressure drops by a factor of n^2, where n represents the number of channel divisions [8].

A major concern in plate heat exchangers is the difficulty to understand flow boiling mechanisms due the complex nature of two-phase flow along the channels [9]. Consequently, quantitative predictions of the flow boiling heat transfer coefficient were obtained through experimental investigations and expressed by empirical correlations such as the ones of Yan Lin for R141a [10], of Ayub [6] for ammonia and R22, of Sterner and Sunden for ammonia [11] and the correlation of Han at al. for R22 and R422A [12]. Research carried out in [13] investigated the effect of refrigerant mass flow rate , saturation temperature (pressure) and fluid properties on pressure drops and heat transfer during the saturated vapor condensation of R236fa, R134a and R410A for a brazed plate heat exchanger that used water as hot stream. The analysis resulted in a great influence of fluid properties and refrigerant mass flow rate and a feeble effect of saturation temperature on the heat transfer coefficient. As concerns pressure losses, it was observed a linear dependence on the kinetic energy per unit volume of refrigerant flow and, consequently, a quadratic dependence on the refrigerant mass flow. According to this study, R410A and R134a have similar heat transfer coefficient, whose values are 10% higher than R236fa for the same device and operating conditions. Moreover, pressure drops for R410A are 50-60% lower than R236fa and 40-50% lower than R134a.

In this paper, a plate heat exchanger used as evaporator of an ORC system for waste heat recovery on an industrial air compressor was tested at different operating conditions using oil as hot stream and R236fa as cold stream. The experimental data acted as a validation baseline for a one-dimensional model of the plate heat exchanger whereas multiple semi-empirical correlations were compared. Off-design conditions of the evaporator were eventually simulated considering the most common regulation strategies in industrial compressors, namely variable speed and throttling of the suction process. The off-design analysis aimed at assessing the influence of oil mass flow rate and temperature on the superheating of the organic working fluid which is a feature strictly required by expansion machines of ORC systems to accomplish a proper conversion of the thermal power recovered into mechanical one.

2. Experimental campaign

The test campaign used as reference for the current study is the one performed on a waste heat recovery application in compressed air industry [14, 15]. In that case, the hot source of the energy recovery system based on Organic Rankine Cycle was the oil of an industrial air compressor that usually heats up during the compression phase because of friction between the machine components it lubricates and the convective heat exchange with air along the compression and discharge processes [16, 17]. The evaporator of the energy recovery system was a off the shelf oil-water plate heat exchanger (PHE) whose features are listed in Table 1. On the cold side of the PHE, a R236fa organic fluid recuperated the thermal energy of the lubricant, that otherwise would be usually dissipated at the compressor radiator, to economize, vaporize and eventually reach a superheated state. Part of the enthalpy of the superheated steam was then converted to mechanical power through a sliding vane rotary expander; the remaining share was rejected to the environment through a water cooled condenser.

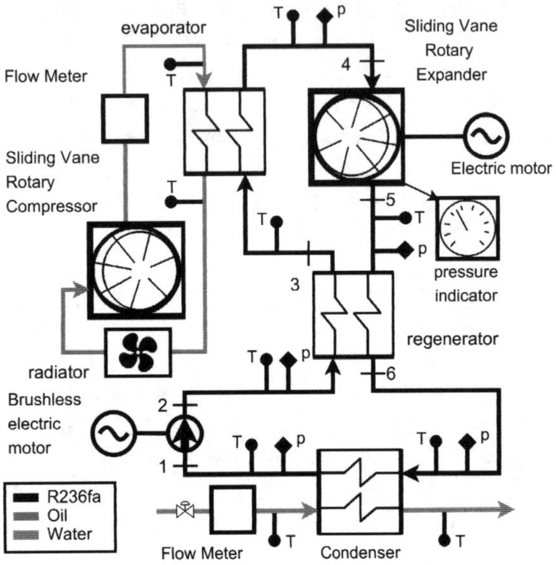

Figure 1: ORC system test bench with sensors layout

plate length	286	mm
plate width	117	mm
number of plates	41	–
nominal heat transfer area	1.2	m^2
plate thickness	1	mm
corrugation amplitude	1.15	mm
plate roughness	1	μm
hydraulic diameter	4.5	mm
weight	6.9	kg
material	copper	

Table 1: Evaporator specifics

temperature	0.1	°C
R236fa mass flow rate	3.57%	
pressure transducers	0.08	bar
thermal power exchanged	2.90%	
oil flow rate	0.6	L/min

Table 2: Direct and derived measurement uncertainties

As reported in Figure 1, the evaporator performances were assessed through measurements across the inlet and outlet ports of both sides of the PHE. In particular, fluids temperatures were measured with T-type thermocouples while a gear flow meter was used to retrieve the oil flow rate. Unlike the oil that stayed in liquid phase while travelling along the evaporator, pressure measurements were additionally needed across the cold side of the heat exchanger to monitor the phase change of the organic fluid. In order to maximize the heat exchanger efficiency, the device was thermally insulated such that the R236fa mass flow rate was calculated from an energy balance. This indirect measurement led to a higher uncertainty on this quantity (Table 2). However, this issue can be overcome using a Coriolis mass flow meter, as did in [2, 18].

Table 3 presents a summary of the experimental campaign performed at working points. For a proper air compressor operation, the oil mass flow rate was kept constant at 1.25 kg/s in all the experimental cases, except for # 13. Furthermore, a by-pass of the oil radiator allowed an overheating of the lubricant up to 120 °C with respect to a conventional operating point at 80-90 °C (# 12). Since mass flow rate of the organic fluid could be varied acting on the revolution speed of the sliding vane rotary pump in the ORC system, this parameter assumes a significant spread and affects the operating pressure on the cold side of the plate heat exchanger. Indeed, high mass flow rates led to higher pressure drops along the circuit and allowed greater upper cycle pressures. On the contrary, the constancy of R236fa inlet temperature is due to fixed operating conditions at the condenser. All these remarks led to a thermal power recovery from 5.6 kW (# 11) up to 19.5 kW (# 7). In this case, the thermal power exchanged at the evaporator was equal the design one with water (20 kW), whose specific heat is almost 3 times than that of liquid R236fa at the inlet conditions of test # 7.

3. Modeling activity

The plate heat exchanger previously tested was further modeled using a steady one-dimensional approach to characterize heat transfer and pressure loss phenomena along fluid passages. In particular, pressures, mass flow rates and inlet temperature of both fluids are the model inputs while outlet temperatures, cold side pressure losses and overall thermal power exchanged are the calculated variables.

case	$T_{hot,in}$	$T_{hot,out}$	m_{hot}	$p_{cold,in}$	$p_{cold,out}$	$T_{cold,in}$	$T_{cold,out}$	m_{cold}	Q
♯	°C	°C	kg/s	bar	bar	°C	°C	kg/s	kW
1	114.8	107.7	1.34	12.4	12.1	42.9	90.6	0.111	18.8
2	118.1	110.9	1.25	12.7	12.4	45.0	96.1	0.111	19.1
3	117.3	110.5	1.25	12.1	11.9	45.1	98.3	0.103	18.0
4	117.1	110.7	1.25	11.4	11.2	44.7	100.1	0.095	16.9
5	121.4	115.4	1.25	10.8	10.6	45.0	106.4	0.086	16.0
6	116.9	111.1	1.25	10.6	10.4	46.6	101.6	0.086	15.3
7	119.7	112.3	1.25	13.4	13.2	48.1	98.3	0.115	19.5
8	118.3	113.4	1.25	9.9	9.8	46.5	104.5	0.071	13.0
9	119.1	115.5	1.25	8.8	8.6	46.5	107.6	0.052	9.80
10	120.6	117.7	1.25	7.9	7.8	46.0	110.8	0.039	7.5
11	120.4	118.2	1.25	6.7	6.5	45.9	112.4	0.029	5.6
12	84.3	80.5	1.28	7.1	7.1	36.0	71.7	0.059	9.9
13	118.9	109.2	0.94	12.4	12.1	48.2	90.5	0.110	19.3

Table 3: Summary of the experimental campaign

3.1. Thermal model

The model is based on the ϵ-NTU method and starts from the geometrical features of the heat exchanger. It follows a 1D approach in which flow channels defined by consecutive plates are divided in a sequence of elementary ducts. Therefore, for each j-th sub-volume, thermal power exchanged Q_j is represented by effectiveness ϵ and NTU values that can be calculated according to Eqns. 1 and 2 respectively.

$$\epsilon_j = \frac{Q_j}{C_{\min,j}(T_{hot,in,j} - T_{cold,in,j})} = f\left(NTU_j, \frac{C_{\min,j}}{C_{\max,j}}\right) \quad (1)$$

$$NTU_j = \frac{U_j \cdot A_j}{C_{\min,j}} \quad (2)$$

where the j-th thermal exchange area A_j is the overall heat transfer area divided by the number of space discretization steps chosen and the minimum heat capacity rate as in Eqn. 3.

$$C_{\min,j} = \min(C_{hot,j}, C_{cold,j}) = \min(m_{hot}c_{hot,j}, m_{cold}c_{cold,j}) \quad (3)$$

The global heat transfer coefficient U_j is calculated by conductive and convective thermal resistances, according to Eqn. 4.

$$\frac{1}{U_j} = \frac{1}{h_{cold,j}} + \frac{t_{wall}}{k_{wall}} + \frac{1}{h_{hot,j}} \quad (4)$$

where t_{wall} and k_{wall} are thickness and thermal conductivity of the PHE's walls. On the other hand, convective heat transfer coefficients h are calculated relying on empirical correlations.

In this regard, considering that heat transfer in PHEs is very often enhanced by turbulent flow [19, 20], for single phase fluids most of the semi-empirical correlations that have been proposed in literature are based on the Dittus-Boelter one [21, 22]. Furthermore, since in the current application flow Reynolds number ($Re \simeq 10^6$) falls in the same range of the above mentioned correlation ($Re \geq 10^4$), convective heat transfer coefficient is calculated according to Eqn. 5.

$$h = 0.023\, Re^{0.8}\, Pr^n\, k_{fluid}/D_h \quad (5)$$

where the Prandtl number exponent n is equal to 0.4 for the cold fluid and to 0.3 for the hot fluid.

As concerns the phase change of the organic fluid, different correlations were tested in order to evaluate the best agreement with experimental data [23, 24].
The first correlation, reported in Eqn. 6, is the Kandlikar one.

$$h = (1.1837\mathrm{Co}^{-0.3} + 225.55\mathrm{Bo}^{2.8})(1-x)^{0.003}h_l \tag{6}$$

where Convection number Co and Boiling number Bo are defined in Eqn. 7.

$$Co = \left(\frac{1-x}{x}\right)^{0.8}\left(\frac{\rho_{sv}}{\rho_{sl}}\right)^{0.5} \qquad Bo = \frac{Q_j}{(H_{sv}-H_{sl})} \tag{7}$$

Since this correlation tends to under predict the heat transfer coefficient due to high values of the Boiling number exponent, a modified Kandlikar correlation [25] was additionally taken into account (Eqn. 8).

$$h = (1.1837\mathrm{Co}^{-0.3} + 225.55\mathrm{Bo}^{0.7})(1-x)^{0.003}h_l \tag{8}$$

h_l is the liquid heat transfer coefficient evaluated according to Eqn. 5.

Another suitable correlation for plate heat exchangers is the Yan-Lin [10] one that it is reported in Eqn. 9.

$$h = 1.1926\, {Bo_{eq}}^{-0.3}\, {Re_{eq}}^{0.5}\, {Pr_l}^{0.33}\left(1 - x + \sqrt{\frac{\rho_{sl}}{\rho_{sv}}}\right)\frac{k_{fluid}}{2b} \tag{9}$$

where equivalent Boiling and Reynolds numbers are evaluated according to Eqn. 10.

$$\mathrm{Re}_{eq} = \frac{m\left(1-x+x\sqrt{\frac{\rho_{sl}}{\rho_{sv}}}\right)2b}{A_s\mu_l} \qquad \mathrm{Bo}_{eq} = \frac{Q_j}{m\left(1-x+x\sqrt{\frac{\rho_{sl}}{\rho_{sv}}}\right)(H_{sv}-H_{sl})} \tag{10}$$

Moreover, two more general heat transfer correlations were considered. Among them, Klimenko's [26] is the unique correlation that accounts for material properties (Eqn. 11).

$$h = 0.087Re^{0.8}\, Pr^{1/6}\left(\frac{\rho_{sv}}{\rho_{sl}}\right)^{0.2}\left(\frac{k_{wall}}{k_{sl}}\right)^{0.09}\frac{k_{sl}}{2b} \tag{11}$$

while Shah and Thome correlation [27] has a better behaviour at lower liquid Reynolds number. In particular, it considers the largest value of Equations 12.

$$\begin{aligned} h &= 230\mathrm{Bo}^{0.5}h_l & h &= 1.8\Big[\mathrm{Co}\Big(0.38{\mathrm{Fr}_l}^{-0.3}\Big)^{\nu}\Big]^{-0.8}h_l \\ h &= F\cdot e^{2.47\left[\mathrm{Co}\left(0.38{\mathrm{Fr}_l}^{-0.3}\right)^{\nu}\right]^{-0.15}}h_l & h &= F\cdot e^{2.74\left[\mathrm{Co}\left(0.38{\mathrm{Fr}_l}^{-0.3}\right)^{\nu}\right]^{-0.1}}h_l \end{aligned} \tag{12}$$

where F is the Shah coefficient that it is equal to 0.064 for Bo>0.0011 and 0.067 otherwise. v is equal to 1 for low liquid Froude number ($Fr_l < 0.04$) and to 0 elsewhere. Liquid Froude number is evaluated according to Eqn. 13.

$$\mathrm{Fr}_l = \frac{m^2}{2b\, g\, A_s^2\, \rho_l^2} \tag{13}$$

3.2. Pressure drops

Measured pressure drops of the organic fluid were correlated to specific kinetic energy according to Eqn. 14.

$$\Delta p_{cold} = f \frac{2\, m_{cold}^2}{A_s^2\, \rho_m} \frac{L}{D_h} \tag{14}$$

where f is the Fanning factor.

4. Model validation

This section presents a comparison between modeling results and experimental measurements on the evaporator at the operating conditions listed in Table 3. In all the simulations, Dittus-Boelter correlation was used for single phase heat exchanges. On the other hand, the literature correlations previously proposed were considered for the phase change of the organic fluid [28, 29]. The goodness of numerical data was evaluated with reference to outlet hot and cold fluids temperatures, which are quantities measured directly as well as output data of the model. Results are presented in Figure 2 for the hot and cold fluids while Figure 3 presents the model validation in terms of thermal power exchanged.

Figure 2a shows a suitable accuracy of all the correlations except for the standard Kandlikar one that underestimates the thermal power exchanged in all cases. Klimenkos formulaiton seems to be quite far from measured values while other correlations have a similar behaviour, with an error always lower than 6 °C.

Thermal powers exchanged (Figure 3) confirmed the same behaviour of Figure 2b. Best fitting of measured values is represented by modified Kandlikar correlation, but also Yan-Lin and Shah-Thome ones have good results. Their mean error is about 3.5 %, with a maximum one of 10 %.

As concerns pressure drops at the cold side of the evaporator, experimental data were used to calculate the Fanning Factor as in Eqn. 14. Figure 4 shows the Fanning factor evaluated with respect to different mass flow rates of the cold fluid; indeed, it represents the link between specific kinetic energy and pressure drops occurred in heat exchanger. The fitting follows a hyperbolic law, confirming a literature result [30].

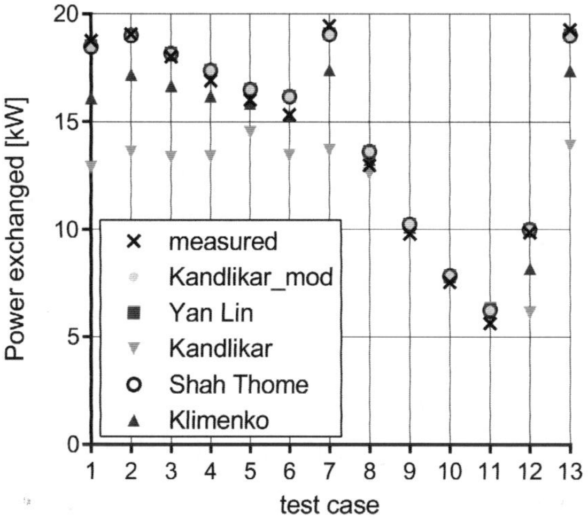

Figure 3: Model validation on thermal power exchanged

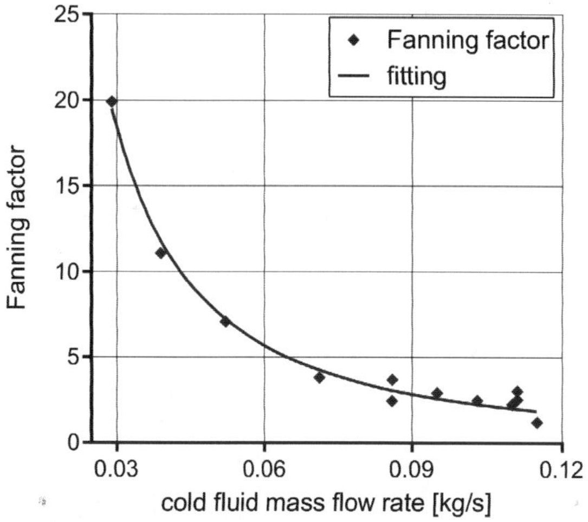

Figure 4: Fanning factor calculated in test cases

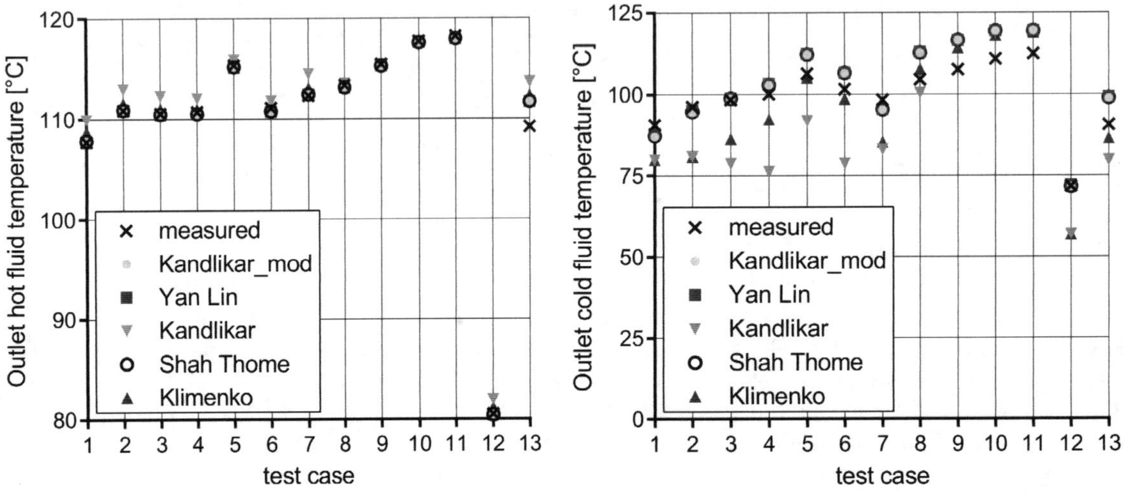

(a) hot side (b) cold side

Figure 2: Comparison between measured outlet temperatures with model results obtained using different correlations for the phase change of the organic fluid

5. Off-design analysis

Even though the modeling methodology developed for plate heat exchangers can be applied to any waste heat recovery application that uses oil and R236fa with phase change, the current study focused on the occurrence of off-design conditions due to an unconventional operation of the sliding vane air compressor whose lubricant was used as upper thermal source of the ORC-based energy recovery system displayed in Figure 1. In an industrial site, compressed air needs are usually variable according to production level. For this reason, the air compressor should always guarantee that air is available at almost the delivery pressure value. Therefore, the most common strategies to vary compressed air supply to match the demand are a regulation on the revolution speed of the compressor through an inverter or throttling the suction valve to reduce the air flow rate.

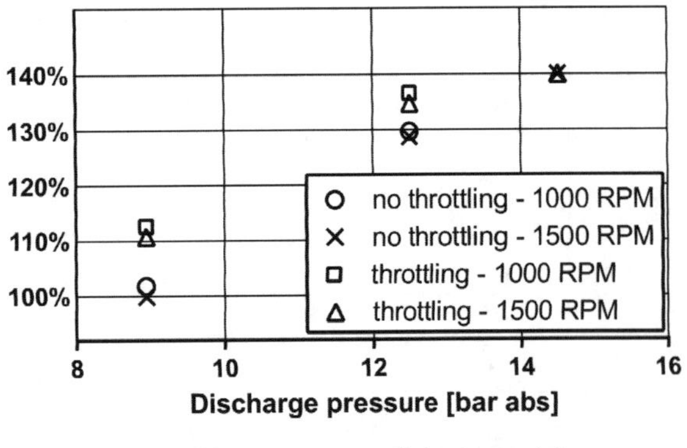

Figure 5: Oil flow rate at off design with respect to test # 7 of Table 3 (8.5 bar at 1500 RPM)

	min	max	
$T_{hot,in}$	100.0	120.0	°C
m_{hot}	1.00	1.50	kg/s
m_{cold}	0.055	0.115	kg/s
p_{cold}	10.0	13.4	bar_a

Table 4: Parameters and simulation ranges for the off-design analysis

As reported in Figure 5, in sliding vane compressors only the latter approach leads to a change in the oil flow rate. Moreover, if flow rate changes, oil temperature at the compressor outlet is also affected. Since both quantities would influence the recovery plate heat exchanger behavior,

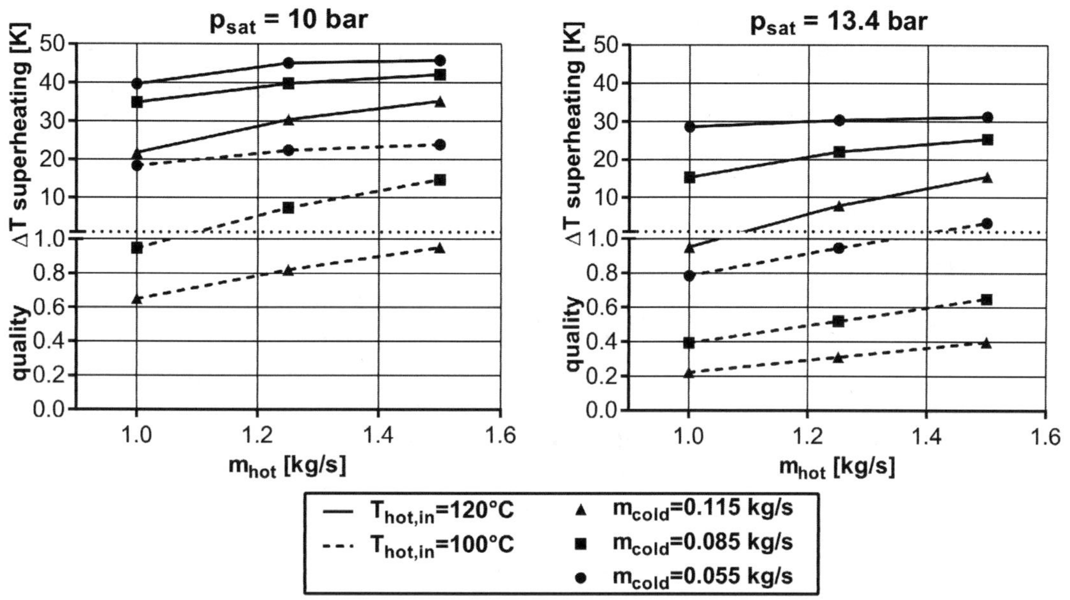

Figure 6: Effects of off-design operating conditions of a plate heat exchanger for waste heat recovery in terms of quality and superheating or subcooling of the organic fluid at the heat exchanger outlet

the off design analysis explored different operating conditions taking into account a variability of organic working fluid mass flow rate and vaporization pressure according to the ranges reported in Table 4.

Figure 6 reports a summary of the off-design analysis carried out at different oil mass flow rate and inlet temperature values. The comparison is made with respect two thermodynamic quantities of the organic fluid, namely quality and degree of superheating (quality>1) or subcooling (quality<0). The choice on these parameters is motivated on the need to ensure a correct operation of the ORC-based power unit even when the compressor operates far from the nominal conditions. In particular, multiple studies on ORC systems confirm that for a proper energy recovery, outlet conditions at the evaporator should consider a 5-10 K of superheating of the working fluid since wet expansion is an unfavourable situation for expansion machines, especially for turbines [31]. Simulations at all pressure levels show that inlet oil temperature highly affect the outlet conditions of R236fa. Indeed, if the thermal energy recovery took place at low temperatures ($T_{hot,in} = 100°\text{C}$), the organic fluid would never reach a complete vaporization, except for a particular case ($m_{cold} = 0.085\, kg/s$, $p_{cold} = 10\, bar$). Moreover, for a given inlet temperature of the cold fluid, vaporization pressure affects the shares of thermal powers needed for pre-heating, vaporization and superheating. For this reason, Figure 6 shows that the higher is the saturation pressure, the lower is the quality of the organic fluid at the evaporator outlet.

According to the trends reported in Figure 6, two control strategies can be outlined to adapt the ORC system at off-design operating conditions imposed by the upper thermal source:

- **variable oil mass flow rate**: if one considers a given inlet temperature of the hot stream ($T_{hot,in}$) and fixed operating conditions of the organic flow circuit (m_{cold}, p_{cold}), a by-pass of the oil mass flow rate at the evaporator could tune the heat gain of the organic fluid to keep it superheated.
- **variable organic fluid mass flow rate**: for given conditions at the hot side of the evaporator ($m_{hot}, T_{hot,in}$), organic fluid mass flow rate might be adjusted to ensure the

superheating. This strategy would be implemented through a feed-forward control on the pump revolution speed of the ORC system. However, this approach should also consider the additional influence of mass flow rate on pressure losses of the organic fluid circuit. Indeed, the higher is the mass flow rate, the higher would be pressure losses that the pump should overcome and, in turn, the vaporization pressure of the organic fluid.

The first strategy is easier to implement (e.g. through a calibrated by-pass valve) but would reduce the energy recovered. On the other hand, in view of a greater complexity, the latter approach would lead the recovery power unit to suitably perform a mechanical conversion of the low grade thermal waste. The optimal regulation technology will definitively depend on economical considerations that can not be forecasted at this development stage of the technology.

6. Conclusions

The heat recovery steam generator is a fundamental component of waste heat recovery systems which are based on an organic Rankine cycle since it has to accomplish an efficient energy recovery with multiple factors that constraint its operation such as space, weight, operating pressures etc.. An even more challenging situation occurs whether the energy recovery system worked at off-design conditions since vaporization and slight superheating of the organic working fluid must be anyway guaranteed for a proper mechanical energy recovery through the expansion machine. To address these issues, this research work analyzed the performances of a 20 kW plate heat exchanger both with numerical and experimental approaches with reference to a waste heat recovery application on industrial air compressors. Experimental data allowed to validate the heat exchanger model in several operating points, whereas the modified Kandlikar semi-empirical correlation turned out to be the most appropriate formulation to represent the phase change of the working fluid R236fa. The off-design analysis that was eventually carried out on the PHE showed that in order to ensure a slight (5-10 K) superheating of the organic working fluid at the heat exchanger outlet, oil mass flow rate might be by-passed or organic fluid mass flow rate could be tuned on the thermal power available at the evaporator. The first control strategy is easier to implement but reduces the energy recovery. On the other hand, a control of the organic fluid mass flow rate acting on the pump revolution speed of the ORC system would maximize the energy recovered but it might be more complex and costly.

References

[1] Tchanche B F, Lambrinos G, Frangoudakis A and Papadakis G 2011 *Renewable and Sustainable Energy Reviews* **15** 3963 – 3979
[2] Bianchi G 2015 *Exhaust Waste Heat Recovery in Internal Combustion Engines - Development of an ORC-based power unit using Sliding Vane Rotary Machines* ISBN 978-8-8871-8269-9
[3] Saidur R, Rahim N and Hasanuzzaman M 2010 *Renewable and Sustainable Energy Reviews* **14** 1135 – 1153
[4] Shah R K and Wanniarachchi A S 1991 *VKI Industrial Heat Exchangers*
[5] Garca-Cascales J, Vera-Garca F, Corbern-Salvador J and Gonzlvez-Maci J 2007 *International Journal of Refrigeration* **30** 1029 – 1041
[6] Ayub Z H 2003 *Heat Transfer Engineering* **24** 3–16
[7] Arie M, Shooshtari A, Dessiatoun S, Al-Hajri E and Ohadi M 2015 *International Journal of Heat and Mass Transfer* **81** 478 – 489
[8] Ohadi M, Choo K, Dessiatoun S and Cetegen E 2013 Force-fed microchannels for high flux cooling applications *Next Generation Microchannel Heat Exchangers* (Springer New York) pp 33–65 ISBN 978-1-4614-0778-2
[9] Huang J, Sheer T J and Bailey-McEwan M 2012 *International Journal of Refrigeration* **35** 325 – 335
[10] Yan Y Y and Lin T F 1999 *Journal of Heat Transfer-transactions of The Asme* **121**(1) 118–127
[11] Han D H, Lee K J and Kim Y H 2003 *Applied Thermal Engineering* **23** 1209 – 1225
[12] Sterner D and Sunden B 2006 *Heat Transfer Engineering* **27** 45–55
[13] Longo G A 2010 *International Journal of Heat and Mass Transfer* **53** 1079 – 1087
[14] Cipollone R, Bianchi G, Battista D D, Contaldi G and Murgia S 2014 *Energy Procedia* **45** 121 – 130

[15] Cipollone R, Contaldi G, Bianchi G and Murgia S 2013 Energy recovery using sliding vane rotary expanders *8th International Conference on Compressors and their Systems* (Woodhead Publishing) pp 183 – 194 ISBN 978-1-78242-169-6
[16] Bianchi G and Cipollone R 2015 *Applied Thermal Engineering* **84** 276 – 285
[17] Bianchi G, Cipollone R, Murgia S and Contaldi G 2015 *International Journal of Refrigeration* **52** 11 – 20
[18] Cipollone R, Bianchi G, Gualtieri A, Di Battista D, Mauriello M and Fatigati F 2015 Development of an organic rankine cycle system for exhaust energy recovery in internal combustion engines *UIT Conference*
[19] Seligman R 1964 *Chemistry & Industry* 1602–1603
[20] Lozano A, Barreras F, Fueyo N and Santodomingo S 2008 *Applied Thermal Engineering* **28** 1109 – 1117
[21] Claesson J 2005 *Thermal and hydraulic performance of compact brazed plate heat exchangers operating as evaporators in domestic heat pumps* Ph.D. thesis KTH Royal Institute of Technology
[22] Ayub Z H 2003 *Heat Transfer Engineering* **24** 3–16
[23] Xiande F, Rongrong S and Zhanru Z 2011 *Energy Science and Technology* **1** 1–15
[24] Zhou Z, Fang X and Li D 2013 *The Scientific World Journal* **2013**
[25] Donowski V D and Kandlikar S G 2000 *Proceedings of Boiling 2000: Phenomena and Emerging Applications, Engineering Foundation*
[26] Klimenko V 1990 *International Journal of Heat and Mass Transfer* **33** 2073 – 2088
[27] Shah M 1982 *Chart correlation for saturated boiling heat transfer: Equations and further study* vol 88
[28] Longo G and Gasparella A 2007 *International Journal of Heat and Mass Transfer* **50** 5194 – 5203
[29] Bennov L, Wronski J, Markussen W B and Haglind F 2015
[30] Huang J, Sheer T J and Bailey-McEwan M 2012 *International Journal of Refrigeration* **35** 325 – 335
[31] Bao J and Zhao L 2013 *Renewable and Sustainable Energy Reviews* **24** 325 – 342

Nomenclature

A	heat transfer area	m^2	t	plate thickness	mm
A_N	PHE nominal heat tr. area	m^2	T	temperature	K
A_s	flow cross sectional area	m^2	U	overall heat tr. coeff.	$W/(m^2\,K)$
b	corrugation amplitude	mm	W	plate width	mm
Bo	boiling number	–	x	vapor quality	–
c	specific heat at const. pressure	$J/kg/K$	Δp	pressure drop	Pa
C	heat capacity rate	W/K	ϵ	efficiency	–
Co	convection number	–	μ	dynamic viscosity	$Pa\,s$
D_h	hydraulic diameter	m	ρ	density	kg/m^3
F	Shah coefficient	–		**subscripts**	
f	Fanning factor	–	hot	hot fluid	
Fr	Froude number	–	cold	cold fluid	
g	gravitational acceleration	m/s^2	in	inlet	
h	convective heat tr. coeff.	$W/(m^2\,K)$	out	outlet	
H	specific enthalpy	J/kg	sat	saturation	
k	thermal conductivity	$W/(m\,K)$	j	discretization index	
L	plate length	mm	max	maximum	
m	mass flow rate	kg/s	min	minimum	
N	number of plates	–	wall	heat exchanger wall	
n	Prandtl number exponent	–	fluid	generic fluid	
NTU	number of thermal units	–	sl	saturated liquid	
p	pressure	Pa	sv	saturated vapor	
Pr	Prandtl number	–	l	liquid	
Q	thermal power	W	eq	equivalent	
Re	Reynolds number	–	m	mean	

Influence of electrolyte concentration on holdup, flow regime transition and local flow properties in a large scale bubble column

Giorgio Besagni[1], **Fabio Inzoli**
Politecnico di Milano, Department of Energy, Via Lambruschini 4a, 20156, Milano

E-mail: giorgio.besagni@polimi.it

Abstract. We experimentally investigate the influence of the electrolyte concentration on holdup, flow regime transition and local flow properties in a large scale bubble column, with air and water as working fluids. The column is 0.24 m inner diameter, 5.3 m height and the air is introduced by a spider sparger up to a superficial gas velocity of 0.2 m/s. The influence of five NaCl concentrations are investigated by using gas holdup and optical probe measurements. The gas holdup measurements are used for analysing the flow regime transition between the homogeneous and the transition regime and the optical probe is used for studying the local flow characteristics at different radial positions. The presence of NaCl - up to a critical concentration - increases the gas holdup. The increase in the gas holdup is due to the inhibition of the coalescence between the bubbles and, thus, the extension of the homogeneous regime. The results are in agreement with the previous literature on smaller bubble columns.

1. Introduction

Bubble columns are widely used as multiphase reactors in the chemical, petrochemical, pharmaceutical and food industries. One of the main hydrodynamic parameters in bubble columns is the gas holdup, ε_G, defined as the gas volume fraction in the mixture. The gas holdup - together with the mean bubble diameter, d_b - enables the computation of the interfacial area for the evaluation of the heat and mass transfer. The values of the gas holdup and the bubble size depend upon the superficial gas velocity, U_G (equal to the ratio of the volumetric flux to the area), and are related to the flow regime: mainly, the homogeneous and the heterogeneous regime. The former is associated with small superficial gas velocities and is characterised by the presence of small, uniform-sized bubbles with negligible interactions. The latter is associated with high superficial gas velocities and is characterised by a wide variety of bubble sizes and high coalescence and breakage phenomena. The transition from the homogeneous to the heterogeneous regime is a gradual process in which a transition flow regime occurs. This flow regime is characterized by large flow macro-structures with large eddies and widened bubble size distribution due to the onset of the bubble coalescence [1].

The correct prediction of the flow regime transition from the homogeneous to the transition regime is very important for the reactor design, effective operation and scale-up. For example, Krishna et al. [2] proved that non-coalescing systems can be used for predicting high-pressure systems. Among the different non-coalescing systems, we consider the electrolytes. It is well known that most electrolytes inhibit bubble coalescence in water [3-8] and, in this respect, a key concept is the transition

[1] To whom any correspondence should be addressed.

concentration, c_t, defined as the concentration above which bubble coalescence is drastically reduced [6, 7]. Depending on the concentration of the electrolyte, we may define a "*coalescent regime*" ($c/c_t \leq 1$) and a "*non-coalescent regime*" ($c/c_t > 1$). In the case of NaCl – which is the subject of this study - the threshold is $c_t = 0.145$ mol/l [9]. It is, therefore, obvious that the use of electrolytes can significantly alter the gas holdup in bubble columns [9-15]. In particular, the increase of ε_G is considered a consequence of the homogeneous regime stabilisation (an increase in the superficial gas velocity of the regime transition onset, U_{trans}) upon the addition of electrolytes into water [11].
Different studies investigated the influence of the electrolytes on the bubble column hydrodynamics and, in the following, we propose a brief literature survey. Thorat and Joshi [16] investigated a 0.385 m inner diameter (between 0.385 and 3.08 m height) bubble column with different gas spargers and using three media (water, 0.2 M NaCl, 1% CMC). The regime transition was investigated by using the Wallis plot and the homogeneous regime was stabilized by using NaCl: the coalescence suppression was seen to be the reason. Grover et al. [17] investigate a 0.1 m inner diameter (1.5 m height) bubble collumn using two electrolytes (NaCl, $CuCl_2$). Using the Wallis plot, they reported a stabilisation of the homogeneous regime. Kelkar et al. [18] investigate a 0.154 m inner diameter (3.25 m height) bubble column using three electrolytes (NaCl, $CaCl_2$ and Na_2SO_4). The authors reported an increase in the gas holdup and a negligible effect of the electrolytes above the critical concentration. Zahradník et al. [9] investigated a 0.14, 0.15 and 0.29 m inner diameter (2.6 m height) bubble columns and studied the influence of nine electrolytes. The gas holdup grew continuously for $c \leqslant c_t$, but little changes in ε_G were observed for $c > c_t$. Moreover, a similar dependence of ε_G vs. U_G was found for all electrolytes at $c = c_t$. Ribeiro and Mewes [11] studied a 0.12 m inner diameter (1.25 m height) bubble column using three electrolytes (NaCl, Na_2SO_4, NaI). Four values of the electrolyte concentration were tested in the "*coalescent regime*" and the gas holdup was found to increase till the critical concentration.
Despite the role of the electrolytes is quite well assessed, there is a lack of studies concerning large scale bubble columns. We contribute to the discussion investigating a 0.24 m inner diameter, 5.3 m height bubble column, which is larger than the columns typically studied. Air is introduced by using a spider sparger up to a superficial gas velocity up of 0.2 m/s and five NaCl concentrations ($0 \leq c/c_t \leq 1.17$) are tested. We investigate the influence of electrolyte concentration over gas holdup, flow regime transition and local flow properties by using gas holdup and optical probe measurements. The gas holdup measurements are used for investigating the flow regime transition. The optical probe measurements were used to study the local flow characteristics at different radial positions. This experimental investigation mainly aim to provide further data for expanding the existing dataset.

2. The experimental setup

The bubble column, made of Plexiglas, is 0.24 m inner diameter and 5.3 m height (Figure 1). The authors have used the same facility for studying the bubble column hydrodynamic in the annular gap configuration [19-20]. The reader may also refer to the dissertation of Carrara [21] for a complete and detailed description of the facility. The compressed air passes through a filter and, then, a pressure valve regulator controls the pressure upstream the rotameters used to set-up the air flow rate (and, then, for computing U_G). The air distributor is a spider sparger with holes ranging from 1 to 4 mm [21]. Due to the nature of the observed flow phenomena and their sensitivity to surface tension forces, clean filtered deionized water and filtered air were used. The air and water temperatures were maintained at room temperature (22 ± 1 °C) during the experiments. Five concentrations of NaCl were tested and are listed in Table 1. The values of gas density, used for computing U_G are based upon the operating conditions existing at the column mid-point. The mid-point column pressure was assumed to be equal to the column outlet pressure plus one-half the total experimental hydrostatic pressure head.

Table 1. NaCl concentrations tested.

c [mol/l]	0	0.02	0.07	0.12	0.145[1]	0.170
c/c_t [-]	0	0.14	0.48	0.84	1	1.17

[1] *Critical concentration, $c_t = 0.145$ mol/l [9]*

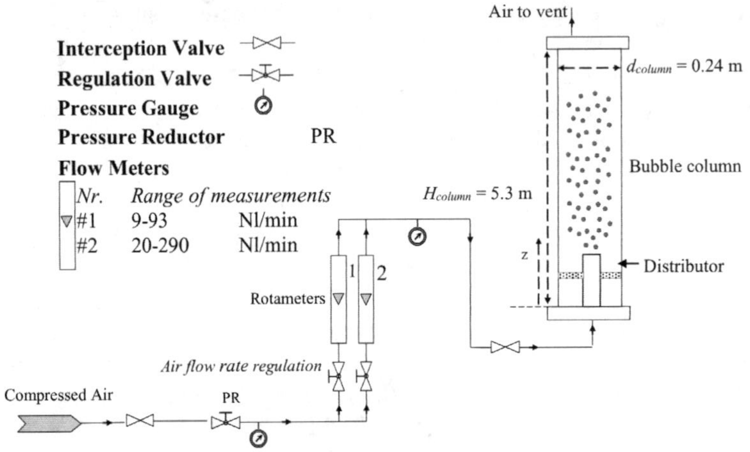

Figure 1. The experimental facility.

3. The measurement techniques

3.1. Holdup measurements

Measurements of the bed expansion allowed the evaluation of the gas holdup ε_G. The procedure involves measuring the location (height) of liquid free surface when air flows in the column. The gas holdup is, then, obtained using the following relation:

$$\varepsilon_G = \frac{(H_D - H_0)}{H_D} \tag{1}$$

Where H_D and H_0 are the heights (measured from the sparger) of the free surface after and before aeration, respectively. The error in the gas holdup measurements is estimated approximately 10%.

3.2. Flow regime transition analysis

Although the flow transition from the homogeneous to the transition regime does not happen instantaneously [1], the definition of an approximate transition point is helpful for modelling the hydrodynamic of bubble columns [2]. In this study, we employ two methods from the literature for investigating the regime transition: (i) the swarm velocity and (ii) the drift flux method.
The swarm velocity method was developed by Zuber and Findlay [22] and is based on the evaluation of the swarm velocity, U_{swarm}:

$$U_{swarm} = U_G / \varepsilon_G \tag{2}$$

In this method, the swarm velocity is plotted against the superficial gas velocity: U_{swarm} is constant in the homogeneous regime and starts to increase - at a certain transition superficial velocity, U_{trans} - as the system enters the transition/heterogeneous regime. The appearance of the first large bubble is responsible for the increase in the swarm velocity and is an indication of flow regime transition. This method has been employed also by Krishna et al. [23], Letzel et al. [24] and Gourich et al. [25], Ribeiro and Mewes [11] and Besagni et al. [19-20].
The drift-flux method was proposed by Wallis [26] and has been widely applied in the literature [11, 26, 27]. This method is based on the drift flux (that represents the gas flux through a surface moving with the speed of the two-phase mixture) and it is experimentally obtained as follows:

$$J_T = U_G(1 - \varepsilon_G) \tag{3}$$

Theoretically, the drift flux is written in terms of a parameter, U_b, whose dependence upon ε_G varies with the prevailing regime:

$$J_E = U_b(1-\varepsilon_G) \tag{4}$$

The idea is to employ a model for U_b valid for the homogeneous regime and to plot J_E and J_T in the same graph as a function of ε_G. In the homogeneous regime J_E is equal to J_T and the transition point is, thus, defined when:

$$J_T \neq J_E \tag{5}$$

The evaluation of U_b is a matter of discussion in the literature, as different models have been proposed and applied. In this study, we follow the approach of Krishna et al. [2, 28], which is based on the empirical model of Richardson and Zaki [29]:

$$U_b = u_\infty(1-\varepsilon_G)^{n-1} \tag{6}$$

where n is fluid-dependent ($n \cong 2$ for water) and u_∞ - the terminal velocity of an isolated bubble - should be fitted with the aid of the experimental data in the determination of the regime transition point. Combining Eq. (4) and Eq. (6) results:

$$J_E = u_\infty \varepsilon_G (1-\varepsilon_G)^n \tag{7}$$

3.3. Local measurements

A double fiber optical probe system (RBI Instrumentation) measures the local flow properties. Similar optical probes have been used in previous studies [19, 20, 30-35] and Besagni et al. [20] proposed a discussion concerning the uncertainty of the optical probe measurements. An optoelectronic module emits the laser to the probe tip and converts the reflected optical signal into a digital signal. From the digital signal, the bubble frequency f (bubble number per unit time) and local void fraction $\varepsilon_{G,Local}$ can be obtained. By cross-correlating the signals of the two tips, the bubble traveling time from one tip to the other can be estimated and bubble velocity u_b can be calculated. Assuming that bubbles are spherical, bubble Sauter mean diameter d_b, is calculated as follow [21, 34]:

$$d_b = 3\varepsilon_{G,Local} u_b / 2f \tag{8}$$

However, the spherical bubbles assumption (Eq. (8)) is approximately valid for small bubbles only (d_b < 3.5 mm [34]). For accounting the non-sphericity of bubbles, the following equation is used [32]:

$$d_{b,corrected} = \alpha^{-2/3} 3\varepsilon_{G,Local} u_b / 2f \tag{9}$$

where α is the aspect ratio. All the measurements have been obtained using a sampling period equal to 1000s, which is large enough to produce reliable time-averaged values and is far above the typical values of 1–5 min for similar optical probes [31, 33, 34, 36]. In this study, the optical probe was used for the following cases: U_G = 0.0037 m/s and c/c_t = 0, c/c_t = 0.48 and c/c_t = 1.17.

4. The experimental results

4.1. Holdup measurements

The gas holdup measurements are presented in Figure 2a. At low superficial gas velocities, the relationship between the gas holdup and the superficial gas velocity is linear, followed by a change in tendency at a transition superficial gas velocity, U_{trans}. The linear trend corresponds to the homogeneous/bubbly flow and the change in tendency is due to flow regime transition toward the transition regime. An example of flow visualisation for the bubbly flow is proposed in Figure 3, whereas the heterogeneous flow is presented in Figure 4. The gas holdup grows continuously while increasing the electrolyte concentration till the critical concentration, as reported in the previous literature [9, 11] (Figure 2a and b). The gas holdup increases also in the homogeneous regime, where the coalescence phenomena are limited, in agreement with Ribeiro and Mewes [11].

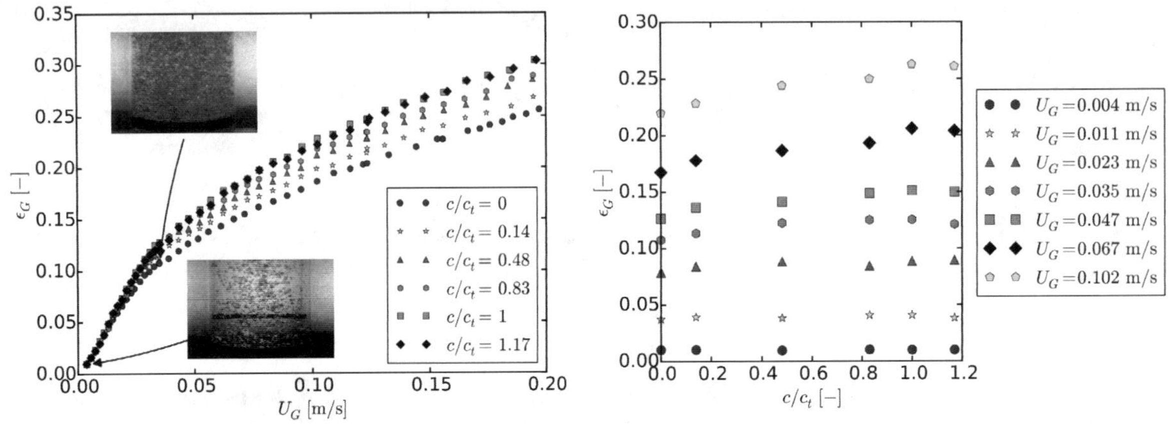

(a) Gas holdup measurements **(b) Influence of the electrolyte concentration**

Figure 2. The gas holdup.

An interesting aspect is the non-linearity of the electrolytes effect upon the gas holdup (Figure 2a and b). The curve for $c/c_t = 0.14$ and 0.48 is already shifted to considerable higher ε_G values in comparison to the curve related to $c/c_t = 0$, while the relative distance between the curves associated with $c/c_t = 0.48$, a 0.83 and 1 is considerable lower. The non-linearity of the electrolytes effect upon the gas holdup was also conclused by Ribeiro and Mewes [11]. Above the critical concentration, there is no remarkable difference in the gas holdup, as expected from the literature [9].

Figure 3. $U_G = 0.0037$ m/s – $c/c_t = 0$.

Figure 4. $U_G = 0.1665$ m/s – $c/c_t = 0$.

4.2. Flow regime transition analysis

The gas holdup increases while increasing the electrolyte concentration till the critical value. Accordingly with the literature, the reason for the increase of the gas holdup lie in the inhibition of the coalescence phenomena; this hypothesis is verified also for our large bubble column by analyzing the flow regime transition using the two methods previously discussed (Section 3.2). The results of the swarm velocity method are presented in Figure 5a and the results of the drift flux method are presented in Figure 6. The values of the transitional gas velocity (Figure 5b and 6) are in agreement between the two methods and, following the proposal of Ribeiro and Mewes [11], the transition points have been evaluated as the mean of the two values and are presented in Figure 7 (where the dash lines are used for giving a better idea of the trend only). The transitional gas velocity, U_{trans} lies in the range between $U_{trans} = 0.0264$ and 0.0338 m/s, depending on the NaCl concentration. Instead, the transitional holdup is between $\varepsilon_{G,trans} = 0.0831$ and 0.1209. Both U_{trans} and $\varepsilon_{G,trans}$ increase while increasing the NaCl concentration, till the critical ratio: U_{trans} and $\varepsilon_{G,trans}$ increase up to 22 and 45%, respectably, if compared with $c/c_t = 0$. This supports the hypothesis that increasing the NaCl concentration, the homogeneous flow regime is stabilized and, thus, ε_G increases by the inhibited coalescence. Beyond the critical concentration, U_{trans} and $\varepsilon_{G,trans}$ do not change anymore.

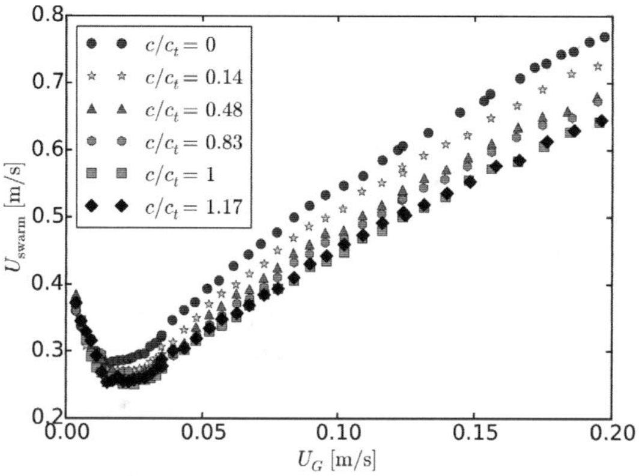

(a) Swarm velocity

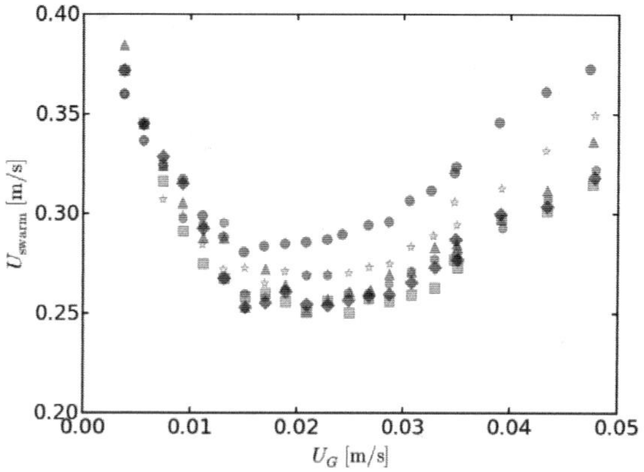

(b) Focus on the transitional velocities

Figure 5. The swarm velocity method for the analysis of the flow regime transition.

The typical values of U_{trans} found in literature for air-water systems (considering $c/c_t = 0$) in bubble columns of diameter higher than 0.15 m range between $U_{trans} = 0.01$ and 0.08 m/s at ambient operating conditions, depending on the distributor geometry [9, 23, 37-42].

We may also compare our data ($c/c_t = 0$) with literature correlations. One of the first correlations proposed is due to Wilkinson et al. [38]:

$$U_{trans} = 0.5U_{b,small}\exp\left(-193\rho_G^{-0.61}\mu_L^{0.5}\sigma^{0.11}\right) \tag{10}$$

where:

$$U_{b,small} = 2.25\frac{\sigma}{\mu_L}\left(\frac{\sigma^3\rho_L}{g\mu_L^4}\right)^{-0.273}\left(\frac{\rho_L}{\rho_G}\right)^{0.03} \tag{11}$$

Another correlation was proposed by Reilly et al. [39] and reads:

$$U_{trans} = \frac{\rho_L}{\rho_G}\left(1-\varepsilon_{G,trans}\right)\left(\frac{B^*}{A^*}\right)^{1.5} \tag{12}$$

where:

$$\varepsilon_{G,trans} = 0.59B^{*1.5}\left(\frac{\sigma^{0.12}\rho_G^{0.96}}{\rho_L}\right)^{0.5} \tag{13}$$

and:

$$A^* = 2.81\rho_L\rho_G^{0.96}\sigma^{-0.12} \tag{14}$$

For water as the liquid phase, $B^* = 4$. The results of the comparison are presented in Table 3: Eq. (10) largely underestimate the transition gas velocity, whereas, Eq. (12) and (13) gives a value for U_{trans} and $\varepsilon_{G,trans}$ in better agreement with the experimental data, as also observed in other studies [20, 40].

Table 2: Flow regime transition: comparison with the literature.

	Exp. $c/c_t = 0$	**Wilkinson et al. [38]**	**Reilly et al. [39]**
U_{trans} [m/s]	0.0241	0.0029	0.0322
$\varepsilon_{G,trans}$ [-]	0.0831	0.01152[1]	0.1484[1]

[1] *The value is given by $U_{trans}/U_{b,small}$ Eq. (11)*

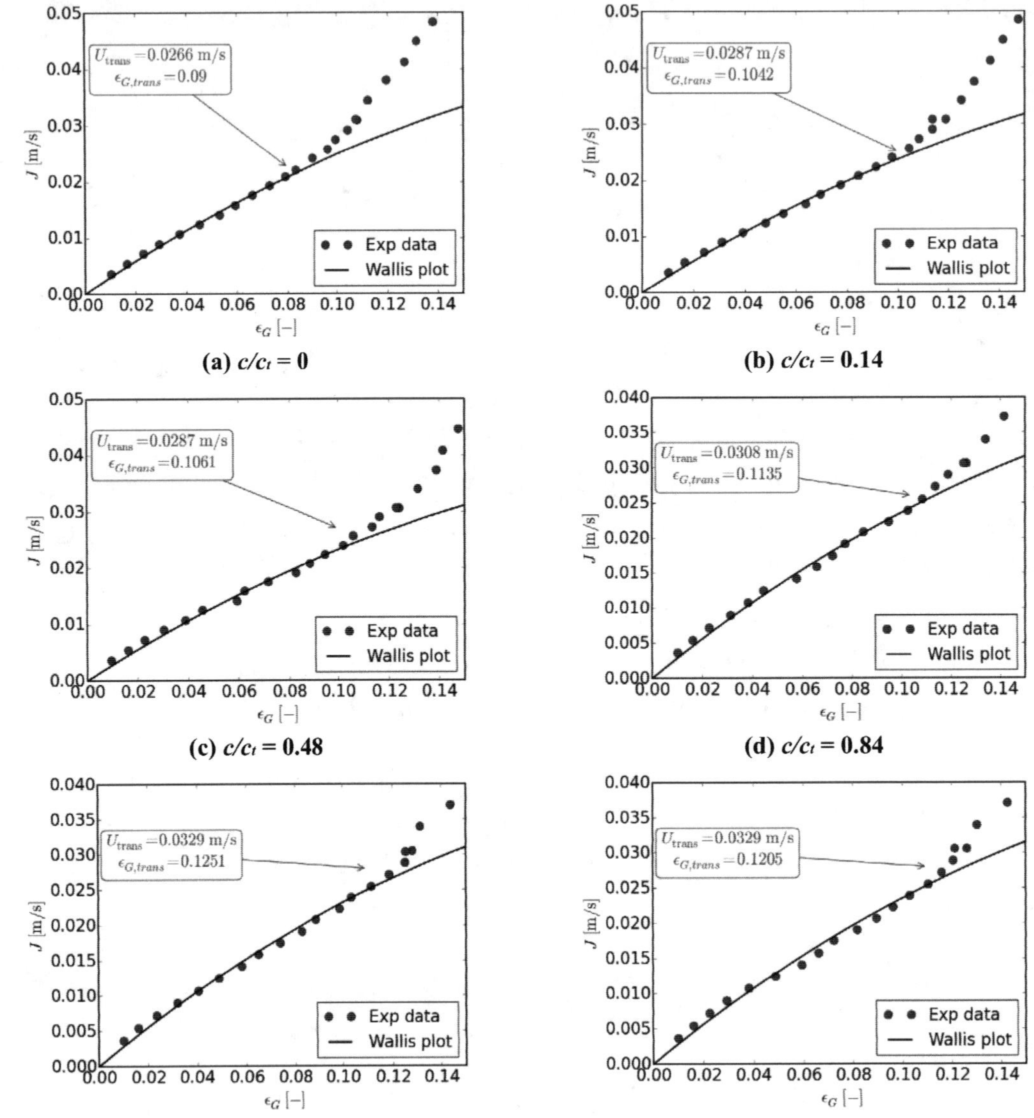

(a) $c/c_t = 0$ **(b)** $c/c_t = 0.14$

(c) $c/c_t = 0.48$ **(d)** $c/c_t = 0.84$

(e) $c/c_t = 1$ **(f)** $c/c_t = 1.17$

Figure 6: The drift flux method for the analysis of the flow regime transition.

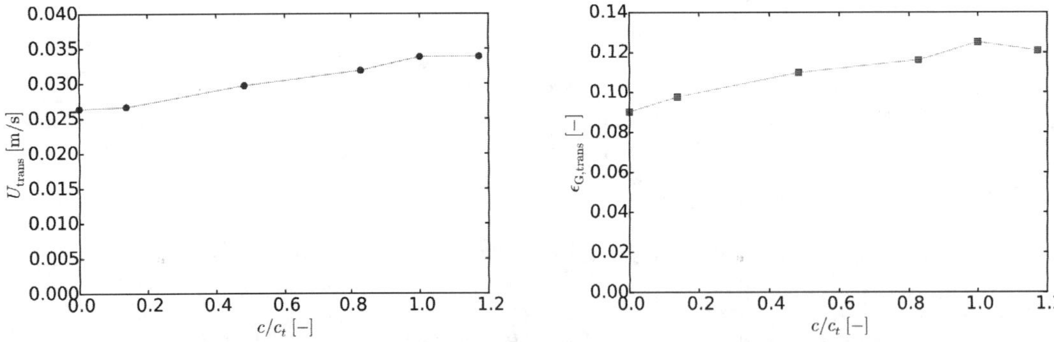

(a) U_{trans} **for the different NaCl concentrations** **(b)** $\varepsilon_{G,trans}$ **for the different NaCl concentrations**

Figure 7: Transitional parameters: U_{trans} and $\varepsilon_{G,trans}$.

4.3. Local measurements

Figure 8 presents the radial optical probe measurements for $c/c_t = 0$, $c/c_t = 0.48$ and $c/c_t = 1.17$.

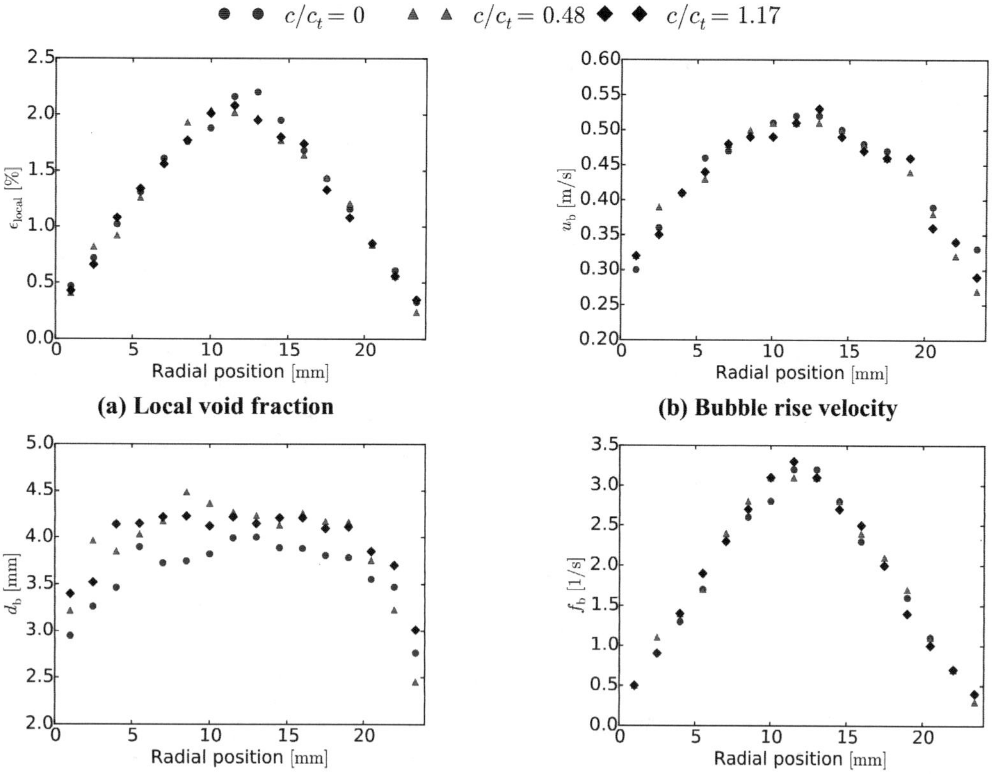

(a) Local void fraction **(b) Bubble rise velocity**

(c) Bubble Sauter mean diameter **(d) Bubble interface frequency**

Figure 8: Optical probe measurements.

The local void fraction profiles are not flat, but are centre peaked. This is probably due to the distributor having large opening, which creates large bubbles in the centre of the column. Furthermore, there is no remarkable difference between the three salt concentrations (Figure 8a). This was expected because - for low U_g - these are no remarkable differences in the gas holdup (Figure 2a and 2b). Local void fraction $\varepsilon_{G,Local}$ obtained with the optical probe were compared to global gas holdups by integrating the radial measurements over the column cross sectional area:

$$< \varepsilon_G > = \frac{1}{\pi R_C^2} \int_0^{R_C} \varepsilon_{G,Local} 2\pi r dr \qquad (15)$$

where R_C, is the radius of the column (R_C = 0.12 m). Table 3 shows that the integrated local measurements were higher if compared to the global value, but they are very similar one another (as for the gas holdup values). The error may derive from the pressure gradient along the column (the volume of the bubbles increases while rising the column) as well as the developing region of the two-phase flow. Further analysis will be devoted to the uncertainty connected to the optical probes.

Table 3. Comparison between local (optical probe) and global gas holdup measurements.

U_G [m/s]	c/c_t [-]	$< \varepsilon_G >$ [%]	ε_G [%]
0.0037	0	1.68	1.02
0.0037	0.43	1.62	0.96
0.0037	1.17	1.61	0.99

Concerning the bubble vertical velocity u_b (Figure 8b), the values are between 0.3 (near wall) and 0.5 (at the center of the column) and there is no remarkable differences between the three curves. Figure 8c presents the bubble mean diameter obtained by using Eq. (9) using mean aspect ratio from the image analysis: $\alpha = 0.649$ for $c/c_t = 0$, $\alpha = 0.808$ for $c/c_t = 0.48$ and $\alpha = 0.816$ for $c/c_t = 1.17$. The image analysis method is the one detailed by Besagni et al. [20] and the sampling was performed considering the bubbles near the wall of the column. The number of bubbles sampled and the operating conditions considered the are the followings:

- $c/c_t = 0$ - 3218 bubbles - $U_G = 0.0037$, 0.0074, 0.111, 0.0149 and 0.188 m/s;
- $c/c_t = 0.48$ - 1481 bubbles - $U_G = 0.0037$ and 0.0074 m/s;
- $c/c_t = 1.16$ - 1341 bubbles - $U_G = 0.0037$ and 0.0074 m/s.

A complete discussion of the number of bubbles sampled, the results concerning the bubbles size distributions and the bubble shape is far beyond the scope of this paper, and detailed results for the image analysis will be presented elsewhere. It is interesting that the mean diameter for $c/c_t = 0.48$ and for $c/c_t = 1.17$ are higher: this is due to the change in the bubble shape, caused by the coalescence inhibition and the modified interaction between the bubbles. Figure 8d presents the bubbly interface frequency and the three curves are similar, as expected (Section 3.3).

5. Conclusions

We have experimentally investigated the influence of electrolyte concentration over holdup, flow regime transition and local flow properties in a large scale bubble column, with air and water as working fluids. The column is 0.24 m inner diameter, 5.3 m height and the air is introduced by using a spider sparger up to a superficial gas velocity up of 0.2 m/s. Five NaCl concentrations were tested and the experimental investigation consists in gas holdup and optical probe measurements. The gas holdup measurements were used for investigating the flow regime transition and the data obtained from the optical probe were used to study the local flow characteristics at different radial positions.

It is found that the presence of NaCl - up to the critical concentration - increases the gas holdup, even for the operation in the homogeneous regime, where bubble coalescence frequencies are not high. A non-linear increase in the gas holdup with the electrolyte concentration was verified, whose extent was greater for small concentrations and progressively levelled off as the electrolyte content in the liquid phase was raised. The gas holdup measurements are used for investigating the flow regime transition between the homogeneous and the transition regime. The transitional gas velocity lies in the range between $U_{trans} = 0.0264$ and 0.0338 m/s, depending on the NaCl concentration. Instead, the transitional holdup is between $\varepsilon_{G,trans} = 0.0831$ and 0.1209. Both the transitional holdup and transitional gas velocity increase while increasing the NaCl concentration, till the critical ratio. This support the hypothesis that increasing the NaCl concentration, the homogeneous flow regime is stabilized and, thus, the gas holdup increase by the inhibited coalescence.

The data obtained from the optical probe were used to study the local flow characteristics at different radial positions. The optical probe data for $c/c_t = 0$, $c/c_t = 0.48$ and $c/c_t = 1.17$ were compared. There is no remarkable difference between the local void fractions, bubbly frequencies and bubble rise velocities between the three cases. However, the bubble Sauter mean diameters increase when considering the cases with NaCl: this is due to the change in the bubble shape, caused by the coalescence inhibition and the modified interaction between the bubbles.

The results - in agreement with previous publication in smaller bubble columns - have extended the existing dataset concerning the effects of electrolytes in bubble columns. Further studies may concern higher NaCl concentrations as well as the study of other surfactants. It would be interesting to study the bubble size distribution and bubble shape for different operating conditions and different NaCl concentrations by means of the image analysis previously presented by the authors [20]. Finally, the present experimental facility may be operated in the counter-current operating condition [19, 20]: the influence of the electrolyte on the counter-current operating condition will be also investigated.

References

[1] Nedeltchev S 2015 *Chem. Eng. Sci.* **137** 436-44
[2] Krishna R, Urseanu M I and Dreher A J 2000 *Chem. Eng. Process.* **39** 371-8
[3] Craig V S J, Ninham B W and Pashley R M 1993 *J. Phys. Chem-US* **97** 10192-7
[4] Deschenes L A, Barrett J, Muller L J, Fourkas J T and Mohanty U 1998 *J. Phys. Chem B* **102** 5115-9
[5] Keitel G and Onken U 1982 *Chem. Eng. Sci.* **37** 1635-8
[6] Lessard R R and Zieminski S A 1971 *Ind. Eng. Chem. Fund.* **10** 260-9
[7] Marrucci G and Nicodemo L 1967 *Chem. Eng. Sci.* **22** 1257-65
[8] Weissenborn P K and Pugh R J 1995 *Langmuir* **11** 1422-6
[9] Zahradnik J, Fialova M, Rruvzivc k, M., Drahovs, J., Kavstanek F and Thomas N H 1997 *Chem. Eng. Sci.* **52** 3811-26
[10] Hikita H, Asai S, Tanigawa K, Segawa K and Kitao M 1980 *Chem. Eng. J.* **20** 59-67
[11] Ribeiro Jr C P and Mewes D 2007 *Chem. Eng. Sci.* **62** 4501-9
[12] Ruthiya K C, van der Schaaf J, Kuster B F and Schouten J C 2006 *Int. J. Chem. React. Eng.* **4**
[13] Kluytmans J H J, Kuster B F M and Schouten J C 2001 *Ind. Eng. Chem. Res.* **40** 5326-33
[14] Kellermann H, Ju¨ttner K and Kreysa G 1998 *J. Appl. Electrochem.* **28** 311-9
[15] Jamialahmadi M and Müller-Steinhagen H 1992 *Chem. Eng. J.***50** 47-56
[16] Thorat B N and Joshi J B 2004 Reg *Ext. Therm. Fluid. Sci.* **28** 423-30
[17] Grover G S, Rode C V and Chaudhari R V 1986 *Can. J. Chem. Eng.* **64** 501-4
[18] Kelkar B G, Phulgaonkar S R and Shah Y T 1983 *Chem. Eng. J.***27** 125-33
[19] Besagni G, Guédon G and Inzoli F 2014 *J. Phys Conf Ser* **547** 012024
[20] Besagni G, Guédon R G and Inzoli F 2016 *ASME J. Fluids Eng* **138** 011302-15
[21] Carrara M 2014 *Experimental investigation of two-phase flows in vertical pipes* M.S. Thesis, Politecnico di Milano
[22] Zuber N and Findlay J A 1965 *J. Heat Transf* **87** 453-68
[23] Krishna R, Wilkinson P M and Van Dierendonck L L 1991 *Chem. Eng. Sci.* **46** 2491-6
[24] Letzel H M, Schouten J C, van den Bleek C M and Krishna R 1997 *Chem. Eng. Sci.* **52** 3733-9
[25] Gourich B, Vial C, Essadki A H, Allam F, Belhaj Soulami M and Ziyad M 2006 *Chem. Eng. Process.* **45** 214-23
[26] Wallis G B 1969 *One-dimensional two-phase flow* (New York)
[27] Passos A D, Voulgaropoulos V P, Paras S V and Mouza A A 2015 *Chem. Eng. Res. Des.* **95** 93-104
[28] Krishna R, Ellenberger J and Maretto C 1999 *Int. Commun. Heat Mass* **26** 467-75
[29] Richardson J F and Zaki W N 1997 *Chem. Eng. Res. Des.***75**, Supplement S82-S100
[30] Kiambi S L, Duquenne A-M, Dupont J B, Colin C, Risso F and Delmas H 2003 *Can. J. Chem. Eng.* **81** 764-70
[31] Chaumat H, Billet-Duquenne A M, Augier F, Mathieu C and Delmas H 2005 *Chem. Eng. Sci.* **60** 6134-45
[32] Simonnet M, Gentric C, Olmos E and Midoux N 2007 *Chem. Eng. Sci.* **62** 858-66
[33] Lima Neto I, Zhu D and Rajaratnam N 2008 *J. Environ Eng.* **134** 283-94
[34] Zhang W and Zhu D Z 2013 *Int. J. Multiphs. Flow* **55** 156-71
[35] Barrau E, Rivière N, Poupot C and Cartellier A 1999 *Int. J. Multiphs. Flow* **25** 229-56
[36] Chang K-A, Lim H-J and Su C B 2003 *Rev. Sci. Intrum.* **74** 3559-65
[37] Schumpe A and Grund G 1986 *Can. J. Chem. Eng.* **64** 891-6
[38] Wilkinson P M, Spek A P and van Dierendonck L L 1992 *A.I.Ch.E. Journal* **38** 544-54
[39] Reilly I, Scott D, Debruijn T and MacIntyre D 1994 *The Can. J. Chem. Eng.* **72** 3-12
[40] Letzel H M, Schouten J C, Krishna R and van den Bleek C M 1999 *Chem. Eng. Sci.* **54** 2237-46
[41] Ruzicka M C, Drahoš J, Fialová M and Thomas N H 2001 *Chem. Eng. Sci.* **56** 6117-24
[42] Hur Y G, Yang J H, Jung H and Park S B 2013 *Int. J. Multiphs. Flow* **50** 89-97

Journal of Physics: Conference Series **655** (2015) 012040 doi:10.1088/1742-6596/655/1/012040

Experimental study of critical heat flux of refrigerant R1234yf in a multi-minichannel heat sink at medium saturation temperatures

W Imparato[1], R Mastrullo[1], A W Mauro[1,2] and L Viscito[1]

[1]Department of Industrial Engineering, Federico II University of Naples, P.le Tecchio 80, 80125, Naples (Italy)

[2]E-mail: alfonsowilliam.mauro@unina.it

Abstract. The main objective of this work is to present experimental results of saturated flow boiling critical heat flux for refrigerant R1234yf at medium reduced pressures. Tests were carried out by taking into account the influence of the saturation temperature and the mass flux. The former was let to vary from 25 up to 45 °C and the latter ranged from 150 up to 300 kg/m^{2}s. The inlet sub-cooling was set from 6 to 10 K. Results are given in the form of boiling curves, from which the CHF was deduced as the heat flux corresponding to an evident decrease in the curve slope and a sudden spike of the wall superheat. The experimental results have shown that the mass flux has a significant influence on the CHF, whilst the latter is almost independent on the saturation temperature.

1. Introduction

New synthetic refrigerants such as R1234yf represent interesting candidates for the substitution of R134a as operating fluids in refrigerating systems, thanks to a low GWP value and similar thermodynamic properties to other HFCs. The wide spread of micro-electronic devices has risen the scientific interests towards small size cooling elements in which the heat is dissipated through boiling processes of refrigerants at high saturation temperatures, thanks to the great efficiency of the heat transfer.

Moreover, new photovoltaic cells are able to concentrate the solar illumination in small spots, and these systems need to be cooled down in order not to damage the electronic equipment and to maintain the cells efficiency at reasonable values. In this regard, heat sinks made up of micro-channels with boiling refrigerants represent an interesting solution, especially for densely packed cells with very high solar concentration (> 150 suns). For these applications, the heat to be dissipated is beyond the traditional sub-cooled liquid water systems and the fans' air forced convection limits.

In this context, the critical heat flux (CHF) refers to the highest thermal load that can be handled in heat flux controlled devices. Exceeding such limit may lead to a lack of liquid in the heated surface and a subsequent sharp increase in the wall surface temperature, with consequent irreversible failures.

Lazarek and Black [1] performed experimental tests on critical heat flux for R113 in a round tube with an internal diameter of 0.31 cm and with two different heated lengths. A correlation was developed to predict critical conditions under low reduced pressures.

Journal of Physics: Conference Series **655** (2015) 012040 doi:10.1088/1742-6596/655/1/012040

Wojtan et al. [2] investigated different conditions of boiling R134a and R245fa to obtain the saturated critical heat flux in 0.5 and 0.8 mm internal diameter tubes with two different heated lengths. In their results, a strong dependence of CHF on mass velocity, heated length and microchannel diameter was detected. On the contrary, no influence of liquid sub-cooling was observed. Furthermore, a new correlation for the prediction of CHF during saturated boiling in microchannels was proposed by the authors.

Park and Thome [3] tested R134a, R236fa, R245fa CHF in two different multi-microchannels copper heat sinks. CHF was observed to increase when increasing the mass velocity, but the increase rate was slower for higher mass fluxes. CHF was instead seen to decrease with increasing inlet saturation temperature. These trends were found to be dependent both on the flow condition and the channel size.

Mauro et al. [4] studied saturated critical heat flux of R134a, R236fa and R245fa in a multi-microchannel copper heat sink, formed by 29 parallel channels that were 199 µm wide and 756 µm deep. The parametric effects of mass velocity, saturation temperature and inlet sub-cooling were investigated. It was found that higher CHFs were obtainable with a split flow system (by means of one inlet and two outlets), providing also a much lower pressure drop.

Tibiriçá et al. [5] obtained CHF results in circular tubes using R134a and R245fa and compared them against experimental data obtained with flattened tube having the same equivalent internal diameter of 2.2 mm, but different aspect ratios of 1/4, 1/2, 2 and 4. The CHF data were found to be independent on the tube aspect ratio when the same heated length was kept.

Although heat transfer characteristics of different synthetic fluids have already been studied for both conventional and mini-channels, quantitative data of critical heat flux for low-GWP fluorinated refrigerants at medium-to-high saturation temperatures are still limited.

Diani et al. [6]-[7] investigated on flow boiling heat transfer of R1234yf and R1234ze inside a 3.4 mm internal diameter microfin tube. The authors obtained CHF values by keeping constant the inlet vapour quality at 0.3. For both fluids, they found that the critical heat flux, defined as the heat flux on the onset of the dryout condition, was increasing with increasing mass velocity.

Anwar et al. [8] performed several tests with R1234yf in a vertical stainless steel test section (1.60 mm inside diameter and 245 mm heated length) under upward flow conditions. They found that signs of dryout first appeared at vapor qualities of 85%, with the values generally increasing with increasing mass flux.

Anwar et al. [9] reported experimental results on dryout of seven refrigerants (R134a, R1234yf, R152a, R22, R245fa, R290 and R600a) in small, single vertical tubes (0.64–1.70 mm and 213–245 mm heated length) under a variety of operating conditions. The authors found that the dryout heat flux increased with increasing mass flux and with increasing tube diameter. No effect of varying saturation temperature was observed. The experimental findings were compared with well-known macro and micro-scale correlations from the literature and a new correlation for prediction of heat flux at dryout conditions was also developed.

These recent works are mainly related to a single channel geometry, which has the advantage of a feasible control of local parameters (such as the vapor quality along the test section). On the other hand a single channel, differently than a multi-minichannels system, is not the realistic representation of a system able to remove high heat fluxes from squared electronic devices, due to greater heated lengths and the absence of possible mal-distribution and back-flows.

The main objective of this paper is to present experimental results of saturated flow boiling critical heat flux for refrigerant R1234yf at medium reduced pressures in an aluminum multi-minichannel heat sink. Differently from other quoted works, the saturation temperature goes by a wider range (25-45 °C) and the heated length is significantly smaller (25 mm), yielding to very different CHF values. Tests were carried out by taking into account the influence both of the saturation temperature and the mass flux. The latter ranged from 150 up to 300 kg/m^{2}s. Results are given in the dedicated section of this paper in the form of boiling curves and diagrams.

2. Test facility and experimental procedure

All the experiments have been carried-out in a closed refrigerant loop where the inlet conditions in terms of pressure, temperature and mass flux were able to be controlled independently. The experimental set-up is located in the Refrigeration Laboratory at the Dipartimento di Ingegneria Industriale of the Università degli Studi di Napoli "Federico II". A schematic diagram of the test facility is shown in Fig. 1. A brief description of the main and cooling loops, as well as of the test section, are given in the following subsections.

2.1. Main and cooling loops

The black line in Fig. 1 refers to the refrigerant closed loop. A magnetic gear pump, powered by a 0.37 kW electrical motor, pushes the liquid refrigerant towards a Coriolis mass flow meter and then it flows into the test section. The pump may elaborate volumetric flow rates within a range of 1.3-2.5 dm^3/min, passing from 1650 rpm to 3400 rpm. An absolute pressure transducer and a resistance thermometer are inserted just before the inlet of the test section in order to determine the R1234yf inlet conditions. A differential pressure transducer measures the pressure drop occurring in crossing the test section. The refrigerant outlet temperature is then procured thanks to another resistance thermometer placed right outside the aluminum multi-minichannel heat sink. The heated organic fluid passes through a throttling valve utilized to manually adjust the system mass flow rate and pressure. The refrigerant then enters a plate heat exchanger in which it condenses thanks to the cold water coming from the secondary loop and finally it flows into a liquid receiver and a double pipe heat exchanger to obtain a slight sub-cooling, before entering the pump again and closing the loop. The main loop is also equipped with a by-pass circuit right after the circulation pump. With the help of a manually-controlled by-pass valve, it is possible to recirculate a certain fraction of the entire organic fluid stream. In this case, very small mass flow rates into the test section can be achieved.

A thermostatic bath supplies demineralized water to the secondary loop. From the suction pipe, the cooling water flows into the double pipe heat exchanger and then it is pumped into the plate heat exchanger to condense the refrigerant in the main loop. An expansion vessel guarantees the restraint of variations of specific volume.

2.2. Description of the test section

In this work, an aluminum multi-minichannel heat sink has been used as test section. Six fins that are 3 mm high, 1.85 mm wide and spaced 2 mm one another are carved into the main block. A photograph of the aluminum heat exchanger is given in Fig. 2. A cogged cover closes the space between fins and provides seven minichannels, that are 2 mm wide, 1 mm high and 35 mm long. In order to prevent mal-distribution due to excessive discrepancies in pressure drop along the channels, two manifolds are provided for the inlet and the outlet section.

The heat sink is electrically heated from the bottom using AC power supply and a ceramic heater put in a dedicated slot carved underneath the test section. In order to guarantee an appropriate thermal contact, a high performance thermal compound is used. According to the manufacturer, the ceramic element is able to provide heating power of 697 (at 25 °C) over a base area of 2.50x2.50 cm^2, thus obtaining a base heat flux up to 110 W/cm^2. The voltage supply is given by a solid state relay, which can provide up to 400 V. The heat power imposed is measured by means of a wattmeter connected between the heater and the solid state relay. Four resistance thermometers (RTDs) are equidistantly placed in the test section to evaluate the wall temperature along the channels.

The insulation of the test rig has been made with an appropriate layer of synthetic rubber, by paying special attention to the test section. According to the manufacturer, the insulating material guarantees a thermal conductivity less than 0.042 W/m K in the whole range of operating conditions.

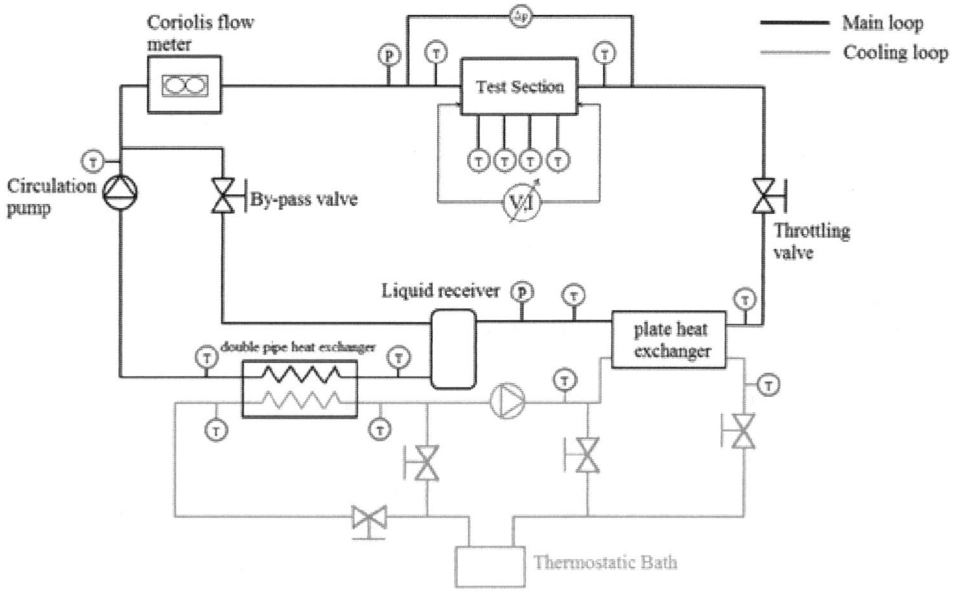

Figure 1 Schematic diagram of the experimental set-up located at the Refrigeration Lab at Università degli Studi di Napoli "Federico II"

Figure 2 Test section

2.3. Experimental procedure

All tests were performed in steady-state conditions. The desired operating conditions in terms of inlet pressure (i.e. saturation temperature), inlet temperature and mass flux were kept constant at the beginning of each test.

The inlet temperature was manually controlled by opening and closing the secondary loop by-pass valve, thus controlling the cooling water flow into the double pipe heat exchanger. However, the refrigerant mass charge in the test facility was kept reasonably low (2.0 kg) not to have high degrees of sub-cooling in the whole range of saturation temperatures tested.

The remaining parameters were remote-controlled via Labview software and Arduino One controller. Specifically, the system pressure was imposed by setting the temperature of the thermostatic bath, whereas the mass flux was mainly controlled by setting the inverter frequency of the electric motor coupled with the circulation pump. Small adjustments were obtained by manipulating the main loop by-pass valve and by using the throttling valve, thus getting a greater or smaller free flow passage for the refrigerant.

Once the test conditions were set, the heat was applied in small increments by controlling the voltage imposed to the ceramic heater via Arduino One within the range of 0-240 V. The boiling mechanism was then triggered and sustained up to the critical condition. The heater temperature was monitored with a K-type thermocouple and the electrical supply was automatically shut off as soon as the device reached 120 °C, in order not to damage the test section irreparably.

During tests, with increasing imposed heat flux the mass flux was gradually adjusted with the by-pass valve and/or the throttling valve, in order to get the chosen nominal conditions again. The system was considered stabilized when the maximum relative deviation of each value from the measured average in the latest 2 minutes was inferior to 3 %. The recorded time was set to two minutes, with a recording frequency of 1 Hz. Thus, each test was made up of 120 measurements and saved into a text file. The nominal value of each data point was assigned to the sample average value.

The post processing was implemented with the software MATLAB and all the thermodynamic and transport properties were calculated using the software REFPROP 9 [10] developed by NIST.

3. Instrument specifications and uncertainty

The test rig (see Fig. 1) is provided with two absolute pressure transducers. One of them is placed at the inlet of the test section and the other has been put right before the liquid receiver. Both devices measure the absolute pressure within the range 0-50 bar. By taking into account the non-linearity, hysteresis and repeatability effects, the instrument accuracy has a typical value of 0.1 %, with a maximum value of 0.3 % (span error). The resulting uncertainty in absolute pressure measurements is ±0.188 bar.

The pressure drop across the test section is evaluated with the differential pressure transducer Rosemount 2051 within the range of 0-60 kPa. According to the manufacturer, the maximum span error is 0.75 % at full scale, which gives an absolute uncertainty of ±0.45 kPa.

Two resistance thermometers (RTDs) are placed at the inlet and the outlet of the test section, and other identical sensors can be found along the main and the secondary loop (see Fig. 1). Each of them is positioned outside the walls and fixed with a nano-aluminum thermal compound to guarantee the equivalence between the fluid and the measured temperature. According to the manufacturer, all RTDs carry an uncertainty of ±0.180 °C.

Four 4-wire Pt100 cylindrical RTDs are instead positioned along the minichannels for the measure of the wall temperature in the test section. Their uncertainty is found to be ±0.154 °C.

A digital wattmeter is used for the heat power measurements. Its uncertainty results to be 1 %, resulting in a maximum uncertainty of ±10 W at full scale.

The mass flow rate occurring in the test section is measured with a Coriolis flow meter (MicroMotion S12S). The instrument has been tested up to 115.17 g/s and the maximum error was found to be 1 % of the measurement, resulting in an absolute uncertainty of ±1.15 g/s at full scale.

For the calculation of the thermodynamic properties obtained with the software REFPROP 9, it has been considered an uncertainty of 2 %.

Tab.1 summarizes the instrument specifications and their uncertainty, obtained by utilizing a coverage factor of $k = 1$.

Table 1 Instruments specifications and their uncertainty

Measurement	Range	B type uncertainty (k=1)
Temperature (4-wire Pt100 RTD)	-80 - +250 °C	±0.180 °C
Temperature (4-wire Pt100 cylindrical RTD)	-80 - +250 °C	±0.154 °C
Absolute pressure	0 – 50 bar	±0.188 bar
Differential pressure	0 – 60 kPa	±0.45 kPa
Flow meter	0.00-115.7 g/s	±1 %
Electrical power	0-8 kW	±1 %
Thermodynamic properties (REFPROP software)	-	± 2 %

4. Data reduction

The sensitivity analysis of the CHF results has been performed with respect to the mass flux G and the saturation temperature T_{sat}. The former parameter has been deduced from the measured mass flow rate by taking into account the minichannels cross sectional area:

$$G = \frac{\dot{m}}{N \cdot w_{ch} \cdot h_{ch}} \ [kg/m^2 s] \tag{1}$$

In the above equation, $\dot{m}$ is the measured mass flow rate [kg/s], whilst N represents the number of minichannels in the aluminum heat sink (N = 7) and w_{ch} and h_{ch} are the minichannels width and height, respectively, expressed in [m].

The saturation temperature, instead, has been evaluated with by knowing the inlet absolute pressure. Since all tests are performed with very low mass fluxes, the measured pressure drop across the test section is insignificant. Therefore, the decrease in the saturation temperature along the minichannels has been considered negligible as well.

The base heat flux supplied from the ceramic heater has been evaluated as:

$$\dot{q}_b = \frac{\dot{Q}}{A_b} \ [W/cm^2] \tag{2}$$

Where $\dot{Q}$ is the measured heat power and A_b is the heater base area of 2.50 x 2.50 cm^2. The wall heat flux presented in the results section is calculated by taking into account the real heat exchange area, by means of the equation below:

$$\dot{q}_w = \frac{\dot{Q}}{(2 \cdot h_{ch} \cdot \eta + w_{ch}) \cdot N \cdot L_h} \ [W/cm^2] \tag{3}$$

In the above equation, L_h represents the heated length (L_h = 25 mm) and also the fin efficiency η has been taken into account, following the same approach of Park and Thome [3]. In this work, an average value of η = 0.9 has been chosen.

For the computation of the average wall superheat, the following equation has been considered:

$$\overline{\Delta T}_{wall} = \overline{T}_{wall} - T_{sat} \ [°C] \tag{4}$$

The average wall temperature has been calculated with the arithmetical average of the four RTDs placed inside the test section, by neglecting the subtractive term due to the aluminum layer between the RTDs placement and the channels wall.

The inlet R1234yf enthalpy has been evaluated by considering the saturated liquid enthalpy h_l at the inlet pressure and the inlet sub-cooling ΔT_{sub}, whilst the outlet enthalpy follows a simple energy balance:

$$h_{out} = h_{in} + \frac{\dot{Q}}{\dot{m}} \ [kJ/kg] \tag{5}$$

Finally, the vapor quality at the outlet is obtained with the following equation, where Δh_{LV} is the latent heat:

$$x_{out} = \frac{h_{out} - h_l}{\Delta h_{LV}} \tag{6}$$

5. Results

5.1. Adiabaticity tests

41 tests in liquid single phase have been performed in order to verify the correct insulation of the experimental set-up and especially of the test section. The imposed heat has been compared to the heat absorbed by the liquid refrigerant, which results in an increase of its temperature. Fig. 3 shows the injected and the absorbed heat power on the horizontal and vertical axes, respectively. The experimental points' trend is about to be linear and it is compared to the bisector y = x. All the data are very close to the expected values, regardless the heat power injected. In relative terms, the calculated heat loss is significant only for low heat rates, whereas for greater injected heat powers the percentage is reasonably inferior to 4 %. Since all the main tests are performed with higher heat fluxes, the small disruptions in the adiabaticity tests can be neglected.

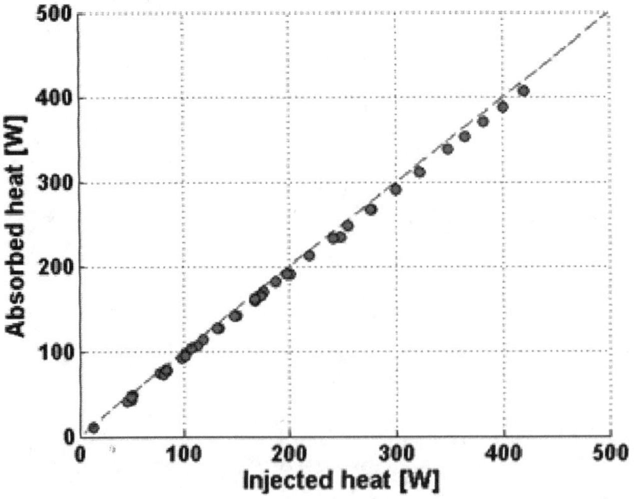

Figure 3 Absorbed versus injected heat power in the test section

5.2. *CHF and sensitivity analysis*

The main tests have been performed with refrigerant R1234yf at saturation temperatures T_{sat} of 25, 35 and 45 °C, with mass fluxes G ranging from 150 to 300 kg/m² s.

The critical condition should be identified when the wall superheat is subjected to a sudden increase with a stable wall heat flux. However, for high mass fluxes, the boiling curves did not exhibit a steep drop in their slope and this behavior has also been observed by Park et al. [3]. In this work the critical heat flux has been defined as the wall heat flux in which the boiling curve decreases its slope $\Delta\dot{q}/\Delta(\Delta T_{wall})$ below a limit value of 2.5 W/cm² K, following the same approach of Mauro et al. [4]. Fig. 4 shows two boiling curves obtained in two different operating conditions. The former is taken at a saturation temperature of 25 °C and a mass flux of 148 kg/m² s, whereas the latter for a saturation temperature of 35 °C and a mass flux of 250 kg/m² s. It is evident that, for the highest mass flux, the thermal crisis is not marked as a sudden flat deviation from the boiling curve trend. This behavior is instead observed for the boiling curve at G = 148 kg/m² s. The mild decrease in the curves' slope might be due to mal-distribution problems occurring in multi-ducts systems. In a previous study, the test section inlet manifold was provided with an orifice insert which reduced the channels inlet to a 20 % of their original size. This solution was taken in order to apply a consistent pressure drop and thus to prevent possible back-flows. However, even with this configuration, the CHF recorded value did not show significant differences from the present solution and the last part of the boiling curve kept its gentle decrease when obtained at high mass velocities. On the contrary, a higher slope was observed for the first part of the boiling curves at any operating condition, meaning a significant increase in the flow boiling heat transfer coefficient. This was probably explained by a flash evaporation through the orifices which instantly turned the flow pattern into an annular regime. In this regard, the recent work of Tibiriçá et al. [11] showed that an inlet positive vapor quality was able to enhance the CHF if compared to a sub-cooled liquid entry. Yet, the effects of the liquid droplets (entrainment) depletion as well as the inlet vapor quality on CHF has to be further investigated.

The parametric effects of mass flux and saturation temperature have been investigated, whilst the inlet sub-cooling effect was not taken into consideration since it has been seen [3] that, except for very high values (e.g. 20 K), it plays a negligible role in determining the critical condition, especially for macro-scale devices.

The parametric effect of the saturation temperature on CHF is shown in Fig. 5. The first diagram refers to G = 149 kg/m² s, whereas the second to a mass flux of G = 300 kg/m² s. In both of them, the saturation temperature varies from 25 to 35 °C. The critical condition is considerably higher for the highest mass flux, but the effect of the saturation temperature is almost negligible. In other words, CHFs are the same for the same mass flux and different saturation temperatures, even if the thermal crisis may occur at slightly inferior wall superheat when augmenting the system pressure, meaning that the flow boiling heat transfer coefficient is affected by the saturation temperature.

The effect of mass velocity is shown in Fig. 6, where the two diagrams displayed refer to the saturation temperatures of 25 and 45 °C, respectively. The four boiling curves in each graph are obtained for mass fluxes equal to 148, 200, 250 and 300 kg/m² s. CHF increases significantly when increasing the mass velocity at a fixed saturation temperature, whereas the first part of the boiling curves blend one another.

The negligible influence of both mass velocity and average vapor quality on the average heat transfer coefficient might be explained by looking at the similar behavior of the local parameters analyzed by Diani et al. [6] in their 3.4 mm internal diameter single tube.

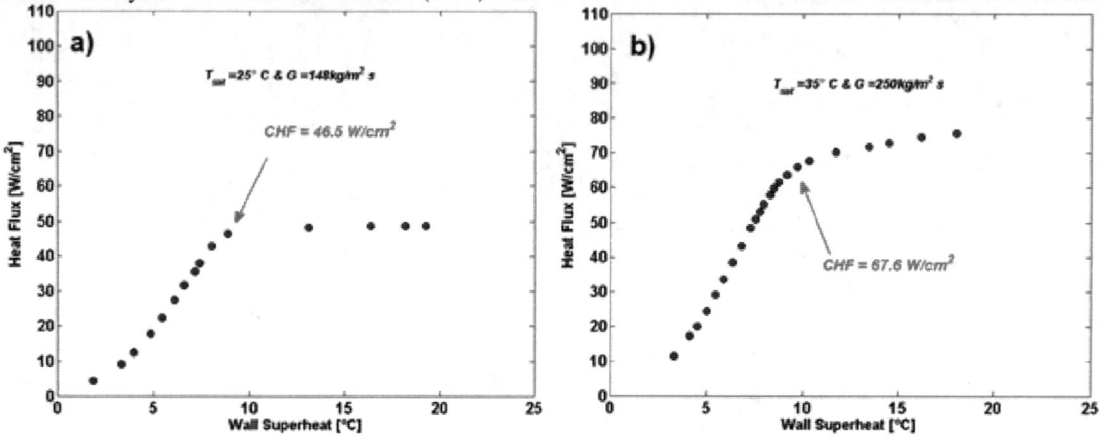

Figure 4 Boiling curves for refrigerant R1234yf at:
a) T_{sat} = 25 °C and G = 148 kg/m² s; b) T_{sat} = 35 °C and G = 250 kg/m² s

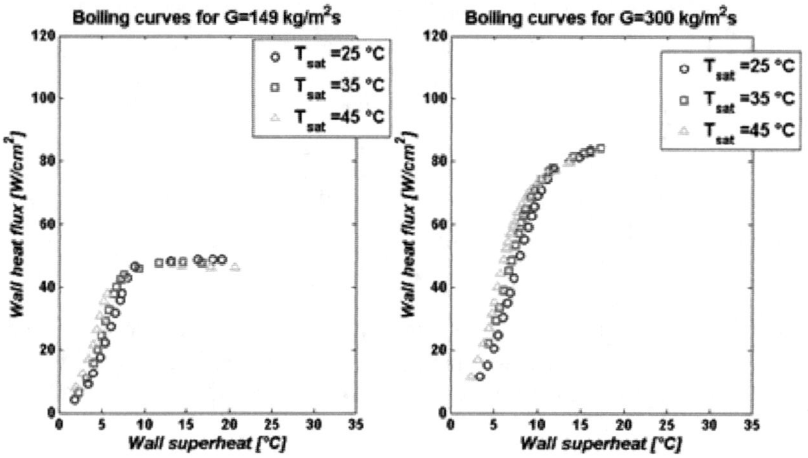

Figure 5 Effect of saturation temperature on CHF for two different mass velocities

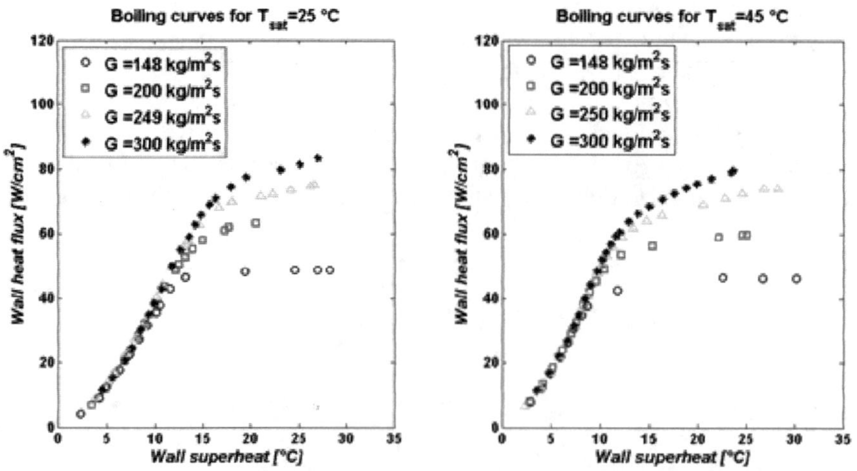

Figure 6 Effect of mass velocity on CHF for two different saturation temperatures

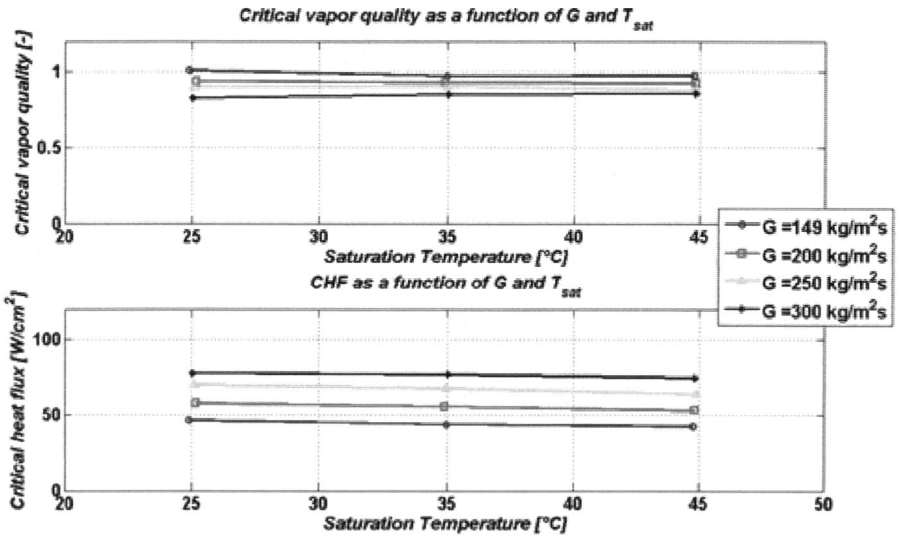

Figure 7 Critical values as a function of the saturation temperature and the mass velocity

Finally, the experimental CHF and critical vapor quality (meaning the outlet vapor quality recorded when the thermal crisis occurred) are plotted in Fig. 7. It is summarized that CHF remarkably increases when increasing mass velocity, whereas for a fixed mass flux it is almost constant within the range of the saturation temperatures investigated. Due to the very small heated length, it is worth noting that experimental CHF values are considerably higher when compared to values obtained in similar studies [8]-[9].

The critical vapor quality does not show a specific trend with the saturation temperature, remaining unchanged from 25 up to 45 °C, at any mass flux displayed. When increasing the mass flux, instead, the critical vapor quality tends to shift towards lower values. Similar results have been obtained in other works available in open literature [2], [3], [8] and [9].

6. Conclusions

Several tests for the investigation of saturated critical heat flux (CHF) of refrigerant R1234yf in a multi-minichannel aluminum heat sink have been proposed in this work. The effect of the mass flux and the system pressure (i.e. the saturation temperature) on CHF and on the boiling curves have been investigated in the range 149-300 kg/m^2 s and 25-45 °C, respectively. The main conclusions are summarized below:

- CHF is significantly affected by the mass flux. When increasing the mass velocity at a fixed saturation temperature, the critical heat flux increases as well.
- The effect of the saturation temperature on CHF is almost negligible. By keeping constant the mass velocity, the thermal crisis occurs approximately at the same heat flux in the range of saturation temperatures investigated, even if a slight decrease is recorded when passing from 25 °C to 45 °C.
- In flow boiling of R1234yf, the average heat transfer coefficient, deduced as the slope in the boiling curves, is moderately affected by the saturation temperature, but when the latter has been fixed it remains almost constant in the whole range of mass velocities investigated.
- The critical vapor quality decreases when increasing the mass flux, whereas it is not affected by a change in the saturation temperature.

Journal of Physics: Conference Series **655** (2015) 012040 doi:10.1088/1742-6596/655/1/012040

However, further studies are required in order to understand the behavior of compact heat sinks handling high thermal loads. More tests at higher mass fluxes and saturation temperatures are needed, as well as a comparison between CHF of different low GWP refrigerants. Finally, experimental CHF values of new refrigerants have to be compared with well-known correlations available in scientific literature.

References

[1] Lazarek G M and Black S H 1982 Evaporative heat transfer, pressure drop and critical heat flux in a small vertical tube with R-113 *Int. J. of Heat and Mass Transfer* **25** 945-960.

[2] Wojtan L, Revellin R and Thome J R 2006 Invetigation of saturated critical heat flux in a single, uniformly heated microchannel *Experimental Thermal and Fluid Science* **30** 765-774.

[3] Park J E and Thome J R 2010 Critical heat flux in multi-microchannel copper elements with low pressure refrigerants *Int. J. of Heat and Mass Transfer* **53** 110-122.

[4] Mauro A W, Thome J R, Toto D and Vanoli G P 2010 Saturated critical heat flux in a multi-microchannel heat sink fed by a split flow system *Experimental Thermal and Fluid Science* **34** 81-92

[5] Tibiriçá C, Ribatski G and Thome J R 2012 Saturated flow boiling heat transfer and critical heat flux in small horizontal flattened tubes *Int. J. of Heat and Mass Transfer* **55**, 7873-7883.

[6] Diani A, Mancin S and Rossetto L 2015 Flow boiling heat transfer of R1234yf inside a 3.4 mm ID microfin tube *Experimental Thermal and fluid Science* **66** 127-136.

[7] Diani A, Mancin S and Rossetto L 2015 Flow boiling heat transfer of R1234ze(E) inside a 3.4 mm ID microfin tube 2014 *Int. J. of Refrigeration* **47** 105-119.

[8] Anwar Z, Palm B and Khodabandeh R 2015 Flow boiling heat transfer, pressure drop and dryout characteristics of R1234yf: Experimental results and predictions *Experimental Thermal and fluid Science* **66** 137-149.

[9] Anwar Z, Palm B and Khodabandeh R 2015 Dryout characteristics of natural and synthetic refrigerants in single vertical mini-channels *Experimental Thermal and fluid Science* **68** 257-267.

[10] Lemmon E W, Mc Linden M O and Huber M L 2009 REFPROP *NIST Standard Reference Database* **23**, *Version 9.0.*

[11] Tibiriçá C, Czelusniak L E and Ribatski G 2015 Critical heat flux in a 0.38 mm microchannel and actions for suppressing of flow boiling instabilities *Experimental Thermal and Fluid Science* – Article in press.

Thin Film Breakup and Rivulet Evolution Modeling

Nicola Suzzi[1], **Giulio Croce**[1] **and Paola D'Agaro**[1]

[1]DIEG - Dipartimento di Ingegneria Elettrica, Gestionale e Meccanica - Universitá di Udine - Via delle Scienze - 33100 - Udine (UD) - Italy

E-mail: suzzi.nicola@spes.uniud.it

Abstract.
The present paper is aimed at the modeling of a continuous film breakup into individual rivulets, leading to the formation of dry patches on the substrate surface. Following an approach already successfully applied to the prediction of still/moving droplet configuration, we attempt to model the details of a single possible film breakup and its evolution over a two-dimensional domain via a phenomenological model. Based on the momentum, energy and mass flow balance of the capillary ridge on the border of the dry patch, the proposed model is validated against both numerical prediction and experimental results from the open literature. Such a detailed prediction may not be practical for the simulation of complex geometrical configuration (which may include, as an example, multiple breakups on the surface of a the whole aircraft subject to icing condition), but can be used to look for statistically significant parameters that can be used to provide proper boundary conditions for fully 3D CFD computations.

1. Introduction

A number of applications of engineering interest are related to the presence and evolution of a thin layer of liquid over a surface, often in presence of a phase change. Such liquid layer may be a collection of tiny droplets, either still moving, driven by gravity or shear stresses, as in fogging and defogging simulations, or can be made up by an ensemble of rivulets, or can be a continuous film, as in several heat transfer devices including de-humidifier or condenser, or may be any combination of these forms.
In most cases, the goal of prediction and simulations are just the evaluation of global average parameters such as percentage of wet surface, or the check of proper critical thresholds such as dry-out limits in a cooling device. Sometimes, however, more details are required. As an example, in-flight icing phenomena on aircraft surfaces are driven by the evolution of the surface water layer fed by a stream of super-cooled droplets from the clouds. Under different conditions of cloud water content, temperature and flow field we may have to deal with all of the above mentioned possible water layer, and we need to know in details such topology, since it will affect the water flow over the surface and, thus, the extent of the area subjected to possible icing condition to be protected.

In previous works [1, 2], the authors addressed the Modeling of the transition between still droplet and moving ones. Here, we present a model for the breakup of a continuous film and the prediction of the evolution of the consequent dry patch.

A first paper on separation of a laminar film flowing vertically under gravity was published by Hartley and Murgatroyd [3], who proposed a critical breakup threshold condition based on a force balance at the apex of the dry patch. Starting from a velocity profile driven by gravity and viscous dissipation only, they defined critical values for both flow rate and film thickness, under which the film may separate into two rivulets, forming a stable dry patch. In the force balance, the pressure increase induced by film inertia compensates the liquid surface tension σ.

Podgorski et al. [4], starting from experimental data on a laminar film flowing down an inclined plane, derived an approximate model for the prediction of the shape of a stable dry patch downstream of a film breakup. The presence of a capillary ridge near the contact line of the dry patch suggested a force balance along the contact line, in which the weight of capillary ridge is balanced by surface tension forces. Practically, Podgorski's model is designed to handle the case where the film inertia is very low and the film thickness is smaller than the critical one defined by Hartley and Murgatroyd.

Also Rio and Limat [5] observed experimentally the presence of this capillary ridge and Wilson et al. [6] obtained its exact shape, solving analytically the governing lubrication equation.

The aim of this paper is to formulate a more general model, based mostly on the basic conservation equation, suited to the prediction of the shape of a stable dry patch, when a laminar liquid film flowing on an inclined plane separates into two rivulets. To enhance the flexibility of the model, both film inertia and capillary ridge weight are taken into account. Furthermore, since we start from the most general form of the balance equations, the shear stress driven case could be easily taken into consideration.

2. Problem Formulation and Model

Consider a thin film of a newtonian liquid with constant and uniform density ρ and viscosity μ flowing down a plane inclined at an angle α to the horizontal. Let h and u_∞ be the film thickness and the film velocity and assume that the external pressure is equal to the atmospheric one, i.e. relative pressure $p_\infty = 0$. Suppose that the film separates into two rivulets forming a stable dry patch and let θ be the static contact angle of the liquid. Following Podgorski [4], it is assumed that the deviated film mass flow is progressively stored into a capillary ridge along the dry patch contact line. Thus, it is possible to derive the differential balances of mass, momentum and energy for an infinitesimal element of liquid inside the capillary ridge.

2.1. Mass, Momentum and Energy Balances

We will solve the continuity, momentum and energy equations in the capillary ridge along a curvilinear coordinate l aligned with the ridge itself. The elementary control volume at a position l is shown in Fig. 1(b): it includes a section of the ridge between l and $l + dl$, and it is bounded by the dry patch contact line and the interface i between the ridge and the film. The flow direction l is inclined at an angle β with respect to the main flow direction y of the undisturbed upstream film. Above the interface i the liquid flow has the same characteristics of the undisturbed film far from the dry patch (u_∞ and p_∞), while below i, in the capillary ridge, the flow is assumed parallel to the contact line (its velocity and pressure become u and p). To let this be true, it is also assumed that the flow deviates instantly and enters the capillary ridge in l direction. The surface S represents a general section of the capillary ridge, see Fig. 1(a). It is not necessary to know its exact shape but only its extension.

From the continuity equation, applied to the control volume of Fig. 1(b), the exit mass flow $(\rho u S)_{l+dl}$ must be equal to the sum of the incoming flow $(\rho u S)_l$ from the upstream section of the capillary ridge and the mass flow δm_i entering through the interface i between the film and the ridge; thus

$$\delta m_i + \rho(uS)_l = \rho(uS)_{l+dl} \tag{1}$$

Rearranging Eq. (1) and neglecting the higher order terms we have

$$\delta m_i = \rho\,(udS + Sdu) = d(\rho u S) \tag{2}$$

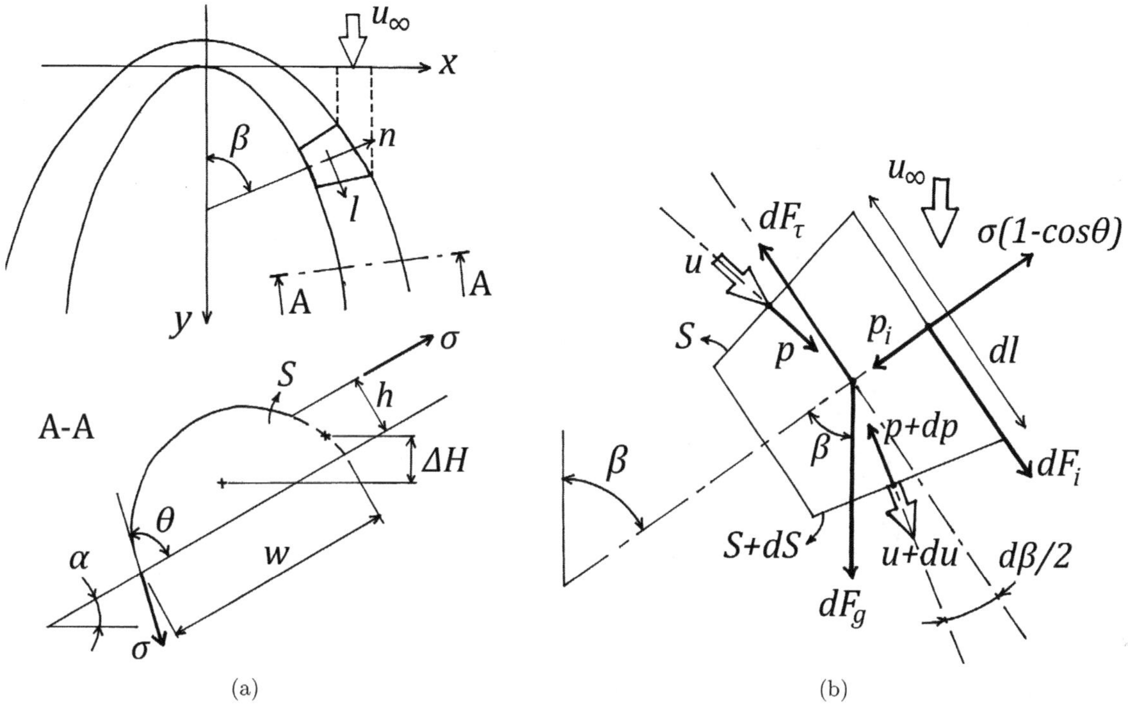

Figure 1. (a) Dry Patch and capillary ridge section. (b) External forces on the control volume

The term δm_i will be defined in section 2.2 as a function of the angle β. The subscript i refers to a property of the liquid crossing the interface i between the ridge and the film.

The momentum equation correlates the momentum fluxes to the external forces acting on the capillary ridge elementary volume. Referring to Fig. 1(b), these external forces are pressure forces on the faces of the volume, the surface tension σ applied along dl, the gravity force dF_g due to the weight of the ridge projected along the inclined plane,

$$dF_g = \rho g \sin\alpha \; S \; dl$$

and the surface shear stress dF_τ along the substrate,

$$dF_\tau = \tau_p \; w \; dl$$

where $w\;dl$ is the base area of the ridge element, where τ_p is applied. Equilibrium along the ridge streamwise coordinate l and along its normal direction n yield respectively

$$\left[\left(\rho u^2 + p\right)_{l+dl} S_{l+dl} - \left(\rho u^2 + p\right)_l S_l\right] \cos\left(\frac{d\beta}{2}\right) - \left(\rho u_i^2 + p_i\right) S_{n,i} = dF_g \sin\beta - dF_\tau \qquad (3)$$

$$dF_g \cos\beta + p_i h \; dl = \; \sigma(1 - \cos\theta) \; dl + \left[\left(\rho u^2 + p\right)_l S_l + \left(\rho u^2 + p\right)_{l+dl} S_{l+dl}\right] \sin\left(\frac{d\beta}{2}\right) \qquad (4)$$

where $S_{n,i}$ is the projection of the interface surface i on the plane normal to the streamwise direction.

$$S_{n,i} = \frac{\delta m_i}{\rho u_i}$$

Rearranging Eqs. (3) and (4) and neglecting the higher order terms yields

$$d\left[\left(\rho u^2+p\right)S\right]-\left(\delta m_i u_i+p_i\frac{\delta m_i}{\rho u_i}\right)=dF_g\sin\beta-dF_\tau \tag{5}$$

$$dF_g\cos\beta+p_i h\ dl=\ \sigma(1-\cos\theta)\ dl+\left(\rho u^2+p\right)S\ d\beta \tag{6}$$

Finally, energy balance can be written as

$$\rho(uS)_l\left(\frac{1}{2}u^2+\frac{p}{\rho}+gH\right)_l+\delta m_i\ e_i=\rho(uS)_{l+dl}\left(\frac{1}{2}u^2+\frac{p}{\rho}+gH\right)_{l+dl}+\rho uS\frac{dR}{\rho} \tag{7}$$

where H is the geodetic height of the corresponding face, dR represents the shear stress losses

$$dR=\tau_p\frac{w\ dl}{S}$$

and e_i is the specific energy of the liquid entering the capillary ridge through its interface i. Assuming energy conservation through the interface:

$$e_i=\frac{1}{2}u_\infty^2+gH_i$$

Applying continuity equation and neglecting the higher order terms, Eq. (7) offers

$$d\left[\rho uS\left(\frac{1}{2}u^2+\frac{p}{\rho}\right)\right]+\rho uS\left(gdH+\frac{dR}{\rho}\right)=\delta m_i\left(\frac{1}{2}u_\infty^2+g\Delta H\right) \tag{8}$$

where the right hand side defines the net energy input from the incoming film mass flow and dH is the geodetic height variation between sections l and $l+dl$ of the elementary volume:

$$dH=H_{l+dl}-H_l=-\sin\alpha\sin\beta\ dl$$

Finally, ΔH is the height difference between the interface i and the ridge centroid

$$\Delta H=\frac{1}{2}\ w\ \sin\alpha\cos\beta=H_i-H_l$$

We can recast the problem in non-dimensional form assuming, as reference quantities, the incoming film height h and velocity u_∞. This allows to define the dimensionless lengths $h^*=1$, dl^*, w^*, ΔH^* velocity u^* and surface S^*. Furthermore, dimensionless pressures p^* are obtained normalizing via the inlet film kinetic energy $\frac{1}{2}\rho u_\infty^2$. Finally, the friction factor and the dimensionless incoming mass flow are defined as:

$$f=\frac{\tau_p}{\frac{1}{2}\rho u_\infty^2}\ ;\ \dot{m}_i^*=\frac{1}{\rho u_\infty h}\frac{\delta m_i}{dl}$$

This normalization lead to the definition of the proper Weber Number and Bond Number

$$\mathrm{We}=\frac{1}{2}\frac{\rho u_\infty^2 h}{\sigma}\ ;\ \mathrm{Bo}=\frac{\rho g h^2}{\sigma}$$

Normalizing Eqs. (2), (5), (6) and (8) a system of four differential equations can be obtained

$$\dot{m}_i^*=\frac{d}{dl^*}\left(u^*S^*\right) \tag{9}$$

$$\frac{d\beta}{dl^*}=-\frac{(1-\cos\theta)-\mathrm{We}\ p_i^*-\mathrm{Bo}\sin\alpha\ S^*\cos\beta}{\mathrm{We}\left(2u^{*2}+p^*\right)S^*} \tag{10}$$

$$\left(\frac{\mathrm{Bo}}{\mathrm{We}}\sin\alpha\ \sin\beta-\frac{w^*}{S^*}f\right)S^*+\dot{m}_i^*\left(2u_i^*+\frac{p_i^*}{u_i^*}\right)=\frac{d}{dl^*}\left[\left(2u^{*2}+p^*\right)S^*\right] \tag{11}$$

$$\left(\frac{\mathrm{Bo}}{\mathrm{We}}\sin\alpha\ \sin\beta-\frac{w^*}{S^*}f\right)u^*S^*+\dot{m}_i^*\left(1+\frac{\mathrm{Bo}}{\mathrm{We}}\Delta H^*\right)=\frac{d}{dl^*}\left[\left(u^{*2}+p^*\right)u^*S^*\right] \tag{12}$$

2.2. Cauchy Problem

Some more hypothesis are needed to solve Eqs. (9), (10), (11) and (12). In particular, from the assumption that the potential energy reduction balances fiction losses, we get

$$gdH + \frac{dR}{\rho} = 0 \Rightarrow \frac{\text{Bo}}{\text{We}} \sin\alpha \sin\beta - \frac{w^*}{S^*} f = 0$$

This is the same condition which is fulfilled in the film flow far from the dry patch. Furthermore, if the film flow is undisturbed up to interface i, the mass flow δm_i entering the control volume through the interface i can be defined as the mass coming from an horizontal section $dx = dl\ \cos\beta$ [4], [5]:

$$\delta m_i = \rho u_\infty h\ dl \cos\beta \Rightarrow \dot{m}_i^* = \cos\beta$$

The velocity u_i and the pressure p_i, that represent the entering condition of the liquid at the interface i, are required to solve Eqs. (10) and (11). Following [4], it is assumed that the relative pressure p_i at the interface is zero, and that the energy transported by incoming mass flow δm_i, see Eq. (8), is fully converted into kinetic energy

$$\frac{1}{2}u_i^2 = \frac{1}{2}u_\infty^2 + g\Delta H \Rightarrow u_i = \sqrt{u_\infty^2 + 2g\Delta H}$$

$$u_i^* = \sqrt{1 + (\text{Bo}/\text{We})\Delta H^*} \tag{13}$$

Finally, assuming that the distance between the centroid and the entering section of the control volume is a linear function of the length S/h, the dimensionless height ΔH^* is estimated as

$$\Delta H^* = \xi\ S^* \sin\alpha \cos\beta \tag{14}$$

where ξ is a constant depending on the shape of capillary ridge section.

Note that from geometrical consideration on Fig.1(a) we can obtain that $\xi = \frac{1}{2}\frac{h}{S/w}$, i.e the average height $\bar{h}$ of the ridge is $1/2\xi$ time the inflow film height, $\bar{h} = \frac{1}{2}\frac{h}{\xi}$. Thus, the maximum possible value for ξ is $1/2$, when the ridge does not exceed the film height and thus has the shape of a simple rectangle of height equal to h, while the minimum value should be limited by the requirement of equilibrium between surface tension and ridge hydrostatic pressure. A reasonable value $\xi = 0.3$ was chosen in the following computations. Although a proper evaluation of ξ would require a 3D formulation, results will show that its actual value doesn't affect considerably the results.

Under these assumptions, we obtain the following Cauchy Problem from Eqs. (9, 10, 11, 12)

$$\dot{u}^* = \frac{du^*}{dl^*} = \frac{\cos\beta}{S^*} \frac{\left[\sqrt{1 + (\text{Bo}/\text{We})\,\Delta H^*} - u^*\right]^2}{p^*} \tag{15}$$

$$\dot{S}^* = \frac{dS^*}{dl^*} = \frac{\cos\beta}{u^*} \frac{p^* - \left[\sqrt{1 + (\text{Bo}/\text{We})\,\Delta H^*} - u^*\right]^2}{p^*} \tag{16}$$

$$\dot{p}^* = \frac{dp^*}{dl^*} = \frac{\cos\beta}{u^* S^*} \frac{\left[1 + (\text{Bo}/\text{We})\,\Delta H^* - \left(u^{*2} + p^*\right)\right] p^* - 2u^{*2}\left[\sqrt{1 + (\text{Bo}/\text{We})\,\Delta H^*} - u^*\right]^2}{p^*} \tag{17}$$

$$\dot{\beta} = \frac{d\beta}{dl^*} = -\frac{(1 - \cos\theta) - \text{Bo} \sin\alpha\ S^* \cos\beta}{\text{We}\,(2u^{*2} + p^*)\,S^*} \tag{18}$$

Journal of Physics: Conference Series **655** (2015) 012041 doi:10.1088/1742-6596/655/1/012041

Ref.	$\dot{V} = u_\infty h$	α	θ_a	m
Rio and Limat [5]	0.01cm^2/s	80°	48°	0.17
Podgorski et al. [4]	0.01cm^2/s	15°	30°	0.33

Table 1. Test case set up

with the initial conditions

$$u_0^* = 0 \; ; \; p_0^* = 1 + \frac{\text{Bo}}{\text{We}} \Delta H_0^* \; ; \; S_0^* = \frac{1 - \cos\theta}{\text{Bo}\sin\alpha - (\text{We} + \text{Bo}\Delta H_0^*)/r_0^*} \; ; \; \beta_0 = 0$$

where $r_0^* = r_0/h = 1/\dot{\beta}_0$ is the dimensionless radius of curvature at the apex of the dry patch.

Differential Eqs. (15), (16), (17) and (18) was solved and the functions $u^*(l^*)$, $S^*(l^*)$, $p^*(l^*)$ and $\beta(l^*)$ obtained. Knowing $\beta(l^*)$ and introducing the dimensionless Cartesian Coordinates $x^* = x/h$ and $y^* = y/h$, see figure 1(a), it's possible to calculate the dry patch shape $y^*(x^*)$

$$y^* = \int_0^{l^*} \sin\beta \; dl^* \quad ; \; x^* = \int_0^{l^*} \cos\beta \; dl^*$$

3. Results

We will compare the model described above with both the theoretical results of Podgorski et al. [4] and the experimental data reported in the same paper and in [5]. Podgorski model is still commonly used as reference benchmark for more sophisticated simulation, such as in [7]. In his model Podgorski et al. use the following equations to describe the dry patch shape

$$x = r_0 \frac{\sin\beta}{\cos^2\beta} \qquad y = -\frac{r_0}{3}\left(\frac{1 - 3\sin^2\beta}{\cos^3\beta} - 1\right) \tag{19}$$

The radius of curvature at the apex of the dry patch is given by an empirical relation [4]

$$r_0 = m \, \frac{(1 - \cos\theta)^4}{\theta - \cos\theta\sin\theta} \, \frac{h}{\text{Ca Bo}\sin\alpha} \tag{20}$$

where $\text{Ca} = \mu u_\infty/\sigma$ is the Capillary number and m is a dimensionless constant depending on the shape of the capillary ridge.

Both Rio and Limat [5] and Podgorski et al. [4] made experimental tests on silicon oil 47V20 sold by Rhodorsil ($\rho = 950 Kg/m^3$, $\mu = 0.019 Pa \cdot s$, $\sigma = 0.021 N/m$). The film flow rate per unit length and the tilt angle of the inclined plane was chosen as input parameters. The advancing contact angle θ_a of the liquid was experimentally measured both by Podgorski et al. and Rio and Limat. The main case parameters are given in Table 1.

Other quantities need to be defined to run our model. The initial condition on S^* is estimated considering the curvature radius r_0 from Eq. (20) with $m = 0.17$ (as in [5]) or $m = 0.33$ (as in [4]) depending on the case, while the mean velocity and the thickness of the film (required to define We, Bo and Ca) are computed assuming that gravity is balanced by viscous shear stress

$$h = \sqrt[3]{\frac{3\mu\dot{V}}{\rho g \sin\alpha}} \; ; \; u_\infty = \frac{\dot{V}}{h}$$

Fig. 2(a) compares our model with the experimental setup of Rio and Limat [5]: predicted dry patch shape shows a fully satisfying agreement with both experimental data and Podgorski model results. The profile of liquid velocity inside the capillary ridge $u(x)$ and its section $S(x)$

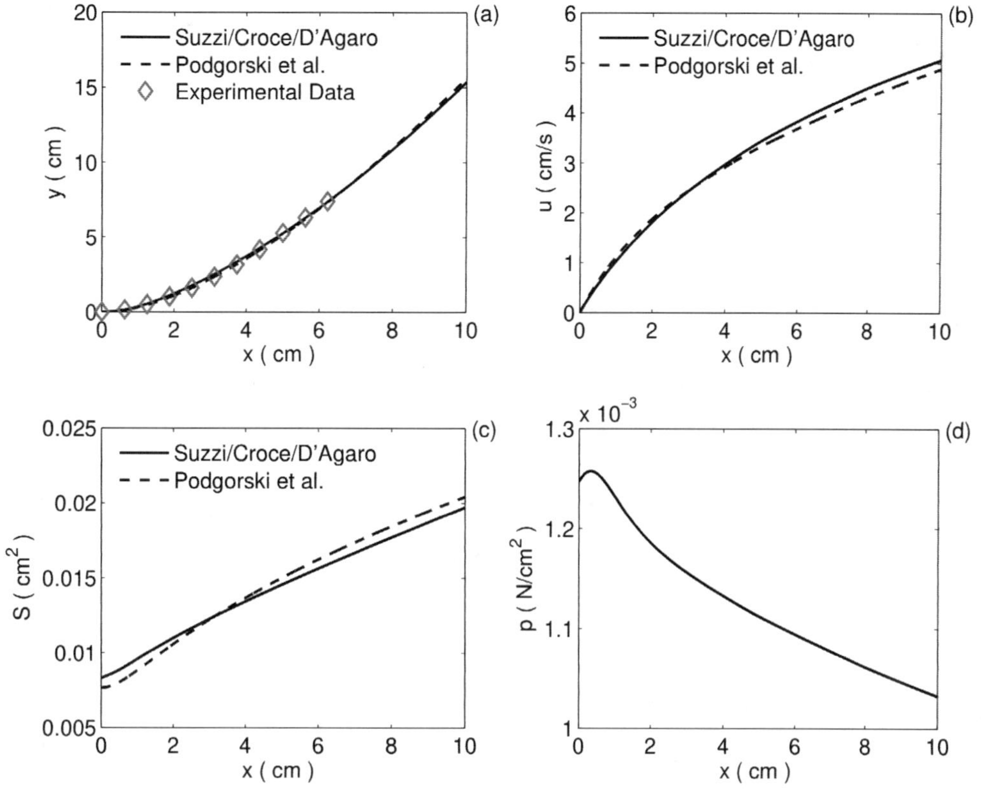

Figure 2. Model validation: $\dot{V} = 0.01$ cm^2/s, $\alpha = 80°$, $\theta = 48°$. (a) Dry patch shapes. (b) Flow velocity along the ridge. (c) Capillary ridge cross section. (d) Pressure inside the ridge.

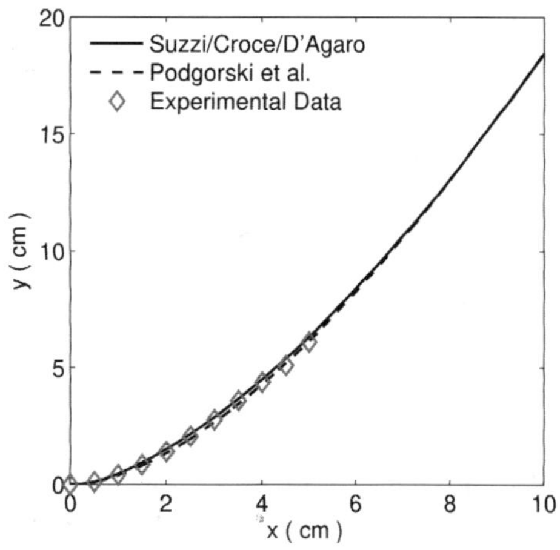

Figure 3. Model validation: $\dot{V} = 0.01$ cm^2/s, $\alpha = 15°$, $\theta = 30°$. Dry patch shapes.

Journal of Physics: Conference Series **655** (2015) 012041 doi:10.1088/1742-6596/655/1/012041

are shown in Figs. 2(b) and 2(c): these profiles replicates fairly well the one predicted by Podgorski et al. model [4], too. It is not possible to compare the liquid pressure inside the capillary ridge because Podgorski et al. didn't take it into account: anyway it can be seen in Fig. 2(d) that $p(x)$ reaches its maximum value of $1.26 \cdot 10^{-3}$ N/cm^2 near the apex of the dry patch ($x \approx 0.335$ cm) and than it starts to decrease for increasing value of x.

The maximum is due to the initial increase of the net potential energy $g\Delta H$, i.e. of the energy that enter the ridge through the interface i.

Fig. 3 shows, again, a fully satisfying matching between our model, Podgorski's model and experimental data [4], for a different set of parameters (see table 1). This confirms the model accuracy and flexibility. The trends of u, S and p are not shown because appears quite similar to the ones of Figs. 2.

It is worth to notice that the details of exact shape of capillary ridge section are not such important: the influence of ξ is showed in Fig. 4(a): the higher is its value, the less it influences the predicted dry patch shape; only at very low values of ξ the predicted dry patch becomes significantly narrower: let remember that $\xi = 0.1$ corresponds to an unrealistic height of the ridge five times taller then the incoming film, while when $\xi = 0$ the centroid of capillary ridge coincides with the ridge interface one, i.e. the capillary ridge has zero width (and thus infinite height).

Also the influence of the empirical parameter m, needed by Eq. (20) to define the radius of curvature at the apex of the dry patch, is analyzed, see Fig. 4(b). The reason why m doesn't affect significantly the shape of our predicted dry patch is that r_0 affects only the initial condition S_0, whilst it scales the dry patch shape obtained by Podgorski, see Eqs. (19). Moreover, being both m and ξ only connected with the ridge shape, their meaning must be similar and it would be desirable to correlate them.

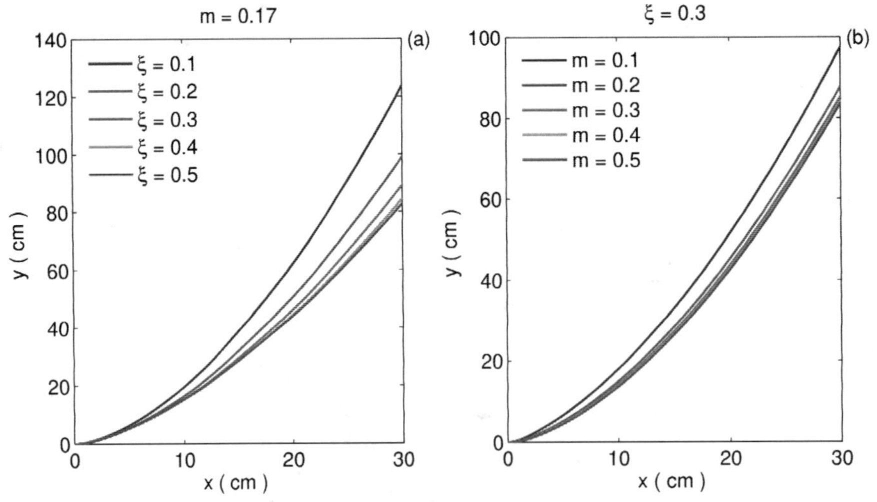

Figure 4. Dry patch shapes: $\dot{V} = 0.01$ cm^2/s, $\alpha = 80°$, $\theta = 48°$: Influence of ξ (a) and m (b).

Fig. 5 illustrates the influence of both film flow rate $\dot{V}$ and tilt angle of the inclined plane α (the two main input parameter defining the general problem) on the dry patch shape. Silicon oil is again considered and the imposed liquid contact angle is $\theta = 48°$. The flow rate ranges from 0.01 to 0.04 cm^2/s, while we consider a tilt angle in the range $15° \div 90°$, where $\alpha = 90°$ is the case of liquid film sliding on a vertical wall. Values of $\dot{V}$ higher than 0.04 cm^2/s are not considered because Rio and Limat observed experimentally that it exists a critical flow rate $\dot{V}_{cr} = 0.03 \div 0.04$ cm^2/s (not depending on α) beyond which the dry patch closes [5].

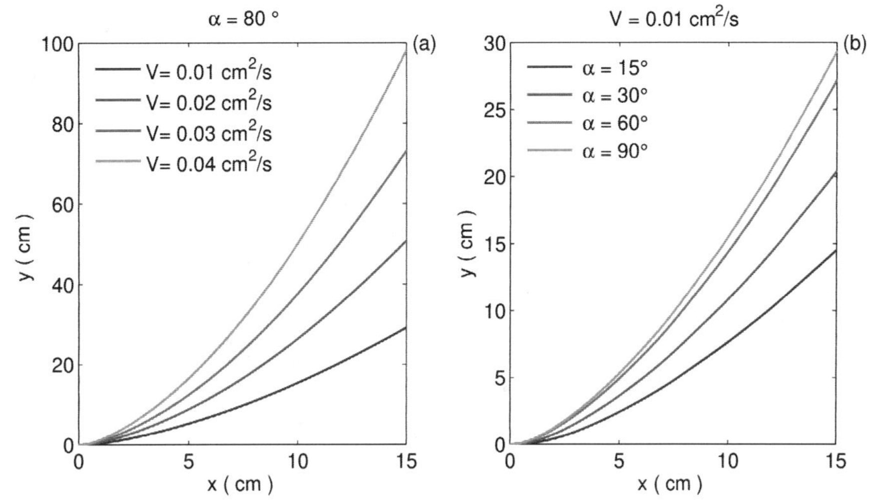

Figure 5. Dry patch shapes: (a) $\theta = 48°$, $\alpha = 80°$, different flow rates: $\dot{V} = 0.01, 0.02, 0.03, 0.04$ cm^2/s; (b) $\theta = 48°$, $\dot{V} = 0.01$ cm^2/s, different tilt angles: $\alpha = 15°, 30°, 60°, 90°$.

Fig. 5(a) shows a progressive reduction of the dry patch width when the film flow rate increases, for a fixed α. This trend is due to the fact that the film inertia (i.e. the Weber number) and the film energy increase with the flow rate; hence the surface tension forces find it harder to widen the dry patch gap. Narrower dry patches are also obtained increasing the tilt angle α at a fixed $\dot{V}$, although this reduction is less noticeable for high value of α, where the curves progressively collapse asymptotically on the limiting case of $\alpha = 90°$, see Fig. 5(b).

It is interesting to notice that the dry patch always tends to a quadratic shape for high values of x, see Fig. 6. This confirms what Wilson et al. [6] obtained from the lubrication approximation. It can be also found that when x is high enough the mean velocity tends to a constant value, the capillary ridge section increases linearly and the liquid pressure decreases proportionally to $x^{-1/2}$:

$$\lim_{x\to\infty} y(x) \propto x^2 \; ; \; \lim_{x\to\infty} u(x) \propto 1 \; ; \; \lim_{x\to\infty} S(x) \propto x \; ; \; \lim_{x\to\infty} p(x) \propto x^{-\frac{1}{2}}$$

Such behavior appears reasonable: if x is large enough the dry patch angle β gets close to $\pi/2$, i.e. the problem reduces to an undisturbed flowing film. In terms of asymptotically solutions, this means that the velocity u of the liquid inside the ridge reaches a constant value (u_∞ should be the expected value), the pressure p is equal to zero and the ridge collapses to a section of the flowing film itself ($S = hx$). From our model, the velocity u tends to a value that is higher than u_∞ but comparable. The only reason is probably due to an inaccurate definition of the height ΔH, see Eq. (14), that is connected with the incoming net potential energy and doesn't reach zero for high values of x (it just tends to a constant value that is lower than the initial one). Podgorski model [4], otherwise, predicts a velocity u inside the ridge proportional to $\sqrt{x}$: this, for $x \to \infty$, yields a non physical infinite velocity, notwithstanding the dry patch angle β tends to $\pi/2$. The expression of $u(x)$ predicted by Podgorski can be obtained rearranging the governing equations of his model [4] and extracting the dry patch angle $\beta(x)$ from Eq. (19).

4. Conclusion

A model based on balance equations was developed with the aim to describe the shape of a stable dry patch on an inclined plane. The existence of a capillary ridge near the dry patch

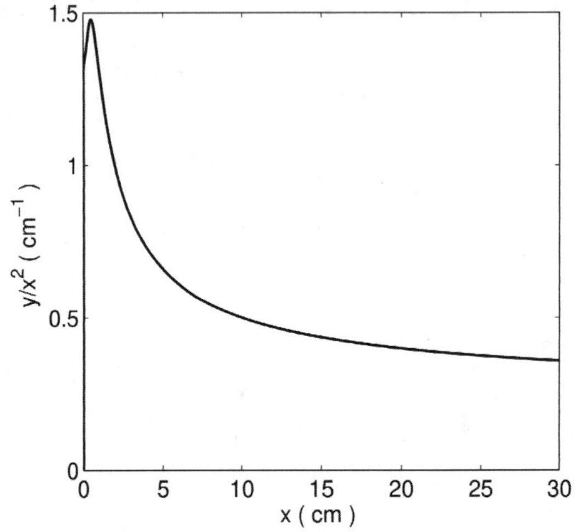

Figure 6. Ratio between dry patch shape $y(x)$ and its parabolic approximation, $\dot{V} = 0.04\text{cm}^2/\text{s}$, $\alpha = 80°$, $\theta = 48°$.

contact line, as proposed by other authors, was assumed. Excellent agreement with the available experimental data was found, although validation was limited to a high viscosity liquid (silicon oil) and the dominant effect was the capillary ridge weight. It would be helpful to validate the model when the liquid inertia is higher (e.g. water), but no experimental data is yet available.

Some empirical information are needed: in particular, in the definition of the height ΔH, function of the ridge shape via the parameter ξ, and in the evaluation of the initial curvature of the dry patch, via a tunable parameter m.

However in the present model the effect of variation of either ΔH and m is limited, and thus reasonable dry patch development predictions are possible without knowing the exact shape of the capillary ridge. Thus, the model seems more flexible than previous literature ones, such as Podgorski's. Furthermore, the proposed model could be able to predict the dry patch shape in a wider range of cases, i.e. when the Weber number is higher, since the film inertia is taken into account.

In perspective, this approach could be extended to the break-up of a shear driven film, adding the shear stress to sum of the external force acting on the capillary ridge, or integrated in more complex problems involving heat and mass transfer.

References

[1] Croce G, DeCandido E, Habashi W, Munzar J, Aube M, Baruzzi G and Aliaga C 2010 *J. of Aircraft* **47** 1283
[2] DeCandido E, Croce G and D'Agaro P 2012 *Heat Transfer Engineering* **33** 1130
[3] Hartley D E and Murgatroyd W 1971 *Int. J. Heat Mass Transfer* **7** 1003
[4] Podgorski T, Flesselles J M and Limat L 1999 *Physics of Fluids* **11** 845
[5] Rio E and Limat L 2006 *Physics of Fluids* **18** 032102
[6] Wilson S K, Duffy B R and Davis S H 2001 *Euro Jnl. of Applied Mathematics.* **12** 233
[7] Sellier M 2015 *Int Jnl. of Multiphase Flow* **71** 66

Pulsating Heat pipe Only for Space (PHOS): results of the REXUS 18 sounding rocket campaign

F Creatini[1], G M Guidi[1], F Belfi[1], G Cicero[1], D Fioriti[1], D Di Prizio[1], S Piacquadio[1], G Becatti[1], G Orlandini[1], A Frigerio[1], S Fontanesi[1], P Nannipieri[1], M Rognini[1], N Morganti[1], S Filippeschi[1], P Di Marco[1], L Fanucci[1], F Baronti[1], M Mameli[1,2], M Manzoni[2], M Marengo[2,3]

[1] University of Pisa, Largo Lucio Lazzarino 2, 56122 Pisa, Italy
[2] University of Bergamo, Viale Guglielmo Marconi 5, 24044 Dalmine (BG), Italy
[3] University of Brighton, Lewes Road, BN2 4GJ Brighton, United Kingdom

E-mail: francescocreatini89@gmail.com

Abstract. Two Closed Loop Pulsating Heat Pipes (CLPHPs) are tested on board REXUS 18 sounding rocket in order to obtain data over a relatively long microgravity period (approximately 90 s). The CLPHPs are partially filled with FC-72 and have, respectively, an inner tube diameter larger (3 mm) and slightly smaller (1.6 mm) than the critical diameter evaluated in static Earth gravity conditions. On ground, the small diameter CLPHP effectively works as a Pulsating Heat Pipe (PHP): the characteristic slug and plug flow pattern forms inside the tube and the heat exchange is triggered by thermally driven self-sustained oscillations of the working fluid. On the other hand, the large diameter CLPHP works as a two-phase thermosyphon in vertical position and doesn't work in horizontal position: in this particular condition, the working fluid stratifies within the device as the surface tension force is no longer able to balance buoyancy. Then, the idea to test the CLPHPs in reduced gravity conditions: as the gravity reduces the buoyancy forces becomes less intense and it is possible to recreate the typical PHP flow pattern also for larger inner tube diameters. This allows to increase the heat transfer rate and, consequently, to decrease the overall thermal resistance. Even though it was not possible to experience low gravity conditions due to a failure in the yo-yo de-spin system, the thermal response to the peculiar acceleration field (hyper-gravity) experienced on board are thoroughly described.

1. Introduction

Pulsating Heat Pipes (PHPs) are novel two-phase passive heat transport devices that seem suitable for the thermal management of electronics components in space related applications. Patented in their most common configurations by Akachi [1-2], PHPs basically consist of a meandering tube of capillary dimensions closed end-to-end to form a closed loop. The tube is first evacuated and then partially filled with a working fluid, which naturally resides in the form of liquid plugs alternated to vapour slugs. When input heat power is provided to the evaporator section, the vapour bubbles expand and push the adjacent liquid plugs to the condenser section, where heat is released to the cold source and condensation process occurs.

In spite of the significant efforts made in the last decades in order to create a comprehensive set of tools able to predict PHPs behaviour and facilitate their design, some issues still remain unsolved [3], such as the effect of the number of turns and the influence of the orientation with respect to the gravity

Journal of Physics: Conference Series **655** (2015) 012042 doi:10.1088/1742-6596/655/1/012042

vector on the thermal performance. Nowadays, the design of PHPs is based on the so-called Bond number confinement criterion: in particular, surface tension can prevail over buoyancy and a capillary slug flow pattern can form inside the tube, if the inner tube diameter is smaller than a critical diameter formulated on Earth gravity conditions, i.e. $d_{cr} = 2\sqrt{\sigma/g\,(\rho_l - \rho_v)}$. If one accounts only for the aforementioned criterion it seems that, under reduced gravity conditions, the inner tube diameter may be increased to the bitter end. However, both Gu et al [4] and Mameli et al [5] asserted that viscous and inertial effects always play a significant role in the definition of the flow pattern within the device that can not be discounted. Considering the recent formulation of the dynamic confinement criterion based on the Weber and Garimella numbers, it seems that a two-phase loop device may works as a PHP under reduced gravity conditions with an inner tube diameter larger than the one evaluated on static Earth gravity conditions. In order to assert the validity of the aforementioned assumption and to decouple completely the inertial effects from buoyancy is therefore fundamental to conduct experiments directly in a reduced gravity environment.

Based on the above considerations, Mangini et al. [6] tested a planar Closed Loop Pulsating Heat Pipe (CLPHP), with an inner tube diameter (3 mm) much larger than the critical diameter value defined on static Earth gravity conditions (1.6 mm at 20 °C for refrigerant FC-72), both on ground and in the micro-gravity conditions experienced during the 61st ESA Parabolic Flight Campaign. On ground, the device works as a closed loop two-phase thermosyphon when gravity assisted and as a pure conductive medium when placed in a horizontal position. When the gravity level decreases, visualization studies reveal a sudden transition of the flow pattern from stratified to slug flow. Then, the oscillations of the working fluid suddenly activate and the device is also able to work as a PHP when placed in horizontal position.

As the time required for the device to reach an operational pseudo-steady-state is about few minutes, the available reduced gravity period in a parabolic flight (approximately 20 s) is not enough to assert anything on the device overall thermal performance. One of the alternative methods in order to obtain a longer reduced gravity period is to exploit longer flight trajectories, such that on board a sounding rocket or a satellite in an orbit around the Earth.

The aim of the present work is to investigate the thermal response of two CLPHPs with an inner tube diameter respectively smaller and larger than the critical one on Earth in the reduced gravity conditions experienced on board REXUS-18 sounding rocket. The reduced gravity period is expected to last around 90 s according to preliminary calculations based on the foreseen overall mass of the payload. The flight test results are compared with the results obtained on ground with the devices both in vertical Bottom Heat Mode (BHM) and horizontal orientations.

2. Experimental set-up

The test-cell and the peripheral facilities are mounted inside a cylindrical module, which constitutes the REXUS 18 sounding rocket casing, according to the architecture sketched in Figure 1.

The geometrical features of the module are given in the rocket manual. The test-cell consists of two CLPHPs and a phase change material that acts as a heat sink. The test-cell is mounted inside an airtight box, also termed as experiment box, in order to avoid leakages that may result in a serious hazard for launch safety. The battery cells that provide the input heat power to the heat sources are integrated inside a pack. Then, the electronic boards for the management of data and power are mounted in the on-board data handling and power management boxes, respectively.

Figure 1. Experimental set-up.

2.1. Test-cell

The geometrical features of both CLPHPs are summarized in Table 1. The basic devices structure, as sketched in Figure 2, consists of an aluminium tube folded in a double-layer configuration with fourteen curves constituting the evaporator section. Two T-junction close the tube in a loop and in addition allow for obtaining two ports: the former is used for the emptying and filling procedure; the latter hosts a pressure transducer (Kulite®, XCQ-093-1.7 bar A). The devices are equipped with T-type thermocouples that have a post-calibration accuracy of ± 0.7 K. The thermocouples are thermally linked to the aluminium tube by means of a high conductive paste. The thermocouples position on the large diameter CLPHP is sketched in Figure 2 and is symmetrical on both devices planes. The thermocouples position on the small diameter CLPHP is the same except in the evaporator section, where the thermocouples are mounted in an alternate fashion. The thermocouples position is selected in order to make assumptions on the circulation of the working fluid and does not influence the experiment in any way being a non-invasive measurement.

Table 1. Geometrical features.

	Small Diameter CLPHP	Large Diameter CLPHP
Internal Diameter (ID)	1.6 mm	3 mm
Outer Diameter (OD)	3.2 mm	5 mm
Radius of Curvature (r_c)	4.8 mm	7.5 mm
Length (L)	192.8 mm	200 mm
Evaporator Length (L_e)	6.4 mm	10 mm
Condenser Length (L_c)	145 mm	145 mm
Width (W)	128 mm	200 mm
Height (H)	12.8 mm	20 mm

The CLPHPs are evacuated by a two-stage vacuum pump (Edwards®, XDS35i and EXT255H 24V) until a pressure level of 10^{-4} Pa is obtained. The selected working fluid, i.e. FC-72, is separately degassed in a secondary loop by continuous boiling: the incondensable gases first accumulate at the top of a tank and then are sucked away by means of several vacuuming cycles. Then, the devices are partially filled with a volumetric ratio of 0.5 ± 0.025 and sealed. The incondensable gases content results in an overpressure included within the margin of accuracy of the pressure transducer.

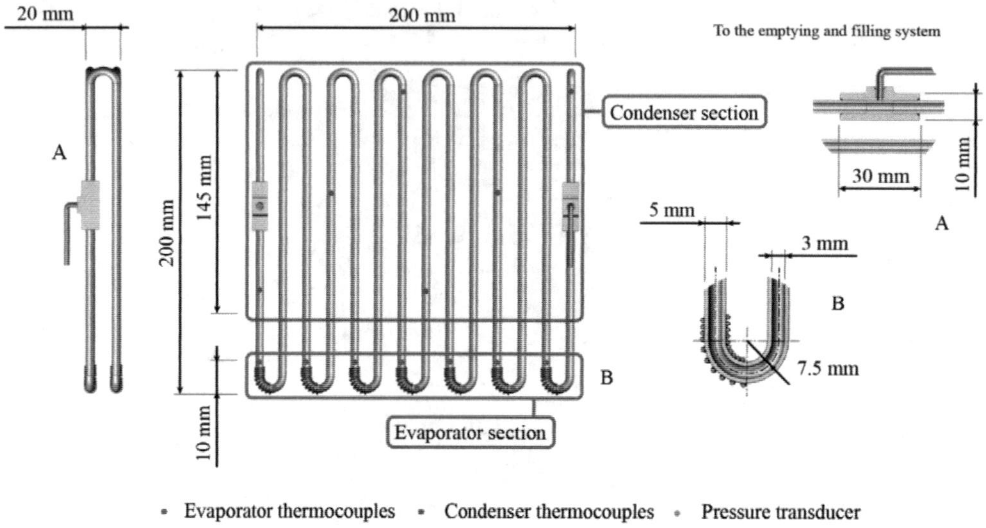

Figure 2. Structure.

2.1.1. Evaporator section. The CLPHPs evaporator sections consist of 2 wire-shaped heating elements (Thermocoax®, 1 Nc I) 1.6 m long with a 1 mm external diameter and an equivalent line resistance of 12 Ω/m. The heating elements are connected in parallel and wrapped around the evaporator curves as shown in Figure 2: the asymmetrical position with respect to the curves is intended to promote the working fluid circulation in a preferential direction. The evaporator sections temperature is limited to 150 °C by two thermal switch (DMP®, 11MP 160L 007E) in contact with the aluminium tube. The heating elements are powered up to 400 W by 14 battery cells (SAFT®, MP176065): 8 DC/DC converters, each that consist of a PRM regulator (Vicor®, P024F048T12AL) and a VTM current multiplier (Vicor®, VTM48EF480T006A00) regulate the voltage and current coming from the cells in order to provide the power profiles shown in Figure 3.

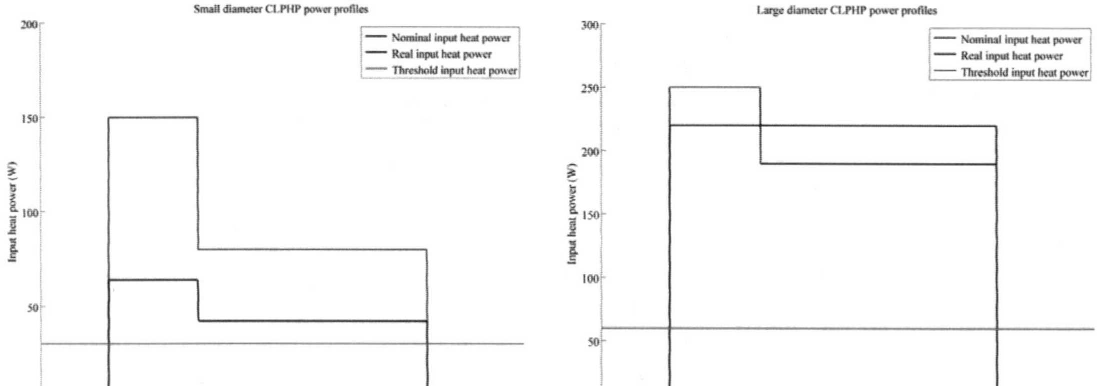

Figure 3. Power profiles: a) small diameter and b) large diameter CLPHPs.

The nominal power values provided to the heating elements are selected in order to supply the evaporator sections of both CLPHPs with an input heat flux larger than a threshold value. The threshold value (dotted line in Figure 3) corresponds to the value beyond which the device shows an unstable behaviour, i.e. it is not able to reach a pseudo-steady-state even if the heat input power is increased [8]. The threshold value is about 5 W/cm^2, thus, the nominal power values are sized so as to result in an input heat flux of 15 W/cm^2. The nominal power values are preceded by peak values intended to rapidly overcome the thermal inertia of the aluminium tubes. The power is evaluated by combining the voltage and current measurements with the imposed duty cycle values. The measurements are carried out by an instrumental amplifier (Analog Devices®, AD8221) with a minimum accuracy of ± 10 % of the imposed power values.

2.1.2. Condenser section. The heat sink consists in a phase change material in contact with the condenser section. The phase change material is an n-octadecane paraffin wax whose thermal properties are summarized in Table 2. Theoretically, the paraffin wax should absorb heat via latent heat of fusion so as to keep constant the condenser temperature. However, as the thermal conductivity of the paraffin wax is small, after a certain time, the melting front stops and the liquid paraffin surrounding the tubes is heated up (sensible heating), thus leading to an increase of the condenser temperature. This phenomenon should be carefully taken into account and can be probably overcome by embedding a conductive structure inside the paraffin (i.e. aluminium honeycomb).

Table 2. Paraffin wax thermal properties.

Phase change material properties	
Melting point temperature	28.0 °C
Boiling point temperature	317.9 °C
Specific heat	2160 J/(kg K)
Latent heat of fusion	244 kJ/kg
Thermal conductivity	0.15 W/(m K)

2.2. Data handling system
The output signals were managed by a microcontroller, mounted on an electronic board specifically designed in order to meet the actual experimental objective. The temperatures and pressure output signals were acquired at 5 and 10 Hz, respectively, while the current and voltage output signals were acquired at 1 Hz.

3. Experimental results

The experimental results are presented in terms of temperatures and pressure temporal trends. In the graphs below, the tubes wall average evaporator and condenser temperatures are sketched together with the local working fluid pressure on the secondary ordinate axis. The input heat power provided to the CLPHPs heating elements is also reported.

The flight results are compared to that obtained on ground with the devices in vertical BHM and horizontal positions in order to provide comparison between the devices thermal performances under different operational conditions. A list of the performed tests is provided in Table 3.

Table 3. List of the performed tests.

	Small Diameter CLPHP	**Large Diameter CLPHP**
Ground Tests -Vertical		
Peak Power	64.5 W	218 W
Nominal Power	42.5 W	218 W
Ground Tests -Horizontal		
Peak Power	64.5 W	218 W
Nominal Power	42.5 W	218 W
Flight Tests		
Peak Power	64.5 W	218 W
Nominal Power	42.5 W	218 W

3.1. Ground tests

The ground tests results relative to the small and large diameter CLPHPs in vertical and horizontal positions are reported in Figure 4 and Figure 5, respectively.

The pressure temporal trends relative to the small diameter CLPHP in vertical (Figure 4a) and horizontal (Figure 4b) positions are different: in the former case, pressure oscillations have smaller amplitude and higher frequency while in the latter larger amplitude and lower frequency. The reason lies in the different orientation of the device with respect to the gravity vector. In vertical position, the gravity vector acts along the flow path direction and assists the working fluid motion. In these conditions, even a small pressure change is sufficient to overcome the inertial effects and to push vapour bubbles and adjoining liquid plugs towards the condenser section: as the vapour bubbles rise rather easily the resulting pressure oscillation frequency is high. In horizontal position, the gravity vector acts on a plane perpendicular to the flow path direction. As a result, a large pressure change is needed in order to displace the liquid plugs and the working fluid motion is least frequent.

The nature of the pressure oscillations results in different start-up behaviours. In vertical position, the start-up occurs suddenly as soon as a sufficient pressure difference is set up between the evaporator and condenser sections. In horizontal position, the start-up takes slightly more time, as gravity has not an influence on the working fluid motion. The working fluid motion inside the CLPHP in different orientations also affects the overall thermal performance. The heat transport capability is higher when gravity assists the working fluid motion as more liquid reaches the condenser section and releases sensible heat. Then, the average evaporator temperature moves towards a smaller value when the device is in vertical than in horizontal position.

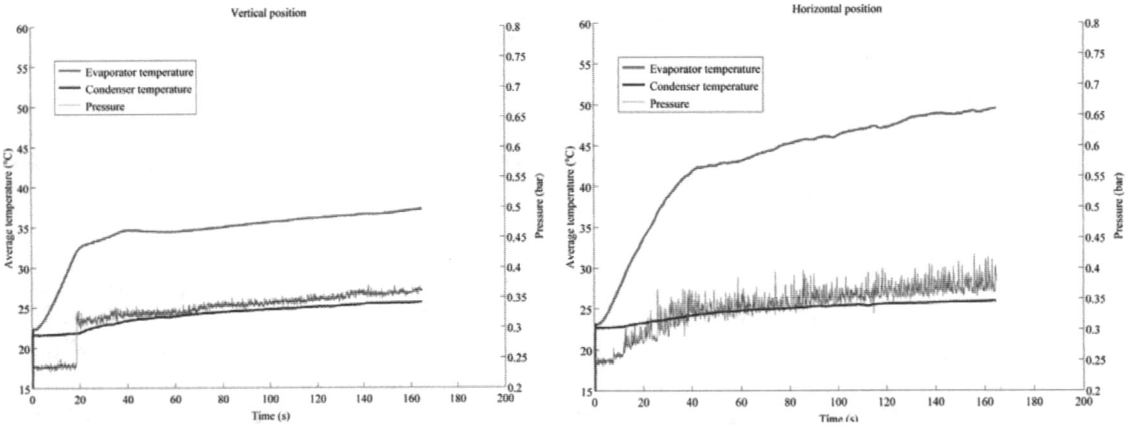

Figure 4. Small diameter CLPHP: a) vertical and b) horizontal ground tests.

The pressure temporal trends relative to the large diameter CLPHP in vertical (Figure 5a) and horizontal (Figure 5b) positions are quite similar to the ones discussed above, although the physical phenomena that occur inside the device have a different nature. In vertical position, the device operates as a two-phase thermosyphon. The liquid phase that accumulates close to the evaporator section boils continuously. The generated vapour rises towards the condenser section and, sometimes, drags on with it on some liquid batches in the so-called "bubble lift" mode thoroughly characterized by Franco et al. [8]. As the working fluid distribution inside the device is optimal, even a small pressure difference between the evaporator and condenser sections is able to trigger the working fluid motion. In horizontal position, gravity restrains the liquid phase in the bottom plane of the device while the vapour phase fills the upper plane. The resulting working fluid distribution within the device is not optimal: beyond gravity, the pressure force that ensue from the vapour expansion on the upper plane, keeps the liquid phase confined to the bottom plane. Therefore, the working fluid motion is completely absent until the vapour phase in the upper plane condenses: the associated pressure decrease in the upper plane leads to a vigorous boiling process in the bottom plane. The vapour expansion process is almost instantaneous and traduces in a water hammer associated with high amplitude pressure oscillations. The heat transport process in such a condition is inefficient and results in a higher average evaporator temperature.

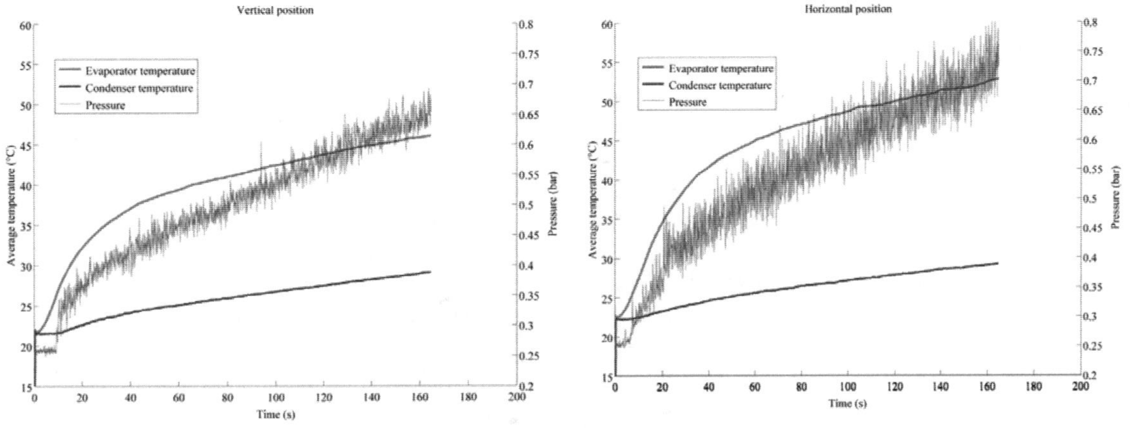

Figure 5. Large diameter CLPHP: a) vertical and b) horizontal ground tests.

3.2. *Flight tests*

The accelerations experienced during REXUS 18 sounding rocket flight are sketched in Figure 6. The red and blue lines represent the accelerations along the longitudinal axis of the rocket and on a plane perpendicular to it, respectively. The black vertical lines represent the timeline main events. The line that corresponds to the ordinate axis represents the rocket Lift Off (LO) and Start of Experiment (SoE). Then, proceeding to right, at LO +28 s, there occurs the Start of Data Storage (SoDS): from this time, power is provided to the heating elements and data are stored in the memory. Finally, power is switched off at LO +160 s and simultaneously the experiment ends.

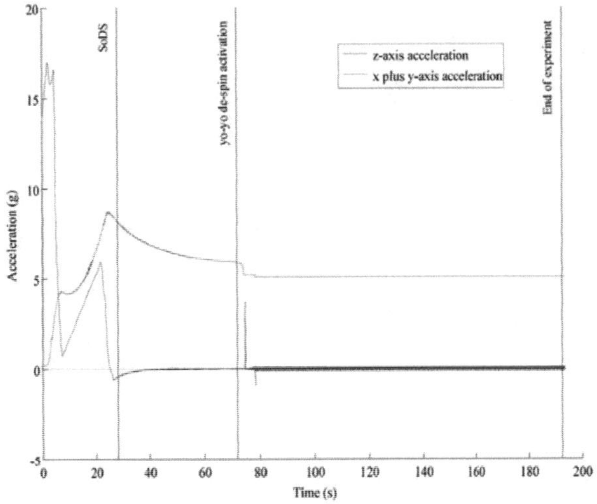

Figure 6. Accelerations.

The black vertical line at LO +70 s corresponds to the instant when the yo-yo system had to act and reduce the rotation of the rocket about its longitudinal axis. However, as can be clearly inferred from Figure 6, the yo-yo system failed and the presence of the residual centrifugal acceleration prevents both CLPHPs to work in the expected reduced gravity environment. For this reason, the flight results are placed in the context of the hyper-gravity experiments documented in literature. An image with a series of concentric circles representing the gravity levels that act on the devices depending on their radial distance from the longitudinal axis of the rocket is sketched for clarity in Figure 7.

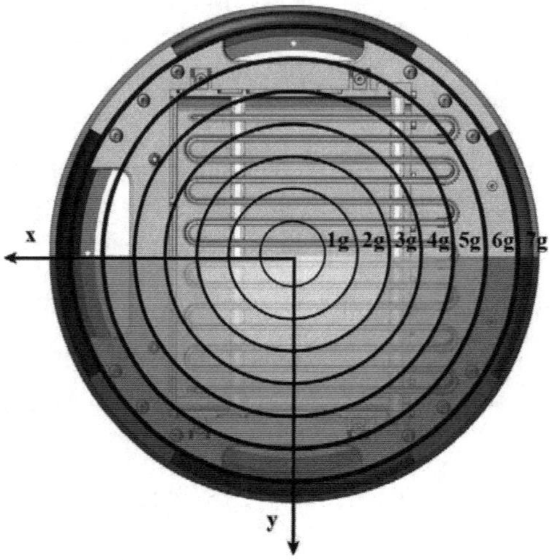

Figure 7. Gravity levels.

The flight test results relative to the small diameter CLPHP are reported in Figure 8. The centrifugal component of the acceleration promotes the working fluid motion and affects the device start-up. However, the acceleration is not oriented from the condenser to the evaporator as in the case with the device oriented in vertical position and it pushes the liquid phase from the central to the peripheral region of the device. The confinement of the liquid phase in the peripheral region of the device results in a detrimental effect on the performance with respect to the vertical ground test but not with respect to the horizontal ground test. The pressure temporal evolution is characterized by the occurrence of random pressure peaks associated with a decrease and increase of the average evaporator temperature, respectively. In fact, when the centrifugal acceleration separates phases, the working fluid motion stops and the heat transport process becomes ineffective. Then, it occurs a decrease of the acceleration level or an increase of pressure in order to allow the working fluid to redistribute properly: the standoff condition is overcome and the device restarts its operation.

The flight test results relative to the large diameter CLPHP are reported in Figure 9. The centrifugal component of the acceleration significantly affects the distribution of the working fluid within the device: the liquid phase is pushed towards the peripheral region while the vapour phase occupies the central region. As such a distribution is not optimal, the heat exchange process is not effective and the average evaporator temperature increases with respect to the vertical case. It is worth to note that, a few moments before the end of the test, a failure in one of the heating elements prevents heat power to be provided to the evaporator section and the average evaporator temperature and pressure decrease.

Journal of Physics: Conference Series **655** (2015) 012042 doi:10.1088/1742-6596/655/1/012042

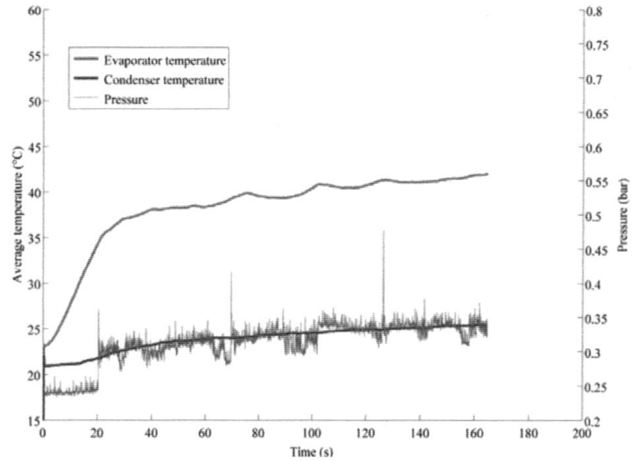

Figure 8. Small diameter CLPHP: flight test.

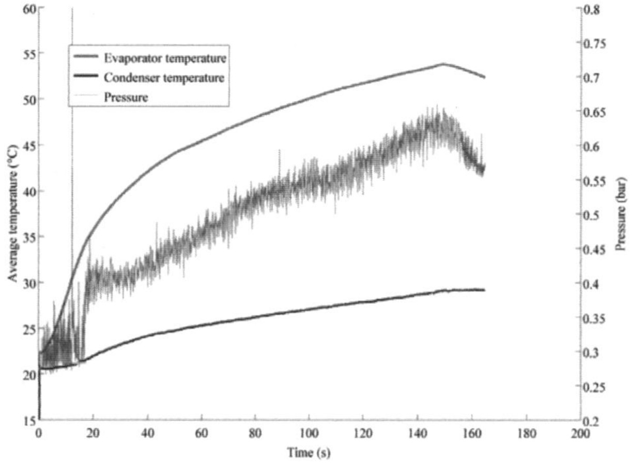

Figure 9. Large diameter CLPHP: flight test.

4. Conclusions

In the present work, the experimental results of two CLPHPs tested on board REXUS 18 sounding rocket are discussed. The devices consist of an aluminium tube folded in a double-layer configuration and filled with refrigerant FC-72. The only remarkable difference between the devices regards the inner tube diameter that is larger (3 mm) and slightly smaller (1.6 mm) than a critical diameter defined on static Earth gravity conditions. The condenser section is embedded in a phase change material, which absorbs heat via latent heat of fusion so as to keep the evaporator temperature to a nearly constant value. The evaporator section consists of 2 wire-shaped heating elements wrapped around the evaporator curves. Due to a malfunction of the rocket de-spin system, the device experienced an augmented gravity environment instead of the desired milli-gravity, so was not possible to observe the expected net transition in the temperatures and pressure temporal evolutions associated with the occurrence of the slug and plug flow pattern within the device. In the light of such considerations, the major outcomes of the present work are listed here below:

- The small diameter CLPHP confirm literature results: the hyper-gravity conditions have a beneficial effect on the overall thermal performance with respect to horizontal orientation as centrifugal accelerations promote the working fluid circulation within the device in every direction. However, the overall thermal performance in vertical position on ground is the best, as gravity acts always from the condenser to the evaporator;

- The large diameter CLPHP on ground is affected by the stratification of the working fluid within the device, as surface tension is not able to balance buoyancy. When in vertical position, the liquid and vapour phases resides in the evaporator and condenser section, respectively and the device behaves like a two-phase thermosyphon. When in horizontal position, the liquid and vapour phases resides in the bottom and upper device planes, respectively, and the device operates intermittently whenever a sufficient temperature gradient arises between the planes to push some liquid in the upper plane;
- The large diameter CLPHP in hyper-gravity conditions is more sensitive to the centrifugal acceleration: the liquid and vapour phases hold relatively in the peripheral and central parts of the device. The overall thermal performance is better than in horizontal position.

References

[1] Akachi H 1990 Structure of a heat pipe *US Pantent* 4.921.041
[2] Akachi H 1993 Structure of a micro heat pipe *US Patent* 5.219.020
[3] Zhang Y and Faghri A 2008 Advances and unsolved issues in pulsating heat pipes *Heat Transfer Eng.* **29** 20-44
[4] Gu J, Kawaji M and Futamata R 2004 Effects of gravity on the performance of pulsating heat pipes *J. Thermophys. Heat Transfer* **18** 371-378
[5] Mameli M, Araneo L, Filippeschi S, Marelli L, Testa R and Marengo M 2014 Thermal response of a closed loop pulsating heat pipe under a varying gravity force *Int. J. Therm. Sci.* **80** 11-22
[6] Mangini D, Mameli M, Georgoulas A, Araneo L, Filippeschi S and Marengo M 2015 A pulsating heat pipe for space applications: ground and microgravity experiments *Int. J. Therm. Sci.* **95** 53-63
[7] Mameli M, Marengo M and Khandekar S 2014 Local heat transfer measurement and thermo-fluid characterization of a pulsating heat pipe *Int. J. of Therm. Sci.* **75** 140-152
[8] Mameli M, Manno V, Filippeschi S and Marengo M 2015 Pulsating heat pipe in hypergravity conditions *Heat Pipe Sci. Tech.*(in press)
[9] Franco A and Filippeschi S 2012 Closed loop two-phase thermosyphon of small dimensions: a review of the experimental results *Microgravity Sci. Technol.* **24** 165–179

Effect of the application of an electric field on the performance of a two-phase loop device: preliminary results

F Creatini[1], P Di Marco[1], S Filippeschi[1], D Fioriti[1] and M Mameli[2]

[1] University of Pisa, Largo Lucio Lazzarino 2, 56122 Pisa, Italy
[2] University of Bergamo, Viale Guglielmo Marconi 5, 24044 Dalmine (BG), Italy

E-mail: francescocreatini89@gmail.com

Abstract. In the last decade, the continuous development of electronics has pointed out the need for a change in mind with regard to thermal management. In the present scenario, Pulsating Heat Pipes (PHPs) are novel promising two-phase passive heat transport devices that seem to meet all present and future thermal requirements. Nevertheless, PHPs governing phenomena are quite unique and not completely understood. In particular, single closed loop PHPs manifest several drawbacks, mostly related to the reduction of device thermal performance and reliability, i.e. the occurrence of multiple operational quasi-steady states. The present research work proposes the application of an electric field as a technique to promote the circulation of the working fluid in a preferential direction and stabilize the device operation. The tested single closed loop PHP is made of a copper tube with an inner tube diameter equal to 2.00 mm and filled with pure ethanol (60% filling ratio). The electric field is generated by a couple of wire-shaped electrodes powered with DC voltage up to 20 kV and laid parallel to the longitudinal axis of the glass tube constituting the adiabatic section. Although the electric field intensity in the working fluid region is weakened both by the polarization phenomenon of the working fluid and by the interposition of the glass tube, the experimental results highlight the influence of the electric field on the device thermal performance and encourage the continuation of the research in this direction.

1. Introduction

Pulsating Heat Pipes (PHPs) are novel promising two-phase passive heat transport devices potentially suitable for the thermal management of power electronics in future space applications. Patented in their most common configuration by Akachi [1-2], Pulsating Heat Pipes (PHPs) consist of a meandering tube of capillary dimensions closed end-to-end to form a closed loop. The loop is first evacuated and then partially filled with a working fluid, which naturally resides in the tube in the form of liquid slugs alternated to vapour plugs. When input heat power is provided to the evaporator section, the vapour bubble expand and push the adjacent liquid slugs to the condenser section, where heat is released to the cold source and condensation process takes place.

Although classified as a sub-class of the family of heat pipes, PHPs governing phenomena are quite unique and not completely understood. In spite of the significant efforts that have been made in last decades, a large number of issues still remain unsolved, such as their characteristic unstable behaviour at low and high heat fluxes [3]. The unstable behaviour seems to be affected by a large number parameters and becomes less and less evident as the number of turns increases. A symmetrically heated single closed loop PHP is therefore the most unstable device that belongs to the PHPs family. Khandekar et al. [4] were the first to address the problem of the existence of multiple operational pseudo-quasi-steady states in a single closed loop PHP. In particular, the experimental tests performed on a single closed loop PHP with an inner tube diameter equal to 2.00 mm and partially filled with pure ethanol (60% filling ratio) highlighted how the device operation evolves towards a thermodynamic equilibrium state characterized by the complete segregation of working fluid phases between the heat power input and output regions [4]. Therefore, the working fluid oscillations stop completely and the device thermal performance rapidly deteriorates. Theoretically, this dynamic equilibrium condition may be disturbed by an externally triggered perturbation or internally generated interface instability. If the applied disturbance is strong enough, it may break the menisci surface tension and cause the redistribution of phases. Then, the system could come back to a previous equilibrium state depending on the extent of spatial phase distribution homogeneity achieved through the mixing of the working fluid. In literature, a large number of techniques have been proposed in order to face the problem and stabilize the device operation.

The techniques range from an asymmetrical arranged diameter configuration [5-6] to the use of flow control check valves [7]. However, none of the techniques listed above result in an enhancement of the thermal performance to justify the major complexity added to the system.

The present research work aims at inquiring the effect of the application of an electric field on the thermal behaviour of a single closed loop PHP. In particular, the electric force that arises from the application of an electric field on a two-phase flow drives the vapour phase (of lower dielectric permittivity) towards a region of weaker electric field intensity and in some circumstances it may break the menisci surface tension and give the opportunity of separating phases [8-9]. The redistribution of phases may renew the level of internal pressure perturbations, which are regarded as the primary working fluid driving mechanism inside the two-phase loop.

2. Experimental set-up

2.1. Test-cell

The basic features of the single closed loop PHP are sketched in Figure 1. The evaporator and condenser are made with copper tubes in order to minimize the thermal resistance between the tubes and the heat source and heat sink while the adiabatic section is made with glass tubes in order to allow the visualization of the working fluid. The tubes have an internal diameter of 2.00 mm and have capillary dimensions for the selected working fluid, i.e. ethanol. The working fluid was selected in order to compare the obtained results with those presented in literature.

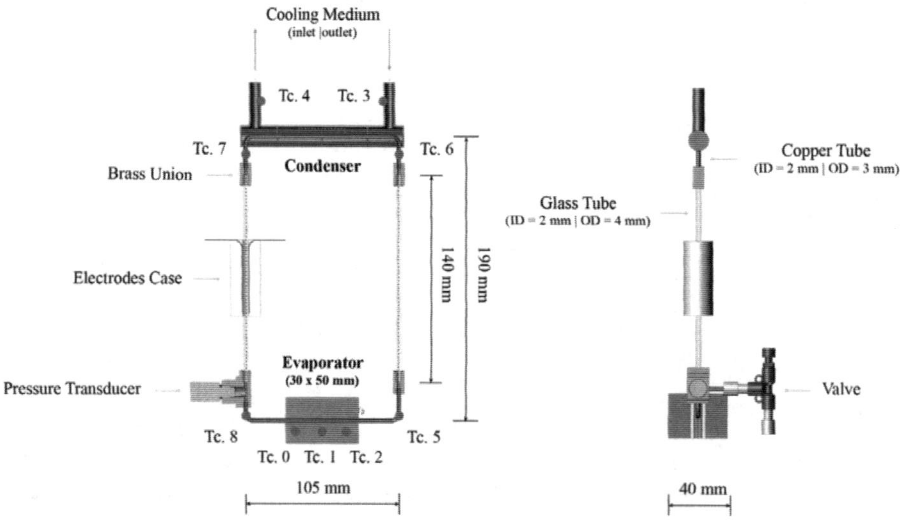

Figure 1. Single closed loop PHP test-cell.

The copper and glass tubes are connected through four brass unions: the copper tubes are brazed to the unions while the glass tubes are bonded with a high-temperature resistant epoxy adhesive (Loctite®, Hysol 9492). The brass T-junction on the left hand side of the evaporator allows to derive two ports: the former joins a copper tube used for the emptying and filling procedure; the latter hosts a pressure transducer (Kulite®, ETL/10T-312M-4BAR-A). The device is equipped with nine T-type thermocouples located as sketched in Figure 1. The thermocouples post-calibration accuracy is 0.1 K.

The single closed loop PHP was first evacuated by a two-stage ultra high vacuum pump (Edwards®, XDS35i and EXT255H) until a pressure level around 10^{-4} Pa was reached and then it was partially filled with a volumetric ratio of 0.5 ± 0.025. The selected working fluid was preventively degassed by continuous boiling: the incondensable gases were vented by several suction cycles as they accumulated in the free volume of the degassing tank. The residual incondensable gases content results in an overpressure above the saturation temperature of ethanol, which is lower than the margin of accuracy of the pressure transducer (0.5% full scale output).

Evaporator section

The evaporator section consists of two copper plates with semi-circular cross-section channels in order to embed the central portion of the copper tube. The copper plates and tube, which are maintained in thermal contact by a high thermal conductivity paste, constitute the evaporator copper block. Two silicone rubber heaters (Minco®, HR5364R125L12A) located at each side of the copper block provide the selected input heat power to the system. The copper block is thermally insulated by two Mica® fiber sheet and two Bakelite® back plates in order to reduce the heat loss to ambient air. The silicone rubber heaters were fed by a dc power supply (Agilent®, 6575A) with a power up to 35 W.

Condenser section

The condenser section is embedded inside a double-pipe heat exchanger, which consists of two concentric copper tubes of different diameter. The working fluid (red line in Figure 1) flows through the inner pipe while the cooling medium (blue line in Figure 1) flows through the annular space in between the tubes. In order to increase the heat exchange area as well as the cooling medium residence time inside the heat exchanger, three copper baffles are brazed to the inner copper pipe.

The condenser section was supplied by water coming from a constant temperature bath and circulator (Hake®, DC10-K20). The water temperature was set at 15 °C. The water mass flow rate was calculated to be equal to 6 g/s after considering the load losses along the circuit.

2.2. *Data acquisition*

The output signals were recorded by a data acquisition system (National Instrument®, cRio-9074) and real-time monitored by a suitable LabView® program. The temperatures and pressure signals were acquired at 10 and 20 Hz, respectively. The ambient temperature was measured by a four-wire resistance temperature detector and acquired at 1 Hz, as well as the heating power.

3. Electric field characteristics

The expression for the electric force per unit volume that acts on a generic medium is [8]:

$$\boldsymbol{f}_e''' = \rho_f \boldsymbol{E} + \frac{1}{2}\varepsilon_0 E^2 grad\varepsilon_r + \frac{1}{2}\epsilon_0 grad(bE^2) \tag{1}$$

where:

$$b = \rho(\frac{d\varepsilon_r}{d\rho})_T \tag{2}$$

The first term in Equation 1, also known as Coulomb force, depends on the sign of the electric field and generally predominates over the other force contributions when free charge is present in the medium. The other force contributions encompass dielectrophoresis and electrostriction: the former is associated with non-homogeneities of the dielectric permittivity of the medium, which in turn are coupled with thermal gradients or phase discontinuities; the latter is associated with non-uniformities of the electric field distribution. The electrical properties of the medium directly influence the formulation and different interfacial and volume force combinations can arise. The matter is widely structured and the interested reader can find additional details in the cited reference [8].

In a dielectric fluid, a non-uniform electric field is required so that a net force can arise at the liquid-vapour interface of a small inclusion: the force drives the phase of lower dielectric permittivity towards the region of higher electric field intensity. However, even the presence of a bubble of different dielectric permittivity than the surrounding medium may be sufficient to originate a gradient in an originally uniform electric field distribution.

3.1. Electrodes design

The electric field is generated by two wire-shaped electrodes laid parallel to the longitudinal axis of the left glass tube and fitted in a Teflon® case in order to prevent the dielectric breakdown phenomenon through the ambient air. The case, as sketched in Figure 2, consists of two symmetric sections, each having a semi-circular cross-section hole in order to embed the glass tube and a longitudinal groove in order to maintain the wire in the correct relative position.

The electrodes were supplied by a power supply (Spellmann®, RHR30P30/220) with constant DC voltage levels up to 20 kV. The voltage was acquired at 1 Hz.

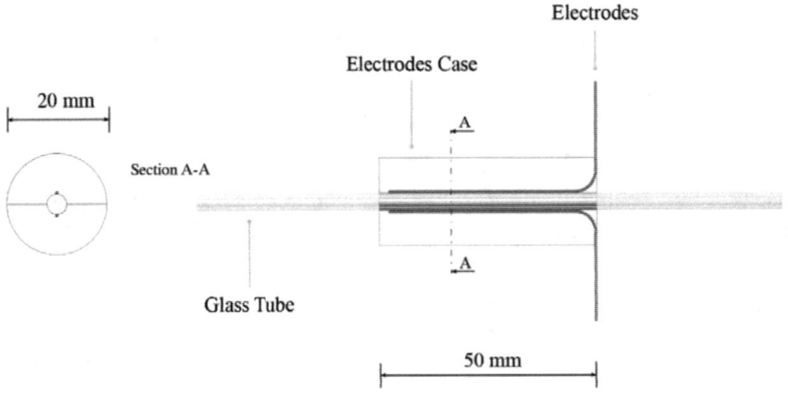

Figure 2. Electrodes case geometrical features.

The optimum electrodes geometry was investigated by performing a two-dimensional electrostatic simulation with the aid of COMSOL Multiphysics® software. The simulation was carried out by ranging the electrodes diameter from 0.25 mm to 1.00 mm with steps of 0.25 mm. The electrodes were connected to ground and to a terminal fed at 10 kV, respectively. The glass tube was considered as completely filled by ethanol. The dielectric permittivity of ethanol is 25 while the dielectric permittivity of the glass tube is 4 when evaluated at ambient temperature.

The electric field trend relative to the 1 mm diameter configuration is sketched in Figure 3. The electric field intensity is largest close to the point of contact between the electrodes and the glass tube while is smallest in the working fluid region. The greater polarization of the working fluid as compared with the glass tube significantly weakens the electric field intensity in the working fluid region regardless of the electrodes diameter. The electric field space distribution with varying electrodes diameter is sketched in Figure 4, where the x and y-axis represent to the line joining the electrodes and the average value of the electric field, respectively. According to Equation 1, the higher the electric field within the working fluid region the higher the electric force per unit volume. Therefore, the 1 mm diameter electrodes configuration was selected.

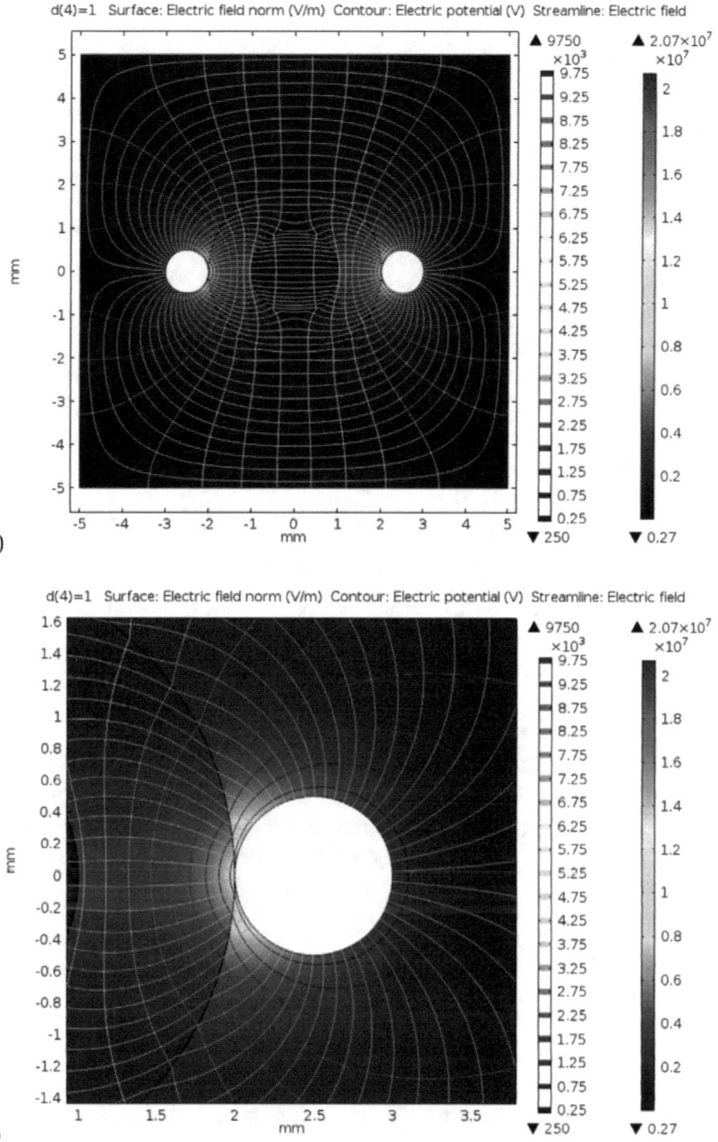

Figure 3. Electric field intensity: a) overall and b) magnified view.

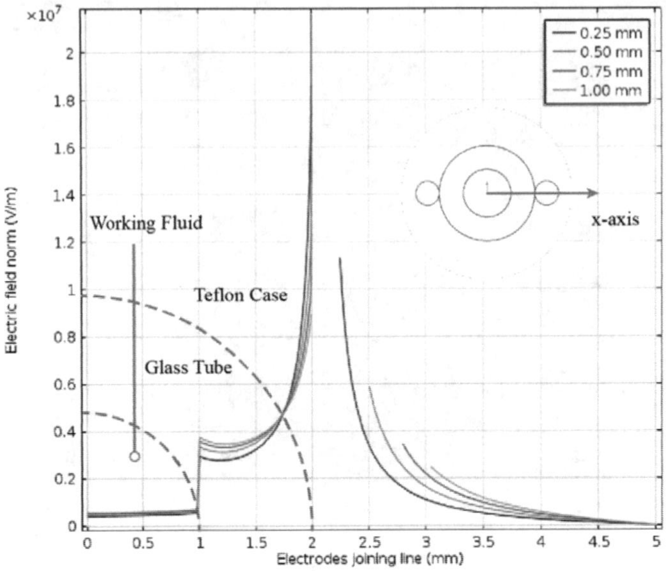

Figure 4. Electric field space distribution along the line joining the center of the electrodes with varying electrodes diameter.

4. Experimental results

The experimental results are presented in terms of temperatures and pressure temporal trend. The warm colors lines represent the temporal evolution of temperatures in the evaporator section while the cool colors lines represent the temporal evolution of temperatures in the condenser section.

The single closed loop PHP thermal performance is evaluated in terms of the equivalent thermal resistance, i.e. $R_{th} = (\bar{T}_e - \bar{T}_c)/\dot{Q}$, where $\bar{T}_e$ and $\bar{T}_c$ are the average evaporator and condenser temperatures in a pseudo-steady-state*, respectively, and $\dot{Q}$ is the input heat power.

4.1. Thermal characterization

A general characterization of the thermal performance of the single closed loop PHP was performed by gradually increasing the input heat power up to 35 W with consecutive steps of 5 W. Before moving to the next step, the input heat power was kept constant for thirty minutes in order to allow the system to reach a new operational pseudo-steady-state. The equivalent thermal resistance associated with each step was evaluated by averaging the evaporator and condenser temperatures. The same procedure was repeated until the evaporator temperatures started to increase abruptly, i.e. until dry-out happened in the evaporator section and the performance limit of the device was reached.

The experimental results without electric field are reported in Figure 5. It is worth to note that three distinct operational regimes can be recognized relative to the input heat power supplied to the evaporator section as already noticed by Khandekar et al [10] and Mameli et al [11].

When the input heat power is set at 5 W, after an initial start-up phase, the single closed loop PHP operation is characterized by alternating periods of working fluid motion and stop-over. The occurrence of such a regime, also termed start and stop regime, relies on the fact that the input heat power is not sufficient to give rise to a stable circulation. The working fluid motion starts after a sufficient temperature gradient is set-up between the evaporator and condenser. However, the onset of oscillations leads to the mixing of hot and cold working fluid portions, which lowers again the temperature gradient and stops the working fluid motion. As the working fluid motion is continuing, the equivalent thermal resistance tends to its highest value, i.e. 8.2 K/W.

* A pseudo-steady state is reached when all temperatures output signals shows an average value constant in time.

When the input heat power is set at 10 W, the single closed loop PHP operation is characterized by successive flow reversals: the vapour pressure is strong enough to establish a net circulation of the working fluid but it is not able to keep it stable all time. As there are no stop periods, the heat exchange is enhanced and the equivalent thermal resistance decreases to 5,5 K/W.

When the input heat power is set at 15 W, the single closed loop PHP operation is characterized by the circulation of the working fluid in the anticlockwise direction. As the circulation is stable, the equivalent thermal resistance moves towards its lowest value, i.e. 2.8 K/W. The occurrence of circulation, in addition to be clearly visible through the transparent section, can be also inferred by observing the temporal evolution of the temperatures measure by the thermocouples: the temperatures in the up-header branch of the device and close to the evaporator (Tc. 5) and in the down-header branch of the device and close to the condenser (Tc. 7) exhibit the highest and lowest values, respectively, as the circulation of the working fluid occurs in a specific direction.

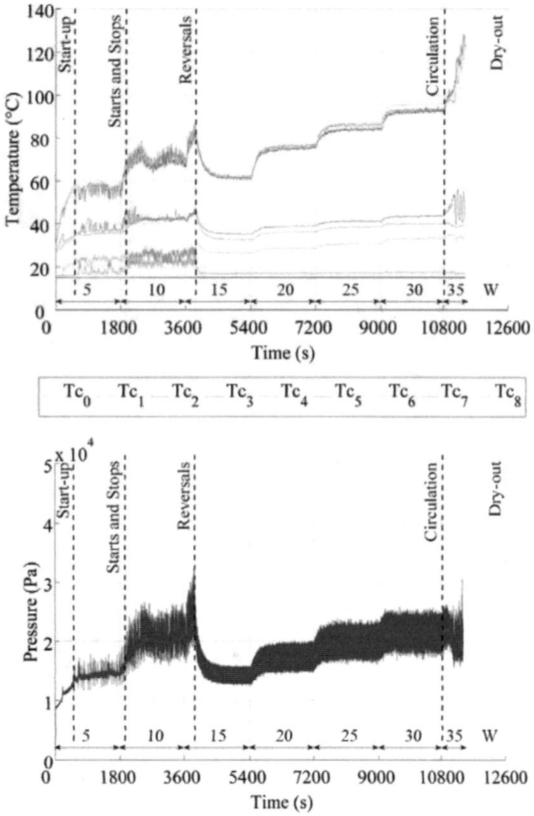

Figure 5. Thermal characterization: temperatures and pressure temporal trend.

4.2. Effect of the electric field

The objective of the present work is to understand if the application of an electric field is able to stabilize the device operation. Then, the attention was focused on the flow regimes that are characterized by an unstable behaviour, i.e. the start and stop and flow reversal regimes. The effect of the application of an electric field was investigated by carrying out long-run experimental tests at a fixed input heat power. The electrodes were powered by increasing the voltage up to 20 kV, with steps of 2.5 kV. Before moving to the next step, the voltage was keep constant for 15 minutes in order to allow the system to reach a new hypothetical operational pseudo-steady state.

The experimental results in terms of temperatures and pressure temporal trend are reported in Figure 6 and summarized in terms of equivalent thermal resistance in Figure 7. The black line represents the equivalent thermal resistance of the device in the absence of an electric field while the colored points correspond to the equivalent thermal resistance of the device in the presence of an electric field, where each color is associated with a specific value of the voltage supply.

It is worth to note that the effect of the application of an electric field is strongly reliant on the input heat power provided to the evaporator section and, accordingly, on the related flow regime. On one hand, for input heat power values associated with an unstable behaviour of the device, the effect of the application of the electric field traduces in an increase of the equivalent thermal resistance, regardless of the voltage supply to the electrodes. On the other, for input heat power values associated with a stable behaviour of the device, the effect of the application of the electric field traduces in a decrease of the equivalent thermal resistance at a high voltage supply. Although the equivalent thermal resistance changes may seem small, they result in a sensible variation of the temperature difference between the evaporator and condenser.

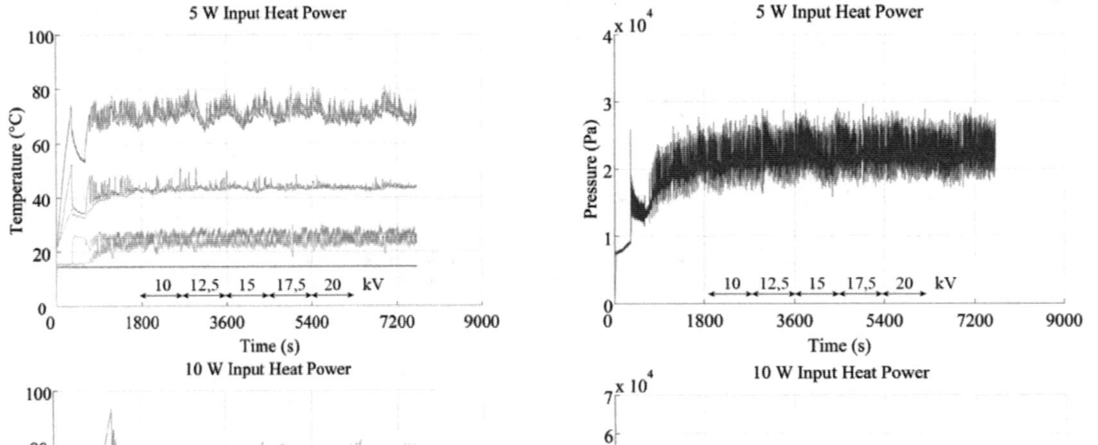

Figure 6. Effect of the electric field: temperatures and pressure temporal evolution.

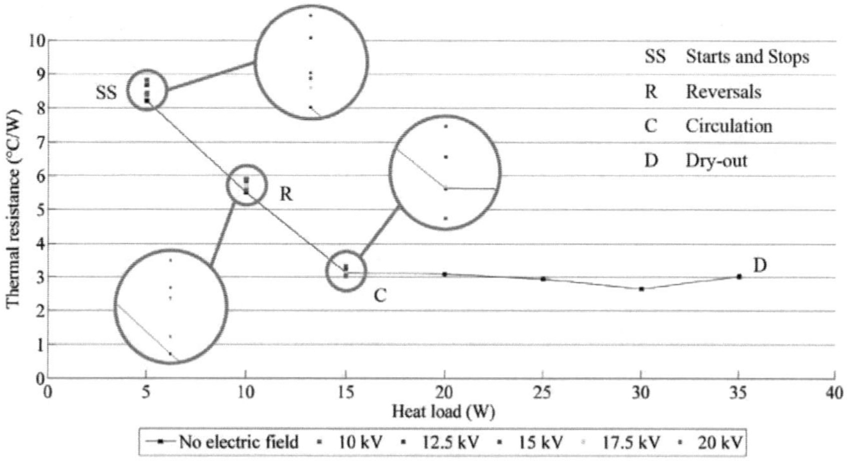

Figure 7. Effect of the electric field: equivalent thermal resistance.

5. Conclusions

In the present work, the behaviour of a single closed loop PHP tested without and with electric field is investigated. The device is made with an inner tube diameter of 2 mm and is partially filled with ethanol in order to provide comparison with the results already discussed in literature.

The results obtained without electric field clearly confirmed that the device operation is characterized by the occurrence of different flow regimes depending on the input heat power. On one hand, a low input heat power was not able to continuously sustain the working fluid motion thus resulting in unstable phenomena such as the "start and stop" fluid oscillations and the continuous flow reversals in the case of circulation. On the other, a high input heat power resulted in a stable circulation of the working fluid in a preferential direction.

The application of an electric field with the actual electrodes geometry and working fluid was not able to stabilize the device operation: the greater polarization of the working fluid than the surrounding glass tube significantly weakened the electric field intensity in the working fluid region and, accordingly, the resultant disturbance is not able to influence the working fluid motion. Nevertheless, the application of an electric field had a different effect on the equivalent thermal performance of the device depending on the input heat power and, therefore, on the related flow regime: for low input heat power, the application of an electric field worsened the thermal performance; for high input heat power, the application of an electric field is mainly traduced in an enhancement of the thermal performance at a high voltage supply (20 kV).

5.1. Future developments

The electric field intensity in the working fluid region could be enhanced by the use either of a couple of electrodes in direct contact with the working fluid or of a working fluid with a lower dielectric permittivity as compared with the glass tube. A working fluid that could be exploited to enhance the electric field intensity in the working fluid region is the refrigerant FC-72: having a dielectric permittivity of 1.6 at ambient temperature it polarizes less than the glass tube, thus resulting in a higher electric field intensity in the working fluid region.

References

[1] Akachi H 1990 Structure of a heat pipe *US Pantent* 4.921.041
[2] Akachi H 1993 Structure of a micro heat pipe *US Patent* 5.219.020
[3] Mameli M, Manno V, Filippeschi S and Marengo M 2014 Thermal instability of a closed loop pulsating heat pipe: combined effect of orientation and filling ratio *Exp. Therm. Fluid Sci.* **59** 222-229

[4] Khandekar S, Gautam A P and Sharma P K 2009 Multiple quasi-steady states in a closed loop pulsating heat pipe *Int. J. Therm. Sci.* **43** 13-20

[5] Holley B and Faghri A 2005 Analysis of pulsating heat pipe with capillary wick and varying channel diameter *Int. J. Heat Mass Transfer* **48** 2635-2651

[6] Chiang C M, Chien K H, Chen H M and Wang C C 2012 Theoretical study of oscillatory phenomena in a horizontal heat pipe with asymmetrical arrayed mini channel *Int. Commun. Heat Mass Reansfer* **39** 923-930

[7] Rittidech S, Pipatpaiboon N and Terdtoon P 2007 Heat transfer of a closed loop oscillating heat pipe with check valves *Appl. Energy* **84** 565-577

[8] Landau L D and Lifsitz E M 1960 *Electrohydrodynamics of Continuous Media* (*Course of Theoretical Physics* vol 8) (New York: Pergamon Press)

[9] Di Marco P 2012 The use of electric force as a replacement of buoyancy in two-phase flow *Microgravity Sci. Tec.* **24** 215-28

[10] Khandekar S and Groll M 2004 An insight into thermo-hydrodynamic coupling in closed loop pulsating heat pipes *Int. J. Therm. Sci.* **43** 13-20

[11] Mameli M, Marengo M and Khandekar S 2014 Local heat transfer measurements and thermo-fluid characterization of a pulsating heat pipe *J. Therm. Sci.* **75** 120-152

Journal of Physics: Conference Series **655** (2015) 012044 doi:10.1088/1742-6596/655/1/012044

Electric field effects on the dynamics of bubble detachment from an inclined surface

P Di Marco[1], N Morganti, G Saccone
DESTEC, University of Pisa, Largo Lucio Lazzarino 1, 56122, Pisa, Italy

Abstract. An experimental apparatus to study bubble detachment from an inclined surface under the action of electric forces is described. It consists of a container filled with FC72 at room temperature and pressure where a train of gas bubbles is injected from an orifice. An electrostatic field can be imposed around the bubble, while the cell can be tilted from 0 to 90°. It is possible to study interface growth with the aid of high-speed cinematography. Since the interface is asymmetrical, a mirror system allowed to acquire, in the same frame, two images at 90° of the bubble. Different inclinations, injection rates and voltages were tested in order to couple the effects of shear gravity and electric field. Curvature and contact angles have been derived with appropriate interpolation methods of the profile. Force balances on the bubble were checked, finding an electric force, which, at first pulls the bubbles from the orifice, then pushes it against the surface. The motion of the center of gravity confirms this behaviour. A power balance has been developed to determine the energy contributions, revealing that surface growth incorporates both the effects of inlet power and electric field.

1. Introduction

Bubble detachment under gravity force of different intensity and direction is encountered in many industrial applications, from electrical power plants based on Rankine - Hirn cycles to the two-phase thermal management for space purposes. In particular, the interest is to predict bubble departure diameter and frequency from a heated wall in order to achieve an increasingly reliable comprehension of two-phase heat transfer. In space applications, the need to reduce size and weight of all thermal management systems (power generation, electronics cooling, cabin temperature control, waste management, and regenerative fuel cells) suggests to replace current single-phase cooling systems with two-phase ones. In fact, this technique can lead to order-of-magnitude enhancement in heat transfer coefficients, giving the possibility to reduce sensibly heat exchangers surface, size and weight 0. Heat transfer performance can be further improved with the application of an electric field (EF); it is particularly in space applications that this technique shows its relevance. It has been assessed also in terrestrial gravity environment that the application of EF enhances boiling heat transfer coefficient [2], [3], [4], [5], [6]; but the most interesting aspect is the possibility to act as a replacement of gravity where it lacks [7], [8].
A relevant part of the current activities in boiling heat transfer research focuses on the influence of a shear flow on bubble detachment from a heated surface. In these conditions, the bubble is subjected to a shear force, deforming the interface from its axisymmetric shape. Studying force balances, different models of bubble detachment in shear flow have been developed [9], [10], [11]. In general, in these

[1] Corresponding author: p.dimarco@ing.unipi.it

models simplifying approximations of the relevant parameters have been adopted, like a spherical or hemispherical bubble shape and a constant contact angle. The more complicated components of force balances are surface tension, drag and lift forces.

A situation similar to shear flow can be reproduced by tilting the detachment surface, in order to have a force component parallel to the surface, due to the buoyancy acting on the bubble. In these conditions, by injecting a controlled amount of gas from an orifice in the liquid, a static or slowly growing interface can be created, which can be more easily characterized and studied in controlled conditions. The results will help in clarifying the role of a transverse force on bubble detachment, and can be compared with existing models in order to check their validity. The experiment constitutes also a ground-based support to RUBI, an experiment developed by an ESA scientific team aimed to studying bubble detachment in microgravity under the action of different forces, which will be installed in the International Space Station in the near future. The contribution of the present work in RUBI experiment is the investigation of the coupled effects of transverse force and electric field on bubble departure. Once the shear effects modify the shape of the bubble, the electric field distribution is, in turn, altered and the way it acts over the bubble can promote or prevent bubble departure and successive trajectory. The present experiment is intended to shed light over the effect of electric field when the interface is modified by a shear force.

2. Experimental setup

The experimental setup used in the present work is the same one, which performed two parabolic flight campaigns carried out by ESA in 2013 and 2014. Its core (see Figure 1 and Figure 2) consists in a bubble generator, comprising a flat stainless steel plate with a circular orifice of 0.55 mm internal diameter drilled on it. The orifice was fed with nitrogen by means of a gas flowrate controller (Bronkhorst F-200, with a maximum flowrate of 100 mm^3/s). In this way, continuous trains of detaching bubbles, or individual bubbles growing very slowly can be generated. In the present experiments, the gas flowrate was limited to a maximum of 7 mm^3/s, in order to remain within the range of the so-called quasi-static detachment regime, where the detachment volume is independent of the gas flowrate, defined by the adimensional Tsuge parameter [12]:

$$\frac{\rho_l u_0 D_{or}^{1.5} g^{0.5}}{\sigma} = N_w < 1 \quad (1)$$

where ρ_l is the liquid density, u_0 is the gas velocity at the orifice and D_{or} the orifice diameter. The test cell (Figure 2) is an aluminum box (volume: about 0.7 L) filled with the test liquid (FC-72). The internal configuration of the cell makes it possible to orient the bubble generator both upwards and downwards. The cell is provided with four Lexan windows on its sides to allow high-speed image recording and illumination. The injected gas is collected in the upper cone of the test cell and, when the collected volume interferes with experiment operation, is vented through an electrovalve. To provide elasticity to the system, the cell is connected to an expansion volume (about 0.05 L) provided with an elastic membrane facing a gas volume where the pressure is controlled by venting and injection valves. The cell was operated at a pressure close to the atmospheric, approximately 100 kPa. The set of measurements is completed by a tri-axial accelerometer, a K-thermocouple for fluid temperature and an absolute pressure transducer.

An electric potential up to 25 kV dc could be applied to a washer-shaped electrode (10x4 mm, 1 mm thickness), located 6 mm above the plate and centered with the orifice axis. Images of the growing and detaching bubble were taken by means of a high-speed camera (Redlake MotionPro Y3) equipped with a microscopic lens allowing for a spatial resolution of 115 pixel/mm. A dedicated mirror system has been developed in order to acquire in the same frame two contemporary images of the interface from different points of view at 90° (see Figure 3a) in order to detect the asymmetries in bubble growth due to the component of gravity parallel to the surface. The acquisition of images in the same

Journal of Physics: Conference Series **655** (2015) 012044 doi:10.1088/1742-6596/655/1/012044

frame avoided the necessity to synchronize frames acquired by different cameras, and made the data processing simpler. Various inclinations have been achieved by tilting the whole experimental rack and checking with the accelerometer when the inclination reached the desired value (see Figure 3b).

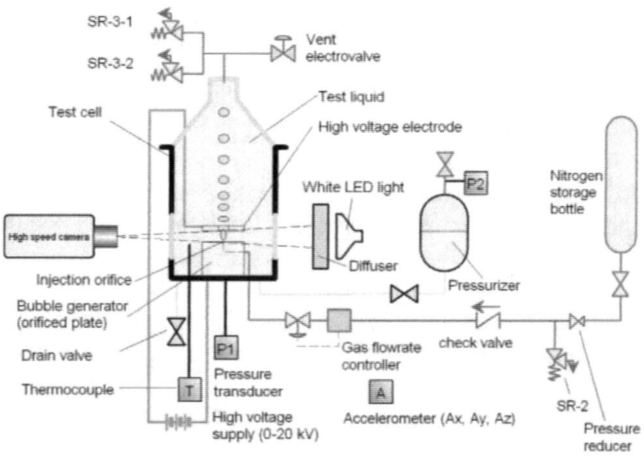

Figure 1: Sketch of the experimental apparatus.

Figure 2: Photo of secondary containment assembled on the rack (left) and detail of the test cell (right) showing the orifice plate and the electrode inside the cell.

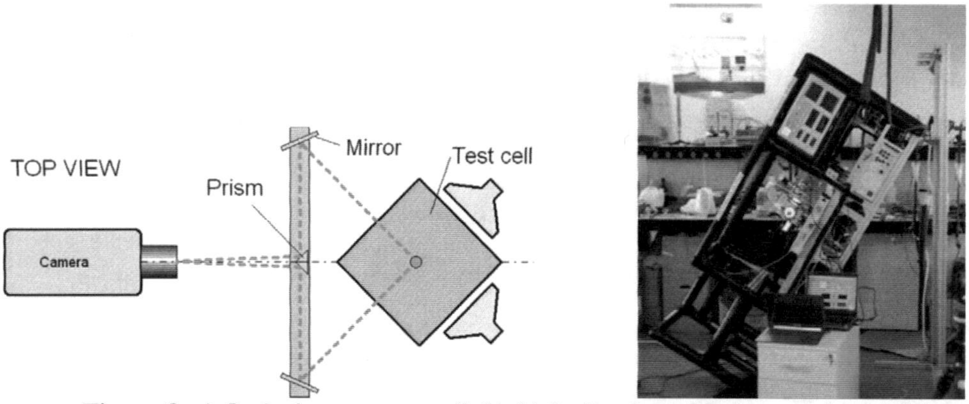

Figure 3: a) Optical arrangement (left). b) Inclination of the test section (right)

3. Force balances

The momentum balance in a control volume surrounding the bubble on the liquid side and the gas side of the base A reads (see Figure 4):

$$\frac{\mathrm{d}}{\mathrm{d}t}\int_V \rho_g \mathbf{v}\,\mathrm{d}V = \int_V \rho_g \mathbf{g}\,\mathrm{d}V + \int_S \left(\bar{\bar{\mathbf{T}}}_{v,f} - p_f \bar{\bar{\mathbf{I}}} + \rho \mathbf{v}_f(\mathbf{v}_f - \mathbf{v}_s) + \bar{\bar{\mathbf{T}}}_{e,f}\right)\cdot \mathbf{n}\,\mathrm{d}S + \\ + \int_A \left(\bar{\bar{\mathbf{T}}}_{v,g} - p_g \bar{\bar{\mathbf{I}}} + \rho \mathbf{v}_g(\mathbf{v}_g - \mathbf{v}_s) + \bar{\bar{\mathbf{T}}}_{e,g}\right)\cdot \mathbf{n}\,\mathrm{d}S + \int_{CL} \sigma_{fg} \mathbf{t}_{fg}\,\mathrm{d}L = 0 \qquad (2)$$

Where, in particular, n and $\underline{t}$ are the normal outward and tangent (in the meridian plane) unit vectors, respectively, p is the pressure, v is the velocity, g is the gravity acceleration, $\bar{\bar{\mathbf{T}}}_e$ is the Maxwell stress tensor, $\bar{\bar{\mathbf{T}}}_v$ is the viscous stress tensor, σ is the surface tension (directed along t, tangent to bubble surface at contact line). The suffixes g, f are referred to gas and liquid sides, respectively. The inertial term in LHS is generally negligible, as only the inertia of the gas is must be taken into account here. Dynamical effects (including the surrounding liquid inertia) are accounted for in the normal and tangential stress distribution around the bubble, which must be evaluated taking into consideration the surrounding fluid motion.

Neglecting dynamical actions and interfacial mass transfer, integration of momentum equation gives a balance of buoyancy and excess of internal pressure, surface tension at the contact line and electric force:

$$\frac{\pi D_{or}^2}{4}\left[2\sigma\left(\frac{1}{R_1}+\frac{1}{R_2}\right)-\rho_f g L\right]\mathbf{j} - V_b(\rho_f-\rho_g)\,\mathbf{g} = \oint_{CL} \sigma \mathbf{t}_{fg}\,\mathrm{d}s + \int_S \bar{\bar{\mathbf{T}}}_{e,f}\cdot \mathbf{n}\,\mathrm{d}S \qquad (3)$$

Where R_1 and R_2 are the curvatures at bubble top, and the local value of contact angle θ has to be considered in the first term of RHS. Projecting the components along local coordinates x (parallel to the surface) and y (perpendicular to the surface), the expressions of vertical and tangential forces are obtained, as reported in Table 1. In particular, the surface tension term has been calculated assuming a linear trend (from β_{max} to β_{min}) of the contact angle versus the polar coordinate along the circumference of the orifice (see Figure 4b).

The force balance without electric field is verified first against the experimental data. Once it has been assessed, the electric contribution is calculated by algebraic sum of the other forces.

Table 1. Force components

	X component (tangential)	Y component (perpendicular)
Buoyancy	$V(\rho_f-\rho_g)g\sin\vartheta$	$V(\rho_f-\rho_g)g\cos\vartheta$
Pressure	none	$\frac{\pi D_{or}{}^2}{4}\left[2\sigma\left(\frac{1}{R_1}+\frac{1}{R_2}\right)-\rho_f g L\right]$
Surface tension	$\pi D_{or}\frac{\beta_{max}-\beta_{min}}{\pi^2-(\beta_{max}-\beta_{min})^2}(\sin\beta_{max}-\sin\beta_{min})$	$\pi D_{or}\frac{\cos\beta_{max}-\cos\beta_{min}}{\beta_{max}-\beta_{min}}$

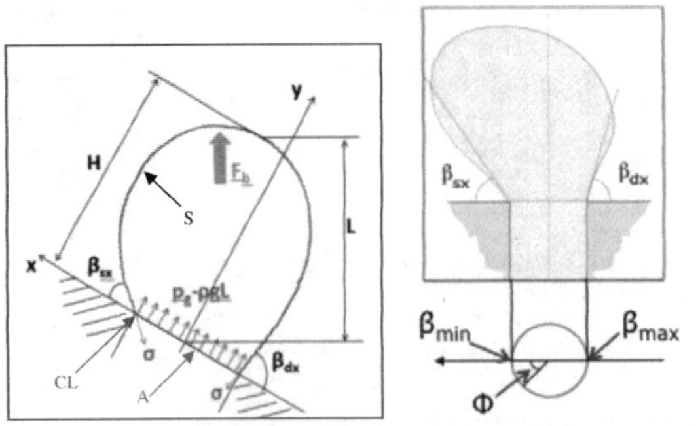

Figure 4: a) Control volume for force calculation. b) Contact angle variation vs. the polar coordinate.

4. Power balance

A last balance will be considered for completeness, i.e. the overall energy balance. It has been presented and ([13]) validated ([14] and [15]) for spreading droplets and has been adapted to study bubble evolution. The phenomena and their contribution to the change in the energy of the bubble are listed below.

- Inlet contribution; this term is mainly due to pressure developed by the injection system, the inlet velocity and by the pressure inside the bubble. It will be labelled as P_{inlet}.

$$P_{inlet} = \frac{dm}{dt}\left[\frac{p_{base} - p_{atm}}{\rho} + \frac{v_{inlet}^2}{2}\right] \tag{4}$$

- Surface potential energy; this term is very effective, it is due to the surface tension energy of the shape and depends on the interface area. It will be labelled as P_{surf}.

$$P_{surf} = \sigma \frac{dA_{fg}}{dt} \tag{5}$$

- Gravitational potential energy; this term is due to change of height of the center of gravity. It will be labelled as P_{grav}.

$$P_{grav} = g\left[m\frac{dz}{dt} + z\frac{dm}{dt}\right] \tag{6}$$

Finally, the overall energy balance in the absence of EF reads as follows:

$$P_{inlet} = P_{surf} + P_{grav} \tag{7}$$

5. Image processing and parameters calculation

The interface shape has been digitized with a dedicated image processing software, based on Matlab. The profile is identified from the grey-scale image by Canny method, with an imposed threshold. The geometrical parameters like curvature and apparent contact angles have been derived with appropriate interpolation methods of the profile [16]. In particular, for what concerns the curvature at the drop top, the 5 highest points are considered (see Figure 5a) and a 2nd order polynomial fitting is performed.

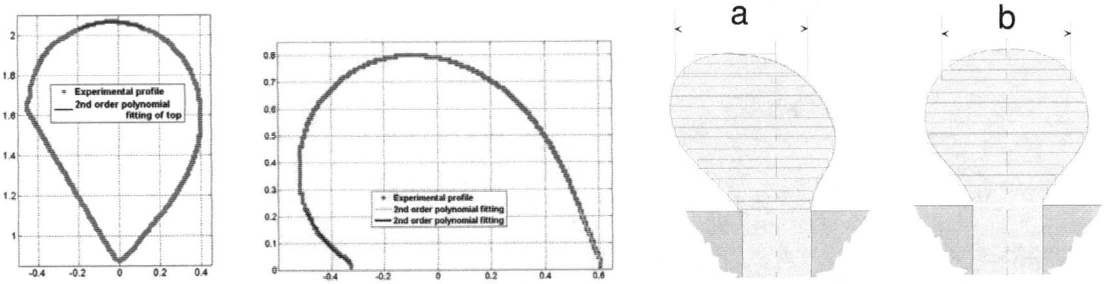

Figure 5: a) Curvature at top calculation. b) Contact angles calculation. c) Volume calculation as a stack of ellipses 1 pixel height; *a* is the major, *b* the minor axis.

When the electric field is not applied, the curvature at bubble top was calculated fitting the highest (in the direction of gravity) points. When the electric field was applied, bubble top turned to be the farthest point in the direction perpendicular to the surface. This is because electric field creates a

pressure contribution inside the droplet, which does not follow the gravity field, but the gradient of the electric potential [8], [17].
The polynomial fitting of contact angles was performed choosing the 25 lowest pixels (see Figure 5b). The bubble volume is determined considering the bubble as a stack of superposed ellipsoidal disks, of 1 pixel height, whose axes are the bubble widths in the two images (see Figure 5c) Once the geometrical parameters of the bubble are known, it was also possible determine the forces acting on it and to check the overall force balance on the inclined bubble.

6. Experimental uncertainty

The two most relevant variables affecting errors in the image processing technique described above are: the pixel size and the threshold chosen to identify the contour of the bubble. Pixel size has been determined *a priori* from the image recording of an object of sharp edges and known size. Error in the estimation of pixels and size of the objects are considered around 5% with a reference value of 115 pixel/mm. In the uncertainty evaluation, a value of 8% has been conservatively adopted in order to check the error induced by a difference of 10 pixel/mm. The threshold is a user-determined dimensionless parameter, discriminating black from white in a grey scale. It is necessary to detect the profile of an object in a grey scale image. The sharper is the contour of the image, the lower is the error induced by its choice. In the present study a nominal value of 0.8 has been chosen and a variation of ±0.05 has been considered. Since these sources of uncertainty are independent and the error distributions can be assumed Gaussian, the total uncertainty for the quantity φ is calculated as:

$$\sigma_\varphi = \sqrt{\left(\frac{\partial\varphi}{\partial ps}\Delta ps\right)^2 + \left(\frac{\partial\varphi}{\partial l}\Delta l\right)^2} \tag{8}$$

Uncertainty in the determination of linear parameters (i.e. bubble height) has been evaluated to be up to 8%, while volume uncertainty can rise to more than 20% for small bubbles. It is worth noting that the larger part of uncertainty is induced by the one in pixel size, and only a small amount is due to the threshold choice. One of the most relevant parameters in the present study is contact angle, whose uncertainty depends in a complex way by the above mentioned parameters. Therefore it has been evaluated with computer simulation, and its variation versus scaling factor and threshold is shown in Figure 6.

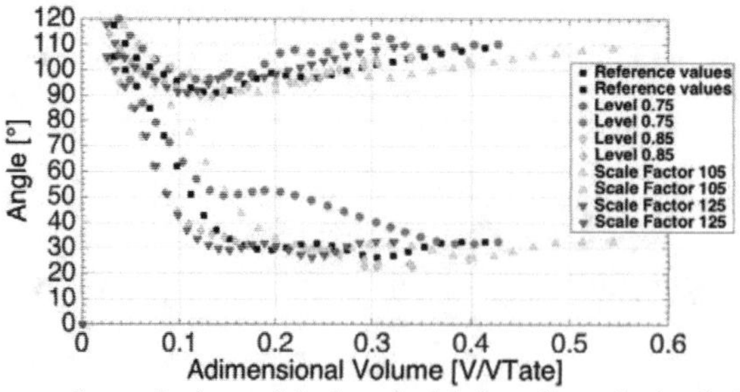

Figure 6: Contact angle variation with threshold (level) and pixel size (scale factor)

7. Results

Once the geometrical parameters of the bubble were known, it was possible to determine the forces acting on it and to check, the overall force balance on the inclined bubble. The overall force balance in vertical direction (inclinations 0 and 180°) has been already assessed and validated [18]. In the present work, different inclinations have been tested, with and without the electric contribution. For space limitations, the overall force balances in the tangential and perpendicular direction with respect to the surface of 75° and 90° are shown comparing the behaviour of the two opposed case: 0kV and 25 kV. As already mentioned, once the force balance was verified in the absence of electric field, the electric force has been calculated as the algebraic sum of the other forces. The abscissa in following figures reports the bubble volume made non-dimensional by dividing by the so-called Tate's volume:

$$V_{Tate} = \frac{\pi D_{or} \sigma}{(\rho_f - \rho_g) g} \tag{9}$$

For what concerns the x-components, as remarked in Table 1, pressure force is absent, so the equilibrium is achieved by the x-components, of gravity and capillary (surface tension) forces. In the absence of electric field (see Figure 7) the resulting unbalance resultant lies below 2 μN, not far from the absolute uncertainty already obtained in previous experiments (1 μN in [18]), but since forces are much smaller, the relative error can grow up to 35% (worst case calculated).

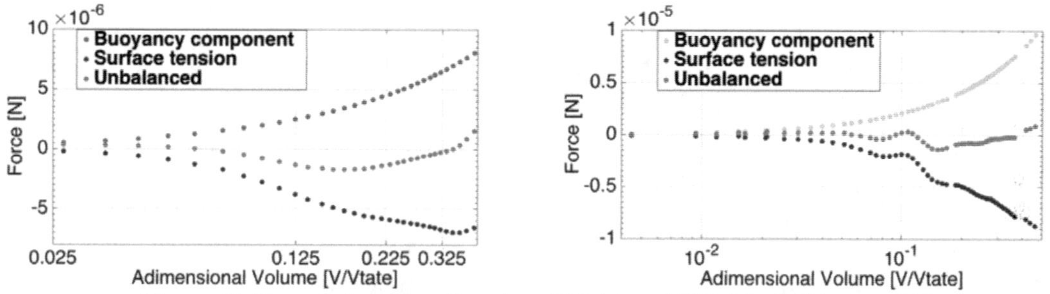

Figure 7: Force balance, x direction, 75° (a) and 90° (b) of tilting, no EF.

After the application of electric field, a force opposing buoyancy can be noted (Figure 8). It seems that the electric field is pulling the bubble downwards, preventing asymmetries in bubble growth. The same behaviour has been assessed also for droplets immersed in an external electric field [13].

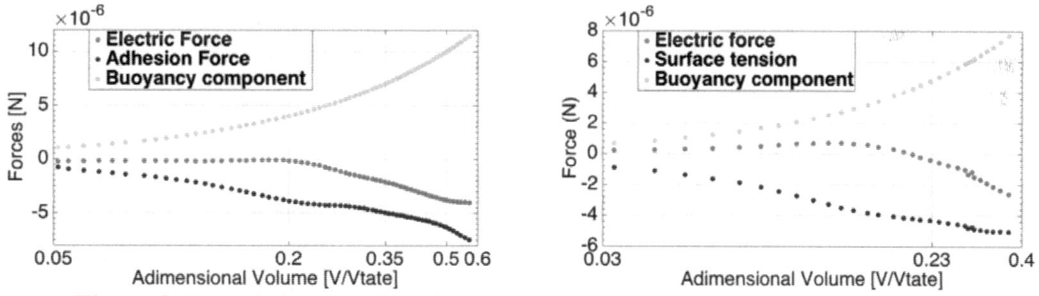

Figure 8: Force balance, x direction, 75° (a) and 90° (b) of tilting, with EF (25 kV).

Concerning the vertical direction, the case without electric field (Figure 9) reveals a lower error with respect to the balance in tangential direction, around 30% in the worst case.

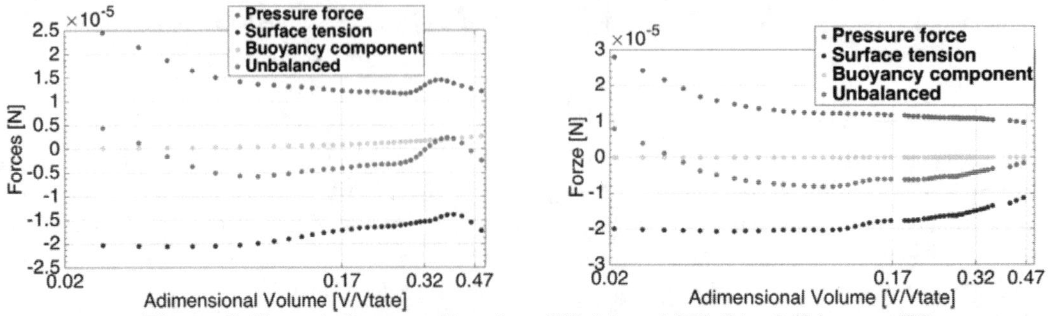

Figure 9: Force balance, *y* direction, 75° (a) and 90° (b) of tilting, no EF.

The electric contribution (Figure 10) results to pull the bubbles away from the orifice. This lift component increases with the tilting angle.

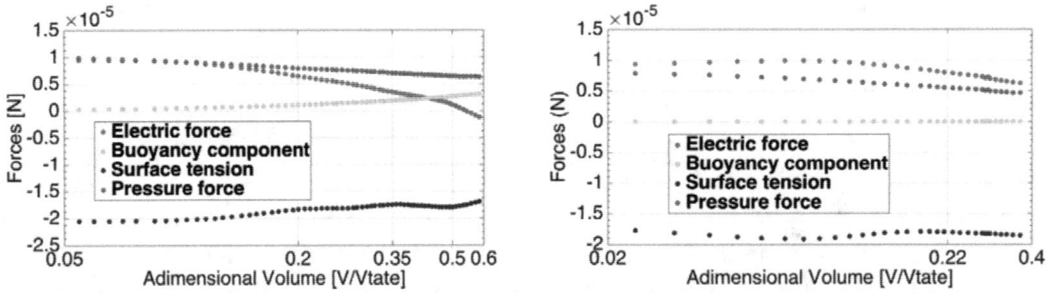

Figure 10: Force balance, *y* direction, 75° (a) and 90° (b) of tilting, with EF (25 kV).

Despite the electric contributions acting over the interface, the contact angle seems not to be altered substantially by the application of electric field until a certain volume value, as can be noticed comparing the history of contact angle for the 0kV and 25 kV cases at the same tilting condition, see Figure 11. This result has also been found in normal gravity and for 180° of tilting in [18].
The motion of the centre of mass of the bubble agrees with the presented force balances. As shown in Figure 12, the electric field pulls the bubble from the orifice towards the electrode and then, close to detachment and after, it pushes the bubble towards the surface.

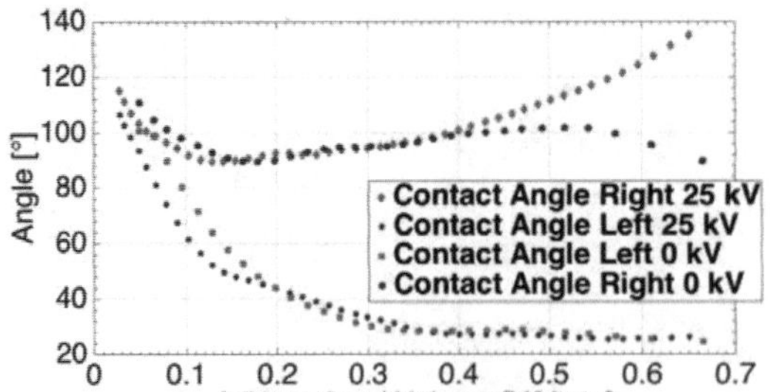

Figure 11: Contact angle, 0kV vs 25kV at 75°

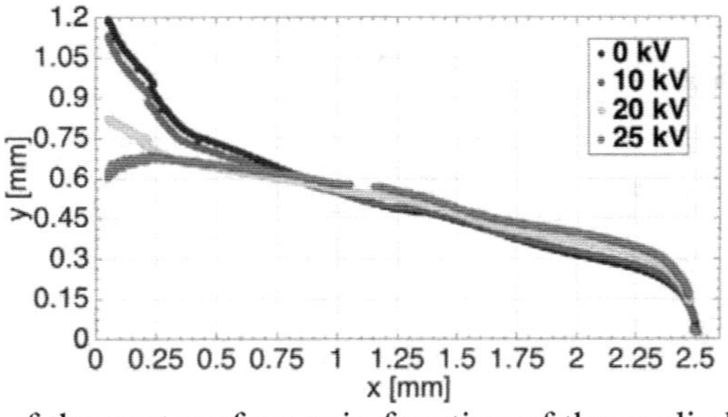

Figure 12: Motion of the centre of mass in function of the applied voltage, 75° of tilting.

For the power balance, the same procedure has been adopted, validating the model first for bubbles (Figure 13) and then determining by difference the contribution of the electric field.
Without electric field, the power spent in the gravitational term is very low compared to the other contributions: surface growth is storing all energy due to pressure and inlet flux.

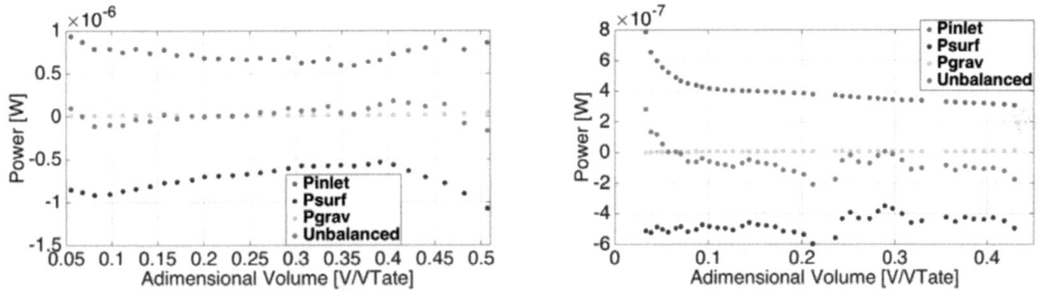

Figure 13: Power balance, 75° (a) and 90° (b) of tilting, no EF.

In the case of applied electric field, an interesting phenomenon can be noticed: in the case of bubbles, differently from droplets, the gravitational term in the power balance is positive (Figure 14). It means that the inlet power doesn't have to withstand it. This condition differs for droplets, where the inlet power is spent both in the growth of the interface and in the rise of the centre of gravity. The interest of this behaviour lies in the fact that when the electric field is applied, it follows the gravity one: for droplets the contribution of the electric field is negative, while for bubbles, the present work shows that it is positive. This is consistent with the fact that the creation of a cavity (with lower dielectric permittivity) causes a reduction of electrostatic energy in the domain. The energy recovered from gravitational and electric terms is used to create the bubble interface.

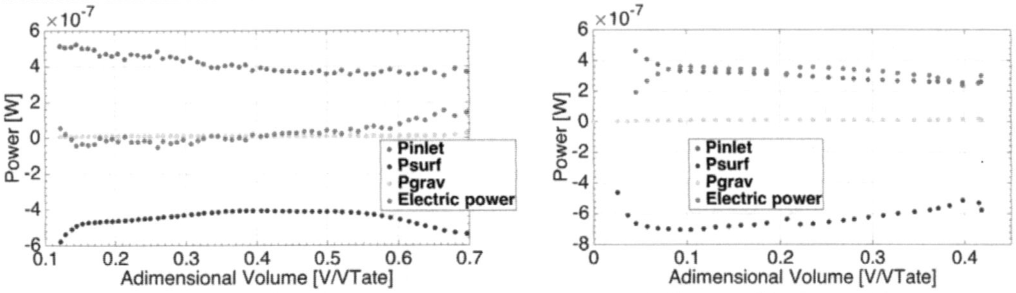

Figure 14: Power balance, 75° (a) and 90° (b) of tilting, with EF (25 kV).

A last figure summarizes all the effects just presented: in Figure 15, the detachment diameter in function of the tilting angle, for different electric potential applied is presented. As can be noticed, the detachment diameter decreases with increasing tilting angle, but it increases with increasing the electric field. In authors' opinion it can be due to the fact that electric forces are contributing in a negative sense to the tangential-x force balance, thus reducing the effects of the shear component of gravity. This detachment diameter increase is negligible at the maximum tilting angle.

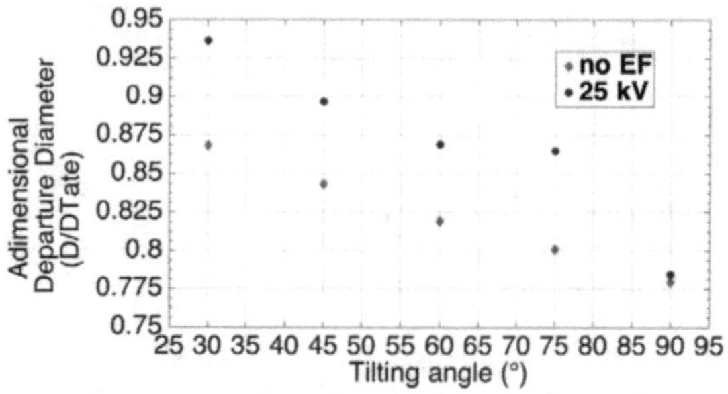

Figure 15: Detachment diameter as function of the tilting angle and of the electric field.

8. Conclusions

Bubbling of FC-72 at different tilting angle and different external electric potential applied has been studied experimentally to determine how tangential forces affect the shape and the behaviour of a single bubble, in order to understand the effect of electric field application to flow boiling.

In general it has been found that the electric field does not affect the contact angle of bubbles, but the resulting perpendicular force it is pulling the bubble away from the orifice during the growth and later, during and after detachment it pushes the bubble against the surface. In this way the bubble is forced to slide along the surface, which may significantly increase the boiling heat transfer coefficient.

Power balances reveal a further increase in the interface growth to compensate the negative electric contribution, thus promoting heat and mass transfer across the interface.

A possible drawback of the application of the electric field is the increase of departure diameter, due to the opposition of electric field to the shear force. In any case, this increase has been found to be very small for all the considered tilting angles (7.4% maximum) and to decrease with increasing tilting angle.

Acknowledgements

This work has been carried out with the financial support of European Space Agency, within the CBC project (MAP AO-2004-111).

References

[1] Zhang H, Mudawar I, Hasan M M 2008 *Thermal and Thermomechanical Phenomena in Electronic Systems ITHERM 11th Intersociety Conference* 949
[2] Sharifi P 2011 *PhD Thesis, Southern Illinois University at Carbondale* UMI Number: 3460461
[3] Zaghdoudi M C and Lallemand M 2005 *Arabian Journal for Science and Eng.* **30** 189
[4] Di Marco P and Grassi W 1993 *J. Enhanced Heat Transfer* **1** 99
[5] Bonjour E, Verdier J, Weil L 1962 *AIChE-ASME 5th Nat. Conference*
[6] Allen P H G and Karayiannis T 1995 *Heat Recovery Systems & CHP* **5** 389
[7] Di Marco P and Grassi W 2007 *Multiphase Sci. Technol.* **19 2** 141
[8] Di Marco P, Kurimoto R, Saccone G, Hayashi K, Tomiyama A 2014 *Proc. 15th Int. Heat Transfer Conference* paper 8960 p 1
[9] Duhar G and Colin C 2006 *Physics of Fluids* **18** 077101
[10] Klausner J F, Mei R, Bernhard M D, Zeng L Z 1993. *Int. J. Heat and Mass Tr.* **36** 651
[11] Thorncroft G E, Klausner J F, Mei R 2001 *Multiphase Science and Technology* **13** 35
[12] Tsuge H 1986 *Encyclopedia of Fluid Mechanics* chap. 9 191

Journal of Physics: Conference Series **655** (2015) 012044 doi:10.1088/1742-6596/655/1/012044

[13] Erickson D, Blackmore B, Li D 2001 *Colloids and Surf. A: Physicochem. Eng. Aspects* **182** 109
[14] Saccone G, Del Seppia G, Rossi C A, Di Marco P 2015 *Proc. Droplet Impact Phenomena and Spray Investigation DIPSI Workshop*, University of Bergamo
[15] Saccone G 2014 *Master Thesis*, University of Pisa
[16] Di Marco P 2009 *7th W. Conf. on Exp. Heat Tr., Fluid Mech., Thermodynamics* 1
[17] Di Marco P and Saccone G, 2012 *23rd International Symposium on Transport Phenomena* 1
[18] Di Marco P, Giannini N, Saccone G 2012 *ECI 8th Int. Conf. on Boiling and Cond. Heat Tr.* 1

Backcoupling of acoustic streaming on the temperature field inside high-intensity discharge lamps

J Schwieger[1], B Baumann[1], M Wolff[1], F Manders[2] and J Suijker[2,3]
[1]Hamburg University of Applied Sciences, Department of Mechanical Engineering and Production, Berliner Tor 21, 20099 Hamburg, Germany
[2]Philips Lighting, Steenweg op Gierle 417, 2300 Turnhout, Belgium
[3]Technical University Eindhoven, Den Dolech 2, 5612AZ Eindhoven, Netherlands

E-mail: joerg.schwieger@haw-hamburg.de

Abstract. Operating high-intensity discharge lamps in the high frequency range (20−300 kHz) provides energy-saving and cost reduction potentials. However, commercially available lamp drivers do not make use of this operating strategy because light intensity fluctuations and even lamp destruction are possible. The reason for the fluctuating discharge arc are acoustic resonances in this frequency range that are excited in the arc tube. The acoustic resonances in turn generate a fluid flow that is caused by the acoustic streaming effect. Here, we present a 3D multiphysics model to determine the influence of acoustic streaming on the temperature field in the vicinity of an acoustic eigenfrequency. In that case a transition from stable to instable behavior occurs. The model is able to predict when light flicker can be expected. The results are in very good accordance with accompanying experiments.

1. Introduction

Approximately one fifth of the world production of electric power is consumed by 30 to 33 billion electric light sources [1–3]. High intensity discharge (HID) lamps represent a main part of the artificial light sources. They are used for street and shop lighting as well as automobile headlights and other applications. A superior color rendering index combined with a sun-like luminance characterizes HID lamps [4]. These advantages over light emitting diodes will guarantee a market for HID lamps in the future. The structure of the low-wattage HID lamp, that is investigated in this article, is shown in figure 1. The distance between the tungsten electrodes is 4.8 mm. The tube from polycrystalline alumina (PCA) contains a mixture of argon, mercury and metal halides. A voltage is applied to the electrodes to establish a plasma arc of high temperature inside the arc tube. In order to minimize material costs of the electronic driver, the lamp should be operated in the high frequency range of 20 kHz to 300 kHz [5]. However, the periodic ohmic heating excites acoustic resonances in this frequency range. The resonances can induce low frequency plasma arc fluctuations of approximately 10 Hz [6] that are visible as light flicker. The acoustic streaming (AS) phenomenon is the link between the high frequency acoustic resonance and the low frequency flicker [7].

To investigate the light flicker phenomenon, a three-dimensional stationary finite element model of this lamp was developed. Initially, it calculates the electric potential, the buoyancy

Journal of Physics: Conference Series **655** (2015) 012045 doi:10.1088/1742-6596/655/1/012045

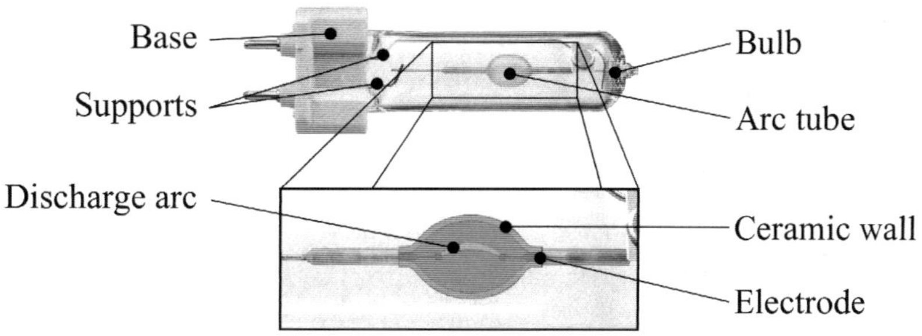

Figure 1. Set-up of the investigated HID lamp (Philips 35 W 930 Elite) in horizontal operation position.

driven velocity field and the temperature field. On the basis of the temperature field the acoustic eigenmodes are determined. Thereafter, the acoustic pressure field and the AS force are determined for the second mode. With the aid of this model, a symmetry breaking transition for the flow field was detected [8]. Here, we present results of the backcoupling of the superimposed AS and buoyancy force on the velocity and the temperature field.

2. Model

The finite element model for the numerical investigation of the horizontally operated HID lamp comprises the arc tube wall, its filling and both electrodes, and it solves several coupled partial differential equations. All simulations were executed with COMSOL Multiphysics and MATLAB.

2.1. Geometry and mesh

The model of the HID lamp was created with slight simplifications. The coils at the electrodes were simplified to cylinders to reduce the number of finite element nodes and, therefore, the computing time. The lamp is tilted by 5° against the horizontal axis to prevent stability problems due to instable paths related to the symmetry breaking transition that was observed in a prior investigation [8]. The model has to be designed in three dimensions in order to consider the upward bending of the plasma arc due to buoyancy. The temperature field is symmetrical to the vertical y-z-plane (figure 2). Therefore, only one half of the physical geometry has to be considered in this simulation step. The temperature field is mirrored at the y-z-plane to map it onto the full geometry and to calculate the acoustic modes.

The finite element mesh (figure 2) used for the simulation of the half model consists of approximately 123 000 elements. In the boundary layer of the arc tube prism elements were used, whereas the rest of the model is composed of tetrahedral elements. The accuracy of the solution was tested with different mesh resolutions. The results obtained with the selected mesh differ from results with a mesh with 164 000 elements by less than 1 %. The computing time needed with the mesh displayed in figure 2 could be reduced by a factor of 10 compared to the simulation with the very fine mesh.

2.2. Temperature Field

The coupled differential equations of mass, momentum, charge and energy conservation were used to obtain the temperature distribution. The differential equation of charge conservation

$$\vec{\nabla} \cdot \left(\sigma \vec{E}\right) = 0 \tag{1}$$

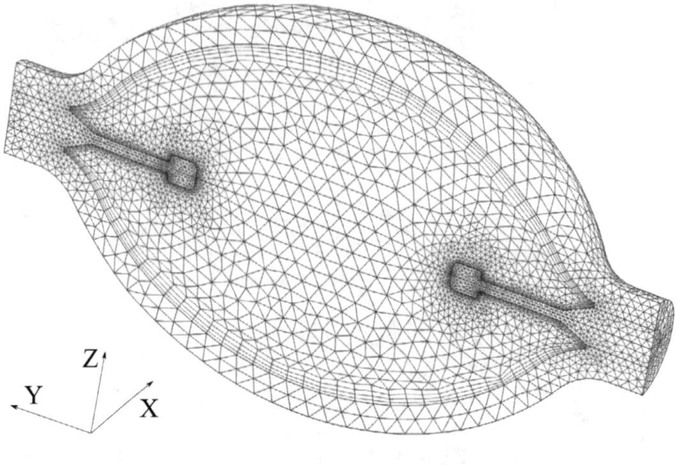

Figure 2. Finite element mesh of the half model. The origin of the coordinate system is located in the center between the electrodes.

enables the calculation of the electric field $\vec{E}$. The electric conductivity σ and other material parameters are listed in a previous publication [8]. Because no essential differences are expected, a direct current (dc) was implemented instead of an alternating current (ac) [1]. The inward current density on the electrodes was obtained by the electrode tip area and the electric current I_{lamp}. With the nominal power of the lamp P_{lamp} and the voltage drop between the electrodes U the electric current $I_{\text{lamp}} = P_{\text{lamp}}/U$ was computed. The tube's wall and the symmetry plane were modeled as electric insulators.

For Newtonian fluids the stationary velocity field $\vec{u}$ is described by the Navier-Stokes equation

$$\rho\left(\vec{u}\cdot\vec{\nabla}\right)\vec{u} = \vec{\nabla}\cdot\left[-P\mathbf{I} + \eta\left(\vec{\nabla}\vec{u} + \left(\vec{\nabla}\vec{u}\right)^{\mathrm{T}}\right) - \frac{2}{3}\eta\left(\vec{\nabla}\cdot\vec{u}\right)\mathbf{I}\right] + \vec{f} \tag{2}$$

and mass conservation

$$\vec{\nabla}\cdot(\rho\vec{u}) = 0 \tag{3}$$

($\mathbf{I}$ identity matrix, ρ density, P pressure inside the arc tube, η dynamic viscosity). Without AS the velocity field is driven by the gravitational force only. Its force density is described by

$$f_l = -\delta_{l3}\rho g \tag{4}$$

(δ_{l3} Kronecker symbol: index 3 corresponds to the Z-direction, $g = 9.81\,\mathrm{m/s^2}$). The boundary condition $\vec{u} = \vec{0}$ was used at the tube's wall and the electrodes. At the symmetry plane $\vec{n}\cdot\vec{u} = 0$ ($\vec{n}$ normal vector) was applied and in the center the static pressure $P = 2.81\,\mathrm{MPa}$ was specified[2].

The Elenbaas-Heller equation

$$\vec{\nabla}\cdot\left(-\kappa\vec{\nabla}T\right) + \rho c_p\vec{u}\cdot\vec{\nabla}T = \sigma|\vec{E}|^2 - q_{\text{rad}} \tag{5}$$

(κ thermal conductivity, c_p specific heat capacity, q_{rad} density of power radiation) determines the temperature field T inside the arc tube, of the wall and of the electrodes. On the external

[1] At ac condition the direction of the carrier wave periodically changes. On time average the system is symmetric. The dc boundary conditions have been set up accordingly.

[2] To preserve model symmetry, the center point was chosen. The position of the static pressure point is of no importance.

surface of the wall heat is emitted by radiation. This is described by the boundary condition $\vec{n} \cdot (-\kappa\vec{\nabla}T) = \epsilon\sigma_{\mathrm{SB}}\left(T_{\mathrm{amb}}^4 - T^4\right)$ (ϵ emissivity, σ_{SB} Stefan-Boltzmann constant, T_{amb} ambient temperature). On the symmetry plane the emissivity is set to zero (no radiation). At the electrode grounds a constant temperature of $T = 1450\,\mathrm{K}$ was defined. The acoustic pressure field in the arc tube can be calculated with the aid of the resulting temperature field and the power density of heat generation.

2.3. Acoustic Streaming

For the calculation of the AS force the acoustic pressure in the arc tube has to be calculated first. To obtain the acoustic pressure p, the Helmholtz equation

$$\vec{\nabla} \cdot \left(\frac{1}{\rho}\vec{\nabla}p\right) + \frac{\omega^2}{\rho c^2}p = \mathrm{i}\omega\frac{\gamma - 1}{\rho c^2}\mathcal{H} \tag{6}$$

has to be solved (c speed of sound, γ heat capacity ratio). The source term of the Elenbaas-Heller equation $\mathcal{H} = \sigma|\vec{E}|^2 - q_{\mathrm{rad}}$ describes the power density of heat generation. The arc tube filling is treated as an ideal gas with the speed of sound field

$$c = \sqrt{\kappa\frac{RT}{M}}. \tag{7}$$

The wall of the arc tube and the electrodes are treated as sound hard. The solution of the Helmholtz equation is expressed by an acoustic mode. It is described elsewhere, how the pressure amplitude, that takes the loss mechanisms of heat conduction and viscosity on the surface as well as in the volume into account, is calculated [9–11].

The AS force arises when a standing sound wave in the arc tube of the HID lamp is excited [7]. In addition to an oscillatory motion, an averaged fluid motion is induced. The motion is a consequence of the viscous force that develops due to the shear viscosity of the fluid particles. It generates vortex-like motions that are distinguished by their location: Outer (or Rayleigh) streaming occurs outside the boundary layer and inner streaming occurs inside the boundary layer [12].

The AS force is implemented into the source term of the Navier-Stokes equation. Instead of the gravitational force only, the force density is now

$$f_l = \frac{\partial\overline{\rho v_k v_l}}{\partial x_k} - \delta_{l3}\rho g \tag{8}$$

($\vec{v}$ sound particle velocity). In this equation Einstein's summation convention is used and the overbar indicates time averaging. In the case of time harmonic waves the force density simplifies to

$$f_l = \frac{1}{2}\frac{\partial\rho\hat{v}_k\hat{v}_l}{\partial x_k} - \delta_{l3}\rho g. \tag{9}$$

$\hat{\vec{v}}$ denotes the amplitude of the sound particle velocity. It can be obtained from the acoustic pressure [13]:

$$\hat{\vec{v}}\left(\vec{r},\omega\right) = \frac{1}{\mathrm{i}\omega\rho}\vec{\nabla}p\left(\vec{r},\omega\right). \tag{10}$$

The velocity drop in the boundary layer had to be considered additionally because the mode calculation does not incorporate this condition. Therefore, the sound particle velocity was multiplied with a factor that drops in the boundary layer from one to zero [14, 15]. The implementation of this function into the model has previously been described [8].

2.4. Simulation procedure

The calculation of the temperature, the acoustic pressure and the AS force field have to be coupled in order to get the stationary temperature field at a certain frequency. Since the discharge arc is excited at this frequency, it is called *excitation frequency.*

Initially, the velocity and temperature fields as well as the acoustic mode and eigenfrequencies are calculated with the gravitational force only. On the basis of the acoustic modes the pressure amplitude is calculated. With the aid of the amplitude, the AS force at a specified excitation frequency is computed. In order to include the back coupling of the AS field on the other fields, a recursion loop was implemented. From the second iteration on, the fluid flow is driven by buoyancy and the AS force. This leads to a different velocity and temperature field compared to the first step. The acoustic eigenmode and the resonance frequency change and lead to an altered AS force that serves as input for the next step. This recursion procedure is continued until the results of the current and the previous step coincide within a certain limit. The resonance frequency is used as a convergence criterion. Once the resonance frequency changes by less than 5 Hz, the solution is considered converged.

The fields obtained in this way are used as initial condition for the next excitation frequency. Since we are mainly interested in results near a resonance, the excitation frequency is increased when the resonance frequency is higher and decreased when the resonance frequency is lower than the excitation frequency. The step size is adopted to the difference of both frequencies. The closer the resonance and the excitation frequency are, the higher the pressure amplitude and AS force. Because the AS force depends quadratically on the acoustic particle velocity (see Equation 8), the force increases significantly near the resonance frequency. Hence, a very small step size of the excitation frequency has to be chosen.

3. Experimental set-up

In the experiment the operating voltage of the lamp consists of two components. The square-wave part with a carrier frequency f_c of 400 Hz is used to operate the lamp at stable conditions. Flicker of the discharge arc is excited by a high frequency sinusoidal part. This excitation frequency f_e is varied in the range of 41 kHz to 44 kHz. The combined voltage input is described by ($\omega_e = 2\pi f_e$, $\omega_c = 2\pi f_c$)

$$U(t) = \begin{cases} A\left[+1+\alpha \sin\left(\omega_e t\right)\right] & 2n < \omega_c t \leq 2n+1 \\ A\left[-1+\alpha \sin\left(\omega_e t\right)\right] & 2n+1 < \omega_c t \leq 2n+2 \end{cases} \qquad n \in \mathbb{N}_0. \tag{11}$$

The amplitude A is adjusted to the nominal power of the lamp. The ratio of the square-wave to the sinusoidal-wave voltage amplitude is described by the modulation depth α. In the experiments it is set to a constant value that is just high enough to stimulate arc flicker. A more detailed description of the experimental set-up can be found elsewhere [16].

Before the measurements the lamp is operated at stable conditions ($\alpha = 0$) for at least 20 minutes to reach a stationary state. Then, the modulation depth is set to a constant value and, simultaneously, the excitation frequency is set to its initial frequency some kHz off the resonance frequency. This initial value is either 41 kHz when the frequency is afterwards increased or 44 kHz when the frequency is decreased. These operating conditions are kept constant for 20 s. During this time the voltage drop between the electrodes is measured 40 times. At the initial step no acoustic resonance is excited so that the mean value of the measured voltages can be defined as the basic voltage. In the next step the excitation frequency is shifted by 50 Hz into the direction of the resonance frequency. To prevent lamp failures, which are caused by too high arc tube temperatures or by temperature oscillations of the arc tube, the voltage measurements are from then on used to control the experiment. If one of the 40 measurements exceeds the basic voltage by 8 V or if the voltage fluctuates by more than 1.5 V, the experiment is terminated. If these criteria are not fulfilled, the experiment is continued until the excitation frequency reaches 41 kHz and 44 kHz, respectively.

4. Results

4.1. Acoustic modes and eigenfrequencies

Initially, the 5° tilted model without AS was solved. The simulation computes a lowest acoustic eigenfrequency of 33 100 Hz which is not apparent in the experiment. At 47 750 Hz and 48 130 Hz two neighboring azimuthal modes appear[3]. In this article the simulations, which include the backcoupling of AS on all other fields, focus on the mode at 47 750 Hz. The experiment shows the influence of both eigenfrequencies.

The excitation frequency was set to 47 750 Hz and the recursion started. Once convergence was reached, the eigenfrequency shifts to 47 400 Hz due to AS. As described above, the excitation frequency for the next step has to be decreased by 50 Hz. The resulting eigenfrequency is 40 Hz below the previous value at 47 360 Hz. Reducing the excitation frequency again by 50 Hz further decreases the difference between excitation and eigenfrequency so that the simulation of the next step is conducted closer to the resonance. This results in a stronger AS force. Continuing in this way the simulations were successful down to an excitation frequency of 46 313 Hz where the difference to the resonance frequency diminishes to 55 Hz. At even lower frequencies the simulation did not converge. We assume that no time-independent solution exists at these frequencies so that stationary simulations are inappropriate[4]. To test this assumption, we plan to perform a linear stability analysis in the near future.

4.2. Voltage drop

A quantity, which is easy to determine in the simulation as well as in the experiment, is the voltage drop between the electrodes. In order to demonstrate the reliability of the FE model, the voltage of the simulations and experiments are compared.

Figure 3 shows the simulation results of the voltage as function of the excitation frequency. The two curves correspond to the behavior of the voltage for increasing and decreasing excitation frequency, respectively. When the frequency is decreased from 47.8 kHz to 46.3 kHz, a significant voltage increase of 7 V occurs. A further lowering of the frequency leads to a voltage jump to values coinciding with the values of the curve representing the increasing frequency simulation. At increasing excitation frequency the voltage stays constant at about 84 V for the frequency range from 45.0 kHz to 47.0 kHz. Then the voltage increases by about 1 V, before it jumps to the upper curve. The data clearly demonstrate the appearance of a hysteresis, which is expected in a nonlinear system. The situation is similar to the one of a Duffing oscillator: The resonance frequency depends on the amplitude. If the oscillation is strong enough, the equation, which couples the amplitude and the excitation frequency, is cubic and its solution includes three real roots in a certain frequency range. The peak of the response curve either tilts to low or high frequencies and overlaps the response curve describing the amplitude far off the resonance [18].

Figure 4 shows the results of the experimental voltage measurements over the excitation frequency. The frequency deviation between the experimental and the simulation result of 4 kHz is related to the simplified geometry and uncertainties in the material properties [16, 19]. Additionally, some boundary conditions, e.g. the constant temperature at the electrode ground or the uniformly distributed electric current density on the electrodes, only approximate the conditions in the lamp. For a decreasing excitation frequency the voltage increases up to 96 V at 42.4 kHz. Here, the experimental results show a voltage jump like the one observed in the

[3] If gravity is switched off, the two modes are degenerate.

[4] A similar situation is known from the Taylor-Couette system, where the stationary Taylor cells undergo a Hopf bifurcation to the time-periodic wavy vortex state [17].

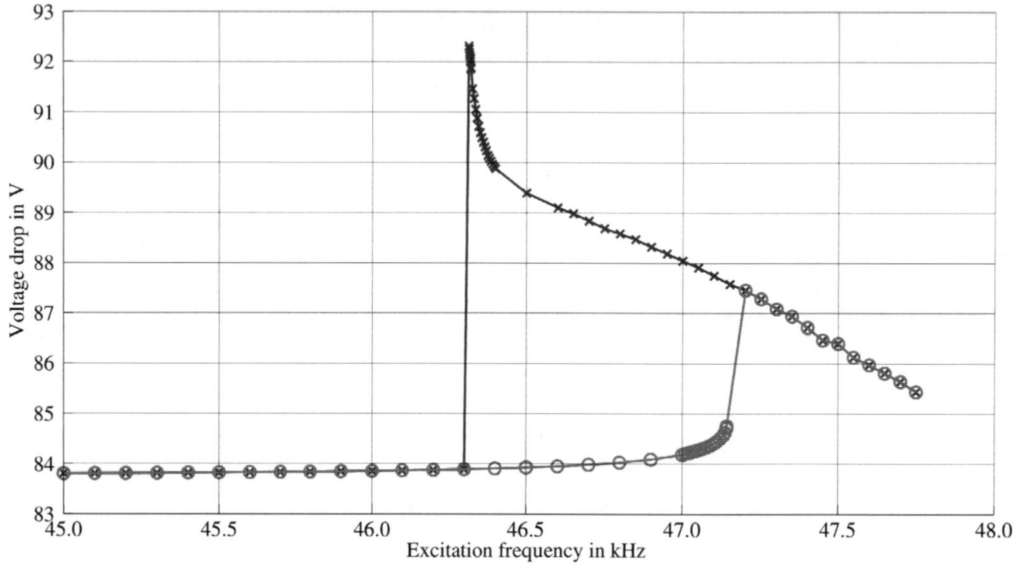

Figure 3. Simulation result of the voltage drop between the electrodes over the excitation frequency. The blue crosses mark voltage drops at decreasing excitation frequency. The red circles indicate voltage drops at increasing excitation frequencies.

simulation. This jump has also been observed in high pressure sodium lamps [20]. The significant voltage increase is related to the excitation of an acoustic mode. In contrast to the simulation, in which only one of the neighboring acoustic modes is considered, figure 4 shows repercussions of both modes. The small voltage peak at the frequency of 43.8 kHz indicates the second mode.

The simplified geometry and uncertainties in the material properties are also responsible for the different voltages: In the simulation the voltage far off an acoustic resonance is 83.8 V and in the experiment approximately 91.5 V. Apart from these differences in the absolute values the graphs in figure 3 and figure 4 show excellent agreement.

The voltage jump at an increasing excitation frequency is 2.5 V compared to 2.3 V in the experiment. At a decreasing excitation frequency the voltage in the experiment increases by 5.1 V and in the simulation by 8.3 V. The frequency range of the hysteresis (voltage peak at an increasing to the voltage peak at a decreasing excitation frequency) is 840 Hz in the simulation and 500 Hz in the experiment. It would be interesting to measure the hysteresis at different modulation depths to determine its influence on the voltage jump and the jump frequencies.

4.3. Temperature and velocity field

Figure 5 right shows the temperature field resulting from the simulation at an excitation frequency of 46 313 Hz, the frequency, where the voltage jump in the blue curve of figure 3 appears (decreasing frequency). A major impact of AS on the temperature field can be observed, when comparing it to the temperature field obtained at 47 750 Hz (figure 5 left). This frequency is far off an acoustic resonance and the AS force accordingly small.

While approaching the resonance, the maximum plasma temperature, located at the hot-spots in front of the electrodes, only increases slightly. Simultaneously, the temperature in the lower part of the arc tube, especially in the vertical center plane of the arc tube, considerably increases. These higher temperatures are caused by a flow that is directed in negative z-direction. The flow in the upper part is directed into the positive z-direction and leads to a steeper temperature gradient near the upper wall.

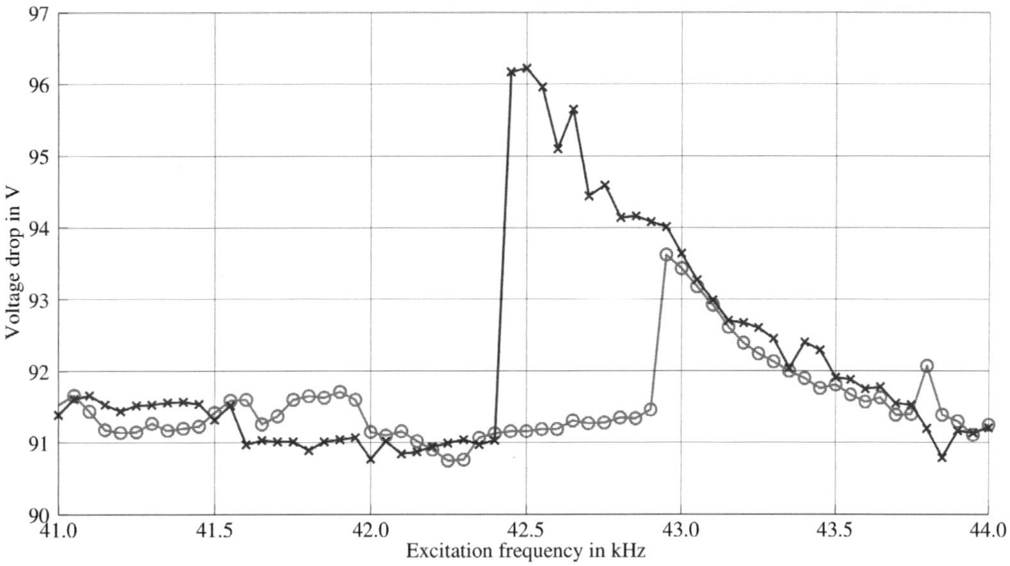

Figure 4. Experimental results of the voltage drop between the electrodes over the excitation frequency at a modulation depth of 2 %. Red circles indicate measurements at increasing and blue crosses at decreasing excitation frequency.

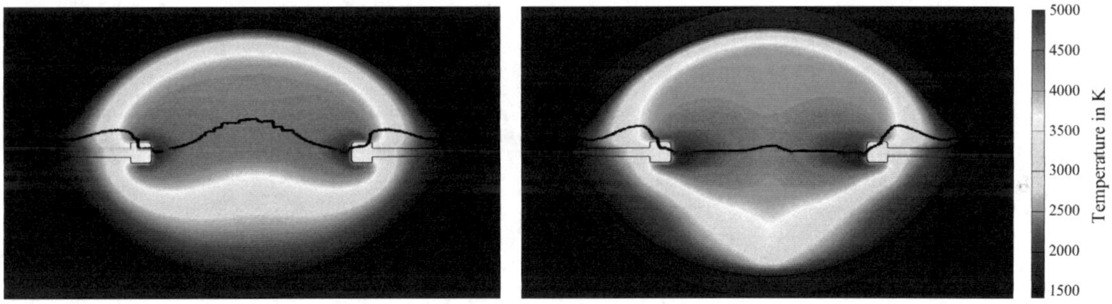

Figure 5. Temperature distribution in the y-z-plane at different excitation frequencies while decreasing the excitation frequency. Left: Far off resonance (47 750 Hz). Right: Near resonance (46 313 Hz). The black line indicates the location of maximal temperature as a function of the y-coordinate.

To compare the temperature field of the simulation with experimental results, the light intensity distribution was recorded by a camera [16]. The left part of figure 6 shows the intensity distribution at stable operation conditions. The buoyancy force bends the plasma arc upward and AS is negligible. The picture resembles the left part of figure 5, where the temperature distribution far off the resonance is depicted.

After tuning the parameters to unstable lamp operation, AS emerges. The intensity distribution on the right shows the discharge arc when it moves towards the lower part of the arc tube. Immediately afterwards, the discharge arc started to rotate around the y-axis and finally the lamp exploded. It highlights that AS acts against buoyancy. The pictures strongly resemble the right part of figure 5 from the simulation.

Figure 7 shows the simulated flow fields corresponding to the temperature fields depicted in figure 5. At an excitation frequency far off the resonance the fluid flow essentially represents

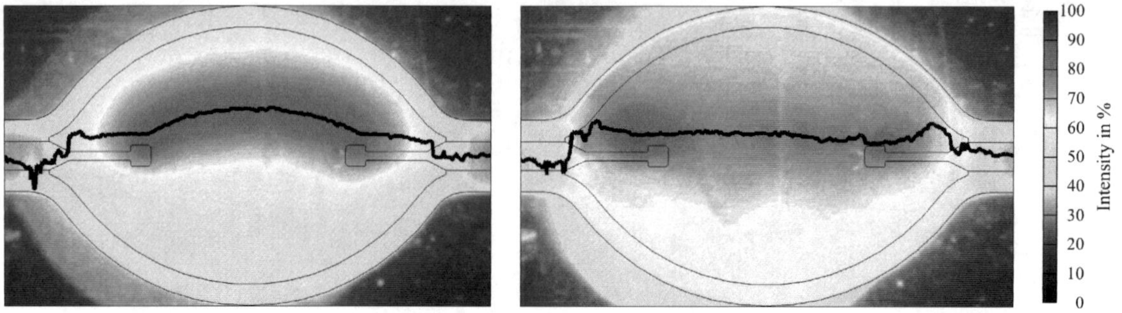

Figure 6. Measurements of the light intensity. The left part shows the plasma arc at stable operation conditions. The right part shows the plasma arc immediately before it starts to rotate violently around the lamp axis. The black line indicates the location of maximal intensity as a function of the y-coordinate.

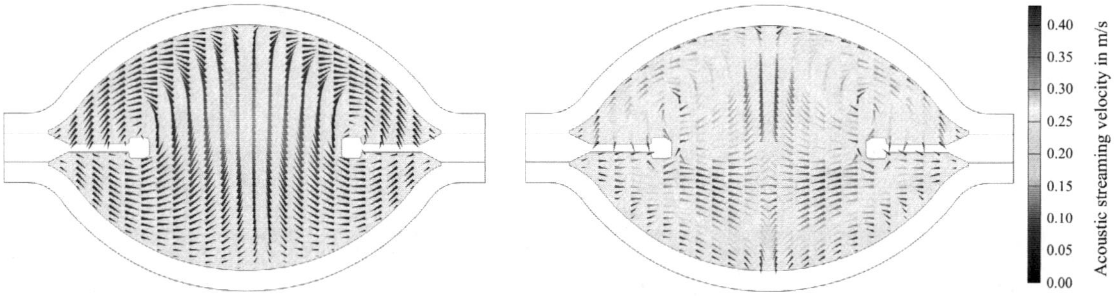

Figure 7. Fluid flow in the y-z-plane at different excitation frequencies while decreasing the excitation frequency: Left: Far off resonance (47 750 Hz). Right: Near resonance (46 313 Hz).

the purely buoyancy-driven flow field. The flow between the electrodes is directed upwards and has its maximum in this region. Instead of a maximal velocity of 0.09 m/s in case of the purely buoyancy-driven flow, the maximal velocity is 0.13 m/s. The flow field is not fully symmetric with respect to the y-z-plane due to the tilting of the lamp.

Near the resonance the AS dominates the buoyancy flow and the asymmetry caused by the lamp tilting is not visible any more because only the buoyancy force is affected by the tilting of the lamp. In the upper part of the arc tube two additional vortices develop (right: clockwise; left: counterclockwise). In the lower part the flow points in the direction of the negative z-axis. The flow continues in the x-direction, which is not visible in the figure. The maximal velocity is 0.42 m/s whereas in the simulation without backcoupling it is 0.61 m/s [8]. The reason for the difference is that in case of the coupled simulation the lamp is excited 55 Hz off the resonance. The AS force is too weak to cause the symmetry breaking transition, which was observed in the uncoupled model [8].

5. Conclusions

In this work a stationary 3D finite element model is used to simulate the backcoupling of acoustic streaming on the temperature field inside a high-intensity discharge lamp. At the investigated acoustic eigenfrequency a hysteresis was detected in the simulation and confirmed by accompanying experimental investigations. The voltage drop when decreasing the excitation frequency shows a different behavior than in case of an increasing excitation frequency. The simulation results show a severe impact of the acoustic streaming on the voltage drop, the temperature field and the flow field. In the simulation as well as in the experiments the voltage drop at the investigated resonance significantly increases compared to a voltage drop far off the acoustic resonance.

AS causes the maximal velocity in the arc tube to become more than 4 times higher compared to the purely buoyancy-driven flow. A buoyancy dominated flow leads to an upward bending of the discharge arc. Acoustic streaming on the contrary induces a straight arc between the electrodes. These results are consistent with measurements of the light intensity distribution. The results will help to understand the physical processes, that lead to flickering of the discharge arc inside high-intensity discharge lamps. They enable development of new lamp systems with improved energy-efficiency.

Acknowledgments

This research was supported by the German Federal Ministry of Education and Research (BMBF) under project reference 03FH025PX2 and Philips Lighting. We are indebted to our colleague Klaus Spohr for discussions.

References

[1] Zissis G and Haverlag M 2010 *J. Phys. D: Appl. Phys.* **43** 230301
[2] Stoffels W W, Baede A H F M, van der Mullen J A M, Haverlag M and Zissis G 2006 *Meas. Sci. Technol.* **17** N67–N70
[3] Waide P and Tanishima S 2006 *Light's labour's lost: Policies for energy-efficient lighting* (Paris: OECD and International Energy Agency)
[4] National Electrical Manufacturers Association 2010 *NEMA standards publication LSD 54-2010: The strengths and potentials of metal halide lighting systems* (Rosslyn: OECD and International Energy Agency)
[5] Trestman G A 2002 Minimizing cost of HID lamp electronic ballast *Proc. IES* ed IEEE (Piscataway: IEEE) pp 1214–1218
[6] Schwieger J, Wolff M, Baumann B, Manders F and Suijker J 2015 *IEEE Trans. Ind. Appl.* **51** 2544–2547
[7] Afshar F 2008 *J. Illum. Eng. Soc.* **5** 27–38
[8] Baumann B, Schwieger J, Wolff M, Manders F and Suijker J 2015 *J. Phys. D: Appl. Phys.* **48** 255501
[9] Kreuzer L B 1977 The physics of signal generation and detection *Optoacoustic spectroscopy and detection* ed Pao Y H (New York: Academic Press) pp 1–25
[10] Baumann B, Wolff M, Kost B and Groninga H 2007 *Appl. Opt.* **46** 1120–1125
[11] Baumann B, Wolff M, Hirsch J, Antonis P, Bhosle S and Barrientos R V 2009 *J. Phys. D: Appl. Phys.* **42** 225209
[12] Boluriaan S and Morris P J 2003 *Int. J. Aeroac.* **2** 255–292
[13] Temkin S 1981 *Elements of acoustics* (New York: Wiley)
[14] Schuster K and Matz W 1940 *Akust. Zeits.* **5** 349–352
[15] Hamilton M F, Ilinskii Y A and Zabolotskaya E A 2003 *J. Acoust. Soc. Am.* **113** 153–160
[16] Schwieger J, Baumann B and Wolff M 2015 *Int. J. Thermophys.* **36** 1327–1335
[17] Cross M and Greenside H 2009 *Pattern formation and dynamics in nonequilibrium systems* (Cambridge: Cambridge University Press)
[18] Thompson J M T and Virgin L N 1986 *Int. J. Nonlinear Mech.* **21** 205–216
[19] Schwieger J, Baumann B, Wolff M, Manders F and Suijker J 2013 *Proc. Comsol Conf.*
[20] Chhun L, Maussion P, Bhosle S and Zissis G 2011 *IEEE Trans. Ind. Appl.* **47** 1071–1076

Numerical validation of a κ-ω-κ_θ-ω_θ heat transfer turbulence model for heavy liquid metals

D Cerroni[1], R Da Vià[1], S Manservisi[1], F Menghini[1], G Pozzetti[2] and R Scardovelli[1]

[1] DIN - Lab. of Montecuccolino, University of Bologna, Via dei Colli 16, 40136 Bologna, Italy
[2] RUES, University of Luxemburg, 6, rue Richard Coudenhove-Kalergi, L-1359 Luxembourg

E-mail: filippo.menghini3@unibo.it

Abstract. The correct prediction of heat transfer in turbulent flows is relevant in almost all industrial applications but many of the heat transfer models available in literature are validated only for ordinary fluids with $Pr \simeq 1$. In commercial Computational Fluid Dynamics codes only turbulence models with a constant turbulent Prandtl number of $0.85 - 0.9$ are usually implemented but in heavy liquid metals with low Prandtl numbers it is well known that these models fail to reproduce correlations based on experimental data. In these fluids heat transfer is mainly due to molecular diffusion and the time scales of temperature and velocity fields are rather different, so simple turbulence models based on similarity between temperature and velocity cannot reproduce experimental correlations. In order to reproduce experimental results and Direct Numerical Simulation data obtained for fluids with $Pr \simeq 0.025$ we introduce a κ-ϵ-κ_θ-ϵ_θ turbulence model. This model, however, shows some numerical instabilities mainly due to the strong coupling between κ and ϵ on the walls. In order to fix this problem we reformulate the model into a new four parameter κ-ω-κ_θ-ω_θ where the dissipation rate on the wall is completely independent on the fluctuations. The model improves numerical stability and convergence. Numerical simulations in plane and channel geometries are reported and compared with experimental, Direct Numerical Simulation results and with results obtained with the κ-ϵ formulation, in order to show the model capabilities and validate the improved κ-ω model.

Keywords: Turbulent Heat Transfer, Turbulence Modeling.

1. Introduction

Heavy liquid metals such as mercury, Lead-Bismuth Eutectic (LBE) or sodium-potassium alloys with low Prandtl number are promising coolant fluids for achieving the necessary requirements for the design of fast nuclear reactors. In ordinary fluids such as water or air, similarity between thermal and dynamical fields holds and a Simple Eddy Diffusivity (SED) model coupled with a two-equation turbulence model seems to be sufficient to obtain reliable thermal and dynamical results. On the contrary, in liquid metals with low Prandtl numbers the temperature and velocity time scales are rather different since heat transfer is due to both molecular diffusion and convection. In these fluids the hypothesis of a constant turbulent Prandtl number fails to reproduce experimental correlations [1, 2, 3]. Algebraic expressions for the turbulent Prandtl number depending on turbulent and velocity fields have been recommended by many authors in order to correctly reproduce experimental data with these fluids, but these correlations are geometry dependent and new expressions have to be defined for each studied case [2]. For this reason the SED model and these algebraic expressions should not be used in Computational

Fluid Dynamics (CFD) simulations over new geometrical configurations where no experimental data are available.

In the last several years many turbulence models taking into account thermal turbulence effects have been proposed. These turbulence models have been successfully employed to compute the turbulent heat transfer in ordinary fluids with $Pr \approx 1$ in complicated geometry and detached flows. All of these models use the Algebraic Flux Model (AFM) in implicit or explicit formulation to calculate the turbulent heat flux. In this paper we deal with an explicit formulation of the AFM and compute the turbulent heat flux using a gradient hypothesis with isotropic eddy heat diffusivity α_t. The eddy or turbulent heat diffusivity has to be defined using appropriate time scales for the turbulence modeling. The thermal turbulent time scale is computed by defining the mean square temperature fluctuation κ_θ, its dissipation ϵ_θ and its specific dissipation rate ω_θ which have to be computed using appropriate transport equations. By coupling a low-Reynolds number κ-ϵ system and a κ_θ-ϵ_θ system one can obtain a four parameter turbulence model which solves the averaged Navier-Stokes system and defines the Reynolds stress tensor and the turbulent heat flux [4, 5, 6, 7, 8]. A four parameter turbulence model based on a κ-ϵ formulation has been already proposed by the authors, see [9, 10]. However it was found that this model shows convergence issues mainly due to the coupling of the variables in the boundary conditions. In this paper we propose a new four parameter κ-ω-κ_θ-ω_θ model based on the already tested κ-ϵ-κ_θ-ϵ_θ to overcome these problems. In this model a fixed Dirichlet boundary condition which does not depend on κ can be imposed for ω on the wall. The model is therefore more stable and robust and numerical convergence is faster.

In the next section the two turbulence models are introduced. They consist both of four transport equations, two for the dynamical turbulence modeling and two for the thermal. The derivation of the κ-ω formulation is presented together with a discussion over the new boundary conditions to be set in this model. In the third section the numerical results obtained with the new turbulence model for the simulation of fully developed turbulent flows in plane and cylindrical geometries are reported. These results are compared with the ones obtained with the κ-ϵ model, with Direct Numerical Simulation (DNS), when they are available, and with experimental heat transfer correlations for the evaluation of the Nusselt number Nu as a function of Peclet number Pe. Finally conclusions and remarks are taken over the usefulness of this turbulence model in nuclear engineering applications when heavy liquid metals are employed.

2. Transport equations and turbulence models

Heavy liquid metals can be considered incompressible in nuclear reactor cores, so the Reynolds-averaged incompressible system of Navier-Stokes can be used. It reads

$$\nabla \cdot \mathbf{u} = 0\,, \tag{1}$$

$$\rho \frac{\partial \mathbf{u}}{\partial t} + \rho(\mathbf{u} \cdot \nabla)\mathbf{u} = \nabla \cdot \sigma - \nabla \cdot \tau + \rho \mathbf{g}\,, \tag{2}$$

$$\rho C_p \left(\frac{\partial T}{\partial t} + (\mathbf{u} \cdot \nabla) T \right) = \nabla \cdot \mathbf{q} - \nabla \cdot \mathbf{q}_\theta + Q\,, \tag{3}$$

where $\mathbf{u}$ is the averaged velocity of the fluid and T is the averaged temperature. The tensors σ and $\mathbf{q}$ are the usual viscous stress and heat flux and they are modeled using Navier-Stokes constitutive law for viscous fluids and Fourier law for heat conduction.

$$\sigma := -p\mathbf{I} + \mu \mathbf{D} \quad \text{with} \quad \mathbf{D} := \nabla \mathbf{u} + \nabla \mathbf{u}^T \tag{4}$$

$$\mathbf{q} := -\lambda \nabla T\,. \tag{5}$$

Two new terms appear after the averaging process, namely the Reynolds stress tensor τ and the turbulent heat flux $\mathbf{q}_\theta$, defined as the averaged product of the fluctuating components of velocity with itself and velocity with temperature. These terms are calculated as

$$\tau = \rho\overline{\mathbf{u}'\mathbf{u}'} \quad , \quad \mathbf{q}_\theta = \rho C_p \overline{\mathbf{u}'T'} \,. \tag{6}$$

Instead of solving the transport equations for the Reynolds stress tensor and for the turbulent heat flux we approximate these terms with the eddy diffusivity model as

$$\tau = -\nu_t \left(\nabla \mathbf{u} + \nabla \mathbf{u}^T\right) + \frac{2\kappa}{3}\mathbf{I} \,, \tag{7}$$

$$\mathbf{q}_\theta = -\alpha_t \nabla T \,, \tag{8}$$

where the eddy diffusivity of momentum ν_t and the heat eddy diffusivity α_t must be properly defined in the turbulence model. We assume them as a function of the turbulence kinetic energy κ and two characteristic time scales, namely τ_{lu}, for dynamical turbulence, and $\tau_{l\theta}$, for thermal turbulence. The eddy viscosity and the eddy diffusivity are then defined as

$$\nu_t := C_\mu \, \kappa \tau_{lu} \,, \quad \alpha_t := C_\theta \, \kappa \tau_{l\theta} \,, \tag{9}$$

where $C_\mu = 0.09$ and $C_\theta = 0.1 = C_\mu/0.9$. The ratio between the two eddy diffusivities is defined as the turbulent Prandtl number, which is $Pr_t = \nu_t/\alpha_t$.

2.1. κ-ϵ-κ_θ-ϵ_θ turbulence model

We can define the turbulence kinetic energy κ and its dissipation ϵ as

$$\kappa := \frac{1}{2}\overline{\mathbf{u}'^2} \qquad \epsilon := \nu\overline{\|\nabla \mathbf{u}'\|^2} \,. \tag{10}$$

The equations for κ and for its dissipation ϵ are obtained taking appropriate moments of the Navier-Stokes equations of motion. A modeling of the terms appearing in the equations is needed and the final equation for κ is

$$\frac{\partial \kappa}{\partial t} + (\mathbf{u}\cdot\nabla)\kappa = \nabla\cdot\left[\left(\nu + \frac{\nu_t}{\sigma_\kappa}\right)\nabla\kappa\right] + P_\kappa - \epsilon \,, \tag{11}$$

where

$$P_\kappa := -\overline{u_i' u_j'}\frac{\partial u_i}{\partial x_j} = \frac{\nu_t}{2}\left|\nabla \mathbf{u} + \nabla \mathbf{u}^T\right|^2 \,. \tag{12}$$

A similar modeling for ϵ gives the equation [7]

$$\frac{\partial \epsilon}{\partial t} + (\mathbf{u}\cdot\nabla)\epsilon = \nabla\cdot\left[\left(\nu + \frac{\nu_t}{\sigma_\epsilon}\right)\nabla\epsilon\right] + C_{1\epsilon}\frac{\epsilon}{\kappa}P_\kappa - C_{2\epsilon}\frac{\epsilon^2}{\kappa}f_\epsilon \,. \tag{13}$$

with $C_{1\epsilon} = 1.5$, $C_{2\epsilon} = 1.9$, $C_\mu = 0.09$, $\sigma_\kappa = 1.4$, $\sigma_\epsilon = 1.4$, P_κ defined by (12) and f_ϵ

$$f_\epsilon = \left(1 - \exp(-0.3226\,R_\delta)\right)^2\left(1 - 0.3\exp(-0.0237\,R_t^2)\right) \,. \tag{14}$$

By defining the characteristic time $\tau_u = \kappa/\epsilon$ one can model the local dynamical time τ_{lu} in many ways. In this work we use the κ-ϵ model as reported in [9]. This model can reproduce the correct near-wall turbulence behavior, that is $\kappa \propto \delta^2$, $\epsilon \propto \delta^0$ and $\nu_t \propto \delta^3$ for $\delta \to 0$ see [7, 8, 9].

We can define also the mean square temperature fluctuation κ_θ and its dissipation ϵ_θ as

$$\kappa_\theta := \frac{1}{2}\overline{T'^2} \,, \qquad \epsilon_\theta := \frac{\nu}{Pr}\overline{\|\nabla T'\|^2} \,. \tag{15}$$

Two transport equations are needed for these variables and they can be obtained as [11]

$$\frac{\partial \kappa_\theta}{\partial t} + (\mathbf{u} \cdot \nabla)\kappa_\theta = \nabla \cdot \left[(\alpha + \frac{\alpha_t}{\sigma_{\kappa\theta}}) \nabla \kappa_\theta \right] + P_\theta - \epsilon_\theta \,, \tag{16}$$

where

$$P_\theta := -\overline{\mathbf{u}'T'} \cdot \nabla T = \alpha_t \, \|\nabla T\|^2 \,. \tag{17}$$

The equation for ϵ_θ can be written as [9, 11]

$$\frac{\partial \epsilon_\theta}{\partial t} + (\mathbf{u} \cdot \nabla)\epsilon_\theta = \nabla \cdot \left[(\alpha + \frac{\alpha_t}{\sigma_{\epsilon\theta}}) \nabla \epsilon_\theta \right] + \frac{\epsilon_\theta}{\kappa_\theta} \left(C_{p1}\, P_\theta - C_{d1}\, \epsilon_\theta \right) + \frac{\epsilon_\theta}{\kappa} \left(C_{p2}\, P_\kappa - C_{d2}\, \epsilon \right), \tag{18}$$

where P_κ is defined by (12) and P_θ by (17). For heavy liquid metals with $Pr \approx 0.025$ we have used the coefficients defined in [5, 6], namely $C_{d1} = 0.9$, $C_{p2} = 0.9$. The coefficient C_{p1} has been set to 0.925 and

$$C_{d2} = 1.9\,(1 - 0.3\,\exp(-0.0237\, R_t^2))(1 - \exp(-0.0308\, R_\delta))^2 \,. \tag{19}$$

For details one can see [4, 11, 12]. As we have done for the dynamical turbulence we can define the thermal time scale as $\tau_\theta = \kappa_\theta / \epsilon_\theta$. The ratio between the turbulent thermal time and the turbulent dynamical time, $R = \tau_\theta / \tau_u$, can be used in order to model the local thermal characteristic time $\tau_{l\theta}$, as reported in [9]. For details one can refer to [4, 7, 8, 11, 13] and references therein.

Having defined the turbulent heat flux and the Reynolds stress tensor the problem is closed and the four parameter system can be solved by assigning appropriate boundary conditions which are discussed in section 2.3.

2.2. κ-ω-κ_θ-ω_θ turbulence model

To obtain a κ-ω formulation of the turbulence model introduced in section 2.1 we define the variables ω and ω_θ, as the specific dissipation rate of κ and κ_θ. These new variables are defined as:

$$\omega = \frac{\epsilon}{C_\mu \kappa} \,, \qquad \omega_\theta = \frac{\epsilon_\theta}{C_\mu \kappa} . \tag{20}$$

By algebraically substituting the definitions (20) in the system of equations (11-13-16-18) we obtain

$$\frac{\partial \kappa}{\partial t} + \mathbf{u} \cdot \nabla \kappa = \nabla \cdot \left[\left(\nu + \frac{\nu_t}{\sigma_k} \right) \nabla \kappa \right] + P_\kappa - C_\mu\, \kappa\, \omega \,, \tag{21}$$

$$\begin{aligned} \frac{\partial \omega}{\partial t} + \mathbf{u} \cdot \nabla \omega = \nabla \cdot \left[\left(\nu + \frac{\nu_t}{\sigma_\epsilon} \right) \nabla \omega \right] + \frac{2}{\kappa} \left(\nu + \frac{\nu_t}{\sigma_\epsilon} \right) \nabla \kappa \cdot \nabla \omega + \\ + (c_{\epsilon 1} - 1)\, \frac{\omega}{\kappa} P_\kappa - C_\mu\, (c_{\epsilon 2} f_\epsilon - 1)\, \omega^2 \,, \end{aligned} \tag{22}$$

$$\frac{\partial \kappa_\theta}{\partial t} + \mathbf{u} \cdot \nabla \kappa_\theta = \nabla \cdot \left[\left(\alpha + \frac{\alpha_t}{\sigma_\theta} \right) \nabla \kappa_\theta \right] + P_\theta - C_\mu\, \kappa_\theta\, \omega_\theta \,, \tag{23}$$

$$\begin{aligned} \frac{\partial \omega_\theta}{\partial t} + \mathbf{u} \cdot \nabla \omega_\theta = \nabla \cdot \left[\left(\alpha + \frac{\alpha_t}{\sigma_\theta} \right) \nabla \omega_\theta \right] + \frac{2}{\kappa_\theta} \left(\alpha + \frac{\alpha_t}{\sigma_\theta} \right) \nabla \kappa_\theta \cdot \nabla \omega_\theta + \\ + (c_{p1} - 1)\, \frac{\omega_\theta}{\kappa_\theta} P_\theta + c_{p2} \frac{\omega_\theta}{\kappa} P_\kappa - (c_{d1} - 1)\, C_\mu \omega_\theta^2 - c_{d2} C_\mu \omega \omega_\theta \,. \end{aligned} \tag{24}$$

For the κ-ω turbulence model the coefficients $c_{\epsilon 1}$, $c_{\epsilon 2}$ and the function f_ϵ are the same used in the κ-ϵ model. For the κ_θ-ω_θ thermal turbulence model the coefficient c_{p2} and the function

c_{d2} are the same used in the κ_θ-ϵ_θ model, while c_{p1} and c_{d1} have been set to 1.025 and 1.1. In a κ-ω formulation the time scales of turbulence can be simply computed as $\tau_u = (C_\mu\,\omega)^{-1}$ and $\tau_\theta = (C_\mu\,\omega_\theta)^{-1}$, so the time ratio becomes $R = \omega/\omega_\theta$. Using this model we can avoid the variables coupling in the boundary conditions which occurs in κ-ϵ formulation, as we prove in the following section.

2.3. Boundary conditions for the turbulence models

We can study the near wall behavior of the turbulent variables by using Taylor series expansion near the wall. By applying this expansion to κ, ϵ, κ_θ and ϵ_θ we obtain the following expressions

$$\kappa_w = \frac{1}{2}\,a\,\delta^2, \qquad \epsilon_w = \nu\frac{\kappa_w}{\delta^2}, \qquad \kappa_{\theta w} = \frac{1}{2}\,a_\theta\,\delta^2, \qquad \epsilon_{\theta w} = \alpha\frac{\kappa_{\theta w}}{\delta^2}\,, \tag{25}$$

where δ is the distance from the wall and a and a_θ are constant values that depend on the velocity and temperature fluctuations. In this case we cannot impose exact Dirichlet boundary conditions on ϵ and ϵ_θ because the values κ_w and $\kappa_{\theta w}$ are not known *a priori*. A Neumann boundary condition is imposed for κ and κ_θ while a Dirichlet boundary condition depending on the value of κ and κ_θ as computed in the previous iteration is employed for ϵ and ϵ_θ. This algorithm can lead to convergence issues if oscillations of the solutions on the wall do generate.

The near wall Taylor series expansion of ω and ω_θ is

$$\omega_w = \frac{2\nu}{C_\mu\delta^2}\,, \qquad \omega_{\theta w} = \frac{2\alpha}{C_\mu\delta^2}\,. \tag{26}$$

As can be seen in (26) it is now possible to impose exact Dirichlet boundary conditions for ω and ω_θ near the wall because the terms on the right hand side are known values. In this case we have no coupling between the variables on the wall boundary condition and the solution of this system is more robust and stable. As it can be seen, in both formulations the time scale ratio R on the wall is equal to the molecular Prandtl number of the fluid while the turbulent kinetic energy κ and the mean square temperature fluctuation κ_θ tend to zero on the wall.

3. Numerical results

In the next sections we report the numerical results of fully developed turbulent flow simulations in different geometries. The physical properties of the fluid are reported in Table 1. These properties are representative of Lead-Bismuth-Eutectic and other heavy liquid metal fluids with a molecular Prandtl number of $Pr = 0.025$.

An in-house finite element code is used to solve all the systems of partial differential equations. The code allows to refine the mesh to obtain better resolution using a multigrid solver. We employ Taylor-Hood finite elements for the system of Navier-Stokes in order to satisfy the inf-sup condition and this system is solved with a fully coupled velocity-pressure method. The two systems of turbulence equations are solved with standard quadratic finite elements. All the transport equations are up-winded by a SUPG algorithm to improve the stability of the solution. The solution of the turbulent fields is obtained in two steps: first the dynamical fields are obtained and then the thermal ones are solved.

Physical properties of LBE fluid

Viscosity	μ	0.00184 Pa s
Density	ρ	10340 Kg/m^3
Thermal conductivity	λ	10.72 W/(m K)
Heat specific capacity	C_p	145.75 J/(Kg K)

Table 1. Physical parameters used in the CFD simulations.

3.1. Plane channel

In this section we report numerical results obtained for a fully developed turbulent flows in plane channel geometry. This geometry is chosen because it is very simple and many DNS data are available for this flow. The physical properties of the fluid are reported in Table 1. We

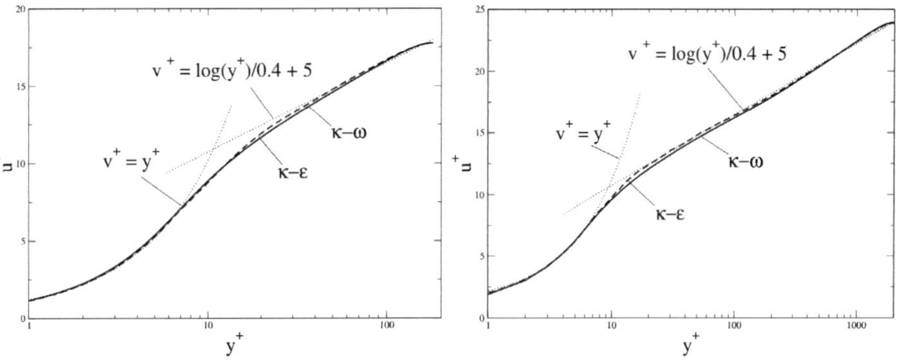

Figure 1. Plane case. Mean velocity distribution for $Re \approx 5500$ (left) and $Re \approx 86200$ (right) as computed with κ-ϵ and κ-ω model.

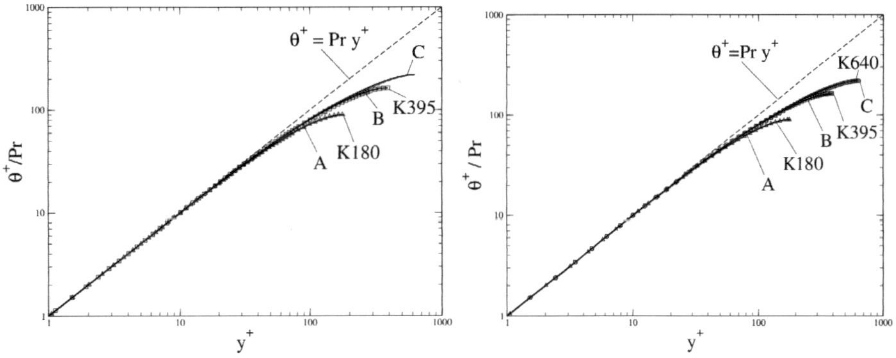

Figure 2. Plane case. Comparison of the temperature distribution θ^+/Pr obtained with κ-ϵ (left) and κ-ω model (right), with DNS data. DNS data are reported for $Re_\tau = 180$ (K180), $Re_\tau = 395$ (K395) and $Re_\tau = 640$ (K640).

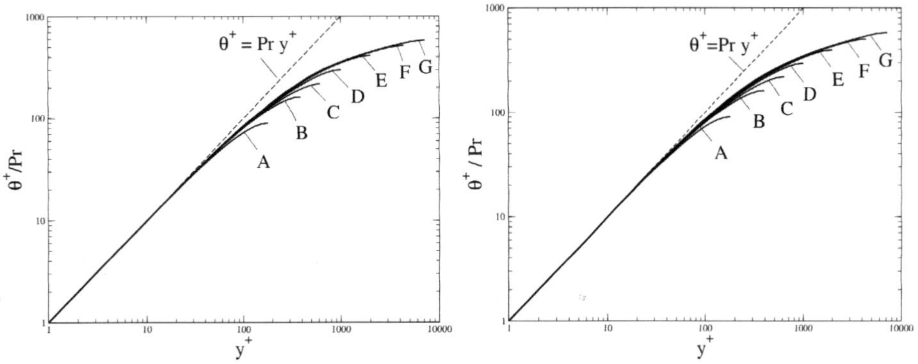

Figure 3. Plane case. Temperature distribution θ^+/Pr for different velocities $Re \approx 5500$ (A), 13500 (B), 23250 (C), 40100 (D), 86200 (E), 203900 (F) and 344800 (G). On the left the results were obtained with the κ-ϵ-κ_θ-ϵ_θ model while on the right the results were obtained with κ-ω-κ_θ-ω_θ model

consider two plates located at a distance $L = 0.0605$ m and with infinite dimensions in the other directions. On the wall a uniform heat flux of 360000 W/m^2 is applied. We can solve this problem in two dimensions with periodic boundary conditions on the inlet and outlet of the channel. The results of the κ-ω-κ_θ-ω_θ model are compared with the ones obtained with the κ-ϵ-κ_θ-ϵ_θ model and with DNS data from Kawamura database, see [3] and references therein. The fields in DNS data are usually reported in non dimensional form. The dimensionless temperature is defined as $\theta^+ = \theta/T_\tau$, where T_τ is the friction temperature calculated as $T_\tau = q/(u_\tau \rho C_p)$ and θ is the difference between the temperature and the linear behavior characteristic of the fully developed flow, $\theta = T - T_{w0} - x\,\Delta T_b$. T_{w0} is the reference inlet wall temperature which is assumed to be zero. By using the temperature θ instead of T we can impose periodic boundary conditions on θ when the flow is fully developed.

Seven simulations have been performed with Reynolds number of $Re \approx 5500$ (A), 13500 (B), 23250 (C), 40100 (D), 86200 (E), 203900 (F) and 344800 (G). They correspond to the friction Reynolds number of 180 (A), 395 (B), 640(C) , 1010 (D), 2000 (E), 4400 (F) and 7200 (G). A full report of these data can be found in [9, 10], where other data such as turbulent heat flux and turbulent Prandtl number are shown and a discussion over the use of different boundary conditions for the four parameter turbulence model is carried on. We remark the differences in convergence and stability of the solver for the two different turbulence models: the κ-ϵ solver needs a dt as time step for the transient solution which is ten to a hundred times smaller than the dt used for the κ-ω solver, otherwise no stable solution can be obtained. In Fig. 1 the non dimensional mean velocity profiles obtained with the κ-ω and with the κ-ϵ model $u^+ = \mathbf{u}/u_\tau$ are reported as a function of the non dimensional wall distance $y^+ = yu_\tau/\nu$ for two test cases, namely the case $Re \approx 5500$ (A) and $Re \approx 86200$ (E). The friction velocity u_τ is defined as $u_\tau = \sqrt{\tau_w/\rho}$ with τ_w the wall shear stress. The κ-ω model well reproduces the linear and logarithmic behaviors of the velocity, namely $u^+ = y^+$ and $u^+ = \log(y^+)/0.4 + 5$. Comparing the results obtained with κ-ω model with the ones obtained with κ-ϵ we can observe only a slight difference between them in the buffer region. In Fig. 2 the non dimensional temperature θ^+ is reported as a function of y^+ and divided by the Prandtl number. The temperature profiles for the cases $Re_\tau = 180$, $Re_\tau = 395$ and $Re_\tau = 640$, are compared with DNS data from Kawamura [3]. The data obtained with the four parameter models agree very well with the DNS ones. The non dimensional temperature profiles of all the simulated cases are reported in Fig. 3. The results obtained with the κ-ω-κ_θ-ω_θ, Fig. 3 (right), show no sensible differences with the κ-ϵ-κ_θ-ϵ_θ ones, Fig. 3 (left). The root-mean-square temperature fluctuation $\theta_{rms} = \sqrt{2\kappa_\theta}$ is an important variable needed to evaluate the time ratio R and the turbulent heat diffusion

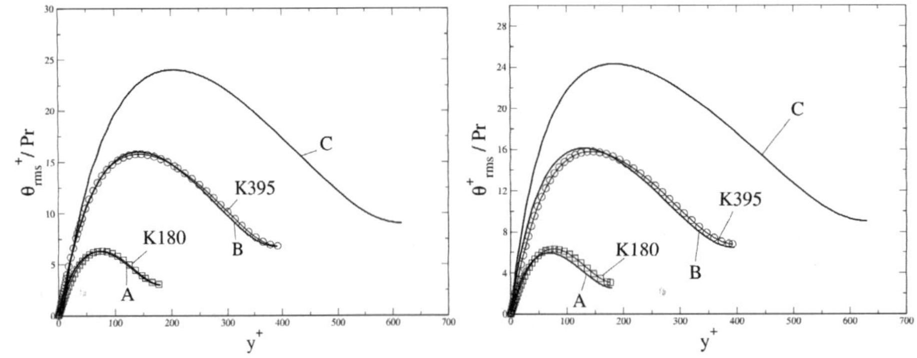

Figure 4. Plane case. The root-mean-square temperature fluctuations θ^+_{rms} for different $Re \approx 5500$ (A), 13500 (B), 23250 (C) and comparison with DNS corresponding data (K180) and (K395). On the left results with the κ-ϵ model, on the right κ-ω model.

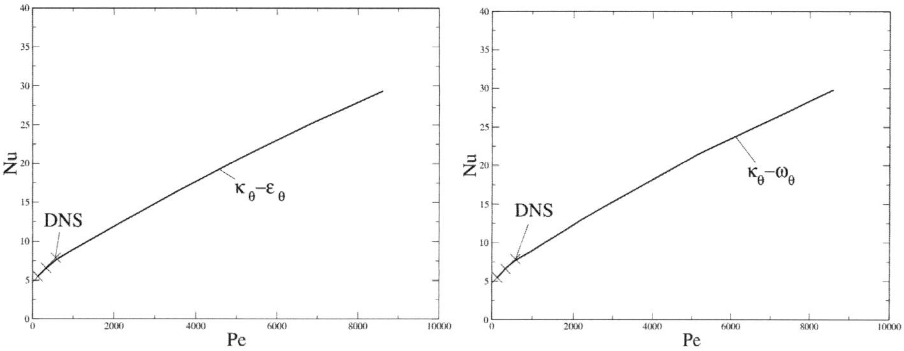

Figure 5. Plane case. Nusselt number (thick line) and DNS values (cross) for different Peclet numbers. On the left results with the κ-ϵ model, on the right κ-ω model.

coefficient α_t. In Fig. 4 the non dimensional $\theta^+_{rms} = \theta_{rms}/T_\tau$ is reported for the κ-ϵ (left) and κ-ω (right) formulation of the model and it is compared with DNS data for $Re_\tau = 180$ and 395. Very similar results to DNS data are obtained with both the turbulence models, a slight difference in the prediction of the position of the peak can be seen between the two models but the overall result is very good. The Nusselt number is the most important non dimensional number in engineering heat transfer calculations and it quantifies the heat transfer effectiveness. It is defined as $Nu = qD_h/(\lambda\Delta T)$ where D_h is the hydraulic diameter of the geometry and ΔT is the difference between the average wall temperature and the bulk temperature, defined as the average temperature on the section of the flow with respect to velocity. In Fig. 5 the Nusselt number calculated with the κ-ϵ (left) and κ-ω (right) four parameter turbulence models is reported as a function of Peclet number $Pe = Re\,Pr$. DNS data are reported as well for comparison. The first two points matches very well while the third (corresponding to $Re \approx 23250$ (C)) seems to slightly overestimate the DNS result.

3.2. *Cylinder channel*

As a second test we simulate a fully developed turbulent flow in a cylindrical pipe with diameter $D = 0.0605$. The physical properties of the simulated flow are reported in Table 1. For this geometrical case we compare our results with DNS data, which are available only for the case

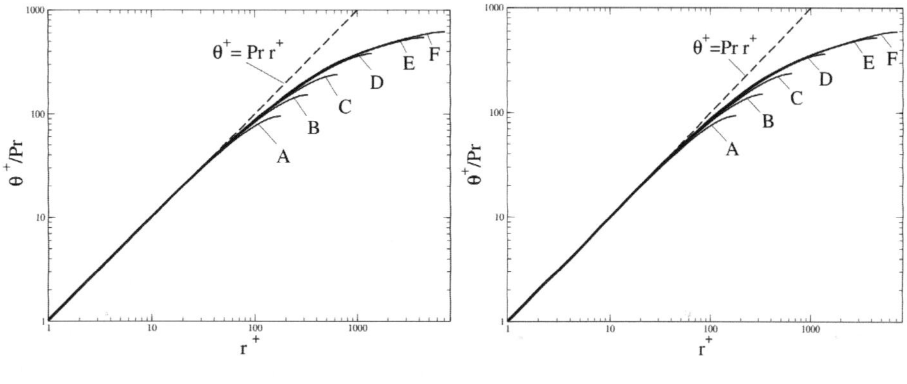

Figure 6. Cylinder case. Temperature distribution θ^+/Pr for $Re \approx 5500$ (A), 11150 (B), 23750 (C), 57500 (D), 213000 (E) and 345000 (F). Results obtained with κ-ϵ-κ_θ-ϵ_θ (left) and κ-ω-κ_θ-ω_θ (right) model.

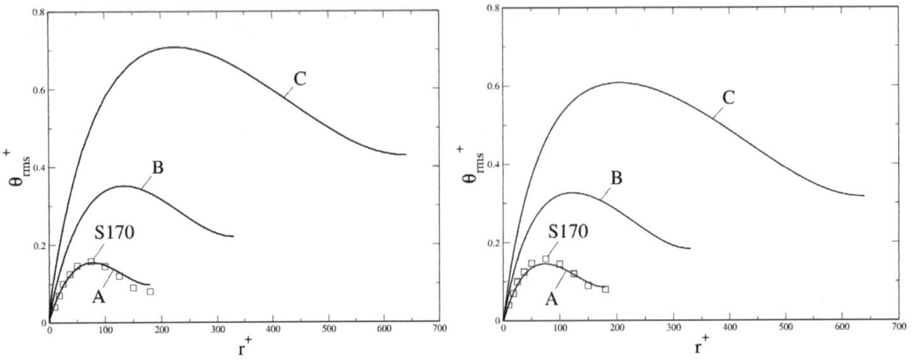

Figure 7. Cylinder case. Non-dimensional root-mean-square temperature fluctuation θ^+_{rms} for $Re \approx 5500$ (A), 11150 (B) and 23750 (C) and comparison with DNS data for $Re_\tau = 170$ (S170). On the left the results were obtained with the κ-ϵ model while on the right they were calculated with the κ-ω model.

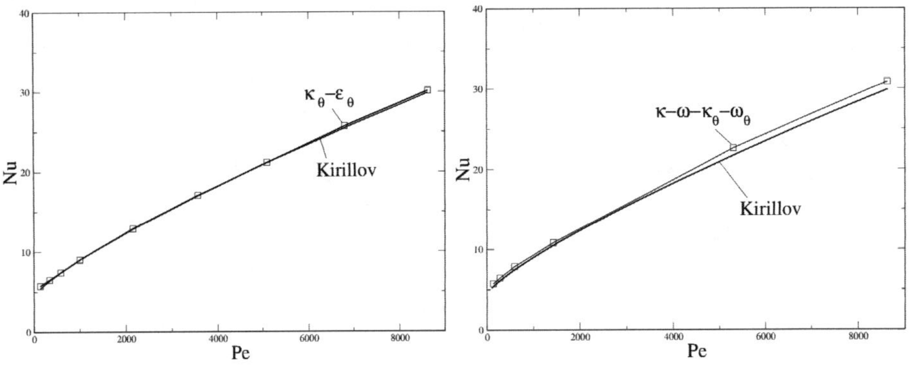

Figure 8. Cylinder case. Nusselt number (square) and Kirillov correlation for cylindrical geometry (thick line). On the left results obtained with the κ-ϵ model while on the right calculated with the κ-ω model.

$Re_\tau = 170$, and with the Kirillov heat transfer correlation

$$Nu = 4.5 + 0.018\, Pe^{0.8}\,. \tag{27}$$

This is the reference correlation for this class of fluids in cylindrical geometry [14, 15] and it is claimed to be valid for $10^4 < Re < 5 \cdot 10^6$.

We performed six simulations over a wide range of Reynolds numbers, namely $Re \approx 5500$ (A), 11150 (B), 23750 (C), 57500 (D), 213000 (E) and 345000 (F) corresponding to $Re_\tau = 180$ (A), 395 (B), 640 (C), 1400 (D), 4500 (E) and 7150 (F). The same behavior of the different solvers for κ-ϵ and κ-ω model as found for the plane cases is observed for the solution in these test cases. In Fig. 6 the non dimensional temperature divided by the Prandtl number θ^+/Pr is reported for all the simulated cases, as a function of the non dimensional wall distance $r^+ = ru_\tau/\nu$. Comparing Fig. 6 (right) with Fig. 6 (left) we can see that the κ-ω results are very close to the κ-ϵ ones. The linear behavior of the non dimensional temperature field is well reproduced in all the simulated cases. The non-dimensional root-mean-square temperature fluctuation θ^+_{rms} for $Re \approx 5500$ (A), 11150 (B) and 23750 (C) is compared with DNS data for $Re_\tau = 170$ (S170) in Fig. 7. A good agreement with the DNS data is obtained with both models. Comparing the κ-ϵ results with the κ-ω ones, for the cases $Re_\tau = 395$ and $Re_\tau = 640$ one can see some differences between the models but there are no reference results to compare with. The Nusselt number for these simulations is reported in Fig. 8. The square dots are the results obtained with the four parameter turbulence model, while the thick line is Kirillov correlation. On the left the results are reported for the κ-ϵ model while on the right for the κ-ω model. The matching between the numerical results and the correlation is almost perfect at any velocity for the κ-ϵ model while little discrepancies can be seen for the κ-ω model at very high velocities.

4. Conclusion

In this work we have proposed and analyzed an improved four parameter turbulence model for heat transfer computation of heavy liquid metal flows with $Pr \approx 0.025$. We have derived a κ-ω formulation from an already tested κ-ϵ turbulence model to obtain a more stable and robust model to compute turbulent heat transfer. The model has been tested in different geometries and a comparison with DNS data, experimental results and numerical results obtained with the κ-ϵ model has been performed. The results obtained show that the new κ-ω-κ_θ-ω_θ four parameter turbulence model is more stable and convergence obtained by using this model is faster. Moreover the comparison with DNS data and experimental results has proven that reliable results can be obtained with the new model, so it can be considered a useful tool in many engineering fields with heavy liquid metal applications.

References

[1] Cheng X and Tak N 2006 Investigation on turbulent heat transfer to lead-bismuth eutectic flows in circular tubes for nuclear applications *Nucl. Eng. Des.* **236** 385-93
[2] Cheng X and Tak N 2006 CFD analysis of thermal-hydraulic behavior of heavy liquid metals in sub-channels *Nucl. Eng. Des.* **236** 1874-85
[3] Kawamura H, Abe H and Matsuo Y 1999 DNS of turbulent heat transfer in channel flow with respect to Reynolds and Prandtl number effects *Int. J. Heat. Fluid. Fl.* **20** 196-207
[4] Hwang C B and Lin C A 1999 A low Reynolds number two-equation $k_t - e_t$ model to predict thermal field *Int. J. Heat. Mass Tran.* **42** 3217-30
[5] Nagano Y and Shimada M 1996 Development of a two equation heat transfer model based on direct simulations of turbulent flows with different Prandtl numbers Phys. Fluids **8** 3379-402
[6] Abe K, Kondoh T and Nagano Y 1995 A new turbulence model for predicting fluid flow and heat transfer in separating and reattaching flows II. Thermal field calculations *Int. J. Heat. Mass Tran.* **38** 1467-81
[7] Nagano Y, Pei C Q and Hattori H 1999 A new low-Reynolds-number one-equation model of turbulence *Flow Turbul. Combust.* **63** 135-51
[8] Abe K, Kondoh T and Nagano Y 1996 A two-equation heat transfer model reflecting second-moment closures for wall and free turbulent flows *Int. J. Heat. Fluid. Fl.* **17** 228-37
[9] Manservisi S and Menghini F 2014 A CFD four parameter heat transfer turbulence model for engineering applications in heavy liquid metals in rod bundle geometries *Int. J. Heat. Mass Tran.* **69** 312-26
[10] Manservisi S and Menghini F 2014 Triangular rod bundle simulations of a CFD κ-ϵ-κ_θ-ϵ_θ heat transfer turbulence model for heavy liquid metals *Nucl. Eng. Des.* **273** 251-70
[11] Hattori H, Nagano Y and Tagawa M 1993 Analysis of turbulent heat transfer under various thermal conditions with two-equation models *Eng. Turb. Mod. Exp.* **2** 43-52
[12] Launder B E, Reece G J and Rodi W 1975 Progress in the development of a Reynolds-stress turbulence closure *J. Fluid Mech.* **37** 537-66
[13] Deng B, Wu W and Xi S 2001 A near-wall two-equation heat transfer model for wall turbulent flow *Int. J. Heat. Mass Tran.* **44** 691-8
[14] Saha S, Chin C, Blackburn H M and Ooi A S H 2011 The influence of pipe length on thermal statistics computed from DNS of turbulent heat transfer *Int. J. Heat. Fluid. Fl.* **32** 1083-97
[15] Kirillov P and Ushakov P A 2001 Heat transfer to liquid metals: specific features, methods of investigation and main relationships *Therm. Eng.* **48** 50-9

33rd UIT (Italian Union of Thermo-fluid-dynamics) Heat Transfer Conference IOP Publishing
Journal of Physics: Conference Series **655** (2015) 012047 doi:10.1088/1742-6596/655/1/012047

Numerical and experimental study of local heat transfer enhancement in helically coiled pipes. Preliminary results.

F Bozzoli[1], L Cattani[1], S Rainieri[1] and A Zachár[2]

[1] Department of Industrial Engineering, University of Parma, Parco Area delle Scienze 181/A, 43124 Parma, Italy.
[2] Institute of Informatics, College of Dunaújváros, Táncsics M. u. 1, Dunaújváros H-2103, Hungary
E-mail address of the corresponding Author (Sara Rainieri): sara.rainieri@unipr.it

Abstract In the last years, the attention of heat transfer equipments manufacturers turned toward helically coiled-tube heat exchangers, especially with regards to applications for viscous and/or particulate products. The recent progress achieved in numerical simulation motivated many research groups to develop numerical models for this kind of apparatuses. These models, intended both to improve the knowledge of the fundamental heat transfer mechanisms in curved geometries and to support the industrial design of this kind of apparatuses, are usually validated throughout the comparison with either theoretical or experimental evidences by considering average heat transfer performances. However, this approach doesn't guarantee that the validated models are able to reproduce local effects in details, which are so important in this kind of non-standard geometries. In the present paper a numerical model of convective heat transfer in coiled tubes for laminar flow regime was formulated and discussed. Its goodness was checked throughout the comparison with the latest experimental outcomes of Bozzoli et al. [1] in terms of convective heat flux distribution along the boundary of the duct, by ensuring the effectiveness of the model also in the description of local behaviours. Although the present paper reports only preliminary results of this simulation/validation process, it could be of interest for the research community because it proposes a novel approach that could be useful to validate many numerical models for non-standard geometries.

1. Introduction

Helically coiled-tube heat exchangers are used in several industrial applications because, for given operating conditions, heat transfer rate in this kind of devices is significantly larger than in straight pipes [1]. The effectiveness of wall curvature in enhancing thermal performances primarily occurs because it gives origin in the fluid to the centrifugal force that induces local maxima in the velocity distribution and, as a consequence, it locally increases the temperature gradients at the wall [2]. This kind of passive heat transfer enhancement is particularly interesting when highly viscous fluids are treated because, with these fluids, the laminar flow regime is usually encountered and convective heat transfer rate is limited.

In the last years the attention of heat transfer equipment manufacturers turned toward helically coiled-tube heat exchangers and the progress in the numerical simulation have motivated many research groups to develop numerical models for this kind of apparatuses. These models are intended both to improve the knowledge of the fundamental heat transfer mechanisms in curved geometries and to support the industrial design of this kind of apparatuses. Some representative examples of numerical models for the laminar fluid flow in curved tubes are proposed by Mitsunobu et al [3], Kozo and Yoshiyuki [4] Yang et al [5] while the turbulent flow regime was deeply investigated by Di Piazza and Ciofalo [6] and Di Liberto and Ciofalo [7].

In [3] a finite-difference solution using a combination of line iterative method and boundary vorticity method is presented for the hydrodynamically and thermally fully developed laminar forced convection in curved pipes. The method was applied to single out the effect of Dean number on local Nusselt for two different fluids having Prandtl number equal to 0.7 and 100 respectively.

Kozo and Yoshiyuki [4] carried out a theoretical study on the effect of the secondary flow on heat transfer in uniformly heated helically coiled tube. Both the centrifugal and buoyancy forces are taken into account in the numerical analysis. The solutions cover a range of Prandtl numbers between 1 and 100. The velocity and temperature profiles, the local friction factor and heat transfer coefficient distribution are obtained.

Yang et al. [5] presented a numerical investigation on the fully developed laminar convective heat transfer in a helicoidal pipe, with particular attention to the effects of torsion on the local heat-transfer coefficient. In particular, the authors reported the Nusselt number distribution varying the coil pitch, and they showed that, due to torsion, the local heat-transfer coefficient, compared to the case of an ideal torus, is increased on half of the tube wall while it is decreased on the other half.

Di Piazza and Ciofalo [6] analysed the heat transfer in turbulent flow regime in curved pipes with curvature ratio, defined as the ratio between the pipe radius and the coil radius, ranging from 0.03 to 0.3. The authors applied to the problem the code ANSYS CFX that employs a coupled technique that, at the same time solves all the transport equations in the whole domain through a false time-step algorithm. The domain under numerical study was a short segment of toroidal pipe of circular cross section. The Authors employed and compared different turbulence models as the standard k-ε with wall functions, the SST k-ω and a second-order RSM. The study was performed by varying the Reynolds number and the Prandtl number in the range 1.4×10^4- 8×10^4 and 0.7-5.6 respectively.

Di Liberto and Ciofalo [7] studied the turbulent heat transfer in curved ducts by numerical simulation considering two different curvature ratios (0.1 and 0.3). The simulations were performed at Prandtl number equal to 0.86 and at Reynolds number varying between 12630 and 17350.

It is universally accepted that numerical models have scientific relevance only if they are validated by the comparison with theoretical or experimental evidences, but, in some cases, the applications of this principle can be ambiguous. Regarding heat transfer models of coiled tubes, they are usually validated only considering average heat transfer performances. However, this approach doesn't guarantee that the so validated models are able to reproduce local effects in details. Moreover most of the available studies regard low Prandtl number fluids, while the experimental data frequently refer to medium and highly viscous fluids. Since the flow velocity and temperature fields in coiled ducts are strongly asymmetrical over the cross-section of the tube, local effects have to be accounted for when a numerical model has to be developed. Moreover, local behavior can affect the performances of many industrial processes. For instance, in food pasteurisation, the irregular temperature field induced by the wall curvature could reduce the bacteria heat-killing or could locally overheat the product. On the other hand, the scarcity of available experimental data regarding local quantities makes this kind of model validation difficult to be performed. Regarding coiled heat exchangers, most of the papers available in the scientific literature did not investigate local thermal behavior, mainly due to the practical difficulty of measuring heat flux on the internal wall surface of a pipe, and they presented the results only in terms of the Nusselt number averaged along the wall circumference. To the Authors's knowledge, only five papers [8-12] present experimental results in terms of local Nusselt number.

Bai et al. [8] experimentally studied the turbulent heat transfer in helically coiled tubes using deionised water as the working fluid. The working fluid was heated by applying alternating current in the tube wall and, in each cross section eight thermocouples were placed on the external surface of the tube wall. The local heat-transfer distribution on the internal wall of the tube was found by solving the two-dimensional inverse heat conduction problem whit the least-square method. As expected, they found that the local heat-transfer coefficient was not evenly distributed along the periphery of the cross section and that, in particular, at the outside surface of the coil, it was three or four times higher than that at the inside surface.

Bozzoli et al. [9] focused their investigation on the thermally fully developed region for the laminar flow regime in the Reynolds number range of 135 to 1050 and in the Prandtl number range of 170 to 200. The temperature distribution maps on the external coil wall were employed as input data of the linear inverse heat conduction problem in the wall under a solution approach based on the Tikhonov regularisation method with the support of the fixed-point iteration technique to determine the proper regularisation parameter. The results showed that, at the outside surface of the coil, the Nusselt number is approximately five times larger than that at the inside surface and this ratio, in the conditions under test, is constant.

Regarding local heat-transfer coefficient, some experimental were discussed by Seban et al. [10] investigating the laminar flow of oil and the turbulent flow of water in tubes coils. These Authors correctly drew the attention to the difference between apparent and true local values: apparent heat transfer coefficient are obtained neglecting the circumferential heat conduction in the tube wall, that means considering the average value of the convective heat flux instead of the punctual value. In terms of true heat transfer coefficient, the ratio of the outside to the inside coefficient found in this experimental campaign was about four for both the laminar flow case and the turbulent one. However, no details about the approach adopted to estimate the punctual convective heat flux were given in this paper.

Both Xin and Ebadian [11] and Janssen and Hoogendoorn [12] developed an extensive experimental campaign with many different fluids on a wide range of curvature ratios and Reynolds numbers. However, these Authors processed their data neglecting the circumferential heat conduction in the tube wall, therefore the reported local heat-transfer coefficient values were the apparent ones and not the real ones.

In the present paper a numerical model of convective heat transfer in coiled tubes for laminar flow regime was formulated and discussed. The validation process was developed through the comparison of the numerical results with the latest experimental outcomes of Bozzoli et al. [1] in terms of convective heat flux distribution along the boundary of the duct. This procedure verifies the robustness of the model in the description of local effects in typical thermal treatments of medium viscous fluids .

2. Mathematical formulation

This section provides the basic equations that must be solved to describe the velocity field and the temperature distribution inside the coiled heat exchanger and the temperature distribution in the coil wall. The physical problem is a time independent three dimensional heat and fluid flow process.

The momentum equation of the fluid inside the heat exchanger coil is expressed by the three dimensional Navier-Stokes equations and the conductive heat transport in the wall of the coil is described by the Laplace equation. A SIMPLE like method is applied to solve the momentum and continuity equations. The dependent variables that describe the present problem are the temperature T, for the fluid and solid wall regions of the studied problem, velocity components U_i in the x_1, x_2 and x_3 directions respectively and the pressure field P.

2.1. Equations

The following set of partial differential equations for U_1, U_2, U_3, P and T as a function of x_1, x_2, x_3 describes the flow and temperature field inside a helically coiled heat exchanger. The conservation equations are formulated in the Cartesian coordinate system because the applied flow solver (Ansys CFX 11.0) uses the Cartesian system to formulate the conservation equations for all quantities (vector U_i and scalar T, P). Temperature dependency of the different physical properties (ρ, c_p, η, λ) of the working fluid are neglected.

Journal of Physics: Conference Series **655** (2015) 012047 doi:10.1088/1742-6596/655/1/012047

In the fluid domain the continuity equation is formulated in the following manner in Cartesian coordinate system:

$$\frac{\partial U_1}{\partial x_1}+\frac{\partial U_2}{\partial x_2}+\frac{\partial U_3}{\partial x_3}=0. \tag{1}$$

The following equation system is the formulation of the momentum equations in Cartesian coordinate system where $i \in \{1,2,3\}$:

$$\rho\left(U_1\frac{\partial U_i}{\partial x_1}+U_2\frac{\partial U_i}{\partial x_2}+U_3\frac{\partial U_i}{\partial x_3}\right)=-\frac{\partial P}{\partial x_i}+\eta\left(\frac{\partial^2 U_i}{\partial x_1^2}+\frac{\partial^2 U_i}{\partial x_2^2}+\frac{\partial^2 U_i}{\partial x_3^2}\right) \tag{2}$$

η is the dynamic viscosity and ρ is the density of the working fluid.
The energy equation, under the assumption of negligible viscous dissipation, is then solved to calculate the temperature field of the studied flow process for the fluid domain

$$\rho c_P\left(U_1\frac{\partial T}{\partial x_1}+U_2\frac{\partial T}{\partial x_2}+U_3\frac{\partial T}{\partial x_3}\right)=\lambda\left(\frac{\partial^2 T}{\partial x_1^2}+\frac{\partial^2 T}{\partial x_2^2}+\frac{\partial^2 T}{\partial x_3^2}\right), \tag{3}$$

where λ and c_p are the thermal conductivity and the specific heat at constant pressure of the fluid respectively. The transport equations have been formulated in a conservative form which is more suitable for numerical treatment.

In the solid domain the steady state energy balance equation is expressed as follows:

$$\left(\frac{\partial^2 T}{\partial x_1^2}\right)+\left(\frac{\partial^2 T}{\partial x_2^2}\right)+\left(\frac{\partial^2 T}{\partial x_3^2}\right)+\frac{q_g}{k}=0 \tag{4}$$

where q_g is the heat generated by Joule effect in the wall, k is the wall thermal conductivity. The solid and the fluid domain are coupled by imposing continuity of temperature and heat flux on the interior wall boundary.
A constant mass flow rate is assumed at the inlet section of the coiled tube. The gradient of the velocity profile and of the temperature field is assumed to be zero at the outlet section. An adiabatic boundary condition (fully insulated) is specified at the outer wall of the helical tube.

2.2. Domain of discretization
The domain of the numerical computation consists of two separate calculation domains: the fluid and the solid pipe wall domain. The geometry has been drawn separately with 3D geometry modeler software (Inventor 11), as shown in figure 2. The developed geometry has been imported into the ICEM CFD software to generate the numerical grids and, after that, the generated grids have been imported by the CFX task preparation environment. A solid-fluid domain interface has been used to connect the fluid and the solid region of the investigated heat and fluid flow problem. So called prismatic cells have been used on the inner surface of the helical pipe and on the near wall region of the fluid domain. Prismatic cells are generally used to discretize boundary layers of different kind of fluid flow problems near solid surfaces. A prismatic cell is a specially constructed finite volume element that is similar to a thin triangular plate. The reason of using this kind of volume element type is to resolve the rapid changing of the normal directional values of the flow variables near the solid surface in the boundary layer.

2.4. Numerical solution of the transport equations

The equations with the appropriate boundary conditions have been solved with a commercially available CFD code (Ansys CFX 11.0). A "High Resolution Up-Wind like" scheme is used to discretize the convection term in the transport equations. The resulting large linear set of equations is solved with an algebraic multi-grid solver.

3. Model validation

The validation process was developed throughout the comparison of the numerical results to the latest experimental outcomes of Bozzoli et al. [1] in terms of convective heat flux distribution along the boundary of the duct. This approach ensures the robustness of the model in the description of local effects.

Bozzoli et al. in [1] tested a helically coiled stainless steel type AISI 304 tube in terms of convective heat flux and heat transfer coefficient distribution along the boundary of the duct. The tube had smooth wall and it was characterised by eight coils following a helical profile along the axis of the tube. The tube internal diameter was 14 mm, and the wall thickness measured 1.0 mm, while the helix diameter and the pitch were approximately 310 mm and 200 mm, respectively.

The working fluid (i.e. ethylene glycol) entered the coiled test section equipped with stainless-steel fin electrodes, which were connected to a power supply. This setup allowed the investigation of the heat transfer performance of the tube under the prescribed condition of uniform heat flux generated by the Joule effect in the wall.

To minimise the heat exchange with the environment, the heated section was thermally insulated. A small portion of the external tube wall, near the downstream region of the heated section, was made accessible to an infrared imaging camera by removing the thermally insulating layer, and it was coated by a thin film of opaque paint of uniform and known emissivity. Therefore, the test section was taken approximately 9 m downstream the inlet section, in the region of the heated section where the laminar boundary layers reached the asymptotic profiles. This condition makes the results obtained for this particular section representative of the thermally fully developed region.

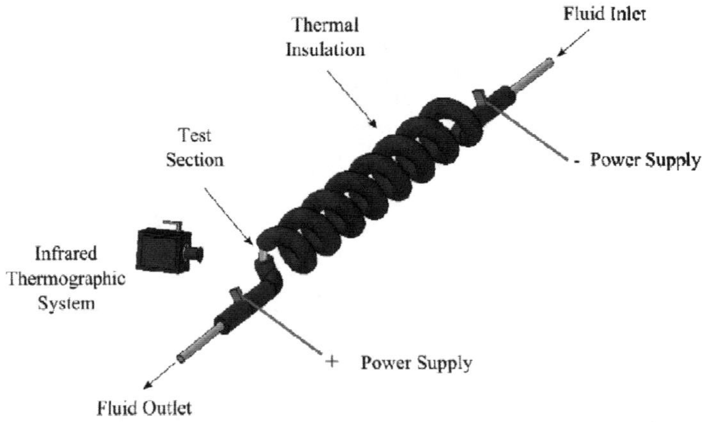

Figure 1. Sketch of the experimental setup.

The surface temperature distribution was acquired by means of a FLIR SC7000 infrared camera. The inlet and the outlet fluid bulk temperatures were measured with type-T thermocouples. The bulk temperature at any location in the heat transfer section was then calculated from the power supplied to the tube wall. The temperature distribution maps on the external coil wall were employed as input data of the linear inverse heat conduction problem in the wall under a solution approach based on the

Tikhonov regularisation method. This procedure is allowed to estimate the local convective wall heat flux and the local convective heat transfer coefficient along the boundary of the duct section.

The whole coiled tube was simulated in ANSYS environment by the numerical model above described. Figure 2 reports a view of the geometrical model of the investigated coiled tube. The applied grids, as shown in figure 3, are strongly non-uniform in order to resolve the wall boundary layers effects. Prismatic cells have been used on the inner surface of the helical pipe and on the near wall region of the helical fluid domain. Fifteen layers of prismatic cells have been generated with 1.15 increment factor to discretize the near wall boundary of the fluid domain inside the helical tube. The inner side boundary region of the solid tube wall (made of stainless steel) domain is created with five layers of prismatic cells. The total sum of finite volumes of the four separately generated grids is 43´664´780 finite volumes.

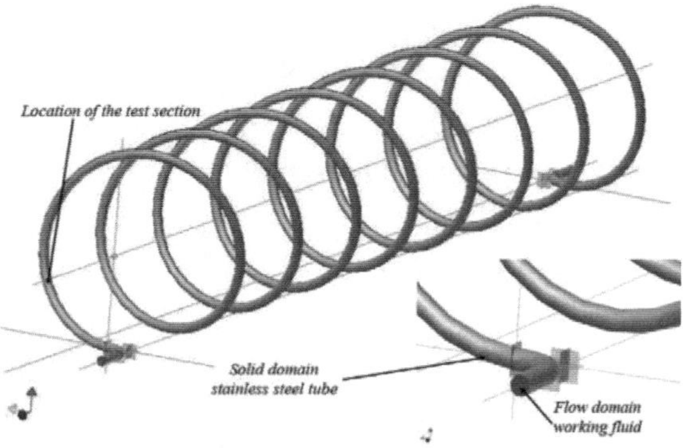

Figure 2. 3D geometrical model of the investigated coiled tube.

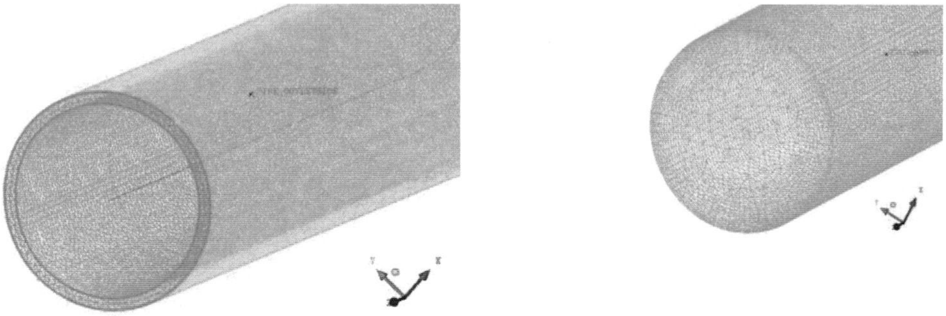

Figure 3. Generetated grid for the coiled tube wall (left) and for the working fluid (right)

To evaluate the local heat transfer performance of the coiled tube, 36 points were defined on the inner surface of the test section.

Numerical values of the local wall heat flux q_w and of the local wall temperature T_w around the test point are collected by sampling the corresponding field variables on each test points.

4. Results

The coiled tube was numerically simulated for two different test conditions for which local measurements are available in [1]. In particular Bozzoli et al. [1] investigated the laminar flow regime in the Reynolds number range of about 130–1100 and they concluded that, by accounting for the experimental uncertainty, for this range the profile of the local Nusselt number is almost independent of the Reynolds number. The two test conditions considered in this paper are representative of the

extreme conditions of the experimental data available in [1] and they correspond to Reynolds number values equal to 135 and 1006, respectively.
The fluid thermo-physical properties considered in the two simulations are reported in Table 1 and they correspond to the values published in [1] for the considered experimental conditions. In particular, they correspond to Prandtl number values of 173 and 190. Following the same approach the values of the heat generated in the wall are taken equal to the experimental values.

Table 1. Test conditions and working fluid thermo-physical properties considered in the numerical simulations.

Re	q_g **(W/m²)**	**λ (W/m·K)**	**ρ (kg/m³)**	**η (Pa·s)**	c_p **(J/Kg·k)**
135	$2.7 \cdot 10^6$	0.252	$1.12 \cdot 10^3$	0.0181	$2.41 \cdot 10^3$
1006	$4.8 \cdot 10^6$	0.251	$1.12 \cdot 10^3$	0.0199	$2.40 \cdot 10^3$

Mesh quality was fully investigated to evaluate the validity of the numerical results. Table 2 compares the results, reported in terms of the higher and the lower temperature values in the test section for different mesh distributions; the data confirm that the adopted mesh ensures a satisfactory solution.

Table 2. Mesh refinement study.

	Re=135		**Re=1006**	
No. of elements	23'758'136	43'664'780	23'758'136	43´664´780
$T_{w,max}$ **(K)**	307.1	307.0	302.5	302.1
$T_{w,min}$ **(K)**	302.3	302.3	297.5	297.3

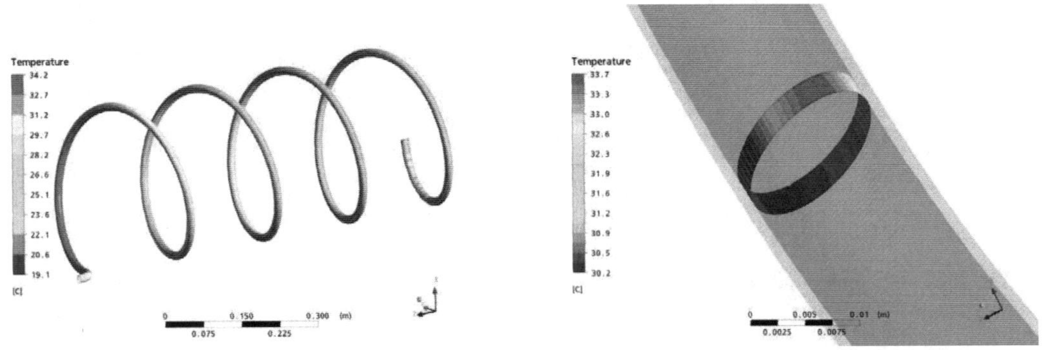

Figure 4. Wall temperature for Re=135 .

Figure 4(left) shows the obtained outer side temperature field on the heated section of the coiled tube. The inner side peripheral temperature distribution is shown in figure 4(right) at the same test section. These data highlight that the temperature is not uniform along the periphery of the coil cross section and that, in particular, at the outside surface of the coil (extrados), it was lower than that at the inside surface (intrados).
In figure 5 the numerical results are compared to experimental ones in terms of heat flux density along the angular coordinate α, whose origin was taken at the extrados.

The data in figure 5 show that the numerical model works quite well in reconstruction of the heat flux at the fluid-wall interface: in particular it is possible to notice that for the case of Reynolds number equal to 1006 the two distributions, the numerical and the experimental one, are very close.
The matching of the two heat flux distributions is less accurate for Reynold number equal to 135: in particular the numerical model tends to overestimate the magnitude of the heat flux in the outside region of the coil. In any case the data confirm that the heat flux presents a great variation along the angular coordinate, by showing in particular values at the outside surface of the coil much higher than the ones at the inside surface as a consequence of the complex flow pattern that establishes due to the wall curvature.

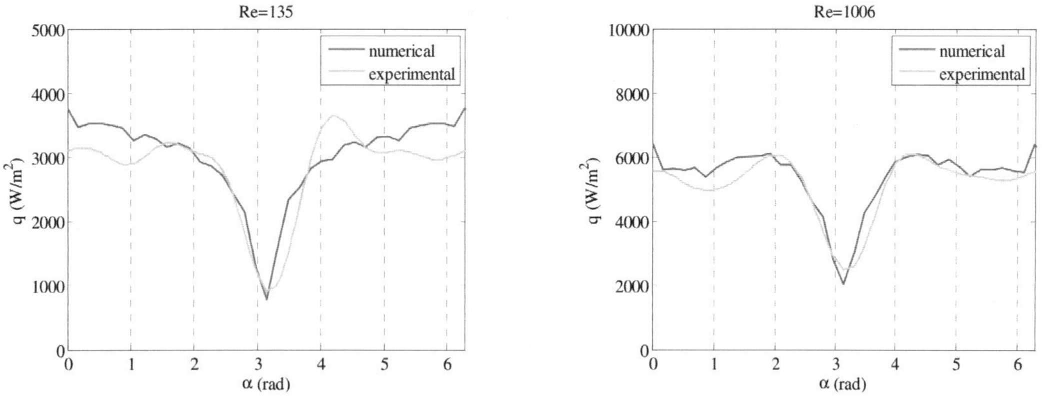

Figure 5. Comparison between experimental and numerical wall heat flux distributions.

In figure 6 the relative wall temperature, defined as the difference between the wall temperature and the fluid bulk temperature T_{bulk} , is reported.
These data show that for both Reynolds number values the numerical model underestimates the relative temperature value with respect to the one obtained experimentally.

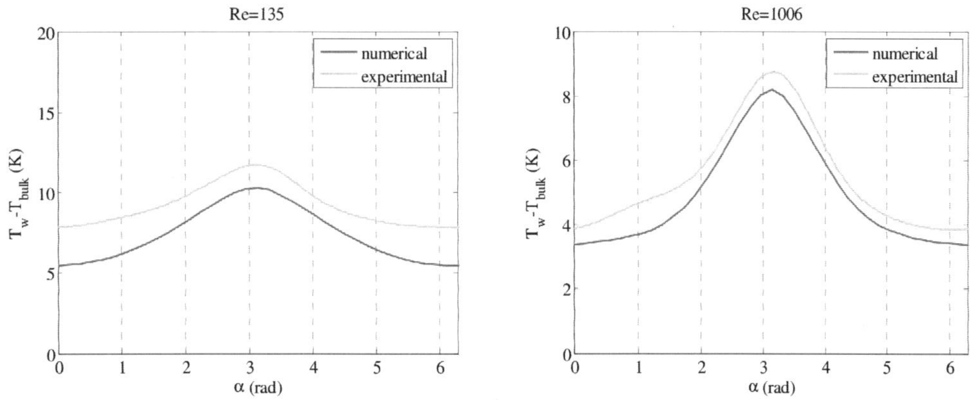

Figure 6. Comparison between experimental and numerical wall relative temperature distributions.

From the quantities obtained by the numerical simulation, the Nusselt number values were calculated as follows:

$$Nu = \frac{q \cdot D_{\text{int}}}{\left(T_w - T_{bulk}\right) \cdot \lambda_f} \tag{5}$$

where D_{int} is the tube internal diameter and λ_f is the fluid thermal conductivity, evaluated at the bulk temperature. Figure 7, reporting Nusselt number distributions, confirms that numerical model tends to overestimate the heat transfer in the outside region of the coil.

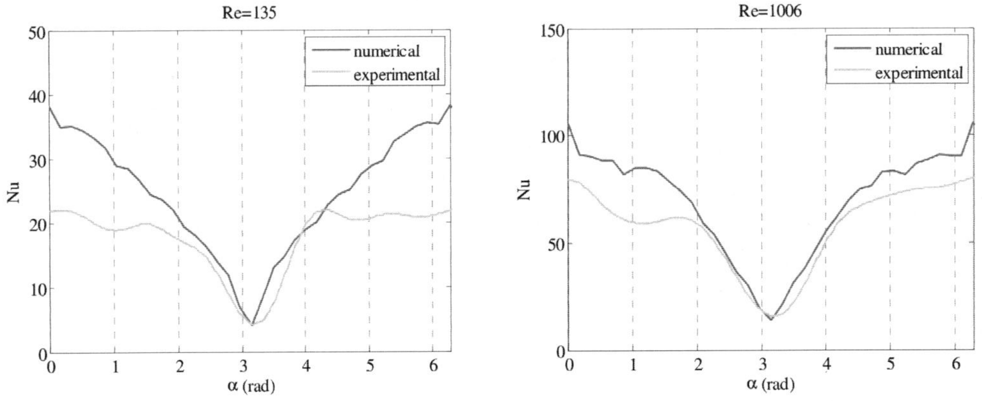

Figure 7. Comparison between experimental and numerical Nusselt number distributions.

Finally, the normalized numerical Nusselt number distribution was compared with the experimental results. The data, reported in figure (8) are in good agreement with the cinclusions of Bozzoli et al. [1]: in this flow regime the local Nusselt number (Nu/Nu_{max}) profile is almost independent on the Reynolds number and its minimum value, located at the inner coil bend, reaches about 0.2.

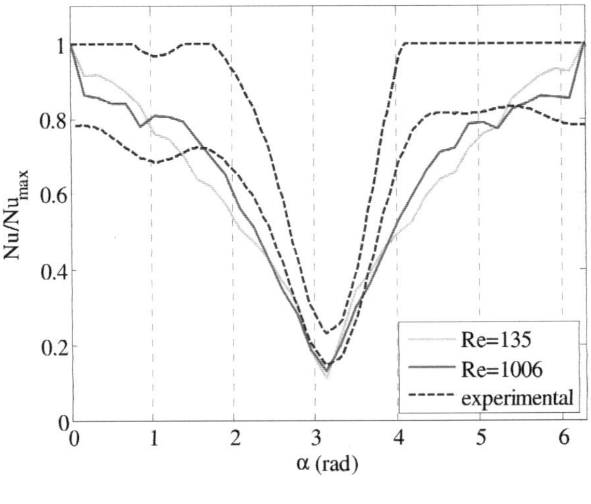

Figure 8. Normalized Nusselt number distribution: comparison between numerical and experimental (95% confidence interval) results.

5. Conclusion

In the present paper a numerical model of convective heat transfer in coiled tubes under the laminar flow regime is presented and discussed with regard to its application to a high Prandtl number fluid. The numerical approach was checked throughout the comparison with the latest experimental outcomes of Bozzoli et al. [1] in terms of convective heat flux distribution along the boundary of the duct. The aim was to verify the effectiveness of the model in the description of the local heat transfer behavior. The preliminary results so far obtained show that the model works well in reconstruction of the heat flux at the fluid-wall interface, by confirming its significant variation along the angular coordinate as a consequence of the complex flow pattern that establishes due to the wall curvature.
The numerical relative temperature does not match the experimental distribution. Possible reasons of this mismatch could be found in the constant fluid properties approximation. In fact, although the fluid adopted in the experimental investigation shows properties that strongly vary with temperature, in the present study the simplified constant properties approach was adopted to limit the computational cost of the simulations. Moreover this simplified approach could affect the thermal development length in comparison to the one observed experimentally. Although this, the local normalized Nusselt number resulting from the numerical simulations is in good agreement with the experimental outcomes [1].
Further analysis is needed to deepen the conjugate heat transfer problem in coiled tubes, where the effect of the wall curvature produce a complex flow behavior. Furthermore, it has to be stressed that the present paper reports only preliminary results of this simulation/validation process but it could be of interest for the research community because it proposes a novel approach that could be useful to validate many numerical models for non-standard geometries.

References

[1] Rainieri S, Bozzoli F, Cattani L and Pagliarini G 2013 Compound convective heat transfer enhancement in helically coiled wall corrugated tubes *Int. J. Heat Mass Tran.* **59** pp 353-362
[2] Naphon P and Wongwises S 2004 A review of flow and heat transfer characteristics in curved tubes *Renew. Sustainable Energy Rev.* **10** pp 463–490
[3] Mitsunobu A and Cheng K C 1971 Boundary vorticity method for laminar forced convection heat transfer in curved pipes *Int. J. Heat Mass Tran.* **14** pp 1659-1675
[4] Kozo F and Aoyama Y 1988 Laminar heat transfer in a helically coiled tube *Int. J. Heat Mass Tran.* **31** pp 387-396
[5] Yang G, and Ebadian M A 1994 Perturbation solution for laminar convective heat transfer in a helix pipe *J. Heat Tran.* **116** pp 771-775
[6] Di Piazza I and Ciofalo M 2010 Numerical prediction of turbulent flow and heat transfer in helically coiled pipes *Int. J. Therm. Sci.* **49** pp 653-663
[7] Di Liberto M and Ciofalo M 2013 A study of turbulent heat transfer incurved pipes by numerical simulation *Int. J. Heat Mass Tran.* **59** pp 112-125
[8] Bai B, Guo L, Feng Z and Chen X 1999 Turbulent Heat Transfer in a Horizontal Helically Coiled Tube *Heat Transf. - Asian Res.* **28** pp 395-403
[9] Bozzoli F, Cattani L, Rainieri S, Viloche Bazán F S and Borges L S 2014 Estimation of the local heat-transfer coefficient in the laminar flow regime in coiled tubes by the Tikhonov regularisation method *Int. J. Heat Mass Tran.* **72** pp 352-361
[10] Seban R A and McLaughlin E F 1963 Heat transfer in tube coils with laminar and turbulent flow *Int. J. Heat Mass Tran.* **6** pp 387-395
[11] Xin R C and Ebadian M A 1997 The effects of Prandtl numbers on local and average convective heat transfer characteristics in helical pipes *J. Heat Tran.* **119** pp 467-473.
[12] Janssen L A M and Hoogendoorn C J 1978 Laminar convective heat transfer in helical coiled tubes *Int. J. Heat Mass Tran.* **21** pp 1197-1206

Parameter estimation approach to the thermal characterization of intumescent fire retardant paints

L Calabrese[1], F Bozzoli[1], G Bochicchio[2], B Tessadri[2], P Vocale[1], S Rainieri[1]

[1]Department of Industrial Engineering - University of Parma
Parco Area delle Scienze 181/A, 43124 Parma, Italy.
[2]National Research Council of Italy (CNR), Trees and Timber Institute – Fire Lab. Via Biasi, 75, 38010 San Michele all'Adige, Italy.

E-mail address of the corresponding Author (Sara Rainieri): sara.rainieri@unipr.it

Abstract Intumescent paints are widely used as passive fire retardant materials in the building sector. They swell on heating to form a highly insulating char, protecting steel members. Intumescent coatings for use in buildings are typically certified according to the standard cellulosic fire resistance test. This test is expensive, often non-representative of realistic fire conditions, and not enough versatile to gather detailed performance information on the response of reactive coatings. A promising approach, that could offer a helpful tool to the engineering community involved in fire safety, is found in the modelling of the behaviour of the intumescent coating. Under this approach, the knowledge of the equivalent thermal conductivity of the intumescent material is a fundamental issue, since it represents the main parameter that allows predicting the thermal protecting capability of the layer. The purpose of this paper is to optimize an estimation procedure intended to the restoration of the equivalent thermal conductivity of intumescent layers. The thermal stress is activated by the action of a cone calorimetric apparatus, while the estimation procedure is based on the inverse heat conduction problem approach under steady state assumption, where the temperature values measured at some locations inside the layer during the expansion process are used as input known data. This procedure was successfully applied to steel samples protected with an intumescent paint; the estimated equivalent thermal conductivity of the layer results to temperature dependent while the initial thickness of the paint does not seem to have a great effect.

1. Introduction

Structural steel is widely used in modern structural construction all over the world due to its high mechanical strength, ductility and execution time. Regarding steel-framed buildings, a sufficient fire resistance is an important and prior requirement of building regulations. A promising approach to achieve this requirement is to apply passive fire protection materials to the structural steel that are inflammable and capable of protecting the steel structure from the effects of high temperatures that are generated during a fire. They can be divided into two categories: non-reactive materials, such as incombustible boards, and reactive materials, such as intumescent coatings. Thin film intumescent coatings have become the dominant choice for passive fire protective materials because of their many advantages including: flexibility and ease of application, light-weight, thin and attractive appearance and high standard finishing. They are typically applied with a thickness ranging from 1 to 8 millimeters and could be solvent or water based [1-4].

Structural steel loses an appreciable part of its load carrying ability when its temperature exceeds 500 °C. The time at which the steel reaches this critical temperature is generally the quantity that has to be considered in the design of a safe evacuation of people from the building and therefore it has become a prior parameter in building regulations in many countries. Intumescent coatings are designed not to burn [5] and to perform under severe conditions with the aim of maintaining the steel integrity between 1 and 3 hours in case of fire [6]. Nowadays, many formulations are available and the basic components used in commercial intumescent paints are: inorganic acids or materials yielding acid at high temperatures, polyhydric materials, organic amines or amides and binders.

The inorganic acids act as a catalyst to speed up the formation of a carbonaceous char; ammonium polyphosphate is commonly used. The polyhydric materials, rich in carbon, form the carbonaceous backbone of the intumescent char; carbohydrates are usually used for this purpose, Pentaerythritol being one of the most common. The organic amines or amides act as a blowing agent; melamine is often used for this purpose because of its low water solubility [7]. These components are linked each other by a polymer matrix which plays a capital role in the intumescent behavior; indeed an increase in the ratio of polymer results in the most noticeable increase in flexibility. Phosphorous polymeric binders are often used as flame retardant.

These innovative coatings, when activated by the heat of a fire, react to form a thick, low-density, polymeric coating or char layer. The parameter that directly reflects the performance in the protection is the pore structure of the char. Water vapor and sulphur dioxide are released during the intumescent reaction. Two fire-protection mechanisms thus become available: the char layer retards the flow of heat, due to the extremely low thermal conductivity and water vapor and sulfur dioxide are released, providing fire-quenching properties. Moreover, if the char, due to the presence of titanium dioxide, reaches a sufficiently high temperature, much of the incident heat load can be reflected [8]. The behavior of the intumescent coating can be described by subsequent physical and chemical steps. Initially, under the effect of the thermal flux, the ablation of a part of the intumescent paint occurs, gases are released and they cause the swelling of a part of the paint. Subsequently, a charring layer appears, which grows with time, until being stabilized in stationary state. The ablative layer goes on regressing and the expanding layer swells. The non reactive char layer grows to the detriment of the reactive layer and the ablative layer. Eventually, the original intumescent paint is totally ablated, the paint stops to swell and a stable char layer establishes [9].

The fire resistance of a commercial coating is usually tested experimentally, by using large-scale tests (see, for instance, EN 13381-4 and EN 1363-1 standards [10,11]). Under this approach, each commercial coating-beam configuration has to be tested one by one.

However, high operating costs, poor repeatability, unrealistic and/or inappropriate boundary conditions, and poor statistical confidence persist as common issues regarding the use of standard fire resistance tests [12].

On the other hand, performance-based structural fire engineering is increasingly common, and it is becoming more widely accepted as a method of structural fire resistance design. According to this approach, it is important to better understand the performance of reactive coatings [2]. Several methods for performance-based fire engineering design have been explored; most of them are based numerical modeling under the assumption of equivalent thermal properties of the intumescent paint. This methodology could provide a helpful tool to simplify the test procedure and to support in the design of fire resistant structure, as well. In literature, several works have been presented on this topic [13-20]. Some researchers proposed to estimate the equivalent thermal conductivity by solving the non stationary energy balance in the intumescent layer. In this procedure, the directly measured parameters are the temperature of steel substrate and the gas temperature of the gas furnace[2,16,21] or the heat flux produced by a cone calorimeter [5,14,15]. However, the determination of the net heat flux absorbed by the coating layer is difficult, because of the uncertainty in parameters which characterize its thermal interaction with the surrounding environment, such as the coefficients of absorption,

reflection and emission of the surface, the view factor towards the environment and surrounding temperature, the convection coefficient. To overcome this difficulties, some researchers [2,15,22,23] have presented simplified models to calculate the effective thermal conductivity of the coating.
However, all these approaches, do not have a sufficient accuracy, due to the weakness correlated to the determination of free boundary variables of the intumescent layer. To overcome this problem, recently Calabrese et al. [3], measured the temperature inside the intumescent layer. In this way, the formulation of the inverse problem does not require the definition of free boundary variables of the intumescent layer and the equivalent thermal conductivity of the intumescent layer becomes the only parameter of the paint that have to be estimated through the solution of the inverse heat conduction problem.
In this paper the approach proposed by Calabrese et al [3] was improved: additional thermocouples were placed inside the foam; a better estimation of the heat lost was obtained by additional thermocouples on the insulating material; finally the numerical model for the thermocouple was enhanced by measuring its thermal properties.

2. Experimental setup and estimation procedure

The paint used in the here presented analysis was a water based intumescent coating (Promapaint by Promat), whose ingredients consist of a PVC polymeric resin with some additives: Titanium Dioxide, Melamine and 2-Butoxyethanol and others. The experimental tests were performed by considering two different dry film thickness (figure 1): 0.8 mm (test case label S08) and 1 mm (test case label S10), both coated on square steel plates, 4 mm thick and 100 mm wide. The specimens were subjected to a heat radiant fluxes of 50 kW/m^2 provided by a cone calorimeter. A specimen holder (figure 2) made of 2 cm of calcium silicate board in order to minimize and control heat losses toward the environment, was used as a cradle for the specimens.

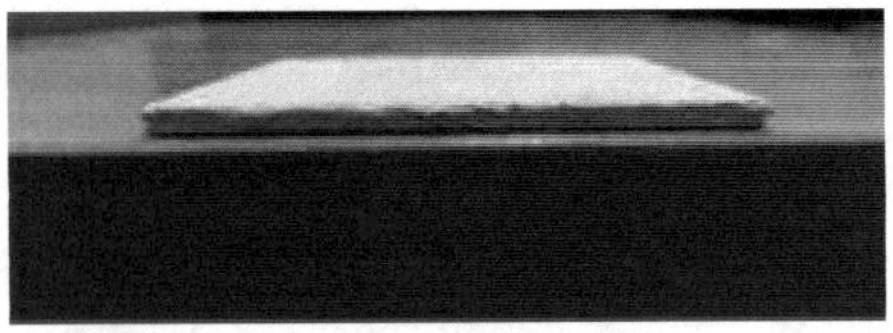

Figure 1. Steel plate with the intumescent paint coating.

Figure 2. Specimen, thermocouples and holder.

The temperature data were acquired by K-type thermocouples sensors, where the sensing element are protected by a metal sheath of circular cross section with a diameter of 1 mm. On the base of data sheet provided by the manufacturers, it was assumed that the sensing element is located at 0.5 mm below the top of metal sheath. The temperature in the foam layer was measured by thermocouples labeled as *T1, T2, T3*. The distance between the surface of steel plate and the sensing element are reported in table 1

Table 1. Distance of the sensing element from the metal surface.

Specimen	Thermocouple *T1* [mm]	Thermocouple *T2* [mm]	Thermocouple *T3* [mm]
S08	9.1	19	28.6
S10	10.6	16.9	25.2

These temperature sensors where installed through a 1.5 mm hole, passing through the sample holder and the coated steel specimen as shown in figure 2. This arrangement prevents the movement of the thermocouples during the growth of the intumescent. The optimized horizontal distances between *T1, T2, T3* were obtained by a Comsol simulation in order to minimize the reciprocal influence of thermocouples and the border effect. When the three sensors are completely covered by the char, the acquired signals start to be processed.

The temperature of steel plate T_s, was acquired by two K-type thermocouples located on the non coated surface of plate, halfway between the edge and the center of the square.

In consideration of the small thickness and the high steel thermal conductivity, the lumped capacity model can be assumed and therefore the temperature inside the plate has a spatially uniform temperature distribution.

Finally, two other K-type thermocouples were placed on at the calcium silicate-metal plate interface and between the calcium silicate and metal holder, with the aim of estimating the amount of thermal energy lost through the base of cradle.

Calabrese et al. [3] have highlighted, that after intumescent layer has completely developed, the heat flux trough the paint will be approximately constant due to the negligible heat capacity of the coating. This implies that the energy balance through the foam layer is almost in a steady state and, in an ideal situation, λ_{eq} could be estimated through direct application of Fourier's law:

$$\lambda_{\mathrm{eq}} = \frac{q \cdot d}{\Delta T} \tag{1}$$

The heat flux q transmitted through the protective layer (figure 3) is estimated by the energy balance of the steel plate:

$$q = q_{\mathrm{s}} + q_{\mathrm{d}} = \rho_{\mathrm{s}}\, c_{\mathrm{s}}\, s\, \frac{dT_{\mathrm{s}}}{dt} + \frac{\lambda_{\mathrm{c}}}{s_{\mathrm{c}}}\,(T_{\mathrm{s}} - T_{\mathrm{c}}) \tag{2}$$

where q_s and q_{d} are the heat absorbed by the steel substrate and the heat lost through its back face; ρ_s, c_s, s are, density, specific heat and thickness of the steel respectively; λ_c, s_c, are the thermal conductivity and the thickness of calcium silicate which accounts for the heat lost from the back face of the specimen.

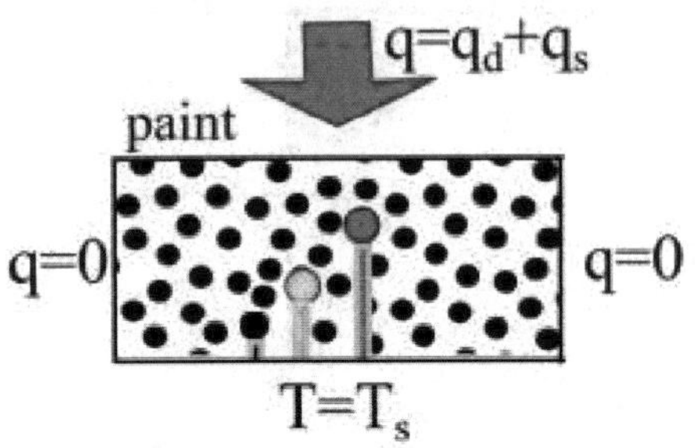

Figure 3. Thermal model of foam for the inverse heat conduction problem.

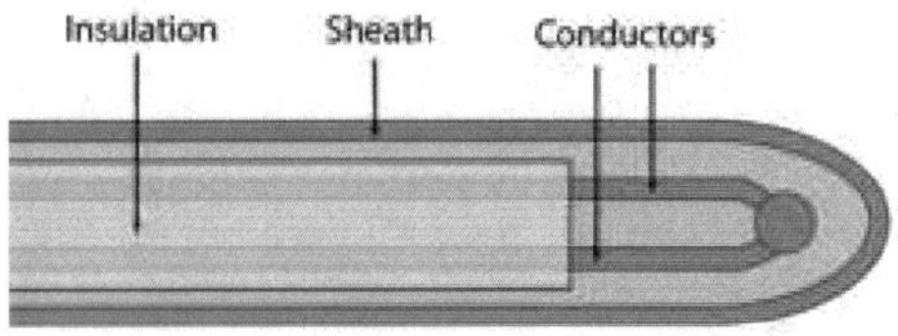

Figure 4. Sketch of the structure of the thermocouple.

Unfortunately, a non-negligible heat flux along the metal sheath of the thermocouple (figure 4), may give an altered temperature field in thickness of foam, compared to the value detectable in its absence. For this reason equation (1) is not directly applicable. To overcome this problem, in this study the inverse heat conduction problem (IHCP) in the intumescent is formulated; the thermal boundary conditions at the boundaries are shown in figure 3, more specifically a Dirichlet (equation (3)) one and a Neumann one (equation (4)) are considered. At the lower boundary:

$$T\big]_{\mathrm{x=0}} = T_{\mathrm{s}} \tag{3}$$

While at the upper one:

$$-\lambda_{\mathrm{eq}} \frac{\partial T}{\partial n}\bigg]_{\mathrm{x=L}} = q \tag{4}$$

The thermocouples were modeled in the inverse problem by following the manufacturer specification, while a sheath thermal conductivity of 3.7 W/m K was estimated. The adiabatic condition on the lateral sides of the specimens was assumed. Because of the temperature field modification induced by the thermocouple sheath, the conduction problem was reformulated in three dimensions. It was then approached within a steady state IHCP solution procedure to estimate the equivalent thermal conductivity of the intumescent layer.

Even if is expected that the temperature across the intumescent layer varies by several hundreds of Kelvin, the corresponding equivalent thermal conductivity of the expanded intumescent char was assumed to be constant [24]. The optimal value of the unknown parameter then result from the minimization of the following target function[25]:

$$F(\lambda_{\mathrm{eq}}) = \sum_{i=1}^{n} [T_{\mathrm{exp}} - T_{\mathrm{sim}}(\lambda_{\mathrm{eq}})]_i^2 \tag{5}$$

Where T_{sim} and T_{exp} are the simulated and experimental temperature, respectively of the three thermocouples placed within the intumescent.

3. Experimental results

The experimental tests under the heat flux of a cone calorimeter was continued until the steel plate temperature reached the value of 600°C that represents the steel critical temperature which was reached in approximately 6000 s (for both specimens). The thermal trend of steel plate, T_s, is shown in figure 5.

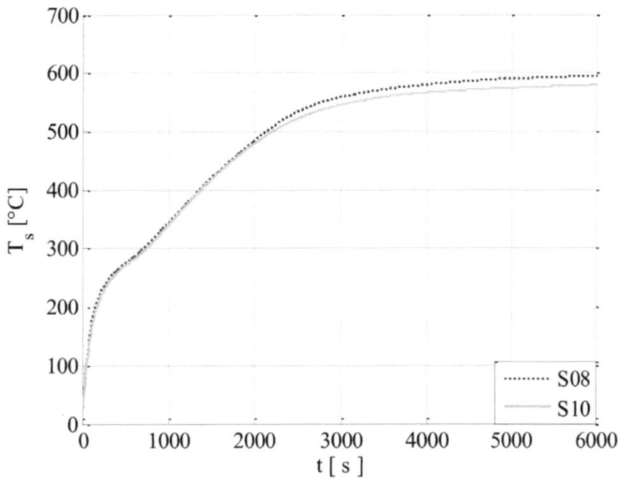

Figure 5. Steel plate temperature histories.

Many researchers measured the histories of the substrate temperature (figure 5) [16, 26-29] to check the protective performance of an intumescent coating. The heating process of the substrate can be divided into three stages [30,31]: preheat, tumescent, and post-tumescent phases. Before the substrate temperature reaches the temperature of intumescence, the heat transfer mode within the coating is dominated by conduction and the heating process is called the preheat phase. The temperatures of the virgin coating and the substrate increase quickly in this stage. When the temperature of intumescence is approached, the rise of the substrate temperature slows because of the energy consumption of the intumescent reaction. The intumescent reaction usually occurs over a temperature range, and the heating process during this stage is called the tumescent phase. At the end of the intumescent reaction, the substrate and the char temperatures rise rapidly again, and this stage is referred to as the post-tumescent phase. Thus, the temperature curve of the substrate at the tumescent phase looks like a bending curve; this kind of bending evidence is one of the characteristics of intumescent coatings.

The decrease of the heating velocity even after the completion of intumescence is due to the increase of the heat loss, q_d as the steel plate temperature increases. The heat lost through the back face of the specimen, q_d was calculated by equation (2); a literature value for calcium silicate thermal conductivity λ_c, was assumed.
For all specimens, the components, q_s and q_d, of total heat flux , q, that reaches the intumescent are shown in figures 6 and 7.

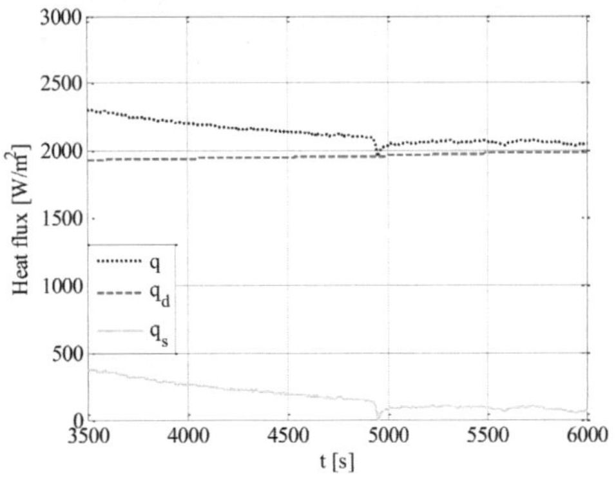

Figure 6 . Heat fluxes for sample S08.

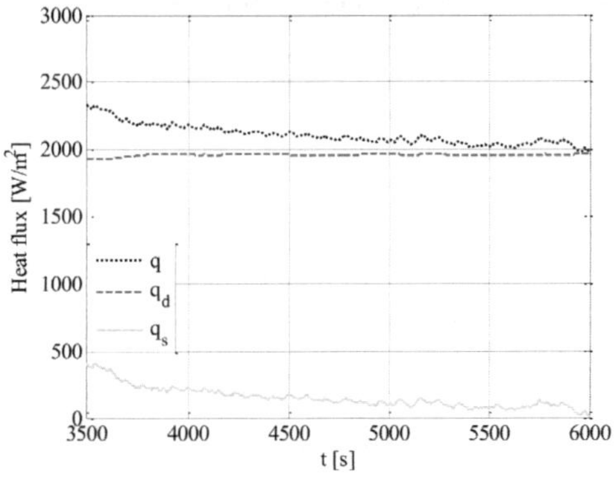

Figure 7. Heat fluxes for sample S10.

Figure 6 and 7 show that q goes to a nearly constant value. To the aim of the present investigation, the steady state condition was assumed to be reached when the percent variation of q was lower than 5%. The assumption of negligible heat losses in the experimental test aimed to the estimation of the equivalent thermal conductivity is a fundamental hypothesis of many researchers, i.e., in , Anderson et al. [16], Mesquita et al. [15]. Disregarding the heat losses in the energy balance equation of the steel plate (equation (2)) may result in an underestimation of the heat flux through the foam layer and, as a consequence, it may result in a lower value of the equivalent thermal conductivity.
The temperature time response by thermocouples *T1, T2* and *T3* is shown in figures 8 and 9, for specimens S08 and S10, respectively.

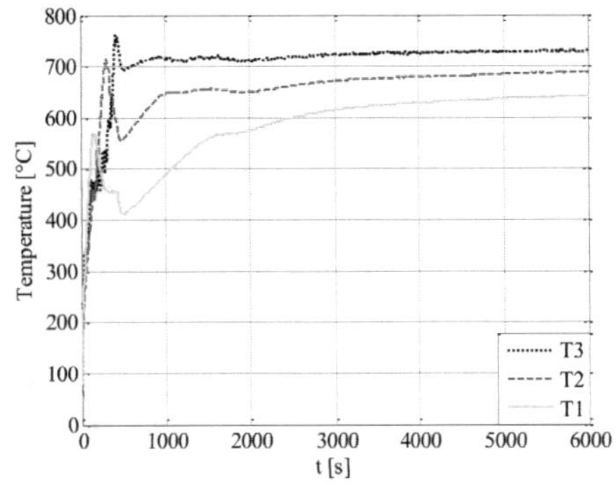

Figure 8. Temperature detected by thermocouples T1, T2, T3 during the intumescence process for sample S08.

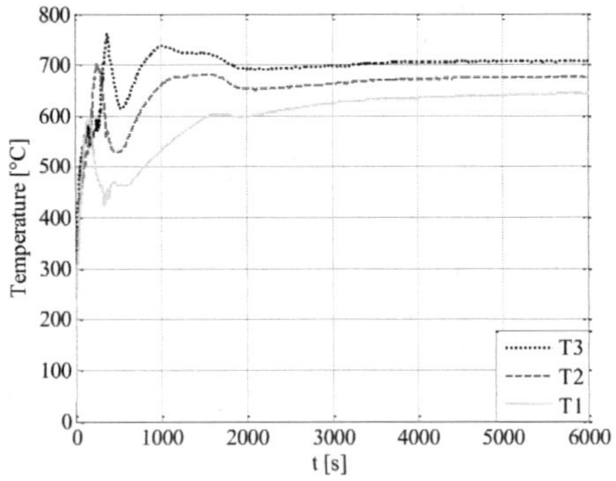

Figure 9. Temperature detected by thermocouples T1, T2, T3 during the intumescence process for sample S10.

The maximum value of temperature is reached when the intumescent layer incorporates the thermocouple, which previously is directly exposed to the heat flux emitted by the cone calorimeter apparatus.

4. Equivalent thermal conductivity estimation

The direct problem as schematized in figure 3 was solved by the numerical finite elements method, implemented inside the Comsol Multiphysics environment. The minimization of the target function (equation (5)) was performed through the Matlab Optimization Toolbox® by using as stopping criterion a relative tolerance on object function lower than $1*10^{-4}$. The Nelder-Mead algorithm was adopted in the minimization process. The results, in terms of equivalent thermal conductivity, obtained for the two specimens are shown in figure 10, as a function of the average temperature recorded by thermocouples installed inside the intumescent.

To identify the main contributions to the uncertainty of the estimated heat flux distribution, the influence coefficient values were calculated:

$$J_{x_i}^{\lambda_{eq}} = \left(\frac{\partial \lambda_{eq}}{\partial x_i} \varepsilon_{x_i}\right)^2 \tag{6}$$

where λ_{eq} is the estimated quantity and x_i is the considered input parameter with an uncertainty equal to ε_{x_i}. For the inverse problem investigated in this work, the partial derivative present in equation (6) are calculated by a finite difference approach:

$$\frac{\partial \lambda_{eq}}{\partial x_i} = \frac{\lambda_{eq}(x_i + \Delta x_i) - \lambda_{eq}(x_i)}{\Delta x_i} \tag{7}$$

Where Δx_i is a small variation of the input parameter x_i. The uncertainty on the unknown parameter was estimated with the equation (8).

$$E_\lambda = \{\sum_{i=1}^{n}[(\partial \lambda_{eq}/\partial x_i) \cdot \varepsilon_{x_i}]^2\}^{\frac{1}{2}} \tag{8}$$

The overall percent uncertainty is $E_\lambda\%$ =12.19%, as reported in figure (10).

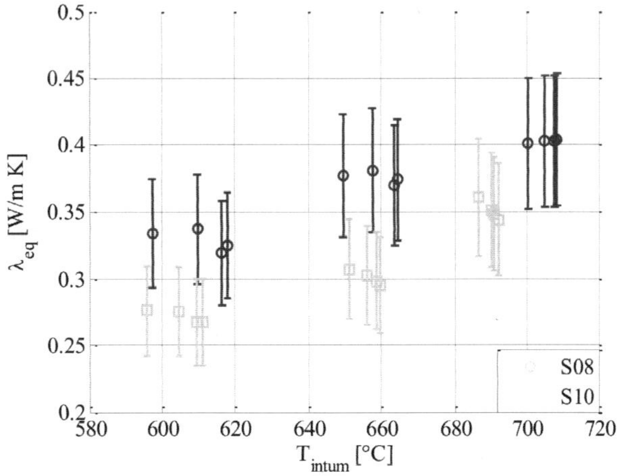

Figure 10. Equivalent thermal conductivity, obtained for the two specimens versus the intumescent temperature.

As expected, the data show that the increase of temperature of intumescence produces the increase of the equivalent thermal conductivity.

Regarding the effect of the initial paint thickness, the results presented in figure 10 show a systematically higher equivalent thermal conductivity for specimen S08. However, this behavior might be due to the above reported dependence on temperature. As a consequence, the initial thickness of the paint seems to have a quite negligible effect on the effective thermal conductivity of the intumescent layer.

5. Conclusion

This paper, starting from the results presented by Calabrese et al. [3], proposes an improved approach for estimating the equivalent thermal conductivity of the char layer generated by the intumescence process. Experimental results were obtained through two specimens, with different initial paint thickness. The methodology is based on the solution of the inverse heat conduction problem within the system under the least-square method with temperature measured in three different locations of the intumescent layer as inputs known data.

The results confirm the conclusion of Calabrese et al. [3]: the equivalent thermal conductivity depends on the temperature, while paint initial thickness, for the experimental condition here investigated, does not seem to have great importance.

6. Acknowledgements

The authors thank Renato Leveghi from National Research Council of Italy (CNR), Trees and Timber Institute for the help in the experimental activity.

References

[1] Di Blasi C, Branca C, 2001, Mathematical Model for the Nonsteady Decomposition of Intumescent Coatings, *AIChE Journal*, **47**, 2359-2370.

[2] Elliot A, Temple A, Maluk C, Bisby L, 2014, Novel Testing to Study the Performance of Intumescent Coatings under Non-Standard Heating Regimes, *Fire Safety Science-Proceedings of the Eleventh International Symposium.*

[3] Calabrese L, Bozzoli F, Bochicchio G, Tessadri B, Rainieri S, Pagliarini G, 2014 Thermal characterization of intumescent fire retardant paints, *IOP Journal of Physics*. **547.**

[4] Zhang Y, Wang Y C, Bailey C G, Taylor A P, 2012 Global modelling of fire protection performance of intumescent coating under different cone calorimeter heating conditions, *Fire Safety Journal*, **50** 51-62.

[5] ISO 5660-1, 2002 Reaction to fire tests-Heat release, smoke reduction and mass loss rate - Part 1: Heat release rate (cone calorimeter method).

[6] Gillet M, Autrique L, Perez L, 2007 Mathematical model for intumescent coatings growth: application to fire retardant systems evaluation, *Journal of Physics D: Applied Physics*. **40** 883-899.

[7] Gillman H D, Nannelli P, 1979 Development of a special purpose spacecraft interior coating, National Aeronautics and Space Administration, Houston.

[8] Parker J A, Fohlen G M, Sawko P M, Fish R H, 1970 Intumescent Coatings as Fire Retardants , National Aeronautics and Space Administration, Houston.

[9] Rodier S, Autrique L, Perez L, Ramdani N, Serra J J, 2005 Fire protective coatings evaluation: tests and modelling based on a nonlinear partial differential equations system, *Imacs -World Congress*. Paper T6-R-00-0406

[10] UNI EN 13381-4, 2013 Test methods for determining the contribution to the fire resistance of structural members - Part 4: Applied passive protection to steel members

[11] UNI EN 1363-1, 2001 Fire resistance tests General requirements

[12] Maluk C, Bisby L, Terrasi G, Krajcovic M, Torero J L, 2012, Novel Fire Testing Metodology:Why, How and what now?, *Proceedings of the Mini Symposium on Perfomance based Fire Safety Engineering of Structures as part of 1 st International Conference on Performance Based land Life Cycle Structural Engineering*. 448-458

[13] Mesquita L, Piloto P, Vaz M, 2009 Efficiency evaluation of intumescent coatings used for fire protection:comparison between numerical method and experimental results, *Integrity Reliability and Failure*. **Chapter XVIII** 537-538
[14] Bartholmai M, Schartel B, 2007 Assessing the performance of intumescent coatings using benchscaled cone calorimeter and finite difference, *Fire and Materials*. **31** 187-205
[15] Mesquita L, Piloto P, Vaz M, 2007 Experimental study of intumescent fire protection coatings, *Fire Retardant Coatings II, Berlin, Germany*.
[16] Anderson C E, Ketchum D E, Mountain W P, 1988 Thermal Conductivity of Intumescent Chars, *Journal of Fire Sciences*. **6** 390-410
[17] Bartholmai M,Schriever R, Schartel B, 2003 Influence of external heat flux and coatings using cone calorimeter and numerical analysis, *Fire and Materials*. **27** 151-162
[18] Omrane A, Wang Y C , Góransson U, Holmstedt G and Aldèn M, 2007 Intumescent coating surface temperature measurement in a cone calorimeter using laser-induced phosphorescence, *Fire Safety Journal* **42** 68–74.
[19] Han Z, Fina A, Malucelli G and Camin G, 2010 Testing fire protective properties of intumescent coatings by in-line temperature measurements on a cone calorimeter, *Progress in Organic Coatings* **69** 475–480.
[20] Wang L L, Wang Y C, Yuan J F and Li G Q, 2013 Thermal conductivity of intumescent coating char after accelerated aging, *Fire and Materials* **37** 440–456.
[21] Wang L, Dong Y, Zhang D, Zhang C, 2015 Experimental Study of Heat Transfer in Intumescent Coatings Exposed to Non-Standard Furnace Curves, *Fire Technology*.
[22] Xianghe D,Wang Y, Bailey C, 2010 A simple method to predict temperatures in steel joints with partial intumescent coating fire protection, *Fire Technology*. **46** 19-35
[23] Li G Q, Lou G B, Zhang C,Wang L L, Wang Y C, 2012 Assess the fire resistance of intumescent coatings by equivalent constant thermal resistance, *Fire Technology*. **48** 529-546
[24] Staggs J E J, 2010 Thermal conductivity estimates of intumescent chars by direct numerical simulation, *Fire Safety Journal*. **45** 228-237
[25] Bcck J V, Blackwcll B, Clair JR. C R ST , Invcrsc hcat conduction, Ill-poscd problcm, Ncw York etc., J. Wiley & Sons 1985.
[26] Anderson C E, Wauters D K, 1984 A Thermodynamic Heat Transfer Model for Intumescent Systems, *International Journal of Engineering Science*. **22** 881-889.
[27] Caliostro D E, Riccitiello S R, Clark K J, Shimizu A B, 1975 Intumescent Coating Modeling, *Journal of Fire & Flammability*. **6** 205-221.
[28] Chaboki A, Kneer M J, Schneider M E, Koo J H, 1991 Experimental and Numerical Results for Thermophysical Properties and Thermal Response of a Fire-Retardant Polymer, ASME, HTD V. 170, Experimental/Numerical. *Heat Transfer in Combustion and Phase Change*, 1-9.
[29] Kneer M J, Koo J H, Miller M J, Schneider M E, 1993 A Cost-Effective Approach to Evaluate Insulative Materials for Low Heat Flux Applications, 31st Aerospace Sciences Meeting & Exhibit, AIAA-93-840, January 11-14, Reno, NV.
[30] Buckmaster, J., Anderson, C. Nachman, A, 1986 A Model for Intumescent Paints, *International Journal of Engineering Science*, **24**, 263-276.
[31] Calabrese L, Cattani L, Vocale P, 2014 Parameter Estimation Approach Applied to the Characterization of an Intumescent Fire Retardant Paints, *JP Journal of Heat and Mass Transfer*. **9**, 101-116.

CFD study on electrolyte distribution in redox flow batteries

S Bortolin[1], P Toninelli, D Maggiolo, M Guarnieri and D Del Col

[1] Corresponding Author: stefano.bortolin@unipd.it

Università degli Studi di Padova, Dipartimento di Ingegneria Industriale
Via Venezia 1, 35131- Padova

Abstract. The most important component in a redox flow battery (RFB) cell is the MEA (membrane electrode assembly), a sandwich consisting of two catalyzed electrodes with an interposed polymeric membrane. In order to allow electrolyte flow toward the electroactive sites, the electrodes have a porous structure that can be obtained with carbon base materials such as carbon felts. The RFB cell is closed by two plates containing the distribution flow channels. Considering that a uniform electrolyte distribution in the reaction region is a prerequisite for high-efficiency operation, the flow pattern is an important parameter to be investigated for the optimization of the cell.
In the present work, the effect of different channels patterns on the electrolyte distribution and on the pressure drop is numerically investigated. Three-dimensional simulations have been carried out with ANSYS Fluent code and four different layouts have been considered. Calculations have been performed both in the distribution channels and in the felt porous region.

1. Introduction

The recent attention to environmental problems has led to an increasing use of renewable sources. However, the intermittent nature of most renewable energy sources has posed a serious challenge for widespread application and for an effective replacement of conventional sources. Therefore, energy storage plays a crucial role in the delivery of electricity from renewable sources, such as solar and wind, providing a solution for the balance problem between the generation and the consumption of the electric power. Nowadays, several energy storage technologies, characterized by different levels of development, have been proposed (e.g. pumped hydro, electrochemical, thermal, compressed air, flywheel...) as reviewed by Alotto *et al.* [1]. Among electrochemical systems, redox flow batteries (RFBs) represent one of the most recent technologies and a highly promising choice for stationary energy storage. Unlike conventional batteries, in which energy is stored in the battery structure, redox flow batteries store energy in two solutions containing different redox couples with proper electrochemical potentials. The electrolyte is stored in two separated tanks. The most appealing features of this technology are: scalability and flexibility, independent sizing of power and energy, long durability, fast responsiveness, and reduced environmental impact. Such features allow for wide ranges of operational power and discharge time, making RFBs an ideal solution for assisting electricity generation from renewable sources. The power size depends on the flow rate of electrolyte and the configuration of the cell stack, whereas the energy stored depends on the reactants chosen, their concentration and the size of the reactant tanks.
The main element of RFB is the membrane-electrode assembly (MEA) where the reduction and oxidation reactions take place in two liquid electrolytes (aqueous and not-aqueous) that contain a

given concentration of metal ions (the active material). The MEA is composed by two catalyzed electrodes (porous media such as carbon felt or metallic foam) with an interposed polymeric membrane. Finally, the RFB is closed by two plates containing the frame flow that can have different configurations to guarantee the best electrolyte distribution inside the porous media. During the charge/discharge operations, the electrons obtained from the redox reactions are collected by one electrode and go to the other one through an external circuit, whereas the ions migrate through the membrane. As compared to the conventional electrochemical batteries, the electrolytes are not permanently sealed, but they are stored in tanks and pumped into the cell stack. Furthermore, if the material of the two electrolytes is the same, as in the case of the vanadium redox flow batteries, then the cross-contamination problem during operation is limited. For these two aspects, the electrolytes have no degradation and a long life of RFB is guaranteed and can be also improved by its configuration, i.e. number of tanks, regulation of flow, type of electrolytes, electrodes, membrane and so on, as analyzed by Cunha *et al.* [2]. The RFB can be categorized by the active species or solvent (aqueous and non-aqueous, respectively), as reported in detail by Weber *et al.* [3] and Ponce de León and coworkers [4].

There are three important aspects related to RFBs design:

- the dimension of whole plant: it is pretty large making their use difficult for a mobile application, as reported by Cunha *et al.* [2];
- the electrolyte temperature in the RFB: it can be controlled through the flow rate avoiding the solution precipitation that occurs outside the allowed operating temperature ranges. For example, a numerical investigation on temperature field inside the MEA has been performed by Wei *et al.* [5];
- the flow field design to achieve uniform distribution of the electrolyte with low pressure drop, minimizing the mass transfer polarization and avoiding problems about solubility limits, as analyzed by Weber *et al.* [3], Xu *et al.* [6] and Wei *et al.* [5].

This paper aims at analyzing the flow field inside the MEA where the electrochemical reactions take place. As mentioned above, the RFB performance is strictly influenced by the electrolyte distribution inside the carbon felt: it is important to investigate the best electrolyte distribution avoiding high pressure drop and stagnation zones inside the porous region. In the following sections, the geometry of different distribution layouts, the setting of CFD simulations and, finally, the results are reported in detail to evaluate the best configuration.

2. Geometry

In the scientific literature, different configurations have been proposed for the distribution of the electrolyte inside the porous felt where electrochemical reactions take place. In particular, two types of solutions can be found: with the first method (indirect feeding), the felt is fed by channels machined in the containing plates of the cell (e.g. parallel channels, serpentine channels), whereas the second solution consists of direct feeding the felt without the use of channels.

In the present work, the authors have focused their attention on four possible geometrical configurations (table 1), with the aim to improve the electrolyte distribution and to evaluate the pressure drop inside the RFB cell. For all the configurations, the geometry of the carbon felt is the same (length 260 mm, width 160 mm, depth 4.3 mm), beside the number and the hydraulic diameter (h.d.) of the square channels is varied. The first three configurations reported in table 1 are referred, as mentioned above, to the indirect feeding: the electrolyte enters the parallel square channels placed above the carbon felt (figure 1, case with 5 inlet parallel channels), then it is sent through the porous zone and it finally exits from the outlet channels (figure 1, case with 6 outlet channels). Instead, in the last configuration, the felt is directly fed with the electrolyte.

Table 1. Selected configurations for the distribution of the electrolyte.

Configuration	Feeding	Hydraulic diameter [mm]	Number inlet channels	Number outlet channels
1	Indirect	1	5	6
2	Indirect	2	5	6
3	Indirect	2	4	4
4	Direct	1	5	6

The four selected configurations reported in table 1 are used to analyze the influence of the following geometrical parameters on the electrolyte distribution inside the carbon felt:

- hydraulic diameter of the square channels forming the frame flow inside the MEA;
- number of inlet/outlet channels and their position on the carbon felt;
- direct/indirect feeding of electrolyte inside the porous zone.

2.1. Mesh

For all the configurations, a tridimensional domain with a structured mesh has been considered. For the purpose of computational time savings, the mesh is finer for regions near the channels walls and in the contact zone between the felt and the channels, whereas it is much coarser in the rest of the volume. In the case of the first three configurations, the mesh is composed of about $7.0 \div 7.6\ 10^6$ hexahedral cells while, for the last configuration, the mesh is composed of about $5.7\ 10^6$ elements. Furthermore, a mesh independence test has been performed for the first case. A new structured mesh composed of about $3.7\ 10^6$ elements has been designed. The maximum deviation between the velocity field obtained with the two different meshes is about 1%.

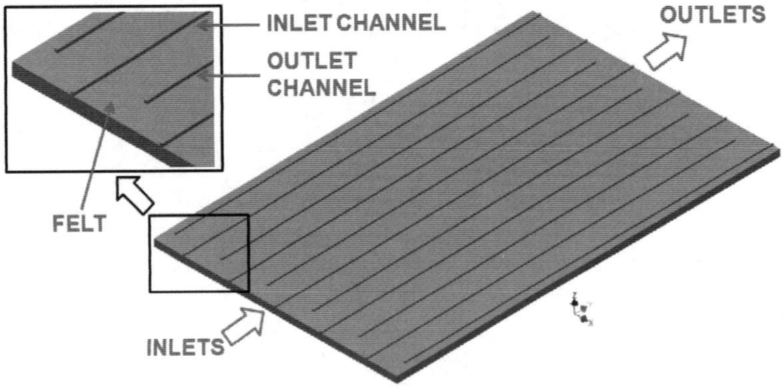

Figure 1. The carbon felt with distribution channels (configuration 1).

3. Numerical Methods

Three-dimensional and steady-state numerical simulations have been performed with ANSYS Fluent 15. The working fluid is a sulfuric acid water solution in which vanadium oxides are dissolved. The fluid is considered incompressible and its density and dynamic viscosity are, respectively, equal to 1354 kg m^{-3} and 0.006 Pa s, as reported by Blanc and Rufer [7] and Tang *et al.* [8].

In all the simulations, the velocity inlet condition has been calculated from an estimation of the cell volumetric flow rate Q_{cell} that was considered to be the same for all the studied cases. The evaluation of volumetric flow rate has been performed at fixed operational conditions of the RFB, as reported in equation (1):

$$Q_{cell} = n_m n_{ch} A_{ch} v_{in} = \frac{J A_m \alpha_Q}{SOC_{\min} c_V F} \tag{1}$$

The cell volumetric flow rate Q_{cell} is a function of electrical density current J, membrane area A_m, flow factor α_Q that defines the operational range of cell volumetric flow rate (Tang *et al.* [9]), minimum state of charge SOC_{min}, vanadium molar concentration c_V and, finally, Faraday constant F. In equation (1), the velocity inlet v_{in} can be obtained from the number of membranes that compose the RFB (n_m), the number of inlet channels (n_{ch}) and the cross section area of the channels (A_{ch}). Besides n_{ch} and A_{ch} already reported in table 1, the other parameters needed in equation (1) are shown in detail in table 2.
As reported by Tang *et al.* [9], the flow factor α_Q can be chosen in the range of 1÷8. All the simulations have been performed with α_Q equal to 1. A pressure outlet condition is imposed for the outlet channels. Being the Reynolds number at inlet lower than 10, the fluid flow is assumed to be laminar.

Table 2. Values of the parameters adopted in equation 1.

n_m	J [A m^{-2}]	A_m [m^2]	α_Q	SOC_{min}	C_V [mol m^{-3}]	F [C mol^{-1}]
2	2000	0.0416	1	0.25	1600	96000

3.1. Governing equations

The porous media model can be used in many applications including flow through packed beds, filter papers, tube banks, felts, etc. In the present CFD simulations, the presence of the porous media is not physically represented, but it has been evaluated adding two source terms to the standard momentum equations: a viscous loss term and an inertial loss term. Only in these additional terms, the effect of the porous media have been considered. In fact, the porosity has not been taken into account in the convection and diffusion terms of transport equations. This numerical procedure is well-known and the fluid velocity obtained from these modified momentum equations is named superficial velocity, i.e. the velocity that the fluid would have if it flowed through the nominal cross section area of the porous media. There is a more accurate option called physical velocity method, where the true fluid velocity inside the porous media can be evaluated and the porosity can be considered inside the convection and diffusion terms, but no significant difference has been noted between the results of the two methods.
All the simulations have been performed with the superficial velocity method and fluid cells have been divided in two domains: the first one without the porous zone, i.e. the flow in the channels, and the second one with the porous zone, i.e. the flow in the carbon felt, where the additional terms have been calculated for a homogeneous porous media.
These momentum sources have been introduced as pressure gradient proportional to the fluid velocity:

$$\nabla p = -\left(\frac{\mu}{\alpha} v_i + \frac{1}{2} C_2 \rho |v| v_i \right) \tag{2}$$

As shown in equation (2), the coefficients of both terms are a function of transport properties of the fluid, i.e. density and dynamic viscosity, and geometrical characteristics of the porous media, i.e. permeability α and inertial loss coefficient C_2, that have been evaluated from the Ergun equation and Blake-Kozeny equations:

$$\alpha = \frac{d_e^2}{150} \frac{\varepsilon^3}{(1-\varepsilon)^2} \; ; \; C_2 = \frac{3.5}{d_e} \frac{(1-\varepsilon)}{\varepsilon^3} \tag{3}$$

The geometrical characteristics reported above in equation 3 are functions of the media porosity ϵ and the hydraulic diameter d_e of cylindrical fibers that compose the carbon felt. For all the simulations, the porosity and the diameter of fibers have been imposed, respectively, equal to 0.84 and 10^{-6} m.

It should be noted that, in the present work, the assumption of a thermal equilibrium between the porous media and the fluid flow has been considered and all the numerical simulations have been run excluding the energy equation.

As a first step, all the numerical simulations have been run considering the whole domain without the porous media to obtain an initial field flow and, in a second time, the porous zone has been enabled in the carbon felt domain. Finally, regarding the solution methods employed, a SIMPLE scheme has been used for pressure-velocity coupling, whereas a standard and second order upwind method have been imposed for the spatial discretization of pressure and momentum.

4. Results

A comparison of velocity fields, velocity distributions and pressure drop has been performed to highlight the different hydraulic performance of the four configurations. This method aims at evaluating the best configuration in terms of velocity field and pressure drop. All the results have been reported in terms of physical velocity v_{ph} on the x-y plane, as defined in equation (4). The physical velocity has been calculated from the porosity ϵ and the superficial velocity v obtained directly from the numerical simulations:

$$v_{ph} = \frac{\sqrt{v_x^2 + v_y^2}}{\varepsilon} \tag{4}$$

The velocity fields are reported in figure 2 showing the difference of the fluid flow behavior in each configuration at 50% of depth inside the carbon felt (the electrolyte enters at the bottom and exits from the top of figure).

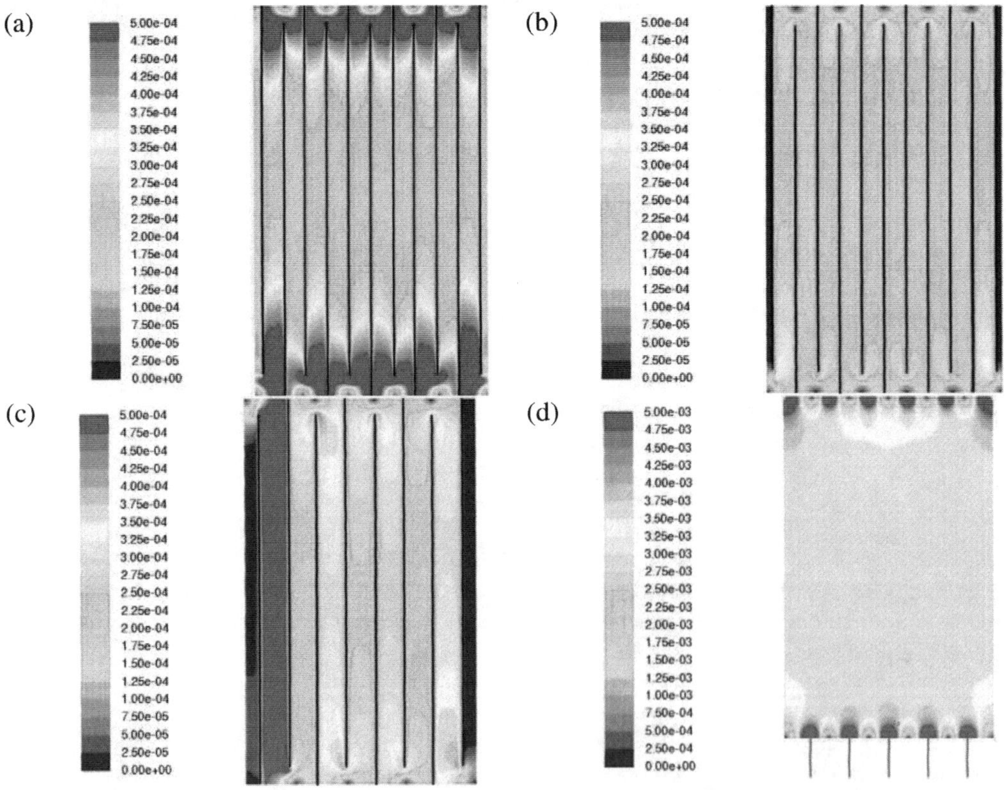

Figure 2. The velocity field ([m s^{-1}]) of the 5x6 1 mm h.d. (a), 5x6 2 mm h.d. (b), 4x4 2 mm h.d. (c) configuration with indirect feeding and the velocity field with direct feeding (d).

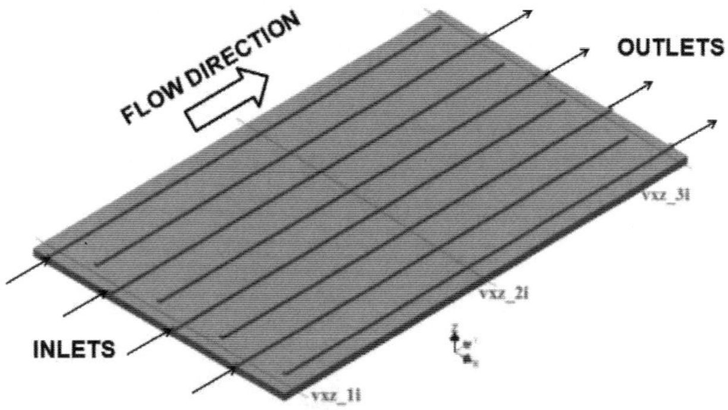

Figure 3. Red lines along which the velocity profiles are calculated. The index *i*, equal to 1, 2 or 3, is referred to 25%, 50%, 75% of depth inside the felt (configuration 3).

As depicted in figures 4-5, the planar velocity profiles along nine lines perpendicularly to the main flow direction (figure 3) have been calculated to investigate the velocity distribution inside the carbon felt. It should be noted that only the *x-y* velocity components have been considered at different depths inside the porous media (index *i*).
Beside the velocity trends, an investigation on velocity distributions has been performed to highlight the uniformity of the velocity field inside the felt. In all the configurations, it can be observed that the velocity profiles at lines *xz_1i* and *xz_3i* are affected by the channels inlets/outlets, while in correspondence of lines *xz_2i* the velocity trend is flatter, especially for the first and last configurations. It is worth noting that the velocities are reported at different depths inside the porous media (the index *i*, equal to 1, 2 or 3, is referred respectively to 25%, 50%, 75% of the depth inside the felt). A non-significant influence of depth on the velocity can be observed: it means that the distribution problem is restricted to the *x-y* plane and the investigation on planar velocity is thus justified.

4.1. Effect of hydraulic diameter

Regarding the influence of the channel hydraulic diameter on the velocity field, a comparison of velocity profiles between the first and the second configuration can be considered. As mentioned above, the volumetric flow rate is the same for all numerical simulations, hence, for case 1, the velocity and pressure drop inside the channel are higher than those for case 2, due to a lower hydraulic diameter. In case 1, the electrolyte solution is mostly driven into the carbon felt and the velocity field is increased, as depicted in figure 2. Instead, for configuration 2, the fluid goes slower into the carbon felt creating stagnation zones below the channels and at lateral sides. The high velocity of case 1 causes a more uniform distribution of electrolyte along the lines *xz_2i.* This aspect is highlighted by the calculation of velocity distribution and it can be observed that, in the second case, the velocity values are more dispersed than those observed in the first case along the lines *xz_2i*.

4.2. Effect of number of inlet/outlet channels

Concerning the influence of the number of channels on the velocity field, a comparison of velocity trends between the second and the third configuration can be considered. Two different behaviors can be observed:

- reducing the number of channels (case 3), the velocity profiles become flatter and the fluid goes through the felt with a longer path, increasing the pressure drop as compared to case 2. Furthermore, it can be observed that the velocity values in case 3 are closer to the mean value as compared to case 2 and thus the velocity field is more uniform.
- when the number of inlet channels is equal to the number of outlet channels (case 3), an asymmetric flow takes place and a maldistribution effect occurs from left to right, as depicted in figure 4. For the third case, this effect occurs because the fluid flowing inside the first inlet

channel on the left side is forced to go only through the outlet channel on the right, whereas the fluid flowing in the other inlet channels can go through its two neighbor outlet channels. For this reason, the electrolyte flow is higher on the left side as compared at the right side. This phenomenon does not occur in case 2, where the fluid in the first inlet channel can flow through the two neighbor outlet channels, providing a more uniform velocity field.

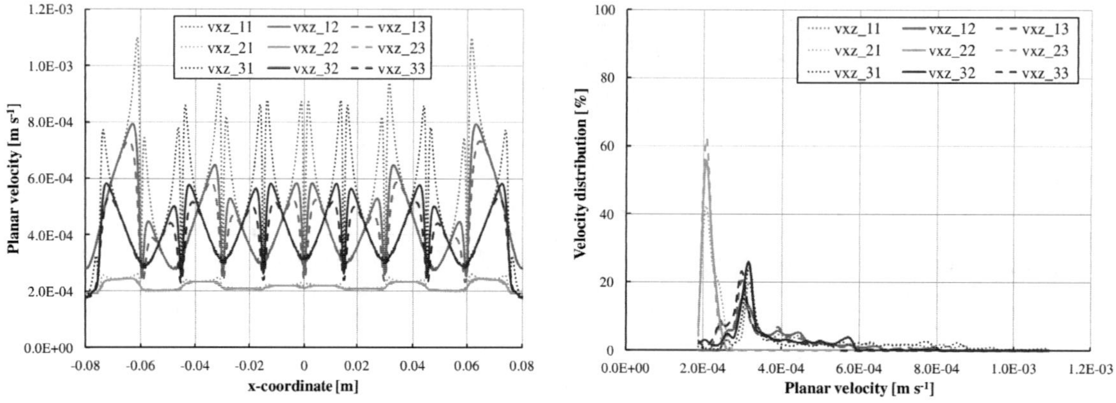

(a)

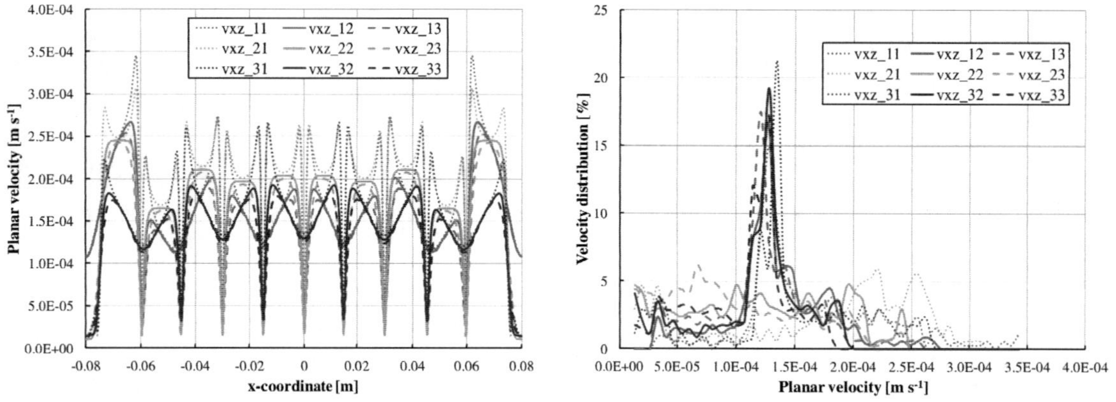

(b)

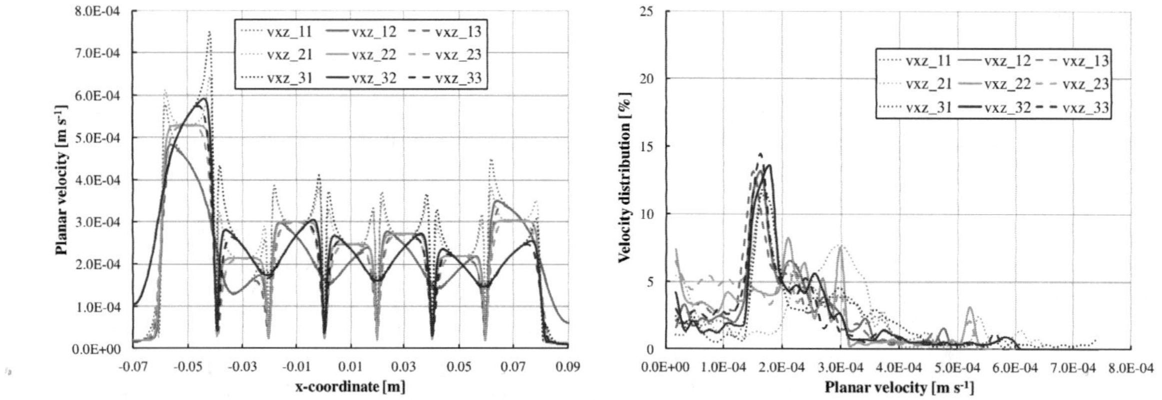

(c)

Figure 4. Planar velocity profiles calculated along the lines of figure 3 (left side) and velocity distribution (right side) for the 5x6 1 mm h.d. (a), 5x6 2 mm h.d. (b) 4x4 2 mm h.d. (c) configuration with indirect feeding.

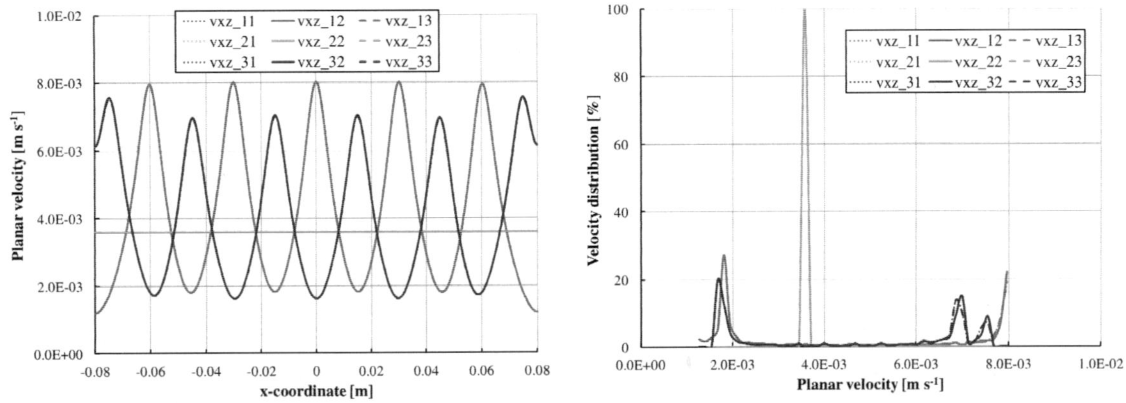

Figure 5. Planar velocity profiles calculated along the lines of figure 3 (left side) and velocity distribution (right side) for the configuration with direct feeding.

4.3. *Effect of feeding type*

With the aim to study the influence of feeding type on the velocity field, a comparison of velocity profiles between the first and the last configuration can be considered. It is important to notice that, in the case of a direct feeding, the velocity field is higher and more uniform as compared to the first configuration, as depicted in figures 4-5. However, using the direct feeding, the pressure drop is higher as compared to the configuration with indirect feeding.

Moreover, from figures 4(a)-5 along lines xz_2i, for both configurations, the velocities are close to the mean values and the velocity trends are similar. For this reason, it is very interesting to compare the four configurations taking into account the pressure drop. Beside the velocity trends along the different lines, the volume-weighted average of planar velocity inside the carbon felt has been calculated as reported in equation 5 to consider, at the same time, the effect of high velocity and stagnation zones.

$$v_{ph}\Big|_V = \frac{1}{V}\int_V v_{ph} dV \tag{5}$$

Table 3. The volume-weighted average of physical planar velocity and pressure drop for the four configurations.

Configuration	$v_{ph}\|_V$ [m s^{-1}]	p_{in}-p_{out} [kPa]
1	3.22E-04	10.61
2	1.98E-04	1.40
3	2.71E-04	2.30
4	3.69E-03	210.50

As reported in table 3, the first configuration is the one that assures the highest volume-weighted average velocity for indirect feeding, whereas considering all the configurations, the case with the highest value of $v_{ph}|_V$ is the case number 4.

4.4. *Pressure Drop*

In the present section, the pressure drop has been calculated in order to evaluate the performance of the RFB cell in terms of pumping consumption. For the estimation of pressure drop, the area-weighted average of static pressure has been calculated at the inlet and outlet of channels for each configuration. The relationship between the pressure drop and the velocity distribution can be observed in table 3. Indeed, the first and the fourth configurations present the highest value of pressure drop. The pressure drop in case 1 and 4 are higher, respectively, up to 7 and 150 times as compared to the configuration 2 that presents the worst distribution. The second and the third case display the lowest pressure drop but both configurations have many stagnation zones and the third one is affected by maldistribution problems. In conclusion, a good compromise would be the first solution since the velocity is high enough to avoid the presence of stagnation zones and to guarantee a velocity field almost uniform. In addition, the total pressure drop of the first case are limited as compared to the fourth case.

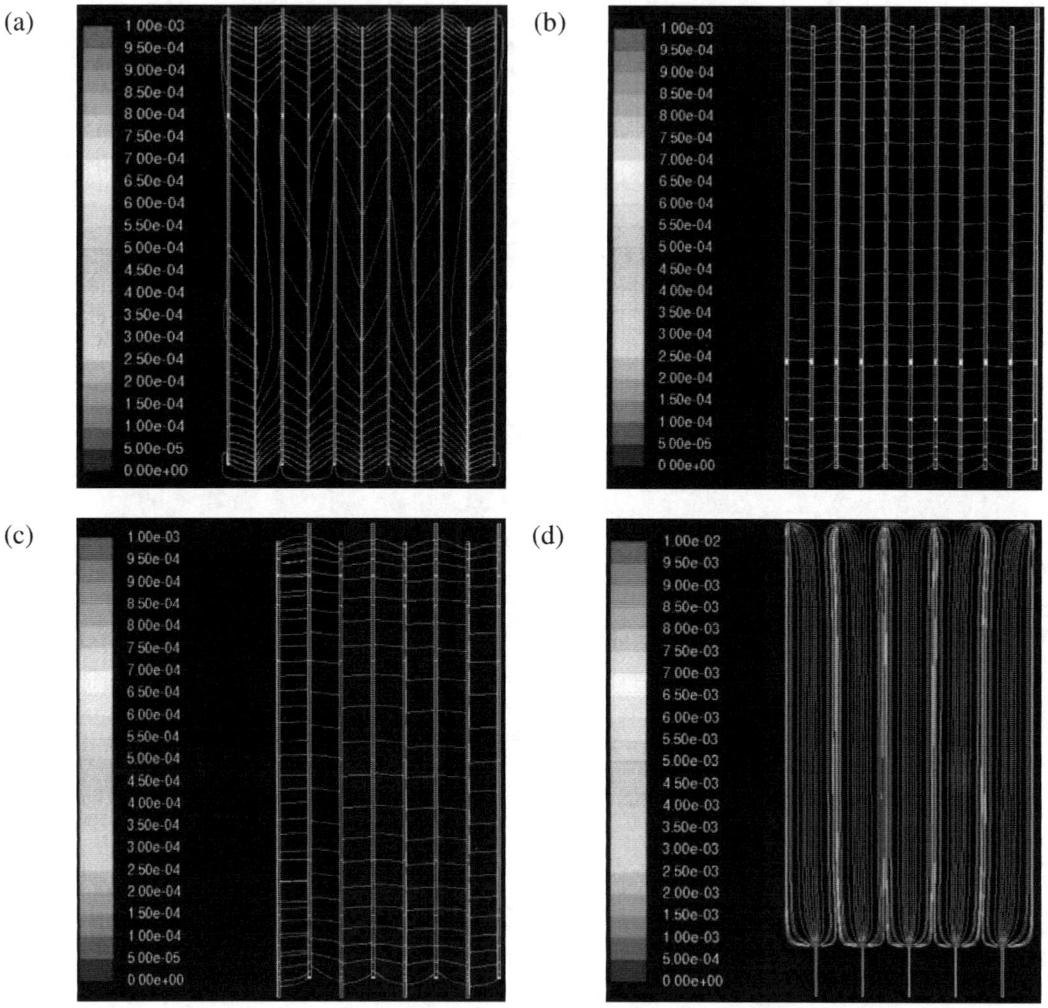

Figure 6. Planar velocity streamlines in [m s^{-1}] for the three configurations with indirect feeding (5x6 1 mm h.d. (a), 5x6 2 mm h.d. (b), 4x4 2 mm h.d. (c)) and for the one with direct feeding (d).

4.5. Streamlines

As mentioned above, to improve the performance of RFB it is fundamental to guarantee:

- a good-distribution of the electrolyte in terms of velocity, avoiding stagnation or dry zones inside the porous felt;
- low pressure drop and low pumping consumption.

For this purpose, the importance of the fluid path inside the porous media must be underlined. In fact, if the fluid has a long path inside the porous felt, the probability of electrochemical reaction occurring is higher and the RFB performance can be improved.

From this point of view, the direct feeding of electrolyte inside the carbon felt is the best configuration since the fluid is constrained to cross the whole felt as compared to the indirect feeding. As depicted in figure 6, it can be observed that, for the second and third configuration, the electrolyte solution flows from inlet to outlet channels through a very short path as compared to the first case where its path inside the felt becomes longer.

5. Conclusions

In this paper, four different geometrical configurations have been considered to evaluate the influence of the electrolyte solution distribution system (inlet/outlet number of channels, channels hydraulic diameter, feeding mode) on velocity field and pressure drop inside a RFB cell. Different parameters have been considered to assess the RFB fluid dynamic performance: planar velocity profiles, pressure drop and velocity streamlines.
The following conclusions can be drawn:

- the first and the fourth configuration present an almost uniform velocity field as compared to the second and third case where stagnation zones have been found below the channels and at the lateral sides of the carbon felt;
- a long fluid path inside the carbon felt can improve the performance of the RFB in terms of electrochemical reactions. For this reason, the second and the third case are the worst configurations due to very short fluid path inside the porous zone;
- the fourth case with direct feeding has the highest pressure drop, up to 150 times higher than the second case. Considering both electrolyte solution distribution and pumping power, the first case results as the best choice according to the present simulations.

Acknowledgements
This work was supported by the University of Padova under the MAESTRA 2011 strategic project.

6. References

[1] Alotto P, Guarnieri M and Moro F 2014 Redox flow batteries for the storage of renewable energy: A review *Renew. Sust. Energy Rev.* **29**, 325-35
[2] Cunha A, Martins J, Rodrigues N and Brito F P 2015 Vanadium redox flow batteries: A technology review *Int. J. Energy Res.* **39**, 889-918
[3] Weber A Z, Mench M M, Meyers J P, Ross P N, Gostick J T and Liu Q 2011 Redox flow batteries: A review *J. Appl. Electrochem.* **41**, 1137-64
[4] Ponce D L C, Frias Ferrer A, Gonzalez Garcia J, Szanto D A and Walsh F C 2006 Redox flow cells for energy conversion *J. Power Sources* **160**, 716-32
[5] Wei Z, Zhao J, Skyllas Kazacos M and Xiong B 2014 Dynamic thermal-hydraulic modeling and stack flow pattern analysis for all-vanadium redox flow battery *J. Power Sources* **260**, 89-99
[6] Xu Q, Zhao T S and Leung P K, 2013 Numerical investigations of flow field designs for vanadium redox flow batteries *Appl. Energy* **105**, 47-56
[7] Blanc C and Rufer A 2008 Multiphysics and energetic modeling of a vanadium redox flow battery, 696-701
[8] Tang Z, Aaron D S, Papandrew A B and Zawodzinski Jr T A 2012b Monitoring the state of charge of operating vanadium redox flow batteries **41**, 1-9
[9] Tang A, Ting S, Bao J and Skyllas-Kazacos M 2012a Thermal modelling and simulation of the all-vanadium redox flow battery J. Power Sources **203**, 165-76

Journal of Physics: Conference Series **655** (2015) 012050 doi:10.1088/1742-6596/655/1/012050

A preliminary investigation of the growth of an aneurysm with a multiscale monolithic Fluid-Structure interaction solver

D Cerroni[1], S Manservisi[1] and G Pozzetti[2]

[1] DIENCA - Lab. di Montecuccolino, Università di Bologna, Via dei Colli, 16, 40136 Bologna
[2] FSTC - University of Luxembourg, 6 rue Richard Coudenhove-Kalergi

E-mail: daniele.cerroni2@unibo.it

Abstract. In this work we investigate the potentialities of multi-scale engineering techniques to approach complex problems related to biomedical and biological fields. In particular we study the interaction between blood and blood vessel focusing on the presence of an aneurysm. The study of each component of the cardiovascular system is very difficult due to the fact that the movement of the fluid and solid is determined by the rest of system through dynamical boundary conditions. The use of multi-scale techniques allows us to investigate the effect of the whole loop on the aneurysm dynamic. A three-dimensional fluid-structure interaction model for the aneurysm is developed and coupled to a mono-dimensional one for the remaining part of the cardiovascular system, where a point zero-dimensional model for the heart is provided. In this manner it is possible to achieve rigorous and quantitative investigations of the cardiovascular disease without loosing the system dynamic. In order to study this biomedical problem we use a monolithic fluid-structure interaction (FSI) model where the fluid and solid equations are solved together. The use of a monolithic solver allows us to handle the convergence issues caused by large deformations. By using this monolithic approach different solid and fluid regions are treated as a single continuum and the interface conditions are automatically taken into account. In this way the iterative process characteristic of the commonly used segregated approach, it is not needed any more.

1. Introduction

In this work the potentialities of a monolithic FEM fluid-structure interaction (FSI) approach for the computational investigation of an abdominal aortic aneurysm are investigated. The development of abdominal aortic aneurysms (AAA) is associated with alterations of the tissue in the aortic wall, in particular collagen degradation, which leads to the loss of elasticity of the vessel. This process can lead to the rupture of the aneurysm with consequent internal bleeding. Accurate numerical simulations could provide large benefits in this field avoiding unnecessary and dangerous surgical removal operations. The simulation of the interaction between blood and blood vessel is not a trivial task because of the softness of the tissues, which leads to very large deformations. As a consequence of this highly non-linear behavior the stability of the numerical problem is not easily ensured. A monolithic approach, which solves the fluid and the structure problems in a coupled way, can bring significant advantages concerning the stability issues. This approach is, however, CPU time expensive and not suitable to simulate a large portion of the circulatory system. The use of mono-dimensional models, for all the components that are not of specific interest, and three-dimensional ones to describe the most critical structures(multi-scale approach) can reduce the computational costs of simulations. Furthermore one can achieve accurate studies of the specific component without loosing the system dynamic. The paper is organized as follows. In a first section the FSI equation of the problem are presented with particular reference to the ALE formulation, the monolithic approach, and the mono-dimensional model for the multi-scale approach. Finally some numerical examples are shown.

2. FSI EQUATIONS

In this section we give a description of the equations that are taken into account in order to solve the fluid structure interaction problem.

2.1. ALE Reference

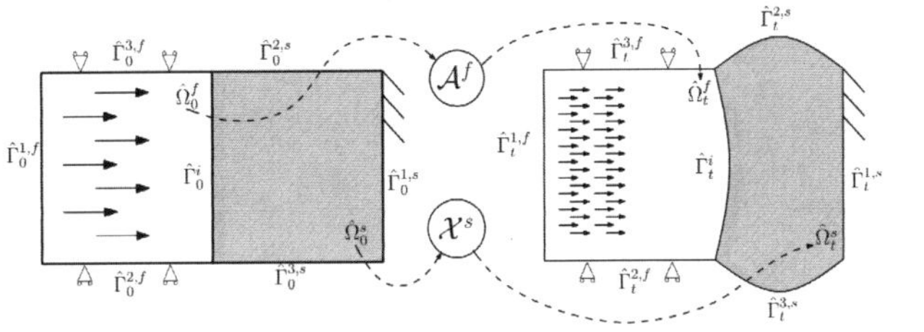

Figure 1. Reference configuration, on the left, and current configuration, on the right, and the operators for the Arbitrary Lagrangian-Eulerian (ALE) formulation.

In this work a Lagrangian reference is adopted for the solid domain while an ALE (Arbitrary Lagrangian Eulerian) reference is introduced in the fluid domain. The Lagrangian viewpoint consists of following the material particles of the continuum in their motion. To this purpose, one can introduce a computational grid, which follows the solid domain. The grid nodes are permanently connected to the same material points. The Lagrangian reference is therefore useful to describe deforming materials, but can only handle (in grid-based approaches) the investigation of systems where the displacement field is limited, such as the solid domain. For this reason in the fluid description the Eulerian reference is very popular. The basic idea of the Eulerian formulation consist on examining, as time evolves, the physical quantities associated with the fluid particles flowing through a fixed region of space. In the Eulerian description, the domain is fixed and the fluid moves and deforms with respect to the computational grid. Obviously these two references can not be coupled in a straightforward way. As previously stated, in this work a mixed Lagrangian-Eulerian description is used in order to combine Lagrangian and Eulerian references. For this purpose let us consider a mechanical system composed by a fluid and solid domain Ω_t as shown in Figure 1. Let Ω_t^f and Ω_t^s be the region occupied by the fluid and the solid respectively at the time $t \in]0, T]$. In the initial configuration fluid and solid regions are therefore defined by Ω_0^f and Ω_0^s, respectively. Let Γ_t^i be the interface between the solid and the fluid and Γ_t^k all the rest of the boundary. The evolution of the domain Ω_t can be described by considering the motion of the solid and fluid domains Ω_0^f and Ω_0^s defined by the mapping

$$\begin{aligned} \chi^s &: \Omega_0^s \times \mathbb{R} \to \mathbb{R}^3 , \\ A_f &: \Omega_0^f \times \mathbb{R} \to \mathbb{R}^3 . \end{aligned} \tag{1}$$

The range of the mapping χ^s is Ω_t^s. χ^s maps the position of the material point $\mathbf{x}_0^s$ to the current solid material configuration Ω_t^s. The solid displacement is then defined as

$$\mathbf{d}^s(\mathbf{x_0}^s, t) = \chi(\mathbf{x_0}^s, t) - \mathbf{d}_0^s . \tag{2}$$

In a similar way for the application A^f we define

$$\mathbf{d}^f(\mathbf{x_0}^f, t) = \chi(\mathbf{x_0}^f, t) - \mathbf{d}_0^f, \tag{3}$$

where $\mathbf{d}^f(\mathbf{x_0}^f, t)$ is as an arbitrary extension over the fluid domain Ω_0^f. This arbitrary extension could be written as

$$\mathbf{d}^f(\mathbf{x_0}^f, t) = Ext(\mathbf{d}^s|_{\Gamma_0^i}) \quad in \quad \Omega_0^f. \tag{4}$$

The harmonic or Laplace operator, which is also adopted in this work, is the most common extension operator. It is now possible to define the velocity $\mathbf{w}^f$ of the points of the fluid domain in the current configuration as

$$\mathbf{w}^f = \frac{\partial \mathbf{d}^f}{\partial t} \circ \mathbf{d}_0^f \quad in \quad \Omega_0^f, \tag{5}$$

where $\partial \mathbf{d}^f / \partial t$ is the velocity in the reference coordinates system.

2.2. Monolithic Approach

In order to couple the fluid and solid equations the liquid-solid interface conditions must be imposed. Two main approaches can be adopted: partitioned and monolithic. The first one aims to solve the fluid and the solid equations with two different dedicated solvers, The stress and displacement functions at the interface are exchanged through the two solvers and imposed as boundary conditions. The second approach solves the solid and fluid equations simultaneously via an unique solver. As we will see in with this approach the equivalence of stresses and displacement at the interface is therefore automatically taken into account. A monolithic approach is generally more computationally expensive than the partitioned ones, but can give some important advantages in therms of stability. In the ALE reference system, the fluid mass and momentum balance equations reads

$$\nabla \cdot \mathbf{u}^f = 0, \qquad \frac{\partial \mathbf{u}^f}{\partial \tau} + (\mathbf{u}^f - \mathbf{w}^f) \cdot \nabla \mathbf{u}^f = -\nabla \cdot p^f \mathbf{I} + \nu \nabla^2 \mathbf{u}^f + f_b, \tag{6}$$

where τ is the time variable, f_b represents a generic body-force, $\mathbf{u}^f$ is the fluid velocity vector, $\mathbf{w}^f$ is the previously defined ALE velocity, p^f is the specific pressure, and ν is the dynamic viscosity which is here taken to be constant. The FSI equation system in strong form then becomes

$$\begin{aligned}
&\nabla \cdot \mathbf{u}^f = 0, \qquad \frac{\partial \mathbf{u}^f}{\partial \tau} + (\mathbf{u}^f - \mathbf{w}^f) \cdot \nabla \mathbf{u}^f = -\nabla \cdot p^f \mathbf{I} + \nu \nabla^2 \mathbf{u}^f + f_b \quad in \quad \Omega_t^f, \\
&\mathbf{w}^f = \frac{\partial \mathbf{d}^f}{\partial t}, \qquad \frac{\partial \mathbf{d}^f}{\partial t} - k \nabla^2 \mathbf{d}^f = 0 \quad in \quad \Omega_t^f, \\
&\rho \frac{\partial \mathbf{u}^s}{\partial t} + \rho (\mathbf{u}^s \cdot \nabla) \mathbf{u}^s = \nabla \cdot \sigma(\mathbf{d}^s) + \mathbf{b}^s, \qquad \mathbf{u}^s = \frac{d\mathbf{d}^s}{dt} \quad in \quad \Omega_t^s,
\end{aligned} \tag{7}$$

where $\mathbf{u}^s$ is the solid velocity vector, $\sigma(\mathbf{d}^s)$ the solid strain tensor as a function of the solid displacement ($\mathbf{d}^s$), and $\mathbf{b}^s$ the external force.

The system (7) must be provided with the following boundary and initial conditions

$$\sigma^s \cdot \mathbf{n} = \sigma^f \cdot \mathbf{n}, \qquad \mathbf{d}^f = \mathbf{d}^s \quad on \quad \Gamma_t^i, \quad \mathbf{u}(\mathbf{x}_0, 0) = \mathbf{u}_0, \quad \mathbf{d}(\mathbf{x}_0, 0) = \mathbf{d}_0 \quad in \quad \Omega_0. \tag{8}$$

In particular the first two conditions set the continuity of stress and velocity at the fluid-solid interface. Other suitable boundary conditions must be set on Γ_t^k. The weak formulation of the problem is obtained by multiplying each equation for a proper test function ϕ and then integrating over the computational domain. For a rigorous treatment of the subject one can refer for example to [2]. For the specific subject of fluid structure interaction one can refer to [1], as well as [5],[8]. For the fluid and solid momentum equations we obtain

$$\int_{\Omega_t^s}(\rho\frac{\partial \mathbf{u}^s}{\partial t}\phi + \rho(\mathbf{u}^s\cdot\nabla)\mathbf{u}^s\phi)d\Omega + \int_{\Omega_t^s}\sigma^s : \nabla\phi d\Omega = \int_{\Gamma_{t,N}^s}\mathbf{g_N^s}\cdot\phi d\gamma - \int_{\Gamma_t^i}(\sigma^f\cdot\mathbf{n})\cdot\phi d\gamma$$
$$\forall\phi\in V_{\Gamma_D}^s\,,\qquad V_{\Gamma_D}^s = [H_{\Gamma_{Dv}}^1(\Omega)^d] = \{\phi\in[H^1(\Omega)^d] : \phi|_{\Gamma_{Dv}} = 0\}\,. \tag{9}$$

$$\int_{\Omega_t^f}(\rho\frac{\partial \mathbf{u}^f}{\partial t}\phi + \rho(\mathbf{u}^f-\mathbf{w})\cdot\nabla\mathbf{u}^f\phi)d\Omega + \int_{\Omega_t^f}\sigma^f : \nabla\phi d\Omega = \int_{\Gamma_{t,N}^f}\mathbf{g_N^f}\cdot\phi d\gamma + \int_{\Gamma_t^i}(\sigma^f\cdot\mathbf{n})\cdot\phi d\gamma$$
$$\phi\in V_{\Gamma_D}^f\,,\qquad V_{\Gamma_D}^f = [H_{\Gamma_{Dv}}^1(\Omega)^d] = \{\phi\in[H^1(\Omega)^d] : \phi|_{\Gamma_{Dv}} = 0\}\,. \tag{10}$$

$H^1(\Omega)^d$ denotes the Sobolev space. It is easy to notice that the last surface integrals of (9) and (10) are equal but with opposite sign due to the equivalence of the stresses at the interface. Such terms are implicitly taken into account in the monolithic approach, while they must be evaluated and substituted in the segregated approach.

2.3. 1D model for the multi-scale approach

The mono-dimensional model is a spatially reduced FSI model of a straight cylinder surrounded by a linear membrane structure. We remark that this simplified model is only used to evaluate the flow rate distributions and pressure wave propagation in order to save computational resources. The physical consistency of the model has being investigated by many authors. On can see reference [1] and citations therein. In particular, referring to a straight cylinder, the integration of the continuum balance equation over a generic section leads to the simplified system which reads

$$\frac{\partial A}{\partial t} + \frac{\partial Q}{\partial x} = 0\,,\qquad \frac{\partial Q}{\partial t} + \frac{\partial}{\partial x}\frac{\alpha Q^2}{A} + \frac{A}{\rho_f}\frac{\partial P}{\partial x} + K_r\frac{Q}{A} = 0\,, \tag{11}$$

where Q is the flow rate over the generic section S, A is its surface and P the average pressure. K_r is a resistance parameter which takes into account the fluid viscosity. The parameter α defines the shape of the velocity profile over the section. The (11) is a system of two equations with three unknowns P, Q, A. For its closure a third equation is needed. For this purpose we use a suitable wall model relating the radial displacement to the mean pressure P. In particular we set

$$P = P_{ext} + \psi(A)\,,\qquad \frac{\partial\psi}{\partial A} > 0\,,\qquad \psi(A_0) = 0\,, \tag{12}$$

where A_0 is the area of the surface S at $t = 0$. A classical choice is to consider an algebraic law of the type

$$\psi(A) = \beta(A)\frac{\sqrt{A}-\sqrt{A_0}}{\sqrt{\pi}}\,,\qquad \beta = \frac{H_s E\pi}{(1-\nu^2)A}\,, \tag{13}$$

where H is the thickness of the structure, E and m are the Young modulus and Poisson ratio, respectively. It is possible to take β constant as a mean value to simplify the treatment.

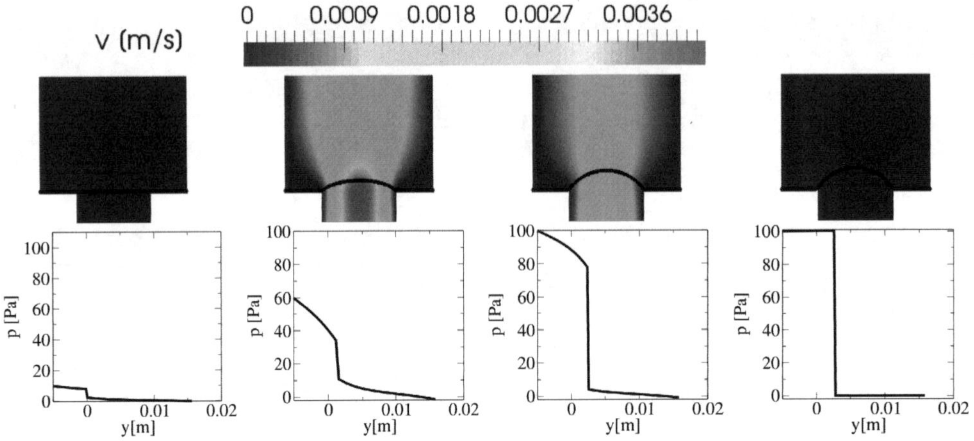

Figure 2. Benchmark 1. Velocity field (on the top) and pressure profile along the axis (on the bottom), from left to right at $t = 0.1$, 0.6, 1 and 2.3s.

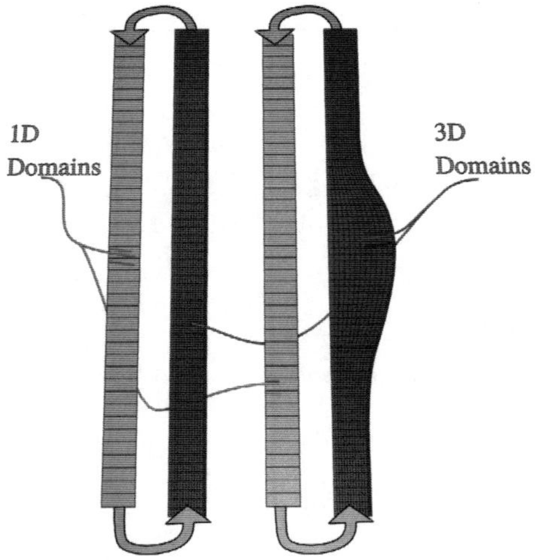

Figure 3. Case A. Configuration of the aneurysm (right) and healthy aorta (left) together with the mono dimensional domain.

3. Numerical Results

3.1. FSI Benchmark

In the first test we validate the fluid structure interaction model. We reproduce the benchmark originally proposed in [4] for large deformation problems and test the stability of the monolithic fluid structure model. In particular the benchmark consider two different fluid regions, divided by a thin soft solid. The lower chamber is loaded with a pressure and the soft material undergoes to large deformation. The interested reader could see all the specific geometrical dimensions of the benchmark in [4]. In figure 2 we show an overview of the obtained results, in particular on the top we can appreciate the continuity of the velocity field at different time steps together with the deformation of the solid structure. On the bottom one can see the evolution of the pressure field at the same time steps.

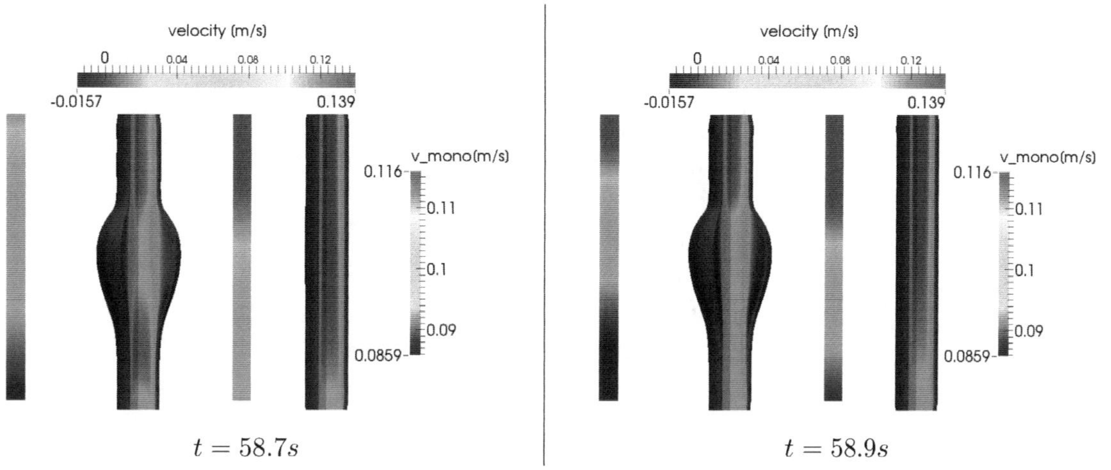

Figure 4. Case A. Velocity along the y-axis in the case of presence of an aneurysm (on the left) and for an healthy aorta (on the right) at $t = 58.7$ (on the left) and $58.9s$ (on the right).

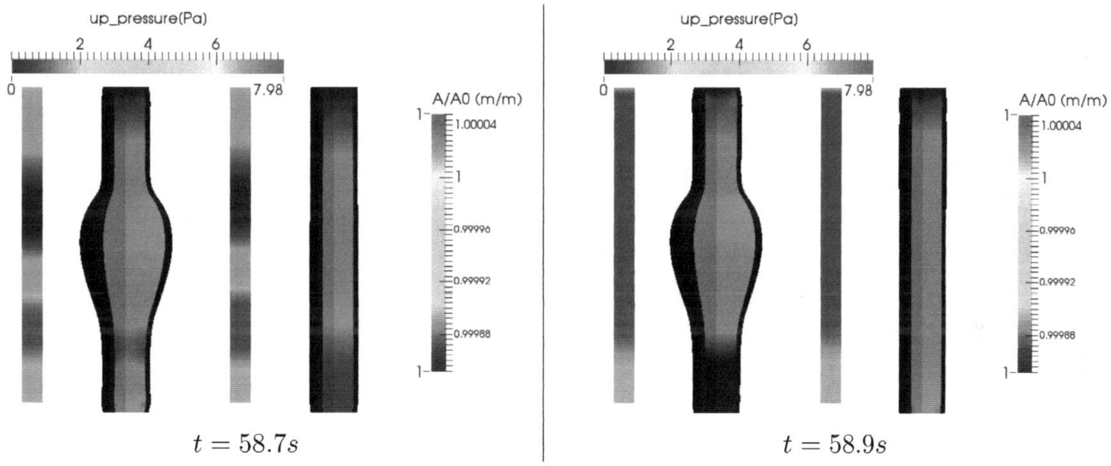

Figure 5. Case A. Pressure in the 3D domain and area variation on the 1D domain in the case of presence of an aneurysm (on the left) and for an healthy aorta (on the right) at $t = 58.7$ (on the left) and $58.9s$ (on the right).

3.2. Stable Aneurysm

A multi-scale, axial symmetric, multi-dimensional simulation is performed with the purpose of studying the behavior of a stable aneurysm inserted in the whole cardiovascular system. We couple the inlet and the outlet surface of the domain with the outlet and the inlet boundary of a mono-dimensional pipe as it is shown in Figure 3. The inlet velocity of the multi-dimensional domain is the one coming from the the simplified system and the outflow of the complex domain is imposed at the inlet region of the 1D system. This coupling represents the first step for a detailed multi-scale approach that could lead to obtain realistic simulations of the whole cardiovascular loop with a reasonable computational cost. The same simulation were performed considering a non-deformed blood vessel in order to compare the obtained results. The Young modulus, as is common in literature for this type of problems, is set to $10^3 Pa$ and the Poisson ratio to 0.4. We also take into account the hearth contraction by adding a periodic momentum

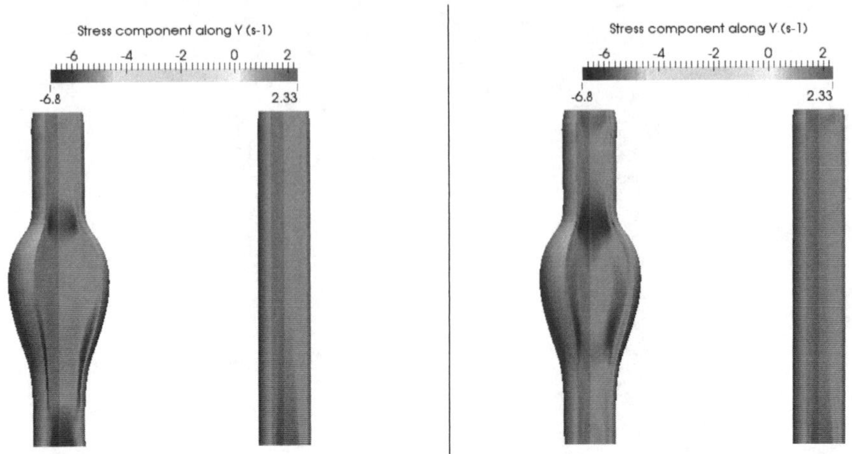

$t = 58.7s$ $t = 58.9s$

Figure 6. Case A. Stress component in the axial direction in the 3D domain in the case of presence of an aneurysm (on the left) and for an healthy aorta (on the right) at $t = 58.7$ (on the left) and $58.9s$ (on the right).

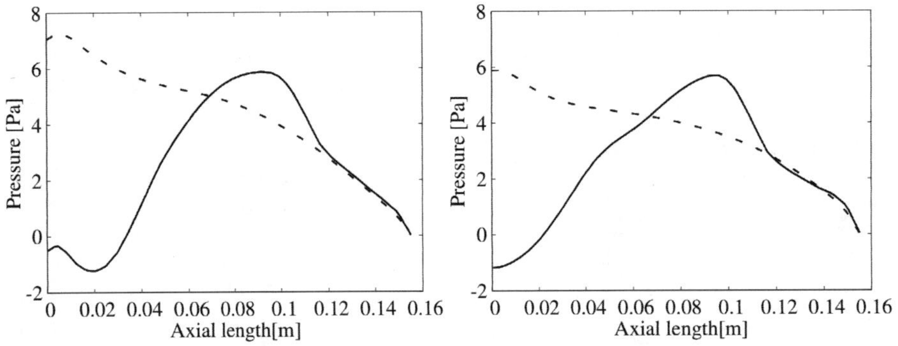

Figure 7. Case A. Pressure along the axis for the 3D domain in the case of presence of an aneurysm (on the left) and for an healthy aorta (on the right) at $t = 58.7$ (on the left) and $58.9s$ (on the right).

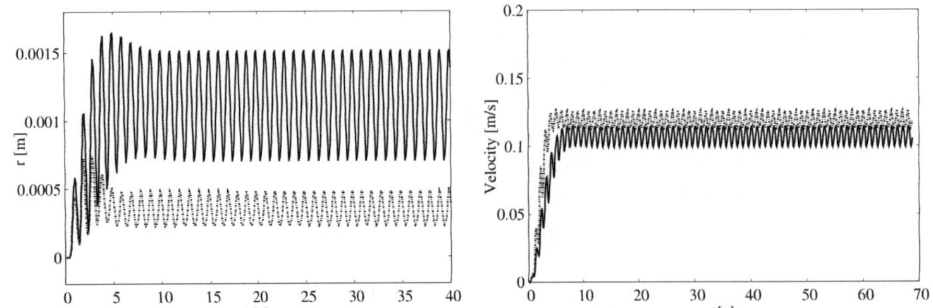

Figure 8. Case A. Evolution of the displacement (on the left) and of the velocity (on the right). in the presence of an aneurysm (solid line) and an healthy aorta (dashed line).

source in the simplified system in the form $Hsin(\pi t)^4$. The multi-scale approach allows us the study of the effects of the presence of the aneurysm on a simplified closed loop. When the flow becomes periodic stabilized a period of $1\,s$ is observed. In Figure 4 the velocity field in the axial direction, for a normal and a deformed domain, is shown at different time steps. In particular we can observe that in the deformed case the inlet velocity undergoes large variations. Such an effect could not be taken into account without using a multidimensional approach. In Figure 5, the pressure field in the three-dimensional domain and the area distribution in the mono-dimensional pipe are shown at different time steps. In Figure 6 one can appreciate the increasing of the axial stress in the presence of an aneurysm at different time steps. In the healthy aorta interface stresses are negligible and, in the presence of an aneurysm, the stresses increase near the inlet and the outlet region of the aneurysm itself with high probability to cause the blood vessel rupture. In Figure 7 the pressure field along the axis at different time steps is shown. We plot, in a dashed line, the values for the healthy aorta and in a solid line the values for the aneurysm. We remark that several differences can be observed between these two cases. Such differences are mostly due to the recirculation phenomena that occur in the presence of the aneurysm. In Figure 8 the radial displacement , at $y = 0.07\,m$ is shown together with the axial velocity. We plot, in a dashed line, the values for the healthy aorta and, in a solid line, the values for the aneurysm. One can notice that in the presence of an aneurysm displacements becomes greater than the ones evaluated in a straight blood vessel. The large deformation that occurs in the presence of the aneurysm reduces the overall magnitude of velocity field and the characteristic frequency of the structure, as we can see on the right of Figure 8.

3.3. Aneurysm formation and growing

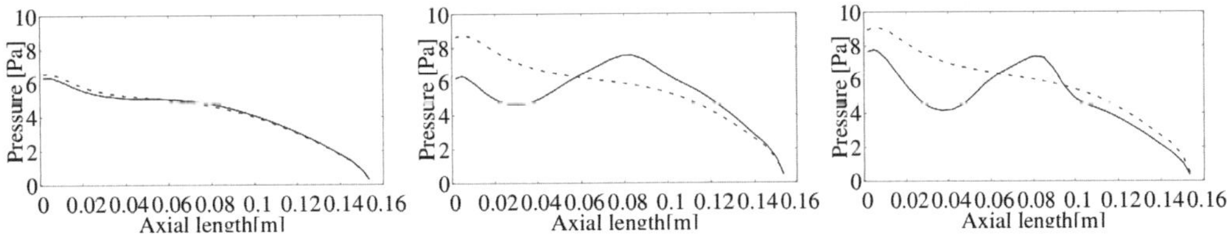

Figure 9. Case B. Formation of the aneurysm pressure as a function of the axial coordinate for the three-dimensional domain. Aneurysm formation (solid line) and healthy aorta (dashed line) for $t = 5$, 35 and $80\,s$.

In this test we simulate the growth of an aneurysm and we show the evolution of the velocity field. The formation of the aneurysm is caused by a progressive, localized loss of elasticity in the aorta wall. The mono-dimensional loop is treated with the model described previously. The boundary and the coupling conditions are the same imposed in the previous section for the healthy aorta in order to compare the velocity profile obtained. A progressive localized loss of elasticity is simulated with a decreasing of the Young modulus in time by using the following model

$$E = E_0(1 - 80t(x - 0.04)(0.1 - x))^4 \quad \forall x \in [0.04; 0.1]\,, \qquad E = E_0 \quad \forall x \notin [0.04; 0.1]\,. \quad (14)$$

The simulation is performed in a time interval of 400 s in which the aneurysm is formed and reaches the steady state. As in the previous example a multi-scale approach is used in order to study the effects of an aneurysm development in the framework of a closed connected loop. This setup may be used to study the evolution of an aneurysm and investigate the behavior of the whole circulatory system when an aneurysm is growing in a certain blood vessel. In Figure

10 the velocity profile and the cross section dimension of the multidimensional and the mono-dimensional domain are shown respectively at different time steps. In Figure 9 we show the pressure distribution along the axis of the multi dimensional domain at different time steps. In Figure 11 we can observe the variation of the inlet velocity in the mono-dimensional domain in the normal and the deformed case. We can notice that the average velocity field is not constant over time. This is a consequence of the huge deformation of the multidimensional domain. Such a feedback could not be taken into account without considering the simplified coupled domain.

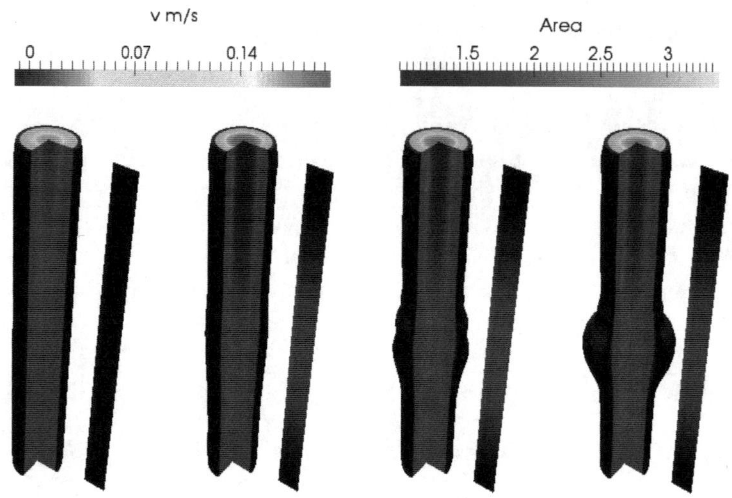

Figure 10. Case B. Formation of the aneurysm, velocity along the axis in the three-dimensional domain and area variation in the mono-dimensional domain for $t = 5$, 35, 65 and 80 s.

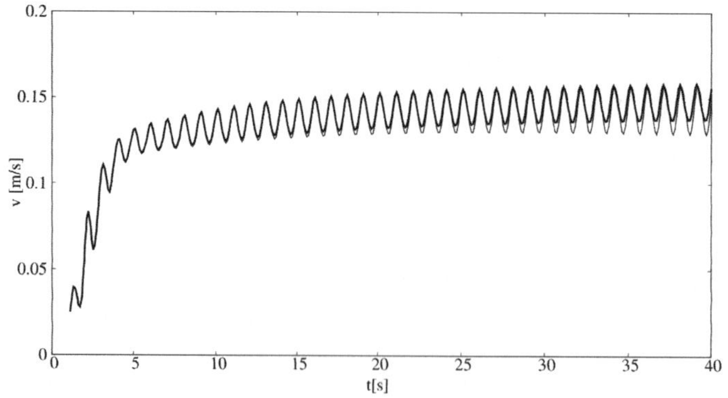

Figure 11. Case B. Formation of the aneurysm. Mono-dimensional velocity as a function of time at the inlet of the mono-dimensional domain. Aneurysm formation (thick line) and healthy aorta (thin line).

4. Conclusion

In this work we have presented a preliminary investigation of an highly unsteady and transient fluid-structure interaction system using a multi-scale, monolithic approach. We have shown how this technique can be used in order to study a complex problem such as an aneurysm growth in a blood vessel. In particular we have used a full scale multi-dimensional model to simulate the growing aneurysm region where and a mono-dimensional simplified model for the rest of the system. Such a coupling scheme allow us to use CFD methodology and still keep into account the whole system response with reasonable computational costs.

References

[1] Kirillov P and Ushakov P A 2001 Heat transfer to liquid metals: specific features, methods of investigation and main relationships *Therm. Eng.* **48** 50-9

[1] Formaggia L Quarteroni A and Veneziani A 2009 Cardiovascular mathematics, Modeling and simulation of the circulatory system *Springer*

[2] Quarteroni A 2008 Numerical Models for Differential Problems *Springer New York*

[3] Hron J and Turek S 2006 A monolithic FEM/multigrid solver for an ALE formulation of fluid-structure interaction with applications in biomechanics *Springer New York*

[4] Bathe K J, Ledzma G A 2007 Benchmark problems for incompressible fluid flows with structural interactions *Computers and Structures* **85** 628-40

[5] Bna S Bornia G and Manservisi S 2012 A Monolithic Fem Multigrid Penalty Projection Solver for Incompressible Fluid-Structure Interaction *Proceedings of the first ECCOMAS-YIC2012 conference on computational method in applied sciences*

[6] Aulisa E Manservisi S and Seshaiyer P 2007 A non-conforming computational methodology for modeling coupled problems *Nonlinear Analysis* **63** 1445-54

[7] Aulisa E Cervone A Manservisi S and Seshaiyer P 2009 A Multilevel domain decomposition approach for studying coupled flow applications *Commun Comput Phys* **6** 319-54

[8] Heil M 2004 An Efficient Solver to the Fully Coupled Solution of Large-Displacement Fluid Structure Interaction Problems *Comput Methods Appl Mech Engrg* **193** 1-2

[9] Tallec P and Mouro J 2012 Fluid structure interaction with large structural displacements *Computer Methods in Applied Mechanics and Engineering* **190** 3039-60

[10] Saks M S Merryman W D and Schmidt D E 2009 On the biomechanics of heart valve function, *Journal of Biomechanics* **42** 1804-24

[11] Sakalihasan N Limet R Defawe O D2005 Abdominal aortic aneurysm *seminar Lancet* **1**

[12] Donea J Huerta A Ponithot J Ph Rodriguez-Ferran A 2004 Arbitrary Lagrangian-Eulerian Methods

[13] Cerroni D Manservisi M and Menghini F Multiscale Techniques for the coupling of 3D-1D FSI equations system in compliant vessels *World Congress on Computational Mechanics* **11**

Journal of Physics: Conference Series **655** (2015) 012051 doi:10.1088/1742-6596/655/1/012051

CFD Simulation of Melting and Solidification of PCM in Thermal Energy Storage Systems of Different Geometry

S Arena, G Cau, C Palomba*
Dept. of Mechanical, Chemical and Materials Engineering, University of Cagliari, Via Marengo 2, 09123 - Cagliari, Italy

* In memory of Chiara, dear friend and estimated researcher, who left us on May 15.

E-mail: simonearena@unica.it

Abstract. At the Department of Mechanical, Chemical and Materials Engineering of the University of Cagliari an experimental and numerical research project has begun with the aim of developing highly efficient thermal energy storage (TES) systems using phase change materials (PCM) of particular interest in concentrating small-medium scale solar power (CSP) applications. The present work aims to simulate the melting and solidification processes in containing boxes and heat transfer devices of different geometrical features which may constitute the elementary cell of a more complex TES system.

Two-dimensional axisymmetric numerical models, developed with COMSOL Multiphysics are considered and used to simulate TES, heat conduction and natural convection. The models are used to determine the temperature profile inside the PCM to identify which configurations are capable of enhancing thermal response between a solid wall and a PCM. The results obtained will be used for comparison with experimental data acquired from a pilot plant under construction in the DIMCM laboratories. At the current stage the laboratory is being brought to completion.

1. Introduction

The growing worldwide energy demand for human activities, mainly based, also in perspective, on fossil fuels, raises the question of sustainability. In particular, as regards the atmospheric concentration of CO_2, renewable energy sources together with power generation efficiency and end-use fuel and electricity efficiency are going to play a fundamental role and will contribute for about 60% to the overall reduction of CO_2 emissions. As is know, renewable energy sources such as solar and wind energy in particular, are not only characterized by discontinuous availability but are also affected by random variations due to local weather conditions. To expand their use, some form of energy storage must be foreseen to permit the deferred use of excess energy produced during periods of high availability or low demand. The development of phase change materials (PCM) for thermal energy storage (TES) is framed in the course of action in which thermal energy at medium-high temperature from, as an example, solar concentration fields or industrial processes, needs to be stored for optimum management. The advantage of this type of storage is that of maintaining a constant temperature of the working fluid preserving its potential to produce work in power generation systems in particular.

However, while it is relatively simple to size the amount of a given material required to store a definite amount of energy, it is more complex to design a phase changing heat exchanger that allows performance of the heat transfer process between PCM and HTF in an efficient way. The heat transfer will influence the time required to store and release the thermal energy and it is therefore a fundamental design parameter when a PCM-TES must interact dynamically with other equipment such as, for example, a solar field and a thermal engine. Poor heat transfer during the charging period will lower solar field efficiency with high re-entry temperatures of the heat transfer fluid. Similar considerations may be drawn for the discharging period when the required thermal power must be supplied, for instance, to a thermal engine.

A number of experimental and theoretical studies on the transient thermal behaviour and performance of a thermal energy storage device during both charging and discharging processes have been conducted by several investigators. Adine and Karma [1] , Akgun et al. [2] and Gil et al.[3] investigated the performance of a shell and tube thermal energy storage device. Gil et al. [4][5][6][7], Solè et al.[8], Medrano et al. [9] studied different types of materials for applications in the field of concentrated solar energy systems at medium temperature. The results show that the selection of operating conditions and geometric parameters depends on the required heat transfer rate and the time in which the energy has to be stored or released. The main disadvantages of PCMs is low thermal conductivity which requires a long time for the melting and solidification processes. Several common techniques have been employed to enhance the heat transfer rate of PCMs. Choi and Kim [10] and Castell et al. [11] used finned tubes, Velraj et al. [12] used rings and bubble agitation, Melhing et al.[13] proposed a graphite compound-material, Agyenim et al. [14] and Al-Abidi et al. [15] proposed a triplex tube heat exchanger (TTHX). This system is composed of three tubes: the heat transfer fluid flows in the inner and outer tubes while the PCM is placed in the middle tube. This configuration increases the heat transfer area, thus improving heat transfer efficiency with respect to that of the double pipe heat exchanger.

The purpose of this analysis is focused on the development of thermal energy storage systems, to be used as back-up devices during transient operations of small and medium scale concentrated solar plants (CSP). Real applications of LHTES for the temperature range chosen (150-220 °C) are typically used in solar technologies associated with Organic Rankine Cycle (ORC) (e.g. REELCOOP EU/FP7 project). In the same temperature range, PCM materials are also used in various types of industrial applications such as in solar cooling systems with absorption chiller [3], related to fuel cell PAFC or heat recovery in metal hydrides hydrogen storage systems.

In the present work, three different geometrical configurations of heat exchangers were investigated numerically. The first configuration is a simple double-pipe, the second is a TTHX and the third is a TTHX with circular fins. The effects of natural convection were considered and the different distributions of heat transfer in the three cases were studied to evaluate the effects of charging and discharging processes on heat transfer coefficients and execution.

2. Material and method

2.1. Phase change material selection

The choice of PCM was mainly made on the basis of a melting temperature range suitable for implementation of a TES system in small-size concentrating solar plants. In this case, the system operates with an upper temperature of 200 °C, which represents the maximum temperature of the HTF leaving the solar collectors. The latent heat is also a significant choice parameter, since on it depends the amount of energy storable per unit mass of PCM. The final choice must also take into consideration other relevant parameter such as chemical properties, thermal expansion and safety.

For the present study, D-Mannitol was selected from among different PCM candidates after a literature review [4][5][9][13][16][17] since it has a phase change temperature range between 164 and 170 °C and a melting enthalpy of 316 kJ/kg [16]. D-Mannitol is classified as a sugar alcohol, molecular formula $C_6H_{14}O_6$: it is derived from a sugar (mannose) by reduction. The properties of D-Mannitol are shown in Table 1. Thermophysical properties of the liquid phase were obtained from other materials belonging to the same family and with similar transition temperatures with an estimated uncertainty of about 10%. Although widely used in other fields (pharmaceuticals, food industry etc.), only in recent years has this kind of materials been studied and used for thermal energy storage. As concerns latent heat, a value of 316 kJ/kg was found in literature [16], obtained for the pure substance and under controlled conditions, but several tests with the DSC [6][18][19][20][20]. detected that this value depends strongly on three important factors, namely purity, cooling/heating rate and air contact. For these reasons, a more "realistic" value of 234 kJ/kg was used in this work.

Table 1. Properties of D-Mannitol.

	Solid	Liquid
Density (kg/m^3)	1520 (20 °C)	1382
Specific heat (kJ/kgK)	1.320 (20 °C) [a]	1.452
Thermal conductivity (W/mK)	0.279 [b]	0.307
Phase change temperature	167°C	
Range of phase transition temperature	±3°C	
Latent heat	234 kJ/kg [c]	

[a] NIST web page (National Institute of Standards and Technology).
[b] Thermal Energy Storage Systems for Solar Cookers, www.ukessays.com
[c] value obtained as the average of the experimental results found in literature [6][18][19][20][21].

2.2. Design of the PCM storage systems
Three cases with different geometrical configurations were considered in characterizing the TES system. Case 1 (Fig. 1a) is based on the configuration of the double-tube heat exchanger, where the outer tube is filled with PCM while the heat transfer fluid (HTF) flows in the inner tube. Case 2 (Fig. 1b) is based on a triplex heat exchanger. It is composed of three tubes: the HTF flows in the outer and inner tubes, while the PCM is placed between them. Case 3 (Fig. 1c) is a triplex with six circular fins. “Syltherm 800” was selected as the heat transfer fluid, as it is the HTF used in the experimental facility under construction.

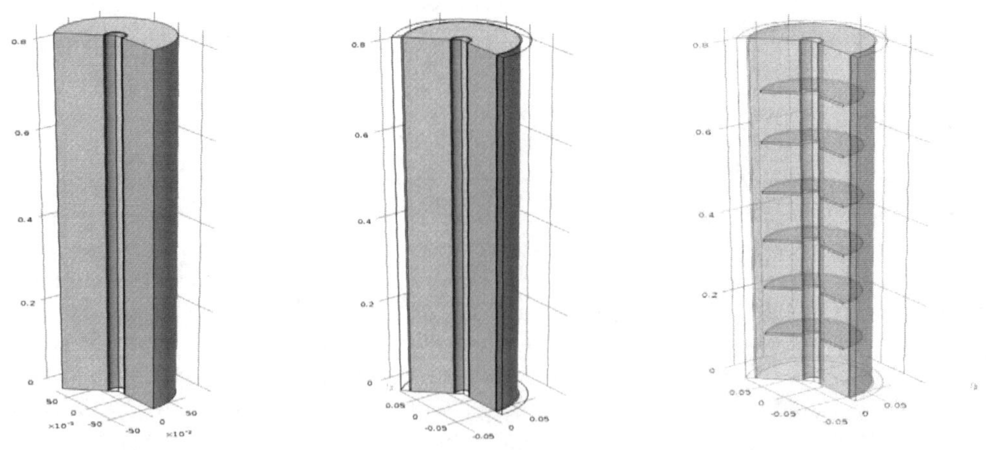

(a) Case 1 - Double tube (b) Case 2 - Triplex (c) Case 3 - Triplex with fins

Figure 1. Geometrical configuration of the three TES systems.

The design of the storage systems is based on the LMTD method and was performed by means of a computer code developed in Matlab environment. The geometrical characteristics of the three TESs are shown in Table 2. The value of HTF mass flow rate was selected as 0.052 kg/s for all three cases analyzed for both charge and discharge phases. The HTF temperature at the inlet of the accumulator was chosen as 180 °C and 100 °C for charge and discharge respectively. The HTF flows from top to bottom during both charge and discharge phases. The PCM mass for Cases 1 and 2 was 23.4 kg, while for Case 3, owing to the space occupied by the fins, the value was slightly lower, 22.8 kg, allowing a thermal energy storage of about 3 kWh.

Table 2. Geometrical characteristics of the storage system

	Case 1	Case 2	Case 3
Heat transfer surface [m^2]	0.077	0.568	0.73
Internal Diameter inner tube [m]	0.0254 (1")	0.0254 (1")	0.0254 (1")
Outside Diameter tube PCM [m]	0.127 (5")	0.127 (5")	0.127 (5")
Outside Diameter outer tube HTF [m]	-	0.2218	0.2218
Length [m]	0.83	0.83	0.83
Number of fins	-	-	6
Fin Length [m]	-	-	0.05
Fin thickness [m]	-	-	0.005
Tube thickness (Steel) [m]	0.002	0.002	0.002

2.3. Mathematical model of the PCM thermal energy storage system

A 2D axisymmetric numerical model was developed with the COMSOL Multiphysics 5.0 platform to simulate the transient and energy behaviour of the entire PCM-TES system during the phase change processes. Three different "physics", described below, were used to create the model [22]. The first and the third physics, "Fluid Flow", were used to simulate the behaviour of the HTF and the liquid PCM, the second, "Heat Transfer", was used to simulate the heat transfer process between the HTF and the PCM; it is composed of three sub-modules: heat transfer by convection, by conduction and phase change heat transfer.

2.3.1. Fluid Flow (HTF)

To simulate the dynamic behaviour of the HTF flowing inside the inner tube (Case 1) and inside the inner and the outer tubes (Case 2-3), the continuity equation (1) and the momentum equation (2) were used. The flow is considered laminar and incompressible and the effect of gravity is negligible (F=0).

$$\frac{\partial \rho}{\partial t} + \nabla \cdot (\rho u) = 0 \tag{1}$$

$$\frac{\partial \rho u}{\partial t} - \nabla \cdot \left[\mu\left(\nabla u + (\nabla u)^{\mathrm{T}}\right)\right] + \rho(u \cdot \nabla)u + \nabla p = F \tag{2}$$

In equations (1) and (2), ρ and μ are the density and the dynamic viscosity of the HTF respectively, u is the velocity vector and p is the pressure.

2.3.2. Heat Transfer in Fluid

Heat transfer from the HTF to the wall of the steel tube takes place by convection. The energy equation (3) was solved using the velocities found from the solution of Eqs. (1) and (2). In equation (3) C_p and k are respectively the specific heat and the thermal conductivity of the material.

$$\rho C_p \frac{\partial T}{\partial t} + \rho C_p u \cdot \nabla T = \nabla \cdot (k \nabla T) \tag{3}$$

2.3.3. Heat Transfer in Solid

Heat transfer from the wall of the steel tube to the PCM takes place by conduction. The energy equation (4) for this case is reported below:

$$\rho C_p \frac{\partial T}{\partial t} = \nabla \cdot (k \nabla T) \tag{4}$$

2.3.4. Heat Transfer with Phase Change

The energy equation (5), in accordance with the apparent heat capacity formulation (6), was used to obtain the temperature field. This method assumes that the transformation occurs in a temperature interval between $(T_m - \Delta T_m/2)$ and $(T_m + \Delta T_m/2)$ where the material phase is modelled by a smoothed function α, representing the fraction of phase that generates during transition. The value of this function is equal to 0 before $(T_m - \Delta T_m/2)$ and to 1 after $(T_m + \Delta T_m/2)$. The equivalent heat conductivity and density are given by equations (7) and (8) in which the parameter $\theta = (1 - \alpha)$ is introduced.

$$\rho C_p \frac{\partial T}{\partial t} + \rho C_p u \cdot \nabla T = \nabla \cdot (k \nabla T) \tag{5}$$

$$C_p = \theta C_{p,phase1} + (1 - \theta) C_{p,phase2} + L \frac{\partial \alpha}{\partial T} \tag{6}$$

$$k = \theta k_{phase1} + (1 - \theta) k_{phase2} \tag{7}$$

$$\rho = \frac{\theta \rho_{phase1} C_{p,phase1} + (1 - \theta) \rho_{phase2} C_{p,phase2}}{\theta C_{p,phase1} + (1 - \theta) C_{p,phase2}} \tag{8}$$

2.3.5. Fluid Flow (liquid PCM)

The continuity and momentum equations (1) and (2) were used to simulate the behaviour of the PCM in liquid phase. To simulate the buoyancy force giving rise to natural convection, a volume force was added. The Boussinesq approximation was introduced to consider these buoyancy forces, according to Eq. (9). In this case ρ, μ and β are respectively the density, the dynamic viscosity and the thermal expansion coefficient of the liquid PCM, g is the gravitational constant and T_m is the melting temperature. The flow is considered laminar and incompressible.

$$F_b = g \rho \beta (T - T_m) \tag{9}$$

In this method, the entire PCM was treated as a liquid, even when its temperature was lower than its melting point. A modified viscosity was used to force this liquid to behave as a solid when required. A step function, continuous second derivative centred at about T_m, was introduced to define the modified viscosity: the value of this step function was 10^6 for the solid PCM and the actual value of μ for the liquid PCM. This approach forces the Navier-Stokes equation to calculate the velocity everywhere, even when the PCM is a solid. This significantly increases the number of calculations and the possibility that the solver will diverge during the simulation. A second volume force, defined by Eq. (10), was added to the momentum equation to calculate velocities equal to zero in the solid phase of PCM.

$$F_v = -A(T) \cdot \mathrm{u} \tag{10}$$

In this equation a step function $A(T)$ was introduced: its value was 10^8 for the solid PCM and zero for the liquid PCM. The role of F_v is to dominate every other force terms in the momentum equation (2) when the PCM is solid, thus improving the time of calculation and effectively forcing a trivial solution of $u = 0$ in the solid. This method is mathematically better defined and allows a reduction of the number of iterations performed by the solver.

3. Results and discussion

In this paragraph the simulation results of the three cases considered are presented. Figures 2 and 3 show the temperature evolution of the PCM evaluated at the bottom of the accumulator (in proximity of the HTF outlet), in function of time during charge (Fig. 2) and discharge (Fig. 3).

In charging mode, the process started from a homogeneous initial state of the entire system (PCM, tubes, and HTF). At the beginning, the entire PCM system was at the set temperature of 100 °C, with an inlet temperature of the HTF of 180 °C. In this condition the heat transfer inside the solid PCM was governed by conduction. Then, when the temperature range of phase transition was reached, the melting process began. During the phase transition both conduction and natural convection in liquid phase occurred. At the end of the transition process, the PCM was in the liquid phase. In this condition, the heat transfer inside the liquid PCM was governed by convection.

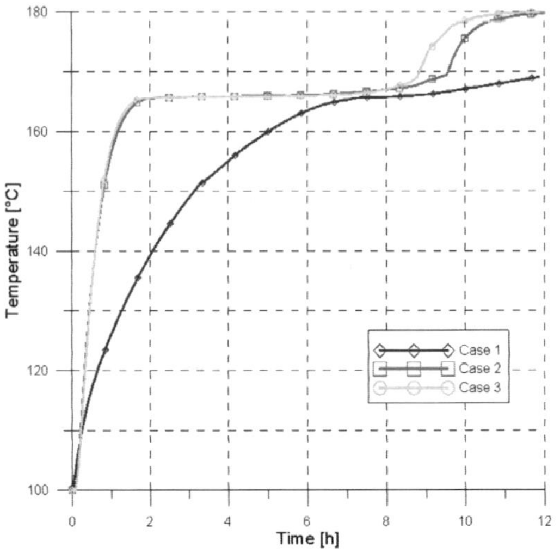

Figure 2. PCM temperature profile, charge.

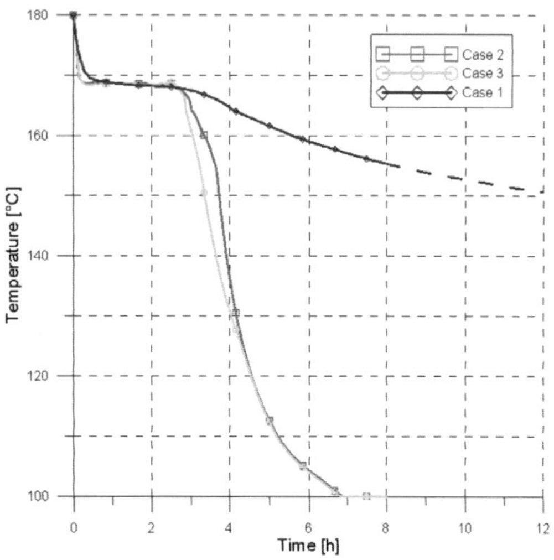

Figure 3. PCM temperature profile, discharge

The discharging process was performed analogously. Initially, the entire PCM system was at the set temperature of 180 °C, with an inlet temperature of the HTF of 100 °C. The time of the charge and discharge processes was imposed as 12 and 8 hours respectively. In the charge phase the transition process finished after approximately 10 hours, while in the discharge phase the complete transition was reached in about 3.5 hours.

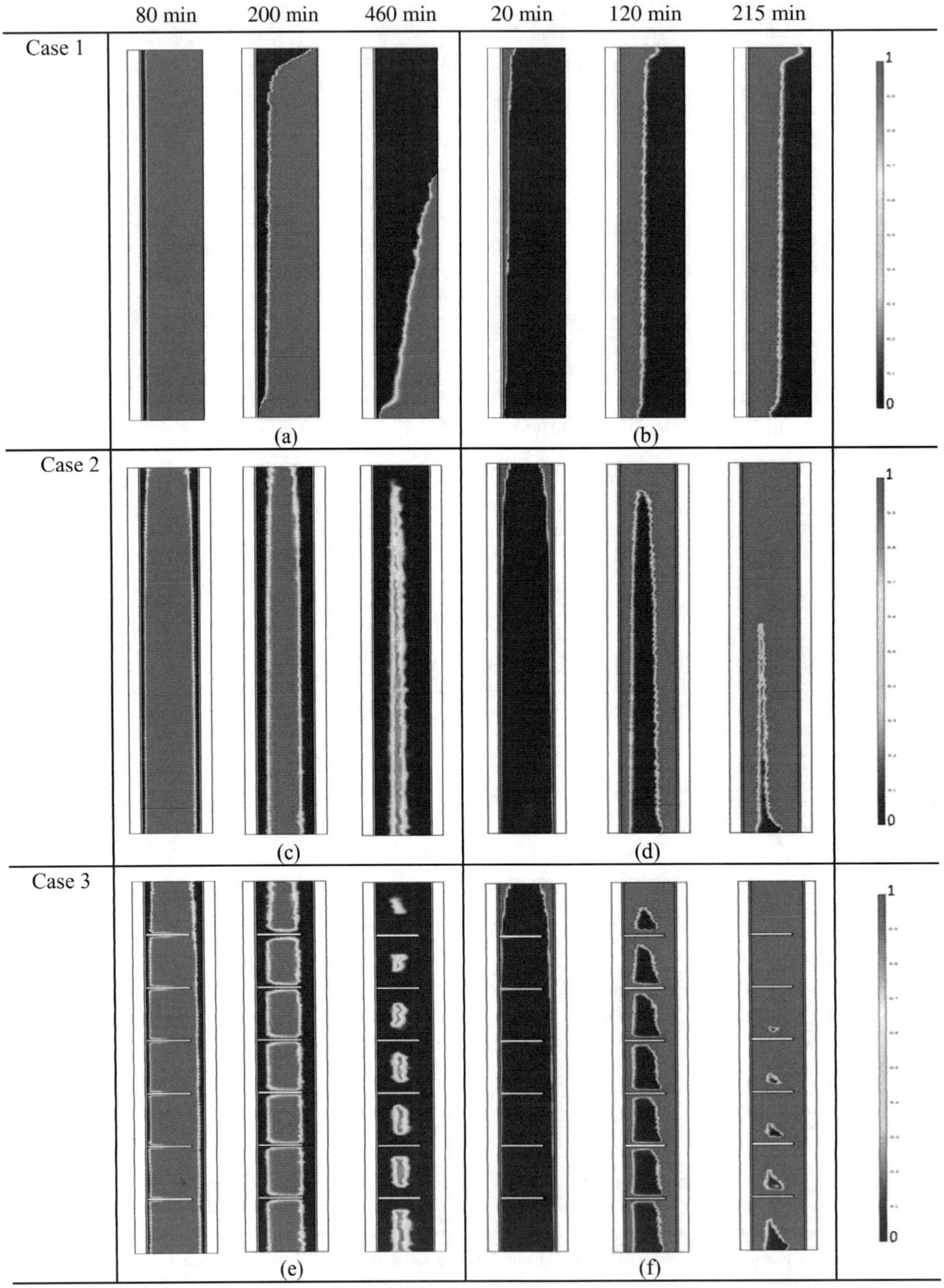

Figure 4. Phase evolution of PCM during charge and discharge process

As expected, the melting process took longer than solidification owing to the lower temperature difference between PCM during transition and HTF at the accumulator inlet, approximately 15 °C during charge and 70 °C during discharge. As expected, Case 3 presents the lowest time of charge and discharge, owing to the enhanced heat transfer surface which allows the exchange of higher and more uniform thermal power with respect to the other cases. Figs. (2-3) illustrate that neither in the charging nor in the discharging phase does Case 1 reach the equilibrium condition for melting/solidification respectively. In fact, after 12 hours of charge and 8 hours of discharge, the PCM shows a lower value of temperature than the HTF. Therefore, in this case the charge and the discharge processes are much slower than in the other two cases. The dashed line refers to the condition where Case 1 continues for 12 hours as for the charge phase.

Fig. (4) shows the phase change evolution of the PCM during charge (a,c,e) and discharge (b,d,f) processes. The figures show a cross section of the TES in which the inner tube is on the left side while the wall (Case 1) or the outer tube (Case 2-3) are on the right side. The red area represents the solid phase of the PCM material, the blue area represents the liquid phase. It can be observed that the solid-liquid interface moves in different time steps depending on the different geometrical configurations.

It is evident that Case 2 and Case 3 have a more homogeneous heat transfer rate with respect to Case 1. After 80 min, the solid-liquid interface starts to form at the top of the TES, where the HTF enters and along the entire length of the inner tube (Case 1) and of the inner and outer tubes (Cases 2-3). Fig. 4e shows that for this time step a film of liquid PCM starts to form also around the fins. After 200 min it is possible to note the different characteristics of the solid-liquid interface for the three cases on account of their different configurations. In Case 1 the solid-liquid interface moves parallel and radially from the centre to the peripheral wall. On the top of the TES device it is possible to observe a larger area of liquid PCM, owing to the proximity of the HTF inlet, while a smaller area appears on the bottom. In Case 2 the interface assumes a kind of parabolic shape (Fig 4c,d), while in Case 3 various cells with liquid PCM form during the process (Fig. 4e,f). In Case 1, after 460 min of charge, a large area of solid material can still be seen (Fig. 4a) while, in Case 2 only a narrow portion in the mid part of TES device remains in solid phase (Fig. 4c). For Case 3, the solid PCM forms small spots into the cells between the fins (Fig. 4e). A similar behaviour can be seen for the discharge phase, but the time to reach complete melting is lower with respect to the charge phase.

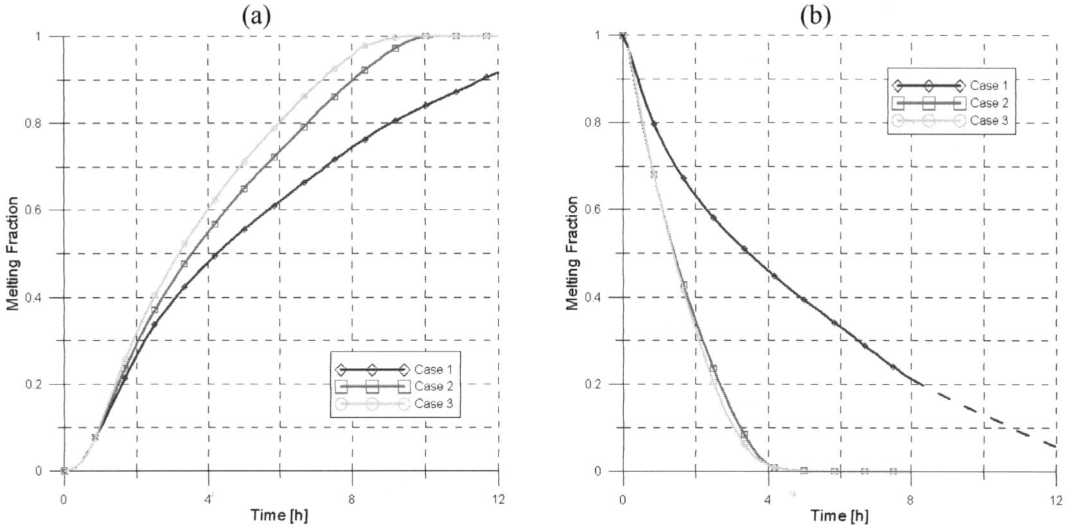

Figure 5. Melting Fraction for charge (a) and discharge (b) processes.

Figure (5) presents the PCM melted fraction versus time for the charge (a) and discharge (b) processes. This parameter is defined as the volume of PCM in liquid phase divided by the total volume occupied by the PCM. By definition, the melted fraction is 0 at the beginning of the charge process and 1 at the beginning of the discharge process. As mentioned, the lowest melting and solidification time was obtained for Case 3. Moreover, it is evident that Case 1 does not reach complete melting and solidification after 12 and 8 hours of charge and discharge respectively. The dashed line again refers to the condition where Case 1 continues for 12 hours as for the charge phase. For Cases 2-3, in the charge phase the melting fraction is equal to 1 after about 9-10 hours and the whole PCM is in the liquid phase. At the same time, for Case 1 the melted fraction is between 0,80-0,84. In the discharge phase, after about 5 hours the melting process is complete for Cases 2-3, while at the same time the value of the melting fraction is about 0.40 for Case 1.

4. Conclusions

Three different geometrical configuration of a TES system were studied to identify the best solution as regards thermal energy storage with PCM materials. A double tube (Case 1), a triplex (Case 2) and a triplex with circular fins (Case 3) were analyzed to evaluate temperature and phase distribution in both charge and discharge phases. It was found that the different configurations of the three systems considered had a strong impact on the performance of the phase change process.

In Case 1 the solid-liquid interface moved parallel and radially from the centre to the wall of the TES. In Case 2 the interface assumed a kind of parabolic shape, while in Case 3 various cells with liquid PCM formed during the process.

As expected, the triplex heat exchanger (Cases 2 and 3) were the solutions that allowed the faster response of the TES device during the charge and discharge processes. The insertion of fins (Case 3) allowed, in particular, not only an increase in the heat exchange surface of the inner HTF tube, but also the creation of cells in the PCM volume that determined faster melting/solidification of the PCM.

A time of 12 and 8 hours was considered as a reference period for charge and discharge respectively. In this period, complete melting and solidification of the PCM was achieved only in Cases 2 and 3. A melted fraction for both phases was analyzed. A value of 1 for the charge process (complete melting) was achieved in about 9-10 hours in Cases 2 and 3, while a value of about 0.80-0,84 was reached in Case 1 in the same time interval. For the discharge process, a value of 0 (complete solidification) was achieved in about 4 hours in Cases 2-3, while a value of about 0.40 was reached in Case 1. The numerical result obtained in this work will be validated with experimental studies to be carried out in the experimental facility under test at the DIMCM.

4. References

[1] Adine H A and El Qarnia H 2009 Numerical analysis of the thermal behaviour of a shell-and-tube heat storage unit using phase change materials *Appl. Math. Model.* **33** 2132–44

[2] Akgün M, Aydin O and Kaygusuz K 2007 Experimental study on melting/solidification characteristics of a paraffin as PCM *Energy Convers. Manag.* **48** 669–78

[3] Gil A, Oró E, Miró L, Peiró G, Ruiz Á, Salmerón J M and Cabeza L F 2014 Experimental analysis of hydroquinone used as phase change material (PCM) to be applied in solar cooling refrigeration *Int. J. Refrig.* **39** 95–103

[4] Gil A, Medrano M, Martorell I, Lázaro A, Dolado P, Zalba B and Cabeza L F 2010 State of the art on high temperature thermal energy storage for power generation. Part 1-Concepts, materials and modellization *Renew. Sustain. Energy Rev.* **14** 31–55

[5] Gil A, Oró E, Peiró G, Álvarez S and Cabeza L F 2013 Material selection and testing for thermal energy storage in solar cooling *Renew. Energy* **57** 366–71

[6] Gil A, Barreneche C, Moreno P, Solé C, Inés Fernández a. and Cabeza L F 2013 Thermal behaviour of d-mannitol when used as PCM: Comparison of results obtained by DSC and in a thermal energy storage unit at pilot plant scale *Appl. Energy* **111** 1107–13

[7] Gil A, Oró E, Castell A and Cabeza L F 2013 Experimental analysis of the effectiveness of a high temperature thermal storage tank for solar cooling applications *Appl. Therm. Eng.* **54** 521–7

[8] Solé A, Neumann H, Niedermaier S, Martorell I, Schossig P and Cabeza L F 2014 Stability of sugar alcohols as PCM for thermal energy storage *Sol. Energy Mater. Sol. Cells* **126** 125–34

[9] Medrano M, Gil A, Martorell I, Potau X and Cabeza L F 2010 State of the art on high-temperature thermal energy storage for power generation. Part 2—Case studies *Renew. Sustain. Energy Rev.* **14** 56–72

[10] Choi J C and Kim S D 1992 Heat-transfer characteristics of a latent heat storage system using MgCl2 · 6H2O *Energy* **17** 1153–64

[11] Castell A, Solé C, Medrano M, Roca J, Cabeza L F and García D 2008 Natural convection heat transfer coefficients in phase change material (PCM) modules with external vertical fins *Appl. Therm. Eng.* **28** 1676–86

[12] Velraj R, Seeniraj R V., Hafner B, Faber C and Schwarzer K 1997 Experimental analysis and numerical modelling of inward solidification on a finned vertical tube for a latent heat storage unit *Sol. Energy* **60** 281–90

[13] Mehling H, Hiebler S and Ziegler F 2000 Latent Heat Storage using a PCM-graphite Composite Material *Terrastock 2000, 8th Int. Conf. Therm. Energy Storage* 375–80

[14] Agyenim F, Eames P and Smyth M 2009 A comparison of heat transfer enhancement in a medium temperature thermal energy storage heat exchanger using fins *Sol. Energy* **83** 1509–20

[15] Al-Abidi A a., Mat S, Sopian K, Sulaiman M Y and Mohammad A T 2013 Internal and external fin heat transfer enhancement technique for latent heat thermal energy storage in triplex tube heat exchangers *Appl. Therm. Eng.* **53** 147–56

[16] Mehling H and Cabeza L F 2008 2 Solid-liquid phase change materials *Heat cold storage with PCM. An up to date Introd. into basics Appl.*

[17] Sharma S D and Sagara K 2005 Latent Heat Storage Materials and Systems: A Review *Int. J. Green Energy* **2** 1–56

[18] Barreneche C, Gil A, Sheth F, Inés Fernández a. and Cabeza L F 2013 Effect of d-mannitol polymorphism in its thermal energy storage capacity when it is used as PCM *Sol. Energy* **94** 344–51

[19] Burger A, Henck J O, Hetz S, Rollinger J M, Weissnicht A a. and Stöttner H 2000 Energy/temperature diagram and compression behavior of the polymorphs of D-mannitol *J. Pharm. Sci.* **89** 457–68

[20] Bruni G, Berbenni V, Milanese C and Girella a 2009 Physico-chemical characterization of anhydrous D-mannitol *J. Therm. nalysis Calorimetrynalysis Calorim.* **95** 871–6

[21] Kumaresan G, Velraj R and Iniyan S 2011 Thermal Analysis of D-mannitol for Use as Phase Change Material for Latent Heat Storage *J. Appl. Sci.* **11** 3044–8

[22] Arena S, Cau G and Palomba C 2014 Development and implementation of a 3D numerical code for designing and predicting performances of PCM thermal energy storage systems *Eurotherm Semin. #99 Adv. Therm. Energy Storage* 1–9

Journal of Physics: Conference Series **655** (2015) 012052 doi:10.1088/1742-6596/655/1/012052

Numerical investigation of the flow in axial water turbines and marine propellers with scale-resolving simulations

Mitja Morgut[1], Dragica Jošt[2], Enrico Nobile[2] and Aljaž Škerlavaj[2]

[1] Turboinštitut d.d., Rovšnikova 7, 1210 Ljubljana (Slovenia)

[2] Dipartimento di Ingegneria e Architettura, Università degli Studi di Trieste, via A. Valerio 10, 34127 Trieste (Italy)

E-mail: `nobile@units.it`

Abstract. The accurate prediction of the performances of axial water turbines and naval propellers is a challenging task, of great practical relevance. In this paper a numerical prediction strategy, based on the combination of a trusted CFD solver and a calibrated mass transfer model, is applied to the turbulent flow in axial turbines and around a model scale naval propeller, under non-cavitating and cavitating conditions. Some selected results for axial water turbines and a marine propeller, and in particular the advantages, in terms of accuracy and fidelity, of Scale-Resolving Simulations (SRS), like SAS (Scale Adaptive Simulation) and Zonal-LES (ZLES) compared to standard RANS approaches, are presented. Efficiency prediction for a Kaplan and a bulb turbine was significantly improved by use of the SAS SST model in combination with the ZLES in the draft tube. Size of cavitation cavity and sigma break curve for Kaplan turbine were successfully predicted with SAS model in combination with robust high resolution scheme, while for mass transfer the Zwart model with calibrated constants were used. The results obtained for a marine propeller in non-uniform inflow, under cavitating conditions, compare well with available experimental measurements, and proved that a mass transfer model, previously calibrated for RANS (Reynolds Averaged Navier Stokes), can be successfully applied also within the SRS approaches.

1. Introduction

CFD has been a useful tool for flow analysis and efficiency prediction of axial turbines and marine propellers for more than 20 years. Regarding water turbines, while the effect of modifications of a runner and other turbine parts on efficiency can be quite accurately estimated, the prediction of absolute efficiency value for different operating regimes is still a challenge. The main problem is an inaccurate flow simulation in a draft tube. RANS models including the SST model, which became a standard in turbo machinery, are not able to model flow in the draft tube accurately. In recent years an intensive research about numerical flow analysis in axial turbines was performed in Turboinštitut [1]. Steady state simulations were done by different turbulence models on grids with different refinements near the walls. At some operating points also transient simulations were done by SST (Shear Stress Transport) and SSG RSM (Reynolds Stress Model) models. On the basis of the results it was concluded, that none of the used turbulence models predicted the efficiency at all operating regimes accurately. With grid refinement some improvements were achieved but the results were still not satisfactory. This was the motivation for time dependent

simulations by more advanced turbulence models, the SAS SST and ZLES. The improvement of results for a Kaplan turbine was presented in [2]. Even greater improvement was achieved in case of a bulb turbine [3]. Recently, in the frame of ACCUSIM project, cavitation prediction and its effect on efficiency of the Kaplan turbine was obtained. Also for the bulb turbine the results were additionally improved by zonal LES model on refined grids [4]. For both turbines a comparison of numerical results to the experimental ones is presented in this paper.
For marine propellers, the predictions of their performances under non-uniform inflow conditions (i.e. wake) and under cavitating conditions is still challenging. In this paper, we will present how a proper combination of a previously calibrated mass transfer model, together with a SAS SST turbulence approach, provides a prediction of the cavitating flow pattern which agrees well with the experimental visualizations.

2. Turbulence models and discretisation schemes

In the analysis reported in this paper, several turbulence models were used: the SST model [5], the SAS SST [6] (hereafter SAS) and the ZLES model [7]. The main idea of the ZLES model is to resolve the flow inside a predefined zone with the LES model, whereas outside the zone a RANS model, either steady or unsteady, is used. At the zone boundaries a Harmonic Flow Generator [8] is used to produce synthetic turbulence. In the simulations of water turbines the zone started just after the interface between the runner and the draft tube, and it is extended to the outlet of the computational domain. Some turbulence models were used in combination with the curvature correction (CC) [9] and with the Kato-Launder limiter (KL) [10] of turbulence production. The CC and KL options were used for water turbines simulations with SST, SAS and ZLES models. In the simulations, we used either 'high resolution' (HRS) or the bounded central difference scheme (BCDS). The high-resolution scheme is a bounded second order upwind biased discretisation, based on the procedure of the Barth and Jespersen's scheme [11]. The bounded central difference scheme (BCDS) is based on the normalised variable diagram and blends from the CDS scheme to the first-order upwind scheme when the convection boundedness criterion [12] is violated. The BCDS was used in combination with SAS and ZLES models, whereas HRS was used in combination with the SST and in order to check the effect of discretisation scheme in some cases also with the SAS turbulence approach.

3. Cavitation and mass transfer models

Cavitating flows can be modelled using several methods. An excellent review of different methods is provided for instance by Koop [13]. In this work we used the homogeneous transport equation based model [14, 15], described in the following.

3.1. Governing equations

As already indicated the cavitating flow is modelled as a mixture of two species i.e. vapour and liquid behaving as a single one, where both phases share the same velocity as well as pressure fields.
In the case of the RANS approach to turbulence, employing the eddy viscosity models and assuming both liquid and vapour phases incompressible, turbulent cavitating flows can be described by the following set of governing equations:

$$\nabla \cdot \mathbf{U} = \dot{m}\left(\frac{1}{\rho_L} - \frac{1}{\rho_V}\right) \tag{1}$$

$$\frac{\partial(\rho\mathbf{U})}{\partial t} + \nabla \cdot (\rho\mathbf{U}\mathbf{U}) = -\nabla P + \nabla \cdot [(\mu + \mu_t)(\nabla\mathbf{U} + (\nabla\mathbf{U})^T)] + S \tag{2}$$

$$\frac{\partial \gamma}{\partial t} + \nabla \cdot (\gamma \mathbf{U}) = \frac{\dot{m}}{\rho_L} \tag{3}$$

which are, in order, the volume continuity and the momentum equation for the liquid-vapour mixture and the volume fraction equation for the liquid phase. In the above equations ρ_L (kg/m^3) and ρ_V (kg/m^3) are the liquid and vapour densities, respectively. $\mathbf{U}$ (m/s) is the averaged velocity and P (Pa) the averaged pressure. S represents the additional momentum sources (for instance the Coriolis and centrifugal forces in the rotating frame of reference). γ is the water volume fraction which is related to the vapour volume fraction α through the volume fraction constraint:

$$\gamma + \alpha = 1 \tag{4}$$

The mixture density ρ (kg/m^3) and laminar dynamic viscosity μ (kg/ms) are evaluated according to:

$$\rho = \gamma \rho_L + (1-\gamma)\rho_V \tag{5}$$

$$\mu = \gamma \mu_L + (1-\gamma)\mu_V \tag{6}$$

where μ_L (kg/ms) and μ_V (kg/ms) are the liquid and vapour dynamic viscosities, respectively. $\dot{m}$ (kg/m^3s) represents the inter-phase transfer rate due to cavitation. In this study $\dot{m}$ was modelled using a mass transfer model briefly presented in the next section.
Finally, in order to close the system of the governing equations, the mixture turbulent viscosity μ_t (kg/ms) was evaluated using one of the turbulence models, developed for the single phase flows, indicated in 2.

3.2. Mass transfer model
The mass transfer model describes/regulates the interphase mass transfer rate due to cavitation. In the past, three different mass transfer models have been considered by the present authors, and their empirical constants have been properly calibrated and validated [16, 17, 18].
Here only the Zwart model, with the calibrated constants, was used, together with a SAS turbulence model.

3.2.1. Zwart model The Zwart model is the native ANSYS CFX mass transfer model. It is based on the the simplified Rayleigh-Plesset equation for bubble dynamics [19]:

$$\dot{m} = \begin{cases} -F_e \dfrac{3r_{nuc}(1-\alpha)\rho_V}{R_B}\sqrt{\dfrac{2}{3}\dfrac{P_V - P}{\rho_L}} & if\ P < P_V \\ F_c \dfrac{3\alpha\rho_V}{R_B}\sqrt{\dfrac{2}{3}\dfrac{P - P_V}{\rho_L}} & if\ P > P_V \end{cases} \tag{7}$$

In the above equations, P_V is the vapour pressure, r_{nuc} is the nucleation site volume fraction, R_B is the radius of a nucleation site, F_e and F_c are two empirical calibration coefficients for the evaporation and condensation processes, respectively.

4. Improvement of performance prediction for axial turbines with advanced turbulence models

4.1. Efficiency and cavitation prediction for Kaplan turbine
The results of a detailed numerical analysis of flow in a 6-blade Kaplan turbine which operates at middle head are presented. Numerical simulations were done for three angles of runner blades at constant head (see table 1).

Table 1. Local best efficiency points for three angles of runner blades of the Kaplan turbine.

Operating point	β°	φ/φ_{BEP}	ψ/ψ_{BEP}
OP1	12	0.64	0.86
OP2	20	0.95	0.86
OP3	28	1.31	0.86

The turbine consists of semi-spiral casing with two vertical piers, 11 stay vanes and a nose, 28 guide vanes, a 6-blade runner and an elbow draft tube with two vertical piers. Tip clearance was modelled while hub clearance was neglected. The grid in the spiral casing with stay vanes was unstructured, while the grids in the other turbine parts were structured. Near the walls the grids were refined to get recommended values of y^+. For the draft tube and the draft tube prolongation, basic and refined grids were used (table 2). For the other turbine parts, our previous studies [1] showed that with additional refinement only a negligible improvement of results can be obtained. The computational grid for the complete computational domain can be seen in Fig. 1.

Table 2. Number of nodes for the Kaplan turbine parts for the basic grid - BG, and the fine grid - FG.

Turbine part	Number of nodes	
Semi spiral casing with stay vanes		1,480,999
Guide vane cascade		2,755,496
Runner		1,858,374
	BG	FG
Draft tube	1,786,432	6,169,935
Draft tube prolongation	398,056	1,681,992
Total	8,279,357	13,946,796

Detailed results of steady state analysis performed on the basic grid are presented in [2]. At OP1 and OP2 all turbulence models predicted the efficiency values rather well. At OP3, the discrepancy between calculated and measured efficiency values was very large, larger than 4 %. We tried to improve the results at OP3 with transient simulations using SST, SAS and ZLES turbulence models (Fig. 2).

The values of calculated efficiency differ mostly due to different values of flow energy losses in the draft tube and different values of torque. The efficiency value calculated from the steady state solution with SST HRS was 4.42% smaller than the measured one. With transient analysis the results improved significantly. With SST HRS and SAS HRS the efficiency values were smaller than the measured ones by 1.01% and 0.24% respectively. The agreement between measured and numerical values is excellent for SAS BCDS and ZLES BCDS, where the discrepancy is 0.09% and 0.05%, respectively. Based on the results it can be concluded that the SAS and ZLES models with BCDS are suitable for flow simulation at operating points with large discharge. Results obtained by SAS HRS are less accurate mostly due to underestimated torque. Comparing the results of ZLES BCDS on fine and basic grids it was seen that the effect of grid density on the efficiency was negligible. Therefore, for transient simulations for different operating regimes the basic grid was used.

Figure 1. Computational domain and grid for the Kaplan turbine.

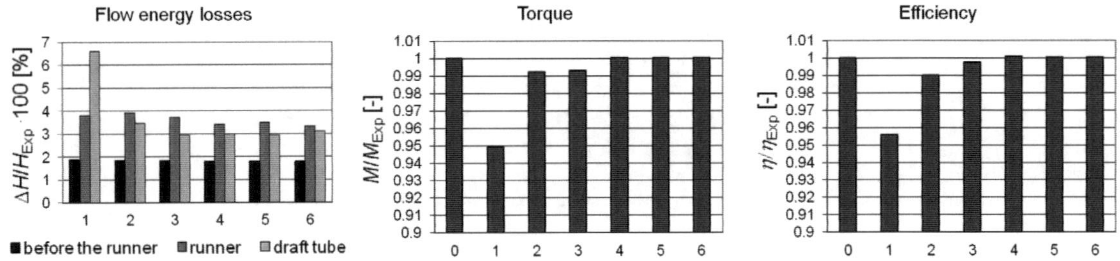

Figure 2. Kaplan turbine - flow energy losses, torque on the shaft and efficiency at OP3. 0 - Experiment; 1 - steady state SST, HRS, FG; 2 - transient SST, HRS, FG; 3 - SAS, HRS, FG; 4 - SAS, BCDS, FG; 5 - ZLES, BCDS, FG; 6 - ZLES, BCDS, BG.

In Fig. 3 flow in the draft tube obtained with steady state SST and transient ZLES is presented. In case of steady state analysis by SST model there is a large swirl at the end of the cone which is by transient simulations nearly completely reduced. Streamlines obtained by the ZLES model are more curled than those obtained by the SST. Vortex structures are represented in Fig. 3(b) and 3(d). By representing viscosity ratio (ratio between eddy and molecular viscosity) it is possible to figure out the amount of modelling in the SST case, and scale-resolving in the ZLES base. By the steady state SST only large vortex structures in the flow were obtained but with the ZLES also small vortex structures were well resolved.

The diagram in Fig. 4 shows a clear distinction between the steady-state simulations and the ZLES, especially at the runner blade angle 28°, where the decrease of calculated flow energy losses in the draft tube was the largest. At the local best efficiency point for runner blade angle 12°(OP1), the calculated and measured values are practically the same. At OP2 (blade angle 20°), the efficiency value obtained with ZLES is about 0.5% smaller than the measured one. The efficiency values at the first three points on the ZLES curve for blade angle 28°agree with the measured results well. At higher discharge, the discrepancy increases. Peak to peak difference

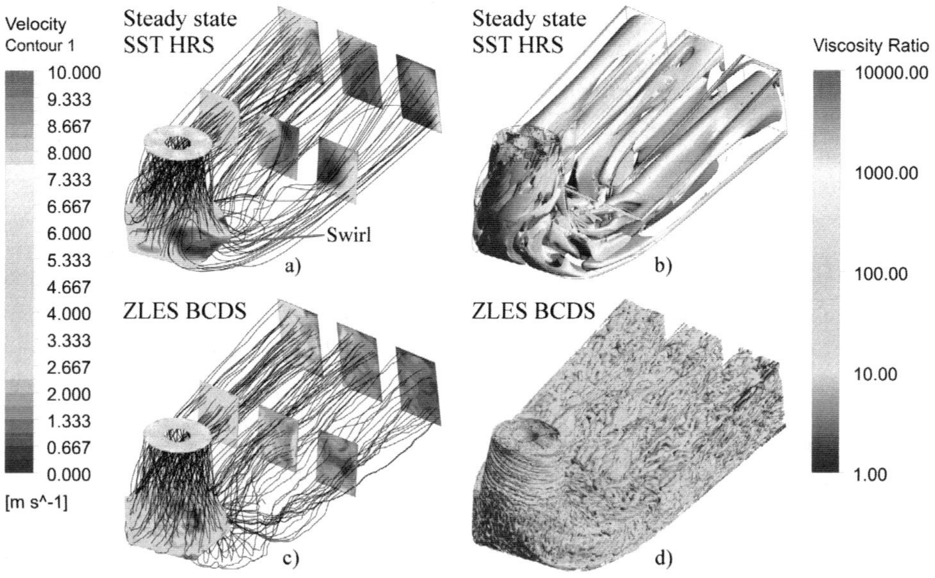

Figure 3. Kaplan turbine - streamlines and velocity contours in the draft tube (left) and isosurfaces of velocity invariant $Q = 0$ (right). Results obtained on fine grid at OP3.

in efficiency values is 0.44% but numerically obtained position of the local best efficiency point is shifted to the left. The tendency that numerical prediction is less accurate at operating points with large discharge still remains, but the improvement with the ZLES model is significant.

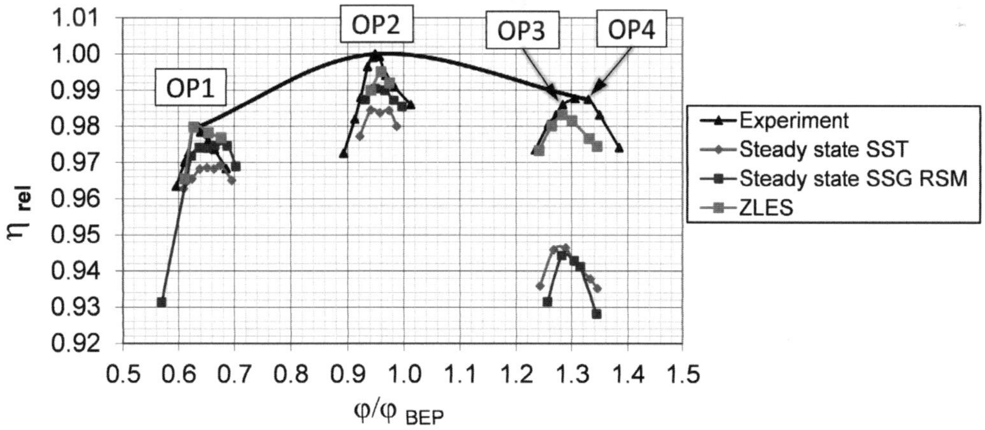

Figure 4. Efficiency diagram for Kaplan turbine - experimental and numerical results.

Cavitation prediction was done at the operating point OP4 (see Fig. 4) because for this point the cavitation was being observed on the test rig and sigma break curve was measured. Two-phase flow simulations are usually less stable, therefore SAS model in combination with robust HR scheme was used instead of ZLES model with BCDS. Cavitating flow was simulated using the homogeneous model. Mass transfer rate due to cavitation was regulated by the Zwart model using vaporization and condensation parameters previously calibrated on a hydrofoil [16, 17].

Journal of Physics: Conference Series **655** (2015) 012052 doi:10.1088/1742-6596/655/1/012052

In order to determine the effect of cavitation on turbine efficiency (sigma-break curve) successive simulations were performed lowering the reference pressure. Steady-state simulations were performed using the MRF (Multiple Reference Frame) approach in combination with the frozen rotor frame/change mixing model. Time-dependent simulations were performed using sliding grids. The first transient simulation that started from steady-state solution was done for non-cavitating conditions. To get the correct flow in the draft tube and stable value of efficiency 20 runner revolutions were needed. For each value of reference pressure three additional runner revolutions were needed.

Numerical results were compared with the observation of cavity size (Fig. 5) on the test rig and with the measured sigma-break curve (Fig. 6). Steady-state simulations did not predict the same extent of cavitation on all blades due to the frozen rotor condition, which somehow preserved differences in circumferential direction. With transient simulation the same amount of cavitation on all runner blades was obtained and the shape and extend of sheet cavitation agreed well with the cavitation observed on the test rig. As already seen in Fig. 4 steady-state simulations significantly under-predicted the efficiency, but the value of sigma, where the efficiency dropped for 1%, agreed well with the experimental result (see Fig. 6). With transient simulations the same amount of cavitation on all runner blades was obtained and the shape and extend of sheet cavitation agreed well with the cavitation observed on the test rig (see Fig. 5). Transient simulations predicted the efficiency more accurately, but the value of sigma where the efficiency dropped for 1% was a bit too large.

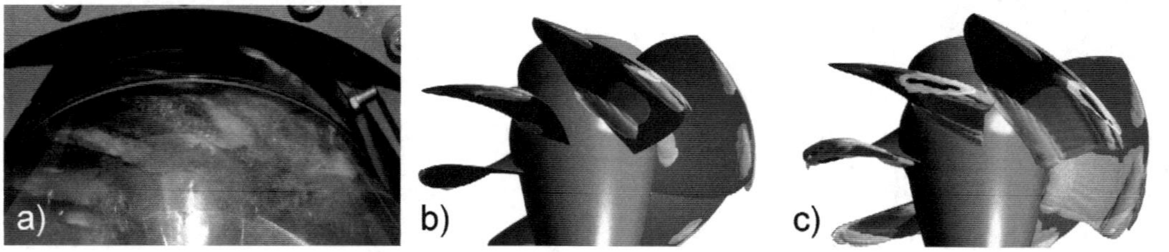

Figure 5. Kaplan turbine - cavity on runner blades: a) observation on test rig, b) steady state simulation, c) time dependent simulation, Th=0.52.

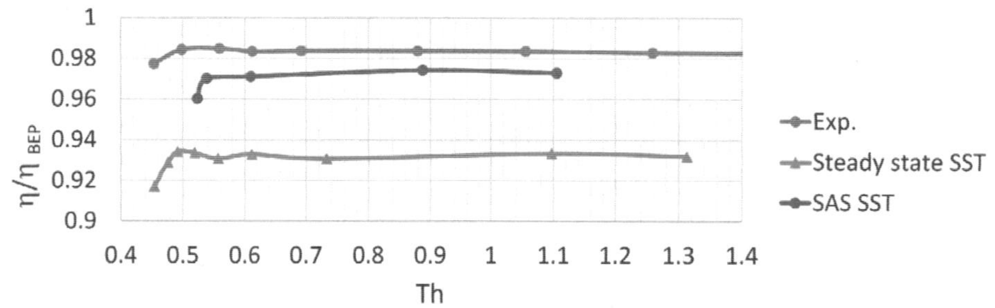

Figure 6. Sigma break curve for Kaplan turbine - experimental and numerical results at OP4.

4.2. Bulb turbine

The bulb turbine considered in this study consists of an inlet part with a vertical pier, 16 guide vanes, 3-blade runner and a draft tube. All simulations were done for a turbine model with runner diameter D=350 mm. The model was tested on a test rig in accordance with international standard IEC 60193 [20].

Numerical simulations were done for three angles of runner blades and two values of head. The complete results obtained by steady state simulations with the SST turbulence model and by transient simulations with ZLES on the basic grid are presented in [3, 4]. This paper is focused on the local best efficiency point for maximal runner blade angle at smaller head ($\beta = 28°$, $\varphi/\varphi_{BEP} = 1.29$, $\psi/\psi_{BEP} = 0.61$). At this operating regime, the difference between numerical and experimental results was the largest.

To see the effect of grid density on results several computational grids were used (see Tab. 3). The basic grid (BG) with about 9×10^6 nodes is like the grids that are usually used in our simulations. Fine grid 1 (FG1) is in comparison to BG refined only in the draft tube and draft tube extension. FG2 is equal to BG everywhere except in the runner where the grid was refined. FG3 has refined grids in the runner, draft tube and draft tube extension. FG4 is equal to FG1 except in the draft tube and draft tube extension where the grids were additionally refined near walls in order to get for ZLES recommended values of y^+. Coarse grid (CG) with about 2.2×10^6 nodes was used to check whether simulation time could be reduced without too large effect on accuracy of results. The basic computational grid on the whole domain can be seen in Fig. 7.

Due to the frozen rotor condition strong wakes behind the runner blades were obtained and in the region where conical shape of the draft tube changed into a rectangular one a large swirl formed (see Fig. 8(a)). Transient simulations started from the steady state results. During the simulations the swirl was moving towards the outlet and after 22 runner revolutions completely vanished (see Fig. 8(b)).

Comparison of numerical results to the measurements is presented in Fig. 9. Steady state analysis failed to predict efficiency for more than 14% due to overestimated losses in the draft tube and underestimated torque on the shaft. Transient simulations with ZLES significantly improved the results. Positive effect of grid refinement in the draft tube is clearly seen. While with grid refinement in the runner only, no improvement was obtained, the best results were obtained when both grids, in the runner and in the draft tube were refined. In this case (FG3), discrepancy between predicted and measured values of torque on the shaft and turbine efficiency were equal to 1.46% and 2.09%, respectively.

Table 3. Number of nodes in computational grids for bulb turbine

	Inlet part	Guide vane	Runner	Draft tube	Draft tube ext.	Total
Coarse grid	223,375	1,032,976	531,603	211,428	170,332	2,169,714
Basic grid	1,394,720	2,059,568	2,076,603	1,936,188	1,573,336	9,040,415
Fine grid 1	1,394,720	2,059,568	2,076,603	16,324,636	13,176,576	35,032,103
Fine grid 2	1,394,720	2,059,568	6,289,611	1,936,188	1,573,336	13,253,423
Fine grid 3	1,394,720	2,059,568	6,289,611	16,324,636	13,176,576	39,245,111
Fine grid 4	1,394,720	2,059,568	2,076,603	22,292,052	17,885,592	45,708,535

5. Cavitation prediction on a propeller in non-uniform inflow

The numerical predictions of the cavitating E779A model propeller in non-uniform inflow were carried out [21] following the experimental/numerical setup described in [22]. In particular, here, we point out that on the domain Inlet the non-homogeneous inflow (nominal wake), kindly

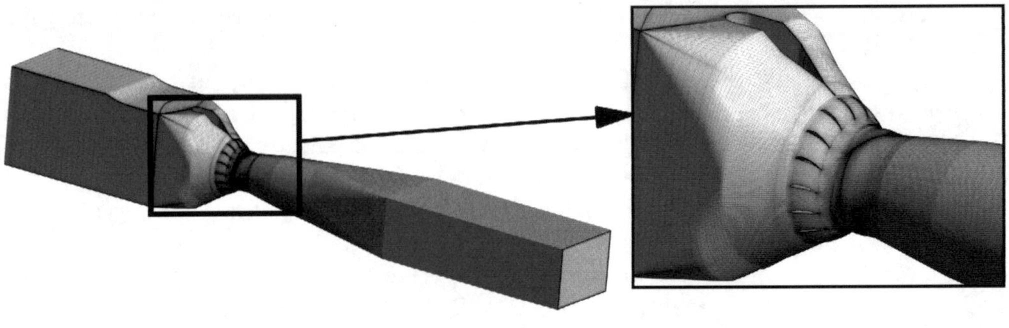

Figure 7. Computational domain and basic grid for bulb turbine.

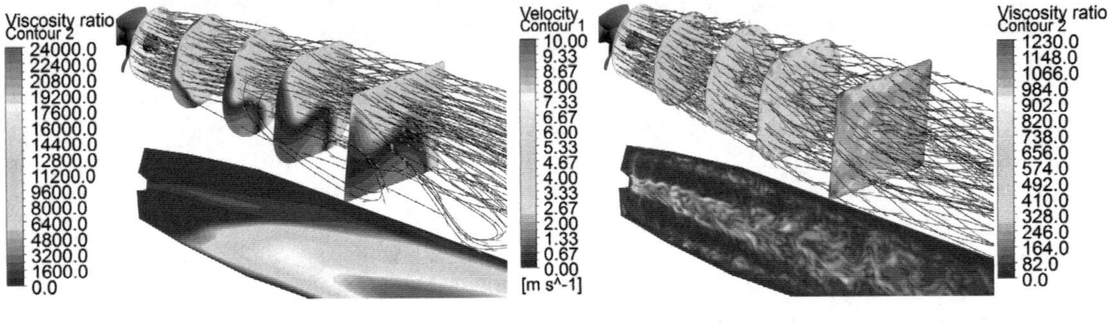

Figure 8. Bulb turbine - flow in the draft tube presented by streamlines and velocity contours, and ratio between eddy viscosity and dynamic viscosity on mid cross-section.

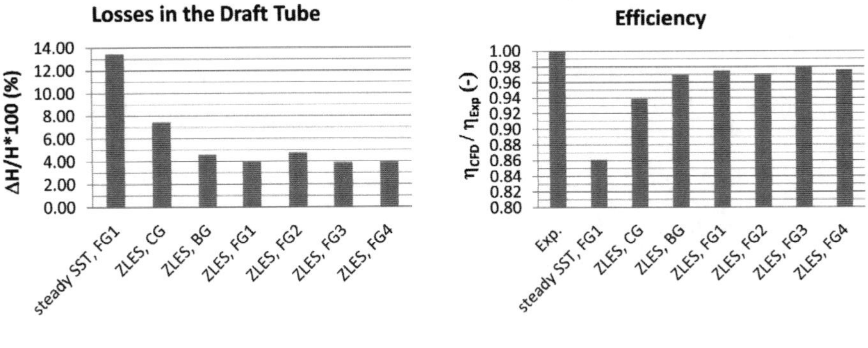

Figure 9. Bulb turbine - comparison of numerical results to the experimental values.

provided by CNR-INSEAN (private communication) was set. On Outlet boundary a fixed value of static pressure was imposed. On the solid surfaces the no-slip wall condition was enforced. The simulated case is schematically presented in Fig. 10. The Fixed domain used a structured numerical grid with 3.3×10^6 nodes, whereas the Rotating domain was meshed as an unstructured hexa-core grid with 7×10^6 nodes. As already described, the cavitation is predicted with the Zwart mass transfer model, with constants previously calibrated on a hydrofoil, using the SAS SST turbulence model. The numerical results agree qualitatively well with the available experimental snapshots describing the cavitation bubble evolution, as reported in figure 11.

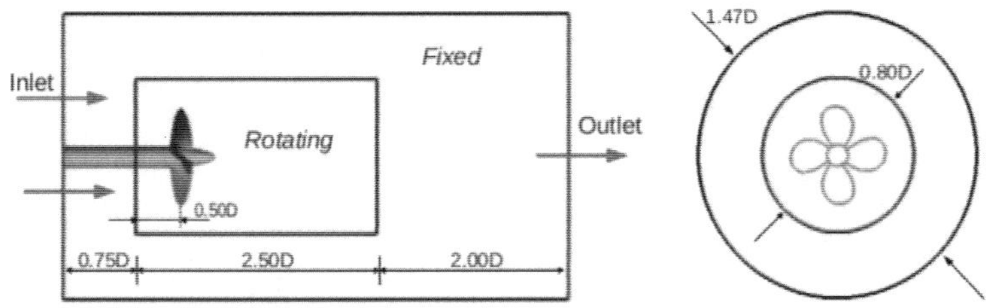

Figure 10. Computational domain for the propeller E779A in non-uniform inflow.

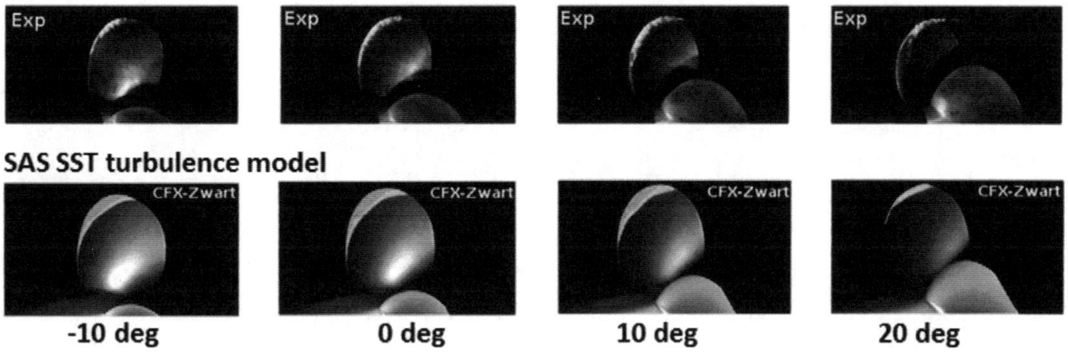

Figure 11. Cavity evolution during propeller rotation. Numerical cavitation patterns depicted using isosurfaces of vapour volume fraction equal to 0.1.

Concluding remarks

On the basis of the detailed analysis of flow in a Kaplan turbine and in a bulb turbine by different turbulence models, and a marine propeller with non-uniform inflow under cavitating conditions, the following conclusions can be drawn:

- For both Kaplan and bulb turbines, it was found out that steady state analysis is not suitable for all operating regimes. While for Kaplan turbine the prediction of efficiency by RANS models was inaccurate only at full rate, for bulb turbine steady state simulations entirely failed. In both cases the main reason for discrepancy between measured and calculated efficiency was the flow in the draft tube, where calculated flow energy losses were overestimated.
- For both turbines the accuracy of results was significantly improved by using ZLES model. For Kaplan turbine the discrepancy was everywhere smaller than 0.8%. For bulb turbine, in spite of large improvement, the discrepancy with measurements was even on the finest grid still 2.09%.
- In the case of the Kaplan turbine the steady-state and time-dependent approaches were compared for the evaluation of the effect of the cavitation on turbine efficiency. The time dependent simulations with SAS SST turbulence model predicted the efficiency more accurately, even though a slightly premature break down of the efficiency was observed.
- In the case of a marine propeller the numerical results, obtained using the SAS SST turbulence approach with a properly calibrated cavitation model, compared qualitatively well with the available experimental data.

Acknowledgments
The research was partly funded by the European Commission - FP7-PEOPLE-2013-IAPP - Grant No. 612279, and by the Slovenian Research Agency ARRS - Contract No. 1000-09-160263.

References
[1] Jošt D and Drešar P 2011 Numerical Analysis of the flow in an axial turbine by different turbulence models (Ljubljana: TurboinÅ¡titut) *Internal report* Nr. 3046 (in Slovene) p 49.
[2] Jošt D Škerlavaj A and Lipej A 2014 Improvement of Efficiency Prediction for a Kaplan Turbine with Advanced Turbulence Models, *J. Mechanical Engineering* **60** 124-34.
[3] Jošt D and Škerlavaj A 2014 Efficiency prediction for a low head bulb turbine with SAS SST and zonal LES turbulence models *IOP Conf. Ser.: Earth Environ. Sci.* **22** 022007.
[4] Jošt D Škerlavaj A Morgut M and Nobile E 2015 Effects of turbulence model and grid resolution on the performance prediction of a bulb turbine. To be presented at *NAFEMS World Congress 2015*, San Diego, USA.
[5] Menter F R Kuntz M and Langtry R 2003 Ten years of industrial experience with the SST turbulence model *Turbulence, Heat and Mass Transfer* 4. ed Hanjalič K Nagano Y and Tummers M (Begell House, New York) pp 625-632.
[6] Egorov Y and Menter F 2008 Development and application of SST-SAS turbulence model in the DESIDER project. *Advances in Hybrid RANS-LES Modelling* ed Peng S -H and Haase W (Springer, Heidelberg), pp 261-270.
[7] Menter, F.R., Gabaruk, A., Smirnov, P., Cokljat, D., Mathey, F 2010 Scale-Adaptive Simulation with Artificial Forcing *Progress in Hybrid RANS-LES Modelling* ed Peng S -H Doerffe P and Haase W (Springer, Berlin) 235-46.
[8] Adamian D and Travin A 2011 An efficient generator of synthetic turbulence at RANS-LES interface in embedded LES of wall-bounded and free shear flows *Computational Fluid Dynamics 2010* ed Kuzmin A (Springer, Berlin) pp 739-744.
[9] Smirnov P E and Menter F 2009 Sensitization of the SST turbulence model to rotation and curvature by applying the Spalart-Shur correction term. *J. Turbomachinery* **131**, no. 4, p 041010.
[10] Kato M and Launder B E 1993 The modelling of turbulent flow around stationary and vibrating square cylinders *Procs 9th Symposium on Turbulent Shear Flows* p 10.4.1-10.4.6.
[11] Barth T J and Jespersen D C 1989 The design and application of upwind schemes on unstructured meshes *27th Aerospace Sciences Meeting* AIAA Paper 89-0366.
[12] Jasak H Weller H.G. and Gosman A D 1999 High resolution NVD differencing scheme for arbitrarily unstructured meshes *Int. J. Num. Meth. Fluids* **31** no. 2 pp 431-449.
[13] Koop A H 2008 Numerical simulation of unsteady three-dimensional sheet cavitation (Ph.D. thesis), University of Twente.
[14] Bouziad Y A 2005 Physical modelling of leading edge cavitation: computational methodologies and application to hydraulic machinery (Ph.D. thesis) Ecole Polytechnique Federale de Lausanne.
[15] ANSYS 2009 *ANSYS CFX-Solver Theory Guide, Release 12.0.*
[16] Morgut M, Nobile E and Biluš I 2010 Comparison of mass transfer models for the numerical prediction of sheet cavitation around a hydrofoil, *Int. J. Multiphase Flow* **37(6)** 620-26.
[17] Morgut M and Nobile E 2012 Numerical Predictions of Cavitating Flow Around Model Scale Propellers by CFD and Advanced Model Calibration, *Int. J. Rotating Machinery* Article ID 618180, 11 pages.
[18] Morgut M and Nobile E 2011 Numerical Predictions of the Cavitating and Non-Cavitating Flow Around the Model Scale Propeller PPTC. *Proc. of the Workshop on Cavitation and Propeller Performance, 2nd International Symposium on Marine Propulsors* (Hamburg, Germany).
[19] Brennen C E 2005 *Fundamentals of Multiphase Flows* (New York, NY, USA Cambridge University Press).
[20] IEC 60193 1999 Hydraulic turbines, storage pumps and pump-turbine - Model acceptance tests. (International Electrotechnical Commission, Geneva).
[21] Morgut M Jošt D Nobile E and Škerlavaj A 2015 Numerical investigations of a cavitating propeller in non-uniform inflow. To appear in *Procs of Fourth International Symposium on Marine Propulsors* (Austin, Texas, USA).
[22] Salvatore F Streckwall H and van Terwisga T 2009 Propeller Cavitation Modelling by CFD-Results from the VIRTUE 2008 Rome Workshop *Proc. of the First International Symposium on Marine Propulsors, smp'09* (Trondheim, Norway).

Scaled models in the analysis of fire-structure interaction

A Andreozzi, N Bianco[1], M Musto and G Rotondo
Dipartimento di Ingegneria Industriale, Università degli studi di Napoli Federico II, P.le Tecchio 80, 80125, Napoli, Italy

[1] E-mail: nicola.bianco@unina.it

Abstract. A fire problem has been scaled both in terms of geometry, boundary conditions and materials thermophysical properties by means of dimensionless parameters. Both the full and scaled models have been solved numerically for two different fire power values. Results obtained by means of the full scale model and the scaled one are compared in terms of velocity and temperature profiles in order to assess the reliability of the scaled model to represent the behavior of the full scale one.

1. Introduction

In fire safety engineering, the prediction of fire evolution and its actions on structures can be very difficult. Several numerical models have been developed during last years, but their validation is difficult due to the lack of experimental data. This is mainly due to the complexity in carrying out experimental tests on real fire. Then, the possibility to operate with scaled experimental models could be very helpful in order to compare numerical and experimental results.

The modeling of fire evolution is very difficult due to the large number of phenomena involved, such as pyrolysis, combustion, conduction, convective and radiative heat transfer. Quintiere [1] proposed a set of dimensionless parameters to be used in fire modeling in order to make the conservation equations and the related boundary conditions dimensionless.

Several authors proposed successful scaled models in fire [2, 3] and Thomas [4, 5] and Quintiere [6] have done reviews of this topic. Chowa and Loa [7] presented a scaled model to predict smoke movement in an atrium. Full scale experiments were carried out for assessing models proposed in literature. It was found that the scaling law for temperatures proposed in literature requires further examination.

The understanding of the behavior of a structure in a fire is crucial in order to assess fire resistance requirements. Wang et al. [8] proposed a method for scaling a compartment fire. Experimental verifications at two scales were performed in order to show the validity of the proposed approach.

In this paper, the dimensionless parameters presented in [1] have been used to compare results obtained for two different models, the first one in full scale and the second one scaled by a given factor (1:2). Results are obtained numerically by means of the commercial code Ansys-Fluent. The proposed model has been scaled both in terms of geometry, boundary conditions and materials thermophysical properties by means of the aforementioned dimensionless parameters. Results obtained by means the full scale model and the scaled one are compared in terms of velocity and

[1] To whom any correspondence should be addressed.

Journal of Physics: Conference Series **655** (2015) 012053 doi:10.1088/1742-6596/655/1/012053

temperature profiles in order to assess the reliability of the scaled model to represent the behavior of the full scale one.

2. Model and methodology

The analyzed problem refers to a room in which a steel beam, located on the room roof, is heated by a fire, located on the room floor. A sketch of the physical domain under investigation is shown in figure 1.

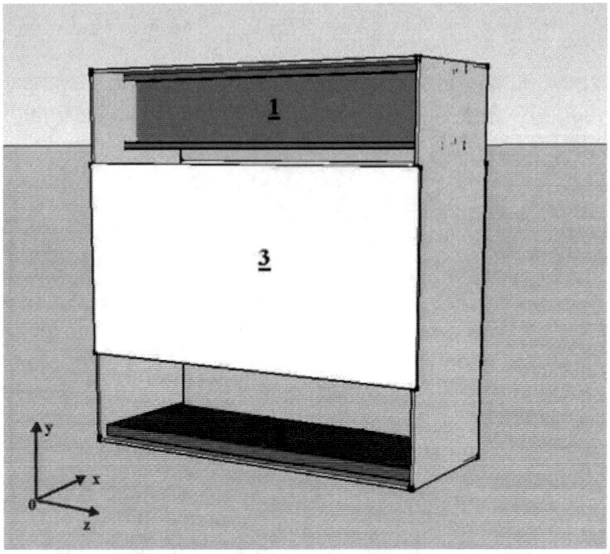

Figure 1. Sketch of the analyzed configuration.
1: IPE beam, 2: burner, 3: plasterboard walls.

A 2.48 m x 1.00 m x 2.50 m (high) parallelepiped enclosure has been taken into account. Four openings 0.60 m high are placed on the two larger vertical walls in proximity of the ceiling and the floor. The walls are made up of plasterboard, 0.020 m thick. A steel IPE beam is placed at the centerline of the ceiling.

The fire is modeled by assigning a heat flux on the portion of floor in which the fire is supposed to be located. The burner dimensions are assumed to be 2.48 m x 0.80 m x 0.08 m. The IPE beam is sketched in figure 2 and its dimensions are reported in table 1. Thermophysical properties of plasterboard, steel and air are reported in table 2.

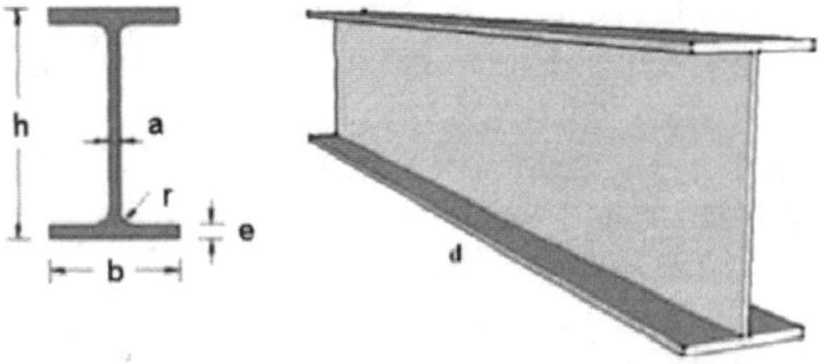

Figure 2. Sketch of the IPE beam.

Table 1. Dimensions of the IPE beam.

h/mm	b/mm	a/mm	e/mm	r/mm	d/mm
500	360	20	50	1	2,480

Journal of Physics: Conference Series **655** (2015) 012053 doi:10.1088/1742-6596/655/1/012053

Table 2. Thermophysical properties of materials.

Material	ρ/(kg m^{-3})	c_p/(J kg^{-1} K^{-1})	k/(W m^{-1} K^{-1})	μ/(Pa s)	β/K^{-1}
Air	1.225	1,006	2.42×10^{-2}	1.789×10^{-5}	3.33×10^{-3}
Steel	8,030	502.5	16.3	-	-
Plasterboard	737	1,423	0.12	-	-

The problem has been solved numerically, by assuming gray surfaces, incompressible flow, constant thermophysical properties for solids and fluid, except for the air density, for which the ideal gas model is assumed. This latter hypothesis can be justified by taking into account that no large air temperature increase is expected, since the predominant heat transfer mechanism between the heater and the walls should be radiative heat transfer.

The governing equations for fluid and solid region, in steady state regime, are time-averaged mass, Navier-Stokes and energy equations combined with k-ε realizable turbulence model [9] and further details on the mathematical model can be found in [10].

2.1 Boundary conditions

An imposed heat flux is assigned on the burning surface depending on the fire power and the scaling factor. The openings are assumed to be black bodies at 300 K, the convective coefficient, h_0, and the temperature, T_0, outside the ceiling are assumed to be 4.0 W K^{-1} m^{-2} and 300 K, respectively. The boundary conditions for the fluid and solid domains are reported in table 3, with reference to figure 3.

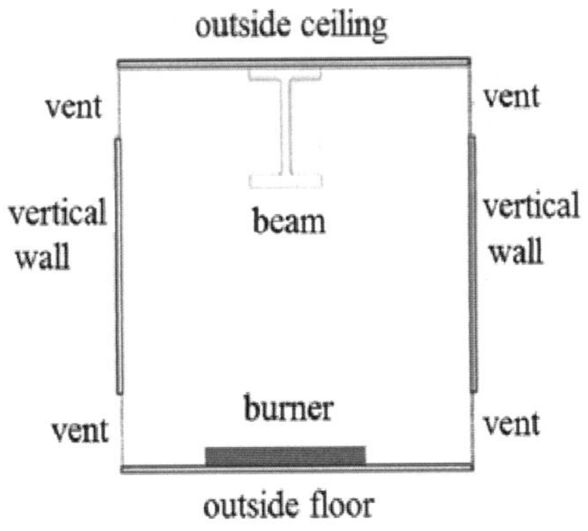

Figure 3. Boundary conditions.

Table 3. Boundary conditions for the solid and fluid domains.

Surface	Boundary condition			
	u	v	w	Temperature
Outside ceiling	0	0	0	$-k_p \frac{\partial T_p}{\partial y} = h_o\left(T_p - T_0\right)$
Vents	$\frac{\partial u}{\partial x}=0$	$\frac{\partial v}{\partial x}=0$	$\frac{\partial w}{\partial x}=0$	300 K
Vertical walls	0	0	0	300 K
Outside floor	0	0	0	300 K

2.2 Scaling and dimensionless parameters

Dimensionless parameter, to be used for scaling models, will be obtained from the conservation equations.

As for the velocity, by making reference to conservation of momentum equation in the hypothesis of natural convection, it is possible to obtain a normalizing factor for velocity equal to [1]:

$$\bar{u} = \sqrt{gl} \tag{1}$$

where g is the acceleration due to the gravity and l is a characteristic length for the analyzed problem. The dimensionless group for scaling firepower is [1]:

$$\dot{Q}^* = \frac{\dot{Q}}{\rho_\infty c_p T_\infty \sqrt{g} l^{5/2}} \tag{2}$$

By denoting with the subscript "s" variables for the scaled model and "f" the ones for the full model, the aforementioned dimensionless group allows evaluating the firepower for the scaled model as:

$$\dot{Q}_s = \dot{Q}_f \left(\frac{l_s}{l_f}\right)^{5/2} \tag{3}$$

where l_s and l_f are characteristic length for scaled and full model and then their ratio represents the geometric scaling factor.

It is worth noticing that temperature values are scaled with respect to the ambient temperature, T_∞, then, by assuming this latter unchanged between the scaled and full models, it follows that expected temperature values will be the same for the two models, for both gaseous and solid domains.

The following dimensionless number allows to scale heat conduction within walls [1]:

$$\dot{Q}_k^* = \frac{\left(k\rho c_p\right)_w^{1/2}}{\rho_\infty c_p g^{1/4} l^{3/4}} \tag{4}$$

The previous parameter allows to evaluate the scaled wall thermal conductivity, k_s, as:

$$k_s = \left(\frac{l_s}{l_f}\right)^{\frac{3}{2}} \cdot k_f \tag{5}$$

2.3 Numerical analysis

The commercial Fluent CFD code was employed to solve the governing equations [9]. The SIMPLE scheme was chosen to couple pressure and velocity. The Discrete Transfer Radiation Model (DTRM), that assumes all surfaces to be diffuse, is chosen. The following convergence criteria were assumed: 10^{-3} for the residuals of the velocity components and 10^{-8} for the residuals of the energy. An analysis of sensitivity was conducted in terms of temperature profiles; a 120 ° polar angle and a 48 ° azimuthal angle were chosen for the radiation model as well as a 0.010 m square grid was chosen for the thermo-fluid-dynamic model.

3. Results

In order to assess the reliability of the aforementioned parameters to be used as scaling parameters for the proposed fire model, a comparison between temperature and velocity profiles for the problem presented in the previous section and the ones obtained for a scaled model has been carried out. A geometric scaling factor equal to 0.5 (1:2) has been assumed. Once the problem has been scaled geometrically, scaled firepower and wall thermal conductivity are evaluated by means of equations (3) and (5). Comparison has been carried out in terms of temperature profiles, not scaled for both models, and velocity profiles; in this latter case velocity values for the scaled model are scaled by means of the normalization factor reported in equation (1). Two firepower values for the unscaled model are assumed, 100 kW and 200 kW. Temperature and velocity profiles have been presented in terms of the dimensionless spatial variables X, Y and Z, by assuming a normalizing factor equal to the computational domain length along the x, y and z direction, respectively.
In figure 4 Z planes and Y values, at which results will be presented, are reported. Moreover, surface beam temperature profiles on the faces indicated with A and B in figure 5 will be presented.

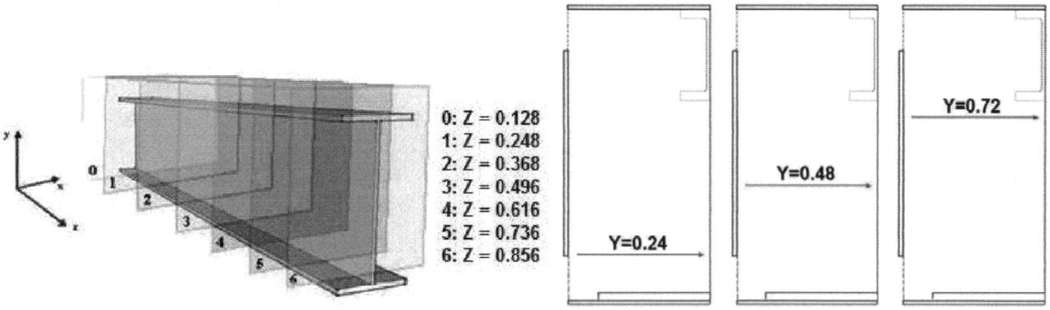

Figure 4. Sketch of the planes at which results will be presented.

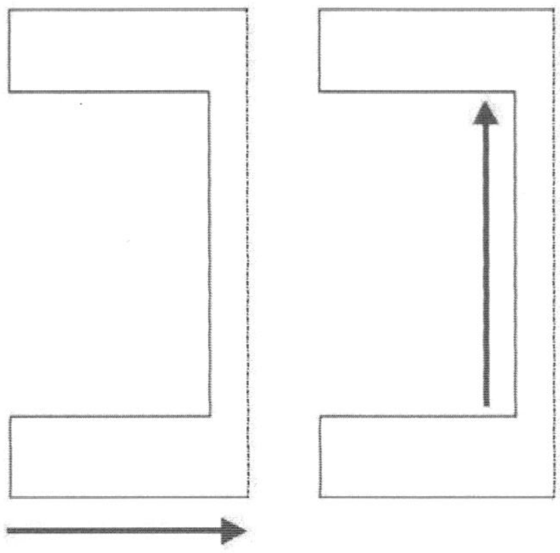

Side A Side B

Figure 5. Sketch of the beam sides at which results will be presented.

The comparison between air temperature profiles along x/L axis for full and scaled models for a firepower value equal to 100 kW is presented in figure 6 for several Z and Y values. For Y=0.24 air temperature differences between full and scaled models are less than 5 K, with a maximum percentage difference of about 15%. Differences between predicted temperature values for the two models increase with Z that is from the boundary of the domain along z axis to the center of the same. For Y=0.72 temperature differences are lower than for Y=0.24, except for Z=0.128 in proximity of the domain walls (x/L values near zero), where differences between temperature predicted by the two analyzed models are about 15 K, with a maximum percentage difference equal to about 30%. For the others analyzed Z values, percentage differences are less than 10%.
In figure 7 beam wall temperature profiles along A and B side showed in figure 5 are reported for the full and scaled models for the same Z values analyzed in the previous figure. As for the horizontal wall, side A, both models predict wall temperature values increasing with Z increase. Large differences between temperature values from full and scaled models are detected, with a maximum percentage difference of about 30%. Similar considerations can be made for the analyzed vertical beam wall, side B, shown in figure 7b. In this case maximum percentage discrepancy between the two models is about 35%. These results suggest that a more accurate scaling procedure has to be accomplished as for the conjugate heat transfer between air and beam, by taking into account scaling factors not only for heat conduction in the walls, but also for convective and radiative heat exchange.

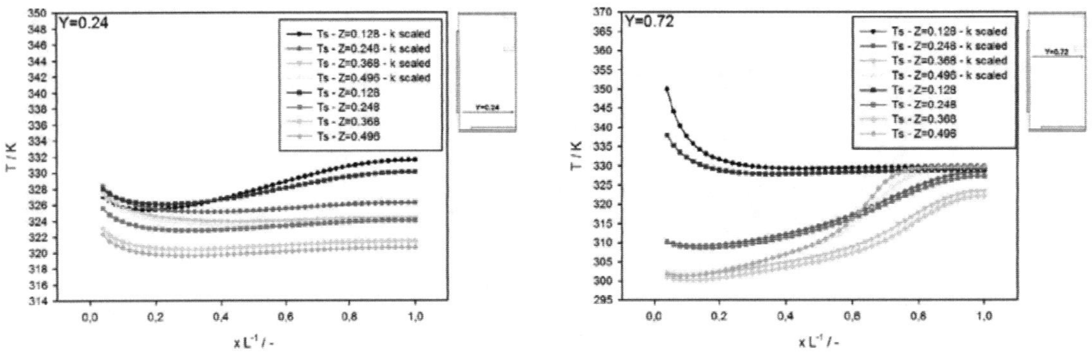

(a) (b)

Figure 6. Air temperature profiles along x/L axis for a firepower equal to 100 kW, for several Z values and for Y=0.24 (a) and Y=0.72 (b).

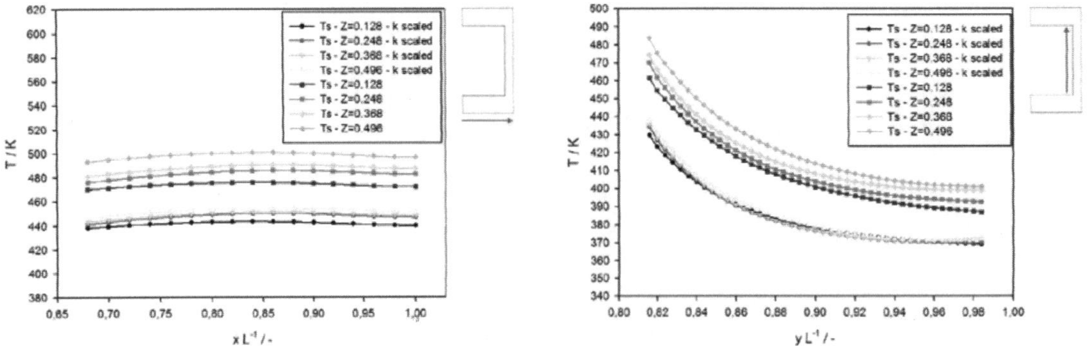

Side A Side B

(a) (b)

Figure 7. Beam wall temperature profiles along x/L axis for a firepower equal to 100 kW, for several Z values on side A (a) and B (b).

The comparison between air velocity profiles predicted by full and scaled models, along x/L axis, for several Y and Z values and for a fire power equal to 100 kW is reported in figures 8 and 9. The y velocity component is reported in figure 8, whereas the velocity magnitude is reported in figure 9. For Y=0.24, figure 8a shows y velocity component profiles almost overlapped for full and scaled models for Z=0.128. Differences between two models increase as Z increases, the maximum percentage difference being about 40% for Z=0.496 and for low x values. Differences between the two models decrease as x increases, where the influence of the domain walls diminishes. A similar behavior is exhibited at Y=0.72, figure 8b. In this case differences are lower than the ones detected for Y=0.24, except for Z=0.128 and in proximity of the domain walls. For all other Z values, maximum percentage difference is about 20%. As for velocity magnitude, figure 9 shows that maximum percentage difference between air velocity magnitude values predicted by the two analyzed models are about 30% and 20% for Y=0.24 and 0.72, respectively. Also in this case maximum differences are attained in proximity of the domain walls.

Similar temperature profiles reported in figures 6-7 for a fire power equal to 100 kW, are reported in figures 10-11 for a fire power equal to 200 kW.

The comparison between air temperature profiles along x/L axis for full and scaled models for a firepower value equal to 200 kW is presented in figure 10 for several Z and Y values. By comparing these temperature profiles with the ones reported in figure 6 for a firepower equal to 100 kW, it can be observed a similar trend, with expected larger temperature values for the 200 kW case. The maximum percentage difference between temperature values predicted by full and scaled models is about 15% for both analyzed Y values.

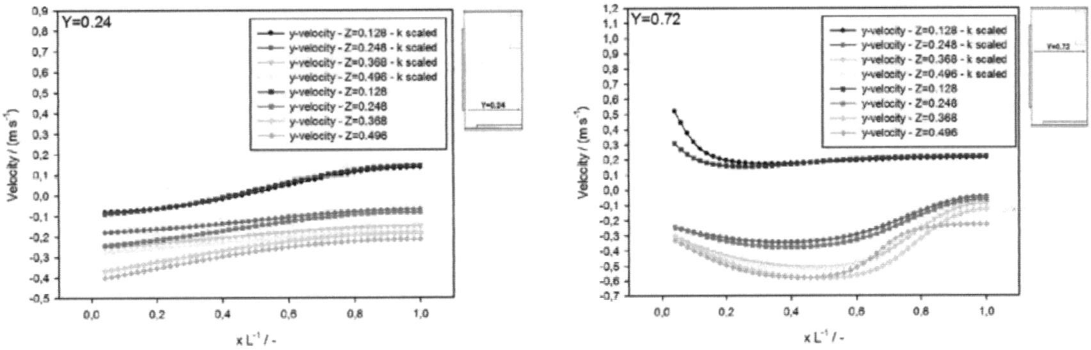

(a) (b)

Figure 8. Air y velocity component profiles along x/L axis for a firepower equal to 100 kW, for several Z values and for Y=0.24 (a) and Y=0.72 (b).

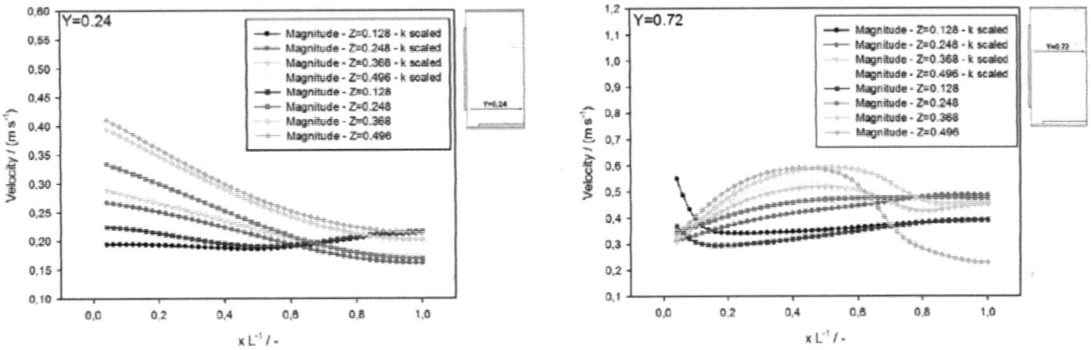

(a) (b)

Figure 9. Air velocity magnitude profiles along x/L axis for a fire power equal to 100 kW, for several Z values and for Y=0.24 (a) and Y=0.72 (b).

In figure 11 beam wall temperature profiles along A and B side showed in figure 5 are reported for the full and scaled models for the same Z values analyzed in the previous figure for a firepower equal to 200 kW. Also in this case similar trends as those observed in figure 7 are observed and temperature values are obviously larger. Maximum percentage differences between the two models are about 30% and 25% for side A and B, respectively.

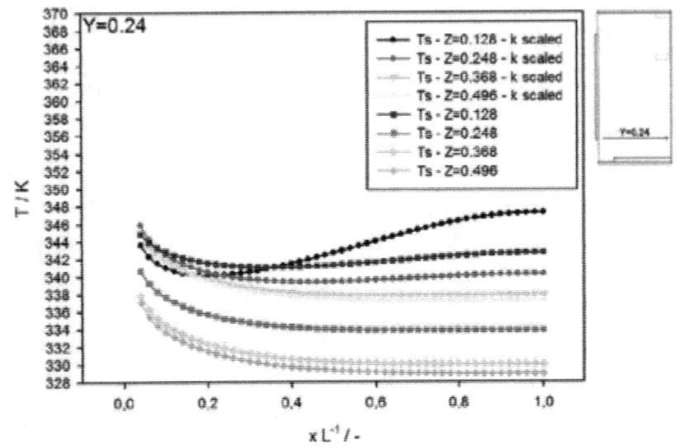

(a)

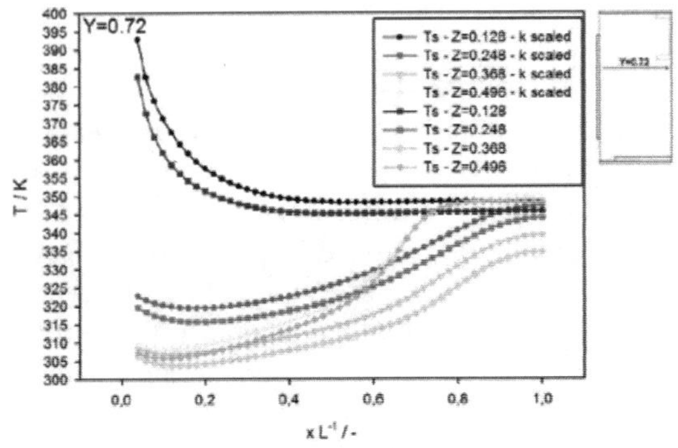

(b)

Figure 10. Air temperature profiles along x/L axis for a firepower equal to 200 kW, for several Z values and for Y=0.24 (a) and Y=0.72 (b).

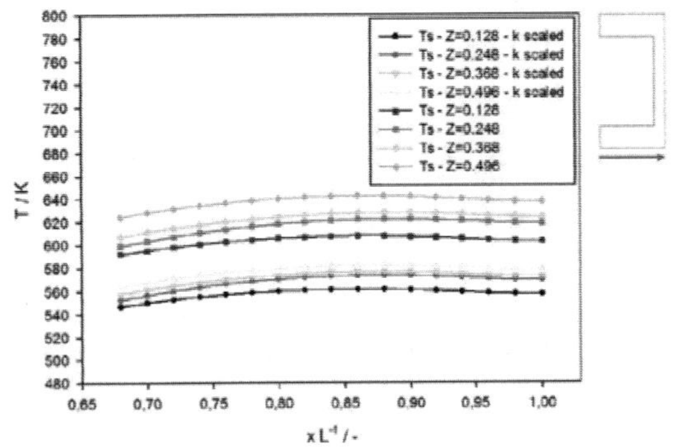

Side A

(a)

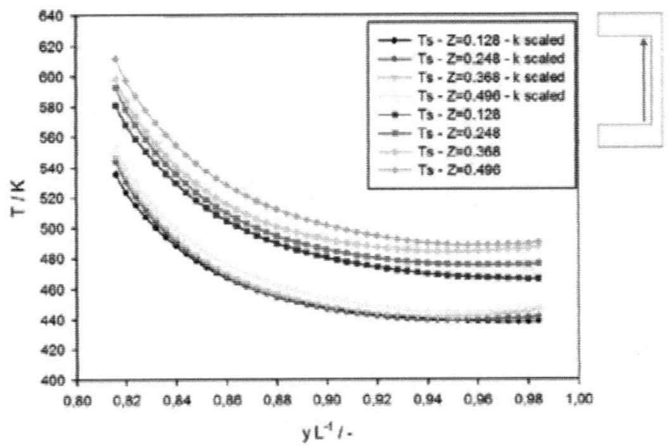

Side B

(b)

Figure 11. Beam wall temperature profiles along x/L axis for a firepower equal to 200 kW, for several Z values on side A (a) and B (b).

The comparison between results for full and scaled models, for all the analyzed cases, shows a better agreement for the air temperature and velocity profiles than for the beam wall temperature profiles. Moreover, the agreement for air temperature and velocity profiles worsens in proximity of the beam walls. This analysis shows that the chosen scaling parameters don't allow a good scaling of the model as for the heat exchange between the beam and the fire compartment, thus indicating that other scaling parameters have to be used, in order to account for convective and radiative heat transfer.

4. Conclusions

A method to scale a compartment fire problem was presented. Dimensionless parameters were identified in order to scale the model in terms of geometry, boundary conditions and materials thermophysical problems. Both the full and scaled models were solved numerically for two different firepower values. Results obtained from the full scale model and the scaled one were compared in terms of velocity and temperature profiles in order to assess the reliability of the scaled model to represent the behavior of the full scale one. Differences between the full and scaled models were large in proximity of the domain walls, whereas they decreased moving away from the walls. Results suggested that a more accurate scaling procedure was to be accomplished as for the conjugate heat transfer between air and domain walls, by taking into account scaling factors not only for heat conduction in the walls, but also for convective and radiative heat exchange.

5. Nomenclature

c_p	Specific heat		
g	Acceleration due to the gravity	μ	Viscosity
h	Convective heat transfer coefficient	**Superscript**	
k	Thermal conductivity	$*$	Dimensionless
l	Characteristic length	**Subscripts**	
$\dot{Q}$	Fire power	0	Ambient
T	Temperature	f	Full
X,Y,Z	Dimensionless Cartesian coordinate	k	Conductive
x, y, z	Cartesian coordinate	P	Ceiling external surface
u, v, w	Velocity component	s	Scaled
Greek symbols		w	Wall
β	Thermal expansion coefficient	∞	Refers to ambient value

6. References

[1] Quintiere J G 2006 *Fundamentals of fire Phenomena*, John Wiley & Sons Ltd.
[2] Quintiere J G, McCaffrey B J and Kashiwagi T 1978 *Combust. Sci. Technol.* **18** 1-19
[3] Emori RI and Saito K1983*Combust. Sci. Technol.* **31** 217-231
[4] Thomas P H 1983 *Fire Safety J.* **5**181-190
[5] Thomas P II 2000 *Fire Safety J.* **34** 111-141
[6] Quintiere J G 1989 *Fire Safety J.* **15** 3-29
[7] Chowa W K and Loa H W 2008 *Heat Transfer Eng.* **29** 76-84
[8] Wang M, Perricone J, Chang P C and Quintiere J G 2008 *J. Fire Prot. Eng.* **18** 223-240
[9] Fluent user's guide, Fluent Inc., 2005
[10] Andreozzi A, Bianco N, Musto M, Naso V and Rotondo G 2014 *Energy Procedia* **45** 385-394

33rd UIT (Italian Union of Thermo-fluid-dynamics) Heat Transfer Conference IOP Publishing
Journal of Physics: Conference Series **655** (2015) 012054 doi:10.1088/1742-6596/655/1/012054

A fast algorithm for Direct Numerical Simulation of natural convection flows in arbitrarily-shaped periodic domains

D Angeli[1], E Stalio[2], M A Corticelli[2], and G S Barozzi[2]

[1] Dipartimento di Scienze e Metodi dell'Ingegneria, Università degli Studi di Modena e Reggio Emilia, Via Amendola 2, 42122 Reggio Emilia (Italy)

[2] Dipartimento di Ingegneria Enzo Ferrari, Università degli Studi di Modena e Reggio Emilia, Via Vivarelli 10, 41125 Modena (Italy)

E-mail: diego.angeli@unimore.it

Abstract. A parallel algorithm is presented for the Direct Numerical Simulation of buoyancy-induced flows in open or partially confined periodic domains, containing immersed cylindrical bodies of arbitrary cross-section. The governing equations are discretized by means of the Finite Volume method on Cartesian grids. A semi-implicit scheme is employed for the diffusive terms, which are treated implicitly on the periodic plane and explicitly along the homogeneous direction, while all convective terms are explicit, via the second-order Adams-Bashfort scheme. The contemporary solution of velocity and pressure fields is achieved by means of a projection method. The numerical resolution of the set of linear equations resulting from discretization is carried out by means of efficient and highly parallel direct solvers. Verification and validation of the numerical procedure is reported in the paper, for the case of flow around an array of heated cylindrical rods arranged in a square lattice. Grid independence is assessed in laminar flow conditions, and DNS results in turbulent conditions are presented for two different grids and compared to available literature data, thus confirming the favorable qualities of the method.

1. Introduction and scope

Direct Numerical Simulation (DNS) is a powerful research tool for the analysis of turbulent flows. As soon as care is taken in the choice of spatial and temporal discretization schemes, and grids or modal expansion truncations are adequate, DNS allows for a full 3D representation of all the turbulent scales of a flow in a set of time steps, from which a large amount of information can be extracted on the nature of the simulated turbulent flow [1].

Although the continuous increase in available computing resources has made it possible to achieve results at very high Reynolds numbers [2] or with extremely refined grids [3], DNSs are often limited to flows in very simple domains, like channels, pipes or flat plates, mainly due to the limitations of certain discretization methods (e.g. spectral methods) or to the numerical errors that might affect the resolution of the discrete equations on unstructured grids, typically involving an excessive amount of unphysical artificial diffusion, which introduces large errors in the representation of the smallest scales of the flow. For this reason, novel parallel and accurate methods for the DNS of flows in complex geometries are still a relevant topic in the field of computational fluid dynamics.

In this context, the approaches based on the representation of arbitrary domains in the frame of a Finite Difference or Finite Volume Cartesian grid, commonly known as Immersed Boundary Methods, are among the most attractive, given the ease in generating the mesh and the possibility of a straightforward parallelization. Stemming from the original development of the Immersed Boundary Method by Peskin [4], a variety of different methodologies have been developed to simulate flow over bodies that, in general, do not conform to the computational grid. These include methods such as the immersed interface method [5], the Cartesian grid method [6] and the cut-cell method [7]; many of these variants are listed and briefly described in the review by Mittal and Iaccarino [8].

Despite the favorable features connected with the adoption of Cartesian grids, those approaches find their main drawback in the treatment of the boundary conditions: since, in fact, in the presence of irregular boundaries the truncation error tends to increase locally, the global accuracy of the spatial discretization can easily be compromised.

In this work, a novel methodology is presented for the Direct Numerical Simulation of fully-developed, buoyancy-induced flow and heat transfer in open or partially confined periodic domains containing immersed cylindrical bodies, aligned with one grid direction (x), of arbitrary cross-section. The continuity, Navier-Stokes and Energy equations are treated in their conservative form, and discretized by means of the Finite Volume method on Cartesian grids which are homogeneous along x and periodic and non-uniform along the two normal directions (y and z). Second-order time and space discretization schemes are used, and pressure-velocity coupling is achieved by means of a Projection Method. The numerical resolution of the discrete equations is carried out by means of efficient and highly parallel direct solvers.

The representation of arbitrarily irregular cylindrical boundaries on Cartesian grids is based on a modification of the original scheme developed by [9] for pure diffusion problems. The technique involves a local modification of the 5-point computational stencil where boundary segments intersect the stencil arms. The variables on the modified stencil are mapped on the global grid, by means of a linear operator determined by geometrical features and boundary conditions. The overall accuracy of the method is virtually preserved, as well as the computational efficiency of the Cartesian approach.

Verification and validation of the numerical procedure is performed. The selected test case is the flow of a fluid with $Pr = 1$ around an array of vertically-mounted, uniformly heated cylindrical rods arranged in a square lattice. A grid sensitivity analysis is performed for the case of laminar flow at $Re_\tau = 17$, for both forced and buoyancy-induced convection. Results in turbulent flow conditions at at $Re_\tau = 170$ are reported, with a focus on grid dependence.

2. Mathematical formulation

The methodology implements a discretization of the the incompressible Navier-Stokes and energy equations. The Boussinesq approximation is enforced, all the fluid properties being consistently assumed as constant, apart from density in the buoyancy term.

2.1. Continuity equation

Under the above hypotheses, the continuity equation reads:

$$\frac{\partial u_i}{\partial x_i} = 0 \tag{1}$$

2.2. Momentum equation

A formulation of momentum equations is

$$\rho\left(\frac{\partial u_i}{\partial t} + u_j\frac{\partial u_i}{\partial x_j}\right) = -\frac{\partial p}{\partial x_i} + \mu\frac{\partial^2 u_i}{\partial x_j \partial x_j} + \rho g_i \tag{2}$$

where u_i indicates the i-th component of the velocity field and the repeated index notation is implied.

The Boussinesq approximation is used in equation (2) to represent the buoyancy force

$$\rho g_i = \rho^o g_i - \rho^o g_i \beta^o (T - T^o) \tag{3}$$

where the superscript "o" denotes reference conditions.

Equation (3) can be conveniently splitted introducing the mean temperature over a y, z plane $\overline{T}_m(x)$

$$\rho g_i = \rho^o g_i - \rho^o g_i \beta^o \left[T - \overline{T}_m(x)\right] - \rho^o g_i \beta^o \left[\overline{T}_m(x) - T^o\right] \tag{4}$$

As in the case of fully developed flow with imposed heat flux, the temperature profile in axial direction is linear, equation (4) can be set as

$$\rho g_i = \rho^o g_i - \rho^o g_i \beta^o \left[T - \overline{T}_m(x)\right] - \rho^o g_i \beta^o a (x - x^o) \tag{5}$$

In equation (5), a denotes the temperature slope in x, which may be evaluated from an energy balance

$$a = \frac{q'' \omega}{\dot{m} c_p} \tag{6}$$

In equation (6), q'' indicates the imposed wall heat flux over perimeter ω, while $\dot{m}$ is the mass flow rate

$$\dot{m} \equiv \int_\Omega \rho^o u \, \mathrm{d}\Omega \tag{7}$$

and Ω is the cross section.

The first and the third terms in equation (5) may be incorporated into the pressure field giving place to a modified pressure[1]

$$p_m \equiv p - \rho^o g_1 x_1 + \rho^o g_1 \beta^o a \frac{(x_1 - x^o)^2}{2} \tag{8}$$

In fully developed conditions, the time averaged streamwise gradient of the modified pressure is constant [10]. As soon as pure natural convection conditions are considered, such a constant equals zero.

Without loss of generality, we can then split the modified pressure in a linear part P_m, depending on the streamwise coordinate only, and a fluctuation p'_m:

$$p_m(x, y, z) = P_m(x) + p'_m(x, y, z) \tag{9}$$

The pressure gradient term becomes then:

$$-\nabla p_m = -\nabla p'_m - \gamma \tag{10}$$

where γ denotes the constant pressure gradient due to external forcing of the flow. If $\gamma = 0$, pure natural convection conditions are recovered, and the fluctuating pressure field p'_m is periodic.

The fluid excess temperature is defined as

$$\theta(x, y, z) = \left[T(x, y, z) - \overline{T}_m(x)\right] \tag{11}$$

In fully developed conditions, $\overline{\theta}$ is independent of the streamwise coordinate [11].

[1] As soon as the first coordinate is directed in streamwise direction $x_1 = x$ and also gravity is directed along x, $(g_1 = -g,\ g_2 = 0,\ g_3 = 0)$

The form of the momentum equations to be solved is then

$$\rho^o \left(\frac{\partial u_i}{\partial t} + u_j \frac{\partial u_i}{\partial x_j} \right) = -\frac{\partial p'_m}{\partial x_i} + \mu \frac{\partial^2 u_i}{\partial x_j \partial x_j} + \rho^o g \beta \theta - \gamma \tag{12}$$

Using the following temperature and velocity scales in terms of the length scale $L_{\rm ref}$

$$T_{\rm ref} = \frac{q''\ L_{\rm ref}}{\lambda} \qquad u_{\rm ref} = u_\tau = \sqrt{\tau_w / \rho^o} \tag{13}$$

the dimensionless momentum equation writes in terms of dimensionless quantities

$$\frac{\partial u_i^*}{\partial t^*} + u_j^* \frac{\partial u_i^*}{\partial x_j^*} = -\frac{\partial p_m^*}{\partial x_i^*} + \frac{1}{\mathrm{Re}_\tau} \frac{\partial^2 u_i^*}{\partial x_j^* \partial x_j^*} + \frac{\mathrm{Gr_q}}{\mathrm{Re}_\tau^2} \theta^* + \frac{\omega^*}{\Omega^*} \tag{14}$$

In equation (14), the friction Reynolds number Re_τ is defined as:

$$\mathrm{Re}_\tau = \frac{u_\tau\, L_{\rm ref}}{\nu} \tag{15}$$

The Grashof number $\mathrm{Gr_q}$ is also introduced, based on the imposed heat flux q'':

$$\mathrm{Gr_q} = \frac{g \beta q''\ L_{\rm ref}^4}{\lambda \nu^2} \tag{16}$$

2.3. Energy equation

As from equations (11) and (6)

$$\frac{\partial T}{\partial x} = a + \frac{\partial \theta}{\partial x}; \qquad \frac{\partial^2 T}{\partial x^2} = \frac{\partial^2 \theta}{\partial x^2} \tag{17}$$

The energy equation becomes

$$\rho^o c_p \left[\frac{\partial \theta}{\partial t} + u_i \frac{\partial \theta}{\partial x_i} \right] = \lambda \frac{\partial^2 \theta}{\partial x_i \partial x_i} - \rho^o c_p a u_1 \tag{18}$$

It is worthy to point out the appearance of the term $\rho^o c_p a u_1$ at the right-hand side of equation (18), emerging from the normalization of temperature [10].

Using the same reference quantities as for the momentum equation the energy equation writes in its dimensionless form

$$\frac{\partial \theta^*}{\partial t^*} + u_i^* \frac{\partial \theta^*}{\partial x_i^*} = \frac{1}{\mathrm{Re}_\tau \mathrm{Pr}} \frac{\partial^2 \theta^*}{\partial x_i^* \partial x_i^*} - \frac{1}{\mathrm{Re}_\tau \mathrm{Pr}} \frac{\omega^*}{\Omega^*} \frac{u_1^*}{u_m^*} \tag{19}$$

where the Prandtl number $\mathrm{Pr} = \nu / \alpha$ expresses the ratio between the diffusivities of momentum and thermal energy.

3. Numerical method

The numerical technique adopted is based on a Finite Volume implementation of a second order Projection Method, following [12]. Time-discretizations of the conservation equations are performed according to a three-level scheme, which is semi-implicit (explicit in the homogeneous direction x) for the diffusive terms, and explicit Adams-Bashfort for the advective terms. Such a practice is second order accurate in time.

Spatial derivatives are approximated with second order central differences on staggered Cartesian grids, which can be non-uniform on the transverse $y-z$ plane. A fast direct resolution of the discrete momentum and energy equations at each time-step is made possible by means of Approximate Factorization, while the Poisson problem associated with the pressure-velocity coupling [12] is solved through a fast Poisson solver, based on Matrix Decomposition [13].

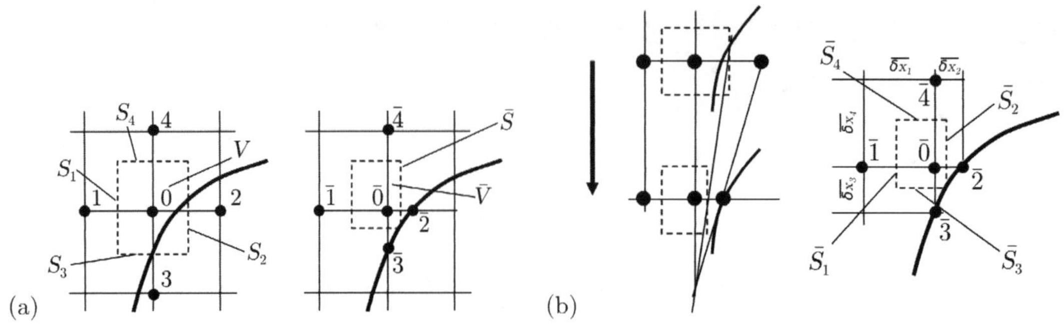

Figure 1. Treatment of irregular boundaries: (a) local stencils g (left) and d (right); (b) local cell resizing (left) and metrics of a boundary cell (right)

3.1. Treatment of irregular boundaries

The methodology employed for the modeling of boundaries non-aligned with the Cartesian grid is briefly outlined here. It represents an extension to convection problems of an algorithm which was originally developed for thermal diffusion problems [9].

On a generic plane along the x direction, in the vicinity of a boundary, two different computational stencils are defined: the stencil g belonging to the global Cartesian grid, and the stencil d, referring to the modified discretization grid, which takes into account the presence of the boundary (Figure 1a). The g stencil is complete only when all its nodes lie inside the domain, otherwise it cannot provide a complete support to the discretization of the equations, because of the "lack" of some of its nodes. Stencil d instead is always complete, since it replaces the missing nodes with the intersection points of the boundary with the grid. If all the nodes of g are internal to the domain, the two stencils coincide.

Two distinct control volumes are identified for the two stencils. The cell V is related to the computational stencil g, and its size depends only on the structure of the grid to which it refers. On the contrary, the size of the discretization cell $\bar{V}$, linked to the d-stencil, is defined coherently with the presence of the geometric boundary: the lenghts of the half-sides of $\bar{V}$ are sized as proportional to those of V, by the ratio between the distance of the cell center from the intersection point, and the length of the associated stencil arm (Figure 1b).

In general, the value of the variables $\bar{f}_k$ on stencil d can be expressed as a linear combination of the values of the nodal unknowns f_k of stencil g, plus a non-homogeneous term accounting for the boundary conditions. Hence, in vector notation, one can write:

$$\bar{\mathbf{f}} = \mathbf{E} \cdot \mathbf{f} + \mathbf{g} \tag{20}$$

If the stencil is completely contained within the domain, it is $\bar{\mathbf{f}} = \mathbf{f} \quad \Rightarrow \quad \mathbf{E} = \mathbf{I},\ \mathbf{g} = \mathbf{0}$:

At the generic intersection node k (Figure 2), the boundary condition for the unknown f_k is expressed by the following general form:

$$a_k f_k + b_k \frac{\partial f_k}{\partial \eta_k} = c_k \tag{21}$$

where η_k denotes the direction normal to the boundary at node k. With reference to Figure 2, the index k' identifies the symmetric of node k with respect to the center of the stencil:

$$\begin{cases} k' = 3 - k & k = 1, 2 \\ k' = 7 - k & k = 3, 4 \end{cases} \tag{22}$$

By performing a second-order Taylor expansion of φ around k, along the direction η, one can

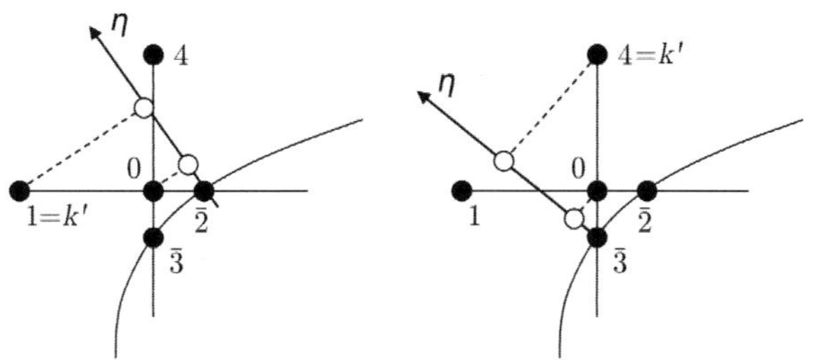

Figure 2. Local discretization stencils for different types of intersections

write:

$$\varphi(\eta) \simeq \varphi_{\eta=0} + \left(\frac{\partial \varphi}{\partial \eta}\right)_{\eta=0} \eta + \left(\frac{\partial^2 \varphi}{\partial \eta^2}\right)_{\eta=0} \eta^2 \tag{23}$$

Here, the node k is taken as the origin of the η axis. The expression (23) holds for any point in the vicinity of k, and, therefore, it can be applied to all the three points 0, k, k' chosen for the approximation of the normal derivative:

$$f_l \simeq f_k + \left(\frac{\partial \varphi}{\partial \eta}\right)_{\eta=0} \eta_l + \left(\frac{\partial^2 \varphi}{\partial \eta^2}\right)_{\eta=0} \eta_l^2 \quad l = 0, k, k' \tag{24}$$

On the basis of the relationship (24), the normal derivative of the variable φ at the boundary can be expressed as a linear combination of the nodal values of the variable itself in the points 0, k, k' of the stencil:

$$\begin{aligned} \left(\frac{\partial \varphi}{\partial \eta}\right)_{\eta=0} &= p_{k,0}\bar{f}_0 + p_{k,k}\bar{f}_k + p_{k,k'}\bar{f}_{k'} = \\ &= \sum_l p_{k,l}\bar{f}_l, \quad l = 0, k, k' \end{aligned} \tag{25}$$

The coefficients $p_{k,l}$ can be determined by substituting equation (24) into equation (25) and then enforcing congruence between the members of the resulting equations.It is then possible to insert the expression (25)into the generic boundary condition (21), thus obtaining:

$$a_k\bar{f}_k + b_k \sum_l p_{k,l}\bar{f}_l = c_k, \qquad l = 0, k, k' \tag{26}$$

Equation (26) expresses a useful constraint to determine the value of f_k, when point k lies on the boundary. If, on the contrary, the computational stencil is completely contained within the domain, one simply has $\bar{f}_k = f_k$. Hence, a general equation can be written as follows:

$$(a_k + b_k p_{k,k})\bar{f}_k + b_k \sum_{l=0, l\neq k}^{4} p_{k,l}\bar{f}_l = c_k + d_k f_k \tag{27}$$

where $d_k = 1$ if the associated node is inside the domain, while $d_k = 0$ if the node lies on the boundary. Equation (27) can be reformulated in matrix notation as:

$$\mathbf{A} \cdot \bar{\mathbf{f}} = \mathbf{D} \cdot \mathbf{f} + \mathbf{c} \tag{28}$$

It is now possible to formulate the expressions for $\mathbf{E}$ and $\mathbf{g}$ introduced by the definition (20):

$$\mathbf{E} = \mathbf{A}^{-1} \cdot \mathbf{D}, \qquad \mathbf{g} = \mathbf{A}^{-1} \cdot \mathbf{c} \tag{29}$$

It should be noted that the values of $\mathbf{E}$ and $\mathbf{g}$ depend only on the coefficients of equation (21) and on the boundary geometry and, therefore, they can be calculated once for all in the pre-processing phase of the numerical procedure.

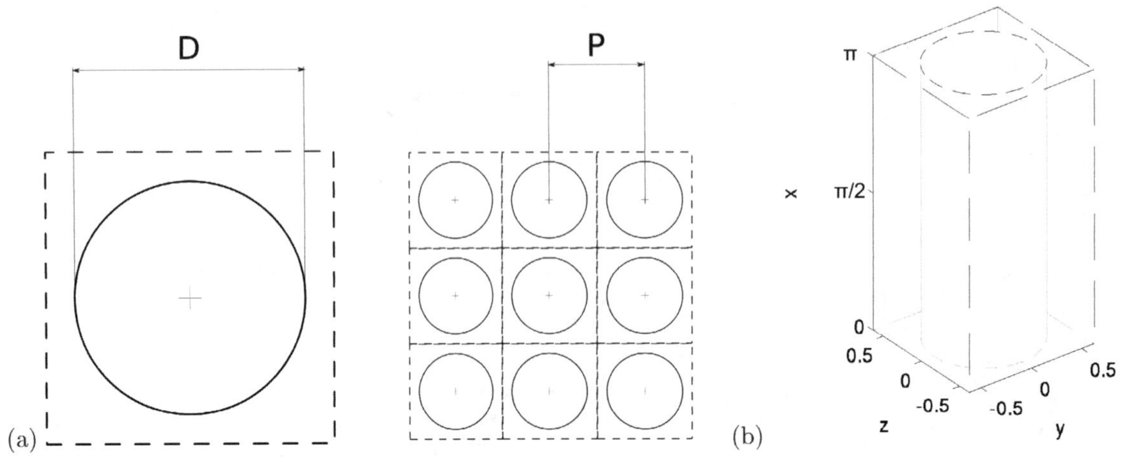

Figure 3. Unit cell of a square array of rods: cross section (a) and vertical development (b)

3.2. Parallel implementation
Parallelism is introduced using the message-passing library MPI. As mentioned above, the momentum and energy equations are made explicit in the homogeneous direction x; for the solution of the momentum and energy equations a one-dimensional domain decomposition is thus implemented in x.

A Fourier decomposition method is applied for solving the Poisson equation, which arises at every time step for the enforcement of mass conservation [12]

$$\nabla^2 \phi = \nabla \cdot \mathbf{u} \tag{30}$$

Thus a one-dimensional Fourier expansion is performed along x followed by the solution of a two dimensional Poisson equation in each y, z plane.

4. Verification and validation
Verification and validation of the numerical method has been performed on the case schematised in Figure 3. A square lattice of cylindrical rods of diameter D, each heated with a uniform heat flux density q'', whose centers are spaced by pitch P, is represented by its unit cell (Figure 3a). In fully-developed conditions, the rod bundle is thought to have an infinite extension along the streamwise, vertical direction x. The computational domain is thus $L_x = \pi$, $L_y = L_z = P$ in the streamwise and the cross-flow directions, respectively (Figure 3b). The physical situation of a bundle of heated rods under the hypothesis of fully developed flow and heat transfer can be represented through periodic conditions, which are enforced on the velocity field, the modified pressure field p_m and the modified temperature-like variable θ along all coordinate directions [14].

As discussed in Ref. [15], from a practical standpoint, the flow induced by buoyancy in a bundle of heated rods is always associated with a non-zero mean flow and a pressure drop, since typically the heat source is of finite height, and it is inserted in a loop together with a suitable cooling section. Such is the situation, for instance, of DHR loops in nuclear applications. Hence, the flow under consideration, although being induced solely by buoyancy, cannot be viewed as a purely natural convection flow, since, in that case, the hypothesis of full development would entail a zero mean flow rate and pressure drop. This apparent paradox is solved by inserting a forcing term in equation (18) which realizes a fully developed mixed convection flow, representative of the actual flow inside a bundle of heated rods.

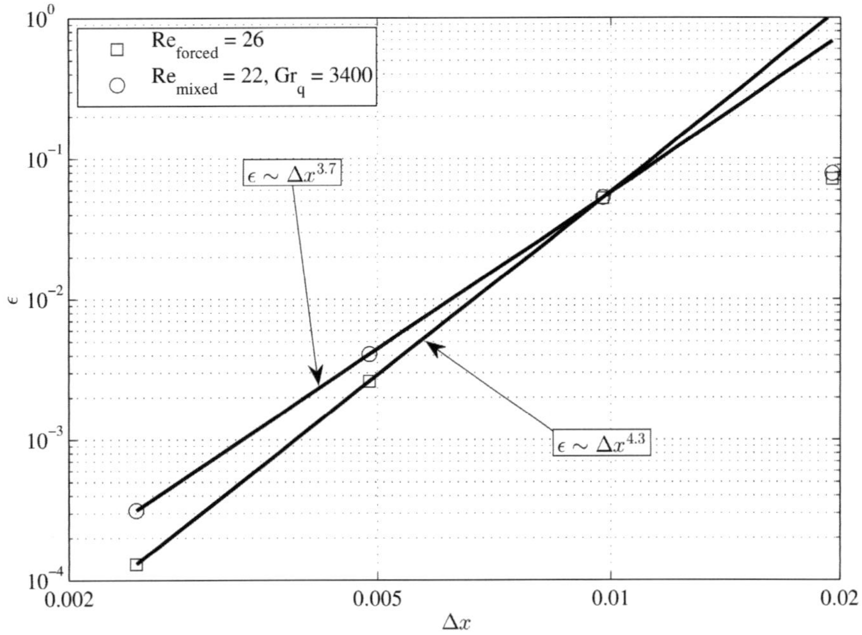

Figure 4. Grid independence test: deviation ϵ from the continuum value vs. Δx.

4.1. Grid independence test
Grid independence is assessed in laminar flow conditions by considering both forced and mixed convection conditions, for $\mathrm{Pr} = 1$. A value $\mathrm{Re}_\tau = 17$ is chosen for the forced convection case, which determines a Reynolds number $\mathrm{Re}_{\mathrm{forced}} = 26$; a value $\mathrm{Gr_q} = 3.4 \times 10^3$ is then set for the mixed convection case, so that $\mathrm{Gr_q}/\mathrm{Re}^2 = \mathcal{O}(1)$, in order to match the experimental findings of El Genk et al. [16]. The resulting Reynolds number for the mixed case is $\mathrm{Re}_{\mathrm{mixed}} = 22$.

Four different grids with constant spacing are used, with a refinement ratio of 2 along the y and z directions (see Table 1). The average Nusselt number on the rod surface is chosen as the reference quantity for assessing grid convergence. Richardson extrapolation [17] was then performed on the results of the three finest grids to determine an estimate of the continuum value, and the order of convergence of the grid. Figure 4 reports the absolute deviation ϵ from the continuum value as a function of the grid spacing Δx, along with an indication of the order of convergence p in the grid independent region. For both cases, $p \sim 4$, indicating an excellent grid convergence.

Table 1. Grid independence test for $\mathrm{Re}_\tau = 17$ and $\mathrm{Gr}_q = 3.4 \times 10^3$

n. of elements	Δx	$\mathrm{Nu_{forced}}$	$\mathrm{Nu_{mixed}}$
64^2	0.0195	4.79	5.68
128^2	0.0098	4.77	5.65
256^2	0.0049	4.73	5.60
512^2	0.0024	4.72	5.60
$\to \infty$	$\to 0$	4.72	5.60

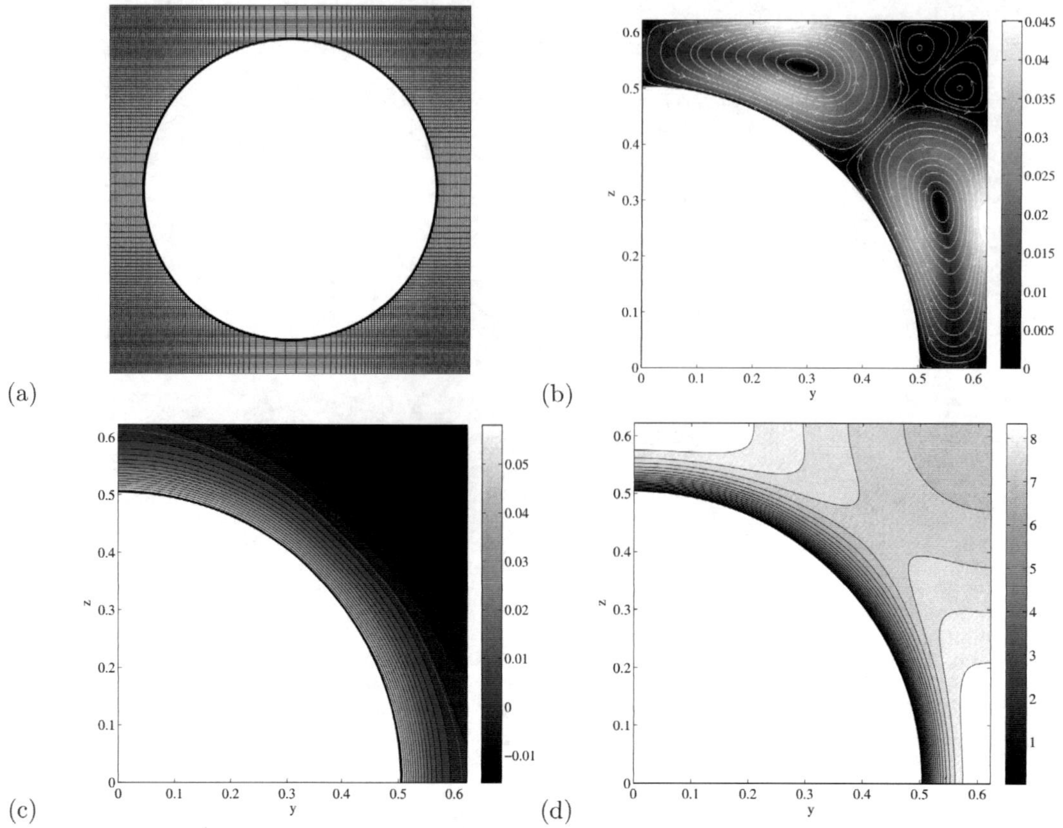

Figure 5. Turbulent flow at $\mathrm{Re}_\tau = 170$, $\mathrm{Gr_q} = 1.32 \times 10^7$, $\mathrm{Pr} = 1$: (a) cross section of the 128^3 computational mesh; (b) streamlines and magnitude of the mean secondary flow components; contours of time-averaged (c) θ and (d) u_x.

4.2. Results in turbulent conditions

Direct Numerical Simulations of fully developed, turbulent, buoyancy-induced flow are performed for $\mathrm{Re}_\tau = 170$, $\mathrm{Gr_q} = 1.32 \times 10^7$, on two different grids. The coarser mesh includes 128^3 elements (see Figure 5a), while the finer counts 256^3 cells.

Non-zero time-averaged secondary flow components are detected. As displayed in Figure 5b, four symmetric vortices are found; two of them are small and very weak $|u_y| \sim |u_z| \sim 0.01 u_{x,\max}$.

Contours of the time-averaged normalized temperature θ and of the time-averaged streamwise velocity component are displayed in Figure 5c and 5d. The peak mean velocity is in $y = P/4, z = 0$ and $y = 0, z = P/4$, while $\overline{u}_x$ decreases towards $y = P/4, z = P/4$, this can be explained by the negative momentum source associated with a negative $\overline{\theta}$. The normalized temperature is negative around $y = P/4, z = P/4$ but becomes positive close to the heated walls. The red line in figure 5 indicates the $\overline{\theta} = 0$ isotherm line. The sign of the time averaged source of momentum changes when

$$\frac{\mathrm{Gr_q}}{\mathrm{Re}_\tau^2}\theta^* = -\frac{\omega^*}{\Omega^*} \tag{31}$$

where $\frac{\mathrm{Gr}}{\mathrm{Re}_\tau^2} \approx 457$ and $\frac{\omega^*}{\Omega^*} \approx 4.04$ in our case.

Profiles of $\overline{u_x}$, $\overline{\theta}$, Reynolds stresses and turbulent heat fluxes across the large $d_l = 0.768D$ gap at a 45° angle with respect to the y and z axes (Figure 6a) show that the velocity field

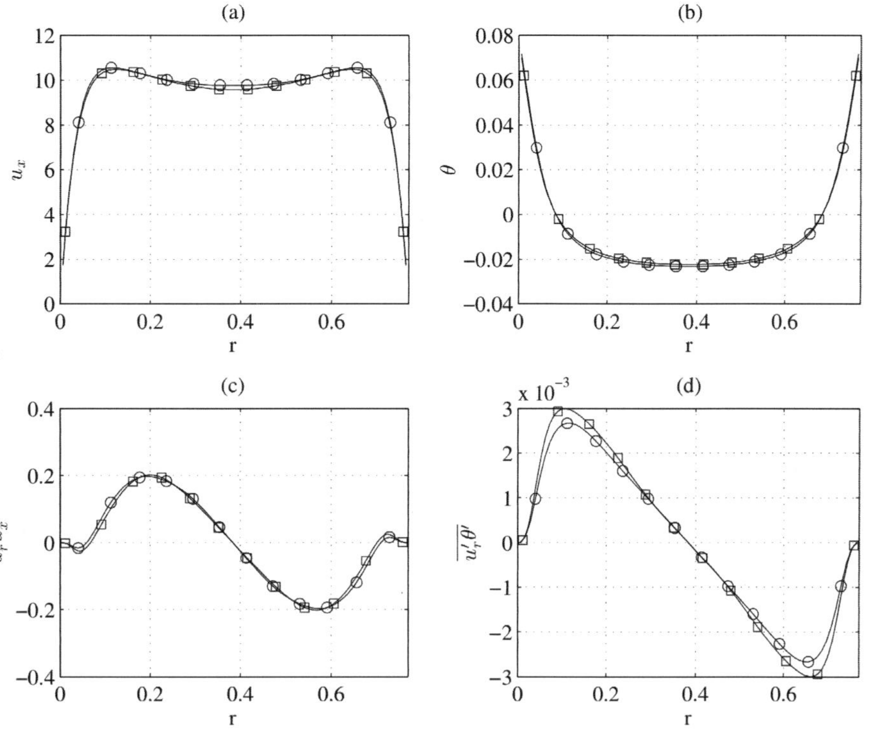

(a)

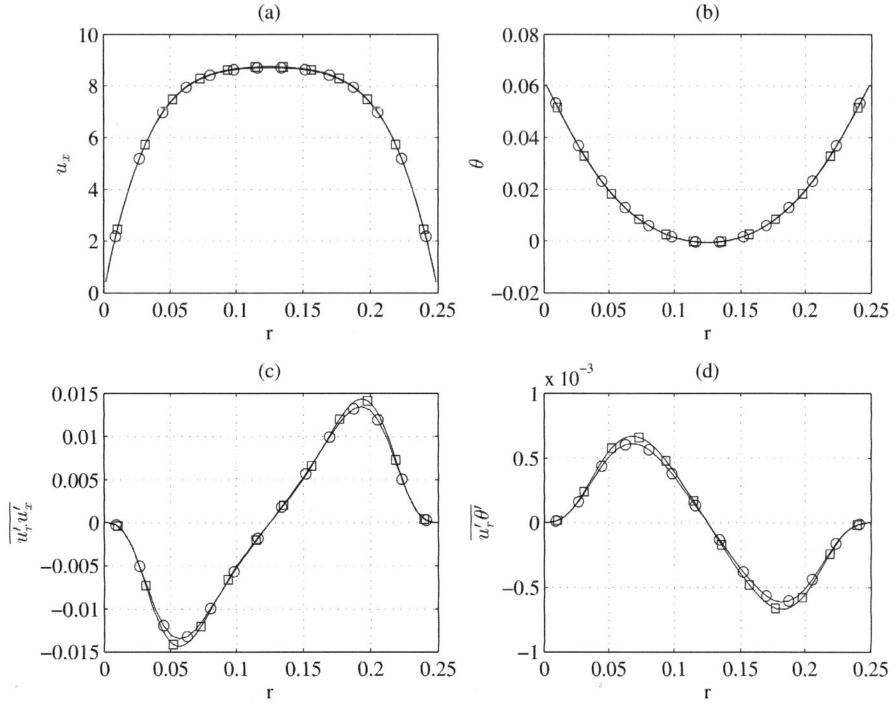

(b)

Figure 6. Turbulent flow at $\mathrm{Re}_\tau = 170$, $\mathrm{Gr_q} = 1.32 \times 10^7$, $\mathrm{Pr} = 1$: profiles of (a) $\overline{u_x}$, (b) $\overline{\theta}$, Reynolds stresses and turbulent heat fluxes across the $d_l = 0.768D$ (a) $d_n = 0.5D$ (b) gaps. Circles and squares indicate results for the 128^3 and 256^3 grid, respectively.

is characterized by two peaks and a central minimum; accordingly Reynolds stresses display two positive peaks and two negative peaks. Across the $d_n = 0.5D$ horizontal and vertical gap (Figure 6b) the profile of the velocity field displays one single, central peak and due to the smaller integral length scale associated with domain dimensions, the turbulent flow characteristics are weaker with respect to Figure 6a.

5. Concluding remarks

A novel technique has been presented, for the Direct Numerical Simulation of buoyancy-induced flows in open or partially confined periodic domains, containing immersed cylindrical bodies. A grid sensitivity analysis has been performed for the case of laminar flow around a square lattice of heated rods. Grid independence tests, based on the average Nusselt number in laminar conditions, suggest that the procedure ensures grid convergence. Results in the weakly turbulent regime display almost laminar characteristics when plotted along the larger, $0.768D$ gap. Across the same gap, the velocity profile displays two peaks; accordingly the Reynolds stress are characterized by four extrema. Across the smaller, $0.5D$ gap the mean velocity is maximum in the center despite the forcing term reaches its minimum in that location. Results in turbulent conditions also highlight that mean quantities and first-order moments have a reduced dependence from the grid. The proposed methodology represents a powerful tool for the analysis of buoyancy-induced turbulent convection in many configurations.

Acknowledgements

The research leading to these results has received funding from the Euratom research and training programme 2014-2018 under grant agreement No 654935 (SESAME).

References

[1] Moin P and Mahesh K 1998 *Annual Review of Fluid Mechanics* **30** 539–578
[2] Lee M and Moser R D 2015 *Journal of Fluid Mechanics* **774** 395–415
[3] Yokokawa M, Itakura K, Uno A, Ishihara T and Kaneda Y 2002 *Procs. ACM/IEEE SC '02 Conf.* pp 1–17
[4] Peskin C 2002 *Acta Numerica* **11** 479–517
[5] Leveque R J and Li Z 1994 *SIAM Journal on Numerical Analysis* **31** 1019–1044
[6] Patankar S V 1978 *Proc. 6th Int. Heat Transfer Conf.* vol 3 p 297
[7] Leveque R 1988 *Journal of Computational Physics* **78** 36–63
[8] Mittal R and Iaccarino G 2005 *Annual Review of Fluid Mechanics* **37** 239–261
[9] Barozzi G S, Bussi C and Corticelli M A 2004 *Numerical Heat Transfer B* **46** 56–77
[10] Piller M and Stalio E 2012 *International Journal of Heat and Mass Transfer* **55** 6506–6513
[11] Patankar S, Liu C and Sparrow E 1978 *International Journal of Heat and Mass Transfer* **21** 557–566
[12] Gresho P M 1990 *International Journal for Numerical Methods in Fluids* **11** 587–620
[13] Babu V and Korpela S 1993 *Journal of Computational Physics* **104** 93–98
[14] Piller M, Nobile E and Hanratty T 2002 **458** 419–441
[15] Hallinan K and Viskanta R 1985 *International Journal of Heat and Fluid Flow* **6** 256–264
[16] El-Genk M, Su B and Guo Z 1993 *International Journal of Heat and Mass Transfer* **36** 2359–2374
[17] Roache P J 1972 *Computational Fluid Dynamics* 1st ed (Hermosa)

Heat transfer along the route to chaos of a swaying thermal plume

D Angeli[1], M A Corticelli[2], A Fichera[3] and A Pagano [3]

[1] Dipartimento di Scienze e Metodi dell'Ingegneria, Università degli Studi di Modena e Reggio Emilia, Via Amendola 2, 42122 Reggio Emilia (Italy)
[2] Dipartimento di Ingegneria Enzo Ferrari, Università degli Studi di Modena e Reggio Emilia, Via Vivarelli 10, 41125 Modena (Italy)
[3] Dipartimento di Ingegneria Industriale, Università degli Studi di Catania, Viale Andrea Doria 6, 95125 Catania (Italy)

E-mail: [1]diego.angeli@unimore.it

E-mail: [2]mauro.corticelli@unimore.it

E-mail: [3]apagano@diim.unict.it

Abstract. Detailed analyses have been recently reported on the low order dynamics of a thermal plume arising from a horizontal cylindrical heat source concentric to an air-filled isothermally cooled square enclosure, together with those of the related flow structures, in the limit of the 2D approximation. In particular, within the range of $0 < Ra < 3Ra_{cr}$, with Ra_{cr} corresponding to the loss of stability of the stationary buoyant plume, the entire evolution from a periodic limit cycle (P_1) to the birth of chaos through a period-doubling cascade has been fully explored. With this respect, special attention has been given to the window of quasiperiodic dynamics onto a T_2-torus that is observed to separate the monoperiodic dynamics from the biperiodic dynamics onto a P_1 and a P_2-limit cycle, respectively. The results of these analyses hint at the bimodal nature of the overall dynamics, in general, and of the subharmonic cascade, in particular, which are still under investigation. Although relevant on a dynamical perspective, a with a main reflection on the laminar-turbulent transition, the observed oscillations appear to be characterised by comparable amplitudes and to be determined by similar evolutions of the flow pattern evolutions, so that their role on the overall heat transfer rate is expected to be marginal. Within this frame, the present study aims at reporting the influence played by the observed dynamics of the thermal plume and of the flow structures on the global heat transfer rate. In particular, the aim is the assessment of the correlation between the Rayleigh number and the average Nusselt number on the cylinder surface, as well as the effect on the latter of the observed series of bifurcations.

1. Introduction

Several studies have addressed the analysis of the thermo-fluid dynamic behaviour of the buoyant flow arising from a horizontal cylindrical heat source confined within a isothermally cooled concentric square cavity. Particular attention has been devoted to the subcritical flows and to steady-state heat transfer, as well as to the first instabilities of the system, both for the limit case of a horizontal line heat source [1] and as a function of the gap width for variable diameter of the heat source [2–6], under the approximation of 2D flow. Similar investigations were also carried out by Bouafia and Daube [7], on an enclosed square-sectioned horizontal cylinder. In a preliminary study [6] with a main focus on the system dynamics, the authors reported the bifurcation path from stationary to chaotic behaviour undergone by the flow for increasing Rayleigh number values. The study was developed under various geometrical configurations, as defined by the aspect ratio $A = H/L$, i.e. the ratio between the maximum gap height and the length of the side walls. For the cavities with aspect ratio $A = 5$ and $A = 10$, the flow asset and early bifurcations were found to be primarily determined by the formation of Rayleigh-Benard-type rolls above the heat source. These steady-state, multicellular patterns were seen to undergo symmetry-breaking pitchfork bifurcations, eventually leading to periodic states. For $A = 3.3$ and $A = 2.5$ the dynamics were found to be governed by the oscillations of a swaying thermal plume originating around the cylindrical heat source. More detailed analyses were hence reported [8] for the behaviour of the cavity characterised by aspect ratio $A = 2.5$. For this case the observed scenario was found to consist of a period doubling cascade, at the beginning of which a window of quasiperiodic behaviour on a T_2-torus is found to separate the period-1 (P_1) orbit from the period-2 (P_2) limit cycle. The subharmonic cascade was followed up to the P_{128}-limit cycle, corresponding to the observation of the seventh period doubling. In order to achieve such a result, together with a very high level of refinement of both the discretisation grid and of the time step adopted for the simulations, a thorough scanning of the Rayleigh number range was performed in order to allow the identification of the whole series of bifurcations leading to chaos.

Even though Squire's theorem [9] ensures that, if the domain is significantly larger along the third dimension with respect to the transverse directions, the first instability takes place always on the transverse plane, this is not generally true for successive transitions; hence, the above-described scenario is not expected to be experimentally observable as a consequence of 3D dynamics, which are not accounted for in the adopted 2D numerical model. Nonetheless, the physical relevance of the scenario under consideration was justified taking into account its qualitative similarity with the experimental results of [10, 11], who reported a similar route to chaos for the Rayleigh-Benard cellular patterns within a rectangular cavity and by Labonia and Guj [12], who reported the route to chaos for a thermal plume confined in a large gap air-filled cylindrical annulus. In the latter, in particular, the influence of three-dimensional effects was shown to emerge well above the critical Ra-value of the first Hopf bifurcation and, hence, in [8] it was inferred that the dynamics of a real 3D square enclosure with the concentric cylindrical heat source can be described with sufficient accuracy by the 2D numerical model, at least for a limited range of Ra above the first Hopf bifurcation.

With respect to the origin of the observed dynamics, the results reported in the previous studies hint at the fundamental bimodal structure of the flow. This structure is, in fact, retained by the modal locking characteristic of the sub-harmonic cascade that follows the window of quasiperiodic behaviour, an observation that was further exploited in [13], with a main focus on the initial transitions, at low values of the Rayleigh number, from the P_1 limit cycle to the quasiperiodic T_2 torus, due to a Neimark-Sacker bifurcation, and from the T_2 torus to the P_2 limit cycle, due to a tangent (flip) bifurcation and corresponding to the locking of the two fundamental modes.

Previous considerations are indeed relevant from a dynamical standpoint in that they play a main role at the beginning of the laminar-turbulent transition. Nonetheless, in any of the observed regimes, the two basic modes are characterised by very similar amplitudes and frequencies and appear to be determined by similar evolutions of the flow patterns within the enclosure. Therefore, their alternate occurrence is not expected to produce relevant variations of the overall heat transfer rate. Within this frame, the present study is devoted to reporting the influence on the global heat transfer rate played by the observed dynamics of the thermal plume and of the flow structures. In particular, the aim is the assessment of the correlation between the Rayleigh number and the average Nusselt number on the cylinder surface, as well as the effect on the latter of the observed series of bifurcations.

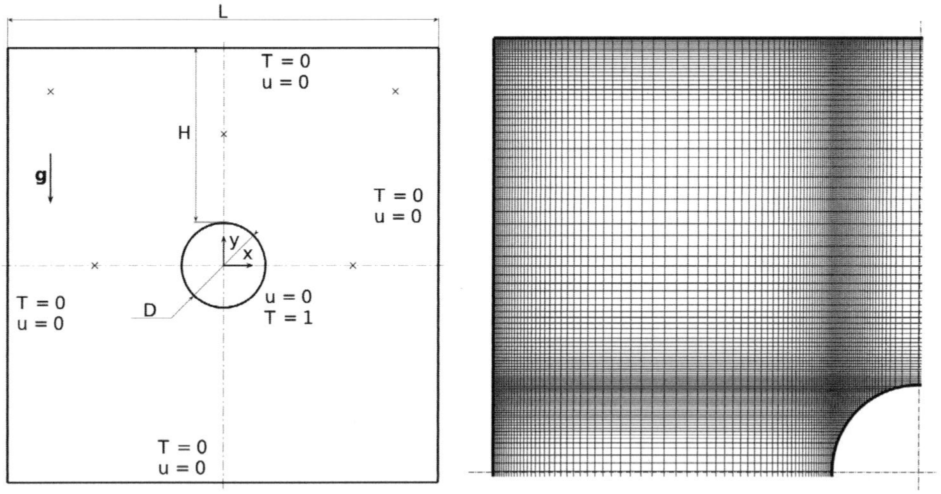

Figure 1. Left: schematic of the system under consideration; (×) symbols indicate locations of the sampling points. Right: quadrant of the computational grid.

2. Problem statement

The problem is stated in terms of the incompressible Navier-Stokes formulation. The Oberbeck-Boussinesq approximation is enforced, all the fluid properties being consistently assumed as constant, apart from density in the buoyancy term.

The governing equations are tackled in their non-dimensional form. The gap between the top of the cylinder and the upper cavity wall is indicated in [5] as the most suitable scale length, $H_{ref} = H$. In fact, it reduces the dependence of the solution ranges on the aspect ratio, in particular, for what concerns the heat transfer rate and the first transitions between different regimes. Moreover, the region above the cylinder is subject to the maximum inverse thermal gradient, *i.e.* to the highest buoyancy force acting on the fluid system, and this is again related to the reference length H.

Temperature is non-dimensionalized according to a reference temperature T_{ref}, *i.e.* the temperature difference between the cylinder and the cavity walls ($T_{ref} = T_S - T_W$), and the following velocity scale is chosen:

$$U_{ref} = \sqrt{g\beta T_{ref} H_{ref}} \tag{1}$$

where g denotes the gravitational acceleration and β is the thermal expansion coefficient of the fluid. The continuity, momentum, and energy equations are given the following form:

$$\nabla \cdot \mathbf{u} = \mathbf{0} \tag{2}$$

$$\frac{\partial \mathbf{u}}{\partial t} + \mathbf{u} \cdot \nabla \mathbf{u} = -\nabla p + \frac{Pr^{1/2}}{Ra^{1/2}} \nabla^2 \mathbf{u} + T\hat{\mathbf{g}} \tag{3}$$

$$\frac{\partial T}{\partial t} + \mathbf{u} \cdot \nabla T = \frac{1}{(RaPr)^{1/2}} \nabla^2 T \tag{4}$$

where t, $\mathbf{u}$, p and T represent the dimensionless time, velocity vector, pressure and temperature, respectively, and $\hat{\mathbf{g}}$ is the gravity unit vector. The Rayleigh and Prandtl numbers are defined as:

$$Ra = \frac{g\beta T_{ref} H_{ref}^3}{\nu\alpha} \quad (5)$$

$$Pr = \nu/\alpha \quad (6)$$

where ν and α represent the momentum and thermal diffusivity, respectively. A constant value $Pr = 0.7$ is assumed for air. With reference to figure 1, the following non-dimensional boundary conditions are imposed:

$$T = 0, \quad \mathbf{u} = \mathbf{0} \quad (7)$$

at the enclosure walls, and:

$$T = 1, \quad \mathbf{u} = \mathbf{0} \quad (8)$$

on the cylinder surface.

3. Numerical methods

The numerical technique adopted is based on a Finite Volume implementation of a second order Projection Method, following [14]. Time-discretizations of the conservation equations are performed according to a three-level scheme, which is fully implicit for the diffusive terms, and explicit Adams-Bashforth for the advective terms. Such a practice is second order accurate in time.

Spatial derivatives are approximated with second order central differences on staggered, non-uniform Cartesian grids. A direct resolution of the discrete momentum and energy equations at each time-step is made possible by means of Approximate Factorization, while the Poisson problem associated with the pressure-velocity coupling [14] is solved through a fast Poisson solver, based on Matrix Decomposition.

The 2D modelling of arbitrarily irregular boundaries on Cartesian grids is achieved thanks to the original scheme developed in [15]. The technique involves a local modification of the 5-point computational stencil where boundary segments intersect the stencil arms. The variables on the modified stencil are mapped on the global grid, by means of a linear operator determined by geometrical features and boundary conditions. The overall accuracy of the method is virtually preserved, as well as the computational efficiency of the Cartesian approach. This represents a significant advantage of the present technique with respect to commercially established, multi-purpose codes. In fact, the possiblity of exploiting fast and direct linear solvers for the discrete equations, even in the presence of non-Cartesian boundaries, entails an extremely low computational cost, which made it possible to perform a large number of long-term prediction of a phenomenon which is typically characterised by very long time scales. Furthermore, a direct solution also implies that no artificial perturbation, associated with the residuals of iterative solutions, is applied to the computed flow regime. Finally, the use of Cartesian grids ensures that second-order accuracy is effectively achieved, and this is not always the case with body-fitted, unstructured grids.

The Cartesian grid employed, made up by 190×190 elements, is shown in figure 1. Along both grid directions, a variable spacing was used, and special care was put on the grid sizing of both near-wall areas and internal domain regions, in view of the work objectives. The criteria adopted for the choice of grid spacing and time step size are outlined in detail in [8], while a grid independency test for a periodic flow is reported in [5].

In order to analyze the system dynamics in the vicinity of bifurcation points, Ra was increased monotonically with suitable steps, each simulation starting from the final frame of the preceeding one. All the simulations were protracted until a fixed dimensionless time span was covered, large enough for an asymptotic flow to be attained [8].

Journal of Physics: Conference Series **655** (2015) 012055 doi:10.1088/1742-6596/655/1/012055

Time histories of dimensionless temperature T, and velocity components along x and y directions, respectively u_x and u_y, were tracked for 5 locations, reported in figure 1, in order to be utilized for phase-space representations of the system dynamics.

4. Results and discussion

Table 1 reports the entire sequence of bifurcations driving the flow in the present system from steady-state to chaos; this was determined by Angeli and Pagano [8] by means of an accurate scanning of the range of the bifurcation parameter (i.e. the Rayleigh number) over which the transition occurred. The bifurcation scenario can be well described by a Feigenbaum cascade, at the beginning of which the first doubling of the fundamental oscillation period Π (P_1 flow) is preceeded by the occurrence of a window of Ra-values for which the flow is quasiperiodic. For such a flow, the trajectories of state-space variables lie on a T_2 torus, and the system dynamics are completely described by two incommensurate frequencies. The peculiarity of the observed behaviour lies in the fact that the period-2 (P_2) flow emerges from the locking of the secondary frequency of the quasiperiodic flow on the exact half of the primary one. A detailed description of the quasiperiodic dynamics has been reported by the authors in [13]. In the following, the overall features of the oscillating thermal and flow fields will be described, and the influence of successive bifurcations on the overall heat transfer will be discussed.

Figure 2 displays a representation of nine snapshots of the evolution of the temperature field in the system over one period Π of oscillation of the P_1 flow, for $Ra = 110000$. The temperature field is represented by means of pseudo-interferograms: dimensionless temperature values is transformed into fringes which represent successive ranges of the variable. The temperature difference between two successive fringes is $\Delta T = 0.1$. The adopted representation is particularly useful since it implies an analogy with experimental visualizations like the ones presented in [16], and allows for a simple but straightforward identification of the main thermal structures.

From the snapshots, it appears evident that the dominating feature is the swaying thermal plume which rises from the heated cylinder. At the beginning of one oscillation period ($t = 0$), the plume stands upright above the cylinder, but it can be noticed how the thermal field is not symmetric with respect to the vertical axis of the enclosure. The plume then starts to bend towards the right-hand side of the cavity, until it gets closest to the lateral enclosure wall at $t \approx Pi/4$. Subsequently, the plume reverts back to its central position for $t \approx Pi/2$: at that time instance, the thermal field is exactly the mirrored image of the field for $t = 0$ with respect to the vertical axis. Then, the plume concludes its oscillation by leaning towards the left-hand

Table 1. Flow regimes and corresponding ranges of Ra throughout the transition to chaos [8].

flow regime		$Ra \times 10^{-5}$
steady-state	S	≤ 0.66200
periodic	P_1	$0.66200 \div 1.73500$
quasiperiodic	T_2	$1.74000 \div 1.79270$
periodic	P_2	$1.79280 \div 1.89750$
periodic	P_4	$1.89800 \div 1.93670$
periodic	P_8	$1.93675 \div 1.94730$
periodic	P_{16}	$1.94735 \div 1.94950$
periodic	P_{32}	$1.94955 \div 1.94990$
periodic	P_{64}	$1.94995 \div 1.95002$
periodic	P_{128}	$1.95002 \div 1.95010$
chaos	N	≥ 1.95025

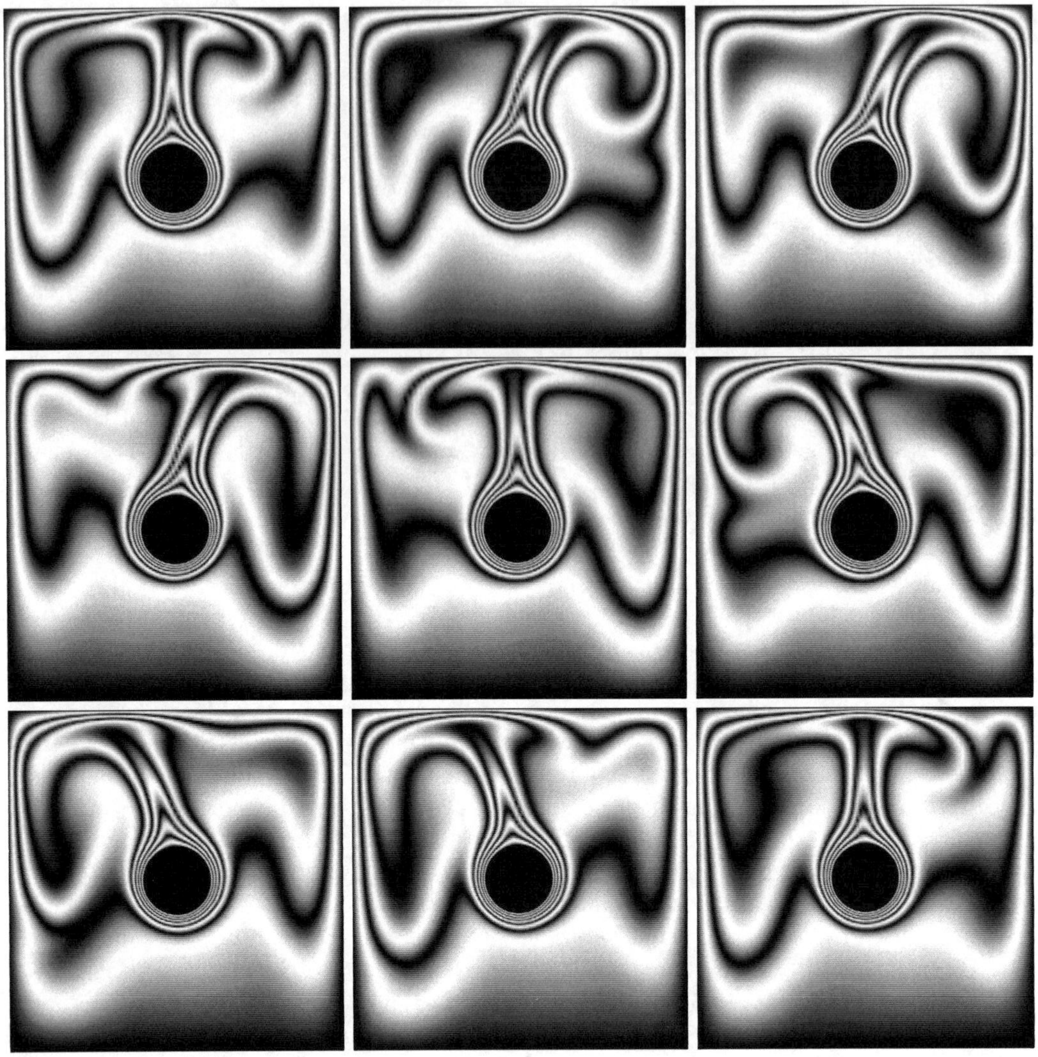

Figure 2. Sequence of pseudo-interferograms portraying the temperature field over one oscillation period Π for P_1 flow at $Ra = 110000$. Each fringe corresponds to a $\Delta T = 0.1$. Time distance between successive snapshots is $\Delta t = \Pi/8$.

cavity wall, and finally returning to the initial position for $t = \Pi$.

Whilst the swaying motion of the plume and its impingement on the enclosure ceiling can be expected to determine large perturbations in the thermal boundary layer of the cavity sidewalls, and a periodic shifting of the peak heat transfer coefficient along the top wall, the boundary layer around the cylinder is observed to be largely uninfluenced by the plume oscillation, with a reduced impact on the time dependence of the Nusselt number along the cylinder itself.

Figure 3 reports a streamline visualization of the flow patterns underlying the temperature fields displayed in Figure 2. The flow field is substantially dominated by the alternate expansion and contraction of two main counter-rotating vortices which alternately sustain the thermal plume. As soon as the plume bends towards the right hand side of the enclosure, the right-hand vortex, which encircles the cylindrical source, contracts, and its center shifts towards the right and the bottom of the cavity. Contemporarily, the left-hand circulation expands and divides in

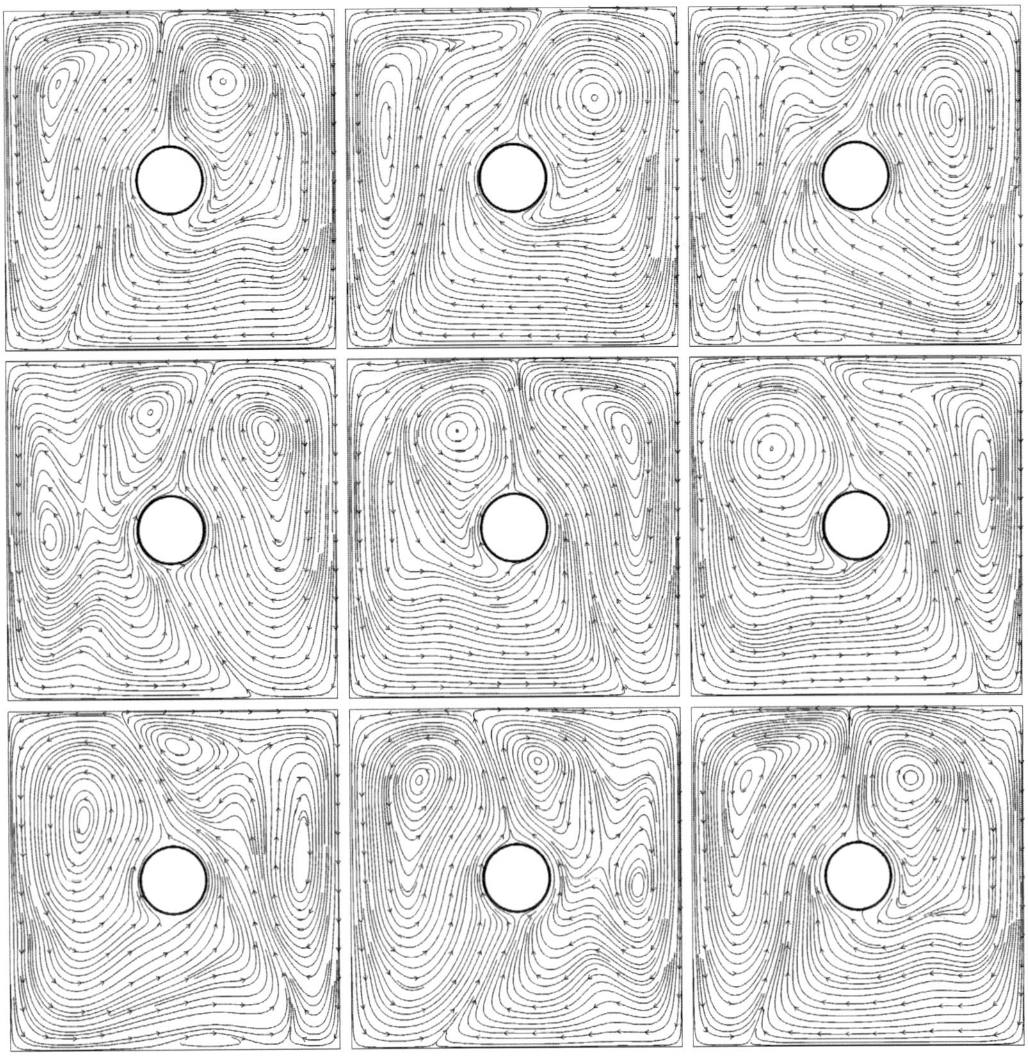

Figure 3. Sequence of streamline visualizations of the flow field over one oscillation period Π for P_1 flow at $Ra = 110000$. Time distance between successive snapshots is $\Delta t = \Pi/8$.

two smaller, distinct cells, which merge back in a bigger circulation when the plume returns in the central position ($t \approx Pi/2$). The same behaviour, albeit with reversed sides, is observed in the second half of the plume cycle.

For the sake of brevity, analogous visualizations for all the flow regimes listed in Table 1 are not included here. Nevertheless, a close inspection of similar sequences of snapshots collected for the T_2 regime revealed that, coherently with the emergence of two independent frequencies f_1 and f_2 in the power spectra of the state variables [13], two distinct "modes" of oscillation could be singled out. These modes alternate almost every other plume oscillation, the ratio f_1/f_2 between the two frequencies being close to 2, and are characterized by the creation and destruction of smaller flow structures aside from the ones described in Figure 3, appearing in slightly but noticeably different manners for each of the two modes. However, such variations in the flow field appeared as hardly distinguishable in the thermal field, whose main pattern never strays significantly from the one depicted in Figure 2. For this reason, the observed evolution of

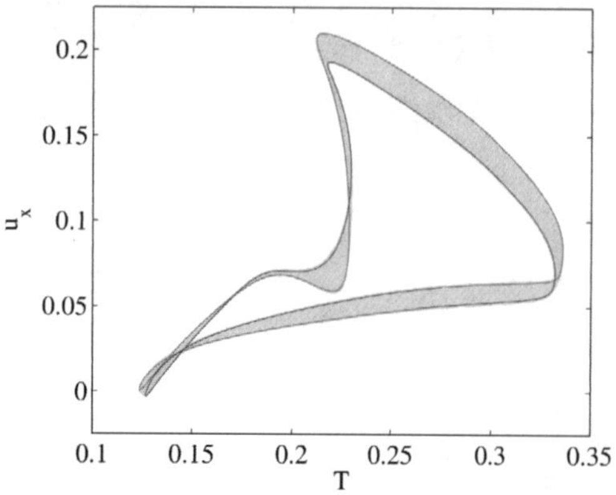

(a)

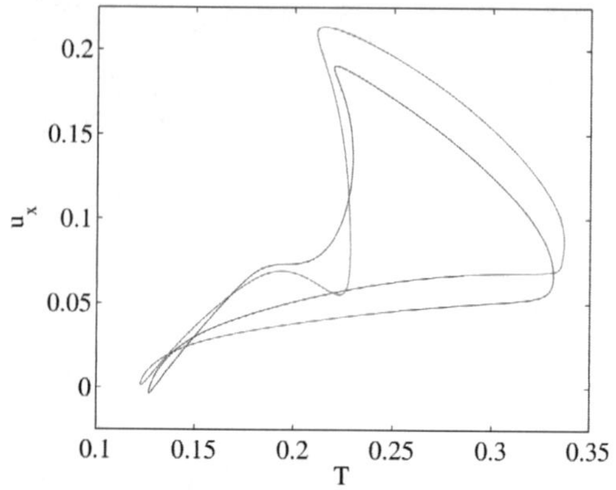

(b)

Figure 4. Attractors at point (0.5,0), projected onto the u_x-T state space for (a) T_2 flow at $Ra = 1.76875 \times 10^5$ and (b) P_2 flow at $Ra = 1.8 \times 10^5$. Trajectories corresponding to the primary (red) and secondary (blue) oscillation modes are highlighted.

the flow patterns and system dynamics are expected to produce marginal effects on the overall heat transfer coefficient. Similar considerations could be inferred for the P_2 regime, when the two frequencies lock at $f_1/f_2 = 2$.

A synthetic but effective visualization of the two basic modes can be drawn in the state space spanned by the variables T and u_x, by plotting the trajectory obtained by the time series sampled at point (0.5,0) (see Figure 4). In the quasiperiodic regime (Figure 4a), the two modes can be highlighted as two successive segments of the trajectory which coils itself onto the T_2 torus. In the P_2 regime (Figure 4b), instead, they form the closed curve that corresponds to the period-2 limit cycle. It can be observed how, for both regimes, the two basic modes are characterised by very similar amplitudes and frequencies, and are determined by similar evolutions of the flow patterns within the enclosure. Moreover, in the subsequent period doublings, the two basic modes remain as the dominant patterns, and are recognizable even when the flow is fully chaotic (up to $Ra = 2 \times 10^5$). For this reason, their alternate occurrence is not expected to produce relevant variations of the overall heat transfer rate throughout the whole transition scenario.

To verify the above hypotheses, the entire database of time- and surface- averaged Nusselt number values on the cylinder has been analyzed in terms of the dependence on Ra. and subdivided into four different macro-groups, based on the observed trend. Such groups are listed in Table 2. For each group, a least-squares fitting has been carried out in order to express the $Nu(Ra)$ dependence via a power-law correlation of the following kind:

$$Nu = C\, Ra^n \tag{9}$$

Table 2. Best-fit coefficients for Equation (9) for different flow types.

flow types		C	n
steady-state flow	S	0.988	0.19
supercritical flow	P_1	1.336	0.16
fully periodic and quasiperiodic flow	$P_1 \to T_2$	0.847	0.20
phase locking and subharmonic cascade	$P_2 \to P_4 \to \ldots \to N$	0.740	0.21
complete transition	$S \to N$	1.021	0.18

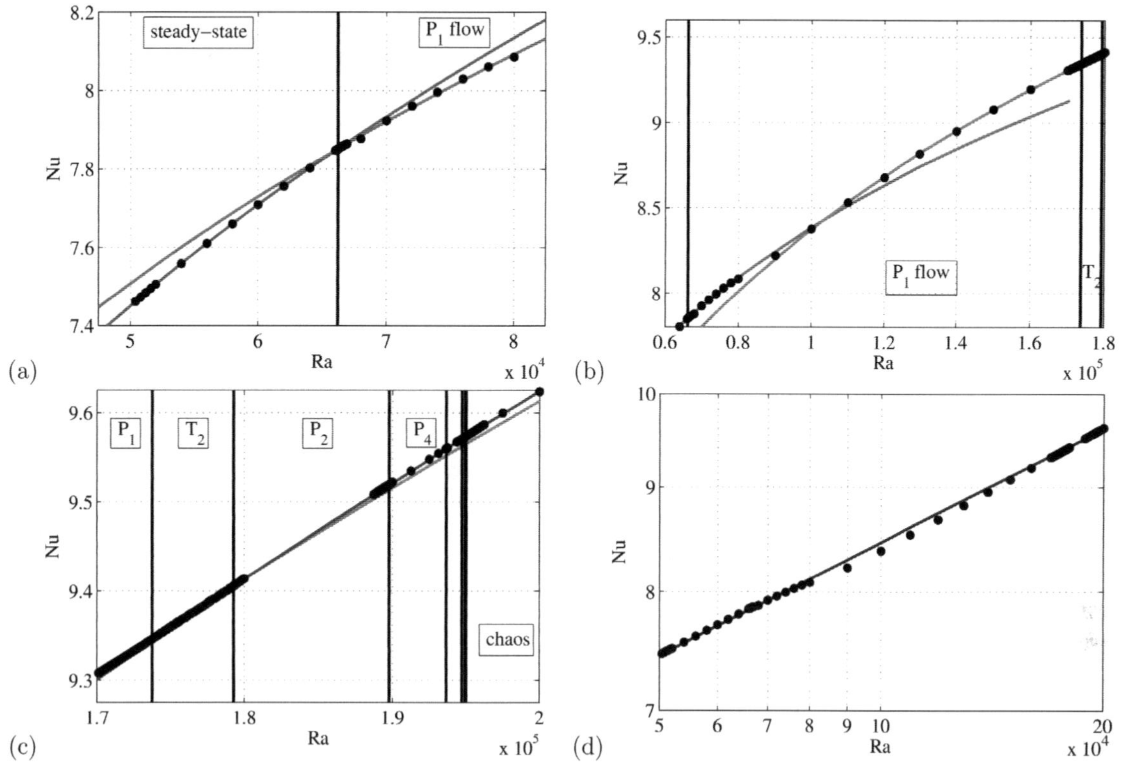

Figure 5. Best-fits (solid lines) and numerical values (dots) of the average Nusselt number on the cylinder for different flow ranges (a)-(c) and for the whole transition range (d).

An additional fitting has been performed for the whole transition range. Such fittings are represented in Figure 5 along with numerical results, in their respective range of interest.

The first variation in the $Nu(Ra)$ dependence can be identified at the transition point between steady-state flow and P_1 flow (Figure 5a), as already suggested in [6]. Quite surprisingly, the second variation occurs in the middle of the P_1 range, around $Ra \approx 10^5$ (Figure 5b): there, the exponent n to be given to Equation (9) $Nu(Ra)$ significantly increases, and remains valid also for the whole T_2 range. Such an unexpected change in behaviour, apparently unlinked with the transitional dynamics, could be due to the full establishment of the regime portrayed in Figures 2 and 3, and to the progressive increase of the oscillation amplitude with Ra with respect to the early stages of the development of the P_1 flow. As long as the T_2 regime disappears, and the period-2 flow emerges, the $Nu(Ra)$ trend changes again, albeit very slightly (Figure 5c). The exponent n increases by a small amount and the observed dependence remains valid up to the chaotic flow regime ($Ra = 2 \times 10^5$).

Figure 5d finally shows the best fit of Equation (9) to the numerical data of the whole transition range: it can be seen how a single exponent is fairly adequate to represent the Nusselt number variation throughout the successive bifurcations. In fact, despite local discrepancies are noticeable (especially in the P_1 range), the maximum deviation of the numerical data from the fit does not exceed 1%, which is indeed a satisfactory accuracy. Hence, it can be concluded that, within the limits of the 2D assumption, the influence of flow transitions on overall heat transfer is not significant.

5. Concluding remarks

Starting from the analysis of an extensive database of numerical simulations already described and discussed in recent works, the influence of successive flow bifurcations on natural convection heat transfer from an enclosed horizontal cylinder was investigated here. In the case under consideration, the scenario leading to chaos consists in a sequence of period-doubling bifurcations, preceeded by a window of quasiperiodic flow.

An analysis of the evolution of the system over one oscillation period revealed that the main feature of the thermal field is the swaying motion of the plume forming above the cylinder, while the flow is characterized by the alternate contraction and expansion of two counter-rotating vortices, and the formation of secondary vortices. As Ra increases, two main oscillation modes emerge in the flow fields and persist throughout the whole transition range.

Despite the peculiarity and complexity of the scenario, flow transitions were observed to have a marginal effect on the overall heat transfer, which could be correlated either with different equations considering different intervals of the leading parameter (Ra), only partially related to the bifurcation pounts, or by a single power-law correlation spanning the entire transition range. The latter matches the numerical data with an accuracy of 1%.

References

[1] Desrayaud G and Lauriat G 1998 *J. Fluid Mech.* **252** 617–646
[2] Moukalled F and Acharya S 1996 *Journal of Thermophys.* **10** 524–531
[3] Shu C and Zhu Y D 2002 *Int. J. Num. Meth. Fluids* **38** 429–445
[4] Kim B S, Lee D S, Ha M Y and Yoon H S 2008 *Int. J. Heat Mass Transf.* **51** 1888–1906
[5] Angeli D, Levoni P and Barozzi G S 2008 *Int. J. Heat Mass Transf.* **51** 553–565
[6] Angeli D, Pagano A, Corticelli M A, Fichera A and Barozzi G S 2011 *Frontiers in Heat and Mass Transfer* **2** 023003
[7] Bouafia M and Daube O 2007 *Int. J. Heat Mass Transf.* **50** 3599–3615
[8] Angeli D and Pagano A 2013 *Int. J. Therm. Sci.* **68** 20–31
[9] Squire H B 1933 *Proceedings of the Royal Society of London. Series A, Containing Papers of a Mathematical and Physical Character* **142** 621–628
[10] Ecke R E, Mainieri R and Sullivan T S 1991 *Physical Review A* **44** 8103–8118
[11] Glazier J A and Libchaber A 1988 IEEE *Transactions on Circuits and Systems* **35** 790–809
[12] Labonia G and Guj G 1998 *J. Fluid Mech.* **375** 179–202
[13] Angeli D, Corticelli M, Fichera A and Pagano A 2014 *Journal of Physics: Conference Series* **501**
[14] Gresho P M 1990 *Int. J. Num. Meth. Fluids* **11** 587–620
[15] Barozzi G S, Bussi C and Corticelli M A 2004 *Numerical Heat Transfer B* **46** 56–77
[16] Cesini G, Paroncini M, Cortella G and Manzan M 1999 *Int. J. Heat Mass Transf.* **42** 1801–1811

Journal of Physics: Conference Series **655** (2015) 012056 doi:10.1088/1742-6596/655/1/012056

Analysis of thermal field within an urban canyon with variable thermophysical characteristics of the building's walls

A De Lieto Vollaro[1], G Galli[1], A Vallati[1,2] and R Romagnoli[1]

[1] Sapienza University of Rome – DIAEE – Via Eudossiana, 18 - 00184 Rome, Italy

Email: andrea.vallati@uniroma1.it

Abstract. In a typical urban configuration, a microclimatic analysis has been carried out. Using a CFD method, a N-S oriented urban street canyon, with a given H/W ratio, has been examined. The standard k–ε turbulence model has been used to simulate a three-dimensional flow field and to calculate the thermo-fluid dynamics parameters that characterize the street canyon. In this study has been analyzed the thermal flow field when the walls of the building change the properties of solar radiation absorption, in particular for α=0.2 and α=0.8. Solar radiation considered is that of 21/07 in Milan in two different hours: at 11:00 a.m. and at 02:00 p.m. The study shows the importance of the thermophysical properties of a wall, in the development of the thermal field and flow field. This is a very important topic, in terms of improvement of well-being and the quality of the air within the cities, through the choice of materials and colors of the facades of buildings.\

1. Introduction

Roughly 25% of the final energy consumption, including all energy delivered to the final consumers (excluding deliveries for transformation and network losses) in the EU is used for residential and 15% for commercial buildings. Heating represents 70% of the residential energy consumption (European Commission Energy, 2009). Therefore there is a great energy saving potential by minimizing the energy demand for space heating and cooling of buildings. Today about 50% of the world population lives in urban areas, increasing to about 70% by 2050.

The urbane climate is strongly influenced by its geometry and surface materials. The urbane temperature increases respect to surrounding areas. The increased heat gains are due to higher absorption of solar radiation on dark surfaces such as asphalt, the reduced energy losses during the night due to buildings that block thermal radiation to the cold sky, the lack of evapotranspiration as well as lower convective heat losses due to wind sheltering by buildings. Waste heat from buildings, industry and transportation further contributes to the urban warming. Besides the higher air temperatures, higher radiative gains, to the building and lower convective heat losses from the building also increase the space cooling.

[2] Author to whom any correspondence should be addressed

This paper studies a particular urban context, an urban canyon, where the characteristics properties of facades are changed trough the absorption coefficients. In effect thermal conditions in street canyons are important topics of urban microclimate, that influences the buildings energy demand and has a large impact on the thermal comfort and health of the people [1]. Surface temperature distribution and air circulation play an important role on heat exchanges between the building and canyon air, that in turn influence pedestrian comfort and the energy demand of buildings [2 -6].

Moreover, the technical characteristics of the used materials determine to a high degree the energy consumption and comfort conditions of individual buildings as well as of open spaces [7, 8]. Many studies have been carried out to understand better the optical and thermal characteristics of materials as well as their impact on the city climate [9-15] and systematic ambient temperature differences above various types of materials have been reported.

The albedo of cities is seriously decreased compared to the surrounding rural areas mainly because of the irregular building structure. Cantat [16] has shown that the albedo in Paris is almost 16% lower than in the surrounding rural areas. Aida and Gotoh [17] have assessed the decrease of the urban albedo because of the geometrical characteristics of urban canyons. It is found that urban configuration with canyon width about twice the width of buildings gives the lowest albedo at all zenith angles. Also, Aida [17] has reported that under clear weather conditions, the absorption of urban structures is increased to about 20% compared to a flat surface of the same material. The use of materials presenting high reflectivity to the solar radiation and high spectral emissivity, cool materials, contribute to increasing the urban albedo and it is considered to be one of the more promising and powerful techniques to mitigate the heat island phenomenon [18].

In this study, using a commercial CFD code (Ansys Fluent), several 3D numerical tests have been performed on an isolated street canyon, to evaluate how the flow field changes within the canyon at different surface properties of facades.

2. CFD numerical model

In this section the CFD numerical model that we have used to perform the simulations, is outlined. Using the commercial CFD code Ansys Fluent 14.0 [19], 3D double precision, pressure based version, the steady RANS equations have been solved in combination with the standard k- ε model.

To evaluate the impact of thermal effects, the natural convection module has been activated by setting incompressible ideal gas model for air density. Based on the best practice guidelines by Franke et al. [21] and Tominaga et al. [23], the computational domain has been extended beyond the built area by 5H in the North, West and South direction, of 15H in the East direction, as can be observed in Figure 1, in the next chapter. The upper boundary condition is located at distance of 5H above the building roofs. These dimensions have been chosen to take into account of the blockage ratio and to ensure the flow re-development behind the building region. The temperature surfaces has been obtained as result of the heat transfer calculations, setting up: the solar load module, the temperature of undisturbed air (303 K), and the internal temperature of buildings (299 K). To simulate the ground influence, the computational domain has been extended 5 m below the ground level. The ground has been simulated setting the following parameters: density = 1000 kg/m^3; specific heat = 1000 J / kg·K; thermal conductivity = 2 W/ m·K; temperature at −5 m = 288 K; emissivity = 0.9; solar radiation absorptivity (direct visible and infrared) = 0.8. The radiation exchanges has been evaluated setting up the S2S radiation model, in which the energy exchange parameters are accounted by a geometric function, i.e., view factor, and activating the solar ray tracing in the solar load model, provided in ANSYS Fluent version 14.0.0. Transient heat conduction in ground and walls has been analyzed in [24]. This study finds that these effects are important to calculate heat fluxes, but not for surface temperatures which influence the natural convection flow fields. Furthermore, the materials characteristics have been reported in [21, 23]: i.e., the building walls have: density = 1000 kg/m^3;

specific heat = 1000 J/ kg·K; thermal conductivity = 0.15 W/ m·K; thickness = 0.30 m; internal air temperature = 299 K; emissivity = 0.9; solar radiation absorptivity (direct visible and near infrared) = 0.8. To ensure a high quality of the computational grid, it is fully structured and the shape of the cells has been chosen hexahedral. The velocity profile has been set giving a uniform velocity magnitude at the velocity inlet boundary, the turbulence intensity at 10% and the roughness length z0 = 0.05 m . As the flow approaches the built area the velocity inlet profile is fully-developed before reaching the buildings [24] and it can be represented by Equation (2.1), where u* is the friction velocity, K is the Von Karman constant (0.4) and z is the height coordinate. This equation represents the wind velocity profile of the inlet flow, when the wind approaches the buildings. In another study [24], it is shown that the profile reaches the asymptotic velocity when it reaches about 4 m in height.

The aerodynamic roughness value z_0 has been set in relation to the roughness parameters in the ground surface boundary conditions (Ramponi and Blocken [25]): the sand-grain roughness height *Ks* and the roughness constant *Cs*. According to the best practices guideline for the CFD simulations by Franke et al. [21] and Tominaga et al. [23], setting Ks = 1 m and Cs = 0.5, we obtain z_0 = 0.05 m (Blocken et al. [26], ANSYS Fluent User's Guide [19]), that has been considered an appropriate value to represent the roughness of the outer region (Blocken et al. [26], Norris and Richards [27]).

$$u(z) = \frac{u^*}{K} \ln\left(\frac{z + z_0}{z_0}\right) \tag{2.1}$$

The friction velocity value has been obtained by the correlation with the calculated value of the turbulent kinetic energy (*k*) at the first node above the ground, as shown in Equation (2.2) [21].

$$u^* = k^{0.5} C_\mu^{0.25} \tag{2.2}$$

Where $C_\mu = 0.09$.

The numerical model used for the computation of the temperature and velocity fields inside the canyon has been validated by reproducing the experimental data reported by Ueara et al. [28], showing a satisfactory degree of agreement, as discussed in Bottillo et al [24].

3. Results

In this study, considering different ambient wind conditions, several simulations have been performed on an isolated urban canyon, in order evaluate the effects on the flow field and heat exchange processes within the canyon. The analyzed urban canyon has the following characteristics: it is placed in Milan, Italy (longitude: 9.18, latitude: 45.47, UTC:+1), it has an aspect ratio H/W = 1 and L/W = 5, the orientation is N–S, the buildings width and height are 20 m, the street width is 20 m and the street length is 100 m. A steady state simulation has been carried out with the ambient temperature and solar radiation at 11.00 a.m. and at 02:00 p.m. of 21 July in Milan, as given by the software meteorological file. As previously mentioned, the computational domain has been extended beyond the built area by 5H in the North, West and South directions, of 15H in the East direction.

To evaluate the relationship between the wind flow field and the temperature field within the canyon, varying the value of the absorption coefficients, we compare the values of heat exchange coefficient h_c, turbulent kinetic energy k, wall temperatures T_{wall} and the intensity of the wind speed u. All these variables are calculate on vertical lines on three different plans, as can be observed in Figure 1, with XZ coordinates: the north plane is placed ten meters from the north limit of the building, the central plane is situated at the middle of the canyon and the south plane is ten meters from the south limit. Specifically the turbulent kinetic energy and wind speed are calculated on lines 0.4 m distant from the façades, while the wall temperatures and the heat exchange coefficients are evaluated on the building façades themselves.

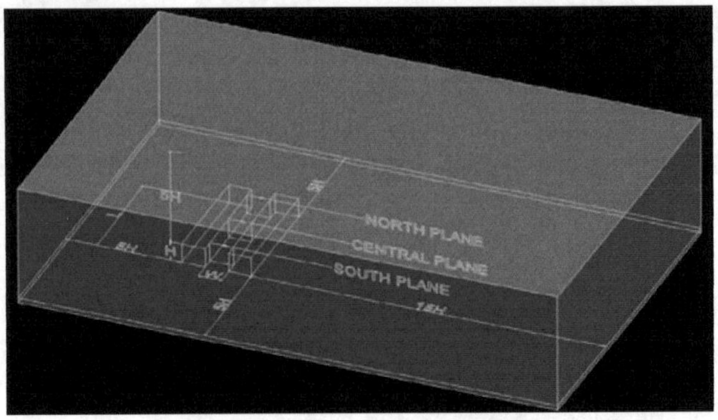

Figure 1. Computational domain.

In previous works [24, 29, 30] we have analyzed the importance of considering a 3D model on the thermal effects and on the formation of the vortices within the canyon. It can be noted in Figure 2, specifically in the central plane, the formation of a rotating vortex, which have effects on the values of the parameters analyzed in these studies and also on the thermal comfort within the canyon. The formation of the vortex within the canyon implies moreover negative effects on the air quality, due to the stagnation of air pollutants in the lower parts of the street canyon. At last, the increase in temperature has a negative effect, in summer, on the thermal comfort for people living inside the building, and on energetic consumption of air conditioning systems.

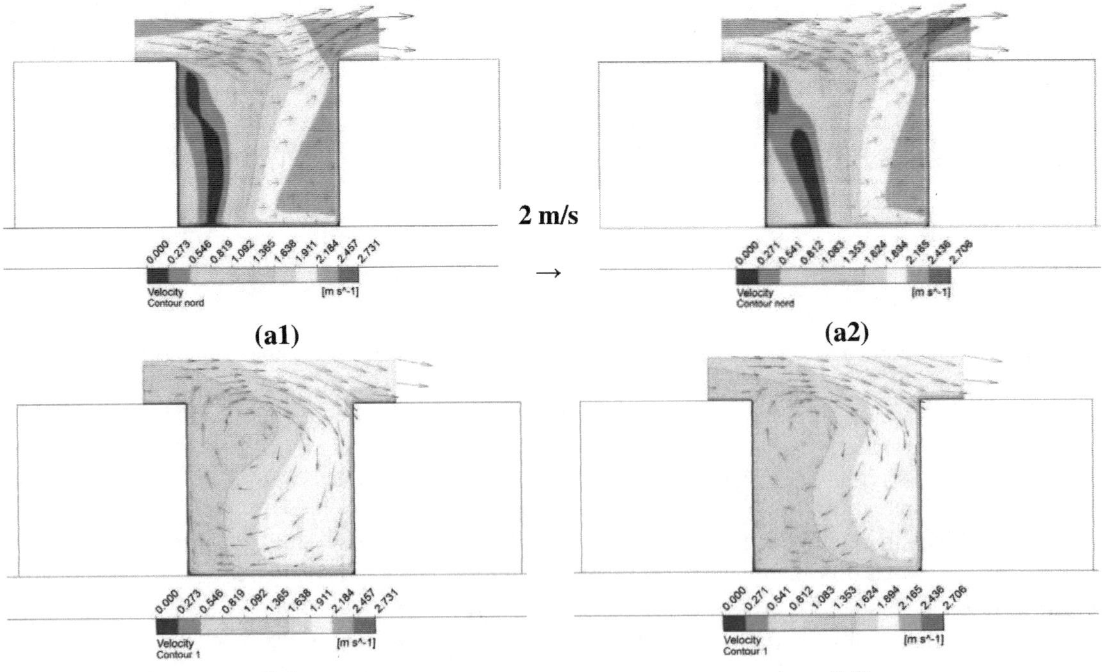

(c1) **(c2)**

Figure 2. XZ velocity vectors on North plane (a); Central plane (b); South plane(c); at 11:00 (subscript 1) and 14:00 (subscript 2).

Considering the various aspects outlined by the previous works, in this study we consider the wind direction of 45° NW and the velocity magnitude of the wind is 2 m/s. Thus the windward façade is always the which ones towards the east direction, i.e. the right façade (WW) in Figure 3. As reported in [24, 26, 29] the values of the convective heat transfer coefficient are not very sensitive to wind direction, so that a wind direction of 45° N can represent a wide range of wind directions between 0° and 60°. This conclusion allow us to perform a parametric analysis on the average values of convective heat transfer coefficient, based on results of 3D CFD simulations characterized by an average wind direction. In Figure 3 it can be observed the configuration of the analyzed urban canyon at 11:00 a.m. and at 02:00 p.m., in particular the direction and the magnitude of the wind and the geometrical parameters H and W.

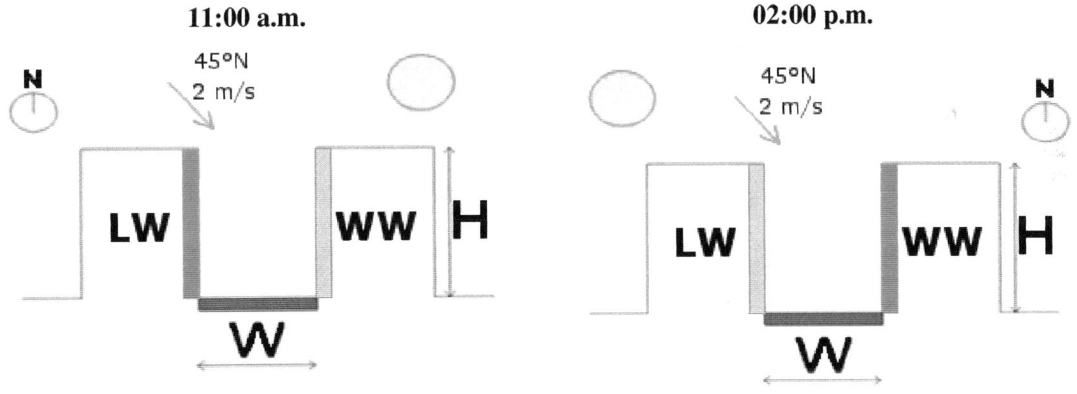

Figure 3. Configuration of the analyzed urban canyon at 11.00 a.m. and at 02:00 p.m.

In this study the parameters, which have been previously described, have been calculated varying the absorption coefficients of the leeward façade (LW) and the windward façade (WW), while the ground absorption coefficients remain always fixed, in different ways: in SIM A have been considered α of both façades equal to 0.8; instead in SIM B $\alpha = 0.2$ for both façades; in SIM C the α of the leeward façade is 0.2, while for windward façade and for the ground are 0.8; at last, in SIM D, has been taken in account $\alpha = 0.2$ for the windward façade and $\alpha = 0.8$ for the leeward façade and for the ground (Table 1).

	α Ground	α LW Façade	α WW Façade
SIM A	0.8	0.8	0.8
SIM B	0.8	0.2	0.2
SIM C	0.8	0.2	0.8
SIM D	0.8	0.8	0.2

Table 1. Presentation of the simulations carried out in this study, varying the absorption coefficients α.

In the next tables, i.e. Table 2, Table 3, Table 4, Table 5, the values of h_c, k, T_{wall}, u, for each simulation are presented. For each simulation carried out, have been considered these parameters on the North, Central and South planes, which have been calculated on the windward façade and on the leeward façade, at 11:00 a.m. and at 02:00 p.m.

SIM A_α_{WW} = 0.8_α_{LW} = 0.8				
WINDWARD FAÇADE at 11:00 a.m.				
	h_c [W/m^{2}K]	**k [m^2/s^2]**	**T_{wall} [K]**	**u [m/s]**
North	13.79	0.33	308.2	1.82
Central	10.03	0.16	309.32	1.61
South	9.02	0.12	310.19	1.39
WINDWARD FAÇADE at 2:00 p.m.				
North	13.72	0.33	321.19	1.81
Central	10.56	0.19	324.67	1.52
South	9.57	0.14	326.64	1.41
LEEWARD FAÇADE at 11:00 a.m.				
North	10.42	0.2	325.04	0.59
Central	10.80	0.21	324.55	1.1
South	9.71	0.15	326.5	1.5
LEEWARD FAÇADE at 2:00 p.m.				
North	8.76	0.13	309.37	0.52
Central	9.44	0.16	310.00	0.88
South	8.53	0.12	310.87	1.29

Table 2. Values of the convective heat transfer coefficient (h_c), turbo-kinetic energy (k), wall temperature (T_{wall}) and velocity magnitude (u) calculated on the windward and leeward façades at 11:00 a.m. and at 02:00 p.m., relatively to the SIM A.

It can be noticed, as expected, the similar values of T_{wall} when the façades are in shadow, else if the absorption coefficients are different. In fact both the windward façade at 11:00 a.m. and the leeward façade at 02:00 p.m., have been calculated a wall temperature comprised between 307 and 311 K, as it's possible to observe in the tables below. Upon the facades in shadow a temperature decrease of 2-3 K has been observed when the absorption coefficient was reduced from 0.8 to 0.2.

As reported in tables 2-5, when the absorption coefficient in the WW façade changes from 0.8 to 02 an important decrease of the T_{wall} temperature can be noticed (11K). The same consideration can be noted about T_{wall} calculated on the leeward façade, when it's exposed to sun radiation. This aspect has important consequences on the street canyon microclimate and, on other hand, also on the thermal fluxes which enter in the buildings.

A further consideration refers to the absence of a mutual thermal interaction between the two façades, when the absorption coefficients are changed.

SIM B_α_{WW} = 0.2_α_{LW} = 0.2				
WINDWARD FAÇADE at 11:00 a.m.				
	h_c [W/m^{2}K]	**k [m^2/s^2]**	**T_{wall} [K]**	**u [m/s]**
North	13.77	0.33	307.08	1.81
Central	9.89	0.15	307.51	1.57
South	8.89	0.12	308.53	1.39
WINDWARD FAÇADE at 2:00 p.m.				
North	13.76	0.33	311.44	1.81
Central	10.14	0.17	313.06	1.53
South	9.23	0.12	314.50	1.40
LEEWARD FAÇADE at 11:00 a.m.				
North	9.83	0.17	313.72	0.54
Central	10.30	0.19	313.35	0.99
South	9.30	0.14	314.82	1.40
LEEWARD FAÇADE at 2:00 p.m.				
North	8.68	0.13	307.90	0.51
Central	9.37	0.15	307.90	0.87
South	8.52	0.12	309.06	1.30

Table 3. Values of the convective heat transfer coefficient (h_c), turbo-kinetic energy (k), wall temperature (T_{wall}) and velocity magnitude (u) calculated on the windward and leeward façades at 11:00 a.m. and at 02:00 p.m., relatively to the SIM B.

Considering SIM A and SIM C at 11:00 (in Table 2 and in Table 4), i.e. when the leeward façade it's exposed to sun radiation, the windward façade, which is in shadow and with α = 0.8 fixed, does not change significantly his wall temperature, when the absorption coefficient on leeward façade is varied from 0.8 (SIM A) to 0.2 (SIM B). In contrast with expectations, the values of T_{wall} calculated on the windward façade does not increase when on leeward façade α = 0.2, but remain quite constants between 307 and 310 K.

SIM C_α_{WW} = 0.8_α_{LW} = 0.2				
WINDWARD FAÇADE at 11:00 a.m.				
	h_c [W/m^{2}K]	**k [m^2/s^2]**	**T_{wall} [K]**	**u [m/s]**
North	13.77	0.33	307.08	1.81
Central	9.89	0.15	307.51	1.57
South	8.89	0.12	308.53	1.39
WINDWARD FAÇADE at 2:00 p.m.				
North	13.73	0.33	321.14	1.81
Central	10.56	0.19	324.64	1.52
South	9.58	0.14	326.63	1.41
LEEWARD FAÇADE at 11:00 a.m.				
North	9.83	0.17	313.72	0.54
Central	10.31	0.19	313.34	0.99
South	9.31	0.14	314.82	1.40
LEEWARD FAÇADE at 2:00 p.m.				
North	8.76	0.13	307.82	0.52
Central	9.44	0.16	307.95	0.88
South	8.53	0.12	309.34	1.29

Table 4. Values of the convective heat transfer coefficient (h_c), turbo-kinetic energy (k), wall temperature (T_{wall}) and velocity magnitude (u) calculated on the windward and leeward façades at 11:00 a.m. and at 02:00 p.m., relatively to the SIM C.

The same consideration can be made between SIM A and SIM D at 02:00 p.m., i.e. when the windward façade it's exposed to sun radiation, and the T_{wall} of the leeward façade are included between about 308 and 311 K.

SIM D_α_{WW} = 0.2_α_{LW} = 0.8				
WINDWARD FAÇADE at 11:00 a.m.				
	h_c [W/m^{2}K]	**k [m^2/s^2]**	**T_{wall} [K]**	**u [m/s]**
North	13.81	0.33	308.15	1.82
Central	10.04	0.16	309.35	1.61
South	9.02	0.12	310.23	1.38
WINDWARD FAÇADE at 2:00 p.m.				
North	13.75	0.33	311.41	1.81
Central	10.14	0.17	313.06	1.53
South	9.23	0.12	314.5	1.40
LEEWARD FAÇADE at 11:00 a.m.				
North	10.42	0.20	325.06	0.59
Central	10.81	0.21	324.53	1.11
South	9.71	0.15	326.50	1.50
LEEWARD FAÇADE at 2:00 p.m.				
North	8.68	0.13	308.32	0.51
Central	9.37	0.15	309.86	0.87
South	8.52	0.12	311.33	1.30

Table 5. Values of the convective heat transfer coefficient (h_c), turbo-kinetic energy (k), wall temperature (T_{wall}) and velocity magnitude (u) calculated on the windward and leeward façades at 11:00 a.m. and at 02:00 p.m., relatively to the SIM D.

4. Discussion

As expected the variations of absorption coefficient from 0.2 to 0.8 have a strong effect on the wall temperatures. This aspect has important consequences on the street canyon microclimate and, on other hand, also on the thermal fluxes which enter in the buildings.

As far as microclimate is concerned, changes in the wall temperature imply changes in the mean radiant temperature and thus in thermal comfort. As regards the thermal loads, diminutions of the absorption coefficient lead to reductions in the solar load. In particular at steady state the reduction of the absorption coefficient from 0.8 to 0.2 brings to a 11 K decrease of the wall temperature and then a decrease of the entering thermal flux.

In fact the thermal fluxes through the wall, in steady case, depend proportionally on the difference of temperature between the inner and the outer surfaces of the façade considered. Thus an increase of about 11 K between the different cases, when α = 0.2 and α = 0.8, produces as consequence an important increase of the thermal fluxes across the wall, with the worsening of the summer cooling load that implants must to counteract and a consequent higher energy consumptions.

According to results obtained, changes in the physical properties of one of the walls do not significantly affect the temperatures of the opposite wall. Moreover, changes in the absorption coefficient have a slight, but not negligible, effect on the CHTC coefficient. For example when the absorption coefficient changes from 0.2 to 0.8, the wall temperature increase of 11K, so that the natural convection influence of the flow field is more important and the CHTC coefficient can increases of about 5%.

5. Conclusions

In this study a numerical simulation method has been used to investigate the values of the convective heat transfer coefficient h_c, the turbo-kinetic energy k, the wall temperature T_{wall} and the velocity magnitude u, within an urban canyon with the geometrical ratio H/W = 1, varying the absorption coefficient α on the windward and leeward façades. These values have been calculated at two different hours: at 11:00 a.m. and at 02:00 p.m. of 21 July, in Milan (Italy). Four simulations have been presented, in which the ground absorption coefficient is fixed and equal to 0.8, instead α_{WW} and α_{LW} varying between 0.8 and 0.2, as it can be observed in Table 1.

Moreover it has been shown that the façades in shadow have similar values of T_{wall}, even if the absorption coefficients are changed. When the façades are sun-heated, i.e. at 11:00 a.m. for the leeward façade and at 02:00 p.m. for the windward façade, it can be noted an important difference between the value of T_{wall} when the absorption coefficient is equal to 0.8 and when it is equal to 0.2 and that causes important different in thermal loads and on the values of the mean radiant temperature.

Regarding to the other parameters considered in this study, it can be observed that the average velocity within the canyon isn't essentially modified from the initial configuration, which is represented from SIM A, but slight but not negligible variation on the CHTC coefficient are produced by the absorption coefficients changes.

References

[1] Moonen P, Defraeye T, Dorer V, Blocken B, Carmeliet J 2012 Urban physics effect of the micro-climate on comfort, health and energy demand. *Front. Archit. Res.* **1**, pp. 197–228.

[2] de Lieto Vollaro A, de Simone G, Romagnoli R, Vallati A, Botillo S 2014 Numerical study of urban canyon microclimate related to geometrical parameters. *Sustainability 2014*, **6**, pp. 7894-7905.

[3] de Lieto Vollaro R, Vallati A, Bottillo S 2013 Differents methods to estimate the mean radiant temperature in an urban canyon. *Advanced Materials Research* **650**, pp. 647-651.

[4] de Lieto Vollaro R, Calvesi M, Battista G, Evangelisti L, Botta F 2014 Calculation model for optimization design of the low impact energy systems for buildings, *Energy Procedia*, **48**, pp. 1459-1467.

[5] Evangelisti L, G. Battista, Guattari C, Basilicata C, de Lieto Vollaro R 2014 Analysis of Two Models for Evaluating Energy Performance of Different Buildings", *Sustainability 2014*, **6**, pp. 5311-5321.

[6] Battista G, Evangelisti L, Guattari C, Basilicata C, de Lieto Vollaro R 2014 Buildings Energy Efficiency: Interventions Analysis under a Smart Cities Approach, *Sustainability* 2014 , **6**, pp. 4694-4705.

[7] Gori P, Bisegna F 2010 Thermophysical parameter estimation of multi-layer walls with stochastic optimization methods, *International Journal of Heat & Technology*, n 1.

[8] Mattoni B, Gugliermetti F, Bisegna F 2015 A multilevel method to assess and design the renovation and integration of Smart Cities, *Sustainable Cities and Society,* **15**, pp. 105-119.

[9] Taha H, Sailor D, Akbari H 1992 High Albedo Materials for Reducing Cooling Energy Use, *Lawrence Berkley Laboratory Report 31721*, UC-350, Berkley CA.

[10] Asaeda T, Ca VT, Wake A 1996 Heat storage of pavement and its effect on the lower atmosphere. *Atmospheric Environment* **30**, **3**, 413– 427.

[11] Doulos L, Santamouris M, Livada I 2001 Passive cooling of outdoor urban spaces. The role of materials. *Solar Energy,* **77**, pp. 231–249.

[12] Niachou A, Livada I, Santamouris M 2008 Experimental study of temperature and airflow distribution inside an urban street canyon during hot summer weather conditions. *Journal of Buildings and Environment* **43** (8), pp. 1383–1392.

[13] Santamouris M, Papanikolaou N, Koronakis I, Livada I, Asimakopoulos D 1999 Thermal and air flow characteristics in a deep pedestrian canyon under hot weather conditions. *Atmospheric Environment* **33**, pp. 4503–4521.

[14] Chen C, Poon C 2009 Photocatalytic construction and building materials: from fundamentals to applications. *Buildings and Environment* **44**, pp. 1899–1906.

[15] White P, Golden JS, Biligiri KP, Kaloush KE 2010 Modeling climate change impacts of pavement production and construction. *Resources, Conservation and Recycling* **54** (11), pp. 776–782.

[16] Cantat O 1989 Contribution a l'etude des variations du bilan d'energie en region parisienne. *PhD. Thesis*, University of Paris Sorbonne.

[17] Aida M 1982 Urban albedo as a function of the urban structure – a model experiment. *Boundary Layer Meteorology* **23**, pp. 405–413.

[18] Akbari H, Menon S, Rosenfeld A 2009 Global cooling: increasing world-wide urban albedos to offset CO2. *Climatic Change* 95.

[19] Ansys Fluent version 14.0.0, 2011. *User's Guide.*

[20] Assimakopoulos VD, Georgakis C, Santamouris M 2006 Experimental validation of a computational fluid dynamics code to predict the wind speed in street canyons for passive cooling purposes. *Solar Energy* **80**, pp. 423-434.

[21] Franke J, Hellsten A, Schlünzen H, Carissimo B 2007 Best practice guideline for the CFD simulation of flows in the urban environment. *COST Action 732.*

[22] Xie X, Liu CH, Leung DYC 2007. Impact of building façades and ground heating on wind flow and pollutant transport in street canyons. *Atmospheric Environment* **41**, pp. 9030-9049.

[23] Tominaga Y, Mochidab A, Yoshiec R, Kataokad H, Nozue T, Yoshikawaf M, Shirasawac T 2008 AIJ guidelines for practical applications of CFD to pedestrian wind environment around buildings. *Journal of Wind Engineering and Industrial Aerodynamics* **96**, pp. 1749-1761.

[24] Bottillo S, de Lieto Vollaro A, Galli G, Vallati A 2013 A. Fluid dynamic and heat transfer parameters in an urban canyon. *Sol. Energy* **99**, pp. 1–10.

[25] Ramponi R, Blocken B 2012 CFD simulation of cross-ventilation for a generic isolated building: impact of computational parameters. *Buildings and Environment* **53**, pp. 34-48.

[26] Blocken B, Stathopoulos T, Carmeliet J 2007 CFD simulation of the atmospheric boundary layer: wall function problems. *Atmospheric Environment* **41**, pp. 238-252.

[27] Richards PJ, Norris SE, 2010 Appropriate boundary conditions for computational wind engineering models revisited. *Journal of Wind Engineering and Industrial Aerodynamics* **99**, pp. 257-266.

[28] Uehara K, Murakami S, Oikawa S, Wakamatsu S 2000 Wind tunnel experiments on how thermal stratification affects how in and above urban street canyons. *Atmospheric Environment* **34**, pp. 1553–1562

[29] Bottillo S, de Lieto Vollaro A, Galli G, Vallati A 2014 CFD modeling of the impact of solar radiation in a tridimensional urban canyon at different wind conditions. *Solar Energy* **102**, pp. 212–222.

[30] Galli G, Vallati A, Recchiuti C, de Lieto Vollaro R, Botta F 2013 Passive cooling design options to improve thermal comfort in an Urban District of Rome, under hot summer conditions. *Int. J. Eng. Technol* **5** , pp. 4495-4500

Characterization of the thermal conductivity for ceramic pebble beds

R Lo Frano[1], **D Aquaro**[1], **L Scaletti**[1] **and N Olivi**[1]
[1] University of Pisa – DICI (Dipartimento di Ingegneria Civile e Industriale)
Largo L. Lazzarino 2, 56122 - Pisa , Italy

E-mail: rosa.lofrano@ing.unipi.it

Abstract. The evaluation of the thermal conductivity of breeder materials is one of the main goals to find the best candidate material for the fusion reactor technology.
The aim of this paper is to evaluate experimentally the thermal conductivity of a ceramic material by applying the hot wire method at different temperatures, ranging from 50 to about 800°C.
The updated experimental facility, available at the Department of Civil and Industrial Engineering (DICI) of the University of Pisa, used to determine the thermal conductivity of a ceramic material (alumina), will be described along with the measurement acquisition system. Moreover it will be also provided an overview of the current state of art of the ceramic pebble bed breeder thermos-mechanics R&D (e.g. Lithium Orthosilicate (Li_4SiO_4) and Lithium Metatitanate (Li_2TiO_3)) focusing on the up-to-date analysis.
The methodological approach adopted is articulated in two phase: the first one aimed at the experimental evaluation of thermal conductivity of a ceramic material by means of hot wire method, to be subsequently used in the second phase that is based on the test rig method, through which is measured the thermal conductivity of pebble bed material. In this framework, the experimental procedure and the measured results obtained varying the temperature, are presented and discussed.

1. Introduction

An open issue for fusion power reactor is the possibility to obtain Tritium in the breeding blanket. The envisaged solution, as known in the scientific literature [1-2-3], foresees the use of pebble beds made of ceramic materials such as lithium orthosilicate (Li_4SiO_4) or lithium metatitanate (Li_2TiO_3), etc. which could also allow to slow down neutrons and extract thermal power.

The pebbles will be loaded into a box-like structure, therefore they must be able to support severe conditions foreseen for the breeding blanket (cyclic mechanical stresses, high temperature and high thermal gradients, irradiation effects) and must exhibit long-term thermo-mechanical stability, and a sufficient compatibility to the adjacent structural materials and low activation levels.

A lot of research effort [1-10] has been spent in developing a thorough understanding and characterization of the pebbles bed, particularly from a thermo-mechanical point of view. Specifically the knowledge of the effective thermal conductivity of pebbles is essential for a proper thermo-mechanical blanket design [6] and assessment of the heat transfer processes.

The research program, developed at the Department of Civil and Industrial Engineering (DICI) of the University of Pisa, aims at characterizing the effective thermal conductivity (λ) of pebbles bed at different temperatures and for compression forces up to 100 kN [8-13].

To the purpose, the methodological approach used, and described extensively in [2], is based on the evaluation of the effective thermal conductivity of pebbles by means of the hot plate with guardian ring method. The effective thermal conductivity (λ) of pebbles bed is therefore calculated in steady state condition by measuring the thermal gradient through an alumina disc (ceramic material) and the pebbles bed [11-12].
Moreover to attain that purpose, the thermal conductivity of the alumina have been experimentally determined by executing hot wire tests on an alumina cylinder for temperatures ranging from 20° to 800°C.

In this paper the experimental activity described refers mainly to the second series of experiments carried out at the Laboratory Scalbatraio of DICI of the University of Pisa.
The results obtained will be presented and discussed in what follows.

2. Description of the experimental facility

The experimental device adopted to execute the hot wire tests consists of (Figure 1):

- Alumina cylinder;
- Oven;
- Thermocouples and electrical resistance;
- Fan-inverter and
- Data Acquisition System (DAS).

The alumina cylinder is made of Alubit 90 (90 wt% Al_2O_3), which has the same composition and was produced with the same technological process of the alumina disk used in the measuring cell of the hot ring method device. It is 280 mm long (L) and has 35 mm radius.
The alumina cylinder has three holes: the first is located in the centre of the cylinder, the second and the third are located at different radial distances from the symmetry axis. Moreover, the central hole allows housing the electrical resistance, while in the other two holes, at 7.75 mm (r_{int}) and 21.2 mm (r_{ext}) from the centre, type K thermocouples are accommodated.
These thermocouples allow to measure the temperature (T_{int} and T_{ext}) inside the cylinder, while the thermocouple (T_{sur}) welded on a thin copper plate, which is placed on the external surface of the cylinder as shown in Figure 1 a, allows to measure the superficial temperature of the specimen itself.
Moreover, a metallic support structure sustains the alumina cylinder and avoids the contact between the specimen and the oven surface.

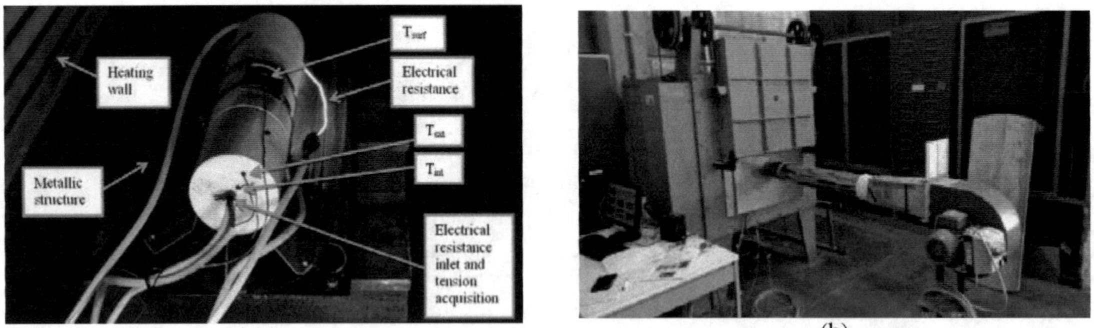

(a) (b)

Figure 1. a) Alumina sample equipped with thermocouples and resistors, and b) the complete experimental apparatus with Fan-Inverter system and DAS.

The oven has all surfaces radiating, during the heating, except the tailgate. It is equipped with a Gefran© GF_Promer, a graphic set-point programmer that provides an accurate configuration of the temperature through a PID controller.

An advanced tuning control allows to check the best PID parameters. The initial conditions of the oven are set up in order to obtain a heating rate of 1°C/1.5 min (assumed increment of oven temperature into the PID) until the desired test temperature was reached.
During tests, the following parameters have been recorded:

- The potential difference across the resistance;
- The voltage drop across the shunt and, based on that, the resulting current;
- The power;
- The alumina temperature through the thermocouples;
- Oven and air temperatures.

Wires and thermocouples are connected to the National Instruments Acquisition System (SCXI1000) that provides directly output data on a pc through LabView software. Additionally, as indicated in [12], to properly determine the thermal conductivity of the alumina specimen, the power generated by the electrical resistance has been removed by forced convection.
The fan (previous Figure 1 b) used to that purpose, has been duly supported to guarantee the parallelism of the tube with the ground; moreover it has been connected to an inverter in order to obtain a more precise regulation. This tube by connecting the fan to the oven conveys the air directly on the outer surface of the alumina.

It is important to remark that the forced convection (operation of the cooling fan-inverter system) has been used for temperatures below 450°C; for higher temperatures (up to 800°C) the radiant heat transfer is considered to be sufficient to remove the power generated by the electrical resistance and to reach accordingly a steady state condition.

2.1 Description of the tests executed

To carry out experimental tests, an already proven procedure has been used [12]. It may be specifically summarised in the following main steps:

1. The alumina specimen is firstly put on the metallic support onto the oven surface;
2. The temperature is set at the desired test value;
3. The oven is turned on and heated up to the set temperature;
4. If the test temperature is below 450°C, the cooling fan is turned on to allow to homogenize the temperature inside the oven chamber;
5. When temperature reaches the set value and stabilises around it (see the yellow behaviours in correspondence of TEMP_EXT_CH1 and TEMP_INT_CH2 indicated in Figure 2), the steady state condition is reached;
6. Data acquisition begins from this instant and lasts for at least 300 s.

Based on the data measured, the thermal conductivity is determined accordingly to the following formula:

$$\lambda_A = \frac{\phi_0}{2 \cdot \pi \cdot \Delta T \cdot L} \ln\left(\frac{r_{ext}}{r_{int}}\right) \qquad (1)$$

where Φ_o is the power of the electric resistance, ΔT is the temperature difference between T_{int} and T_{ext} measured at the two radii r_{int} and r_{ext}, L is the length of the electric resistance.

The thermal conductivity resulted so dependent on Φ_o and ΔT being fixed the two radii and the resistance length L (the geometrical parameters may be considered constant).

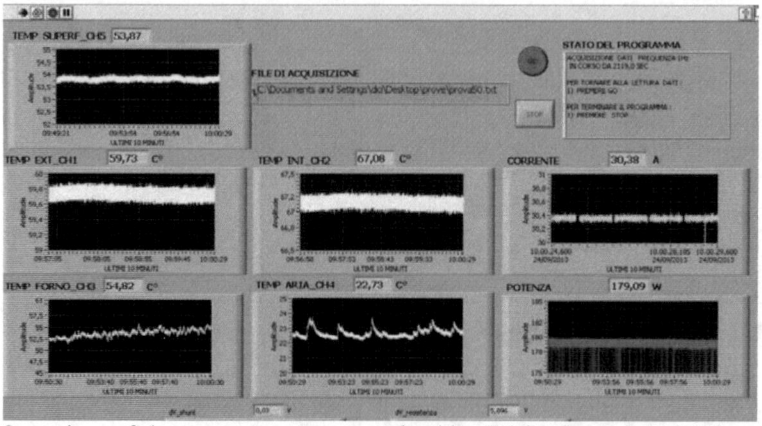

Figure 2. Overview of the test parameters acquired by DAS (through Lab View interface).

2.2 Discussion

Several tests have been performed into the range from 50°C to 800°C, with increment of 50°C. In the following Figure 3 and Figure 4 the plots of the temperatures vs. time obtained carrying out wire tests respectively at 200°C and 500°C, are showed. Moreover in Figure 4 is indicated the instant at which the oven (indicated as electrical resistance) and the fan are turned on.

As shown the steady state condition is achieved by means of the fan regulation (by inducing a forced convection) in the first case, while in the second one, as previously mentioned, the radiant heat transfer is sufficient to remove the amount of heat generated in the alumina centre. In this latter case (like in Figure 4) the steady state condition is reached with a longer time interval, due to the energy balance between the radiant heat transfer and the heat of the resistance.

Specifically, the results show that at an average temperature of 200°C (Figure 3), with an average power of 162 W and a differential temperature (between the internal and external thermocouple) of 7.3°C, the thermal conductivity of alumina is 12.67 W/m°C. Furthermore, at an average temperature of 523°C, with an average power of 181 W and with a differential temperature of 13°C, the thermal conductivity of alumina is 7.96 W/m°C (Figure 4).

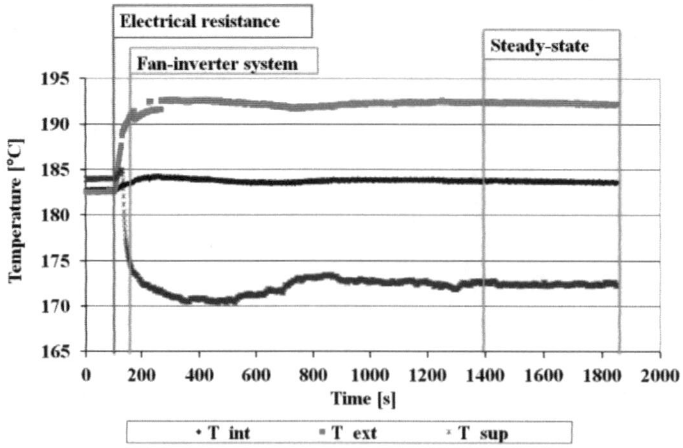

Figure 3. Hot wire test results at 200°C.

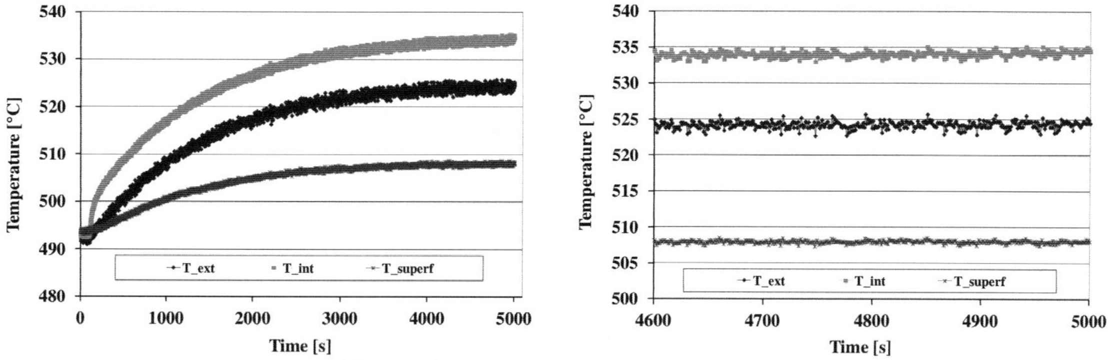

Figure 4. Hot wire test results at 500°C.

The wire tests were repeated several times; for each test campaign the calibration of thermocouples was also repeated: the mean error related to the measure resulted about 4 %.

The trend of the alumina thermal conductivity versus temperature, as experimentally calculated, is shown in Figure 5.

Analysing this behaviour, it is possible to observe that the conductivity decreases along with the increase of the test temperature. The maximum standard deviation is 5 %, as clearly recognizable from the bars of the error depicted in Figure 5: the experimental data have been also interpolated with a fifth-order polynomial curve (indicated with the dark line in Figure 5).

Analysing the results fit very well the interpolation curve even if some spreading appears at low temperature. Moreover analysing the λ behaviour, it is possible to observe that the thermal conductivity decreases along with the increase of the test temperature.

This outcome, in line with the expected results, confirms the trend observed in the first experimental campaign [12] carried out at the DICI of the University of Pisa. The thermal conductivity decreases when temperature increases, and this result agrees with the data available in literature.

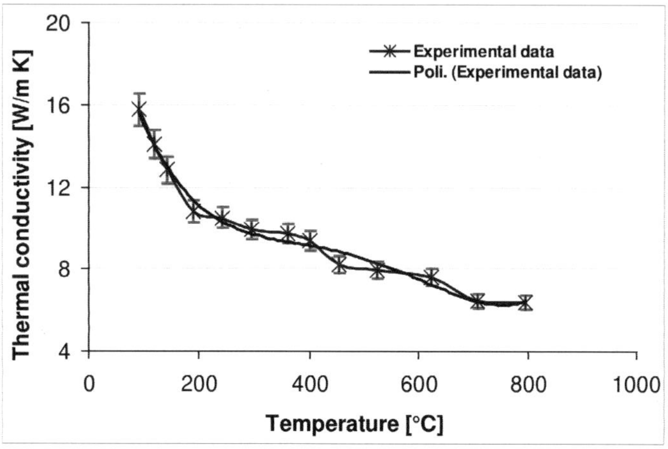

Figure 5. Thermal conductivity vs. temperature.

Figure 7 shows the comparison of the alumina thermal conductivity vs. temperature obtained in both the first (2014) and second experimental campaign: λ values in the former are indicated with the red triangle squared point, while in the latter (2015) with blue point.

The comparison of these curves, represented with a proper interpolation curve, highlights a good agreement despite some singularity at lower temperature (at 60°C the discrepancy is about 9%).

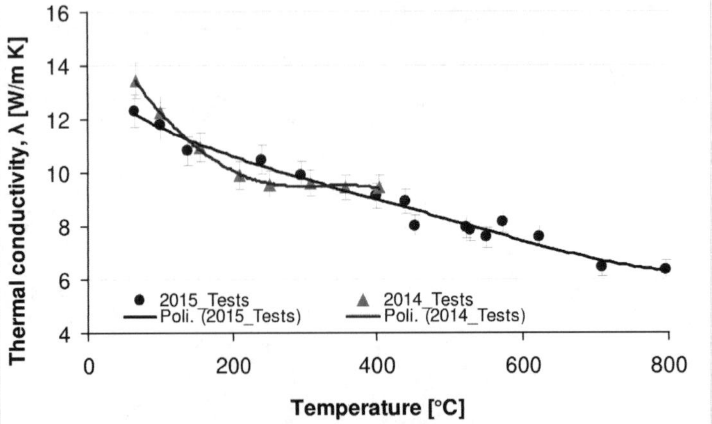

Figure 6. Experimental λ values.

Finally the values of the alumina thermal conductivity have been compared to those available in literature. Figure 8 shows that λ values have a similar trend to that of Bansal and Zhu [14] with a quite better agreement at higher temperature (T > 350°C). Once more, even if the discrepancy observed is about 10-15%, it could be explained by considering the different material composition of the specimen used to execute tests (alumina 100%wt compared to Alubit90) and could also rely on the different test conditions for which no accurate description is given in [14](leak in literature data).

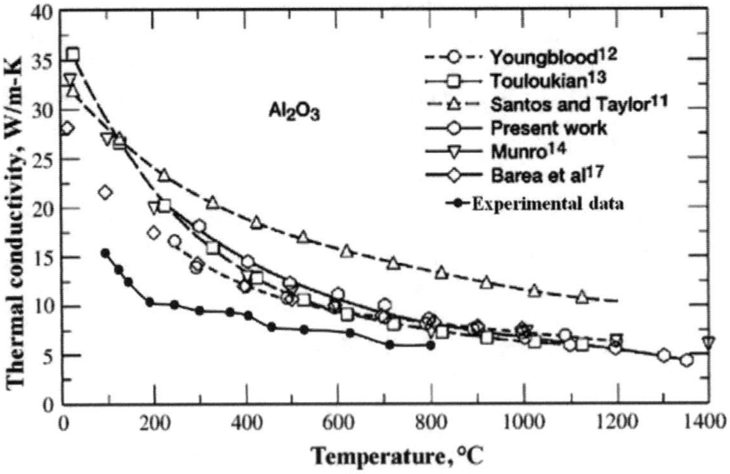

Figure 7. Comparison between literature and experimental alumina thermal conductivity.

3. Conclusions

The determination of the thermal conductivity of breeder materials is one of the main goals to find/select the best candidate material for the fusion reactor technology.

In this paper the thermal conductivity of the alumina, to be used as reference thermal conductivity in the hot plate with guardian ring method, has been determined by performing hot wire tests at different temperatures.

The thermal conductivity of an alumina cylindrical specimen has been calculated at stationary conditions (lasting at minimum 5 minutes) on the basis of the temperature gradient measured by thermocouples positioned at two different radii from the central axis.

The experimental activity (second test campaign) was carried out at the Lab. Scalbatraio of the Dep. DICI of the University of Pisa by using a dedicated device and an adequate DAS.

Hot wire tests have been executed in the temperature range 50÷800°C, with increment of 50°C.

The results show, as an example, that at an average temperature of 200°C, the thermal conductivity of alumina is equal to 12.67 W/m°C, whereas at about 550° it is 7.96 W/m°C.

Analysing the λ behaviour, it is possible to observe that the conductivity decreases as the test temperature increases: the calculated maximum deviation is less than 5%. Results fit very well the interpolation curve even if some spreading appears at low temperature.

Moreover analysing the λ behaviour, it is possible to observe that the results confirms the trend observed in the first experimental campaign.

A further comparison with the data available in literature (e.g. Bansal, D. Zhu, 2005) highlights a similar trend, despite some discrepancy at lower temperature that could be probably due to the different material composition and test condition.

The alumina thermal conductivity will be used subsequently as reference value for the investigation of the thermo-mechanical characterization of ceramic pebble beds of interest for the breeding blanket (this activity is currently ongoing at the DICI- University of Pisa).

References

[1] Zaccari N and Aquaro D, 2007 Mechanical characterization of Li2TiO3 and Li4SiO4 pebble beds: Experimental determination of the material properties and of the pebble bed effective values, *Fusion Engineering and Design*, **82** 2375-2382

[2] Lo Frano R et al. 2014 Thermo-mechanical test rig for experimental evaluation of thermal conductivity of ceramic pebble beds, *Fusion Engineering and Design*, **89**, Issue **7-8**, 1309-1313

[3] Aquaro D and Zaccari N, 2006 Experimental and numerical analysis on pebble beds used in an ITER Test Module Blanket, *Fusion Engineering and Design*, **81 A**, **Issue 1-4,** 707-712

[4] Enoeda M et al. 2001 Effective Thermal Conductivity Measurement of the Candidate Ceramic Breeder Pebble Beds by the Hot Wire Method *Fusion Tech.* **39** 612

[5] Abou-Sena A, Ying A and Abdou M 2003 Experimental Investigation and Analysis of the Effective Thermal Properties of Beryllium Packed Beds *Fusion Science and Techology* **44** 79

[6] Reimann J Hermsmeyer S 2002 Thermal conductivity of compressed ceramic breeder pebble beds *Fusion Engineering and Design* **61-62** 345-351

[7] Enoeda M et al. 1998 Effective thermal conductivity measurements of the binary pebble beds by hot wire method for the breeding blanket *Fusion Technology* **34** November

[8] Ying A et al. 2012 Status of ceramic breeder pebble bed thermo-mechanics R&D and impact on breeder material mechanical strength *Fusion Engineering and Design* **87** 1130-1137

[9] Aquaro D and Zaccari N, 2006 Experimental and numerical analysis on pebble beds used in an ITER Test Module Blanket *Fus. Eng. Des.* **81** 707-712

[10] Aquaro D and Zaccari N 2007 Experimental and numerical analyses on Li4SO4 and Li2TiO3 Pebble bed used in a ITER test module blanket *J. Nucl. Mat.* **367-370** 1293-1297

[11] Aquaro D and Orlandi F 2014 Thermal conductivity of ceramic pebble beds used as breeder in fusion nuclear reactors *Proc. Of XXXI UIT Conference*

[12] Aquaro D and Lo Frano R 2014 Experimental evaluation of thermal conductivity of ceramic pebble beds *J. Phys.: Conf. Ser.* **501** 012001

[13] Lo Frano R, Moscardini M and Aquaro D 2014 Numerical-experimental analyses by Hot-Wire method of an alumina cylinder for future studies on thermal conductivity of the fusion breeder materials *Proc. Of XXXI UIT Conference*

[14] Bansal N P and Zhu D 2005 Thermal conductivity of zirconia-alumina composites *Ceramic International* **31** 911-916

Journal of Physics: Conference Series **655** (2015) 012058 doi:10.1088/1742-6596/655/1/012058

An anisotropic numerical model for thermal hydraulic analyses: application to liquid metal flow in fuel assemblies

F Vitillo, D Vitale Di Maio, C Galati, G Caruso[1]

"Sapienza" University of Rome – Dep. of Astronautical, Electrical and Energy Engineering (DIAEE) – Corso Vittorio Emanuele II, 244 – 00186 Rome, IT

E-mail: gianfranco.caruso@uniroma1.it

Abstract. A CFD analysis has been carried out to study the thermal–hydraulic behavior of liquid metal coolant in a fuel assembly of triangular lattice. In order to obtain fast and accurate results, the isotropic two-equation RANS approach is often used in nuclear engineering applications. A different approach is provided by Non-Linear Eddy Viscosity Models (NLEVM), which try to take into account anisotropic effects by a nonlinear formulation of the Reynolds stress tensor. This approach is very promising, as it results in a very good numerical behavior and in a potentially better fluid flow description than classical isotropic models. An Anisotropic Shear Stress Transport (ASST) model, implemented into a commercial software, has been applied in previous studies, showing very trustful results for a large variety of flows and applications. In the paper, the ASST model has been used to perform an analysis of the fluid flow inside the fuel assembly of the ALFRED lead cooled fast reactor. Then, a comparison between the results of wall-resolved conjugated heat transfer computations and the results of a decoupled analysis using a suitable thermal wall-function previously implemented into the solver has been performed and presented.

1. Introduction

In order to obtain fast and accurate results, the isotropic eddy-viscosity RANS approach is often used in nuclear engineering applications: in particular, the k–ε model, k–ω model and SST model are examples of isotropic two-equations models widely used. As known from the literature, these models can fail when dealing with complex flows, especially when secondary flows are important. Another approach is then provided by Reynolds Stress Models (RSM), in which all the components of the Reynolds stress tensor are directly computed: this makes them suitable for anisotropic flows treatment but, as main drawback, it requires an expensive computational effort and potentially suffers numerical stability problems.

A different approach is provided by Non-Linear Eddy Viscosity Models (NLEVM), which try to take into account anisotropic effects by a nonlinear formulation of the Reynolds stress tensor. This approach is very promising, as it results in a very good numerical behaviour and in a potentially better fluid flow description if compared with classical isotropic models. An Anisotropic Shear Stress Transport (ASST) model, developed in a previous study [1] and implemented into the commercial software ANSYS FLUENT ®, has been applied showing very trustful results for a large variety of flows and applications.

[1] To whom any correspondence should be addressed.

As liquid metals do not meet the so-called Reynolds analogy (which assumes a complete similarity between the momentum and the thermal boundary layers via the use of the turbulent Prandtl number close to one), a suitable expression for the turbulent Prandtl number has been implemented into the solver; it allows at performing a complete thermal-hydraulic analysis using the Simple Gradient Diffusion Hypothesis to model the turbulent heat flux. A comparison between the results of wall-resolved conjugated heat transfer computations and the results of a decoupled analysis using a suitable thermal wall-function which had been previously implemented into the solver has been performed and presented [2].

In the present paper, a CFD study focused on the application of the abovementioned approaches has been carried out on the liquid metal flow within the fuel assembly of the ALFRED lead cooled fast reactor.

2. Wall function approach

To be able to keep a Simple Gradient Diffusion Hypothesis in our analysis and to take into account the enhanced heat transfer in liquid metals (due to their high thermal conductivity) a new correlation to link turbulent viscosity to turbulent heat diffusivity by means of the turbulent Prandtl number $\left(\mathrm{Pr}_t = \nu_t/\alpha_t\right)$ is needed. The needing of assessment is evidenced by the numerous correlations proposed in the past through experimental campaigns or analytical studies. Among them, the model of Kays [3] provides very good results [2]:

$$Pr_t = 0.85 + \frac{0.7}{Pr}\frac{\nu}{\nu_t} \tag{1}$$

In the wall-resolved approach the first node along the y axis need to be positioned in such a way that $y_1^+ \leq 1$, leading to very time-consuming simulations, because this method devotes a substantial fraction of the whole computational time to the near wall region, where velocity and temperature profiles are steep due to viscous effects.

The wall function approach, on the contrary, allows at using a more uniform mesh with y_1^+ greater than one and using semi empirical formulas to link the viscosity affected region between the wall and the fully-turbulent region.

These functions are based on the universal linear and log laws for velocity; the conventional temperature wall functions have been built for conventional fluids (air, water) and hence they depend on the velocity field according to the Reynolds analogy. Since no logarithmic law is visible in the temperature profile dealing with liquid metals, a new wall function has to be used.

The law proposed by Bricteux et al. [4] as temperature wall function in liquid metals is:

$$\theta^+ = \frac{Pr_t}{\kappa} log\left(1 + \frac{\kappa}{Pr_t} Pr\, y^+\right) \tag{2}$$

DNS and LES data, for a range of Péclet numbers (Pr=0.025-0.01 and Re_τ=180-2000), were compared with this law showing a good agreement in the range y^+=70-300, where the logarithmic law for the velocity profile is applicable.

Since ANSYS FLUENT does not allow the user to define an user wall function (UDF) for the temperature field, a strategy to overcome this limit has been previously proposed and consists in a step-by-step solution of the flow field, as explained below [2].

First, the velocity field is computed using the enhanced wall treatment. Once the solution is converged, a user defined wall function for the velocity is hooked up in the solver, while are freezed, the resolution of the momentum and turbulence equations and is activated the resolution of the energy transport equation. When the calculation is run again, the user defined wall function will not affect the velocity field, but it will be used by the solver to compute the temperature field near the wall.

Few iterations (i.e. five or six), alternating the enhanced wall treatment and the user wall function computations, in which the energy and the momentum and turbulence equations are alternatively freezed, are needed to quickly obtain a converged solution with temperature-dependent thermo-physical properties, as in the present analyzed case.

The UDF employed is based on the pre-existing standard wall functions (since these wall functions are replaced by the new one). Let's consider the temperature wall function as a linear variation of the velocity one:

$$T^* = \begin{cases} \Pr y^* = \Pr U^* & for\ y^* < y_T^* \\ Pr_t\left[\frac{1}{\kappa}\ln\left(Ey^*\right)+P\right] = Pr_t\left[U^*+P\right] & for\ y^* > y_T^* \end{cases} \quad (3)$$

Where y_t^* is the thermal sub-layer thickness. Since the first y^* is such as it is outside of the thermal sub-layer (i.e. in the velocity logarithmic sub-layer), only the second law is applied (i.e. for $y^* > y_t^*$). Thus, remembering that in the log-region it holds $U^* = \frac{1}{\kappa}\ln\left(Ey^*\right)$, it results that:

$$T^* = \frac{(T_w - T_P)\rho c_p C_\mu^{0.25} k_P^{0.5}}{q_w} = Pr_t\left[U^* + P\right] \quad (4)$$

It can be imposed:

$$Pr_t\left[U^*+P\right] = T^* = \theta^+ \frac{C_\mu^{0.25} k_P^{\ 0.5}}{u_\tau} = \frac{C_\mu^{0.25} k_P^{\ 0.5}}{u_\tau} \cdot \frac{Pr_t}{\kappa} log\left(1 + \frac{\kappa}{Pr_t} Pr\, y^+\right) \quad (5)$$

Thus, defining:

$$U^* = \frac{C_\mu^{0.25} k_P^{\ 0.5}}{u_\tau} \cdot \frac{1}{\kappa} log\left(1 + \frac{\kappa}{Pr_t} Pr\, y^+\right) - P \quad (6)$$

And setting a value for Pr_t at the wall, one can have:

$$T^* = Pr_t\left[\frac{C_\mu^{0.25} k_P^{0.5}}{u_\tau} \cdot \frac{1}{\kappa} log\left(1 + \frac{\kappa}{Pr_t} Pr\, y^+\right)\right] \quad (7)$$

$$\Rightarrow \theta^+ = \frac{(T_w - T_P) c_p \rho u_\tau}{q_w} = \frac{u_\tau T^*}{C_\mu^{0.25} k_P^{\ 0.5}} = \frac{\Pr_t}{\kappa} log\left(1 + \frac{\kappa}{Pr_t} Pr\, y^+\right) \quad (8)$$

as required from Eq. (2). The new temperature wall function is employed in a Realizable k-ε model [5] simulation.

3. A non-linear eddy viscosity model: the ASST model

Non-linear eddy viscosity (NLEV) models offer a potentially powerful solution to improve numerical model ability to describe anisotropic characteristics of complex flows. In particular they are supposed to be capable of better predicting secondary motion magnitude as well as flow turbulent features. The Anisotropic Shear Stress Transport (ASST) model has been developed and implemented into the solver ANSYS FLUENT via User-Defined Functions. The model is based on a second-order formulation of the Reynolds stress tensor, written according to the Cayley-Hamilton theorem as follows:

$$\rho\overline{u_i' u_j'} = \frac{2}{3}\rho k - \mu_t S_{ij} + \rho C_1 k\tau^2\left(S_{ik}S_{kj} - \frac{1}{3}\delta_{ij}S_{kl}S_{kl}\right) + \rho C_2 k\tau^2\left(\Omega_{ik}S_{kj} + \Omega_{jk}S_{ki}\right)$$
$$+\rho C_3 k\tau^2\left(\Omega_{ik}\Omega_{kj} - \frac{1}{3}\delta_{ij}\Omega_{kl}\Omega_{kl}\right) \quad (9)$$

$S_{ij} = \left(\frac{\partial U_i}{\partial x_j} + \frac{\partial U_j}{\partial x_i}\right)$ being the strain tensor, $\Omega_{ij} = \left(\frac{\partial U_i}{\partial x_j} - \frac{\partial U_j}{\partial x_i}\right)$ being the rotation tensor, k being the turbulent kinetic energy and ω the specific dissipation rate. τ is the turbulent time scale, which, for an ω-based model is typically equal $\tau = \frac{1}{\beta^* \omega}$,with $\beta^* = 0.09$.

The model uses the k-ω transport equations of the well-known SST model of Menter [6]. However a modification of the eddy viscosity formulation is necessary to take into account the Bradshaw's assumption inside the boundary layer and the new anisotropic formulation of the Reynolds stresses. The proposed expression for the eddy viscosity is then:

$$\mu_t = \frac{\rho \mathcal{B}}{max\left[\frac{\mathcal{B}}{C_\mu k}\omega, F_2 S\right]} \tag{10}$$

Where $C_\mu = \frac{7.4}{A_1 + \tau S}$, $\mathcal{B} = (a_1 + C_1\tau^2 S^2 + C_2\tau^2 S^2 + C_3\tau^2 S^2) \cdot k$ and $S = \left(\frac{1}{2} S_{ij} S_{ij}\right)^{1/2}$. Closure coefficients are A_1=3.9 and a_1=0.31.

The model is fully realizable thanks to some constraints on the closure coefficients. In particular, retaining the same approach presented in literature [7] and [8], the following relations must hold:

$$C_1 = \frac{C_{NL1}}{C_{NL4} + C_{NL5} \cdot (\tau S)^3} \tag{11}$$

$$C_2 = \frac{C_{NL2}}{C_{NL4} + C_{NL5} \cdot (\tau S)^3} \tag{12}$$

$$C_3 = \frac{C_{NL3}}{C_{NL4} + C_{NL5} \cdot (\tau S)^3} \tag{13}$$

Reference values [8] are used for all the constants, i.e. C_{NL1}=0.8, C_{NL2}=11, C_{NL3}=4.5, C_{NL4}=1000 and C_{NL5}=1. The C_{NL2} value is of particular interest for a rod bundle flow analysis since it allows improved predictions of secondary motions in this kind of flow [8].

The model is closed through the SST k and ω transport equations [6]:

$$\frac{\partial}{\partial t}(\rho k) + \frac{\partial}{\partial x_j}(\rho k u_j) = \frac{\partial}{\partial x_j}\left[\left(\mu + \frac{\mu_t}{\sigma_k}\right)\frac{\partial k}{\partial x_j}\right] + \\ + \widetilde{G_k} - \rho\beta^* k\omega \tag{14}$$

$$\frac{\partial}{\partial t}(\rho \omega) + \frac{\partial}{\partial x_j}(\rho \omega u_j) = \frac{\partial}{\partial x_j}\left[\left(\mu + \frac{\mu_t}{\sigma_\omega}\right)\frac{\partial \omega}{\partial x_j}\right] + \\ + G_\omega - \rho\beta\omega^2 + D_\omega \tag{15}$$

To be consistent with the NLEV formulation, the turbulence kinetic energy production term can be expressed in the proper form:

$$G_k = -\rho\overline{u_i' u_j'}\frac{\partial U_i}{\partial x_j} \tag{16}$$

Where $\overline{u_i' u_j'}$ is given by Eq. (9). Note that, due to the realizability condition given by the eddy viscosity modification and the model coefficient expressions, no turbulence kinetic energy production limiter is needed. Finally the specific dissipation rate is expressed in the original form i.e.

$$G_\omega = \frac{\rho\alpha}{\mu_t} G_k \tag{17}$$

The other parameters not mentioned so far maintains the same definition as the original SST model formulation. Closure constants are:

$$\sigma_{k,1} = 1.176, \quad \sigma_{k,2} = 1.0, \quad \sigma_{\omega,1} = 2.0, \quad \sigma_{\omega,2} = 1.168,$$
$$\beta^* = 0.09, a_1 = 0.31, \beta_{i,1} = 0.075, \beta_{i,2} = 0.0828$$

4. The ALFRED core and nodalization

The simulations performed are based on the ALFRED (Advanced Lead Fast Reactor European Demonstrator) triangular core lattice geometry. This lead cooled reactor is the third considered within the LEADER (Lead - cooled European Advanced DEmonstration Reactor) project, and is thought to be realized in the short term, relying on presently available technology. This scaled down reactor is thus designed using solutions close to the lead fast reactor reference conceptual design as much as possible, but still considering the essential need to proceed to construction in a short time frame. Many design details of this reactor can be found in [9]; the main parameters are reported in Table 1.

Table 1. ALFRED core parameters.

Parameter (Core)	Value	Unit	Parameter (Fuel pin)	Value	Unit
Thermal power	300	MW_{th}	Cladding material	15 - 15 Ti	-
Coolant mass flow rate	25460	kg/s	Clad rugosity	$5 \cdot 10^{-6}$	m
Coolant average velocity, $\bar{u}$	1.4	m/s	Relative rugosity, ε	$5.17 \cdot 10^{-4}$	-
Number of FAs/Total assemblies	171/295	-	Cladding outer radius, r	$5.25 \cdot 10^{-3}$	m
Pins per fuel assembly	127	-	Pitch to diameter ratio, p/D	1.32	-
Coolant inlet temperature, T_{in}	400	°C	Hydraulic diameter, D_h	0.009673	m
Coolant outlet temperature, T_{out}	480	°C	Total length/ Active height	1.42/0.6	m

The central assembly power is 2.063 MW and the maximum heat flux in the central fuel assembly is 9.92 10^5 W/m^2; these values have been estimated considering the radial and axial peak factors, 1.2 and 1.209 respectively, and the extrapolated length equal to 150% of the active length. The average flow rate in the central assembly is 176.5 kg/s, with a FA inlet velocity of 1.54 m/s. A bypass flow of 0.373 kg/s associated to each FA (1.7 kW of thermal power in each FA bypass) has been considered.

In the near-wall approach, the estimation of the y_1^+ is crucial to create the mesh to employ.

As known, the requirement to be fulfilled is $y_1^+ < 1$, thus:

$$y_1^+ = \frac{y u_\tau \rho}{\mu} < 1 \tag{18}$$

For the geometry described in Table 1, a preliminary evaluation of the flow field in the active zone provided that the condition $y_1^+ = 1$ will be satisfied with distances from the wall in the range 2.22 μm and 2.6 μm. The domain used in the simulations consists of 1/12 of the fuel element 1 m long (composed by the asset length of 0.3 m at the inlet (more than 30 equivalent diameters), 0.6 m for the active length and 0.1 m in the upper zone).

A sensitivity analysis on the nodalization to be used in the near-wall approach and the ASST model has been performed, by using two different boundary layers, $y_1^+ = 1$ (Case A) and $y_1^+ = 0.5$ (Cases B and C), for 16 and 18 layers with a growth rate (which is the ratio of a layer's height and the previous one) equal to 1.3 and two geometrical elements (thetraedrical – Cases A and C - and hexaedrical elements – Case B). The axial size of the mesh is 5 mm for the whole length of the fuel element. Table 2 summarizes the mesh sensitivity and the total number of nodes. Due to the negligible differences in the results obtained with the A, B and C nodalizations, mainly the Cases B and D will be discussed in the paper.

For the wall function approach, a coarser hexaedrical mesh has been prepared, without boundary layer and a $y_1^+ = 170$ (Case D), with the same axial discretization as in the previous cases.

Table 2. Nodalization of the computational domain

Case	Elements type	y_1^+	Approach	Number of nodes
A	Thetra	1	Wall resolved	≈2.6M
B	Hexa	0.5	Wall resolved	≈3.5M
C	Thetra	0.5	Wall resolved	≈2.9M
D	Hexa	170	User Wall function	≈0.8M

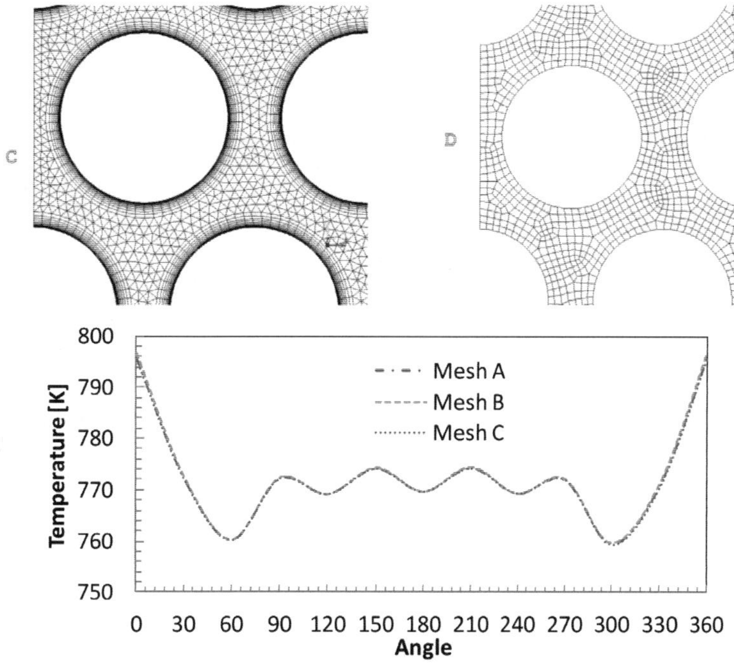

Figure 1. Domain nodalizations and convergence test

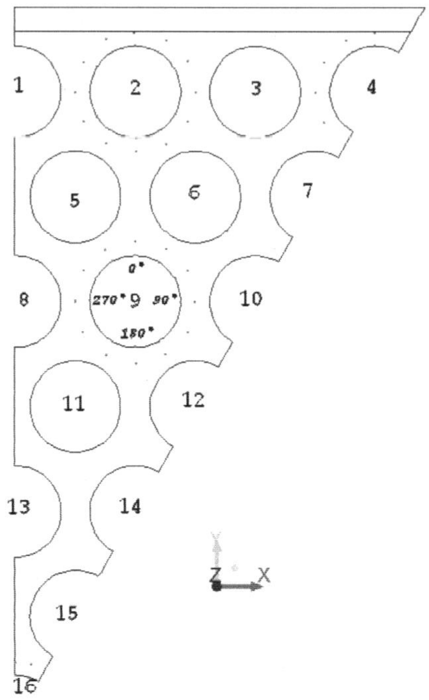

Figure 2. Geometry (transversal section of the 1/12 FA) and temperature lines location

An example of the analyzed nodalizations is shown in Figure 1, where case C and D of Table 2 are shown. Also, an example of the mesh convergence tests for the three wall-resolved configurations is given in the same figure for the angular wall temperature distribution in pin #2.

The domain includes 16 pins, as shown in Figure 2, where the pins are numbered for reference with respect to the result presentation and discussion. In the same figure the positions of the axial lines where pin wall and fluid temperatures were extracted (to evaluate their axial and circumferential profiles) are evidenced. With reference each fluid and wall temperatures couple, the Nusselt numbers have been evaluated from locally imposed the heat flux, the hydraulic diameter and the temperature-dependent thermal conductivity of lead. The distance of the fluid temperature line from the corresponding wall temperature one is half of the minimum distance between pins ($0.16\ d_{pin} = 1.68$ mm $\approx 650\ y^+$).

Journal of Physics: Conference Series **655** (2015) 012058 doi:10.1088/1742-6596/655/1/012058

5. Results and discussion

First of all it is of primary importance to verify that the used models are able to correctly reproduce the main features of the triangular lattice flow. In this sense, observe in Figure 3 and Figure 4 the velocity and temperature field respectively calculated by the realizable k-ε model with wall-resolved (WR, with mesh B) and wall-function (WF, with mesh D) approaches.

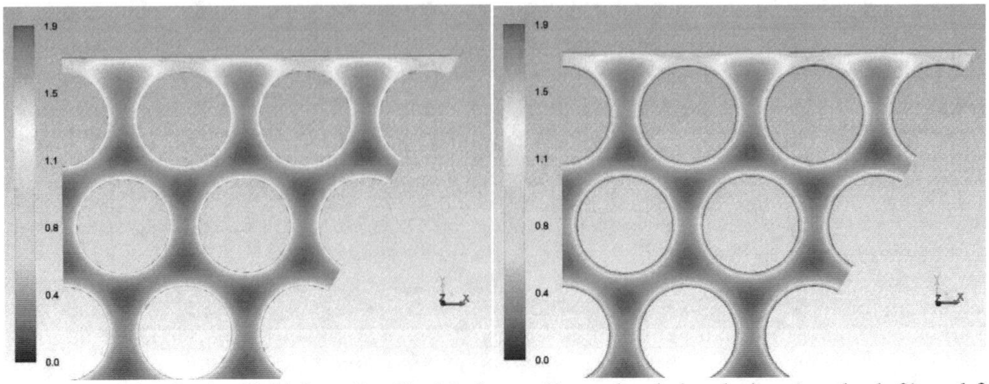

Figure 3. Axial velocity field for a Realizable k-ε wall-resolved simulation (on the left) and for a k-ε simulation using the new temperature wall function (on the right) [m/s]

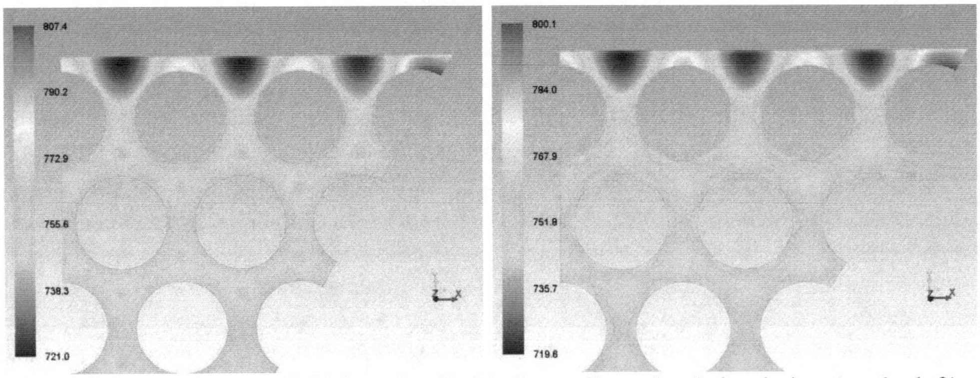

Figure 4. Axial temperature field for a Realizable k-ε wall-resolved simulation (on the left) and for a k-ε simulation using the new temperature wall function (on the right) [K]

The two different approaches provide very similar results, both qualitatively and quantitatively. The velocity field presents a peak in the region where hydraulic resistance is lower (i.e. in the central region of the triangular lattice). The main differences between the two approaches are mainly located in the boundary layer region, which is what it is expected due to the different near-wall approach. In particular, the momentum boundary layer is thicker in the k-ε wall-function computation than in the Realizable k-ε wall-resolved case. This is quite interesting, since the standard wall functions for velocity are developed for a one-dimensional flow when secondary motions are practically negligible compared to the principal motion. This is not the case for rod bundle flow, where secondary motions are responsible for a strong homogenization of the flow inside the different subchannels. Moreover, the observed thicker boundary layer results in a thicker thermal boundary layer as well: the flower-shaped thermal boundary layer is much more pronounced for the wall-function calculations than for the wall-resolved case.

Due to the observed differences, it is important to verify whether the two approaches can reproduce the correct trend of the Nusselt number predicted by the currently used correlations found in the literature, specifically the Ushakov's [10] and Mikityuk's [11] correlations. Note that these correlations are evaluated with respect to the theoretical temperature profile of the coolant. Figure 5 shows the comparison of the computed Nusselt number as a function of core elevation (centered in the center of the active core height), with that obtained using the two aforementioned correlations, for the 0° position in pin #9 and #16. The same comparison has been also done on the angular distribution of the Nusselt number for pin #9 at the middle point of the active core height. These two pins #9 and #16 are supposed to be the more appropriate to be compared to the experimental correlations, since the effects of the assembly shroud are lower and the fluid flow is closer to the theoretical flow inside an infinite triangular lattice geometry. Note that the quite constant value of the Nusselt number obtained through correlations is used, since it is evaluated by means of the local non-dimensional numbers of the studied case (taking into account the limited effects of the lead temperature variation on the thermophysical properties). The Mikityuk's correlation is presented with its root-mean-squared error bars, trying to quantify eventual differences between numerical and experimental data.

All the different tested models and approaches seem to be able to satisfactory reproduce values of the Nusselt number. In particular, the axial Nusselt number approaches the experimental value once the inlet effects become negligible. The computed Nusselt number varies as function of the angular position (i.e. the already mentioned flower-shape observed in Figure 4), but still the average value remains within the experimental uncertainty of the Mikityuk's correlation. Therefore the straightforward conclusion is that, when dealing with global feature of the fluid flow, commonly used CFD is capable of reproducing the correct trends, no matter the turbulence model used. However, differences are expected when looking at local quantities, (i.e. wall temperature distributions) due to the different type of model and approach to resolve the boundary layer. In this sense, Figure 6 shows the axial evolution of the wall temperature at the 0° position and Figure 7 shows the angular distribution of the wall temperature at the center of the core active length for pin #2 and #9. The pin #9 is used again as the effects of the assembly shroud are negligible, whereas the pin #2 is chosen since it could provide information on the changes induced by the shroud itself.

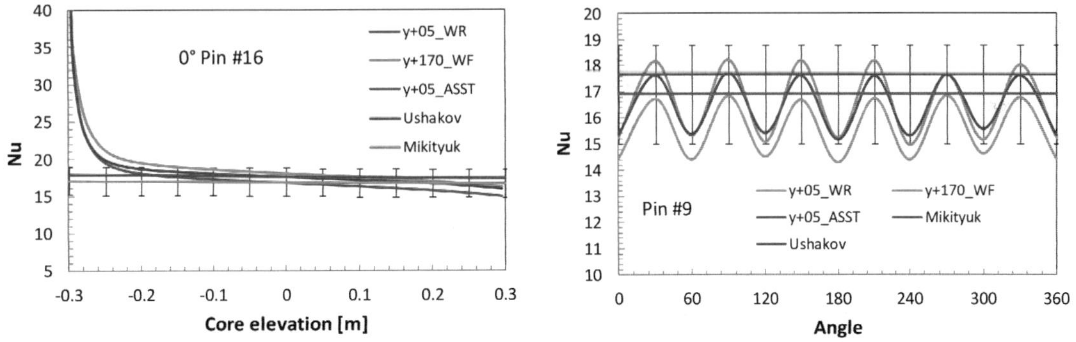

Figure 5. CFD model validation with regard to Nusselt number correlations available in the literature

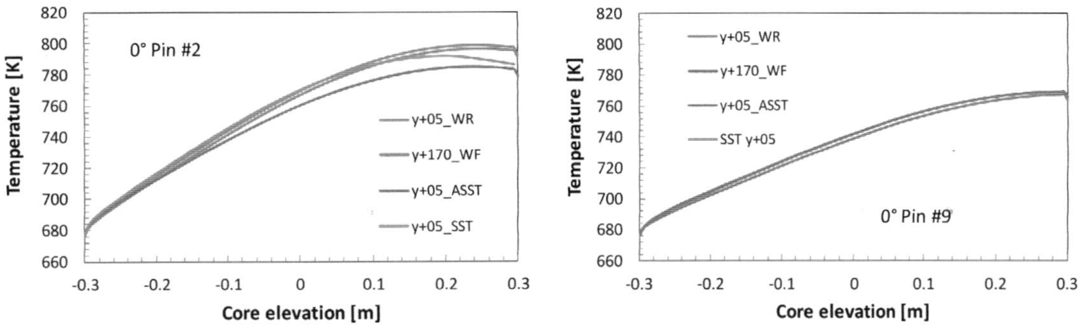

Figure 6. Comparison of the axial wall temperature distribution

As Figure 6 and Figure 7 show, differences do exist: in particular, if the axial wall temperature evolution is practically the same no matter the turbulence model and the near-wall approach used for pin #9, the assembly shroud changes this distribution so that the ASST compute wall temperature at the top of the core active height is 10-15°C lower than that of the k-ε based calculations. Differences are even clearer when looking at the angular distribution of the pin wall temperature in pin #2 in Figure 7.

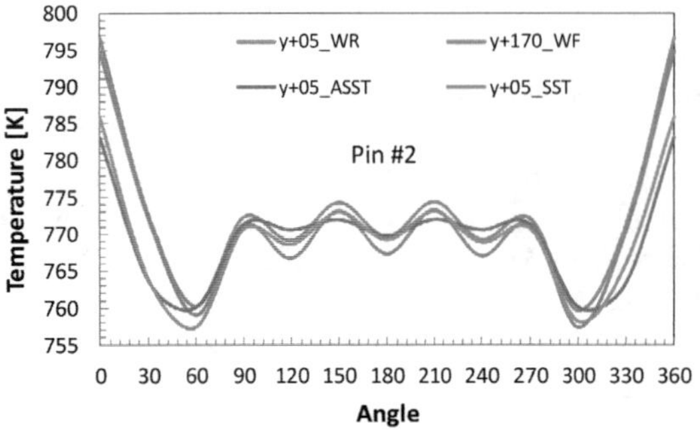

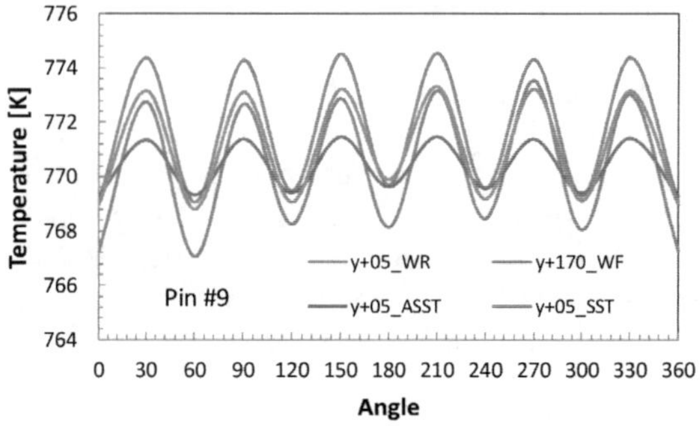

Figure 7. Comparison of the angular wall temperature distribution

Since it could be stated that it is difficult to understand the contribution of the several modeling differences between the wall-resolved ASST results and the wall-function k-ε results (i.e. the near-wall approach, the dissipation equation, the expression of the Reynolds stress tensor), the results of an isotropic SST [6] calculations are shown as well in Figure 7. This helps a lot the comprehension of the shown comparison, since the SST model employs the same dissipation equation as the ASST model but its isotropic nature is the same as that of the k-ε model. Therefore, note that when the ω-dissipation equation is used, the wall temperature is underestimated of around 10°C compared to that of a k-ε simulation (by using either the user defined wall function for temperature with the coarse mesh D or the wall-resolved approach with mesh B) in the region right in front of the assembly shroud (from 270° to 90° in Figure 2).

This difference is likely due to the fact that this region presents a mesh with a y^+ (= 170) in the logarithmic region of the boundary layer (for sake of feasibility of the computations with available means). Therefore, for a given imposed solution of the velocity field in the first cell next to the wall, the temperature field is somehow influenced by the turbulence dissipation occurring in such region, which is modeled differently. Remember that the turbulence dissipation equation is a major source of uncertainty in CFD calculations, since no exact equation exist for this kind of quantity. However, difference are much lower on the other side of pin #2 (i.e. 90° to 270°) oriented towards the inner part of the fuel assembly. By the way, observe that the angular fluctuations are practically the same for the isotropic models, whereas the ASST models have a smoother distribution due the effect of anisotropy: in particular, its favorable characteristics of better reproducing secondary motions can enhance the heat transfer mixing among the different subchannels.

The same conclusion is valid for wall temperature angular distribution in pin #9, when the fluctuations around the mean value are of the order of 2-3°C for the isotropic models and of around 0,5°C for the ASST model.

6. Conclusions

A CFD analysis has been carried out to study the thermal–hydraulic behavior of liquid metal coolant in the fuel assembly of triangular lattice of the lead-cooled ALFRED reactor. Different turbulence models (namely the realizable k-ε, the SST and the ASST models) and different near wall approaches (i.e. wall-resolved and wall-functions) have been tested and compared.

Results show that all the tested models and near wall approaches can predict quite well the global flow characteristics: the results are qualitatively the same and quantitatively very similar. In particular, the Nusselt number predicted by all the tested models lies within the experimental uncertainty range of well-known correlations found in the literature (Ushakov's and Mikityuk's correlations). Differences are observed when looking at very local quantities such as the wall temperature. In this case, the isotropic models predict larger spatial fluctuations compared to the ASST model. The former can in fact damp the fluctuations thanks to a better prediction of the inter-subchannel mixing due to secondary motions. The Simple Gradient Diffusion Hypothesis approach for the turbulent heat flux seems to be able to capture the global heat transfer inside the fuel assembly, when a proper correlation for the turbulent Prandtl number is used. This is likely due to the fact that for such flows the Péclet number is low, hence the conduction heat transfer prevails on the convection (i.e. turbulence-driven) heat transfer.

As a conclusion, the tested models and approaches are able to reproduce the main features of the rod-bundle flow and to provide a good first estimation of the flow characteristics. However, they can potentially be untruthful when performing a finer analysis looking at very local quantities. Therefore care must be taken when performing a CFD analysis of such complex flows.

List of Symbols

k	turbulent kinetic energy [m^2/s^2]
Nu	Nusselt number
p/D	pitch to diameter ratio
Pe	Peclet number
Pr	Prandtl number
Re	Reynolds number
T^*	dimensionless temperature
T_b	bulk temperature [K]
T_w	channel wall temperature [K]
u, U	streamwise velocity [m/s]
u^+, U^*	dimensionless velocity
u_τ	shear stress velocity [m/s]
y^+, y^*	dimensionless wall-normal coordinate
y_1^+	dimensionless first grid point at wall boundary in wall normal direction

Greek symbols

ε	energy dissipation [m^2/s^3]
θ	mean transformed temperature [K]
θ^+	dimensionless θ
κ	Von Kàrman constant
μ	fluid dynamic viscosity [kg/ms]
μ_t	fluid turbulent viscosity [kg/ms]
ν	fluid kinetic viscosity [m^2/s]
ν_t	turbulent momentum diffusivity [m^2/s]
ρ	fluid density [kg/m^3]
τ_w	wall shear stress [kg/ms^2]
ω	specific energy dissipation [s^{-1}]

References

[1] Vitillo F, Galati C, Cachon L, Laroche E and Millan P, "An Anisotropic Shear Stress Transport (ASST) Model Formulation", *Comput. Math. Appl.*, http://dx.doi.org/10.1016/j.camwa.2015.08.023
[2] Vodret S, Vitale Di Maio D and Caruso G 2014 *J. Phys.: Conf. Ser.* **547** 012033
[3] Kays W 1994 *J. Heat Trans-T. ASME* **2** (116) 284-95
[4] Bricteux L, Duponcheel M, Winckelmans G, Tiselj I and Bartosiewicz Y 2012 *Nucl. Eng. Des.* **246** 91-97
[5] Shih T H, Liou W W, Shabbir A, Yang Z and Zhu J 1995 *Comput. Fluids* **24**(3), 227-238
[6] Menter F R, Kuntz M and Langtry R, "Ten years of industrial experience with the SST Turbulence model", Turbulence, Heat and Mass Transfer 4, Begell House, Inc., 2003
[7] Shih T H, Zhu J and Lumley J L, "A Realizable Reynolds Stress Algebraic Equation Model", NASA TM 105993
[8] Baglietto E, Ninokata H, and Misawa T 2006 *Nucl. Eng. Des.* **236** 1503-1510
[9] Grasso G, Petrovich C, Mattioli D, Artioli C, Sciora P, Gugiu D, Bandini G, Bubelis E and Mikityuk K 2014 *Nucl Eng. Des.* **278** 287–301
[10] Ushakov P, Zhukov A and Matyukhin M 1977 *High Temp.* **5** (15) 868-73
[11] Mikityuk K 2009 *Nucl. Eng. Des.* **239** 680-7

Journal of Physics: Conference Series **655** (2015) 012059 doi:10.1088/1742-6596/655/1/012059

Transient analysis of "2 inch Direct Vessel Injection line break" in SPES-2 facility by using TRACE code

S D'Amico[1], C Lombardo[2], I Moscato[1], M Polidori[2], G Vella[1]
[1] University of Palermo, Dipartimento di Energia, Ingegneria dell'Informazione e Modelli Matematici, Building 6, Viale delle Scienze, 90128 Palermo, Italy
[2] ENEA-UTFISSM/SICSIS, Via Martiri di Monte Sole 4, 40129 Bologna, Italy

E-mail: giuseppe.vella22@unipa.it

Abstract. In the past few decades a lot of theoretical and experimental researches have been done to understand the physical phenomena characterizing nuclear accidents. In particular, after the Three Miles Island accident, several reactors have been designed to handle successfully LOCA events. This paper presents a comparison between experimental and numerical results obtained for the "2 inch Direct Vessel Injection line break" in SPES-2. This facility is an integral test facility built in Piacenza at the SIET laboratories and simulating the primary circuit, the relevant parts of the secondary circuits and the passive safety systems typical of the AP600 nuclear power plant. The numerical analysis here presented was performed by using TRACE and CATHARE thermal-hydraulic codes with the purpose of evaluating their prediction capability. The main results show that the TRACE model well predicts the overall behaviour of the plant during the transient, in particular it is able to simulate the principal thermal-hydraulic phenomena related to all passive safety systems. The performance of the presented CATHARE noding has suggested some possible improvements of the model.

1. Introduction

After the Three Miles Island accident the International Community has changed the way to design nuclear power plants starting the growth of new advanced reactors which employ several safety systems that are able to mitigate a wide spectrum of accidents such as Loss of Coolant Accidents (LOCAs) or Station Black Out (SBO). Therefore, throughout last decades, lots of theoretical and experimental researches have been done to deeply understand the physical phenomena characterizing the innovative devices introduced in new nuclear designs and their performances during a postulated accident.

Westinghouse has designed and developed an advanced type of Pressurized Water Reactor (PWR) which is known in its two variants: the AP600 and its updated version AP1000. This kind of PWR employs a series of passive safety systems to mitigate the accidental transients. To achieve the design approval by U.S. NRC (United States Nuclear Regulatory Commission), from the early '90, Westinghouse has strictly collaborated with some partners in an experimental programme which aim was to build several facilities simulating the thermal-hydraulic behaviour of AP600 reactor during different accident scenarios. One of these plants is SPES-2 which is an integral test facility built in Piacenza at SIET (Società Italiana per Esperienze Termofluidodinamiche) laboratories and simulating the primary circuit, the relevant parts of the secondary circuits and the passive safety systems typical of the AP600 [1]. The aforementioned programme has demonstrated the effectiveness of the AP600 safety systems and the results obtained have been also used to provide data for system codes validation in the framework of International Code Assessment and Applications Program (ICAP). In fact code validation on qualified experimental data is a fundamental issue in the design and safety analyses of nuclear reactors.

This paper presents the numerical results that has been obtained by DEIM (Dipartimento di Energia, Ingegneria dell'Informazione e Modelli Matematici) of the University of Palermo and ENEA (Agenzia nazionale per le nuove tecnologie, l'energia e lo sviluppo economico sostenibile) for the "2 inch Direct Vessel Injection line break" with both TRACE and CATHARE best estimate thermal-hydraulic codes and their comparison with experimental data collected by SIET in the SPES-2 facility. The aim of this work is to study the differences on the code predictions considering the different modelling approach with one and/or three-dimensional components and to compare the capability of these codes. The main results show that the TRACE model well predicts the overall behaviour of the plant during the transient, in particular it is able to simulate the principal thermal-hydraulic phenomena related to all passive safety systems. The performance of the presented CATHARE model has suggested some possible improvements of the models.

2. SPES-2 facility description

The SPES-2 experimental plant, Fig. 1, is a full height and full pressure facility with an overall volume scale of 1/395 [2]. In addition, components and lines have been designed considering others scaling criteria to better duplicate the AP600 behaviour such as: conservation of thermodynamic conditions, preservation of fluid transit time, heat flux conservation in heat transfer components and conservation of elevation in lines and components.

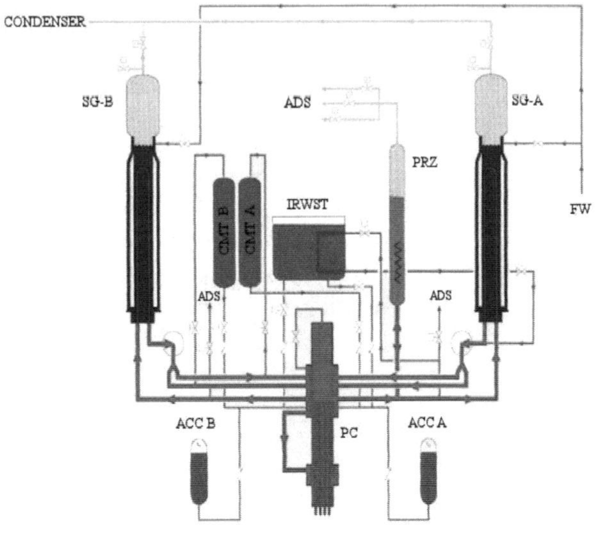

Figure 1. SPES-2 layout.

2.1. Power channel

The power channel (PC) is composed of the following components: lower plenum, riser (including the core), downcomer (DC), upper plenum, upper head (UH) and downcomer-upper head by-pass (DC-UH by-pass). Ninety seven electrically heated rods are located inside the core, they reproduce the same geometry (rod pitch, diameter and length) as the AP600 rod bundle and the scaled full power used for the simulated transients is 4.99 MW. The downcomer is composed of an annular section in which the four cold legs and two direct vessel injection (DVI) nozzles for safety injection enter and a tubular (TB) section that connects annular downcomer to the lower plenum.

2.2. Loop piping and pumps
The primary piping consists of two loops (loop A and B) each one including one hot leg (HL) and two cold legs (CLs). The hot leg reproduces the geometry of the AP600 up to surge line nozzle (same L/D in the horizontal section and same angle in the inclined section). The AP600 two cold legs per loop design is reproduced as well, however, they detach from a single coolant pump vertical discharge. The split from single pump discharge into the two cold legs is positioned at the elevation of the AP600 steam generator channel head in order to reproduce the geodetic flow path during a cold leg break transient: the broken cold leg must take the fluid from the unbroken one. The pumps (one per loop) are centrifugal, their suction line is horizontal while the delivery is directed downwards discharging in a pipe common to the two cold legs.

2.3. Pressurizer
The pressurizer (PRZ) consists of a cylindrical vessel with flanged ends and contains two immersion type electrical heaters each having a maximum controlled power of 16 kW. The pressurizer heat losses are compensated by means of six electrical external heaters. The pressurizer volume is scaled and the bottom elevation is preserved. The level swelling is preserved by ensuring that the average void fraction in the test is equal to AP600 for similar thermo-hydraulic conditions using Wilson bubble rise models [3].

2.4. Steam generators
The plant has two identical steam generators (SGs) that allow the transfer of thermal power from the primary to the secondary circuit. The steam generator primary side consists of a tube bundle and inlet/outlet plena. Its bundle includes 13 inconel 600 U-tubes assembled in a square array. The secondary side volumes are scaled by 1/395 and all vertical elevations are preserved up to the steam separator.

2.5. Passive safety systems
The passive safety systems are: two Core Make-Up Tanks (CMTs), an In-Containment Refueling Water Storage Tank (IRWST), a Passive Residual Heat Removal (PRHR) system, two Accumulators (ACCs), an Automatic Depressurization System (ADS) and all their connection lines to the primary system. The two CMTs are connected to the cold legs of loop B by two balance lines and they discharge cold water in the DVI lines through their injection lines. The volumetric scaling and the elevation change preservation have defined the component which is cylindrical with hemispherical heads. Their design has been developed so that the CMT metal mass is scaled to the AP600 CMT in order to preserve the rate of steam condensation on the walls. The IRWST provides low pressure gravity injection to each of the two DVI connections. The tank has been simulated conserving the water volume overall scaling, the water level and the bottom elevation. The IRWST also provides a heat sink for the PRHR. The PRHR consists in a full height C shaped heat exchanger with friction pressure drops maintained and the heat transfer area is scaled such that the natural circulation behaviour of the AP600 PRHR is simulated. The PRHR top is connected by a supply line with the hot leg A while its bottom is connected by a line to the pump suction A. The two accumulators in SPES-2 facility are not spherical as in AP600 design but they have been scaled preserving the volume of the accumulators in the reference plant. The gas (air) to water volume ratio is preserved. Each of their injection lines are connected to the respective DVI. The four stages of ADS consist of ball valves (one per stage) with an orifice in series to achieve the proper scaled flow area. The two sets of piping connected to the steam space of the pressurizer in the AP600 are combined into single set with the first, second and third stage valves in SPES-2. The ADS fourth stage is located on the hot legs of the primary piping.

3. TRACE and CATHARE codes brief description

Best-estimate thermal-hydraulic system codes are complex tools developed to simulate both normal and transient conditions in nuclear power plants. The general objective of the best estimate approach is to get an evaluation of the plant behaviour which will be as realistic as possible.

3.1. TRACE code

The TRACE code is an advanced best-estimate code, two-fluid, thermal-hydraulic computer code designed for use in realistic studies of light water reactors. The capability exists to model thermal-hydraulic phenomena in both one and three dimensional space. In addition, it contains not only a point kinetics model but also the capability to be coupled with external 3D neutronic codes-and the possibility to be managed through the graphical user interface SNAP [4], [5].

3.2. CATHARE code

The Code for Analysis of THermalhydraulics during an Accident of Reactor and safety Evaluation (CATHARE) is a system code for PWR safety analysis, accident management, definition of plant operating procedures and for research and development. The CATHARE code is based on a 2-fluid 6-equation model including non-condensable gas equations and additional equations for radio-chemical components transport. The code would be utilized by means of the graphical user interface GUITHARE [6], [7].

4. TRACE model

The SPES-2 TRACE nodalization, Fig. 2, has been made by using TRACE V5.0 patch 3 and it models in detail the experimental facility [8], [9].

The total number of thermal-hydraulic components presented in TRACE model is 432, being 83 PIPEs, 13 VESSELs, 26 VALVEs, 11 BREAKs, 2 FILLs, 4 SINGLE JUNCTIONs, 11 TEEs, 2 PUMPs, 276 HEAT STRUCTUREs and 4 POWERs. In addition 42 TRIPs, 124 CONTROL BLOCKs, 505 SIGNAL VARIABLEs and 144 TABLEs complete the model. In particular, where the 1D approach is not sufficient (i.e. secondary flows are not negligible) as in the power channel, in the steam generators and in the IRWST pool, the 3D component "VESSEL" has been used.

4.1. Principal physical models adopted

In order to better simulate the behavior of the various systems during a SBLOCA, several physical models made available by the code have been used [10].

The "level tracking" model have been adopted in all vertical components to localize, as well as possible, the liquid-gas interfaces. Models of counter-current flow limitation (CCFL) have been used at the cell junctions where this event was expected during flooding and reflux condensation phases. The "offtake" model has been applied at the side junctions of ducts where a horizontal stratified two-phase flow was expected to predict the correct offtake flow quality that would have come out from the main tube; this has allowed to better simulate the void fraction distribution downstream of the junctions that connect the primary circuit to some passive safety systems which mass flow rates are deeply influenced by the buoyancy force. At cell edges that simulate the break and the ADS orifices the "critical flow" model has been used.

4.2. Power channel, steam generators and IRWST

The lower plenum, the riser and the upper head have been described using the VESSEL component: its nodalization consists of 38 axial levels, 4 radial rings and 8 azimuthal sectors.

The annular downcomer has been simulated apart by means a VESSEL component with 12 axial levels, 2 radial rings and 10 azimuthal sections to better simulate the circumferential position of the nozzles and fillers with the purpose of reproducing as best as possible the azimuthal pressure drops. Fig. 3 shows the noding of power channel. Four PIPE components have been used to describe the tubular downcomer and upper head-downcomer by-pass.

The upper part of each steam generator has been described using three VESSEL components which have been needed to better reproduce the geometrical features and the complex flow paths taking place inside steam separator and steam dome.
The IRWST pool nodalization has been made through the application of a cartesian VESSEL that is divided in 24 axial levels, 3 X-sections and 6 Y-sectors.

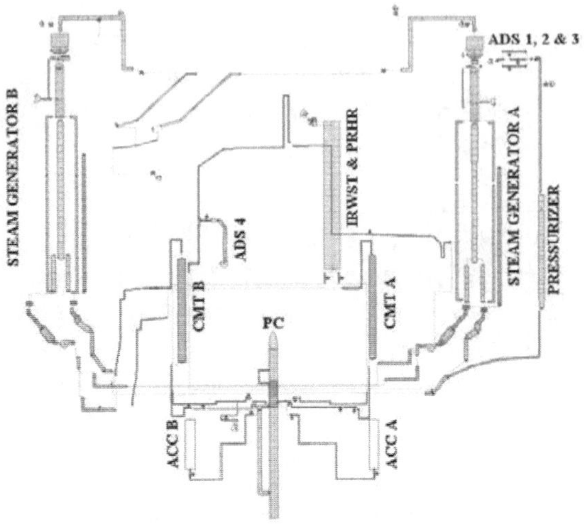

Figure 2. SPES-2 TRACE-SNAP nodalization.

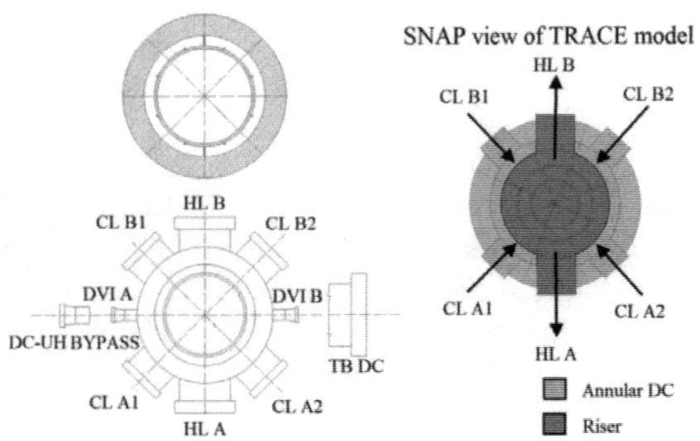

Figure 3. Power channel cross sections.

5. CATHARE model

The latest version V2.5_2 of CATHARE 2 has been adopted to simulate the SPES-2 facility, the related nodalization has been developed by respecting the geometrical dimensions of different parts and components as well as the topology of the circuits. In particular, the modular structure of the code has allowed to find a good compromise to preserve heights, flow areas and fluid volumes. It is worth to remind that the facility is scaled 1/1 in height and 1/395 in volume respect to the AP600 reactor. The annular downcomer has been simulated by means a VOLUME. In Figs. 4 and 5 are reported the CATHARE nodalization schemes of the primary vessel and of the loop A that includes the pressurizer.

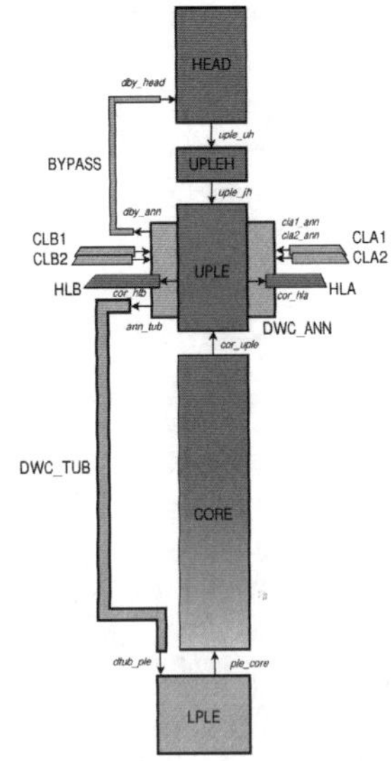

Figure 4. CATHARE PC nodalization.

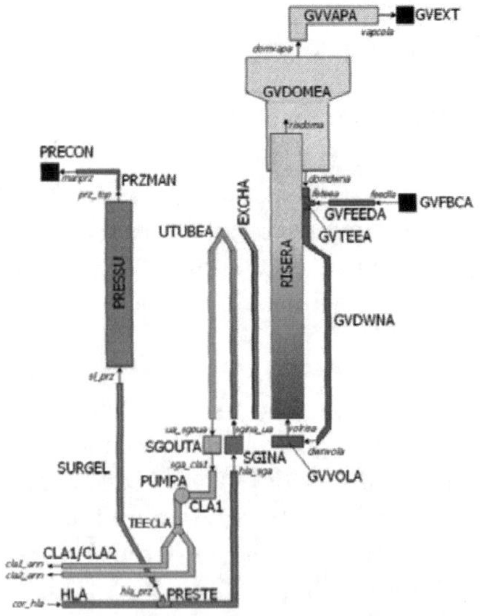

Figure 5. CATHARE loop A nodalization

6. Small break LOCA phenomenology in AP600

The AP600 Small Break LOCA (SBLOCA) transient can be subdivided into four different phases that characterize thermal-hydraulic phenomena, namely the blow-down phase, the natural circulation phase, the ADS blow-down phase and the IRWST injection phase [11]. Fig. 6 shows the pressure transient for a typical SBLOCA event. Superimposed on the graph is the relative timing of events as described in the following sections.

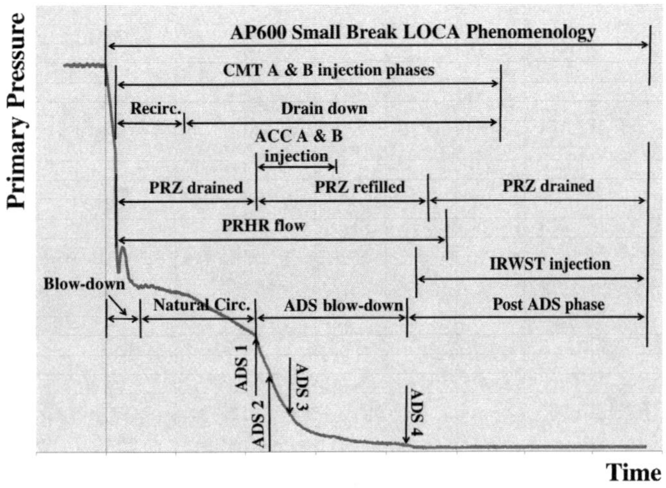

Figure 6. AP600 typical SBLOCA pressure transient.

6.1. Blow-down phase

During this phase the primary system pressure decreases from the operating pressure to the secondary-side pressure of steam generator, then reaches stable primary pressure. After the break is opened the coolant mass is lost from the primary system and the pressure falls continuously as consequence. When the pressurizer pressure reaches the setpoint value of 12.41 MPa, the reactor scram and steam generator main steam line isolation valves (MSLIVs) closure are actuated. Quickly the pressure falls below 11.72 MPa: "safety signal" condition is reached. This signal causes the opening of the CMT and PRHR isolation valves (CMTIVs, PRHRIVs) and, after a short delay, the reactor coolant pumps coastdown. Meanwhile the core power decrease according to decay heat model and the reactor system starts to be cooled by natural circulation that occurring in CMT and PRHR lines; moreover the energy is removed from reactor system by heat up of steam generators and fluid loss through the break. At the end of the blow-down phase the pressurizer is empty of water but full of steam. The liquid in the upper plenum and upper head may flash and the upper head may begin to drain.

The phenomena that take place in this phase are common to all PWR reactors except that behaviour of CMTs and PRHR that are installed only in AP600 nuclear power plant.

6.2. Natural circulation phase

This phase covers the period between the stabilization of the primary pressure with the secondary-side pressure and the ADS stage 1 opening. The primary system is cooled down by various modes of natural circulation that depend on the system inventory; that is, single-phase natural circulation, two-phase natural circulation and reflux condensation.

The decay heat is removed through the break, steam generators, PRHR and CMTs recirculation flow. The steam generator U-tubes drain relatively quickly and, as consequence, the IRWST becomes the main heat sink for the reactor coolant system. During the time between the steam generators isolation and ADS activation, the PRHR and the break flow provide a large heat removal rate. The IRWST is heated up by the two-phase mixture which flows into PRHR tubes. The heat transfer in the pool occurs

Journal of Physics: Conference Series **655** (2015) 012059 doi:10.1088/1742-6596/655/1/012059

either by free convection or boiling depending on the tube outer wall temperature and tank water temperature near the tubes. The CMTs continue the recirculation mode, providing heat removal and a small net injection to the primary system until the cold legs and balance lines drain: when this occurs the steam can flow up to the CMT top, after which CMT injection grows and its level drops continuously. When CMT level falls below ADS actuation setpoint the third phase of the SBLOCA transient begins.

6.3. ADS blow-down phase

At the beginning of the ADS blow-down phase, the primary system is depressurized through the ADS stage 1 valve opening. After ADS stage 1 valve opening, to ensure a controlled depressurization of primary system, the other two stages are opened via progressive delay within few minutes. When either CMTs water level falls below 20% of the nominal, the fourth-stage valves attached to both hot legs open and discharge water directly to the containment.

Since ADS stage 1 actuation results in a reduction in pressure at the top of the pressurizer, the pressurizer two-phase fluid level increases markedly. Thus, pressurizer level and surge line phenomena are important factors in the depressurization behaviour following stages actuation. Fluid flashing occurs again in the reactor coolant system due to increased depressurization rate after the ADS is actuated.

The accumulators start to inject cold water once the pressure drops below 4.83 MPa. This reduces the discharge flow from the CMTs, which may be stopped temporarily due to increased backpressure in the DVI lines, caused by the high accumulator flow rate. ADS stage 4 performance is affected by the nature of flow in the hot legs and by drainage from the pressurizer.

The flow through the ADS valves is the major factor that determine the instant when the primary system pressure drops such as allowing the gravity injection of water from IRWST.

6.4. IRWST injection phase

When this phase starts the pressure of the primary circuit is near the containment pressure. The injection from the IRWST signs the end of the SBLOCA transient and the beginning of the long-term cooling.

During IRWST injection phase, the core coverage with liquid or two-phase mixture depends upon the decay heat level and the IRWST injection flow. At this time the CMT flow can be stopped again owing to the increased IRWST gravity drain rate, anyway the phenomena that affect the CMT behaviour are less important compared to those of IRWST.

7. Two inch Direct Vessel Injection line break in SPES-2

The aim of the test was to investigate the response of the passive safety systems during the whole transient. The test is characterized by a single failure event: one of the two ADS stage 4 valves had to remain closed.

7.1. Experimental test description

The test starts at time 0 s by opening the break valve from full power steady-state condition. The pressurizer pressure and water level quickly decrease owing to the mass lost from the break. About 62 seconds after the beginning of the transient the pressure reaches the "R-signal" (12.41 MPa) so the reactor scram is actuated, 2 seconds later the main steam isolation valves are closed. The "S-signal" setpoint (11.72 MPa) is crossed at 70 s, 2 seconds delay the CMTs and the PRHR isolation valves are opened, at the same time the main feedwater isolation valves (MFWIVs) close. The primary pumps starts coastdown 16.2 second after S-signal. Meanwhile the primary pressure falls to the secondary-side pressure of steam generator and it reaches a quasi-stable condition. Soon after S-signal actuation the CMTs start injection in recirculation mode; when the steam occurs in cold legs, at about 400 s, it can flow through the balance line reaching the CMTs which begin to drain. Moreover after presence of steam into the top of the CMTs the primary pressure restarts to drop slowly owing to expansion of

steam in large volumes. The CMT flow rates, as expected, are slightly asymmetric during both recirculation and drain down phases.
ADS stage 1 opens at 809 s, 30 second after CMT A level reaches 67% of nominal value. ADS stage 2 and 3 open with a delay of 95 and 215 seconds, respectively, from the actuation of the first stage. At this time primary system depressurization becomes more rapid, increasing the presence of steam in the system.
Accumulator injection begins contemporaneously ADS stage 1 opening, once the pressure drops below about 4.87 MPa and the flow rate released reaches its maximum value after the ADS stage 3 actuation. The pressurizer level rises after ADS stage 1 opening and it continues to increase until accumulators injection stops.
At about 30 minutes since the beginning of the transient, CMT A level falls under 20% of its nominal value and, 60 s after, the ADS stage 4 opens. Since ADS stage 4 actuation the pressurizer goes back to emptying again.
IRWST injection begins about 100 seconds later: this signs the end of the SBLOCA transient and the beginning of the long-term cooling phase.

7.2. Comparison between experimental data and numerical results
The SPES-2 transient analysis here presented has been performed by the research groups of DEIM of the University of Palermo and ENEA of Bologna. In order to improve the prediction capability of the SPES-2 TRACE and CATHARE models, several nodalizations of the experimental plant have been studied and they have been used to perform different simulations of steady-state conditions comparing the results with data available in literature [12], [13]. A great importance was given to the correct simulation of the heat losses that affected the behaviour of these kinds of facilities especially during the analysed transient and to the simulation of the pressure distribution and pressure drops along all circuits.
In the following all figures will show the values normalized: Y-axes are normalized over the maximum value of plot scales, while X-axes are divided by a constant value.
The simulation of the transient by using TRACE code has been performed until the beginning of the long-term cooling phase, since the main phenomena were investigated. The TRACE numerical results show a good agreement with experimental data. The calculated primary pressure well reproduces the facility behaviour during the main phases of the transient though some mismatch is present in timing and value when a slow depressurization occurs after the beginning of the CMTs drain down phase, Fig. 7(a). The CMT A flow rate predicted by TRACE code is quite similar to experimental data in trend and values, Fig 8(a), as well as the calculated PRHR flow rate until the onset of ADS blow-down phase when numerical results show an unstable flow rate which did not appear in the experimental test, Fig. 9(a). These oscillations are a result of the calculated ADS flow rate which has more fluctuations than the experimental one although its mean value is very similar to experimental data. Anyway the overall behaviour of plant is well predicted by TRACE code providing a confirmation of the model capability to simulate the thermal hydraulic phenomena that occur in passive systems during the transient.
The main results obtained by a preliminary CATHARE calculation are compared with the experimental trends in Figs. 7(b) to 9(b). The Fig. 7(b) shows the fast depressurization of the primary system, this behaviour is well predicted by CATHARE in the first seconds of the transient while in the following the experimental depressurization results faster than in the simulation. The Figs. 8(b) and 9(b) report respectively the flow rate discharged by the CMT A in the primary side of the plant and the mass flowing in the PRHR after the break opening. As can be seen, the flow rates predicted by CATHARE are not in good agreement with the experimental data. This could be a consequence of the CATHARE nodalization: the annular downcomer representation by means of a volume (0D) module of CATHARE is not able to describe the real behaviour of this part of plant. In fact considering the complex phenomena that occur into this component the noding should be updated considering the possibility to introduce a 3D element to describe the behaviour of the annular downcomer.

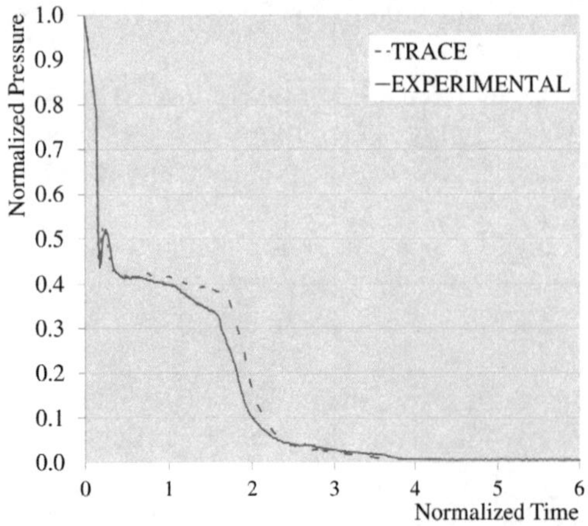

Figure 7(a). TRACE vs EXP primary pressure.

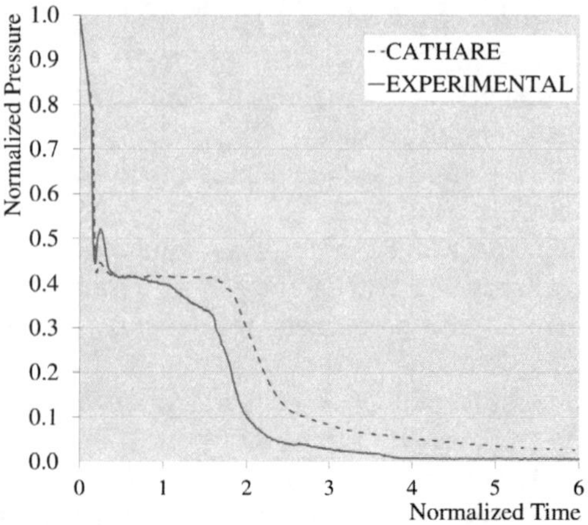

Figure 7(b). CATHARE vs EXP primary pressure.

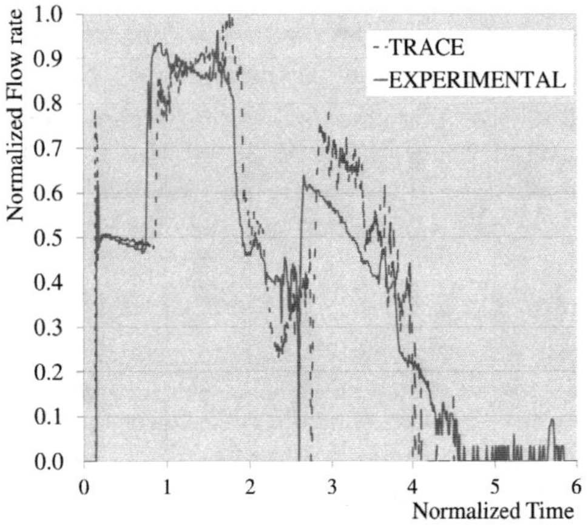

Figure 8(a). TRACE vs EXP CMT A flow rate.

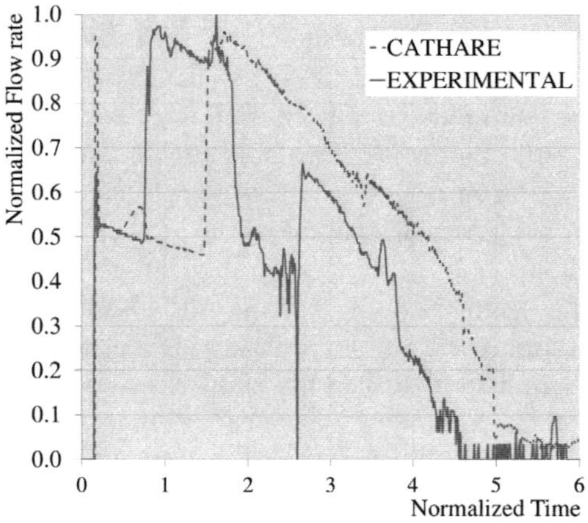

Figure 8(b). CATHARE vs EXP CMT A flow rate.

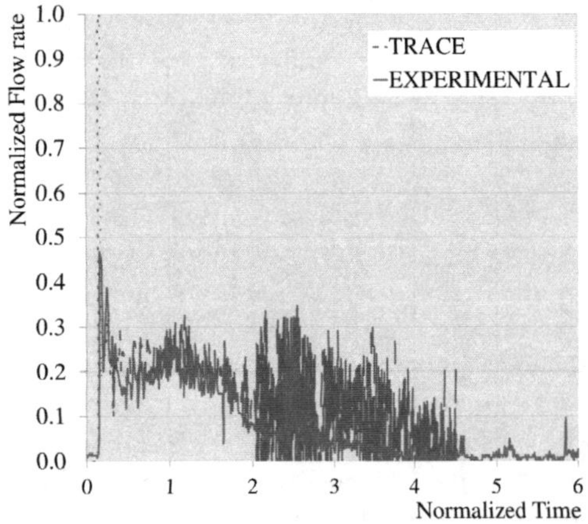

Figure 9(a). TRACE vs EXP PRHR flow rate.

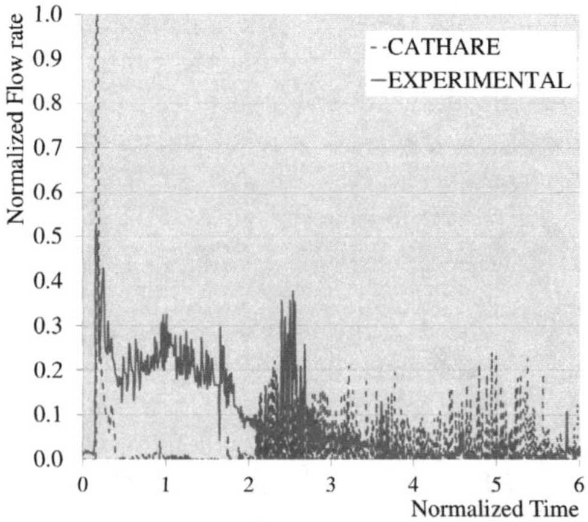

Figure 9(b). CATHARE vs EXP PRHR flow rate.

Journal of Physics: Conference Series **655** (2015) 012059 doi:10.1088/1742-6596/655/1/012059

8. Conclusions

The present work concerns a comparison between experimental and numerical results obtained for the "2 inch Direct Vessel Injection line break" in SPES-2. This facility is an integral test facility built in Piacenza at the SIET laboratories and simulating the primary circuit, the relevant parts of the secondary circuits and the passive safety systems typical of the AP600 nuclear power plant. The numerical analysis here presented was performed by using TRACE and CATHARE thermal-hydraulic codes with the purpose of evaluating their prediction capability. The main results show that the TRACE model well predict the overall behaviour of the plant during the transient and it is able to simulate the principal thermal-hydraulic phenomena related to all passive safety systems; in particular, physical models such as CCFL and "offtake" manage to respectively catch the upper-plenum\upper-head pool formations and the two-phase flows that move toward CMTs and PRHR. The performance of the presented CATHARE model has suggested a possible improvement of the geometrical nodalization to predict the behaviour of the facility during the transient that is the application of 3D module (instead of 0D module) with the aim of describing as well as possible multidimensional effects in the annular downcomer which is a key component for steam displacement around the primary circuit during the two-phase natural circulation.

9. References

[1] Alemberti A, Bacchiani M, Conway L E and Rigamonti M 1995 *Proc of the 7th Int. Meeting on Nuclear Reactor Thermal-Hydraulics (Saratoga Springs, USA, 10–15 September 1995)* vol 4 (American Nuclear Society) pp 2869-2883.

[2] Rigamonti M 1995 *SPES-2 facility description* (*SIET laboratory report)* (Piacenza: SIET).

[3] Wilson J F 1962 The velocity of rising steam in a bubbling two phase mixture *ANS-Translation* *5* **151**.

[4] NRC 2010 *TRACE V5.0 User's Manual Volume 1: Input Specification* (Washington: United States Nuclear Regulatory Commission).

[5] NRC and Applied Programming Technology 2012 *SNAP 2.2.1 User's manual* (Bloomsburg: Applied Programming Technology).

[6] Lavialle G 2005 *CATHARE 2 V2.5: 1 User's Guidelines* SSTH/LDAS/EM/2005-034.

[7] CEA 2006 *CATHARE2 V2.5_1: User's Guidelines* DER/SSTH/LDAS/EM/2005-034.

[8] D'Amico S and Moscato I 2014 *Analisi di test sperimentali nell'impianto SPES-2 e di progetto nell'impianto SPES-3 mediante codici di sistema best-estimate* (*Master Thesis*) (Palermo: University of Palermo Press).

[9] Chiovaro P D'Amico S Lombardo C Moscato I Polidori and M Vella G 2014 *Validazione e Verifica (V&V) di CATHARE2 e TRACE sul Programma Sperimentale SPES2* (*CIRTEN-ENEA report)* (Bologna: ENEA).

[10] NRC 2010 *TRACE V5.0 Theory Manual: Field Equations, Solution Methods, and Physical Models* (Washington: United States Nuclear Regulatory Commission).

[11] Wright R F 2007 Simulated AP1000 response to design basis small-break LOCA events in APEX-1000 test facility *Nuclear Engineering Technology* **39** 287-298.

[12] Ferri R 1999 *SPES-99: 10" IB-LOCA in Cold Leg Experimental Data Report* (*SIET report*) (Piacenza: SIET).

[13] D'Amico S Lombardo C Moscato I Polidori M and Vella G 2015 Transient analysis of "2 inch cold leg to core make-up tank pressure balance line break" in SPES-2 facility by using TRACE code *Proc. Of ASME-ATI-UIT on Thermal Energy Systems: Production, Storage, Utilization and the Environment* (Napoli: Enzo Albano Editore).

Heat transfer characteristics in forced convection through a rectangular channel with broken V-shaped rib roughened surface

D Fustinoni, P Gramazio, L Colombo and A Niro
Politecnico di Milano, Department of Energy,
Campus Bovisa, Via Lambruschini 4, 20156 Milano, Italy

Corresponding author email: alfonso.niro@polimi.it

Abstract. In this paper we present experimental results for a rectangular channel with the lower and upper walls configured with V-shaped broken ribs pointing both upward and downward the flow stream, as well as in a mixed configuration. Results for V-shaped ribs upward pointing are also reported for comparison. The duct cross-section is 120-mm wide and 12-mm height; the rib-roughened walls are operated at fixed temperature whereas the channel side walls are adiabatic. The ribs have square cross section of 2 mm in side, a V-apex angle of 60°, and a pitch-to-side ratios of 10 and 40. Reynolds number has been varied between 700 and 7500. The results show that the flow regime seems to be turbulent even at the lowest tested value of Re, and the friction factor is quite independent of Re, i.e., the typical trend for k-roughened surfaces. Broken and continued ribs seems to perform with no evident differences between them, except at the lowest values of Re while prevailing the former or the latter according to *p/e*. Anyway, adopting ribbed surface is always advantageous and allows a maximum heat transfer enhancement factor of 4.9.

1. Introduction

In many leading-edge engineering components the best performances entail high heat transfer rates, frequently in order to avoid the temperature overcoming limits in some their critical points, such as in gas-turbine blade cooling. Thus, very efficient and stable cooling systems are required to ensure a good operation and durability of these devices. To this end, many passive techniques to enhance heat transfer have been studied [1], [2] and, among them, the rib-roughened surfaces because they are reliable, robust, effective and quite inexpensive. As the possible shapes and arrangements of ribs are countless, starting from the '70s, a continuously growing number of studies have been dealing with heat transfer over this kind of enhanced surfaces.

From the '90s several studies, such as Han et al. [3], Kukreja et al. [4] and Gao and Sunden [5], have been devoted to V-shaped ribs as they are supposed to be effective in enhancing secondary flows. Han et al. in [3] found that the V-shaped ribs with apex angle of 45° and 60° are more effective than the parallel-tilted ribs with the same angles; besides, they also found the arrangements with the V-apex pointing upward the main flow stream are more effective than the contrary ones. However, Gao and Sunden in [5] asserted just the opposite.

From the same '90s, some studies also started proposing V-shaped broken rib configurations, namely, V-shaped ribs with the last part of their branches detached and staggered, because this kind of ribs was expected to promote more cells of secondary flows than the V-shaped "continued ribs", and that in turn should increase local turbulence and, subsequently, heat transfer.

Actually, Han and Zhang [6] found that the V-shaped broken ribs, inside a channel with a square cross-section and uniformly heated walls, promoted a higher heat transfer augmentation than the V-shaped "continued ribs". Liou et al. in [7] reported the results of a study on twelve different configurations, with both continuous and broken transverse-, tilted-, V-shaped-ribs mounted only on one wall of a channel with a square cross-section; however, they found the highest heat transfer increase is generated by the use of continuous V-shaped ribs.

Again, also in this case, we can't draw a definitive conclusion whether V-shaped broken ribs perform better or not than continued one, because the performances depend on many factors. For this reason, we started an experimental campaign aimed at investigating heat transfer characteristics and pressure drops inside a rectangular channel, with the lower and upper walls configured with a large variety of ribs.

2. Experimental setup

As schematically shown in figure 1(a), the experimental setup consists of two independent circuits, namely, the air-circuit containing the test section, and the heating-water circuit used to control the temperature of the channel walls.

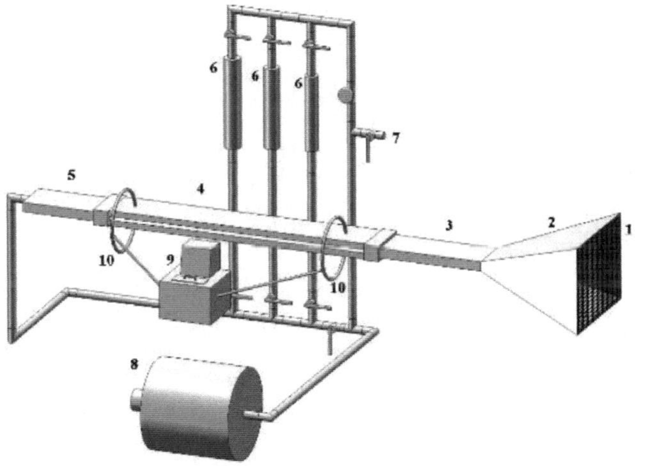

Figure 1(a). Schematics of the experimental setup: 1. fine mesh screen cover; 2. convergent air inlet; 3. entry-section; 4. test-section; 5. exit-section; 6. rotameters; 7. by-pass; 8. blower; 9. thermostatic bath; 10. water circuit piping.

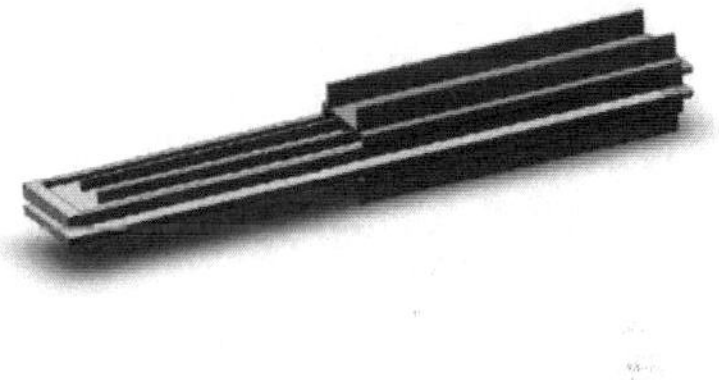

Figure 1(b). Test section with the duct on the wall backside.

Through a convergent duct, room air flows into the circuit; a fine mesh screen covers the convergent entrance, whereas at its exit there is a flow straightener consisting in a matrix of staggered 4-mm-diameter, 40-mm long thin polystyrene tubes. Inside the convergent a PTR is mounted to measure the air temperature. From the straightener air flows into an entry-section that is a rectangular duct with the same cross-section dimensions as the tested channel; entry-section is 0.8-m long, i.e., near 40 times the channel hydraulic diameter, and its walls are made with 10-mm thick plexiglass plates and are not heated. At the end of this section, air enters the test-section that is a 120-mm wide, 12-mm height, 880-mm long duct; its lower and upper walls are two aluminium plates of 10-mm minimum thickness with a rib configured face.

As shown in figure 1(b), the backside of each plate is covered by a cap strongly tightened to the plate, so that they form a jacket where the heating water flows. Inside and outside the water jacket there are ribs which prevent the plate buckling. To check if temperature is uniform over the heated walls, four thermocouples are embedded in the lower wall and three in the upper one; each thermocouple is cemented into a 1.8-mm-wide, 0.5-mm-depth groove cut in the flat face. Eventually, the test-section sides are closed by 4-mm-thick glass plates to allow optical access inside the channel.

At the test-section exit, there is a short exit-section equipped with two turbolizer rows followed by a convergent that conveys air through a 20-mm-wide, 6-mm-high channel partially filled by a fine-meshed plastic net; after the screen, a PTR and a thermocouple are positioned on the centreline of this channel to measure the air bulk-temperature. Finally, inside the couplings between the entry-section and the test-section, and between this one and the exit-section there are two pressure taps, each consisting in a 1.5-mm-diameter hole (great care was devoted in eliminating any burr). Downstream the exit-section there are three rotameters, i.e., float-type flow-meters, connected in parallel with full scale of 6, 23.5 and 40 m^3/h respectively, a metering valve, and a 7-stage, 30-kPa-head, 5.5-kW-power blower operating in suction mode. The exhausted air is discharged outside the laboratory. Air temperature is also measured upstream the flow-meters by means of a thermocouple plugged in the pipe.

The heating circuit is mainly composed by a heat bath which provides high mass flow-rate of water at constant temperature, and by the channels built into the upper and lower test-section walls (water and air stream in counterflow); two thermocouples are placed inside each water channel, near the inlet and outlet ports, respectively. The heat bath is the ThermoHaake B12 with a tank of 12-dm^3, a 3-kW heater and a high precision controller; water temperature inside the tank is kept constant within 0.01 K.

3. Measurements, data processing and error analysis

All the thermocouples here used are T-type with 0.5-mm-diameter wires, whereas the PTR are 4-wires, 100-ohm, Platinum type, i.e., PT100, with dimensions of 2 mm x 4 mm. Both thermocouples and PTR were all preliminary calibrated over five points within the temperature range from 22 to 60 °C, by means of the ThermoHaake heat bath. The probes were immersed all together into the bath while devoting great care in their positioning; for each calibration point, 160 readings per probe were collected; the resulting standard deviation is of 0.02 K for the thermocouples, and of 0.01 K for the thermo-resistances. All temperature measurements are performed by means of an Agilent 34970A data logger equipped with a relay multiplexer and a 6½ digit multimeter. The channels are sequentially read, by waiting a 0.5-s settling time after each channel-locking, with an integration time of 400 ms which ensures a standard deviation of 0.01 K that is less than or equal to the probe uncertainties; consequently, reading cycles are performed every 20 s. Pressure drops are measured by connecting the two pressure taps, placed at the inlet and outlet of the test channel, to a differential micromanometer with a full scale of 250 Pa and a 0.125-Pa sensitivity. Air volume-flow-rate is measured by means of the aforementioned rotameters which have a 2% nominal accuracy; however, by means of a calibration performed by measuring pressure drops of laminar air-flows through a smooth circular tube, we found that their accuracy is better of 1%. Finally, in order to guarantee repeatability to measurements, we adopted a precise test procedures described in the following. First, we power the bath heater and water starts to circulate through the entire circuit included the ducts on the test-section wall backside. After 30 minutes, that is the time to stabilize the channel wall temperature to a prefixed value within a 0.01-K band, we power the blower and air starts to flow through the test section. Then we need to wait for other 30 minutes in order to attain regime conditions with all temperature time-fluctuations within 0.02-K band; in this conditions, channel wall-temperature is uniform within 0.1 K over the entire heated length. Eventually, we start to collect 20 reading-cycles which take 400 s; during this time, measurements of the air-volume-flow-rate and of reading-cycles was chosen first by collecting data for N=200 for some of the most representative regime conditions, and then by calculating their average standard deviation as a function of N; as a result, we found that at N=20 the average standard deviation becomes less than 0.01 K, and thus we assumed that 20 is the minimum reading-cycle number to be collected.

The average Nusselt over the test section and the Darcy-Weisbach friction factor are calculated as follows

$$Nu = \frac{hD_H}{k}, \ \mathrm{Re} = \frac{\rho u D_H}{\mu} \tag{1}$$

$$h = \frac{\dot{Q}/A_S}{\Delta T_{m,Log}} = \frac{\dot{m}c_p(\theta_i - \theta_o)}{A_S[(\theta_i - \theta_o)/\ln(\theta_i - \theta_o)]} = \frac{\rho \dot{V} c_P}{A_S}\ln\frac{\theta_i}{\theta_o} \tag{2}$$

$$Nu = \frac{D_H}{k}\frac{\rho \dot{V} c_P}{A_S}\ln\frac{\theta_i}{\theta_o}, \ \mathrm{Re} = \frac{\rho \dot{V} D_H}{A\mu} \tag{3}$$

$$f = \frac{\Delta p}{\rho u^2/2}\frac{D_H}{\ell_{taps}} = \frac{2\Delta p A^2 D_H}{\rho \dot{V}^2 \ell_{taps}} \tag{4}$$

where h the average convective coefficient, $\dot{Q}$ the total heat power exchanged, A and A_S the cross-section area and the total heated area, respectively, of the test section, $D_H=2A/(w+H)$ the channel hydraulic diameter, the log-mean-temperature difference, θ_i and θ_o the wall-to-air-bulk-temperature difference at the entrance and exit, respectively, of the test section, and $\dot{V}$ the mass-flow-rate and the volume-flow-rate of air, respectively, ρ the density calculated at the temperature where is measured, c_P the specific heat at constant pressure and k the thermal conductivity of air, Δp the pressure drop, and ℓ_{taps} the distance between the pressure taps.

The error analysis performed according to Moffat [8] gives an uncertainty less than 3% on the convective coefficient, and less than 7% on the apparent Darcy friction factor.

4. Results

All data here reported refer to a rectangular channel with aspect ratio of 0.1, and to air-flows entering at room temperature, namely, about 22 °C. Heat transfer characteristics refer to the test section operated with the lower and upper wall at uniform and fixed temperature at the nominal value of 40 °C, while friction factors are obtained with channel walls at room temperature. Table 1 shows the main dimension of the channel geometry.

Table 1. Main channel parameters.

Channel height, H	[mm]	12.00
Channel length, L	[mm]	880.00
Channel width, w	[mm]	120.00
Hydraulic diameter, D_H	[mm]	21.82
Height to width ratio	[-]	0.10

Tests have been carried out on broken V-shaped ribs pointing both upward and downward the main stream, referred to as 3D_Up and 3D_down respectively, and furthermore in a mixed configuration, with ribs on the lower wall pointing downward while upward the ones on the upper wall referred to as 3D_mix. In addition, results for V-shape continuous ribs pointing upward (V_up) are also here presented in order to evaluate differences between these two configurations. Figure 2 shows a schematic of the different tested configurations. All ribs have a square cross section, with a side dimension of 2 mm and a pith-to-side ratio equal to 10 and 40.

Journal of Physics: Conference Series **655** (2015) 012060 doi:10.1088/1742-6596/655/1/012060

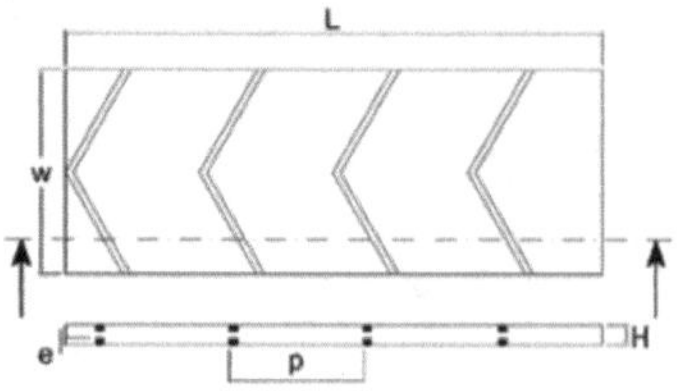

Figure 2(a). V_up: V-shaped ribs in upstream configuration.

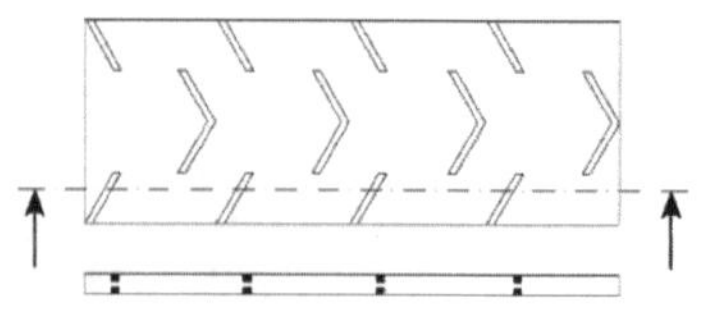

Figure 2(b). 3D_up and 3D_down: broken V-shaped ribs in upstream or downstream configuration.

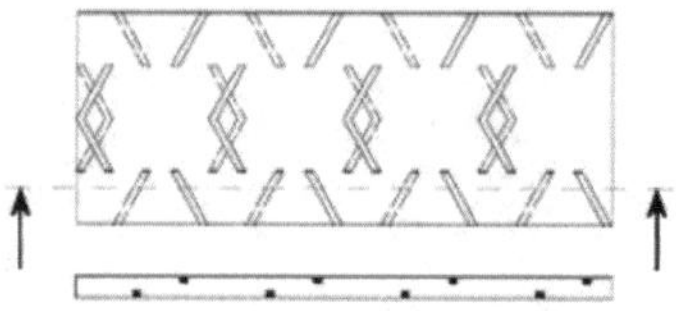

Figure 2(c). 3D_mix: broken V-shaped ribs in mixed up-stream-downstream configura-tion.

In figures 3(a) and 3(b) the average Nusselt numbers over the test section, evaluated according to equation (3), are plotted versus Reynolds number for all ribs arrangements. As reference values, data for a flat channel with the same dimensions and operating conditions are also reported. Finally, as solid lines, the values predicted for a flat rectangular-channel by the Shah-London correlation, with the Wibulsas correction, for Re<2300, and by the Gnielinski correlation for Re≥2300 [9] are plotted in the same graph. All ribs configurations show an increasing trend from the lower investigated Reynolds number to the upper ones, and a power-law dependence on Re with an exponent of about 0.9. The absence of a region where Nusselt numbers are independent by Reynolds numbers indicates that flow regime is turbulent since the lower investigated Reynolds number. The highest heat transfer increase is for the broken V-shaped with downward apex and *p/e*=10, as it displays a mean value of the ratio Nu/Nu_0 equal to 3.0 (Nu_0 refers to the flat channel operated at the same conditions) with a maximum of 4.9 at Re=3250. Among the broken ribs, the arrangements with the downward apex always show values higher than the upward one, or the mixed configuration. With regard to the pitch-to-side ratio, the arrangements with *p/e*=10 perform better than *p/e*=40. It is noteworthy that broken V-shape with upward apex displays values of *Nu* higher than the mixed (broken) configuration for *p/e*=10 but lower than the latter for *p/e*=40. By comparing thermal performances of broken and continued ribs with upward apex, seemingly there are no differences between them except at the lowest Reynolds number where a opposite behaviour is displayed according to *p/e*. In particular, for *p/e*=10 continued ribs perform better than broken ones, but for *p/e*=40 the latter configuration is winning.

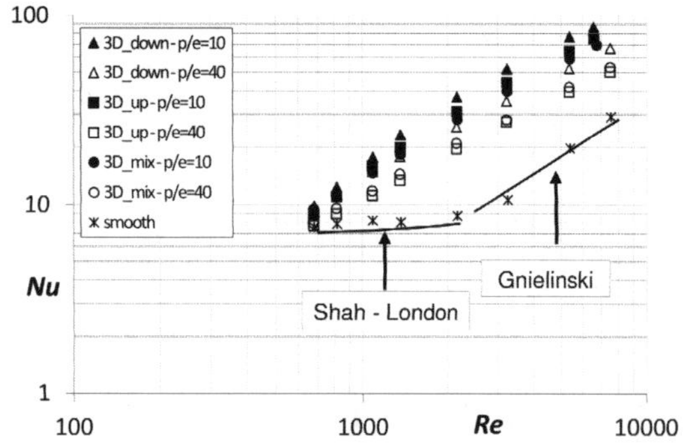

Figure 3(a). Average Nusselt number versus Reynolds number for the broken V-shaped arrangements.

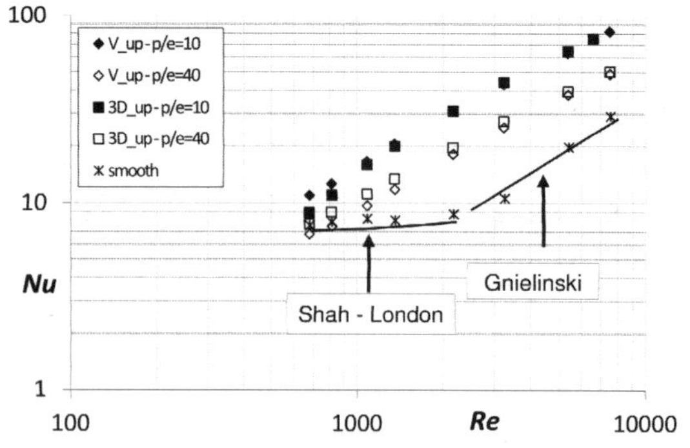

Figure 3(b). Average Nusselt number versus Reynolds number for the broken and continuous V-shaped pointing upward arrangements.

Figures 4(a) and 4(b) show experimental values of the apparent Darcy friction factor *f* plotted versus *Re* for all the rib configurations. For comparison, the figures also report as solid lines the values f_0 predicted for a flow through the flat rectangular channel, both in laminar regime and turbulent regime, by the Shah-London correlation [9]. All configurations show higher values of *f* than the smooth channel. In particular, the highest value of ratio f/f_0 is 16 for broken V-shaped ribs, pointing both upward and downward, for *p/e*=10. Figures 4(a) shows that V-shaped ribs with both upward and downward apex display pressure drops higher than the mixed configuration. That always holds for *p/e*=10 whereas for *p/e*=40 it is true only at high Reynolds numbers. Comparison of data for broken and continued ribs shows the former has always values higher than the latter for *p/e*=10, while for *p/e*=40 there is a dependence on the Reynolds number, namely, *f* of broken ribs is higher than continued ones for *Re*<2100 while becoming lower in the fully-turbulent regime.

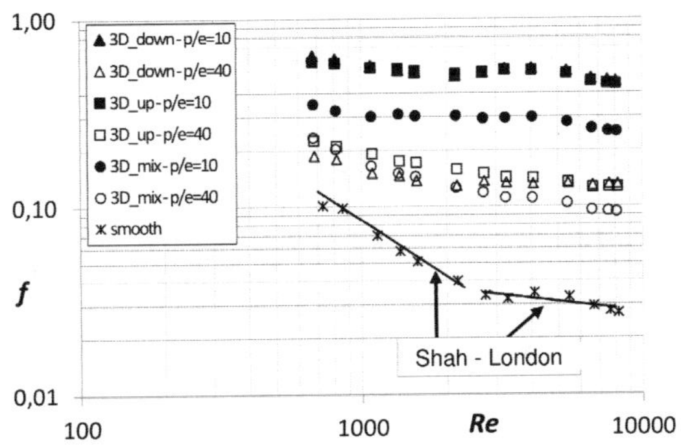

Figure 4(a). Darcy friction factor versus Reynolds number for the broken V-shaped arrangements.

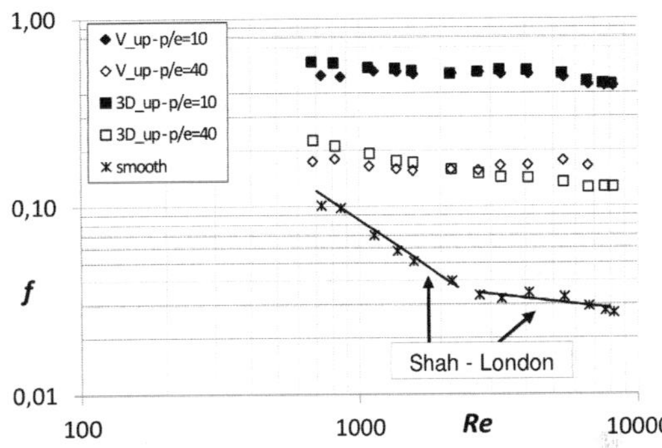

Figure 4(b). Darcy friction factor versus Reynolds number for the broken and continuous V-shaped pointing upward arrangements.

As a concluding remark, from an engineering point of view it is more interesting to compare performances of different rib configurations by taking into account simultaneously both heat transfer enhancement and pressure-drop penalization at constant pumping power, equation (5).

$$\frac{Nu/Nu_0}{(f/f_0)^{1/3}} \tag{5}$$

Figure 5 shows data obtained according to this criterion plotted versus the Reynolds number; all rib arrangements have similar trends with a maximum at Reynolds number of 3250. Broken V-shaped with apex pointing downward for *p/e*=40 seems to be the best configuration with the highest value of 2.07.

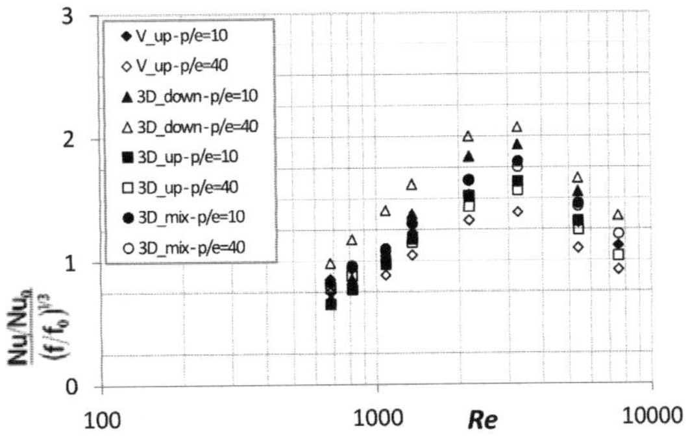

Figure 5. Heat exchangers comparison at constant pumping power.

On the other side the worst one is the continuous V-shaped pointing upward for *p/e*=40. According to this criterion, broken ribs perform better than continued ones over all the investigated Reynolds number but only for *p/e*=40; it can be seen that the maximum values are 1.57 and 1.39, respectively. At the lowest pitch-to-side ratio, the difference seems to decrease.

5. Conclusions

Within the investigated range of Reynolds numbers, i.e., 700 – 7500, the configuration with broken V-shaped pointing downward for *p/e*=40 shows the best compromise between heat transfer enhancement and pressure drop penalization at constant pumping power. From the collected data, the following general conclusions can be drawn.

Average Nusselt number:

- data for all ribbed tested configurations show the flow regime seems to be turbulent even at the lowest tested value of *Re*, and a power law dependence on *Re* with an exponent of 0.9;
- the highest heat transfer enhancement factor is 4.9 at *Re*=3250 for V-shaped broken ribs pointing downward with *p/e*=10;
- V-shaped pointing downward arrangements seem to have larger values than other investigated configurations;
- data for broken and continued ribs seem to have no evident difference between them except at the lowest values of *Re*.

Pressure drop:

- data confirm the flow regime seems to be turbulent even at the lowest tested value of *Re*
- the highest value of the ratio f/f_0 is 16 and is attained by the V-shaped broken ribs, pointing both upward and downward, with *p/e*=10;
- mixed configuration is characterized by values lower than the others except for *p/e*=40 at the lowest *Re*.

6. Nomenclature

Symbol	Quantity	Unit
A	Channel cross section area	m^2
A_S	Channel total heated area	m^2
AR	Channel aspect ratio	-
c_p	Air specific heat at constant pressure	$J\ kg^{-1}\ K^{-1}$
D_H	Channel hydraulic diameter	m
e	Rib side dimension	mm
f	Darcy-Weisbach friction factor	-
H	Channel height	mm
h	Average convective coefficient	$W\ K^{-1}\ m^{-2}$
k	Air thermal conductivity	$W\ K^{-1}\ m^{-1}$
ℓ_{taps}	Distance between the pressure taps	m
$\dot{m}$	Air mass flow-rate	$kg\ s^{-1}$
Nu	Average Nusselt number	-
p	Rib pitch	mm
Δp	Pressure drop over ℓ_{taps}	Pa
Pr	Prandtl number	-
$\dot{Q}$	Total heat power exchanged	W
Re	Reynolds number	-
$\Delta T_{m,Log}$	Log-mean-temperature difference	K
$\dot{V}$	Air volume flow-rate	$m^3\ s^{-1}$
w	Channel width	mm
Greek symbols		
α	Rib angle respect to main stream	deg

θ	Wall-to-air-bulk-temperature difference	°C
ρ	Air density	kg m^{-3}
μ	Dynamic viscosity	kg m^{-1} s^{-1}

Subscripts

i	Inlet
o	Outlet
0	Performance of smooth channel

7. References

[1] Webb R L 1994 *Principles of enhanced heat transfer* (New York: Wiley-Interscience)
[2] Gupta S, Chaube A and Verma P 2012 Review on Heat Transfer Augmentation Techniques: Application in Gas Turbine Blade Internal Cooling *J. Eng. Sci. Tech. Rev.* **1** 57-62
[3] Han J C and Park J S 1985 Heat transfer enhancement in channels with turbulence promoters *J. Eng. Gas Turbines Power* **107** 628-35
[4] Kukreja R T, Lau S C and McMillin R D 1993 Local heat/mass transfer distribution in a square channel with full and V-shaped ribs *Int. J. Heat Mass Tran.* **36** 2013-20
[5] Gao X and Sunden B 2001 Heat transfer and pressure drop measurements in rib-roughened rectangular ducts *Exp. Therm. Fluid Sci.* **24** 25-34
[6] Han J C and Zhang Y M 1992 High performance heat transfer ducts with parallel broken and V-shaped broken ribs *Int. J. Heat Mass Tran.* **35** 513-23
[7] Liou T M, Chen C C and Tsai T W 1999 Heat transfer and fluid flow in a square duct with 12 different shaped vortex generators *J. Heat Transf.* **122** 327-35
[8] Moffat R J 1988 Describing the uncertainties in experimental results *Exp. Therm. Fluid Sci.* **1** 3-17
[9] Kakaç S, Shah R K and Aung W 1987 *Handbook of single-phase convective heat transfer* (New York: Wiley-Interscience)

Journal of Physics: Conference Series **655** (2015) 012061 doi:10.1088/1742-6596/655/1/012061

Passive and active infrared thermography: An overview of applications for the inspection of mosaic structures.

P Theodorakeas, E Cheilakou, E Ftikou, M Koui

National Technical University of Athens, NDT Lab, Materials Science & Engineering Department, School of Chemical Engineering, Zografou Campus, 157 80 Athens, Greece.

corresponding author's e-mail: pantheod@mail.ntua.gr

Abstract. Infrared Thermography is a non destructive testing and evaluation (NDT&E) technique, which has been widely used for the investigation of cultural heritage and art objects. The main purpose of this study is to present the capabilities of both passive and active thermography for the inspection of mosaic structures, evaluating the performance of each testing approach through its application in representative mosaic structures. In situ passive thermography was applied on mosaic pavements in an attempt to acquire knowledge about their preservation state, while the active approach was used in order to study plastered mosaics and characterise the tesserae layer beneath the plaster. The results from this study revealed that passive approach can be efficiently applied as a moisture detection tool and a rapid monitoring technique of the mosaic condition, while the active thermographic investigation showed much more potentiality as quantitative information for the detected feature was further retrieved.

1. Introduction

Among the different NDT&E techniques, thermal imaging can accurately obtain subsurface features information without compromising the structural integrity of the inspected target. The above along with the non contact nature of this technique have resulted to its wide use in the cultural heritage conservation field [1,2]. The two basic configurations for implementing an infrared thermographic survey are passive and active approach respectively. In passive thermography, the monitoring of the thermal radiation, emitted by the surface of the test body under natural conditions, is used and it is widely applied as a standard quality control technique of historic structures since many years ago [3,4].

The active configuration is deployed through the generation of a heat flow such that the thermophysical properties of the test object can be made to enhance or impede this flow. The thermographic investigation during and/or after a heating process, has been proven to present much more potentiality than the conventional passive approach [5], while recent advances in thermographic signal processing and the development of efficient numerical algorithms have made it practical to implement active thermal imaging for quantitative purposes as well [6,7].

The main objective of this study was to evaluate the performance of both passive and active thermography as inspection methods and monitoring techniques of mosaic artefacts. In particular, passive thermography was applied in order to test three mosaic pavements having received different conservation interventions, while the active approach was used to test a plastered mosaic in order to reveal the hidden artefact and gain information regarding its location, its thickness and its characterisation based on the estimation of its thermal properties.

2. In situ passive thermography for the inspection of mosaic pavements

2.1. Early Christian mosaic floor at Delphi Museum

The first mosaic tested was a stone-glass tesserae consisted pavement, discovered at the archaeological site of Delphi and transferred on the roof of the Delphi Museum, in Greece. Its origin is dated back to the Early Byzantine Period (5th century A.D) and as can be seen from Figure 1, its decoration schematises a tress as a framework and the main scene by geometrical motives and by human and animal representations. Initially, the mosaic was conserved in 1962, where the *tessellatum* layer was detached and placed on reinforced cement mortar, while in 2006 the mosaic was transferred in pieces and reattached on the roof of the Museum. In the latter conservation procedure, the infrastructure was built from bricks arranged in a square formation, within which polystyrene foam covered by metal armatures and cement was fitted. The sections were placed on the infrastructure and the gaps were filled with mortar. The thickness of the overall construction is about 25cm, while some parts of the mosaic carry inner tubes for drainage of water around the base.

(a) (b)

Figure 1. a) Panoramic view of the Delphi floor mosaic's eastern area and (b) mosaic area of animal representations on the *tessellatum* layer.

The main purpose of this survey was to apply thermographic testing in order to investigate for moisture accumulation on the mosaic pavement. Visual observations revealed local regions with detached tesserae, while the climate conditions where the mosaic is located, are also harsh during the winter time. Here, it shall be mentioned that the above mosaic's thermographic investigation was performed during the noon when the solar energy was directly heating up the inspected structure, producing a transient in a manner thermal regime. Representative thermal images from the inspection of the Delphi mosaic are illustrated in Figure 2, where temperature variations were observed on the surface decorative layer.

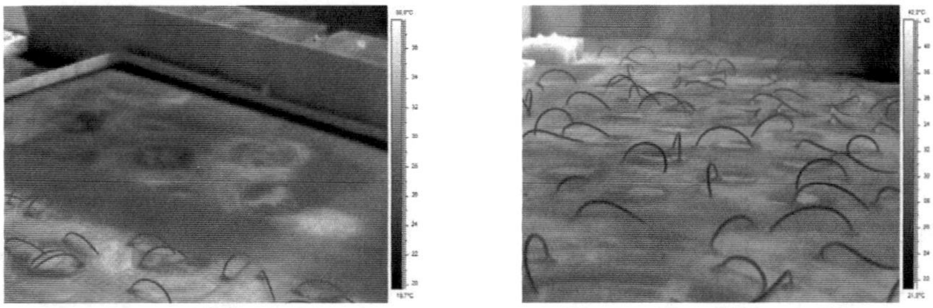

(a) (b)

Figure 2. Thermal image from (a) the mosaic region without drainage system and (b) with drainage system.

More specifically, Figure 2a presents the surface temperature map from an area without drainage system and Figure 2b presents the respective thermal image from a region having the drainage system discussed above. Comparing the two images and considering that the testing scenario was applied in a "naturally active" mode, temperature uniformity was observed in the latter case, indicating the better preserved state of the mosaic part with the drainage system. In other words, deterioration phenomena are more aggressive along the eastern area of the mosaic, where the water accumulation occurred due to the lack of the drainage system.

2.2. Mosaic pavement at the Ancient Agora of Athens

The second mosaic inspected it is ruins of an ancient house, revealed during the excavation of *Dörpfeld* at the Ancient Agora of Athens. The mosaic floor dates back to the 2nd century B.C. and it is decorated with geometric patterns. In 1960, the mosaic was conserved using cement mortar, while in 2002, a second conservation procedure was applied during which a lower level of pebble floor was appeared (dated possibly at the Hellenistic period), that was covered with geotextile and sand. Photographs from the current state of the mosaic are presented in Figure 3, where visual observations revealed local tiles detachments, few biological deposits and surface cracks on the mosaic surface.

(a) (b) (c)

Figure 3. Mosaic pavement from the ruins of an ancient house located at the Ancient Agora of Athens with (a) the state of the structure after the 2002 conservation intervention, (b) the preservation state during the survey in 2014 and (c) a region with surface cracks obvious through visual inspection.

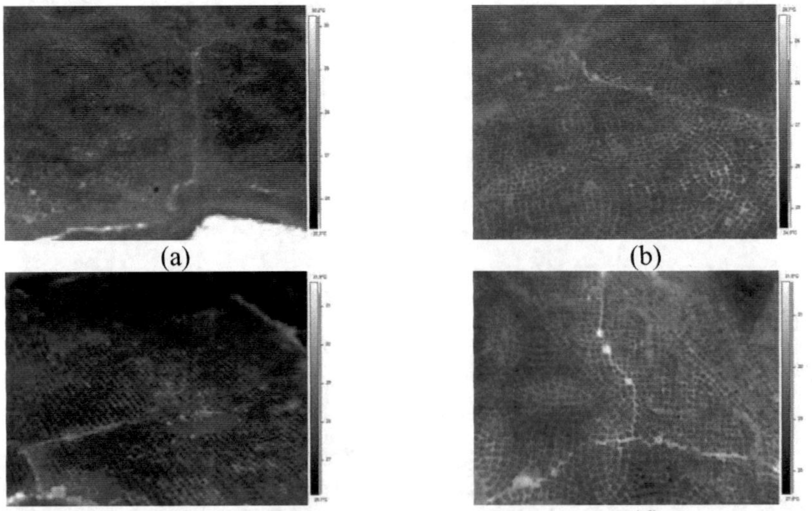

(a) (b) (c) (d)

Figure 4. Thermal images revealing surface cracks with (a), (b) being invisible and (c), (d) being partially visible from visual inspection.

Apart from the above mentioned deterioration phenomena, thermographic inspection revealed some further surface cracks, which were either difficult to be detected through visual testing or undetectable in some instances. Nevertheless, through infrared vision it was very easy to detect and map the cracking as presented on the indicative thermal images of Figure 4, acquired at different parts on the mosaic surface. As the cracks were randomly detected at several areas, this phenomenon can be possibly attributed to the conservation procedure performed by placing cement mortar as a substrate layer, enhancing the incompatibility of the structure in terms of materials lay-up and mechanical behaviour. In this point, it is interesting to note that cracks propagation obvious to the naked eye was mainly detected on regions made of glass tesserae (such that in Fig. 3c), while cracks of smaller thickness and not easily detected though visual testing were mainly observed on the stone tesserae constructed regions.

2.3. Mosaic pavement at the Sanctuary of Pan

The third in situ passive thermographic inspection was performed on a mosaic consisted of lime mortar substrate and a coloured stone tesserae decorative layer, dated between the 2nd century B.C. and 1st century A.D. The Sanctuary of Pan was discovered in 2001, where outside the caved chamber a rectangular room with a floor mosaic of geometric patterns was found (Figure 5a).

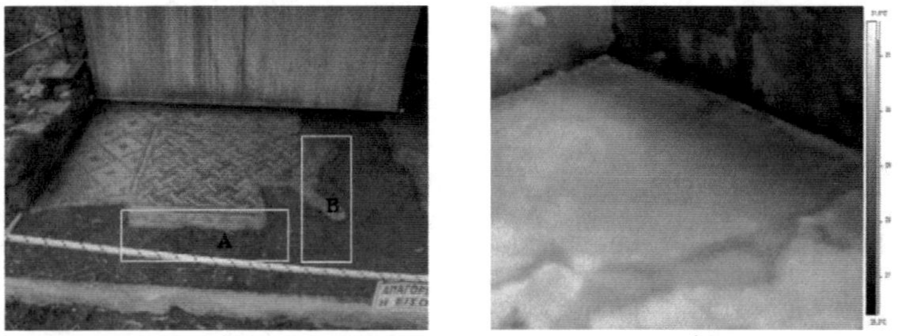

(a) (b)

Figure 5. (a) Mosaic pavement at the *Sanctuary of Pan* and (b) thermal image confirming the good preservation state of the mosaic.

The room into which the mosaic was found should be a villa of the early Imperial period and its relationship with the adjacent "*Sanctuary of Pan*" has not been fully elucidated. Between 2001 and 2004, a complex program of restoration works was implemented by the Ministry of Culture of Greece. The conservation interventions focused on the reattachment of two scattered mosaic pieces on new mortar (areas A and B in Figure 5a), while cleaning and consolidation procedures were also applied on the structure. Moreover, a shelter was built in order to protect the artefact from its direct exposure to rain and solar radiation. Visual inspection verified that the mosaic structure was in a good preservation state, something that was also confirmed through the aid of thermographic testing. In particular, as presented on Figure 5b, surface temperature uniformity was detected on the decorative layer of the mosaic. Nevertheless, humidity was detected at the consolidated edges mainly next to the lateral cave wall even if a tent that was placed vertically on the cave wall.

3. Active thermographic investigation of plastered mosaics

3.1. Samples preparation and testing methodology description

The aim of this study was to experimentally and numerically demonstrate the potential of active thermography on revealing and quantitatively characterising a hidden mosaic. Images seeing-through the mortar on a plastered mosaic surface were obtained using the principles of cooling down thermography, while quantitative information regarding the hidden artefact was retrieved through the simultaneous monitoring of the surface temperature decay over a reference (just plaster) and a mosaic-consisted area, and through the conduction of numerical parametric studies for the evaluation of specific parameters alterations (i.e. covering thickness, mosaic layer thickness, mosaic thermophysical properties) on the thermal response of the structure. For this study, two (2) panels -one mosaic sample consisted of marble tesserae and covered with two different layers of plaster and one blank sample with just plaster and no tiles underneath it- were prepared in the laboratory. The cross section of the investigated panel is shown in Figure 6b, while the description of the tested panels is presented in Table 1.

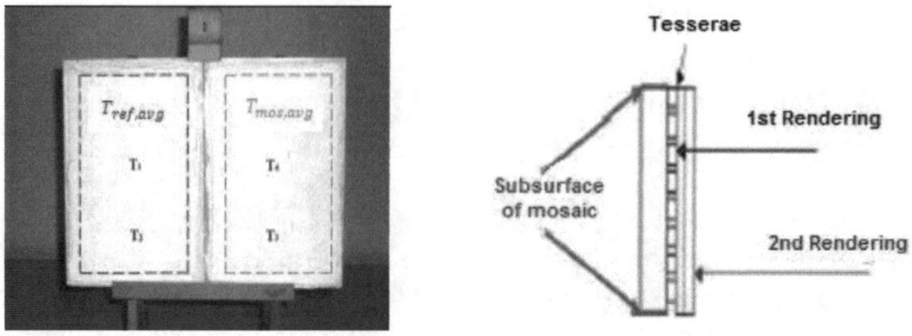

(a) (b)

Figure 6. a) Real image of the tested surfaces. The panel on the left side is the reference sample and the panel on the right side is the plastered mosaic. b) Cross-section schematic of the plastered mosaic panel.

For the cooling down thermographic approach, the inspected specimens were heated with the use of an external 1500 W heat source (infrared lamp of INFRATECH type), placed at the distance of 40 cm from the inspected surfaces. The thermal excitation process was performed for 90 min, while the recording of the transient cooling phase was performed for 60 min, through the aid of an infrared camera placed always vertically to the surface of the panels. The thermographic system used for this study (ThermaCAM SC640) was operating at the wavelength region from 7.5 to 13.5 μm and it was a focal plane array device with an image resolution of 640 x 480 pixels. For the monitoring of the cooling down process, thermal images were recorded sequentially with a time step of 30 s. The concept of the measurements performed in the present study, was based on the assumption that infrared thermography will be able to detect the hidden mosaic, presented with temperature variations on the surface, due to the dissimilar thermal diffusion that each layer renders. Thus, the measurement was performed by comparing the thermal behaviour of the covered mosaic with respect to the behaviour of the corresponding in terms of plastering reference sample. For quantitative analysis, the monitoring of surface temperature variation as a function of time from areas above the plastered mosaic and above the blank sample were compared and the produced temperature differences were further plotted as a function of time.

Table 1. Test samples description

Sample No.	Tesserae material	Lay-up description (starting from the top surface layer)	Lateral dimensions
M2	Marble	1cm lime mortar + 1 cm of cement mortar + 0.5 cm tesserae layer + 2cm lime mortar	20 cm x 33cm
M5	No tesserae	1cm lime mortar + 1 cm of cement mortar+ 2cm lime mortar	20 cm x 33cm

The purpose of ΔT vs. time plotting was to monitor how this quantity changes as a function of time and to identify the maximum ΔT_{max} and the distinct time of this occurrence t_{max}. In order to compute the above quantity, two different options were investigated in the sequential analysis of the cooling down process, selecting initially a single pixel above the plastered mosaic and a single pixel above the mosaic-free surface. Additionally, the same analysis was repeated after computing the average temperature values on the two aforementioned areas. As the first option raises the crucial issue of how the control points should be selected (the feature of interest occupies a large volume of pixels in the thermal image), four different points were selected –two on each sample surface- accommodated in diametrically opposed positions between them (pixels T1 and T4 were selected at the centre of each panel, while pixels T2 and T3 were selected near the bottom edges, Figure 6a).

Along with the experimental testing, parametric studies were conducted in a simulation environment, modelling transient thermal regimes. Through numerical simulations, the aim was to evaluate the mosaic detectability by altering parameters such as the covering intervention thickness, the mosaic layer thickness and the thermophysical properties of the hidden mosaic. The above parametric studies were implemented through the aid of *ThermoCalc-3D*^{TM} heat transfer model [8], calculating the thermal response of the covered mosaics after the application of a heat flux. The dimensions of the physical models were 40cm x 33cm x 4 cm with the half area referring to the blank panel (20cm x 33cm), whilst the other half was designed to consist of a simulated subsurface mosaic area (20cm x 33 cm). For the creation of the physical models, the materials properties were obtained by the literature [8,9] and the complete description of the layers specifications and the simulated testing parameters are summarised in Table 2. For numerical analysis and data interpretation, transient curves were calculated in a similar manner as in the case of experimental testing. Nevertheless, in order to measure the temperature decay and the thermal contrast evolution curves above the mosaic-free and covered mosaic regions, only two points accommodated in diametrically opposed positions and placed at the centres of each respective area, were selected due to the thermal stimulation uniformity that the simulation regime provided.

Table 2. Input parameters for the description of the simulated samples' layers and testing procedure

	Thermal conductivity ($Wm^{-1}K^{-1}$)	Specific heat ($Jkg^{-1}K^{-1}$)	Density (kgm^{-3})	Thermal effusivity ($Ws^{1/2}m^{-2}K^{-1}$)	Thermal diffusivity ($x10^{-7}$ m^2s^{-1})
Lime plaster	0.87	800	1440	1001	7.55
Concrete plaster	1.4	1100	2000	1755	6.36
Marble tesserae layer	3.14	870	2700	2715	13.4
Geometrical characteristics					
Concrete cover (cm)	length = 40	width = 33	thickness= 1		
Lime cover (cm)	length = 40	width = 33	thickness= 1		
Subsurface lime layer (cm)	length = 40	width = 33	thickness= 2		
Marble mosaic layer (cm)	length = 20	width = 33	thickness= 0.5		
Simulation testing parameters					
Power input (W/m^2)	1500	Duration of energy input (min)		90	
Duration of observation in min (including heating)	150	Time interval (s)		30	

3.2. *Experimental and numerical results*

The figure of Table 3 presents an indicative thermal image from the inspected set, after thermally exciting simultaneously the reference panel and the covered mosaic. As can be seen from the thermogram, increased thermal energy was deposited on the right surface (plastered mosaic) with respect to the thermal energy deposited on the left surface (blank sample), revealing the presence of the hidden mosaic layer beneath the coating. In other words, the different type of material embedded into the structure is altering the cooling down behaviour on the respective area. Additionally, the thermal contrast levels measured using the different control points are also summarised in Table 3, providing the temperature difference values at the beginning, at the time point of maximum thermal contrast occurrence and at the end of the cooling down process. In particular, from the ΔT values monitored through the T4-T1 spots analysis, it was observed that a local negative ΔT value of -5.7 °C was initially detected, having an upward trend, until reaching its maximum value ΔT_{max} of 12.4 °C at t_{max} = 31 min. The signal after that slowly decays to a ΔT value of 8.8 °C at the end of the cooling down recording. The thermal characterisation obtained from the second spot testing (T3/T2) and the measurement of the average temperature over the investigated areas, produced similar results as in the case of T4/T1 analysis. However, some small variations in terms of ΔT and t_{max} values were observed, indicating that the quantity of heat delivered to the sample surface was not uniform directly influencing their temperature decays and thus the resulting thermal contrast evolutions. This can be further confirmed if consider that the surface emissivity was constant on the two panels due to the same covering intervention that they received.

Table 3. Summary of experimental results

Set of samples	Indicative thermal image at t=30 min after the beginning of the cooling down process	Temperature differences (^{0}C)		
M2-M5 marble/cement, lime		$\Delta T_{4,1}$ t=0 (min)	$\Delta T_{4,1}$ max	$\Delta T_{4,1}$ t=60 (min)
		-5.7	12.4/ t= 31	8.8
		$\Delta T_{3,2}$ t=0 (min)	$\Delta T_{3,2}$ max	$\Delta T_{3,2}$ t=60 (min)
		-2.3	8.8/ t=30.5	5.8
		ΔT_{avg} t=0 (min)	ΔT_{avg} max	ΔT_{avg} t=60 (min)
		-4.4	10.9/ t=32	7.7

In order to conduct the parametric studies discussed above and to extract results regarding the influence of specific parameters variations on the hidden mosaic identification, a series of simulations were also carried out. Particularly, in order to study the influence of covering layer thickness on hidden mosaic detectability, simulations were performed for a mosaic layer of stable thickness (5mm) covered with plaster varying from 0.5 to 3.5 cm. The thickness step increase was of 0.5 cm and each layer was increased equally from 0.25 to 1.75 cm. The numerical investigation was fitted to simulate the experimental testing conditions and in each individual model setup, the characteristic ΔT_{max} and the respective t_{max} were calculated. The above procedure aimed to correlate the variation of these physical quantities with respect to the covering thickness increase, as illustrated in Figure 7a. As can be seen from the plot, for thicker covering interventions the hidden mosaic layer can be still identified, despite the reduced thermal contrasts produced on the modeled surfaces, while the simulated thermographic monitoring of thinner interventions produced a faster seeing-though condition of the hidden feature.

(a) (b)

Figure 7. Variation of ΔT_{max} and of the correspontent t_{max} with respect to (a) the plaster cover thickness increase and (b) the mosaic layer thickness increase.

Along with the covering influence, the thickness of the tesserae layer is another parameter affecting the identification of the covered features. In a similar manner, the influence of mosaic layer thickness on the thermal response of the structure was studied in a 2 cm covering model with tiles thickness varying from 2 to 14 mm. The mosaic thickness was increased by a 2 mm step and Figure 7b presents the variations of the peak thermal contrast values and of the correspondent maximum times associated with the mosaic thickness increase. The numerical results from this parametric study indicate that an enhanced visibility can be achieved for more resistive subsurface features. In other words, as higher was the thickness of the hidden tiles the more enhanced was its visibility under the plastering regimes, as hidden mosaics react better in heat accumulation.

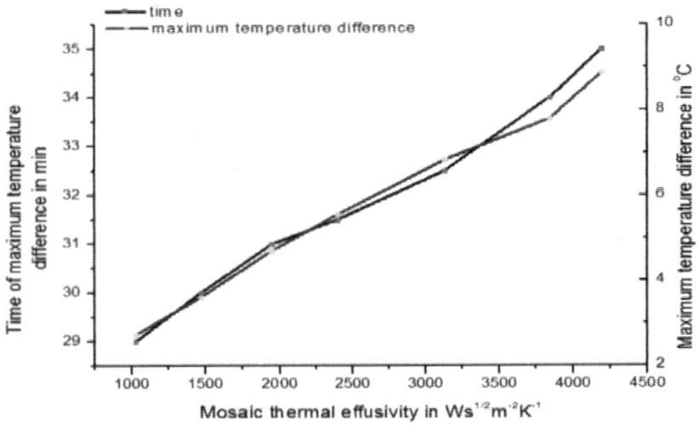

Figure 8. Variation of ΔT_{max} and of the correspontent t_{max} with respect to the mosaic thermal effusivity for the lime/cement covering intervention.

Finally, the thermal response of the covered mosaic was evaluated in terms of its thermophysical properties, taking into consideration the material properties of thermal effusivity and thermal diffusivity, respectively. In a model having the physical dimensions of the real samples, the influence of the mosaic thermophysical characteristics was investigated by varying simultaneously the thermal conductivity of the mosaic surface from 1 to 4.5 W/(mK), the mass density from 1500 to 3000 kg/m^3 and the specific heat from 700 to 1300 Jkg^{-1}K^{-1}. The above variations produced 7 models, where the thermal effusivity of the investigated feature was varied from 1024 to 4189 Ws$^{1/2}$m^{-2}K^{-1} and its respective thermal diffusivity from 9.5 to 11.5 x10^{-7} m^{2}s^{-1}. Figure 8 illustrates the variations of the informative parameters ΔT_{max} and t_{max} as a function of the mosaic thermal effusivity (a similar behaviour would be exhibited if the x-axis had the values of thermal diffusivity), where as can be seen as the effusivity of the covering layer remained constant, as greater was the thermal effusivity of the hidden feature the lower was its reaction in the conductive heat transfer back to the surface due to the produced temperature difference.

3.3. Quantitative analysis of plastered mosaic

Along with the qualitative characterisation of the plastered mosaic structures, quantitative information were retrieved correlating numerical and experimental results and solving the inverse problem through the parameter estimation approach. As a result, three different inverse problems were produced, where in the first one the location of the mosaic layer was estimated considering mosaic's thermophysical properties and thickness known, the second one was conducted for the estimation of mosaic thickness being known its location and its thermophysical properties and the last one predicted the thermophysical properties of the hidden mosaic being known the geometrical characteristics of the structure. As a result, least square fitting was used in order to find the best fitting function describing the evolution of the dependent variable of t_{max} with respect to the parameter estimated (independent variable) in each individual inverse problem. An example of this analysis is presented in Figure 9, where the fit function of t_{max} related to the depth at which the mosaic is located to, is presented. On the aforesaid plot, the experimental values of maximum times from the three different analysis procedures can be connected to particular mosaic depths.

The results obtained by solving the inverse problem for mosaic depth and tiles thickness prediction are summarised in Table 4. The measured depths determined by the inverse solution differ from 0.11 mm to 0.17 mm to the nominal depth, which in terms of error estimation can be deduced to a range from 5.5% to 10%. These deviations can be characterised as relatively short, as their production can be possible attributed to a non uniform plastering during the sample manufacturing. As regards the comparison of depth information retrieval through the different analysis procedures, variations were detected when using the experimental data derived from the correspondent pixels' analyses, which was fixed using the average surface temperature.

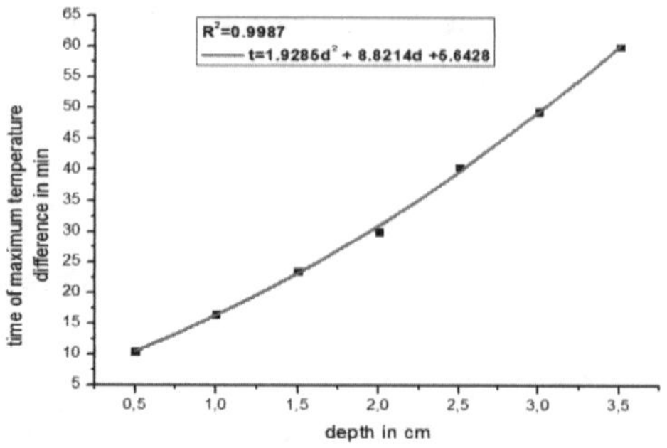

Figure 9. Time of maximum temperature difference as a function of mosaic depth, with the curve following a fitting function of a second order polynomial.

The second inverse problem investigated the correlation of numerical and experimental data in order to acquire quantitative information regarding the thickness of the hidden mosaic layer. In general, the acquired results were able to provide estimations regarding the tesserae thickness, however larger prediction errors were produced in this case with respect to the accuracy produced in the former one. Nevertheless, a comparison of the produced results through the different data analysis procedures confirmed as well that the temporal monitoring of the average surface temperature produces more accurate results with respect to these derived from the pixel analyses.

Table 4. Determination of mosaic depth and thickness through the inverse solution

Actual depth (cm)	Estimated depth (cm)		Error (%)	Actual thickness (mm)	Estimated thickness (mm)		Error (%)
	Centre	1.83	-8.5		Centre	2.55	-49
2	Edge	1.80	-10	5	Edge	2.06	-58.8
	Average	1.89	-5.5		Average	3.54	-29.2

Finally, the third inverse problem was created in order to fit the maximum temporal characteristic with respect to the mosaic's thermal effusivity and thermal diffusivity changes. The classical thermophysical properties of a material that are characterising its response to a heat flux are the thermal conductivity k, the specific heat cp and its mass density ρ. The latter two can be combined to the heat capacity which is a material property, derived from the product of specific heat and density (cp·ρ). These thermophysical properties in a transient thermal regime can be considered in two other material properties, the thermal effusivity ($e = \sqrt{k\rho c_p}$) and the thermal diffusivity ($\alpha = k/\rho c_p$) respectively. Thus, by determining the aforementioned thermal properties, estimations regarding the thermal conductivity and the heat capacity of the investigated feature can be obtained [10], according to $k = \sqrt{\alpha}e$ and $\rho c_p = e/\sqrt{a}$. In this point, it should be mentioned that the indirect characterisation of the mosaic material though the estimation of its thermophysical properties was performed taking into consideration the temporal information derived from the average temperature data and the results produced are summarised in Table 5.

From the results produced, it can be observed that a rough characterisation of the hidden mosaic layer can be performed. By defining the properties of thermal conductivity and heat capacity, these can be combined searching in the literature to match both values with a specific material. In this point it should be mentioned that the nominal values of the thermal properties, are actually the values that have been used in order to create the numerical simulations and as stated above these have been retrieved from the literature. However a bibliographic search exhibited that the several types of marble existed have different thermophysical properties. For instance according to the type of marble used its thermal conductivity can vary from 2.08 to 3.34 [11].

Table 5. Determination of mosaic thermophysical properties

Nominal e ($W\ s^{1/2}\ m^{-2}\ K^{-1}$)	Estimated e ($W\ s^{1/2}\ m^{-2}\ K^{-1}$)	Nominal α ($x10^{-7}\ m^2s^{-1}$)	Estimated α ($x10^{-7}\ m^2s^{-1}$)
2715	2616	13.4	10.97
Nominal k ($Wm^{-1}K^{-1}$)	Estimated k ($Wm^{-1}K^{-1}$)	Nominal c_p ($KJm^{-3}K^{-1}$)	Estimated c_p ($KJm^{-3}K^{-1}$)
3.14	2.74 (-12.7%)	2349	2515 (7%)

4. Conclusions

From the results produced through this study, it can be concluded that infrared thermography either through the passive configuration or through the active one can act as a useful tool for the evaluation of mosaic structures. In particular, through the aid of passive thermography, results concerning the condition of the structure can be extracted in real time, while the surface temperature readings can be correlated with physical and chemical phenomena, such as the influence of humidity in the structure and the evaluation of incompatible conservation interventions to the structure's reaming life. On the other hand, the thermal excitation in active thermography and the sequential monitoring of the surface transient cooling can produce a complete set of qualitative and quantitative information. A limitation characterising the active approach could be the fact that it can be applied in cases where only an unknown parameter is missed. In other words, in order to create the relevant models and follow the above described methodology, only one parameter each time can be altered and can be defined by the inverse problem (i.e. in order to estimate the mosaic depth, its thermophysical properties and its thickness have to be known and so on). However, this approach can be efficiently applied in cases of cultural heritage investigations, where a detail is missing from the archives documenting the artefact, or can be overcome by using experimental procedures to gain missing quantitative information, so that these can be inputted into the model.

ACKNOWLEDGEMENTS

Acknowledgements are attributed to the Doc-Culture research project entitled "Development of an Integrated Information Environment for assessment and documentation of conservation interventions to cultural works/objects with Non Destructive Techniques (NDTs)", which is coordinated and managed by NTUA. MIS: 379472. The 3-year project began in April 2012 and is co-financed by the European Union (European Social Fund – ESF) and Greek national funds through the Operational Program "Education and Lifelong Learning" of the National Strategic Reference Framework (NSRF) - Research Funding Program: THALES Investing in knowledge society through the European Social Fund.

References

[1] Grinzato E, Bressan C, Peron F, Romagnoni P, Stevan A G 2000, *Proc SPIE* **4020** 314.
[2] Theodorakeas P, Ibarra-Castanedo C, Sfarra S, Avdelidis N P, Koui M, Maldague X P V, Paoletti D, Ambrosini D 2012 *NDT&E Int.* **47** 150.
[3] Grinzato E , Bison P G, Marinetti S 2002, *J Cult Herit* **3** 21.
[4] Paoletti D, Ambrosini D, Sfarra S, Bisegna F 2013 *J Cult Herit* **14** 116.
[5] Grinzato E 2012 *Proc. 18th World Conf on Nondest Test- 18WNDT*, Durban, South Africa.
[6] Avdelidis N P, Koui M, Ibarra-Castanedo C, Maldague X 2007 *Infrared Phys Technol* **49** 254.
[7] Theodorakeas P, Avdelidis N P, Cheilakou E, Koui M 2014 *J Constr Build Mater* **73** 417.
[8] ThermoCalc 3D Software operating manual. Version 2.0, Innovation Inc., 1998.
[9] Almond D P, Patel P M 1996 *Photothermal Science and Techniques*(London: Chapman & Hall)
[10] Bison P, Bortolin A, Cadelano G, Ferrarini G, Grinzato E 2012 *Proc 11th Int Conf on Quant Infrared Thermography - QIRT2012*, Naples, Italy.
[11] The engineering toolbox: http://www.engineeringtoolbox.com/thermal-conductivity-d_429.html

Journal of Physics: Conference Series **655** (2015) 012062 doi:10.1088/1742-6596/655/1/012062

The use of infrared thermography to detect the skin temperature response to physical activity

G Tanda[1]
DIME, Università degli Studi di Genova, Genova, Italy

E-mail: giovanni.tanda@unige.it

Abstract. Physical activity has a noticeable effect on skin blood flow and temperature. The thermal regulatory and hemodynamic processes during physical activity are controlled by two conflicting mechanisms: the skin vasoconstriction induced by the blood flow demand to active muscles and the skin vasodilation required by thermoregulation to increase warm blood flow and heat conduction to the skin. The time-evolution of skin temperature during exercise can give useful information about the adaptation of the subject as a function of specific type, intensity and duration of exercise. In this paper, infrared thermography is used to investigate the thermal response of skin temperature during running exercise on treadmill for a group of seven healthy and trained runners. Two different treadmill exercises are considered: a graded load exercise and a constant load exercise; for both exercises the duration was 30 minutes. Within the limits due to the relatively small size of the sample group, results typically indicate a fall in skin temperature during the initial stage of running exercise. As the exercise progresses, the dynamics of the skin temperature response depends on the type of exercise (graded versus constant load) and probably on the level of training of the subject.

1. Introduction

Human body temperature comprises the temperatures of the core and skin. The core temperature refers to the temperatures of the abdominal, thoracic and cranial cavities, the skin temperature refers to the temperatures of the skin and subcutaneous tissues. Core temperature is essentially the temperature of the blood in the circulation and is regulated by the brain, whereas skin temperature is influenced more by skin blood flow and environmental conditions [1].

The thermo-regulatory system of the human body is aimed at maintaining a constant core temperature against a wide range of environmental and/or physical work conditions. This constant temperature (about 37°C during rest) is the result of a balance between the metabolic heat production and the heat dissipation to the environment. During intense exercise, metabolic heat production can increase by 10- to 20 fold with respect to heat production at rest. Even if core temperature elevation up to 40°C may be tolerated during long-distance running races or training workouts, most of the extra metabolic heat has to be carried away from the body by conduction, convection, radiation and evaporation [2]. These heat transfer processes are mainly controlled by the skin temperature; indeed, the higher the temperature difference between the skin and the external environment, the higher the heat transfer rate dissipated to the environment. At the same time, intense physical activity activates

[1] To whom any correspondence should be addressed

compensatory vasoregulation through a reduction of blood flow in the skin compartment (skin vasoconstriction) and an increase demand of blood flow to active muscles. Therefore, the thermal regulatory and hemodynamic processes during physical activity are controlled by two conflicting mechanisms: the skin vasoconstriction induced by the blood flow demand to active muscles and the skin vasodilation required by thermoregulation to increase warm blood flow and heat conduction to the skin [3].

Consider the circulatory system model shown in Fig.1. The scheme on the left-hand, taken from Bejan's book [4], consists of three main parts: the heart, the pulmonary circulation, and the systemic circulation; blood is forced away from the heart to flow into the systemic and pulmonary circuits and then returns to the heart. In the right-hand picture, the system circulation has been here modified in order to split the cutaneous blood flow circulation from the internal body flow circulation.

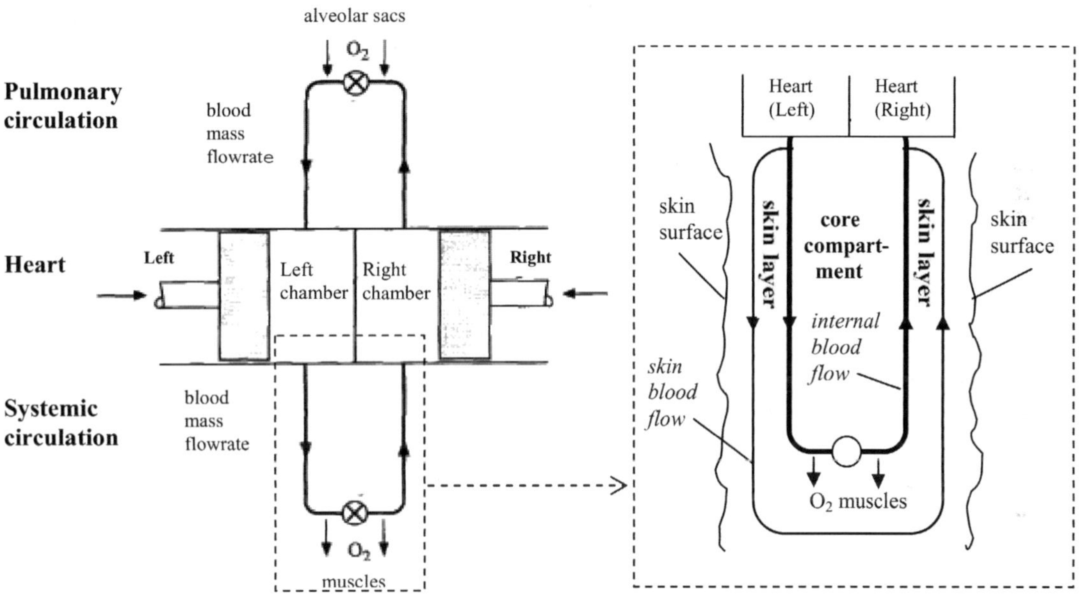

Figure 1. Simple scheme of circulatory system taken from [4] (left-hand) and implementation of the circulatory system by taking into account cutaneous blood circulation (right-hand).

The role of skin blood flow, vital to the human thermoregulatory control [5], can be inferred in combination with the energy balance of body during exercise shown in Figure 2. At the beginning of exercise, the increased oxygen supply to muscles implies a larger demand of blood flow to active muscles and thus a reduction of cutaneous blood flow (skin vasoconstriction). If the environmental air temperature and mean radiant temperature are lower than body temperature, skin vasoconstriction induces a reduction of the skin temperature due to the exposition to a cooler environment. At the same time, heat starts to accumulate into the body due the intense metabolic heat production, leading to an increase in core temperature. Beyond a "set-point temperature", the hypothalamic thermoregulatory centers generate a "thermal command signal" to activate the heat loss response consisting in cutaneous vasodilation and sweating. Cutaneous vasodilation increases blood flow to the skin several-fold: the warmer blood redirected from the body core to the surface (blood perfusion) substantially increases the convective heat transfer from the core to the periphery; as a result the skin temperature tends to increase. Well trained and/or elite runners able to tolerate high core temperature or produce a reduced amount of metabolic heat by the oxidation process (as compared with less trained and/or recreational runners) delay skin vasodilation without compromising the blood (and oxygen) supply to active muscles. In these conditions, they may sustain a great effort for the whole duration of the exercise. Conversely, runners whose increasing metabolic production, at a certain point, induces a cutaneous vasodilation will face a reduction in performance during the exercise due to decreased blood (and oxygen) supply to muscles.

It is apparent from above considerations that for exercising subjects the knowledge of skin temperature during an intense activity like long-distance running is of great importance to understand the sustainability of the required muscular work as well as to infer a possible association of the thermoregulatory response with the athletic performance. Modern thermal imaging devices like infrared thermography are particularly suitable to precisely map the cutaneous temperature distribution and its evolution during exercise. Infrared thermography has been widely used to visualize skin temperature in biomedical sciences (see, for instance, refs. [6,7]); however, a limited number of investigations [8-10] are devoted to dynamic thermographic imaging during running exercise. The time-evolution of skin temperature during running activity can give useful information about the adaptation of the subject as a function of specific type, intensity and duration of exercise. Moreover, recent studies performed on cyclists [11,12] suggested the possibility to use thermal imaging as an indirect method to evaluate the level of training efficiency.

The main purpose of this study was to evaluate the cutaneous temperature response to running exercise under controlled laboratory conditions, in order to gain an insight into the involved thermal regulatory and hemodynamic processes and to provide the basis for subsequent studies aimed at investigating the possible association between the skin temperature changes and the athletic performance.

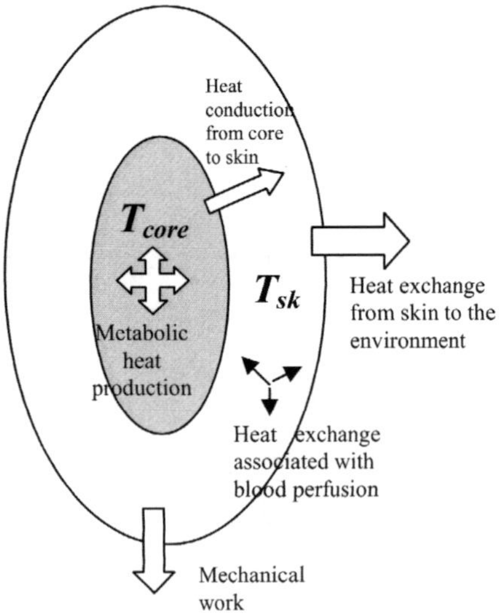

Figure 2. Energy balance during physical activity for core and skin compartments.

2. Materials and methods

The study was designed to measure the skin temperature response to different levels of physical activity, consisting of running exercise on a treadmill with graded or constant load.

2.1 Subjects

Seven healthy and active subjects (six males, one female) gave informed consent to participate in this investigation. All of them regularly trained from 3 to 5 sessions per week and were used to run in middle-distance and long-distance races. Some of them took part to Italian youth and absolute championships in middle-distance track competitions (5000m, 3000m steeplechase), others regularly run half-marathons and marathons with PB (Personal Best) under 1hour, 25minutes and 3hours, respectively. Their age, weight, and height ranged from 18 to 57 years (mean: 33 years), from 50 to 71 kg (mean: 62 kg), and from 1.69 to 1.9 m (mean: 1.75 m), respectively. The body mass index was in the 17.2-24.3 kg/m^2 range (mean: 20.2 kg/m^2).

2.2 Procedures

Treadmill running exercise was performed in an internal room with controlled air temperature and humidity conditions (temperature = 22°C±0.3°C, relative humidity 60%±2%, no direct ventilation). During tests, the athletes wore a light running kit or exercised bare-chested. Each experiment was conducted at the same time of day (late morning) to avoid variation due to circadian rhythm of body temperature.

Two different treadmill protocols were used: a graded load exercise and a constant load exercise. During the graded load exercise, from an initial value of 6 km/h the treadmill velocity was incremented by 1.5 km/h every 5 minutes up to the maximum value of 13.5 km/h. Throughout each step the speed was kept constant and thermal images of anterior and posterior body were typically taken after 4-4.5 minutes from the initiation of the respective speed step. The constant load exercise consisted of 25 min-running at the constant speed of 12 km/h, preceded by a warm-up period (5 min at 6 km/h) with thermal images (front and back of the body) typically taken every 5 minutes. The duration of both running sessions was 30 minutes. The standardized exercise programs are summarized in the scheme reported in Table 1.

For both protocols, the treadmill was set at zero inclination and no airflow was present during the exercise; heart rate was continuously monitored by means of a standard ECG device. Before each exercise, the participants stood motionless for a 10 minute acclimatisation period. At the end of each 30-minute exercise period, subjects stood motionless for a further 5-10 minute rest period.

All seven subjects performed the graded load treadmill test. Four of them performed the constant load test at a few days distance from the previous test. For each subject, the highest heart rate recorded during each exercise was reached at the end of the test. It ranged from 72% to 91% of respective age-predicted maximal heart rate for the graded load exercise and from 71% to 90% for the constant load exercise.

Table 1. Graded and constant load exercise scheme.

duration (min)	**graded load**	**constant load**
	velocity (km/h)	velocity (km/h)
acclimatisaton (10 min)	0	0
0-5	6	6
5-10	7.5	12
10-15	9	12
15-20	10.5	12
20-25	12	12
25-30	13.5	12
recovery (10 min)	0	0

2.3 Estimation of total-body skin temperature

Thermal images of subjects during running exercise were taken by using a digital infrared camera (FLIR T335, Flir Systems Inc, USA, 320 x 240 pixels, Thermal sensitivity/NETD < 0.05°C). All the thermographic measurements were performed on the nude skin. Infrared emissivity of skin was set at

0.98 [7]; it is worth noting that the sweat, when present on the skin, does not significantly affect the surface emissivity as reported in [8].

The accuracy of infrared temperature measurements was assessed by a calibration test carried out by gradually heating an aluminium plate (110 mm × 150 mm × 8 mm) equipped with a plane heater on the back face and coated with a high-emissivity black paint on the frontal face. The calibration plate was surrounded, on lateral and bottom sides, by thermally insulating material (extruded polystyrene) to promote uniform temperature conditions. Ten fine-gauge, pre-calibrated, thermocouples, housed in small holes drilled in the material at different positions as close to the surface as possible, were used to measure the plate temperature and to verify the temperature uniformity. During the test, the plate was heated in order to reach a desired plate temperature, measured by the thermocouples, and the power was adjusted to maintain the equilibrium. At the steady state, the infrared image of the plate was captured and thermocouple readings were recorded. The maximum difference in temperature values given by thermocouples was 0.1°C, thus proving the high degree of temperature uniformity in the plate. This procedure was repeated for plate temperatures from 25 to 42°C. For an emissivity setting of 0.95 on the infrared camera, a typical agreement within 0.2°C was found between the temperatures detected by infrared camera and thermocouples.

The major limitation of the total body, infrared thermal imaging is that each acquisition is a two-dimensional representation of a three-dimensional structure. Ideally, acquisitions would provide a three-dimensional representation of the surface temperature of the entire three-dimensional structure by using a multiple camera system and suitable post-processing techniques. This procedure largely increases cost and time to perform the investigation. In this study, the total body skin temperature was calculated by averaging the thermographic measurements over points encompassing a significant surface area of each region of interest (e.g., anterior thigh, abdomen, posterior calf, etc.). In order to minimize the errors associated with the calculation procedure of the total-body skin temperature, different literature relationships, proposed for the mean skin temperature evaluation, were considered. They are typically structured in the following form

$$T_{sk} = \left(\sum_{i=1}^{n} f_i T_{sk,i}\right)/f_0 \tag{1}$$

where T_{sk} is the calculated total-body mean skin temperature, $T_{sk,i}$ are local (or regionally averaged) skin temperatures, f_i are the corresponding weighting factors (with f_0=1, unless differently specified) and n is the number of skin temperature measurement sites. A survey of literature relationships is given in [13] where total-body skin temperature calculations were based on a number of skin locations n ranging from 3 to 15. Table 2 reports the weighting factors f_i (for each region of the body surface) and the value of f_0 to be used in Eq.(1) for four different formulas, based on n=3, 8, 12 and 15. Each formula is named according to the notation introduced in [13]: the number of local skin temperature sites (8, 10, 12 or 15) is followed by a letter (W or U) indicating a weighted or unweighted (same weighting factor for all the skin sites) formula. The apices denote formulas (12W' and 15U') modified here to exclude the contribution of foot skin temperature, whose value can not be extracted from infrared thermographic images of runners. Some measurement sites were changed in [13] from those adopted in the original studies since accurate thermographic measurements revealed, for the new sites, a better agreement with the true mean skin temperature, measured for ten nude subjects as the arithmetic mean of local temperatures (from 10841 pixels) provided by infrared images of the total body surface. The individual sites subjected to local (or regionally averaged, as shown in the picture close to the table) temperature measurements were: cheek **a**, chest **b**, abdomen **c**, scapula **d**, subscapula **e**, lumbar **f**, posterior upper arm **g**, anterior forearm **h**, posterior forearm **i**, hand **j**, anterior thigh **k**, posteromedial thigh **l**, posterolower thigh **m**, anterior calf **n**, posterior calf **o**, foot **p**. Choi et al [13] found that formulas based on a number of at least eight locations (n≥8) were in close agreement (within ±0.2°C) with the true mean skin temperature for 80 to 90% of the performed measurements on their subjects, staying at rest under ambient temperature conditions ranging from 4 to 40°C.

Journal of Physics: Conference Series **655** (2015) 012062 doi:10.1088/1742-6596/655/1/012062

Table 2. Body surface regions (**a**→**p**), corresponding weighting factors f_i and value of f_0 used in the literature relationships: 3W (n=3), 8W (n=8), 12W (n=12), 15U (n=15). The apices denote formulas 12W' (n=11) and 15U' (n=14) modified here to exclude the contribution of the foot skin temperature. The measurement sites have been modified from the original ones according to Choi et al [13].

	3 W	8 W	12 W	15 U	12 W'	15 U'
a		0.07	0.07	0.0666	0.07	0.0666
b			0.0875	0.0666	0.0875	0.0666
c	0.5	0.175	0.0875	0.0666	0.0875	0.0666
d				0.0666		0.0666
e		0.175	0.0875	0.0666	0.0875	0.0666
f			0.0875	0.0666	0.0875	0.0666
g		0.07		0.0666		0.0666
h		0.07		0.0666		0.0666
i	0.14		0.14		0.14	
j		0.05	0.05	0.0666	0.05	0.0666
k		0.19	0.095	0.0666	0.095	0.0666
l			0.095	0.0666	0.095	0.0666
m				0.0666		0.0666
n			0.065	0.0666	0.065	0.0666
o	0.36	0.2	0.065	0.0666	0.065	0.0666
p			0.07	0.0666	/	/
f_0	1	1	1	1	0.93	0.9333

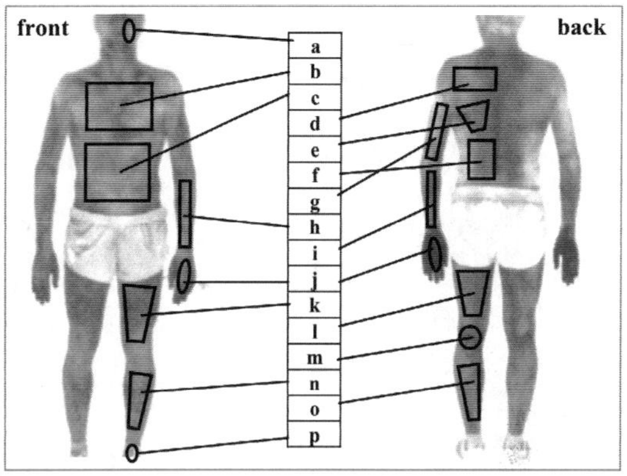

Despite their results were obtained for resting nude subjects, literature relationships implemented by Choi et al [13] were used in the present study to process the thermographic measurements performed on running subjects, due to the lack of specific formulas devoted to the mean skin temperature evaluation during exercise. Figure 3 shows the mean skin temperature, evaluated from the combination of regionally averaged thermographic measurements, for an athlete during a treadmill running exercise at constant load. Mean skin temperatures have been evaluated during the warm-up (0-5 minutes) following the acclimatisation period, the running period at constant velocity V=12km/h (5-30 minutes), and the resting period (30-35 minutes). Inspection of figure shows that temperature distributions obtained by literature relationships have a common trend, except for results given by the relationship based on only three measurement points. Similar results were found by processing the individual measurements performed for a large cohort of athletes. Based on the suggestions reported in [13], the formula involving the largest number of measurement sites (15U', n=14) was deemed as the most accurate and thus was adopted to process the present thermographic measurements.

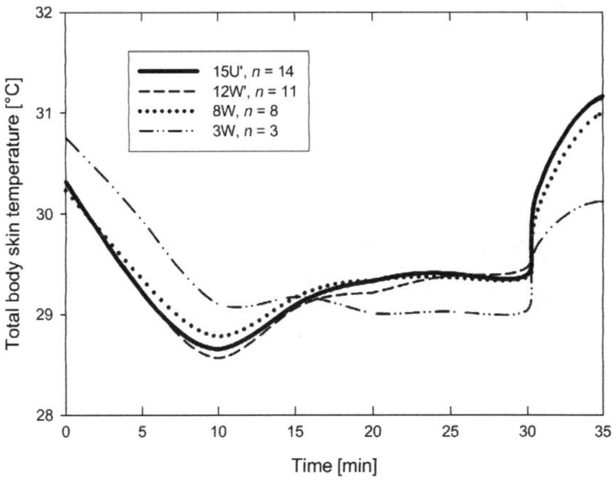

Figure 3. Mean skin temperature, evaluated as a combination of regionally averaged thermographic measurements by using different literature formulas, for a subject during a constant load treadmill exercise.

Journal of Physics: Conference Series **655** (2015) 012062 doi:10.1088/1742-6596/655/1/012062

3. Results and discussion

Examples of thermographic images, with color-coded temperature maps, are presented in Figures 4 and 5. Each group of thermographic images has the same thermal scale in order to facilitate the comparison. Images, taken for three different subjects under conditions of graded load, show the typical skin temperature modifications on the front of the body during different phases of exercise.

Figure 4 shows the changes in temperature for the female athlete involved in this study: from the baseline distribution on left-hand side, the skin temperature decreases as the exercise progresses. Temperature reduction is not uniform over the body; it seems to be less pronounced over calves, neck and cutaneous projections of liver and spleen. A rapid increase in skin temperature is observed during motionless recovery from exercise (right-hand side); hyperthermal spots, probably associated with cutaneous vasodilation occurring at rest, are clearly visible. Similar features of skin temperature modifications for two male athletes during the grade load exercise are documented in Fig.5.

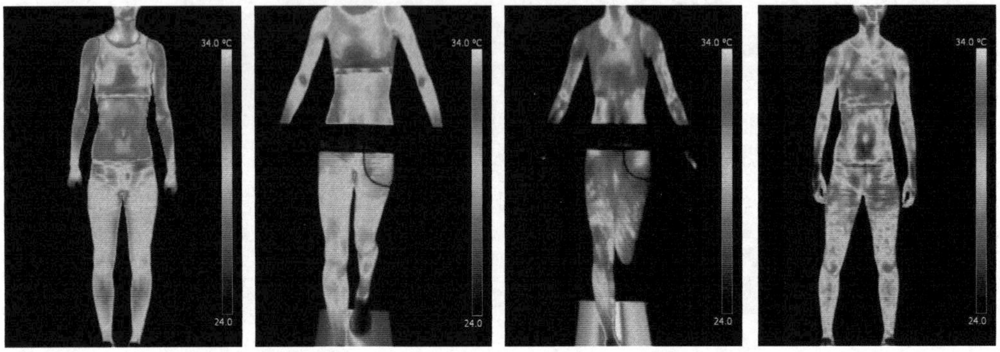

Figure 4. Infrared thermal images of the anterior body for an athlete during graded load exercise. From left to right: before exercise, after 15 min exercise period, after 29 min exercise period, immediately after the end of the exercise. The same thermal scale is adopted.

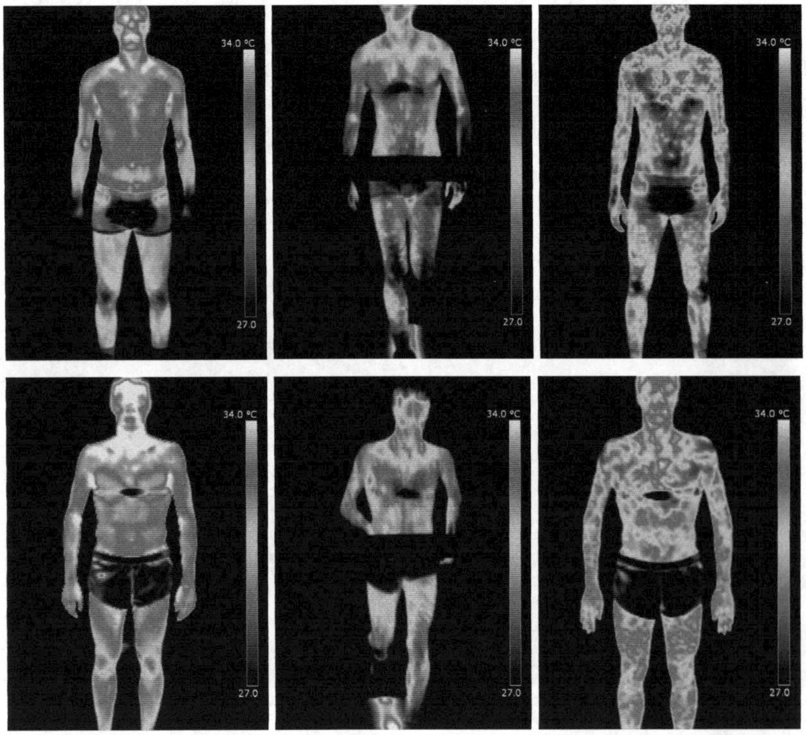

Figure 5. Infrared thermal images of the anterior body for two athletes during graded load exercise. From left to right: before exercise, after 15 min exercise period, immediately after the end of the exercise. The same thermal scale is adopted.

Total-body skin temperature distributions during graded and constant load exercise, averaged among all the subjects (7 for the graded load exercise, 4 for the constant load exercise), are shown in Figure 6. The skin temperature distribution for the graded load exercise, averaged only among the same four subjects participating in the constant load exercise, has been reported too. The same figure reports the velocity profiles set for the two treadmill protocols.

During the graded load exercise, where the load on the body is progressively increasing (and the blood demand to the working organs is increasing too), the mean skin temperature of the subjects decreases throughout the exercise. Time-distributions of skin temperature for the majority of the participants were similar in shape to that averaged among the subjects; thus individual data are not reported for the sake of brevity. The reduction of skin temperature occurs immediately on starting to run, even during the first steps at moderately low velocities, without the appearance of sweat on the skin and thus it is not related to thermal factors such as evaporation due to skin sweat. This finding is consistent with results found in [10] under similar experimental conditions: a continuous vasoconstrictor response, leading to a progressive reduction of blood flow in the tegumentary apparatus as the exercise intensity is increased, was deemed to justify the decrease in the mean skin temperature. The mean skin temperature distribution for a reduced group of subjects (4 out of 7, those participating in both treadmill protocols) has a similar trend but a different slope in the initial and final steps of exercise, while for both sample groups a nearly flat distribution in the central phase of exercise is noticed. These observations suggest the influence of individual characteristics (genetic and/or anthropometric and/or training factors) on the time-evolution of skin temperature during exercise. At the end of exercise, during recovery, mean skin temperatures increase toward pre-exercise values.

In the case of constant load exercise, the skin temperature, averaged among four subjects, still decreases at the beginning of the work due to the initial skin vasoconstriction and then attains a minimum value, followed by a little and gradual increase over time. It is argued that the initial decline and the subsequent slight rise in skin temperature is the net result of the competition between the vasoconstrictor response, which lasts as long as the exercise is continued, and the vasodilator response, induced by the increasing body temperature. All subjects exhibited the same time-distribution, but the start and the extent of the temperature rise during the last phase of exercise seem to be related to the individual grade of vasoregulation. No literature data are available, to the author's knowledge, for the dynamic response of skin temperature to a constant load running exercise; however, a behaviour similar to that here reported, i.e., an initial descending trend (reflecting vasoconstriction) followed by an ascending trend (reflecting vasodilation) of regional skin temperatures, was found in [14] (hands) and [15] (hands, thighs and legs) during a steady-load exercise on a bycycle ergometer.

Further information can be inferred from Figs.7-8, showing the regional skin temperature response to running exercise. Figure 7 reports, for the female athlete, the skin temperature distributions over total body, upper limbs (upper arms and forearms) and calves, for the graded load exercise. Inspection of the figure reveals that the skin temperature decrease is more pronounced over the most peripheral regions (e.g., upper limbs) less involved in running. Conversely, the skin over exercising muscle mass (e.g., calves) exhibits only a moderate reduction in temperature and tends to be warmer than skin over other structures as the exercise progresses; this result is probably related to the increasing delivery of blood flow to the working muscles for metabolic needs and to the heat conduction from the muscles to the skin surface.

Comparisons of total body and regionally averaged skin temperature distributions, for two athletes (one female and one male) under constant load exercise are reported in Figure 8. Solid lines represent the total body skin temperature distributions, whereas the dashed lines indicate the skin temperature over upper limbs (top) and calves (bottom). Different colours refer to different athletes. It can be inferred from Fig.8 that skin temperature over upper arms and forearms shows a time-evolution similar to that for total body but at a lower level, as outlined in the comment to Fig.7. Temperature of skin over active muscles (calves) decreases in a similar manner for both athletes; for one subject the skin over calves is warmer than total body almost throughout the exercise, for the other subject the temperature of skin over calves starts from a lower value than total body and becomes higher in the second part of exercise. During the last part of the constant load exercise, skin temperature over calves decreases up to a value close to that registered for total body, probably due to skin vasodilation and the consequent reduction of warm blood flow to active muscles.

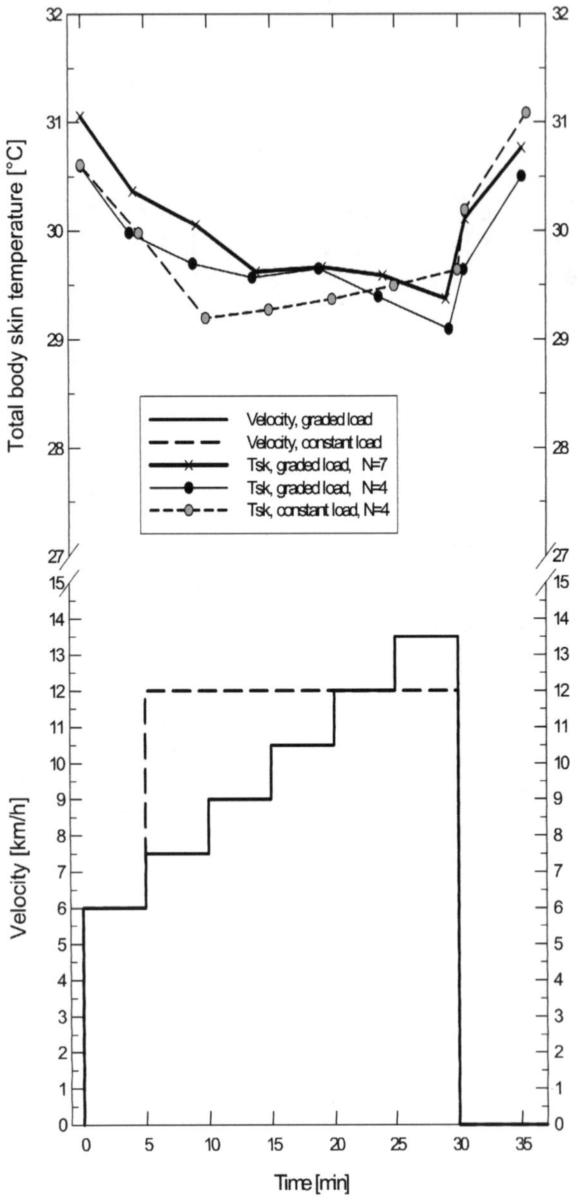

Figure 6. Total body skin temperature distributions averaged among the N subjects during graded and constant load exercise. Treadmill velocities versus time are reported too.

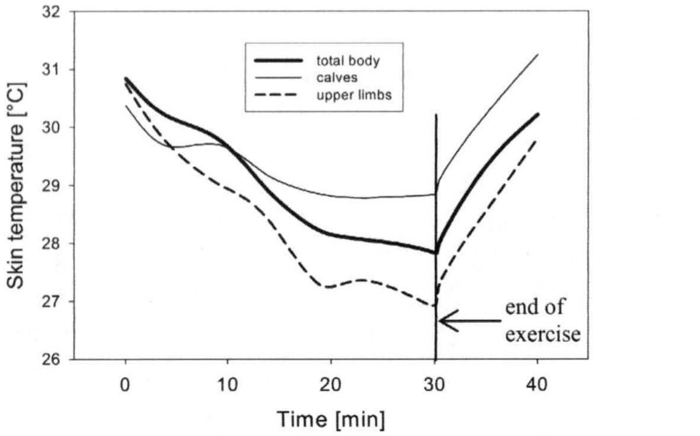

Figure 7. Distributions of the total-body skin temperature and the regionally averaged (over upper limbs and calves) skin temperature for an athlete during graded and constant load exercise

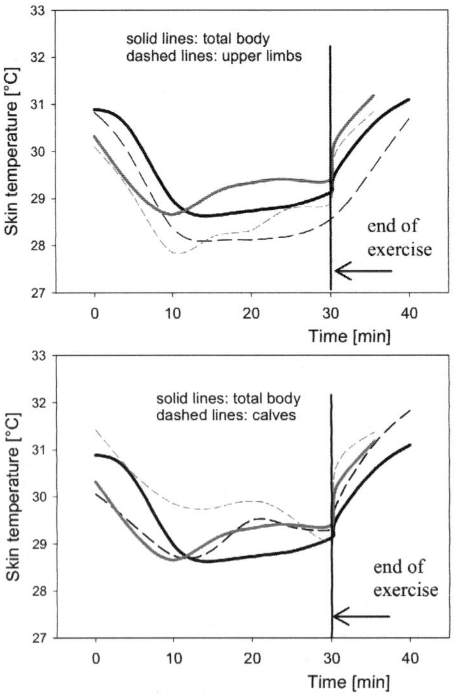

Figure 8. Distributions of the skin temperature over the total-body, over upper limbs (top) and over calves (bottom) for two athletes during constant load exercise. Different colours refer to different athletes.

Journal of Physics: Conference Series **655** (2015) 012062 doi:10.1088/1742-6596/655/1/012062

4. Conclusions

Human thermoregulation during running exercise is ruled by a dynamic energy balance, involving environmental and physical work conditions, where the skin temperature represents the main variable controlling the heat exchanges at the body/environment interface. Infrared thermal imaging is currently the only method for real-time, non-invasive monitoring of cutaneous temperature.

In this work infrared thermography was used to visualize skin temperature of athletes during indoor treadmill running. For all the investigated subjects a fall in skin temperature during the initial stage of running exercise was recorded, regardless of the type of work (with graded or constant load). This skin temperature decrease is probably associated with the cutaneous vasoconstrictor response to exercise. A continuous increase in load intensity was found to produce further reductions in skin temperature; conversely, a relative minimum of skin temperature, followed by a little rise, was attained during a constant load running exercise, probably reflecting the competition between the vasoconstriction induced by exercise and vasodilation induced by thermoregulation. Hyperthermal spots were observed during the recovery following the running exercise, probably due to peripheral vasodilation enabling a progressive transfer of warmer blood from the body core to the surface.

This preliminary study, performed on a limited number of subjects, provided the basis upon which a subsequent massive investigation of skin temperature response to running exercise, according to specifically designed protocols and controlled laboratory conditions, could attempt to infer an indirect estimation of the physical efficiency and/or the training level of athletes.

References

[1] Lim CL, Byrne C and Lee JKW 2008 Human thermoregulation and measurement of body temperature in exercise and clinical settings *Ann. Acad. Med. Singapore* **37** 347-353

[2] Kenefick RW, Cheuvront SN and Sawka MN 2007 Thermoregulatory function during the marathon *Sports Med.* **37** 312-315

[3] Johnson JM 1992 Exercise and the cutaneous circulation *Exercise Sport Sci. Rev.* **20** 59-97

[4] Bejan A 2000 *Shape and Structure, from Engineering to Nature* (Cambridge Press)

[5] Charkoudian N 2003 Skin blood flow in adult human thermoregulation: how it works, when it does not, and why *Mayo Clin. Proc.* **78** 603-612

[6] Lahiri BB, Bagavathiappan S, Jayakumar T and Philip J 2012 Medical applications of infrared thermography: A review *Infrared Phys.Technol.* **55** 221–235

[7] Fernández-Cuevas I, Bouzas Marins JC, Arnáiz Lastras J, Gómez Carmona PM, Piñonosa Cano S, García-Concepción MA and Sillero-Quintana M 2015 Classification of factors influencing the use of infrared thermography in humans: A review *Infrared Phys. Technol.* **71** 28-55

[8] Clark RP, Mullan BJ and Pugh LGCE 1977 Skin temperature during running - A study using infra-red colour thermography *J. Physiol.* **267** 53-62

[9] Roberts BC, Waller TM and Caine MP 2007 Thermoregulatory response to base-layer garments during treadmill exercise *Int. J. Sports Sci. Eng.* **1** 29-38

[10] Merla A, Mattei PA, Di Donato L and Romani GL 2010 Thermal imaging of cutaneous temperature modifications in runners during graded exercise *Annals Biomed. Eng.* **38** 158–163

[11] Akimov EB and Son'kin VD 2011 Skin temperature and lactate threshold during muscle work in athletes *Hum. Physiol.* **37** 621-628

[12] Cholewka A, Kasprzyk T, Stanek A, Sieroń-Stołtny K and Drzazga Z 2015 May thermal imaging be useful in cyclist endurance tests? *J Therm. Anal. Calorim.* (in press) DOI 10.1007/s10973-015-4662-5

[13] Choi JK, Miki K, Sagawa S and Shiraki K 1997 Evaluation of mean skin temperature formulas by infrared thermography *Int. J. Biometeorol.* **41** 68–75

[14] Zontak A, Sideman S, Verbitsky O and Beyar R 1998 Dynamic thermography: analysis of hand temperature during exercise *Annals Biomed. Eng.* **26** 988–993

[15] Nakayama T, Ohnuki Y, and Niwa K 1977 Fall in skin temperature during exercise, *Jap. J. Physiol.* **27** 423-437

Stream temperature estimated in situ from thermal-infrared images: best estimate and uncertainty

F Iezzi[1] and M T Todisco[2]

[1] Department of Engineering and Geology, University "G. d'Annunzio" of Chieti-Pescara, Viale Pindaro, 42 – 65127 Pescara, Italy
[2] Department I.C.E.A.A., University of L'Aquila, Piazzale Pontieri, 1 – 67100 Monteluco di Roio, L'Aquila, Italy

E-mail: fabrizio.iezzi@unich.it

Abstract. The paper aims to show a technique to estimate in situ the stream temperature from thermal-infrared images deepening its best estimate and uncertainty. Stream temperature is an important indicator of water quality and nowadays its assessment is important particularly for thermal pollution monitoring in water bodies. Stream temperature changes are especially due to the anthropogenic heat input from urban wastewater and from water used as a coolant by power plants and industrial manufacturers. The stream temperatures assessment using ordinary techniques (e.g. appropriate thermometers) is limited by sparse sampling in space due to a spatial discretization necessarily punctual. Latest and most advanced techniques assess the stream temperature using thermal-infrared remote sensing based on thermal imagers placed usually on aircrafts or using satellite images. These techniques assess only the surface water temperature and they are suitable to detect the temperature of vast water bodies but do not allow a detailed and precise surface water temperature assessment in limited areas of the water body. The technique shown in this research is based on the assessment of thermal-infrared images obtained in situ via portable thermal imager. As in all thermographic techniques, also in this technique, it is possible to estimate only the surface water temperature. A stream with the presence of a discharge of urban wastewater is proposed as case study to validate the technique and to show its application limits. Since the technique analyzes limited areas in extension of the water body, it allows a detailed and precise assessment of the water temperature. In general, the punctual and average stream temperatures are respectively uncorrected and corrected. An appropriate statistical method that minimizes the errors in the average stream temperature is proposed. The correct measurement of this temperature through the assessment of thermal-infrared images obtained in situ via portable thermal imager is confirmed by the direct measurement of stream temperature using an ordinary technique based on an appropriate thermometer.

1. Introduction

Today large quantities of water are artificially brought into urban watersheds via water works and hydraulic systems and consumed to maintain the various and common urban activities. Therefore, vast quantities of urban wastewater are discharged into the local water system, in general at different temperatures from the naturally occurring water. Stream temperature is an important indicator of water quality. Into water bodies, most chemical and biological activities are a function of temperature. The fish habitat in particular is sensitive to water temperature. Water bodies temperature is affected by various environmental process, both natural and anthropogenic. Human activities can affect stream temperature through global climate change [1], regional land-use alteration [2-3], heated effluents from power generation plants [4-5] and summertime urban stormwater runoff [6]. Disentangling the various effects is not easy. Stream temperature changes are especially due to the anthropogenic heat input from urban wastewater and from water used as a coolant by power plants and industrial manufacturers. Wastewater from municipal treatment plants is probably the largest anthropogenic heat source for urban streams. Its effect on stream temperature depends on the temperature and volume of wastewater added to the stream [7]. The increase in stream temperature related to anthropogenic heat input from urban wastewater is described in [8].

Nowadays, the water temperature assessment is important particularly for thermal pollution monitoring of the water bodies (including streams, rivers, lakes and reservoirs). In Italy, [9] regulates the stream temperature changes due to heat input from a discharging. The regulation regards the limited area of the water body near the discharging. Therefore, it is necessary an appropriate and suited technique to estimate the water temperature. It is note that stream temperatures assessment using ordinary techniques (e.g. appropriate thermometers) is limited by sparse sampling in space due to a spatial discretization necessarily punctual. These techniques are limited due to the necessary entrance of the temperature measuring instruments into the water body. On the contrary of the ordinary techniques, latest and most advanced techniques assess the water temperature in distant way from the water body. These innovative techniques estimate the stream temperature using thermal-infrared remote sensing based on thermal imagers placed on aircrafts [10-12] or using satellite images [13-15]. Recently, portable thermal imager is used to estimate in situ the stream temperature from thermal-infrared images [16-17]. These techniques are based on infrared (IR) thermography. In fact, they assess the thermal images based on the radiant power output of the surface of the water body; it is possible, therefore, to estimate only the surface water temperature. On the contrary of the thermal images taken over in situ via portable thermal imager, those taken over from aircrafts or satellites are suitable to detect the temperature of vast water bodies but do not allow a detailed and precise surface water temperature assessment in limited areas of the water body. In fact, the thermal images taken over from aircraft or satellite have a low resolution, so they do not distinguish the thermal fields of stream water respect anomalous thermal fields generate by foreign matter to the water. On the contrary, the thermal images taken over in situ via portable thermal imager have an high resolution, therefore the thermal anomalies are evident. The thermographic technique in situ is based on the sampling, from appropriate thermal images, of measuring points of the water temperature and to process, with suitable statistical methods, the temperature data to minimized the negative effects of thermal anomalies. The paper aims to review the technique to estimate in situ the stream temperature from thermal-infrared images via portable thermal imager deepening its best estimate and uncertainty and to show a method to assess the thermal pollution of the stream itself. A stream with the presence of a surface discharge of urban wastewater is proposed as case study to validate the technique and to show its limits. Since the technique analyzes limited areas in extension of the water body, it allows a detailed and precise assessment of the water temperature. In general, the punctual and average stream temperatures are respectively uncorrected and corrected. An appropriate statistical method that minimizes the errors in the average stream temperature, due to the thermal anomalies, is proposed. The correct measurement of the average stream temperature through the assessment of thermal-infrared images obtained in situ via portable thermal imager is confirmed by the direct measurement of stream temperature via an ordinary technique based on an appropriate thermometer.

2. Approach

2.1. *Case study*

For the IR thermographic applications in situ, the studied surface water body is the Aterno River in urban area of L'Aquila Municipality (Italy). A surface discharge of urban type is present in the

Journal of Physics: Conference Series **655** (2015) 012063 doi:10.1088/1742-6596/655/1/012063

analyzed situ. The discharge is characterized by an average wastewater hydraulics flow rate of 0,06 m^3/s during the thermographic experimental period. In the vicinity of the input point of the discharge wastewater, the Aterno River is characterized by an average hydraulics flow rate of 3,8 m^3/s and a river channel width of 20 m about. Thermal-infrared images are taken over via portable thermal imager in no. 3 areas of limited extension: close to the input point of discharge, upstream and downstream areas of discharge (60 m about is the distance from each other).

2.2. *Thermographic experimentation: equipment and boundary conditions*

Thermal-infrared images are taken over by a FLIR Systems S65 HS portable thermal imager that measures a range of the electromagnetic radiation spectrum between 7,5 and 13 μm wavelengths. The portable thermal imager has an average incidence angle, respect to surface of the water body, of 30° close to the input point area and in the downstream area of the discharge, of 90° in the upstream area of the discharge. The measurement days have been 21st, 22nd and 23rd November 2011. The measurement hours have been 9 am, 1 pm and 8 pm. During the measurement days, the maximum and minimum air temperatures have been 0 °C and 15 °C respectively. The average inclination of the solar rays to the ground has been 27,60° about. The water of the Aterno River is physically modelled as an opaque body having a constant emissivity coefficient 0,96 [16].

2.3. *Image processing*

Image processing was conducted using ThermaCAM Researcher Pro 2.8 software, distributed by FLIR Systems. Thermal-infrared images taken over by a portable thermal imager can take over only an area of the river channel with a very limited extension. Therefore, in order to have thermographic images that can estimate the stream temperature of an area with suitable extension of the water body, it is necessary to take over a suitable number of thermal images and to combine them with an appropriate and specific software. Visually, some cross-sections of the river channel are studied (no. 8 in total: no. 2 upstream, no. 2 in the vicinity and no. 4 downstream of the discharge). From these fluvial cross-sections, punctual temperatures are extrapolated from thermal images through a specific software. These punctual thermal data are analyzed for each measurement carried out in a specific day and in a specific hour. For each fluvial cross-section: the spatial distribution of punctual temperatures is determined; the arithmetic average temperatures and their variances are calculated. In figures 1, 2 and 3 are reported some examples of the taken over and assessed thermal-infrared images.

3. Sources of uncertainty

In figures 1, 2 and 3 the thermal images present several anomalous thermal fields (i.e. uncertainties) generated by foreign matters to the water and contamination from atmospheric effects. Sources of uncertainty include thermal scattering from the near-bank environment, the role of shade, surface effects such as surface roughness and viewing geometry, i.e. the observation angle. The near-bank environment includes objects with a wide variety of temperatures and emissivities, including bark, branches, dead grasses, leaves, soil, sand and rocks, that are exposed during the period of the thermographic measurements due to the low water levels of the river, and that have commonly different temperatures from those of the river water. Some of the IR radiation emitted by objects in the near-bank environment will reach the stream surface, either directly or through multiple scattering from objects near and in the water river. IR radiation emitted from near-bank objects may also pass directly into the path of the sensor of the portable thermal imager or be multiply scattered or reflected from other surfaces into the sensor, both of which will increase the observed temperature. However, this effects is small compared to the temperature of the water. The IR radiation emitted by objects depends by the observation angle; in fact this angle can be a source of uncertainty [18].

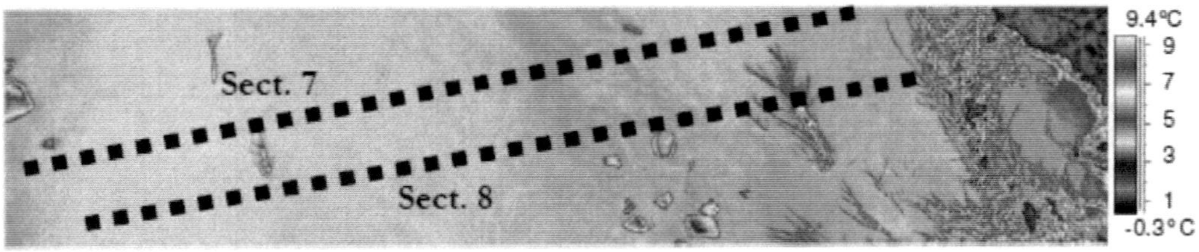

Figure 1. Example of a thermographic image of the upstream area of the discharge.

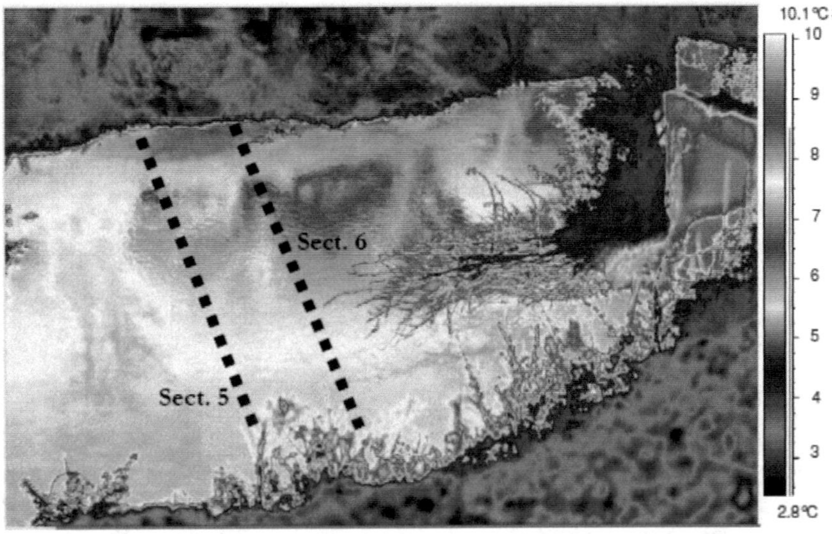

Figure 2. Example of a thermographic image of the area in the vicinity of the discharge.

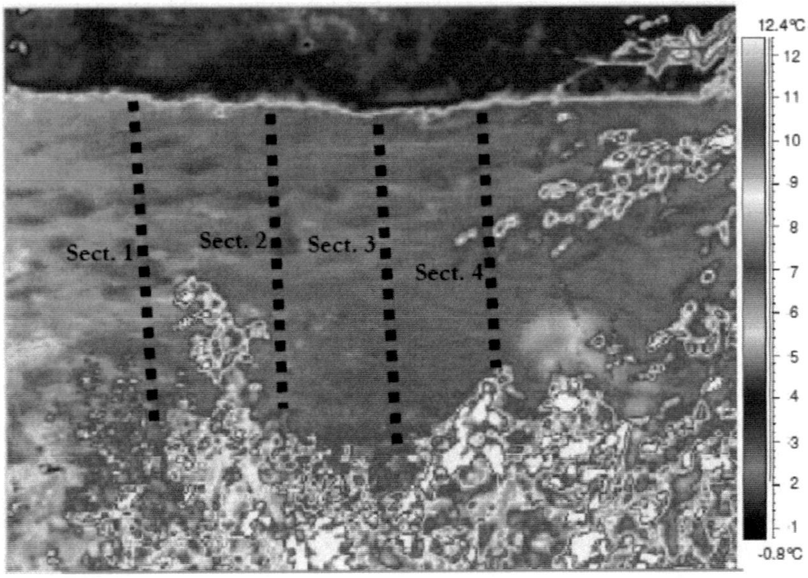

Figure 3. Example of a thermographic image of the downstream area of the discharge.

4. Best estimate of temperature

As mentioned in section 2, it is necessary to assess the average stream temperature along the fluvial cross-sections in order to limit the effects of the sources of uncertainty, i.e. to minimize the contribution of the anomalous thermal fields.

The best estimate of the average stream temperature (T_w) along a generic fluvial cross-section is assessed using the Gauss weighted average [19]. Therefore, the fluvial cross-section is idealized as discrete, i.e. composed by a series of temperature points and divided in ranges. These temperature points are grouped in a suitable number (i) of ranges and the i-th range has a determined number of points (n). For each i-th range, the arithmetic average temperature and its variance (σ^2_i) are calculated. The contribution of the quality of the i-th range to the temperature of the generic fluvial cross-section is therefore assessed by assigning a normalized weight (W_i):

$$W_i = \frac{\frac{1}{\sigma_i^2}}{\sum_{i=1}^{n} \frac{1}{\sigma_i^2}} \quad (1)$$

where ($1/\sigma^2_i$) is used in the calculation of the weight for each range so that the ranges with a great variance contribute little to the calculation of T_w. These normalized weights are used to calculate, in each fluvial cross-section, T_w as the weighted sum of arithmetic average temperatures of each i-th range (T_i):

$$T_w = \sum_{i=1}^{n} W_i T_i \quad (2)$$

5. Validation of results

In order to validate the average surface stream temperature (based on thermal-infrared images via portable thermal imager in situ) along fluvial cross-section, estimated using the Gauss weighted average (equations (1) and (2)), an ordinary technique is adopted: use of an appropriate portable thermometer. The measures using the portable thermometer are carried out in a point of the river localized at 8 m about downstream from the input point of the discharge, in each of the no. 3 hours of measurement of the no. 3 experimental days. So there are no. 9 measures with an arithmetic average of 7,6 °C. The comparison of the thermal results of equations (1) and (2) with those of the ordinary technique permits to validate the IR thermographic technique based on the use of portable thermal imager in situ.

6. Uncertainty quantification

6.1. On total thermal power taken over

In general, the total thermal power emitted by a detected object by the portable thermal imager is [20]:

$$W_{tot} = \varepsilon\tau W_{obj} + (1-\varepsilon)\tau W_{refl} + (1-\tau)W_{atm} \quad (3)$$

where ε is the emission coefficient of the detected object, τ is the atmospheric transmission coefficient, W_{obj} is the thermal power emitted by the detected object, W_{refl} is the thermal power reflected by other sources (i.e. foreign matters and bodies to the detected object) present in the environment and W_{atm} is the thermal power emitted by the atmosphere. So that, $\varepsilon\tau W_{obj}$ is the object emission, $(1-\varepsilon)\,\tau W_{refl}$ is the environmental sources emission and $(1-\tau)W_{atm}$ is the atmospheric emission.

In equation (3) it is assumed that the generic thermal power (W) emitted by a generic actual source is [21]:

$$W = \varepsilon\sigma T^4 \quad (4)$$

where ε is the emission coefficient of the source, σ is the Stefan-Boltzmann constant equal to $5{,}67 \cdot 10^{-8}$ W/m^2K^4 and T^4 fourth power of the source temperature. The equation (4) is used to estimate W_{obj}, W_{refl} and W_{atm}. In this study, the generic detected object above mentioned is the water body (i.e. Aterno River). In the present research, $\varepsilon = 0{,}96$ and $T_{obj} = 7{,}6$ °C are assumed respectively for the emission coefficient and temperature of the river water; the atmospheric transmission coefficient is assumed $\tau = 1$ [21]. In the period of the thermographic applications, there is a atmospheric (i.e. environmental) temperature T_{atm} varying between 0 °C and 15 °C. Since in the analyzed areas of the river there are not punctual heat sources, it is assumed $T_{atm} = T_{refl}$ (T_{refl} is reflected temperature) [22]. Therefore, using the equations (3) and (4) it is possible to assess the contribution of the radiation emitted by the water body, by the atmosphere and by the reflected radiation from the river respect to the total radiation taken over by the portable thermal imager. The synthetic graphic results are shown in figure 4.

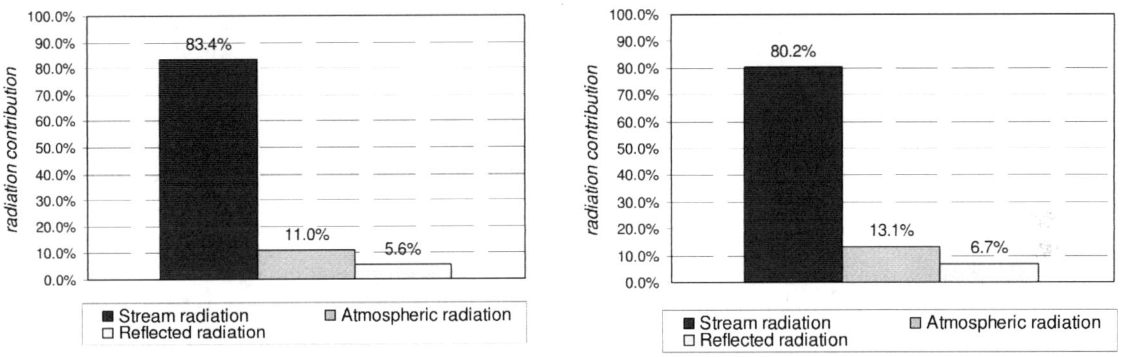

Figure 4. Radiation contribution emitted by the surface water body, by the atmosphere and by the reflected radiation on the total radiation recorded by thermal imager for emissivity and temperature of the water of the water body respectively of 0,96 and 7,6 °C, an atmospheric transmission coefficient of 1 and a reflected temperature of 0 °C (left graph) and 15 °C (right graph).

Observing the figure 4 it is possible to note that the weight, i.e. the contribution, of the water body radiation (W_{obj}), respect to the total radiation measured by the portable thermal imager (W_{tot}), decreases with the increase of the environment (T_{atm}) and reflected (T_{refl}) temperatures. Therefore, the increase of T_{atm} and T_{refl} generates consequently a growth of the noise, due to respectively atmospheric and solar radiations, on the values of stream temperature. As is shown in figure 4, the object radiation, i.e. the river water radiation, is included between the 80,2% and 83,4%, therefore the uncertainty on thermal power of the river water taken over by the portable thermal imager is 4,0%.

6.2. On stream temperature estimated

Considering the equation (4), it is possible to formulate the relation between the uncertainty on total thermal power taken over by the portable thermal imager (ΔW) and the uncertainty on the generic source temperature (ΔT):

$$\frac{\Delta W}{W} = 4\varepsilon\sigma T^3 \frac{\Delta T}{T} \tag{5}$$

Then for the Aterno River, since $\Delta W/W = 4{,}0\%$ and $T = 7{,}6$ °C (see section 6.1), the uncertainty on the stream temperature estimated in situ from thermal-infrared images via portable thermal imager is $\Delta T = 2{,}4$ °C, i.e. $T = \pm 1{,}2$ °C.

7. Stream temperature resulting

Punctual stream temperatures oscillate between 5 °C and 10 °C along upstream and downstream fluvial cross-sections, whereas in the vicinity of the discharge between 2,5 °C and 9,5 °C. Thermal peaks at low temperatures of the river water around the discharge are due to the greater presence of foreign matters to the water of the water body transported in surface by the river itself and outcropped from the river surface, with a different emission coefficient respect to that of the river water.

Average temperatures along fluvial cross-sections at 8 pm (about 8 °C) are higher and less oscillating than those at 9 am and 1 pm because of the absence of "cold" solar reflections in the infrared range.

By averaging, between all measures of each fluvial cross-section, all the weighted and arithmetic average temperatures, the Aterno River, in the detected area portion, is characterized by temperature values respectively equal to 7,2 °C and 7,3 °C. The small difference between the two temperature values is due to the preliminary choice of fluvial cross-sections little thermally disturbed, i.e. with a limited presence of anomalous thermal fields, carried out at view on the thermal images.

The thermal plume (figure 5) in the input point of the discharge in the river has a surface temperature of about 11 °C. This temperature becomes about 8 °C, i.e. equal to that of the equilibrium surface temperature of the river, through a distance of about 6 m, from the input point, in all directions including between that orthogonal and parallel at the discharge axis. In more detail, this thermal decrease has almost exponential law. Furthermore, it is maximum for the orthogonal direction at the discharge axis and minimum for that coincident with this axis. Moreover, the plume has not a symmetric spatial configuration due to the non-coincidence between the direction of the water flow of the river and the wastewater flow from the discharge, and due to the water push of the river during its natural propagation in the space.

By averaging all the weighted average temperatures in the no. 2 upstream and in the no. 4 downstream fluvial cross-sections, the stream temperatures are respectively equal to 7,7 °C and 7,3 °C. Therefore, the limits for the thermal pollution laid down by [9] are widely respected.

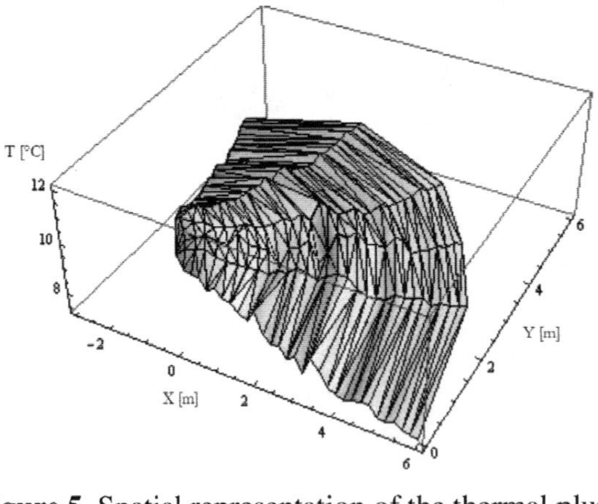

Figure 5. Spatial representation of the thermal plume.

8. Conclusions

The technique to estimate the stream temperature from thermal-infrared images in situ via portable thermal imager con be not suitable for the punctual temperatures. On the contrary, it is acceptable for the average temperatures along fluvial cross-sections; it is necessary to depurate from the calculation of these temperatures the potential errors generates, in general, by different experimental conditions of individual measurements, by the noise due to the foreign matters to the water of the water body, transported in surface by the river itself and outcropped from the river surface, highlighted from the registrations of anomalous thermal fields in the thermal images.

In according to the results and the experimental conditions characterizing this research, it is recommended to carry out the measurements with the portable thermal imager in position as possible orthogonal to the surface of the surface water body, to carry out the measurements in the evening and/or night or with overcast sky, to choose to view the sections most suitable and to use appropriate statistical methods to estimate the average surface temperatures (e.g. Gauss weighted average).

The technique analyzed in this paper allows to carry out the temperature analysis of the surface water body at a distance, without direct contact, on fairly extensive areas and to obtain rapidly numerous data readily available for subsequent and more evolved elaborations (e.g. calculation of the temperature gradients and relative heat flows).

This technique is suitable only for surface discharges and not for those submerged, since it does not allow to analyze the temperatures within the volume of the water body but only that at the surface. It allows, in addition to the assessment and monitoring of thermal pollution, also to identify sites with probable chemical pollution in place by recording of anomalous thermal fields. From the above, this technique may be of aid for government departments and public bodies responsible of the environment and land protection, for the monitoring of authorized discharges, i.e. censused, and for the identification of those abusive especially if they are not visible to the eye.

References

[1] Mohseni O and Stefan H G 1999 Stream temperature/air temperature relationship: a physical interpretation *Journal of Hydrology* **218** 128–141

[2] Chen Y D, Carsel R F, McCutcheon S C and Nutter W L 1998a Stream temperature simulation of forested riparian areas: I Watershed-scale model development *Journal of Environmental Engineering ASCE* **124**(4) 304–315

[3] Chen Y D, McCutcheon S C, Norton D J and Nutter W L 1998b Stream temperature simulation of forested riparian areas: II. Model application *Journal of Environmental Engineering, ASCE* **124**(4) 316–328

[4] Stefan H G and Chau V N 1976 River cooling water temperature fluctuations *Journal of Hydraulics Division ASCE* **102**(11) 1613–1627

[5] Chen C Q, Shi P and Mao Q W 2003 Application of remote sensing techniques for monitoring the thermal pollution of cooling water discharge from nuclear power plant. *Journal of Environmental Science and Health Part A* **38**(8) 1659–1668

[6] Van Buren M A, Watt W E, Marsalek J and Anderson B C 2000 Thermal enhancement of stormwater runoff by paved surfaces *Water Research* **34**(4) 1359–1371

[7] Kinouchi T 2007 Impact of long-term water and energy consumption in Tokyo on wastewater effluent: implication for thermal degradation of urban streams *Hydrological Processes* **21**(9) 1207–1216

[8] Kinouchi T, Yagi H and Miyamoto M 2007 Increase in stream temperature related to anthropogenic heat input from urban wastewater *Journal of Hydrology* **335** 78–88

[9] Italian Legislative Decree 3rd April 2006 no. 152 – *Environmental Regulations* (in Italian)

[10] Torgersen C E, Faux R N, McIntosh B A, Poage N J and Norton D J 2001 Airborne thermal remote sensing for water temperature assessment in rivers and streams *Remote Sensing of Environmental* **763**(3) 386–398

[11] Hedger R D, Malthus T J, Folkard A M and Atkinson P M 2007 Spatial dynamics of estuarine water surface temperature from airborne remote sensing *Estuarine, Coastal and Shelf Science* **71** 608–615

[12] Marmorino G O and Smith G B 2008 Thermal remote sensing of estuarine spatial dynamics: effects of bottom-generated vertical mixing *Estuarine, Coastal and Shelf Science* **78** 587–591

[13] Bolgrien D Q and Brooks A S 1992 Analysis of thermal features of Lake-Michigan from AVHRR satellite images *Journal of Great lakes Research* **18**(2) 259–166

[14] Parkinson C L 2003 Aqua: An earth-observing satellite mission to examine water and other climate variables *IEEE Transactions on Geoscience and Remote Sensing* **41**(2) 173–183

[15] Cherkauer K A, Burges S J, Handcock R N, Kay J E, Kampf S K and Gillespie A R 2005 Assessing satellite- and aircraft-based thermal infrared remote sensing for monitoring Pacific

Northwest rivers *Journal of the American Water Resources Association* **41** 1149–1159

[16] Cardenas M B, Harvey J W, Packman A I and Scott D T 2008 Ground-based thermography of fluvial systems at low and high discharge reveals potential complex thermal heterogeneity driven by flow variation and bioroughness *Hydrological Processes* **22** 980-986

[17] Iezzi F and Todisco M T 2012 Measures of thermal pollution of the surface water bodies via IR thermographic technique *Proceedings of the XXXIII National Congress on Hydraulics and Hydraulic Constructions*, Brescia, Italy (in Italian)

[18] Handcock R N, Gillespie A R, Cherkauer K A, Kay J E, Burges S J and Kampf S K 2006 Accuracy and uncertainty of thermal-infrared remote sensing of stream temperature at multiple spatial scales *Remote Sensing of Environment* **100** 427–440

[19] Taylor J R 2000 *Introduction to Error Analysis* (Zanichelli, ed. 2, in Italian)

[20] Flir Systems 2006 *User Manual*

[21] Çengel Y A 2005 *Thermodynamics and Heat Transmission* (McGraw-Hill, ed. 2, in Italian)

[22] Testo SpA 2008 *Pocket Guide on the Thermography* (in Italian)

Journal of Physics: Conference Series **655** (2015) 012064 doi:10.1088/1742-6596/655/1/012064

Spectral BRDF measurements of metallic samples for laser processing applications

L Vitali, D Fustinoni, P Gramazio and A Niro
Politecnico di Milano, Department of Energy,
Campus Bovisa, Via Lambruschini 4, 20156 Milano, Italy

Corresponding author email: alfonso.niro@polimi.it

Abstract. The spectral bidirectional reflectance distribution function (BRDF) of metals plays an important role in industrial processing involving laser-surface interaction. In particular, in laser metal machining, absorbance is strongly dependent on the radiation incidence angle as well as on finishing and contamination grade of the surface, and in turn it can considerably affect processing results. Very recently, laser radiation is also used to structure metallic surfaces, in order to produce many particular optical effects, ranging from a high level polishing to angular color shifting. Of course, full knowledge of the spectral BRDF of these structured layers makes it possible to infer reflectance or color for any irradiation and viewing angles. In this paper, we present Vis-NIR spectral BRDF measurements of laser-polished metallic, opaque, flat samples commonly employed in such applications. The resulting optical properties seem to be dependent on the atmospheric composition during the polishing process in addition to the roughness. The measurements are carried out with a Perkin Elmer Lambda 950 double-beam spectrophotometer, equipped with the Absolute Reflectance/Transmittance Analyzer (ARTA) motorized goniometer.

1 Introduction

Laser surface processing techniques on metals are a subject of great interest from both a scientific and industrial standpoint, since they can allow faster machining times and yield more precise finishing properties than the mechanical counterpart. Moreover, they can be used to obtain several number of interesting features, like controlled microstructures for enhancing glues adhesion [1], coloring via oxidation phenomena, and surface polishing. However, laser machining requires the fine adjustment of several parameters, i.e., beam power, velocity and frequency, atmosphere composition and angle of incidence. Hence most of the works in open literature are about finding general correlations between the aforementioned parameters and the experimental results. In this context, the paper by Veiko et al. [2] proposes a correlation between a chromaticity coefficient and laser treatment regimes for stainless steel and titanium coloring, while the work by Bordatchev et al. [3] is an interesting review of laser polishing experiments on various metal surfaces carried out by several research groups, but the comparison takes into account only the starting and the achieved average roughness. The assessment of laser polishing processes requires a detailed optical surface characterization to define the visual appearance of the finished product, as well as to evaluate its radiative properties in engineering applications. Although the BRDF is only one of the parameters considered in [4] to evaluate visual appearance, where texture properties have to be taken into account, it still is the main measurable feature in characterizing surfaces, and it can give a full insight on the optical properties of a surface finish.

In this paper, Vis-NIR BRDF measurements are carried out on four laser polished steel samples, machined with two different sets of laser parameters in two different atmospheric compositions. The achieved optical properties are compared with those of the untreated steel, to show the effect of the laser surface processing.

2 Experiments

2.1 Spectral BRDF

The definition of BRDF, or f_r, for an element of area dA, has been given by Nicodemus et al. [5] as

$$f_r(\theta_i,\varphi_i;\theta_r,\varphi_r) = \frac{dL_r(\theta_i,\varphi_i;\theta_r,\varphi_r)}{dE_i(\theta_i,\varphi_i)} \quad [sr^{-1}] \tag{1}$$

where dL_r is the reflected radiance, dE_i is the incident irradiance, the spherical coordinates (θ_i,φ_i) and (θ_r,φ_r) are the generic directions of the incoming and the reflected radiation, respectively. This definition is valid under the assumption of geometrical (ray) optics, flat reference surface uniformly irradiated, and scattering properties uniform and isotropic across the reference plane.

Although BRDF is defined as a differential quantity, the use of finite quantities is needed for a definition of BRDF that can be applied in real measurement situations, typically with the rotation of the sample along one or two axes and a movable detector with an aperture area A_d that can revolve around the test object. Under these approximations, we get [6]

$$f_r(\theta_i,\varphi_i;\theta_r,\varphi_r) \approx \frac{L_r(\theta_i,\varphi_i;\theta_r,\varphi_r)}{E_i(\theta_i,\varphi_i)} = \frac{\Phi_r}{\Phi_i}\frac{r^2}{\cos(\theta_r)A_d} \quad [sr^{-1}] \tag{2}$$

where Φ_i and Φ_r are the incident and reflected fluxes, respectively, r is the center to center distance from the sample to detector aperture. However, f_r is a total quantity whereas for our purposes, where fluorescence effects are absent and the wavelength of the incident and reflected radiation is the same, we introduce a spectral BRDF $f_{r,\lambda}$, defined by similarity with equation (2) as

$$f_{r,\lambda}(\theta_i,\varphi_i;\theta_r,\varphi_r;\lambda) \approx \frac{L_{r,\lambda}(\theta_i,\varphi_i;\theta_r,\varphi_r,\lambda)}{E_{i,\lambda}(\theta_i,\varphi_i,\lambda)} = \rho(\lambda)\frac{r^2}{\cos(\theta_r)A_d} \quad [sr^{-1}] \tag{3}$$

where $\rho(\lambda)$ is the measured spectral reflectance for a given direction of incident radiation and for a given detector position, i.e. the $\Phi_{r,\lambda}(\lambda)$ to $\Phi_{i,\lambda}(\lambda)$ ratio.

2.2 Geometric definitions

With respect to the geometry of the experimental apparatus that will be introduced in the next paragraph, let us define a few relevant angles, which are also visualized in figures 1 and 2. Given a specimen whose test surface lays on the x-y plane, and being the z axis coincident with its normal direction, in our configuration the incident and reflected beams lay always on the x-z plane, thus the azimuth angle φ is always 0. Three angles define the geometry of the setup for a single spectrophotometric measure, namely the incidence angle θ_i, between the incident beam and the sample normal, positive clockwise, the detector angle θ_d, between the incident beam and the line joining the detector and sample centers, positive clockwise, and the aspecular angle θ_a, defined as the angle from direct specular direction, positive towards incident illumination. In particular, the relationship between these three angles is expressed in equation 4:

$$\theta_a = 2\theta_i - \theta_d \tag{4}$$

being $2\theta_i$ the angle between incident and specular reflected radiation. θ_d is the angular position of the detector that collects the reflected radiation, hence it corresponds to θ_r in the BRDF definition given by equation (3). The range of the angles is from 10° to 350° for θ_d, although the extremities shouldn't be reached because the beam periscope blocks part of the detector viewing angle, and from -90° to 90° for θ_i in case of reflectance measurements.

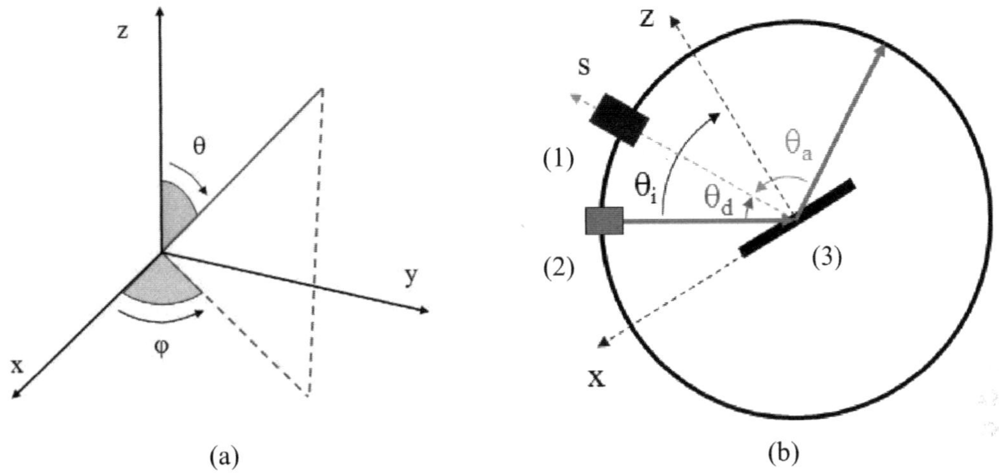

Figure 1. Generic spherical coordinates (a), and top-view scheme of the ARTA with moving detector (1), beam periscope (2), and sample (3), (b).

2.3 Experimental setup

The measurements are carried out with a Perkin Elmer Lambda 950 double beam spectrophotometer equipped with the ARTA goniophotometer accessory by OMT Solutions BV, presented in [7] and shown in figure 2. The sample is positioned on a motorized rotation stage in the center, whereas the PMT and PbS sensors, mounted in a 25 mm integrating sphere with a 30 mm wide and 17 mm high entrance port, can rotate around the same axis as the sample holder.

Figure 2. Sample compartment of the ARTA with illustrated axes and angles: sample beam periscope (1), sample with alignment beam (2).

The accessory has two polarizing filters, and every measurement is repeated twice, in *s* and *p* polarization, and results are then averaged. The beam size with normal incidence is 10 mm x 5 mm, and its width is increased with the sample rotation angle θ_i by the factor $1/\cos(\theta_i)$. The distance between the sample center and the entrance port is 91.5 mm. For this experimental campaign, the sample port width has been reduced to 14 mm with black paperboard in order to improve angular accuracy, giving the sensor a horizontal view angle $\Delta\theta_v = 8.8°$. The vertical view angle is $\Delta\varphi_v = 10.6°$. The errors introduced by using a finite sized detection solid angle in benchtop goniophotometers are well analyzed in [6], and they depend on both the aperture size and the variation of the measured radiation over the aperture, thus they can vary with the instrument setup, the measurement sample, and the sample properties. An in-depth discussion of these effect is beyond the purpose of this work, however it should be pointed out that measurement errors of about 3% are expected.

In this work, measurements are taken in the Vis-NIR wavelength range, i.e., from 350 to 2000 nm, with steps of 10 nm. A preliminary analysis has been carried out with an angle of incidence of incoming radiation $\theta_i = 45°$, and sensing angles θ_d ranging from 15° to 105° with 15° increments. Since very low reflected radiation values have been detected outside aspecular angles θ_a of ±15°, the main measurement campaign has been carried out with θ_a ranging from -15° to 15°, with 5° increments. Two angles of incidence, $\theta_i = 45°$ and $\theta_i = 15°$ are considered. The measurement geometries are listed in tables 1 and 2 and illustrated in figure 3.

Table 1. Preliminary test geometries

	$\theta_i = 45°$							
θ_d [°]	345	15	30	45	60	75	90	105
θ_a [°]	105	75	60	45	30	15	0	-15

Table 2. Test geometries

	$\theta_i = 15°$						
θ_d [°]	15	20	25	30	35	40	45
θ_a [°]	15	10	5	0	-5	-10	-15
	$\theta_i = 45°$						
θ_d [°]	75	80	85	90	95	100	105
θ_a [°]	15	10	5	0	-5	-10	-15

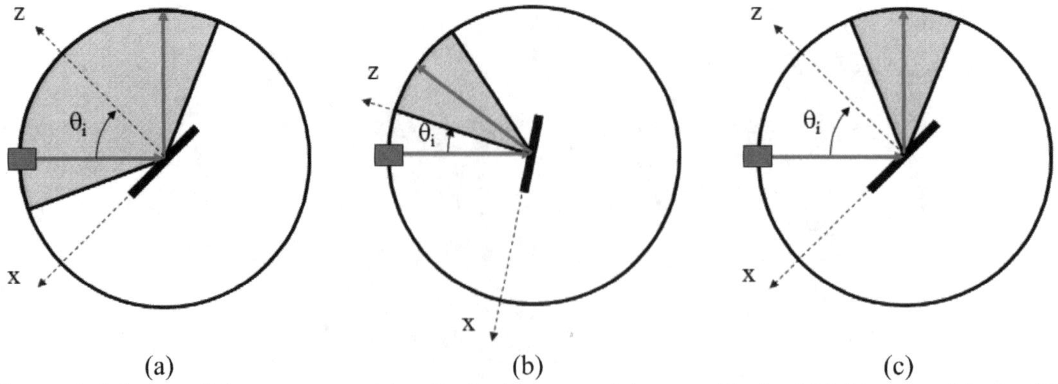

Figure 3. Scheme of the test geometries. Preliminary test (a), test with θ_i =15° (b), test with θ_i =45°. The highlighted (grey) region is swept by the detector with discrete angular intervals.

Journal of Physics: Conference Series **655** (2015) 012064 doi:10.1088/1742-6596/655/1/012064

2.4 Samples
The samples are 0.5 mm thick steel plates with 16 mm x 12 mm surface polished areas, prepared with a pulsed active fiber laser source, operating at a wavelength of 1064 nm. The setup is fully presented in [8]. Four finishes are considered representing two different roughness classes and processing environments. Finishes N1A1 and N2A2 were obtained in Ar atmosphere, exhibiting low and high roughness classes respectively. Analogously, N1 and N2 were obtained under N_2, and represent low and high surface roughness cases. The different textures of the laser polished surfaces are highlighted in figure 4. The "untreated steel" measurements have been taken on the backside of the specimen with N1 and N2 treatments.

Figure 4. Pictures of the laser-treated samples.

3 Results

3.1 Angle of incidence effect
A slight increase of the specular component of the reflection with the increase of the angle of incidence θ_i from 15° to 45° is shown in figures 5 and 6. In particular, sample N1A1 shows lesser sensitivity to variations of θ_i than the other samples. It is common behavior between all samples that reflectance differences peak at the specular direction.

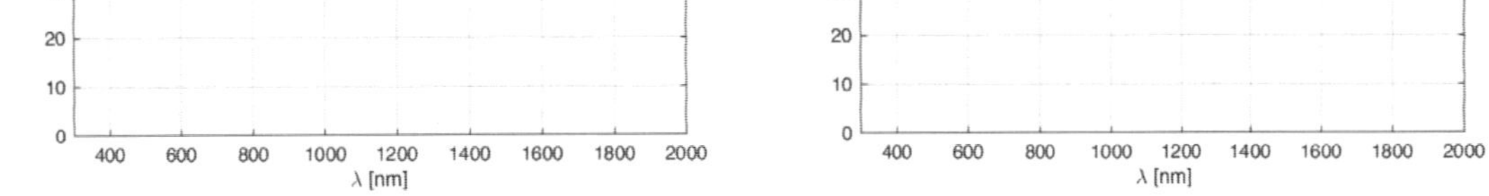

Figure 5. Spectra in specular directions, angle of incidence $\theta_i = 15°$ and $\theta_i = 45°$, samples N1 and N1A1

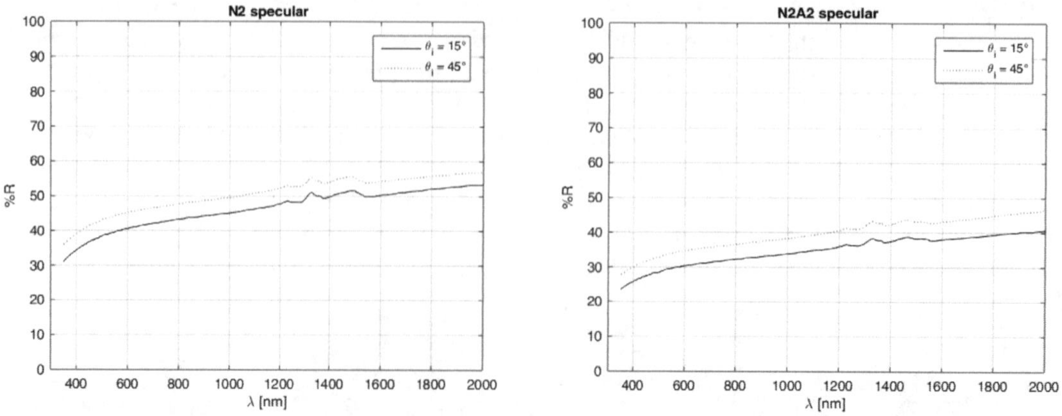

Figure 6. Spectra in specular directions, angle of incidence $\theta_i = 15°$ and $\theta_i = 45°$, samples N2 and N2A2

3.2 BRDF

The BRDFs of the laser-treated surfaces are shown for $\theta_i = 45°$ in figure 7. It should be noted that the four different finishes show different angular and spectral variations: sample N1 is the only one that presents high BRDF values for negative aspecular angles θ_a, whereas sample N1A1 shows a perfectly symmetric behavior. The peculiarity of sample N2A2 is that it's the only one whose BRDF doesn't peak at $\theta_a = 0°$.

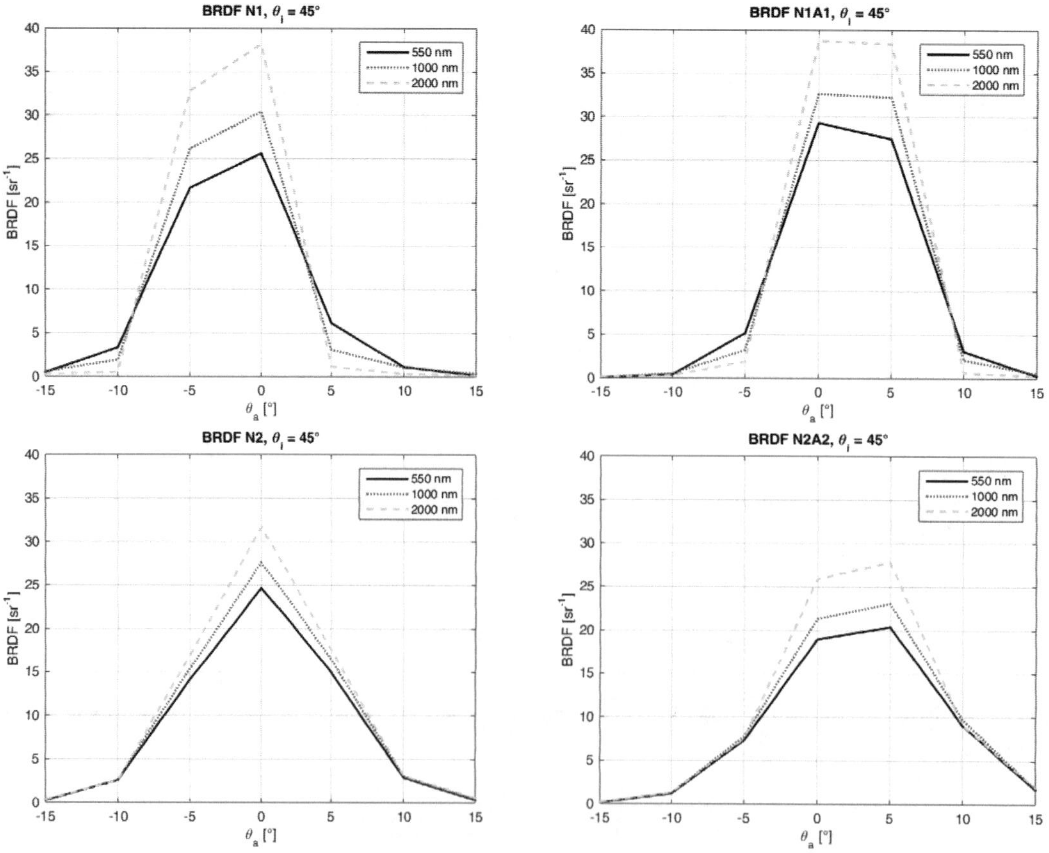

Figure 7. BRDF of the four laser-treated surfaces at three wavelegths, angle of incidence $\theta_i = 45°$.

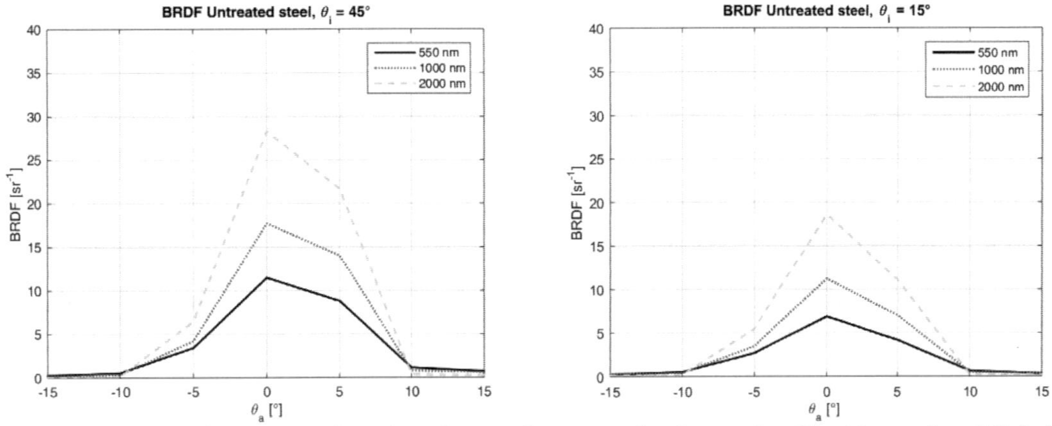

Figure 8. BRDF of the untreated steel surface at three wavelegths, angle of incidence $\theta_i = 45°$ (left) and $\theta_i = 15°$ (right).

The curves show similar trends for $\theta_i = 15°$, as shown in figure 9, but with smaller values, and as pointed out in paragraph 3.1. The reflectance increase of the laser polished surfaces with respect to the untreated bare steel is easily noticeable in figure 8 by comparison. Further considerations on this point are drawn in section 3.3.

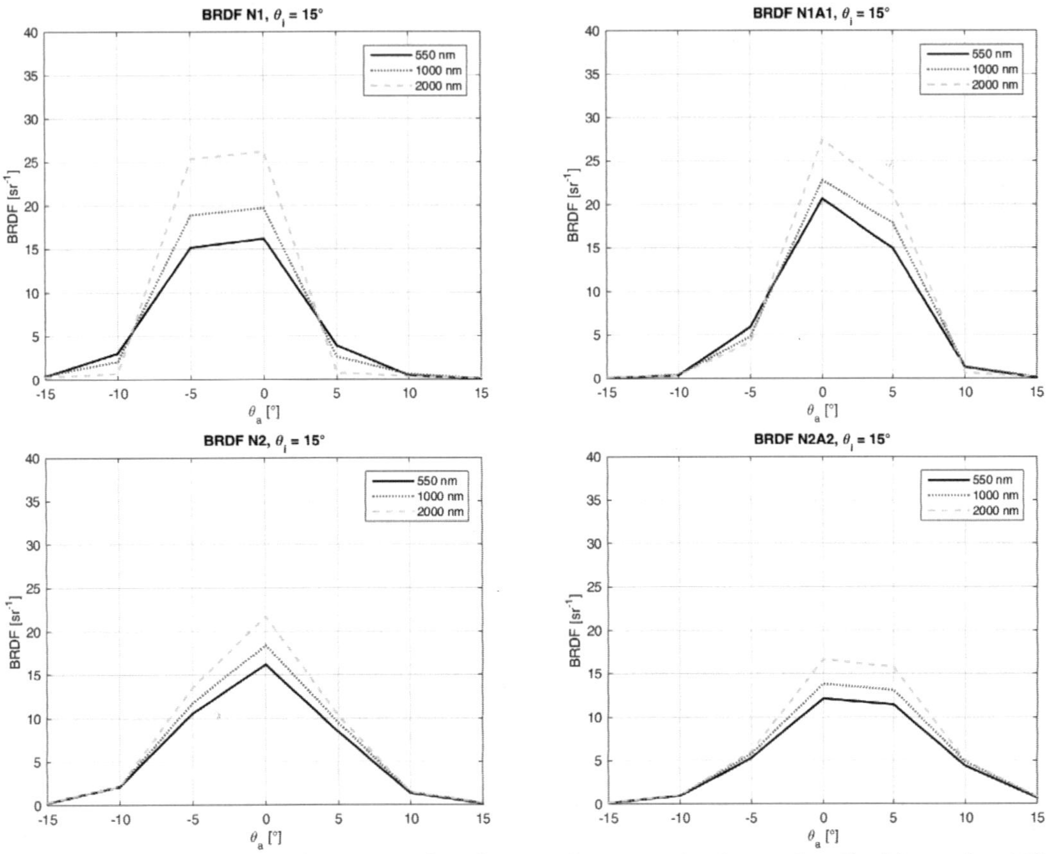

Figure 9. BRDF of the four laser-treated surfaces at three wavelegths, angle of incidence $\theta_i = 15°$.

Journal of Physics: Conference Series **655** (2015) 012064 doi:10.1088/1742-6596/655/1/012064

3.3 Reflectance differences

The surface polishing process causes an increase of the mirror-like behavior in steel, by factors listed in table 4, while considering the visible wavelength range. The largest specular reflectance increase is shown by sample N1A1, while the lowest by sample N2A2. However, the most interesting behavior is the change of the angular reflection properties, which are very specific for each of the four samples. Figure 10 shows, indeed, that the four angular shapes of reflectance increase peak at different angles. The only sample that also shows a reflection decrease, for $\theta_a = 5°$, is sample N1.

Table 3. Average reflectance increase in specular direction due to laser polishing, Vis range

	N1	N1A1	N2	N2A2
$\theta_i = 15°$ [%R]	22.7	32.9	22.2	12.2
$\theta_i = 45°$ [%R]	25.1	31.2	23.0	12.7

Figure 11 shows a 3D plot of the angular variation of the spectra with θ_a, where the differences in shape between the sample N1 and the untreated bare steel surface can be highlighted.

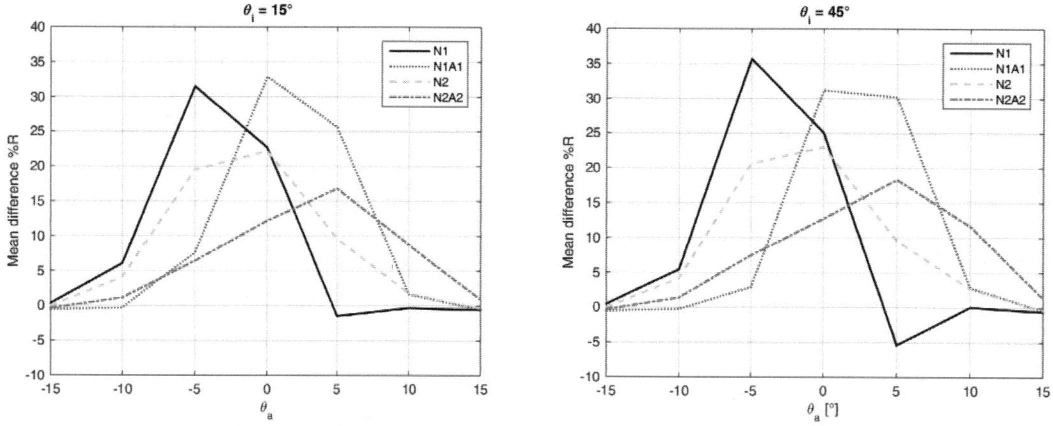

Figure 10. Average reflectance increase, Vis range, for $\theta_i = 15°$ and $\theta_i = 45°$, at seven aspecular angles.

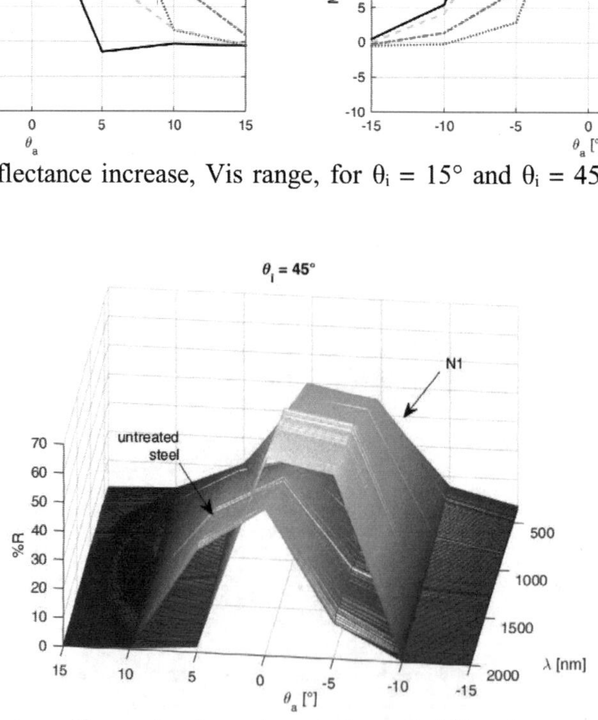

Figure 11. Full spectra with angular dependency of sample N1 and untreated steel, $\theta_i = 45°$

4 Conclusions

A first attempt to characterize the angular-dependent optical properties of steel laser polished surfaces has been carried out by means of the benchtop spectrogoniometer ARTA. The analyses allow to characterize the reflectance increase due to the laser machining. The BRDF has been calculated for two different angles of incidence, showing spectral and angular properties of the four considered samples. In addition, the results show that different machining parameters lead to different angular-dependent values of reflectance increase. Further studies are needed to verify the link between the angular and spectral variations of the surface reflectance, and the laser machining parameters, as well as the connection of spectral properties to the superficial roughness and texture.

Acknowledgments

The authors wish to thank prof. Barbara Previtali and dr. Ali Gökhan Demir of SITEC - Laboratory for Laser Applications of the Department of Mechanical Engineering at the Politecnico di Milano, to have provided the laser-treated samples analysed in this work.

References

[1] Maressa P, Anodio L, Bernasconi A, Demir A G and Previtali B 2014 *J of Adhes.* **91** 518-37
[2] Veiko V, Odintsova G, Ageev E, Karlagina Y, Loginov A, Skuratova A and Gorbunova E 2014 *Opt. Express* **22** 24342-7
[3] Bordatchev E V, Hafiz A M K and Tutunea-Fatan O R 2014 *Int. J. Adv. Manuf. Tech.* **73** 35-52
[4] Désage S F, Pitard G, Pillet M, Favrelière H, Maire J L, Frelin F, Samper S and Le Goïc G 2015 Extended visual appearance texture features *Proc. of SPIE* **9398**
[5] Nicodemus F E, Richmond J C, Hsia J J, Ginsberg I W and Limperis T 1977 *Geometrical Considerations and Nomenclature for Reflectance* (U.S. Dept. of Commerce, National Bureau of Standards)
[6] Johansson N, Neuman M, Andersson M and Edström P 2014 *App. Opt.* **53** 1212-20
[7] van Nijnatten P A 2003 *Thin Solid Films* **442** 74-9
[8] Franceschini F, Demir A G, Griffiths J D, Dowding C and Previtali B 2014 *Proc. of the 10th Conferenza del Colore (Genova, IT)* vol X B (Maggioli Editore) 20-31

Journal of Physics: Conference Series **655** (2015) 012065 doi:10.1088/1742-6596/655/1/012065

Accuracy of lumped-parameter representations for heat conduction modeling in multilayer slabs

Paola Gori, Claudia Guattari, Roberto de Lieto Vollaro, Luca Evangelisti

Università Roma Tre, Department of Engineering, Mechanical and Industrial Engineering Section, via della Vasca Navale 79, 00146 Rome, Italy

E-mail: paola.gori@uniroma3.it

Abstract. Heat conduction in homogeneous solids can be studied by resorting to one-dimensional schemes, as is often done, e.g., for building construction elements. In such situations, a simple model often employed makes use of an electrical analogy between temperature and heat flux, on one side, and voltage and electrical current on the other side. Within this framework, a few lumped-parameter representations have been described in literature to describe the thermal behavior of a single homogeneous slab or of multilayer slabs. Such models have the advantage of providing some physical insight into the phenomenon of one-dimensional heat conduction, by conveying the concepts of thermal resistance and thermal capacitance, the latter related to heat storage ability. There is, however, a certain degree of approximation in such models. The simplifying assumptions and approximations underlying these approaches will be reviewed and discussed in this contribution.

The accuracy of some lumped-parameter model will be analyzed in order to show under which circumstances the approximate solutions can be satisfactorily employed. In particular, the focus will be on the comparison of the predictions that approximate and accurate methods provide when studying the influence of layer order and distribution on the thermal performance of multilayer structures.

1. Introduction

The modeling of heat conduction in multilayer structures is an essential part for the evaluation of thermal loads in buildings. The accuracy of such modeling reflects in the proper sizing of the HVAC systems and as such can be the source of cost-effective energy use [1, 2, 3, 4, 5]. Equivalent electric circuits have been widely employed to model heat conduction [6, 7]. Their use has a double value. First, they offer a simpler mathematical treatment to be used as compared to exact solutions. Second, they have a considerable appeal because they trace thermal problems back to familiar circuit behavior using equivalent electric circuits essentially based on (single or chains of) simple RC cells. Such equivalence has been used extensively both for time sinusoidal variation regime and for transient phenomena [8]. The validity of lumped representations has been investigated by several authors [6, 7, 9]. Assessments lead to markedly different results for the two regimes cited above (steady state sinusoidal and transient). Even for single slabs, transient regime is considerably more demanding than sinusoidal regime. It has been shown [9] that good approximations for short times (0.1 RC) may require up to 50 lumped elements. Sinusoidal cases can be treated with simpler circuitry, the basic reason for this being that the fundamental frequency to be taken into account for daily variation phenomena is about 12 μHz while the spectrum required to account for functions affected by discontinuities may contain much higher frequencies.

In recent times, multilayer walls came into play, and analyses of their performances in sinusoidal regime were investigated, mainly by using approximate methods. In particular, a resurgence of electric analogs took place [10, 11, 12, 13, 14, 15]. Available results show some discrepancies, and the need for an analysis of RC model reliability in the realm of layered structures is evident.

In this context, an issue that has been frequently investigated in the design of multilayer walls is that of the influence of layers order and distribution on the wall's thermal performances [13, 16, 17, 18, 19, 20]. This subject will be studied in this work using the full analytical model of a multilayer structure in sinusoidal regime and certain relevant results will be derived. After that, the comparison with the predictions provided by the approximate lumped-parameter models will be presented.

The analysis will concern summer weather conditions, as the most demanding in typical Mediterranean climates, and on the decrease effect that the building envelope produces on the inner temperature variations. Of course, similar evaluations can be similarly carried out when the issue is to limit as much as possible the dispersion of internal heat toward the outer environment in winter conditions.

2. Harmonic temperature variation on a multilayer wall

We recall here some basic equations that are used to describe, in a one-dimensional model, the behavior of a multilayer wall subject to harmonic temperature variation. We focus in particular on the exact model of the system and on lumped-parameter models based on electrical analogy.

2.1. Exact model

The one-dimensional heat conduction problem in a homogeneous wall of thickness L, thermal conductivity λ and thermal diffusivity D, subject to harmonic temperature variation on its boundary faces can be analytically described [21] by linking temperatures and heat fluxes on its faces by a transfer matrix in the following way:

$$\begin{bmatrix} T_0 \\ q_0 \end{bmatrix} = \begin{bmatrix} \cosh(KL) & \sinh(KL)/(\lambda K) \\ \lambda K \sinh(KL) & \cosh(KL) \end{bmatrix} \begin{bmatrix} T_L \\ q_L \end{bmatrix} = \begin{bmatrix} C & \frac{S}{Y} \\ YS & C \end{bmatrix} \begin{bmatrix} T_L \\ q_L \end{bmatrix}, \qquad (1)$$

where $K = (1+j)\sqrt{\omega/(2D)}$ and $\omega = 2\pi f$ is the angular frequency (typically considered to be that of the daily temperature variation: $\omega = 2\pi/86400 = 7.27 \cdot 10^{-5}$ rad/s). A shorthand notation to be used in the following has also been introduced, with $Y = \lambda K$ denoting a thermal admittance. It is worth noting that the transfer matrix in equation (1) has unity determinant, a property that reduces to three the independent matrix elements. This of course is still valid for the matrix product that is employed for the description of a multilayer wall.

It is possible to introduce a temperature decrement factor as the ratio $F_{\text{dec}} = T_0/T_L$ that quantifies the effect that the massive wall has on the inner temperature variations. If an adiabatic boundary condition is selected ($q_L = 0$), $F_{\text{dec}} = \cosh(KL)$. The decrement factor is a complex quantity that takes into account both an amplitude reduction and a phase lag (illustrated in the following by a corresponding time delay between outer and inner temperatures). Similarly, a heat flux decrement factor $F^q_{\text{dec}} = q_0/q_L = \cosh(KL)$ can be defined if an isothermal boundary condition is selected at the inner face ($T_L = 0$).

For small values of $|KL|$, a second-order Taylor expansion of the cosh(.) function gives in the adiabatic case $F_{\text{dec}} \simeq 1 + K^2L^2/2 = 1 + j\omega RC/2$, where $R = L/\lambda$ and $C = \rho c_p L$ have been introduced, with ρ the volume density and c_p the specific heat of the homogeneous wall. At this level, R and C just define a thermal resistance and a thermal capacitance (per unit surface in both cases) of the wall but they still do not imply a specific lumped-parameter scheme that will be introduced instead in the following. In the general case (no restriction to adiabatic boundary condition), by approximating both terms in the expression of T_0 through their leading terms in the Taylor expansion (using $\sinh(x) \simeq x + x^3/6$), one obtains

$$T_0 = T_L \left(1 + j\frac{\omega RC}{2}\right) + q_L \left(R + j\frac{\omega R^2 C}{6}\right) \tag{2}$$

For multilayer walls subject to sinusoidal temperature variations, assuming continuity of temperature and heat flux at the interfaces, the description is simply based on the sequential application of transfer matrices of the type given in equation (1), each one depending on the thickness and thermophysical properties of the pertaining layer.

2.2. Two simple lumped electrical models

Two widely used models [6] of a homogeneous wall based on the electro-thermal analogy are recalled in the following.

The 2R1C or T model, is a first-order model with two equal resistances and a capacitance connected as sketched in figure 1a. The lumped parameters are the thermal resistance per unit surface $R_L = L/\lambda$ and the thermal capacitance per unit surface $C_L = \rho c_p L$. In this case

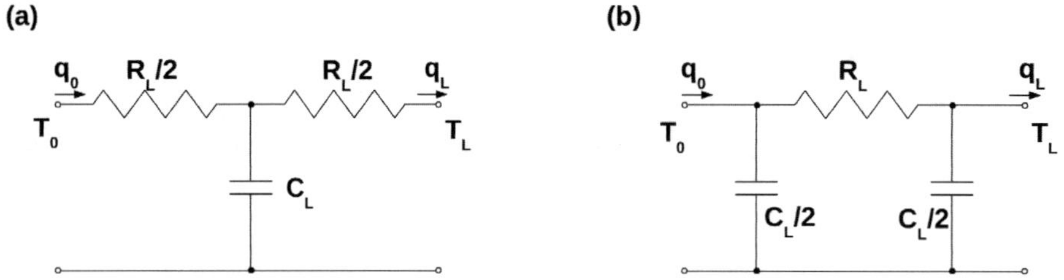

Figure 1. Equivalent electric circuits of a homogeneous wall: (a) T-model; (b) Π-model.

temperatures and heat fluxes on the two boundary faces can still be related by a matrix that reads

$$\begin{bmatrix} T_0 \\ q_0 \end{bmatrix} = \begin{bmatrix} 1 + Z_R/Z_C & 2Z_R + Z_R^2/Z_C \\ 1/Z_C & 1 + Z_R/Z_C \end{bmatrix} \begin{bmatrix} T_L \\ q_L \end{bmatrix}, \tag{3}$$

with $Z_R = R_L/2$ and $Z_C = 1/(j\omega C_L)$.

By explicitly writing the frequency dependence in equation (3), it is easily seen that, in the adiabatic case, the lumped-parameter value of F_{dec} coincides with the small-value expression found in section 2.1, whereas a small difference occurs in the general case where one obtains

$$T_0 = T_L \left(1 + j\frac{\omega R_L C_L}{2}\right) + q_L \left(R_L + j\frac{\omega R_L^2 C_L}{4}\right). \tag{4}$$

This shows that the lumped-parameter T representation of a wall gives very similar but not exactly identical results to a small-value approximated expression of the full analytical representation.

Another employed model is a 1R2C model, or Π model, that is a second-order model including two equal capacitances and a resistance connected as sketched in figure 1b.

$$\begin{bmatrix} T_0 \\ q_0 \end{bmatrix} = \begin{bmatrix} 1 + Z_R/Z_C & Z_R \\ 2/Z_C + Z_R/Z_C^2 & 1 + Z_R/Z_C \end{bmatrix} \begin{bmatrix} T_L \\ q_L \end{bmatrix}, \quad (5)$$

where now $Z_R = R_L$ and $Z_C = 2/(j\omega C_L)$. As in the exact case, both the determinant of the transfer matrix in equation (3) and in equation (5) are one. Also for the Π model the value of the decrement factor in the adiabatic case is $F_{\text{dec}} = 1 + j\omega RC/2$.

2.3. *Higher-order lumped electrical models*

Higher-order Taylor expansions of the hyperbolic functions appearing in the transfer matrix in equation (1) give rise to more accurate lumped-parameter approximations of the behavior of the multilayer wall. Correspondingly, more complex electrical circuits can be devised to model one-dimensional heat conduction (see, e.g., refs. [10, 22]).

3. Comparison of model predictions

The accuracy of the approximated models introduced in the previous section will be now analyzed by comparing their predictions with the ones of the full matrix treatment. As anticipated, the comparison will deal with the effect that the order of layers and their distribution have on the overall wall performance. At first, a few significant results will be derived by means of the exact analytical model. Then the issue will be addressed within the lumped-parameter schemes.

3.1. *Exact model*

3.1.1. Layer order Let us consider a double-layer wall composed by a conducting layer (identified by the subscript c in the following) and an insulating layer (subscript i). The matrix treatment of the problem immediately shows that changing the order of the layers affects the resulting inner temperature for a given outer temperature, being the matrix product non commutative. Indeed, for the two alternate arrangements, using equation (1) it is possible to write

$$\begin{bmatrix} T_0 \\ q_0 \end{bmatrix} = \begin{bmatrix} C_c C_i + \frac{Y_i}{Y_c} S_i S_c & \frac{C_c S_i}{Y_i} + \frac{C_i S_c}{Y_c} \\ C_c S_i Y_i + C_i S_c Y_c & C_c C_i + \frac{Y_c}{Y_i} S_i S_c \end{bmatrix} \begin{bmatrix} T_L \\ q_L \end{bmatrix}, \quad (6)$$

or

$$\begin{bmatrix} T_0 \\ q_0 \end{bmatrix} = \begin{bmatrix} C_c C_i + \frac{Y_c}{Y_i} S_i S_c & \frac{C_c S_i}{Y_i} + \frac{C_i S_c}{Y_c} \\ C_c S_i Y_i + C_i S_c Y_c & C_c C_i + \frac{Y_i}{Y_c} S_i S_c \end{bmatrix} \begin{bmatrix} T_L \\ q_L \end{bmatrix}. \quad (7)$$

It can be observed that the layer arrangement affects only the diagonal elements. If one is interested in the temperature decrement factor, under adiabatic conditions $F_{\text{dec}} = C_c C_i + \frac{Y_c}{Y_i} S_i S_c$ or $F_{\text{dec}} = C_c C_i + \frac{Y_i}{Y_c} S_i S_c$ and therefore the different complex ratios Y_c/Y_i and Y_i/Y_c are at the origin of the different wall behaviors in the two situations. A few examples are shown in the following. Let us consider typical values of the thermophysical parameters of concrete ($\lambda_c = 0.81\text{W}/(\text{m}\cdot\text{K})$, $D_c = 5.96 \cdot 10^{-7}\text{m}^2/\text{s}$) and polystyrene, ($\lambda_i = 0.034\text{W}/(\text{m}\cdot\text{K})$, $D_i = 11.55 \cdot 10^{-7}\text{m}^2/\text{s}$). Figure 2a shows the ratio $|T_L/T_0| = |F_{\text{dec}}|^{-1}$ for the two alternate arrangements of an insulating and of a conducting layer (CI, conductor in the outer layer, or IC, insulator in the outer layer) when they have equal thickness ($\alpha = 0.5$) or when the conducting layer thickness is 90% ($\alpha = 0.9$) of the total wall extension L. This comparison shows that data reported in literature on the layer order influence [13, 17] are easily explained within our model that indicates how arranging an insulating layer closer to the heat source (red curves) is always more effective than starting with a conducting layer (black curves) to reduce temperature oscillations. It is also apparent that order is more important than thickness and that using a thin insulating layer in IC configuration (red dashed line in figure 2a) is more effective than using a five times larger insulating layer in CI configuration (black full line in figure 2a). It is possible

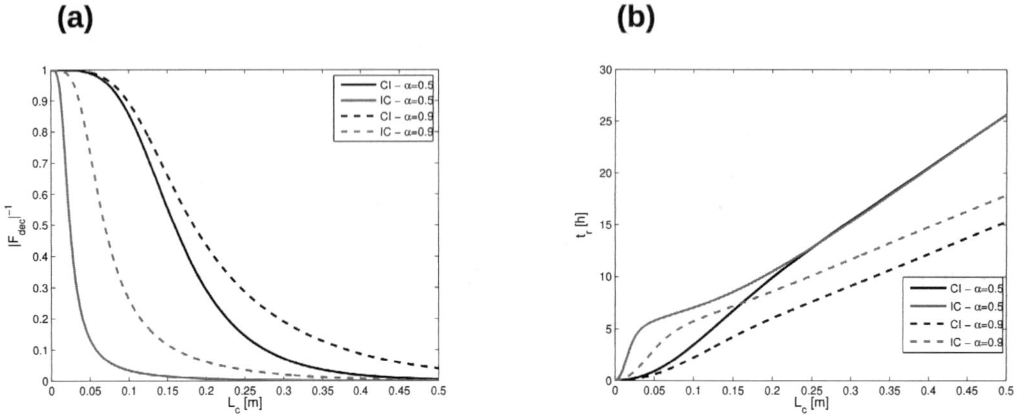

Figure 2. (a) Modulus of inverse of the decrement factor for the two alternate arrangements of the bilayer and two values of $\alpha = L_c/L$ as a function of the conductor thickness L_c; (b) Time delay in hours for the two alternate arrangements of the bilayer and two values of $\alpha = L_c/L$ as a function of the conductor thickness L_c.

to calculate the time delay introduced by the bilayer wall in the same conditions: this is shown in figure 2b suggesting that time delay is also significantly affected by layer order. In this case a more critical role of the insulation thickness is observed. For smaller values of L_c, however, IC configuration provide increased time delays in addition to larger temperature attenuation.

3.1.2. Layer splitting Recent works [13, 17] have shown the possibility that alternating insulator and conducting layers by reducing their thicknesses, while keeping their total amount constant, could improve the insulation capabilities of a wall. We investigate here this issue. Figure 3a shows that halving the thickness of individual layers in an IC configuration, employing an ICIC arrangement, can be effective to significantly increase the wall attenuation. However this is true only above a given critical thickness of the conducting layer, below which halving layer thickness results instead in a worsening of the wall insulating properties. For the example shown in figure 3a (where $L_i = 0.2L_c$) this critical value is L_c=13.4 cm.

Also the time delay between T_L and T_0 is affected by the layer distribution, as shown in figure 3b, mainly above L_c=6.0 cm for the example shown here.

The critical thickness that makes layer halving detrimental for insulation is plotted as a function of the ratio L_i/L_c in figure 4 that shows how this value becomes larger as long as smaller percentages of insulation are employed.

3.2. T and Π models

The effect that layer order and distribution have on temperature reduction and delay will be investigated in this paragraph in the cases in which the approximate T and Π models are employed. The adiabatic hypothesis is again applied.

It is generally found that the qualitative behavior is reproduced with the lumped-parameter models, but that there are quantitative discrepancies that can be large. For this reason, in

 doi:10.1088/1742-6596/655/1/012065

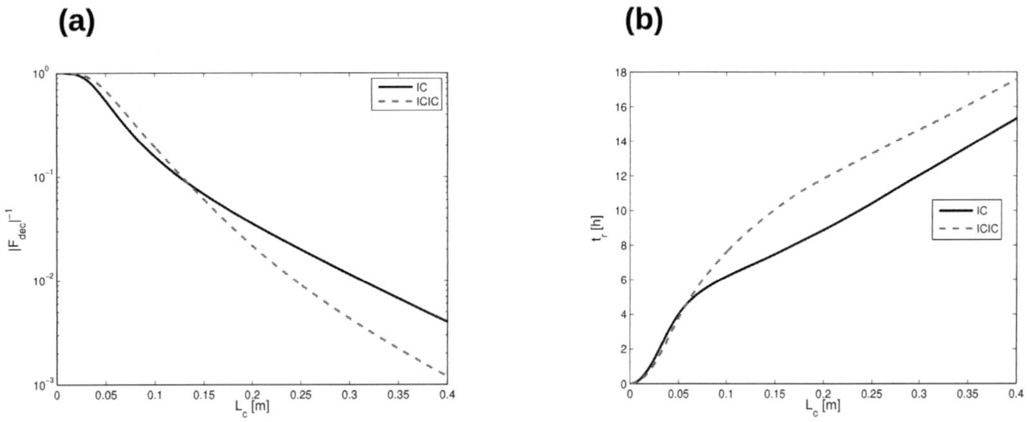

Figure 3. (a) Modulus of the inverse of the decrement factor versus the conducting layer thickness for IC or ICIC configuration, where the second configuration refers to layers with halved thickness; (b) Time delay in hours versus the conducting layer thickness for IC or ICIC configuration where the second configuration refers to layers with halved thickness.

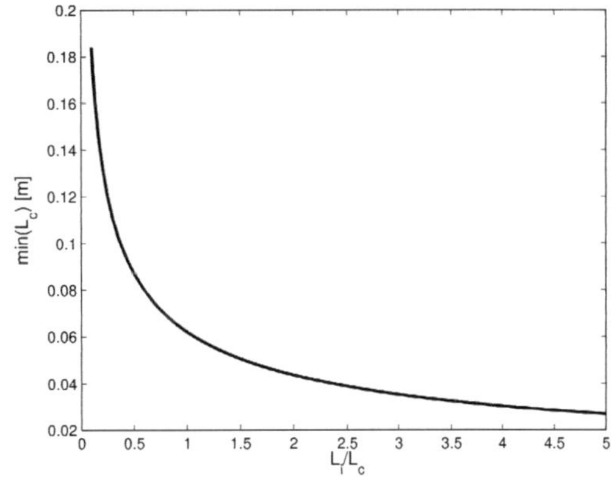

Figure 4. Value of the minimum thickness of the conducting layer below which halving it worsens the wall insulation capability. The plot is made as a function of the relative amount of insulating and conducting layer. IC and ICIC configurations are compared.

the following only error plots will be presented, omitting graphs similar to the ones shown in figures 2-4, as approximate models provide trends similar to those indicated by the exact model.

3.2.1. Layer order Figures 5a and b display the percent error on the inverse modulus of the decrement factor arising from the use of the lumped-parameter models: it can be observed that, while both models provide very similar results for CI configurations, where the attenuating behavior of the wall is less effective, they largely differ in the case where the decrement factor is more significant, i.e. in IC configurations. In such cases, the T model can be satisfactory up to $L_C = 0.2 - 0.3$ m (depending on the relative ratio L_i/L_C), while the Π model validity ceases for thinner walls. If also the delay estimation is considered, however (see figures 6a and 6b) the indication is that the use of approximate models can be acceptable if the layers are even thinner

Journal of Physics: Conference Series **655** (2015) 012065 doi:10.1088/1742-6596/655/1/012065

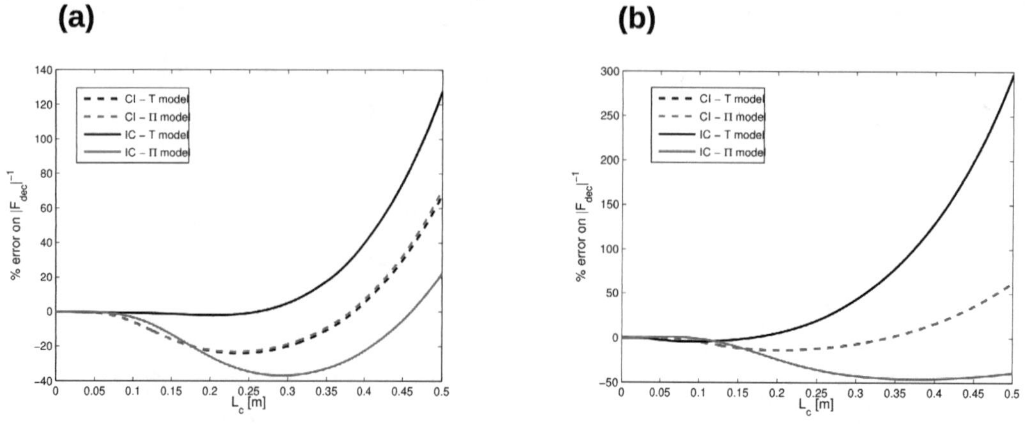

Figure 5. Percent error on the prediction of the inverse modulus of the decrement factor obtained with T and Π models for (a) $L_c/L = 0.5$ and (b) $L_c/L = 0.9$. In (b) black dashed and red dashed curves are almost completely superimposed.

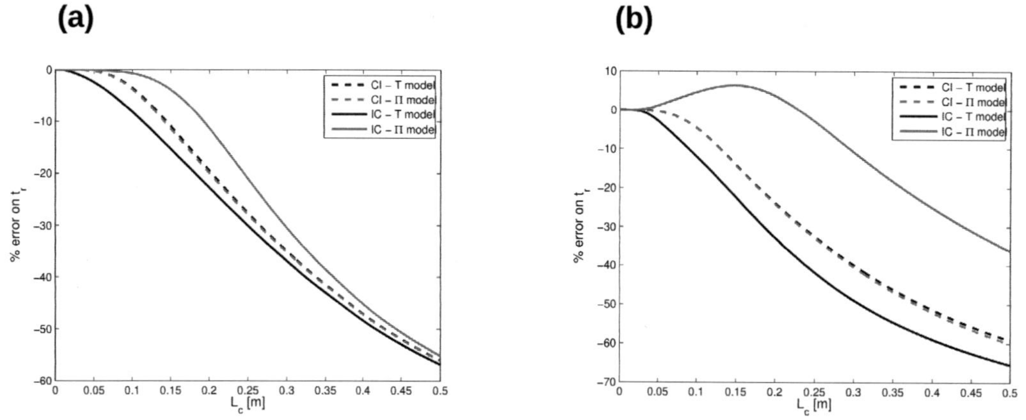

Figure 6. (a) Percent error on the prediction of the delay obtained with T and Π models for (a) $L_c/L = 0.5$ and (b) $L_c/L = 0.9$.

(e.g. approximately smaller than 0.10-0.15 m if the accepted error threshold is set at 10%).

3.2.2. Layer splitting A relevant result derived in section 3.1.2 is the determination of a critical conductor thickness below which the performances of a double-layer wall (of IC type) worsen if both of them are split in two equal parts and alternated. It is shown in figures 7a and 7b that, using lumped-parameter models, such critical value is determined with percent errors that can be large, especially when using the Π model. In both cases the largest errors arise for small values of the ratio L_i/L_c that imply, as shown in figure 3, larger values of the critical thickness L_c. In this latter cases, as expected, less accurate descriptions of the wall behavior are obtained in the lumped-parameter schemes.

4. Conclusions

This work analyzed the behavior of multilayer walls in sinusoidal regime as far as layer order and distribution are concerned. Both exact and approximate equivalent electrical models were

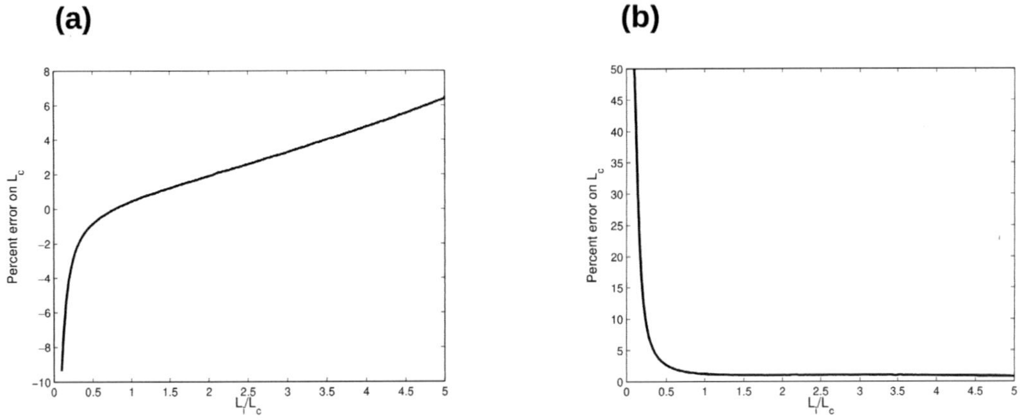

Figure 7. (a) Percent error on the prediction of the T model for the estimation of the minimum thickness of the conducting layer below which halving the thickness worsens the wall insulation capability; (b) The same as in (a) but for the Π model.

employed and their predictions were compared. The conclusion turns out to be that approximate models are usually sufficient to capture the trends of the different behaviors of multilayer walls. Concerning quantitative comparisons, however, they can largely fail especially for larger thicknesses. In such cases more than one electrical equivalent network can be necessary to model a single layer, and the question can be how useful lumped parameter models still prove being the exact treatment not only more accurate but, paradoxically enough, even simpler than approximate electrical models.

References

[1] Kossecka E and Kosny J 2002 Influence of insulation configuration on heating and cooling loads in a continuously used building *Energy Build.* **34** 321-31
[2] Foucquier A, Robert S, Suard F, Stéphan L and Jay A 2013 State of the art in building modelling and energy performances prediction: a review *Renew. Sustain. Energy Rev.* **23** 272-88
[3] Fumo N 2014 A review on the basics of building energy estimation *Renew. Sustain. Energy Rev.* **31** 53-60
[4] Pisello A L, Cotana F, Nicolini A and Buratti C 2014 Effect of dynamic characteristics of building envelope on thermal-energy performance in winter conditions: in field experiment *Energy Build.* **80** 218-30
[5] Stazi F, Bonfigli C, Tomassoni E, Di Perna C and Munafó P 2015 The effect of high thermal insulation on high thermal mass: is the dynamic behaviour of traditional envelopes in Mediterranean climates still possible? *Energy Build.* **88** 367-83
[6] Letherman K M 1977 A rational criterion for accuracy of modelling of periodic heat conduction in plane slabs *Build. Environ.* **12** 127-30
[7] Davies M G 1983 Optimum design of resistance and capacitance elements in modelling a sinusoidally excited building wall *Build. Environ.* **18** 19-37
[8] Davies M G 1997 Wall transient heat flow using time-domain analysis *Build. Environ.* **32** 427-46
[9] Barbaro S, Giaconia C and Orioli A 1986 Analysis of the accuracy in modelling of transient heat conduction in plane slabs *Build. Environ.* **21** 81-7
[10] Fraisse G, Viardot C, Lafabrie O and Achard G 2002 Development of a simplified and accurate building model based on electrical analogy *Energy Build.* **34** 1017-31
[11] Ciampi M, Leccese F and Tuoni G 2004 Multi-layered walls design to optimize building-plant interaction *Int. J. Therm. Sci.* **43** 417-29
[12] Peng C and Wu Z 2008 Thermoelectricity analogy method for computing the periodic heat transfer in external building envelopes *Appl. Energy* **85** 735-54
[13] Bond D E M, Clark W W and Kimber M 2013 Configuring wall layers for improved insulation performance *Appl. Energy* **112** 235-45

[14] Ramallo-González A P, Eames M E and Coley D A 2013 Lumped parameter models for building thermal modelling: an analytic approach to simplifying complex multi-layered constructions *Energy Build.* **60** 174-84
[15] Oliveti G, Arcuri N, Mazzeo D and De Simone M 2015 A new parameter for the dynamic analysis of building walls using the harmonic method *Int. J. Therm. Sci.* **88** 96-109
[16] Asan H 2000 Investigation of walls optimum insulation position from maximum time lag and minimum decrement factor point of view *Energy Build.* **32** 197-203
[17] Al-Sanea S A and Zedan M F 2011 Improving thermal performance of building walls by optimizing insulation layer distribution and thickness for same thermal mass *Appl. Energy* **88** 3113-24
[18] Kontoleon K J, Theodosiou T G and Tsikaloudaki K G 2013 The influence of concrete density and conductivity on walls thermal inertia parameters under a variety of masonry and insulation placements *Appl. Energy* **112** 325-37
[19] Evangelisti L, Battista G, Guattari C, Basilicata C and de Lieto Vollaro R 2014 Influence of the thermal inertia in the european simplified procedures for the assessment of buildings energy performance *Sustainability* **6** 4514-24
[20] Ozel M 2014 Effect of insulation location on dynamic heat-transfer characteristics of building external walls and optimization of insulation thickness *Energy Build.* **72** 288-95
[21] Carslaw H S and Jaeger J C 1959 *Conduction of Heat in Solids* (Oxford: Clarendon Press)
[22] Naveros I and Ghiaus C 2015 Order selection of thermal models by frequency analysis of measurements for building energy efficiency estimation *Appl. Energy* **139** 230-44